# ALGAL PHYSIOLOGY AND
# BIOCHEMISTRY

# BOTANICAL MONOGRAPHS

BOTANICAL MONOGRAPHS · VOLUME 10

# ALGAL PHYSIOLOGY AND BIOCHEMISTRY

EDITED BY

## W. D. P. STEWART

PhD, DSc

Professor, Department of
Biological Sciences,
University of Dundee

UNIVERSITY OF CALIFORNIA PRESS

BERKELEY AND LOS ANGELES 1974

UNIVERSITY OF CALIFORNIA PRESS
Berkeley and Los Angeles, California

ISBN: 0–520–02410–9
Library of Congress Catalog Card Number: 72–97731

Printed in Great Britain

# CONTENTS

Contributors      ix

Preface      xi

1   Biochemical Taxonomy      1
    R. A. LEWIN

2   Cell Wall and Intercellular Region Polysaccharides      40
    W. MACKIE and R. D. PRESTON

3   Cytoplasmic Organelles      86
    L. V. EVANS

4   Plastids      124
    T. BISALPUTRA

5   Chlorophylls      161
    J. C. MEEKS

6   Carotenoids and Biliproteins      176
    T. W. GOODWIN

7   Storage Products      206
    J. S. CRAIGIE

8   Fatty Acids and Saponifiable Lipids      236
    B. J. B. WOOD

9   Sterols      266
    T. W. GOODWIN

10   Nucleic Acids and their Metabolism      281
    BEVERLEY R. GREEN

11   Nuclear and Cytoplasmic Inheritance in Green Algae      314
    RUTH SAGER

12   Light Absorption, Emission and Photosynthesis                            346
     GOVINDJEE and BARBARA ZILINSKAS BRAUN

13   Photosynthetic Electron Flow and Photophosphorylation                    391
     J. A. RAVEN

14   Mutant Studies on Photosynthetic Electron Transport                      424
     R. P. LEVINE

15   Carbon Dioxide Fixation                                                  434
     J. A. RAVEN

16   Hydrogenase, Photoreduction and Anaerobic Growth                         456
     E. KESSLER

17   Photorespiration                                                         474
     N. E. TOLBERT

18   Dark Respiration                                                         505
     D. LLOYD

19   Heterotrophy of Carbon                                                   530
     M. R. DROOP

20   Nitrogen Fixation                                                        560
     G. E. FOGG

21   Nitrogen Assimilation and Protein Synthesis                              583
     I. MORRIS

22   Inorganic Nutrients                                                      610
     J. C. O'KELLEY

23   Phosphorus                                                               636
     A. KUHL

24   Silicification and Calcification                                         655
     W. M. DARLEY

25   Ion Uptake                                                               676
     E. A. C. MACROBBIE

26   Physico-Chemical Factors affecting Metabolism and Growth Rate            714
     C. J. SOEDER and E. STENGEL

27  Vitamins and Growth Regulators                      741
    L. PROVASOLI and A. F. CARLUCCI

28  Morphogenesis                                       788
    BETTY MOSS

29  Reproduction                                        814
    M. J. DRING

30  Extracellular Products                              838
    J. A. HELLEBUST

31  Movements                                           864
    W. NULTSCH

32  Synchronous Cultures                                894
    H. LORENZEN and M. HESSE

    General Appendix 1: A Note on Taxonomy              909
    W. D. P. STEWART

    Species Index                                       919

    Subject Index                                       956

# CONTRIBUTORS

T. BISALPUTRA, Department of Botany, University of British Columbia, Vancouver 8, Canada

BARBARA ZILINSKAS BRAUN, Department of Botany, University of Illinois, 297 Morrill Hall, Urbana, Illinois 61801, U.S.A.

A. F. CARLUCCI, Institute of Marine Resources, University of California, La Jolla, California, U.S.A.

J. S. CRAIGIE, Atlantic Regional Laboratory, National Research Council of Canada, Halifax, Nova Scotia, Canada

W. M. DARLEY, Department of Botany, University of Georgia, Athens, Georgia 30601, U.S.A.

M. J. DRING, Department of Botany, Queen's University of Belfast, Belfast, Northern Ireland, U.K.

M. R. DROOP, Scottish Marine Biological Association, Oban, Scotland, U.K.

L. V. EVANS, Department of Plant Sciences, University of Leeds, Leeds, England, U.K.

G. E. FOGG, Marine Sciences Laboratory, University College of North Wales, Menai Bridge, Anglesey, Wales, U.K.

T. W. GOODWIN, Department of Biochemistry, University of Liverpool, P.O. Box 147, Liverpool, England, U.K.

GOVINDJEE, Department of Botany, University of Illinois, 297 Morrill Hall, Urbana, Illinois 61801, U.S.A.

BEVERLEY R. GREEN, Botany Department, University of British Columbia, Vancouver 8, Canada

J. A. HELLEBUST, Department of Botany, University of Toronto, Toronto, Canada

M. HESSE, Pflanzenphysiologisches Institut der Universität, 34 Göttingen, Untere Karspüle 2, Germany

E. KESSLER, Botanisches Institut der Universität, 852 Erlangen, Germany

A. KUHL, Pflanzenphysiologisches Institut der Universität, 34 Göttingen, Untere Karspüle 2, Germany

R. P. LEVINE, Biological Laboratories, Harvard University, Cambridge, Massachusetts 02138, U.S.A.

R. A. LEWIN, University of California, Scripps Institution of Oceanography, La Jolla, California 92037, U.S.A.

D. LLOYD, Department of Microbiology, University College, Cardiff, Wales, U.K.

H. LORENZEN, Pflanzenphysiologisches Institut der Universität, 34 Göttingen, Untere Karspüle 2, Germany

W. MACKIE, Astbury Department of Biophysics, University of Leeds, Leeds, England, U.K.

E. A. C. MACROBBIE, Botany School, University of Cambridge, Cambridge, England, U.K.

J. C. MEEKS, Department of Biological Sciences, University of Dundee, Dundee, Scotland, U.K.

I. MORRIS, Department of Botany and Microbiology, University College London, Gower Street, London W.C.1, England, U.K.

BETTY MOSS, Department of Botany, University of Newcastle upon Tyne, Newcastle upon Tyne, England, U.K.

W. NULTSCH, Department of Botany, University of Marburg, Marburg, Germany

J. C. O'KELLEY, Department of Biology, University of Alabama, University, Alabama, U.S.A.

R. D. PRESTON, Astbury Department of Biophysics, University of Leeds, Leeds, England, U.K.

L. PROVASOLI, Haskins Laboratories, Biology Department, Yale University, New Haven, Connecticut, U.S.A.

J. A. RAVEN, Department of Biological Sciences, University of Dundee, Dundee, Scotland, U.K.

R. SAGER, Department of Biological Sciences, Hunter College of the City University of New York, 695 Park Avenue, New York, U.S.A.

C. SOEDER, Kohlenstoffbiologische, Forschungsstation, e.V. Dortmund, Germany

E. STENGEL, Kohlenstoffbiologische, Forschungsstation, e.V. Dortmund, Germany

N. E. TOLBERT, Department of Biochemistry, Michigan State University, East Lansing, Michigan 48823, U.S.A.

B. J. B. WOOD, Department of Applied Microbiology, University of Strathclyde, Glasgow, Scotland, U.K.

# PREFACE

Over the years plant physiologists and biochemists have made increasing use of the algae for the study of physiological mechanisms in photosynthetic plants. Many of the studies have used the algae simply as convenient physiological tools, while others have been concerned with the peculiar and fascinating physiology of particular algal groups. This book is an attempt to provide within one cover an up-to-date picture of the current state of various aspects of algal physiology and biochemistry. The hope is that it will be of use to phycologists, to physiologists and biochemists, irrespective of the organisms with which they work, and in general to those whose work involves algae, for example limnologists, oceanographers, those interested in seaweed utilization, teachers etc.

Ten years ago it was probably possible to include in a book of this type, over 90 per cent of the relevant aspects of algal physiology and biochemistry but this is no longer the case. I was therefore faced with the problem of what to include and what to leave out. What I tried to do therefore was to include chapters on the basic processes of algal metabolism such as photosynthesis, respiration and nitrogen assimilation. I then asked my graduate students who are 'physiologically orientated phycologists' what topics they would like to see in a book of this type. There were many suggestions (graduate students are not reticent!) and eventually I arrived at a list of chapters rather similar to those on pp. v–vii (the original list had to be modified slightly during the preparation of the volume due to the failure of certain of the original contributors to turn in chapters). There are many other topics which could usefully have been included, but at the expense of these I have given each author sufficient space (within reason) to develop his, or her, own particular theme. Reference to the blue-green algae has been kept to a minimum as they are dealt with in detail in a companion volume to this one.

My main thanks must, of course, go to the contributors, each a recognized authority in his or her particular field of algal physiology and biochemistry, who have responded so well (some with gentle nudging!) to my editorial scissors. I also acknowledge the generous help of the following phycologists who have helped me with various problems: Dr M.Parke, F.R.S., Mr N.I.Hendey, Dr G.T. Boalch, Mr H.T.Powell and Dr D.Hibberd. I thank Professor J.H.Burnett for his interest in the book and for his friendly advice, and Blackwell Scientific Publications, Oxford, for seeing the book through the press. Finally I must thank especially Miss Gail Alexander and Mrs B.Cunningham of my own department for all their help in Dundee.

*August 1973*                                                                         W.D.P.Stewart

CHAPTER 1

# BIOCHEMICAL TAXONOMY

## R. A. LEWIN

University of California,
Scripps Institution of Oceanography,
La Jolla, California 92037, U.S.A.

1   Introduction  2

2   A note on nomenclature  2

3   Biochemistry  3
3.1  Pigments  3
    (a) Chlorophylls  3
    (b) Carotenoids  4
    (c) Biliproteins  6
3.2  Carbohydrates  9
    (a) More or less soluble
        metabolites  9
    (b) Alginic acids  10
    (c) Fucoidins  10
    (d) Agaroses  10
    (e) Mannans and xylans  12
    (f) Cellulose  13
    (g) Intercellular plugs  14
    (h) Sulphate groups  14
    (i) Inconsistencies  15
3.3  Lipids  15
    (a) Fatty acids  16
    (b) Hydrocarbons  16
    (c) Plastiquinones and toco-
        pherols  17
    (d) Steroids  17
    (e) Lipids of blue-green al-
        gae  18
3.4  Proteins  19
    (a) Hydroxyproline  20
    (b) Enzymes  20
    (c) Amino acid sequences 21

3.5  Nucleic acids  21
    (a) Deoxyribonucleic acid
        21
    (b) Ribonucleic acid  22
3.6  Other compounds of low
     molecular weight  23
3.7  Cell surfaces  24

4   Taxonomy  25
4.1  Eukaryote genera of uncertain
     taxonomic status  25
4.2  Specific distinctions among the
     Chlorococcales  26
    (a) Chlorella  26
    (b) Ankistrodesmus and
        Scenedesmus  27
    (c) Other genera  29
4.3  Problems with the taxonomy
     of blue-green algae  29
    (a) Generic and specific dis-
        tinctions  29
    (b) Class distinctions  30

5   Epilogue and acknowledge-
    ments  32

6   References  32

## 1  INTRODUCTION

To write a really good chapter under this title one would have to be a generalist and a specialist in two different fields. On the one hand one would need to be familiar with algal taxonomy, both its classical history and its recent revelations. There are so many uncertainties here; the field is in such a state of flux that even the major subdivisions are a source of debate and are subject to frequent re-arrangements. On the other hand, one would need to be familiar with many classical methods of biochemical analysis and with at least as many of the newer techniques which are fast superseding them. I am certainly not competent to write a comprehensive survey of that vast and diffuse subject. Instead, I shall touch on a few aspects that have caught my attention in recent years, and make reference to publications which may open up special fields to the more enquiring minds among experimental phycologists.

The earlier literature was surveyed in a book I edited (Lewin 1962) and there are more recent reviews: on diatoms (Lewin & Guillard 1963), dino-flagellates (Loeblich 1966), coccolithophorids (Paasche 1968) and blue-green algae (Holm-Hansen 1968, Fogg *et al.* 1973). In 1967 I reviewed aspects of algal biochemistry in relation to taxonomy; and here I shall recapitulate some of the latter, fleshing them out with more recent items including several of the contribu-tions to two symposia, held in 1969, which dealt in large measure with this subject (Harborne 1970, Fredrick & Klein 1970).

But before going on, let me interpose a word of caution. I believe we have progressed beyond the stage where one asserts that 'algae differ from land plants (say) in such-and-such a feature' on the basis of studies of a single undesignated strain of a single unspecified species of 'chlorella' (sometimes set in lower-case letters, like sodium chloride). There is still a tendency—not confined to physi-ologists—to over-generalize on the basis of investigations on no more than one or two examples. Some workers even today tend to propound such statements as 'Lysine synthesis in the Euglenophyceae goes thus and so' although they may have examined only one strain of one species of one genus grown under only one set of known conditions; or, worse, they may have analysed only a single tuft of seaweed, grown under unspecified and probably unknown conditions, and identified merely to genus by whichever passing botanist was rash enough to do so (see, for example, De Souza & Nes 1969). Only the future will show how wrong they may have been.

## 2  A NOTE ON NOMENCLATURE

In the interests of brevity I have referred to many algae by only their generic names; for the specific names and authorities the reader should consult the original publications.

# 3 BIOCHEMISTRY

## 3.1 *Pigments*

### (a) *Chlorophylls*

The presence of chlorophylls *a* and *b*, together with various carotenoids, is a general feature of all photosynthetic land plants. Taken with other biochemical data this suggests that pigment composition may be a phylogenetically conservative feature. The Chlorophyceae resemble these land plants in this as in other biochemical features, so that we have reason to think of some early green algae as ancestors of higher plants.

Despite many biochemical similarities between green algae and land plants, some clear-cut cytological, physiological and biochemical differences are apparent. First, land-plant chloroplasts generally contain clearly recognizable grana, which are not evident in algal plastids. Second, land-plant plastids lack pyrenoids, which are found in many classes of algae. Third, the 'Hatch and Slack' pathway of $CO_2$-fixation (to C-4 dicarboxylic acids), now known to occur in many angiosperms, has not been recorded in any alga (Gibbs *et al.* 1970). (This *may* be because physiologists have concentrated their efforts on unicellular or filamentous algae with a relatively short diffusion path for $CO_2$ between the medium and the chloroplast lamellae.) Fourth, land plants seem to lack a hydrogenase system (fixation of $CO_2$ using the energy of $H_2$ oxidation), which has been demonstrated in certain green, brown, red, and blue-green algae (see Gibbs *et al.* 1970, and Chapter 16, p. 456).

The Euglenophyceae, which differ from the Chlorophyceae in many basic biochemical features (cell walls and carbohydrate reserves, for instance), have a chlorophyll complement like that of land plants. The Prasinophyceae, a class originally set off from other green flagellates by their hairy flagella and other features of fine structure, also have chlorophylls *a* and *b* in amounts approximately similar to those of other green algae, but in this class they are associated with an unusual pigment resembling protochlorophyll (Ricketts 1967).

Even among algae which have hitherto been considered as *bona fide* members of the Chlorophyceae, some unusual features are being discovered. The volvocalean *Chlamydobotrys stellata* produces very little chlorophyll *b* when grown photoheterotrophically with acetate instead of $CO_2$ as its main carbon source (Wiessner & French 1970), indicating that future investigators might profitably pay more attention to specifying the conditions under which their plants are grown. In a little marine planktonic alga with the general appearance of a *Stichococcus* (Chlorophyceae, order Ulotrichales), chlorophyll *b* seems to be absent (Yentsch & Guillard 1969), which suggests its affinity with members of the Xanthophyceae. The absence of chlorophyll *b* was among the decisive features which led Soma (1960) to assign *Vaucheria* to the Xanthophyceae. (The absence of demonstrable cellulose in the walls of such algae is no longer of itself a sound argument for this choice, as will be seen from consideration of the chemical composition of the walls of other siphonous algae: see below).

Chlorophyll $a$ is the sole chlorophyll in the Cyanophyceae where it is associated with phycobilins. This may, however, not be invariably so. Allen and Smith (1969) have reported that in the much studied *Anacystis nidulans* (a rod-like alga of shape and size very similar to *Synechococcus*) cells depleted of nitrogen lose all detectable phycocyanin, although they survive well and retain their chlorophyll content and their ability to resynthesize a full pigment complement in a few hours when a suitable source of nitrogen is resupplied.

The status of chlorophyll $d$, reported to date solely from the Rhodophyceae, is still in doubt. Perhaps it is merely a breakdown product of chlorophyll $a$, but, if so, why does this decomposition occur only among the red algae? On the other hand, recent work has suggested that chlorophyll $c$, which differs from chlorophylls $a$ and $b$ chiefly in lacking a phytyl group, actually embraces two pigments, distinguished as chlorophylls $c_1$ and $c_2$ (Jeffrey 1969). They appear to have restricted phyletic distributions. They have been found together in all of the Phaeophyceae (5 spp.), diatoms (3 spp.) and chrysomonads (2 spp.) examined, generally in a ratio $c_2/c_1$ between 1 and 2; but only $c_2$ was found in the dinoflagellates (3 spp.) and in the single member of the Cryptophyceae examined. All of these algae in Jeffrey's analyses were marine forms; it might be worthwhile to examine critically the chlorophylls of a few freshwater diatoms and dinoflagellates before accepting her general conclusions. For further details of algal chlorophylls see Chapter 5, p. 161 and Chapter 12, p. 346.

## (b) *Carotenoids*

Next to the chlorophylls, the main lipid pigments of algae are the carotenoids; of these, several have been considered to be of diagnostic taxonomic importance. The Phaeophyceae, like Chrysophyceae, Haptophyceae and Bacillariophyceae, derive their brown colour from the presence of the carotenoid fucoxanthin, a photosynthetically active accessory pigment that can be dissociated from its protein matrix, presumably on the thylakoids, by the action of heat or organic solvents. In most of the dinoflagellates, peridinin, another much-oxygenated xanthophyll, plays a similar physiological role. Members of the Xanthophyceae, however, lack both fucoxanthin and accessory chlorophylls, the main xanthophylls, in the nine species examined, being 5,6 epoxides. Vaucheriaxanthin, which has been recorded only in Xanthophyceae (*Vaucheria, Botrydium* and *Tribonema*) may constitute, in *Tribonema*, as much as 40% of the total carotenoid content (Falk & Kleinig 1968).

In most members of the Chlorophyceae, as in those few of the Euglenophyceae and Prasinophyceae which have been examined, the predominant carotene (hydrocarbon) is β-carotene, and the predominant xanthophyll (with two oxygen atoms per molecule) is lutein (Strain 1965). Lutein and lutein epoxide were found

**Table 1.1.** Distribution of major carotenoids in algae (after Goodwin 1971). Footnotes indicating minor variations, etc., have been omitted.

| Pigment | Charo-phyceae | Chloro-phyceae | Prasino-phyceae | Xantho-phyceae | Bacillario-phyceae | Chryso-phyceae* | Phaeo-phyceae | Rhodo-phyceae | Dino-phyceae | Eugleno-phyceae | Cyano-phyceae | Crypto-phyceae |
|---|---|---|---|---|---|---|---|---|---|---|---|---|
| α-Carotene | | | | | | | | | | | | + |
| β-Carotene | + | ++ | ++ | + | + | ++ | + | ++ | | ++ | ++ | |
| Echinenone | | + | | | | | | | | | | |
| Lutein | ++ | + | + | | | | | | | | | |
| Zeaxanthin | | | | | | | | +++ | | | | |
| Antheraxanthin | | | | | | | | | | | | |
| Neoxanthin | + | + | + | | | | | | | ++ | | |
| Fucoxanthin | | | | | +++ | +++ | + | | | | | |
| Diatoxanthin | | | | | | | | | | | | |
| Diadinoxanthin | | | | + | | | | | | | | |
| Peridinin | | | | | | | | | ++ | | | |
| Alloxanthin | | | | | | | | | | | | + |
| Myxoxanthophyll | | | | | | | | | | | ++ | |
| Oscillaxanthin | | | | | | | | | | | ++ | |
| Violaxanthin | + | + | + | | | | + | | | | | |

* Including Haptophyceae

to be generally present among the 50 species of siphonous green algae examined by Kleinig (1969), although they were not found in any of the five species of *Codium*. Also, in the siphonous green algae, e.g. in *Halimeda* and *Codium*, α-carotene predominates over β-carotene, and various less common xanthophylls are also found, notably siphonein and siphonoxanthin (Strain 1958, 1965). The presence of siphonein in *Ostreobium* (an unusual siphonous alga with the ability to bore into the skeletons of marine molluscs and corals) indicates that this genus is probably correctly assigned to the siphonous Chlorophyceae (Jeffrey 1968). Since *Bryopsis* contains lutein, which is absent from *Codium* (Kleinig 1969), and forms cell walls of xylan, whereas those of *Codium* consist largely of mannan (see below), one might question the advisability of assigning these two genera of plants, so different in thallus construction, to the same order.

The only cellular algae reported to contain siphonein are two out of nine species of flagellates belonging to the Prasinophyceae and examined by Ricketts (1967). However, the occurrence of this pigment is not regularly correlated with the presence or absence of an unusual protochlorophyll derivative or of a new xanthophyll (micronene), and it is hard to draw taxonomic conclusions from their distribution. Ricketts (1967) concluded from his study of these organisms 'The pigment groupings do not correspond with the orders of the class Prasinophyceae suggested by Christensen (1966).'

In the Rhodophyceae, as in most siphonous green algae, α-carotene seems generally to predominate over β-carotene. Here, too, we have a need for caution in extrapolating from limited data. β-carotene constitutes 70 to 80% of the total carotenoids of the blue-green alga *Spirulina platensis*, and cryptoxanthin only 5%; whereas the almost indistinguishable species *S. geitleri* has been reported to lack β-carotene, and to have instead some 95% of cryptoxanthin (Palla *et al.* 1970).

Among the blue-green algae, two of the characteristic xanthophylls, oscillaxanthin and myxoxanthophyll, have been shown to bear rhamnose residues, thereby revealing one more affinity between these prokaryotic algae and certain bacteria (from which other glycosidic carotenoids have been isolated; Jensen 1970). As pointed out by Goodwin (1971) (see also Chapter 6, p. 176), the carotenoids of blue-green algae are apparently all derivatives of β-carotene, which, he considers, is evidence for their phylogenetic primitiveness. Table 1.1 was taken from Goodwin (1971) who, in this paper and in Chapter 6, reviews the more biochemical aspects of the subject. It provides a useful survey of the major carotenoids. Bisalputra (Chapter 4, p. 124) also considers the relationship between thylakoid type and carotenoid content of the various algal classes.

## (c) *Biliproteins*

The phycobilin pigments, which, among land plants, are apparently represented only by the elusive phytochromes, play a major role in the photosynthetic apparatus of algae in three groups: Cyanophyceae, Rhodophyceae and Cryptophyceae. When differences were first noted among blue and red phycobilin pigments of blue-green algae and red algae, they were respectively designated as C-phycocyanin, C-phycoerythrin, R-phycocyanin and R-phycoerythrin. As more of these bilin pigments from different sources were examined, it emerged that the situation may not be so simple. Surveys of phycoerythrins from 44 red algae and one blue-green alga (Hirose & Kumano 1966) and of phycocyanins from 14 red algae and 31 blue-green algae (Hirose *et al.* 1969) revealed a wider range of pigments, which I have tried to summarize in Table 1.2. The more biochemical aspects of the biliproteins are dealt with in Chapter 6, p. 194, and more details of their absorption spectra are given in Chapter 12, p. 353.

We should be cautious about accepting the results in Table 1.2 uncritically, since in some cases, at least, what was claimed to be a single pigment of a new kind may have been merely a mixture of two or more known phycobiliproteins. Furthermore, the pigment composition of an algal species is not always a constant feature and may be subject to physiological and genetic modifications. For instance, Hirose *et al.* (1969) reported that algae identified as *Oscillatoria limosa* from Kobe, and *Phormidium autumnale* from various locations in Japan, contained both phycocyanin and phycoerythrin, whereas in samples from Ainu *O. limosa* lacked phycocyanin, and *P. autumnale* lacked phycoerythrin. In view of results reported with other blue-green algae, notably in the work of Hattori and Fujita (1959a,b) on *Tolypothrix tenuis*, these variations should be critically re-examined with unialgal, if not pure, cultures under physiologically controlled conditions. Castenholz (1970) has provided indisputable evidence for such variation within a single clone in that a culture of *Oscillatoria terebriformis*, which normally contains both phycocyanin and phycoerythrin, spontaneously gave rise to a phycocyanin-less mutant in culture.

The presence of similar phycobiliproteins in the photosynthetic apparatus of such different organisms as the prokaryotic blue-green algae and the eukaryotic rhodophytes and cryptophytes has raised various phylogenetic problems. Did these algal classes independently evolve biosynthetic pathways for red and blue bilin pigments? At least three lines of evidence point away from this possibility. First, whatever their sources, there are only two chemical species of bilins: phycoerythrobilin obtainable from phycoerythrins, and phycocyanobilins from most phycocyanins and allophycocyanins. One molecule of each is, apparently, present in R-phycocyanin (Siegelman *et al.* 1968). Second, the protein complexes show antigenic cross-reactions, indicating much in common among the protein moieties of (a) the phycoerythrins from all three algal classes and (b) the phycocyanins (and allophycocyanins) from blue-green algae and red algae (though not from the Cryptophyceae, Berns 1967). Third, the protein moieties of phycoerythrins from all members of the Cryptophyceae and Rhodophyceae so far examined have an N-terminal methionine and a C-terminal alanine residue (O'Carra 1965, Raftery & O'hEocha 1965).

Admittedly, red algae, like blue-green algae, lack flagella; and their thylakoids likewise occur singly, not in stacks. In other respects, however, algae of these

**Table 1.2.** Phycoerythrin types in various Cyanophyceae and Rhodophyceae (compiled from Hirose & Kumano 1966, Hirose et al. 1969)

| Phycoerythrin type | Absorption maxima (nm) | Cyanophyceae | Rhodophyceae Order | Genus and species |
|---|---|---|---|---|
| I (= C-type) | ca. 570, 615[2] | 18 spp. | None | |
| II (= B-type) | 548, 567 | Chroococcus minutus[1] | Bangiales | Porphyridium cruentum<br>Smithora naiadum |
| III | 550, 570, 615[2] | Lyngbya confervoides<br>Oscillatoria princeps<br>Phormidium ectocarpi | Bangiales | Compsopogon oishii |
| IV | 495, 565 | Oscillatoria irrigua[1] | Bangiales<br><br>Nemaliales | Bangia (2 spp.)<br>Porphyra suborbiculata<br>Batrachospermum (6 spp.)<br>Sirodotia suecica<br>Thorea okadai |
| V (= R-type) | 497, 540, 564 | None | Nemaliales<br><br><br>Gelidiales[3]<br>Cryptonemiales<br>Gigartinales<br>Rhodymeniales<br>Ceramiales | Bonnemaisonia hamifera<br>Delisia pulchra<br>Galaxaura fastigiata<br>2 spp.<br>10 spp.<br>9 spp.<br>2 spp.<br>8 spp. |
| None detected | | 9 spp. | | |

[1] No phycocyanin detected  [2] Probably contaminated with C-phycocyanin  [3] Parke and Dixon (1968) consider the Gelidiales as a family (Gelidiaceae) of the Nemaliales, and that Porphyridium is a member of the Porphyridiales

two classes differ widely. Compare the apparently sexless life of the blue-green algae with the elaborate 3-generation life-cycle of most red algae, for instance, and it is hard to believe in a close phylogenetic relationship between these two algal classes. (The subject has been extensively discussed by Klein, 1970). Perhaps that leaves as the only reasonable alternative explanation the possible symbiotic origin of plastids propounded in the nineteenth century (see Mereschkowsky 1905), shelved for decades, and recently reconsidered seriously by Margulis (1970).

### 3.2 *Carbohydrates*

The chemistry of algal polysaccharides is a vast and specialized field with extensive applications to algal taxonomy. Percival and McDowell (1967) published an authoritative work on the subject and a later paper by Percival (1968) has provided an invaluable supplement. Algal carbohydrates are discussed further in Chapter 2, p. 40 (cell wall polysaccharides), Chapter 7, p. 206 (reserve carbohydrates) and Chapter 30, p. 838 (extracellular carbohydrates).

### (a) *More or less soluble metabolites*

The occurrence of mannitol and laminaran (or chrysolaminaran) in such structurally dissimilar algae as the Phaeophyceae, Chrysophyceae and Bacillariophyceae supports earlier indications of affinity among these three classes based on their content of fucoxanthin and chlorophyll *c*. However, mannitol has also been found to be the principal product of photosynthetic metabolism among the Prasinophyceae (22 spp. examined) and in at least one presumed member of the Chlorophyceae, a *Stichococcus* sp. (McLachlan & Craigie 1967, Gooday 1970). Species of marine unicellular and multicellular green algae generally produce sucrose, although species of the green flagellate *Dunaliella* chiefly form glycerol (McLachlan & Craigie 1967).

The word *laminaran* covers a range of linear β-1,3-linked glucans. In a few species, some of the molecules are terminated at the reducing end by a mannitol residue. The laminarans of certain brown algae, e.g. *Eisenia bicyclis* and *Ishige okamurai*, have also some 1,6 inter-residue linkages in the chains; and indeed most samples of laminaran have a small degree of branching at C-6 (Maeda *et al.* 1966a,b, 1968, Maeda & Nisizawa 1968a,b). These substances appear to play a role in the metabolism of brown algae similar to that of the starches in the Chlorophyceae and Rhodophyceae.

In addition to starch, green algae also synthesize a wide range of more soluble polysaccharides, including complex sulphated water-soluble polysaccharides which comprise different monosaccharide residues depending on the genus (Percival & McDowell 1967). In the Cladophorales, sucrose, various fructans and galactoarabans are commonly found (Percival & Young 1971); in at least some of the siphonous green algae the major sugar moieties in the soluble polysaccharides are D-galactose and L-arabinose; while in the Ulvales they are L-rhamnose and D-glucuronic acid. In some siphonous green algae such as *Batophora oerstedii* (Meeuse 1963) the major water-soluble polysaccharide is of the inulin type (1,2-linked fructan), but some sulphated polysaccharides comprising residues of D-galactose, L-rhamnose and D-glucuronic acid are also present (E. Percival pers. comm.). Small proportions of D-xylose appear to be ubiquitous in the sulphated polysaccharides of all green algae so far examined.

### (b) *Alginic acids*

Alginic acid, a linear polysaccharide comprising β-1,4-linked D-mannuronic acid and 1,4-linked L-guluronic acid units, apparently randomly arranged along the macromolecules, is produced only by brown algae (and certain bacteria). A number of values for the ratio of mannuronic to guluronic acid are presented in Table 1.3. By the use of a specific polymannuronidase and a specific polyguluronidase, followed by chromatography of the resulting molecular fragments, Nisizawa *et al.* (1969) found that such ratios differ not only from species to species, but also, considerably, from one part of a thallus to another, indicating another *caveat* to the would-be algal chemotaxonomist!

### (c) *Fucoidins*

Certain highly branched sulphated polysaccharides based mainly on L-fucose units, but containing also varying proportions of D-glucuronic acid, D-xylose and D-galactose (Percival & Mian 1971), are also present in the Phaeophyceae, and are apparently confined to this class.

### (d) *Agaroses*

Agar, carrageenan and porphyran are characteristic polysaccharides of the Rhodophyceae. They are all sulphated, essentially linear galactans, consisting mostly of alternating 1,3- and 1,4-linked D- and L-galactose units, though some of the units may be present in the polysaccharide as the D- or L-3,6-anhydro-derivative. According to Yaphe and Duckworth (1972), the ability to convert galactose-6-sulphate to 3,6 anhydrogalactose is confined to the members of the Rhodophyceae. Depending on the species, varying proportions of methoxyl and sulphate groups substitute the hydroxyl groups on C-2, C-4 and C-6 of the galactose residues.

The compositions of some representative agaroses and related compounds are summarized in Table 1.4 (Percival & McDowell 1967). Attempts to correlate the taxonomy of red algae with the agar-like gels extractable from their cell walls, or, perhaps more correctly stated, from their intercellular matrices (McCully 1970, see Chapter 2, p. 40) proved of only limited success (Stoloff &

**Table 1.3.** Composition of alginates from different species of algae (Percival & McDowell 1967)

| Species | mannuronic acid/ guluronic acid | | |
|---|---|---|---|
| | (a) | | (b) |
| Ectocarpales | | | |
| *Ectocarpus confervoides* | 0·4 | (0·3) | — |
| *Ectocarpus* sp. | — | | 0·45 |
| Sphacelariales | | | |
| *Sphacelaria bipinnata* | 0·6 | (0·4) | — |
| Dictyotales | | | |
| *Dictyota dichotoma* | 0·6 | (0·4) | 1·05 |
| *Dictyopteris polypodioides* | 0·6 | (0·4) | |
| Chordariales* | | | |
| *Mesogloia vermiculata* | — | | 0·25 |
| *Chordaria flagelliformis* | — | | 0·90 |
| *Spermatochnus paradoxus* | — | | 1·30 |
| Dictyosiphonales | | | |
| *Scytosiphon lomentaria** | — | | 1·15 |
| *Dictyosiphon foeniculaceus* | — | | 0·85 |
| Desmarestiales | | | |
| *Desmarestia aculeata* | — | | 0·85 |
| Laminariales | | | |
| *Chorda filum* | 1·1 | (0·8) | — |
| *Laminaria digitata* | 3·1 | (2·1) | 1·45–1·6 |
| *Laminaria digitata* (f) | | | 1·2–1·85 |
| *Laminaria digitata* (nf) | | | 2·35 |
| *Laminaria hyperborea* | 1·6 | (1·1) | |
| *Laminaria hyperborea* (s) | — | | 0·4–1·0 |
| *Laminaria hyperborea* (f) | — | | 1·05–1·65 |
| *Laminaria hyperborea* (nf) | — | | 1·90 |
| *Laminaria saccharina* (f) | — | | 1·25–1·35 |
| *Alaria esculenta* | | | 1·20–1·70 |
| Fucales | | | |
| *Ascophyllum nodosum* | 2·6 | (1·7) | 1·40–2·25 |
| *Fucus vesiculosus* | 1·3 | (0·9) | 0·75–1·20 |
| *Fucus serratus* | 2·7 | (1·8) | 1·15 |
| *Pelvetia canaliculata* | 1·5 | (1·0) | 1·30–1·50 |
| *Himanthalia elongata* | 2·7 | (1·8) | 1·00–1·80 |
| *Halidrys siliquosa* | 1·1 | (0·8) | 0·75 |
| *Cystoseira barbata* | 0·7 | (0·5) | — |
| *Cystoseira abrotanifolia* | 1·9 | (1·2) | — |
| *Sargassum linifolium* | 0·8 | (0·6) | — |

(a) Fischer and Dörfel (1955). Figures in brackets are corrected by Haug's factor for the higher rate of destruction of guluronic acid.

(b) Haug (1964).

Whole plants were used for analysis except in the cases indicated as follows: (f) = fronds; (nf) = new fronds; (s) = stipes.

* The order Chordariales is not recognized by Parke and Dixon (1968). They place *Mesogloia* and *Chordaria* in the Chordariaceae and *Spermatochnus* in the Spermatochnaceae, both families of the Ectocarpales; they place *Scytosiphon* in the Scytosiphonales.

Silva 1957, Yaphe 1959). Yaphe (1971) later questioned the value of such earlier studies, in view of the relatively unsophisticated methods of polysaccharide analysis available at the time. In this later paper he divided agaroids into neutral agaroses, pyruvated (only slightly sulphated) agaroses, and sulphated galactans.

**Table 1.4.** Percentage composition of different agaroses (Percival & McDowell 1967)

| Source | D-Galactose | 6-O-Methyl-D-galactose | 3,6-Anhydro-L-galactose | L-Galactose | D-Xylose |
|---|---|---|---|---|---|
| Gelidium amansii | 51·0 | 1·4 | 44·1 | 1·9 | 0·3 |
| Gelidium subcostatum | 45·1 | 7·3 | 43·8 | 1·8 | 1·3 |
| Gelidium japonicum | 50·8 | 1·6 | 44·0 | 1·9 | 0·3 |
| Pterocladia tenuis | 51·7 | 0·8 | 44·2 | 1·4 | 0·6 |
| Acanthopeltis japonica | 49·4 | 3·2 | 44·5 | 2·0 | 0·4 |
| Campylaephora hypnaeoides | 50·0 | 0·8 | 43·7 | 4·0 | 1·8 |
| Gracilaria verrucosa | 36·3 | 16·3 | 44·0 | 2·1 | 0·2 |
| Ceramium boydenii | 31·9 | 20·8 | 44·1 | 1·0 | 0·7 |

More refined techniques, involving the controlled fragmentation of agaroid molecules with specific enzymes, followed by chromatographic fractionation of the resulting oligosaccharides, have reopened this field (Duckworth & Yaphe 1971a,b, Young et al. 1971). Hopefully, simplified methods, perhaps applicable even in the field or in the herbarium (as are many of the empirical tests now extensively used by lichenologists), will eventually come into use.

### (e) Mannans and xylans

β-1,4-linked mannans have been found in the cell-wall fractions of certain Bangiales, namely *Porphyra umbilicalis* and *Bangia fuscopurpurea*, and constitute the main components of the cell wall of some of the siphonous green algae (see Table 1.5). A xylomannan has been extracted from the green alga *Prasiola japonica* (Takeda et al. 1968).

In the red algae two types of xylans are found. In one type, either as a homo-xylan or combined as a heteropolysaccharide, both 1,3- and 1,4-xylosidic linkages occur in a structure which may be linear or branched. In the second type of xylan the molecule is essentially linear, with only 1,3 links or only 1,4 links (Turvey & Williams 1970). Xylans are also the main structural constituents of many siphonous green algae and their allies, which exhibit a number of other interesting physiological and biochemical features. Most of them are marine and are confined to sub-tropical sea shores, indicating relatively high minima of temperature and salinity. Many are calcified. Their cell walls, unlike those of most cellular green algae, generally lack cellulose (β-1,4-glucan), its place being taken by a more or less comparable microfibrillar β-1,4-mannan or β-1,3-xylan. Some of the genera studied in this connexion (usually only one species of each, examined by only one or two investigators) are listed in Table 1.5. It would seem safe to predict that comprehensive surveys of the carbohydrate metabolism of algae in these orders would reveal a variety of features not present in that 'ordinary' alga *Chlorella*.

Table 1.5. Structural cell-wall polysaccharides of coenocytic green algae (after Miwa *et al.* 1961, Parker 1970)

| Polysaccharide type | Genera | |
|---|---|---|
| | Non-calcified | Calcified |
| Cellulose | *Chaetomorpha* | |
| (β-1,4-glucan) | *Cladophora* | |
| | *Rhizoclonium* | |
| | *Spongomorpha* | |
| | *Urospora* | |
| | *Blastophysa* | |
| | *Anadyomene* | |
| | *Apjohnia* | |
| | *Cladophoropsis* | |
| | *Dictyosphaeria* | |
| | *Siphonocladus* | |
| | *Struvea* | |
| | *Valonia* | |
| | *Valoniopsis* | |
| β-1,4-mannan | *Batophora* | *Acetabularia* |
| | *Halicoryne* | *Dasycladus* |
| | *Derbesia* | |
| | *Codium* | |
| β-1,3-xylan | *Bryopsis* | |
| | *Avrainvillea* | *Halimeda* |
| | *Caulerpa* | *Penicillus* |
| | *Chlorodesmis* | *Udotea* |
| | *Dichotomosiphon* | |
| | *Pseudodichotomosiphon* | |

## (f) *Cellulose*

Critical examination and chemical analysis of the wall components of certain green algae have shown that some, but certainly not all (see Chapter 2, p. 40), lack cellulose. Those of *Spongiococcum*, for example, are demonstrably not cellulose (Deason & Cox 1971). Consequently, the general presence of this polysaccharide in other algal groups may now be questioned. So far, however, we have no reason to doubt the ubiquity of cellulose among the brown algae. The presence of cellulose as a component of the cell walls of all brown algae was indicated, on the basis of birefringence and staining reactions, and its presence has now been chemically established in *Laminaria hyperborea* and *L. digitata, Fucus vesiculosus, Ascophyllum nodosum* and *Pelvetia canaliculata* (Percival & McDowell 1967) and in

*Himanthalia lorea, Bifurcaria bifurcata* and *Padina pavonia* (E. Percival pers. comm.). Although cellulose has been reported as a general feature of the walls of red algae too, it has been critically characterized only in *Iridaea laminarioides, Gelidium amansii* and *Rhodymenia pertusa* (Whyte & Englar 1971). The main structural component of the walls of *Vaucheria* (Xanthophyceae) has been chemically identified as cellulose (Maeda *et al.* 1966b).

## (g) *Intercellular plugs*

A feature of special interest in the walls of most red algae is the intercellular plug, which apparently stoppers the duct connecting cells that have recently separated after cell division. Its precise role is unknown. Hitherto it was believed that the presence of such plugs was confined to (indeed, was almost diagnostic of) the Florideophycidae, but we know now from the work of several investigators that these structures occur even among the Bangiales, though only between cells of the conchocelis stages. Species which form plugs in this microscopic phase lack them in the macroscopic phase (presumably the gametophyte). Why? Could one correlate this difference with any detectable change in the cellular, or intercellular, biochemistry? So far, it seems, no-one has done so. The chemical composition of such plugs, too, presents a challenge. Those of *Griffithsia pacifica* were shown by Ramus (1971) to consist of protein and carbohydrate moieties, the latter with both sulphate and carboxyl groups. If intercellular plugs of red algae in general have the refractory nature of such *Griffithsia* plugs, which resist treatment with 10 N sulphuric acid at 100° for 12 hours, then we might expect to find them associated with fossilized sediments, offering us a chance ultimately to unearth palaeontological evidence for the origin of the Florideophycidae.

## (h) *Sulphate groups*

Sulphated polysaccharides are characteristic of most marine algae, being of general occurrence not only among the red and brown algae but also in the intercellular matrical substance of marine green algae such as *Ulva* and *Enteromorpha* and in the mucilage of a marine diatom, *Gomphonema olivaceum* (Huntsman & Sloneker 1971). It is noteworthy that the polysaccharides (a water-soluble xylogalactan, an insoluble xylan, and cellulose) of certain freshwater red algae, namely two species of *Batrachospermum*, seem to lack sulphate residues (Iriki & Tsuchiya 1963). On the other hand, the pectin produced by a marine angiosperm, *Zostera marina*, differs from that of land plants in having some 70% of ester sulphate (Maeda *et al.* 1966a). The incorporation of radioactive sulphate into carrageenan was demonstrated in *Chondrus* (Wagner *et al.* 1971), and one may anticipate that enzyme systems controlling such reactions will prove to be common at least among marine red algae.

## (i) *Inconsistencies*

Sugar residues in polysaccharides of the three main algal groups are summarized in Table 1.6. On the whole, data on cell-wall composition are consistent with other criteria considered of taxonomic value: e.g. most green algae with xylan walls contain leucoplasts as well as chloroplasts (Feldmann 1946). However certain species produce different kinds of wall at different stages of their life

**Table 1.6.** Sugars present in some plant polysaccharides (Percival & McDowell 1967)

|  | Phaeo-phyceae | Rhodo-phyceae | Chloro-phyceae | Flowering plants |
|---|---|---|---|---|
| D-Glucose | ** | ** | ** | ** |
| D-Galactose | * | ** | ** | ** |
| D-Mannose | * | ** | ** | ** |
| L-Galactose |  | ** |  | * |
| D-Fructose |  |  | ** | ** |
| D-Xylose | * | ** | ** | ** |
| L-Arabinose |  |  | ** | ** |
| D-Glucuronic acid | * | * | ** | ** |
| D-Galacturonic acid |  | * |  | ** |
| D-Mannuronic acid | ** |  |  |  |
| L-Guluronic acid | ** |  |  |  |
| L-Fucose | ** |  |  | * |
| L-Rhamnose |  |  | ** | * |
| Mannitol | * |  |  |  |
| Sulphate ester | ** | ** | ** |  |

** Sugars forming a major part of polysaccharide.
* Sugars found in small quantity in polysaccharide.

cycles. For example, *Spongomorpha* and *Urospora* (presumably haploid plants) have cellulosic walls, whereas the walls of the (presumably diploid) phase hitherto identified as *Codiolum* spp. appear to be purely pectic (Hanic, quoted by Parker 1970). The walls of *Derbesia* consist largely of mannan, whereas those of a species of *Halicystis* (which may be the haploid phase of *Derbesia*) contain cellulose and possibly xylan The chemical components of the walls of such plants should be checked critically.

### 3.3 *Lipids*

The lipids of algae comprise photosynthetic pigments—chlorophylls and carotenoids, discussed above—and other types of compound which may be saponifiable (such as wax esters, glycerides, phospholipids, sulpholipids and glycolipids) or not saponifiable (hydrocarbons, steroids, etc.).

## (a) *Fatty acids*

In a comparative study of the lipids of 12 species of marine unicellular algae, Ackman *et al.* (1968) found that the 4 diatoms examined were distinguished from the other algae by a virtual absence of unsaturated C-18 acids, leading to an exceptionally low iodine value. Shortly afterwards, Chuecas and Riley (1969) presented a review of the literature on the lipids of planktonic (and a few other) algae, together with the results of their own survey of 27 species of marine plankton grown in unialgal (though presumably not pure) culture. In all, they detected and determined the proportions of some 40 fatty acids, ranging from C-12 to C-24 in chain length. They found the fats of diatoms (7 species) to be distinguished by the virtual absence of C-18:2, 3 and 4 acids (i.e. with respectively 2, 3 and 4 double bonds), and those of cryptophytes (4 species) by a large proportion (10–18%) of the C-20:1 acid. The lipids of some 39 algae and phytoflagellates, mostly marine and most of them grown in pure culture, were analysed by Lee and Loeblich (1971). Their species comprised 2 blue-green algae, 3 rhodophytes, 2 xanthophytes, 6 chlorophytes, 3 euglenoids, 8 dinoflagellates, 2 cryptomonads, 7 diatoms, 3 chrysophytes and 3 phaeophytes—an impressive range. Of these, the fatty acids of 10 unicellular and 3 filamentous types were subjected to more detailed analysis, revealing a range from C-12:0 to C-22:6. As shown also by Schneider *et al.* (1970), the C-12 to C-16 acids were mostly saturated and the C-18 to C-22 acids were mostly unsaturated. Among the latter, the C-20 fraction exceeded the C-18 fraction in fats from the 2 cryptomonads and the 2 filamentous brown algae, data which refute the generalization (indicated by Lewin 1968) that a high C-20/C-18 ratio is a special feature of red algae. Docosohexaenoic acid (C-22:6) was generally present in small amounts (less than 10% of the total), but in two of the three chrysophytes examined it exceeded 10%, and in the three dinoflagellates it exceeded 20%. Similar results were obtained by Jamieson and Reid (1972), who examined in detail the fatty acids of 12 red, 17 brown and 5 green seaweeds from Scottish shores. A special feature of the lipids of chlorophytes was the high proportion of a C-16:4 acid, largely combined as monogalactosyl diglycerides. This acid was present in only small amounts in the other algae they examined.

## (b) *Hydrocarbons*

Inversely correlated with the abundance of the long-chain, highly unsaturated fatty acid (C-22:6) is the corresponding hydrocarbon, *n*-heneicosahexaene (C-21:6). In many of the algae examined by Lee and Loeblich (1971) it constituted 80–90% of the hydrocarbon fraction and in some cases more than 10% of the total lipid, i.e. more than 1% of the total dry weight of some of the diatoms analysed. It is significant that, in its taxonomic distribution among brown algae and diatoms, a high content of heneicosahexaene tends to go along with chlorophyll *c*, fucoxanthin and laminarin.

At about the same time two somewhat similar surveys were carried out by Blumer and his colleagues. A C-19 isoprenoid, pristane, was found in relatively large amounts in a diatom, a chrysophyte and a cryptophyte, but was barely detectable in the brown, green and red algae examined (Clark & Blumer 1967). Blumer et al. (1971), who studied the hydrocarbons of some 23 marine planktonic algae among 9 classes, obtained results rather similar to those of Lee and Loeblich (1971). The taxonomic generalizations that emerged from their studies are discussed at length in that paper and are not recapitulated here. Youngblood et al. (1971) examined the saturated and unsaturated hydrocarbons of 24 Atlantic coast seaweeds, comprising 4 green, 14 brown and 6 red algae. Their data indicated that n-pentadecane (C-15) predominates in brown algae, n-heptadecane (C-17) in red algae; that olefines predominate in red algae more than in brown or green algae; and that, although C-19 poly-unsaturated olefines occur in all three of these classes, the C-21 homologues seem to be absent from the red algae. According to Youngblood et al. (1971), 'No consistent correlation between hydrocarbon content and evolutionary position can be made', and 'The variability in hydrocarbon chemistry with growth stage . . . cautions against premature chemotaxonomic considerations.' If by 'considerations' they mean 'conclusions', this seems very wise. Nichols (1970) also made references to the considerable effects of cultural conditions on the lipid composition of various algae, and Gelpi et al. (1970), in presenting similar data on the hydrocarbons of 7 blue-green algae, 4 green algae and a xanthophyte, Botryococcus braunii, concluded, understandably, 'One is hard pressed to make phylogenetic conclusions solely on the basis of hydrocarbon distribution. . . .'

### (c) Plastoquinones and tocopherols

Several classes of algae have been shown to contain various plastoquinones and tocopherols. These include; green algae (5 species), yellow-green algae (1 species), red algae (3 species) and brown algae (3 species), in addition to a Euglena and several blue-green algae (Carr et al. 1967, Powls & Redfearn 1967, Sun et al. 1968, Antia et al. 1970). Apparently most algae produce the same types of quinones as are found in higher plants, including phylloquinone (vitamin $K_1$), plastoquinone-9, and α-tocopherolquinone. The blue-green algae however, generally form neither ubiquinones nor menaquinones, and Anacystis nidulans also lacks α-tocopherolquinone.

### (d) Steroids

The presence of steroids is characteristic of eukaryotes, and the distribution of different steroids seems to follow taxonomic lines. Algae in most divisions produce fucosterol or sitosterol, but in the red algae, as in animals, cholesterol is the usual type (Gibbons et al. 1967), while ergosterol has been reported in euglenophytes (Aaronson & Baker 1961). Earlier reports had led us to believe that steroids were not formed by blue-green algae or other prokaryotes, but this is apparently not the case: sitosterol, cholesterol and certain other steroids have now been found in *Anabaena, Anacystis, Fremyella,* and *Phormidium* (see Carr & Craig 1970).

### (e) *Lipids of blue-green algae*

Special interest attaches to the lipids of the blue-green algae, since these plants may be expected to exhibit closer affinities to the bacteria than to eukaryotic algae, and since demonstration of their presence in early geological strata may provide information on the origins of the Prokaryota. A comparative study of the fatty acid components of 4 green algae and 7 blue-green algae, all grown in pure culture, revealed no consistent differences between representatives of these two algal classes (Schneider *et al.* 1970). Other studies, however, have shown indications of considerable differences between the lipid components of blue-green

**Table 1.7.** Occurrence of certain lipid components among blue-green algae, eukaryotic algae and higher green plants (compiled chiefly from Han & Calvin 1969, Nichols 1970)

| | Cyanophyceae | Eukaryotic algae | Higher plants |
|---|:---:|:---:|:---:|
| Monogalactosyl diglyceride<br>Digalactosyl diglyceride<br>Sulphoquinovosyl diglyceride<br>Phosphatidyl glycerol | + | + | + |
| Phosphatidyl choline<br>Phosphatidyl ethanolamine<br>Phosphatidyl inositol<br>Arachidonic acid<br>Trans-3-hexadecenoic acid<br>Lecithin* | − | + | + |
| Cerebroside<br>Sterol glycoside<br>Phytane<br>Pristane<br>Squalene, etc.<br>High-M.W. hydrocarbons (C-25, 27, 29) | − | − | + |
| Branched C-18 hydrocarbons<br>7 and 8 Methyl heptadecanes | + | − | − |
| n-C-17 Hydrocarbons | + | + | − |

* James & Nichols 1966.

Note: Only limited numbers of representative types were examined; these generalizations should be regarded as tentative only.

algae and those of eukaryotic algae and higher plants; many of these are summarized in Table 1.7.

All blue-green algae seem to contain mono- and digalactosyl diglycerides, which are also commonly associated with the chloroplasts of eukaryotic plants. The presence of these galactolipids is apparently correlated with the ability of blue-green algal cells, like chloroplasts, to evolve oxygen under conditions suitable for photosynthesis. The galactolipids, however, are absent from photosynthetic bacteria. Indeed, Kenyon and Stanier (1970) felt that the correspondence in lipid composition between certain *Chroococcus* spp. and angiosperm plastids supports the hypothesis that the latter organelles arose from symbiotic 'protocyanophytes'.

**Table 1.8.** Distribution of fatty acids in various Cyanophyceae (Kenyon & Stanier 1970)

| | Number of strains | Poly-unsaturated fatty acids | Ratio of 0- or 1-enic to 2-enic fatty acids |
|---|---|---|---|
| Nostocales | 15 | +++ | 1 to 3 |
| *Spirulina* | 2 | + or − | >59 |
| Chroococcales | | | |
| Type I | 6 | +++ | ~1 |
| Type II | 12 | + or − | >13 |

From data of Winters *et al.* (1969), Kenyon and Stanier (1970), and Holton and Blecker (1970), the blue-green algae appear to fall roughly into two biochemical groupings, as indicated in Table 1.8. It would be of interest if one could culture for comparative studies the marine blue-green alga *Trichodesmium erythraeum*, in which as much as 25–50% of the fatty acids is represented by *n*-decanoic acid (C-10:0). This is an unusually small molecule not found in any of the other blue-greens examined (Parker *et al.* 1967). (The alga has recently been reclassified as a species of *Oscillatoria*, Sournia 1968.)

### 3.4 *Proteins*

Although one would not expect bulk proteins of algae to differ much in their overall proportions of amino acids, specific proteins present in large amounts (such as the storage proteins of leguminous seeds) can exhibit specific differences. Perhaps proteins of this kind occur in resting spores of certain algae, but to date none seems to have been characterized. Some algal cell walls contain appreciable proportions of proteins, especially in the outermost layers (Thompson & Preston 1967, Hanic & Craigie 1969), which may prove to be of taxonomic value. Some of the values reported are, for *Cyanidium*, 50 to 55% (Bailey & Staehelin 1968); *Derbesia*, 5 to 10% (Parker 1970); *Chaetomorpha*, *Codium*, *Nitella*, etc., 5 to 10% (Thompson & Preston 1967).

## (a) *Hydroxyproline*

Gotelli and Cleland (1968), who surveyed the occurrence of hydroxyproline in proteins from 50 species of algae belonging to 7 classes, discovered some rather significant taxonomic correlations. In most of the 16 green algae analysed, this amino acid was found generally to be concentrated in the cell walls, being especially abundant in those of *Codium* and *Bryopsis*. Two exceptional members of the Chlorophyceae were *Chlorella pyrenoidosa*, in which hydroxyproline occurred chiefly in the intracellular protein fraction, and a *Nitella* species, from which it appeared to be absent altogether. In the Phaeophyceae (12 species), it occurred chiefly in the cytoplasm; in the Rhodophyceae (14 species), it was not found at all. The blue-green algae (5 species) contained little or no hydroxyproline.

## (b) *Enzymes*

Special enzymes distinguish special kinds of algae. One immediately thinks of those involved in the fixation of molecular nitrogen (see Chapter 20, p. 560), or in the biosynthesis of α-ε diaminopimelic acid (DAPA) as a cell-wall component, in blue-green algae.

A special feature of the red algae is a particle-bound amino-acid decarboxylase, found by Hartmann (1972) in most of the 25 species of marine Florideophycidae that he examined (though not in any of the 4 Bangiales included in his survey). This enzyme is apparently absent from the brown and green seaweeds, the diatoms, and the blue-green algae.

Two blue-green algae have been shown to synthesize lysine via DAPA, thereby resembling bacteria (but so do two species of *Chlorella*, thereby confusing the picture!); whereas in *Euglena gracilis* lysine is synthesized via α-amino adipic acid as in many fungi (Vogel 1965). When lysine pathways in a few dozen more algae have been similarly studied, we may begin to be in a better position to draw a few tentative taxonomic generalizations. Among 33 species of microscopic algae examined (including a *Euglena*, a *Platymonas*, and several red algae, cryptophytes and pennate diatoms) the only ones found capable of evolving $H_2$ when illuminated anaerobically were members of the Chlorophyceae: *Chlamydomonas* (3 species), *Chlorella* (2 species) and *Scenedesmus* (2 species) (Healey 1970). The aldolase of *Anacystis nidulans* is of the metal-dependent type characteristic of other prokaryotes (Willard & Gibbs 1968).

At lower taxonomic levels, too, we may hope for help from protein chemists, who have already provided simplified techniques for the separation of isoenzymes. However, until we know how much intraspecific variation to expect in such studies, their interpretation must be open to question. A survey such as that carried out by Thomas and Brown (1970) on 19 species of *Chlorococcum* and *Tetracystis*, each represented by only a single strain, can tell us hardly more taxonomically than we already knew.

## (c) *Amino-acid sequences*

Reference has already been made to the terminal amino-acid residues of phyco-biliproteins, from which some phylogenetic conclusions can be drawn. Looking further into the future, one may predict that it may ultimately be possible to establish phylogenetic trees for algae with a reasonable degree of confidence, based on objective data. This might be possible if we had as much information about the proteins associated with chlorophylls from various sources as we have already begun to collect about the amino-acid sequences of that other, related, tetrapyrrol-protein complex, haemoglobin. '. . . Computer programs applied to data from phylogenetic trees and protein polymorphism can be used to find the rate of molecular evolution of different species . . .' (Goodman *et al.* 1971). If this can be done for globins of insects, fishes, birds and mammals, why not for the protein moieties of algal chlorophylls?

## 3.5    *Nucleic acids*

With regard to the nucleic acids of algae, we may consider separately the DNAs of their nuclei (both eukaryons and prokaryons), of chloroplasts and of mito-chondria, and the RNAs in the various ribosomal fractions. Nucleic acid metabo-lism in general is dealt with in more detail in Chapter 10, p. 282.

## (a) *Deoxyribonucleic acid*

The unequivocal demonstrations of deoxyribonucleic acid (DNA) in chloroplasts support the hypothesis, already mentioned, that these organelles may have descended from autonomous blue-green algae. Indeed, in molecular hybridiza-tion experiments, some degree of homology has been found between the DNA of blue-green algae (7 species examined) and that of *Euglena* chloroplasts (Pigott & Carr 1972). Both in plastids and in blue-green algae (as in bacteria), the DNA fibrils seem not to be associated with basic proteins such as the histones that characterize the nuclei of most eukaryotes. This lends further weight to the symbiont hypothesis (Makino & Tsuzuki 1971). However, a similar condition also exists in the dinoflagellates (Kowallik 1971), suggesting that this class has a long phylogenetic history apart from that of other eukaryotic algae. Both in plastids and in blue-green algae, the DNA appears densitometrically homo-geneous, generally exhibiting only a single peak in gradient centrifugation (Iwamura 1970), although an accountable second peak has been reported for one blue-green alga, *Plectonema boryanum* (Kaye *et al.* 1967). The unexpected dis-covery that, among eukaryotic algae, the total amount of DNA per cell is more or less proportional to the cell volume (Holm-Hansen 1969) suggests that per-haps only a small fraction of it carries the full complement of the genotype. In this respect, algae may differ from higher plants, but adequate data are not yet available.

B

Iwamura (1970) also pointed out that the chloroplast DNA of algae tends to have a lower percentage of guanine *plus* cytosine than does their nuclear DNA, whereas in higher plants, i.e. the terrestrial counterparts of the Chlorophyceae, the reverse is generally the case.

Endosymbionts, of course, may be expected to retain their own DNA characteristics. In preparations of zooxanthellae isolated from sea anemones, for example, the DNA can be physico-chemically distinguished from that of their host cells (Franker 1970). If this is true also of the symbiotic bacteria that Leedale (1969) detected, by electron microscopy, living within the nuclei of certain strains of euglenoid flagellates, it might be worth looking for similar cases of symbiosis among other algae with satellite DNAs that cannot be otherwise accounted for.

## (b) *Ribonucleic acid*

Ribosomes of blue-green algae, like those of bacteria, have a Svedberg centrifugation value of 70-S, while those of eukaryotic algae typically are 80-S (Carr & Craig 1970). Evidence from studies of ribonucleic acid (RNA) fractions seems to support the hypothesis that plastids are phylogenetically derived from blue-green algae (see Table 1.9). Molecular weights of the ribosomal RNAs of two species

**Table 1.9.** Physical and physiological features of ribosomes from Cyanophyceae and eukaryotic plants. (For references see text.)

| Source of ribosomes | Ultracentrifugal fractions | | | Inhibition of protein synthesis by: | |
|---|---|---|---|---|---|
| | | | | chloramphenicol | cycloheximide |
| Cyanophyceae | 70-S | 23-S | 16-S | + | — |
| Chloroplasts (of algae and higher plants) | 70-S | 23-S | 16-S | + | — |
| Cytoplasm (of algae and higher plants) | 80-S | 25-S | 18-S | — | + |

of red algae (a *Porphyr dium* sp. and a *Griffithsia* sp.), determined by electrophoresis in polyacrylamide gel, indicate that the affinity between the Rhodophyceae and blue-green algae may be closer than that between other algae and blue-green algae. The molecular weight of the ribosomal RNA from chloroplasts was found to be similar to that of blue-green algae, while that of the heavy cytoplasmic ribosomal RNA (28-S) was only $1 \cdot 21 \times 10^6$ (Howland & Ramus 1971), closer to that of blue-green algae (4 species, $1 \cdot 07 \times 10^6$) and lower than that of any other eukaryote ($1 \cdot 27$ to $1 \cdot 75 \times 10^6$) so far studied (Loening 1968). Although these results do not provide as direct evidence for affinities between the Rhodophyceae and the blue-green algae as do comparative studies of their biliprotein pigments (mentioned earlier), they certainly point in the same direction.

### 3.6 Other compounds of low molecular weight

In addition to compounds of major metabolic significance, there are several kinds of substances whose distribution may prove of some taxonomic interest. Poly-β-hydroxybutyric acid, for instance, a characteristic storage reserve material of many bacteria, has been reported in the blue-green alga *Chlorogloea fritschii* (Carr 1966), but in no eukaryotic algae.

The occurrence of halogenated compounds appears to follow strict taxonomic lines. Mono- and di-iodotyrosines have long been known to occur in certain families of brown algae (Tong & Chaikoff 1955). Iodine (probably iodide, though this is far from clear in the literature) is also concentrated in special cells (called 'ioduques') of *Bonnemaisonia, Falkenbergia* and certain other red algae (Feldmann 1961). Since these algae have complicated anisomorphic alternations of generations, it would be of interest to compare iodine accumulation and metabolism in the different stages of their life cycles.

A class of unusual brominated and chlorinated compounds has been identified during the last few years in red algae of another family, the Rhodomelaceae, including *Laurencia* (7 species), *Polysiphonia* (2 species), *Chondria* (1 species) and *Rhodomela* (1 species) (see, for example, Irie *et al*. 1970a,b, Sims *et al*. 1971, W. Fenical pers. comm.). Their metabolic role is unknown, although they may be distasteful and a deterrent to potential browsing animals. Attempts to use them as aids to the specific identification of *Laurencia* spp. have been only partially successful, since these substances tend to be interconverted or destroyed *in vivo* as well as *in vitro* (W. Fenical pers. comm.). Although not yet chemically identified, substances with characteristic absorption spectra in the ultra-violet range have been found among the extracellular products of 5 species of *Acrochaetium* grown in laboratory culture. These may prove to be of taxonomic value in future studies of this genus (Boney 1972).

More information is becoming available on the content of choline and of various inositols in algae (Ikawa *et al*. 1968), e.g. C-methyl inositol (laminitol) in *Macrocystis* and other Phaeophyceae and Rhodophyceae (Schweiger 1967, Nagashima *et al*. 1969).

A survey of volatile amines in 5 green, 11 brown and 12 red marine algae revealed rather more variety in the red algae than among seaweeds of other classes (Steiner & Hartmann 1968). Another volatile compound, dimethyl propiothetin, was recognized early as one of the products of red algal metabolism contributing, *via* dimethyl sulphide, to the fishy odour of *Polysiphonia lanosa*. Acrylic acid, the other product of dimethyl propiothetin hydrolysis, has more recently been found as an extracellular product of *Phaeocystis* (Sieburth 1964) and other kinds of phytoplankton (Ackman *et al*. 1966), where it may serve as a kind of antibiotic.

The distribution of caulerpin and caulerpacin in several species of *Caulerpa* (Aguilar-Santos & Doty 1968) suggests that one should seek similar neurotoxic compounds among other siphonous green algae. (Indeed, it is almost certain that

a search of this sort is now under way in more than one pharmaceutical laboratory).

### 3.7   Cell surfaces

Long before there was any algal taxonomy on our part, fellow members of algal
species were recognizing and distinguishing one another by biochemical means.
As they have done for aeons, motile male gametes home in on prospective female
partners by following concentration gradients of pheromones, and pairing is
promoted or miscegenation blocked according to the compatibilities of iso-
agglutinins on flagella or other gamete surfaces. These are obviously the first
sites where we may expect to find interspecific biochemical differences but, since
they must depend on a large variety of subtle differences, they present special
problems for whose solutions we have only recently developed adequate
analytical techniques. The revelation that the male-attracting pheromone of
*Ectocarpus siliculosus* is a butenyl cycloheptadiene (Müller 1967, Müller *et al.*
1971) doubtless presages an enthusiastic survey of such substances from other
species and genera, substances which have direct, almost causal, relationships
with their taxonomy. Flagellar iso-agglutinins of various species of *Chlamydomonas* and *Volvox* are being studied by techniques of ever increasing refine-
ment. So far, they all appear to be muco-polysaccharides (Wiese 1969). Possibly
the variety of extracellular polysaccharides of *Chlamydomonas* spp. (Lewin 1956)
may have some metabolic relationships to substances controlling the compatibilities of their gametic cells. For, whatever we may think about the factors that
establish specific distinctions, these are the chemicals that facilitate or block
copulation and thereby permit or inhibit exchanges of information between
genotypes. Presumably similar agglutinins control the adhesion of red algal
spermatia to the trichogynes of conspecific females, but I know of no biochemical
work on this subject.

Cryptic species have been demonstrated in several genera of Volvocales in
which sexual reactions have been critically studied, but the various clones do not
necessarily fall neatly into syngen pairs. For instance, among the isolates from
33 natural populations of *Gonium pectorale*, Stein (1965, 1966a) found quite a
complicated pattern of differential sexual compatibilities. Physiological differences among the various strains appeared to be relatively trivial, although from
an evolutionary standpoint they may be vital. The 15 syngen pairs of *Pandorina
morum* examined by Palmer and Starr (1971) responded differently to acetate,
although within each pair the two clones agreed in this respect.

How can so many cryptic species continue to look alike? Why have they not
diverged, morphologically and physiologically? Perhaps a clue to this enigma
lies in the discovery (Stein 1966b) that genetic barriers between these syngens are
not absolute: challenged at different temperatures or pH values, some of them
prove permeable, as it were, permitting mating and gene exchange between
members of different syngens, and thereby re-uniting genetic sub-pools into one
recognizable species (see also Wiese 1969, and Chapter 29, p. 814).

Another kind of biochemical recognition, this one involving organisms which are not conspecific, is that between an epiphyte, symbiont, or parasite and its host. Here we have almost no experimental data to account for the highly specific relationships that are often observed. The fact that some red algae parasitize only members of their own families (e.g. *Erythrocystis* on *Laurencia* species) suggests that the selective biochemical mechanisms involved here may be similar to those controlling conspecific reactions in fertilization.

In studies of algae capable of forming symbiotic relationships with the flat-worm *Convoluta roscoffensis*, Provasoli *et al.* (1968) showed that a variety of species and strains are acceptable and can be incorporated alive into the cells of the worm, but that only the normally symbiotic *Platymonas convolutae* is able to continue thriving there. L. Muscatine (pers. comm.) has some pertinent observations on the acceptance or rejection of different kinds of *Chlorella* by cells of a *Hydra* species normally characterized by endosymbiotic zoochlorellae. Here, too, the acceptance of algal cells is highly selective, but it is still not known how the surface chemistry differs between potential food and potential symbiont. One might start to look into such phenomena experimentally by seeking specific surface antigens, using such immunological techniques as were employed in studies of *Chlamydomonas* species by Brown and Walne (1967) and by Sato and Tomochika (1969).

## 4 TAXONOMY

### 4.1 *Eukaryote genera of uncertain taxonomic status*

A few organisms have defied the attempts of orderly minded taxonomists to fit them neatly into the usual classificatory pigeonholes. For instance, although no-one seriously doubts that species like *Nitzschia alba* are apochlorotic diatoms, there is considerably less certainty about the classes to which one should assign other non-photosynthetic organisms with more-or-less algal appearances.

*Prototheca*, with at least three species, is generally accepted as the apochlorotic equivalent of *Chlorella*. Largely on the basis of physiological similarity, *P. zopfii* and *P. chlorelloides* appear to be conspecific (Casselton & Stacey 1969). The activity of mitochondria isolated from *P. zopfii* indicates that the respiratory chain of this organism is essentially similar to that of green algae and higher plants (Webster & Hackett 1965). Nevertheless, *Prototheca* differs from almost all photosynthetic algae in its inability to utilize nitrate as a source of nitrogen (Casselton & Stacey, 1969), and its main cell-wall component is embarassingly unlike cellulose (Lloyd & Turner 1968). The discovery, on human skin, of a *Prototheca*-like organism which is capable of producing multicellular filaments, and which can grow by assimilating various disaccharides and sugar alcohols that are not used by other species of *Prototheca* (Arnold & Ahearn 1972), casts further doubt on the algal nature of this genus.

*Saprochaete*, which at first sight looks like an apochlorotic *Cladophora*, and which certainly does not fit readily into any of the recognized classes of algae, proved to have cell walls which lack cellulose but contain some chitin, and reserve granules of a material more like glycogen than starch. Taken together with evidence from electron microscopy, these data led Dawes (1969) to conclude that this strange organism is best relegated to the fungi.

*Cyanidium*, which looks like a blue-green *Chlorella*, has been bandied about from the blue-green algae (though it is not prokaryotic) to the Rhodophyceae, the Chlorophyceae, and the Cryptophyceae, each new analysis of its biochemistry serving only to complicate the picture further. The pigments of this alga seem to resemble those of blue-green algae (Allen 1959), its lipids those of green algae and higher plants (Allen *et al.* 1970), and its tolerance of acidity (including the ability to grow at pH values below 1) and heat (growth occurs at temperatures up to 56°C; Allen, 1959, Doemel & Brock 1971) are unique among eukaryotic organisms. Certain despairing phylogenists are inclined to assign it to limbo, or in a subclass by itself (Schnepf *et al.* 1966, Klein & Cronquist 1967).

With the blue-green eukaryote *Glaucocystis*, we are in similar trouble, and not even with the help of the electron microscope have we been able to establish unequivocally whether its pigmented bodies are blue-green algae or plastids (Schnepf *et al.* 1966, Schnepf & Brown 1969). In some ways they resemble blue-green algae: their pigments comprise chlorophyll *a* (but not *b* or *c*), β-carotene, zeaxanthin (but no echinenone or myxoxanthophyll), C-phycocyanin and some allophycocyanin (Chapman 1966).

### 4.2   *Specific distinctions among Chlorococcales*

Whereas in the taxonomy of bacteria the main groups are distinguished largely on the basis of morphological features (branching or unbranching filaments, single or catenate cells, presence or absence of flagella, etc.), the major subdivisions of algae have been distinguished for the last 150 years on biochemical grounds (primarily on the nature of the photosynthetic pigments, and secondarily on the chemical nature of the cell walls and food reserves). Conversely, whereas bacterial species are separated, *faute de mieux*, by biochemical differences such as their diverse abilities to dissimilate or ferment various carbon sources, algal species are generally distinguished morphologically. Only when evident morphological differences are slight or absent, as in *Chlorella* spp., or exhibit an inconveniently protean plasticity of form, as in *Scenedesmus* spp., have we felt the need to examine the possibilities of using biochemical features as additional specific criteria.

### (a)  *Chlorella*

Morphological features seem insufficient for *Chlorella* taxonomy. Several cultural approaches to the problem of species delimitation, some more and some less successful, have been attempted, but it is only within the last few years that a satisfactory classification of the commoner species of *Chlorella* has begun to emerge. What makes it seem relatively satisfactory, as compared with earlier schemes, has been the high degree of agreement between taxonomies based, respectively, on microscopic (morphological) and cultural (physiological and biochemical) features. A synthesis of the system now generally agreed on by Fott, Kessler and their colleagues is summarized in Table 1.10 (see also Chapter 16, p. 456).

It should be noted that most of these *Chlorella* species are represented here by a number of separately isolated clones. This is an essential feature of such a combined approach, and it gives us some confidence to believe that species, in this case, have objective reality, i.e. that there exist in Nature real interspecific discontinuities. The fewer the strains examined, the shakier are the grounds on which we base our taxonomy. When we have only one strain of each putative species, and examine such single strains biochemically or physiologically, we have no way of knowing whether the correlations we find are of specific value or whether they are merely fortuitous. For instance, *Chlorella vulgaris*—but perhaps only a single strain—has been reported to differ from *C. ellipsoidea* and certain other species in producing 1-pentacosene and 1-heptacosene in darkness, and in being unable to convert Δ-7 sterols (chondrillastenol and ergostenone) to Δ-5 sterols (Patterson 1967a,b). How typical are such biochemical features of the species in general? This remains to be studied.

### (b) *Ankistrodesmus and Scenedesmus*

As shown by Trainor (1970) and by Schlichting and Bruton (1970), in some cultures of *Scenedesmus* and *Ankistrodesmus* pluricellular colonies may give rise to unicells in media of high pH, and spines may be suppressed by high concentrations of iron, etc. It is evident that for the taxonomy of many genera of smaller algae, 'natural' cell structure is not enough. As we learn more about cultural behaviour in such genera, we find ourselves less dependent on purely morphological criteria for specific distinctions and, incidentally, in a better position to combine with them certain algae now masquerading under other generic names. Thus there is little doubt that many of the named species of *Scenedesmus* (Uherkovich 1966), together with certain unicellular algae presently assigned to other genera such as *Franceia* and *Chodatella*, will eventually be recognizable as growth forms of a limited number of *Scenedesmus* spp. It seems a pity that *Ankistrodesmus* (22 strains) and *Scenedesmus* (27 strains) have both proved so remarkably homogeneous in respect to the three features studied in comparative cultures: hydrogenase activity, gelatin digestion, and the accumulation of secondary carotenoids under conditions of nitrogen starvation (Kessler & Czygan 1967). Maybe other biochemical features will prove more helpful.

**Table 1.10.** Correlations of morphological, biochemical and physiological characters of *Chlorella* species

| Species | 1[a] | 2[a] | 3[a] | 4[a] | 5[a] | 6[a] | 7 | 8[b] | 9[b] | 10[b] | 11[b] | 12[b] | 13[b] |
|---|---|---|---|---|---|---|---|---|---|---|---|---|---|
| *C. fusca* Shihira et Krauss | – | 16 | Parietal | + | + | – | 12 | + | + | + | 3·5–4·0 | + | – |
| *C. fusca* var. *rubescens* (Dangeard) | ± | 13 | Parietal | + | + | – | 1 | + | + | – | 4·5 | + | – |
| *C. homosphaera* Skuja | – | 9 | Mantle | – | – | – | 1 | + | + | – | 6·0 | + | – |
| *C. kessleri* Fott et Nováková | – | 10 | Mantle | – | + | – | 7 | + | – | – | 3·0 | + | – |
| *C. luteoviridis* Chodat | – | 14 | Dish/Band | – | + | + | 6 | – | – | – | 3·0 | – | – |
| *C. minutissima* Fott et Nováková | – | 3 | Cup | – | – | – | 1 | – | – | – | 5·5 | + | – |
| *C. protothecoides* Krüger | – | 9 | Cup/Band | – | – | – | 14 | – | – | – | 3·5–4·0 | – | + |
| *C. saccharophila*[c] (Krüger) Nadson | + | 15 | Dish/Band | – | + | – | 10 | – | – | – | 2·0–3·0 | + | – |
| *C. vulgaris* Beijerinck | – | 10 | Girdle | – | + | – | 21 | – | – | – | 4·0–4·5 | + | – |
| *C. vulgaris* f. *tertia*[d] Fott et Nováková | – | 12 | Girdle | – | + | – | 17 | + | – | – | 4·0–4·5 | + | –[e] |
| *C. zofingiensis* Dönz. | – | 17 | Mantle | – | – | – | 4 | – | + | – | 4·5–5·5 | + | – |

1. Shape: elongate?
2. Cell diameter (max., in μm)
3. Chloroplast-form
4. Chloroplast-slits?
5. Pyrenoid
6. Walls—thickened?
7. Number of strains in physiological study
8. Hydrogenase activity
9. Secondary carotenoids formed under nitrogen-deficient conditions
10. Liquefaction of gelatin
11. Lower pH limit of growth
12. Nitrate reduction
13. Thiamine requirement

a. for morphological characters, see Fott and Nováková (1969)
b. for physiological characters, see Kessler (1968) and Chapter 16, p. 456
c. including *C. ellipsoidea* Gerneck
d. thermophilic
e. 2 strains require thiamine

(c) *Other genera*

Biochemical features, like any others, can be used for taxonomic purposes only when we have some idea how they may vary within specific limits. Analyses of data on available sources of nitrogen (including several amino acids and bases) utilizable by *Chlamydomonas* spp. (Cain 1965), for instance, are of questionable value since for each 'species' only one strain was examined. How much intra-specific variation may be found in such genera is illustrated by a comparative study of 14 clones of *Pseudochlorococcum typicum* subjected to a variety of tests, including their ability to use six carbon sources and their sensitivity to five antibiotics (Archibald 1970).

Many other genera or families might lend themselves to comparable investigations, notably the freshwater cryptomonads of which even Pringsheim (1968) despaired. And, if all else fails, one solution to some of our taxonomic difficulties might be pyrolysis and gas-liquid chromatography (cf. Sprung & Wujek 1971), although it is still too early to expect that meaningful results will come from such drastic measures. But although biochemical and physiological approaches to the taxonomy of different groups provide new data, it remains to be demonstrated that they are of intrinsically greater taxonomic value than, say those based on morphological features.

### 4.3   Problems with the taxonomy of blue-green algae

(a) *Generic and specific distinctions*

The taxonomy of the Cyanophyceae is a Pandora's box and I hesitate even to lift the lid here. My task could be evaded if, like some strict phycologists, we were to exclude the blue-green algae from the algae altogether, since, like the bacteria, they are prokaryotic. But I will venture here a few rash comments on this difficult topic.

The taxonomy of the blue-green algae has been, until recently, in much the same chaotic state as was the taxonomy of the bacteria before cultural methods had been invented. Until recently many phycologists, not only physiologists and biochemists, were unaware that *Hapalosiphon laminosus* is a synonym of *Mastigocladus laminosus; Anacystis marina* may be conspecific with *Synechococcus cedrorum*; apparently the same alga in different laboratories may be called a *Phormidium*, a *Lyngbya* or a *Plectonema*. That favourite of algal physiologists, *Anacystis nidulans*, has been assigned to four different genera by as many 'authorities' in the last four years (Komarek 1970). Fortunately, bacteriologists progressed beyond such chaos long ago, but phycologists are trying only now to do likewise. Until we study blue-green algae in culture, we cannot know to what extent environmental conditions may affect the form and colour of the trichomes, the widths of the cells or their degree of cohesion, the intercellular constrictions and terminal tapers, the presence of false or true branching of filaments, etc. Consider sheaths, for example. Species of *Phormidium* and *Oscillatoria* are supposed to be naked, species of *Lyngbya* are characterized by separate sheaths, and species of *Plectonema* by diffluent, more or less confluent sheaths. Cultural studies such as those of Drouet (1963) indicated that several differential features of this sort are subject to influence by such factors as light intensity, available sources of nitrogen, and even turbulence. In two considerable but controversial works, Drouet proposed revisions of the whole structure of blue-green algal taxonomy, reducing more than 2,500 taxa of coccoid forms to 25 species (Drouet & Daily 1956) and a comparable number of filamentous blue-green algae to 23 species (Drouet 1968)! In view of the present limitations of our knowledge, this is surely carrying 'lumping' too far. In the travails of uncertainty, we should not throw out on this account the many tedious man-years of labour, leading to so many apparently useful specific distinctions, which we owe to taxonomists working with blue-green algae during the past century.

Probably the most convenient taxonomy, if not absolute truth, will emerge when students will have studied one group at a time, critically evaluating variations both in Nature and in culture under different conditions, and taking into account biochemical and physiological features no less than morphological and structural ones. Baker and Bold (1970) and Kenyon *et al.* (1972) have made a good start with experimental studies of filamentous species, as have, *a fortiori*, Stanier *et al.* (1971) with unicellular forms. An example of the monographic treatment of one group of unicellular blue-green algae, those which divide in only one plane (i.e. *Synechococcus* as redefined), is illustrated in Table 1.11. It is hoped that other workers on blue-green algae will follow such examples, for it seems only appropriate to deal with these microorganisms, at long last, by the methods of modern microbiology (see also Fogg *et al.* 1973, Carr & Whitton 1973).

Ultimately, we may be able to take into consideration potentialities of recombination between gene pools, as one can in higher plants. An encouraging paper for the prospects of an objectively based taxonomy of blue-green algae is that of Shestakov and Nguyen (1970), who presented convincing evidence that these organisms can be transformed by DNA preparations in much the same way as can such bacteria as *Salmonella* spp. This opens the door to paragenetic studies, even with algae such as these which lack manifest sexuality.

### (b) *Class distinctions*

Lastly, what unequivocally sets the Cyanophyceae apart from the other prokaryotes? *Beggiatoa* may be a colourless counterpart of *Oscillatoria*—but how are we to distinguish other apochlorotic blue-green algae from, say, flexibacteria (Soriano & Lewin 1965)? Studies of the composition of their cell walls, lipids, carotenoid pigments and intermediate metabolites have left us still uncertain. The discovery of a filamentous, gliding organism which normally contains only carotenoids but which develops bacteriochlorophyll when grown anaerobically

**Table 1.11.** Characteristics of strain clusters of typological group I (*Synechococcus*) (Stanier *et al.* 1971)

| Cluster | Cell width (μm) | Polar granules | Motility | Temp. max. (°C) | Content of poly-unsaturated fatty acids | Requirement for Vitamin $B_{12}$ | Requirement for 1% NaCl | Phyco-erythrin | Peak ratio of phycocyanin: chlorophyll *a* | Deoxyribo-nucleic acid base composition[a] | Strains included |
|---|---|---|---|---|---|---|---|---|---|---|---|
| 1 | 0·8–1·5 | – | – | 35–37 | Low | – | – | – | 1·19–1·57 | 66–71 | 6307, 6603, 6706, 6707, 6708, 6709, 6710, 6713, 6907, 6911 |
| 2 | 0·8 | – | – | 41 | High | – | – | – | 1·03 | 69 | 7001 |
| 3 | 1·5 | – | – | >53 | Low | – | – | – | 0·76–0·83 | 52–54 | 6715, 6716, 6717 |
| 4 | 1·0 | – | – | 43 | Low | – | – | – | 0·80–1·08 | 55–56 | 6301, 6311, 6908 |
| 5 | 1·5 | – | – | 37 | Low | – | – | – | 0·88 | 50 | 6312 |
| 6a | 1·5 | – | – | 43 | High | + | – | – | 0·96 | 49 | 7002 |
| 6b | 2·0 | – | – | 39 | High | + | + | – | 1·17 | 49·5 | 7003 |
| 7 | 3 | – | – | <35 | High | – | – | + | 0·87 | 47 | 6605 |
| 8 | 3 | – | – | 41 | High | – | – | – | 0·78 | 48 | 6910 |
| 9a | 2·5–3·0 | + | + | 37 | High | – | – | + or – | 0·79–0·98 | 46 | 6802, 6903 |
| 9b | 1·5 | + | + | 37 | High | – | – | – | 0·98 | 45 | 6901 |

[a] Expressed as moles % guanine plus cytosine.

in the light (Pierson & Castenholz 1971) is particularly perplexing to prokaryote taxonomists.

As a conclusion to this discussion of the phylogenetic aspects of algal biochemistry, I would like to recall Mereschkowsky's fanciful corollary of the theory deriving chloroplasts from endosymbiotic blue-green algae. 'If only animals had such intracellular symbionts', he postulated, 'then lions would have no need to prey on antelopes, but could relax on the banks of streams, photosynthesizing as peacefully as palm trees' (Mereschkowsky 1905).

## 5 EPILOGUE AND ACKNOWLEDGEMENTS

To try to follow progress in two disciplines as disparate as taxonomy and biochemistry is like trying to ride two bicycles at once. The pure algal physiologist and the strict algal taxonomist, each from his own side of the street, may decry the inadequacy of these clumsy strivings to straddle two saddles. But if my efforts serve in some measure to point out some ways, and to indicate some pitfalls, then they will not have been altogether wasted. I could not have written this chapter without generous help from colleagues who contributed expertise from several specialized fields. Their comments and criticisms have been invaluable. Obviously, errors and omissions must be my sole responsibility. Those who have helped in various ways include Drs. A. A. Benson, L. Cheng, W. Fenical, B. Fott, D. L. Fox, F. T. Haxo, O. Holm-Hansen, R. B. Johns, E. Kessler, E. Marques, E. Percival, H. Ramus and P. C. Silva.

This article was prepared while my research was supported by National Science Foundation grants GRA-650 and GB-8724. This is a Contribution from the Scripps Institution of Oceanography.

## 6 REFERENCES

AARONSON S. & BAKER H. (1961) Lipid and sterol content of some Protozoa. *J. Protozool.* **8**, 274–7.

ACKMAN R.G., TOCHER C.S. & MCLACHLAN J. (1966) Occurrence of dimethyl-β-propiothetin in marine phytoplankton. *J. Fish. Res. Bd. Can.* **23**, 357–64.

ACKMAN R.G., TOCHER C.S. & MCLACHLAN J. (1968) Marine phytoplankter fatty acids. *J. Fish. Res. Bd. Can.* **25**, 1603–20.

AGUILAR-SANTOS G. & DOTY M.S. (1968) Chemical studies on three species of the marine algal genus *Caulerpa*. In *Trans. Symp. Drugs from the Sea*, ed. Freudenthal H.O., pp. 173–6. Marine Technology Society.

ALLEN C.F., GOOD P. & HOLTON R.W. (1970) Lipid composition of *Cyanidium. Pl. Physiol., Lancaster*, **46**, 748–51.

ALLEN M.B. (1959) Studies with *Cyanidium caldarium*, an anomalously pigmented chlorophyte. *Arch. Mikrobiol.* **32**, 270–7.

ALLEN M.M. & SMITH A.J. (1969) Nitrogen chlorosis in blue-green algae. *Arch. Mikrobiol.* **69**, 114–20.

ANTIA N.J., DESAI I.D. & ROMILLY M.J. (1970) The tocopherol, vitamin $K_1$ and related isoprenoid quinone composition of the unicellular red alga *Porphyridium cruentum*. *J. Phycol.* **6**, 305–12.

ARCHIBALD P.A. (1970) *Pseudochlorococcum*, a new Chlorococcalean genus. *J. Phycol.* **6** 127–32.

ARNOLD P. & AHEARN D.G. (1972) The systematics of the genus *Prototheca* with a description of a new species *P. filamenta*. *Mycologia*, **64**, 265–75.

BAKER A.F. & BOLD H.C. The genus *Chlorococcum* Meneghini. *Phycol. Studies*. (U. Texas Publ. 7015) 105pp.

BERNS D.S. (1967) Immunochemistry of biliproteins. *Pl. Physiol., Lancaster* **42**, 1569–86.

BAILEY R.W. & STAEHELIN L.A. (1968) The chemical composition of isolated cell walls of *Cyanidium caldarium. J. gen. Microbiol.* **54**, 269–76.

BLUMER M., GUILLARD R.R.L. & CHASE T. (1971) Hydrocarbons of marine phytoplankton. *Mar. Biol.* **8**, 183–9.

BONEY A.D. (1972) Fluorescent substances from *Acrochaetium* species (Nemaliales, subclass Florideophycidae, Rhodophyceae). *J. nat. Hist.* **6**, 47–53.

BROWN R.L. & WALNE P.L. (1967) Comparative immunology of selected wild types, varieties and mutants of *Chlamydomonas. J. Protozool.* **14**, 365–73.

CAIN J. (1965) Nitrogen utilization in 38 freshwater chlamydomonad algae. *Can. J. Bot.* **43**, 1367–78.

CARR N.G. (1966) The occurrence of poly-β-hydroxybutyrate in the blue-green alga, *Chlorogloea fritschii. Biochim. biophys. Acta* **120**, 308–10.

CARR N.G. & CRAIG I.W. (1970) The relationship between bacteria, blue-green algae and chloroplasts. In *Phytochemical Phylogeny*, ed. Harborne J.B., pp. 119–43. Academic Press, New York & London.

CARR N.G., EXELL G., FLYNN V., HALLAWAY M. & TALUKDAR S. (1967) Minor quinones of some Myxophyceae. *Archs Biochem. Biophys.* **120**, 503–7.

CARR N.G. & WHITTON B.A. (1973) eds. *The Biology of Blue-Green Algae*. Blackwell, Oxford.

CASSELTON P.J. & STACEY J.L. (1969) Observations on the nitrogen metabolism of *Prototheca* Krüger. *New Phytol.* **68**, 731–49.

CASTENHOLZ R.W. (1970) Laboratory culture of thermophilic cyanophytes. *Schweiz. Z. Hydrol.* **32**, 538–51.

CHAPMAN D.J. (1966) The pigments of the symbiotic algae (cyanomes) of *Cyanophora paradoxa* and *Glaucocystis nostochinearum* and two Rhodophyceae, *Porphyridium aerugineum* and *Asterocytis ramosa. Arch. Mikrobiol.* **55**, 17–25.

CHRISTENSEN T. (1966) Botanik. II. Systematisk Botanik 2. 128–46. Munksgaard, Copenhagen.

CHUECAS L. & RILEY J.P. (1969) Component fatty acids of the total lipids of some marine phytoplankton. *J. mar. biol. Ass. U.K.* **49**, 97–116.

CLARK R.C. & M. BLUMER (1967) Distribution of n-paraffins in marine organisms and sediments. *Limnol. Oceanogr.* **12**, 79–87.

DAWES C.J. (1969) *Saprochaete saccharophila*: ultrastructure, X-ray diffractions and chitin assay of cell walls as aids in evaluating taxonomic position. *Trans. Am. microsc. Soc.* **88**, 572–81.

DEASON T.R. & COX E.R. (1971) The genera *Spongiococcum* and *Neospongiococcum*. 2. Species of *Neospongiococcum* with labile walls. *Phycologia* **10**, 255–62.

DE SOUZA N.J. & NES W.R. (1969) The presence of phytol in brown and blue-green algae and its relationship to evolution. *Phytochem.* **8**, 819–22.

DROOP M.R. & McGILL S. (1966) The carbon nutrition of some algae: the inability to utilize glycollic acid for growth. *J. mar. biol. Ass. U.K.* **46**, 679–84.

DROUET F. (1963) Ecophenes of *Schizothrix calcicola* (Oscillatoriaceae). *Proc. Acad. nat. Sci. Philad.* **115**, 261–81.

DROUET F. (1968) Revision of the classification of the Oscillatoriaceae. *Acad. nat. Sci. Philad.* Monograph 15, 370pp.

DROUET F. & DAILY W.A. (1956) Revision of the coccoid Myxophyceae. *Butler Univ. Bot. Studies* **12**, 218pp.

DUCKWORTH M. & YAPHE W. (1971a, b) The structure of agar. I. Fractionation of a complex mixture of polysaccharides. II. The use of a bacterial agarase to elucidate structural features of the charged polysaccharides in agar. *Carbohyd. Res.* **16**, 189–97, 435–45.

FALK H. & KLEINIG H. (1968) Feinbau und Carotenoide von *Tribonema* (Xanthophyceae). *Arch. Mikrobiol.* **61**, 347–62.

FELDMANN J. (1946) Sur l'hétéroplastie de certaines Siphonales et leur classification. *C. r. hebd. Séanc. Acad. Sci. Paris* **222**, 752–3.

FELDMANN J. (1961) Sur quelques problèmes biochimiques posés par l'étude cytologique des algues rouges. *Colloques Int. du C. N. R. S.* (No. 103), ed. Heim R., pp. 17–24.

FOGG G.E., STEWART W.D.P., FAY P. & WALSBY A.E. (1973) *The Blue-green Algae*. Academic Press, London & New York.

FOTT B. & NOVÁKOVÁ M. (1969) A monograph of the genus *Chlorella*: the fresh water species. In *Studies in Phycology*, ed. Fott B., pp. 10–74. Academia, Prague.

FRANKER C.K. (1970) Some properties of DNA from zooxanthellae harbored by an anemone *Anthopleura elegantissima*. *J. Phycol.* **6**, 299–305.

FREDRICK J.F. & KLEIN R.M. (1970) (eds.) Phylogenesis and morphogenesis in the algae. *Ann. N.Y. Acad. Sci.* **175**, 413–781.

GELPI E., SCHNEIDER H., MANN J. & ORÓ J. (1970) Hydrocarbons of geochemical significance in microscopic algae. *Phytochem.* **9**, 603–12.

GIBBONS G.F., GOOD L.J. & GOODWIN T.W. (1967) The sterols of some marine red algae. *Phytochem.* **6**, 677–83.

GIBBS M., LATZKO E., HARVEY M.J., PLAUT Z. & SHAIN Y. (1970) Photosynthesis in the algae. *Ann. N.Y. Acad. Sci.* **175**, 541–45.

GOODAY G.W. (1970) A physiological comparison of the symbiotic alga *Platymonas convolutae* and its free-living relatives. *J. mar. biol. Ass. U.K.* **50**, 199–208.

GOODMAN M., BARNABAS J., MATSUDA G. & MOORE G.W. (1971) Molecular evolution in the descent of man. *Nature, Lond.* **233**, 604–13.

GOODWIN T.W. (1971) Algal carotenoids. In *Aspects of Terpenoid Chemistry and Biochemistry* ed. Goodwin T.W., pp. 315–56. Academic Press, New York & London.

GOTELLI I.B. & CLELAND R. (1968) Differences in the occurrence and distribution of hydroxyproline proteins among the algae. *Am. J. Bot.* **55**, 907–14.

HAN J. & CALVIN M. (1969) Hydrocarbon distribution of algae and bacteria and microbiological activity in sediments. *Proc. natn. Acad. Sci. U.S.A.* **64**, 436–43.

HANIC L.A. & CRAIGIE J.S. (1969) Studies on the algal cuticle. *J. Phycol.* **5**, 89–102.

HARBORNE J.B. (1970) Phytochemical Phylogeny. *Proc. Phytochem. Soc. Sym.*, Bristol, 1969. Academic Press, New York & London.

HARTMANN T. (1972) Leucin-carboxylase aus marinen Rhodophyceae: Vorkommen, Verbreitung und einige Eigenschaften. *Phytochem.* **11**, 1327–36.

HATTORI A. & FUJITA Y. (1959a) Formation of phycobilin pigments in a blue-green alga, *Tolypothrix tenuis*, as induced by illumination with colored lights. *J. Biochem., Tokyo* **46**, 521–4.

HATTORI A. & FUJITA Y. (1959b) Effect of pre-illumination on the formation of phycobilin pigments in a blue-green alga, *Tolypothrix tenuis*. *J. Biochem., Tokyo* **46**, 1259–61.

HEALEY F.P. (1970) Hydrogen evolution by several algae. *Planta* **91**, 220–6.

HIROSE H. & KUMANO S. (1966) Comparison of phycobilins of 44 Rhodophyceae and *Phormidium*. *Bot. Mag. Tokyo* **79**, 105–13.

HIROSE H., KUMANO S. & MADONO K. (1969) Spectroscopic studies on phycoerythrins from cyanophycean and rhodophycean algae with special reference to their phylogenetical relations. *Bot. Mag. Tokyo* **82**, 197–203.

HOLM-HANSEN O. (1968) Ecology, physiology, and biochemistry of blue-green algae. *A. Rev. Microbiol.* **22**, 47–70.

HOLM-HANSEN O. (1969) Algae: amounts of DNA and organic carbon in single cells. *Science, N.Y.* **163**, 87–8.

HOLTON R.W. & BLECKER H.H. (1970) Fatty acids of blue-green algae. In *Properties and Products of Algae*, ed. Zajic J.E., pp. 115–27. Plenum Press, New York.

HOWLAND G.P. & RAMUS J. (1971) Analysis of blue-green and red algal ribosomal RNA's by gel electrophoresis. *Arch. Mikrobiol.* 76, 292–8.

HUNTSMAN S.A. & SLONEKER J.H. (1971) An exocellular polysaccharide from the diatom *Gomphonema olivaceum. J. Phycol.* 7, 261–4.

IKAWA M., BOROWSKI P.T. & CHAKRAVARTI A. (1968) Choline and inositol distribution in algae and fungi. *Appl. Microbiol.* 16, 620–3.

IRIE T., IZAWA M. & KUROSAWA E. (1970a) Laureatin and isolaureatin, constituents of *Laurencia nipponica* Yamada. *Tetrahedron* 26, 851–70.

IRIE T., SUZUKI M., KUROSAWA E. & MASAMUNE T. (1970b) Laurinterol, debromolaurinterol and isolaurinterol, constituents of *Laurencia intermedia* Yamada. *Tetrahedron* 26, 3271–7.

IWAMURA T. (1970) DNA species in algae. *Ann. N.Y. Acad. Sci.* 175, 488–510.

IRIKI Y. & TSUCHIYA Y. (1963) Constituents of the cell wall of *Batrachospermum virgatum* from Shiga Heights. *Bull. Inst. Nat. Educ. (Shinshu Univ.)* 2, 1–8.

JAMES A.T. & NICHOLS B.W. (1966) Lipids of photosynthetic systems. *Nature*, 210, 372–5.

JAMIESON G.R. & REID E.H. (1972) The component fatty acids of some marine algal lipids. *Phytochem.* 11, 1423–32.

JEFFREY S.W. (1968) Pigment composition of Siphonales algae in the brain coral *Favia. Biol. Bull. mar. biol. Lab., Woods Hole* 135, 141–8.

JEFFREY S.W. (1969) Properties of 2 spectrally different components in chlorophyll c preparations. *Biochim. biophys. Acta* 177, 456–67.

JENSEN S.L. (1970) Developments in the carotenoid field. *Experientia* 26, 697–710.

KAYE A.M., SALOMON R. & FRIDLENDER B. (1967) Base composition and presence of methylated bases in DNA from a glue-green alga *Plectonema boryanum. J. mol. Biol.* 24, 479–83.

KENYON C.N., RIPPKA R. & STANIER R.Y. (1972) Fatty acid composition and physiological properties of some filamentous blue-green algae. *Arch. Mikrobiol.* 83, 216–36.

KENYON C.N. & STANIER R.Y. (1970) Possible evolutionary significance of polyunsaturated fatty acids in blue-green algae. *Nature, Lond.* 227, 1164–6.

KESSLER E. (1967) Physiologische und biochemische Beiträge zur Taxonomie der Gattung *Chlorella*. Merkmale von 8 autotrophen Arten. *Arch. Mikrobiol.* 55, 346–57.

KESSLER E. & CZYGAN F.-C. (1967) Physiologische und biochemische Beiträge zur Taxonomie der Gattungen *Ankistrodesmus* und *Scenedesmus*. I. Hydrogenase, Sekundär-Carotinoide und Gelatine-Verflüssigung. *Arch. Mikrobiol.* 55, 320–6.

KESSLER E. & CZYGAN F.-C. (1970) Physiologische und biochemische Beiträge zur Taxonomie der Gattung *Chlorella*. IV. Verwertung organischen Stickstoff-verbindungen. *Arch. Mikrobiol.* 70, 211–16.

KESSLER E. & ZWEIER I. (1971) Physiologische und biochemische Beiträge zur Taxonomie der Gattung *Chlorella*. V. Die auxotrophen und mesotrophen Arten. *Arch. Mikrobiol.* 79, 44–8.

KLEIN R.M. (1970) Relationships between blue-green and red algae. *Ann. N.Y. Acad. Sci.* 175, 623–33.

KLEIN R.M. & CRONQUIST A. (1967) A consideration of the evolutionary and taxonomic significance of some biochemical, micromorphological and physiological characters in the thallophytes. *Q. Rev. Biol.* 42, 105–295.

KLEINIG H. (1969) Carotenoids of siphonous green algae: a chemotaxonomic study. *J. Phycol.* 5, 281–4.

KOMAREK J. (1970) Generic identity of the '*Anacystis nidulans*' strain Kratz-Allen/Bloom. 625 with *Synechococcus* Näg. 1849. *Arch. Protistenk.* 112, 343–64.

KOWALLIK K.V. (1971) The use of proteases for improved presentations of DNA in chromosomes and chloroplasts of *Prorocentrum micans. Arch. Mikrobiol.* 80, 154–65.

LEE R.F. & LOEBLICH A.R. (1971) Distribution of 21:6 hydrocarbon and its relationship to 22:6 fatty acid in algae. *Phytochem.* 10, 593–602.

LEEDALE G.F. (1969) Observations on endonuclear bacteria in euglenoid flagellates. *Öst. bot. Z.* **116**, 279–94.

LEWIN J. & GUILLARD R.R.L. (1963) Diatoms. *A. Rev. Microbiol.* **17**, 373–414.

LEWIN R.A. (1956) Extracellular polysaccharides of green algae. *Can. J. Microbiol.* **2**, 665–72.

LEWIN R.A. (1962) *Physiology and Biochemistry of Algae*. Academic Press, New York & London.

LEWIN R.A. (1968) Biochemistry and physiology of algae: taxonomic and phylogenetic considerations. In *Algae, Man and the Environment*, ed. Jackson D.F., pp. 15–26. Syracuse University Press.

LLOYD D. & TURNER G. (1968) The cell wall of *Prototheca zopfii*. *J. gen. Microbiol.* **50**, 421–7.

LOEBLICH A.R. III. (1966) Aspects of the physiology and biochemistry of the Pyrrophyta. *Phykos* **5**, 216–55.

LOENING U.E. (1968) Molecular weights of ribosomal RNA in relation to evolution. *J. molec. Biol.* **38**, 355–65.

MAEDA M., KOSHIKAWA M., NISIZAWA K., & TAKANO K. (1966a) Cell wall constituents, especially pectic substance of a marine phanerogam *Zostera marina*. *Bot. Mag. Tokyo* **79**, 422–6.

MAEDA M., KURODA K., IRIKI Y., CHIHARA M., NISIZAWA K. & MIWA T. (1966b) Chemical nature of major cell wall constituents of *Vaucheria* and *Dichotomosiphon* with special reference to their phylogenetic positions. *Bot. Mag. Tokyo* **79**, 634–43.

MAEDA M. & NISIZAWA K. (1968a) Fine structure of laminaran of *Eisenia bicyclis*. *J. Biochem., Tokyo* **63**, 199–206.

MAEDA M. & NISIZAWA K. (1968b) Laminaran of *Ishige okamurai*. *Carbohyd. Res.* **7**, 97–9.

MAKINO F. & TSUZUKI J. (1971) Absence of histone in the blue-green alga *Anabaena cylindrica*. *Nature, Lond.* **231**, 446–7.

MARGULIS L. (1970) *Origin of Eukaryotic Cells*. Yale Univ. Press.

McCULLY M.E. (1970) The histological localization of the structural polysaccharides of seaweeds. *Ann. N.Y. Acad. Sci.* **175**, 702–11.

McLACHLAN J. & CRAIGIE J.S. (1967) Photosynthesis in algae containing chlorophyll b. *Br. phycol. Bull.* **3**, 408–9.

MEEUSE B.J.D. (1963) Inulin in the green alga, *Batophora*. J. Ag. (Dasycladales). *Acta. bot. neerl.* **12**, 315–18.

MERESCHKOWSKY C. (1905) Ueber Natur und Ursprung den Chromatophoren in Pflanzenreich. *Biol. Centralbl.* **25**, 593–604.

MIWA T., IRIKI Y. & SUZUKI T. (1961) Chimie et physico-chimie des principes immediats tirés des algues. *Colloques Internationaux du C. N. R. S.* **103**, 135–44.

MÜLLER D.G. (1967) Ein leicht flüchtiges Gyno-Gamon der Braunalge *Ectocarpus siliculosus*, *Naturwissenschaften* **18**, 496–7.

MÜLLER D.G., JAENICKE L., DONIKE M. & AKINTOBI T. (1971) Sex attractant in a brown alga: chemical structure. *Science, N.Y.* **171**, 815–17.

NAGASHIMA H., NAKAMURA S. & NISIZAWA K. (1969) Isolation and identification of low molecular weight carbohydrates from a red alga, *Serraticardia maxima*. *Bot. Mag. Tokyo* **82**, 379–81.

NICHOLS B.W. (1970) Comparative lipid biochemistry of photosynthetic organisms. IV. Algae. In *Phytochemical Phylogeny*, ed. Harborne J.B., pp. 109–15. Academic Press, New York & London.

NISIZAWA K., FUJIBAYASHI S. & HABE H. (1969) On the fine structure of alginic acid. *Proc. 6th Int. Seaweed Symp.* pp. 553–65.

O'CARRA P. (1965) Purification and N-terminal analyses of algal biliproteins. *Biochem. J.* **94**, 171–4.

PAASCHE E. (1968) Biology and physiology of coccolithophorids. *A. Rev. Microbiol.* **22**, 71–86.

PALLA J-C., MILLE G. & BUSSON F. (1970) Étude comparée des carotenoïdes de *Spirulina*.

*platensis* (Gom.) Geitler et de *Spirulina geitleri* J. de Toni (Cyanophycées). *C. r. hebd. Séanc. Acad. Sci., Paris* **270**, 1038–41.

PALMER E.G. & STARR R.C. (1971) Nutrition of *Pandorina morum. J. Phycol.* **7**, 85–9.

PARKER B.C. (1970) Significance of cell wall chemistry to phylogeny in the algae. *Ann. N.Y. Acad. Sci.* **175**, 417–28.

PARKER P.L., VAN BAALEN C. & MAURER L. (1967) Fatty acids in 11 species of blue-green algae: geochemical significance. *Science, N.Y.* **155**, 707–8.

PATTERSON G.W. (1967a) The effect of culture conditions on the hydrocarbon content of *Chlorella vulgaris. J. Phycol.* **3**, 22–3.

PATTERSON G.W. (1967b) Sterols in *Chlorella*. II. The occurrence of an unusual sterol mixture in *Chlorella vulgaris. Plant Physiol., Lancaster* **42**, 1457–9.

PERCIVAL E. (1968) Marine algal carbohydrates. *Oceanogr. Mar. Biol. A. Rev.* **6**, 137–161.

PERCIVAL E. & MCDOWELL R.H. (1967) *Chemistry and Enzymology of Seaweed Polysaccharides*. Academic Press, New York & London.

PERCIVAL E. & MIAN A.J. (1971) Fucose-containing polysaccharides of the Phaeophyceae. *Abs. VII Int. Seaweed Symp.* p. 98.

PERCIVAL E. & YOUNG M. (1971) Low-molecular-weight carbohydrates and water-soluble polysaccharides metabolized by the Cladophorales. *Phytochem.* **10**, 807–12.

PIERSON B.K. & CASTENHOLZ R.W. (1971) Bacteriochlorophylls in gliding filamentous prokaryotes from hot springs. *Nature, Lond.* **233**, 25–7.

PIGOTT G.H. & CARR N.G. (1972) Homology between nucleic acids of blue-green algae and chloroplasts of *Euglena gracilis. Science, N.Y.* **175**, 1259–61.

POWLS R. & REDFEARN E.R. (1967) The tocopherols of the blue-green algae. *Biochem. J.* **104**, 24c–6c.

PRINGSHEIM E.G. (1968) Zur Kenntnis der Cryptomonaden des Süsswassers. *Nova Hedwigia* **16**, 367–401.

PROVASOLI L., YAMASU T. & MANTON I. (1968) Experiments on the resynthesis of symbiosis in *Convoluta roscoffensis* with different flagellate cultures. *J. mar. biol. Assoc. U.K.* **48**, 465–79.

RAFTERY M.A. & O'HEOCHA C. (1965) Amino acid composition and C-terminal residues of algal biliproteins. *Biochem. J.* **94**, 166–70.

RAMUS J. (1971) Properties of septal plugs from the red alga *Griffithsia pacifica. Phycologia* **10**, 99–103.

RICKETTS T.R. (1967) Further investigations into the pigment composition of green flagellates possessing scaly flagella. *Phytochem.* **6**, 1375–86.

SATO C.T. & TOMOCHIKA K. (1969) Immunological relationships among strains of *Chlamydomonas. Jap. J. Genet.* **44**, 241–6.

SCHLICHTING H.E. & BRUTON B.A. (1970) Some problems of pleomorphism in algal taxonomy. *Lloydia* **33**, 472–6.

SCHNEIDER H., GELPI E., BENNETT E.O. & ORÓ J. (1970) Fatty acids of geochemical significance in microscopic algae. *Phytochem.* **9**, 613–17.

SCHNEPF E. & BROWN R.M. (1969) On relationships between endosymbionts and the origin of plastids and mitochondria. In *Origin and development of cell organelles*, eds. Ursprung W. & Reinert C., pp. 299–322, Springer Verlag.

SCHNEPF E., KOCH W. & DEICHGRÄBER G. (1966) Zur Cytologie und taxonomischen Einordnung von *Glaucocystis. Arch. Mikrobiol.* **55**, 149–74.

SCHWEIGER R.G. (1967) Low molecular weight compounds in *Macrocystis pyrifera*, a marine alga. *Arch. Biochem. Biophys.* **118**, 383–7.

SHESTAKOV S.V. & NGUEN T.K. (1970) Evidence for genetic transformation in blue-green alga *Anacystis nidulans. Molec. gen. Genetics* **107**, 372–5.

SIEBURTH J.McN. (1964) Antibacterial substances produced by marine algae. *Devs. ind. Microbiol.* **5**, 124–34.

SIEGELMAN H.W., CHAPMAN D.J. & COLE W.J. (1968) The bile pigments of plants. *Biochem. Soc. Symp.* **28**, 107–20, ed. Goodwin T.W., Academic Press, New York & London.

SIMS J.J., FENICAL W., WING R.M. & RADLICK W. (1971) Marine natural products. I. Pacifenol, a rare sesquiterpene containing bromine and chlorine from the red alga *Laurencia pacifica. J. am. Chem. Soc.* **93**, 3774–5.

SOMA S. (1960) Chlorophyll in *Vaucheria* as a clue to the determination of its phytogenetic position. *J. Fac. Sci. Toyko Univ.* Sec. 3 (Bot.) **7**, 535–42.

SORIANO S. & LEWIN R.A. (1965) Gliding microbes: some taxonomic reconsiderations. *Antonie van Leeuwenhoek* **31**, 66–80.

SOURNIA P.A. (1968) La Cyanophycée *Oscillatoria* (= *Trichodesmium*) dans le plancton marin : Taxinomie, et observations dans le Canal de Mozambique. *Nova Hedwigia.* **15**, 1–12.

SPRUNG D.C. & WUJEK D.E. (1971) Chemotaxonomic studies of *Pleurastrum* Chodat by means of pyrolysis-gas-liquid chromatography. *Phycologia* **10**, 251–4.

STANIER R.Y., KUNISAWA R., MANDEL M. & COHEN-BAZIRE G. (1971) Purification and properties of unicellular blue-green algae (order Chroococcales). *Bact. Rev.* **35**,171–205.

STEIN J.R. (1965) Sexual populations of *Gonium pectorale* (Volvocales). *Am. J. Bot.* **52**, 379–88.

STEIN J.R. (1966a) Growth and mating of *Gonium pectorale* (Volvocales) in defined media. *J. Phycol.* **2**, 23–8.

STEIN J.R. (1966b) Effect of temperature on sexual populations of *Gonium pectorale* (Volvocales). *Am. J. Bot.* **53**, 941–4.

STEINER M. & HARTMANN T. (1968) Ueber Vorkommen und Verbreitung flüchtiger Amine bei Meeresalgen. *Planta* **79**, 113–21.

STOLOFF L. & SILVA P. (1957) An attempt to determine possible taxonomic significance of the properties of water-extractable polysaccharides in red algae. *Econ. Bot.* **11**, 327–30.

STRAIN H.H. (1958) *Chloroplast Pigments and Chromatographic Analysis.* Penn. Univ. Press

STRAIN H.H. (1965) Chloroplast pigments and the classification of some Siphonalean green algae of Australia. *Biol. Bull.* **129**, 365–70.

SUN E., BARR R. & CRANE F.L. (1968) Comparative studies on plastoquinones. IV. Plasto-quinones in algae. *Pl. Physiol., Lancaster* **43**, 1935–40.

TAKEDA H., NISIZAWA K. & MIWA T. (1968) A xylomannan from the cell wall of *Prasiola japonica* Yatobe. *Sci. Rept. Tokyo Kyoiku Daigaku.* Sec. B, **13**, 183–98.

THOMAS D.L. & BROWN R.M. (1970) New taxonomic criteria in the classification of *Chlorococcum* species. III. Isozyme analysis. *J. Phycol.* **6**, 293–9.

THOMPSON E.W. & PRESTON R.D. (1967) Proteins in the cell walls of some green algae. *Nature, Lond.* **213**, 684–5.

TRAINOR F.C. (1970) Algal morphogenesis: nutritional factors. *Ann. N.Y. Acad. Sci.* **175**, 749–56.

TONG W. & CHAIKOFF I.L. (1955) Metabolism of I[131] by the marine alga, *Nereocystis luetkeana. J. biol. Chem.* **215**, 473–84.

TURVEY J.R. & WILLIAMS E.L. (1970) The structures of some xylans from red algae. *Phytochem.* **9**, 2383–8.

UHERKOVICH G. (1966) Die *Scenedesmus*-Arten Ungarns. Akad. Kiado, Budapest, 173pp.

VOGEL H.J. (1965) Lysine biosynthesis and evolution. In *Evolving Genes and Proteins*, eds. Bryson V. & Vogel H.J., pp. 25–40. Academic Press, New York & London.

WAGNER G., TSANG M.L-S., SCHIFF J.A. & LOEWUS F. (1971) Studies on the incorporation of sulphate into carrageenan in *Chondrus. Biol. Bull. mar. biol. Lab., Woods Hole* **141**, 405.

WEBSTER D.A. & HACKETT D.O. (1965) Respiratory chain of colorless algae. I. Chlorophyta and Euglenophyta. *Pl. Physiol., Lancaster* **40**, 1091–100.

WHYTE J.N.C. & ENGLAR J.R. (1971) Polysaccharides of the red alga *Rhodymenia pertusa.* II. Cell-wall glucan; proton magnetic resonance studies on permethylated polysaccharides. *Can. J. Chem.* **49**, 1302–5.

WIESE L. (1969) Algae. In *Fertilization*, eds. Metz C.B. & Monroy A., pp. 135–88. Academic Press, New York & London.

WIESSNER W. & FRENCH C.S. (1970) The forms of native chlorophyll in *Chlamydobotrys stellata* and their changes during adaptation from heterotrophic to autotrophic growth. *Planta* **94**, 78–90.

WILLARD J.J. & GIBBS M. (1968) Purification and characterization of the FUDP aldolases from *Anacystis nidulans* and *Saprospira thermalis*. *Biochim. biophys. Acta* **151**, 438–48.

WINTERS K., PARKER P.L. & VAN BAALEN C. (1969) Hydrocarbons of blue-green algae: geochemical significance. *Science, N.Y.* **163**, 457–68.

YAPHE W. (1959) The determination of K-carrageenan as a factor in the classification of the Rhodophyceae. *Can. J. Bot.* **37**, 751–7.

YAPHE W. & DUCKWORTH M. (1972) The relationship between structures and biological properties of agars. *Proc. 7th Int. Seaweed Symp.* 15–22.

YENTSCH C.S. & GUILLARD R.R.L. (1969) The absorption of chlorophyll b *in vivo*. *Photochem. Photobiol.* **9**, 385–8.

YOUNG K., DUCKWORTH M. & YAPHE W. (1971) The structure of agar. III. Pyruvic acid, a common feature of agars from different agarophytes. *Carbohydrate Res.* **16**, 446–8.

YOUNGBLOOD W.W., BLUMER M. GUILLARD R.R.L. & FIORE F. (1971) Saturated and unsaturated hydrocarbons in marine benthic algae. *Mar. Biol.* **8**, 190–201.

CHAPTER 2

# CELL WALL AND INTERCELLULAR REGION POLYSACCHARIDES

W. MACKIE and R. D. PRESTON

Astbury Department of Biophysics,
University of Leeds,
Leeds, U.K.

1    Introduction  40

2    Organization of cell walls  42
2.1  General  42
2.2  *Valonia*-type walls  44
2.3  Other wall types  46

3    Cellulose  48

4    Mannan  53

5    Xylans  55

6    Alginic acid  58

7    Sulphated polysaccharides  63
7.1  Galactans  of  the  Rhodo-
     phyceae  63
       (a) Agar  64
       (b) Porphyran  65
       (c) Carrageenan  66
       (d) Other galactans  69

7.2  Sulphated  polysaccharides  of
     the Phaeophyceae  72
7.3  Sulphated  polysaccharides  of
     the Chlorophyceae  73
       (a) Galactoarabinoxylans
           73
       (b) D-Glucuronoxylorham-
           nans  74

8    Miscellaneous polysaccharides
     75

9    Conclusion  75

10   References  76

## 1  INTRODUCTION

Polysaccharide products, on account of their physical properties, have long been used by man in a variety of practical applications, for example as structural materials, protective coverings, adhesives, hydrated gels and viscous solutions, and more recently in the laboratory, for ion-exchange and gel filtration chromatography. Undoubtedly, these are the same properties which are so important in the biological functions of polysaccharides. Marine algae are particularly interesting with regard to the range and diversity of polysaccharide structures (many of which are rarely, if ever, encountered elsewhere) which are used for these purposes. Some of their most common monosaccharide constituents are shown in Fig. 2.1.

In this review, we have not attempted to provide a comprehensive catalogue of algal polysaccharides for many of their commercial uses, their chemical structures and the methods used for elucidating them have been extensively covered elsewhere (Kreger 1962, O'Colla 1962, Peat & Turvey 1965, Percival & McDowell 1967, Percival 1970). Instead, we have tried mainly to bring together the fruits of some recent botanical, chemical and physical advances in the study

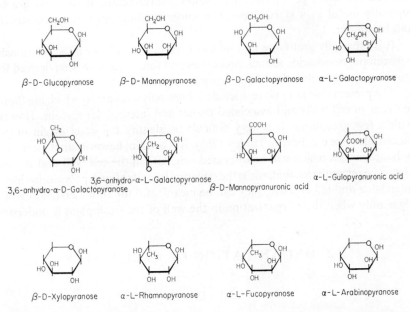

Fig. 2.1. Some monosaccharide constituents in algal polysaccharides. Variations include the occurrence of Ester Sulphate ($-OSO^-_3$) and Methyl Ethers ($-OCH3$) on some Hydroxyl groups.

of algae to show how these are helping us to understand not only the structure and properties but also the functions of polysaccharide molecules and for this purpose we have restricted the discussion almost exclusively to members of the Rhodophyceae, Phaeophyceae and Chlorophyceae.

Nowadays it is recognized that polysaccharide molecules have hierarchies of structure analogous to the primary, secondary, tertiary and quaternary structures of other macromolecules, and it is accepted that the complete understanding of their physical properties and biological functions depends on understanding the structures of polysaccharides at all these levels.

A great deal of effort has gone (and is still going) into the elucidation of polysaccharide primary structure, that is the nature, sequence and linkage of the monosaccharides. It has also become generally recognized, through experimental and theoretical methods, that the secondary structures or shapes of polysaccharides are principally determined by the geometry of their linking regions— the glycosidic linkages (Rees 1969a, Rees & Scott 1971 and references within). To a first approximation, polysaccharides may be flat and ribbon-like, buckled, crumpled or helically coiled, according to their linking geometry, and it has also been pointed out that, to some extent, there appears to be a correlation between the biological functions and molecular shapes of polysaccharides. This is not to say that the shapes of polysaccharides are immutably fixed and do not vary between the solid, gel and solution states. Nevertheless, these ideas are conceptually useful as a starting point in understanding polysaccharide structure and behaviour.

It is especially gratifying that many of the current ideas on the relationships between polysaccharide shapes, properties and functions have been derived from studies of polysaccharides of algal origin.

The present discussion is restricted to those polysaccharides which are thought to occur in cell walls and associated tissues and intercellular regions. However, with a few exceptions, it is very difficult to identify the exact origin of polysaccharides (see e.g. Peat & Turvey 1965). It is certain, however, that many of the polysaccharides which will be discussed originate in the cell wall and it should always be remembered that, while the structural details of such polysaccharides are interesting and informative, they can be meaningful for an understanding of the algae only when their organization in the wall of the fresh plant is understood.

## 2   ORGANIZATION OF CELL WALLS

### 2.1   *General*

In the context of cell wall organization, only a few algae have so far been examined in any depth and even with these many details remain to be resolved.

In general, cell walls are composite materials and at least two components, a fibrillar one and a mucilaginous one can be identified. The latter is usually regarded as forming a non- or para-crystalline matrix in which the former (as a set, or sets, of microfibrils) is embedded, and there is evidence that, in some cases, these components occur in part as alternating layers as in a sandwich (see e.g. Hanic & Craigie 1969). The microfibrils form the most inert and resistant part of the cell wall and are left behind in conditions which remove the greater proportion of the mucilaginous components. The best known and most common of these skeletal components is cellulose, a polymer of 1,4 linked β-D-glucose, although this fibrillar cellulose component is replaced in certain instances by polymers of 1,4 linked β-D-mannose and 1,3 linked β-D-xylose respectively (Frei & Preston, 1961a, Parker 1964). The occurrence of these different cell walls provides an interesting illustration of the principle that different chemical means may achieve similar physical results. This principle will be encountered again in other algal polysaccharides.

It is obvious that the derived structures of skeletal components such as cellulose depend entirely on observations made on material removed from its natural environment, and at least air-dried, and the question may properly be raised how far the derivation applies to the fresh wall. This question was answered for *Cladophora* (Preston *et al.* 1948) by showing that the normal X-ray diagram can be recorded from fresh material maintained in running sea water in the X-ray beam; and at the same time that the contrary finding for cotton hairs (Berkeley & Kerr 1946) was due to an artefact depending upon the construction of the hairs. There seems no reason to doubt that, in all algae, the cellulose is crystalline in the fresh wall and that it has the organization revealed in the electron microscope. Since fresh walls always show intrinsic birefringence there is in any case no doubt about the presence of crystallinity in some form. It follows that the demonstration that non-cellulosic substances, other than those firmly attached to the microfibrils, can be removed without effect upon the X-ray diagram of the constituent cellulose (see e.g. Cronshaw *et al.* 1958) implies that these substances surround the microfibrils but do not penetrate them. It seems likely that the microfibrils are bonded to the matrix, although the nature and distribution of the bonds is speculative. There is evidence that protein is usually also present in cell walls and so disulphide bonds may occur and some weight is currently being thrown on these as the bonds which may be important in controlling cell growth. Stress relaxation measurements in *Nitella, Penicillus* and *Acetabularia*, however (Haughton *et al.* 1968, Haughton & Sellen 1969, Cleland & Haughton 1971), seem to imply that the bonding may be, in the main, hydrogen bonding.

In those algae for which the relevant observations have been made, the microfibrils of the wall are more or less oriented, the orientation varying from a very dispersed preferred orientation to one almost exactly parallel. Different microfibrillar orientations lie in different lamellae and the lamellation may be fine or coarse. It seems to be the case that the walls with the finer lamellation are often those with the better orientation. In terms of wall organization, therefore, the algae will be considered in two groups separated by these criteria. It is a common occurrence among algae that the outer lamellae of the wall are much more encrusted and more tightly bound together than are inner lamellae. This is a feature which develops as the algae grow. The suggestion is clearly that non-skeletal polysaccharides or their precursors pass out through the wall and accumulate in the outer lamellae, in intercellular spaces or even diffuse into the aqueous environment. An especially clear case occurs in the rhizoids of *Porphyra*, whose walls are based upon a β-1,3 linked xylan, which becomes enveloped in a cylindrical sheath of mannan as they pass downward through the base of the plant (Frei & Preston 1964). This infiltration of outer lamellae may account for the so-called 'cuticle' of many algae.

## 2.2 *Valonia-type walls*

These are so named here because *Valonia* was the first of the type to be discovered. It was proved by polarization microscopy and by X-ray diffraction analysis, long before the advent of electron microscopy, that in the walls of *Valonia* (Preston & Astbury 1937), *Chaetomorpha* (Nicolai & Frey-Wyssling 1938) and *Cladophora* (Astbury & Preston 1940) the crystallites of cellulose lie mostly in two directions lying mutually at less than a right-angle and that the two orientations occurred in different lamellae. This has been amply confirmed by electron microscopy (Frei & Preston 1961b) (Fig. 2.2). The lamellation visible in the light microscope does not, however, correspond to this difference in orientation. The lamellae are alternately cellulosic and non-cellulosic and within each cellulosic lamella both orientations occur, in laminations visible in the electron microscope. In *Valonia* the two orientations detectable in any piece of wall, manifested in the light microscope by two striations, form parts of two helices, one shallow and one steep, running around the vesicle (Fig. 2.3). A third orientation was known to exist at that time, but its components were less abundant (Cronshaw & Preston 1958) and this is included in the model presented in Fig. 2.3. The walls of the Cladophorales have since that time been studied intensively (Frei & Preston 1961b) and it is clear that all the algae in this order possess the same type of wall. Recently it has been shown that two unicellular algae *Glaucocystis* (Schnepf 1965, Robinson & Preston 1971a) and *Oocystis* (Robinson & White 1972) though taxonomically unrelated to either *Valonia* or the Cladophorales, also fall into this group.

*Glaucocystis* and *Oocystis* resemble *Valonia* moreover in that the structural helices curl into two 'poles' at opposite ends of a diameter, defining an axis. The same applies also to the wall newly formed about the swarmers of *Chaetomorpha* (Nicolai 1957) and *Cladophora* (Frei unpubl.) in which the axis becomes a morphological axis on the outgrowth of the rhizoid from one 'pole'. In vegetative *Cladophora* and *Chaetomorpha* the cells are cylindrical with two flat ends and, although the helices on the side walls swirl in toward the centre from the edge of the cross walls, the orientation is then lost and over most of the cross wall the microfibrils lie at random. In the side walls there may be only two helices crossing at somewhat less than a right angle (*Chaetomorpha melagonium, Cladophora rupestris* in which, however, there is an occasional lamella with the third orientation) or three helices may be present (*Chaetomorpha princeps* and *Cladophora prolifera*) in which the third orientation, as with *Valonia*, is not so abundant. The walls constitute basically, therefore, a two-lamella repeat system with occasionally a third lamella with the third orientation.

The succession of deposition by the cytoplasm is slow, fast, followed sometimes by a third lamella, the signs of the helices being left-right-left-hand in *Chaetomorpha princeps* and in *Chaetomorpha melagonium* and *Cladophora rupestris* (where, however, the third orientation is missing) and right-left-left-hand in *Cladophora prolifera*, compared with left-left-right-hand in *Valonia*. The

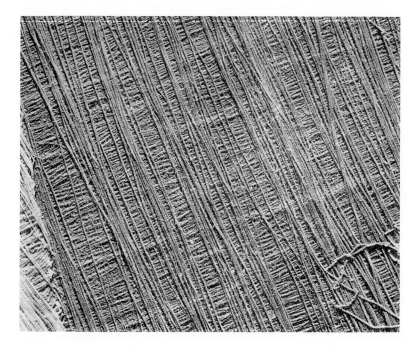

**Fig. 2.2.** Electron micrograph of inner lamellae of a side wall of *Chaetomorpha melagonium*, viewed from the outside. Shadowed Pd/Au, magnification 30,000. Cell axis ↕ (Frei & Preston 1961).

rhythm of deposition is fixed for each species and, in *Cladophora* and *Chaeto-morpha* persists through several cell divisions. The descendants of any one cell, however, eventually fall out of step.

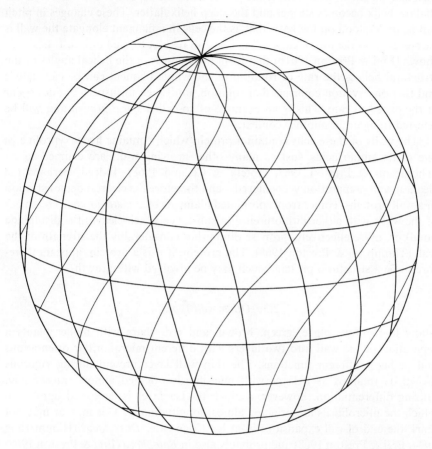

**Fig. 2.3.** Schematic representation of the run of the microfibrils in the wall of a vesicle of *Valonia ventricosa*. The vesicle is considered spherical, the helical lines representing microfibril directions.

In none of these plants is there a differentiation between primary and secondary wall layers such as occurs in higher plants. The first wall lamella laid down in swarmers of *Cladophora* and *Chaetomorpha* and in the aplanospores of *Valonia* contains sparse randomly oriented microfibrils, and is preceded in *Cladophora*, *Chaetomorpha*, *Glaucocystis* and *Oocystis* by a fibrous layer of unknown composition.

The walls of growing filaments of *Cladophora* and *Chaetomorpha* present excellent examples of the kind of growth called by Roelofsen multi-net growth

(Frei & Preston 1961b). The microfibrils of the steeper helix become progressively more nearly parallel and straight while those of the flatter set, if anything, show less order as the cells elongate. The outside lamellae are commonly torn and shrunken to collars at the positions of the septa. At the same time the shallow helix becomes steeper and the steep helix flatter. These changes in pitch can be understood on the basis that as the cells in a filament elongate the wall is 'stretched' and the top of each cell twists with respect to the bottom. It has been shown (Frei & Preston 1961b) that from the changes in the helical angles of the structural helices the rate of rotation of the whole filament can be calculated; and that observation confirms that rotation, until then unsuspected, does occur at the expected rate. This is an example of so-called *spiral growth* and will be referred to again in another context.

The walls of these cells contain a protein which contains hydroxyproline as one of its amino acids, just as many other growing walls are claimed to do (Thompson & Preston 1967, Gotelli & Cleland 1968). Indeed, perhaps the cleanest wall preparation yet achieved—and therefore the clearest demonstration of a wall protein—comes from sporulated filaments of *Chaetomorpha*. Treatment of the walls with either dithiothreitol (a substance which cleaves the disulphide bond) or, more effectively, pronase diminishes considerably the strength of the wall (Thompson & Preston 1968). The protein, if it is a protein, appears therefore to be a structural protein which may be involved with growth.

## 2.3   *Other wall types*

Many filamentous algae, green, brown, and red, share with higher plants a separation of the wall into two layers with different microfibrillar orientations and perhaps different functions. The thin wall layer present during vigorous growth is retained as a layer with microfibrils oriented almost transversely. During differentiation, however, a much thicker layer is deposited upon it in which the microfibrils are oriented almost longitudinally. This may or may not mark the end of cell expansion; in at least one case, *Dasycladus* (Hämmerling 1944, Frei & Preston 1968) and probably also in *Batophora* (Frei & Preston 1968) the cell proceeds to shrink. The implication is clear, however, that the primary/secondary layer differentiation of higher plants is foreshadowed in the algae.

The structure of the 'primary' wall and its relation to growth has been dealt with extensively with one alga, *Nitella*, in the internodal cells. The cellulosic nature of the wall of this plant was not firmly established until the work of Probine and Preston (1961) but, before that, Green (1958), assuming the presence of cellulose, had shown that tritium in the medium is incorporated only at, or very close to, the inner surface of the wall. Wall deposition is therefore, at least in part, by apposition. Subsequently Probine and Preston (1961) showed that the detailed architecture of the wall conforms with the multi-net growth hypothesis. The microfibrils of the innermost layer lie distributed about a slow helix, those of the outermost layer tend toward an axial alignment, while intermediate layers show intermediate orientation. Growth is distributed uniformly over the wall of these internodal cells and these show the phenomena of spiral growth already referred to above. These cells furnish elegant material for a study of the relation between cell growth and wall structure. In a series of papers (Probine & Preston 1962) a clear connexion has been worked out between cell growth and 'creep' in the wall under constant stress, the cylindrical shape of the cell resulting from anisotropy of creep stemming from the almost transverse orientation in the bulk of the wall. Both this work and the converse study of stress relaxation already referred to (p. 43) make it clear that the bonds which must be broken during both wall creep and cell growth must reside in the inter-microfibrillar polysaccharides. Comparison of the rate and duration of creep with the rate and duration of growth, point toward an involvement of the bio-chemical machinery of the cell in the yielding of the growing wall.

In many respects, the constituents of the amorphous regions present a more difficult problem, especially since there are few ways of studying them without removing them from the plant. Firstly, there is no evidence to suggest that their chemical composition throughout cell wall layers is constant. In fact, the available results suggest that this is not the case. Secondly, the methods of preparation of these components usually means that they are obtained in mixtures with polysaccharides not properly belonging to the cell wall such as those of inter-cellular regions and the middle lamella which themselves appear to have similar compositions. Thirdly, these substances are invariably far more chemically complex than are the skeletal components. They are built up from a wide range of sugar units and linkages and are often characterized by the presence of carboxyl groups (uronic acids) and/or sulphate half-esters. Their biological functions are also less apparent but, as with the skeletal compounds, these are more likely to be dependent on their physical properties than their chemical reactivity. The balance of evidence suggests that, together with the substances of the intercellular regions, they may well be multi-functional molecules. For example, they may serve directly as structural materials and also as cementing substances and it may be that they can be synthesized or structurally modified during growth in response to changes in local environmental requirements; for example, to prevent desiccation, to protect against tidal action or to control selectively ion and solute flow pathways. Current ideas suggest that these poly-saccharides may be far from 'structureless' in native cell walls and there is accumulating *in vitro* evidence that in some cases their biological functions may depend to a great extent on the formation of well defined hydrated networks which are built up in the gel state (see pp. 62, 70–2).

Finally, the involvement of algae in recent developments of ideas concerning the mechanisms of the biosynthesis of cell walls necessitates at least a brief con-sideration. A more detailed account of this subject appears elsewhere (Preston 1974a, b).

In both higher plants and algae the locus of synthesis differs between cellulose (and perhaps the other skeletal polysaccharides) and the other polysaccharides of the wall. As first proposed on theoretical grounds (Preston 1964) cellulose is now known to be synthesized by granules lying on the surface of the plasmalemma. This has recently been supported by observation of freeze-etched swarmers of *Cladophora* and *Chaetomorpha* (Robinson & Preston 1971b, Robinson *et al.* 1972) and *Oocystis* (Robinson & Preston 1972a). There is now strong evidence that the other site at one time proposed, namely in the location of the cytoplasmic microtubules, is no longer acceptable and most workers reject this possibility. In the case of the non-skeletal polysaccharides, the early claim (Mollenhauer *et al.* 1961) that synthesis occurs in Golgi bodies has been substantiated (Northcote & Pickett-Heaps 1966, Harris & Northcote 1971). Only with one organism, *Pleurochrysis scherffelii*, has a claim been made that cellulose is also synthesized in Golgi bodies (Brown *et al.* 1970). It is not clear, however, that the synthesized product, possibly 1,4 linked, is any more than a short chain glucan.

The biochemistry of polysaccharide synthesis (see Hassid 1970) has received a good deal of attention in recent years and it seems certain that nucleoside diphosphate sugars are involved as precursors. It also appears that the enzymes concerned are particulate, and the possible involvement of a lipid (Pinsky & Ordin 1969, Villemez & Clark 1969) suggests that the particles may be membrane bound. It had been concluded that GDP-D-glucose (guanosine-5-diphosphoglucose) is the precursor for β-1,4 linked polysaccharides including cellulose, and UDP-D-glucose (uridine-5-diphosphoglucose) for 1,3 linked polysaccharides (e.g. callose and laminaran), but other studies (Villemez & Heller 1970) have cast doubt on the chemical nature of the substances which are produced in some of these experiments, so that these points cannot be regarded as being settled. It has also been claimed that the *in vivo* synthesis of cellulose in cotton hairs from UDP-D-glucose agrees well with the *in vitro* experiments (Meier 1969). A more recent examination of the synthesized products by X-ray diffraction analysis (Robinson & Preston 1972b, in which references to the earlier work will be found) has failed to reveal any differences between the products from GDP-glucose and UDP-glucose and the evidence suggests that only low molecular weight oligosaccharides are being synthesized by the cell-free systems being used.

## 3  CELLULOSE

During the past few years the celluloses of marine algae have proved to be critical for a re-examination of the structure of cellulose. This is equally true for the structure at the level of the molecular lattice as for structure at the electron microscope level.

The unit cell of Meyer and Misch (1937) has dimensions $a = 0.834$nm, $b = 1.03$nm, $c = 0.79$nm, $\beta = 84°$ and contains two cellobiose residues. These occur as four residues along four of the '$b$' edges (each shared by four unit cells and therefore counting in total as one) and one lying centrally also parallel to the *b* axis. These are all, of course, segments of long cellulose chains which can be built up by stacking one unit cell above the other. The central chain was visualized by Meyer and Misch, for no very convincing reason, as lying upside down and translated along the *b* axis by a distance *b*/4 relative to the corner chains (which by definition must be identical). However, not only the size of the unit cell but also the details of chain ordering and conformation have repeatedly been called into question. Although it has long been believed that the 1,4 linked β-D-glucose units, which make up the cellulose polymer, have the C1 chair form so that they are diequatorially (1e, 4e) linked (see Sundaralingham 1968, for a discussion of this nomenclature), the conformation of the chain itself has frequently been disputed. It is now considered unlikely that the original conformation proposed by Meyer and Misch is correct and most workers now

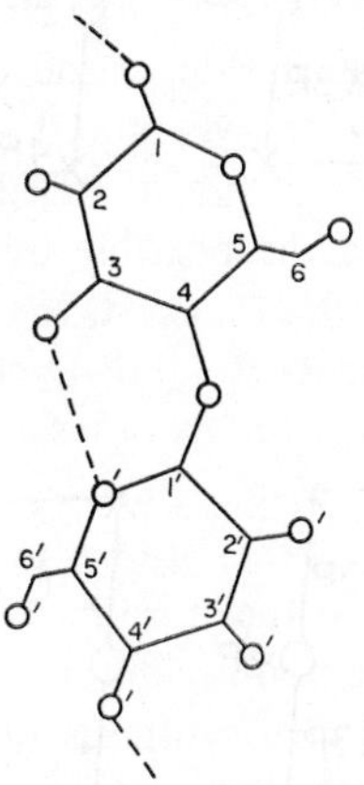

**Fig. 2.4.** Cellulose repeating segment in Hermans Bent Chain Conformation showing *intra*-molecular ($O_3 \ldots O'_5$) hydrogen bond. (Carbon atoms are numbered, hydrogen atoms not shown.)

favour a 'bent' chain conformation (Hermans 1949). This conformation incorporates both a two-fold screw axis (which is not strictly necessary) and an *intra* molecular hydrogen bond between —$O_3H$ and the ring oxygen atom ($O'_5$) of the next unit in the chain (see Fig. 2.4) for which there is polarized infra-red evidence (Mann & Marrinan 1958, Liang & Marchessault 1959). Conformational analysis calculations also favour a conformation close to the Hermans type (Rao *et al.* 1967, Rees & Skerrett 1968).

Using the cellulose of *Valonia* sp. it was early suggested (Honjo & Watanabe 1958, Fischer & Mann 1960) that the unit cell is four times larger than the Meyer-Misch cell and this was supported by the finding (Frei & Preston 1961b) in *Chaetomorpha melagonium* cellulose of a first layer line spacing in the X-ray diagram which could be indexed only on the larger cell. Such a cell would contain 8 cellobiose residues so that in the whole structure there should be 8, and not 2, chains, which in some way or other differ from each other. For this reason, and on account of conceptual difficulties in biosynthesis, the simple anti-parallel arrangement of the Meyer-Misch cell has also been re-examined. In spite, however, of more recent investigations, both theoretical and practical, it has turned out to be extraordinarily difficult to be sure whether all the chains lie parallel or whether some of them lie antiparallel.

More recently the X-ray data referring to the cellulose of *Chaetomorpha melagonium* in comparison with that of ramie (*Boehemeria nivea*) have been reassessed (Nieduszynski & Atkins 1970). Since cell planes lying parallel to chain

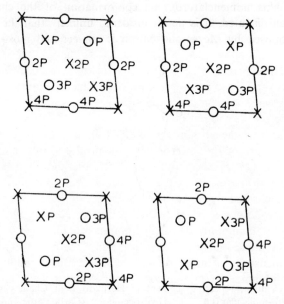

**Fig. 2.5.** Four eight-chain unit cells with positive translation (*plus* four cells with negative translations ($p = b/4$)). Circles and crosses represent chains in antiparallel arrangement.

length and yielding an X-ray reflection can be indexed on the one chain unit cell originally conceived by Sponsler and Dore, it was taken as a fixed point that all the chains have the same projection laterally. Using all the spacings available from this alga in a computer programme a unit cell with parameters $a = 1.643$nm; $b =$ (fibre axis) $= 1.033$nm; $c = 1.570$nm; $\beta = 96°\ 58'$ was deduced in conformity with the earlier workers (but, in harmony with crystallographic practice, using the complement of the angle in the Meyer-Misch cell). The placing of the 8 chains within this unit cell was then considered in terms of the above 'fixed point' so that rotation of chains about the chain axis is excluded, and assuming that each chain lies in the same environment. On these grounds 8 possible unit cells were found, all 8-chain unit cells, of which the basal projection is presented in Fig. 2.5. All of them contain sheets of cellulose chains lying in the same sense alternating with sheets of chains of opposite sense, so that the antiparallel structure is retained. No cell could be found with parallel chains which would yield a match for the known X-ray spacings and intensities.

Unfortunately, even with antiparallel chains the calculated intensities do not give a good overall fit with observed intensities for any of the 8 unit cells. Apart, therefore, from the confirmation that the unit cell must be an 8-chain unit cell, this leaves the structure of cellulose still uncertain. It seems likely that further progress may be made only by relaxing the restrictions covering chain rotation and chain environment, but then an X-ray diagram much richer than any so far obtained would be necessary to differentiate between the multitudinous structures that would then be possible. It remains, indeed, possible that cellulose chains are units which can form a whole range of structures between which the differences are rather subtle. The X-ray spacings of algal cellulose vary appreciably from one species to another particularly in the red algae (Frei & Preston unpubl.) just as similar variations are known with higher plants (Wellard 1954). This is perhaps why the diagrams are not richer. The search for a unique structure could therefore be without point.

The celluloses of algae have proved of equal significance in attempts to define the internal structure of the electron microscopically visible microfibrils (Fig. 2.2), and are central to the current dispute concerning the existence of subunits, 3·5 nm in diameter, first called by Frey-Wyssling *elementary fibrils*. When algal celluloses are separated from the walls by currently accepted extraction procedures the hydrolysates of the extracted celluloses commonly contain sugars other than glucose, commonly xylose, in amounts ranging up to 50% (Cronshaw *et al.* 1958). The algae *Valonia*, *Cladophora* and *Chaetomorpha* are unusual in yielding only glucose under these circumstances. The consideration that the polysaccharides containing the non-glucose sugars cannot be removed without destroying the microfibrils, taken together with the undoubted fact that the microfibrils are in part crystalline and are the wall component yielding the X-ray diagram, led immediately to the concept of microfibril structure shown in Fig. 2.6 (Preston 1959). This was not at that time at variance with views expressed among others, by Rånby and Frey-Wyssling. The demonstrations (Rånby 1949, 1951, Ribi & Rånby 1950) that under certain conditions extracted cellulose microfibrils can be broken down into rodlets which still give the X-ray diagram of native cellulose (cellulose I) were clearly in harmony with a central crystalline core of 'pure' cellulose. Final confirmation came from the demonstration (Dennis & Preston 1961) that when the celluloses of *Rhodymenia*, *Ulva* and *Laminaria* (and of the xylem of *Pinus* and *Ulmus* which also contain about 15% xylose or mannose) are treated by Rånby's methods the resulting rodlets, which still give the diagram of cellulose I, hydrolyse to give glucose only.

A natural consequence of the model of Fig. 2.6 is that in all microfibrils, irrespective of their width, there is a single crystalline core. In *Valonia*, *Cladophora* and *Chaetomorpha* cellulose, therefore, with microfibrils about 20 nm wide, the central crystallite is considered to be broader than with other algal celluloses (and higher plant celluloses) where the microfibrils are about 10 nm wide. Visual inspection of the corresponding X-ray diagrams suggests that this is the case. The matter has, however, been put on a sure footing by the quantitative findings (Nieduszynski & Preston 1970, Caulfield 1971) that both X-ray line broadening and X-ray low-angle scattering can be accounted for by the presence of crystallites 17 nm wide for *Chaetomorpha* compared with about 5 nm for ramie cellulose. There is therefore sound reason to believe that microfibrils are the biological units of cellulose.

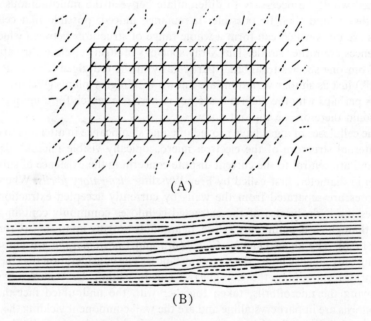

(A)

(B)

**Fig. 2.6.** Diagrammatic representation of sections of a cellulose microfibril.

(A) Transverse section. Oblique lines represent the trace in the plane of the paper of the sugar rings; full lines represent cellulose, broken lines other polysaccharides. The central lattice represents the crystalline core, the broader face tending to lie parallel to wall surface.

(B) Longitudinal section. Full lines represent cellulose chains, broken lines chains of other polysaccharides.

During recent years, however, entirely electron-microscopic evidence has repeatedly been quoted as suggesting that the basic unit of cellulose is an *elementary microfibril* 3·5 nm wide and that larger microfibrils are composed of aggregates of these units (Mühlethaler 1960, Frey-Wyssling & Mühlethaler 1963, Heyn 1965, 1966). Observations of this kind have been criticized whether referring to single microfibrils (Colvin 1963) or more recently to negatively-stained wall sections (Preston 1971). In the latter case it is abundantly clear that overlapping of images ensures that fibril diameter has been underestimated by a factor of 2 or 3 at least. Even with isolated microfibrils, phase-contrast effects can lead to spuriously low values and these always occur except when the electron microscope is at precise focus. Until firm evidence is presented for the existence of elementary fibrils as subunits pre-existing in microfibrils the whole idea must be discarded. This applies with even greater force to the concept of elementary fibrils with folded chains envisaged some time ago by Manley and recently restated in modified form (Manley 1971). Both the size of the native cellulose chain and its biosynthesis are controversial subjects (see pp. 47–8). For example, it has been claimed that cellulose in *Gossypium* (cotton) (Marx-Figini & Schulz 1966) and in *Valonia* (Marx-Figini 1969) is synthesized according to a template mechanism and is monodisperse (at least in the secondary cell wall of *Gossypium*) having degrees of polymerization (DP) of about 14,000 and 18,500 respectively. However, more recent investigations of the molecular sizes and distributions of *Gossypium* cellulose have not confirmed either such high values of DP or the monodispersity of the cellulose (Holt *et al*. 1972).

## 4   MANNAN

The cell walls of a number of green seaweeds including species of *Codium*, *Dasycladus*, *Batophora*, *Neomeris* and *Acetabularia* utilize a partially crystalline mannan as the skeletal component (Frei & Preston 1961a; 1968). A mannan of similar structure occurs in association with a xylan in the cuticle of the red algae *Porphyra umbilicalis* and *Bangia fuscopurpurea* (Frei & Preston 1964). Chemical studies suggested that these mannans are essentially linear and contain a predominance of β-1,4 linkages (Love & Percival 1964, Iriki & Miwa 1960). (Small amounts of glucose found in hydrolysates of these mannans may be spurious.) The best X-ray fibre diffraction photographs of mannan have been obtained from the central siphon of *Dasycladus*, *Batophora* and *Cymopolia* and show that there the mannan crystallites are arranged parallel to the siphon axis. In outer parts of the walls of *Dasycladus* and *Batophora*, however, the crystallites have been found to lie more or less transversely with respect to the siphon axis (Frei and Preston 1968). These conclusions have been supported by studies of cell wall lamellae by electron microscopy, and of cell wall sections by polarization microscopy which show that inner wall layers exhibit positive birefringence whereas outer layers appear to be negatively birefringent (the mannan chain is itself intrinsically positively birefringent). Although the distinction between inner and outer wall layers is not precisely that between secondary and primary wall layers in higher plants, it is clear that here and elsewhere in the thallophyta the secondary/primary layer differentiation is foreshadowed. It is interesting to note, in this context, that some of these mannan-containing seaweeds, such as *Dasycladus*, appear to be unique in that as the plant grows the constituent cells become shorter. No explanation of this phenomenon is yet forthcoming.

C

On treatment with alkali the native mannan I structure undergoes a transformation to the so-called mannan II form. The fibre repeat value obtained for mannan I was 1·028 nm and so it has been concluded that the chain conformation is basically similar to that of cellulose and contains two β-D-mannose units in the C1 chair form (1e, 4e linked) in a molecular repeat. It was also suggested that the unit cell is orthorhombic and contains two chain segments (Frei & Preston 1968).

These conclusions have been supported by further consideration of the X-ray data and computer-aided chain packing calculations, in conjunction with the results of polarized infra-red studies. The mannan chains are arranged in anti-parallel fashion and it has been proposed that, in addition to *intra*molecular hydrogen bonds between $-O_3H$ and the ring oxygen atom $(O'_5)$ (see Fig. 2.4), there are weaker *inter*molecular hydrogen bonds formed between the axial $-O_2H$ and ring oxygen atom in adjoining chains (Nieduszynski & Marchessault 1972).

Surprisingly, the original electron microscopic examination of mannan cell wall layers failed to produce evidence for the existence of well formed microfibrils. The layers all appeared granular or at best had a short rod-like appearance even after efforts to remove structure-concealing, encrusting substances. This anomaly was linked with another, namely the value of 16 which had been suggested for the degree of polymerization (DP) of mannan chains in *Codium* (Iriki & Miwa 1960). This corresponds to a chain length of about 8 nm, and this result, together with the electron microscopic evidence, seemed to be inconsistent with the function of mannan as a skeletal substance like cellulose which has a contour length of 2 to 5 μm (DP = 5–10,000). These anomalies were resolved in subsequent studies. In the first of these, it proved possible to obtain mannan microfibrils from cell wall preparations of *Codium fragile* and *Acetabularia crenulata*, providing these were not subjected to unduly harsh physical and chemical cleaning procedures (Mackie & Preston 1968). It also appeared likely that these microfibrils as well as being less stable to normal treatments than cellulosic microfibrils, might form a relatively minor component of the cell wall, the major proportion being made up of the granular or non-microfibrillar component. However, although an investigation of the degrees of polymerization and polydispersity of the corresponding nitrated mannan from *C. fragile* indicated that 90% of the mannan has DP's in the range 100–2,500 which is sufficiently great to account for both types of ultrastructure, no real evidence for two mannan components has been obtained (Mackie & Sellen 1969).

Further investigations of the cell wall protein of *Codium fragile* (Mackie unpubl.) have indicated that a protein component enriched in hydroxyproline and hydroxylysine is firmly bound to the skeletal mannan. It also appears that this complex is, in turn, linked by alkali labile (possibly disulphide) bonds to another protein-carbohydrate complex. This latter complex, however, contains little hydroxyproline, hydroxylysine or mannose, but has high proportions of galactose and arabinose—the characteristic monosaccharides of *Codium* sulphated polysaccharide (see p. 74).

## 5 XYLANS

Some members of the Rhodophyceae and Chlorophyceae contain polymers based on β-D-xylose as major polysaccharide constituents. These are of different types and appear to vary in their proportions of 1,3 and 1,4 linkages and in the degree of branching. There is no doubt that the best characterized and most interesting of algal xylans are those which form the skeletal component of several siphonous green algae and in the walls of the order Bangiales in the red algae (Frei & Preston 1961a; 1964). These plants again show the differentiation in microfibrillar orientation between inner and outer wall layer precisely as mentioned with the mannan-containing seaweeds.

Early chemical studies of the xylan from *Caulerpa filiformis* and other species indicated that they all have essentially linear chains of β-1,3 linked D-xylose (Mackie & Percival 1959, Iriki *et al.* 1960). (A small and variable proportion of D-glucose units was also present in acid hydrolysates of the xylans but its role is still unsettled.)

Early model building trials indicated that the molecular chain has a different conformation from the cellulose/mannan type, a consequence of the di-equatorially (1e, 3e) linked D-xylose units. This evidence, together with that of X-ray fibre diffraction, led to the proposition that in the native xylan two chains were intertwined into double helices, which in turn were hexagonally packed and organized into microfibrils. This model accounted for many of the features of xylan cell walls including their unusual negative birefringence in the polarizing microscope (Frei & Preston 1964).

However, a more recent reconsideration of the X-ray diffraction results, supplemented by polarized infra-red observations has led to a new model being proposed (Atkins *et al.* 1969). The basic structure has been shown not to be one of double but triple helices (Fig. 2.7). Various multi-strand helices were considered and the available data is in accord with a three-stranded structure. It was considered that the individual xylan chains are right-handed helices containing six D-xylose units in a pitch of 1·836 nm. This conformation allows the formation of *inter*-strand hydrogen bonds between the $-O_2H$ hydroxyl groups. Thus the triple helices are stabilized by cyclic triads of hydrogen bonds every 0·306 nm along the helix axis (see Fig. 2.8). This arrangement was supported by the occurrence of a band in the hydroxyl-stretching region of the infra-red spectrum, which exhibited perpendicular dichroism and which was deuterated with difficulty. In accord with density measurements, it was proposed that the unit cell contains one such triple helix with water molecules bound between the remaining hydroxyl group ($-O_4H$) and the ring oxygen atom in an adjacent chain (Atkins & Parker 1969).

It is interesting that this structure which is so different from the cellulose and mannan polymers also occurs in microfibrillar form. The microfibrils are distinct from those of cellulose and mannan in being negatively birefringent, a consequence of the helical chain conformation. Like mannan, the DP's of xylan had been reported to fall in the range of 42 to 67 (Iriki *et al.* 1960), which, even at full extension, seemed far too short (20 to 35 nm) to account for the observed microfibrils. Once more this anomaly was resolved through preparation and investigation of the nitrated xylan. The xylan, in this case (cf. mannan) was

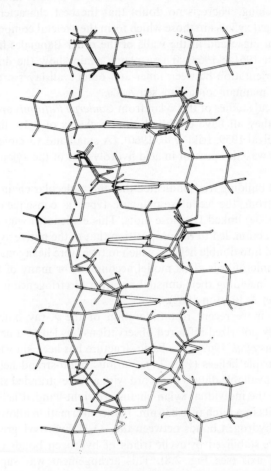

**Fig. 2.7.** Three-stranded helical structure of β-1,3 linked xylan.

nitrated very rapidly and undoubtedly was degraded. Even so, it was shown that $DP_n$ has a minimum value of about 1,000 (Mackie 1969) and that xylan chains having DPs much greater than that may be present in the native cell wall (Mackie & Sellen 1971).

Of the other xylans, the best studied is the water soluble xylan from *Rhodymenia palmata*. The original studies by methylation and periodate oxidation showed that the xylan chains are probably linear and composed of mixed β-1,3 (ca. 20%) and β-1,4 (ca. 80%) linkages (Percival & Chanda 1950, Barry *et al.* 1950). Other preparations containing higher proportions (30 to 40%) of 1,3 linkages have also been obtained (Björndahl *et al.* 1965). Both chemical (Manners & Mitchell 1963) and enzymic (Howard 1957, Björndahl *et al.* 1965) degradations resulted in the isolation of fragments which confirm that both linkages are present in the original molecule and also show that very few adjacent β-1,3 linkages are present. From an analysis of the oligomeric products of one enzymic degradation, it was concluded that the arrangement of the two kinds of links is not regular (Björndahl *et al.* 1965). Xylans, which appear to have a similar structure as judged from methylation analysis, have been isolated from a

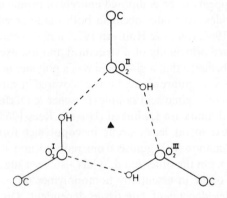

**Fig. 2.8.** Cyclic triads of hydrogen bonds in β-1,3 linked xylan, viewed down helix axis. Each –COH group is contributed to by one chain of the three-strand helix. The individual chains are denoted by the symbols I, II and III. Hydrogen bonds are represented by dotted lines. The arrangement is drawn with all nine atoms coplanar.

chlorite extract of *Porphyra umbilicalis* (which also contains a wholly β-1,3 linked xylan) and a dilute alkaline extract of *Laurencia pinnatifida* (Turvey & Williams 1970). A more detailed investigation of the water soluble xylan which is a major component of *Chaetangium fastigiatum* showed that it, too, has a similar constitution (Cerezo *et al.* 1971, Cerezo 1972).

On the other hand, extraction of the residual components of *Rhodymenia palmata* with dilute alkali, produced a xylan, which from methylation analysis, is also linear, but contains only 1,4 linked D-xylose units (cf. hemicellulosic xylans of higher plants such as esparto grass). The same methods showed that the xylan from *Caulerpa racemosa*, though mainly 1,3 linked, contains about 10% of branch points, while the water and chlorite soluble, xylose rich, polysaccharides from *Rhodochorton floridulum* are similar to *Rhodymenia* xylan in the proportions of 1,3 and 1,4 linked units, but also contain a considerable number of branch points. It was considered possible that these xylans are attached as comparatively short chains to another polysaccharide (possibly glucan) backbone (Turvey & Williams 1970).

## 6  ALGINIC ACID

Alginic acid, together with its salts (alginates) is one of the most extensively studied polysaccharide preparations (see Percival and McDowell 1967 for a comprehensive review). Until recently it was thought to be restricted to members of the Phaeophyceae but polymers of similar chemical constitution are now known to be synthesized extracellularly by certain strains of bacteria (Linker & Jones 1966, Gorin & Spencer 1966, Larsen & Haug 1971). It is an unusual substance in that it appears to be composed entirely of uronic acid units with no neutral monosaccharides. Alginates occur in both the intercellular regions and cell walls (Baardseth 1966, Evans & Holligan 1972) and it is considered that their biological functions are principally of a structural and ion-exchange type.

Originally, it was believed that alginic acid was a polymer of β-D-mannuronic acid only. Subsequently, L-guluronic acid was discovered in variable proportions in hydrolysates of several alginic acid samples (Fischer & Dörfel 1955). Evidence was obtained that both units are 1,4 linked (Hirst & Rees 1965, Rees & Samuel 1967) and to some extent, at least, occur in copolymer form, following the isolation of 4-O-β-D-mannosyl-L-gulose from reduced and hydrolysed alginic acid (Hirst *et al.* 1963). On the other hand, it has also been suggested that alginic acid may sometimes occur in essentially homopolymer form, the composition being not only species dependent but tissue dependent. On the basis of the examination (primarily by X-ray powder diffraction) of several alginic acid preparations obtained by selective extraction from several species, it has been proposed that alginic acid enriched in polymannuronic acid is characteristic of young cell wall tissue and/or intercellular regions, whereas polyguluronic-rich alginic acid appears to be located in the cell wall proper (Frei & Preston 1962). These results were later supported by an investigation of the composition of alginic acids from various parts of *Ascophyllum*, *Fucus* and *Laminaria* species. It was clearly demonstrated that alginic acid could be prepared ranging from 97% mannuronic acid (in the intercellular fluid of *Ascophyllum*) to about 30% mannuronic acid (in the cortex of *L. hyperborea*) (Haug *et al.* 1969). Thus it is to be expected that 'alginic acid', as normally prepared from whole plants, will show wide ranges in the proportions of mannuronic and guluronic acids. It has also emerged from a series of partial hydrolysis studies of alginates that to a large extent, they are block copolymers, in which regions of 1,4 linked D-mannuronic acid $(-M-)_n$, 1,4 linked L-guluronic acid $(-G-)_n$ and alternating D-mannuronic and L-guluronic acid $(-MG-)_n$ occur (Haug *et al.* 1966, 1967a) (Fig. 2.9) and it is known that native alginates vary in their block content according to source, etc. This then is the currently held view of the primary structures of alginic acids and we shall now explore to what extent this structure provides a basis for understanding the physical and biological properties of alginates.

The characteristic physical properties of alginates such as formation of gels, viscous solutions, and selective interaction with cations are well catalogued but are less well understood in molecular terms. It is well known that alginates which have high contents of guluronic acid have correspondingly high affinities for divalent ions such as $Ca^{2+}$, $Sr^{2+}$ and $Ba^{2+}$, (Haug & Smidsrød 1965, 1967a, Smidsrød & Haug 1968a), and it has also been shown that the relative amounts of ions such as $Mg^{2+}$, $Ca^{2+}$ and $Sr^{2+}$ accumulated by brown algae are closely correlated with the uronic acid composition of the alginate in the plants (Haug & Smidsrød 1967b). This specificity of guluronic acid-rich alginates for divalent ions has also formed the basis of medical applications in attempts to remove toxic cations such as radioactive $Sr^{2+}$ (see e.g. Sutton *et al.* 1971, Van der Borght *et al.* 1971).

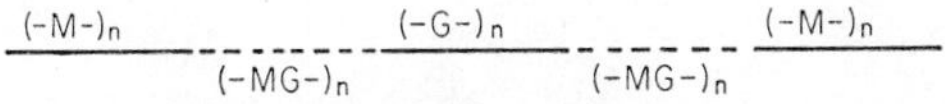

**Fig. 2.9.** Possible block structure of alginic acid.

These ion selectivities are more complicated than hitherto believed and, from more recent studies of the calcium-magnesium selectivity exchange, it has been proposed that alginates enriched in guluronic acid exhibit two types of $Ca^{2+}$ binding depending on the equivalent fraction ($X_{Ca}$) of $Ca^{2+}$ which is available (Smidsrød & Haug 1972a). The first type occurs in solution at low values of $X_{Ca}$ appears to involve isolated alginate chains and does not depend on the presence of contiguous guluronic acid units. This type of selectivity is removed by modification of the hydroxyl groups (for example by acetylation) in the sugar units. The second type involves a higher selectivity for $Ca^{2+}$ and occurs at higher values of $X_{Ca}$ with concomitant formation of the gel state. It has been suggested that for this type, autoco-operative effects are important in forming inter-chain bridges through binding of $Ca^{2+}$. The formation of these bridges does require the presence of several contiguous guluronic acid units and is eliminated on modification of the carboxyl groups. It has also been suggested that, at low values of $X_{Ca}$, the order of selectivity of alginate blocks for $Ca^{2+}$ is $(-M-)_n < (-MG-)_n < (-G-)_n$ (Smidsrød *et al.* 1972). Selectivity effects have also been observed in changes in the circular dichroism spectra of alginates which take place during gel formation. However, these experiments suggest that in gel formation, as calcium ions are diffused into the solution G blocks are first to be affected, then M blocks, while MG blocks appear to be hardly affected at all (Rees 1972).

The principal techniques for investigating the shapes of molecules have been applied to alginates. In addition, more recent theoretical calculations of molecular shape by conformational analysis have been made (Whittington 1971).

The original investigators of alginates by X-ray diffraction are now known to have been misled, in that the X-ray diffraction patterns they studied arose

from the then unknown polyguluronic acid, and not as supposed from polymannuronic acid (Astbury 1945). The relationships between the crystalline
components of alginic acid have been established recently (Frei & Preston 1962,
Atkins *et al.* 1970). Investigations of samples of alginic acid enriched in polymannuronic acid and polyguluronic acid respectively, by X-ray methods, supplemented by polarized infra-red results, showed that stereochemically the molecular
chains are quite different (Atkins *et al.* 1971). In the acid form, both appear to
form twofold helices, but whereas polymannuronic acid is a flat ribbon-like
chain, whose molecular repeat of 1·035 nm contains two diequatorially (1e, 4e)
linked β-D-mannuronic acid units in the C1 chair form (Fig. 2.10), polyguluronic

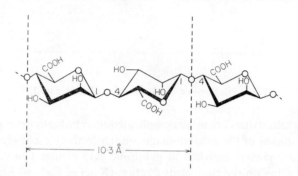

**Fig. 2.10.**   Segment of polymannuronic acid chain.

acid contains two diaxially (1a, 4a) linked α-L-guluronic acid units in the 1C
chair form, producing a rod-like conformation with a molecular repeat of
0·87 nm (Fig. 2.11). The polymannuronic chain conformation is close to that
proposed for the other 1,4 linked hexosans, cellulose and mannan, and, like
these, has a stabilizing *intra*-residue hydrogen bond between $-O_3H$ and the ring
oxygen atom $(O'_5)$ of the next unit in the chain (see Fig. 2.4). It is also proposed
that the chains are packed in a similar fashion to β-1,4 mannan (p. 54) and are
bonded into sheets by hydrogen bonds formed between the carboxyl groups and
hydroxyl groups of both parallel and antiparallel chains (Atkins *et al.* 1973).

In the proposed polyguluronic acid model, the rodlike chain may be stabilized
by an *intra*-residue hydrogen bond between the equatorial $-O_2H$ and the
carbonyl oxygen atom in adjacent units. The *inter*-chain bonds are more complicated than in the case of polymannuronic acid and probably involve water
molecules. A study of some of the salts of these substances by X-ray techniques
has revealed some interesting differences (Mackie 1971). In the polymannuron-
ates (Na$^+$, Li$^+$, K$^+$ and Ca$^{2+}$ salts) the fibre repeating value is changed to about
1·5 nm, indicating that the chain has relaxed to a threefold helix containing three
C1 chairs per turn. Similar conformations have been reported for β-1,4 D-xylan
(Nieduszynski & Marchessault 1971), soda cellulose (Nieduszynski unpubl.) and
mannan triacetate (Bittiger & Marchessault 1971). In contrast, the twofold helix

and molecular repeat of 0·87 nm is retained in the $NH_4^+$ and $K^+$ salts of poly-guluronic acid. The X-ray photographs of other salts of polyguluronic acid are poorly resolved at present but they show similar intensity distributions to poorly crystalline samples of $NH_4^+$ and $K^+$ salts, so it is likely that they have similar overall chain conformations. It is interesting to recall that the building blocks of alginic acid, β-D-mannuronic acid and α-L-guluronic acid, differ only in their stereochemistry at $C_5$. The preceding results illustrate the important influence which the conformations of the monomer units have on the overall molecular shape since it is basically these conformations which determine the linking geometry. The same monosaccharide conformations have been reported to be

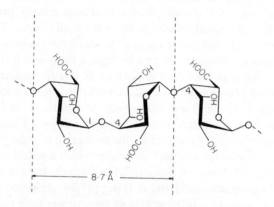

Fig. 2.11. Segment of polyguluronic acid chain.

retained in solution conditions (Rees 1972) and so it is almost certain that they are maintained in the gel state—the biologically important form.

The preceding experimental observations are also supported by theoretical calculations based on conformational analysis, in which the potential energies of macromolecular conformations are plotted as functions of torsional angles. Calculations have also been made of characteristic ratios (which describe the flexibility of polymer chains) of $(-M-)_n$, $(-G-)_n$ and $(-MG-)_n$ segments and these suggest that $(-G-)_n$ is more stiff than $(-M-)_n$ which in turn is more stiff than $(-MG-)_n$ (Whittington 1971). Unfortunately, there is still a lack of definitive experimental observations on pure, high molecular weight alginate blocks or homopolymers. An investigation of alginate of mixed composition by light scattering indicated that it was polydisperse and behaved as a highly extended flexible coil in 0·1M salt solution (Smidsrød & Haug 1968b). From a related investigation, it has been shown that like other polyelectrolytes, the intrinsic viscosity of alginate solutions varies linearly with the reciprocal of the square root of the ionic strength of the solution, and it was concluded that the chain extension is due mainly to the intrinsic chain rigidity (Smidsrød 1970). More recent estimates of the unperturbed dimensions have indicated that in solution $(-G-)_n$ is more extended than $(-M-)_n$ which is in turn more extended than $(-MG-)_n$ which is in accord with the theoretical calculations (Brucker *et al.* 1971, Smidsrød *et al.* 1972a, b). It is also worth noting that the X-ray evidence also indicates that polymannuronate may be more flexible than poly-guluronate, in the sense that rotations can take place about the glycosidic bonds in the former, whereas the units appear to be more rigidly held in twofold helices in the latter. This result is also reflected in the very small number of allowed conformations calculated for the dimer of guluronic acid (in the 1C chair form) as compared with that of mannuronic acid (in the C1 chair form).

The prime importance of all these investigations is that they are bringing us closer to understanding the nature of the cross-links which enable coherent gel structures to be built up. For example, the structures already proposed for the partially crystalline fibrous alginic acids show the kind of cross-links, involving certain hydrogen bonds and water molecules, which are probably important in the acid gel structures. However, from a biological viewpoint it is obviously the gels formed by their salts which are important. It was recognized early on that correlations existed between the block structures of alginates and their physical properties (Haug *et al.* 1967) and the affinity of guluronic acid sequences for divalent ions such as $Ca^{2+}$ has already been mentioned. There is now clear evidence that this affinity is paralleled by the formation of rigid gels, in contrast to the behaviour of mannuronic acid sequences which are said to show little or no $Ca^{2+}$ selectivity (cf. however, Rees 1972) and which, instead of gelling, tend to form slurried aggregates (Smidsrød & Haug 1972b). However, the elucidation of the exact molecular cross-links in gels of the calcium type must await more reliable experimental data, although possible types of structure have been proposed from consideration of molecular models. These suggest how the buckled, diaxially linked chains such as polyguluronic acid could possibly co-ordinate divalent ions in particular geometrical arrangements, involving both carboxyl and hydroxyl groups (Rees 1972, Smidsrød *et al.* 1972a, b).

Altogether the available experimental evidence suggests that in many respects the observed physical properties can be correlated not only with cation selectivities but also with flexibilities of alginate block segments. It may be, therefore, that one of their principal biological functions is to control the flexibility requirements of tissues throughout different species and according to environment. In this respect, it will be interesting to see whether enzyme preparations (such as have already been prepared from bacterial sources) can be detected in living algae, which can convert D-mannuronic into L-guluronic acid units in the polymer chains (Haug & Larsen 1971, Larsen & Haug 1971). Such an enzyme would allow metabolic control of polysaccharide conformation and therefore of physical properties *via* primary structure as has already been suggested for the sulphated galactans of the Rhodophyceae (pp. 70–2).

# 7 SULPHATED POLYSACCHARIDES

Polysaccharides in which some of the monosaccharide units carry sulphate half-ester groups appear to have been found in most, if not all of the marine algae which have been investigated. They appear to originate from both the cell walls and intercellular regions and exist as diverse and chemically complex substances. To some extent, each division of the algae appears to synthesize its own characteristic type of polysaccharide, but within each group there are important variations in structure.

## 7.1 *Galactans of the Rhodophyceae*

Polysaccharides of a distinct structural kind are widespread throughout the Rhodophyceae (Rees 1965, Anderson & Rees 1966). Included in this group are superficially different polysaccharide preparations such as agar, carrageenan, porphyran, furcelleran and funoran.

Although great variations occur in the chemical nature of the building units in these polysaccharides, particularly in the occurrence and distribution of sulphate ester and methyl ether groups, it is apparent that, in many cases, these variations obscure or mask an underlying repeating structure of a galactan in which the building units are alternately β-1,3 and α-1,4 linked (Anderson *et al.* 1965). Whereas the 1,3 linked units usually have the D-configuration, the 1,4 linked units may have the D- or L-configuration, often as the 3,6-anhydride.

Polysaccharides which contain both D- and L-galactose units in alternation are best exemplified by agarose and porphyran, and structures of this type are thought to occur in the following families: Gelidiaceae (*Gelidium, Pterocladia, Acanthopeltis, Suhria*), Gracilariaceae (*Gracilaria*), Ceramiaceae (*Ceramium*), Rhodomelaceae (*Laurencia*), Bangiaceae (*Porphyra*), Endocladiaceae (*Endocladia, Gloiopeltis*), and Solieraceae (*Anatheca*). Polysaccharides containing sugar units of the D-configuration only are illustrated by carrageenan and furcelleran. Similar polysaccharides are thought to occur in the Gigartinaceae (*Chondrus, Gigartina, Iridaea, Rhodoglossum*), Solieraceae (*Eucheuma, Agardhiella*), Phyllophoraceae (*Ahnfeltia, Gymnogongrus*), Hypneaceae (*Hypnea*), Furcellariaceae (*Furcellaria*), Grateloupiaceae (*Grateloupia, Aeodes, Phyllymenia, Pachymenia*). Within this group of polysaccharides it is now apparent that in many ways they are more usefully thought of as continuous spectra of related molecular types (see also alginates and mucilaginous polysaccharides of brown and green algae) of which the proportions and structure can vary during growth and development as well as according to factors such as species, season and geographical origin. As with alginates, the characteristic physical property of this group of polysaccharides is the ability to form gels and viscous solutions under appropriate conditions and, following a discussion of their chemical structures, it will be seen how, in some cases, certain structural features have been correlated with their characteristic physical and biological properties.

(a) *Agar*

Agar (Araki 1966 and references within) is probably the most familiar of all the red algal polysaccharide preparations. It is usually prepared from *Gelidium* and *Gracilaria* (Duckworth *et al.* 1970) species, and the gelling properties of the products vary accordingly. It is atypical in having a very low sulphate content and for many years it was considered that agar was a mixture of two poly-saccharides. The major component was called agarose, shown to be built up on the basis of a repeating unit of alternating 1,4 linked 3,6-anhydro-α-L-galactopyranose and 1,3 linked β-D-galactopyranose (Fig. 2.12). This neutral fraction was separated from the minor acidic component agaropectin after acetylation of the mixture. Agaropectin was a more complex structure and contained residues of sulphuric, pyruvic (as 4,6-carboxyethylidene-D-galactose) and uronic acids, in addition to D-galactose and 3,6-anhydro-L-galactose.

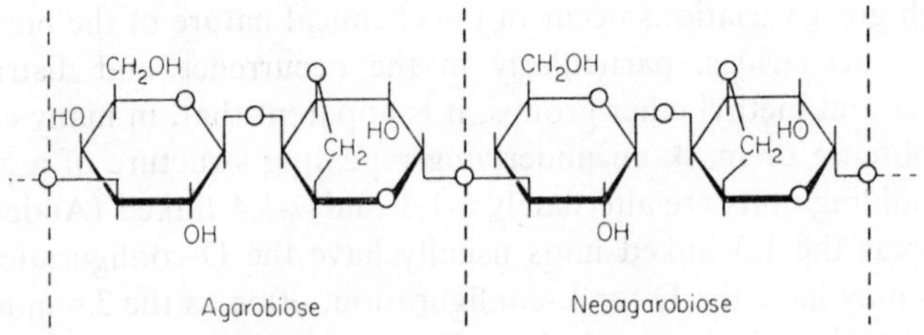

Fig. 2.12. Repeating sequence of Agarose.

From more recent results involving fractionation of the agars from *Gelidium amansii* (Izumi 1971a) and a commercial source (Duckworth & Yaphe 1971) by ion exchange chromatography, it has been suggested that these ideas are an oversimplification. It seems more likely that, as usually prepared, agar is better regarded as a complex mixture or continuous distribution of polysaccharides, which, though basically similar, are substituted to a variable degree with charged groups (Young *et al.* 1971). For example, in commercial ('Difcobacter') agar three extreme molecular types, neutral agarose, pyruvated agarose and a sulphated galactan, have been recognized (Duckworth & Yaphe 1971).

Originally, most of the structural work was done on agar from *Gelidium amansii*. The main evidence for the structure of agarose rests on the isolation in high yield of 4-O-β-D-galactopyranosyl-3,6-anhydro-L-galactose (agarobiose) and its derivatives after partial hydrolysis (Araki 1944), methanolysis (Araki & Hirase 1954), mercaptolysis (Hirase & Araki 1954) and methylation, in addition to characterization of the hydrolysis products of agarose and its methylated derivative. The alternative repeating disaccharide segment, 3-O-(3,6-anhydro-α-L-galactopyranosyl)-D-galactose (neoagarobiose) and the corresponding tetrasaccharide (neoagarotetraose) have also been isolated and characterized following enzymic digestion of agarose (Araki & Arai 1956, 1957).

In another investigation large proportions of neoagarotetraose (50%) and neoagarohexaose (42%) were produced from neutral agarose. The same enzyme produced a tetrasaccharide ($4^363$-carboxyethylidine neoagarotetraose) and a hexasaccharide ($4^565$-carboxyethylidine neoagarohexaose) from pyruvated agarose (Duckworth & Yaphe 1971).

## (b) *Porphyran*

This polysaccharide preparation from *Porphyra umbilicalis* appears to be closely related to agarose in its basic structure but it has more variation in monosaccharide composition (Anderson & Rees 1966). For example, the 1,3 linked β-D-galactose units sometimes occur as the 6-O-methyl derivative, while 3,6-anhydro-L-galactose units may be replaced by L-galactose 6-sulphate as shown by the identification of the disaccharides 4-O-β-D-galactopyranosyl-L-galactose 6-sulphate, 3-O-(α-L-galactopyranosyl 6-sulphate)-D-galactose, 4-O-(6-O-methyl-β-D-galactopyranosyl)-L-galactose 6-sulphate and 3-O-(α-L-galacto-pyranosyl 6-sulphate)-6-O-methyl-D-galactose in the products of partial acid hydrolysis (Turvey & Williams 1963). However, in spite of the variations in the proportions of the constituents in different samples, the sum of the L-galactose 6-sulphate and 3,6-anhydro-L-galactose units was always equal to the sum of the D-galactose and 6-O-methyl-D-galactose proportions. It has been estimated that at least 90% of the polymer possesses the ideal repeating structure. The basic similarity to agarose was shown by alkaline conversion of the L-galactose 6-sulphate units to the 3,6-anhydride, followed by methylation. The product was indistinguishable from methylated agarose (Anderson & Rees 1965).

As indicated previously, there are other polysaccharide preparations related to the agarose/porphyran type. They contain agarobiose segments and presumably have a masked repeating structure, sometimes including other variations in addition to those already described. For example, the polysaccharide from *Gloiopeltis furcata* (funoran) has been shown to be highly heterogeneous (Izumi 1971b) and in one case has been fractionated to give two polysaccharide sulphates which differ in composition (Hirase & Watanabe 1971). The major component contained D-galactose, 6-O-methyl-D-galactose, 3,6-anhydro-L-galactose, 3,6-anhydro-2-O-methyl-L-galactose and produced the dimethylacetal of agarobiose and its 6-O-methyl derivative on partial methanolysis. It was also concluded that all of the 1,3 linked β-D-galactose units are sulphated or methylated at $C_6$ while the 1,4 linked 3,6-anhydro-L-galactose units are non-sulphated, sulphated or methylated at $C_2$.

The polysaccharide from *Laurencia pinnatifida* is more complex and contains units of D-galactose and its 6-O-methyl derivative along with L-galactose, 3,6-anhydro-L-galactose, and their 2-O-methyl ethers. There is also evidence for the presence of sulphate groups on $C_6$ of both L-galactose and 2-O-methyl-L-galactose, and on $C_2$ of D-galactose units, and the polysaccharide may also contain 1,3 linked D-galacturonic acid (Bowker & Turvey 1968). The polysaccharide from *Ceramium boydenii* yielded the dimethylacetal of both agarobiose and 4-O-(6-O-methyl-β-D-galactopyranosyl)-3,6-anhydro-L-galactose (6-O-methyl agarobiose) on methanolysis (Araki 1966) while the polysaccharide from *Polysiphonia fastigiata* contains D- and L-galactose, their 6-O-methyl ethers, 3,6-anhydrides and possibly 6-sulphates (Peat & Turvey 1965). Other constituents which have been detected include 2-O-methyl-L-galactose, 3,6-anhydro-2-O-methyl-L-galactose and 4-O-methyl-D-galactose in acid hydrolysates of the polysaccharide from *Grateloupia elliptica* (Hirase *et al.* 1967) and 4-O-methyl-L-galactose from the agar of *Gelidium amansii* (Araki *et al.* 1967). The polysaccharides from *Rhodomela larix* (Usov *et al.* 1971) and *Laingia pacifica* (Kochetkov *et al.* 1970) also contain D- and L-galactose, 3,6-anhydro-L-galactose and its 2-O-methyl derivative. The latter polysaccharide has also been reported to contain galactose 6- and 4-sulphates in addition to all the mono-methyl ethers of galactopyranose.

All of the preceding polysaccharides contain variable but appreciable proportions of 3,6-anhydro-L-galactose, but there are other instances in which the sulphated galactans are virtually devoid of 3,6-anhydrogalactose. For example, the sulphated polysaccharide from *Anatheca dentata* (Nunn *et al.* 1971) is considered to have a substantial proportion of alternating 1,3 linked β-D- and 1,4 linked α-L-galactose units as shown by the isolation, after partial hydrolysis, of the disaccharides 4-O-β-D-galactopyranosyl-L-galactose and 3-O-α-L-galacto-pyranosyl-D-galactose, the trisaccharide O-β-D-galactopyranosyl-(1 → 4)-O-α-L-galactopyranosyl-(1 → 3)-D-galactose and the tetrasaccharide O-β-D-galactopyranosyl-(1 → 4)-O-α-L-galactopyranosyl-(1 → 3)-O-D-galactopyranosyl-(1 → 4)-L-galactose. However, the excess proportions of D-galactose relative to L-galactose and the isolation of the disaccharide 4-O-β-D-galactopyranosyl-D-galactose indicates that the molecule also contains regions of at least two contiguous 1,4 linked D-galactose units. The hydrolysis products also contain small amounts of D-xylose, 3-O-methylgalactose and uronic acid.

The sulphated galactan obtained from *Corallina officinalis* contains high proportions of D-xylose as well as D- and L-galactose, very few of which seemed to be 1,3 linked. It is also devoid of 3,6-anhydrogalactose units and ester sulphate occurs on $C_6$ of L-galactose and $C_4$ of D-galactose (Turvey & Simpson 1966).

## (c) *Carrageenan*

Originally, two distinct components were recognized in carrageenan preparations the proportions of which ranged widely according to species, geographical origin and season. These were designated as '$\kappa$' or '$\lambda$' according to their solubility in potassium chloride solutions. However, since carrageenan is now considered to include a wide range or family of structural types (cf. preceding discussion on agar/porphyran group), this simple classification is considered to be unsatisfactory, and recently the nomenclature of carrageenans has undergone substantial modification. In accord with the presently available evidence, four additional structural types have been defined. These are *idealized* structures and are shown in three pairs (Fig. 2.13) of which the left-hand member can be regarded as a precursor of the right-hand member, in the sense that the latter are formed from the former on alkaline or enzymic desulphation (Stanicoff & Stanley 1969). All of these structures conform to the masked repeating type and differ in the

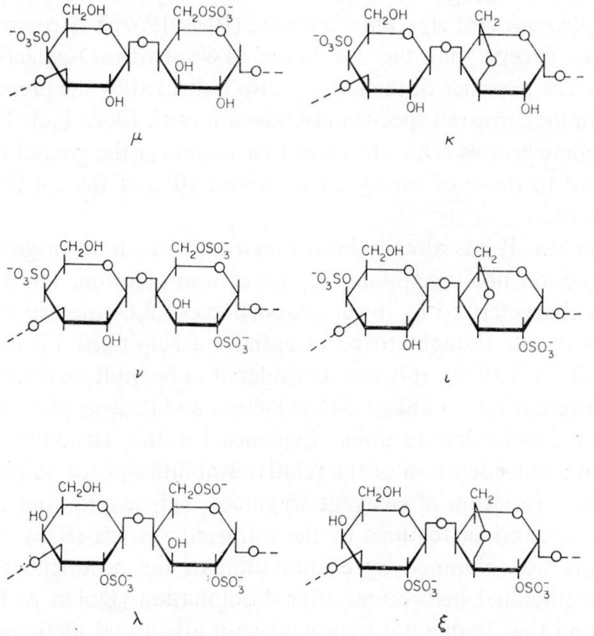

**Fig. 2.13.** Idealised Carrageenan repeating sequences (after Stanicoff & Stanley 1969, Rees 1969a).

occurrence and distribution of sulphate groups. As usually prepared, carrageenans appear to contain varying proportions of these extreme components, as independent molecules and/or within the same molecule.

κ-*carrageenan*. This carrageenan fraction is precipitated by potassium chloride. All the evidence of chemical investigations indicates that it largely consists of the repeating sequence (Fig. 2.13) of 1,3 linked β-D-galactose 4-sulphate and 1,4 linked 3,6-anhydro-α-D-galactose (O'Neill 1955, Araki & Hirase 1956, Anderson *et al.* 1968c). Variations in native polymer preparations include the occurrence of 1,4 linked α-D-galactose 6-sulphate and 2,6-disulphated units, which undergo alkaline elimination with formation of the corresponding anhydride. Alkaline treatment of native κ-carrageenan from *Chondrus crispus* followed by methanolysis produced the dimethylacetal of 4-O-β-D-galacto-pyranosyl-D-galactose (carrabiose), in a yield close to that calculated for a perfectly alternating copolymer (Anderson & Rees 1966). The fragment 3-O-(3,6-anhydro-α-D-galactopyranosyl)-D-galactose 4-sulphate (neocarrabiose sulphate) and other homologues have been isolated from enzymic digests of κ-carrageenan (Weigl et al. 1966). The results of periodate oxidation, methylation analysis and infra-red spectroscopy are also consistent with the above structural variations (Anderson et al. 1968b,c). The significance of some of these variations will be considered later.

ι-*carrageenan*. Carrageenan enriched in this component can be prepared from *Eucheuma spinosum* and *Agardhiella tenera* (Rees 1969b). Structurally it is like κ-carrageenan except that the 1,4 linked 3,6-anhydro-D-galactose units are 2-sulphated. The presence of the latter is also indicated by the presence of a band at 805cm$^{-1}$ in the infra-red spectrum (Anderson et al. 1968a,b,c). The occurrence of these sulphate groups is also thought to account for the greater elasticity of its gels compared to those of carrageenan. About 10% of the 1,4 linked units are 2,6-disulphated.

λ-*carrageenan*. It has already been noted that this term originally described the entire fraction not precipitated by potassium chloride. Unlike the κ component, it is characterized by lower proportions of 3,6-anhydro units and from early studies it was thought to be essentially a sulphated 1,3 linked galactan (Morgan & O'Neill 1959). It is now considered to be built up from the repeating segment of alternating 1,3 linked β-D-galactose and its 2-sulphate and 1,4 linked α-D-galactose 2,6-disulphate units. Evidence for this structure was obtained initially from a consideration of the relative stabilities of the sulphate groups to acid and alkali, isolation of cleavage fragments before and after alkaline treatment, and characteristic features of the infra-red spectra (Rees 1963a). Subsequent support was obtained by examination of the products from the polysaccharide methylated before and after desulphation (Dolan & Rees 1965). It was also found that sequential treatment with alkali and methanolic hydrogen chloride results in almost theoretical production of carrabiose dimethyl acetal (cf. κ-carrageenan) (Anderson & Rees 1966). Further support has been obtained from the isolation, from the acetolysis products of λ-carrageenan, of the fragments 4-O-β-D-galactopyranosyl-D-galactose, 3-O-α-D-galactopyranosyl-D-galactose, O-α-D-galactopyranosyl-(1 → 3)-O-β-D-galactopyranosyl-(1 → 4)-D-galactose, O-β-D-galactopyranosyl-(1 → 4)-O-α-D-galactopyranosyl-(1 → 3)-D-galactose and O-α-D-galactopyranosyl-(1 → 3)-O-β-D-galactopyranosyl-(1 → 4)-O-α-D-galactopyranosyl-(1 → 3)-D-galactose (Lawson & Rees 1968).

μ-*carrageenan*. This carrageenan component has the idealized structure shown in Fig. 2.13 and occurs to a variable extent with λ-carrageenan in the potassium chloride soluble fraction (Anderson et al. 1968a). As yet, no means has been found of separating native λ- and μ-carrageenans. The procedure followed to identify the μ-fraction was that of alkali-borohydride treatment and precipitation of the modified μ-fraction by potassium chloride. Examination of this modified fraction by methylation, infra-red spectroscopy and identification of methanolysis products indicates its overall similarity to κ-carrageenan. Variations from the depicted ideal structure include some replacement of D-galactose 6-sulphate by 3,6-anhydro-D-galactose and the presence of 2-sulphate on a small proportion of the 1,4 linked units. It has also been suggested that μ-carrageenan rather than λ-carrageenan, as originally thought, is a likely biological precursor of κ-carrageenan.

### (d) Other galactans

The polysaccharide from *Furcellaria fastigiata* (furcelleran) appears to be chemically similar to κ-carrageenan. Enzymic and chemical degradation by partial mercaptolysis indicate the existence of κ-carrageenan-like segments and D-galactose, 2-, 4- and 6-sulphates and 3,6-anhydro-D-galactose 2-sulphate have been obtained by partial acid hydrolysis (Yaphe 1959, Painter 1966). Examination of the hydrolysis products of the methylated, desulphated polysaccharide suggests the occurrence of branching, at $C_6$, in galactose units (Painter 1960).

The sulphated polysaccharide from *Hypnea specifera* appears also to be of this type. It yielded 3,6-anhydro-4-O-β-galactopyranosyl-D-galactitol on partial methanolysis, followed by hydrolysis and reduction, while methylation indicated that D-galactose is also linked 1,3 and sulphated on $C_4$ (Clingman & Nunn 1959).

Sulphated D-galactans have also been isolated from members of the Grateloupiaceae, *Aeodes orbitosa, A. ulvoidea, Phyllymenia cornea* and *Pachymenia carnosa* (Allsobrook *et al.* 1969). These contain little or no 3,6-anhydrogalactose or alkali labile ester sulphate but they do contain high proportions of monomethylated galactose units in addition to D-galactose (cf. porphyran). It is considered likely that these polysaccharides also consist largely of alternating sequences of β-1,3 and α-1,4 linkages.

The polysaccharide from *Aeodes orbitosa* (aeodan) (Nunn & Parolis 1968) yielded 4-O-β-D-galactopyranosyl-D-galactose and 3-O-D-galactopyranosyl-D-galactose on partial hydrolysis and from the results of methylation and periodate oxidation of the sulphated and desulphated polysaccharides, it was concluded that most of the 1,3 linked units are sulphated at $C_2$ and some at $C_2$ and $C_6$. The latter accounts for the small proportion of 3,6-anhydrogalactose formed on alkali treatment. Evidence has also been obtained for the occurrence of 1,3 linked galactose 4-sulphate units and the polysaccharide also gave rise to 2-O-methyl-D-galactose and small amounts of 6-O-methyl-D-galactose, 4-O-methyl-L-galactose and D-xylose on acid hydrolysis.

The polysaccharide from *Aeodes ulvoidea* appears to be similar to that of *A. orbitosa* but contains higher proportions of 4-O-methyl-L-galactose, probably as non-reducing end group (Allsobrook *et al.* 1971).

The sulphated polysaccharide obtained from *Phyllymenia cornea* (phyllymenan) contains higher proportions of 2-O-methyl-D-galactose as well as

D-galactose, 6-O-methyl-D-galactose, 4-O-methyl-L-galactose and D-xylose. The ester sulphate was stable to alkali and this result, together with the results of methylation analysis and the observed immunity of the desulphated poly-saccharide to periodate oxidation, indicates that sulphate groups are probably present on $C_2$ of 1,3 linked units in the native polymer. Partial acid hydrolysis produced 4-O-β-D-galactopyranosyl-D-galactose, 2-O-methyl-4-O-(β-D-galacto-pyranosyl)-D-galactose and possibly 3-O-(2-O-methyl-α-D-galactopyranosyl)-D-galactose as well as other disaccharides, composed of 6-O-methyl and 2-O-methyl-D-galactose, which may have been 1,4 and 1,3 linked (Nunn & Parolis 1969, 1970).

The polysaccharide from *Pachymenia carnosa* has also been subjected to partial hydrolysis resulting in the isolation of 4-O-β-D-galactopyranosyl-D-galactose, 4-O-β-D-galactopyranosyl-2-O-methyl-D-galactose, 3-O-(2-O-methyl-D-galactopyranosyl)-D-galactose, 4-O-(6-O-methyl-β-D-galactopyranosyl)-D-galactose, 2-O-methyl-4-O-(6-O-methyl-β-D-galactopyranosyl)-D-galactose and 6-O-methyl-(2-O-methyl-D-galactopyranosyl)-D-galactose (Farrant *et al.* 1971).

Less extensively studied sulphated galactans have also been reported from *Dilsea edulis* (Barry & McCormick 1957) and *Dumontia incrassata* (Dillon & McKenna 1950). The former is considered to possess a 1,3 linked backbone in which some of the galactose units are sulphated at $C_4$. It is also considered to contain xylose, glucuronic acid, 3,6-anhydrogalactose and some 1,4 linked galactose 6-sulphate (Rees 1961a). The polysaccharides from *Tichocarpus crinitus* have also been separated into κ- and λ-like fractions, of which the latter produced O-α-D-galactopyranosyl-(1 → 3)-O-β-D-galactopyranosyl-(1 → 4)-D-galactose among other fragments, on partial acetolysis (cf. λ-carrageenan). However, some deviation from a regular sequence of 1,3 and 1,4 links was indicated by the isolation of 2-O-β-D-galactopyranosyl-D-threitol after periodate oxidation, reduction and hydrolysis (Usov *et al.* 1969, 1970, Kochetkov *et al.* 1971).

The preceding sections indicate the complexity of primary structure which exists in the agar/carrageenan system of polysaccharides. We can now consider the extent to which these structural variations have been rationalized through the study of their secondary and tertiary structures.

In the partially crystalline fibrous form, several salts of κ- and ι-carrageenan (after treatment with alkali) gave similar X-ray diffraction patterns (Anderson *et al.* 1969). With this evidence, supplemented by polarized infra-red spectro-scopic results, it has been proposed that both κ- and ι-carrageenans contain pairs of polysaccharide chains arranged in double helices joined by hydrogen bonds (perpendicular to the helix axis) between $O_2$ and $O_6$. It is considered that each chain is a right-hand helix containing three disaccharide segments in a turn. In the potassium salt of ι-carrageenan, the observed fibre repeat of 1·30 nm is consistent with two parallel chains, each of pitch 2·60 nm, staggered half way with respect to each other. In the unit cell, it has been proposed that three double helix rods are arranged in a hexagonal lattice of side 2·26 nm. In κ-carrageenan, the fibre repeat of 2·4 nm suggests that there has been a contraction in the helices and a rearrangement of the strands with respect to each other. It has been proposed that the double helix model provides a basis for understanding the melting and setting of carrageenan gels, and it has also been suggested that this mechanism of gel formation, *via* a network of double helices, provides a means for regulating biological polysaccharide texture by controlling helix content *via* polysaccharide primary structure (Rees 1969b, Rees *et al.* 1969). These ideas received further support following the isolation of an enzyme preparation from *Gigartina stellata* which converts the D-galactose 2,6-disulphate units in ι-carrageenan into the 3,6-anhydride (Lawson & Rees 1970). Enzyme preparations of a similar kind have previously been isolated from *Porphyra umbilicalis* extracts and shown to convert L-galactose 6-sulphate units into the 3,6-anhydride in porphyran (Rees 1961b). The sequence of events during this conversion can be visualized as follows with reference to ι-carrageenan. In native ι-carrageenan, the regular sequence of $(-A-B-)_n$ (A = 3,6-anhydro-D-galactose 2-sulphate,

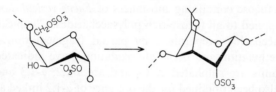

**Fig. 2.14.** Desulphation of 1a,4a linked D-galactose 2,6-disulphated (A*) units to form 1e,4e linked 3,6 anhydro-D-galactose 2-sulphate units in ι-carrageenan.

B = D-galactose 4-sulphate), is occasionally interrupted by 'λ-like' segments of D-galactose 2,6-disulphate (A*). These latter units undoubtedly will be in the C1 chair conformation (Fig. 2.14) whereas the A units will have the 1C conformation. The consequence of desulphation therefore is to replace diaxially linked (1a,4a) units by diequatorially linked (1e,4e) units and thus it can be seen that the inclusion of 1a, 4a (A*) units in the polymer chain will cause abrupt local changes in the chain direction (Rees 1969b, 1972). These irregularities in the structure have been described as kinks and it is evident that desulphation will result in their removal while, at the same time, it is observed that the gelling capacity is greatly increased. Solutions of κ- and ι-carrageenan gel on cooling, and it has been observed by optical rotary dispersion techniques that optical rotation shifts, characteristic of coil-helix transitions occur during cooling and gel setting (Rees *et al.* 1969). A better interpretation of the magnitude of these optical rotation shifts has been obtained by examination of solutions of specifically degraded ι-carrageenan in which coil-helix transformations are observed to take place, uncomplicated by gel formation (McKinnon *et al.* 1969).

Not surprisingly, it has been speculated that double helix formation may be important for the other related polysaccharides in the group including some glycosaminoglycans, which also have a basic repeating pattern of 1,3 and 1,4 linkages (Rees 1969a). Although this suggestion is supported to some extent by calculations of various stereochemical parameters by conformational analysis methods and, in the case of agarose solutions, by spectroscopic (N.M.R.) changes (Child *et al.* 1970) and optical rotation shifts (Rees 1972), so far only hyaluronates, among the glycosaminoglycans, have been shown to sometimes form double helices (Atkins & Sheehan 1973, Dea *et al.* 1973). It is worth emphasizing (Rees 1969b) that the texture of carrageenan gels is also dependent on the disposition of sulphate groups and the nature of their associated cations, so that changes in these factors provide additional scope for the modification of physical properties in the biological situation.

## 7.2   *Sulphated polysaccharides of the Phaeophyceae*

In addition to alginic acid, water soluble extracts of brown algae usually contain polysaccharides characterized by the presence of sulphate ester and residues of L-fucose (2-deoxy-L-mannose). Originally the name fucoidin (fucoidan) was coined for the fucose containing substances of *Fucus vesiculosus*, but this term tended to be applied to all fucose-rich polysaccharides. Chemical investigations of this polysaccharide preparation by partial fragmentation (O'Neill 1954, Côté 1959), methylation and periodate oxidation, have indicated that some of the L-fucose units are sulphated at $C_4$ and are $\alpha$-1,2, $\alpha$-1,3 and $\alpha$-1,4 linked. Evidence has also been obtained for the presence of $\alpha$-1,2 linked and 4-sulphated L-fucose in fucoidan from *Pelvetia wrightii* (Anno *et al.* 1969).

In addition to L-fucose, acid hydrolysates of fucoidan contain various proportions of D-xylose, D-galactose and uronic acid and in the past these were usually considered to be impurities.

However, an examination of the fucose containing polysaccharides from *Ascophyllum nodosum* has resulted in the separation of three distinct components, containing different proportions of L-fucose, D-xylose, hexuronic acid (mainly D-glucuronic acid) and sulphate ester attached to a polypeptide moiety (Larsen *et al.* 1966). The major component has been called 'ascophyllan' and, at the same time, the use of the term 'fucoidan' has been questioned. The fragment 3-O-$\beta$-D-xylopyranosyl-L-fucose has been isolated (Larsen 1967) and from this and other chemical studies, it has been suggested that ascophyllan contains a polyuronide backbone to which side chains composed of D-xylose and L-fucose and sulphate are attached. The polypeptide is considered to be linked through L-fucose units. The other two components were found to be intermediate in composition between ascophyllan and a fucan sulphate, but possessed similar structural features.

In a subsequent investigation of extracts of *Ascophyllum nodosum*, another complex containing fucose, xylose, galactose, mannose, glucuronic acid, mannuronic acid, guluronic acid, ester sulphate and polypeptide was isolated. Complexes of similar composition have been obtained from *Pelvetia wrightii* and *Sargassum pallidum* (Ovodov & Pavlenko 1970). It was found that the complex from *Ascophyllum* can be specifically degraded, under acidic conditions, to a sulphated glucuronoxylofucan (ascophyllan), alginic acid and a small amount of a polymer of L-fucose sulphate (fucoidan). The same treatment of *Fucus vesiculosus* produced a similar complex but contained much greater proportions of material which could be regarded as fucoidan (Larsen *et al.* 1970). It has also been shown that the fucose-rich polysaccharides from *Pelvetia wrightii* contain both a galactofucan and a fucan sulphate (Anno & Uemura 1971) and it has been concluded that native fucoidan mostly occurs as a building element of a much more complex macromolecule which may be of general occurrence in the Fucaceae (Larsen *et al.* 1970).

An extensive investigation of another fucose-rich polysaccharide also obtained from *Ascophyllum nodosum* has been reported (Percival 1968, 1971). It is considered to originate from the cell wall proper and contains different proportions from ascophyllan of L-fucose, D-xylose and uronic acid. It has resisted all attempts at fractionation and contains 3-O-β-D-glucopyranosyluronic acid-L-fucose as a major structural feature and lesser amounts of 3-O-β-D-xylopyranosyl-L-fucose and 4-O-α-L-fucopyranosyl-D-xylose. With this and other evidence it was concluded that the structure is dissimilar to that of ascophyllan in that it is probably highly branched and does not possess a polyuronide backbone.

It was also concluded that the D-xylose is 1,4 linked and present as end groups and that the L-fucose is present as end groups and 1,2 linked units, as 1,2 and 1,3 linked units sulphated at $C_4$ and as 1,2,3,4 linked and/or sulphated units.

Other species of brown algae such as *Himanthalia lorea*, *Bifurcaria bifurcata* and *Padina pavonia* also yield glucuronoxylofucans which appear to be of similar structure (Percival & Jabbar-Mian 1971).

### 7.3 *Sulphated polysaccharides of the Chlorophyceae*

The sulphated polysaccharides of several green algae have now received attention. These are invariably highly complex substances and at the moment it is extremely difficult to see any structural patterns of the Rhodophyceae-type emerging.

### (a) *Galactoarabinoxylans*

Polysaccharides containing mainly D-galactose, L-arabinose and D-xylose as well as ester sulphate have been prepared from species of *Cladophora*, *Chaetomorpha*, *Codium* and *Caulerpa* species. The preparations from *Cladophora* and *Chaetomorpha* undoubtedly have many structural features in common (Hirst *et al.* 1965). Partial hydrolysis produced β-1,3 and β-1,6 linked galactobioses and 1,4 or 1,5 linked L-arabinobiose in addition to D-galactose 6-sulphate and L-arabinose 3-sulphate. The latter sulphate was alkali labile in the polysaccharide and on treatment with sodium methoxide gave substantial proportions of 2-O-methyl-L-xylose. Other studies involving methylation (Fisher & Percival 1957) and successive oxidations with sodium metaperiodate, reduction and hydrolysis (Smith degradation) (Johnson & Percival 1969) have confirmed its highly branched character and indicate that L-arabinose and D-galactose are linked together (Bourne *et al.* 1970), that some of the galactose exists as D-galactofuranose and that D-xylose is end group or 1,4 linked.

The sulphated polysaccharide obtained from *Codium fragile* is distinctly different in its structural details, although it is composed of D-galactose and L-arabinose with a little D-xylose (Love & Percival 1964). Partial hydrolysis produced the disaccharides 3-O-β-D-galactopyranosyl-D-galactose and 3-O-α-L-arabinopyranosyl-L-arabinose together with D-galactose 6- and 4-sulphates. No evidence was obtained for the occurrence of ester sulphate on L-arabinose units.

A preliminary investigation of the water soluble polysaccharides from *Caulerpa* species showed that these contain D-galactose, D-xylose, D-mannose and variable proportions of L-arabinose. No definite conclusions were reached concerning the glycosidic linkages and site of ester sulphate (Mackie & Percival 1961).

### (b) *D-glucuronoxylorhamnans*

Polysaccharides of similar structure containing D-glucuronic acid, D-xylose and L-rhamnose and ester sulphate have been isolated from *Acrosiphonia centralis* (*Spongomorpha arcta*) (O'Donnell & Percival 1959), *Enteromorpha compressa* (McKinnell & Percival 1962a) and *Ulva lactuca* (McKinnell & Percival 1962b, Percival & Wold 1963). In each case a major structural component was found to be 4-O-β-D-glucopyranosyluronic acid-L-rhamnose. The results of periodate oxidation, before and after desulphation are consistent with the presence of some L-rhamnose 2-sulphate residues and D-xylose 2-sulphate units. The latter were confirmed by isolation of D-arabinose and 2-O-methyl-D-xylose following treatment of the polysaccharide with sodium hydroxide and sodium methoxide respectively, followed by hydrolysis. Partial hydrolysis of the carboxyl reduced polysaccharide led to the isolation of O-L-rhamnopyranosyl-(1 → 4)-O-D-xylopyranosyl-(1 → 3)-D-glucose, while partial hydrolysis of the native polysaccharide established the occurrence of O-D-glucopyranosyluronic acid-(1 → 4)-L-rhamnose 2-sulphate as a major structural feature, and also produced evidence for the occurrence of the fragments 3-O-D-glucopyranosyluronic acid-D-xylose, 4-O-D-glucopyranosyluronic acid-D-xylose and O-D-glucopyranosyluronic acid-(1 → 4)-L-rhamnopyranosyl-(1 → 3)-D-glucopyranosyluronic acid-(1 → 3)-D-xylose (Haq & Percival 1966a). Methylation analysis and degradation of the polymer by successive periodate oxidations and hydrolysis confirmed the highly branched nature of the substance (Haq & Percival 1966b).

A polysaccharide containing the same monosaccharides in addition to D-galactose has been prepared and examined from *Acetabularia crenulata* (Bourne *et al.* 1972). This polysaccharide also appears to be highly branched and its structural features include 1,3 linked D-galactose 4- and 6-sulphates, 1,2 linked L-rhamnose, while D-glucuronic acid is thought to be linked to both galactose and rhamnose units.

## 8 MISCELLANEOUS POLYSACCHARIDES

In comparison with the algae of the Rhodophyceae, Chlorophyceae and Phaeophyceae, the structural polysaccharides of other algae, with few exceptions, remain virtually unknown and most reports simply provide information about their constituent monosaccharides. Polysaccharides such as arabino-galactans and acidic xylans have been isolated from the fresh water algae *Nitella translucens* (Anderson & King 1961) and *Apjohnia laetevirens* (Stewart *et al.* 1969) and these appear to be closely related to the pectins and hemicelluloses of higher plants.

The polysaccharides from some members of the Cyanophyceae (e.g. *Anabaena, Amorphonostoc, Calothrix, Nostoc, Rivularia*) (O'Colla 1962, Moore & Tischer 1965, Kirillova *et al.* 1967) contain a wide range of monosaccharides including glucuronic acid, glucose, galactose, mannose, arabinose, xylose, fucose and rhamnose. In this field, particularly, there is scope for further investigation. From the Bacillariophyceae, the diatom *Phaeodactylum tricornutum* has been shown to contain a sulphated β-1,3 linked D-mannan to which side chains of O-D- glucopyranosyluronic acid-(1 → 3)-O-D-mannopyranosyl-(1 → 2)-O-D-mannose are attached (Ford & Percival 1965a). The exocellular polysaccharide from *Gomphonema olivaceum* is considered to contain predominantly 1,6 and 1,3 linked D-galactose, 1,3 and 1,4 linked D-xylose and ester sulphate which is stable to alkali and may be on $C_4$ of D-galactose residues (Huntsman & Sloneker 1971). In other studies, xylose, mannose and rhamnose have been found to be constituents of the tubes of *Amphipleura rutilans* while the principal component of the capsule of *Navicula pelliculosa* may be a polyuronide (Lewin 1955, 1958, O'Colla 1962). In contrast, it has been shown that the extracellular spines or fibres of *Thalassiosira fluviatilis* and *Cyclotella cryptica* are composed of highly oriented 1,4 linked 2-acetamido-2-deoxy-β-D-glucopyranose (chitin) (Blackwell *et al.* 1967, Dweltz *et al.* 1968), while the principal cell wall polysaccharide of *Monodus subterraneus*, of the Xanthophyceae, is a β-D-glucan containing both 1,3 and 1,4-linked units (Ford & Percival 1965b).

## 9 CONCLUSION

From the preceding discussions, it is clear that the reconstitution of the cell wall is a formidable problem. It should also be borne in mind that all together the algae mentioned in this discussion constitute less than 10% of all known species.

It is therefore apparent that our overall knowledge of algal polysaccharides is still very meagre so that generalizations must be accompanied with caution.

Nevertheless, it is evident that substantial progress in certain aspects has been accomplished in the last decade. For example, three of the principal types of skeletal polysaccharides, cellulose, mannan and xylan have been described in great detail at the molecular and ultrastructural level. In these instances the correlations between primary structure, molecular shape and biological functions are fairly obvious; and they provide excellent examples of the ways in which different chemistry is utilized for similar physical purposes.

Equally impressive are the successes which have been achieved in determining not only the primary but also the secondary and tertiary structures of some of the mucilaginous polysaccharides, particularly the alginates and carrageenans. In spite of great complexity we are beginning to see why, in terms of shape, molecules such as alginates and carrageenans have the properties they do and how, like proteins and nucleic acids, they may fulfil particular biological functions through their tertiary structures.

## 10   REFERENCES

ALLSOBROOK A.J.R., NUNN J.R. & PAROLIS H. (1969) Some Grateloupiaceae polysaccharides. *Proc. VIth Int. Seaweed Symp.* pp. 417–20.

ALLSOBROOK A.J.R., NUNN J.R. & PAROLIS H. (1971) A polysaccharide from *Aeodes ulvoidea. Carbohyd. Res.* 8, 361–2.

ANDERSON D.M.W. & KING N.J. (1961) Polysaccharides of the Characeae. Part IV. A Non-esterified pectic acid from *Nitella translucens. J. chem. Soc.* 5333–8.

ANDERSON N.S., CAMPBELL J.W., HARDING M.M., REES D.A. & SAMUEL J.W.B. (1969) X-Ray diffraction studies of polysaccharide sulphates: double helix models for $\varkappa$- and $\iota$-carrageenans. *J. molec. Biol.* 45, 85–99.

ANDERSON N.S., DOLAN T.C.S., LAWSON C.J., PENMAN A. & REES D.A. (1968a) Carrageenans. Part V. The masked repeating structure of $\lambda$- and $\mu$-carrageenans. *Carbohyd. Res.* 7, 468–73.

ANDERSON N.S., DOLAN T.C.S., PENMAN A., REES D.A., MUELLER G.P., STANICOFF D.J. & STANLEY N.F. (1968b) Carrageenans. Part IV. Variations in the structure and gel properties of $\varkappa$-carrageenan and the characterisation of sulphate esters by infrared spectroscopy. *J. chem. Soc. C.* 602–6.

ANDERSON N.S., DOLAN T.C.S. & REES D.A. (1968c) Carrageenans. Part III. Oxidative hydrolysis of methylated $\varkappa$-carrageenan and evidence for a masked repeating structure. *J. chem. Soc. C.* 596–601.

ANDERSON N.S., DOLAN T.C.S. & REES D.A. (1965) Evidence for a common structural pattern in the polysaccharide sulphates of the Rhodophyceae. *Nature, Lond.,* 205, 1060–2.

ANDERSON N.S. & REES D.A. (1965) Porphyran: A polysaccharide with a masked repeating structure. *J. chem. Soc.* 5880–7.

ANDERSON N.S. & REES D.A. (1966) The repeating structure of some polysaccharide sulphates from red seaweeds. *Proc. Vth Int. Seaweed Symp.* pp. 243–9.

ANNO K., SENO N. & OTA M. (1969) Structural studies on fucoidan from *Pelvetia wrightii. Proc. VIth Int. Seaweed Symp.* pp. 421–6.

ANNO K. & UEMURA K. (1971) Heterogeneity of fucoidan obtained from *Pelvetia wrightii*. *Abs. VIIth Int. Seaweed Symp.* p. 97.

ARAKI C. (1944) Chemical studies on agar-agar. XIII. Separation of agarobiose from the agar-agar like substance of *Gelidium amansii* by partial hydrolysis. *J. chem. Soc. Japan*, 65, 533–8, 627–32.

ARAKI C. (1966) Some recent studies on the polysaccharides of agarophytes. *Proc. Vth Int. Seaweed Symp.* pp. 3–17.

ARAKI C. & ARAI K. (1956) Studies on the chemical constitution of agar-agar. XVIII. Isolation of a new crystalline disaccharide by enzymatic hydrolysis of agar-agar. *Bull. chem. Soc. Japan*, 29, 339–45.

ARAKI C. & ARAI K. (1957) Studies on the chemical constitution of agar-agar. XX. Isolation of a tetrasaccharide by enzymatic hydrolysis of agar-agar. *Bull. chem. Soc. Japan*, 30, 287–93.

ARAKI C., ARAI K. & HIRASE S. (1967) Studies on the chemical constitution of agar-agar. XXIII. Isolation of D-xylose, 6-O-methyl-D-galactose, 4-O-methyl-L-galactose and O-methylpentose. *Bull. chem. Soc. Japan*, 40, 959–62.

ARAKI C. & HIRASE S. (1954) Studies on the chemical constitution of agar-agar. XVII. Isolation of crystalline agarbiose dimethylacetal by partial methanolysis of agar-agar. *Bull. chem. Soc. Japan*, 27, 109–12.

ARAKI C. & HIRASE S. (1956) Partial methanolysis of the mucilage of *Chondrus ocellatus* Holmes. *Bull. chem. Soc. Japan*, 29, 770–5.

ASTBURY W.T. (1945) Structure of alginic acid. *Nature, Lond.*, 155, 667.

ASTBURY W.T. & PRESTON R.D. (1940) The structure of the cell wall in some species of the filamentous green alga *Cladophora*. *Proc. R. Soc. B* 129, 54–76.

ATKINS E.D.T., MACKIE W., PARKER K.D. & SMOLKO E.E. (1971) Crystalline structures of poly-β-D-mannuronic and poly-α-L-guluronic acids. *J. Polymer Sci. B* 9, 311–16.

ATKINS E.D.T., MACKIE W. & SMOLKO E.E. (1970) Crystalline structures of alginic acids. *Nature, Lond.*, 225, 626–8.

ATKINS E.D.T., MACKIE W., NIEDUSZYNSKI I.A., PARKER K.D. & SMOLKO E.E. (1973) Structural components of alginic acid. Part I. The crystalline structure of poly-β-D-mannuronic acid. Results of X-ray diffraction and polarised infra-red studies. Part II. The crystalline structure of poly-α-L-guluronic acid. Results of X-ray diffraction and polarised infra-red studies. *Biopolymers (in press)*.

ATKINS E.D.T. & PARKER K.D. (1969) The helical structure of a β-1,3 xylan. *J. Polymer Sci. C* 28, 69–81.

ATKINS E.D.T., PARKER K.D. & PRESTON R.D. (1969) The helical structure of the β-1,3-linked xylan in some siphoneous green algae. *Proc. R. Soc. B* 173, 209–21.

ATKINS E.D.T. & SHEEHAN J. (1973) Hyaluronates: Relation between molecular conformation. *Science, N.Y.* 179, 563–4.

BAARDSETH E. (1966) Localization and structure of alginate gels. *Proc. Vth Int. Seaweed Symp.* pp. 19–28.

BARRY V.C., DILLON T., HAWKINS B. & O'COLLA P. (1950) Xylan of *Rhodymenia palmata*. *Nature, Lond.*, 166, 788.

BARRY V.C. & McCORMICK J.E. (1957) Properties of periodate oxidised polysaccharides. Part VI. The mucilage from *Dilsea edulis*. *J. chem. Soc.*, 2777–83.

BERKELEY E.E. & KERR T. (1946) Structure and plasticity of undried cotton fibers. *Ind. Eng. Chem.* 38, 304–9.

BITTIGER H. & MARCHESSAULT R.H. (1971) Fiber structure of mannan triacetate. *Carbohyd. Res.* 18, 469–70.

BJÖRNDAHL H., ERIKSSON K.-E., GAREGG P.J., LINDBERG B. & SWAN B. (1965) Studies on the xylan from the red seaweed *Rhodymenia palmata*. *Acta chem. scand.* 19, 2309–15.

BLACKWELL J., PARKER K.D. & RUDALL K.M. (1967) Chitin fibres of the diatoms *Thalassiosira fluviatilis* and *Cyclotella cryptica*. *J. molec. Biol.* 28, 383–5.

BOURNE E.J., JOHNSON P.G. & PERCIVAL E. (1970) Water soluble polysaccharides of *Cladophora rupestris*. IV. Autohydrolysis methylation of the partly desulphated material and correlation with the results from Smith degradation. *J. chem. Soc.*, 1561–9.

BOURNE E.J., PERCIVAL E. & SMESTAD B. (1972) Carbohydrates of *Acetabularia* species. Part I. *A. crenulata. Carbohyd. Res.* **22**, 75–82.

BOWKER D.M. & TURVEY J.R. (1968) Water soluble polysaccharides of the red alga *Laurencia pinnatifida*. Part I. Constituent units. *J. chem. Soc.* 983–5.

Part II. Methylation analysis of the galactan sulphate. *Ibid.* 989–91.

BROWN R.M., FRANCKE W.W., KLEINIG H., FALK H. & SITTE P. (1970) Scale formation in Chrysophycean algae. *J. Cell Biol.* **45**, 246–71.

BRUCKER R.F., WORMINGTON C.M. & NAKADA H.I. (1971) Comparison of some physiochemical properties of alginic acids of differing composition. *J. Macromol. Sci. Chem.* **5**, 1169–85.

CAULFIELD D.F. (1971) Crystallite size in wet and dry *Valonia ventricosa. Textile Research J.* **41**, 267–74.

CEREZO A.S., LEZEROVICH A., LABRIOLA R. & REES D.A. (1971) A xylan from the red seaweed *Chaetangium fastigiatum. Carbohyd. Res.* **19**, 289–96.

CEREZO A.S. (1972) The fine structure of *Chaetangium fastigiatum* xylan: Studies of the sequence and configuration of the 1 → 3 linkages. *Carbohyd. Res.* **22**, 209–11.

CHILD T.F., PRYCE N.E., TAIT M.J. & ABLETT S. (1970) Proton and deuteron magnetic resonance studies of aqueous polysaccharides. *J. chem. Soc. D* **18**, 1214–15.

CLELAND R. & HAUGHTON P.M. (1971) The effect of auxin on stress relaxation in isolated *Avena* coleoptiles. *Pl. Physiol., Lancaster,* **47**, 812–15.

CLINGMAN A.L. & NUNN J.R. (1959) Red seaweed polysaccharides. Part III. Polysaccharides from *Hypnea specifera. J. chem. Soc.* 493–6.

COLVIN J.R. (1963) The size of the cellulose microfibril. *J. Cell Biol.* **17**, 105–9.

CÔTÉ R.H. (1959) Disaccharides from fucoidin. *J. chem. Soc.* 2248–54.

CRONSHAW J., MYERS A. & PRESTON R.D. (1958) A chemical and physical investigation of the cell walls of some marine algae. *Biochim. biophys. Acta,* **27**, 89–103.

CRONSHAW J. & PRESTON R.D. (1958) A re-examination of the fine structure of the walls of vesicles of the green alga *Valonia. Proc. R. Soc. B* **148**, 137–48.

DEA I.C.M., MOORHOUSE R., REES D.A., ARNOTT S., GUSS J.M. & BALAZS E.A. (1973) Hyaluronic acid: A novel double helical model. *Science N.Y.* **179**, 560–2.

DENNIS D.T. & PRESTON R.D. (1961) Constitution of cellulose microfibrils. *Nature, Lond.* **191**, 667–8.

DILLON T. & McKENNA J. (1950) Mucilage of *Dumontia incrassata. Nature, Lond.* **165**, 318.

DOLAN T.C.S. & REES D.A. (1965) The Carrageenans. Part II. The positions of the glycosidic linkages and sulphate esters in λ-carrageenan. *J. chem. Soc.* 3534–9.

DUCKWORTH M., HONG K.C. & YAPHE W. (1970) The agar polysaccharides of *Gracilaria* species. *Carbohyd. Res.* **18**, 1–9.

DUCKWORTH M. & YAPHE W. (1971) The structure of agar. Part I. Fractionation of a complex mixture of polysaccharides. *Carbohyd. Res.* **16**, 189–97.

Part II. The use of a bacterial agarase to elucidate structural features of the charged polysaccharides in agar. *Ibid.* **16**, 434–45.

DWELTZ N.E., COLVIN J.R. & McINNES A.G. (1968) Studies on chitin (β-(1 → 4)-linked 2-acetamido-2-deoxy-D-glucan) fibers of the diatom *Thallassiosira fluviatilis* Hustedt. III. The structure of chitin from X-ray diffraction and electron microscope observations. *Can. J. Chem.* **46**, 1513–21.

EVANS L.V. & HOLLIGAN M.S. (1972) Correlated light and electron microscope studies on brown algae. I. Localisation of alginic acid and sulphated polysaccharides in *Dictyota. New Phytol.* **71**, 1161–72.

FARRANT A.J., NUNN J.R. & PAROLIS H. (1971) A polysaccharide from *Pachymenia carnosa. Carbohyd. Res.* **19**, 161–8.

FISCHER F.G. & DÖRFEL H. (1955) Die Polyuronsäuren der Braunalgen. *Hoppe-Seyler's Z. Physiol. Chem.* **302**, 186–203.

FISCHER D.G. & MANN J. (1960) Crystalline modifications of cellulose. Part VI. Unit cell and molecular symmetry of cellulose I. *J. Polymer Sci.* **42**, 189–94.

FISHER I.S. & PERCIVAL E. (1957) The water soluble polysaccharides of *Cladophora rupestris*. *J. chem. Soc.* 2666–75.

FORD C.W. & PERCIVAL E. (1965a) The carbohydrates of *Phaeodactylum tricornutum*. Part II. A sulphated glucuronomannan. *J. chem. Soc.* 7042–6.

FORD C.W. & PERCIVAL E. (1965b) Polysaccharides synthesised by *Monodus subterraneus*. *J. Chem. Soc.* 3014–3016.

FREI E. & PRESTON R.D. (1961a) Variants in the structural polysaccharides of algal cell walls. *Nature, Lond.* **192**, 939–43.

FREI E. & PRESTON R.D. (1961b) Cell wall organisation and wall growth in the filamentous green algae *Cladophora* and *Chaetomorpha*. I. The basic structure and its formation. *Proc. Roy. Soc. B* **154**, 70–94. II. Spiral structure and spiral growth. *Ibid. B* **155**, 55–77.

FREI E. & PRESTON R.D. (1962) Configuration of alginic acid in marine brown algae. *Nature, Lond.* **196**, 130–4.

FREI E. & PRESTON R.D. (1964) Non-cellulosic structural polysaccharides in algal cell walls. I. Xylan in siphoneous green algae. *Proc. Roy. Soc. B* **160**, 293–313. II. Association of xylan and mannan in *Porphyra umbilicalis*. *Ibid. B* **160**, 314–27.

FREI E. & PRESTON R.D. (1968) Non-cellulosic structural polysaccharides in algal cell walls. III. Mannan in siphoneous green algae. *Proc. R. Soc. B* **169**, 127–45.

FREY-WYSSLING A. & MÜHLETHALER K. (1963) Die Elementarfibrillen der Cellulose. *Makromol. Chemie.* **62**, 25–30.

GORIN P.A.J. & SPENCER J.F.T. (1966) Exocellular alginic acid from *Azotobacter vinelandii*. *Can. J. Chem.* **44**, 993–8.

GOTELLI I.B. & CLELAND R. (1968) Differences in the occurrence and distribution of hydroxyproline-proteins among the algae. *Am. J. Bot.* **55**, 907–14.

GREEN P.B. (1958) Concerning the site of the addition of new wall substances to the elongating *Nitella* cell walls. *Am. J. Bot.* **45**, 111–16.

HÄMMERLING J. (1944) Zur Lebensweise, Fortpflanzung und Entwicklung verschiedener Dasycladaceen. *Arch. Protist.* **97**, 7–56.

HANIC L.A. & CRAIGIE J.S. (1969) Studies on the algal cuticle. *J. Phycol.* **5**, 89–102.

HAQ Q.N. & PERCIVAL E. (1966a) Structural studies on the water-soluble polysaccharide from the green seaweed, *Ulva lactuca*. *Some Contemporary Studies in Marine Sciences*, ed. Barnes H., pp. 355–63. Allen and Unwin, London.

HAQ Q.N. & PERCIVAL E. (1966b) Structural studies on the water soluble polysaccharide from the green seaweed, *Ulva lactuca*. Part IV. Smith degradation. *Proc. Vth Int. Seaweed Symp.* pp. 261–9.

HARRIS P.J. & NORTHCOTE D.H. (1971) Polysaccharide formation in plant Golgi bodies. *Biochim. biophys. Acta,* **237**, 56–64.

HASSID W.Z. (1970 Biosynthesis of sugars and polysaccharides. *The Carbohydrates. Chemistry and Biochemistry Vol. IIA*, eds. Pigman W. & Horton D., pp. 301–73. Academic Press, New York & London.

HAUG A. & LARSEN B. (1971) Biosynthesis of alginate. Part II. Polymannuronic C-5-epimerase from *Azotobacter vinelandii* (Lipman). *Carbohyd. Res.* **17**, 297–308. Part III. Tritium incorporation with polymannuronic acid 5-epimerase from *Azotobacter vinelandii* (Lipman). *Ibid.* **20**, 225–32.

HAUG A., LARSEN B. & BAARDSETH E. (1969) Comparison of the constitution of alginates from different sources. *Proc. VIth Int. Seaweed Symp.* pp. 443–51.

HAUG A., LARSEN B. & SMIDSRØD O. (1966) A study of the constitution of alginic acid by partial acid hydrolysis. *Acta chem. scand.* **20**, 183–90.

HAUG A., LARSEN B. & SMIDSRØD O. (1967) Studies on the sequence of uronic acid residues in alginic acid. *Acta chem. scand.* **21,** 691–704.

HAUG A., MYKLESTAD S., LARSEN B. & SMIDSRØD O. (1967) Correlation between chemical structure and physical properties of alginates. *Acta chem. scand.* **21,** 769–78.

HAUG A. & SMIDSRØD O. (1965) The effect of divalent metals on the properties of alginate solutions. *Acta chem. scand.* **19,** 341–51.

HAUG A. & SMIDSRØD O. (1967a) Strontium-calcium selectivity of alginates. *Nature, Lond.* **215,** 757.

HAUG A. & SMIDSRØD O. (1967b) Strontium, calcium and magnesium in brown algae. *Nature, Lond.* **215,** 1167–8.

HAUGHTON P.M. & SELLEN D.B. (1969) Dynamic mechanical properties of the cell walls of some green algae. *J. exp. Bot.* **20,** 516–35.

HAUGHTON P.M., SELLEN D.B. & PRESTON R.D. (1968) Dynamic mechanical properties of the cell wall of *Nitella opaca. J. exp. Bot.* **19,** 1–12.

HERMANS P.H. (1949) *Physics and Chemistry of Cellulose Fibres.* Elsevier, New York.

HEYN A.N.J. (1965) Observations on the size and shapes of the cellulose microcrystallite in cotton fiber by electron staining. *J. Applied Phys.* **36,** 208.

HEYN A.N.J. (1966) The microcrystalline structure of cellulose in cell walls of cotton, ramie and jute fibers as revealed by negative staining of sections. *J. Cell Biol.* **29,** 181–97.

HIRASE S. & ARAKI C. (1954) Studies on the chemical constitution of agar-agar. XVI. Isolation of crystalline agarobiose diethylmercaptal by mercaptolysis of agar-agar. *Bull. Chem. Soc. Japan,* **27,** 105–9.

HIRASE S., ARAKI C. & WATANABE K. (1967) Component sugars of the polysaccharide of the red seaweed *Grateloupia elliptica. Bull. Chem. Soc. Japan,* **40,** 1445–8.

HIRASE S. & WATANABE K. (1971) Fractionation and structural investigation of funoran. *Abs. VIIth Int. Seaweed Symp.* p. 100.

HIRST E.L., MACKIE W. & PERCIVAL E. (1965) The water soluble polysaccharides of *Cladophora rupestris* and *Chaetomorpha* spp. Part II. The site of ester sulphate groups and the linkage between the galactose residues. *J. chem. Soc.* 2958–67.

HIRST E.L., PERCIVAL E. & WOLD J.K. (1963) Structural studies of alginic acid. *Chemy. Ind.* 257.

HIRST E.L. & REES D.A. (1965) The structure of alginic acid. Part V. Isolation and unambiguous characterisation of some hydrolysis products of the methylated polysaccharide. *J. chem. Soc.* 1182–7.

HOLT C., MACKIE W. & SELLEN D.B. (1972) Degree of polymerization and polydispersity of native cellulose. I.U.P.A.C. *Int. Symp. on Macromolecules, Helsinki,* **5,** Section V, 37–41.

HONJO G. & WATANABE M. (1958) Examination of cellulose fibre by the low-temperature specimen method of electron diffraction and electron microscopy. *Nature, Lond.* **181,** 326–8.

HOWARD B.H. (1957) Hydrolysis of the soluble pentosans of wheat flour and *Rhodymenia palmata* by ruminal microorganisms. *Biochem. J.* **67,** 643–51.

HUNSTMAN S.A. & SLONEKER J.H. (1971) An exocellular polysaccharide from the diatom *Gomphonema olivaceum. J. Phycol.* **7,** 261–4.

IRIKI Y. & MIWA T. (1960) Chemical nature of the cell wall of the green algae, *Codium, Acetabularia* and *Halicoryne. Nature, Lond.* **185,** 178–9.

IRIKI Y., SUZUKI T., NISIZAWA K. & MIWA T. (1960) Xylan of siphonaceous green algae. *Nature, Lond.* **187,** 82–3.

IZUMI K. (1971a) Chemical heterogeneity of the agar from *Gelidium amansii. Carbohyd. Res.* **17,** 227–30.

IZUMI K. (1971b) Heterogeneity of anhydrogalactose containing polysaccharides from *Gloiopeltis furcata. Agr. Biol. Chem.* **35,** 651–5.

JOHNSON P.G. & PERCIVAL E. (1969) The water soluble polysaccharides of *Cladophora rupestris*. Part III. Smith degradation. *J. chem. Soc. C*, 906–9.

KIRILLOVA V.S., KOSENKO L.V. & RATUSHNAYA M.Y. (1967) Chemical composition of two nitrogen fixing blue-green algae. *Chem. Abstr.* **72**, 39945Y.

KOCHETKOV N.K., USOV A.I. & MIROSHNIKOVA C.I. (1970) Polysaccharides of algae IV. Fractionation and methanolysis of a sulphated polysaccharide from *Laingia pacifica*. *Zh. Obsch. Khim.* **40**, 2469–78 (*Chem. Abstr.* **75**, 58759u, 58761p).

KOCHETKOV N.K., USOV A.I. & REKHTER M.A. (1971) Polysaccharides of algae VIII. Acetolysis of λ-polysaccharides from *Tichocarpus crinitus*. *Zh. Obsch. Khim.* **41**, 1160–5 (*Chem. Abstr.* **75**, 772126).

KREGER D.R. (1962) Cell Walls. In *Physiology and biochemistry of algae*, ed. Lewin R.A., pp. 315–35. Academic Press, New York & London.

LARSEN B. (1967) Sulphated polysaccharides in brown algae. II. Isolation of 3-O-β-D-xylopyranosyl-L-fucose from ascophyllan. *Acta chem. scand.* **21**, 1395–6.

LARSEN B. & HAUG A. (1971) Biosynthesis of alginate. Part I. Composition and structure of alginate produced by *Azotobacter vinelandii* (Lipman). *Carbohyd. Res.* **17**, 287–96.

LARSEN B., HAUG A. & PAINTER T.J. (1966) Sulphated polysaccharides in brown algae. I. Isolation and preliminary characterisation of three sulphated polysaccharides from *Ascophyllum nodosum* (L.) Le Jol. *Acta chem. scand.* **20**, 219–30.

LARSEN B., HAUG A. & PAINTER T.J. (1970) Sulphated polysaccharides in brown algae. III. The native state of fucoidan in *Ascophyllum nodosum* and *Fucus vesiculosus*. *Acta chem. scand.* **24**, 3339–52.

LAWSON C.J. & REES D.A. (1968) Carrageenans. Part VI. Reinvestigation of the acetolysis products of λ-carrageenan. Revision of the structure of 'α-1,3-galactotriose' and a further example of the reverse specificities of glycoside hydrolysis and acetolysis. *J. chem. Soc. C*, 1301–4.

LAWSON C.J. & REES D.A. (1970) An enzyme for the metabolic control of polysaccharide conformation and function. *Nature, Lond.* **227**, 343.

LEWIN J.C. (1955) The capsule of the diatom *Navicula pelliculosa*. *J. gen. Microbiol.* **13**, 162–9.

LEWIN R.A. (1958) The mucilage tubes of *Amphipleura rutilans*. *Limnol. Oceanog.* **3**, 111–13.

LIANG C.Y. & MARCHESSAULT R.H. (1959) Infrared spectra of crystalline polysaccharides. I. Hydrogen bonds in native celluloses. *J. Polymer Sci.* **37**, 385–95.

LINKER A. & JONES R.S. (1966) A new polysaccharide resembling alginic acid isolated from pseudomonads. *J. biol. Chem.* **241**, 3845–51.

LOVE J. & PERCIVAL E. (1964) The polysaccharides of the green seaweed *Codium fragile*. Part II. The water soluble sulphated polysaccharides. *J. chem. Soc.* 3338–45. Part III. A β-1,4-linked mannan. *Ibid.* 3345–50.

MACKIE I.M. & PERCIVAL E. (1959) The constitution of xylan from the green seaweed *Caulerpa filiformis*. *J. chem. Soc.* 1151–6.

MACKIE I.M. & PERCIVAL E. (1961) Polysaccharides from the green seaweeds of *Caulerpa* spp. Part III. Detailed study of the water soluble polysaccharides of *C. filiformis*: Comparison with the polysaccharides synthesised by *C. racemosa* and *C. sertularoides*. *J. chem. Soc.* 3010–15.

MACKIE W. (1969) The degree of polymerization of xylan in the cell wall of the green seaweed *Penicillus dumetosus*. *Carbohyd. Res.* **9**, 247–9.

MACKIE W. (1971) Conformations of crystalline alginic acids and their salts. *Biochem. J.* **125**, 89P.

MACKIE W. & PRESTON R.D. (1968) The occurrence of mannan microfibrils in the green algae *Codium fragile* and *Acetabularia crenulata*. *Planta*, **79**, 249–53.

MACKIE W. & SELLEN D.B. (1969) The degree of polymerization and polydispersity of mannan from the cell wall of the green seaweed, *Codium fragile*. *Polymer*, **10**, 621–32.

MACKIE W. & SELLEN D.B. (1971) The degree of polymerization and polydispersity of xylan from the cell wall of the green seaweed *Penicillus dumetosus*. *Biopolymers*, **10**, 1–9.

MANLEY R.ST.JOHN (1971) The molecular morphology of cellulose. *J. Polymer Sci.* **9** B, 1025–59.

MANN J. & MARRINAN H.J. (1958) Polarized infrared spectra of cellulose I. *J. Polymer Sci.* **27**, 595–6.

MANNERS D.J. & MITCHELL J.P. (1963) The fine structure of *Rhodymenia palmata* xylan. *Biochem. J.* **89**, 92P.

MARX-FIGINI M. (1969) Untersuchungen zur Biosynthese der Cellulose in der Alga *Valonia*. *Biochim. biophys. Acta*, **177**, 27–34.

MARX-FIGINI M. & SCHULZ G.V. (1966) Über die Kinetik und den Mechanismus der Biosynthese der Cellulose in den höheren Pflanzen (Nach Versuchen an den Samenhaaren der Baumwolle). *Biochim. biophys. Acta*, **112**, 81–101.

MCKINNELL J.P. & PERCIVAL E. (1962a) Structural investigations on the water soluble polysaccharides of the green seaweed *Entermorpha compressa. J. chem. Soc.* 3141–8.

MCKINNELL J.P. & PERCIVAL E. (1962b) The acid polysaccharide from the green seaweed *Ulva lactuca. J. chem. Soc.* 2082–3.

MCKINNON A.A., REES D.A. & WILLIAMSON F.B. (1969) Coil to double helix transition for a polysaccharide. *Chem. Commun.* 701–2.

MEIER H. (1969) Biosynthesis of cellulose in growing cotton hairs. *Phytochemistry*, **8**, 579–83.

MEYER K.H. & MISCH L. (1937) Positions des atomes dans le nouveau modèle spatiale de la cellulose. *Helv. Chim. Acta*, **20**, 232–44.

MOLLENHAUER H.H., WHALEY W.G. & LEECH J.H. (1961) A function of the Golgi apparatus in outer rootcap cells. *J. Ultrastr. Res.* **5**, 193–200.

MOORE B.G. & TISCHER R.G. (1965) Biosynthesis of extracellular polysaccharides by the blue-green alga *Anabaena flos-aquae. Can. J. Microbiol.* **11**, 877–85.

MORGAN K. & O'NEILL A.N. (1959) Degradative studies on λ-carrageenin. *Can. J. Chem.* **37**, 1201–9.

MÜHLETHALER K. (1960) Der Feinstruktur der Zellulosemikrofibrillen. *Beih. Z. Schweiz. Forstver.* **30**, 55.

NICOLAI E. (1957) Wall deposition in *Chaetomorpha melagonium* (Cladophorales). *Nature, Lond.* **180**, 491–3.

NICOLAI M.F.E. & FREY-WYSSLING A. (1938) Über den Feinbau der Zellwand von *Chaetomorpha. Protoplasma*, **30**, 401–13.

NIEDUSZYNSKI I.A. & ATKINS E.D.T. (1970) Preliminary investigation of algal cellulose. I. X-ray intensity data. *Biochim. biophys. Acta*, **222**, 109–18.

NIEDUSZYNSKI I.A. & MARCHESSAULT R.H. (1971) Structure of β-D-(1 → 4′) xylan hydrate. *Nature, Lond.* **232**, 46–7.

NIEDUSZYNSKI I.A. & MARCHESSAULT R.H. (1972) Structure of β-D-(1 → 4) mannan. *Can. J. Chem.* **50**, 2130–2138.

NIEDUSZYNSKI I.A. & PRESTON R.D. (1970) Crystallite size in natural cellulose. *Nature, Lond.* **225**, 273–4.

NORTHCOTE D.H. & PICKETT-HEAPS J.D. (1966) A function of the Golgi apparatus in polysaccharide synthesis and transport in root cap cells of wheat. *Biochem. J.* **98**, 159–67.

NUNN J.R. & PAROLIS H. (1968) A Polysaccharide from *Aeodes orbitosa. Carbohyd. Res.* **6**, 1–11. Sulphated polysaccharides of the Grateloupiaceae family. Part II. Isolation of 4-O-methyl-L-galactose, 6-O-methyl-D-galactose and two disaccharides from hydrolysates of aeodan. *Ibid.* **8**, 361–2.

NUNN J.R. & PAROLIS H. (1969) A polysaccharide from *Phyllymenia cornea. Carbohyd. Res.* **9**, 265–76.

NUNN J.R. & PAROLIS H. (1970) Methylation analysis of phyllymenan and desulphated phyllymenan. *Carbohyd. Res.* **14**, 145–50.

NUNN J.R., PAROLIS H. & RUSSELL I. (1971) Sulphated polysaccharides of the Solieraceae family. Part I. A polysaccharide from *Anatheca dentata. Carbohyd. Res.* **20**, 205–15.

O'COLLA P.S. (1962) Mucilages. In *Physiology and biochemistry of algae*, ed. Lewin R.A., pp. 337–56. Academic Press, New York & London.

O'DONNELL J.J. & PERCIVAL E. (1959) Structural investigations on the water soluble poly-saccharides from the green seaweed *Acrosiphonia centralis* (*Spongomorpha arcta*). *J. chem. Soc.* 2168–78.

O'NEILL A.N. (1954) Degradative studies on fucoidin. *J. Am. chem. Soc.* **76**, 5074–6.

O'NEILL A.N. (1955) Derivatives of 4-O-β-D-galactopyranosyl-3,6-anhydro-D-galactose from x-carrageenan. *J. Am. chem. Soc.* **77**, 6324–6.

OVODOV Y.S. & PAVLENKO A.F. (1970) Polysaccharides from brown algae. III. Pelvetian from *Pelvetia wrightii. Khim. Prir. Soedin.* **6**, 400–2 (*Chem. Abstr.* **74**, 10339j).

PAINTER T.J. (1960) The polysaccharides of *Furcellaria fastigiata.* I. Isolation and partial mercaptolysis of a gel fraction. *Can. J. Chem.* **38**, 112–18.

PAINTER T.J. (1966) The location of sulphate half ester groups in furcelleran and x-carra-geenan. *Proc. Vth Int. Seaweed Symp.* pp. 305–11.

PARKER B.C. (1964) The structure and composition of cell walls of three chlorophycean algae. *Phycologia*, **4**, 63–74.

PEAT S. & TURVEY J.R. (1965) Polysaccharides of marine algae. *Fortschritte der Chemie Organischer Naturstoffe*, **23**, 1–45.

PERCIVAL E. (1968) Glucuronoxylofucan, a cell wall component of *Ascophyllum nodosum.* Part I. *Carbohyd. Res.* **7**, 272–83.

PERCIVAL E. (1970) Algal polysaccharides. In *The Carbohydrates. Chemistry and Biochemistry. Vol. IIB*, eds. Pigman W. & Horton D., pp. 537–68. Academic Press, New York & London.

PERCIVAL E. (1971) Glucuronoxylofucan, a cell wall component of *Ascophyllum nodosum* Part II. *Carbohyd. Res.* **17**, 121–6.

PERCIVAL E.G.V. & CHANDA S.K. (1950) Xylan of *Rhodymenia palmata. Nature, Lond.* **166**, 787–8.

PERCIVAL E. & JABBAR MIAN A. (1971) Fucose containing polysaccharides of the Phaeo-phyceae. *Abs. VIIth Int. Seaweed Symp.* p. 98.

PERCIVAL E. & MCDOWELL R.H. (1967) *Chemistry and Enzymology of Marine Algal Poly-saccharides.* Academic Press, New York & London.

PERCIVAL E. & WOLD J.K. (1963) The acid polysaccharide from the green Seaweed *Ulva lactuca.* Part II. The site of ester sulphate. *J. chem. Soc.* 5459–68.

PINSKY A. & ORDIN L. (1969) Role of lipid in the cellulose synthetase enzyme system from oat seedlings. *Pl. Cell Physiol., Tokyo*, **10**, 771–85.

PRESTON R.D. (1959) Wall organisation in plant cells. *Int. Rev. Cytol.* **8**, 33–60.

PRESTON R.D. (1964) Structural and mechanical aspects of plant cell walls with particular reference to synthesis and growth. In *The Formation of Wood in Forest Trees*, ed. Zimmerman M.H., pp. 169–88. Academic Press, New York & London.

PRESTON R.D. (1971) Negative staining and cellulose microfibril size. *J. Micros.* **93**, 7–13.

PRESTON R.D. (1974a) Plant cell walls. In *Dynamic aspects of plant ultrastructure*, ed. Ro-bards A.W. In press.

PRESTON R.D. (1974b) *Physical biology of plant cell walls.* Chapman & Hall, Lond. In press.

PRESTON R.D. & ASTBURY W.T. (1937) The structure of the wall of the green alga *Valonia ventricosa. Proc. R. Soc. B*, **122**, 76–97.

PRESTON R.D., WARDROP A.B. & NICOLAI M.F.E. (1948) The fine structure of cell walls in fresh plant tissues. *Nature, Lond.* **162**, 957–9.

PROBINE M.C. & PRESTON R.D. (1961) Cell growth and the structure and mechanical proper-ties of the wall in internodal cells of *Nitella opaca.* I. Wall structure and growth. *J. Exp. Bot.* **12**, 261–82.

PROBINE M.C. & PRESTON R.D. (1962) Cell growth and the structure and mechanical proper-ties of the wall in internodal cells of *Nitella opaca.* II. Mechanical properties of the wall. *J. exp. Bot.* **13**, 111–27.

RÅNBY B.G. (1949) Aqueous colloidal solutions of cellulose micelles. *Acta chem. scand.* **3**, 649–50.

RÅNBY B.G. (1951) The colloidal properties of cellulose micelles. *Disc. Faraday Soc. No. 11*, 158–64.

RAO V.S.R., SUNDARARAJAN P.R., RAMAKRISHNAN C. & RAMACHANDRAN G.N. (1967) In *Conformation of Biopolymers*, ed. Ramachandran G.N. Academic Press, New York.

REES D.A. (1961a) Estimation of the relative amounts of isomeric sulphate esters in some sulphated polysaccharides. *J. chem. Soc.* 5168–71.

REES D.A. (1961b) Enzymic synthesis of 3,6-anhydro-L-galactose within porphyran from L-galactose 6-sulphate units. *Biochem. J.* **81**, 347–52.

REES D.A. (1963a) The carrageenan system of polysaccharides. Part I. The relation between the ϰ- and λ-components. *J. chem. Soc.* 1821–32.

REES D.A. (1963b) A note on the characterisation of carbohydrate sulphates by acid hydrolysis. *Biochem. J.* **88**, 343–5.

REES D.A. (1965) Carbohydrate sulphates. *Ann. Rep. Chem. Soc.* **62**, 479–84.

REES D.A. (1969a) Conformational analysis of polyscaaharides. Part II. Alternating copolymers of the agar-carrageenan-chondroitin type by model building in the computer with calculation of helical parameters. *J. chem. Soc. B*, 217–26.

REES D.A. (1969b) Structure, conformation and mechanism in the formation of polysaccharide gels and networks. *Advances in Carbohydrate Chemistry and Biochemistry*, **24**, 267–332.

REES D.A. (1972) Shapely polysaccharides. *Biochem. J.* **126**, 257–73.

REES D.A. & SAMUEL J.W.B. (1967) The structure of alginic acid. Part VI. *J. chem. Soc. C*, 2295–8.

REES D.A. & SCOTT W.E. (1971) Polysaccharide conformation. Part VI. Computer model building for linear and branched pyranoglycans. Correlations with biological function. Preliminary assessment of inter-residue forces in aqueous solution. Further interpretation of optical rotation in terms of chain conformation. *J. chem. Soc. B*, 469–79.

REES D.A. & SKERRETT R.J. (1968) Conformational analysis of cellobiose, cellulose and xylan. *Carbohyd. Res.* **7**, 334–48.

REES D.A., STEELE I.W. & WILLIAMSON F.B. (1969) Conformational analysis of polysaccharides. Part III. The relationship between stereochemistry and properties of some natural polysaccharide sulphates. *J. Polymer Sci. C*, **28**, 261–76.

RIBI E. & RÅNBY B.G. (1950) Zur elektronmikroskopischen Präparaten von Kolloiden. *Experientia*, **6**, 27–8.

ROBINSON D.G. & PRESTON R.D. (1971a) Studies on the fine structure of *Glaucocystis nostochinearum* Itzigs. I. Wall structure. *J. exp. Bot.* **22**, 635–43.

ROBINSON D.G. & Preston R.D. (1971b) The fine structure of swarmers of *Cladophora* and *Chaetomorpha*. I. The plasmalemma and Golgi apparatus in naked swarmers. *J. Cell Sci.* **9**, 581–601.

ROBINSON D.G. & PRESTON R.D. (1972a) Plasmalemma structure in relation to microfibril biosynthesis in *Oocystis*. *Planta*, **104**, 234–46.

ROBINSON D.G. & PRESTON R.D. (1972b) Polysaccharide synthesis in mung bean roots— An X-ray investigation. *Biochim. biophys. Acta*, **273**, 336–45.

ROBINSON D.G. & WHITE R.K. (1972) The fine structure of *Oocystis apiculata* W. West with particular reference to the cell wall. *Br. Phycol. J.* **7**, 109.

ROBINSON D.G., WHITE R.K. & PRESTON R.D. (1972) Fine structure of swarmers of *Cladophora* and *Chaetomorpha*. III. Wall synthesis and development. *Planta*, **107**, 131–144.

SCHNEPF E. (1965) Struktur der Zellwände und Cellulosefibrillen bei *Glaucocystis*. *Planta*, **67**, 213–24.

SMIDSRØD O. (1970) Solution properties of alginate. *Carbohyd. Res.* **13**, 359–72.

SMIDSRØD O. & HAUG A. (1968a) Dependence upon uronic acid composition of some ion-exchange properties of alginates. *Acta chem. scand* **22**, 1989–97.

SMIDSRØD O. & HAUG A. (1968b) A light scattering study of alginate. *Acta chem. scand.* **22**, 797–810.

SMIDSRØD O. & HAUG A. (1972a) Dependence upon the gel-sol state of the ion exchange properties of alginates. *Acta chem. scand.* **26**, 2063–74.

SMIDSRØD O. & HAUG A. (1972b) Properties of poly (1,4-hexuronates) in the gel state. II. Comparison of gels of different chemical composition. *Acta chem. scand.* **26**, 79–88.

SMIDSRØD O., HAUG A. & WHITTINGTON S.G. (1972a) The molecular basis for some physical properties of polyuronides. Presented at the I.U.P.A.C. Int. Symp. Macromolecules, Helsinki, Finland.

SMIDSRØD O., HAUG A. & WHITTINGTON S.G. (1972b) Molecular basis for some physical properties of polyuronides. *Acta chem. scand.* **26**, 2563–6.

STANICOFF D.J. & STANLEY N.F. (1969) Infrared studies on algal polysaccharides. *Proc. VIth Int. Seaweed Symp.* pp. 595–609.

STEWART C.M., DAWES C.J., DICKINS B.M. & NICHOLLS J.W.P. (1969) Cell wall constituents of *Apjohnia laetevirens. Aust. J. Mar. Freshwater Res.* **20**, 143–55.

SUNDARALINGHAM M. (1968) Some aspects of stereochemistry and hydrogen bonding of carbohydrates related to polysaccharide conformations. *Biopolymers,* **6**, 189–213.

SUTTON A., HARRISON G.E., CARR T.E.F. & BARLTROP D. (1971) Reduction in the absorption of dietary strontium in children by an alginate derivative. *Int. J. Radiat. Biol.* **19**, 79–85.

THOMPSON E.W. & PRESTON R.D. (1967) Proteins in the cell walls of some green algae. *Nature, Lond.* **213**, 684–5.

THOMPSON E.W. & PRESTON R.D. (1968) Evidence for a structural role of protein in algal cell walls. *J. exp. Bot.* **19**, 690–7.

TURVEY J.R. & SIMPSON P.R. (1966) Polysaccharides from *Corallina officinalis. Proc. Vth Int. Seaweed Symp.* pp. 323–7.

TURVEY J.R. & WILLIAMS T.P. (1963) Sugar sulphates from the mucilage of *Porphyra umbilicalis. Proc. IVth Int. Seaweed Symp.* pp. 370–3.

TURVEY J.R. & WILLIAMS E.L. (1970) The structures of some xylans from red algae. *Phytochemistry,* **9**, 2383–8.

USOV A.I., LOTOV R.A. & KOCHETKOV N.K. (1971) Polysaccharides of algae. VII. Preliminary study of polysaccharides of the red alga *Rhodomela larix. Zh. Obsch. Khim.* **41**, 1154–60 (*Chem. Abstr.* **75**, 72782c).

USOV A.I., REKHTER M.A. & KOCHETKOV N.K. (1969) Polysaccharides from algae III. Separation and preliminary study of a γ-polysaccharide from *Tichocarpus crinitus. Zh. Obsch. Khim.* **39**, 905–11. (*Chem. Abstr.* **71**, 56809d).

USOV A.I. REKHTER M.A. & KOCHETKOV N.K. (1970) Polysaccharides from algae. Part VI. α-Polysaccharide from *Tichocarpus crinitus. Zh. Obsch. Khim.* **40**, 2732–7. (*Chem. Abstr.* **75**, 1289m).

VAN DER BORGHT O., VAN PUYMBROECK S. & COLARD J. (1971) Intestinal absorption and body retention of $^{226}$Ra and $^{47}$Ca in mice. *Health Phys.* **21**, 181–96.

VILLEMEZ C.L. & CLARK A.F. (1969) A particle bound intermediate in the biosynthesis of plant cell wall polysaccharides. *Biochem. biophys. Res. Comm.* **36**, 57–63.

VILLEMEZ C.L. & HELLER J.S. (1970) Is guanosine diphosphate-D-glucose a precursor of cellulose? *Nature, Lond.* **227**, 80–1.

WEIGL J., TURVEY J.R. & YAPHE W. (1966) The enzymic hydrolysis of ϰ-carrageenan. *Proc. Vth Int. Seaweed Symp.* pp. 329–32.

WELLARD H.J. (1954) Variation in the lattice spacing of cellulose. *J. Polymer Sci.* **13**, 471–6.

WHITTINGTON S.G. (1971) Conformational energy calculations of alginic acid. I. Helix parameters and flexibility of the homopolymers. *Biopolymers,* **10**, 1481–9. II. Conformational Statistics of the Copolymer. *Ibid.* **10**, 1617–23.

YAPHE W. (1959) The determination of ϰ-carrageenan as a factor in the classification of the Rhodophyceae. *Can. J. Bot.* **37**, 751–7.

YOUNG K., DUCKWORTH M. & YAPHE W. (1971) The structure of agar. Part III. Pyruvic acid, a common feature of agars from different agarophytes. *Carbohyd. Res.* **16**, 446–8.

D

# CHAPTER 3

# CYTOPLASMIC ORGANELLES

L. V. EVANS

Department of Plant Sciences,
University of Leeds, Leeds, U.K.

1  **Introduction**  86

2  **Organelles in unicellular forms**
   87
2.1  Flagella  87
2.2  Haptonemata  89
2.3  Trichocysts  89
2.4  Cell coverings  89
2.5  Lysosomes  90
2.6  Microbodies  91
2.7  Contractile vacuoles  91

3  **Organelles in multicellular forms**
   92
3.1  Walls  92

3.2  Intercellular connections  92

4  **Organelles in unicellular and
   multicellular forms**  95
4.1  Nuclei  95
4.2  Microtubules  96
4.3  Nuclear division  97
4.4  Mitochondria  101
4.5  Golgi apparatus  103

5  **References**  113

## 1  INTRODUCTION

In the algae, as in other plant groups, electron microscopy is providing detailed knowledge of the structure of cells and their organelles. Biochemical and bio-physical techniques, on the other hand, contribute much information on their possible function. Such observations, considered together, provide a basis for an improved understanding of the inter-relationship between organelle structure and function in the integrated cell or organism. Here, a brief account of more important recent findings on the fine structure of algal cell organelles will be given and an attempt made to relate this, wherever possible, to what is known of the functioning of such organelles (see also Dodge 1973, 1974).

Despite the tremendous variation in form, organization and size shown by the algae, algal cells essentially possess the same organelles as are present in eukaryotic organisms, except for the Cyanophyceae, the ultrastructure of which is described by Lang (1968) and Fogg et al. (1973). Nuclei, mitochondria and plastids are ubiquitous under natural conditions as also are Golgi bodies, endoplasmic reticulum, ribosomes and vacuoles. Additional structures may also be present, e.g. pyrenoids, contractile vacuoles, trichocysts and haptonemata.

## ORGANELLES IN UNICELLULAR FORMS

### 2.1 *Flagella*

Unicellular algae may be non-motile and without external appendages, as in the red algae *Porphyridium* (Gantt & Conti 1965) and *Rhodella* (Evans 1970). The majority of unicellular algae however, are typically motile in the vegetative condition, having one, two, four or many flagella. These may be inserted terminally or laterally, and if there is a pair they may be of equal length (the isokont condition) as in many Chlorophyceae, or of unequal length (the hetero-kont condition) as in the Chrysophyceae, Xanthophyceae and motile stages of the Phaeophyceae. Certain Euglenophyceae and dinoflagellates have a flagellar condition unique to these groups.

Flagella consist of a ring of nine double peripheral fibres enclosing two single axial fibres, the whole being surrounded by a membrane which is continuous with the plasma membrane of the cell. In *Chlamydomonas* (Rosenbaum *et al.* 1969) assembly of the flagellum, particularly the microtubules, occurs at the elongating distal tip. Inhibitors of protein synthesis showed that flagellar development depended partly on pre-existing flagellar precursors and partly on new protein synthesis. There was, however, little turnover of flagellar proteins in full grown flagella (Gorovsky *et al.* 1970). Subsidiary components have been observed associated with the peripheral and axial fibres in the flagella of *Chlamy-domonas reinhardtii* (Hopkins 1970). The $9+2$ pattern, first seen by Manton and Clarke (1951) is common to the cilia and flagella of most animal and plant cells. It has been suggested that the axial filaments are essential for movement, since their absence in e.g. *Chlamydomonas* (Witman 1972) gives rise to paralysed mutants. However, both axial filaments are absent in the flagellum of the male gametes of the diatoms *Lithodesmium* (Manton & von Stosch 1966) and *Biddul-phia* (Heath & Darley 1972) and there is only one in the green alga *Golenkinia* (Moestrup 1972), but in all three movement is normal suggesting the motility mechanism is not in the axial components. Flagellar movement in all eukaryotes is dependent on ATP-ase activity but it is unlikely that this alone accounts for flagellar movement (see Warner 1972).

In the typical $9+2$ flagellum, as the peripheral fibres run into the cell cyto-plasm their arrangement changes becoming first the well-known 'stellate pattern', after which the fibres assume a triplet structure and end at the lowest level in the 'cartwheel pattern' of the flagellar base or blepharoplast (see e.g. Manton 1964a, Ringo 1967a). In the biflagellate motile cells of many brown algae, e.g. spermato-zoids of *Halidrys* and *Cystoseira* (Manton 1964b) and zoospores of *Scytosiphon* (Manton 1957) the flagella possess a two-limbed root consisting of six to nine fibres, often running in close association with mitochondria. It is thought that the root, as well as being mechanical, has some metabolic or conducting function

connected with respiration. In quadriflagellate motile cells of green algae, e.g.
*Stigeoclonium* and *Draparnaldia* (Manton 1964c) there are four flagellar roots,
one pair being five-stranded and the other pair two-stranded. No explanation is
yet available, in developmental or functional terms, for the structural differences
which occur between the two kinds of roots.

Flagella may be smooth, i.e. devoid of hairs (*Peitschengeissel* or whiplash
flagella) or bear hairs which in the case of heterokont algae are called Flimmer
or mastigonemes (*Flimmergeissel* or hairy flagella) (see Bouck 1972). The Flimmer
on motile cells of heterokont algae such as members of the Phaeophyceae are very
different in structure from the hairs on the flagella of non-heterokont algae such as
members of the Prasinophyceae. The former are tripartite in structure, having a
tapering basal region 200nm in length from which extends a closed microtubular
shaft ending in a group of two to three fine, thread-like structures (Bouck 1969).
The bases of such Flimmer do not penetrate the flagellar membrane. During de-
elopment of, for example, male gametes of *Fucus* and *Ascophyllum* (Bouck, 1969)
and the diatom *Biddulphia* (Heath & Darley 1972), in the Xanthophycean flagellate
*Olisthodiscus*, in zoospores of the filamentous algae *Bumilleria*, *Heterococcus* and
*Tribonema* (Leedale *et al.* 1970), and also in the fungi *Saprolegnia* and *Dictyuchus*,
and the algae *Synura* and *Cryptomonas* (Heath *et al.* 1970) Flimmer-like structures are found in membrane-lined sacs in the cytoplasm and within the dilated
region between the two membranes of the nuclear envelope. They are thought to
be transported extracellularly to the sites of attachment on the anterior flagellum.
In *Ochromonas* (Bouck, 1971) Flimmer development appears to begin by
assembly of the basal region and shaft in the perinuclear continuum, but addition
of lateral filaments to the shaft and Flimmer discharge is mediated by the Golgi
apparatus. Mignot *et al.* (1972) also implicated the Golgi apparatus in the
discharge of glycoprotein Flimmer elements in *Ochromonas*. Heywood (1972)
showed that in *Vacuolaria* hairs on the anterior flagellum consist of a tapered
base superficially attached to the flagellar membrane, and a microtubular shaft.
Dilated regions of endoplasmic reticulum contain many aligned microtubules
identical to the latter and it is suggested that microtubular proteins are synthesized
by ribosomes on the endoplasmic reticulum and then secreted into the cisternal
space. Membranes of the vesicles containing flagellar hairs could originate from
the endoplasmic reticulum and the Golgi does not appear to be involved. There is
little information on the mode of deposition, chemical composition or functional
significance of Flimmer and flagellar hairs. Bouck (1971) has isolated a single
polypeptide (different from microtubular proteins) and several carbohydrates
from preparations of Flimmer. Flagellar hairs, it is suggested (Heywood 1972),
may reverse the direction of locomotion.

In addition to Flimmer, flagella may bear appendages such as spines. Thus
the anterior flagellum of the spermatozoid of the brown alga *Himanthalia*
(Manton 1956) bears a large spine, over a micron long, near the distal end, borne
on one fibril of the peripheral series. Spermatozoids of the brown alga *Dictyota*
have a row of about 12 smaller spines placed in a median position between the two
lateral rows of Flimmer on the front end of the anterior flagellum (Manton 1956).

In *Sphaleromantis* (Manton 1966a), a heterokont alga belonging to the Chrysophyceae, and members of the Prasinophyceae e.g. *Pyramimonas* spp. (Manton *et al.* 1963), *Heteromastix* (Manton *et al.* 1965) and *Mesostigma* (Manton & Ettl 1965) flagellar scales are present. These will be dealt with again later in more detail in a consideration of Golgi activity.

## 2.2 Haptonemata

In addition to flagella, many flagellates (classified together in the Haptophyceae) possess another sort of filamentous appendage called a haptonema. This may be short and peg-like as in *Prymnesium parvum* (Manton 1964d), where it is about 5µm long and arises between the bases of the two flagella and projects rigidly forward. Cells are able to attach themselves temporarily to glass slides, for example, by means of the suctorial haptonemal tip, and they then rotate about a fixed point. In *Chrysochromulina* spp. (Manton 1968) the haptonema is long and thread-like and can coil up in a fraction of a second. A haptonema is anatomically and functionally entirely different from a flagellum. In *Chrysochromulina chiton* it is composed of a ring of seven fibres, increasing to nine in the haptonemal base inside the cell. The fibres are enclosed in three concentric unit membranes, the two inner of which collectively bound a space, the haptonemal cavity, which is continuous with peripheral endoplasmic reticulum in the cell and thought to be important in haptonemal movement.

## 2.3 Trichocysts

Flagellates of the class Cryptophyceae, e.g. *Chilomonas paramecium* (Dragesco 1951), *Cryptomonas* and *Hemiselmis* (Wehrmeyer 1970) possess slender, tapered, needle-like trichocysts. Summaries of the work on trichocysts in this and other groups may be found in Hovasse (1965), and Hovasse *et al.* (1967). Certain dinoflagellates also discharge complex trichocysts in the form of cross-banded ribbons (Bouck & Sweeney 1966, Leadbeater & Dodge 1966). Some members of the Chrysophyceae possess an entirely different type of ring-shaped trichocyst (Hovasse 1948, Hibberd 1970). In *Pyramimonas grossii* (Manton 1969), a green flagellate belonging to the Prasinophyceae, the discharged trichocyst resembles a slender, hollow, tapered needle about 0·1µm wide and some 35µm long, i.e. over six times the length of the flagella. In the undischarged state, the trichocysts are coiled ribbons, up to four in number, in membrane-bounded cavities at the anterior end of the cell. The functional significance of trichocysts is unknown.

## 2.4 Cell coverings

Unicellular algae may be enclosed, or naked i.e. surrounded only by a plasma membrane which may be enveloped in mucilage. There is a wide range of form and composition of cell coverings. In the red alga *Rhodella*, the mucilage is composed of sulphated polysaccharide chiefly of xylose and glucuronic acid (Evans *et al.* 1973b). The mucilage of the red alga *Porphyridium* has been shown to be a high molecular weight, sulphated, acidic polysaccharide (Ramus 1972, Ramus &

Groves 1972). In members of the Prasinophyceae, e.g. *Micromonas* (Manton & Parke 1960) and *Pyramimonas* spp. (Manton *et al.* 1963, Manton 1966b) the cells (and often the flagella) are covered externally with scales; in *Heteromastix* (Manton *et al.*1965) and *Mesostigma* (Manton & Ettl 1965) the scales form a coherent periplast, whilst in *Platymonas* (Manton & Parke 1965) the cell is covered in a theca formed by coalescence of minute stellate particles. Members of the Haptophyceae, e.g. *Chrysochromulina* (Figs. 3.6 & 3.7) (Manton & Leedale 1961a, Manton & Parke 1962, Manton 1967a,b) and *Prymnesium* (Manton 1966c) are also covered in scales, often of more than one type. In addition to scales, the cells of some members, e.g. *Crystallolithus* (Manton & Leedale 1963a), *Coccolithus* and *Cricosphaera* (Manton & Leedale 1969) bear calcified bodies known as coccoliths. Scales exhibit a wide range of size, shape and surface pattern and often there are a number of different types on the same cell. Little is known of the chemical composition of scales and thecae. In *Platymonas subcordiformis*, Lewin (1958) showed that in a preparation of discarded thecae the major components were galactose and a uronic acid. Recent work on *Platymonas tetrathele* (Manton *et al.* 1973) using electron microscope X-ray microanalysis (EMMA) has shown the presence of calcium in thecal preparations. This is thought to be present as the calcium salt of galacturonic acid (see also Gooday 1971). In *Chrysochromulina chiton*, Green and Jennings (1967) have shown that the main sugars present in the scales are ribose and galactose with no discernible glucose. Brown *et al.* (1970), however, record the presence of a microfibrillar cellulosic moiety in the scales of the closely-related alga *Pleurochrysis scherffelii*, in addition to galactose and ribose.

Some unicellular algae possess cell walls consisting largely of cellulose and hemicellulose, e.g. *Chlorella pyrenoidosa* (Northcote *et al.* 1958). Work on *Chlorella* and certain other algae, however (Atkinson *et al.* 1972), shows that in some strains the carotenoid polymer sporopollenin is present in the outer trilaminar wall layer. Barnett and Preston (1969) found that in freeze-etch preparations of *Chlamydomonas*, orientated cellulose microfibrils were present in the wall. Preliminary studies by Horne (1971), however, suggest that the wall is composed of a lattice of protein or lipoprotein. Roberts *et al.* (1972) have shown the presence of seven main wall layers, hydroxyproline-containing glycoprotein lattices being sandwiched between microfibrillar layers. Some members of the Dinophyceae are covered by a wall composed of plates thought to consist of cellulose (see e.g. Dodge 1971a), whilst in the Euglenophyceae many organisms are surrounded by a pellicle consisting of helically arranged, flexible overlapping strips composed largely of protein (Barras & Stone 1965, Leedale 1967). Bacillariophyceae (diatom) cells are surrounded by perforated frustules composed of silica, though chemical studies on two diatoms *Thalassiosira fluviatilis* and *Cyclotella cryptica* revealed the presence of chitin (Falk *et al.* 1966).

## 2.5  *Lysosomes*

Although lysosomes, first described by DeDuve, have been intensely studied and ascribed a variety of functions in animal cells, the whole question of their status in plant cells is a matter of some doubt and often flimsy analogy. They are poly-morphic organelles lacking internal structure and enclosed in one unit membrane. They are individually defined only by their biochemical properties, principally their enzyme content (hydrolases). Lysosomes are not well known in algal cells. Quiescent, non-flagellated, aged cells of *Euglena granulata*, however, contain lysosome-like structures with pigmented bodies and membrane fragments (Palisano & Walne 1972). These lysosome-like bodies show acid phosphatase activity, as also does the maturing face of dictyosomes and associated vesicles. Acid phosphatase is associated with multivesicular bodies of *Hymenomonas* (Pienaar 1971b), and ageing cells of *Ochromonas* (Grusky & Aaronson 1969). Lysosomes may also be seen to contain fragments of cytoplasmic organelles, e.g. lysosomes of certain Cryptophyceae contain trichocyst material (Wehremeyer 1970). Since acid phosphatase activity had been demonstrated in the Golgi, e.g. in *Euglena* (Brandes 1965) it was suggested (Mollenhauer & Morré 1966) that lysosomes are derived from the Golgi apparatus. This has led Holtzman *et al.* (1967), on the basis of common acid phosphatase activity, to introduce the name GERL to indicate the relationship between the maturing face of the Golgi apparatus, the endoplasmic reticulum and the lysosomes (see also Chapter 4, p. 133).

## 2.6 *Microbodies*

Microbody is a general term used to describe a cytoplasmic organelle with granular content, bounded by a single membrane thought to be derived from the endoplasmic reticulum and of uncertain function. Such organelles are also referred to as peroxisomes or glyoxysomes as a consequence of their different biochemical composition and metabolic function. (For a review, see Tolbert 1971.) Micro-bodies were found in *Chlorella* (Gergis 1971) and cell-free fractions of these were shown by Codd *et al.* (1972) to contain glycolate oxidase and catalase, enzymes associated with photorespiration in higher plants (Kisaki & Tolbert 1969, Yamasaki & Tolbert, see also Chapter 17 p. 474). On these grounds the micro-bodies have been called algal peroxisomes. There is cytochemical evidence for catalase activity in *Micrasterias* (Tourte 1972) and in *Klebsormidium* (Stewart *et al.* 1972). Microbodies were once reported in brown algae (Bouck 1965).

## 2.7 *Contractile vacuoles*

Contractile vacuoles occur in certain unicellular algae such as *Euglena* (Leedale *et al.* 1965), *Vacuolaria* (Schnepf & Koch 1966b), *Chlamydomonas* (Johnson & Porter 1968) and *Ochromonas* (Hibberd 1970), and in zoospores of green algae such as *Stigeoclonium* (Manton 1964c). These are regarded as organelles of osmoregulation. The mechanism by which this occurs is not understood. The intake of water could be purely osmotic or due to active secretion. Likewise water could be expelled by active contraction of the membrane or simply due to a build-up of hydrostatic pressure in the cytoplasm forcing the contractile vacuole to burst at a weak spot. In *Euglena* (Leedale *et al.* 1965), the vacuole appears as a membrane-limited cavity. After discharge, the surrounding cyto-plasm contains collapsed accessory vesicles and numerous spherical vesicles with a well-defined alveolate pattern on their walls. Whether such vesicles have any functional association with the contractile vacuole is not known. In *Glaucocystis* and *Vacuolaria* (Schnepf & Koch 1966a,b) contractile vacuoles arise from the Golgi. In *Glaucocystis* the vacuoles discharge their watery contents by bursting and their membranes are incorporated into the plasmalemma. In *Vacuolaria* a volume of water equal to the cell volume can be discharged in this way within 30 min. In dinoflagellates an organelle known as the pusule is thought to have an excretory or osmoregulatory function (Dodge 1972, 1973).

# 3 ORGANELLES IN MULTICELLULAR FORMS

## 3.1 *Walls*

These are bounded by cell walls of a wide ranging composition. Cell walls are dealt with in Chapter 2 p. 40, and only passing reference will be made to the subject here. In the larger multicellular green, brown and red algae, highly orientated polysaccharides such as cellulose, some xylans and mannans con-stitute the basic cell wall structure (Percival & McDowell 1967). In addition, compounds such as sulphated polysaccharides are present, contributing to mechanical flexibility whilst allowing for environmental stresses. Calcification of the walls also occurs in some members of the green, brown and red algae (see e.g. Bailey & Bisalputra 1970, and Chapter 24, p. 662).

## 3.2 *Intercellular connections*

Cytoplasmic continuity in many multicellular algae is considered to be provided by pores or pits in the walls between adjacent cells. In red algae such as *Lomentaria* (Bouck 1962), *Laurencia* (Bisalputra *et al.* 1967) *Pseudogloeophloea* (Ramus 1969) and *Rhodymenia* (Fig. 3.1) (Evans unpubl.) apertures in the walls between neighbouring cells contain a discrete, membrane-bounded plug. These structures

---

**Fig. 3.1.** A pit-connection between two vegetative cells of the red alga *Rhodymenia*. A plug (P), bounded by a unit membrane traverses the wall (W) between the two cells and the plasmalemma is continuous from one cell to the other (black arrows). The plug is capped on each side by endoplasmic reticulum-like material (white arrows) and further endoplasmic reticulum (ER) is present in the cell cytoplasm near the plug. Also visible are part of a nucleus (N), vacuole (V), chloroplast (C), floridean starch grains (S) and Golgi bodies (G). × 45,000 (Evans, unpubl.)

**Fig. 3.2.** A cell wall of the brown alga *Dictyota* showing it to be perforated by pores each with a thin strand of cytoplasm. The pores are bounded by membranes continuous with the plasmalemma of the adjacent cells. Part of a chloroplast (C) is also visible. × 40,000 (From Evans & Holligan 1972.)

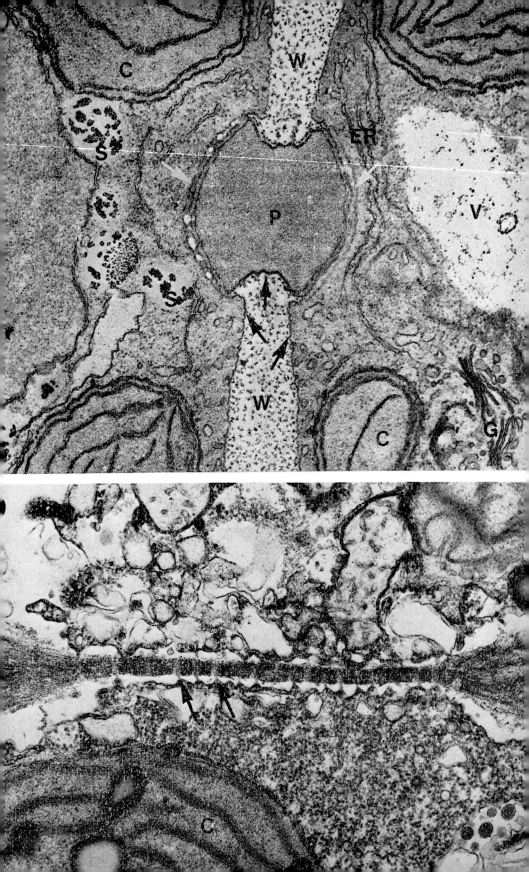

are classically termed pit connections. Ramus (1971) has shown the plugs in *Griffithsia pacifica* to be composed of an acid polysaccharide-protein complex. Although there is an accumulation of endoplasmic reticulum in the cytoplasm near the plugs, there would not appear to be cytoplasmic connection between adjacent cells across the plugs. However, translocation has been shown to occur in the red algae *Delesseria sanguinea* and *Cystoclonium purpureum* (Hartmann & Eschrich 1969). In *Delesseria*, leucine-(U)-$^{14}$C has been found to be transported (the highest velocity calculated being 63cm h$^{-1}$) and the pathway is thought to be through conducting elements, up to 540 μm long, in the 'veins'. These conducting cells contain several nuclei and a few small chloroplasts surrounded by starch grains. They are interconnected by 'pit fields' in transverse and lateral walls and the pit fields are largely occluded by slime-like material.

Translocation between one alga and another has also been shown to occur. $^{14}$C-labelled metabolites are transferred from the red alga *Gracilaria* to the parasitic, chloroplast-free red alga *Holmsella* (Evans *et al.* 1973a). Such translocation, from one plant to another may prove an important and widespread algal phenomenon since metabolite movement probably also occurs between brown algae and 'epiphytic' red algae growing on them (Evans unpub., Citharel 1972).

In multicellular brown algae cell wall pits contain protoplasmic connections, e.g. *Dictyota* (Fig. 3.2) (Evans & Holligan 1972). In *Himanthalia* (Berkaloff 1963), walls of adjacent meristematic cells are perforated by pores about 50nm in diameter, each with a thin cytoplasmic strand passing through it. In the epidermal and outer cortex cell walls of *Egregia* and *Fucus evanescens* (Bisalputra 1966) there are pits 0·5 to 2μm in diameter. Several small pores averaging 40nm in diameter traverse the pit membrane and protoplasmic connections (plasmodesmata) run through these. The protoplasmic connections are bounded by limiting membranes continuous with the plasma membranes of the adjacent cell protoplasts. Endoplasmic reticulum has not yet been seen passing through the plasmodesmata as may occur in some higher plant cells (see e.g. Robards 1971).

Very large brown algae such as *Macrocystis* and *Nereocystis*, members of the Laminariales, possess cells in the inner cortex and medulla resembling the sieve tubes of vascular plants. The sieve tubes in *Macrocystis* are divided by simple transverse sieve plates into sieve elements and callose is often associated with sieve plates (Ziegler & Ruck 1967). Parker (1965, 1966) showed that in *Macrocystis* $^{14}$C-labelled assimilates moved with velocities of 65 to 78 cm h$^{-1}$ while in *Nereocystis* (Nicholson & Briggs 1972) rates of 37cm h$^{-1}$ are reported. Girdled blades in which the mucilage ducts were interrupted and the medulla left intact continued to translocate $^{14}$C-labelled photosynthates (Nicholson & Briggs 1972) indicating that conduction occurs in the medulla. The major $^{14}$C-labelled substances in sieve tube exudate of *Macrocystis* (Parker 1966) were D-mannitol and some amino acids. The exudate also had a high protein and K$^+$ content and in general is similar to that obtained from higher plants. Electron microscope examination (Parker & Huber 1965) showed the absence of plasmodesmata in longitudinal walls of sieve tubes. The sieve pores in freeze killed material were not occluded but the cell lumens were found to be filled with a fine granular material and aggregated slime. The sieve elements lacked nuclei and contained a peripheral layer of plastids, many vesicles, numerous mitochondria with well-developed cristae, endoplasmic reticulum and occasional Golgi bodies. Although in many ways similar to the sieve tubes of higher plants, the greater complement of organelles might indicate a greater degree of metabolic independence by *Macrocystis* sieve tubes which lack companion cells (Parker & Huber 1965). Using $H^{14}CO_3^-$, translocation through medullary conductive tissue has also been shown in *Laminaria hyperborea* and *Laminaria saccharina* by Schmitz *et al.* (1972). Mannitol and amino acids were the main substances translocated but transport was much slower (5–10cm $h^{-1}$) than in either *Macrocystis* or *Nereocystis*. Likewise, slow translocation (1cm $h^{-1}$) of photosynthate from old portions to new portions of the blade was reported in *Laminaria digitata* (Hellebust & Haug 1972).

Sieve tubes are not homologous with the well-known trumpet hyphae found in the medulla of members of the Laminariales. Trumpet hyphae have numerous plasmodesmata in their cross walls and are thought not to function in translocation (Parker & Huber 1965), although recent work on *Laminaria digitata* (van Went & Tammes 1972) suggests they might have a conducting function.

In the filamentous green alga *Bulbochaete*, plasmodesmata lack internal components but consist of a cylindrical connection between the plasma membranes of adjacent cells (Fraser & Gunning 1969). The cylinder is constricted at each end and it is suggested that the constrictions act as a control mechanism regulating the flow of materials through the groundplasm. Plasmodesmata are also present in *Oedogonium* (Hill & Machlis 1968) but in *Tribonema*, a less advanced filamentous type, they are not found (Falk & Kleinig 1968). In *Ulothrix* and *Stigeoclonium* plasmalemma-lined plasmodesmata with an electron-dense core occur (Floyd *et al.* 1971, Floyd 1972) although plasmodesmata do not occur in the related genus *Microspora* (Pickett-Heaps 1973). In *Chara*, plasmodesmata have been observed with (Pickett-Heaps 1968b), and without (Pickett-Heaps 1967a, b), a strand of endoplasmic reticulum, and in *Volvox* cytoplasmic connections have also been found to contain elements of endoplasmic reticulum (Bisalputra & Stein 1966, Pickett-Heaps 1970a). Adjacent elements of endoplasmic reticulum are only rarely associated with plasmodesmata in *Nitella* (Spanswick & Costerton 1967) and, although there is some restriction of ion diffusion, the plasmodesmata have been shown, by measurements of electrical resistance, to provide an efficient means of intercellular transport.

# 4 ORGANELLES IN UNICELLULAR AND MULTICELLULAR FORMS

## 4.1 *Nuclei*

As in other eukaryotes, algal nuclei (excepting the nuclear-like materials of the blue-green algae) are enclosed in a limiting nuclear envelope consisting of two unit membranes separated by a perinuclear space (see Feldhen 1972). The nucleoplasm often contains one or two dense nucleolar regions rich in RNA. The nuclear envelope is frequently in continuity with endoplasmic reticulum, so that there may be free passage from the perinuclear space into the lumen of the endoplasmic reticulum. In some algae, such as members of the Phaeophyceae, the nuclear envelope is characteristically continuous with the envelope of endoplasmic reticulum that surrounds the chloroplast (Bouck 1965, Evans 1966).

In brown algae (see e.g. Bouck 1965, Cole 1970, Baker & Evans 1973), amongst others, e.g. the Xanthophyceae (Falk 1967, Falk & Kleinig 1968, Massalski & Leedale 1969), the Golgi apparatus is also closely associated with the nuclear envelope and there is evidence of vesicular transfer from the nuclear envelope to the Golgi bodies (Fig. 3.11). The nuclear envelope is commonly perforated by pores (Fig. 3.15) e.g. in *Bumilleria*, a member of the Xanthophyceae, (Massalski & Leedale 1969), the red alga *Membranoptera* (McDonald 1972) and in certain brown algae (Neushul & Dahl 1972). They are typically 80 to 90nm in diameter, and in some dinoflagellates, e.g. *Prorocentrum micans* (Wecke & Giesbrecht 1970) the pores are arranged in a symmetrical hexagonal pattern. Nuclear pores are believed to be structural entities permitting passage of molecules such as RNA and protein in to and out of the nucleus (Northcote 1971a). In the desmid *Micrasterias* (Kiermayer 1971) rod-like vesicles 100 to 120nm in diameter may be seen in the cytoplasm near the nuclear pores, and these may function in nuclear-cytoplasmic exchange. A similar function has been attributed to openings in the peripheral fibrillar nucleoplasmic reticulum which in *Bryopsis* germlings are found adjacent to nuclear pores (Burr & West 1971). In *Asteromonas gracilis*, a member of the Prasinophyceae, the outermost unit membrane of the nuclear envelope is extended into a membrane-lined channel in the pyrenoid of the chloroplast (Peterfi & Manton 1968). Likewise in *Prasinocladus marinus* (Parke & Manton 1965), another member of the Prasinophyceae, and in *Rhodella maculata* (Fig. 3.3) (Evans 1970), a unicellular red alga, branched nuclear protrusions enclosed in both membranes of the nuclear envelope enter branched, membrane-lined pyrenoid channels. The physiological significance of this close relationship between the nucleus and chloroplast-pyrenoid complex is not yet known.

### 4.2    *Microtubules*

Before considering the ultrastructure of nuclear division, brief account must be given of microtubules, widely distributed and important elements involved in a number of subcellular structures and functions (see e.g. Newcomb 1969, Tilney 1971). They occur in cell cytoplasm, are involved in nuclear division (as meiotic and mitotic spindle fibres) as well as being present in flagella (constituting the 9+2 fibres discussed earlier) and the flagellar basal bodies and centrioles. Microtubule walls are composed of a variable number of globular subunits each about 4nm in diameter; in the fibres of the flagella and basal body of *Chlamydomonas reinhardtii* (Ringo 1967b) the number of subunits seen in cross section has been found to be 13, whilst in the Flimmer of *Fucus* flagella (Bouck 1969) the number of subunits is 10 and in *Ochromonas* Flimmer about 6 (Barton *et al.* 1970). Microtubules are built up of protein subunits (see e.g. Northcote 1971a), and their diameters fall between 18 and 30nm (average 24nm). Biochemical analysis of isolated doublets of 9+2 flagella from *Chlamydomonas* (Witman *et al.* 1972a,b) has shown that there are two microtubule proteins (tublins 1 and 2) with molecular weights of 56,000 and 53,000 respectively. Microtubules appear to be somewhat stiff rigid structures and when they occur in the cytoplasm appear to carry out a structural or skeletal function (see e.g. Bouck & Brown 1973, Brown & Bouck 1973). They may also aid organized movement of materials in the cytoplasm and nucleoplasm (see e.g. Wilson *et al.* 1973) and the orientation of deposited microfibrils during cell wall formation. Pickett-Heaps (1967b) observed microtubules aligned to wall microfibrils in *Chara* and Green (1963) showed that destruction of microtubules by colchicine caused microfibrils normally deposited transversely to become laid down randomly in *Nitella*. In the developing spermatozoid of *Fucus* there is evidence that microtubules participate in the initial alignment of eyespot granules within the chloroplast (Bouck 1970). Brown and Franke (1971) record a characteristic hexagonal array of microtubules associated with the Golgi body in *Pleurochrysis scherffelii* and suggestions for their functional significance include a role in secretion and extrusion of scales. In *Acetabularia* microtubules in the cytoplasm serve to anchor the nuclei (Woodcock 1971).

Another organelle which must be considered briefly at this point is the centriole. It is a complex cylindrical body composed of a ring of nine evenly-spaced triplet microtubules. Centrioles serve as organizing centres for the growth of spindle microtubules and also, in the form of basal bodies (bodies with a centriolar structure which are generally considered to be centrioles) produce the 9+2 microtubules found in flagella. DNA is thought to be present in centrioles, although this is not yet definitely established (see Fulton 1971, Wolfe 1972). Among the algae there is considerable variation in the part played by centrioles or bodies resembling centrioles in the mitotic process.

### 4.3   *Nuclear division*

The process of nuclear division has been studied with the electron microscope in only a very small number of algae. Most observations have been made on unicellular forms. Nuclear division in the dinoflagellates, e.g. *Woloszynskia* (Leadbeater & Dodge 1967) and *Gyrodinium cohnii* (Kubai & Ris 1969) is of an unusual sort and probably represents one of the most primitive known in plants. Following longitudinal division of the chromosomes at prophase in *Woloszynskia* a number of cytoplasmic inpushings become continuous across the nucleus. These are surrounded by the nuclear envelope and each contains 8 to 20 extra-nuclear microtubules (of similar size to normal spindle fibres) which pass completely through the nucleus from one side to the other. The microtubules are not connected to the chromosomes so that they cannot be involved in chromosome separation at anaphase and their function is not understood. There are no centrosomes or centrioles, and the nuclear envelope remains continuous throughout the mitotic process. The chromosomes differ chemically from normal chromosomes in lacking histone protein and probably consist entirely of DNA fibrils. *Glenodinium foliaceum* has (Dodge 1971b), in addition to a single meso-karyotic nucleus (a typical dinoflagellate nucleus with characteristic 'visible' chromosomes), a second nucleus, termed the eukaryotic nucleus because of its similarity to eukaryotic nuclei of higher organisms. It is a polymorphic body which varies in shape, contains several nucleoli and is surrounded by a per-forated envelope. Its function is unknown.

As in dinoflagellates, the mitotic process in the euglenoids differs consider-ably from that encountered in higher plants (see e.g. Leedale 1970). Electron microscope studies on mitosis in *Euglena gracilis* show that the nuclear envelope remains intact throughout the whole process and the flagellar basal bodies have no centriolar activity. Bundles of nucleoplasmic microtubules run from pole to pole between the chromosomes but there are no centrosomes or chromosomal microtubules. The microtubule bundles may provide guide-lines to the poles, and chromosome movement is thought to be autonomous. The nucleolus persists and divides.

Unusual features have also been revealed in detailed and elegant studies by Manton *et al.* (1969a,b, 1970a,b) on the mitotic and meiotic processes observed during gametogenesis in the diatom *Lithodesmium undulatum*. Non-dividing cells possess a highly characteristic body, the spindle precursor, closely pressed to the nuclear envelope. This rectangular body is composed of parallel plates, the enlarged end ones of which are regarded as centrosomal. During prophase the microtubular spindle is laid down between the precursor and the nuclear envelope, and at metaphase, following breakdown of the nuclear envelope, the spindle sinks into the crowded mass of chromosome material. In the spindle equator 16 to 20 bundles of microtubules occur, perhaps one bundle for each chromosome. Detailed analysis suggests that the spindle is composed of two half-spindles with many microtubules running only from one pole to the spindle equator where they cohere laterally with their counterparts from the opposite pole. The remaining tubules run uninterruptedly from one pole to the other. The meiotic spindle is much more massive than the mitotic spindle but the number of microtubules produced is smaller. There is a numerical relationship between chromosome mass and microtubule number and when chromatids are moving there are only about half the total number of microtubules present per bundle as there are during chromosome movement.

In other algae where nuclear division has been examined ultrastructurally, although there are variations, the process resembles that encountered in higher plants and the same organelles are involved. For example in *Prymnesium parvum*, a member of the Haptophyceae, organelle replication precedes nuclear division and there is proliferation of fibres from the flagellar base region (Manton 1964e). The flagellar basal bodies probably act as centrioles. Spindle microtubules and an equatorial plate become visible during metaphase and the nuclear envelope breaks down into tubular cristae, often still carrying ribosomes, which remain in the spindle region. Spindle microtubules, 20 to 24nm in diameter, pass between chromosomes but there is no evidence of attachment. During metaphase, cell elongation occurs and the nuclear envelope becomes reconstituted, and during telophase cleavage of the daughter cells takes place. Mitosis in the chrysophycean alga *Ochromonas* is somewhat similar to that in *Prymnesium* except that in the former the spindle microtubules converge on a rhizoplast (a fibrous striated root attached to the basal body of the long flagellum) at each pole (Slankis & Gibbs 1972).

In *Chlamydomonas reinhardtii* cells in an early stage of division, the basal bodies of the two flagella replicate and the two original basal bodies become detached from their flagella (Johnson & Porter 1968). However, although free to move in the cytoplasm and morphologically indistinguishable from centrioles, the basal bodies are not found at the poles of the spindle and do not appear to participate in nuclear division. Rather, they are involved with formation of cytoplasmic microtubules which become orientated along the cleavage plane during cytokinesis. These microtubules are thought to provide a framework against which cytoplasmic furrowing can take place. The four classical stages of mitosis occur. Spindle microtubules traverse the nucleoplasm but the nuclear envelope does not break down. However, discontinuities appear at the poles and spindle microtubules arise from (or end in) a specialized cytoplasmic region devoid of ribosomes and other structures which occurs just beyond the polar openings.

In *Chlorella pyrenoidosa*, a non-motile alga, centrioles occur at all stages of the cell cycle and their main function appears to be as microtubule-orientating centres, both in the cytoplasm and in intranuclear spindle formation (Atkinson *et al.* 1971). In *Kirchneriella lunaris*, another non-motile green alga, rudimentary centriolar complexes and associated microtubules migrate into the nucleus through polar discontinuities prior to intranuclear spindle formation (Pickett-Heaps 1970b). Such rudimentary centrioles suggest that the organelles are non-functional relics in non-motile algae (Atkinson *et al.* 1971).

In the coenobial alga *Hydrodictyon* (Marchant & Pickett-Heaps 1970) persistent centrioles replicate and migrate to the poles of the prophase spindle, as microtubules appear between them. A perinuclear envelope of endoplasmic reticulum encloses the prophase nucleus, centrioles and cytoplasmic microtubules, and this remains intact until telophase. Extranuclear microtubules invade the nucleus through large polar discontinuities in the nuclear envelope. During cytoplasmic cleavage (Marchant & Pickett-Heaps 1971) the centrioles move into a characteristic position near the cleavage furrows and during cleavage transversely orientated microtubules occur.

In *Ulva mutabilis* (Løvlie & Bråten 1970), as in *Chlamydomonas*, the nuclear envelope remains during mitosis but with polar discontinuities. Spindle microtubules associated with centrioles pass through these into the nucleus. In *Oedogonium* (Pickett-Heaps & Fowke 1969, 1970a, Pickett-Heaps 1971a) the nuclear envelope again remains intact during mitosis, but with polar discontinuities through which microtubules pass. Complex paired chromosomal centromeres become attached to bundles of intranuclear microtubules by metaphase, and at anaphase the centromeres separate and the chromosomes move apart. No recognizable centrioles or other polar structures have been observed in mitotic vegetative cells of *Oedogonium*. Since these structures presumably occur in large numbers and give rise to flagella during zoospore production, Hoffman (1966) and Pickett-Heaps (1971a) believe they arise *de novo*. In *Chara* also, no centrioles were observed in vegetative cells or young cells of antheridia (Pickett-Heaps 1967a,b, 1968a). Centrioles do occur however, in spermatogenous filaments (Pickett-Heaps 1968b) where they become associated with the mitotic spindle, and are later involved in flagellum formation. Pickett-Heaps (1968b) suggests they may be an appendage to, and not a vital component of, the mitotic apparatus. In *Spirogyra*, Fowke and Pickett-Heaps (1969a,b) have shown that the nucleolus fragments during mitosis and the chromosomes become coated with nucleolar material (see also Pickett-Heaps 1972). Cytoplasmic microtubules penetrate the nucleus through polar discontinuities in the nuclear envelope. By metaphase many microtubules enter the nucleus through discontinuities in the envelope and as the spindle enlarges the nuclear envelope finally disintegrates. No localized centomeres or centrioles were observed. In *Stigeoclonium*, *Ulothrix* and *Klebsormidium* (Floyd *et al.* 1972a,b) and *Microspora* (Pickett-Heaps 1973) centrioles occur and the nuclear envelope becomes at least partially disrupted during mitosis. In *Vaucheria* (Ott & Brown 1972) nuclear division occurs within a completely closed nuclear envelope, there being no gaps or discontinuities at the poles. This is the first alga known where the nuclear envelope remains completely intact during division.

Nuclear division has not yet been studied in detail with the electron microscope in any brown alga, as far as is known. However, it is reported (Dr. G. B. Bouck pers. comm., cited in Leedale 1970) that during mitosis in the antheridium of *Fucus vesiculosus*, spindle microtubules arise from paired centrioles and radiate in to the nucleus where they pass between chromosomes through polar gaps in an otherwise intact nuclear envelope. A similar situation exists in some members of the Dictyotales (Neushul & Dahl 1972).

There is so far only one ultrastructural study of mitosis in a red algal nucleus. In a study of mitosis in *Membranoptera platyphylla*, McDonald (1972) showed that the spindle-shaped prophase nucleus is surrounded by microtubules and a concentric sheath of endoplasmic reticulum. A unique organelle, the polar ring, is associated with the early stages of mitosis. This is a short hollow cylinder, superficially resembling a centriole but lacking a cartwheel substructure. The nuclear membrane remains intact throughout mitosis except for polar discontinuities, through which spindle microtubules pass into the cytoplasm. Spindle microtubules are attached to kinetochores (centromeres) on the chromosomes and continuous microtubules run from pole to pole. The nucleolus disperses but its granular components remain in the nucleoplasm to be reassembled in the daughter nuclei.

Synaptonemal complexes (structures thought to correspond to the axes of homologous chromosomes in the meiotic process) have been shown in young tetraspore mother cell nuclei of the red algae *Janczewskia gardneri, Levringiella gardneri, Gonimophyllum skottsbergii* and *Polycoryne gardneri* (Kugrens & West 1972). They correspond in structure and dimensions with similar structures found in other organisms (see Wettstein & Sotelo 1971), being composed of a less densely stained central region from which fibrils extend to two lateral more heavily stained elements.

### 4.4  *Mitochondria*

Mitochondria range in length from less than 1μm to several μm, are sometimes branched and may be of variable shape. There may be only one mitochondrion per cell, as in unicellular algae such as *Micromonas* spp. (Manton & Parke 1960) and *Chromulina pusilla* (Manton 1959), but generally there are more. Since mitochondria may branch profusely (Arnold *et al.* 1972), they can reach overall lengths far in excess of the diameter of the cell, and three-dimensional models show that the number per cell may be very much less than might appear from examining single sections.

Mitochondria of all eukaryotic cells have the same basic structure with an outer smooth membrane surrounding a highly infolded inner membrane which encloses a central space, the lumen or matrix. The infoldings of the inner membrane lie in the matrix as microvilli or cristae and these may be swollen at their tips as in *Acetabularia* (Bouck 1964). Evenly-spaced bristle-like structures, 22 to 31nm in length and 4nm in width, have been found projecting from some mitochondrial cristae in *Oedogonium* cells undergoing zoosporogenesis (Pickett-Heaps 1971b). Rather similar particles attached to mitochondrial membranes have been seen by Wigglesworth *et al.* (1970) and are of the order of magnitude expected for cytochromes and dehydrogenases. In carpospores and tetraspores of the red alga *Ceramium* (Chamberlain & Evans 1973) continuity between the outer mitochondrial membrane and adjacent profiles of endoplasmic reticulum are occasionally observed. This has recently been reported for several organisms, e.g. *Pythium* (Bracker & Grove 1971), rat liver and onion stem (Morré *et al.* 1971a). Franke and Kartenbeck (1971) speculate that such continuity in the ciliate *Tetrahymena* may provide a route for transfer of special proteins and possibly lipid from rough endoplasmic reticulum to mitochondria (see also Lord *et al.* 1973).

Algal mitochondria are frequently seen with constrictions suggestive of an impending division by fission. However, a lot more positive evidence is required to prove this. In *Chrysochromulina brevifilum* (Manton 1961a) mitochondrial multiplication may result from growth leading to clusters of attached mitochondria which later separate. Giant mitochondria formed during the growth phase of the cell cycle in *Chlamydomonas reinhardtii* later divide to give smaller forms (Osafune *et al.* 1972). Formation of giant mitochondria results in decrease in oxygen uptake activities, whilst normal oxygen uptake is restored on reformation of smaller forms.

In the mitochondria of the brown alga *Egregia menziesii* one or two areas containing 1·0 to 2·5nm fibrils with the characteristics of DNA have been described by Bisalputra and Bisalputra (1967) (Fig. 3.4). These areas are removed by DNase treatment (Fig. 3.5) but they are not membrane-bound and it is not

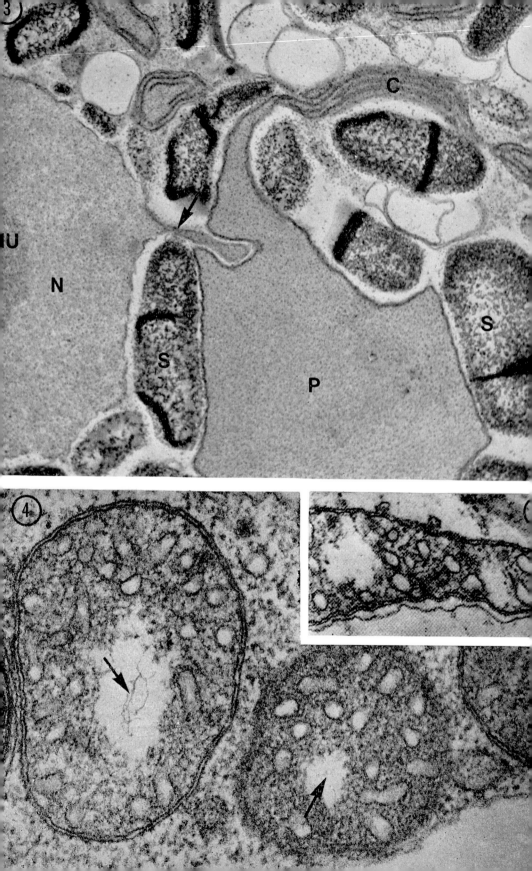

known whether the fibrils penetrate the surrounding mitochondrial matrix and become associated with cristae. In the red alga *Polysiphonia* electron transparent areas in the mitochondrial matrix may show, in addition to DNA fibrils, twisted structures interpreted as stages in replication of the mitochondrial DNA (Tripodi *et al.* 1972). DNA extracted from isolated mitochondria of strepto-mycin-bleached cells of *Euglena gracilis* has been shown by Edelman *et al.* (1966) to be double-stranded. Each mitochondrion contained 9 to $22 \times 10^6$ daltons of DNA. In *Ochromonas*, Gibbs (1968) describes mitochondrial ribosomes, 15 to 17 nm in diameter (cf. cytoplasmic ribosomes, 21 to 23 nm), both free in the matrix and attached to the cristae. Although there is no information on the extent to which algal mitochondria are autonomous there is considerable evidence that they are partially so in fungi and animal cells (e.g. Woodward *et al.* 1970, Kroon & DeVries 1970).

## 4.5 *The Golgi apparatus*

The Golgi apparatus has a key position in the complex membrane-bound trans-port system within the cell (see e.g. Whaley *et al.* 1971). The perinuclear space, the lumen of the endoplasmic reticulum and the vesicles and cisternae of the Golgi complex are all linked, and connected to the outside of the cell across the plasma membrane, either by direct connection or by functional continuity. The Golgi apparatus consists of individual stacks of cisternae, the Golgi bodies or dictyo-somes. The number of Golgi bodies per cell varies from only one (Manton 1961b) to c. 25,000 in the rhizoid apex of *Chara* (Sievers pers. comm., cited in Mollen-hauer & Morré 1966). Although not usually within the limit of resolution of the light microscope, Golgi bodies several microns in length have been observed in living cells of the diatom *Pinnularia* (Jarosch 1962), the desmid *Micrasterias* (Drawart & Mix 1962), the chrysophycean alga *Paraphysomonas* (Manton & Leedale 1961b) and the haptophycean alga *Crystallolithus* (Manton & Leedale 1963a). The stack of parallel, disc-shaped cisternae constituting a Golgi body are often slightly curved so as to give the organelle concave (maturing or secreting) and convex (forming) faces (Fig. 3.15). Material sometimes visible between the cister-nae may form part of an intercisternal structure. An anastomosing network of

---

**Fig. 3.3.** A projection (arrowed) from the nucleus (N), lined by the two membranes of the nuclear envelope, entering a channel in the central pyrenoid (P) of the uni-cellular red alga *Rhodella*. The pyrenoid is connected to lobes of the much-branched chloroplast (C) by narrow isthmuses and the pyrenoid channel is lined by the two membranes of the chloroplast/pyrenoid envelope. Floridean starch grains (S) and part of the nucleolus (NU) are also present. $\times$ 30,000 (From Evans 1970.)

**Fig. 3.4.** Mitochondria of the brown alga *Egregia* showing electron transparent areas containing DNA fibrils (arrowed) $\times$ 75,000 Courtesy Drs. T. & A.A. Bisalputra. (From Bisalputra & Bisalputra 1967.)

**Fig. 3.5.** Mitochondria of *Egregia* after 6h treatment with DNase showing the removal of the DNA by this treatment. $\times$ 35,000 Courtesy Drs. T. & A.A. Bisalputra. (From Bisalputra & Bisalputra 1967.)

tubules arises from the cisternal edges and these swell in places to form different types of vesicles, so that the cisternae may be fenestrated (Fig. 3.16). Fenestration becomes more pronounced from forming to maturing Golgi faces, so that at the forming face there are complete membranes whilst at the other face the membranes are extensively pierced by pores which break up the cisternae into the tubule network (Northcote 1971b). Analysis of Golgi body membranes shows that those on the forming face are similar to membranes of the nucleus and endoplasmic reticulum, and those on the maturing face resemble the plasma membrane, the central ones being intermediate in composition (Keenan & Morré 1970). A function of the Golgi system is therefore in the modification of lipoprotein membranes.

As well as serving as part of an internal transport system in the cell, the Golgi complex is involved in the formation and packaging of substances for extracellular transport. A variety of substances are known to be packaged in this way in animal and plant cells (e.g. Mollenhauer & Morré 1966, Beams & Kessel 1968, Schnepf 1969, Morré et al. 1971b). These are largely polysaccharides (Northcote & Pickett-Heaps 1966) or are composed of carbohydrate, lipid or protein moieties conjugated to give, for example, glycoprotein/mucopolysaccharide (Neutra & Leblond 1966a,b) or lipoproteins (Claude 1970a,b). There are also less frequently encountered Golgi functions. For example, in the green alga *Vacuolaria* interconnected Golgi bodies form vesicles which coalesce to become the transient contractile vacuoles responsible for osmoregulation (Schnepf & Koch 1966b). Similarly, in *Chara* (Pickett-Heaps 1967a) Golgi-derived vesicles discharge into the vacuole or contractile vacuole, and in *Oedogonium* (Pickett-Heaps & Fowke 1969) similar vesicular discharge is suggested to cause the increased turgor pressure which leads to wall rupture. In *Ulva* the ring-shaped Golgi complex in vegetative cells may be responsible for ion transport and in particular the selective flux of $K^+$ ions across the cytoplasm into the vacuole (West & Pitman 1967). In certain Cryptophyceae (Wehremeyer 1970) trichocysts are believed to be derived from the Golgi system, and in *Ochromonas* (Chrysophyceae) the Golgi apparatus is thought to be involved in production of Flimmer (Bouck 1971). Acid phosphatase activity has been demonstrated in the Golgi apparatus of vegetative cells of *Ulva* (Micalef 1972).

Since cell division does not generally result in reduction in numbers of Golgi bodies (Mollenhauer & Morré 1966), multiplication must occur, although the way in which this takes place is far from certain. Doubling of Golgi bodies accompanies cell division in *Botrydium granulatum* (Nagy & Fridvalsky 1968) and similar replication by division of cisternae and fission was proposed in *Micrasterias* by Drawert and Mix (1963) and Kiermayer (1967, 1970). However, the positional relationships of paired Golgi bodies in some algae suggests that multiplication could not have occurred in this way (Drum 1966, Manton 1967a, Pickett-Heaps 1968b). For a discussion on other possible multiplication mechanisms see Morré et al. (1971b).

The extent of molecular assembly that takes place within the Golgi apparatus varies. In the case of polysaccharide products, the carbohydrate is acquired in the membrane system of the Golgi apparatus and its vesicles. Thus in the extensive and well-documented work, of Manton and co-workers on scale production in unicellular algae there is good evidence to show that the polysaccharide scales develop progressively and completely within Golgi cisternae, and are subsequently transported in vesicles to the exterior to form a covering to the cell and sometimes also the flagella. The one or more types of scale produced by a cell are of characteristic and distinctive shape, size and surface markings. For example, in *Mesostigma* (Manton & Ettl 1965), a prasinophycean alga, there are large spectacular basket-shaped scales overlying several layers of small naviculoid plate scales which in turn overlie an array of tiny subrhomboid plate scales. All these types can be traced to Golgi-derived vesicles within the cell. In *Prymnesium parvum* (Manton 1966c) and *Chrysochromulina chiton* (Manton 1967a,b) scale production occurs only on the side of the Golgi body away from subtending endoplasmic reticulum (Fig. 3.8) and cisternae from the centre to the maturing face of the Golgi contain progressively more mature scales. In *Chrysochromulina* different scale types (Fig. 3.6 and 3.7) may also be seen within adjacent cisternae (Fig. 3.8). The scales have a dorsoventral assymetry and scales of one type are all similarly orientated with respect to the inner and outer faces of the cisternae containing them (Fig. 3.8). The very large scales become displaced into T-shaped cisternae, the scale occupying the crosspiece (Manton 1967b). In the marine coccolithophorids *Coccolithus* and *Cricosphaera* (Manton & Leedale 1969) and *Hymenomonas* (Pienaar 1971a), coccoliths, as well as ordinary body scales, are produced within cisternae of the Golgi body.

The theca of *Platymonas tetrathele* was shown by Manton and Parke (1965) to consist of coalesced particles originating from Golgi cisternae and to be composed of a pectin-like material (Gooday 1971). Using the technique of electron microscope autoradiography Gooday (1971) has confirmed the observations of Manton and Parke. He demonstrated the movement of tritiated glucose into cells of *Platymonas tetrathele* and followed its incorporation into Golgi polysaccharides and finally into the theca surrounding the cell. The results together with those from an analysis of the carbohydrate composition of the theca are consistent with the idea of Northcote and Pickett-Heaps (1966) that a function of the Golgi apparatus is the synthesis of pectin-like material from a pool of precursors that contains galactose, galacturonic acid and arabinose. The products, in the form of discrete stellate particles, are then exported from the cell and coalesce to form a new theca, beginning next to the pyrenoid. As Gooday points out, the evidence that starch provides material for theca formation also supports the idea that formation of the theca is intimately connected with chloroplast metabolism.

Following the findings of Green and Jennings (1967) where the scales of the haptophycean flagellate *Chrysochromulina chiton* were found to be pectin-like in nature, Brown *et al.* (1970), in a study of the related alga *Pleurochrysis scherffelii*, produced evidence that the scales consist of a layer of concentrically arranged microfibrils resembling cellulose, densely-coated with particles assumed to be pectic in nature and identical with the gelatinous mass in which the scales are

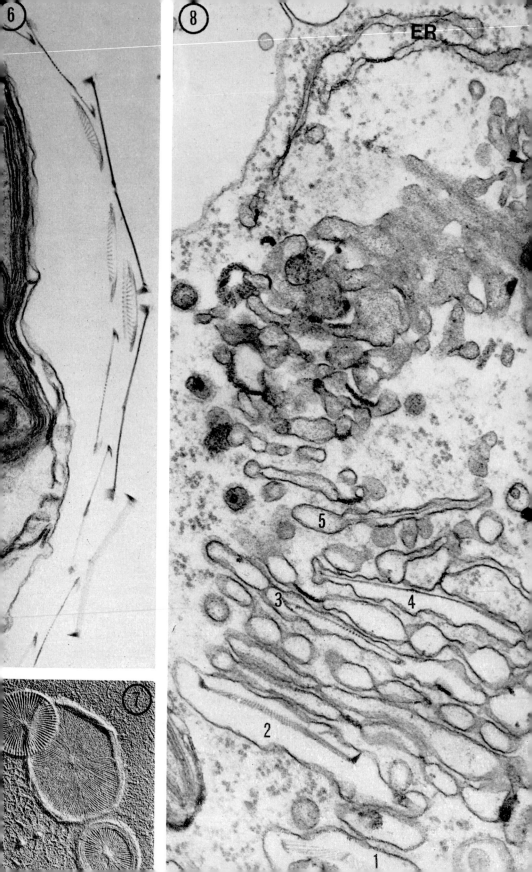

embedded. Detailed analyses of isolated scales (Herth *et al.* 1972) indicate that peptide material is covalently linked to the cellulosic structural polysaccharide to give a cellulosic 'glycoprotein'. Using electron microscopy and freeze etch techniques, the formation of the scales has been followed from the dictyosome cisternae to their secretion by exocytosis. Scales were observed to be initiated in distinct cisternae characterized by a central dilation, often with a dense layer on its membranous surface. The dilations are tentatively suggested in functional terms as 'centers of polymerization' of the cellulosic microfibrils. Franke (1970) discusses the presence of similar cisternae in other organisms and suggests they may be a morphological indicator of certain stages in polysaccharide formation. Since cellulose-like material was found in isolated scales, Brown *et al.* (1970) interpret their results as evidence for cellulose synthesis by the Golgi apparatus. This hypothesis contrasts with results of other workers (see Preston & Goodman 1968) which suggest that cellulose synthesis takes place at the plasma membrane surface. However, in certain higher plants, enzymes involved in cellulose synthesis are found in association with Golgi membranes (Ray *et al.* 1969) and, since it is likely that membranes from the Golgi complex fuse with and contribute to the plasma membrane, it is possible that cellulose-synthesizing enzymes attached to the insides of Golgi membranes are transported from there to other sites at the outer plasma membrane surface (see Northcote 1971a).

There is also evidence of Golgi involvement in polysaccharide synthesis in larger algae. Light microscope histochemical studies on the brown algae *Fucus* (McCully 1968) and *Dictyota* (Evans & Holligan 1972) suggest that the polyuronide alginic acid and sulphated polysaccharides such as fucoidan are Golgi-synthesized. In the brown alga *Pelvetia* (Evans *et al.* 1973c), light microscope autoradiography using labelled inorganic sulphate $^{35}SO_4^{--}$ shows heavy labelling in the perinuclear Golgi region, towards the bases of epidermal cells, an area which histochemical studies show to be rich in sulphated materials. After a chase period the label finally appears in the layer of sulphated material on the thallus surface. *Laminaria* species (Evans *et al.* 1973c), have been shown to have a system of secretory cells intimately associated with canals (Figs. 3.9 and 3.10), the content of which is composed entirely of sulphated polysaccharide. Labelled inorganic

---

**Fig. 3.6.** Section through part of the cell surface of the mutant form of *Chrysochromulina chiton* showing the arrangement of the two types of scales. × 48,000 Courtesy Prof. I. Manton. (From Manton 1967b.)

**Fig. 3.7.** Isolated scales of the mutant form shown in Fig. 3.6 illustrating a large scale and the two faces of a small scale. Reversed print. × 30,000 Courtesy Prof. I. Manton. (From Manton 1967a.)

**Fig. 3.8.** Transverse section of a Golgi system active in scale production on the side away from the subtending endoplasmic reticulum (ER). The scales increase in maturity from the Golgi centre outwards, scale 1 being the most mature and scale 5 the least mature. Scales 1 and 3 are small scales and 2 and 4 large ones. Both types are orientated so that the morphologically outer surface is away from the Golgi centre. × 52,000 Courtesy Prof. I. Manton. (From Manton 1967a.)

sulphate $^{35}SO_4^{--}$ can be shown autoradiographically to be incorporated into the secretory cells and passed into the secretory canals (Figs. 3.12 and 3.13) and eventually onto the thallus surface. Histochemical techniques show that the location of sulphated material within the secretory cells is around the nucleus, an area occupied almost entirely by large Golgi bodies (up to $3\mu$m long) and associated vesicles (Fig. 3.10). Electron microscope autoradiography indicates that polysaccharide sulphation takes place within the Golgi cisternae. In addition, secretory cells of some species (e.g. *Laminaria saccharina*) are characterized by a complex anastomosing system of concentric profiles of endoplasmic reticulum (Fig. 3.11). Such structures are typical of animal secretory cells (see e.g. Jamieson & Palade 1967a), but their role in the secretory process in *Laminaria* is unknown, since protein cannot be detected in the secreted mucilage. However, the proposed mechanism for synthesis of sulphated polysaccharide by the Golgi would necessitate a rapid turnover of Golgi cisternal membranes, and it is possible that the endoplasmic reticulum is synthesising Golgi membrane proteins. The haptera of *Laminaria* spp. lack secretory cells but here the epidermal cells have well-developed perinuclear Golgi complexes which appear to synthesise sulphated polysaccharides (Davies *et al.* 1973).

Reference to the close association which often occurs between the nuclear envelope and Golgi bodies has already been made. There is, in addition, evidence that adjacent endoplasmic reticulum may also be involved (e.g. Manton 1961b, Manton & Leedale 1961b, Berkaloff 1963, Drum & Pankratz 1964, Bouck 1965, Schötz *et al.* 1972). This is often found to subtend and to be in close association with the forming face of the Golgi (Fig. 3.15), but there is no absolute evidence of direct continuity between the two (see Pickett-Heaps 1971a for an example of

---

**Fig. 3.9.** Light micrograph of four small secretory cells of *Laminaria saccharina* adjacent to a larger secretory canal. The cells have large nuclei and sulphated polysaccharide is discharged from the cells into the canal. × 1,000 (Evans *et al.* 1973c).

**Fig. 3.10.** A secretory cell of *Laminaria hyperborea* showing the central nucleus (N) with large perinuclear Golgi bodies (G). Numerous vesicles (arrowed) believed to be of Golgi origin occur in the cytoplasm and these are thought to empty their contents into the canal. Other organelles visible include mitochondria (M). × 24,000 (Evans *et al.* 1973c.)

**Fig. 3.11.** Part of a secretory cell of *Laminaria saccharina* showing anastomosing, swollen endoplasmic reticulum (arrowed). A Golgi area (G) is also visible and a proliferation of small vesicles at the Golgi/nuclear envelope interface. N = nucleus. × 22,000 (Evans *et al.* 1973c).

**Fig. 3.12.** Light micrograph of a secretory canal and secretory cells from a lamina of *Laminaria hyperborea* incubated in inorganic sulphate $^{35}SO_4^{--}$ for 8h showing heavy labelling of the secretory cells surrounding the small subsidiary secretory canals. No labelling of main canal (C). × 375 (Evans *et al.* 1973c.)

**Fig. 3.13.** Light micrograph of material incubated in inorganic sulphate $^{35}SO_4^{--}$ for 3h and chased in unlabelled medium for 41h. Silver grains are associated with the main canal (C) in addition to the subsidiary canals (arrowed). × 225 (Evans *et al.* 1973c.)

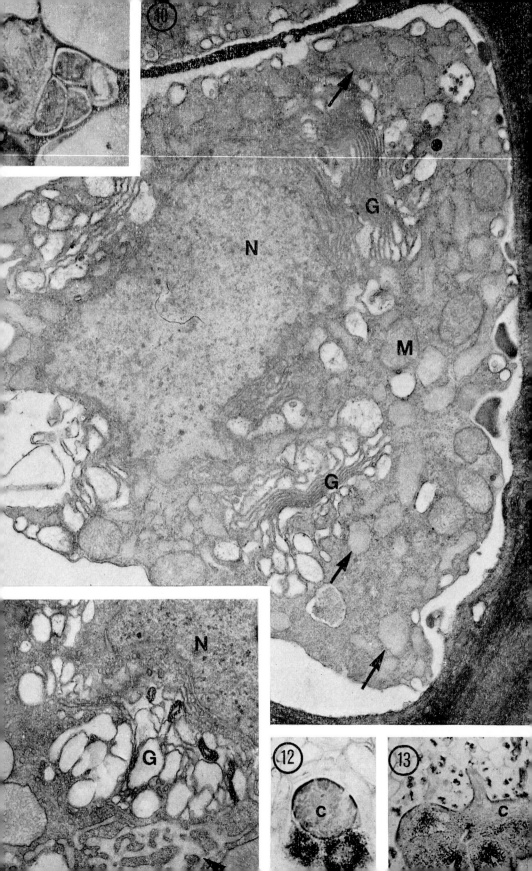

possible direct connection). Material has, however, been shown by several workers to pass from the endoplasmic reticulum to the Golgi cisternae. For example, Caro and Palade (1964) and Jamieson and Palade (1967a,b) have investigated the transport of protein in relation to the Golgi complex. In pancreatic exocrine cells, enzymes synthesized by ribosomes attached to rough endoplasmic reticulum have been shown to pass into the lumen of the endoplasmic reticulum and later to be transferred via small vesicles to the forming face of the Golgi apparatus. Here the protein is concentrated into membrane-bounded zymogen granules which are secreted by reverse pinocytosis. Investigations on the production of adhesive by motile zoospores of the green seaweed *Enteromorpha* (Evans & Christie 1970, Callow & Evans 1973), indicate that a similar series of events occurs here also (Figs. 3.14 and 3.15). Electron microscope cytochemistry for the detection of carbohydrate-complexed substances (Thiéry 1967) and electron microscope autoradiography using L-leucine-4,5-3H, show the Golgi bodies to be sites of synthesis of vesicles containing carbohydrates and protein moieties. The vesicle contents are formed in association with adjacent rough endoplasmic reticulum through the medium of small transfer vesicles. The glycoprotein material is secreted from the cells by exocytosis and forms the fibrillar extracellular substance responsible for attaching the spores to a surface during settlement. The Golgi bodies of zoospores of the filamentous brown alga *Ectocarpus* (Baker & Evans 1973) and of carpospores and tetraspores of the red alga *Ceramium* (Chamberlain & Evans 1973) have also been shown to be involved in the synthesis of adhesive material which is secreted during settlement. The adhesive in both cases appears to be composed of two or more moieties. In the case of *Ectocarpus* a complex molecule probably composed of a carbohydrate and possibly a protein is involved, and in the case of *Ceramium* the adhesive is basically proteinaceous with probably a phospholipid and possibly a polysaccharide component as well. Release and attachment of monospores of the red alga *Smithora naiadum* are also aided by Golgi-derived substances of unknown chemical composition (McBride & Cole 1971). In *Bulbochaete hiloensis* acid mucopolysaccharide secreted from vesicles present in the zoospore head is suggested to function as an adhesive (Retallack & Butler 1972).

It seems that materials such as polysaccharides are both synthesised and

---

**Fig. 3.14.** Part of a motile zoospore of the green alga *Enteromorpha* showing 3 Golgi bodies (G) the upper of which illustrates a developmental sequence in vesicle formation in cisternae from the forming to the maturing face. Abstricted vesicles (V) lie in the cytoplasm and these provide the glycoprotein adhesive secreted during spore settlement. × 84,000 (From Evans & Christie 1970.)

**Fig. 3.15.** A Golgi body in an *Enteromorpha* zoospore subtended by a characteristic arc of endoplasmic reticulum (ER). From the smooth inner surface of the latter small vesicles are abstricted (arrow) and these are believed to transport material across to the Golgi complex. Also visible is part of a nucleus (N) with a pore containing dense material in the nuclear envelope. × 50,000 (From Evans & Christie 1970.)

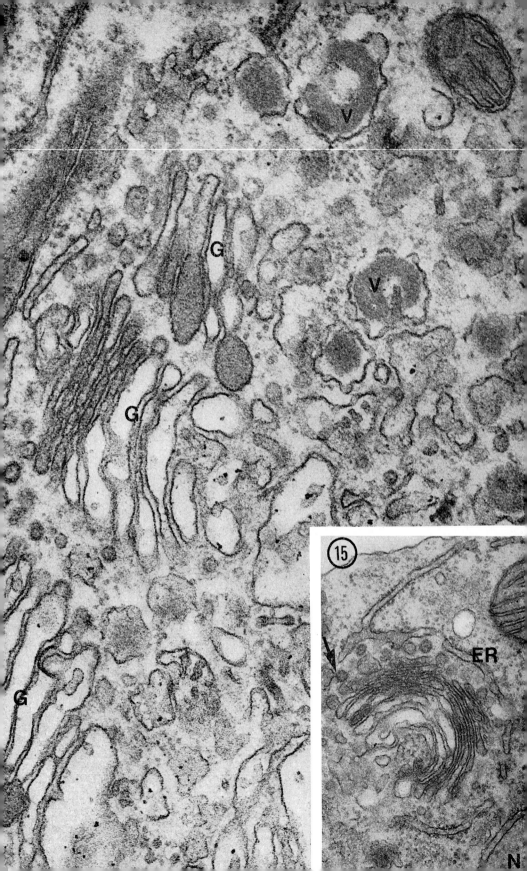

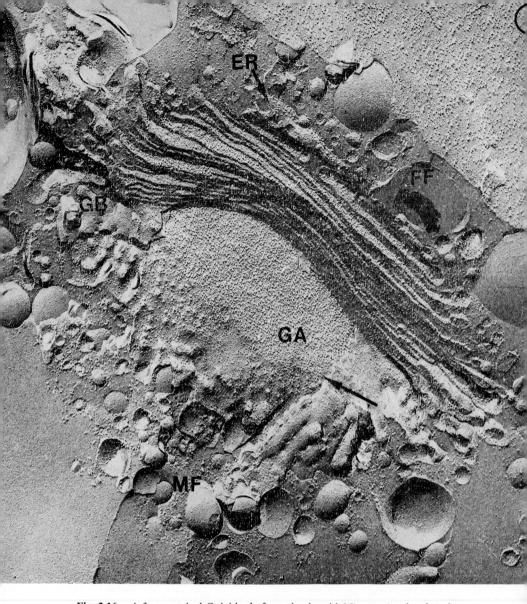

**Fig. 3.16.** A freeze-etched Golgi body from the desmid *Micrasterias* showing cisternae in cross-sectional (upper half) and surface (lower half) views. At the forming face (FF) there is association with membranes of the endoplasmic reticulum (ER). Membrane faces GA and GB are parts of cisterna of the maturing face (MF). Particle density on face GA declines sharply (arrow) where fenestration starts to occur. × 60,000 Courtesy Drs. L. A. Staehelin & O. Kiermayer. (From Staehelin & Kiermayer 1970.)

modified within the Golgi cisternae and vesicles. Enzymes necessary for synthesis such as nucleotide diphosphate sugar transglycosylases should therefore be contained on the membranes, as well as enzymes being modified and transported through the system for use elsewhere in the cell (Northcote 1971b). Cytochemical techniques reveal enzymes such as nucleoside phosphatases and also that activity of a certain enzyme may be confined to some Golgi cisternae only; that is, not all cisternae necessarily contain the same enzyme (Novikoff *et al.* 1962, Dauwalder *et al.* 1969, Wise & Flickinger 1970). Freeze-etch studies on the Golgi of the desmid *Micrasterias* (Staehelin & Kiermayer 1970) show a gradual increase in particle density on the cisternal membranes from the forming to the maturing face (Fig. 3.16). Such particles could be membrane-bounded globular enzymes or multienzyme-complexes (Branton 1969). It also seems probable that not all the Golgi body populations of a cell are necessarily carrying out identical functions simultaneously. Pickett-Heaps and Fowke (1970b) speculate that Golgi bodies in *Oedogonium* may be forming vesicles with at least two different contents, thus perhaps performing two functions at once, viz. contributing material to the ring and being involved, through fusion with the contractile vacuole, in the control of turgor pressure. Pickett-Heaps (1971a) also provides evidence which suggests the concurrent derivation of two characteristic wall layers from vesicles of two apparently differentiated populations of Golgi within the cell during zoosporogenesis in *Oedogonium*.

An indication of the rate at which Golgi-associated activities may take place may be had from the findings of Brown (1969, 1971) who reports that in *Pleurochrysis* a Golgi-derived vesicle ejects its scale content extracellularly every 1 to 2 min. The activity of the Golgi complex, as a dynamic membrane system involved in synthesis and organized transportation of soluble and insoluble molecules, varies with metabolic conditions and with the stage of cell development. A wide variety of materials is elaborated at different stages during growth and differentiation and this results in a continued changing pattern in the synthetic activities of the Golgi system.

## 5 REFERENCES

ARNOLD C.G., SCHIMMER O., SCHÖTZ F. & BATHELT H. (1972) Die Mitochondrien von *Chlamydomonas reinhardii. Arch. Microbiol.* **81**, 50–67.

ATKINSON A.W., GUNNING B.E.S. & JOHN P.C.L. (1971) Centrioles and microtubules in *Chlorella. Nature New Biol.* **234**, 24–5.

ATKINSON A.W., GUNNING B.E.S. & JOHN P.C.L. (1972) Sporopollenin in the cell wall of *Chlorella* and other algae: Ultrastructure, chemistry, and incorporation of $^{14}$C-acetate, studied in synchronous cultures. *Planta,* **107**, 1–32.

BAILEY A. & BISALPUTRA T. (1970) A preliminary account of the application of thin-sectioning, freeze-etching and scanning electron microscopy to the study of coralline algae. *Phycologia,* **9**, 83–101.

BAKER J.R.J. & EVANS L.V. (1973) The ship fouling alga *Ectocarpus* I. Ultrastructure and cytochemistry of plurilocular reproductive stages. *Protoplasma,* **77**, 1–13.

BARNETT J.R. & PRESTON R.D. (1969) The fine structure of *Chlamydomonas* as judged by freeze etching using an apparatus of novel design. *Proc. Roy. Microscop. Soc.* **4**, 135.

BARRAS D.R. & STONE B.A. (1965) The chemical composition of the pellicle of *Euglena gracilis* var. *bacillaris*. *Biochem. J.* **97**, 14P.

BARTON R., DAVIS P.J. & THOMAS S.R. (1970) Globular subunits in negatively stained hairs (mastigonemes) of *Ochromonas* flagella. *Microscopie Électronique* **3**, ed. Favard P. pp. 433–4. Grenoble 1970.

BEAMS H.W. & KESSEL R.G. (1968) The Golgi apparatus, structure and function. *Int. Rev. Cytol.* **23**, 209–76.

BERKALOFF G. (1963) Les cellules méristématiques d'*Himanthalia lorea* (L.) S.F. Gray. Étude au microscope-elecktronique. *J. Microscopie*, **2**, 213–28.

BISALPUTRA T. (1966) Electron microscope study of the protoplasmic continuity in certain brown algae. *Can. J. Bot.* **44**, 89–93.

BISALPUTRA T. & BISALPUTRA A.A. (1967) Chloroplast and mitochondrial DNA in a brown alga *Egregia menziesii*. *J. Cell Biol.* **33**, 511–20.

BISALPUTRA T., RUSANOWSKI P.C. & WALKER W.S. (1967) Surface activity, cell wall, and fine structure of pit connections in the red alga *Laurencia spectabilis*. *J. Ultrastruct. Res.* **20**, 277–89.

BISALPUTRA T. & STEIN J.R. (1966) The development of cytoplasmic bridges in *Volvox aureus*. *Can. J. Bot.* **44**, 1697–702.

BOUCK G.B. (1962) Chromatophore development, pits and other fine structure in the red alga *Lomentaria baileyana* (Harv.) Farlow. *J. Cell Biol.* **12**, 553–69.

BOUCK G.B. (1964) Fine structure in *Acetabularia* and its relation to protoplasmic streaming. In *Primitive Motile Systems in Cell Biology*, pp. 7–18. Academic Press, New York & London.

BOUCK G.B. (1965) Fine structure and organelle associations in brown algae. *J. Cell Biol.* **26**, 523–37.

BOUCK G.B. (1969) Extracellular microtubules. The origin, structure and attachment of flagellar hairs in *Fucus* and *Ascophyllum* antherozoids. *J. Cell Biol.* **40**, 446–60.

BOUCK G.B. (1970) The development and post fertilization fate of the eyespot and the apparent photoreceptor in *Fucus* sperm. *Ann. N.Y. Acad. Sci.* **175**, 673–85.

BOUCK G.B. (1971) The structure, origin, isolation and composition of the tubular mastigonemes of the *Ochromonas* flagellum. *J. Cell Biol.* **50**, 362–84.

BOUCK G.B. (1972) Architecture and assembly of mastigonemes. In *Advances in Cell and Molecular Biology*, Vol. 2, ed. Du Praw E.J. pp. 237–71. Academic Press, New York.

BOUCK G.B. & BROWN D.L. (1973) Microtubule biogenesis and cell shape in *Ochromonas* I. The distribution of cytoplasmic and mitotic microtubules. *J. Cell Biol.* **56**, 340–69.

BOUCK G.B. & SWEENEY B.M. (1966) The fine structure and ontogeny of trichocysts in marine dinoflagellates. *Protoplasma*, **61**, 205–23.

BRACKER C.E. & GROVE S.N. (1971) Continuity between cytoplasmic endomembranes and other mitochondrial membranes in fungi. *Protoplasma*, **73**, 15–34.

BRANDES D.J. (1965) Observations on the apparent mode of formation of 'pure' lysosomes. *J. Ultrastruct. Res.* **12**, 63–80.

BRANTON D. (1969) Membrane Structure. *A. Rev. Pl. Physiol.* **20**, 209–38.

BROWN D.L. & BOUCK G.B. (1973) Microtubule biogenesis and cell shape in *Ochromonas* II. The role of nucleating site in shape development. *J. Cell Biol.* **56**, 360–78.

BROWN R.M. (1969) Observations on the relationship of the Golgi apparatus to wall formation in the marine chrysophycean alga *Pleurochrysis scherffelii*. Pringsheim. *J. Cell Biol.* **41**, 109–23.

BROWN R.M. (1971) Plasmabewegungen und Funktion des Golgi-Apparates in Algen. Film V 1401. Institut für den Wissenschaftlichen Film, Göttingen.

BROWN R.M. & FRANKE W.W. (1971) A microtubular crystal associated with the Golgi field in *Pleurochrysis scherffelii*. *Planta*, **96**, 354–63.

BROWN R.M., FRANKE W.W., KLEINIG H., FALK H. & SITTE P. (1970) Scale formation in Chrysophycean algae I. Cellulosic and non-cellulosic wall components made by the Golgi apparatus. *J. Cell Biol.* **45**, 246–71.

BURR F.A. & WEST J.A. (1971) Comparative ultrastructure of the primary nucleus in *Bryopsis* and *Acetabularia*. *J. Phycol.* **7**, 108–13.

CALLOW M.E. & EVANS L.V. (1973) Studies on the ship-fouling alga *Enteromorpha* III. Cytochemistry and autoradiography of adhesive production. *Protoplasma.* **80** 15–27.

CARO L.G. & PALADE G.E. (1964) Protein synthesis, storage, and discharge in the pancreatic exocrine cell. An autoradiographic study. *J. Cell Biol.* **20**, 473–95.

CHAMBERLAIN, A.H.L. & EVANS L.V. (1973) Aspects of spore production in the red alga *Ceramium*. *Protoplasma* **76**, 139–59.

CITHAREL J. (1972) Contribution à l'etude du métabolisme azoté des Algues marines. Utilisation métabolique d'acide glutamique-$^{14}$C par *Ascophyllum nodosum* (Linne) Le Jolis et *Polysiphonia lanosa* (Linne) Tandy. *Bot. Mar.* **15**, 157–61.

CLAUDE A. (1970a) Growth and differentiation of cytoplasmic membranes in the course of lipoprotein granule synthesis in the hepatic cell. I. Elaboration of elements of the Golgi complex. *J. Cell. Biol.* **47**, 745–66.

CLAUDE A. (1970b) Origin and differentiation of membrane structures of the Golgi apparatus in the hepatic cell. *Microscopie Electronique* 3, ed. Favard P., pp. 85–6. Grenoble, 1970.

CODD G.A., SCHMID G.H. & KOWALLIK W. (1972) Enzymic evidence for peroxisomes in a mutant of *Chlorella vulgaris*. *Arch. Mikrobiol.* **81**, 264–72.

COLE K. (1970) Ultrastructural characteristics in some species in the order Scytosiphonales. *Phycologia*, **9**, 275–83.

DAUWALDER M., WHALEY W.G. & KEPHART J.E. (1969). Phosphatases and differentiation of the Golgi apparatus. *J. Cell Sci.* **4**, 455–97.

DAVIES, J.M., FERRIER N.C. & JOHNSTON C.S. (1973) The ultrastructure of the meristoderm cells of the hapteron of *Laminaria*. *J. mar. biol. Ass. U.K.* **53**, 237–46.

DODGE J.D. (1971a) Fine structure of the Pyrrophyta. *Bot. Rev.* **37**, 481–508.

DODGE J.D. (1971b) A dinoflagellate with both a mesocaryotic and a eucaryotic nucleus. I. Fine structure of the nuclei. *Protoplasma*, **73**, 145–57.

DODGE J.D. (1972) The ultrastructure of the Dinoflagellate pusule: a unique osmo-regulatory organelle. *Protoplasma*, **75**, 285–302.

DODGE J.D. (1973) *The fine structure of algal cells*. Academic Press, London & New York.

DODGE J.D. (1974) Fine structure and phylogeny in the algae. *Sci. Progr.* **61**, 257–74.

DRAGESCO J. (1951) Sur la structure des trichocystes du Flagellé Cryptomonadine *Chilomonas paramecium*. *Bull. Micr. appl.* **1**, 172–5.

DRAWERT H. & MIX M. (1962) Zur Funktion des Golgi-Apparates in der Pflanzenzelle. *Planta*, **58**, 448–52.

DRAWERT H. & MIX M. (1963) Licht- und elektronenmikroskopische Untersuchungen an Desmidiaceen. XI. Mitteilung: Die Struktur von Nucleolus und Golgi-Apparat bei *Micrasterias denticulata* Breb. *Port. Acta Biol.* **7**, 17–28.

DRUM R.W. (1966) Electron microscopy of paired Golgi structures in the diatom *Pinnularia nobilis*. *J. Ultrastruct. Res.* **15**, 100–7.

DRUM R.W. & PANKRATZ H.S. (1964) Pyrenoids, raphes and other fine structure in diatoms. *Am. J. Bot.* **51**, 405–18.

EDELMAN M., EPSTEIN H.T. & SCHIFF J.A. (1966) Isolation and characterization of DNA from the mitochondrial fraction of *Euglena*. *J. molec. Biol.* 463–9.

EVANS L.V. (1966) Distribution of pyrenoids among some brown algae. *J. Cell Sci.* **1**, 449–54.

EVANS L.V. (1970) Electron microscopical observations on a new red algal unicell *Rhodella maculata* gen. nov., sp. nov. *Br. phycol. J.* **5**, 1–13.

EVANS L.V. & CHRISTIE A.O. (1970) Studies on the ship-fouling alga *Enteromorpha*. I. Aspects of the fine-structure and biochemistry of swimming and newly settled zoospores. *Ann. Bot.* **34**, 451–66.

EVANS L.V. & HOLLIGAN M.S. (1972) Correlated light and electron microscope studies on brown algae. I. Localisation of alginic acid and sulphated polysaccharides in *Dictyota*. *New Phytol.* **71**, 1161–72.

EVANS L.V., CALLOW J.A. & CALLOW M.E. (1973a) Structural and physiological studies on the parasitic red alga *Holmsella*. *New Phytol.* **72**, 393–402.

EVANS L.V., CALLOW M.E., PERCIVAL E. & FAREED V. (1974) Studies on the synthesis and composition of extracellular mucilage in the unicellular red alga *Rhodella*. *J. Cell Sci.* **15**. (In press.)

EVANS L.V., SIMPSON M. & CALLOW M.E. (1973c) Sulphated polysaccharide synthesis in brown algae. *Planta*, **110**, 237–52.

FALK H. (1967) Zum feinbau von *Botrydium granulatum* Grev. (Xanthophyceae). *Arch. Mikrobiol.* **58**, 212–27.

FALK H. & KLEINIG H. (1968) Feinbau und Carotinoide von *Tribonema* (Xanthophyceae). *Arch. Mikrobiol.* **61**, 347–62.

FALK M., SMITH D.G., MCLACHLAN J. & MCINNES A.G. (1966) Studies on chitan (β-(1 → 4)- linked 2- acetamido-2-deoxy-D-glucan) fibres of the diatom *Thalassiosira fluviatilis* Hustedt. *Can. J. Chem.* **44**, 2269–81.

FELDHEN C.M. (1972) Structure and function of the nuclear envelope. In *Advances in Cell and Molecular Biology*, Vol. 2, ed. Du Praw E.J. pp. 273–307. Academic Press, New York.

FLOYD G.L. (1972) Comparative cytology of *Ulothrix* and *Stigeoclonium*. *J. Phycol.* **8**, 68–81.

FLOYD G.L., STEWART K.D. & MATTOX K.R. (1971) Cytokinesis and plasmodesmata in *Ulothrix*. *J. Phycol.* **7**, 306–9.

FLOYD G.L., STEWART K.D. & MATTOX K.R. (1972a) Cellular organization, mitosis and cytokinesis in the Ulotrichalean alga, *Klebsormidium*. *J. Phycol.* **8**, 176–84.

FLOYD G.L., STEWART K.D. & MATTOX K.R. (1972b) Comparative cytology of *Ulothrix* and *Stigeoclonium*. *J. Phycol.* **8**, 68–81.

FOGG G.E., STEWART W.D.P., FAY P. & WALSBY A.E. (1973) *The blue-green algae*. Academic Press, London & New York.

FOWKE L.C. & PICKETT-HEAPS J.D. (1969a) Cell division in *Spirogyra*. I. Mitosis. *J. Phycol.* **5**, 240–59.

FOWKE L.C. & PICKETT-HEAPS J.D. (1969b) Cell division in *Spirogyra*. II. Cytokinesis. *J. Phycol.* **5**, 273–81.

FRANKE W.W. (1970) Central dilations in maturing Golgi cisternae—a common feature among plant cells? *Planta*, **90**, 370–3.

FRANKE W.W. & KARTENBECK J. (1971) Outer mitochondrial membrane continuous with endoplasmic reticulum. *Protoplasma*, **73**, 35–41.

FRASER T.W. & GUNNING B.E.S. (1969) The ultrastructure of plasmodesmata in the filamentous green alga *Bulbochaete hiloensis* (Nordst.) Tiffany. *Planta*, **88**, 244–54.

FULTON C. (1971) Centrioles. In *Origin and Continuity of Cell Organelles*, eds. Reinert J. & Ursprung Z. pp. 170–221. Springer-Verlag, Berlin.

GANTT E. & CONTI S.F. (1965) The ultrastructure of *Porphyridium cruentum*. *J. Cell Biol.* **26**, 365–81.

GERGIS M.S. (1971) The presence of microbodies in three strains of *Chlorella*. *Planta*, **101**, 180–4.

GIBBS S.P. (1968) Autoradiographic evidence for the *in situ* synthesis of chloroplast and mitochondrial RNA. *J. Cell Sci.* **3**, 327–40.

GOODAY G.W. (1971) A biochemical and autoradiographic study of the role of the Golgi bodies in thecal formation in *Platymonas tetrathele*. *J. exp. Bot.* **23**, 959–71.

GOROVSKY M.A., CARLSON K. & ROSENBAUM J.L. (1970) Simple method for quantitative densitometry of polyacrylamide gels using fast green. *Analyt. Biochem.* **35**, 359–370.

GREEN J.C. & JENNINGS D.H. (1967) A physical and chemical investigation of the scales produced by the Golgi apparatus within and found on the surface of the cells of *Chrysochromulina chiton* Parke *et* Manton. *J. exp. Bot.* **18**, 359–70.

GREEN P.B. (1963) On mechanisms of elongation. In *Cytodifferentiation and Macromolecular Synthesis*, ed. Locke M., pp. 203–34. Academic Press, New York and London.

GRUSKY G.E. & AARONSON S. (1969) Cytochemical changes in aging *Ochromonas*; evidence for an alkaline phosphatase. *J. Protozool.* **16**, 686–9,

HARTMANN T. & ESCHRICH W. (1969) Stofftransport in Rotalgen. *Planta*, **85**, 303–12.

HEATH I.B. & DARLEY W.M. (1972) Observations on the ultrastructure of the male gametes of *Biddulphia levis* Ehr. *J. Phycol.* **8**, 51–9.

HEATH I.B., GREENWOOD A.D. & GRIFFITHS H.B. (1970) The origin of flimmer in *Saprolegnia, Dictyuchus, Synura* and *Cryptomonas*. *J. Cell Sci.* **7**, 445–61.

HELLEBUST J.A. & HAUG A. (1972) Photosynthesis, translocation and alginic acid synthesis in *Laminaria digitata* and *Laminaria hyperborea*. *Can. J. Bot.* **50**, 169–76.

HERTH W., FRANKE W.W., STADLER J., BITTIGER H., KEILICH G. & BROWN R.M. Jr. (1972) Further characterization of the alkali-stable material from the scales of *Pleurochrysis scherffelii*: A cellulosic glycoprotein. *Planta*, **105**, 79–92.

HEYWOOD P. (1972) Structure and origin of flagellar hairs in *Vacuolaria virescens*. *J. Ultrastruct. Res.* **39**, 608–23.

HIBBERD D.J. (1970) Observations on the cytology and ultrastructure of *Ochromonas tuberculatus* sp. nov. (Chrysophyceae), with special reference to the discobolocysts. *Br. phycol. J.* **5**, 119–43.

HILL G.T.C. & MACHLIS L. (1968) An ultrastructural study of vegetative cell division in *Oedogonium borisianum*. *J. Phycol.* **4**, 261–71.

HOFFMAN L.R. (1966) The fine structure of zoospore development in *Oedogonium cardiacum*. *J. Phycol.* **2**, Suppl. p. 5.

HOLTZMAN E., NOVIKOFF A.B. & VILLAVERDE H. (1967) Lysosomes and GERL in normal and chromatolytic neurons of the rat ganglion nodosum. *J. Cell Biol.* **33**, 419–35.

HOPKINS J.M. (1970) Subsidiary components of the flagella of *Chlamydomonas reinhardii*. *J. Cell Sci.* **7**, 823–39.

HORNE R.W. (1971) Electron microscopy applied to the study of macromolecular components assembled to form complex biological structures. In *Control Mechanisms of Growth and Differentiation*. XXV Symp. S.E.B., eds. Davies D.D. & Balls M. pp. 71–92. Cambridge University Press.

HOVASSE R. (1948) Le discobolocyste, organite lanceur de projectile, chez la Chrysomonadine *Cyclonexis annularis* Stokes. *C. r. hebd. Séanc. Acad. Sci. Paris*, **226**, 1038–9.

HOVASSE R. (1965) Trichocysts, corps trichocystoïdes, cnidocystes et colloblastes. *Protoplasmatologia*, **3**, 1–57.

HOVASSE R., MIGNOT J.P. & JOYON L. (1967) Nouvelles observations sur les trichocystes des Cryptomonadines et les 'R bodies' des particules Kappa de *Paramecium aurelia* Killer. *Protistologica*, **3**, 241–55.

JAMIESON J.D. & PALADE G.E. (1967a) Intracellular transport of secretory proteins in the pancreatic exocrine cell. I. Role of the peripheral elements of the Golgi complex. *J. Cell Biol.* **34**, 577–96.

JAMIESON J.D. & PALADE G.E. (1967b) Intracellular transport of secretory proteins in the pancreatic exocrine cell II. Transport to condensing vacuoles and zymogen granules. *J. Cell Biol.* **34**, 597–615.

JAROSCH R. (1962) Die 'Doppelplättchen' der Diatomeen als Golgi-Apparate. *Protoplasma*, **55**, 552–4.

JOHNSON U.G. & PORTER K.R. (1968) Fine structure of cell division in *Chlamydomonas reinhardii*. *J. Cell Biol.* **38**, 403–25.

KEENAN T.W. & MORRÉ D.J. (1970) Phospholipid class and fatty acid composition of Golgi apparatus isolated from rat liver and comparison with other cell fractions. *Biochemistry. N.Y.* **9**, 19–25.

KIERMAYER O. (1967) Dictyosomes in *Micrasterias* and their 'division'. *J. Cell Biol.* **35**, 68A.

KIERMAYER O. (1970) Elektronmikroskopische Untersuchungen zum Problem der Cytomorphogenese von *Micrasterias denticulata* Breb. I. Allgemeiner Überblick. *Protoplasma*, **69**, 97–132.

KIERMAYER O. (1971) Elektronenmikroskopischer Nachweis spezieller cytoplasmatischer Vesikel bei *Micrasterias denticulata* Bréb, *Planta*, **96**, 74–80.

E

KISAKI T. & TOLBERT N.E. (1969) Glycolate and glyoxylate metabolism by isolated peroxisomes or chloroplasts *Pl. Physiol., Lancaster*, **44**, 242–50.

KROON A.M. & DE VRIES H. (1970) Antibiotics: A tool in the search for the degree of autonomy of mitochondria in higher animals. In *Control of Organelle Development.* XXIV Symp. S.E.B., ed. Miller P.L. pp. 181–99. Cambridge University Press.

KUBAI D.F. & RIS H. (1969) Division in the dinoflagellate *Gyrodinium cohnii* (Schiller). *J. Cell Biol.* **40**, 508–28.

KUGRENS P. & WEST J.A. (1972) Synaptonemal complexes in red algae. *J. Phycol.* **8**, 187–91.

LANG N.J. (1968) The fine structure of blue-green algae. *A. Rev. Microbiol.* **22**, 15–46.

LEADBEATER B.S.C. (1971) The intracellular origin of flagellar hairs in the dinoflagellate *Woloszynskia micra* Leadbeater & Dodge. *J. Cell Sci.* **9**, 443–51.

LEADBEATER B.S.C. & DODGE J.D. (1966) The fine structure of *Woloszynskia micra* sp. nov., a new marine dinoflagellate. *Br. phycol. Bull.* **3**, 1–17.

LEADBEATER B.S.C. & DODGE J.D. (1967) An electron microscope study of nuclear and cell division in a dinoflagellate. *Arch. Mikrobiol.* **57**, 239–54.

LEEDALE G.F. (1967) Euglenida/Euglenophyta. *A. Rev. Microbiol.* **21**, 31–48.

LEEDALE G.F. (1970) Phylogenetic aspects of nuclear cytology in the algae. *Ann. N.Y. Acad. Sci.* **175**, 429–53.

LEEDALE G.F., LEADBEATER B.S.C. & MASSALSKI A. (1970) The intracellular origin of flagellar hairs in the Chrysophyceae and Xanthophyceae. *J. Cell. Sci.* **6**, 701–19.

LEEDALE G.F., MEEUSE B.J.D. & PRINGSHEIM E.G. (1965). Structure and physiology of *Euglena spirogyra*. I and II. *Arch. Mikrobiol.* **50**, 68–102.

LEWIN R.A. (1958) The cell walls of *Platymonas*. *J. gen. Microbiol.* **19**, 87–90.

LORD J.M., KAGAWA T., MORRÉ T.S. & BEEVERS H. (1973) Endoplasmic reticulum as the site of lecithin formation in castor bean endosperm. *J. Cell Biol.* **57**, 659–67.

LØVLIE A. & BRÅTEN T. (1970) On the mitosis in the multicellular alga *Ulva mutabilis. J. Cell Sci.* **6**, 109–29.

MCBRIDE D.L. & COLE K. (1971) Electron microscopic observations on the differentiation and release of monospores in the marine red alga *Smithora naiadum. Phycologia* **10**, 49–61.

MCCULLY M.C. (1968) Histological studies on the genus *Fucus* III. Fine structure and possible functions of the epidermal cells of the vegetative thallus. *J. Cell Sci.* **3**, 1–16.

MCDONALD K. (1972) The ultrastructure of mitosis in the marine red alga *Membranoptera platyphylla. J. Phycol.* **8**, 156–66

MANTON I. (1956) Plant cilia and associated organelles. In *Cellular Mechanisms in Differentiation and Growth*, ed. Rudnick D., pp. 61–71. Princeton University Press.

MANTON I. (1957) Observations with the electron microscope on the internal structure of the zoospore of a brown alga. *J. exp. Bot.* **8**, 294–303.

MANTON I. (1959) Electron microscope observations on a very small flagellate; the problem of *Chromulina pusilla* Butcher. *J. mar. biol. Ass. U.K.* **38**, 319–33.

MANTON I. (1961a) Some problems of mitochondrial growth. *J. exp. Bot.* **12**, 421–9.

MANTON I. (1961b) Plant cell structure. In *Contemporary Botanical Thought*, eds. McLeod A.M. & Cobley L.S., pp. 171–97. Oliver & Boyd, Edinburgh.

MANTON I. (1964a) The possible significance of some details of flagellar bases in plants. *J.R. microsc. Soc.* **82**, 279–85.

MANTON I. (1964b) A contribution towards understanding of 'The Primitive Fucoid'. *New Phytol.* **63**, 244–54.

MANTON I. (1964c) Observations on the fine structure of the zoospore and young germling of *Stigeoclonium. J. exp. Bot.* **15**, 399–411.

MANTON I. (1964d) Further observations on the fine structure of the haptonema in *Prymnesium parvum. Arch. Mikrobiol.* **49**, 315–30.

MANTON I. (1964e) Observations with the electron microscope on the division cycle in the flagellate *Prymnesium parvum* Carter. *Jl. R. microsc. Soc.* **83**, 317–25.

MANTON I. (1966a) Observations on the microanatomy of the brown flagellate *Sphalero-*

*mantis tetragona* Skuja with special reference to the flagellar apparatus and scales. *J. Linn. Soc. (Bot.).* **59**, 397–403.

MANTON I. (1966b) Observations on scale production in *Pyramimonas amylifera* Conrad. *J. Cell Sci.* **1**, 429–38.

MANTON I. (1966c) Observations on scale production in *Prymnesium parvum. J. Cell Sci.* **1**, 375–80.

MANTON I. (1967a) Further observations on scale formation in *Chrysochromulina chiton. J. Cell Sci.* **2**, 411–18.

MANTON I. (1967b) Further observations on the fine structure of *Chrysochromulina chiton* with special reference to the haptonema, 'peculiar' Golgi structure and scale production. *J. Cell Sci.* **2**, 265–72.

MANTON I. (1968) Further observations on the microanatomy of the haptonema in *Chrysochromulina chiton* and *Prymnesium parvum. Protoplasma,* **66**, 33–53.

MANTON I. (1969) Tubular trichocysts in a *Pyramimonas (P. grosii* Parke). *Osterr. Bot. Z.* **116**, 378–92.

MANTON I. & CLARKE B. (1951) An electron microscope study of the spermatozoid of *Fucus serratus. Ann. Bot.* **15**, 461–71.

MANTON I. & ETTL H. (1965) Observations on the fine structure of *Mesostigma viride* Lauterborn. *J. Linn. Soc. (Bot.).* **59**, 175–84.

MANTON I. & LEEDALE G.F. (1961a) Further observations on the fine structure of *Chrysochromulina ericina* Parke & Manton. *J. mar. biol. Ass. U.K.* **41**, 145–55.

MANTON I. & LEEDALE G.F. (1961b) Observations on the fine structure of *Paraphysomonas vestita,* with special reference to the Golgi apparatus and the origin of scales. *Phycologia,* **1**, 37–57.

MANTON I. & LEEDALE G.F. (1963a) Observations on the microanatomy of *Crystallolithus hyalinus* Gaarder & Markali. *Arch. Mikrobiol.* **47**, 115–36.

MANTON I. & LEEDALE G.F. (1963b) Observations on the fine structure of *Prymnesium parvum* Carter. *Arch. Mikrobiol.* **45**, 285–303.

MANTON I. & LEEDALE G.F. (1969) Observations on the microanatomy of *Coccolithus pelagicus* and *Cricosphaera carterae,* with special reference to the origin and nature of coccoliths and scales. *J. mar. biol. Ass. U.K.* **49**, 1–16.

MANTON I. & PARKE M. (1960) Further observations on small green flagellates with special reference to possible relatives of *Chromulina pusilla. J. mar. biol. Ass. U.K.* **39**, 275–298.

MANTON I. & Parke M. (1962) Preliminary observations on scales and their mode of origin in *Chrysochromulina polylepis* sp. nov. *J. mar. biol. Ass. U.K.* **42**, 565–78.

MANTON I. & PARKE M. (1965) Observations on the fine structure of two species of *Platymonas* with special reference to flagellar scales and the mode of origin of the theca. *J. mar. biol. Ass. U.K.* **45**, 743–54.

MANTON I. & VON STOSCH H.A. (1966) Observations on the fine structure of the male gamete of the marine centric diatom *Lithodesmium undulatum. Jl. R. microsc. Soc.* **85**, 119–34.

MANTON I., KOWALLIK K. & VON STOSCH H.A. (1969a) Observations on the fine structure and development of the spindle at mitosis and meiosis in a marine centric diatom (*Lithodesmium undulatum*). I. Preliminary survey of mitosis in spermatogonia. *J. Microscopy.* **89**, 295–320.

MANTON I., KOWALLIK K. & VON STOSCH H.A. (1969b) Observations on the fine structure and development of the spindle at mitosis and meiosis in a marine centric diatom (*Lithodesmium undulatum*). II. The early meiotic stages in male gametogenesis. *J. Cell Sci.* **5**, 271–98.

MANTON I., KOWALLIK K. & VON STOSCH H.A. (1970a) Observations on the fine structure and development of the spindle at mitosis and meiosis in a marine centric diatom (*Lithodesmium undulatum*). III. The later stages of meiosis I in male gametogenesis. *J. Cell Sci.* **6**, 131–57.

MANTON I., KOWALLIK K. & VON STOSCH H.A. (1970b) Observations on the fine structure and development of the spindle at mitosis and meiosis in a marine centric diatom (*Lithodesmium undulatum*). IV. The second meiotic division and conclusion. *J. Cell Sci.* 7, 407–44.

MANTON I., OATES K. & GOODAY G. (1973) Further observations on the chemical composition of thecae of *Platymonas tetrathele* West (Prasinophyceae) by means of the X-ray microanalyser electron microscope (EMMA). *J. exp. Bot.* **24**, 223–9.

MANTON I., OATES K. & PARKE M. (1963) Observations on the fine structure of the *Pyramimonas* stage of *Halosphaera* and preliminary observations on three species of *Pyramimonas*. *J. mar. biol. Ass. U.K.* **43**, 225–38.

MANTON I., RAYNS D.G., ETTL H. & PARKE M. (1965) Further observations on green flagellates with scaly flagella: the genus *Heteromastix* Korshikov. *J. mar. biol. Ass. U.K.* **45**, 241–55.

MARCHANT H.J. & PICKETT-HEAPS J.D. (1970) Ultrastructure and differentiation of *Hydrodictyon reticulatum* I. Mitosis in the coenobium. *Aust. J. biol. Sci.* **23**, 1173–86.

MARCHANT H.J. & PICKETT-HEAPS J.D. (1971) Ultrastructure and differentiation of *Hydrodictyon reticulatum* II. Formation of zooids within the coenobium. *Aust. J. biol. Sci.* **24**, 471–86.

MASSALSKI A. & LEEDALE G.F. (1969) Cytology and ultrastructure of the Xanthophyceae. I. Comparative morphology of the zoospores of *Bumilleria sicula* Borzi and *Tribonema vulgare* Pascher. *Br. phycol. J.* **4**, 159–80.

MICALEF H. (1972) Mise en évidence de phosphatase acide dans l'appareil de Golgi des cellules végétatives d'*Ulva lactuca* (Chlorophycées, Ulvales). *C. R. Acad. Sc. Paris*, **275**, sér. D, 2481–4.

MIGNOT J.P., BRUGEROLLE G. & METENIER G. (1972). Compléments a l'étude des mastigonème des protistes flagellés. Utilisation de la technique de Thiéry pour la mise en évidence des polysaccharides sur coupes fines. *J. Microscopie* **14**, 327–42.

MOESTRUP O. (1972) Observations on the fine structure of spermatozoids and vegetative cells of the green alga *Golenkinia*. *Br. phycol. J.* **7**, 169–83.

MOLLENHAUER H.H. & MORRÉ D.J. (1966) Golgi apparatus and plant secretion. *A. Rev. Pl. Physiol.* **17**, 27–46.

MORRÉ D.J., MERRITT W.D. & LEMBI C.A. (1971a) Connections between mitochondria and endoplasmic reticulum in rat liver and onion stem. *Protoplasma*, **73**, 43–9.

MORRÉ D.J., MOLLENHAUER H.H. & BRACKER C.E. (1971b) Origin and continuity of Golgi apparatus. In *Origin and Continuity of Cell Organelles*, eds. Reinert J. & Ursprung Z., pp. 82–126. Springer-Verlag, Berlin.

NAGY J. & FRIDVALSKY L. (1968) Dictyosome-nuclei relationships in *Botrydium granulatum* (Chrysophyta). In *Electron Microscopy 1968*, ed. Cocciarelli D.J., **2**, 423–4. Fourth European Conference on Electron Microscopy. Rome: Tipografia Poliglotta Vaticana.

NEUSHUL M. & DAHL A.L. (1972) Ultrastructural studies of brown algal nuclei. *Am. J. Bot.* **59**, 401–10.

NEUTRA M. & LEBLOND C.P. (1966a) Synthesis of the carbohydrate of mucus in the Golgi complex as shown by electron microscope radioautography of goblet cells from rats injected with glucose-$H^3$. *J. Cell Biol.* **30**, 119–36.

NEUTRA M. & LEBLOND C.P. (1966b) Radioautographic comparison of the uptake of galactose-$H^3$ and glucose-$H^3$ in the Golgi region of various cells secreting mucoproteins or mucopolysaccharides. *J. Cell Biol.* **30**, 137–50.

NEWCOMB E.H. (1969) Plant microtubules. *A. Rev. Pl. Physiol.* **20**, 253–88.

NICHOLSON N.L. & BRIGGS W.R. (1972) Translocation of photosynthate in the brown alga *Nereocystis*. *Am. J. Bot.* **59**, 97–106.

NORTHCOTE D.H. (1971a) Organisation of structure, synthesis and transport within the plant during cell division and growth. In *Control Mechanisms of Growth and Differentiation.* XXV Symp. S.E.B., eds. Davies D.D. & Balls M., pp. 51–69. Camb. Univ. Press.

NORTHCOTE D.H. (1971b) The Golgi apparatus. *Endeavour*, **30**, 26–33.

NORTHCOTE D.H., GOULDING K.J. & HORNE R.W. (1958) The chemical composition and structure of the cell wall of *Chlorella pyrenoidosa*. *Biochem. J.* **70**, 391–7.

NORTHCOTE D.H. & PICKETT-HEAPS J.D. (1966) A function of the Golgi apparatus in polysaccharide synthesis and transport in the root-cap cells of wheat. *Biochem. J.* **98**, 159–67.

NOVIKOFF A.B., ESSNER E., GOLDFISCHER S. & HEUS M. (1962). Nucleosidephosphatase activities of cytomembranes. In *The Interpretation of Ultrastructure*, **1**, Symp. Int. Soc. Cell. Biol., ed. Harris R.J.C., pp. 149–92. Academic Press, New York & London.

OSAFUNE T., MIHARA S., HASE E. & OHKURO I. (1972) Electron microscope studies on the vegetative cellular life cycle of *Chlamydomonas reinhardii* Dangeard in synchronous culture I. Some characteristics of changes in subcellular structures during the cell cycle, especially in formation of giant mitochondria. *Pl. Cell Physiol., Tokyo*, **13**, 211–227.

OTT D.W. & BROWN R.M. Jr. (1972) Light and electron microscopical observations on mitosis in *Vaucheria litorea* Hofman ex C. Agardh. *Br. phycol. J.* **7**, 361–74.

PALISANO J.R. & WALNE P.L. (1972) Acid phosphatase activity and ultrastructure of aged cells of *Euglena granulata*. *J. Phycol.* **8**, 81–8.

PARKE M. & MANTON I. (1965) Preliminary observations on the fine structure of *Prasinocladus marinus*. *J. mar. biol. Ass. U.K.* **45**, 525–36.

PARKER B.C. (1965) Translocation in the giant kelp *Macrocystis*. I. Rates, direction, quantity of $C^{14}$-labeled products and fluorescein. *J. Phycol.* **1**, 41–6.

PARKER B.C. (1966) Translocation in *Macrocystis*. III. Composition of sieve tube exudate and identification of the major $C^{14}$-labeled products. *J. Phycol.* **2**, 38–41.

PARKER B.C. & HUBER J. (1965) Translocation in *Macrocystis*. II. Fine structure of the sieve tubes. *J. Phycol.* **1**, 172–9.

PERCIVAL E. & McDOWELL R.H. (1967) *Chemistry and enzymology of marine algal polysaccharides*. Academic Press, New York & London.

PETERFI L.S. & MANTON I. (1968) Observations with the electron microscope on *Asteromonas gracilis* Artari emend. (*Stephanoptera gracilis* (Artari) Wisl.) with some comparative observations on *Dunaliella* sp. *Br. Phycol. Bull.* **3**, 423–40.

PICKETT-HEAPS J.D. (1967a) Ultrastructure and differentiation in *Chara* sp. I. Vegetative cells. *Aust. J. biol. Sci.* **20**, 539–51.

PICKETT-HEAPS J.D. (1967b) Ultrastructure and differentiation in *Chara*. II. Mitosis. *Aust. J. biol. Sci.* **20**, 883–94.

PICKETT-HEAPS J.D. (1968a) Ultrastructure and differentiation in *Chara*. III. Formation of the antheridium. *Aust. J. biol. Sci.* **21**, 255–74.

PICKETT-HEAPS J.D. (1968b) Ultrastructure and differentiation in *Chara* (*fibrosa*). IV. Spermatogenesis. *Aust. J. biol. Sci.* **21**, 655–90.

PICKETT-HEAPS J.D. (1970a) Some ultrastructural features of *Volvox*, with particular reference to the phenomenon of inversion. *Planta*, **90**, 174–90.

PICKETT-HEAPS J.D. (1970b) Mitosis and autospore formation in the green alga *Kirchneriella lunaris*. *Protoplasma*, **70**, 325–47.

PICKETT-HEAPS J.D. (1971a) Reproduction by zoospores in *Oedogonium*. I. Zoosporogenesis. *Protoplasma*, **72**, 275–314.

PICKETT-HEAPS J.D. (1971b) 'Bristly' cristae in algal mitochondria. *Planta*, **100**, 357–9.

PICKETT-HEAPS J.D. (1972) Cell division in *Cosmarium botrytis*. *J. Phycol.* **8**, 343–60.

PICKETT-HEAPS J.D. (1973) Cell division and wall structure in *Microspora*. *New Phytol.* **72**, 347–55.

PICKETT-HEAPS J.D. & FOWKE L.C. (1969) Cell division in *Oedogonium*. I. Mitosis, cytokinesis, and cell elongation. *Aust. J. biol. Sci.* **22**, 857–94.

PICKETT-HEAPS J.D. & FOWKE L.C. (1970a) Cell division in *Oedogonium*. II. Nuclear division in *O. cardiacum*. *Aust. J. biol. Sci.* **23**, 71–92.

PICKETT-HEAPS J.D. & FOWKE L.C. (1970b) Cell division in *Oedogonium*. III. Golgi bodies, wall structure, and wall formation in *O. cardiacum*. *Aust. J. biol. Sci.* **23**, 93–113.

PIENAAR, R.N. (1971a) Coccolith production in *Hymenomonas carterae*. *Protoplasma*, **73**, 217–24.

PIENAAR R.N. (1971b) Acid phosphatase activity in the cells of the coccolithophorid *Hymenomonas*. *Proc. S.A. Electron Microscopy Society, Pretoria, 1971*.

PRESTON R.D. & GOODMAN R.N. (1968) Structural aspects of cellulose microfibril biosynthesis. *Jl. R. microsc. Soc.* **88**, 513–27.

RAMUS J. (1969) Dimorphic pit connections in the red alga *Pseudogloiophloea*. *J. Cell Biol.* **41**, 340–5.

RAMUS J. (1971) Properties of septal plugs from the red alga *Griffithsia pacifica*. *Phycologia*, **10**, 99–103.

RAMUS J. (1972) The production of extracellular polysaccharide by the unicellular red alga *Porphyridium aerugineum*. *J. Phycol.* **8**, 97–111.

RAMUS J. & GROVES S.T. (1972) Incorporation of sulfate into the capsular polysaccharide of the red alga *Porphyridium*. *J. Cell Biol.* **54**, 399–407.

RAY, P.M., SHININGER T.L. & RAY M.M. (1969) Isolation of β-glucan synthetase particles from plant cells and identification with Golgi membranes. *Proc. natn. Acad. Sci. U.S.A.* **64**, 605–11.

RETALLACK B. & BUTLER R.D. (1972) Reproduction in *Bulbochaete hiloensis* (Nordst.) Tiffany. I. Structure of the zoospore. *Arch. Mikrobiol.* **86**, 265–80.

RINGO D.L. (1967a) Flagellar motion and fine structure of the flagellar apparatus in *Chlamydomonas*. *J. Cell Biol.* **33**, 543–71.

RINGO D.L. (1967b) The arrangement of subunits in flagellar fibres. *J. Ultrastruct. Res.* **17**, 266–77.

ROBARDS A.W. (1971) The ultrastructure of plasmodesmata. *Protoplasma*, **72**, 315–23.

ROBERTS K., GURNEY-SMITH M. & HILLS G.J. (1972) Structure, composition and morphogenesis of the cell wall of *Chlamydomonas reinhardi*. *J. Ultrastruct. Res.* **40**, 599–613.

ROSENBAUM J.L., MOULDER J.E. & RINGO D.L. (1969). Flagellar elongation and shortening in *Chlamydomonas*. The use of cyclohexamide and colchicine to study the synthesis and assembly of flagellar proteins. *J. Cell Biol.* **41**, 600–19.

SCHMITZ K., LÜNING K. & WILLENBRINK J. (1972) $CO_2$-Fixierung und Stofftransport in benthischen marinen Algen. II. Zum Ferntransport $^{14}C$-markierter Assimilate bei *Laminaria hyperborea* und *Laminaria saccharina*. *Z. Pflanzenphysiol.* **67**, 418–29.

SCHNEPF E. (1969) Sekretion und Exkretion bei Pflanzen. *Protoplasmatologia VIII* 8. Springer-Verlag. Wein. New York.

SCHNEPF E. & KOCH W. (1966a) Golgi-Apparat und Wasserausscheidung bei *Glaucocystis*. *Z. Pflanzenphysiol.* **55**, 97–109.

SCHNEPF E. & KOCH W. (1966b) Über die Entstehung der pulsierenden Vacuolen von *Vacuolaria virescens* (Chloromonadophyceae) aus dem Golgi-Apparat. *Arch. Mikrobiol.* **54**, 229–36.

SCHÖTZ F., BATHELT H., ARNOLD C. & SCHIMMER O. (1972) Die Architektur und Organisation der *Chlamydomonas* Zelle. Ergebrisse der Elektronenmikroskopie von Serienschnitten und der darans resultierenden dreidimensionalen Rekonstruktion. *Protoplasma*, **75**, 229–54.

SLANKIS T. & GIBBS S.P. (1972) The fine structure of mitosis and cell division in the Chrysophycean alga *Ochromonas danica*. *J. Phycol.* **8**, 243–56.

SPANSWICK R.M. & COSTERTON J.W.F. (1967) Plasmodesmata in *Nitella translucens*: structure and electrical resistance. *J. Cell Sci.* **2**, 451–64.

STAEHELIN L.A. & KIERMAYER O. (1970) Membrane differentiation in the Golgi complex of *Micrasterias denticulata* Bréb. visualized by freeze-etching. *J. Cell Sci.* **7**, 787–92.

STEWART K.L., FLOYD G.L., MATTOX K.R. & DAVIES M.E. (1972) Cytochemical demonstration of a single peroxisome in a filamentous green alga. *J. Cell Biol.* **54**, 431–4.

THIÉRY J.P. (1967) Mise en évidence des polysaccharides sur coupes fines en microscopie électronique. *J. Microscopie*. **6**, 987–1017.

TILNEY L.G. (1971) Origin and continuity of microtubules. In *Origin and Continuity of Cell Organelles*, eds. Reinert J. & Ursprung Z, pp. 222–60, Springer-Verlag, Berlin.

TOLBERT N.E. (1971) Microbodies—peroxisomes and glyoxysomes. *Ann. Rev. Pl. Physiol.* **22**, 45–74.

TOURTE M. (1972) Mise en evidence d'une activité catalasique dans les peroxysomes de *Micrasterias fimbriata* (Ralfs). *Planta*, **105**, 50–9.

TRIPODI G., PIZZOLONGO P. & GIANNATTASIO M. (1972) A DNase-sensitive twisted structure in the mitochondrial matrix of *Polysiphonia* (Rhodophyta). *J. Cell Biol.* **55**, 530–2.

WARNER F.D. (1972) Macromolecular organisation of eukaryotic cilia and flagella. In *Advances in Cell and Molecular Biology*. Vol. 2, ed. Du Praw E.J. pp. 193–235. Academic Press, New York and London.

WECKE J. & GIESBRECHT P. (1970) Freeze-etching of the nuclear membrane of dinoflagellates. In *Microscopie électronique*, 3, ed. Favard P., pp. 233–4. Grenoble 1970.

WEHRMEYER W. (1970). Struktur, Entwicklung und Abbau von Trichocysten in *Cryptomonas* and *Hemiselmis* (Cryptophyceae). *Protoplasma*, **70**, 295–315.

VAN WENT J.L. & TAMMES P.M.L. (1972) Experimental fluid flow through plasmodesmata of *Laminaria digitata*. *Acta Bot. Neerland.* **21**, 321–6.

VAN WENT J.L. & TAMMES P.M.L. (1973) Trumpet filaments in *Laminaria digitata* as an artefact. *Acta Bot. Neerland.* **22**, 112–19.

VAN WENT J.L., VAN AELST A.C. & TAMMES P.M.L. (1973) Open plasmodesmata in sieve plates of *Laminaria digitata*. *Acta Bot. Neerland.* **22**, 120–3.

WEST K.R. & PITMAN M.G. (1967) Ionic relations and ultrastructure in *Ulva lactuca*. *Aust. J. biol. Sci.* **20**, 901–14.

WETTSTEIN R. & SOTELO J.R. (1971) The molecular architecture of synaptonemal complexes. In *Advances in Cell and Molecular Biology*. Vol. 1, ed. Du Praw E.J. pp. 109–52. Academic Press, New York and London.

WHALEY W.G., DAUWALDER M. & KEPHART J.E. (1971) Assembly, continuity and exchanges in certain cytoplasmic membrane systems. In *Origin and Continuity of Cell Organelles*, eds. Reinert J. & Ursprung Z. pp. 1–45, Springer-Verlag, Berlin.

WILSON H.J., WANKA F. & LINSKENS H.F. (1973) The relationship between centrioles, microtubules and cell plate initiation in *Chlorella pyrenoidosa*. *Planta*, **109**, 259–63.

WISE G.E. & FLICKINGER C.J. (1970) Cytochemical staining of the Golgi apparatus in *Amoeba proteus*. *J. Cell Biol.* **46**, 620–6.

WITMAN G.B., CARLSON K., BERLINER J. & ROSENBAUM J.L. (1972a) *Chlamydomonas* flagella. I. Isolation and electrophoretic analysis of microtubules, matrix, membranes and mastigonemes. *J. Cell Biol.* **54**, 507–39.

WITMAN G.B., CARLSON K., BERLINER J. & ROSENBAUM J.L. (1972b) *Chlamydomonas* flagella. II. The distribution of tubulins 1 and 2 in the outer doublet microtubules. *J. Cell Biol.* **54**, 540–55.

WOLFE J. (1972) Basal body fine structure and chemistry. In *Advances in Cell and Molecular Biology*. Vol. 2, ed. Du Praw E.J. pp. 151–92. Academic Press, New York and London.

WOODCOCK C.L.F. (1971) The anchoring of nuclei by cytoplasmic microtubules in *Acetabularia*. *J. Cell Sci.* **8**, 611–21.

WOODWARD D.O., EDWARDS D.L. & FLAVELL R.B. (1970) Nucleocytoplasmic interactions in the control of mitochondrial structure and function in *Neurospora*. In *Control of organelle development*. XXIV Symp. S.E.B., ed. Miller P.L., pp. 55–69. Cambridge University Press.

WRIGGLESWORTH J.M., PACKER L. & BRANTON D. (1970) Organisation of mitochondrial structures as revealed by freeze-etching. *Biochim. biophys. Acta.* **205**, 125–35.

YAMAZAKI R.K. & TOLBERT N.E. (1970) Enzymic characterization of leaf peroxis' .nes. *J. biol. Chem.* **245**, 5137–44.

ZIEGLER H. & RUCK I. (1967) Untersuchungen über die Feinstruktur des Phloems III. Die 'Trompetenzellen' von *Laminaria*-Arten. *Planta* **73**, 62–73.

# CHAPTER 4

# PLASTIDS

T. BISALPUTRA

Department of Botany,
University of British Columbia,
Vancouver 8, Canada

1   Introduction   124

2   Basic terminologies   125

3   The photosynthetic apparatus in blue-green algae   126

4   Structure of the algal chloroplast   128
4.1   The chloroplast envelope and related membranes   128
  (a) Relationship between the envelope and the thylakoid   133
  (b) Relationship with the external membrane system—the chloroplast endoplasmic reticulum   133
4.2   Arrangement of the thylakoids   140
  (a) Type I   140
  (b) Type II   141
  (c) Type III   141
  (d) Type IV   142
  (e) Type V   142

4.3   Thylakoid membranes   143
  (a) Nature of the associations between thylakoids   144
4.4   The pyrenoid and location of storage polysaccharides   144
4.5   Microtubular elements   146
4.6   Plastoglobuli and the 'eyespot'   147
4.7   Ribosomes   147

5   Chloroplast division and the transmission of the genophore   148

6   Correlation of chloroplast ultrastructure with chloroplast pigments   152

7   References   155

## 1   INTRODUCTION

Algal plastids, particularly their colour, shape, size, number and distribution within the cells, have been used along with other important characters such as pigments and reserve products in the classification of algae. Since the advent of the electron microscope the study of chloroplast ultrastructure has become one of the focal points of interest to taxonomists, physiologists and biochemists.

Fine structure work of importance dates back to the early 1950's with the studies of chloroplasts of a chrysomonad, *Poteriochromonas*, and *Euglena* (Wolken & Schwertz 1953); *Fucus* (von Wettstein 1954, Leyon & von Wettstein 1954); *Chlorella* (Albertsson & Leyon 1954) and the algal pyrenoid (Leyon 1954). The study of *Chlamydomonas* by Sager and Palade (1957), however, laid the foundation for the modern description of chloroplast ultrastructure. The fundamental approach during these early years of electron microscopy had been mainly morphological and in relation to its usefulness in taxonomy. Although the number of species of algae investigated is extremely small, a pattern emerged very early that the ultrastructure of chloroplasts, particularly the arrangement of their photosynthetic membranes is characteristic of the major classes. A number of reviews on the comparative ultrastructure of algal chloroplasts have appeared (Granick 1961, Jónsson 1963, Manton 1966b, Kirk & Tilney-Basset 1967, Echlin 1970, Gibbs 1970, Dodge 1974). The present chapter will summarize the background of this and other aspects of chloroplast cytology including recent developments such as membrane structure and biogenesis, chloroplast growth and division as related to the chloroplast genophores, and possible evolutionary trends of algal chloroplasts.

## 2  BASIC TERMINOLOGIES

The *photosynthetic apparatus* can be defined as the entire complex of structures and components responsible for the photosynthetic processes of an organism. In algae and higher plants this would include the pigment-containing structures, together with components associated with the photochemical reactions on one hand and a set of components of relevance to $CO_2$ fixation on the other. From a morphological point of view the basic structures of the photosynthetic apparatus as revealed by electron microscopy consist of a system of flattened membranous vesicles and a surrounding matrix. Each of the former has been termed a '*disc*' (Sager & Palade 1957, Gibbs 1960) or '*thylakoid*' (Menke 1962). The latter term is now widely employed and will be adopted here. The thylakoids are chlorophyll-containing and considered to be the sites of photochemical reactions. $CO_2$ fixation on the other hand, is thought to occur within the surrounding matrix (see Branton 1968, Kirk 1971, Park & Sane 1971). In prokaryotic blue-green algae the photosynthetic apparatus does not segregate into a discrete organelle, and the matrix surrounding the thylakoids is the cytoplasmic matrix.

The condition is different in eukaryotic algae. Here the thylakoids are partitioned off from the cytoplasm by a pair of limiting membranes, the '*chloroplast envelope*' giving rise to a discrete cell organelle the '*chloroplast*' with its own matrix or '*stroma*'. The proteinaceous stroma may contain besides *chloroplast ribosomes*, and the electron-translucent DNA-containing areas or '*genophores*', *plastoglobuli* (Lichtenthaler 1968), *eyespots, pyrenoids* and *crystalline inclusions*

Within the chloroplast the thylakoids are arranged in various ways. They may remain separated from each other along their length (i.e. in the Rhodophyceae); become closely approximated with one another to form '2 or 3-thylakoid bands' (Gibbs 1970), such as those in the Cryptophyceae and Phaeophyceae respectively; or exist as 'stacked thylakoids' (Goodenough & Staehelin 1971) whereby a number of them become closely adpressed so that the spacings between neighbouring thylakoids are eliminated. This last type of apposition will result in the formation of 'partitions' (Weier et al. 1966) between the adjoining thylakoids and therefore each of these stacks is considered as representing a 'granum' (Weier et al. 1966, Gibbs 1970). Such a structure is characteristic of the chloroplasts of the Euglenophyceae and Chlorophyceae.

Further terminology related to each aspect of the chloroplast will be given and discussed in appropriate sections.

## 3 THE PHOTOSYNTHETIC APPARATUS IN BLUE-GREEN ALGAE

Among the photosynthetic prokaryotes only the blue-green algae have the ability to produce $O_2$ from water. Photosynthetic bacteria depend upon electron donors from sources other than water. Detailed discussions on the mechanism and evolution of photosynthesis in these organisms may be found in recent reviews by Olson (1970), Evans and Whatley (1970), and Fogg et al. (1973). Prokaryotes do not possess true chloroplasts in that the photosynthetic apparatus in these organisms does not segregate into discrete organelles. Only the structures related with the photochemical reactions, the thylakoids, can be visualized with electron microscopy. In blue-green algae, as well as in purple bacteria, the thylakoids show a direct connexion with the plasma membranes of the cell or arise from them during ontogeny (Lang 1968, Echlin 1968, 1970, Whitton et al. 1971, Cohen-Bazire 1971). The components involved in the dark reactions cannot be distinguished morphologically from other cytoplasmic constituents.

Results from electron microscopic studies have shown that thylakoids of blue-green algae are, in general, of the lamellar type. According to Echlin (1970) the disposition and structure of the thylakoids within the cells depends not only on the species but also on physiological conditions, and on the methods of handling the material for electron microscopy. In some species the individual thylakoids remain separated from each other along their length but branching and elongation of the membrane may occur during rapid cell growth. In others the thylakoids may form complicated interlocking plates. During the development of heterocysts the thylakoids tend to become greatly distorted (Lang 1965).

Earlier works often describe the thylakoids of blue-green algae as being similar to unit membranes in appearance and dimensions. Freeze-etching work clearly shows subunits in these lamellae (Jost 1965). The particles are of two distinct sizes, 5 to 7 nm and 10 to 20 nm. It has now been established that

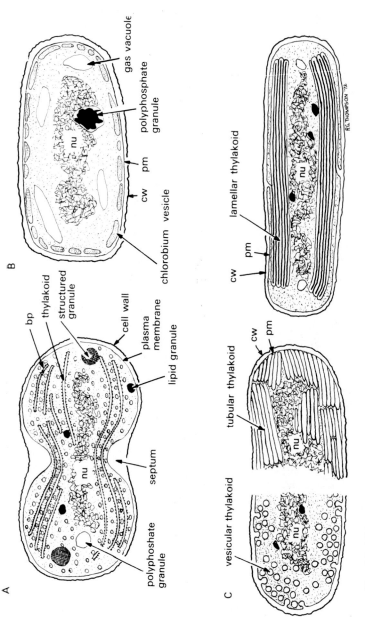

**Fig. 4.1.** Diagrams to show the arrangement of the thylakoids in A. Blue-green algae, B. Green bacteria and C. Purple bacteria, illustrating the three major types of thylakoids in these groups. From left to right: vesicular, tubular and stacked lamellar types (cw = cell wall; nu = nuclear region; pm = plasma membrane).

aggregates of phycobiliproteins, in the form of 35 nm granules, are attached to the outer surfaces of the thylakoids of blue-green algae (Gantt & Conti 1969) in a similar way to those found earlier in the red alga *Porphyridium cruentum*. While the real structure of the thylakoid remains obscure, its nature must be considered in terms of a mosaic of functional complexes. There is some evidence that, in addition to photosynthesis, these membranes may also be the sites of $N_2$ fixation and respiration (Lang 1968, Fogg *et al.* 1973).

The photosynthetic apparatus of blue-green algae and of the major types of photosynthetic bacteria are illustrated in Fig. 4.1. Purple bacteria are similar to blue-green algae in that their chlorophylls, bacteriochlorophyll *a* and *b*, are located within the thylakoid membranes. However, the morphology of these thylakoids is very much diversified and the three major types are shown in Fig. 4.1C. In both purple sulphur and purple non-sulphur bacteria the thylakoid can be either of vesicular, tubular or lamellar types (Cohen-Bazire 1971, Echlin 1970). In the last type, the lamellae may show multiple foldings and appear as grana-like stacks (Fig. 4.1C). In all cases the thylakoid of the purple bacteria originates by invagination of the cell membrane.

The photosynthetic apparatus of green bacteria (Fig. 4.1B), differs from other prokaryotic organisms in certain aspects. Their pigments, mainly bacterio-chlorophyll *c* or *d*, are located within the 'chlorobium vesicles' which are situated along the periphery of the cell (Cohen-Bazire 1971). These small elongated vesicles are bounded by a thin, single, electron dense membrane of 2 to 3 nm thick, similar to the membrane of the gas vacuoles of both bacterial and blue-green algal cells.


## 4   STRUCTURE OF THE ALGAL CHLOROPLAST

The chloroplast is a discrete cell organelle which represents the photosynthetic apparatus of eukaryotic cells. There are distinct differences in the architecture of the chloroplasts belonging to various classes of algae. In subsequent sections important characteristics of the ultrastructure of this organelle and its relationship with other cell organelles will be discussed.


### 4.1   *The chloroplast envelope and related membranes*

The envelope of algal chloroplasts has received little attention despite its importance in the functional and structural integrity of the organelle. The chloroplast envelope is semi-permeable, and work on isolated chloroplasts of higher plants has led to a better understanding of its permeability (Walker 1969). It has been shown that the survival of isolated chloroplasts depends on the presence of the intact envelope (Ridley & Leech 1969). Chloroplasts isolated from *Caulerpa sedoides* have been shown not only to survive for several weeks *in vitro*, but also to retain the ability to divide repeatedly (Giles & Sarafis 1972). The survival of algal chloroplasts in animal tissue under symbiotic conditions (see Taylor 1970) can be again attributed to the retention of their envelopes (Trench *et al.* 1969, Taylor 1973).

The chloroplast envelope has generally been thought of as a double membrane structure (Figs. 4.2, 4.3). The envelopes of the euglenoid and dinoflagellate chloroplasts, however, are unique in that they have three membrane layers (Fig. 4.4). It is likely that this is a derived condition and it will be discussed at the end of this section. The two membranes of the basic type of envelope were described

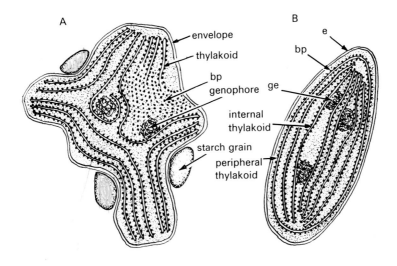

**Fig. 4.2.** Diagrams of chloroplasts of Rhodophyceae showing the arrangement of thylakoids. A. Type 1*a*, lobed chloroplast, peripheral thylakoid absent, thylakoid terminates close to envelope, arrangement of phycobiliprotein granules as seen from surface view are also shown (bp). B. Type 1*b* discoid chloroplast, peripheral thylakoid present (e = envelope; bp = phycobiliprotein granule; ge = genophore).

in early works as a pair of parallel unit membranes. Each is approximately 5 to 6 nm wide, separated from the other by a narrow gap of approximately 6 nm. The double nature of the chloroplast envelope has often been cited as one of the main lines of evidence in support of the endosymbiotic origin of the chloroplast (Schnepf & Brown 1971, Raven 1970, Taylor 1970, 1973). Under this assumption the outer and inner membranes of the envelope are regarded as belonging to different systems, i.e. the host's and the endosymbiont's respectively. If this is so one could expect certain functional and morphological differences between the two membranes. Even though there are not extensive data comparable to that of mitochondrial membranes, biochemical differences between the outer and inner membranes of the chloroplast envelope are known (Racker 1970). Recently morphological differences between the outer and inner membranes of the chloroplast envelopes of a red alga *Bangia* and a brown alga *Sphacelaria*, were recognized in freeze-etched specimens (Bisalputra & Bailey 1973). In *Bangia* the fracture face of the outer membrane is smooth whereas the inner membrane

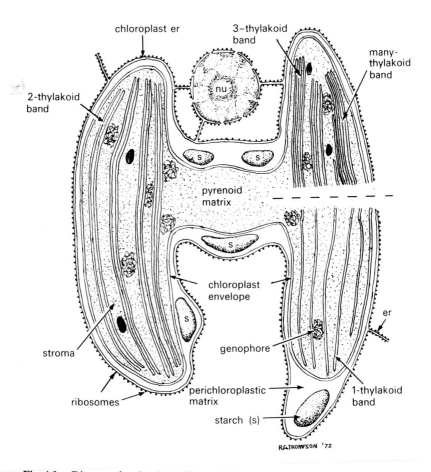

**Fig. 4.3.** Diagram showing 3 possible variations of the arrangement of thylakoids in different cells of the Cryptophyceae. Left, 'typical' Type II arrangement, with 2-thylakoid bands. Top right, a variation from the typical with 3-thylakoid bands. Lower right, single-thylakoid (er = endoplasmic reticulum; nu = nucleus).

shows extensive formation of folded ridges (Fig. 4.5). Analysis of the density of particles on fracture faces of both membranes indicates differences in the make-up of the two membranes and perhaps their activities; of the two, the inner membrane is the more active.

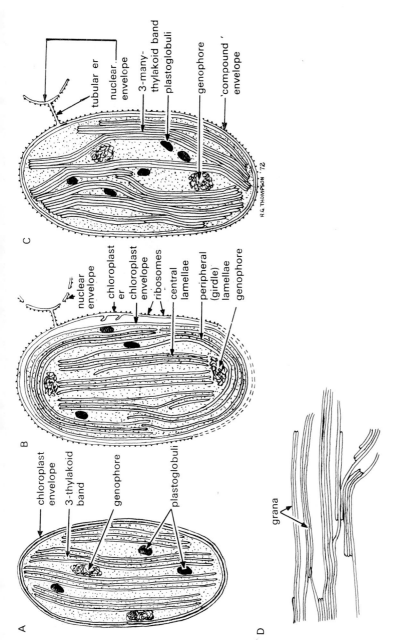

**Fig. 4.4.** Diagram of thylakoid arrangements. A. Dinoflagellate chloroplast, showing the triple layered envelope, 3-thylakoid bands without peripheral or girdle lamellae (Type IIIa). B. 3-thylakoid bands with girdle lamellae, 3-thylakoid bands from chloroplast ER as in Xanthophyceae, Chrysophyceae, Phaeophyceae (Type IIIb), are shown. Haptophyceae and Eustigmatophyceae are similar except for the lack of girdle lamellae. C. Euglenoid chloroplast. Note the compound envelope of three membrane layers, elongated grana of three fused thylakoids. D. Grana of the Chlorophyceae.

**Fig. 4.5.** Freeze-etched replica of a red alga *Bangia fuscopurpurea* chloroplast (Type I*a*) showing the arrangement of the thylakoid and the differences in morphology between the inner (IE) and outer membranes (OE) of the envelope (PM = plasma membrane; T = thylakoid; W = cell wall). × 25,000.

(a) *Relationship between the envelope and the thylakoid*

As indicated previously the thylakoids of prokaryotic cells are derived from the invagination of plasma membranes. In higher plants it is well known that the inner membrane of the proplastid envelope plays an important role in contributing material for the differentiation of the thylakoids. Direct connexions between the inner envelope membrane and the thylakoid are maintained in some fully developed chloroplasts. In algae, proplastids are found in the euglenoids and in apical cells of the more advanced red algae. In the former, the thylakoids are also derived from material invaginated from the proplastid envelope (Ben-Shaul *et al.* 1964). In the case of red algae with proplastids, the inner thylakoids are derived from the peripheral thylakoids which are always present in the proplastid (Bouck 1962). No direct connexion between the envelope membrane and the thylakoids has been reported in the functional chloroplasts of these two groups of algae. Freeze etching observations of the fracture faces of the inner membrane of the envelope and the thylakoids of a red alga *Bangia*, have shown that the two membranes are of a vastly different nature (Bisalputra & Bailey 1973).

(b) *Relationship with the external membrane system—the chloroplast endoplasmic reticulum*

In the Rhodophyceae and Chlorophyceae the chloroplasts are bounded by the basic double membrane envelope. Although relationships between the chloroplasts and other cellular membranes are often reported, there is not extensive development or elaboration of other membranes around the chloroplast envelope in these two groups of algae.

Chloroplasts of the Cryptophyceae, Euglenophyceae, Haptophyceae, Chrysophyceae, Bacillariophyceae, Xanthophyceae and certain Phaeophyceae illustrate intimate relations with the endoplasmic reticulum and to a lesser extent the nuclear membrane. Gibbs (1962b) was the first to describe an 'outer double-membrane envelope' in *Ochromonas danica* and *Cryptomonas* chloroplasts. She pointed out that in the area where the chloroplasts lie adjacent to the nucleus the continuity between the 'outer envelope' and the nuclear membrane can be seen clearly. Bouck (1965) found similar sets of membranes around the chloroplasts of brown algae and appropriately termed this system of membranes the 'chloroplast endoplasmic reticulum'. It is clear now that the chloroplast endoplasmic reticulum (ER) is an extension of the extensive endoplasmic reticular channels which link up directly with the nucleus, the Golgi and lysosomes (GERL).

In general, chloroplast ER can be recognized by the presence of ribosomes on the surface facing the cytoplasm. The surface facing the chloroplast is devoid of ribosomes (Figs. 4.3, 4.4B, 4.7). Gibbs (1970) has reported that in *Ochromonas*, the ribosomes are arranged in polysomes. There are, however, some variations in the structure of the chloroplast ER in various algal groups. For example, in

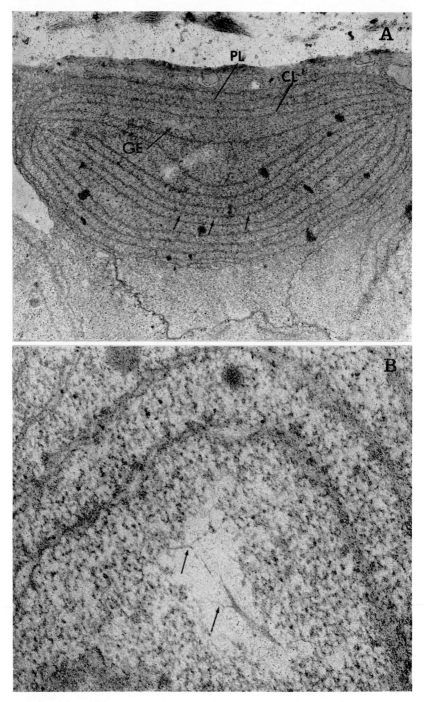

**Fig. 4.6.** A. Longitudinal section of a chloroplast of a red alga *Delesseria* (Type I*b*), showing the peripheral thylakoid (PL), central thylakoid (CL) and scattered genophore (GE). Arrows indicate phycobilisomes on the surface of a thylakoid. × 4,000. B. A portion of the genophore showing DNA fibrils (arrowed), in a comparatively electron lucent area. × 65,000.

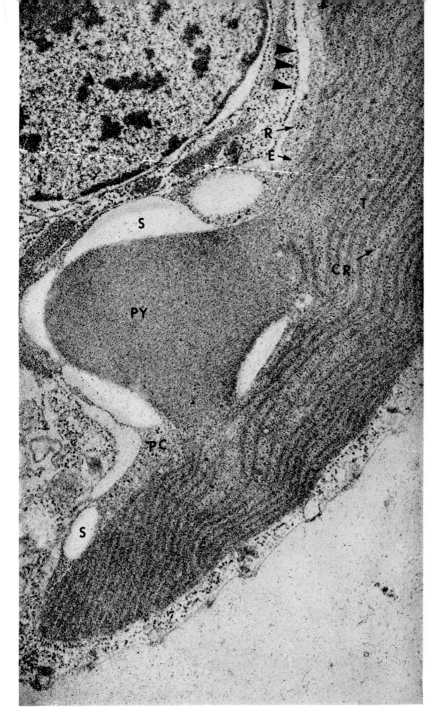

**Fig. 4.7.** *Cryptomonas stigmatica.* Ribosomes on the chloroplast ER are indicated by large arrow-heads; the opposite side of the chloroplast ER is devoid of ribosomes. Note, however, a few groups of ribosomes (R) and the perichloroplastic matrix (PC) within which starch (S) is also located. This species shows a departure from the typical Type II thylakoid arrangement in that the thylakoids (T) remain un-associated. E = envelope, PY = pyrenoid. × 25,000.

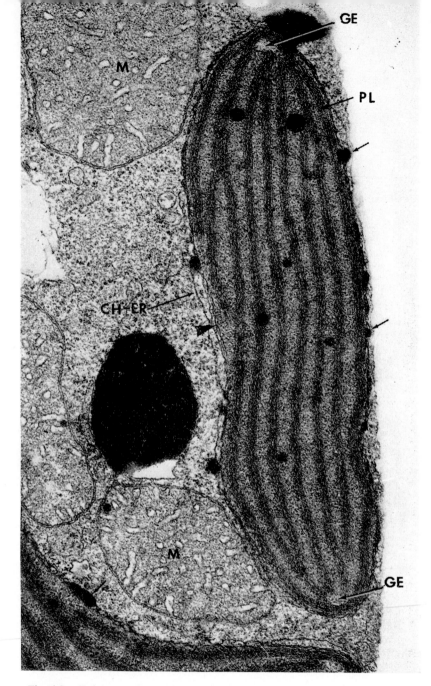

**Fig. 4.8.** *Tribonema minus* chloroplast illustrating Type III*b* arrangement of the thylakoids, i.e. 3-thylakoid bands, with peripheral or girdle lamellae also of 3-thylakoids, no grana. Note the chloroplast ER (CH-ER) and lipid materials within it (arrows). Some lipid materials are of considerable size and extend from the chloroplast; storage oil granules of similar density are seen in the cytoplasm and seem to be limited by a membrane. Parts of the genophore (GE) are seen at both ends inside the peripheral bands (PL). M = mitochondrion. × 36,000.

the Cryptophyceae the space between the chloroplast ER and the chloroplast
envelope may become greatly distended (Fig. 4.3). Starch grains are located in
this space (Gibbs 1962b, Dodge 1969b, Lucas 1970). It has been suggested that
the content of this space is probably cytoplasmic in origin (Lucas 1970). From
our observations it is filled with a very fine granular matrix and appears to be
slightly denser than the cytoplasmic matrix. A few ribosome-like structures are
found, but their numbers are very much lower than those in the surrounding
cytoplasm (Fig. 4.7). It seems likely that this space is completely separated from
the remainder of the cytoplasm and should be termed the 'perichloroplastic
space', and its matrix the 'perichloroplastic matrix'.

In the Chrysophyceae, Xanthophyceae, Bacillariophyceae, Haptophyceae
and to a certain extent the Phaeophyceae, the chloroplast ER is somewhat
similar in that the perichloroplastic space is very much narrower than that
of the Cryptophyceae (Figs. 4.4B, 4.8, 4.10A). In the area where the perichloroplastic space is formed by the nuclear membrane and the envelope, small tubules
are found (Gibbs 1962b, Hibberd & Leedale 1971). These are probably extensions
from the nuclear membrane. Tubular networks from the chloroplast ER towards
the chloroplast envelope were also reported (Bouck 1965, Falk & Kleinig 1968,
Massalski & Leedale 1969). Recently Oliviera and Bisalputra (1972) found a
direct connexion between the chloroplast ER and the outer membrane of the
chloroplast envelope in *Ectocarpus*. Hence, the GERL system has now been
shown to extend into the compartment between the two envelope membranes.
The continuity of outer mitochondrial membranes with the ER membrane has
been previously established in both plant and animal cells (Morré *et al.* 1971).

It should be noted that in brown algae, chloroplast ER is best developed
when the number of chloroplasts per cell is relatively low. In those genera where
the number is high, such as *Sphacelaria*, *Egregia*, and *Dictyota*, the chloroplast
ER is poorly developed. The chloroplast is not completely enclosed and in some
cases the degree of association and contact between the elements of the ER and
the chloroplast envelope is small and random. It is questionable, in such cases,
whether these ER elements should be classified as chloroplast ER.

As indicated previously, the chloroplast envelope of the euglenoids and dinoflagellates consists of a triple layer of membrane (Fig. 4.4). The envelope of
*Euglena* has been termed a 'compound' envelope (Leedale *et al.* 1965). Owing to
the prevalence of the chloroplast ER- envelope association in algae as described
above, it is not surprising to find that the triple-layered envelope is considered to
result from the fusion of these two membrane systems (Dodge 1968, 1971,
Gibbs 1970). This view seems to be well supported in the case of the euglenoids,
by the fact that there the outer layer and the nuclear envelope are connected by
a tubular endoplasmic reticulum (Leedale 1968). Furthermore, ribosomes are
associated with the outer surface of the chloroplast envelope (Fig. 4.9). In the
dinoflagellates the central membrane of the chloroplast envelope is occasionally
found to be thicker than the outer ones (Dodge 1968), suggesting the possibility
of the fusion between two pairs of membranes. However, the picture is far from

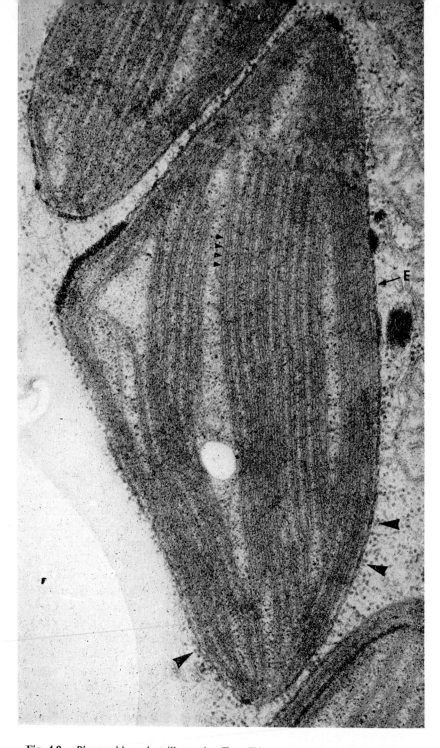

**Fig. 4.9.** *Phacus* chloroplast illustrating Type IV arrangement of the thylakoids. Large arrow heads indicate ribosomes, observed on the outer surface of the 'compound' envelope (E), whose three layers of membranes can be seen in places (small arrows). × 40,000.

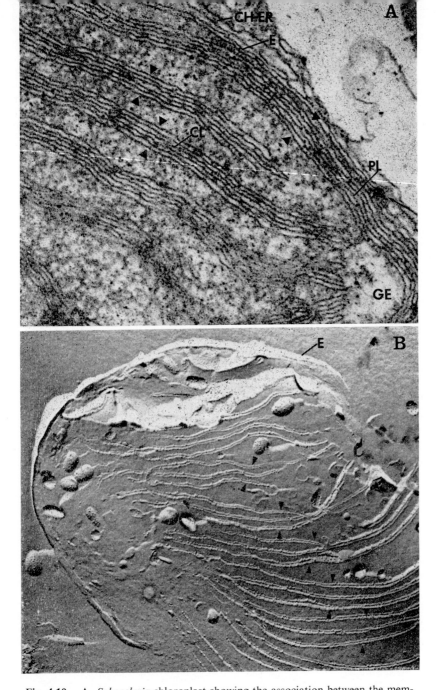

**Fig. 4.10.** A. *Sphacelaria* chloroplast showing the association between the members of the 3-thylakoid band, opposite pairs of arrows. Gaps between neighbouring thylakoids can be observed. (CH-ER = chloroplast ER; E = envelope; GE = genophore; PL = peripheral lamellae; CL = central lamellae). × 86,000.
B. Freeze-etched micrograph of an isolated *Sphacelaria* chloroplast with intact envelope (E). Note the separation of lamellar bands. The three members of each band can still be accounted for (between pairs of opposite arrow heads). Disruption of envelope usually results in vesiculation of the thylakoid. × 5,400.

clear since ribosomes are not found on the outer surface of the envelope. In addition, Taylor (1969) has described a direct connexion between the chloroplast envelope and the nuclear membrane which appears to be of a unique type. The result is that it provides an open channel between the chloroplast stroma and the nucleoplasm.

From the foregoing, it is clear that chloroplast ER exemplifies the close functional relationship between the chloroplast at one end and the GERL system which can be traced back to the nuclear membrane at the other. It is well known that one of the functions of the GERL system is concerned with transport of cellular products. Bouck's (1965) speculation that the chloroplast ER may serve as a channelling system for carbohydrate reserves and secretory precursors seems justified. The extent of this function would depend upon the degree of development of the chloroplast ER around the chloroplasts. Of the algae which store reserve polysaccharides outside the chloroplasts, all have well developed chloroplast ER except members of the Rhodophyceae. As for the Rhodophyceae it now seems certain that associations between the chloroplasts and the Golgi, a member of the GERL system, do exist. The deposition of starch grains has been reported to involve the Golgi apparatus (Tripodi 1971). There is also evidence that other products, such as lipid material originating in the chloroplasts, may find their way out of the chloroplasts through the chloroplast ER. Lipid granules are found almost universally in algal cytoplasm with high concentrations in the Chrysophyceae, Xanthophyceae, and Bacillariophyceae. The appearance of lipid droplets of various sizes around the chloroplast envelope and also in the cisternae of the chloroplast ER (Fig.4.8), is indicative of the secretion of this material from the chloroplast. The release by budding off, of these lipid materials and their subsequent coalescence into larger droplets outside the chloroplasts has been reported (Bisalputra *et al.* 1971). Besides being channels for the transport of products, the chloroplast ER provides a direct passage between the chloroplasts and other organelles including the nucleus, a factor which may be of importance in connexion with the mechanisms for cellular control and regulation.

### 4.2   *Arrangement of the thylakoids*

The arrangements of the thylakoids in the chloroplasts of various groups of algae can be classified from electron microscopic investigations as follows (other pertinent information is also included):

### (a) *Type I*

No association between individual thylakoids, i.e. thylakoids not banded or stacked, but inter-connexions between neighbouring thylakoids may occur; no grana; phycobilisomes located on the surfaces of the thylakoids; genophore interlamellar, scattered; chloroplast ER absent (Fig. 4.2). *Note:* The shape of 'phycobilisomes' (phycobiliprotein aggregates) is apparently determined by the phycoerythrin/phycocyanin content. They appear as 35 nm granules when phycoerythrin predominates (Gantt & Conti 1965) or aggregate into discs or tubules when phycocyanin is predominant (Gantt *et al.* 1968, Lichtlé & Giraud 1970).

*a.* No peripheral thylakoids and most, or a certain number, of the thylakoids terminate close to the chloroplast envelope (Fig. 4.2A). This type is found in the lower Rhodophyceae, i.e. *Porphyridium, Rhodella* and *Bangia* (Fig. 4.5). (Giraud 1962, Gantt & Conti 1965, Evans 1970). *Note:* This type of arrangement is found usually in stellate or deeply lobed chloroplasts; a pyrenoid is usually present.

*b.* Peripheral thylakoids are present and completely, or almost completely, enclose the internal set of thylakoids. (Fig. 4.2B, 4.6). This type is found in the higher Rhodophyceae i.e. *Lomentaria, Laurencia, Batrachospermum, Pseudogloeophloea*, etc. (Bouck 1962, Gibbs 1962a, Bisalputra & Bisalputra 1967a, Brown & Weier 1970). *Note:* This type of arrangement is usually associated with discoid chloroplasts; a pyrenoid is usually absent. *Smithora* (McBride & Cole 1969) appears to be an intermediate form in that its lobed chloroplast contains a pyrenoid. It often shows peripheral thylakoids and the inner thylakoids do not terminate at the envelope.

## (b) *Type II*

Association of thylakoids into 2-thylakoid bands; members of each thylakoid pair are separated usually by a gap of 4 nm along their length; partitions do not form and hence there are no grana; no peripheral thylakoids; genophores interband and are scattered; chloroplast ER present (Fig. 4.3). This type is found in the Cryptophyceae, i.e. *Cryptomonas, Chroomonas, Hemiselmis, Rhodomonas*, etc. (Gibbs 1962b, Dodge 1969b, Lucas 1970, Gantt *et al.* 1971). *Note:* Exceptions to the 2-thylakoid band condition are found frequently. Members of this class of algae may occasionally show unassociated thylakoids (Fig. 4.7) or 3- or 4-thylakoid bands; phycobiliproteins do not appear as aggregates but are reported to be within the intra-thylakoid spaces (Gantt *et al.* 1971).

## (c) *Type III*

Association of thylakoids into 3-thylakoid bands; degree of association variable and gaps of 2 to 4 nm usually exist between adjacent thylakoids within the band; partitions not formed and hence no grana; genophores scattered or single, ring shaped; chloroplast ER present.

*a.* No peripheral or 'girdle band(s)'; genophores are scattered.
Dinophyceae (Fig. 4.4A), i.e. *Amphidinium, Peridinium, Woloszynskia, Gonyaulax, Gymnodinium, Prorocentrum, Aureodinium*, etc. (Gibbs 1962b, Dodge 1968,

Dodge & Crawford 1971). *Note:* Arrangements of 2-thylakoids, or multiples of 3-thylakoid bands are found in this group of algae; chloroplast ER fuses with envelope ( ?).

Haptophyceae, i.e. *Chrysochromulina, Prymnesium, Coccolithus* (Manton 1966a,b, Manton & Leedale 1969). *Note:* Interband communication frequent in Haptophyceae.

Eustigmatophyceae, i.e. *Polyedriella, Vischeria* (Hibberd & Leedale 1970). *Note:* This group of algae has been separated from the Xanthophyceae on the basis of the absence of girdle lamellae, eyespots, flagella, etc.

*b.* Peripheral or girdle bands of 3-thylakoids present; single, ring genophore inside the peripheral bands (Fig. 4.4C).

Xanthophyceae (including the chloromonads) i.e. *Tribonema* (Fig. 4.8), *Botrydium, Vaucheria, Ophiocytium.* (Massalski & Leedale 1969, Falk & Kleinig 1968). *Note:* Of all the members of this group reported so far only *Bumilleria sicula* lacks a girdle band. In *Tribonema* the thylakoids of each band are reported to often fuse together (Falk & Kleinig 1968).

Chrysophyceae, i.e. *Ochromonas, Synura* (Gibbs 1962a, Giraud 1962).

Bacillariophyceae, i.e. *Lithodesmium, Gomphonema, Nitzschia, Amphipleura* (Drum & Pankratz 1964, Manton *et al.* 1969).

Phaeophyceae (Fig. 4.10), i.e. *Fucus, Chorda, Giffordia, Scytosiphon, Laminaria, Sphacelaria,* etc. (Gibbs 1962a, Bouck 1965, Evans 1966, Bisalputra *et al.* 1971, Cole 1970). *Note:* Interband communication is frequent in the above three groups.

(d) *Type IV*

Bands of 3-fused thylakoids; forming partitions and hence grana are present. Grana much elongated; interconnexions between grana are common. Genophores are usually scattered (Figs. 4.4, 4.9).

Euglenophyceae, i.e. *Euglena, Phacus* (Gibbs 1960, Leedale 1967, Ben-Shaul *et al.* 1964).

(e) *Type V*

Bands of 2-6-many fused thylakoids, partitions and grana are present; grana of variable sizes (Fig. 4.4D), interconnexions between grana are common; scattered genophores.

Chlorophyceae, Prasinophyceae and Charophyceae, i.e. *Chlorella, Chlamydomonas, Volvox, Carteria, Acetabularia, Chara, Bryopsis, Trebouxia, Closterium, Scenedesmus, Platymonas, Pyramimonas* (Gibbs, 1962a, Lang 1963, Burr & West 1970, Puiseux-Dao 1970, Johnson & Porter 1968, Weier *et al.* 1966, Pickett-Heaps & Fowke 1970, Hoffman 1967, Lembi & Lang 1965. Manton & Parke 1965).

Although some variations are found within each type of thylakoid arrangement as noted, this feature of algal ultrastructure has already proved to be useful in taxonomic and possibly in phylogenetic determinations. One must be cautious in the sense that only a few representatives of each class of algae have been investigated to date and the range and distribution of the variations are not fully known. Furthermore, little attention has been paid in the literature to the effects of factors which may cause changes in the arrangement of the thylakoids.

### 4.3   Thylakoid membranes

Recent high resolution electron microscopy and freeze-etching studies indicate that the membranes of the thylakoids of algae contain subunits similar to their counterparts in the higher plants. With conventional thin sectioning methods the thylakoid membrane of *Scenedesmus* is visualized as being comprised of spherical subunits of approximately 10 nm in diameter. In the partition area there are two rows of these spherical units appressed to each other (Weier *et al.* 1966). Freeze-etching studies (Guérin-Dumartrait 1968, Brown & Weier 1970, Neushul 1970, Goodenough & Staehelin 1971, Bisalputra & Bailey 1973) show that the fractured faces of the algal thylakoids are similar to those described by Branton and Park (1967) for higher plant chloroplasts. As is well known, membranes subjected to freeze-etching are cleaved in half and both fracture faces can be observed. In unassociated thylakoids such as those of the Rhodophyceae, only two fracture faces, i.e. face B and C (C being complementary) are recognized (Branton & Park 1967). In stacked lamellae two additional faces, termed Bs and Cs are also found (Goodenough & Staehelin 1971). In most cases the B face is reported to contain 17·5 nm particles, whereas on the C face there appear smaller 11 nm particles of relatively higher density. The Bs' face in *Chlamydomonas* has a dense population of 16 nm particles, while the Cs' face contains a number of large holes of 5 to 16 nm diameter and a population of smaller particles ranging from 3 to 10 nm in size. The 16 nm particles (termed S particles, Goodenough & Staehelin 1971) are considered to be an important factor in the stacking process.

The nature of the 11 nm and 17·5 nm particles is far from clear although attempts have been made to link them with photosystem I and photosystem II respectively. The reader is referred to a recent review by Kirk (1971). The problems are further complicated in the algal thylakoids as these contain a vast range of pigments (see Table 4.1) and their photosynthetic systems vary from one group to the next.

The growth of the chloroplast thylakoid has been studied in some detail in *Chlamydomonas reinhardtii* (Hoober *et al.* 1969, de Petrocellis *et al.* 1970, Goldberg & Ohad 1970). By using mutants, appropriate inhibitors of protein synthesis, and radioautographic and fractionation methods, the chloroplast thylakoids have been shown to develop neither from the chloroplast envelope or the pyrenoidal membranes. New membrane material is incorporated into the pre-existing membrane by a multistep assembly. In such a process the membrane is built up from a small number of components while other constituents are incorporated at later stages. There is a time lapse before the new membrane becomes fully functional. It appears that new material is added perhaps to the centre portion of the thylakoid rather than at its edge (Kirk 1971).

(a) *Nature of the associations between thylakoids*

In the cases where the thylakoids become closely appressed to form partitions' the association established is usually strong and able to survive osmotic shock and drastic treatment such as sonication (Weier & Benson 1967). The bonding between the thylakoids in these partition regions was thought, at first, to be due to hydrophobic bonding. Kirk (1971), however, has pointed out that the junctions between the thylakoids along the partition are, in fact, aqueous regions and the electrostatic bridges formed by divalent ions, localized protein-protein interaction, or localized hydrophobic bonding may be responsible for the adhesion between the thylakoids. Whether 'S-particles' represent these is questionable.

In other associations where grana are not formed and the thylakoids within the band are separated by narrow gaps (i.e. thylakoid arrangements Types II, IIIa and IIIb), the associations have been described as loose or variable (Gibbs 1970). Osmotic pressure and the type of fixation are known to influence the appearance of the thylakoid stacks and may increase or reduce the width of the gaps. Work conducted in our laboratory has indicated that in the brown alga, *Sphacelaria*, the associations between members of the 3-thylakoid bands will not survive the aqueous media used in the isolation technique, even although the chloroplast envelope remains intact (Fig. 4.10B). A similar effect can be induced in material fixed in glutaraldehyde by sonication, heat, or adding detergent to the washing solution. If the treatment is not too drastic the separated thylakoids will not vesiculate. The bonding between the thylakoids is probably due to $H^+$ bonds.

Staehelin (1967), using the freeze-etching technique, demonstrated the presence of 4 nm fibrils within the chloroplasts of *Cyanidium caldarium*. He reported that the thylakoids are linked together by these fibrils and the spacing between any pair of neighbouring thylakoids is constantly maintained. The constancy of the spacing between thylakoid bands is common in other types of chloroplasts, particularly the Type IIIb arrangement. It is possible that these 4 nm fibrils may play a similar role in these chloroplasts.

### 4.4    *The pyrenoid and the location of storage polysaccharides*

Pyrenoid-containing species are widely distributed in every class of algae, except the Cyanophyceae. The pyrenoid is a differentiated region of the chloroplast. It may be embedded within the chloroplast or occur in the form of a projection from the chloroplast. The embedded pyrenoid is surrounded by a starch sheath, as in the Chlorophyceae and Prasinophyceae, or by a membrane as in certain diatoms (Drum & Pankratz 1964) and dinoflagellates (Dodge & Crawford 1971).

In other groups of algae the pyrenoid is recognizable by the density of its matrix, which consists mainly of closely packed granular or fibrillar protein-aceous material. Cytochemical tests do not establish conclusively either the absence or presence of DNA and RNA (Brown & Arnott 1970). Projection or stalked pyrenoids have been described for the members of the Phaeophyceae, Euglenophyceae, Dinophyceae and Haptophyceae. These are characterized by a narrow stalk (or stalks) as in the case of the multistalked pyrenoid of certain dinoflagellates (Dodge & Crawford 1971). The matrix of this type of pyrenoid is never penetrated by the thylakoids and is bordered by the chloroplast envelope. Outside this boundary is a 'pyrenoid cap' (Evans 1966), which is a swollen endoplasmic reticulum-like membrane sac.

The variability in the distribution, structure and development of the pyrenoid can be seen in the dinoflagellates in which five main types of pyrenoids are found (Dodge & Crawford 1971). In some algae, pyrenoids are present only in certain phases of the algal life cycle (Evans 1968, Bourne & Cole 1968, Bisalputra *et al.* 1971). In *Laminaria* for example, pyrenoid-like structures occur in the germina-ting meiospores, gametophytes, gametes and zygotes and disappear during the sporophyte generation. In general the presence of the pyrenoid seems to occur with high cellular activity and the build-up of reserve products. Changes in the pyrenoid structure may be induced experimentally. Griffith (1970), concluded that the pyrenoid was involved in the conversion and translocation of early photo-synthates in chloroplasts and the differences in pyrenoid structure are possibly a reflection of the way in which different algal types have dealt with this aspect. The matrix of the pyrenoid can, therefore, be looked upon as a localized con-centration of enzymes and material necessary for such functions. Its position and appearance is conceivably determined also by other factors such as the architec-ture of the chloroplast, the site for the storage of photosynthetic products, and the role of other cell components such as the endoplasmic reticulum.

In many algae, the spatial relationship between the pyrenoid and storage polysaccharide can be taken as indicative of the role of the pyrenoid discussed above. In others, where no relationship is evident, its role remains unclear. The storage of starch grains within the chloroplast is restricted to two classes of algae, the Prasinophyceae, and the Chlorophyceae. When the pyrenoid is present, starch grains are deposited as a sheath around the pyrenoid. In species without the pyrenoid the deposition of starch occurs at random as in higher plants.

In other classes of algae, polysaccharides are stored outside the chloroplast. In the Cryptophyceae, reserve starch grains are found in the perichloroplastic matrix (Fig. 4.3, 4.7). Other reserve polysaccharides, i.e. cyanophycean starch, floridean starch of red algae, laminaran of the brown algae, leucosin of the chrysophytes and diatoms, paramylon of the euglenoids, and starch of the dino-flagellates, are found in the cytoplasm and may or may not show a spatial relationship with the pyrenoids. When an association between the storage poly-saccharide and pyrenoid exists, the pyrenoid is usually one of the stalked types.

Examples of these are the polysaccharide deposits within the pyrenoid cap of the brown algae, and the starch platelets found adjacent to the various types of stalked pyrenoids in the dinoflagellates (Evans 1966, Dodge & Crawford 1971). Polysaccharide caps of similar type are also common among the Euglenophyceae, Haptophyceae, and certain Rhodophyceae such as *Rhodosorus marinus* (Giraud 1962). Recently, *Peridinium westii*, a fresh-water dinoflagellate, has been reported to possess starch grains within the chloroplast (Messer & Ben-Shaul 1969). Dodge and Crawford (1971) however, indicate that this flagellate possesses an interlamellar pyrenoid which, as a rule, has extrachloroplastic starch. The pyrenoid has also been proposed as a site for the storage of proteins. Crystalline structures have also been found within the pyrenoid matrix (Kowallik 1969, Retellack & Butler 1970).

Electron microscopic studies on the mode of multiplication of the pyrenoid have provided evidence to support the classical concept that there are two ways by which this can happen: *de novo* formation subsequent to cell division, and the division of the pre-existing pyrenoids by constriction prior to, or during, cell division. The former process is described in detail in *Scenedesmus* (Bisalputra & Weier 1964), *Tetracystis* (Brown & Arnott 1970) and *Oedogonium* (Hoffman 1968) and involves the dissolution of the pyrenoid matrix prior to cell division and the laying down of a new pyrenoid ground substance in the daughter cell, followed by the formation of starch sheets. Examples of pyrenoid division are found in red algae, euglenoids, haptophytes and several green algae including *Chlorella ellipsoidea* and *Chlamydomonas reinhardtii* (Griffith 1970). This process, which occurs concurrently with chloroplast division, simply involves the elongation, constriction and separation of daughter halves. The paucity of biochemical information on the pyrenoid appears to be due mainly to the difficulty in the isolation of this structure. The real function and activities of the pyrenoid remain poorly understood.

## 4.5    *Microtubular elements*

Microtubule-like structures have been reported to occur either singly or in groups within the stroma of chloroplasts of *Oedogonium cardiacum* (Hoffman 1967), *Chara fibrosa* and *Volvox* sp. (Pickett-Heaps 1968). In all accounts, chloroplast microtubules have larger diameters than those found in the cytoplasm (27 to 32 nm vs. 24 nm). The walls of chloroplast microtubules are found to have banded or helical striations. Hoffman (1967) has constructed models illustrating the structure of the chloroplast microtubule. It appears to be made up of two ribbon-like units wound in a helical manner. Each of these units may consist of three filamentous subunits similar to those commonly found in cytoplasmic microtubules. Chloroplast microtubules are found not only in vegetative cells but also in zoospores, germlings and eggs. Pickett-Heaps (1968) concluded that they were generally found in greatest numbers in elongating and enlarging chloroplasts. Their roles remain obscure.

### 4.6   *Plastoglobuli ('osmiophilic' granules) and the 'eyespot'*

The occurrence of lipid granules in plastids has been reviewed by Lichtenthaler (1968). He introduced the term 'plastoglobuli' to denote these osmiophilic granules. The size of plastoglobuli varies from 30 to 100 nm. They usually lack bounding membranes and are found scattered within the stroma. The amount present in any one chloroplast seems to depend on the rate of lipid synthesis. In higher plants where more extensive work has been done, their number increases during the period of rapid growth in the light and decreases during the differentiation of prolamellae into the thylakoids. It appears that these materials serve as a pool of lipid reserve for the synthesis and growth of the lipoprotein membranes within the chloroplast (Sprey & Lichtenthaler 1966). During senescence of the cells of higher plants, there is also a marked increase in the accumulation of plastoglobuli. Whitton and Peat (1967) reported a similar increase in plastoglobuli during heterocyst formation in the blue-green alga *Chlorogloea fritschii*. Greenwood *et al.* (1963) reported that plastoglobuli contain carotenoids, plastoquinones, and a small amount of chlorophyll.

In motile algae, groups of tightly packed carotenoid globules may become photosensitive and serve as 'eyespots' or 'stigma'. These structures are often regarded as primitive photoreceptors. It has been reported that eyespot granules may originate from the plastoglobuli (Lembi & Lang 1965, Arnott & Brown 1967). Dodge (1969a) classified eyespots into five different types. Of these only two, type *A* and type *B* are found within the chloroplast. Type *A* eyespots show no relation with the flagella, but this relation is established in type *B*. Type *A* eyespots are found in the Chlorophyceae, Prasinophyceae and Cryptophyceae and type *B* in the Chrysophyceae, Xanthophyceae and Phaeophyceae. Of the remaining three types, type *C* and *D* may have originated from the chloroplasts. Type *C* is typical of the euglenoids and type *D* of the dinoflagellates. The last type (*E*), is also found in the dinoflagellates but the structure is not likely to be derived from the chloroplast. Bartlett *et al.* (1971) have devised a method for the isolation and purification of eyespots from *Euglena*, thus opening the door for biochemical investigations on this structure.

### 4.7   *Ribosomes*

Chloroplast ribosomes have been shown, e.g. in *Chlamydomonas*, to be smaller and have lower sedimentation constants than cytoplasmic ribosomes (Hoober & Blobel 1969) and they resemble prokaryotic ribosomes, not only in size and sedimentation value, but also in the size of their RNA components, the presence of acidic proteins and their sensitivity to antibiotics (Ellis 1970). Activities of chloroplast and mitochondrial ribosomes are sensitive to inhibitors such as chloramphenicol, tetracycline, spiromycin, erythromycin and lincomycin at concentrations which do not affect cytoplasmic ribosomes. On the other hand protein synthesis in the cytoplasm is sensitive to cycloheximide while the chloroplast and mitochondrial ribosomes are not, suggesting that the protein synthesis of these organelles is more similar to that of prokaryotic cells than to the cytoplasmic systems of the cell.


## 5  CHLOROPLAST DIVISION AND THE TRANSMISSION OF THE GENOPHORE

It is now clear that chloroplasts are derived from pre-existing chloroplasts, and among the algae there are two ways by which new chloroplasts may be formed. In the majority of algae the chloroplasts divide by fission, producing two smaller daughter chloroplasts. Chloroplast fission in *Nitella* has been examined and described by Green (1964).

The development of the chloroplasts from proplastids has been observed in *Euglena* (Ben-Shaul *et al.* 1964, Schiff 1971) and in red algae such as *Lomentaria* (Bouck 1962) and *Batrachospermum* (Brown & Weier 1968, Burton 1971). There are certain differences between the differentiation of proplastids of *Euglena* and the red algae, in that in the former, the thylakoid materials are probably derived from the inner proplastid envelope (Ben-Shaul *et al.* 1964) and prolamellar bodies are found. In the red algae, the inner thylakoids differentiate from the peripheral thylakoids by means of invaginations and no connexion between the thylakoids and the envelope membrane is found (Bouck 1962, Brown & Weier 1968).

The genophore, as defined by Ris (1961), is a gene-containing body similar to that in viruses and bacteria, or any genetic linkage group. Under the electron microscope the chloroplast genophore appears as an electron translucent region containing fibrils with DNA properties. It was first identified in *Chlamydomonas* (Ris & Plaut 1962) and later it was reported in a variety of algal groups (Bouck 1965, Bisalputra & Bisalputra 1967a,b, 1969a,b, Falk & Kleinig 1968, Gibbs 1968, Green & Burton 1970, Kowallik & Haberkorn 1971). It may appear as a ring-shaped structure inside the girdle lamella (Gibbs 1968, Bisalputra & Bisalputra 1969a). This type of genophore is found strictly in those chloroplasts with a Type III thylakoid arrangement, i.e. the Xanthophyceae, Chrysophyceae, Bacillariophyceae and Phaeophyceae (see p. 141). Other algal groups contain a number of separated chloroplast genophores with no specific localization within the stroma. Kowallik and Haberkorn (1971) have shown by serial sectioning that each chloroplast of *Prorocentrum* contains no fewer than 80 to 100 individual genophores of all sizes and shapes. It has been shown both by thin-sectioning and by isolation methods that the DNA molecules within the chloroplast genophores are attached to the thylakoid membranes (Bisalputra & Bisalputra 1969b, Woodcock & Fernandez-Moran 1968, Herrmann & Kowallik 1970). Such an attachment is an important factor in the replication and segregation of chloroplast DNA during the division of the chloroplast (Bisalputra & Burton 1970).

In *Antithamnion subulatum* both types of chloroplast continuity occur (Burton 1971). Figure 4.11 illustrates this fact. The *A. subulatum* chloroplast may multiply by elongation and constriction (Fig. 4.11, A → B → C → E → A),

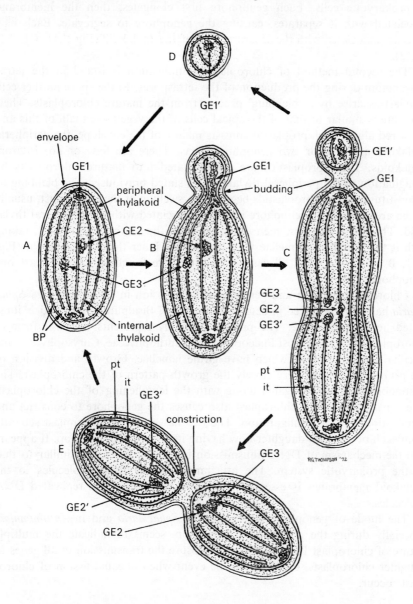

**Fig. 4.11.** Diagram showing the continuity of the chloroplast of red algae possesing proplastids. A → B → C → D, regeneration of proplastid. D → A, differentiation of proplastid into mature chloroplast. C → E → A, division of chloroplast by fission. The continuity of the genophore is explained in the text.

F

resulting in two daughter chloroplasts each with complete membranes and other components. The genophores are also found to have undergone division and are transmitted to the daughter chloroplasts in a rather similar manner to that found in prokaryotic cells. Each genophore first elongates, then the membrane associated with it separates, causing the genophore to segregate. Each half becomes associated with the separated end of the thylakoid (Fig. 4.11, GE 2 & GE 3).

The second method of chloroplast multiplication is found in the tetrasporangium during the production of the tetraspores. In the spore mother cell, proplastids arise by a 'blebbing' process from the mature chloroplasts. Their structure is similar to that of the apical cells of the vegetative thalli of this and other red algae. Each proplastid consists mainly of the envelope, the peripheral thylakoid and one or two genophore areas. There are few, or no, internal thylakoids. These proplastids are differentiated into mature chloroplasts by invagination of the peripheral thylakoids as stated previously. The blebbing or reconstitution of the proplastids begins by a swelling of the chloroplast, usually in the area where the genophore is found associated with the peripheral thylakoid. The 'bleb' enlarges, then separates from the mother chloroplast, taking with it the peripheral lamellae with little or no inner thylakoid material (Fig. 4.11, B → C → D). In all cases proplastids are found to have at least one genophore.

Chloroplast division and genophore transmission in the brown alga *Sphacelaria* has been described in detail (Bisalputra & Bisalputra 1970) and is illustrated here in Fig. 4.12. The process probably represents the basic form of chloroplast division in most Phaeophyceae, Xanthophyceae, Chrysophyceae and Bacillariophyceae, all of which have girdle lamellae. Growth and division of the genophore follows very closely the growth pattern of the chloroplast. The genophore becomes elongated along with the lengthening of the chloroplast. The constriction of the chloroplast also causes the genophore to constrict and form a double loop configuration. The two halves of the chloroplast separate from each other, each daughter now having its own genophore loop. It appears that the mechanism of DNA transmission in the chloroplast is similar to that of the prokaryotic system. The attachment of the DNA molecules to the thylakoid membranes is essential for the segregation of the replicated DNA molecules.

The mode of genophore segregation in *Sphacelaria* and in *Antithamnion*, especially during the proplastid regeneration, seems to indicate the multiple nature of chloroplast DNA. This would ensure the transmission of all genes to daughter chloroplasts, especially in the event when unequal fission of chloroplasts occur.

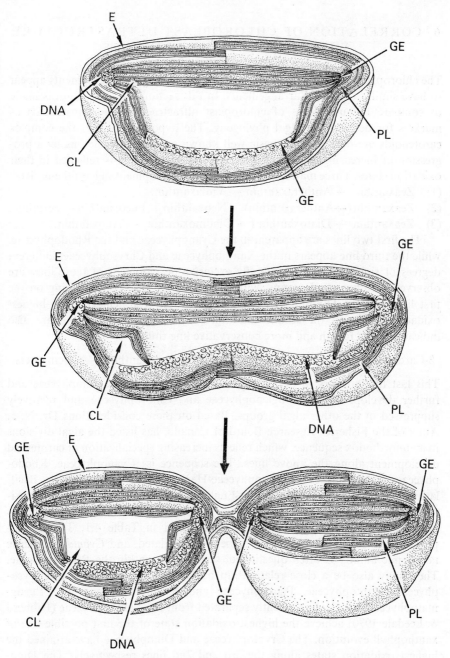

**Fig. 4.12.** The transmission of chloroplast DNA in algae with a ring-shaped geno-phore involves the elongation of the chloroplast together with its membranes and the genophore; constriction then leads to a figure 8 formation and the separa-tion of the daughter halves of the chloroplast, together with daughter genophores (CL = central lamellae; DNA = DNA fibrils with genophore; E = envelope; GE = genophore; PL = peripheral lamellae).

## 6　CORRELATION OF CHLOROPLAST ULTRASTRUCTURE WITH CHLOROPLAST PIGMENTS

The chloroplasts of different classes of algae containing similar pigments appear to have similar thylakoid arrangements. In this section a brief attempt is made to compare the usefulness of chloroplast ultrastructure and carotenoids as markers for determining algal phylogeny. The hypothesis is that the complex carotenoids are derived from basic well known carotenoid skeletons by a progression of increasing levels of chemical complexity which is reflected in their oxidation states. Three major lines of molecular development suggest themselves:

(1)　Zeaxanthin → Antheraxanthin → Violaxanthin.
(2)　Zeaxanthin (→Antheraxanthin) → Neoxanthin → Fucoxanthin → Peridinin.
(3)　Zeaxanthin → Diatoxanthin (→ Diadinoxanthin) → Alloxanthin.

The first two lines are apparent in the Cyanophyceae and the Rhodophyceae, while the third line appears in the Xanthophyceae and Chrysophyceae. Different degrees of simultaneous or mutually exclusive development of these lines are observed in other algal groups, with the Chlorophyceae concentrating on the first line, and the Dinophyceae and the Cryptophyceae achieving the highest oxidation levels on the second and third lines respectively. There are also indications of a fourth and more conservative line thus:

β-Carotene → Echinenone → Canthaxanthin → Astaxanthin → Astacin esters.

This last line appears to be already well advanced in the Cyanophyceae and further developed in the Euglenophyceae and Chlorophyceae, but relatively suppressed in the other algal groups. Based on these considerations Dr. N. J. Antia of the Fisheries Research Board of Canada, has listed the algal divisions in an unorthodox sequence, which reflects increasing specialization of carotenoid development along the above lines. The sequence is: Cyanophyceae, Rhodophyceae, Xanthophyceae, Chrysophyceae-Haptophyceae, Phaeophyceae, Bacillariophyceae, Eustigmatophyceae, Euglenophyceae, Chlorophyceae, Prasinophyceae, Cryptophyceae and Dinophyceae. (The insertion of Eustigmatophyceae is my own.) The scheme is summarized in Table 4.1. The Cyanophyceae and Rhodophyceae seem to be closely related, and *Cyanidium*, whose taxonomic affinity is often questioned, seems to fit within the Rhodophyceae. There may also be a close relationship between the Xanthophyceae, Chrysophyceae, Haptophyceae, Phaeophyceae and Bacillariophyceae. The Eustigmatophyceae, which were recently separated from the Xanthophyceae (Hibberd & Leedale 1970) achieve the highest oxidation state of the first possible line of xanthophyll evolution. The Cryptophyceae and Dinophyceae have attained the highest oxidation states along the 3rd and 2nd lines respectively. The Dinophyceae also maintain the 3rd line of xanthophyll development. Chloroplasts of the dinoflagellates, especially the nature of their envelopes appear to be the more advanced. The Euglenophyceae have affinities with both the Cryptophyceae and

**Table 4.1.** Thylakoid arrangement and photosynthetic pigment composition in various classes of algae (the pigments are adapted from a list compiled by Dr. N. J. Antia, Fisheries Research Board of Canada; the predominant xanthophylls are italicized)

| Class | Thylakoid type | Chlorophyll* | Carotenoids* | Phycobilins* | Comments |
|---|---|---|---|---|---|
| Cyanophyceae | I | $a$ | β-carotene, *zeaxanthin*, echinenone, canthaxanthin, mutatochrome, antherax-anthin, β-cryptoxanthin, *myxoxanthophyll*, aphanizophyll, oscillaxanthin | C-phycocyanin C-phycoerythrin allophycocyanin | |
| Rhodophyceae | I | $a$ (+ rarely $d$) | β-carotene, *zeaxanthin*, antheraxanthin, β-cryptoxanthin, lutein, neoxanthin. | R-phycocyanin R-phycoerythrin, C-phycocyanin C-allophycocyanin C-phycoerythrin | *Cyanidium caldarium* contains chlorophyll $a$, β-carotene, *zeaxanthin*, lutein, C-phycocyanin and allophycocyanin |
| Xanthophyceae | III$b$ | $a, c$ | β-carotene, *diadinoxanthin*, diatoxanthin, *heteroxanthin*, *vaucheriaxanthin ester*, neoxanthin, β-cryptoxanthin 5′, 6′ monoepoxide and 5 6, 5′6′-diepoxide | | |
| Chrysophyceae and Haptophyceae | III$b$ | $a, c_1, c_2$ | β-carotene, *fucoxanthin*, diatoxanthin, diadinoxanthin, echinenone | | *Ochromonas* is exceptional in containing chlorophyll $a$ only and *violaxanthin*, zeaxanthin, antheraxanthin, neoxanthin, β-cryptoxanthin (and its 5′, 6′-monopoxide) in addition to β-carotene and *fucoxanthin* |
| Phaeophyceae | III$b$ | $a, c_1, c_2$ | β-carotene, (±α-carotene, + rarely ε-carotene) *fucoxanthin*, violaxanthin, zeaxanthin, antheraxanthin, mutatochrome | | |
| Bacillariophyceae | III$b$ | $a, c_1, c_2$ | β-carotene, (±α-carotene, + rarely ε-carotene) *fucoxanthin*, diaoxanthin, diadinoxanthin, neoxanthin | | |

* Further details of algal chlorophylls are given in Chapter 5, and algal carotenoids and phycobilins are detailed in Chapter 6.

**Table 4.1.**—continued

| Class | Thylakoid type | Chloro-phyll* | Carotenoids* | Phycobilins* | Comments |
|---|---|---|---|---|---|
| Eustigmatophyceae | IIIa | $a$, $c$ | β-carotene, *violaxanthin* (±zeaxanthin and antheraxanthin), diatoxanthin, heteroxanthin, vaucheriaxanthin ester, neoxanthin, β-cryptoxanthin 5′6′ monoepoxide, and 5 6, 5′6′-diepoxide | | |
| Euglenophyceae | IV | $a$, $b$ | β-carotene (± γ-carotene), *diadinoxanthin*, diatoxanthin, neoxanthin, β-cryptoxanthin (and its 5′6′ monoepoxide), echinenone, 3-hydroxy-echinenone, 3-hydroxy-echinenone, astaxanthin ester | | |
| Chlorophyceae and Prasinophyceae | V | $a$, $b$ | β-carotene (±α-carotene, rarely γ-carotene and lycopene), *lutein*, violaxanthin, zeaxanthin, antheraxanthin, neoxanthin, β-cryptoxanthin (or its 5′6′-monoepoxide), lutein-5, 6-epoxide, loroxanthin, pyrenoxanthin, echinenone, c anthaxanthin, 3-hydroxy-echinenone ester, adonirubin ester, adonixanthin ester, crustaxanthin ester, astaxanthin ester, phoenicopterone | | Several species of Prasinophyceae and genera of siphoneaceous Chlorophyceae e.g. *Codium*, *Derbesia* and *Caulerpa* are exceptional in containing *siphonaxanthin* and its ester *siphonein*, as additional major xanthophylls, sometimes totally replacing lutein. These species of Prasinophyceae also show a chlorophyll $c$ type pigment, in addition to chlorophylls $a$ and $b$. |
| Cryptophyceae | II | $a$, $c_2$ | α-carotene (±β-carotene, rarely + ε-carotene), *alloxanthin*, crocoxanthin, monadoxanthin | 3 spectral types of phycoerythrin absorbing at 544nm, 555nm and 568nm; 3 spectral types of phycocyanin absorbing at 615nm, 630nm and 645nm | |
| Dinophyceae | IIIa | $a$, $c_2$ | β-carotene, *peridinin*, diadinoxanthin, diatoxanthin, dinoxanthin | | *Gymnodinium veneficum* and *Glenodinium foliaceum* are exceptional in containing *fucoxanthin* in place of peridinin, the same may be true for *Woloszynskia* species |

Chlorophyceae. In general, details of ultrastructure of the chloroplasts, when used in conjunction with other characteristics such as eye-spots, flagella, etc. may prove useful in phylogenetic studies.

The origin of chloroplasts is not discussed here, except to say that there is increasing evidence to support the endosymbiotic hypothesis. There are numerous reviews dealing with this concept (Carr & Craig 1970, Cohen 1970, Echlin 1970, Raven 1970, Margulis 1970, Taylor 1970, Schnepf & Brown 1971, Whitton *et al.* 1971, Flavell 1972). In summary it gains support from: the double nature of the chloroplast envelope, the similarity of many aspects of the ultrastructure and nucleic acid biochemistry of modern prokaryotes and chloroplasts, and the numerous present-day examples of similar associations showing various degrees of host-dependability. The alternative view, that is, that the origin of various organelles of the eukaryotic cells resulted from accumulative and progressive transformations of prokaryotic cells is considered by Allsopp (1969).

## ACKNOWLEDGEMENTS

The assistance of Dr N. J. Antia of the Fisheries Research Board of Canada, Vancouver, my wife Alice-Ann and David Hamilton of the Australian National University in the preparation of the manuscript is gratefully acknowledged.

## 7  REFERENCES

ALBERTSSON P.A. & LEYON H. (1954) The structure of chloroplasts. V. *Chlorella pyrenoidosa* Pringsheim studied by means of electron microscopy. *Expl. Cell. Res.* **7**, 288–90.

ALLSOPP A. (1969) Phylogenetic relationships of the Procaryota and the origin of the eucaryotic cell. *New Phytol.* **68**, 591–612.

ARNOTT H.J. & BROWN R.M. (1967) Ultrastructure of the eyespot and its possible significance in phototaxis of *Tetracystis excentrica*. *J. Protozool.* **14**, 529–39.

BARTLETT C.J., WALNE P.L., SCHWARY O.J. & BROWN D.H. (1971) Isolation and purification of eyespot granules from *Euglena gracilis* var. *bacillaris* by combined rate zonal and isopycnic centrifugation. *J. Phycol.* **7** (suppl.), 3–4.

BEN-SHAUL Y., SCHIFF J.A. & EPSTEIN H.T. (1964) Studies of chloroplast development in *Euglena* VII. Fine structure of the developing plastid. *Pl. Physiol., Lancaster*, **39**, 231–40.

BISALPUTRA T. & BISALPUTRA A.A. (1967a) The occurrence of DNA fibrils in chloroplasts of *Laurencia spectabilis*. *J. ultrastruct. Res.* **17**, 14–22.

BISALPUTRA T. & BISALPUTRA A.A. (1967b) Chloroplast and mitochondrial DNA in a brown alga *Egregia menziesii*. *J. Cell. Biol.* **33**, 511–20.

BISALPUTRA T. & BISALPUTRA A.A. (1969a) The ultrastructure of chloroplast of a brown alga *Sphacelaria* sp. I. Plastid DNA configuration—the chloroplast genophore. *J. Ultrastruct. Res.* **29**, 151–70.

BISALPUTRA T. & BISALPUTRA A.A. (1969b) The ultrastructure of chloroplast of a brown alga *Sphacelaria* sp. II. Association between the chloroplast DNA and photosynthetic lamellae. *J. Ultrasrtuct. Res.* **29**, 224–35.

BISALPUTRA T. & BISALPUTRA A.A. (1970) The ultrastructure of the chloroplast of a brown alga *Sphacelaria* sp. III. Replication and segregation of chloroplast genophores. *J. Ultrastruct. Res.* **32**, 417–29.

BISALPUTRA T. & BAILEY A. (1973) The fine structure of the chloroplast envelope of a red alga *Bangia fusco-purpurea. Protoplasma*, **76**, 443–54.

BISALPUTRA T. & BURTON H. (1970) On the chloroplast DNA-membrane complex in *Sphacelaria* sp. *J. Microscopie*, **9**, 661–6.

BISALPUTRA T., SHIELD C.M. & MARKHAM J.W. (1971) *In situ* observation of the fine structure of *Laminaria* gametophytes and embryos in culture. I. Methods and the ultrastructure of the zygote. *J. Microscopie*, **10**, 83–98.

BISALPUTRA T. & WEIER T.E. (1964) The pyrenoid of *Scenedesmus quadricauda. Am. J. Bot.* **54**, 881–92.

BOUCK G.B. (1962) Chromatophore development, pits, and other fine structure in the red alga, *Lomentaria baileyana* (Harv.) Farlow. *J. Cell. Biol.* **12**, 553–64.

BOUCK G.B. (1965) Fine structure and organelle associations in brown algae. *Ibid.*, **26**, 523–37.

BOURNE V.L. & COLE K. (1968) Some observations on the fine structure of the marine brown alga *Phaeostrophion irregulata* S. et. G. *Can. J. Bot.* **46**, 1369–75.

BRANTON D. (1968) Structure of photosynthetic apparatus. In *Photophysiology*. III, ed. Giese A.C. pp. 197–224. Academic Press, New York & London.

BRANTON D. & PARK R.B. (1967) Subunits in chloroplast lamellae. *J. Ultrastruct. Res.* **19**, 283–303.

BROWN D.L. & WEIER T.E. (1968) Chloroplast development and ultrastructure in the fresh water red alga *Batrachospermum. J. Phycol.* **4**, 199–206.

BROWN D.L. & WEIER T.E. (1970) Ultrastructure of the fresh water alga *Batrachospermum.* I. Thin-section and freeze-etch analysis of juvenile and photosynthetic filament vegetative cells. *Phycologia*, **9**, 217–35.

BROWN R.M. & ARNOTT H.J. (1970) Structure and function of the algal pyrenoid. I. Ultrastructure and cytochemistry during zoosporogenesis of *Tetracystis excentrica J. Phycol.* **6**, 14–22.

BURR F.A. & WEST J.A. (1970) Light and electron microscopic observations on the vegetative and reproductive structures of *Bryopsis hypnoides. Phycologia*, **9**, 17–37.

BURTON A.H.S. (1971) Chloroplast continuity during the formation of the tetraspore in *Antithamnion subulatum.* M.Sc. Thesis, University of British Columbia, Canada.

CARR N.G. & CRAIG I.W. (1970) The relationship between bacteria, blue-green algae and chloroplasts. In *Phytochemical Phylogeny*, ed. Harborne J.B., pp. 119–40. Academic Press, London & New York.

COHEN S.S. (1970) Are/were mitochondria and chloroplasts micro-organisms? *Amer. Scientist*, **58**, 281–9.

COHEN-BAZIRE G. (1971) The photosynthetic apparatus of procaryotic organisms. In *Biological Ultrastructure: The origin of cell organelles*, ed. Harris P., pp. 65–90. Oregon State Univ. Press, Corvallis.

COLE K. (1970) Ultrastructural characteristics in some species in the order Scytosiphonales. *Phycologia*, **9**, 275–83.

DODGE J.D. (1968) The fine structure of chloroplasts and pyrenoids in some marine dinoflagellates. *J. Cell Sci.* **3**, 41–8.

DODGE J.D. (1969a) A review of the fine structure of algal eyespots. *Br. Phycol. J.* **4**, 199–210.

DODGE J.D. (1969b) The ultrastructure of *Chroomonas mesostigmatica* Butcher (Cryptophyceae). *Arch. Mikrobiol.* **69**, 266–80.

DODGE J.D. (1971) Fine structure of the Pyrrophyta. *Bot. Rev.* **37**, 481–508.

DODGE J.D. (1974) Fine structure and phylogeny in the algae. *Sci. Progr.* **61**, 257–74.

DODGE J.D. & CRAWFORD R.M. (1971) A fine structural survey of dinoflagellate pyrenoids and food reserves. *Bot. J. Linn. Soc.* **64**, 105–15.

DRUM R.W. & PANKRATZ H.S. (1964) Pyrenoids, raphes, and other fine structure in diatoms. *Am. J. Bot.* **51**, 405–18.

ECHLIN P. (1968) The fine structure and taxonomy of unicellular blue-green algae. *Proc. 4th European Regional Conference on Electron Microscopy*, **2**, 389.

ECHLIN P. (1970) The photosynthetic apparatus in prokaryotes and eukaryotes. In *Organization and Control in Prokaryotic and Eukaryotic Cells*, eds. Charles H.P. & Knight B.C.J.G. pp. 221–48. Cambridge Univ. Press, Cambridge.

ELLIS R.J. (1970) Further similarities between chloroplast and bacterial ribosomes. *Planta*, **91**, 329–32.

EVANS M.C.W. & WHATLEY F.R. (1970) Photosynthetic mechanisms in prokaryotes and eukaryotes. In *Organization and Control in Prokaryotic and Eukaryotic Cells*, eds. Charles H.P. & Knight B.C.J.G. pp. 203–20 Cambridge Univ. Press, Cambridge.

EVANS L.V. (1966) Distribution of pyrenoids among some brown algae. *J. Cell Sci.* **1**, 449–54.

EVANS L.V. (1968) Chloroplast morphology and fine structure in British fucoids. *New Phytol.* **67**, 173–8.

EVANS L.V. (1970) Electron microscopical observations on a new red algal unicell, *Rhodella maculata* Gen. Nov., Sp. Nov. *Br. Phycol. J.* **5**, 1–13.

FALK H. & KLEINIG H. (1968) Feinbau und Carotinoide von *Tribonema* (Xanthophyceae). *Arch. Mikrobiol.* **61**, 347–62.

FLAVELL R. (1972) Mitochondria and chloroplasts as descendants of prokaryotes. *Biochem. Genetics*, **6**, 275–91.

FOGG G.E., STEWART W.D.P., FAY P. & WALSBY A.E. (1973) *The Blue-green Algae*. Academic Press, London & New York.

GANTT E. & CONTI S.F. (1965) The ultrastructure of *Porphyridium cruentum*. *J. Cell. Biol.* **26**, 305–38.

GANTT E. & CONTI S.F. (1969) Ultrastructure of blue-green algae. *J. Bact.* **97**, 1486–1493.

GANTT E., EDWARDS M.R. & CONTI S.F. (1968) Ultrastructure of *Porphyridium aerugineum*, a blue-green colored Rhodophytan. *J. Phycol.* **4**, 65–71.

GANTT E., EDWARDS M.R. & PROVASOLI L. (1971) Chloroplast structure of the Cryptophyceae. Evidence for phycobiliproteins within intrathylakoid spaces. *J. Cell. Biol.* **48**, 280–90.

GIBBS S.P. (1960) The fine structure of *Euglena gracilis* with special reference to the chloroplasts and pyrenoids. *J. Ultrastruct. Res.* **4**, 127–48.

GIBBS S.P. (1962a) The ultrastructure of the chloroplasts of algae. *Ibid.*, **7**, 418–35.

GIBBS S.P. (1962b) Nuclear envelope-chloroplast relationships in algae. *J. Cell. Biol.* **14**, 433–44.

GIBBS S.P. (1968) Autoradiographic evidence for the *in situ* synthesis of chloroplast and mitochondrial RNA. *J. Cell Sci.* **3**, 327–40.

GIBBS S.P. (1970) The comparative ultrastructure of the algal chloroplast. *Ann. N.Y. Acad. Sci.* **175**, 454–73.

GILES K.L. & SARAFIS V. (1972) Chloroplast survival and division *in vitro*. *Nature New Biol.* **236**, 56–8.

GIRAUD G. (1962) Les infrastructures de quelques Algues et leur physiologie. *J. Microscopie*, **1**, 251–74.

GOODENOUGH U.W. & STAEHELIN L.A. (1971) Structural differentiation of stacked and unstacked chloroplast membranes. Freeze-etch electron microscopy of wild-type and mutant strains of *Chlamydomonas*. *J. Cell. Biol.* **48**, 594–619.

GOLDBERG I. & OHAD I. (1970) Biogenesis of chloroplast membranes. V. A radioautographic study of membrane growth in a mutant of *Chlamydomonas reinhardi* y-I. *J. Cell Biol.* **44**, 572–91.

GRANICK S. (1961) The chloroplasts: Inheritance, structure and function. In *The Cell, Biochemistry, Physiology, Morphology*, eds. Brachet J. & Mirsky A.E., pp. 489–602. Academic Press New York & London.

GREEN B.R. & BURTON H. (1970) *Acetabularia* chloroplast DNA: electron microscopic visualization. *Science, N. Y.* **168**, 981–2.

GREEN P.B. (1964) Cinematic observation on the growth and division of chloroplast in *Nitella. Am. J. Bot.* **51**, 334–42.

GREENWOOD A.D., LEECH R.M. & WILLIAMS J.P. (1963) The osmiophilic globules of chloroplasts. I. Osmiophilic globules as a normal component of chloroplasts and their isolation and composition in *Vicia faba* L. *Biochim. biophys. Acta.* **78**, 148–62.

GRIFFITH D.J. (1970) The pyrenoid. *Bot. Rev.* **36**, 29–58.

GUÉRIN-DUMARTRAIT E. (1968) Etude, en cryodécapage, de la morphologie des surfaces lamellaires chloroplastiques de *Chlorella pyrenoidosa* en cultures synchrones. *Planta*, **80**, 96–109.

HERRMANN R.G. & KOWALLIK K.V. (1970) Selective presentation of DNA-region and membranes in chloroplasts and mitochondria. *J. Cell. Biol.* **45**, 198–202.

HIBBERD D.J. & LEEDALE G.F. (1970) Eustigmatophyceae—a new algal class with unique organization of the motile cell. *Nature, Lond.* **225**, 758–60.

HIBBERD D.J. & LEEDALE G.F. (1971) Cytology and ultrastructure of the Xanthophyceae. II. The zoospore and vegetative cell of coccoid forms, with special reference to *Ophiocytium majus* Naegeli. *Br. Phycol. J.* **6**, 1–23.

HOFFMAN L.R. (1967) Observations on the fine structure of *Oedogonium* III. Microtubular elements in the chloroplasts of *Oedogonium cardiacum. J. Phycol.* **3**, 212–21.

HOFFMAN L.R. (1968) Observations on the fine structure of *Oedogonium*. V. Evidence for the de novo formation of pyrenoids in zoospores of *O. cardiacum. J. Phycol.* **4**, 212–18.

HOOBER J.K. & BLOBEL G. (1969) Characterization of the chloroplastic and cytoplasmic ribosomes of *Chlamydomonas reinhardi. J. molec. Biol.* **41**, 121–38.

HOOBER J.K., SIEKEVITZ P. & PALADE G.E. (1969) Formation of chloroplast membranes in *Chlamydomonas reinhardi* y-I. Effects of inhibitors of protein synthesis. *J. biol. Chem.* **144**, 2621–31.

JOHNSON U. & PORTER K.R. (1968) Fine structure of cell division in *Chlamydomonas reinhardi. J. Cell. Biol.* **38**, 403–25.

JÓNSSON S. (1963) Données récentes sur l'ultrastructure des plastes, des mitochondries et de l'appareil le Golgi chez les végétaux eucaryotes Chlorophylliens. *Ann. Biol.* **2**, 207–55.

JOST M. (1965) Die Ultrastruktur von *Oscillatoria rubescens* D.C. *Arch. Mikrobiol.* **50**, 211–45.

KIRK J.T.O. (1971) Chloroplast structure and biogenesis. *A. Rev. Biochem.* **40**, 161–96.

KIRK J.T.O. & TILNEY-BASSETT R.A.E. (1967) *The Plastids: their chemistry, structure, growth and inheritance.* W.H. Freeman & Co., London.

KOWALLIK K. (1969) The crystal lattice of the pyrenoid matrix of *Prorocentrum micans. J. Cell. Sci.* **5**, 251–69.

KOWALLIK K. & HABERKORN G. (1971) The DNA-structure of the chloroplast of *Prorocentrum micans* (Dinophyceae). *Arch. Mikrobiol.* **80**, 252–61.

LANG N.J. (1963) Electron microscopy of the Volvocaceae and Astrephomenaceae. *Am. J. Bot.* **50**, 280–300.

LANG N.J. (1965) Electron microscopic study of heterocyst development in *Anabaena azollae* Strasburger. *J. Phycol.* **1**, 127–31.

LANG N.J. (1968) The fine structure of blue-green algae. *A. Rev. Microbiol.* **22**, 15–46.

LEEDALE G.F. (1967) *Euglenoid Flagellates* Prentice-Hall, Inc. Englewood Cliffs, New Jersey.

LEEDALE G.F. (1968) The nucleus in *Euglena.* In *The Biology of Euglena.* Vol. I, ed. Buetow D.E. pp. 185–242. Academic Press, New York & London.

LEEDALE G.F., PRINGSHEIM E.G. & MEEUSE B.J.D. (1965) Structure and physiology of *Euglena spirogyra.* II. Cytology and fine structure. *Arch. Mikrobiol.* **50**, 70–102.

LEMBI C.A. & LANG N.J. (1965) Electron microscopy of *Carteria* and *Chlamydomonas. Am. J. Bot.* **52**, 464–77.

LEYON H. (1954) The structure of chloroplasts. III. A study of pyrenoids. *Expl. Cell. Res.* **6**, 497–505.

LEYON H. & VON WETTSTEIN D. (1954) Der Chromatophoren-Feinbau bei den Phaeophyceen. *Z. Naturf.* **9**, 471–5.

LICHTENTHALER, H.K. (1968) Plastoglobuli and the fine structure of plastids. *Endeavour*, **27**, 144–9.

LICHTLÉ C. & GIRAUD G. (1970) Aspects ultrastructureous particuliers au plasti du *Batrachospermum virgatum* (Sirdt) Rhodophycée—Nemalionale. *J. Phycol.* **6**, 281–9.

LUCAS I.A.N. (1970) Observations on the fine structure of the Cryptophyceae. I. The genus *Cryptomonas*. *J. Phycol.* **6**, 30–8.

MANTON I. (1966a) Further observations on the fine structure of *Chrysochromulina chiton*, with special reference to the pyrenoids. *J. Cell. Sci*, **1**, 187–92.

MANTON I. (1966b) Some possible significant structural relations between chloroplasts and other cell components. In *Biochemistry of Chloroplasts*. Vol. I., ed. Goodwin T.W. 23–47. Academic Press, London & New York.

MANTON I., KOWALLIK K. & VON STOSCH H.A. (1969) Observations on the fine structure and development of the spindle at mitosis and meiosis in a marine centric diatom (*Lithodesmium undulatum*). II. Early meiotic prophases in male gametogenesis. *J. Cell. Sci.* **5**, 271–98.

MANTON I. & LEEDALE G.F. (1969) Observations on the microanatomy of *Coccolithus pelagicus* and *Cricosphaera carterae*, with special reference to the origin and nature of coccoliths and scales. *J. mar. biol. Assoc. U.K.* **49**, 1–16.

MANTON I. & PARKE M. (1965) Observations on the fine structure of two species of *Platymonas* with special reference to flagellar scales and the mode of origin of the theca. *Ibid.*, **45**, 743–54.

MARGULIS L. (1970) *The origin of Eukaryotic cells*. Yale University Press, New Haven.

MASSALSKI A. & LEEDALE G.F. (1969) Cytology and Ultrastructure of *Bumilleria sicula* Borzi and *Tribonema vulgare* Pascher. *Br. phycol. J.* **4**, 159–80.

MCBRIDE D.L. & COLE K. (1969) Ultrastructural characteristics of the vegetative cells of *Smithora naiadum* (Rhodophyta). *Phycologia*, **8**, 177–86.

MENKE W. (1962) Structure and chemistry of plastids. *A. Rev. Pl. Physiol.* **13**, 27–44.

MESSER G. & BEN-SHAUL Y. (1969) Fine structure of the *Peridinium westii* Lemm., a fresh water dinoflagellate. *J. Protozool.* **16**, 272–80.

MORRÉ D.J., MERRITT W.D. & LEMBI C.A. (1971) Connections between mitochondria and endoplasmic reticulum in rat liver and onion stem. *Protoplasma*, **73**, 43–9.

NEUSHUL M. (1970) A freeze-etching study of the red alga *Porphyridium*. *Am. J. Bot.* **57**, 1231–9.

OLIVIERA L. & BISALPUTRA T. (1972) Studies on the brown alga *Ectocarpus* in culture. I. General ultrastructure of the sporophytic vegetative cells. *J. Submicros. Cytol.* (in press).

OLSON J.M. (1970) The evolution of photosynthesis. *Science, N.Y.* **168**, 438–46.

PARK R.B. & SANE P.V. (1971) Distribution of function and structure in chloroplast lamellae. *A. Rev. Pl. Physiol.* **22**, 395–430.

DE PETROCELLIS B., SIEKEVITZ P. & PALADE G.E. (1970) Changes in chemical composition of thylakoid membranes during greening of the y-1 mutant of *Chlamydomonas reinhardi*. *J. Cell. Biol.* **44**, 618–34.

PICKETT-HEAPS J.D. (1968) Microtubule-like structure in the growing plastids or chloroplasts of two algae. *Planta*, **81**, 193–200.

PICKETT-HEAPS J.D. & FOWKE L.C. (1970) Mitosis, cytokinesis and cell elongation in the desmid, *Closterium littorale*. *J. Phycol.* **6**, 189–215.

PUISEUX-DAO S. (1970) *Acetabularia and cell biology*. Springer-Verlag, Inc. New York.

RACKER E. (1970) *Membranes of mitochondria and chloroplasts*. van Nostrand, Reinhold, New York.

RAVEN, P.H. (1970) A multiple origin for plastids and mitochondria. *Science, N.Y.* **169**, 641–6.

RETELLACK B. & BUTLER R.D. (1970) The development and structure of pyrenoids in *Bulbochaete hiloensis. J. Cell Sci.* **6**, 229–41.

RIDLEY S.M. & LEECH R.M. (1969) Chloroplast survival *in vitro*. In *Progress in Photosynthesis Research*, Vol. I, ed. Metzner H., pp. 229–44. H. Laupp, Jr. Tubingen.

RIS H. (1961) Ultrastructure and molecular organization of genetic systems. *Canad. J. Genet. Cytol.* **3**, 95–120.

RIS H. & PLAUT W. (1962) Structures of DNA-containing areas in the chloroplast of *Chlamydomonas. J. Cell. Biol.* **13**, 383–91.

SAGER R. & PALADE G.E. (1957) Structure and development of the chloroplast in *Chlamydomonas*. I. The normal green cell. *J. Biophys. Biochem.* **3**, 463–88.

SCHIFF J.A. (1971) Developmental interaction along cellular compartments in *Euglena*. In *Autonomy and biogenesis of mitochondria and chloroplasts*, eds. Boardman N.K., Linnane, A.W. & Smillie R.M., pp. 98–118. North Holland, Amsterdam.

SCHNEPF E. & BROWN R.M. (1971) On the relationship between endosymbiosis and the origin of plastids and mitochondria. In *Origin and Continuity of Cell Organelles*, eds. Reinert J. & Ursprung H. Springer-Verlag, New York.

SPREY B. & LICHTENTHALER H. (1966) Zur Frage de Beziehungen zwichen Plastoglobuli und Thylakoid Genese in Gerstenkeim Lingen. *Z. Naturf.* **21**b, 697–9.

STAEHELIN A. (1967) Chloroplast fibrils linking the photosynthetic lamellae. *Natu^ e, Lond.* **214**, 1158.

TAYLOR D.L. (1969) Identity of zooxanthellae isolated from some Pacific Tirdacnidae. *J. Phycol.* **5**, 336–40.

TAYLOR D.L. (1973) Algal symbionts of invertebrates. *A. Rev. Microbiol.* **27**, 171–87.

TAYLOR D.L. (1970) Chloroplasts as symbiotic organelles. *Int. Rev. Cytol.* **27**, 24–64.

TRENCH R.K., GREEN R.W. & BYSTROM B.G. (1969) Chloroplasts as functional organelles in animal tissues. *J. Cell. Biol.* **42**, 404–17.

TRIPODI G. (1971) Some observations on the ultrastructure of the red alga *Pterocladia capillacea* (Gmel.). Born. et Thur. *J. Submicros. Cytol.* **3**, 63–70.

WALKER O.A. (1969) Permeability of the chloroplast envelope. In *Progress in Photosynthesis Research*, Vol. I, ed. Metzner H., pp. 250–7. H. Laupp, Jr. Tubingen.

WEIER T.E. & BENSON A.A. (1967) The molecular organization of chloroplast membranes. *Am. J. Bot.* **54**, 389–402.

WEIER T.E., BISALPUTRA T. & HARRISON A. (1966) Subunits in chloroplasts of *Scenedesmus quadricauda. J. Ultrastruct. Res.* **15**, 38–56.

VON WETTSTEIN D. (1954) Formwechsel und Teilung der Chromatophoren von *Fucus vesiculosus. Z. Naturf.* **9**b, 476–81.

WHITTON B.A., CARR N.G. & CRAIG I.W. (1971) A comparison of the fine structure and nucleic acid biochemistry of chloroplasts and blue-green algae. *Protoplasma*, **72**, 325–57.

WHITTON B.A. & PEAT A. (1967) Heterocyst structure in *Chlorogloea fritschii. Arch. Mikrobiol.* **58**, 324–38.

WOLKEN J.J. & SCHWERTZ F.A. (1953) Chlorophyll monolayers in chloroplasts. *J. gen. Physiol.* **37**, 111–20.

WOODCOCK C.L.F. & FERNANDEZ-MORAN H. (1968) Electron microscopy of DNA conformations in spinach chloroplasts. *J. mol. Biol.* **31**, 627–31.

CHAPTER 5

# CHLOROPHYLLS

## J. C. MEEKS

Department of Biological Sciences,
University of Dundee, Dundee, U.K.

1  **Introduction**  161

2  **Distribution and
   properties**  161

3  **Analysis**  164
3.1 Extraction  164
3.2 Identification  164
3.3 Estimation  165

4  **Biosynthesis of
   chlorophylls**  167
4.1 The biosynthetic
   chain  167

4.2 Control of chlorophyll biosynthesis and plastid
   development  168

5  **Conditions affecting chlorophyll content**  169
5.1 Nutrition  169
5.2 Light intensity, temperature
   and cell age  170

6  **References**  171

## 1  INTRODUCTION

Chlorophylls are the basic pigments involved in light absorption and photo-chemistry in higher plants, algae, and photosynthetic bacteria. The aim of this chapter is to present a brief general review of algal chlorophylls with relatively equal treatment being given to the distribution, *in vitro* properties, control of biosynthesis, and physiological and environmental conditions which affect the cellular content of algal chlorophylls. For detailed discussions of the physical and biological properties of chlorophylls, the reader should consult the review articles cited in the text and the comprehensive volume edited by Vernon and Seely (1966) (see also Chapter 12, p. 346).

Four major *in vitro* algal chlorophylls (Chl) have been described: Chl *a*, Chl *b*, Chl *c* (Chl $c_1$ and Chl $c_2$) and Chl *d*. The structural formula of Chl *a* is shown in Fig. 5.1. Variations of this basic structure which give the other chlorophylls are also listed.

## 2  DISTRIBUTION AND PROPERTIES

The current distribution of chlorophylls in the algae is listed in Table 5.1.

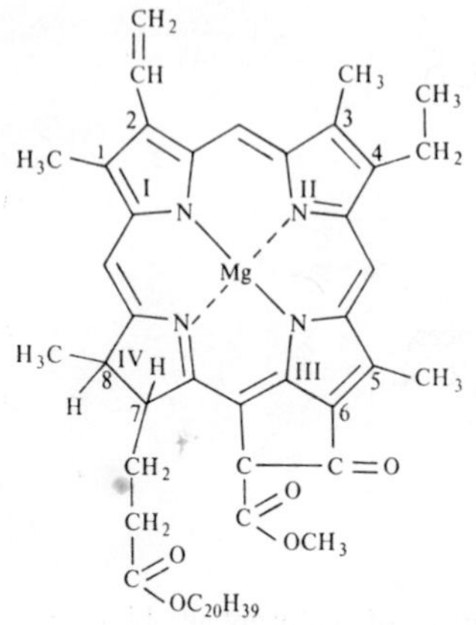

Chl $a$ : as above
Chl $b$ : II - 3 = CHO
Chl $d$ : I  - 2 = CHO
Chl $c_1$ : IV- 7 = CH = CHCOOH; double bond at IV - 7, 8
Chl $c_2$ : IV- 7 = CH = CHCOOH; double bond at IV - 7, 8;
　　　　　II - 4 = CH = CH$_2$

**Fig. 5.1.** Chemical structure of chlorophyll $a$.

Chlorophyll $a$ is the primary photosynthetic pigment in all oxygen evolving photosynthetic organisms investigated but the other algal chlorophylls have a limited distribution and are considered as accessory or secondary photosynthetic pigments. The Chl $a$ content of algal cells can range from 0·3 to 2·0 per cent of the dry weight (Rabinowitch 1945) and there are indications that it may exist in multiple states *in vivo* (see Chapter 12, p. 346). *In vitro* Chl $a$ is soluble in alcohols, diethyl ether, benzene, and acetone and is insoluble in water. Pure Chl $a$ is only slightly soluble in petroleum ether and must be dissolved in a small amount of acetone or methanol prior to transfer to petroleum ether (see Smith & Benitez 1955). The pigment shows two main absorption bands *in vitro*: one band is in the red light region near to 660 and 665nm and the other is in the Soret

**Table 5.1.** Distribution of chlorophylls among the algae

| Algal group | Chlorophyll | | | |
|---|---|---|---|---|
| | Chl $a$ | Chl $b$ | Chl $c$ ($c_2 \pm c_1$) | Chl $d$ |
| Cyanophyceae | + | — | — | — |
| Rhodophyceae | + | — | — | + |
| Cryptophyceae | + | — | + | — |
| Dinophyceae | + | — | + | — |
| Rhaphidophyceae | + | — | + | — |
| Chrysophyceae | + | — | + | — |
| Haptophyceae | + | — | + | — |
| Bacillariophyceae | + | — | + | — |
| Xanthophyceae[1] | + | — | + | — |
| Phaeophyceae | + | — | + | — |
| Prasinophyceae | + | + | — | — |
| Euglenophyceae | + | + | — | — |
| Chlorophyceae[2] | + | + | — | — |

[1] includes Eustigmatophyceae, [2] includes Charophyceae.
+ means the chlorophyll is present in at least some members; — means the chlorophyll has not been recorded in any member. See text for discussion.

region near to 430nm. The ratio of the Soret to red absorption maximum ranges from 1.11:1 (Goedheer 1966) to 1.32:1 (Holt 1965).

Chlorophyll $b$ is found in the Prasinophyceae (Ricketts 1966), Euglenophyceae, and Chlorophyceae (Allen 1966). These classes form the 'green plant line' which extends in structural complexity to the higher plants (see Chapter 1, p. 3). Chl $b$ functions photosynthetically as a light harvesting pigment, transferring absorbed light energy to Chl $a$ for primary photochemistry (see Chapter 12, p. 369). The Chl $a$:Chl $b$ ratio generally ranges from 2:1 to 3:1 (Strain, Cope & Svec 1971). Chl $b$ may be detected in low temperature *in vivo* absorption spectra (Chapter 12, p. 357). Its solubility characteristics are similar to those of Chl $a$ (Smith & Benitez 1955) and *in vitro* shows two main absorption maxima. The red band maximum is near to 645nm in acetone or methanol and the Soret maximum near to 435nm. The Soret:red band ratio is about 2·85.

Chlorophyll $c$ is found in those organisms which constitute the 'brown plant line', i.e. the Cryptophyceae, Dinophyceae, Rhaphidophyceae, Chrysophyceae, Haptophyceae, Bacillariophyceae, Xanthophyceae and Phaeophyceae. Chl $c$ preparations contain two spectrally distinct components: Chl $c_1$ and Chl $c_2$ (Jeffrey 1969, Doughetry *et al.* 1970). Both pigments have been isolated (Jeffrey 1969, 1972, Strain *et al.* 1971) and their structural formulas determined (Strain *et al.* 1971). Chl $c_2$ is always present but Chl $c_1$ is absent in the Dinophyceae (free-living and symbiotic) and Cryptophyceae (Jeffrey 1969). Chl $c_1$ and $c_2$ both occur in the Bacillariophyceae, Phaeophyceae and Chrysophyceae (Jeffrey 1969). Some controversy exists as to the presence of Chl $c$ in the Chrysophyceae in that those Chrysophyceae reported to contain Chl $c$ by Ricketts (1965) and Jeffrey (1969) are now classified in the Haptophyceae (see Parke & Dixon 1968). Ricketts (1965) found no Chl $c$ in *Ochromonas*, a member of the Chrysophyceae (Parke & Dixon 1968), but suggests that the lack of pigment may have been due to its heterotrophic mode of growth. However, Guillard and Lorenzen (1972) found Chl $c$ in a freshwater *Synura* sp. which is classified in the Chrysophyceae. Moreover, Guillard and Lorenzen (1972) reported the presence of Chl $c$ in members of the Xanthophyceae, including species of *Tribonema*, *Botrydium* and *Ophiocytium*. They also found Chl $c$ in *Vaucheria* which also argues in favour of its placement in the Xanthophyceae (see De Greef & Cauberg 1970, Cauberg & De Greef 1971). Chl $c$ was not present in two related species of Xanthophyceae (Guillard & Lorenzen 1972) which are now presumed to belong to the new class, Eustigmatophyceae (Hibberd & Leedale 1970, 1972). Guillard and Lorenzen (1972) also reported that two species (*Vacuolaria virescens* and *Gonyostomum semen*) of the class Rhaphidophyceae contain Chl $c$.

Chl $c$ may function as an accessory pigment to photosystem II in diatoms (Mann & Myers 1968) and brown algae (Fork 1963, Goedheer 1970). The Chl $a$:Chl $c$ ratio ranges from 1·2:1 to 5·5:1 (Jeffrey 1972). Because of its low absorption in the red light region, Chl $c$ is difficult to detect in *in vivo* absorption spectra. The pigment is soluble in ether, acetone, methanol and ethyl acetate but insoluble in petroleum ether (Smith & Benitez 1955). However, crystalline preparations of Chl $c_1$ and Chl $c_2$ are not fully soluble in these solvents and must first be dissolved in pyridine (Jeffrey 1972). Extracted Chl $c_1$ has main absorption maxima at 634nm, 583nm and 444nm in methanol, while Chl $c_2$ shows maxima at 635nm, 586nm and 452nm (Jeffrey 1969). The main absorption band for both $c$ type chlorophylls is in the Soret region. The Soret to red (628nm) band ratios in methanol are 7·3:1 for Chl $c_1$ and 9·65:1 for Chl $c_2$.

Chlorophyll $d$ is a minor component in extracts of many Rhodophyceae. It is now accepted as a genuine pigment and no longer appears to be an artifact of extraction (O'hEocha 1971). Nevertheless, it is not seen in all species and has yet to be detected in the Bangiophycidae. The pigment has not been observed *in vivo* and has no known photosynthetic function. It is soluble in ether, acetone, alcohol and benzene, and very slightly soluble in petroleum ether (Smith & Benitez 1955). It has three main absorption maxima at 696nm, 456nm and 400nm (Manning & Strain 1943). The main absorption band is in the red light region and the ratio of 'blue' (456nm) to red bands is about 0·70.

Chlorophyll $e$ was reported in methanol extracts from two members of the Xanthophyceae, *Tribonema bombycinum* and *Vaucheria hamata* (see Allen 1966). Both isolates were from feral populations and it has not been detected in cultured species. It is possible that Chl $e$ is an alteration product of Chl $c$.

Accessory chlorophylls appear to be absent in the Cyanophyceae, in the enigmatic phycocyanin-containing *Cyanidium caldarium* (Allen 1959), and in some Rhodophyceae, Xanthophyceae (probably Eustigmatophyceae), and Chrysophyceae (*Ochromonas* sp.).

## 3   ANALYSIS

### 3.1   *Extraction*

For algal material, 80 to 90 per cent acetone or 90 per cent methanol is used most frequently but sometimes methanol gives a more quantitative extraction (Steeman Nielsen 1961) and has been recommended over acetone for extraction, but not for estimation (Marker 1972). To prevent alteration products fresh material should be extracted quickly under alkaline conditions in the cold and dark. Larger seaweeds may be boiled for 1 to 2 minutes in water to facilitate extraction, but there will be some alteration products as a result (Strain, Cope & Svec 1971); Jeffrey (1972) recommended freezing marine phytoplankton in water to ease extraction. Several extractions with hot methanol are required to quantitatively remove the chlorophylls from some green algae such as *Scenedesmus* (Bishop 1971). Quantitative extraction of Chl $a$ from *Cyanidium* is only possible with N,N-dimethyl formamide (Volk & Bishop 1968). Extraction methods are reviewed by Smith and Benitez (1955), Holden (1965), Strain and Svec (1966), Subba Rao and Platt (1969) and Strain, Cope and Svec (1971).

### 3.2   *Identification*

For definite identification of the chlorophylls in an extract, the individual pigments must be isolated and purified. Extracts in methanol or acetone may be transferred to petroleum ether or used directly for chromatographic separation. Alternatively, the pigments may be quantitatively transferred to diethyl ether and concentrated prior to separation by chromatography (see Smith & Benitez 1955, Holden 1965, Strain & Svec 1966, Jeffrey 1969, 1972, Doughetry *et al.* 1970, Strain, Cope & Svec 1971, Strain & Sherma 1972). The chlorophylls and their pheophytins are initially identified in pure form by their characteristic absorption spectra. Spectra of the chlorophylls in diethyl ether were given by Smith and Benitez (1955), Holt 1965), Strain and Svec (1966) and Govindjee (see Chapter 12, p. 353). Smith and Benitez (1955) discuss additional methods, such as phase tests and HCl number, for the identification of individual chlorophylls.

### 3.3 *Estimation*

The concentration of chlorophylls is generally determined spectrophotometrically using the specific absorption (extinction) coefficient of the chlorophyll in the particular solvent. The specific absorption coefficient ($\alpha$) is defined as $\alpha = D/dC$; where $D$ is the optical density or absorbancy of the extract, $d$ is the inside pathlength of the spectrophotometer cuvette in cm, and $C$ is the concentration of pigment in grams per litre. Table 5.2 gives the wavelengths of the near maximum red absorption peak and specific absorption coefficients of chlorophylls $a$, $b$, $c$ ($c_1$ and $c_2$), and $d$ in various solvents.

For algae containing only Chl $a$, the specific absorption coefficient in the extraction solvent may be used directly. However, methanol and acetone

**Table 5.2.** Wavelengths of maximum red absorption peaks and specific absorption coefficients of chlorophylls in various solvents

| Solvent | Wavelength of maximum red absorption peak (nm)* | | | | | |
|---|---|---|---|---|---|---|
| | Chl $a$ | Chl $b$ | Chl $c$ | Chl $c_1$ | Chl $c_2$ | Chl $d$ |
| 80–90% Acetone | 663[1] (84·0) | 645[1] (51·8) | 630[2] (19·5) | 630·6[3] (44·8) | 630·9[3] (40·4) | — |
| 90% Methanol | 665[4] (76·07) | 650[5] (36·4) | 635[2] (15·2) | — | — | — |
| Diethyl Ether | 662[1] (100·9) | 644[1] (62·0) | 628[1] (15·8) | — | — | 688[1] (110·4) |
| N,N-dimethyl formamide | 665[6] (72·114) | — | — | — | — | — |

* Figures in brackets are specific absorption coefficients (l.g$^{-1}$ cm$^{-1}$).
References: (*1*) Smith and Benitez, 1955; (*2*) Jeffrey, 1963; (*3*) Jeffrey, 1972, plus 1% pyridine; (*4*) Marker, 1972. Also reported in 100% methanol as 74·5 at 665nm (Mackinney, 1941); (*5*) Mackinney, 1941; (*6*) Volk and Bishop, 1968.

extracts are often turbid; if the turbidity is excessive transfer to diethyl ether is recommended. In order to facilitate estimation and circumvent time-consuming separation, a number of equations for determining chlorophyll concentrations have been derived from the specific absorption coefficients of chlorophylls in mixed samples. The most popular equations for chlorophylls $a$ and $b$ have been derived from the specific absorption coefficients of Mackinney (1941) which are probably low (S. W. Jeffrey *pers. comm.*). A re-analysis of the specific absorption coefficients of purified chlorophylls $a$, $b$, $c_1$ and $c_2$ in various solvents has been completed and new equations for mixed samples are forthcoming (S. W. Jeffrey *pers. comm.*, but see Jeffrey 1972). The following equations are now most frequently used for Chl $a$ and Chl $b$ in 90 per cent methanol (Holden 1965):

$$\text{total Chl} \quad (\text{mgl}^{-1}) = 25 \cdot 5 D_{650nm} + 4 \cdot 0 D_{665nm}$$
$$\text{Chl } a \ (\text{mgl}^{-1}) = 16 \cdot 5 D_{665nm} - 8 \cdot 3 D_{650nm}$$
$$\text{Chl } b \ (\text{mgl}^{-1}) = 33 \cdot 8 D_{650nm} - 12 \cdot 5 D_{665nm}$$

80 to 90 per cent acetone (Arnon 1949):

$$\text{total Chl} \quad (\text{mgl}^{-1}) = 20 \cdot 2 D_{645nm} + 8 \cdot 02 D_{663nm}$$
$$\text{Chl } a \ (\text{mgl}^{-1}) = 12 \cdot 7 D_{663nm} - 2 \cdot 69 D_{645nm}$$
$$\text{Chl } b \ (\text{mgl}^{-1}) = 22 \cdot 9 D_{645nm} - 4 \cdot 64 D_{663nm}$$

diethyl ether (Strain, Cope & Svec 1971):

$$\text{total Chl} \quad (\text{mgl}^{-1}) = 7 \cdot 12 D_{660nm} + 16 \cdot 8 D_{642 \cdot 5nm}$$
$$\text{Chl } a \ (\text{mgl}^{-1}) = 9 \cdot 92 D_{660nm} - 0 \cdot 77 D_{642 \cdot 5nm}$$
$$\text{Chl } b \ (\text{mgl}^{-1}) = 17 \cdot 60 D_{642 \cdot 5nm} - 2 \cdot 81 D_{660nm}$$

A method for determining Chl $c$ concentrations which utilizes absorbancy changes at 450nm on conversion to pheophytin $c$ by acidification was developed by Parsons (1963). Ricketts (1967) discussed the difficulties involved with this method. Chl $c$ concentration may also be reasonably estimated using the trichromatic equations calculated by Parsons and Strickland (1963). These equations are the most commonly applied in analysis of Chl $a$, $b$ and $c$ concentrations from phytoplankton (see Strickland & Parsons 1968). This method is most generally used despite possible errors which may arise due to alteration products (Strickland 1972). Marker (1972) has discussed problems due to alteration products which may arise in the analysis of pigments from mixed planktonic and soil algae. The Parsons and Strickland (1963) equations in 90 per cent acetone are:

$$\text{Chl } a \ (\text{mgl}^{-1}) = 11 \cdot 6 D_{665nm} - 0 \cdot 14 D_{630nm} - 1 \cdot 31 D_{645nm}$$
$$\text{Chl } b \ (\text{mgl}^{-1}) = 20 \cdot 7 D_{645nm} - 4 \cdot 34 D_{665nm} - 4 \cdot 42 D_{630nm}$$
$$\text{Chl } c \ (\text{mgl}^{-1}) = 55 \cdot 0 D_{630nm} - 16 \cdot 30 D_{645nm} - 4 \cdot 64 D_{665nm}$$

Smith and Benitez (1955) gave the following equation for Chl $a$ and Chl $d$ in diethyl ether:

$$\text{Chl } a \ (\text{mgl}^{-1}) = 9 \cdot 93 D_{663nm} - 1 \cdot 020 D_{688nm}$$
$$\text{Chl } d \ (\text{mgl}^{-1}) = 9 \cdot 07 D_{688nm} - 0 \cdot 135 D_{663nm}$$

# 4  BIOSYNTHESIS OF CHLOROPHYLLS

## 4.1  *The biosynthetic chain*

The general biosynthetic chain of chlorophyll *a* formation is fairly well understood and is shown in Fig. 5.2. The evidence for this pathway has been detailed
by Bogorad (1966). Kirk (1970) discussed in depth the origins of δ-aminolevulinic
acid (ALA), Granick (1971) has outlined the work with *Chlorella* mutants blocked
in various synthetic steps, and Goodwin (1971) has recently reviewed chlorophyll
biosynthesis in plants in general. Current concepts concerning the biosynthesis
of chlorophyll *b* have been reviewed by Shlyk (1971), who suggests a sequential
oxidative step via an alcohol from Chl *a* to Chl *b* rather than a parallel synthetic
pathway. Details of the origins of the unphylated *c* type chlorophylls and of
Chl *d* are not presently known. Since the general synthetic pathway is presumed
similar in algae and higher plants, the enzymology of chlorophyll biosynthesis
will not be discussed here and the reader should consult the above reviews.

  However, algae in general differ from higher plants in that most higher plants
grown in the dark accumulate protochlorophyllide (PChl) and the conversion to
Chl *a* is a photodependent process involving several spectral changes (Shibata
shift transitions) (see Shibata 1957, Boardman 1966, Kirk 1970, Goodwin 1971).
On the other hand most, but not all, algae synthesize chlorophyll while growing
heterotrophically in the dark and yet little is known about this dark, or non-
photosensitized reduction of PChl. Algae that require light for chlorophyll
biosynthesis include strains of *Euglena*, *Ochromonas* and *Cyanidium caldarium*;
mutants of *Chlamydomonas*, *Chlorella* and *Scenedesmus*; and glucose grown
*Chlorella protothecoides*. The photoreceptor for the conversion in most algae and
higher plants appears to be protein bound protochlorophyllide (holochrome)
(Goodwin 1971). Dark grown *Euglena gracilis* accumulates PChl-635 (635nm
absorption maximum) which, upon illumination, is reduced to Chl-673 without
the Shibata shift transitions (Butler & Briggs 1966). Dark grown mutants of
*Chlorella* accumulate PChl absorbing at 632 nm (Bryan & Bogorad 1963) to
625nm (Bryan *et al.* 1967). From the kinetics of Chl *a* synthesis by dark grown
*Chlamydomonas reinhardtii y-l*, Ohad *et al.* (1967a,b) assume that low levels
of an unidentified PChl accumulate. Spectroscopic changes have not been
reported during the greening of these mutants.

  Not all dark grown algae accumulate PChl. Glucose bleached *Chlorella
protothecoides* contains no detectable PChl in the presence or absence of exogenous ALA but accumulates small quantities of chlorophyll(ide) (Ochiai &
Hase 1970). Blue light rather than red light is also more effective in enhancing
chlorophyll formation in these cells. A recently isolated mutant of *Scenedesmus
obliquus* forms only traces of chlorophyll when grown heterotrophically and
does not appear to accumulate PChl (Senger & Bishop 1972). The initial stages
of greening in this mutant are not dependent on the typical photoconversion of

PChl. The photoreceptor, as in glucose bleached *C. protothecoides*, appears to be a carotenoid or flavoprotein rather than PChl.

## 4.2 *Control of chlorophyll biosynthesis and plastid development*

Primary control of the chlorophyll and haem biosynthetic pathway in algae is generally assumed to occur at the level of ALA formation (Bogorad 1966, Granick 1967, Goodwin 1971). Organisms which do not form chlorophyll in the dark accumulate only small quantities of PChl ($10^{-3}$ of the normal Chl content) unless supplied with exogenous ALA (Granick 1967). Concurrent with the observable spectroscopic shifts at the onset of illumination, a lag phase of varying duration occurs prior to a constant rate of chlorophyll accumulation. Alternating light and dark periods during the lag phase and the kinetics of initial synthesis both suggest the necessity of ALA synthesis (Granick 1967). Phytochrome may be involved in regulating the lag phase and the rate of chlorophyll synthesis in higher plants (De Greef *et al.* 1971, Masoner *et al.* 1972) but although the lag phase in *Euglena* may be reduced by a brief period of pre-illumination (Holowinsky & Schiff 1970) there is no evidence that phytochrome is involved (Boutin & Klein 1972). The phytochrome control in higher plants is inhibited by chloramphenicol, an inhibitor of organelle ribosomes (Beridze *et al.* 1966). Bogorad (1967) suggests that the phytochrome response and inhibitor sensitivity reflect the constant need to synthesize ALA synthetase. Beale (1971) concludes that the enzyme responsible for the synthesis of ALA in *Chlorella* has a half life of about 30 minutes. Goldberg (1971, cited by Ohad *et al.* 1972) however, calculates the half life of the RNA and enzyme system forming ALA in *C. reinhardtii y*-1 to be about 90 minutes.

The regulation of chlorophyll biosynthesis is additionally complex since the accumulation of the pigment appears to be intimately associated with thylakoid development and photosynthetic activity. Chloroplast biogenesis has been studied extensively in *Euglena* (see Schiff 1971), in wild type and mutant strains of *C. reinhardtii* (see Goodenough *et al.* 1971, Ohad *et al.* 1972), and in glucose bleached *C. protothecoides* (Hase 1971).

Much work has been done on the effects of inhibition of protein synthesis on the control of chloroplast development and on chlorophyll biosynthesis, but the results must be interpreted cautiously since the degree of inhibition is frequently concentration dependent (Smillie *et al.* 1971). Chlorophyll biosynthesis is generally reported to be sensitive to inhibitors of both cytoplasmic (cycloheximide) and organelle (chloramphenicol) ribosomes (Kirk & Tilney-Bassett 1967). Cycloheximide completely inhibits chlorophyll synthesis in *C. reinhardtii* whereas chloramphenicol, spectinomycin, and rifampicin, differing inhibitors of organelle and prokaryote RNA translation and transcription, produce less than 50 per cent inhibition (Surzycki *et al.* 1970, Eytan & Ohad 1970, 1972, Jennings & Ohad 1972). In a careful study and review of protein synthesis inhibitors, Surzycki *et al.* (1970) concluded, like Ohad *et al.* (1972), that chlorophyll biosynthesis is not a chloroplast directed activity in *C. reinhardtii*. Schiff (1971) has tentatively come to the same conclusion with respect to the initial steps in porphyrin biosynthesis in *Euglena*. Interaction between the plastid and cytoplasm is well illustrated in *C. reinhardtii y*-1 mutant where continuous illumination is necessary for cytoplasmic directed synthesis of a structural protein(s) and continued chlorophyll synthesis (Eytan & Ohad 1972, Hoober & Stegeman 1973). Cytoplasmic synthesis of the structural protein(s) may be controlled at the transcriptional level by the intra-plastid conversion of PChl to Chl (Eytan & Ohad 1972) and the control mediated by a protein synthesized on choroplast ribosomes (Hoober & Stegeman 1973).

The general trend thus emerging in some eukaryotic algae is that chlorophyll biosynthesis, although it may occur in the plastids (Carell & Kahn 1964, Rebeiz & Castelfranco 1971), is initially under nuclear control (Surzycki *et al.* 1970, Ohad *et al.* 1972) and that later steps (Schiff 1971) require cooperation and interaction between nuclear and plastid genomes for the control and synthesis of both the apoprotein of protochlorophyllide holochromes (Goodwin 1971) and the structural components of the photosynthetic apparatus (Hase 1971, Eytan & Ohad 1972, Hoober & Stegemen 1973).

## 5 CONDITIONS AFFECTING CHLOROPHYLL CONTENT

### 5.1 *Nutrition*

Mineral nutrition affects a number of growth and metabolic parameters including the chlorophyll content of algae (see Chapter 22, p. 610). Predictably, deficiencies of iron, nitrogen and magnesium, essential constituents of haem and chlorophyll, have pronounced effects on chlorophyll synthesis and content (Kirk & Tilney-Basset 1967, O'Kelley 1968). However, the blue-green alga *Anacystis nidulans* maintained normal levels of Chl *a* for a while but rapidly lost phycocyanin under conditions of nitrogen starvation (Allen & Smith 1969).

Utilizable sources of organic carbon or energy in the growth medium in the light retard chlorophyll formation by some algae. For example, acetate reversibly inhibits chlorophyll synthesis in *Euglena gracilis* (App & Jagendorf 1963, Buetow 1967), *Chlorella variegata* (Fuller & Gibbs 1959) and *Golenkinia* sp. (Ellis 1970); previous growth in ethanol inhibits the light induced synthesis in *E. gracilis* (Kirk & Keylock 1967), and glucose inhibits light or dark synthesis in *Chlorella protothecoides* (Ochiai & Hase 1970). *Chlamydomonas stellata*, when grown photoheterotrophically on acetate, loses photosystem II activity (Merrett 1969) and forms little Chl *b* but contains the same amount of Chl *a* as photoautotrophically grown cells (Wiessner & French 1970). However, the relative proportions of the *in vivo* forms of Chl *a* in *C. stellata* appear to vary with photoheterotrophic growth conditions (Wiessner & French 1970). Hutch (1967) indicates that there are differing glucose effects on chlorophyll formation under heterotrophic (dark plus glucose), mixotrophic (light plus glucose) and photoautotrophic (light minus glucose) growth conditions depending on the organism. Mixotrophic conditions do not inhibit chlorophyll synthesis in *Chlorella pyrenoidosa*, *E. gracilis* or *Chlamydomonas eugametos*, but do so in

*Chlorella luteoviridis*. However, *C. luteoviridis* grew poorly under mixotrophic conditions. There is no chlorophyll synthesis in *E. gracilis* or *C. eugametos* under dark heterotrophic conditions. However, *C. eugametos* does not grow heterotrophically. *C. luteoviridis*, which shows poor heterotrophic growth, forms only small quantities of chlorophyll. In general, putative obligate phototrophs, such as certain Cyanophyceae, contain similar amounts of chlorophyll when grown photoheterotrophically (dim light) or dark heterotrophically as when grown photoautotrophically (Hoare *et al.* 1971). An exception is *Plectonema boryanum* which grows heterotrophically in the dark on glucose in a continuous dialysis apparatus (Pan 1972). Under these conditions the alga is unable to synthesize Chl *a* or phycobilins although Chl *a* accumulates when the filaments are returned to the light. Hase (1971) found that the ratio of glucose to nitrogen source in the medium regulates the cell type of *Chlorella protothecoides* produced. Glucose has a dual suppressive effect on chlorophyll accumulation, one being the loss of pigment at high concentrations and the other repression of synthesis. Disintegration of plastid ribosomes accompanies the loss of chlorophyll. Chlorophyll is resynthesized in the bleached cells in the dark (slowly) and light in the presence of a nitrogen source and in the complete, or near, absence of glucose. However, low concentrations of glucose, in the absence of a nitrogen source, inhibit the greening process by interfering with synthetic reactions prior to ALA formation and the cyotplasmic assembly of proteins essential for maintenance of plastid structure.

## 5.2   *Light intensity, temperature and cell age*

Within certain limits, the chlorophyll content of numerous algae is inversely proportional to the light intensity during growth (see e.g. Kirk & Tilney-Bassett 1967, Brown & Richardson 1968, Sheridan 1972). Beale and Appleman (1971) investigated the physiological basis of light regulation of chlorophyll biosynthesis in *Chlorella vulgaris*. Their findings suggest that the degree of light limitation of growth is the primary controlling factor. When light is limiting for growth the chlorophyll content increases and it decreases when light is not limiting. Regulation was hypothesized as being due to the accumulation of a photosynthetic product under non-limiting light conditions which in some manner inhibited chlorophyll synthesis. Photo-destruction of chlorophyll may occur in *Chlorella* during prolonged exposure to high light intensities (Kok 1956). A similar situation occurs with a thermophilic alga during exposures to subminimal growth temperatures (Castenholz 1972).

    Cross gradients of light intensities and growth temperatures indicate that moderate light intensities (3,000 lux) and optimal growth temperatures (30 to 40°C) result in maximal chlorophyll concentrations in *Anacystis nidulans* (Halldal & French 1958). Although prolonged exposure to supra-optimal temperatures (34·5°C) in *E. gracilis* results in bleached cells, chlorophyll synthesis at these temperatures is only 25 per cent lower by dark-grown cells taken from

optimal temperatures (Kirk & Tilney-Bassett 1967). Suboptimal growth temperatures for thermophilic algae result in lower chlorophyll concentrations than those obtained at optimal temperatures (Castenholz 1969).

Synchronizing cell division in *A. nidulans* by a temperature cycle of 8 hours at 26°C and 6 hours at 32°C in continuous light results in cessation of chlorophyll synthesis during exposure to the suboptimal temperature (26°C), but synthesis begins again after the shift to optimal temperature (32°C) (Venkataraman & Lorenzen 1969). The chlorophyll content of light-dark synchronized *Chlorella ellipsoidea* cells varies during the cell cycle (Tamiya 1966). The percentage of chlorophyll per cell is lowest immediately after the dark division of the mother cells, increases to a maximum as the young cells grow in the light, and then decreases gradually as these cells mature prior to cell division (light followed by dark). *E. gracilis* cells synchronized in a 16 and 8 hour light-dark cycle synthesize chlorophyll at a constant rate in the light but reach a plateau 2 hours prior to the dark period (Edmunds 1965). Following an initial lag phase, light-dark synchronized *Scenedesmus obliquus* cells synthesize chlorophyll at a constant rate in the light up to the dark period (Senger & Bishop 1969). Although randomly dividing *S. obliquus* forms chlorophylls in continuous darkness (Senger & Bishop 1972), it is of interest that synthesis stops when synchronized cultures are in the dark phase (Senger & Bishop 1969) (see also Chapter 32, p. 894).

# 6 REFERENCES

ALLEN, M.B. (1959) Studies with *Cyanidium caldarium* an anomalously pigmented chlorophyte. *Arch. Mikrobiol.* **32**, 270–7.

ALLEN M.B. (1966) Distribution of the chlorophylls. In *The Chlorophylls*, eds. Vernon L.P. & Seely G.R., pp. 511–19. Academic Press, New York & London.

ALLEN M.B. & SMITH A.J. (1969) Nitrogen chlorosis in blue-green algae. *Arch. Mikrobiol.* **69**, 114–20.

APP A. & JAGENDORF A.T. (1963) Repression of chloroplast development in *Euglena gracilis* by substrates. *J. Protozool.* **10**, 340–3.

ARNON D.I. (1949) Copper enzymes in isolated chloroplasts. Polyphenoloxidase in *Beta vulgaris. Pl. Physiol.*, Lancaster, **24**, 1–15.

BEALE S.I. (1971) Studies on the biosynthesis and metabolism of δ-aminolevulinic acid in *Chlorella. Pl. Physiol.*, Lancaster, **48**, 316–19.

BEALE S.I. & APPLEMAN D. (1971) Chlorophyll synthesis in *Chlorella*. Regulation by degree of light limitation of growth. *Pl. Physiol.*, Lancaster, **47**, 230–5.

BERIDZE G., ODINTSOVA M.S., CHERKASHINA N.A. & SIDDAKIAN N.M. (1966) The effect of nucleic acid synthesis inhibitors on the chlorophyll formation by etiolated bean leaves. *Biochem. Biophys. Res. Commun.* **23**, 683–9.

BISHOP N.I. (1971) Preparation and properties of mutants: *Scenedesmus*. In *Methods in Enzymology*, ed. San Pietro A., pp. 130–43, Vol. XXIII A. Academic Press, New York & London.

BOARDMAN N.K. (1966) Protochlorophyll. In *The Chlorophylls*, eds. Vernon L.P. & Seely G.R., pp. 437–79. Academic Press, New York & London.

BOGORAD L. (1966) The biosynthesis of chlorophylls. In *The Chlorophylls*, eds. Vernon L.P. & Seely G.R., pp. 481–510. Academic Press, New York & London.

BOGORAD L. (1967) Biosynthesis and morphogenesis in plastids. In *Biochemistry of Chloroplasts*, Vol. II, ed. Goodwin T.W., pp. 615–32. Academic Press, New York & London.

BOUTIN M.E. & KLEIN R.M. (1972) Absence of phytochrome participation in chlorophyll synthesis in *Euglena*. *Pl. Physiol., Lancaster* **49**, 656–7.

BROWN T.E. & RICHARDSON F.L. (1968) The effect of growth environment on the physiology of algae: light intensity. *J. Phycol.* **4**, 38–54.

BRYAN G.W. & BOGORAD L. (1963) Protochlorophyll and chlorophyll formation in response to iron nutrition in a *Chlorella* mutant. In *Studies on Microalgae and Photosynthetic Bacteria*. Japan Soc. Plant Physiol., pp. 399–405. Tokyo University Press, Tokyo.

BRYAN G.W., ZADYLAK A.H. & EHRET C.F. (1967) Photoinduction of plastids and of chlorophyll in a *Chlorella* mutant. *J. Cell Sci.* **2**, 513–28.

BUETOW D.E. (1967) Acetate repression of chlorophyll in *Euglena gracilis*. *Nature, Lond.* **213**, 1127–8.

BUTLER W.L. & BRIGGS W.R. (1966) The relation between structure and pigments during the first stages of proplastid greening. *Biochim. biophys. Acta* **112**, 45–53.

CARELL E.F. & KAHN J.S. (1964) Synthesis of porphyrins by isolated chloroplasts of *Euglena*. *Archs. Biochem.* **108**, 1–6.

CASTENHOLZ R.W. (1969) Thermophilic blue-green algae and the thermal environment. *Bact. Revs.* **33**, 476–504.

CASTENHOLZ R.W. (1972) Low temperature acclimation and survival in thermophilic *Oscillatoria terebriformis*. In *Taxonomy and Biology of Blue-Green Algae*, ed. Desikachary T.V., pp. 406–18. Univ. of Madras, Madras.

CAUBERG R. & DE GREEF J.A. (1971) Pigment studies in the genus *Vaucheria*. *Bot. Mag. Tokyo* **84**, 222–30.

DE GREEF J.A. & CAUBERG R. (1970) Chlorophyll *c* in *Vaucheria*. *Naturwissenschaften* **12**, 673–4.

DE GREEF J., BUTLER W.L. & ROTH T.F. (1971) Greening of etiolated bean leaves in far red light. *Pl. Physiol., Lancaster* **47**, 457–64.

DOUGHETRY R.C., STRAIN H.H., SVEC W.A., UPHAUS R.A. & KATZ J.J. (1970) The structure, properties and distribution of chlorophyll *c*. *J. Amer. chem. Soc.* **92**, 2826–33.

EDMUNDS L.N. JR. (1965) Studies on synchronously dividing cultures of *Euglena gracilis* Klebs (Strain Z) II. Patterns of biosynthesis during the cell cycle. *J. Cell and Comp. Physiol.* **66**, 159–82.

ELLIS R.J. (1970) Effects of acetate on the growth and chlorophyll content of *Golenkinia*. *J. Phycol.* **6**, 364–8.

EYTAN G. & OHAD I. (1970) Biogenesis of chloroplast membranes VI. Co-operation between cytoplasmic and chloroplast ribosomes in the synthesis of photosynthetic lamellar proteins during the greening process in a mutant of *Chlamydomonas reinhardi y*-1. *J. biol. Chem.* **245**, 4297–07.

EYTAN G. & OHAD I. (1972) Biogenesis of chloroplast membranes VIII. Modulation of chloroplast lamellae composition and function induced by discontinuous illumination and inhibition of ribonucleic acid and protein synthesis during greening of *Chlamydomonas reinhardi y*-1 mutant cells *J. biol. Chem.* **247**, 122–9.

FORK D.C. (1963) Observations on the function of chlorophyll *a* and accessory pigments in photosynthesis. In *Photosynthetic Mechanisms of Green Plants*, eds. Kok B. & Jagendorf A.T., pp. 352–61, N.A.S.-N.R.C. Washington DC, 1145.

FULLER R. & GIBBS M. (1959) Intracellular and phylogenetic distribution of ribulose-1,5-diphosphate carboxylase and D-glyceraldehyde-3-phosphate dehydrogenases. *Pl. Physiol., Lancaster* **34**, 324–9.

GOEDHEER J.C. (1966) Visible absorption and fluorescence of chlorophyll and its aggregates in solution. In *The Chlorophylls*, eds. Vernon L.P. & Seely G.R., pp. 147–85. Academic Press, New York & London.

GOEDHFER J.C. (1970) On the pigment system of brown algae. *Photosynthetica* **4**, 97–106.

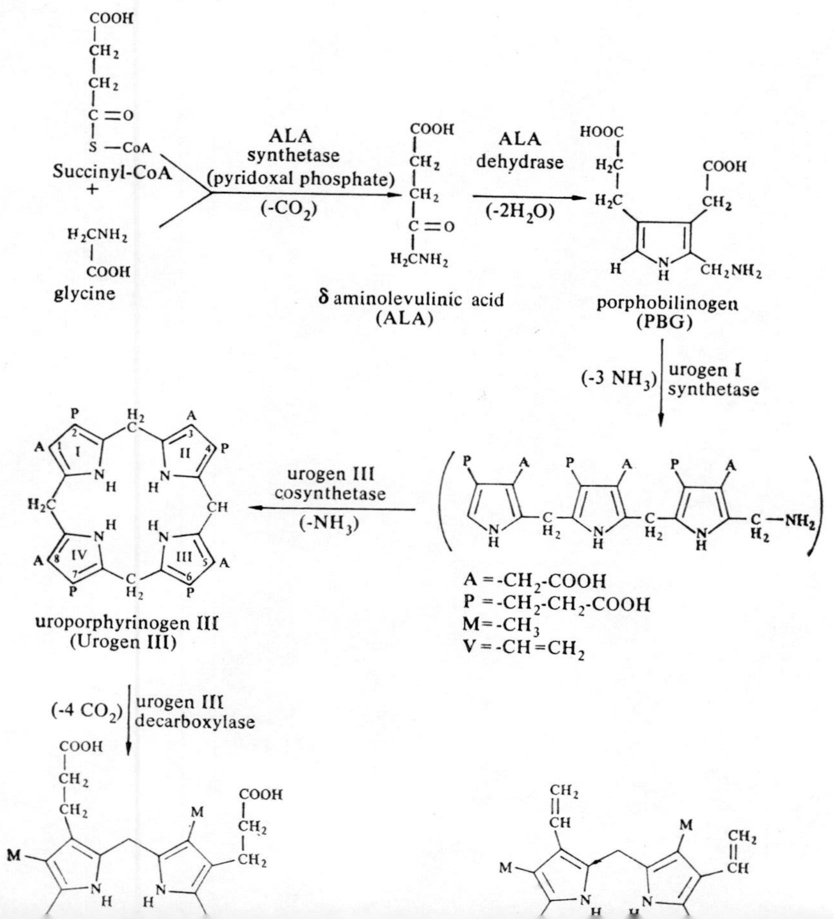

COOH
|
CH₂
|
CH₂
|
C=O
|
S—CoA

Succinyl-CoA

+

H₂CNH₂
|
COOH

glycine

ALA
synthetase
(pyridoxal phosphate)

(-CO₂)

COOH
|
CH₂
|
CH₂
|
C=O
|
CH₂
|
H₂CNH₂

δ aminolevulinic acid
(ALA)

ALA
dehydrase

(-2H₂O)

porphobilinogen
(PBG)

(-3 NH₃) urogen I
synthetase

A =-CH₂-COOH
P =-CH₂-CH₂-COOH
M=-CH₃
V =-CH=CH₂

urogen III
cosynthetase

(-NH₃)

uroporphyrinogen III
(Urogen III)

(-4 CO₂) urogen III
decarboxylase

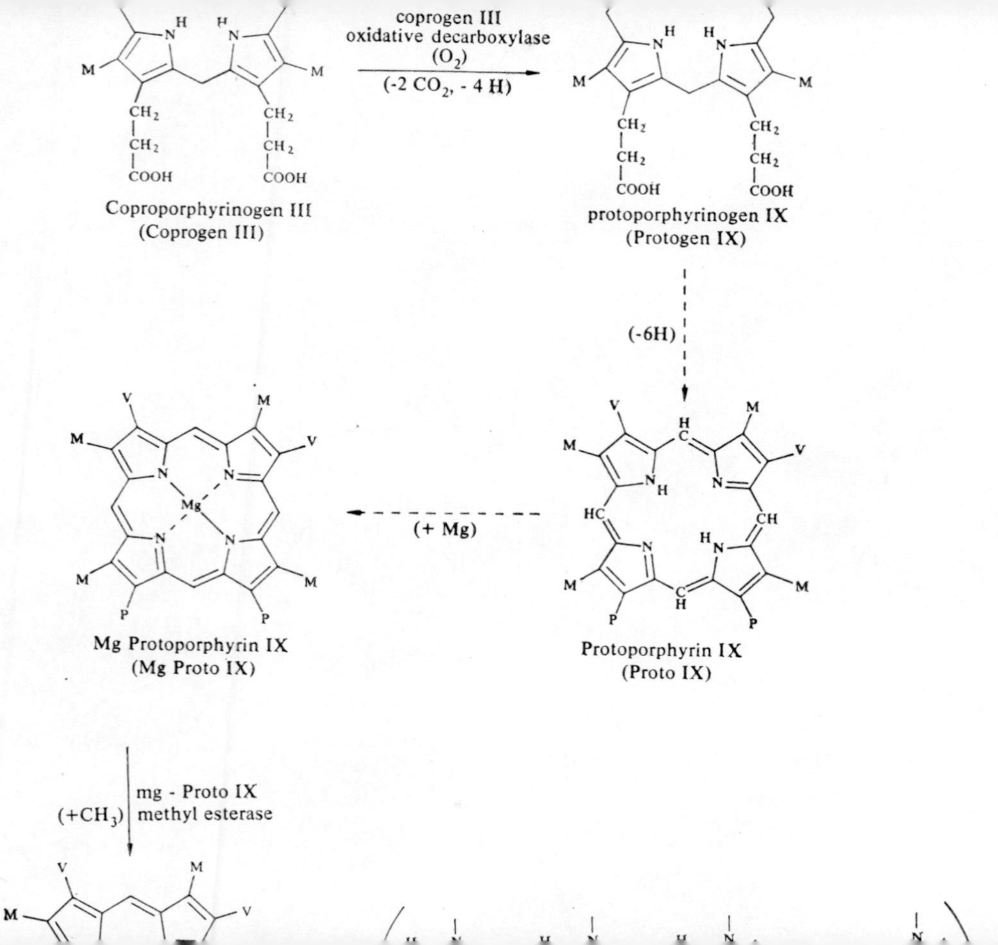

protoporphyrinogen IX
(Protogen IX)

Coproporphyrinogen III
(Coprogen III)

coprogen III
oxidative decarboxylase
$(O_2)$
$(-2\ CO_2,\ -4\ H)$

$(-6H)$

Protoporphyrin IX
(Proto IX)

$(+ Mg)$

Mg Protoporphyrin IX
(Mg Proto IX)

mg - Proto IX
methyl esterase
$(+CH_3)$

M
$CH_2$
$CH_2$—$CH_2$—COOH
$CH_2$—$CH_2$—COOH
M

N
H
H
N

CH
V
M
P
N
H
N
V
M
P
HC
M
M
HC

N
Mg
N
N
N
V
M
P
M
M
V

M
V
M
V

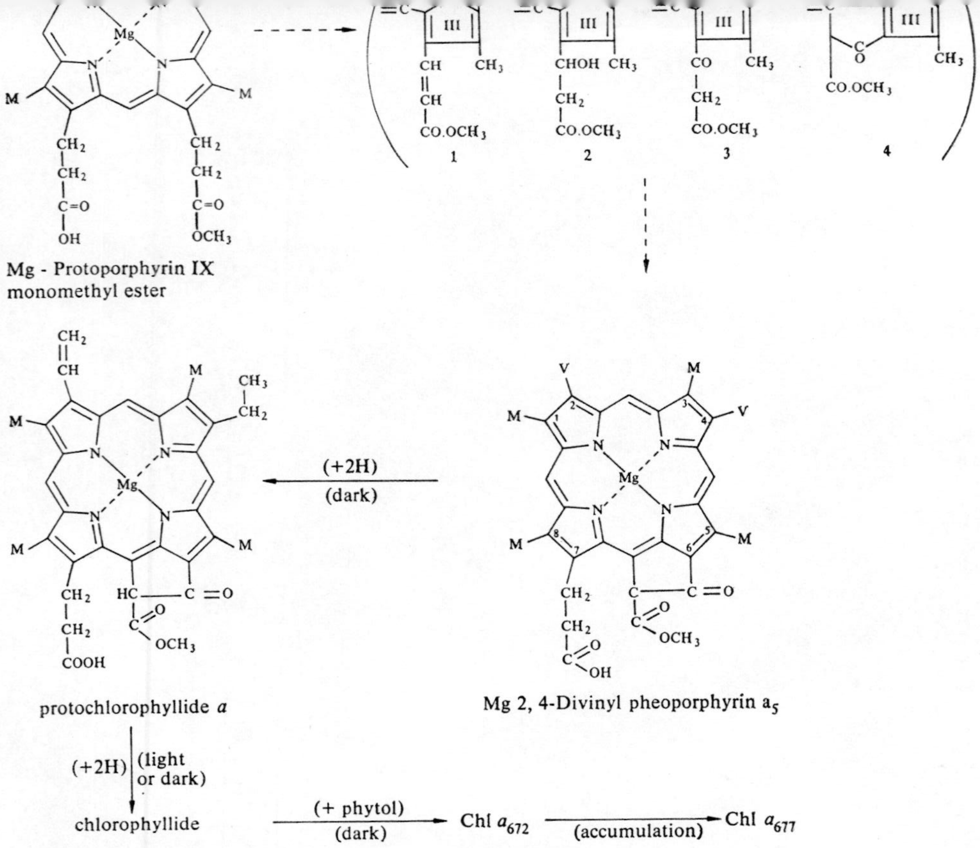

**Fig. 5.2.** Pathway of chlorophyll biosynthesis (after Bogorad 1966).

GOODENOUGH U.W., TOGASAKI R.K., PASZEWSKI A. & LEVINE R.P. (1971) Inhibition of chloroplast ribosome formation by gene mutation in *Chlamydomonas reinhardi*. In *Autonomy and Biogenesis of Mitochondria and Chloroplasts*, eds. Boardman N.K., Linnane A.W. & Smillie R.M., pp. 224–34. North Holland, Amsterdam.

GOODWIN T.W. (1971) Biosynthesis by chloroplasts. In *Structure and Function of Chloroplasts*, ed. Gibbs M., pp. 215–76. Springer-Verlag, Berlin, Heidelberg & New York.

GRANICK S. (1967) The heme and chlorophyll biosynthetic chain. In *Biochemistry of Chloroplasts*. Vol. II, ed. Goodwin T.W., pp. 373–410. Academic Press, New York & London.

GRANICK S. (1971) Preparations and properties of *Chlorella* mutants in chlorophyll bio-synthesis. In *Methods in Enzymology*, XXIII, A. ed. San Pietro A., pp. 162–8. Academic Press, New York & London.

GUILLARD R.R.L. & LORENZEN C.J. (1972) Yellow-green algae with chlorophyllide *c*. *J. Phycol.* **8**, 10–14.

HALLDAL P. & FRENCH C.S. (1968) Algal growth in crossed gradients of light intensity and temperatures. *Pl. Physiol., Lancaster* **33**, 249–52.

HASE E. (1971) Studies on the metabolism of nucleic acid and protein associated with the process of de- and re-generation of chloroplasts in *Chlorella prototothecoides*. In *Autonomy and Biogenesis of Mitochondria and Chloroplasts*, eds. Boardman N.K., Linnane A.W. & Smillie R.M., pp. 434–46. North Holland, Amsterdam.

HIBBERD D.J. & LEEDALE G.F. (1970) Eustigmatophyceae—a new algal class with unique organization of the motile cell. *Nature, Lond.* **225**, 758–60.

HIBBERD D.J. & LEEDALE G.F. (1972) Cytology and ultrastructure of Eustigmatophyceae. *Ann. Bot.* **36**, 49–71.

HOARE D.S., INGRAM L.O., THURSTON E.L. & WALKUP R. (1971) Dark heterotrophic growth of an endophytic blue-green alga. *Arch. Mikrobiol.* **78**, 310–21.

HOLDEN M. (1965) Chlorophylls. In *Chemistry and Biochemistry of Plant Pigments*, ed. Goodwin T.W., pp. 461–88. Academic Press, New York & London.

HOLOWINSKY A. & SCHIFF J.A. (1970) Events surrounding the early development of *Euglena* chloroplasts I. Induction by pre-illumination. *Pl. Physiol., Lancaster* **45**, 339–47.

HOLT A.S. (1965) Nature, properties and distribution of chlorophylls. In *Chemistry and Biochemistry of Plant Pigments*, ed. Goodwin T.W., pp. 3–28. Academic Press, New York & London.

HOOBER J.K. & STEGEMAN W.J. (1973) Control of the synthesis of a major polypeptide of chloroplast membranes in *Chlamydomonas reinhardi*. *J. Cell Biol.* **56**, 1–12.

HUTCH W. VON (1967) Enzyme in grünen einzellern in abhängigkeit won der kohlenstoffversorgung. *Flora, Jena, Abt A.* **158**, 58–87.

JEFFREY S.W. (1963) Purification and properties of chlorophyll *c* from *Sargassum flavicans*. *Biochem. J.* **86**, 313–18.

JEFFREY S.W. (1969) Properties of two spectrally different components in chlorophyll *c* preparations. *Biochim. biophys. Acta* **177**, 456–67

JEFFREY S.W. (1972) Preparation and some properties of crystalline chlorophyll $c_1$ and $c_2$ from marine algae. *Biochim. biophys. Acta* **279**, 15–33.

JENNINGS R.C. & OHAD I. (1972) Biogenesis of chloroplast membranes XI. Evidence for the translation of extrachloroplast RNA on chloroplast ribosomes in a mutant of *Chlamydomonas reinhardi*, *y*-1. *Archs. Biochem. Biophys.* **153**, 79–87.

KIRK J.T.O. (1970) Biochemical aspects of chloroplast development. *Ann. Rev. Plant Physiol.* **21**, 11–42.

KIRK J.T.O. & KEYLOCK M.J. (1967) Control of chloroplast formation in *Euglena gracilis*: dependence of rate of chlorophyll synthesis on previous nutritional history of cells. *Biochem. Biophys. Res. Commun.* **28**, 927–31.

KIRK J.T.O. & TILNEY-BASSETT R.A.E. (1967) *The Plastids*. W.H. Freeman & Co., London.

KOK, B. (1956) On the inhibition of photosynthesis by intense light. *Biochim. Biophys. Acta* **21**, 234–44.

MACKINNEY G. (1941) Absorption of light by chlorophyll solutions. *J. Biol. Chem.* **140**, 315–22.

MANN J.E. & MYERS J. (1968) Photosynthetic enhancement in the diatom *Phaeodactylum tricornutum*. *Pl. Physiol.*, Lancaster **43**, 1991–3.

MANNING W.M. & STRAIN H.H. (1943) Chlorophyll D, a green pigment of red algae. *J. Biol. Chem.* **151**, 1–19.

MARKER A.F.H. (1972) The use of acetone and methanol in the estimation of chlorophyll in the presence of phaeophytin. *Freshwater Biol.* **2**, 361–85.

MASONER M., UNSER G. & MOHR H. (1972) Accumulation of protochlorophyll and chlorophyll *a* as controlled by photomorphogenically effective light. *Planta* **105**, 267–72.

MERRETT M. (1969) Observations on the fine structure of *Chlamydobotrys stellata* with particular reference to its unusual chloroplast structure. *Arch. Mikrobiol.* **65**, 1–11.

OCHIAI S. & HASE E. (1970) Studies on chlorophyll formation in *Chlorella protothecoides*. I. Enhancing effect of light and added ALA and suppressive effect of glucose on chlorophyll formation. *Plant and Cell Physiol.* **11**, 663–73.

OHAD I., SIEKEVITZ P. & PALADE G.E. (1967a) Biogenesis of chloroplast membranes I. Plastid dedifferentiation in a dark-grown algal mutant (*Chlamydomonas reinhardi*). *J. Cell Biol.* **35**, 521–52.

OHAD I., SIEKEVITZ P. & PALADE G.E. (1967b) Biogenesis of chloroplast membranes II. Plastid differentiation during greening of a dark-grown algal mutant (*Chlamydomonas reinhardi*). *J. Cell Biol.* **35**, 553–84.

OHAD I., EYTAN E., JENNINGS R.C., GOLDBERG I., BAR-NUN S. & WALLACH D. (1972) Biogenesis of chloroplasts membranes in *Chlamydomonas reinhardi*. In *Proceedings of the Second Annual International Congress on Photosynthetic Research*, eds. Forti G., Avron M. & Melandri A., pp. 2563–84. Dr. W. Junk, N.V., The Hague.

O'HEOCHA C. (1971) Pigments of the red algae. *Oceanogr. Mar. Biol. Ann. Rev.* **9**, 61–82.

O'KELLEY J.C. (1968) Mineral nutrition of algae. *A. Rev. Pl. Physiol.* **19**, 89–112.

PAN P. (1972) Growth of a photoautotroph, *Plectonema boryanum*, in the dark on glucose. *Can. J. Microbiol.* **18**, 275–80.

PARKE M. & DIXON P.S. (1968) Check-list of British marine algae—second revision. *J. mar. biol. Ass. U.K.* **48**, 783–832.

PARSONS T.R. (1963) A new method for the micro-determination of chlorophyll *c* in sea water. *J. mar. Res.* **21**, 164–71.

PARSONS T.R. & STRICKLAND J.D.H. (1963) Discussion of spectrophotometric determination of plant pigments, with revised equations for ascertaining chlorophylls and carotenoids. *J. mar. Res.* **21**, 155–63.

RABINOWITCH E.I. (1945) *Photosynthesis and Related Processes*, Vol. I, Interscience Publishers Inc. New York.

RABEIZ C.A. & CASTELFRANCO P.A. (1971) Chlorophyll biosynthesis in a cell-free system from higher plants. *Pl. Physiol.*, Lancaster **47**, 33–7.

RICKETTS T.R. (1965) Chlorophyll *c* in some members of the Chrysophyceae. *Phytochem.* **4**, 725–30.

RICKETTS T.R. (1966) Magnesium 2,4-divinyl-phaeoporphyrin a₅ monomethyl ester, a protochlorophyll-like pigment present in some unicellular flagellates. *Phytochem.* **5**, 223–9.

RICKETTS T.R. (1967) A note on the estimation of chlorophyll *c*. *Phytochem.* **6**, 1353–4.

SCHIFF J.A. (1971) Developmental interactions among cellular components in *Euglena*. In *Autonomy and Biogenesis of Mitochondria and Chloroplasts*, eds. Boardman N.K., Linnane A.W. & Smillie R.M., pp. 98–118. North Holland, Amsterdam.

SENGER H. & BISHOP N.I. (1969) Changes in the photosynthetic apparatus during the synchronous life cycle of *Scenedesmus obliquus*. In *Progress in Photosynthesis Research*, Vol. I, ed. Metzner H., pp. 425–34. Internat. Union Biol. Sci. Tübingen.

SENGER H. & BISHOP H.I. (1972) The development of structure and function in chloroplasts of greening mutants of *Scenedesmus* I. Formation of chlorophyll. *Plant and Cell Physiol.* **13**, 633–49.

SHERIDAN R.P. (1972) A qualitative and quantitative study of plastoquinone A in two thermophilic blue-green algae. *J. Phycol.* **8**, 47–50.

SHIBATA K. (1957) Spectroscopic studies on chlorophyll formation in intact leaves. *J. Biochem.* **44**, 147–73.

SHLYK A.A. (1971) Biosynthesis of chlorophyll *b*. *A. Rev. Pl. Physiol.* **22**, 169–84.

SMILLIE R.M., BISHOP D.G., GIBBONS G.C., GRAHAM D., GRIEVE A.M., RAISON J.K. & REGER B.J. (1971) Determination of the sites of synthesis of proteins and lipids of the chloroplast using chloramphenicol and cycloheximide. In *Autonomy and Biogenesis of Mitochondria and Chloroplasts*, eds. Boardman N.K., Linnane A.W. & Smillie R.M., pp. 422–33. North Holland, Amsterdam.

SMITH J.H.C. & BENITEZ A. (1955) Chlorophylls: analysis in plant materials. In *Modern Methods of Plant Analysis*, eds. Paech K. & Tracey M.V., pp. 142–96. Springer-Verlag, Berlin.

STEEMAN NIELSEN E. (1961) Chlorophyll concentration and the rate of photosynthesis in *Chlorella vulgaris*. *Physiologia. Pl.* **14**, 868–76.

STRAIN H.H. & SVEC W.A. (1966) Extraction, separation, estimation and isolation of the chlorophylls. In *The Chlorophylls*, eds. Vernon L.P. & Seely G.R., pp. 22–66. Academic Press, New York & London.

STRAIN H.H., COPE B.T. & SVEC W.A. (1971) Procedures for the isolation, identification, estimation and investigation of the chlorophylls. In *Methods in Enzymology*, ed. San Pietro A., pp. 452–76, XXIII A. Academic Press, New York & London.

STRAIN H.H., COPE B.T. JR., MCDONALD G.N., SVEC W.A. & KATZ J.J. (1971) Chlorophylls $c_1$ and $c_2$. *Phytochem.* **10**, 1109–14.

STRAIN H.H. & SHERMA H. (1972) Investigations of the chloroplast pigments of higher plants, green algae, and brown algae and their influence upon the invention modifications, and applications of Tsweet's chromatographic method. *J. Chromat.* **73**, 371–97.

STRICKLAND J.D.H. (1972) Research on the marine planktonic food web at the Institute of Marine Resources: a review of the past. *Oceanogr. Mar. Biol. Ann. Rev.* **10**, 349–414.

STRICKLAND J.D.H. & PARSONS T.R. (1968) *A Practical Handbook of Seawater Analysis*. Bull. 167, Fish. Res. Bd. Canada.

SUBBA RAO D.V. & PLATT T. (1969) Optimal extraction conditions of chlorophylls from cultures of five species of marine phytoplankton. *J. Fish. Res. Bd. Canada* **26**, 1625–30.

SURZYCKI S.J., GOODENOUGH U.W., LEVINE R.P. & ARMSTRONG J.J. (1970) Nuclear and chloroplast control of chloroplast structure and function in *Chlamydomonas reinhardi*. In *Control of Organelle Development*, ed. Miller P.L., pp. 13–35. XXIV Symp. S.E.B. Cambridge Univ. Press, London.

TAMIYA H. (1966) Synchronous cultures of algae. *A. Rev. Pl. Physiol.*, **17**, 1–26.

VENKATARAMAN G.S. & LORENZEN H. (1969) Biochemical studies on *Anacystis nidulans* during its synchronous growth. *Arch. Mikrobiol.* **69**, 34–9.

VERNON, L.P. & SEELY G.R. (1966) *The Chlorophylls—Physical, Chemical, and Biological Properties*. Academic Press, New York & London.

VOLK S.L. & BISHOP N.I. (1968) Photosynthetic efficiency of a phycocyanin-less mutant of *Cyanidium*. *Photochem. Photobiol.* **8**, 213–21.

WIESSNER W. & FRENCH C.S. (1970) The forms of native chlorophyll in *Chlamydobotrys stellata* and their changes during adaptation from photo-heterotrophic to autotrophic growth. *Planta* **94**, 78–90.

# CHAPTER 6

# CAROTENOIDS AND BILIPROTEINS

## T. W. GOODWIN

Department of Biochemistry,
University of Liverpool,
P.O. Box 147, Liverpool, U.K.

1    Introduction   176

2    Carotenoids   176
2.1  Nature and nomenclature   176
2.2  Chlorophyceae   178
2.3  Rhodophyceae   183
2.4  Xanthophyceae and Eustigma-
     tophyceae   183
2.5  Chrysophyceae and Hapto-
     phyceae   185
2.6  Phaeophyceae   185
2.7  Bacillariophyceae   185
2.8  Dinophyceae   186
2.9  Cryptophyceae   187
2.10 Euglenophyceae   187
2.11 Cyanophyceae   187

2.12 Non-photosynthetic algae   192

3    Biliproteins   194
3.1  Introduction   194
3.2  Phycoerythrins   196
     (a) General   196
     (b) C-phycoerythrins   196
     (c) R-phycoerythrin and B-
         phycoerythrin   197
3.3  Phycocyanins   197
     (a) C-phycocyanins   197
     (b) Allophycocyanins   198
     (c) R-phycocyanins   199
3.4  Effect of light on phycobilin
     synthesis   199

4    References   200

## 1  INTRODUCTION

In Chapter 4 Bisalputra has listed briefly some of the carotenoids and biliproteins of various algal groups, and how these relate to chloroplast ultrastructure. In this chapter a survey of the chemical composition of the various carotenoids and biliproteins is given.

## 2  CAROTENOIDS

### 2.1  *Nature and nomenclature*

Carotenoids are tetraterpenes, that is they are made up of eight isoprenoid (ip) (branched 5-carbon, $\overset{c\text{-}c\text{-}c\text{-}c}{\underset{c}{\diagup}}$,) units. They can be considered to be formed by the tail to tail condensation of two 20 carbon units, themselves formed by head to tail condensation of four isoprenoid (ip) residues, thus: ipipipippipipipi. The first $C_{40}$ polyene formed biosynthetically is phytoene (1) which is stepwise desaturated to form lycopene (2) which is probably the precursor of all carotenoids found in algae. Hydrocarbon derivatives (carotenes) of lycopene important in

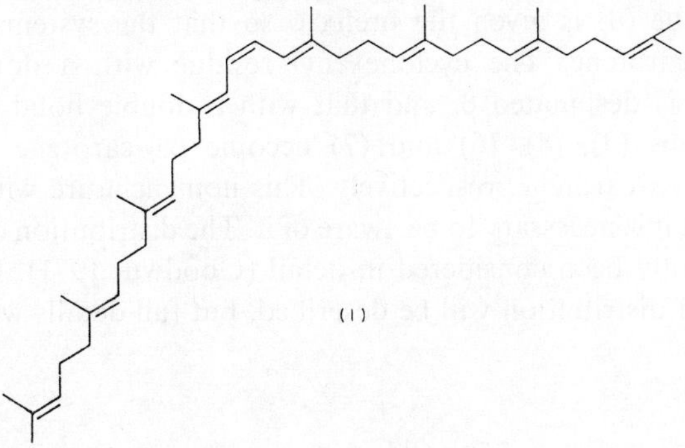

(1)

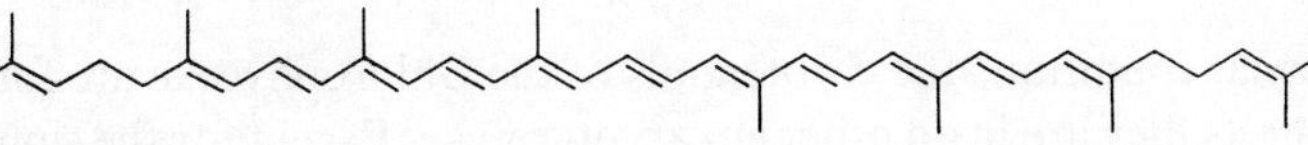

(2)

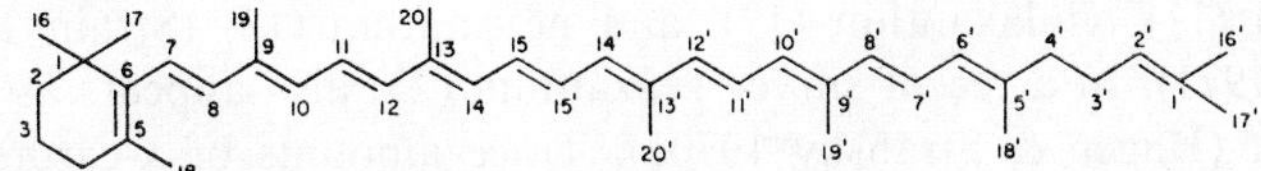

(3)

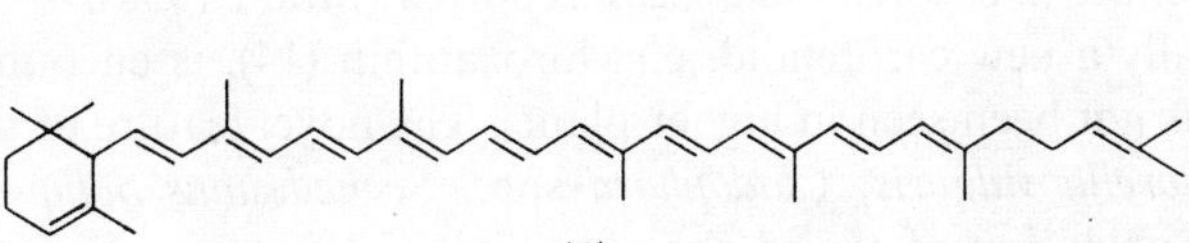

(4)

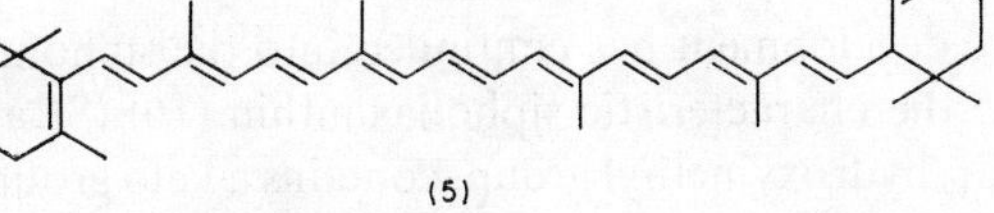

(5)

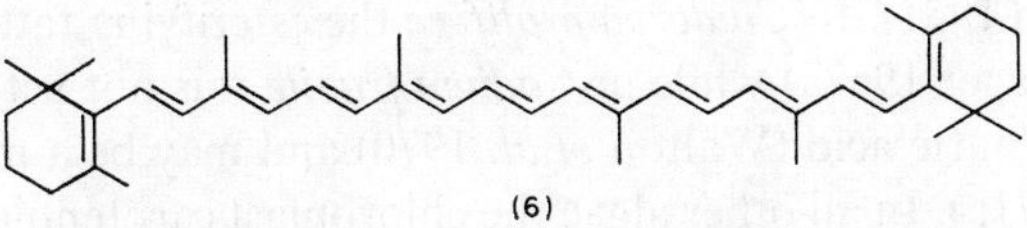

(6)

the present context are γ-carotene (3), δ-carotene (4), α-carotene (5), β-carotene (6) and ε-carotene (7). The structures of all algal carotenoids including the xanthophylls (oxygen-containing carotenoids) can be related to one of these carotenes. The numbering of the carotenoid molecule is indicated for γ-carotene.

Recently the International Union of Pure and Applied Chemistry and the International Union of Physics tentative rules on carotenoid nomenclature have been published. In brief, the 9 carbon end groups are taken as reference points. The acyclic residue (8) is given the prefix ψ so that the systematic name for lycopene is ψ,ψ-carotene. The cyclohexenyl residue with a double bond at position 5, 6 (9) is designated β, and that with a double bond at 4,5 (10) is designated ε. Thus (3), (4), (6) and (7) become β,ψ-carotene, ε,ψ-carotene, β,β-carotene and ε,ε-carotene, respectively. This nomenclature will not be used in this chapter but it is necessary to be aware of it. The distribution of carotenoids in algae has recently been considered in detail (Goodwin 1971). Here only the general pattern of distribution will be described, but full details will be given of newer material.

## 2.2 *Chlorophyceae*

Under normal conditions the carotenoids of the Chlorophyceae are found in the chloroplasts as they are in all other eukaryotic algae. Except for the siphonaceous green algae and Prasinophyceae the quantitative pattern is similar to that found in higher plants, the main pigments being β-carotene (6), lutein (3,3'-dihydroxy-α-carotene) (11), violaxanthin (12), and neoxanthin (13) (Strain 1958, 1966, Goodwin 1971). In a recent survey zeaxanthin (23) also appears as a constant component (Hager & Stransky 1970a). Trace amounts of α-carotene (5) are sporadically encountered as in higher plants, but in one case, *Chlamydomonas agloeformis*, 'much α-carotene' has been reported (Strain 1958).

Occasionally a new carotenoid, e.g. loroxanthin (14), is encountered which up to now has not been seen in higher plants. The novel feature of this pigment, noted in *Chlorella vulgaris*, *Cladophora* spp., *Scenedesmus obliquus* and *Ulva rigida*, is the oxidation of the in-chain methyl (C-19) group to hydroxymethyl (Aitzetmüller *et al.* 1969). Pyrenoxanthin (15), isomeric with loroxanthin, is present in *Chlorella pyrenoidosa* (Yamamoto *et al.* 1969) and the possibility exists that it is identical with loroxanthin.

The oxidative development has continued into the siphonaceous green algae which synthesize the characteristic siphonaxanthin (16) (Strain 1951) which, in addition to the $C_{19}$ hydroxymethyl group, contains a keto group at C-8 (Walton *et al.* 1970, Strain *et al.* 1971a, Ricketts 1971a). Apart from this, another characteristic of these algae is that siphonaxanthin frequently also exists as an ester (siphonein) (Strain 1951). In *Caulerpa prolifera* the esterifying fatty acid is lauric acid (Kleinig & Egger 1967a) while in *Codium fragile* it is not yet identified, but is apparently not lauric acid (Walton *et al.* 1970) and may be a mixture of fatty acids (Ricketts 1971a). In all other algae the chloroplast carotenoids are unesterified. The usual higher plant xanthophylls are present alongside siphonaxanthin and siphonein and in many cases lutein epoxide (17) is also present (Kleinig 1969). With one exception, all contain more α-carotene than β-carotene (Strain 1965).

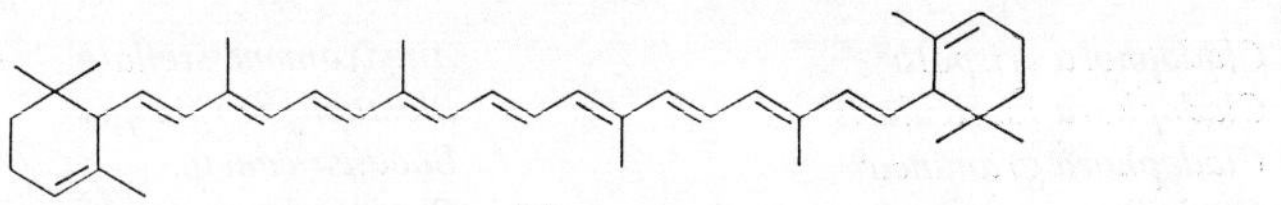

(7)

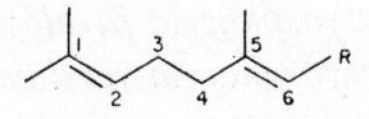

(8) 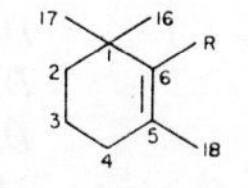  (9) 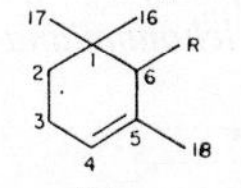  (10)

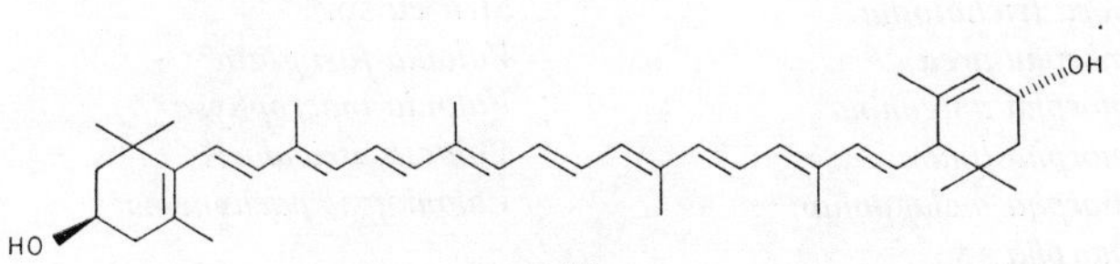

(11)

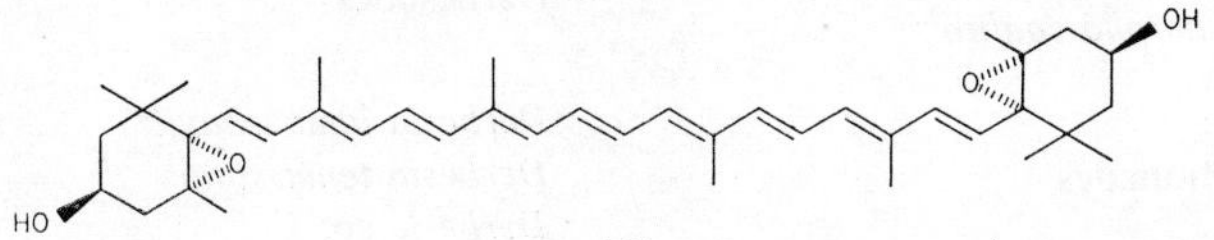

(12)

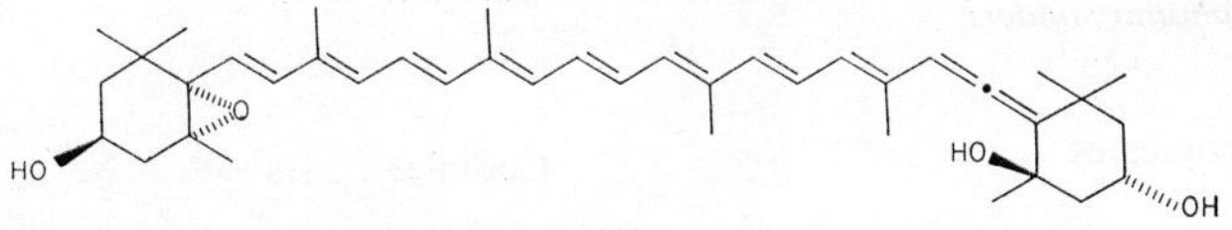

(13)

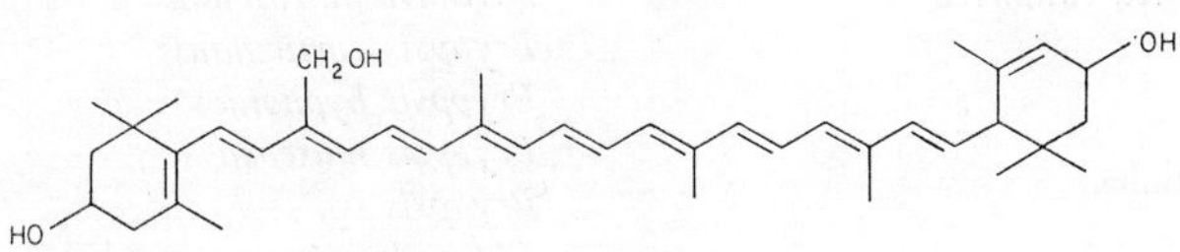

(14)

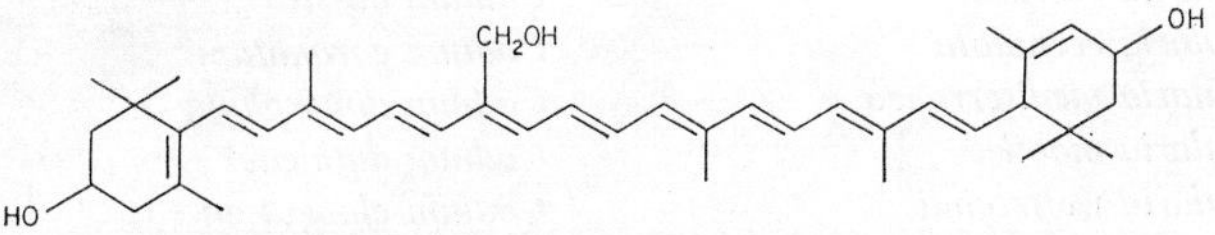

(15)

The exception, *Dichotomosiphon tuberosus*, is the only freshwater specimen so far examined (Strain 1958). Kleinig (1969) carried out a detailed study of the occurrence of siphonaxanthin and siphonein in a wide variety of algae (see Table 6.1). The findings confirm and extend the earlier distribution studies of Strain (1958,

**Table 6.1.** Distribution of siphonaxanthin and siphonein in certain green algae. (From Kleinig 1969, except where otherwise stated *, **).

| Cladophorales | | Siphonocladales | |
|---|---|---|---|
| *Cladophora crispata** | — | *Anadyomene stellata* | — |
| *Cladophora fascicularis** | — | *Blastophysa rhizopus* | 1 |
| *Cladophora graminea** | — | *Boodlea coacta* | 1 |
| *Cladophora membranacea** | — | *Boodlea kaeneana** | — |
| *Cladophora rupestris* | — | *Cladophoropsis herpestica*** | — |
| *Cladophora* cf. *prolifera* | 1 | *Cladophoropsis zollingeri* | — |
| *Cladophora* cf. *lehmanniana* | 1 | *Dictyosphaeria cavernosa* | — |
| *Cladophora* sp. | — | *Dictyosphaeria favulosa** | — |
| *Cladophora* sp.* | — | *Dictyosphaeria versluysii** | — |
| *Cladophora* sp. | 1, 2 | *Microdictyon setchellianum** | — |
| *Cladophora trichotoma* | — | *Struvea* sp.* | — |
| *Chaetomorpha area* | — | *Valonia fastigiata** | — |
| *Chaetomorpha antennina** | — | *Valonia macrophysa* | 1 |
| *Chaetomorpha linum* | — | *Valonia utricularis* | 1 |
| *Chaetomorpha melagonium* | — | *Valoniopsis pachynema*** | — |
| *Chaetomorpha* sp. | — | | |
| *Rhizoclonium implexum** | — | | |
| *Rhizoclonium tortuosum* | — | Derbesiales | |
| *Spongomorpha coalita** [a] | — | | |
| | | *Derbesia lamourouxii** | 1, 2 |
| Acrosiphoniales | | *Derbesia tenuissima* | 1, 2 |
| | | *Derbesia* sp. | 1, 2 |
| | | *Derbesia vaucheriaeformis** | 1, 2 |
| *Acrosiphonia arcta* | — | *Halicystis ovalis** | 1, 2 |
| *Acrosiphonia sonderi* | — | | |
| | | | |
| Sphaeropleales | | Codiales | |
| | | | |
| *Sphaeroplea annulina* | — | *Avrainvillea nigricans* | 1, 2 |
| *Sphaeroplea cambrica* | — | *Avrainvillea rawsoni* | 1, 2 |
| | | *Bryopsis corticulans** | 1, 2 |
| | | *Bryopsis hypnoides** | 1, 2 |
| Dasycladales | | *Bryopsis muscosa* | 1, 2 |
| | | *Bryopsis* sp. | 1, 2 |
| | | *Chlorodesmis comosa** [b] | 1, 2 |
| *Acetabularia clavata** | — | *Codium bursa* | 1, 2 |
| *Acetabularia crenulata* | — | *Codium coronatum** | 1, 2 |
| *Acetabularia mediterranea* | — | *Codium dimorphum* | 1, 2 |
| *Acetabularia mobii* | — | *Codium duthieae** | 1, 2 |
| *Acetabularia wettsteinii* | — | *Codium elongatum* | 1, 2 |
| *Acicularia schenkii* | — | *Codium fragile* | 1, 2 |
| *Batophora oerstedi* | — | *Codium lucasii** | 1, 2 |
| *Bornetella sphaerica* | — | *Codium muelleri** | 1, 2 |
| *Cymopolia barbata* | — | *Codium spongiosum* | 1, 2 |
| *Dasycladus clavaeformis* | — | *Codium tomentosum* | 1, 2 |
| *Neomeris annulata** | — | *Penicillus capitatus** [b] | 1, 2 |

| Caulerpales | | Caulerpales (contd.) | |
|---|---|---|---|
| *Caulerpa cupressoides** | 1, 2 | *Halimeda discoidea** | 1, 2 |
| *Caulerpa distichophylla** | 1, 2 | *Halimeda opuntia** | 1, 2 |
| *Caulerpa filiformis** | 1, 2 | *Halimeda tuna* | 1, 2 |
| *Caulerpa lentillifera** | 1, 2 | *Udotea flabellum** | 1, 2 |
| *Caulerpa prolifera* | 1, 2 | *Udotea petlolata* | 1, 2 |
| *Caulerpa racemosa** | 1, 2 | | |
| *Caulerpa serrulata** | 1, 2 | | |
| *Caulerpa sertularioides* | 1, 2 | Dichotomosiphonales | |
| | | *Dichotomosiphon* sp. | 2 |
| | | *Dichotomosiphon tuberosus* | 2 |

\* Data of Strain (1958, 1965, 1966); \*\* data of de Nicola (1961).

1, siphonaxanthin; 2, siphonein; —, absence of both pigments.

*a* Classified in Acrosiphoniales by Round (1971); *b* Classified in Caulerpales by Round (1971).

1965, 1966), but Strain did not find siphonein or siphonaxanthin in *Caulerpa filiformis*. Both pigments have been found in the green alga *Ostreobium* which lives symbiotically with the brain coral *Favia* (Jeffrey 1968). The general picture of carotenoid distribution in siphonous green algae is summarized in Table 6.1.

Many of the Prasinophyceae contain the usual carotenoids but some produce in addition siphonaxanthin (and siphonein) (e.g. *Asteromonas propulsa*) (Ricketts 1970) or a keto-carotenoid, micronone. The latter has not been fully characterized but has similarities with siphonaxanthin (Ricketts 1966, 1967). Others. (e.g. *Pterosperma* sp., *Heteromastix* sp.) are characterized by the absence of lutein and the presence of siphonaxanthin and siphonein (Ricketts 1970, 1971b)

Carotenoids accumulate in the gametes of some colonial green algae such as *Ulva* spp. (Strain 1951, Haxo & Glendenning 1953). The major pigment involved in *U. lactuca* is γ-carotene (3) (Haxo & Glendenning 1953) which also accumulates along with lycopene (2) in the antheridia of *Chara ceratophylla* and *Nitella syncarpa* (Charophyceae) (Karrer *et al.* 1943). Extra-plastidic carotenoids also appear in many algae under unfavourable cultural conditions, particularly nitrogen deficiency. The accumulation is such that the cultures often appear orange or red. The pigments responsible are generally β-carotene and/or its keto-derivatives echinenone (18), canthaxanthin (19) and astaxanthin (20). For example, β-carotene is the major extra-plastidic pigment in *Trentepohlia aurea* (Czygan & Kalb 1966) and *Dunaliella salina* (Aasen *et al.* 1969); echinenone (18) preponderates in *Scenedesmus brasiliensis* (Czygan, 1964) and is found together with canthaxanthin (19) and astaxanthin (20) in *Protosiphon botryoides* (Kleinig & Czygan 1969). Other keto derivatives are also found in trace amounts (see Goodwin 1971). In addition, crustaxanthin (21) and phoenicopterone (22) are present in *Haematococcus pluvialis* (Czygan 1970).

G

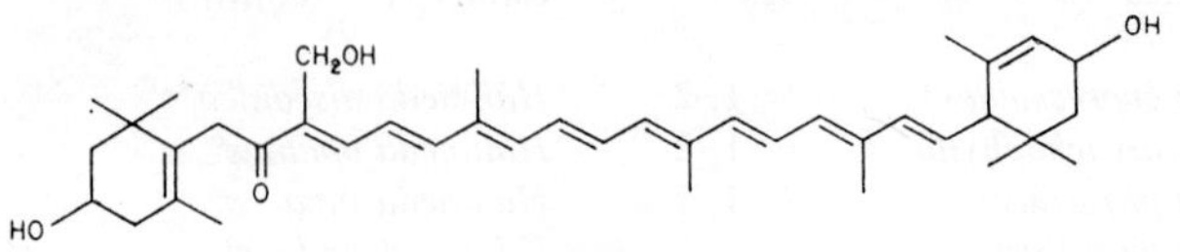

(16)

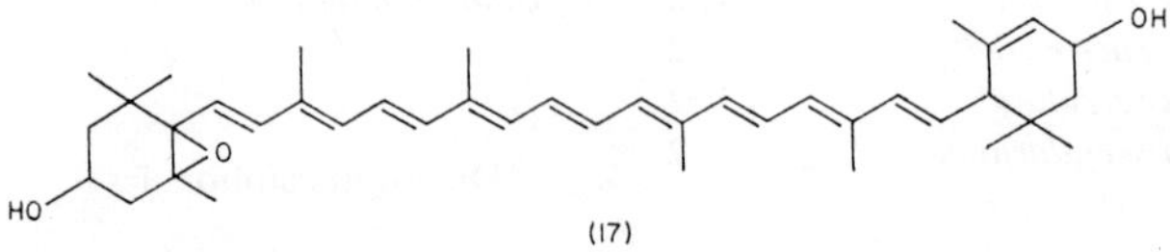

(17)

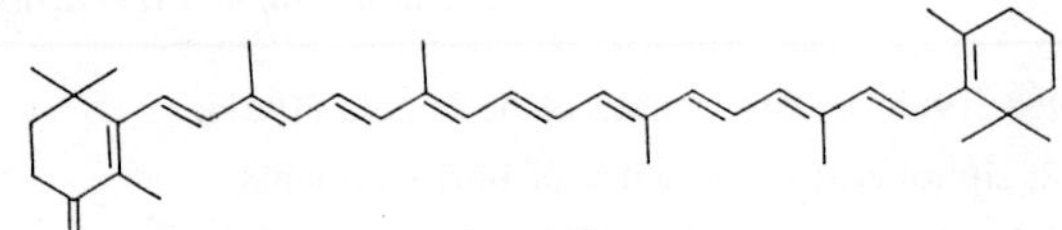

(18)

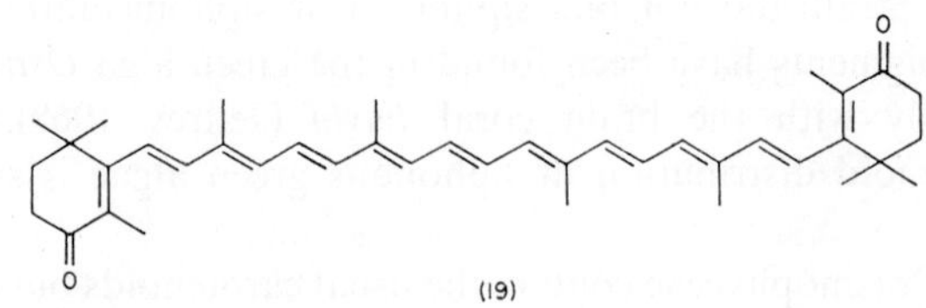

(19)

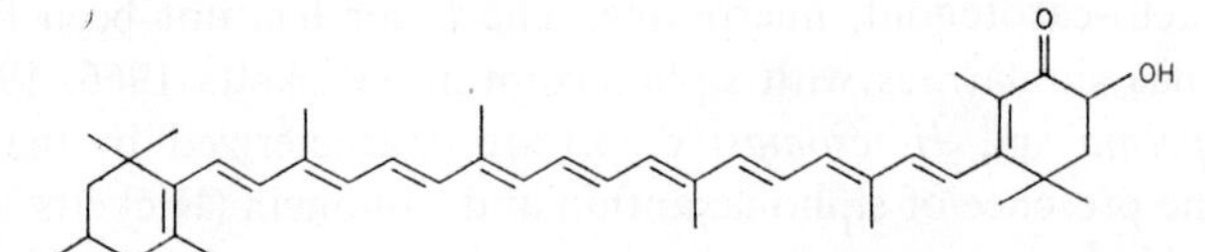

(20)

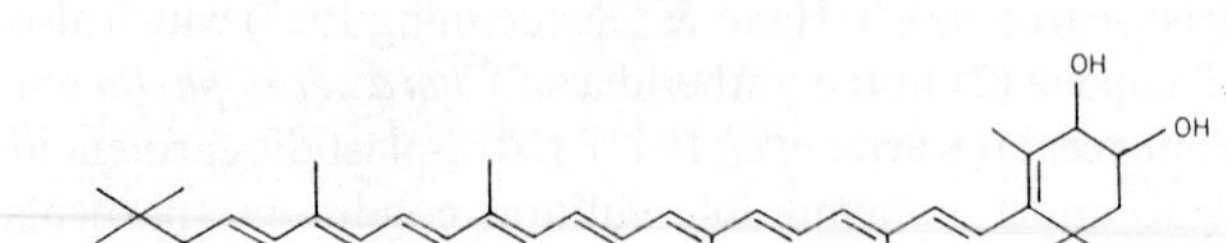

(21)

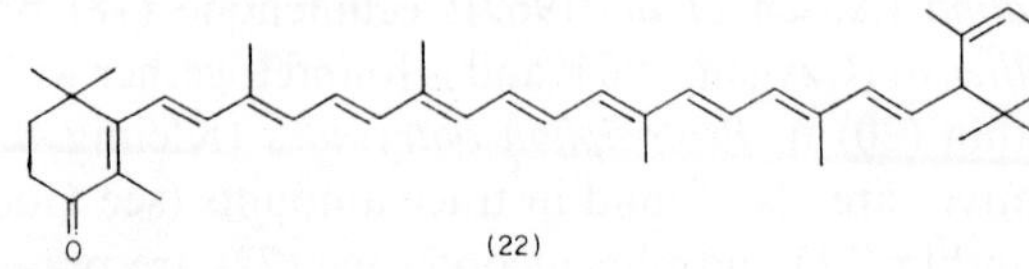

(22)

The location of these extra-plastidic carotenoids appears to vary according to the alga examined. They are found in intracytoplasmic deposits with no limiting membrane in *Protosiphon botryoides* (Berkaloff 1967), in lipid vacuoles in *Ankistrodesmus braunii* (Mayer & Czygan 1969) and in plastoglobuli in *Haematococcus pluvialis* (Sprey 1970). In *Haematococcus* sp., however, they have been reported in the aplanospores (Czygan & Kessler 1967, Czygan 1970) and are said by Lang (1968) not to exist within any organelle or vesicle.

### 2.3 Rhodophyceae

The hydrocarbons α- and β-carotenes (5,6) together with the corresponding xanthophylls, lutein (11) and zeaxanthin (23), are widely distributed in red algae (Strain 1958, 1966, Allen *et al.* 1964). Thus the general pattern is relatively simple compared with all other classes and in some cases is made even more simple by the occasional absence of lutein (11), e.g. in *Porphyridium aerugineum* (Chapman 1966a) and *P. cruentum* (Stransky & Hager 1970a) and the frequent absence of α-carotene (5). In one case, *Phycodrys sinuosa*, β-carotene (6) is absent (Larsen & Haug 1956). The monohydroxy derivative, β-cryptoxanthin (24), has been found in two Hawaiian algae, *Acanthophora spicifera* and *Gracilaria lichenoides* (Aihara & Yamamoto 1968) whilst α-cryptoxanthin (25) is present in *Lenormandia prolifera* (Saenger *et al.* 1968).

Epoxides have only rarely been encountered, but in the red algae from Hawaii, just mentioned, *Acanthophora spicifera* and *Gracilaria lichenoides*, antheraxanthin (26) is the main pigment found (Aihara & Yamamoto 1968). However in *G. sjoestedtii* from California (Strain 1958) and *G. edulis* from Australia (Strain 1966) no antheraxanthin was present. Violaxanthin (12) has been reported in *Halosaccion glandiforma* (Strain 1958) and traces of what appeared to be neoxanthin (13) were present in *Nemalion multifidum* (Allen *et al.* 1964). It appears difficult to find *Cyanidium caldarium* a taxonomic home. Some consider that it resides in the Rhodophyceae (e.g. D. J. Chapman quoted by Goodwin 1971). From the carotenoid point of view this is acceptable; the components are β-carotene, zeaxanthin and probably lutein (Allen *et al.* 1960).

### 2.4 Xanthophyceae and Eustigmatophyceae

β-carotene is present in all members of the Xanthophyceae so far examined and three quite distinct xanthophylls are present in all these except one. These pigments, all characterized by the presence of acetylenic bonds at positions 7,8 and/or 7′,8′, are diadinoxanthin (27), diatoxanthin (28) and heteroxanthin (29) (Strain *et al.* 1968, Strain *et al.* 1970, Egger *et al.* 1969, Stransky & Hager 1970a). In early investigations diadinoxanthin (27) was confused with antheraxanthin (26) (Thomas & Goodwin 1965, Kleinig & Egger 1967b). An alternative structure for heteroxanthin (Nitsche 1970) is probably incorrect. Vaucheriaxanthin (30), in the form of a partial ester, is often but not always present (Strain *et al.* 1968,

Nitsche & Egger 1970). Neoxanthin (13) is also reported to be present (Stransky & Hager 1970a).

The one recorded exception to this distribution pattern is *Pleurochloris commutata* which contains violaxanthin (12), antheraxanthin (26) and zeaxanthin

(23)

(24)

(25)

(26)

(27)

(28)

(23) in addition to free and esterified vaucheriaxanthin (30) (Stransky & Hager 1970a). A closely related alga *Botrydiopsis alpina*, contains the expected carotenoids (Thomas & Goodwin 1965, Stransky & Hager 1970a).

## 2.5   *Chrysophyceae and Haptophyceae*

The characteristic xanthophyll in the Chrysophyceae and Haptophyceae is fucoxanthin (31). It represents 75% of the total pigment in *Ochromonas danica* and *Prymnesium parvum* (Allen *et al.* 1960). It is accompanied by diatoxanthin (28) in these organisms and in *Isochrysis galbana* and *Sphaleromantis* sp. (Jeffrey 1961). Diadinoxanthin (27) is present in the latter, but the situation in the former is still unclear (Dales 1960, Jeffrey 1961) although it is present in an *Isochrysis* sp. (Hager & Stransky 1970a). Dinoxanthin (structure still unknown) is probably present in *I. galbana* (Jeffrey 1961). β-Carotene (6) is either the only carotene present (Allen *et al.* 1960) or the major component of a mixture which includes traces of α- and γ-carotenes (5,3) (Dales 1960).

The concentration of carotenoids, especially the xanthophylls, is high (e.g. $10 \cdot 57 \text{mg.g}^{-1}$ of dry matter in *P. parvum*) (Allen *et al.* 1960) and in one organism, *Hymenomonas huxleyii*, they appear to be localized in a finely coiled lamellar system which is distinct from the chloroplast (Olson *et al.* 1967).

## 2.6   *Phaeophyceae*

An exhaustive investigation revealed that the main pigments of all Phaeophyceae examined are β-carotene (6), violaxanthin (12) and fucoxanthin (31) (Jensen 1966). No α-carotene (5) or lutein (11) has been detected in recent studies (Jensen 1966, Strain 1966), although traces of diatoxanthin (28) and diadinoxanthin (27) are occasionally encountered (Jensen 1966). β-Carotene (6) accumulates extra-plastidically in the male gametes of *Fucus* spp. and *Ascophyllum nodosum*, which imparts to them the characteristic bright-orange colour; the olive-green of the ova is due to a mixture of fucoxanthin (31) and chlorophylls (Carter *et al.* 1948).

## 2.7   *Bacillariophyceae*

Carotenoid distribution here is very similar to that in the Phaeophyceae, in that the major pigments are β-carotene (6), diatoxanthin (28), diadinoxanthin (27) and fucoxanthin (31) (Strain 1951, 1958, 1966). The rather rare ε-carotene (7) is present in *Nitzschia closterium* (*Phaeodactylum tricornutum*) (Strain *et al.* 1944) and in *Navicula pelliculosa* (Hager & Stransky 1970b).

## 2.8 *Dinophyceae*

The characteristic pigment of this class of algae is peridinin (32) (Strain *et al.* 1971b). Small amounts of diadinoxanthin (27) and dinoxanthin (unknown structure) are associated with peridinin in *Prorocentrum micans* (Pinckard *et al.* 1953), *Gymnodinium* sp. (Jeffrey 1961) and *Amphidinium* spp. (Bunt 1964, Parsons & Strickland 1963). Peridinin is also the major pigment in the symbiotic

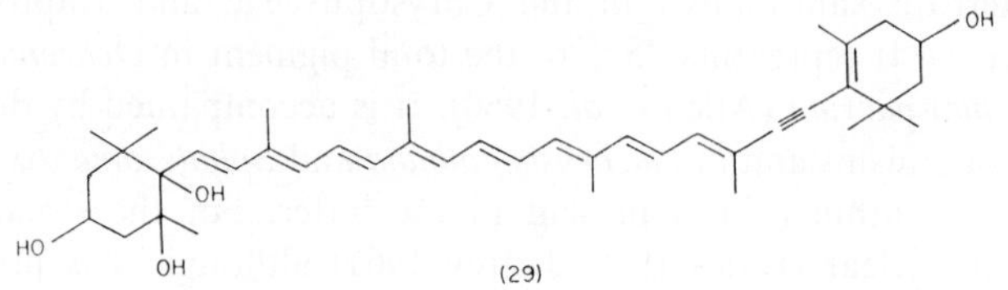

(29)

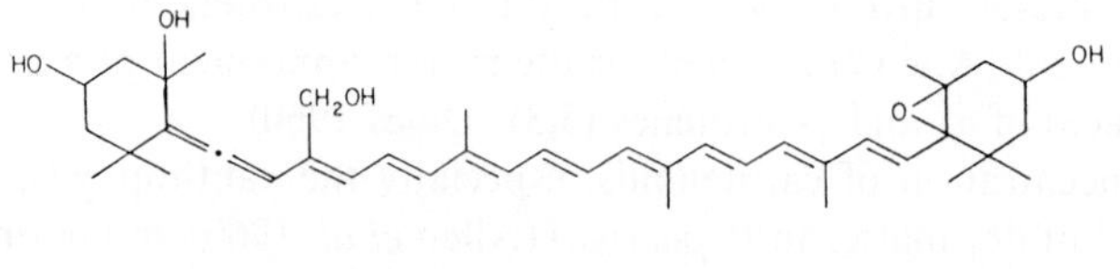

(30)

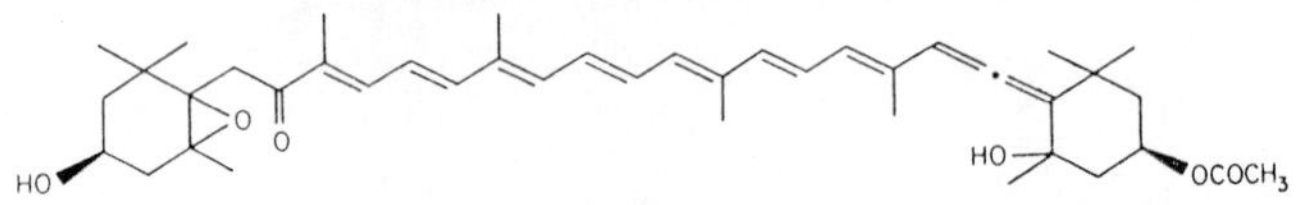

(31)

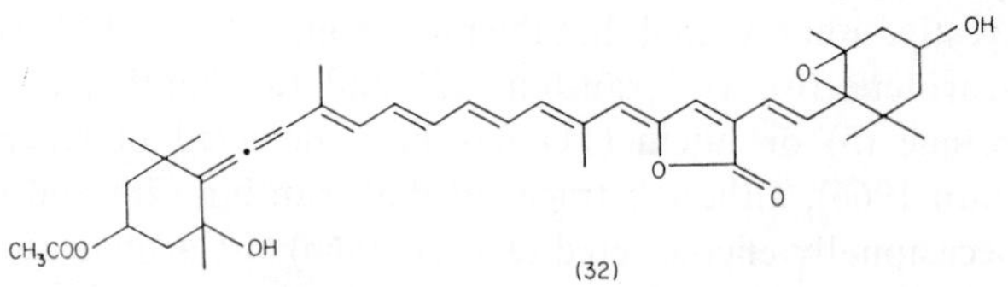

(32)

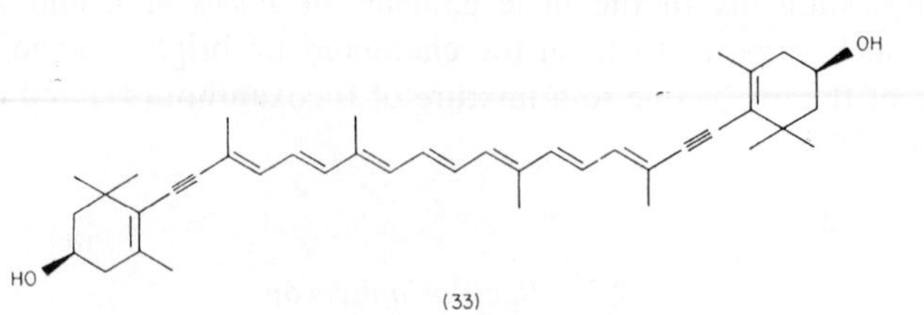

(33)

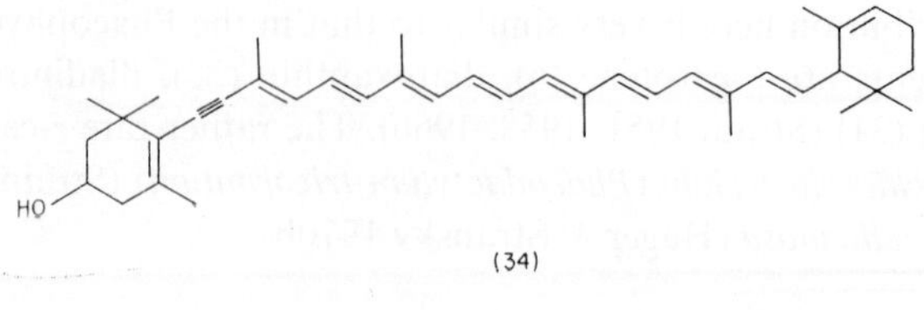

(34)

dinoflagellates in sea anemones, clams and corals (Jeffrey & Haxo 1968). Fucoxanthin (31) is said to be the main pigment in *Gymnodinium veneficum* (Riley & Wilson 1965) and *Glenodinium foliaceum* (Mandelli 1968).

### 2.9 *Cryptophyceae*

The characteristics of those algae so far examined in this class are that α-carotene (5) predominates over β-carotene (6) (Allen *et al.* 1964, Chapman & Haxo 1963) and that the constituent xanthophylls are acetylene derivatives; indeed it was in these algae that acetylenic carotenoids were first discovered. The major xanthophyll which has been noted in *Cryptomonas* spp., *Hemiselmis viridis* and *Rhodomonas* sp. is alloxanthin (33). Crocoxanthin (34) is present in the last two and monadoxanthin (35) in the last named (Chapman 1966b, Mallams *et al.* 1967). ε-Carotene (7) is found in *Cryptomonas ovata* (Chapman & Haxo 1963).

### 2.10 *Euglenophyceae*

The carotenoids in the chloroplasts of various *Euglena* spp. are somewhat similar to the green leaf carotenoids in consisting of β-carotene (6), zeaxanthin (23) and neoxanthin (13), but differ in that the main xanthophyll is the acetylenic diadinoxanthin (27) (Aitzetmüller *et al.* 1968) and not antheraxanthin (26) (Krinsky & Goldsmith 1960) or lutein (11) (Goodwin & Jamikorn 1954), as previously suggested. Diatoxanthin (28) has also recently been reported (Hager & Stransky 1970b, Johannes *et al.* 1971) as has a pigment considered to be diepoxyneoxanthin (36) (Nitsche *et al.* 1969). Traces of the keto carotenoids, echinenone (18), 3-hydroxyechinenone (37) and canthaxanthin (euglenanone) (19) are also present (Goodwin & Gross 1958, Krinsky & Goldsmith 1960). Astaxanthin (20) is apparently unique to *E. heliorubescens* from which it was isolated as the oxidative artefact astacin (38) (Kuhn *et al.* 1939, Tischer 1941). The pigment was originally called euglenarhodone (Tischer 1936). The keto-carotenoids are said to be found in the eye-spots of *Euglena* spp.

### 2.11 *Cyanophyceae*

The characteristic carotenoids of the blue-green algae are β-carotene (6), echinenone (18) (= myxoxanthin, aphanin) (Goodwin & Taha 1951, Hertzberg & Liaaen-Jensen 1966b) and zeaxanthin (23) (Hertzberg & Liaaen-Jensen 1966b). Canthaxanthin (19) (= aphanicin) and 4-keto-3'-hydroxy-β-carotene (39) are encountered less frequently; for example the former is the major pigment in *Aphanizomenon flos-aquae* and the latter occurs in *Arthrospira* sp. (Hertzberg & Liaaen-Jensen 1966a,b). Blue-green algae characteristically synthesize carotenoid glycosides, a property they share with certain non-photosynthetic bacteria. The predominant pigments of this type are myxoxanthophyll, oscillaxanthin and aphanizophyll. As originally isolated, myxoxanthophyll was a mixture of glycosides, but the rhamnoside of myxol (the aglycone) is by far the major component and is now termed myxoxanthophyll (40) (Hertzberg & Liaaen-Jensen 1969a). Two other components are myxol 2'-O-methylpentoside (41) and myxol-2'-glucoside (see 40) (Francis et al. 1970). Myxoxanthophyll is present in

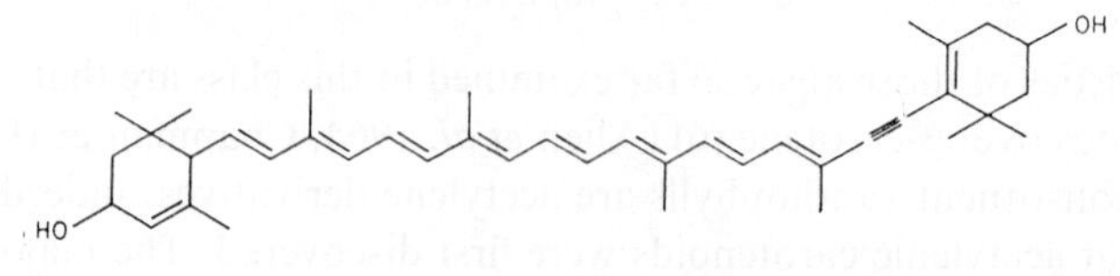

(35)

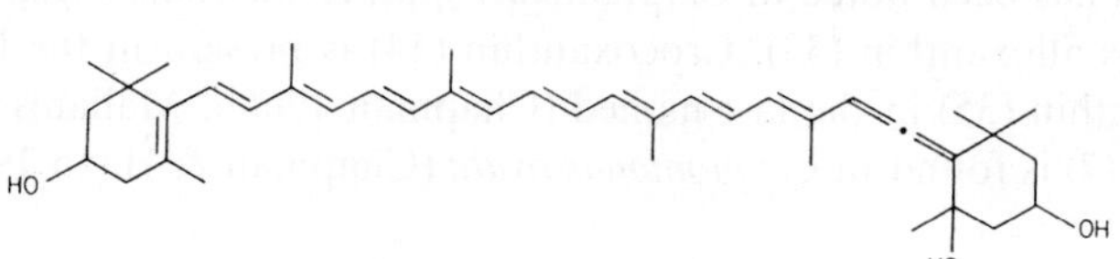

(36)

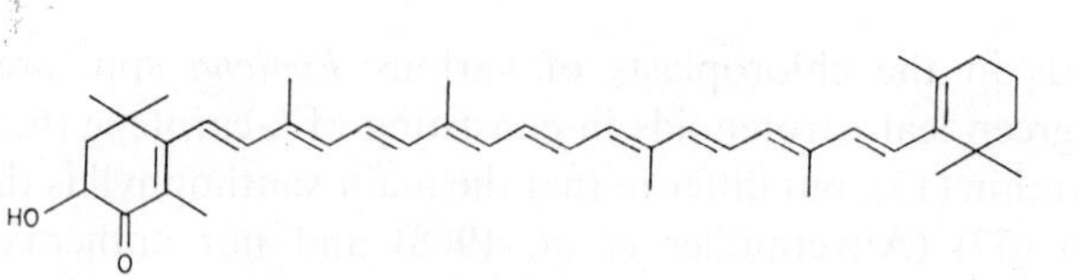

(37)

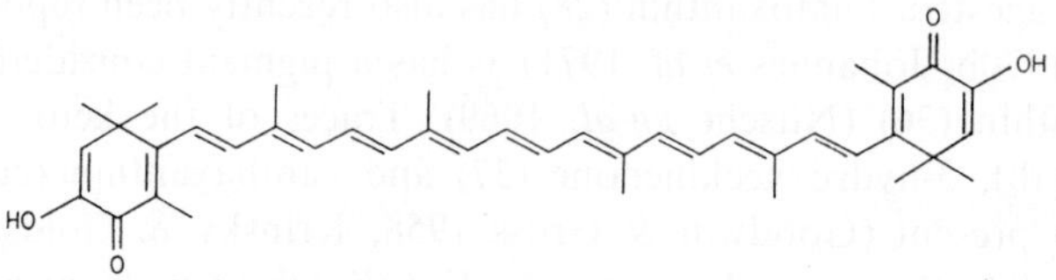

(38)

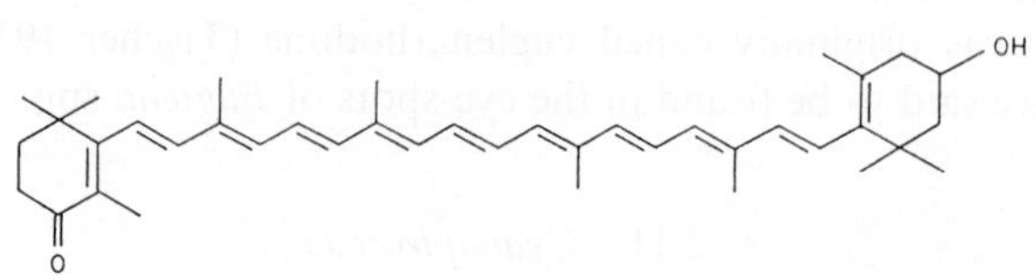

(39)

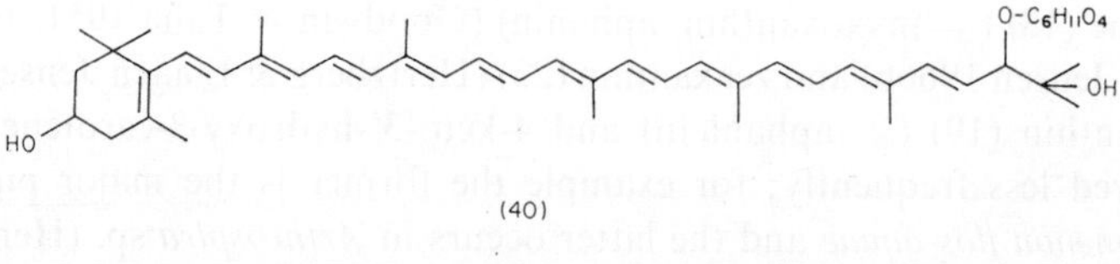

(40)

all blue-green algae so far examined except *Phormidium ectocarpi*, *P.foveolarum* (Hertzberg et al. 1971), *P.persicinum* (Healey 1968) and *Anacystis nidulans* (Stransky & Hager 1970a,b). However, a recent investigation of *A. nidulans* revealed the presence of myxoxanthophyll (Halfen & Francis 1972).

Oscillaxanthin (42), first isolated from *Oscillatoria rubescens* (Karrer & Rutschmann 1944) but also present in *Arthrospira* sp. (Hertzberg & Liaaen-Jensen 1969b), is a diglucoside derivative of lycopene (2). Aphanizophyll (43),

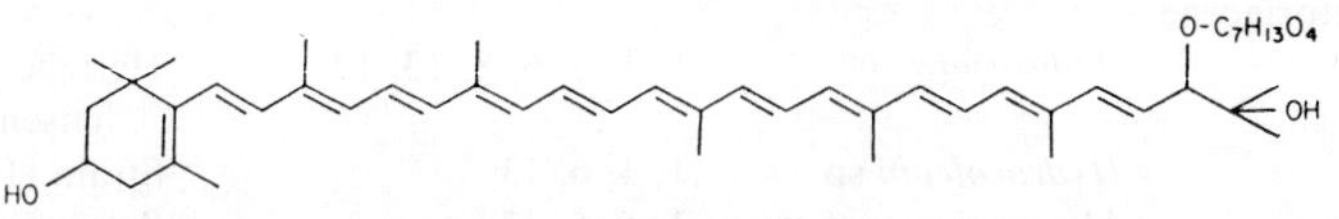

(41)

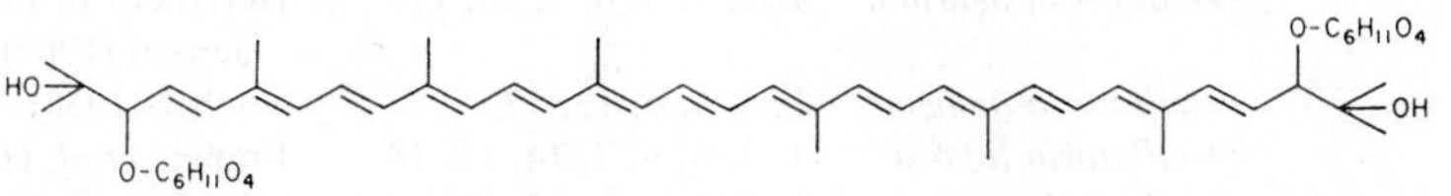

(42)

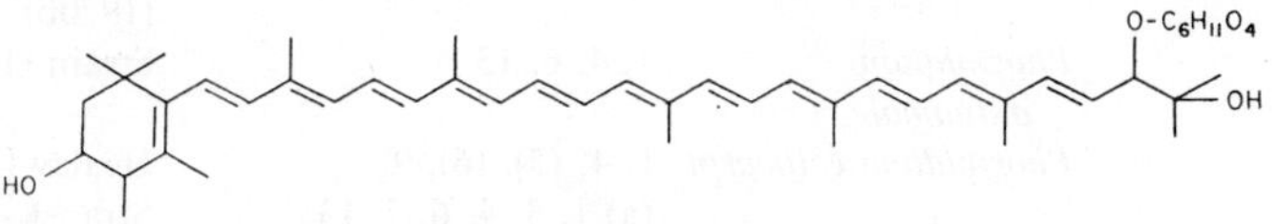

(43)

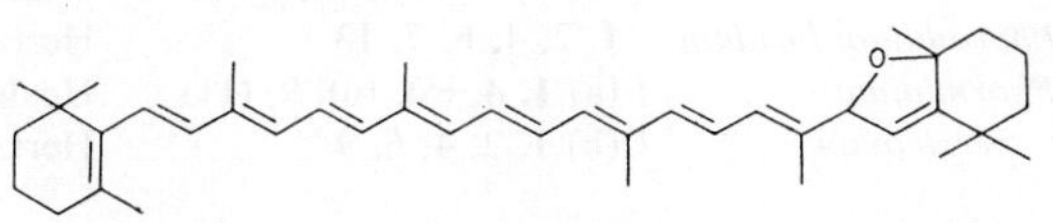

(44)

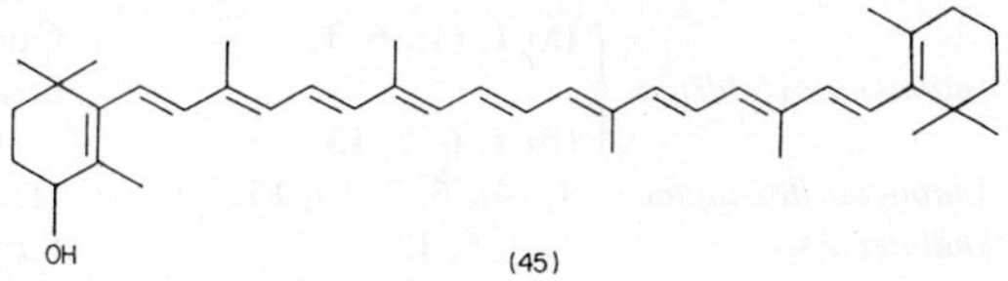

(45)

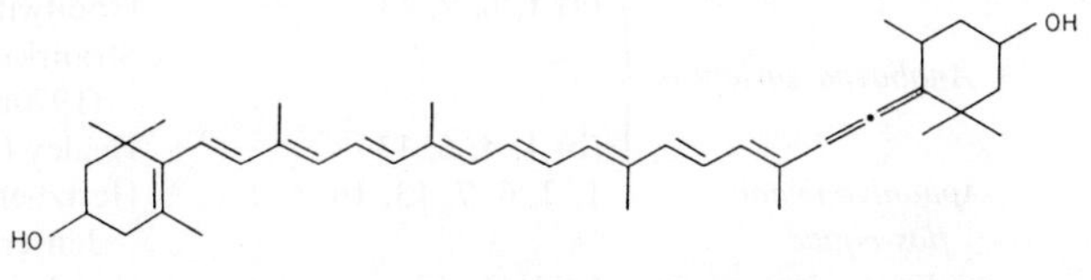

(46)

from *Aphanizomenon flos-aquae* (Goodwin 1957), is 4-hydroxymyxoxanthophyll (Hertzberg & Liaaen-Jensen 1971). Traces of γ-carotene and lycopene have been recently found in a hot spring *Oscillatoria* sp. (Francis & Halfen 1972). Mutato-chrome (= flavacin) (44) (Hertzberg & Liaaen-Jensen 1967) is occasionally

**Table 6.2.** Distribution of carotenoids in blue-green algae.

| Family | Species | Pigments | References |
|---|---|---|---|
| Oscillatoriaceae | | | |
| | *Arthrospira* sp. | 1, 3, 4  6, 8, **13**, 17 | Hertzberg & Liaaen-Jensen (1966a) |
| | *Hydrocoleum* sp. | 1, 4, 6, 13 | Strain (1958) |
| | *Microcoleus paludosus* | **1**, 5, 6, 13 | Stransky & Hager (1970b) |
| | *Microcoleus vaginatus* | **1**, 4, 6, 13 | Goodwin (1957) |
| | *Oscillatoria agardhii* | **1**, 2, 3, 4, 6, 8, **13**, 17 | Hertzberg & Liaaen-Jensen (1967) |
| | *Oscillatoria amoena* | 1, 3, 4, 6, 13, 17 | Tischer (1958) |
| | *Oscillatoria limosa* | 1, 3, 4, 6, 7, **14**, 15, 18 | Francis *et al.* (1970) |
| | *Oscillatoria rubescens* | 1, 3, 4, 6, 8, **13**, 17 | Hertzberg & Liaaen-Jensen (1966a) |
| | *Oscillatoria tenuis* | **1**, 6, 7, 13, 16 | Stransky & Hager (1970b) |
| | *Phormidium autumnale* | 1, 4, 6, 13 | Strain (1958) |
| | *Phormidium ectocarpi* | 1, 4, (5), (6), 9 | Healey (1968) |
| | *Phormidium foveolarum* | (a) 1, 3, 4, **6**, 7, 13 | Stransky & Hager (1970b) |
| | | (b) **1**, 2, 4, 6, 13, 15 | Hertzberg *et al.* (1971) |
| | *Phormidium luridum* | **1**, 2, 4, 6, 7, 13 | Hertzberg *et al.* (1971) |
| | *Phormidium persicinum* | (a) **1**, **4**, (5), (6), 9, (11) | Healey (1968) |
| | | (b) **1**, 2, 4, 6, 9 | Hertzberg *et al.* (1971) |
| Nostocaceae | | | |
| | *Anabaena aerulosa oscillatorioides* | 1, 4, 6, 7, **13**, 16 | Stransky & Hager (1970b) |
| | *Anabaena cylindrica* | (a) **1**, (4), 6, 13 | Goodwin (1957) Stransky & Hager (1970b) |
| | | (b) **1**, **6**, 7, **13** | |
| | *Anabaena flos-aquae* | **1**, (4), 6, 7, 13, 15 | Hertzberg *et al.* (1971) |
| | *Anabaena* sp. | 1, 4, 6, 13 | Strain (1958) |
| | *Anabaena variabilis* | (a) **1**, 6, 7, 13 | Goodwin (1957), Stransky & Hager (1970b) |
| | | (b) **1**, 4, 6, 13 | Healey (1968) |
| | *Aphanizomenon flos-aquae* | 1, 2, 6, **7**, 13, 16 | Hertzberg & Liaaen-Jensen (1966b) |
| | *Cylindrospermum* sp. | **1**, (4), 6, 13 | Goodwin (1957) |
| | *Hormothamnion enteromorphoides* | 1, 4, 6, 13 | Strain (1958) |
| | *Nostoc commune* | **1**, 4, 5, 6, 7, 11, 12, 13 | Stransky & Hager (1970b) |
| | *Nostoc muscorum* | **1**, 4, 6, 13 | Strain (1958) |
| | *Nostoc* sp. | 1, (4), **6**, 13 | Goodwin (1957) |

| Family | Species | Pigments | References |
|--------|---------|----------|------------|
| Scytonemataceae | | | |
| | *Tolypothrix tenuis* | 1, 4, **6**, 7, 13 | Stransky & Hager (1970b) |
| Rivulariaceae | | | |
| | *Calothrix parietina* | **1**, 4, 5, **6**, 7, 11, 12, 13 | Stransky & Hager (1970b) |
| Mastigocladaceae | | | |
| | *Mastigocladus* sp. | **1**, 4, **6**, 13 | Goodwin (1957) |
| Chroococcaceae | | | |
| | *Anacystis nidulans* | 1, 3, **4**, 11, 12 | Stransky & Hager (1970b) Rotfarb (1970) |
| | | **1**, 6, 7, 13, 15 | Halfen & Francis (1972) |
| | *Chroococcus* sp. | 1, 4, 6, 13 | Strain (1958) |
| | *Coccochloris elabens* | **1**, (4), **6**, **13** | Goodwin (1957) |
| | *Merismopedia punctata* | **1**, 3, 4, 5, 6, 7, 13 | Stransky & Hager (1970b) |
| | *Microcystis aeruginosa* | **1**, 3, 4, 5, 6, 7, 9, 13, 16, 17 | Stransky & Hager (1970b) |
| | *Synechococcus elongatus* | 1, 4, 6, (9), 11, 12, 13 | Stransky & Hager (1970b) |

\* Figures in bold type indicate the major pigment present; figures in brackets represent pigments present in traces.

Key: 1, β-Carotene; 2, Mutatochrome; 3, Cryptoxanthin; 4, Zeaxanthin; 5, Isozeaxanthin; 6, Echinenone; 7, Canthaxanthin; 8, 4-Keto-3′-hydroxy-β-carotene; 9, Trihydroxy-β-carotene; 10, Unknown 3; 11, Caloxanthin; 12, Nostoxanthin; 13, Myxoxanthophyll; 14, Myxol-2′-O-methyl-methyl pentoside; 15, 4-Ketomyxol-2′-methyl pentoside; 16, Aphanizophyll; 17, Oscillaxanthin; 18, Oscillol-2,2′-di(O-methyl) methyl pentoside.

encountered, as in *Oscillatoria agardhii*. Similarly isocryptoxanthin (45) turns up spasmodically as in *Oscillatoria* spp. (Francis & Halfen 1972). Two allenic xanthophylls, caloxanthin (46) and nostoxanthin (47) have been reported in *Nostoc commune*, *Calothrix parietina*, *Anacystis nidulans* and *Synechococcus elongatus* (Stransky & Hager 1970a,b) but the structures are still in some doubt (Hertzberg *et al.* 1971). They may be related to the trihydroxy-β-carotene found in *Phormidium* spp. (Hertzberg *et al.* 1971) and 'unknown 3' from *P. persicinum* (Healey 1968). The relationship between the pigments of *Calothrix scopulorum* described by Kylin (1927) as carotene, myxorhodin-α, myxorhodin-β and those just described in the other blue-green algae has not yet been clarified. In *Anabaena cylindrica* the pigments are quantitatively and qualitatively the same in heterocysts and vegetative cells (Winkenbach *et al.* 1972). Earlier reports had suggested differences (Fay 1969, Wolk & Simon 1969). Calorhodin α and calorhodin β may be echinenone (18) and canthaxanthin (19), respectively.

The surface algal mat of a Californian sub-tropical desert, which consists mainly of filamentous blue-green algae (*Schizothrix* spp., *Nostoc muscorum, Scytonema hoffmanii*) together with a smaller number of Chroococcales (*Anacystis* spp., *Coccochloris peniocystis*) and a few Chlorophyceae and algae-containing protozoa, accumulated mainly echinenone (18) and canthaxanthin (19) but no β-carotene (6) (Bauman *et al.* 1971). As already indicated, related species in culture contain considerable amounts of β-carotene.

Carotenoids with α-rings are probably absent from blue-green algae. The early report of lutein (11) in *Oscillatoria rubescens* (Heilbron 1942) has not been substantiated in later work (Hertzberg & Liaaen-Jensen 1966a).

Two symbiotic algae (cyanelles) which exist with the colourless algae *Cyanophora paradoxa* and *Glaucocystis nostochinearum* synthesize only β-carotene (6) and zeaxanthin (23); the pigments characteristic of the free living forms echinenone (18) and myxoxanthophyll (40) were not present (Chapman 1966a).

The well-authenticated distribution of carotenoids in blue-green algae is summarized in Table 6.2.

## 2.12   *Non-photosynthetic algae*

The heterotrophic phytoflagellate *Polytoma uvella* synthesizes a keto carotenoid of unknown structure named polytomaxanthin (Links *et al.* 1960) whilst *Astasia ocellata* probably forms phoenicopterone (22) (Thomas *et al.* 1967).

The Flexibacteria, which may be considered as non-photosynthetic blue-green algae, synthesize carotenoids reminiscent of those produced by blue-green algae. For example, *Flexibacter* sp. produces the keto carotenoid flexixanthin (49) (Aasen & Liaaen-Jensen, 1966a) and *Saprospira grandis*, saproxanthin (49); both have similarities to myxol, the aglycone of myxoxanthophyll (Aasen & Liaaen-Jensen 1966b).

Mention should be made here of the carotenoids of the recently investigated 'gliding filamentous organism F-2' (Halfen *et al.* 1972). This organism was once thought to be a colourless flexibacterium but it has now been shown to contain chlorophylls characteristic of the green sulphur bacteria. Its carotenoids represent characteristic pigments of both blue-green algae and photosynthetic bacteria; of the former we have β-carotene (6) and echinenone (18) and of the latter, carotenoids characterized by hydration across the 1,2 double bond; for example 1'-hydroxy-1',2'-dihydro-γ-carotene (50), and by desaturation at 3,4, for example O-glycosyl-1'-hydroxyl-1',2'-dihydro-3',4'-didehydro-γ-carotene (51). γ-Carotene itself is not very characteristic of photosynthetic bacteria, being found only in small amounts in some green bacteria. Finally the characteristic xanthophylls of flexibacteria are absent from this organism.

(47)

(48)

(49)

(50)

(51)

(52)

(53)

(54)

## 3 BILIPROTEINS

### 3.1 *Introduction*

The intense red and blue colours of some algae are due to the presence of biliproteins. These are chromoproteins in which the prosthetic group is a bile pigment tightly bound by co-valent linkage(s) to its apoprotein. They occur generally in Rhodophyceae, Cryptophyceae and the Cyanophyceae. Table 6.3 indicates that there are two well defined prosthetic groups phycoerythrobilin (50) and phycocyanobilin (51). The existence of a third prosthetic group, phycourobilin, has been claimed (O'Carra *et al.* 1964) but this is still in dispute (Chapman *et al.* 1968). Phytochrome, the biliprotein which mediates photomorphogenesis in higher plants (see e.g. Mohr 1970) and perhaps also in certain algae (see Chapter 28, p. 790 and Chapter 29, p. 817), has a prosthetic group (52) similar to that of phycocyanin and phycoerythrobilin.

**Table 6.3.** Algal biliproteins and their prosthetic groups

| Biliprotein | Bile Pigment (prosthetic group) |
|---|---|
| C-Phycoerythrin | Phycoerythrobilin |
| C-Phycocyanin | Phycocyanobilin |
| Allophycocyanin | Phycocyanobilin |
| R-Phycoerythrin | { Phycoerythrobilin <br> Phycourobilin |
| R-Phycocyanin | { Phycoerythrobilin <br> Phycocyanobilin |

The general classification of biliproteins is based on their absorption spectra. There are, for example, three types of phycoerythrin: R-phycoerythrin and B-phycoerythrin found in the Rhodophyceae (Fig. 6.1) and C-phycoerythrin present in Cyanophyceae (Fig. 6.2). There are also three types of phycocyanin: R-phycocyanin, from red algae (Fig. 6.3), C-phycocyanin and allophycocyanin (Fig. 6.2). Originally the letters C-, B- and R- designated the source of the pigments, the Cyanophyceae, the Bangiales, and Rhodophyceae other than the Bangiales, respectively. Further investigations showed that this relationship was not entirely valid (see O'hEocha 1971) and, as stated above, the prefixes now refer to pigments with different characteristic absorption spectra.

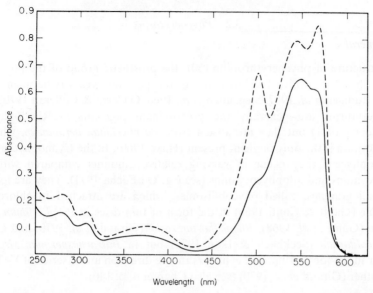

**Fig. 6.1.** Absorption spectra of (i) R-phycoerythrin (— —) from *Ceramium rubrum*; (ii) B-phycoerythrin (——) from *Rhodochorton floridulum*; both in aqueous solution at pH 6–7 (O'hEocha 1971).

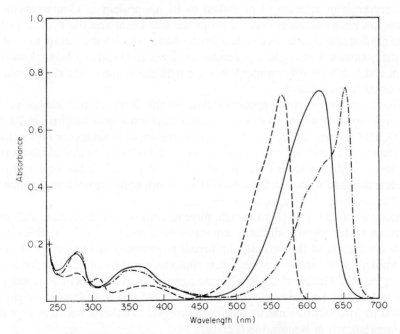

**Fig. 6.2.** Absorption spectra of purified C-phycoerythrin (— —), C-phyco-cyanin (——) and allophycocyanin (—·—·) from *Fremyella diplosiphon*: all three in 0.1M sodium phosphate (pH 7·0) (Bennett & Bogorad 1971).

## 3.2  *Phycoerythrins*

### (a) *General*

The structure of phycoerythrobilin (50), the prosthetic group of the phycoery-thrins, has been fully elucidated during the past few years (Chapman *et al.* 1967, Rudiger *et al.* 1967, Siegelman *et al.* 1968, O'Carra & Colleran 1970). The phycoerythrins are generally the predominant pigments in Rhodophyceae (O'hEocha 1971) but they are absent from *Porphyridium aerugineum* where a phycocyanin is the only pigment present (Haxo 1960). In the Cyanophyceae and Cryptophyceae they occur in varying relative amounts compared with the phycocyanins and allophycocyanins (see e.g. O'hEocha 1971). They are located in small granules, called phycobilisomes, which are attached to chloroplast lamellae (Gantt & Conti 1966) in the form of thin discs in *Porphyridium aeru-gineum* (Gantt *et al.* 1968), *Rhodochorton rothii* (Giraud *et al.* 1970) and *Cyani-dium caldarium* (Seckbach & Ikan 1972) but in *Batrachospermum virgatum* appear in the form of parallel cylinders in one direction and a series of V-shapes in another (Giraud *et al.* 1970) (see also Chapter 4, p. 140).

### (b) *C-Phycoerythrins*

Phycoerythrobilin appears to be linked to its apoprotein in C-phycoerythrin through an ester linkage between a propionic side chain and the hydroxyl of a serine residue and also through a less strong bond between the lactam in ring A and the γ-carboxyl of a glutaryl residue (Killilea & O'Carra 1968). However, Crespi and Smith (1970) consider that the ethylidene group of the chromophore is involved in the binding.

The C-phycoerythrins appear to exist in the form of two sub-units. In *Fremyella diplosiphon* and *Tolypothrix tenuis* their molecular weights are 20,000 and 18,300 (Bennett & Bogorad 1971); the corresponding values for *Phormidium persicinum* are 22,000 and 19,700 (O'Carra & Killilea 1971). In *F. diplosiphon* two phycoerythrobilin molecules are linked covalently to each heavy unit and one molecule is linked with each light unit. The two sub-units are only weakly bound together and no SH linkages are involved.

Because of the ease with which phycoerythrins can aggregate and dis-aggregate under slightly differing environmental conditions, it is difficult to decide on the size of the naturally occurring polymers. Some years ago Hattori and Fujita (1959) reported a maximum molecular weight of 226,000 for an *in vitro* aggregate. More recently the largest aggregates in extracts of *T. tenuis* and *F. diplosiphon* had a sedimentation constant of 10S and a molecular weight of 180,000 to 210,000; this means that the aggregate is either a decamer or a do-decamer (Bennett & Bogorad 1971).

## (c) R- and B-phycoerythrin

The absorption spectra of these two forms of phycoerythrin have already been given (Fig. 6.1). They apparently both contain two prosthetic groups, phycoery-throbilin (51) and phycourobilin (structure unknown). These pigments are present in the ratio 2:1 in R-phycoerythrin and about 6:1 in B-phycoerythrin (P. O'Carra, quoted by O'hEocha 1966). The intensity of absorption in the region 495 to 500nm (Fig. 6.1) reflects the amount of phycourobilin present. R-Phycoerythrins are antigenically related to phycoerythrin from other sources but not to phycocyanins or allophycocyanins (Berns 1967).

## 3.3  Phycocyanins

### (a) C-Phycocyanins

Analysis of phycocyanins on calibrated sodium dodecyl sulphate gels indicates that the pigments consist of two sub-units. In *Fremyella diplosiphon* these have molecular weights of 17,600 and 16,300 (Bennett & Bogorad 1971); the cor-responding values for the pigment in *Plectonema boryanum* are 17,200 and 15,100 (H. Rice, quoted by Bennett & Bogorad 1971), in *Anacystis nidulans* and *Aphanocapsa* sp. 20,000 and 16,000 (Glazer *et al.* 1971), in *Nostoc punctiforme*, *Anacystis nidulans*, *Anabaena variabilis A. cylindrica* and *Calothrix scopulorum* 18,500 and 20,500 (O'Carra & Killilea 1971), in *Mastigocladus laminosus* 14,000 and 14,000 (Binder *et al.* 1972) and in *Oscillatoria agardhii* 12,200 and 14,100 (Torjesen & Sletten 1972). From their results O'Carra and Killilea (1971) incline to the view that there is no species variation. It remains to be seen whether this is generally true and whether the varying molecular weights just quoted from different laboratories are real or due to vagaries in technique. Identical antigenic reactions were observed with phycocyanins from *A. variabilis* and *N. muscorum* (Bogorad 1965) which would support the view of O'Carra and Killilea. Bogorad also found that the pigments from *Porphyra laciniata* and *Cyanidium caldarium* are immunologically very similar.

There are reports that phycocyanin exists as a single sub-unit of molecular weight about 30,000 (Kao & Berns 1968), about 46,000 (Hattori *et al.* 1965) and 15,000 (Bloomfield & Jennings 1969).

If one assumes the presence of one phycocyanobilin chromophore per sub-unit then the pigment content of phycocyanin from *Synechococcus lividus* and *Phormidium luridum* indicates a minimum molecular weight of 14,600 (Crespi *et al.* 1968). Thus it can be concluded that each sub-unit contains one molecule of phycocyanobilin. Similar conclusions can be reached for the chromoprotein from *Cyanidium caldarium*; two bands have been observed with molecular weights between 15,000 and 17,000 (Bennett & Bogorad, unpubl., quoted by Bennett & Bogorad 1971) and the minimum molecular weight calculated on phycocyanobilin content is 16,200 (Troxler & Lester 1968). The two sub-units of *M. laminosus*, each of which consists of a single polypeptide chain, differ in amino acid composition, amino terminal residues and absorption spectra (Binder *et al.* 1972). Both sub-units of *O. agardhii* have amino terminal methionine (Torjeson & Sletten 1972).

The association of the basic sub-units into aggregates does not involve disulphide linkages in *F. diplosiphon* (Bennett & Bogorad 1971) but does in *O. agardhii* (Torjeson & Sletten 1972). The largest aggregate observed *in vitro* is one with a molecular weight of 226,000 (Hattori & Fujita 1959).

### (b) *Allophycocyanins*

The allophycocyanins from *Fremyella diplosiphon* and *Plectonema boryanum* consist of single sub-units of molecular weight 16,000 (Bennett & Bogorad 1971), and 15,300 (Rice quoted by Bennett & Bogorad 1971) respectively. On the other hand, the phycocyanins from *Anacystis nidulans* and *Aphanocapsa* sp. consist of two sub-units of 17,500 and 15,500 molecular weight (Glazer *et al.* 1971). The difference in absorption spectra of C-phycocyanin and allophycocyanin (Fig. 6.2), both of which contain phycocyanobilin as prosthetic group, is probably due to different means of attachment of the pigment to the constituent apoproteins. Certainly the two proteins from *Arthrospira maxima* differ significantly in their amino acid composition (Raftery & O'hEocha 1965) and the two pigments from other sources are antigenically unrelated (Bennett & Bogorad 1971, Glazer *et al.* 1971).

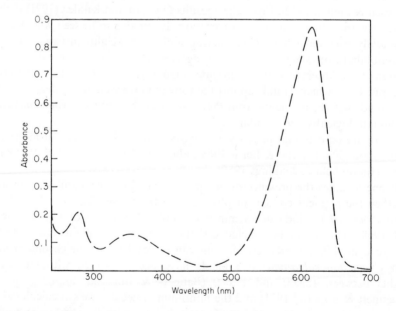

**Fig. 6.3.** Absorption spectrum of R-phycocyanin from *Porphyra laciniata* (after O'hEocha 1971).

## (c) R-phycocyanins

R-phycocyanin occurs in red algae but apparently never together with C-phyco-cyanin (O'hEocha 1965). It has been crystallized (O'Carra 1965, O'Carra & O'hEocha 1965, Chapman *et al.* 1967). It contains both phycoerythrobilin and phycocyanobilin and its absorption band with a maximum at 553nm is due to phycoerythrobilin whilst that with a maximum at 615nm is due to phycocyan-obilin (Chapman *et al.* 1967) (Fig. 6.3). Apparently there are three times as many phycocyanobilin as phycoerythrobilin residues present in one molecule of the chromoprotein (O'Carra quoted by O'hEocha 1966).

## 3.4  *Effect of light on phycobilin synthesis*

Although this chapter is not concerned with the biosynthesis of biliproteins it is important to point out that light has profound quantitative and qualitative effects on their formation.

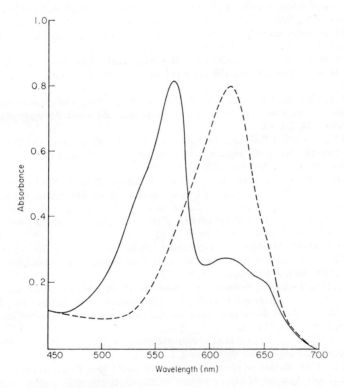

**Fig. 6.4.** Absorption spectra of crude extracts of *Fremyella diplosiphon* grown under fluorescent illumination (—), and incandescent illumination (wavelengths above 595nm) (— —) (after Bennett & Bogorad 1971).

*Tolypothrix tenuis* can produce biliproteins in the dark under appropriate conditions (see e.g. Hattori & Fujita 1959) whilst *Cyanidium caldarium* does not do so (Bogorad 1965). The quality of light is important; *Tolypothrix tenuis* grown in red light is relatively rich in phycocyanin and poor in phycoerythrin while the reverse situation exists in cultures grown in green light (Hattori & Fujita 1959). *Fremyella diplosiphon* behaves similarly, and the absorption spectrum of the extracts of the cells grown under fluorescent and incandescent lamps is given in Fig. 6.4.

# 4 REFERENCES

AASEN A.J., EIMHJELLEN K.E. & LIAAEN-JENSEN S. (1969) An extreme source of β-carotene. *Acta chem. scand.* **23**, 2544–5.

AASEN A.J. & LIAAEN-JENSEN S. (1966a) The carotenoids of flexibacteria. III. The structures of flexixanthin and deoxyflexixanthin. *Acta chem. scand.* **20**, 1970–88.

AASEN A.J. & LIAAEN-JENSEN S. (1966b) The carotenoids of flexibacteria. II. A new xanthophyll from *Saprospira grandis*. *Acta chem. scand.* **20**, 811–19.

AIHARA M.S. & YAMAMOTO H.Y. (1968) Occurrence of antheraxanthin in two Rhodophyceae *Acanthophora spicifera* and *Gracilaria lichenoides*. *Phytochem.* **7**, 497–9.

AITZETMÜLLER K., STRAIN H.H., SVEC W.A., GRANDOLFO M. & KATZ J.J. (1969). Loroxanthin, a unique xanthophyll from *Scenedesmus obliquus* and *Chlorella vulgaris*. *Phytochem.* **8**, 1761–70.

AITZETMÜLLER K., SVEC W.A., KATZ J.J. & STRAIN H.H. (1968) Structure and chemical identity of diadinoxanthin, the principal xanthophyll of *Euglena*. *Chem. Comm.* **32**, 32–3.

ALLEN M.B., FRIES L., GOODWIN T.W. & THOMAS D.M. (1964) The carotenoids of algae: Pigments from some cryptomonads, a heterokont, and some Rhodophyceae. *J. gen. Microbiol.* **34**, 259–67.

ALLEN M.B., GOODWIN T.W. & PHAGPOLNGARM S. (1960) Carotenoid distribution in certain naturally occurring algae and in some artificially induced mutants of *Chlorella pyrenoidosa*. *J. gen. Microbiol.* **23**, 93–103.

BAUMAN A.J., BOETTGER H.G., KELLY A.M., CAMERON R.E. & YOKOYAMA H. (1971) Isolation and characterization of keto-carotenoids from the neutral extract of algal mat communities of a desert soil. *Eur. J. Biochem.* **22**, 287–93.

BENNETT A. & BOGORAD L. (1971) Properties of sub-units and aggregates of blue-green algae biliproteins. *Biochemistry, N.Y.* **10**, 3625–34.

BERKALOFF C. (1967) Modifications ultrastructurale du plaste et de divers autres organites cellulaires au cours de developpment et de l'encystement du *Protosiphon botryoides* (Chlorophycées). *J. Microscopie*, **6**, 839–52.

BERNS D.S. (1967) Immunochemistry of biliproteins. *Pl. Physiol., Lancaster*, **42**, 1569–86.

BINDER A., WILSON K. & ZUBER H. (1972) C-Phycocyanin from the thermophilic blue-green alga *Mastigocladus laminosus*. Isolation, characterization and sub-unit composition. *FEBS. Letters*, **20**, 111–16.

BLOOMFIELD V.A. & JENNINGS B.R. (1969) Molecular weight and shape of phycocyanin monomer and aggregates. *Biopolymers*, **8**, 297–9.

BOGORAD L. (1965) Studies of phycobiliproteins. *Record Chem. Progress*, **26**, 108–19.

BUNT J.S. (1964) Analysis of algal pigments by thin-layer chromatography. *Nature, Lond.* **203**, 1261–3.

CARTER P.W., CROSS L.C., HEILBRON I.M. & JONES E.R.H. (1948) The lipochromes of the male and female gametes of some species of the Fucaceae. *Biochem. J.* **43**, 349–52.

CHAPMAN D.J. (1966a) The pigments of symbiotic algae of *Cyanophora paradoxa*, *Glaucocystis nostochinearum* and two Rhodophyceae, *Porphyridium aerugineum* and *Asterocytis ramosa*. *Arch. Mikrobiol.* **55**, 17–25.

CHAPMAN D.J. (1966b) Three new carotenoids isolated from algae. *Phytochem.* **5**, 1331–3.

CHAPMAN D.J. & HAXO F.T. (1963) Identity of $\epsilon$-carotene and $\epsilon_1$-carotene. *Pl. Cell. Physiol.*, *Tokyo*, **3**, 57–63.

CHAPMAN D.J. & HAXO F.T. (1966) Chloroplast pigments of Chloromonadophyceae. *J. Phycol.* **2**, 89–91.

CHAPMAN D.J., COLE W.J. & SIEGELMAN H.W. (1967) Chromophores of allophycocyanin and R-phycocyanin. *Biochem. J.* **105**, 903–5.

CHAPMAN D.J., COLE W.J. & SIEGELMAN H.W. (1968) A comparative study of the phycoerythrin chromophore. *Phytochem.* **7**, 1831–5.

COLE W.J., CHAPMAN D.J. & SIEGELMAN H.W. (1967) The structure of phycocyanobilin. *J. Am. chem. Soc.* **89**, 3643–5.

CRESPI H.L. & SMITH U. (1970) The chromophore-protein bonds in phycocyanin. *Phytochem.* **9**, 205–12.

CRESPI H.L., SMITH U. & KATZ J.J. (1968) Phycobilins. Structure and exchange studies by nuclear magnetic resonance and its mode of attachment in phycocyanin. A model for phytochrome. *Biochemistry N.Y.* **7**, 2232–42.

CZYGAN F.C. (1964) Canthaxanthin als Sekundar-Carotinoid einiger Grunalgen. *Experientia*, **20**, 573–4.

CZYGAN F.C. (1970) On the occurrence of crustaxanthin and phoenicopterone in aplanospores of *Haematococcus pluvialis* Flowtow em. Wille. *Flora Abt. A. Physiol. Biochem.* **159**, 339–45.

CZYGAN F.C. & KALB K. (1966) Untersuchungen zur Biogenese der Carotinoide in *Trentepohlia aurea*. *Z. Pfl. Physiol.* **55**, 59–64.

CZYGAN F.C. & KESSLER E. (1967) Nachweis von 3-Oxi-4,4'-dioxo-carotin in den Grunalgen *Chlorococcum wimmeri* und *Haematococcus* sp. *Z. Naturf.* **22b**, 1085–6.

DALES R.P. (1960) On the pigments of the Chrysophyceae. *J. mar. biol. Ass. U.K.* **39**, 693–9.

DE NICOLA M. (1961) Nota preliminare sui carotenoidi in *Chaetomorfa aerea* e in *Ulva latissima*. *Boll. Ist. Univ. Catania*, **2**, 35–9.

DRING M.J. (1967) Phytochrome in red alga, *Porphyra tenera*. *Nature, Lond.* **215**, 1411–12.

EGGER K., NITSCHE H. & KLEINIG H. (1969) Diatoxanthin und diadinoxanthin—Bestandteile des xanthophyllgemisches von *Vaucheria* und *Botrydium*. *Phytochem.* **8**, 1583–6.

FAY P. (1969) Cellular differentiation and pigment composition in *Anabaena cylindrica*. *Arch. Mikrobiol.* **67**, 62–70.

FRANCIS G.W. & HALFEN L.N. (1972) $\gamma$-Carotene and lycopene in *Oscillatoria princeps*. *Phytochem.* **11**, 2347–8.

FRANCIS G.W., HERTZBERG S., ANDERSEN K. & LIAAEN-JENSEN S. (1970) New carotenoid glycosides from *Oscillatoria limosa*. *Phytochem.* **9**, 629–35.

GANTT E. & CONTI S.F. (1966) Granules associated with the chloroplast lamellae of *Porphyridium cruentum*. *J. Cell. Biol.* **29**, 423–34.

GANTT E., CONTI S.F. & EDWARDS M.R. (1968) Ultrastructure of *Porphyridium aerugineum* a blue-green coloured rhodophyte. *J. Phycol.* **4**, 65–71.

GIRAUD G., LICHTLÉ C. & THOMAS J.C. (1970) Localization of phycobiliproteins in the Rhodophyceae and the Cyanophyceae. *Biochem. J.* **119**, 15P–16P.

GLAZER A.N., COHEN-BAZIRE G. & STANIER R.Y. (1971) Comparative immunology of algal biliproteins. *Proc. natn. Acad. Sci. U.S.A.* **68**, 3005–8.

GOODWIN T.W. (1957) The nature and distribution of carotenoids in some blue-green algae. *J. gen. Microbiol.* **17**, 467–73.

GOODWIN T.W. (1971) Algal carotenoids. In *Aspects of Terpenoid Chemistry and Biochemistry*, ed. Goodwin T.W., pp. 315–56. Academic Press, New York & London.

GOODWIN T.W. & GROSS J.A. (1958) The carotenoid distribution in various bleached substrains of *Euglena gracilis. J. Protozool.* **5**, 292–6.

GOODWIN T.W. & JAMIKORN M. (1954) Studies on carotenogenesis. Some observations on carotenoid synthesis in two varieties of *Euglena gracilis. J. Protozool*, **1**, 216–19.

GOODWIN T.W. & TAHA M.M. (1951) A study of the caretenoids echinenone and myxoxanthin with special reference to their probable identity. *Biochem. J.* **48**, 513–14.

HAGER A. & STRANSKY H. (1970a) Das carotenoidmuster und die Verbreitung des lichtinduzierten Xanthophyllcyclus in Verschiedenen Algenklassen. III. Grünalgen. *Arch. Mikrobiol.* **72**, 68–83.

HAGER A. & STRANSKY H. (1970b) Das carotenoidmuster und die Verbreitung des lichtinduzierten Xanthophyllcyclus in Verschiedenen Algenklassen. V. Einzelne Vertreter der Cryptophyceae, Euglenophyceae, Bacillariophyceae, Chrysophyceae und Phaeophyceae. *Arch. Mikrobiol.* **73**, 77–89.

HALFEN L.N. & FRANCIS G.W. (1972) The influence of culture temperature on the carotenoid composition of the blue-green alga *Anacystis nidulans. Arch. Mikrobiol.* **81**, 25–35.

HALFEN L.N., PIERSON B.K. & FRANCIS G.W. (1972) Carotenoids of a gliding organism containing bacteriochlorophylls. *Arch. Mikrobiol.* **82**, 240–6.

HATTORI A. & FUJITA Y. (1959) Crystalline phycobilin chromoproteins obtained from blue-green alga, *Tolypothrix tenuis. J. Biochem., Tokyo,* **46**, 633–44.

HATTORI A., CRESPI H.L. & KATZ J.J. (1965) Association and dissociation of phycocyanin and the effects of deuterium substitution on the processes. *Biochemistry, N.Y.* **4**, 1225–38.

HAXO F.T. (1960) The wavelength dependence of photosynthesis and the role of accessory pigments. In *Comparative Biochemistry of Photoreactive Pigments,* ed. Allen M.B., pp. 339–60. Academic Press, New York & London.

HAXO F.T. & GLENDENNING K.A. (1953) Photosynthesis and phototaxis in *Ulva lactuca* gametes. *Biol. Bull. Woods Hole,* **105**, 103–14.

HEALEY F.P. (1968) The carotenoids of four blue-green algae. *J. Phycol.* **4**, 126–9.

HEILBRON I.M. (1942) Some aspects of algal chemistry. *J. Chem. Soc.* 79–89.

HERTZBERG S. & LIAAEN-JENSEN S. (1966a) The carotenoids of blue-green algae. I. The carotenoids of *Oscillatoria rubescens* and an *Arthrospira* sp. *Phytochem.* **5**, 557–63.

HERTZBERG S. & LIAAEN-JENSEN S. (1966b) The carotenoids of blue-green algae. II. The carotenoids of *Aphanizomenon flos-aquae. Phytochem.* **5**, 565–70.

HERTZBERG S. & LIAAEN-JENSEN S. (1967) The carotenoids of blue-green algae. III. A comparative study of mutatochrome and flavacin. *Phytochem.* **6**, 1119–26.

HERTZBERG S. & LIAAEN-JENSEN S. (1969a) The structure of myxoxanthophyll. *Phytochem.* **8**, 1259–80.

HERTZBERG A. & LIAAEN-JENSEN S. (1969b) The structure of oscillaxanthin. *Phytochem.* **8**, 1281–92.

HERTZBERG S. & LIAAEN-JENSEN S. (1971) The constitution of aphanizophyll. *Phytochem.* **10**, 3251–2.

HERTZBERG S., LIAAEN-JENSEN S. & SIEGELMAN H.W. (1971) The carotenoids of blue-green algae. *Phytochem.* **10**, 3121–7.

JEFFREY S.W. (1961) Paper chromatographic separation of chlorophylls and carotenoids from marine algae. *Biochem. J.* **80**, 336–42.

JEFFREY S.W. (1968) Pigment composition of Siphonales algae in the brain coral *Favia. Biol. Bull. mar. biol. Lab. Woods Hole,* **135**, 141–8.

JEFFREY S.W. & HAXO F.T. (1968) Photosynthetic pigments of symbiotic dinoflagellates (Zooxanthellae) from corals and clams. *Biol. Bull. mar. biol. Lab. Woods Hole,* **135**, 149–65.

JENSEN A. (1966) Report No. 31 *Norwegian Institute of Seaweed Research.*

JOHANNES B., BRZEZINKA H. & BUDZIEKIEWICZ H. (1971) Photosynthesis in green plants. VI. Isolation of diatoxanthin from *Euglena gracilis. Z. Naturf.* **26**, **B**, 377–8.

KAO O. & BERNS D.S. (1968) The monomer molecular weight of C-phycocyanin. *Biochem. biophys. Res. Commun.* **33,** 457–62.

KARRER P., FATZER W., FAVARGER M. & JUCKER E. (1943) Die Antheridien farbstoffe von *Chara*-Arten. *Helv. Chim. Acta,* **26,** 2121–2.

KARRER P. & RUTSCHMANN J. (1944) Beitrag zur Kenntnis der Carotinoide aus *Oscillatoria rubescens. Helv. Chim. Acta,* **27,** 1691–5.

KILLILEA S.D. & O'CARRA P. (1968) Amino acids involved in chromophore-protein linkages in phycoerythrins. *Biochem. J.* **110,** 14–15P.

KLEINIG H. (1969) Carotenoids of siphonous green algae: a chemotaxonomic study. *J. Phycol.* **5,** 281–4.

KLEINIG H. & CZYGAN F.C. (1969) Lipids of *Protosiphon* (Chlorophyta). I. Carotenoids and carotenoid esters of five strains of *Protosiphon botryoides. Z. Naturf.* **24b,** 927–30.

KLEINIG H. & EGGER K. (1967a) Zur struktur von siphonaxanthin und siphonein den hauptcarotinoiden siphonaler grünalgen. *Phytochem.* **6,** 1681–6.

KLEINIG H. & EGGER K. (1967b) Carotinoide der Vaucheriales *Vaucheria* und *Botrydium. Z. Naturf.* **22b,** 868–72.

KRINSKY N.I. & GOLDSMITH T.H. (1960) The carotenoids of the flagellated alga, *Euglena gracilis. Archs. Biochem. Biophys.* **91,** 271–9.

KUHN R., STENE J. & SÖRENSEN N.A. (1939) Astaxanthin im Tier und Pflanzenreich. *Ber. dt. Chem. ges.* **72,** 1688–701.

KYLIN H. (1927) Über die karotinoiden Farbstoffe der höherein Pflanzen. *Hoppe-Seyler's Z. physiol. Chem.* **163,** 229–60.

LANG N.J. (1968) Electron microscopic studies of extraplastidic astaxanthin in *Haematococcus. J. Phycol.* **4,** 12–19.

LARSEN B. & HAUG A (1956) Carotene isomers in some red algae. *Acta chem. scand.* **10,** 470–2.

LINKS J., VERLOOP A. & HAVINGA E. (1960) The carotenoids of *Polytoma uvella. Arch. Mikrobiol.* **36,** 306–24.

MALLAMS A.K., WAIGHT E.S., WEEDON B.C.L., CHAPMAN D.J., HAXO F.T., GOODWIN T.W. & THOMAS D.M. (1967) A new class of carotenoids. *Proc. chem. Soc.* 301–2.

MANDELLI E.F. (1968) Carotenoid pigments of the dinoflagellate *Glenodinium foliaceum* Stein. *J. Phycol.* **4,** 347–8.

MAYER F. & CZYGAN F.C. (1969) Ünderungen der Ultrastrukturen in den Grünalgen *Ankistrodesmus braunii* und *Chlorella fusca* var. *rubescens* bei Stickstoffmangel. *Planta,* **86,** 175–85.

MOHR H. (1970) Mechanism of action of phytochrome. *Biochem. J.* **119,** 3–4.

NITSCHE H. (1970) Die Struktur von *Vaucheria*-Heteroxanthin. *Tetrahedron Letters,* 3345–8.

NITSCHE H. & EGGER K. (1970) Die Struktur von Vaucheriaxanthin. *Tetrahedron Letters,* 1435–8.

NITSCHE H., EGGER K. & DABBAGH A.G. (1969) Diepoxyneoxanthin, das Hauptcarotinoid in Bluten von *Mimulus guttatus. Tetrahedron Letters,* 2999–3002.

O'CARRA P. (1965) Purification and N-terminal analyses of algal biliproteins. *Biochem. J.* **94,** 171–4.

O'CARRA P. & COLLERAN E. (1970) Separation and identification of biliverdin isomers and isomer analysis of phycobilins and bilirubin. *J. Chromat.* **50,** 458–68.

O'CARRA P. & KILLILEA S.D. (1971) Sub-unit structures of C-phycocyanin and C-phycoerythrin. *Biochem. biophys. Res. Commun.* **45,** 1192–7.

O'CARRA P. & O'hEOCHA C. (1965) R-phycocyanin, a distinct type of biliprotein. *Phytochem.* **4,** 635–8.

O'CARRA P., O'hEOCHA C. & CARROL D.M. (1964) Spectral properties of the phycobilins II. Phycoerythrobilin. *Biochemistry, N.Y.* **3,** 1343–50.

O'hEOCHA C. (1965) Biliproteins of Algae. *A. Rev. Pl. Physiol.* **16,** 415–32.

O'HEOCHA C. (1966) Biliproteins. In *Biochemistry of Chloroplasts*, Vol. I, ed. Goodwin T.W., pp. 407–21. Academic Press, New York & London.

O'HEOCHA C. (1971) Pigments of the red algae. *A. Rev. Oceanogr. Mar. Biol.* **9**, 61–82.

OLSON R.A., JENNINGS W.R. & ALLEN M.B. (1967) Spectral properties of a pigmented body in *Hymenomonas* sp. An extra-chloroplast organelle containing chlorophyll. *J. Cell. Physiol.* **70**, 133–40.

PARSONS T.R. & STRICKLAND J.D.H. (1963) Discussion of spectrophotometric determination of marine plant pigments, with revised equations for ascertaining chlorophylls and carotenoids. *J. mar. Res.* **21**, 155–63.

PINCKARD J.H., KITTREDGE J.S., FOX D.L., HAXO F.T. & ZECHMEISTER L. (1953) Pigments from a marine 'red-water' population of the dinoflagellate *Prorocentrum micans*. *Archs. Biochem. Biophys.* **44**, 189–99.

RAFTERY M.A. & O'HEOCHA C. (1965) Amino acid composition and C-terminal residues of algal biliproteins. *Biochem. J.* **94**, 166–70.

RICKETTS T.R. (1966) The carotenoids of the phytoflagellate, *Micromonas pusilla*. *Phytochem.* **5**, 571–80.

RICKETTS T.R. (1967) Further investigations into the pigment composition of green flagellates possessing scaly flagella. *Phytochem.* **6**, 1375–86.

RICKETTS T.R. (1970) The pigments of the Prasinophyceae and related organisms. *Phytochem.* **9**, 1835–42.

RICKETTS T.R. (1971a) The structures of siphonein and siphonaxanthin from *Codium fragile. Phytochem.* **10**, 155–60.

RICKETTS T.R. (1971b) Identification of xanthophylls KI and KIS of the Prasinophyceae as siphonein and siphonaxanthin. *Phytochem.* **10**, 161–4.

RILEY J.P. & WILSON T.R.S. (1965) The use of thin-layer chromatography for the separation and identification of phytoplankton pigments. *J. mar. biol. Ass. U.K.* **45**, 583–91.

ROTFARB R.M. (1970) Isolation and purification of zeaxanthin from the blue-green alga *Anacystis nidulans. Fiziol Biokhim. Issled Rast.* **3–5**.

ROUND F.E. (1971) The taxonomy of the Chlorophyta. II. *Br. Phycol. J.* **6**, 235–64.

RÜDIGER W., O'CARRA P. & O'HEOCHA C. (1967) Structure of phycoerythrobilin and phycocyanobilin. *Nature, Lond.* **215**, 1477–8.

SAENGER P., ROWAN K.S. & DUCKER S.C. (1968) Lipid soluble pigments of marine red alga *Lenormandia prolifera. Helgolander Wiss. Meersunter,* **18**, 549.

SĔCKBACH J. & IKAN R. (1972) Sterols and chloroplast structure of *Cyanidium caldarium. Plant Physiol. Lancaster,* **49**, 457–9.

SIEGELMAN H.W., CHAPMAN D.J. & COLE W.J. (1968) The bile pigments of plants. In *Porphyrins and related compounds*, ed. Goodwin T.W., pp. 107–20. Academic Press, New York and London.

SPREY B. (1970) Die Lokalisierung von Sekundärcarotinoiden von *Haematococcus pluvialis*. Flotow em. Wille. *Protoplasma,* **71**, 235–50.

STRAIN H.H. (1951) The pigments of algae. In *Manual of Phycology*, ed. Smith G.M., pp. 243–62. Ronald Press Co., New York.

STRAIN H.H. (1958) Chloroplast pigments and chromatographic analysis. *32nd Annual Priestley Lectures.* Penn. State University.

STRAIN H.H. (1965) Chloroplast pigments and the classification of some siphonalean green algae of Australia. *Biol. Bull. mar. biol. Lab. Woods Hole,* **129**, 366–70.

STRAIN H.H. (1966) Fat soluble chloroplast pigments: Their identification and distribution in various Australian plants. In *Biochemistry of Chloroplasts*, Vol. I, ed. Goodwin T.W., pp. 387–406. Academic Press, New York & London.

STRAIN H.H., BENTON F.L., GRANDOLFO M.C., AITZETMÜLLER K., SVEC W.A. & KATZ J.J. (1970) Heteroxanthin, diatoxanthin and diadinoxanthin from *Tribonema aequale. Phytochem.* **9**, 2561–5.

STRAIN H.H., MANNING W.M. & HARDEN G.J. (1944) Xanthophylls and carotenes of diatoms, brown algae, dinoflagellates and sea anemones. *Biol. Bull. mar. biol. Lab. Woods Hole*, **86**, 169–91.

STRAIN H.H., SVEC W.A., AITZETMÜLLER K., COPE B.T., HARKNESS A.L. & KATZ J.J. (1971a) Mass fragmentation and structure of siphonaxanthin, siphonein, and derivatives. *Org. Mass. Spectrom.* **5**, 565–72.

STRAIN H.H., SVEC W.A., AITZETMÜLLER K., GRANDOLFO M. & KATZ J.J. (1968) Molecular weights and empirical formulas of the xanthophylls of *Vaucheria*. *Phytochem.* **7**, 1417–18.

STRAIN H.H., SVEC W.A., AITZETMÜLLER K., GRANDOLFO M.C., KATZ J.J., KJØSEN H., NORGARD S., LIAAEN-JENSEN S., HAXO F.T., WEGFAHRT P. & RAPOPORT H. (1971b) The structure of peridinin the characteristic dinoflagellate carotenoid. *J. Am. chem. Soc.* **93**, 1823–5.

STRANSKY H. & HAGER A. (1970a) Das carotenoidimuster und die Verbreitung des lichtinduzierten Xanthophyllcyclus in Verschiedenen Algenklassen. II. Xanthophyceae. *Arch. Mikrobiol.* **71**, 164–90.

STRANSKY H. & HAGER A. (1970b) Das carotenoidimuster und die Verbreitung des lichtinduzierten Xanthophyllcyclus in Verschiedenen Algenklassen. IV. Cyanophyceae und Rhodophyceae. *Arch. Mikrobiol.* **72**, 84–96.

THOMAS D.M. & GOODWIN T.W. (1965) Nature and distribution of carotenoids in the Xanthophyta. *J. Phycol.* **1**, 118–21.

THOMAS D.M., GOODWIN T.W. & RYLEY J.F. (1967) Nature and distribution of carotenoid pigments in *Astasia ocellata*. *J. Protozool.* **14**, 654–7.

TISCHER J. (1936) Über das Euglenarhodon und andere Carotinoide einer roten Euglena (Carotinoide der Süsswasseralgen, I. Teil). *Hoppe-Seyler's Z. physiol. Chem.* **239**, 257–69.

TISCHER J. (1941) Uber die Identität von Euglenarhodon mit Astacin. *Hoppe-Seyler's Z. physiol. Chem.* **267**, 281–4.

TISCHER J. (1958) Carotinoide der Süsswasseralgen. XI. Über die carotinoide aus *Oscillatoria amoena*. *Hoppe-Seyler's Z. physiol. Chem.* **311**, 140–7.

TORJESEN P.A. & SLETTEN K. (1972) C-Phycocyanin from *Oscillatoria agardhii*. *Biochim. biophys. Acta*, **263**, 258.

TROXLER R.F. & LESTER R. (1968) Formation, chromophore composition, and labelling specificity of *Cyanidium caldarium* phycocyanin. *Pl. Physiol., Lancaster*, **43**, 1737–9.

WALTON T.J., BRITTON G., GOODWIN T.W., DINER B. & MOSHIER S. (1970) The structure of siphonaxanthin. *Phytochem.* **9**, 2545–52.

WINKENBACH F., WOLK C.P. & JOST M. (1972) Lipids of membranes and of the cell envelope in heterocysts of a blue-green alga. *Planta*, **107**, 69–80.

WOLK C.P. & SIMON R.D. (1969) Pigments and lipids of heterocysts. *Planta*, **86**, 92–7.

YAMAMOTO H., YOKOYAMA H. & BOETTGER H. (1969) Carotenoids of *Chlorella pyrenoidosa*. Pyrenoxanthin, a new carotenol. *J. org. Chem.* **34**, 4207–8.

# CHAPTER 7

# STORAGE PRODUCTS

## J. S. CRAIGIE

Atlantic Regional Laboratory,
National Research Council of Canada,
Halifax, Nova Scotia, Canada

1   Introduction   206

2   α-(1,4)-linked glucans   207
2.1   Floridean starch   207
2.2   Myxophycean starch   208
2.3   Chlorophycean and other starches   210

3   β-(1,3)-linked glucans   212
3.1   Laminaran   212
3.2   Chrysolaminaran   214
3.3   Paramylon   215

4   Fructosans and inulin   216

5   Low molecular weight compounds   217
5.1   Sugars and simple glycosides
217
5.2   Polyols   220
5.3   Cyclitols   223
5.4   Function in osmoregulation
225

6   References   226

## 1   INTRODUCTION

The definition of a storage product is rather arbitrary and this chapter follows an earlier review (Meeuse 1962) which treated starches, β-(1,3)-glucans, fructans, and low-molecular weight carbohydrates. A description of algal cyclitols is presented because they are prominent photosynthetic products of some algae, and are directly derived from simple carbohydrates. Only the enzymes immediately involved in the biosynthesis or degradation of the storage products are discussed. Recent general reviews are available on saccharide biosynthesis (Nikaido & Hassid 1971), carbohydrate interconversions (Loewus 1971), enzyme regulation (Preiss & Kosuge 1970), and algal photosynthesis (Gibbs *et al*. 1970). Evolutionary aspects of metabolism are discussed by Klein and Cronquist (1967) and Fredrick (1970).

## 2  α-(1,4)-LINKED GLUCANS

### 2.1  *Floridean starch*

Historical aspects, the X-ray diffraction spectra, and the results of amylolytic and chemical degradation of floridean starches were reviewed by Meeuse (1962), and excellent accounts of the chemical characterization of this starch are available (Peat & Turvey 1965, Percival & McDowell 1967, Percival 1968).

Greenwood and Thompson (1961) prepared pure *Dilsea edulis* starch by ultracentrifugation and grouped it with amylopectins rather than glycogens on the basis of its average chain length 18·6, internal chain length 7, molecular weight $7 \times 10^8$, limiting viscosity No. 160, $\lambda_{max}$ for the iodine complex of 550nm, and $(\alpha)_D + 190°$. Broad bean R-enzyme will not debranch animal glycogen but acts on both *D. edulis* and waxy maize starches (Peat *et al.* 1959). Further support for the amylopectin structure is provided by the failure of floridean starch to precipitate in the presence of a globulin from jack-bean meal. Some 47 α-(1 → 4)-glucans were tested for their ability to precipitate in the presence of the globulin, concanavalin-A, and the low reactivity of floridean starch in this test clearly placed it with the amylopectins rather than the glycogens (Manners & Wright 1962).

*Corallina officinalis* starch was purified by chromatography on ion-exchange cellulose and shown to be chemically similar to other floridean starches viz. an iodine complex with an absorption maximum at 500nm, a β-amylolysis of 44%, a rotation $(\alpha)_D + 171·2°$, and a chain length of 11·8 as determined by periodate oxidation (Turvey & Simpson 1966). Extracts of another calcareous alga, *Joculator maximus*, were fractionated with benzalkonium chloride and yielded a starch with $(\alpha)_D + 195·5°$, an iodine complex absorption maximum at 550nm, and a chain length of 17 (Ozaki *et al.* 1967). These authors reported a 3% yield of starch from the dried *Joculator*, and remarked that this was one of the highest values observed in some 60 sources they examined.

*Rhodymenia pertusa* was the source of starch used by Whyte (1971) who demonstrated how the absorption maximum of the iodine complex and the optical rotation changed with increasing purity of the glucan. Repeated chromatography on DEAE-Sephadex resulted in a preparation with an optical rotation $(\alpha)_D + 177°$, an absorption maximum at 500nm for the iodine complex, and a chain length of 12–13 D-glucose units for each one bearing a C-6 branch point.

Reports of the absence of starch from *Bangia fuscopurpurea*, *Porphyra perforata*, *P. naiadum* (= *Smithora*) (Meeuse *et al.* 1960), and *Rhodymenia palmata* (Serenkov & Zlochevskaya 1962) are exceptional and may reflect the physiological state of the specimens as starch is formed in *P. umbilicalis* (Peat *et al.* 1961) and in both *R. palmata* f. *palmata* and *R. pertusa* (Meeuse *et al.* 1960). Iodine staining reactions should be interpreted cautiously as α-(1 → 4)-linked polysaccharides such as rhodymenian xylan will afford a positive reaction (Gaillard & Bailey 1966).

Nagashima *et al.* (1968) describe chloroplast preparations from *Serraticardia maxima* capable of biosynthesizing floridean starch from ADP- or UDP-glucose and a starch primer. Physiological studies with intact plants confirmed that floridean starch was readily formed from $H^{14}CO_3^-$ in the light, and that it declined markedly during respiration (Nagashima *et al.* 1969a). In a later study they demonstrated that ADP-glucose: α-1,4-glucan α-4-glucosyltransferase was bound only to the floridean starch granules, and was not part of the enzyme complement of the chloroplast (Nagashima *et al.* 1971). The enzyme would transfer ADP-glucose, and to a lesser extent, UDP- and GDP-glucose to a floridean starch primer.

Fredrick (1967) fractionated extracts of *Rhodymenia pertusa* by polyacryla-mide gel electrophoresis and demonstrated one phosphorylase, branching isoenzymes, and two glucosyltransferases each capable of using either ADP- or UDP-glucose. Further examination revealed that two of these isoenzymes would introduce α-(1 → 6)-linkages into amylose, and thus act as classical Q enzymes. A third branching enzyme was distinguished by its ability both to branch amylose, and to introduce additional (1 → 6)-linkages into amylopectin (Fredrick 1971a). The single phosphorylase from *R. pertusa* required neither AMP nor primer, and appeared to be a glycoprotein capable of forming a linear glucan by adding glucose from glucose-1-phosphate to its own glycosyl moiety (Fredrick 1971b).

Five sources of amylase were exploited by Meeuse and Smith (1962) who compared the amylolysis of raw floridean starch from *Constantinea subulifera*, *Laurencia spectabilis*, *Plocamium pacificum*, and *Rhodymenia pertusa* with starches from other plant sources. Algal and moss starches were more readily degraded than higher plant starch, but the authors were unable to detect any optical differences in the corrosion pattern between floridean starch and native grains of wheat starch.

Briefly, the starch of red algae is an amylopectin synthesized on particles outside the chloroplasts in a form ranging from bowl-shaped to irregular, and from 0·5 to 25μm in size. Amylose has not been encountered in red algae.

## 2.2   *Myxophycean starch*

The early literature convinced Fritsch (1945) that the first visible product of myxophycean photosynthesis was glycogen. This conclusion was based upon histological staining reactions and microscopical examination of small granules which appear between, but not attached to the thylakoids of many blue-green algae (Lang 1968). Diastase digestion of the granules in *Oscillatoria amoena* indicated that they were polyglucans (Fuhs 1963), and the manipulation of physiological variables such as available nitrogen, light intensity, and culture age of *O. chalybia* suggested that the particles functioned as photosynthetic reserves (Giesy 1964). The morphology of the granules varies with the algal source, and they may be rod-shaped (Giesy 1964), 25nm particles (Ris & Singh

1961), or elongated bodies of 31 × 65nm in size (Chao & Bowen 1971). The rapidity of formation of these polyglucans is demonstrated by the fact that radioactive glucose is recovered from polysaccharides after 5 seconds of photosynthesis in $^{14}CO_2$ (Kandler 1961, Kindel & Gibbs 1963).

The biosynthesis of cyanophycean polysaccharides was studied by Fredrick (1951) who prepared an extract of *Oscillatoria princeps* which synthesized a water-soluble glucan resembling glycogen. An independent structural analysis of the polysaccharides of *Oscillatoria* sp. led Hough *et al.* (1952) to conclude that the food reserve was an amylopectin-like polyglucan. The apparent disparity in these results presumably reflects strain or species variability as a variant of *O. princeps* also produces a sparsely branched amylopectin (Fredrick 1952). Recent direct evidence has established that the intracellular particles may indeed be glycogen-like. Chao and Bowen (1971) purified the α-granules of *Nostoc muscorum* and describe them as uniform elongated particles having a sedimentation coefficient of 265S, and consisting of more than 95% glucose together with 2 to 3% protein. Although further structural investigations are obviously required, their α-amylolysis studies showed that the glucan is highly branched. The iodine complex exhibited a maximum absorption at 410nm, and is consistent with the presence of short external chains similar to those of shellfish glycogen.

The absorption spectrum of the polyglucan iodine-complex from *Cyanidium caldarium* resembles that from *Oscillatoria princeps*, and both differ from that for starch from *Spirogyra setiformis* with absorption maxima at 540, 550 and 580nm respectively (Fredrick 1968d).

The immunological relationship of the phosphorylases and branching enzymes of various blue-green algae has been demonstrated (Fredrick 1961) and additional research has shown the same two phosphorylases, two ADP-/UDP-glucosyltransferases, and two branching enzymes in three genera of blue-green algae (Fredrick 1962, 1968b,c, 1971b,d). *Cyanidium caldarium* fits the cyanophycean pattern except that the AMP sensitive phosphorylase is lacking (Fredrick 1968d).

The two cyanophycean phosphorylases were distinguished on the basis of their electrophoretic mobilities and by the observation that one functioned only with AMP (Fredrick 1962, 1963, 1967) while the other required neither cofactor nor primer (Fredrick 1971b). A single protein apparently possessing both phosphorylase and Q enzyme activity was detected in the variant LTV strain of *Oscillatoria* (Fredrick 1962). The formation of a functional phosphorylase was shown to depend upon the presence of vitamin $B_6$ in the growth medium of several species (Fredrick 1971d).

The addition of either $Mn^{2+}$ or $Fe^{3+}$ will overcome the *in vitro* inhibition of phosphorylase caused by the herbicide amitrole (Fredrick & Gentile 1960). Amitrole inhibits the formation of the AMP sensitive phosphorylase, but is without effect on the AMP insensitive enzyme in all algae tested (Fredrick 1969).

The ADP-/UDP-glucosyltransferases appear to be quite similar regardless of the algal source (Fredrick 1967, 1971c). The branching isoenzymes of the cyanophycean algae and *C. caldarium* possess dual functions in that each has the capacity to increase the branching of amylopectin in addition to its usual function as a classical Q enzyme (Fredrick 1968b,d, 1971a).

### 2.3   Chlorophycean and other starches

It is generally conceded (Meeuse 1962) that green seaweeds form starch but it remained for Mackie and Percival (1960) to chemically establish the presence of amylopectin in *Caulerpa filiformis*. The starches of *Enteromorpha compressa*, *Ulva lactuca*, *Cladophora rupestris*, and *Codium fragile* were separated into amylose and amylopectin fractions (Love *et al.* 1963). Approximately 16 to 22% of these starches was composed of an amylose that showed an iodine complex with an absorption maximum at 610 to 635nm, and a degree of polymerization some 4 to 8 times less than potato amylose. Amylopectin was also recovered from *Chaetomorpha rupestris* but amylose that may have been present was destroyed. Except for the low intrinsic viscosity of these algal amylopectins, they were similar in optical rotation, absorption maxima, and α- and β-amylolysis limits to the corresponding preparation from potatoes.

The reserve polysaccharides of several freshwater members of the Chlorophyceae and Charophyceae appear similar to those of green seaweeds. A detailed fractionation of *Chlorella pyrenoidosa* led to the isolation of a pure starch consisting of 7% amylose and an amylopectin (Olaitan & Northcote 1962). Two heterotrophic chlorophytes were examined and a typical glycogen was recovered from *Prototheca zopfii* (Manners *et al.* 1967) whereas *Polytoma uvella* produced a relatively insoluble starch consisting of 16% amylose, and an amylopectin similar to that of higher plants (Manners *et al.* 1965). Both *Chara australis* and *Nitella translucens* have starches but only the latter species was examined chemically (Anderson & King 1961a,b). It contained 16% of an amylose with an average chain length of 21 D-glucose units.

Raw starch from both freshwater and marine green algae, red algae, a moss and several higher plants was treated with amylases of diverse origin (Meeuse & Smith 1962). The general vulnerability of cryptogamic starch to enzymic attack is in good accord with data from X-ray diffraction studies which indicate that algal starch grains are less 'crystalline' than those of higher plants (Meeuse 1962).

The enzymes involved in the metabolism of algal starch appear to be similar to those of higher plants. α-Glucan phosphorylase activity was demonstrated in *Hydrodictyon* (Richter & Pirson 1957), in nucleate and enucleated *Acetabularia mediterranea* (Clauss 1959), and in *Ulva pertusa* extracts (Kashiwabara *et al.* 1965). Two distinct phosphorylases were resolved in both *Spirogyra setiformis* (Fredrick 1967) and *Chlorella pyrenoidosa* (Fredrick 1969). One of these isoenzymes is unusual in that it utilizes either glucose-1-phosphate or ADP-glucose as a substrate (Fredrick 1968a), and the author speculates that this reflects an evolutionary advance in the development of the enzyme (Fredrick 1969). In other respects the chlorophycean phosphorylases are similar to those of blue-green algae as one requires both AMP and a primer, and is suppressed by amitrole, while the other requires no additives and is unaffected by the herbicide (Fredrick 1969, 1971b). Algae grown in vitamin $B_6$ deficient culture lack both phosphorylase isoenzymes (Fredrick 1971d).

Partial purification of ADP-glucose: α-1,4-glucan transferase from *Chlorella pyrenoidosa* was reported by Preiss and Greenberg (1967). The enzyme had a pH optimum of 9·0, was sensitive to *p*-chloromercuribenzoate, and was not activated by glucose-6-phosphate or other glycolytic intermediates. In this respect it resembled the corresponding enzyme in bacteria and other plants excluding yeast. In contrast to the substrate specificity shown by the *Chlorella* transferase, the enzyme from *Spirogyra setiformis* can use either ADP- or UDP-glucose (Fredrick 1967).

A periodicity in starch production and utilization may be observed with synchronous algal cultures maintained on either alternating light and dark regimes (Müller 1961, Ruppel 1962), or under continuous illumination (Duynstee & Schmidt 1967).

Badenhuizen and his collaborators studied the effects of agitation, temperature and culture age on the starch composition and enzymes of starch synthesis in *Polytoma uvella* (McCracken & Badenhuizen 1970, Mangat & Badenhuizen 1970, 1971). They observed that ADP- and UDP-glucose concentrations, and phosphorylase and ADP-glucose transferase activities passed through simultaneous maxima soon after the cells entered the stationary phase of growth. Later the Q enzyme activity rose and the ratio of amylose to amylopectin declined. No changes in the ratio of UDP- to ADP-glucose were detected. Cells grown at 30°C lost much transferase activity but continued to produce starch of a lower amylose content and this appeared to reflect changes in the phosphorylase to Q enzyme ratio rather than the reduced level of transferase activity (Mangat & Badenhuizen 1971). Mature *Polytomella agilis* cells respond differently by forming more amylose at 18°C than at 9°C (Sheeler *et al.* 1968).

Short term photosynthetic experiments using $^{14}CO_2$ indicated that ADP-glucose was a likely precursor of starch in *Chlorella pyrenoidosa* (Kauss & Kandler 1962), and it has been suggested that starch synthesis in *Chlorella* may be controlled by the influence of ADP-glucose pyrophosphorylase on the synthesis of ADP-glucose (Sanwal & Preiss 1967). Experiments with *Polytoma*, however, indicate that starch synthesis is not rigorously coupled to the amount of ADP-glucose available, nor is it strictly dependent upon the levels of ADP-glucose transferase activity (Mangat & Badenhuizen 1970, 1971).

Enzymes of starch mobilization other than phosphorylase have received relatively little attention. Both maltase and α-amylase activities were noted in protein preparations of *Cladophora rupestris* (Manners 1964). An amylase was reported in *Ulva pertusa* extracts (Kashiwabara *et al.* 1965), and was found as an extracellular product of *Chlorella vulgaris* (Niimi *et al.* 1969). Investigations with light and dark synchronized *Chlorella* have established that α-amylase activity is greatest during the darkness while polyglucan phosphorylase is most active during the latter half of the light period (Wanka *et al.* 1970). The phosphorylase activity of *Hydrodictyon* differs in reaching its maximum during the dark period (Richter & Pirson 1957).

Starch is readily available for energy production and biosynthesis of proteins (Müller 1961, Ruppel 1962), chloroplast and pigment synthesis (Matsuka *et al.* 1966, Ohad *et al.* 1967), cell division, and cell wall or thecal formation (Duynstee & Schmidt 1967, Gooday 1971a,b). The degradation of starch for the biosynthesis of chloroplast structures during greening of a *Chlamydomonas reinhardtii* mutant is both initiated by and dependent upon continued illumination (Ohad *et al.* 1967).

Only one chemical examination of starch from a cryptomonad appears to have been reported (Archibald *et al.* 1960). Aside from its unusually high content of amylose (45%), there were no unique features found in either starch component from this species. Starchy reserves appear to characterize the Dinophyceae and Prasinophyceae but their exact chemical nature remains uncertain. Bursa (1968) described the diversity in form of the native dinoflagellate starch grains.

## 3 β-(1,3)-LINKED GLUCANS

### 3.1 *Laminaran*

Excellent descriptions of the early investigations on laminaran are available (Meeuse 1962, Peat & Turvey 1965, Bull & Chesters 1966, Percival & McDowell 1967, Percival 1968), and a suitably abbreviated account is presented below.

Despite the wide distribution of this reserve polysaccharide in the Phaeophyceae (Meeuse 1962, Powell & Meeuse 1964), our knowledge of laminaran structure rests upon samples prepared from a few representatives of the Ectocarpales, Laminariales, and Fucales. Generally it is extracted with dilute aqueous acid and in some cases, e.g. *Laminaria hyperborea* it retrogrades from solution as insoluble laminaran which may be recrystallized in high yield. Other species contain a soluble laminaran which is recovered by alcohol precipitation (Fleming & Manners 1965).

The Edinburgh school re-examined the chemical fine structure of both the non-reducing, mannitol-terminated molecule (M-chain), and the glucan terminated by a reducing glucose residue (G-chain). Annan *et al.* (1962) investigated the model compound 1-O-β-laminaritetraosyl mannitol and several samples of M-chain laminaran, and established that all yielded 3 moles of formic acid when oxidized with periodate at 2°C. This observation, together with the demonstration of ethylene glycol in the hydrolytic products of periodate oxidized, borohydride reduced M-chains show that the mannitol was monosubstituted at the C-1 position (Annan *et al.* 1962, 1965a). Mannitol ranges from 2·4 to 3·7% of both soluble and insoluble laminarans (Fleming *et al.* 1966). The mannose levels were less than 0·2% and were not significant structurally (Annan *et al.* 1965a). A

Smith degradation of insoluble laminaran revealed that a $1 \to 6$ branch point is present in only about 30% of the molecules, and the similarity in average chain length ($\overline{CL}$) and degree of polymerization ($\overline{DP}$) of 15–19 and 16–21 respectively also suggests an essentially unbranched structure. Samples of soluble laminaran from *L. digitata*, *L. saccharina* and *Fucus serratus*, however, appear to bear 2 or 3 $1 \to 6$ branch points per molecule, and this branching is reflected in $\overline{CL}$ values of 7–10 and $\overline{DP}$ values of 26–31 (Fleming *et al.* 1965, Annan *et al.* 1965b). The presence of occasional $1 \to 6$ inter-residue linkages in these laminaran chains should be considered as both possible and probable.

Evidence for a slightly different molecular fine structure is given by Handa and Nisizawa (1961) who examined the soluble laminaran from another laminarian, *Eisenia bicyclis*. The preparation had an ($\alpha$) $_D$ of $-45 \cdot 5°$, lacked mannitol end groups, and appeared to be an unbranched $\beta$-glucan of about 20 residues. There were about three times as many $1 \to 3$- as $1 \to 6$-linkages, and evidence from alkaline digestion together with the release of gentio-oligosaccharides by partial hydrolysis, indicates that most of the $1 \to 6$-links occur as a block within the molecule (Maeda & Nisizawa 1968a). Laminaran obtained from *Ishige okamurai* resembles that of *Eisenia* in being an unbranched chain of about 18 glucose units. Both $1 \to 6$- and $1 \to 3$-linkages were present in a ratio of 1:6, and mannitol was again absent (Maeda & Nisizawa 1968b). A mannitol containing $\beta$-glucan of undefined structure was noted in *Monodus subterraneus*, a member of the Xanthophyceae (Beattie & Percival 1962).

Laminaran, therefore, consists of a related group of predominantly $\beta$-1 $\to$ 3-linked glucans usually containing 16 to 31 residues. Variation in the molecule is introduced by the number of $1 \to 6$-linkages, the degree of branching, and the presence or absence of a terminal mannitol molecule. As yet, no systematic relationships have emerged with respect to these variables.

Powell and Meeuse (1964) found laminaran contents of $<2$ to 34% of the algal dry weight in 19 species representing 16 genera of brown algae. The authors did not examine seasonal fluctuations, but reported that sporophylls of *Alaria* sp. were richer in laminaran than the frond, and basal portions of *Laminaria saccharina* frond contained less laminaran than the more distal parts. Lamina and mid-ribs of *Ecklonia radiata* are better sources of laminaran than stipes especially if they are harvested in summer (Stewart *et al.* 1961). Laminaran analyses are reported for the following genera: *Saccorhiza* (LaBruto & Bruno 1960), *Fucus*, *Dictyopteris* and *Cystoseira* spp. (Lokar 1967, Shah & Rao 1969), *Laminaria* spp. (Bukhryakova 1965), *Arthrothamnus*, *Cymathaera* and *Laminaria* spp. (Bukhryakova & Levanidov 1969), *Cystoseira* spp. and *Sargassum* (Abdel-Fattah & Hussein 1970). Pellegrini and Pellegrini (1971) report a comprehensive analysis of *Cystoseira stricta* at 14 monthly intervals and observed an increase in laminaran in June. In general the reports reinforce the idea that laminaran increases during spring and summer and declines to minimal values during winter (Meeuse 1962). *Alaria*, *Ecklonia* and *Laminaria* spp. are richer sources of laminaran than are other genera.

H

The soluble glucan of *Eisenia bicyclis* is analogous to starch as it is rapidly formed in photosynthesis and is metabolized to other compounds in darkness (Yamaguchi *et al.* 1966). Bidwell and Ghosh (1962) showed that exogenously supplied $^{14}$C-mannitol enters the laminaran fraction of *Fucus vesiculosus* better in darkness than in light, and even although the laminaran pool is small in this species it functions as a reserve product (Bidwell *et al.* 1972).

### 3.2   *Chrysolaminaran*

Soluble polyglucans of low optical rotation have been isolated from *Hydrurus foetidus* and *Ochromonas malhamensis*, members of the Chrysophyceae and from freshwater diatoms (cf. Meeuse 1962). Archibald *et al.* (1963) reinvestigated *Ochromonas* and succeeded in preparing a soluble glucan free from mannose containing polymers. Under the action of fungal laminarinase it yielded principally glucose and laminaribiose, and because formaldehyde was not produced during low temperature periodate oxidation, the molecule appears to lack a mannitol end group. A $\overline{\text{DP}}$ of 34 was recorded and evidence was presented for two $1 \rightarrow 6$-glucosidic bonds per molecule which the authors suggest are inter-chain rather than inter-residue linkages.

Chemical and enzymatic investigation of a soluble polyglucan from the diatom *Phaeodactylum tricornutum* revealed that it, too, consisted of $\beta$-$(1 \rightarrow 3)$-linked D-glucose chains with a small degree of branching at C-6. Again no terminal mannitol residues were observed (Ford & Percival 1965). A natural mixed population of *Biddulphia* sp. and *Coscinodiscus* sp. was recently shown to contain an extractable $\beta$-$(1 \rightarrow 3)$-linked glucan with a $\overline{\text{DP}}$ of 28 (Handa & Tominaga 1969); a similar polymer occurs in *Skeletonema costatum* (Handa 1969). Glucose-containing polysaccharides are also the major water soluble carbohydrates extracted from *Isochrysis galbana*, *Prymnesium parvum* (Marker 1965), and from the benthic alga *Phaeosaccion collinsii* (Craigie *et al.* 1971). They probably account for much of the carbohydrate reported in a variety of diatoms and chrysophytes (Parsons *et al.* 1961, Ricketts 1966).

*Ochromonas* chrysolaminaran becomes strongly radioactive during photosynthesis in $^{14}$CO$_2$ and therefore it is an assimilatory product (Kauss 1962). As such, its accumulation would depend upon physiological and environmental parameters, and this may explain the wide variations observed in the chrysolaminaran content of *Phaeodactylum tricornutum* (Ford & Percival 1965). Some evidence suggesting the utilization of carbohydrate reserves has been presented for *Prymnesium parvum* (Ricketts 1966). The $\beta$-$(1 \rightarrow 3)$-glucan of *Skeletonema costatum* is consumed when the diatom is placed in darkness and apparently it serves as a respiratory substrate (Handa 1969). Such circumstantial evidence together with structural information and data on laminaran metabolism would support the idea that chrysolaminaran is a photosynthetic reserve in both the Bacillariophyceae and the Chrysophyceae.

Little definitive information is available on the biosynthesis of the laminarans but it has been suggested that nucleoside diphosphate glucose and corresponding transferases will be involved (Fleming et al. 1965). An irresistible parallel exists with paramylon formation in the euglenoids where UDP-glucose pyrophosphorylase, β-1,3-glucan synthetase, and laminaribiose phosphorylase have been shown to be active (Smillie 1968). Additional enzymes would, however, be required to introduce the 1 → 6-inter-chain and inter-residue linkages found in laminaran and chrysolaminaran. Kauss and Kriebitzsch (1969) have discovered in *Ochromonas malhamensis* a β-(1 → 3)-glucan phosphorylase which differs from the euglenoid enzyme in substrate specificity, but which adds glucose residues to laminaran of various origins. The enzyme requires a laminaran primer and seems to be specific for glucose-1-phosphate. The $K_m$ values show that the enzyme has a greater affinity for laminaran than glucose-1-phosphate and it may function in the mobilization of the reserve. AMP stimulates the enzyme activity but is without effect on the $K_m$ values measured (Albrecht & Kauss 1971).

### 3.3 *Paramylon*

The reserve polysaccharide of the euglenoids occurs as water insoluble, single membrane bound cellular inclusions of widely variable shapes and dimensions. The paramylon of three genera, including both green and colourless forms, has been chemically examined and consists solely of β-(1 → 3)-linked D-glucose residues. The $\overline{DP}$ values range from 50 to more than 150 and indicate that paramylon is at least twice as large a molecule as either laminaran or chrysolaminaran. Paramylon can no longer be considered an exclusive product of euglenoid algae as *Pavlova mesolychnon*, a member of the Chrysophyceae produces water-insoluble granules which give X-ray powder patterns identical to those of *Euglena* paramylon (Kreger & van der Veer 1971). Details of the occurrence and physical and chemical nature of paramylon are available in a comprehensive review (Barras & Stone 1968). Enzymes with β-(1 → 3)-glucosyl transferase, hydrolase, and phosphorylase activities have been demonstrated in *Euglena* extracts and these reports are summarized by Smillie (1968).

A soluble β-(1 → 3)-glucanase capable of degrading alkali disrupted paramylon was partially purified from streptomycin bleached *E. gracilis* var. *bacillaris* (Vogel & Barber 1968). The enzyme would not attack starch or glycogen but released glucose sequentially from paramylon and thus appears to be an exohydrolase. Its activity increased sharply during the stationary phase of growth and corresponded with an observed decline in cellular paramylon content. Both exo- and endo-hydrolase activities have now been detected and fractionated (Barras & Stone 1969a,b). One of the two exo-hydrolase isoenzymes, β-(1,3)-glucan glucohydrolase, was shown to be highly specific for β-(1 → 3)-glucans and to release glucose with inversion of configuration at the anomeric position. The endo-hydrolase, β-(1 → 3)-glucan glucanohydrolase, was partially bound to the native paramylon granules.

Heterotrophically grown *Euglena* shows an abrupt decline in paramylon synthetase activity which coincides with the depletion of exogenous carbon sources and the cessation of net paramylon synthesis. Active β-(1 → 3)-glucanase and hydrolase enzymes were demonstrated but only a slow utilization of the glucan reserve occurred until the cells were exposed to light (Dwyer *et al.* 1970). Rapid paramylon utilization was triggered by light and sustained by continuous illumination. During this period, the synthetase activity was low whereas the level of both degradative enzymes, β-(1 → 3) glucan phosphorylase and β-(1 → 3)-glucanase, was high (Dwyer & Smillie 1970). In light, some 58% of the re-mobilized paramylon was respired to $CO_2$ with the remainder being converted to other cell components, chiefly lipids which were localized in the regenerating chloroplasts (Dwyer & Smillie 1971). A similar effect of light on the reserve glucan of a permanently bleached *Euglena* was reported by Mitchell (1971).

Euglenoid β-(1 → 3) glucan thus can be utilized for respiration and as a carbon supply for the formation of new cellular material in much the same way that starch is metabolized by green algae. In heterotrophically grown algae of both groups, light appears to be indispensable for initiating and sustaining the rapid utilization of reserves that takes place during the formation of new cell structures.

## 4  FRUCTOSANS AND INULIN

Early evidence for the occurrence of fructose containing oligosaccharides and polymers in the Dasycladales has been summarized by Meeuse (1962). Examination of another member of this order, *Batophora oerstedi*, resulted in the recovery of a polysaccharide having many of the properties of inulin. It possessed an $(\alpha)_D$ −40° and analysis of the hydrolyzates established that fructose was the major constituent. X-ray diffraction patterns of the purified fructan were closely similar to those of authentic inulin (Meeuse 1963). The inulin-like structure of these oligosaccharides was established chemically with the recent demonstration in *Acetabularia crenulata* of a series of (1 → 2)-linked fructose units terminated by a glucose end group. The caps of these cells contained 25 times as much hot water soluble fructan as the stalks, and this was characterized as an inulin with a glucose to fructose ratio of 1:33 (Bourne *et al.* 1972).

The localization of the intracellular sites of fructan synthesis is clearly of interest, but conflicting reports have been published. A Seliwanoff positive material recovered from disrupted *Acetabularia* chloroplasts was reported to be inulin (Vanden Driessche & Bonotto 1967). The nature of this polysaccharide was reinvestigated in chloroplasts from three *Acetabularia* spp., and a glucan was found which was degraded by α-amylase and appeared to be a starch and not inulin (Werz & Clauss 1970). Reasons for these conflicting reports are unknown, but the latter result is in keeping with evidence that *Acetabularia* possesses enzymes associated with starch biosynthesis (Clauss 1959).

Fructose containing oligosaccharides are not restricted to the Dasycladales. A survey of nine members of the Cladophorales and *Urospora* sp. revealed that only the two *Chaetomorpha* spp. and *Urospora* sp. failed to produce fructosides other than sucrose (Percival & Young 1971a). A series of compounds ranging from a trisaccharide to polymeric fructans was observed in *Cladophora rupestris*, and the hexasaccharide and some polymeric material were isolated. Hydrolysis of the methylated derivatives and recovery of 3,4,6-tri-O-methylfructose with smaller amounts of 1,3,4,6-tetra-O-methylfructose, and 2,3,4,6-tetra-O-methyl-glucose, conclusively established that the glycosidic linkage was 1 → 2. These oligosaccharides are terminated by sucrose and belong to the inulin series. An oligosaccharide composed of fructose and glucose was also reported in a chlorophycean unicell (Craigie *et al.* 1967), and glucofructans occur in blue-green algae (Tsusue & Fujita 1964).

The metabolic role of these fructans has not been demonstrated in algae but it is presumed that they are photosynthetic reserves. A prolonged light period is reported to favour the accumulation of inulin in *Acetabularia* (Vanden Driessche 1969). No evidence of a seasonal variation was detected in the fructosan content of several *Cladophora* samples (Percival & Young 1971a).

# 5   LOW-MOLECULAR WEIGHT COMPOUNDS

## 5.1   *Sugars and simple glycosides*

Green algae were excluded from Table 7.1 as representatives of most orders characteristically produce sucrose. Some recent reports on its distribution are those of Craigie *et al.* (1966, 1967) and Percival and Young (1971a). Sucrose is synthesized by intact chloroplast preparations from *Acetabularia mediterranea* (Bidwell *et al.* 1969, 1970), and its formation may be stimulated by the presence of external osmoticants (Hiller & Greenway 1968), isonicotinyl hydrazide (Pritchard *et al.* 1963), and decreased by blue light (Hauschild *et al.* 1962, Laudenbach & Pirson 1969) or ammonium salts (Kanazawa *et al.* 1970, Bassham 1971). A novel derivative of sucrose, 4-O-lactyl β-D-fructofuranosyl α-D-glucopyranoside, occurs in *Cladophora laetevirens* and *Rhizoclonium* sp., while these and several additional members of the Cladophorales metabolize the oligosaccharides 6-O-D-glucosylmaltotriose, maltotriose, maltotetraose, and panose (Percival & Young 1971b). *Acetabularia crenulata* was the source of the rare ketose, *ribo*-hexulose or allulose (Bourne *et al.* 1972).

Sucrose is listed in Table 7.1 as a constituent of five freshwater red algae but the quantities are low and it cannot be considered as a true reserve product (Quillet 1965). It has not been reported in *Batrachospermum*, and there is no critical evidence to establish its presence in marine Rhodophyceae. The demonstration of sucrose in the flagellates *Euglena* and *Prymnesium* deserves further investigation, and it is noteworthy that sucrose is produced by mature *Euglena* whereas juvenile cells form maltose (Codd & Merrett 1971). The recovery of sucrose from *Fucus* is unusual and its presence may be due to associated Chlorophyceae.

**Table 7.1.** Soluble sugars and simple glycosides reported from algae other than the Chlorophyceae[1]

| | | |
|---|---|---|
| **Bacillariophyceae** | | |
| *Cyclotella nana* | G[2] | Wallen & Geen (1971) |
| *Phaeodactylum tricornutum* | G, Laminari-biose & -triose | Ford & Percival (1965) |
| *Skeletonema costatum* | G, O | Handa (1969) |
| **Chrysophyceae** | | |
| *Ochromonas malhamensis* | S, I, Fr, G, Ga, R, X, Allose | Kauss (1965,[3] 1966,[3] 1967a, 1967b, 1969a) |
| **Haptophyceae** | | |
| *Prymnesium parvum* | S, Fr, G | Gooday (1970b) |
| **Cyanophyceae** | | |
| *Anabaena flos-aquae* | F | Moore & Tischer (1965) |
| *Calothrix* sp. | G | Richardson *et al.* (1968) |
| *Nostoc* sp. | G | Richardson *et al.* (1968) |
| *N. commune* | S, Fr, Fr-osides, Ga, G, R, X | Quillet (1967) |
| *Oscillatoria amphibia* | S, G, O | Tarchevskii (1969) |
| *Rivularia bullata* | S, T, Fr, Fr-osides, Ga, G, R | Quillet (1967) |
| *Scytonema* sp. | G | Richardson *et al.* (1968) |
| *Tolypothrix tenuis* | G, Fr, Fr-osides | Tsusué & Fujita (1964), Tsusué & Yamakawa (1965) |
| **Rhodophyceae** | | |
| *Ahnfeltia plicata* | F | Majak *et al.* (1966) |
| *Batrachospermum* sp. | F, T | Feige (1970), Majak *et al.* (1966) |
| *Bornetia secundiflora* | Na-M, S | Quillet & Priou (1962), Quillet (1965) |
| *Bostrychia scorpioides* | F, S | Quillet (1965) |
| *Callithamnion tetricum* | S, T | Quillet (1965) |
| *Ceramium rubrum* | F | Majak *et al.* (1966) |
| *Chondrus crispus* | F, I | Craigie *et al.* (1968) |
| *Corallina officinalis* | F, I, T | Majak *et al.* (1966), Craigie *et al.* (1968) |
| *Cystoclonium purpureum* | F, I | Majak *et al.* (1966), Craigie *et al.* (1968) |
| *Delesseria sanguinea* | Na-M, T | Gervais (1959) |
| *Dumontia incrassata* | F, I | Craigie *et al.* (1968) |
| *Furcellaria fastigiata* | F, I | Craigie *et al.* (1968) |
| *Gigartina stellata* | F, I | Majak *et al.* (1966), Craigie *et al.* (1968) |
| *Iridophycus flaccidum* | F, I | Kauss (1968) |
| *Lemanea fluviatilis* | F | Feige (1970) |
| *L. nodosa* | F, S, T | Quillet (1965) |
| *Lithothamnion* sp. | F, T | Craigie *et al.* (1968) |
| *Phycodrys rubens* | F | Majak *et al.* (1966) |
| *Phyllophora membranifolia* | F, I | Craigie *et al.* (1968) |
| *Phymatolithon compactum* | F, T | Craigie *et al.* (1968) |

Rhodophyceae—*contd.*

| | | |
|---|---|---|
| *Polyides rotundus* | F, I | Majak *et al.* (1966), |
| | | Craigie *et al.* (1968) |
| *Porphyra leucosticta* | F, I | Craigie *et al.* (1968) |
| *P. linearis* | F, I | McLachlan *et al.* (1972) |
| *P. perforata* | F, I | Su & Hassid (1962a), |
| | | Kauss (1968) |
| *P. miniata* | F, I | Craigie *et al.* (1968) |
| *P. umbilicalis* | F, I, G | Turvey (1961), Peat & Rees |
| | | (1961), Majak *et al.* (1966), |
| | | Craigie *et al.* (1968) |
| *Porphyridium* sp. | F, G | Majak *et al.* (1966), |
| | | Craigie *et al.* (1968), |
| | | Feige (1970) |
| *Rhodomela larix* | Na-M | Whyte (1970) |
| *Rhodymenia palmata* | F, I | Craigie *et al.* (1968) |
| *Sacheria fluviatilis* | F, S, T | Quillet (1965) |
| *Serraticardia maxima* | F, I, M, T | Nagashima *et al.* (1969) |
| *Trailliella intricata* | F | Majak *et al.* (1966) |

Others

| | | |
|---|---|---|
| *Euglena gracilis* | M, S, T | Codd & Merrett (1971), |
| | | Marzullo & Danforth (1969) |
| *Fucus virsoides* | S | Coassini *et al.* (1967) |
| *Monodus subterraneus* | G, O | Beattie & Percival (1962) |

[1] See also Meeuse (1962).
[2] F = floridoside, Fr = fructose, G = glucose, Ga = galactose, I = isofloridoside, M = maltose, Na-M = sodium mannoglycerate, O = oligosaccharides, R = rhamnose, S= sucrose, T =trehalose, and X = xylose.
[3] Chromatograms also show spots thought to be altrose, gulose, idose, lyxose, ribose and sorbose.

Trehalose is the next most frequently identified algal disaccharide and ranges from 0·55% of the dry weight of *Rivularia* to the very low levels encountered in red algae. A trehalose phosphorylase from *Euglena gracilis* catalyzes the reversible reaction trehalose $\leftrightarrows$ β-glucose-1-phosphate + glucose (Belocopitow & Maréchal 1970).

Maltose occurs in several unrelated species and occasionally may be an important photosynthetic product of some green algae such as *Zoochlorella* (Muscatine 1965), and *Ulva lactuca* (Joshi 1967, Patil & Joshi 1970). Maltose, however, was not among the photosynthetic products of *U. lactuca* found by Percival and Smestad (1972).

The glycerol glycosides, floridoside and isofloridoside are widely distributed in the Rhodophyceae except in the Rhodomelaceae where they are largely replaced by sodium mannoglycerate and mannitol. Floridoside is generally present at several times the concentration of isofloridoside but occasionally the reverse occurs as in *Porphyra linearis* and *P. miniata*. It was thought that these α-galactosides were peculiar to red algae, but this belief must be modified as isofloridoside occurs in the chrysophyte, *Ochromonas malhamensis* (Kauss 1967a), and floridoside is rather common in the Cryptophyceae (Craigie unpubl.). Isofloridoside phosphate has been demonstrated in *Ochromonas* and probably is formed from α-glycerophosphate and UDP-galactose (Su & Hassid 1962b, Kauss 1969b).

Both D- and L-glycerol-(1 → 1)-O-α-D-galactopyranose (isofloridoside) have been reported from red algae either as a mixture (cf. Meeuse 1962, Craigie *et al.* 1968), or as the D-isomer alone (Peat & Rees 1961, Su & Hassid 1962a).

### 5.2   *Polyols*

Lewis and Smith (1967) reviewed the distribution and metabolism of polyols in various plants including algae. Table 7.2 supplements and updates this information.

Mannitol is the commonest polyol reported and is conspicuous by its almost universal presence in the Phaeophyceae, and in the Prasinophyceae where it replaces sucrose as a photosynthetic product. Various crystalline forms of mannitol are reported in the white efflorescence arising on the surface of desiccated brown algae of five genera (Walter-Levy & Strauss 1969). Mannitol is the translocation product in *Macrocystis* (Parker 1966). Levels of up to 0·004% of 1-mannitol-β-glucoside and 1,6-mannitol di-β-glucoside occur together with 1-mannitol acetate in *Fucus virsoides* (Coassini Lokar & Baradel 1967).

Numerous red algae also metabolize mannitol, but rarely does it accumulate in significant quantity except in the Ceramiales where floridoside is a minor reserve product. There are but few reports of mannitol in blue-green, green and other algae. *Nannochloris* sp. and *Stichococcus* sp. are unusual members of the Chlorophyceae in that they photosynthesize mannitol and not sucrose. Paper chromatographic evidence suggests that mannitol is an extracellular product of several other phytoplankton species including *Coccolithus huxleyi*, *Monochrysis lutheri*, and *Thalassiosira fluviatilis* (Hellebust 1965). Extracts of the latter three species have been examined on several occasions by paper and vapour phase chromatography but mannitol does not accumulate (Craigie unpubl.).

Exogenously supplied mannitol is poorly utilized by *Fucus* (Bidwell & Ghosh 1963, Drew 1969) whereas endogenous mannitol is readily respired, and is interconvertible with laminaran in both *Eisenia* and *Fucus* (Yamaguchi *et al.* 1966, Bidwell 1967). Drew (1969) investigated the metabolism of several [14]C-labelled sugars supplied to 11 species of brown algae. No species would convert fructose or galactose to mannitol, and most formed a glucan, presumably laminaran, from glucose. *Ascophyllum nodosum* and *Pelvetia canaliculata* transformed both glucose and mannose to mannitol. Enzymes that may be involved in mannitol biosynthesis were examined in five species of brown algae and the presence of aldolase, hexosediphosphatase, glucose phosphate isomerase and mannitol-1-phosphatase supports the suggestion that mannitol is formed from mannitol-1-

**Table 7.2.** Recent reports of polyols in algae[1]

| | | |
|---|---|---|
| Chlorophyceae | | |
| *Chlamydomonas* spp. | G[2] | Craigie *et al.* (1967) |
| *Cladophora* sp. | M | Craigie *et al.* (1966) |
| *Coccomyxa* sp. | R | Richardson *et al.* (1968) |
| *Dunaliella* spp. | G | Craigie & McLachlan (1964), Craigie *et al.* (1966), Wegmann (1969, 1971), Wallen & Geen (1971) |
| *Hyalococcus* sp. | S | Richardson *et al.* (1968) |
| *Myrmecia* sp. | R | Richardson *et al.* (1968) |
| *Nannochloris* sp. | M | Craigie *et al.* (1967) |
| *Stichococcus* sp. | M | Craigie *et al.* (1967) |
| *Trebouxia* sp. | R | Richardson *et al.* (1967, 1968), Richardson & Smith (1968a, 1968b), Komiya & Shibata (1971) |
| *Trentepohlia* sp. | E | Richardson *et al.* (1968) |
| Chrysophyceae | | |
| *Monochrysis lutheri* | G | Craigie (unpublished) |
| *Ochromonas malhamensis* | G, threitol | Kauss (1966) |
| Haptophyceae | | |
| *Prymnesium parvum* | G | Gooday (1970b) |
| Cyanophyceae | | |
| *Nostoc commune* | M, S, G, heptitol | Quillet (1967) |
| *Oscillatoria amphibia* | G, glycol | Tarchevskii (1969) |
| *Rivularia bullata* | M, S, G, heptitol | Quillet (1967) |
| Phaeophyceae | | |
| *Ascophyllum nodosum* | M | Munda (1964a, 1967) |
| *Eisenia bicyclis* | M | Yamaguchi *et al.* (1966, 1969) |
| *Fucus* sp. | M | Pompowski & Trokowicz (1966) |
| *F. ceranoides* | M | Munda (1967) |
| *F. serratus* | M | Munda (1967) |
| *F. vesiculosus* | M, G | Bidwell & Ghosh (1963), Bidwell (1967), Munda (1967) |
| *F. virsoides* | M | Coassini Lokar & Baradel (1967) |
| *Macrocystis pyrifera* | M, G | Parker (1966), Schweiger (1967) |
| *Sargassum natans* | M | Martinez Nadal *et al.* (1963) |
| *S. wightii* | M | Rao (1969) |
| *Stilophora rhizodes* | M | Craigie (unpublished) |
| *Turbinaria conoides* | M | Rao (1969) |
| Prasinophyceae | | |
| *Micromonas* spp. | M | Craigie *et al.* (1967) |
| *Prasinocladus* spp. | M | Craigie *et al.* (1967) |
| *Pyramimonas* sp. | M | Craigie *et al.* (1967) |
| *Tetraselmis* spp (*Platymonas*) | M | Craigie *et al.* (1966, 1967), Gooday (1970a) |
| Unidentified '*Carteria*' | M | Craigie *et al.* (1967) |

**Table 7.2.**—continued

| Rhodophyceae | | |
|---|---|---|
| *Ahnfeltia plicata* | M | Majak *et al.* (1966) |
| *Batrachospermum* sp. | G | Feige (1970) |
| *Bostrychia scorpoides* | D, S | Quillet (1965) |
| *Ceramium rubrum* | M | Majak *et al.* (1966) |
| *Corallina officinalis* | M | Majak *et al.* (1966), Craigie *et al.* (1968) |
| *Cystoclonium purpureum* | M | Majak *et al.* (1966) |
| *Gigartina stellata* | M | Majak *et al.* (1966) |
| *Lemanea fluviatilis* | M, G | Feige (1970) |
| *Lithothamnion* sp. | M | Craigie *et al.* (1968) |
| *Phycodrys rubens* | M | Majak *et al.* (1966) |
| *Phymatolithon compactum* | M | Craigie *et al.* (1968) |
| *Polyides rotundus* | M | Majak *et al.* (1966) |
| *Polysiphonia fastigiata* | | |
| (= *lanosa*) | M | Wickberg (1957) |
| *P. nigrescens* | M | Majak *et al.* (1966) |
| *Rhodomela larix* | M | Whyte (1970) |
| *Serraticardia maxima* | M | Nagashima *et al.* (1969) |
| *Trailliella intricata* | M | Majak *et al.* (1966) |
| Xanthophyceae | | |
| *Monodus subterraneus* | M | Beattie & Percival (1962) |

[1] See also Meeuse (1962), Lewis & Smith (1967).
[2] D = dulcitol (galactitol), E = erythritol, G = glycerol, M = mannitol, R = ribitol, and S = sorbitol.

phosphate which may arise from the reduction of fructose-6-phosphate (Yamaguchi *et al.* 1969).

Free glycerol occurs fairly widely in algae and is an important photosynthetic product in several zooxanthellae (Muscatine 1967, Trench 1971) and in some marine Volvocales, especially *Dunaliella* spp. (Craigie & McLachlan 1964, Craigie *et al.* 1967). Short-term experiments with $H^{14}CO_3^-$ show that glycerol is an early product of photosynthesis and probably arises from α-glycerophosphate derived from the reduction of dihydroxyacetone-phosphate (Wegmann 1968). More glycerol is produced in white light than in blue (Wallen & Geen 1971). Glycerol promotes the growth, in light, of a variety of phytoplankton species, especially members of the Chrysophyceae and Cryptophyceae (Cheng & Antia 1970), and it will serve as a carbon source for the heterotrophic growth of *Prymnesium parvum* (Rahat & Spira 1967, Gooday 1970b) and *Chroomonas salina* (Cheng & Antia 1970).

Sorbitol is reported from two blue-green algae, and sorbitol, ribitol, and erythritol occur in green phycobionts (Table 7.2). Only small amounts of these polyols may be produced by a free-living phycobiont but this does not necessarily reflect the potential of the lichenized alga (Richardson & Smith 1968b, Richardson *et al.* 1968).

## 5.3   Cyclitols

Five classes of algae are now known to contain free cyclitols as indicated in Table 7.3 and Fig. 7.1. Generally they do not accumulate in significant quantities but a notable exception is the 1,4/2,5-cyclohexanetetrol (X) of *Monochrysis lutheri* which acquires radioactivity within 60 seconds of photosynthesis in $H^{14}CO_3^-$ (Craigie unpubl.), and can account for 7% of the dry weight of the cells (Laycock & Craigie 1970). In *Porphyridium* sp. this compound is second only to floridoside in importance as a soluble photosynthetic product. A survey of 20 species of algae shows that green algae and *Euglena* are richer (up to 0·38% of the dry weight) in total *myo*-inisitol (I) than are red or brown algae while the cyclitol was undetectable in blue-green algae (Ikawa *et al.* 1968).

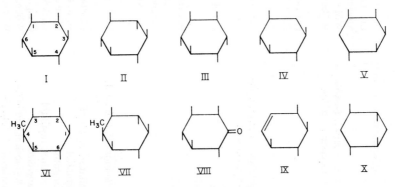

**Fig. 7.1.** Free cyclitols reported as algal constituents. I = *myo*-Inositol; II = *scyllo*-inositol; III = D-*chiro*-inositol; IV = 2-deoxy-*myo*-inositol (1,3,5/2,4-cyclohexanepentol); V = 1,2,3,4/5-cyclohexanepentol; VI = laminitol (1D-4-C-methyl-*myo*-inositol); VII = mytilitol; VIII = D-2,3,5/4,6-pentahydroxycyclo-hexanone (1-oxo-*myo*-inositol); IX = L-leucanthemitol (L-1,2,4/3-cyclohexene-tetrol); X = D(+)-1,4/2,5-cyclohexanetetrol. Vertical lines indicate the position of hydroxyl groups.

Kinetic $^{14}CO_2$ uptake studies with *Chlorella fusca* have shown that *myo*-inositol is formed by cyclization of D-glucose and is later epimerized to give D-*chiro*-inositol (III). This pathway of cyclitol biosynthesis is common to other organisms but does not account for the formation of the C-methyl cyclitols, laminitol (VI) or mytilitol (VII) (Wöber & Hoffmann-Ostenhof 1969). The methyl group does not arise from methionine, nor does the C-skeleton originate with *myo*-inositol (Wöber & Hoffmann-Ostenhof 1970b). Further experiments with $^{14}C$-labelled potential precursors of the C-methyl cyclitols revealed that *Porphyridium* sp. formed $^{14}C$-laminitol most readily from L-*gluco*-heptulose, glucose and arabinose in decreasing order of efficiency. Laminitol is isomerized to mytilitol by *C. fusca*, and the intermediate in the enzymatic reaction is D-2,3,5/4,6-pentahydroxycyclohexanone (VIII) (Wöber & Hoffmann-Ostenhof

**Table 7.3.** Cyclitols reported from algae

| | Cyclitols | References |
|---|---|---|
| **Bacillariophyceae** | | |
| *Phaeodactylum tricornutum* | *myo*-Inositol, *scyllo*-inositol, laminitol | Ford & Percival (1965) |
| **Chlorophyceae** | | |
| *Acetabularia crenulata* | *myo*-Inositol, 1,2,3,4/5-cyclohexanepentol[1] | Bourne *et al.* (1972) |
| *Chlorella* sp. | *myo*-Inositol | Lindberg (1955a) |
| *C. ellipsoidea* | *myo*-Inositol | Morimura (1959) |
| *C. fusca* | *myo*-Inositol, D-*chiro*-inositol, 1-oxo-*myo*-inositol, laminitol, L-leucanthemitol, mytilitol | Wöber & Hoffmann-Ostenhof (1969, 1970a), Wöber *et al.* (1971) |
| *C. pyrenoidosa* | *myo*-Inositol | Pratt & Johnson (1966) |
| *C. vulgaris* | *myo*-Inositol | Pratt & Johnson (1966) |
| **Chrysophyceae** | | |
| *Monochrysis lutheri* | D-(+)-1,4/2,5-Cyclohexanetetrol, *myo*-inositol, 2-deoxy-*myo*-inositol | Ramanathan *et al.* (1966), Laycock & Craigie (1970) |
| **Haptophyceae** | | |
| *Prymnesium parvum* | *myo*-Inositol | Gooday (1970b) |
| **Phaeophyceae** | | |
| *Desmarestia aculeata* | Laminitol | Bouveng & Lindberg (1955) |
| *Fucus spiralis* | Laminitol | Bouveng & Lindberg (1955) |
| *Laminaria cloustonii* | Laminitol | Lindberg & McPherson (1954) |
| *Macrocystis pyrifera* | Laminitol | Schweiger (1967) |
| **Rhodophyceae** | | |
| *Furcellaria fastigiata* | *myo*-Inositol, *scyllo*-inositol, laminitol, mytilitol | Lindberg (1955b), Wickberg (1957) |
| *Gelidium cartilagineum* | *myo*-Inositol, laminitol, mytilitol[2] | Lindberg (1955d) |
| *Porphyra perforata* | *scyllo*-Inositol, laminitol | Su & Hassid (1962a) |
| *P. umbilicalis* | *scyllo*-Inositol, laminitol | Lindberg (1955c) |
| *Polysiphonia fastigiata* (= *lanosa*) | *scyllo*-Inositol, laminitol, mytilitol | Wickberg (1957) |
| *Porphyridium* sp. Lewin strain | 1,4/2,5-Cyclohexanetetrol | Craigie *et al.* (1968) |
| *P. cruentum* | *myo*-Inositol, *scyllo*-inositol, laminitol, mytilitol, L-leucanthemitol, 1,4/2,5-cyclohexanetetrol | Wöber & Hoffmann-Ostenhof (1970b), Wöber (personal comm.) |
| *Serraticardia maxima* | Laminitol | Nagashima *et al.* (1969) |

1970a, Wöber *et al.* 1971). The epimerization requires both NAD+ and NADPH. $^{14}$C-Laminitol is also formed from photo-assimilated H$^{14}$CO$_3^-$ by *Porphyridium* sp. (Wöber & Hoffmann-Ostenhof 1970a), but is not readily synthesized by *Serraticardia* during photosynthesis (Nagashima *et al.* 1969).

### 5.4 Function in osmoregulation

That organic solutes function in osmotic control is thoroughly established with animals where amino acids and amines are recognized as important osmo-regulators. Physiologists have long employed sugars and polyols to adjust the osmotic environment for plant cells. It seems obvious, when one considers that some algae, e.g. Phaeophyceae, accumulate mannitol up to 50% of their dry weight, that organic solutes must contribute significantly to the osmotic balance in algal cells. Direct evidence that the intracellular level of simple carbohydrates could be affected by the salinity of the medium was provided by experiments with the euryhaline flagellate *Dunaliella*. Cells grown in 2·5 M NaCl contained 133 times more glycerol than those maintained at 0·025 M NaCl (Craigie & McLachlan 1964). *Dunaliella* responds similarly when osmoticants other than NaCl are used in the medium, and part of the excess glycerol may be formed through a fermentation process (Wegmann 1971).

Kauss (1967a,b) studied the response of *Ochromonas malhamensis* to various osmotic regulators and determined that the cellular concentration of isoflori-doside was directly dependent upon the external osmotic pressure. He further reported that the response was reversible and involved a reciprocal transforma-tion between isofloridoside and a polysaccharide. These experiments have been extended to the red macrophytes *Iridophycus flaccidum* and *Porphyra perforata* in which it was shown that floridoside and isofloridoside formation were markedly enhanced by elevated salinities (Kauss 1968, 1969a).

The 1,4/2,5-cyclohexanetetrol of *Monochrysis lutheri* is an example of another class of organic solute, the levels of which are governed by the external osmotic environment (Craigie 1969). Both steady-state and kinetic experiments were conducted and a reversible response was again observed. This time dilution of the medium caused the rapid expulsion of tetrol from the cells while increased salinity resulted in the slower photosynthetic accumulation of new tetrol. In a similar manner, sucrose accumulates in *Chlorella* when the cells are stressed with external osmoticants (Hiller & Greenway 1968).

It now seems reasonable to assume that some algae regulate the level of organic solutes as part of their osmotic control mechanism. Field data to support this hypothesis are few. Munda (1964a,b, 1967) measured the mannitol content of *Fucus* spp. and *Ascophyllum nodosum* plants taken from localities of varying salinities and did reciprocal transplanting experiments. The mannitol levels declined in plants moved to a lower salinity and increased in those moved to higher salinities. Although unknown factors may influence these results, the plants are probably responding to osmotic change.

# 6 REFERENCES

ABDEL-FATTAH A.F. & HUSSEIN M.M. (1970) Biochemical studies on marine algal constituents. I. Composition of some brown algae as influenced by seasonal variation. *Phytochem.* **9**, 721–4.

ALBRECHT G.J. & KAUSS H. (1971) Purification, crystallization and properties of a β-(1 → 3)- glucan phosphorylase from *Ochromonas malhamensis*. *Phytochem.* **10**, 1293–8.

ANDERSON D.M.W. & KING N.J. (1961a) Polysaccharides of the Characeae. I. Preliminary examination of a starch-type polysaccharide from *Nitella translucens*. *J. chem. Soc.* 2914–19.

ANDERSON D.M.W. & KING N.J. (1961b) Polysaccharides of the Characeae. III. The carbohydrate content of *Chara australis*. *Biochim. biophys. Acta*, **52**, 449–54.

ANNAN W.D., HIRST E.L. & MANNERS D.J. (1962) The position of mannitol in laminarin. *Chem. Ind.* 984–5.

ANNAN W.D., HIRST E.L. & MANNERS D.J. (1965a) The constitution of laminarin. IV. The minor component sugars. *J. chem. Soc.* 220–6.

ANNAN W.D., HIRST E.L. & MANNERS D.J. (1965b) The constitution of laminarin. V. The location of 1,6-glucosidic linkages. *J. chem. Soc.* 885–91.

ARCHIBALD A.R., CUNNINGHAM W.L., MANNERS D.J. & STARK J.R. (1963) Studies on the metabolism of the protozoa. X. The molecular structure of the reserve polysaccharides from *Ochromonas malhamensis* and *Peranema trichophorum*. *Biochem. J.* **88**, 444–51.

ARCHIBALD A.R., HIRST E.L., MANNERS D.J. & RYLEY J.F. (1960) Studies on the metabolism of the protozoa. VIII. The molecular structure of a starch-type polysaccharide from *Chilomonas paramecium*. *J. chem. Soc.* 556–60.

BARRAS D.R. & STONE B.A. (1968) Carbohydrate composition and metabolism in *Euglena*. In *The Biology of Euglena II*, ed. Buetow D.E., pp. 149–91. Academic Press, New York & London.

BARRAS D.R. & STONE B.A. (1969a) β-1,3-Glucan hydrolases from *Euglena gracilis*. I. The nature of the hydrolases. *Biochim. biophys. Acta*, **191**, 329–41.

BARRAS D.R. & STONE B.A. (1969b) β-1,3-Glucan hydrolases from *Euglena gracilis*. II. Purification and properties of the β-1,3-glucan exo-hydrolase. *Biochim. biophys. Acta* **191**, 342–53.

BASSHAM J.A. (1971) The control of photosynthetic carbon metabolism. *Science, N.Y.* **172**, 526–34.

BEATTIE A. & PERCIVAL E. (1962) The polysaccharides synthesized by *Monodus subterraneus* when grown on artificial media under bacteria-free conditions. *Proc. R. Soc. Edinb.* **68B**, 171–85.

BELOCOPITOW E. & MARÉCHAL L.R. (1970) Trehalose phosphorylase from *Euglena gracilis*. *Biochim. biophys. Acta*, **198**, 151–4.

BIDWELL R.G.S. (1967) Photosynthesis and metabolism in marine algae. VII. Products of photosynthesis in fronds of *Fucus vesiculosus* and their use in respiration. *Can. J. Bot.* **45**, 1557–65.

BIDWELL R.G.S. & GHOSH N.R. (1962) Photosynthesis and metabolism in marine algae. IV. The fate of $C^{14}$-mannitol in *Fucus vesiculosus*. *Can. J. Bot.* **40**, 803–7.

BIDWELL R.G.S. & GHOSH N.R. (1963) Photosynthesis and metabolism of marine algae. V. Respiration and metabolism of $C^{14}$-labelled glucose and organic acids supplied to *Fucus vesiculosus*. *Can. J. Bot.* **41**, 155–63.

BIDWELL R.G.S., LEVIN W.B. & SHEPHARD D.C. (1969) Photosynthesis, photorespiration, and respiration of chloroplasts from *Acetabularia mediterranea*. *Pl. Physiol., Lancaster*, **44**, 946–54.

BIDWELL R.G.S., LEVIN W.B. & SHEPHARD D.C. (1970) Intermediates of photosynthesis in *Acetabularia mediterranea* chloroplasts. *Pl. Physiol., Lancaster*, **45**, 70–5.

BIDWELL R.G.S., PERCIVAL E. & SMESTAD B. (1972) Photosynthesis and metabolism of marine algae. VIII. Incorporation of $^{14}$C into the polysaccharides metabolized by *Fucus vesiculosus* during pulse labelling experiments. *Can. J. Bot.* **50**, 191–7.

BOURNE E.J., PERCIVAL E. & SMESTAD B. (1972) Carbohydrates of *Acetabularia* species. I. *A. crenulata. Carbohyd. Res.* **22**, 75–82.

BOUVENG H. & LINDBERG B. (1955) Low-molecular carbohydrates in algae. VII. Investigation of *Fucus spiralis* and *Desmarestia aculeata. Acta chem. scand.* **9**, 168–9.

BUKHRYAKOVA L.K. (1965) Chemical composition of *Laminaria* on the southwest coast of Sakhalin Island. *Rybn. Khoz.* **41**, 62–4. *C.A.* **63**, 16777h.

BUKHRYAKOVA L.K. & LEVANIDOV I.P. (1969) Chemical composition of the Laminariaceae of the Sakhalin-Kuril Islands area. *Rast. Resur.* **5**, 183–7. *C.A.* **72**, 9883j.

BULL A.T. & CHESTERS C.G.C. (1966) The biochemistry of laminaran and the nature of laminarinase. *Adv. Enzymol.* **28**, 325–64.

BURSA A.S. (1968) Starch in the oceans. *J. Fish. Res. Bd. Canada,* **25**, 1269–84.

CHAO L. & BOWEN C.C. (1971) Purification and properties of glycogen isolated from a blue-green alga, *Nostoc muscorum. J. Bact.* **105**, 331–8.

CHENG J.Y. & ANTIA N.J.(1970) Enhancement by glycerol of phototrophic growth of marine planktonic algae and its significance to the ecology of glycerol pollution. *J. Fish. Res. Bd. Canada,* **27**, 335–46.

CLAUSS H. (1959) Das Verhalten der Phosphorylase in kernhaltigen und kernlosen Teilen von *Acetabularia mediterranea. Planta,* **52**, 534–42.

COASSINI LOKAR L.C. & BARADEL P. (1967) Sugli oligosaccaridi e sui polisaccaridi delle alghe del'alto Adriatico. III. Determinazione del contenuto di mannitolo nel *Fucus virsoides. Univ. Stude Trieste, Fac. Econ. Commer. Ist. Merceol.* No. 32, 5–21.

CODD G.A. & MERRETT M.J. (1971) Photosynthetic products of division synchronized cultures of *Euglena. Pl. Physiol., Lancaster,* **47**, 635–9.

CRAIGIE J.S. (1969) Some salinity-induced changes in growth, pigments, and cyclohexanetetrol content of *Monochrysis lutheri. J. Fish. Res. Bd. Canada,* **26**, 2959–67.

CRAIGIE J S., LEIGH C., CHEN L.C.-M. & MCLACHLAN J. (1971) Pigments, polysaccharides, and photosynthetic products of *Phaeosaccion collinsii. Can. J. Bot.* **49**, 1067–74.

CRAIGIE J.S. & MCLACHLAN J. (1964) Glycerol as a photosynthetic product in *Dunaliella tertiolecta* Butcher. *Can. J. Bot.* **42**, 777–8.

CRAIGIE J.S., MCLACHLAN J., ACKMAN R.G. & TOCHER C.S. (1967) Photosynthesis in algae. III. Distribution of soluble carbohydrates and dimethyl-β-propiothetin in marine unicellular Chlorophyceae and Prasinophyceae. *Can. J. Bot.* **45**, 1327–34.

CRAIGIE J.S., MCLACHLAN J., MAJAK W., ACKMAN R.G. & TOCHER C.S. (1966) Photosynthesis in algae. II. Green algae with special reference to *Dunaliella* spp. and *Tetraselmis* spp. *Can. J. Bot.* **44**, 1247–54.

CRAIGIE J.S., MCLACHLAN J. & TOCHER R.D. (1968) Some neutral constituents of the Rhodophyceae with special reference to the occurrence of the floridosides. *Can. J. Bot.* **46**, 605–11.

DREW E.A. (1969) Uptake and metabolism of exogenously supplied sugars by brown algae. *New Phytol.* **68**, 35–43.

DUYNSTEE E.E. & SCHMIDT R.R. (1967) Total starch and amylose levels during synchronous growth of *Chlorella pyrenoidosa. Archs. Biochem. Biophys.* **119**, 382–6.

DWYER M.R. & SMILLIE R.M. (1970) A light-induced β-1,3-glucan breakdown associated with the differentiation of chloroplasts in *Euglena gracilis. Biochim. biophys. Acta,* **216**, 392–401.

DWYER M.R. & SMILLIE R.M. (1971) β-1,3-Glucan: a source of carbon and energy for chloroplast development in *Euglena gracilis. Aust. J. biol. Sci.* **24**, 15–22.

DWYER M.R., SMYDZUK J. & SMILLIE R.M. (1970) Synthesis and breakdown of β-1.3-glucan in *Euglena gracilis* during growth and carbon depletion. *Aust. J. biol. Sci.* **23**, 1005–13.

FEIGE B. (1970) Beiträge zur Physiologie einheimischer Algen. I. $^{14}$C-Markierungsprodukte von drei Süsswasserrotalgen. *Z. Pfl. Physiol.* **63**, 288–91.

FLEMING M. & MANNERS D.J. (1965) A structural difference between the soluble and insoluble forms of laminarin. *Biochem. J.* **94**, 17P.

FLEMING M., HIRST E. & MANNERS D.J. (1966) The constitution of laminarin. VI. The fine structure of soluble laminarin. *Proc. Int'l. Seaweed Symposium*, **5**, 255–60.

FORD C.W. & PERCIVAL E. (1965) The carbohydrates of *Phaeodactylum tricornutum*. I. Preliminary examination of the organism and characterization of low molecular weight material and of a glucan. *J. chem. Soc.* 7035–41.

FREDRICK J.F. (1951) Preliminary studies on the synthesis of polysaccharides in the algae. *Physiologia Pl.* **4**, 621–6.

FREDRICK J.F. (1952) Preliminary studies on the synthesis of polysaccharides in the algae. II. A polysaccharide variant of *Oscillatoria princeps*. *Physiologia Pl.* **5**, 37–40.

FREDRICK J.F. (1961) Immunochemical studies of phosphorylases of Cyanophyceae. *Phyton, B. Aires*, **16**, 21–6.

FREDRICK J.F. (1962) Multiple molecular forms of 4-glucosyltransferase (phosphorylase) in *Oscillatoria princeps*. *Phytochem.* **1**, 153–7.

FREDRICK J.F. (1963) An algal α-glucan phosphorylase which requires adenosine-5-phosphate as coenzyme. *Phytochem.* **2**, 413–15.

FREDRICK J.F. (1967) Glucosyltransferase isozymes in algae. *Phytochem.* **6**, 1041–6.

FREDRICK J.F. (1968a) Biochemical evolution of glucosyltransferase isozymes in algae. *Ann. N. Y. Acad. Sci.* **151**, 413–23.

FREDRICK J.F. (1968b) Multiple forms of polyglucoside-branching enzyme in the algae. *Physiologia Pl.* **21**, 176–82.

FREDRICK J.F. (1968c) Glucosyltransferase isozymes in algae. II. Properties of branching enzymes. *Phytochem.* **7**, 931–6.

FREDRICK J.F. (1968d) Glucosyltransferase isozymes in algae. III. The polyglucoside and enzymes of *Cyanidium caldarium*. *Phytochem.* **7**, 1573–6.

FREDRICK J.F. (1969) Effects of amitrole on biosynthesis of phosphorylases in different algae. *Phytochem.* **9**, 725–9.

FREDRICK J.F. (1970) Evolution of polyglucoside synthesizing isozymes in the algae. *Ann. N.Y. Acad. Sci.* **175**, 524–30.

FREDRICK J.F. (1971a) Polyglucan branching isoenzymes of algae. *Physiologia Pl.* **24**, 55–8.

FREDRICK J.F. (1971b) *De novo* synthesis of polyglucans by a phosphorylase isoenzyme in algae. *Physiologia Pl.* **25**, 32–4.

FREDRICK J.F. (1971c) Storage polyglucan-synthesizing isozyme patterns in the Cyanophyceae. *Phytochem.* **10**, 395–8.

FREDRICK J.F. (1971d) Requirement for pyridoxal in the biosynthesis of 4-glucosyltransferases (phosphorylases) in the algae. *Phytochem.* **10**, 1025–9.

FREDRICK J.F. & GENTILE A.C. (1960) The effect of 3-amino-1,2,4-triazole on phosphorylase of *Oscillatoria princeps*. *Archs. Biochem. Biophys.* **86**, 30–3.

FRITSCH F.E. (1945) *The structure and reproduction of the algae*. II. Cambridge University Press.

FUHS G.W. (1963) Cytochemish-elektronenmikroscopische Lokalisierung der Ribonukleinsäure und des Assimilates in Cyanophyceen. *Protoplasma*, **56**, 178–87.

GAILLARD B.D.E. & BAILEY R.W. (1966) Reaction with iodine of polysaccharides dissolved in strong calcium chloride solution. *Nature, Lond.* **212**, 202–3.

GERVAIS J. (1959) Les glucides de la Floridée *Delesseria sanguinea* (L.) Lamouroux. *Rev. gén. Bot.* **66**, 395–401.

GIBBS M., LATZKO E., HARVEY M.J., PLAUT Z. & SHAIN Y. (1970) Photosynthesis in the algae. *Ann. N.Y. Acad. Sci.* **175**, 541–54.

GIESY R.M. (1964) A light and electron microscope study of interlamellar polyglucoside bodies in *Oscillatoria chalybia*. *Am. J. Bot.* **51**, 388–96.

GOODAY G.W. (1970a) A physiological comparison of the symbiotic alga *Platymonas convolutae* and its free-living relatives. *J. mar. biol. Ass. U.K.* **50**, 199–208.

GOODAY G.W. (1970b) Aspects of the carbohydrate metabolism of *Prymnesium parvum*. *Arch. Mikrobiol.* **72**, 9–15.

GOODAY G.W. (1971a) Control by light of starch degradation and cell-wall biosynthesis in *Platymonas tetrathele. Biochem. J.* **123**, 3P.

GOODAY G.W. (1971b) A biochemical and autoradiographic study of the role of the golgi bodies in thecal formation in *Platymonas tetrathele. J. exp. Bot.* **22**, 959–71.

GREENWOOD C.T. & THOMPSON J. (1961) Physicochemical studies on starch. XXIII. Physical properties of floridean starch and the characterization of structure-type of branched α-1,4-glucans. *J. chem. Soc.* 1534–7.

HANDA N. (1969) Carbohydrate metabolism in the marine diatom *Skeletonema costatum. Mar. Biol.* **4**, 208–14.

HANDA N. & NISIZAWA K. (1961) Structural investigation of a laminaran isolated from *Eisenia bicyclis. Nature, Lond.* **192**, 1078–80.

HANDA N. & TOMINAGA H. (1969) A detailed analysis of carbohydrates in marine particulate matter. *Mar. Biol.* **2**, 228–35.

HAUSCHILD A.W.H., NELSON C.D. & KROTKOV G. (1962) The effect of light quality on the products of photosynthesis in green and blue-green algae, and in photosynthetic bacteria. *Can. J. Bot.* **40**, 1619–30.

HELLEBUST J.A. (1965) Excretion of some organic compounds by marine phytoplankton. *Limnol. Oceanogr.* **10**, 192–206.

HILLER R.G. & GREENWAY H. (1968) Effects of low water potentials on some aspects of carbohydrate metabolism in *Chlorella pyrenoidosa. Planta*, **78**, 49–59.

HOUGH L., JONES J.K.N. & WADMAN W.H. (1952) An investigation of the polysaccharide components of certain fresh-water algae. *J. chem. Soc.* 3392–9.

IKAWA M., BOROWSKI P.T. & CHAKRAVARTI A. (1968) Choline and inositol distribution in algae and fungi. *Appl. Microbiol.* **16**, 620–3.

JOSHI G.V. (1967) Studies in photosynthesis in marine plants of Bombay. In *Proc. Seminar Sea, Salt & Plants*, ed. Krishnamurthy V. pp. 256–64. Catholic Press, Ranchi, India.

KANAZAWA T., KIRK M. & BASSHAM J.A. (1970) Regulatory effects of ammonia on carbon metabolism in photosynthesizing *Chlorella pyrenoidosa. Biochim. biophys. Acta*, **205**, 401–8.

KANDLER O. (1961) Verteilung von $C^{14}$ nach photosynthese in $^{14}CO_2$ von *Anacystis nidulans. Náturwissenschaften*, **48**, 604.

KASHIWABARA Y., SUZUKI H. & NISIZAWA K. (1965) Some enzymes relating to the metabolism of starch in a sea-lettuce. *Pl. Cell. Physiol., Tokyo*, **6**, 537–46.

KAUSS H. (1962) Der Einfluss von Thiamin- und Biotin-mangel auf den Stoffwechsel von *Ochromonas malhamensis. Vortr. Gesamtbeg. Bot.* No. 1, 129–32.

KAUSS H. (1965) Isolierung und Identifizierung von Allose aus der chrysomonadalen Alge *Ochromonas. Z. Pflanzenphysiol.* **53**, 58–63.

KAUSS H. (1966) Isolierung und identifizierung von Threit. *Z. Pfl. Physiol.* **55**, 85–8.

KAUSS H. (1967a) Isofloridosid und Osmoregulation bei *Ochromonas malhamensis. Z. Pflanzenphysiol.* **56**, 453–65.

KAUSS H. (1967b) Metabolism of isofloridoside (O-α-D-galactopyranosyl-(1 → 1)-glycerol) and osmotic balance in the fresh water alga *Ochromonas. Nature, Lond.* **214**, 1129–30.

KAUSS H. (1968) α-Galaktosyl-glyzeride und Osmoregulation in Rotalgen. *Z. Pfl. Physiol.* **58**, 428–43.

KAUSS H. (1969a) Osmoregulation mit α-Galaktosylglyzeriden bei *Ochromonas* und Rotalgen. *Ber. dt. bot. Ges.* **82**, 115–25.

KAUSS H. (1969b) Isofloridosidephosphat, ein Zwischenstoff bei der Osmoregulation in *Ochromonas. Z. Naturf.* **24**b, 363–4.

KAUSS H. & KANDLER O. (1962) Adenosindiphosphatglucose aus *Chlorella*. *Z. Naturf.* **17b**, 858–60.

KAUSS H. & KRIEBITZSCH CH. (1969) Demonstration and partial purification of a β-(1 → 3)-glucan phosphorylase. *Biochem. biophys. Res. Comm.* **35**, 926–30.

KINDEL P. & GIBBS M. (1963) Distribution of carbon-14 in polysaccharide after photosynthesis in carbon dioxide labelled with carbon-14 by *Anacystis nidulans*. *Nature, Lond.* **200**, 260–1.

KLEIN R.M. & CRONQUIST A. (1967) A consideration of the evolutionary and taxonomic significance of some biochemical, micromorphological, and physiological characters in the thallophytes. *Q. Rev. Biol.* **42**, 105–296.

KOMIYA T. & SHIBATA S. (1971) Polyols produced by the cultured phyco- and mycobionts of some *Ramalina* species. *Phytochem.* **10**, 695–9.

KREGER D.R. & VEER J. VAN DER (1971) Paramylon in a chrysophyte. *Acta Bot. neerl.* **19**, 401–2.

LABRUTO G. & BRUNO E. (1960) Constituents of *Saccorhiza bulbosa*, a phaeophyte inhabiting the straits of Messina. *Ann. Chim. Rome*, **50**, 1349–56. *C.A.* **55**, 6610i.

LANG N.J. (1968) The fine structure of blue-green algae. *A. Rev. Microbiol.* **22**, 15–46.

LAUDENBACH B. & PIRSON A. (1969) Über den Kohlenhydratumsatz in *Chlorella* unter dem Einfluss von blauem und rotem Licht. *Arch. Mikrobiol.* **67**, 226–42.

LAYCOCK M.V. & CRAIGIE J.S. (1970) *Myo*-inositol and 2-deoxy-*myo*-inositol in the chrysophycean alga *Monochrysis lutheri* Droop. *Can. J. Biochem.* **48**, 699–701.

LEWIS D.H. & SMITH D.C. (1967) Sugar alcohols (polyols) in fungi and green plants. I. Distribution, physiology and metabolism. *New Phytol.* **66**, 143–84.

LINDBERG B. (1955a) Low-molecular carbohydrates in algae. VIII. Investigation of two green algae. *Acta chem. scand.* **9**, 169.

LINDBERG B. (1955b) Low-molecular carbohydrates in algae. X. Investigation of *Furcellaria fastigiata*. *Acta chem. scand.* **9**, 1093–6.

LINDBERG B. (1955c) Low-molecular carbohydrates in algae. XI. Investigation of *Porphyra umbilicalis*. *Acta chem. scand.* **9**, 1097–9.

LINDBERG B. (1955d) Methylated taurines and choline sulphate in red algae. *Acta chem. scand.* **9**, 1323–6.

LINDBERG B. & MCPHERSON J. (1954) Low-molecular carbohydrates in algae. VI. Laminitol, a new C-methyl inositol from *Laminaria cloustoni*. *Acta chem. scand.* **8**, 1875–6.

LOEWUS F. (1971) Carbohydrate interconversions. *A. Rev. Pl. Physiol.* **22**, 337–64.

LOKAR A. (1967) Sugli oligosaccaridi e sui polisaccaridi delle alghe dell' alto Adriatico. IV. Determinazione del contenuto di laminarina di alcune Phaeophyceae. *Univ. Studi Trieste, Fac. Econ. Commer. Ist. Merceol.* No. 33, 5–15.

LOVE J., MACKIE W., MCKINNELL J.W. & PERCIVAL E. (1963) Starch-type polysaccharides isolated from the green seaweeds, *Enteromorpha compressa, Ulva lactuca, Cladophora rupestris, Codium fragile*, and *Chaetomorpha capillaris. J. chem. Soc.* 4177–82.

MACKIE I.M. & PERCIVAL E. (1960) Polysaccharides from the green seaweed *Caulerpa filiformis*. II. A glucan of amylopectin type. *J. chem. Soc.* 2381–4.

MAEDA M. & NISIZAWA K. (1968a) Fine structure of laminaran of *Eisenia bicyclis. J. Biochem.* **63**, 199–206.

MAEDA M. & NISIZAWA K. (1968b) Laminaran of *Ishige okamurai. Carbohyd. Res.* **7**, 94–7.

MAJAK W., CRAIGIE J.S. & MCLACHLAN J. (1966) Photosynthesis in algae. I. Accumulation products in the Rhodophyceae. *Can. J. Bot.* **44**, 541–9.

MANGAT B.S. & BADENHUIZEN N.P. (1970) Studies on the origin of amylose and amylopectin in starch granules. II. Changes in amylose content and enzyme activities in cultures of *Polytoma uvella. Stärke*, **22**, 329–33.

MANGAT B.S. & BADENHUIZEN N.P. (1971) Studies on the origin of amylose and amylopectin in starch granules. III. The effect of temperature on enzyme activities and amylose content. *Can. J. Bot.* **49**, 1787–92.

MANNERS D.J. (1964) Carbohydrates in marine algae. *Proc. Int. Seaweed Symp.* **4**, 348–51.

MANNERS D.J., MERCER G.A., STARK J.R. & RYLEY J.F. (1965) Studies on the metabolism of the protozoa. The molecular structure of a starch-type polysaccharide from *Polytoma uvella*. *Biochem. J.* **97**, 530–2.

MANNERS D.J., PENNIE I.R. & RYLEY J.F. (1967) The reserve polysaccharides of *Prototheca zopfii*. *Biochem. J.* **104**, 32P.

MANNERS D.J. & WRIGHT A. (1962) α-1,4-Glucosans. XIV. The interaction of concanavalin-A with glycogens. *J. chem. Soc.* 4592–5.

MARKER A.F.H. (1965) Extracellular carbohydrate liberation in the flagellates *Isochrysis galbana* and *Prymnesium parvum*. *J. mar. biol. Ass. U.K.* **45**, 755–72.

MARTINEZ NADAL N.G., RODRIGUEZ L.V. & CASILLAS C. (1963) Low molecular weight carbohydrates in *Sargassum natans* from Puerto Rico. *J. Pharm. Sci.* **52**, 498.

MARZULLO G. & DANFORTH W.F. (1969) Ethanol-soluble intermediates and products of acetate metabolism by *Euglena gracilis* var. *bacillaris*. *J. gen. Microbiol.* **55**, 257–66.

MATSUKA M., OTSUKA H. & HASE E. (1966) Changes in contents of carbohydrate and fatty acid in the cells of *Chlorella protothecoides* during the processes of de- and regeneration of chloroplasts. *Pl. Cell. Physiol., Tokyo*, **7**, 651–62.

MCCRACKEN D.A. & BADENHUIZEN N.P. (1970) Studies on the origin of amylose and amylopectin in starch granules. I. The use of *Polytoma uvella* as a source of NDPG-glucosyltransferases and some properties of the enzymes. *Stärke*, **22**, 289–91.

MCLACHLAN J., CRAIGIE J.S., CHEN L.C.-M. & OGATA E. (1972) *Porphyra linearis* Grev. An edible species from Nova Scotia. *Proc. Int. Seaweed Symp.* **7**, 473–6.

MEEUSE B.J.D. (1962) Storage products. In *Physiology and Biochemistry of Algae*, ed. Lewin R.A. pp. 289–311. Academic Press, New York & London.

MEEUSE B.J.D. (1963) Inulin in the green alga *Batophora oerstedi* J. Ag. (Dasycladales). *Acta bot. neerl.* **12**, 315–18.

MEEUSE B.J.D., ANDRIES M. & WOOD J.A. (1960) Floridean starch. *J. exp. Bot.* **11**, 129–40.

MEEUSE B.J.D. & SMITH B.N. (1962) A note on the amylolytic breakdown of some raw algal starches. *Planta*, **57**, 624–35.

MITCHELL J.L.A. (1971) Photoinduced division synchrony in permanently bleached *Euglena gracilis*. *Planta*, **100**, 244–57.

MOORE B.G. & TISCHER R.G. (1965) Biosynthesis of extracellular polysaccharides by the blue-green alga *Anabaena flos-aquae*. *Can. J. Microbiol.* **11**, 877–85.

MORIMURA Y. (1959) Synchronous culture of *Chlorella*. II. Changes in content of various vitamins during the course of the algal life cycle. *Pl. Cell. Physiol., Tokyo*, **1**, 63–9.

MÜLLER H.-M. (1961) Über die Veränderung der chemischen zusammensetzung von *Scenedesmus obliquus* bei synchroner Kultur im Licht-Dunkel-Wechsel. *Planta*, **56**, 555–74.

MUNDA I. (1964a) The quantity and chemical composition of *Ascophyllum nodosum* along the coast between the rivers Olfusa and Thjorsa (Southern Iceland). *Botanica mar.* **7**, 76–89.

MUNDA I. (1964b) Observations on variation in form and chemical composition of *Fucus ceranoides* L. *Nova Hedwigia*, **8**, 403–14.

MUNDA I. (1967) Der Einfluss der Salinität auf die chemische Zusammensetzung, das Wachstum und die Fruktifikation einiger Fucaceen. *Nova Hedwigia*, **13**, 471–508.

MUSCATINE L. (1965) Symbiosis of hydra and algae. III. Extracellular products of the algae. *Comp. Biochem. Physiol.* **16**, 77–92.

MUSCATINE L. (1967) Glycerol excretion by symbiotic algae from corals and *Tridacna* and its control by the host. *Science, N.Y.* **156**, 516–19.

NAGASHIMA H., NAKAMURA S. & NISIZAWA K. (1968) Biosynthesis of floridean starch by chloroplast preparations from a marine red alga *Serraticardia maxima*. *Bot. Mag. Tokyo*, **81**, 411–13.

NAGASHIMA H., NAKAMURA S., NISIZAWA K. & HORI T. (1971) Enzymic synthesis of floridean starch in a red alga *Serraticardia maxima*. *Pl. Cell Physiol., Tokyo,* **12**, 243–53.

NAGASHIMA H., OZAKI H., NAKAMURA S. & NISIZAWA K. (1969) Physiological studies on floridean starch, floridoside and trehalose in a red alga, *Serraticardia maxima. Bot. Mag. Tokyo,* **82**, 462–73.

NIIMI S., ISHIHARA H., IKEZAWA H. & ISHIZAKA O. (1969) Utilization of *Chlorella vulgaris* components. V. Externally secreted carbohydrase. *Nagoya Shiritsu Daigaku Yakugakuba Kenkyu Nempo,* **15**, 225–31. *C.A.* **72**, 62957.

NIKAIDO H. & HASSID W.Z. (1971) Biosynthesis of saccharides from glycopyranosyl esters of nucleoside pyrophosphates ('sugar nucleotides'). *Adv. Carbohyd. Chem. Biochem.* **26**, 351–483.

OHAD I., SIEKEVITZ P. & PALADE G.E. (1967) Biogenesis of chloroplast membranes. II. Plastid differentiation during greening. *J. Cell. Biol.* **35**, 553–84.

OLAITAN S.A. & NORTHCOTE D.H. (1962) Polysaccharides of *Chlorella pyrenoidosa. Biochem. J.* **82**, 509–19.

OZAKI H., MAEDA M. & NISIZAWA K. (1967) Floridean starch of a calcareous red alga, *Joculator maximus. J. Biochem., Tokyo,* **61**, 497–503.

PARKER B.C. (1966) Translocation in (the giant kelp) *Macrocystis*. III. Composition of sieve tube exudate and identification of the major [14]C-labelled products. *J. Phycol.* **2**, 38–41.

PARSONS T.R., STEPHENS K. & STRICKLAND J.D.H. (1961) On the chemical composition of eleven species of marine phytoplankters. *J. Fish. Res. Bd. Canada,* **18**, 1001–16.

PATIL B.A. & JOSHI G.V. (1970) Photosynthetic studies in *Ulva lactuca. Botanica mar.* **13**, 111–15.

PEAT S. & REES D.A. (1961) Carbohydrase and sulphatase activities of *Porphyra umbilicalis. Biochem. J.* **79**, 7–12.

PEAT S. & TURVEY J.R. (1965) Polysaccharides of marine algae. *Fortschr. Chem. Org. Naturstoffe,* **23**, 1–45.

PEAT S., TURVEY J.R. & EVANS J.M. (1959) The structure of floridean starch. II. Enzymic hydrolysis and other studies. *J. chem. Soc.* 3341–4.

PEAT S., TURVEY J.R. & REES D.A. (1961) Carbohydrates of the red alga, *Porphyra umbilicalis. J. chem. Soc.* 1590–5.

PELLEGRINI L. & PELLEGRINI M. (1971) Contribution à l'étude biochimique des Cystoseiracées Mediterranéennes. I. *Cystoseira stricta* (Mont.) Sauvageau. *Botanica mar.* **14**, 6–16.

PERCIVAL E. (1968) Marine algal carbohydrates. *Oceanogr. Mar. Biol. Ann. Rev.* **6**, 137–61.

PERCIVAL E. & MCDOWELL R.H. (1967) *Chemistry and enzymology of marine algal polysaccharides*. Academic Press, London & New York.

PERCIVAL E. & SMESTAD B. (1972) Photosynthetic studies on *Ulva lactuca. Phytocem.* **11**, 1967–72.

PERCIVAL E. & YOUNG M. (1971a) Low molecular weight carbohydrates and water-soluble polysaccharide metabolized by the Cladophorales. *Phytochem.* **10**, 807–12.

PERCIVAL E. & YOUNG M. (1971b) Characterization of sucrose, lactate and other oligosaccharides found in the Cladophorales. *Carbohyd. Res.* **20**, 217–23.

POMPOWSKI T. & TROKOWICZ D. (1966) Determination of mannitol in the Baltic *Fucus* (changes in the mannitol content depending on the season). *Przem. Spozyw.* **20**, 491–2. *C.A.* **66**, 92236v.

POWELL J.H. & MEEUSE B.J.D. (1964) Laminarin in some Phaeophyta of the Pacific Coast. *Econ. Bot.* **18**, 164–6.

PRATT R. & JOHNSON E. (1966) Production of pantothenic acid and inositol by *Chlorella vulgaris* and *C. pyrenoidosa. J. Pharm. Sci.* **55**, 799–802.

PREISS J. & GREENBERG E. (1967) Biosynthesis of starch in *Chlorella pyrenoidosa*. I. Purification and properties of the adenosine diphosphoglucose; α-1,4-glucan, α-4-glucosyl transferase from *Chlorella. Archs. Biochem. Biophys.* **118**, 702–8.

PREISS J. & KOSUGE T. (1970) Regulation of enzyme activity in photosynthetic systems. *A. Rev. Pl. Physiol.* **21**, 433–66.

PRITCHARD C.G., WHITTINGHAM C.P. & GRIFFIN W.J. (1963) The effect of isonicototinyl-hydrazide on the photosynthetic incorporation of radio-active carbon dioxide into ethanol-soluble compounds of *Chlorella. J. exp. Bot.* **14**, 281–9.

QUILLET M. (1965) Sur le saccharose produit par les Rhodophycees. *C. r. hebd. Séanc. Acad. Sci. Paris*, **260**, 6192–4.

QUILLET M. (1967) Sur le saccharose et les glucides vascuolaires de deux espèces d'Algues bleues: *Rivularia bullata* (Berk.) et *Nostoc commune* (Vauch.) *C. r. hebd. Séanc. Acad. Sci. Paris*, **264**, 1718–20.

QUILLET M. & PRIOU M.L. (1962) Sur le sucre soluble et les glucides membranaires de l'algue rouge *Bornetia secundiflora* (J. Ag.) Thuret. *C. r. hebd. Séanc. Acad. Sci. Paris*, 2210–12.

RAHAT M. & SPIRA Z. (1967) Specificity of glycerol for dark growth of *Prymnesium parvum. J. Protozool.* **14**, 45–8.

RAMANATHAN J.D., CRAIGIE J.S., MCLACHLAN J., SMITH D.G. & MCINNES A.G. (1966) Occurrence of D-(+)-1,4/2,5-cyclohexanetetrol in *Monochrysis lutheri* Droop. *Tetrahedron Letters*, No. 14, 1527–31.

RAO M.U. (1969) Seasonal variations in growth, alginic acid, and mannitol contents of *Sargassum wightii* and *Turbinaria conoides. Proc. Int. Seaweed Symp.* **6**, 579–84.

RICHARDSON D.H.S., HILL D.J. & SMITH D.C. (1968) Lichen physiology. XI. The role of the alga in determining the pattern of carbohydrate movement between lichen symbionts. *New Phytol.* **67**, 469–86.

RICHARDSON D.H.S., SMITH D.C. & LEWIS D.H. (1967) Carbohydrate movement between the symbionts of lichens. *Nature, Lond.* **214**, 879–82.

RICHARDSON D.H.S. & SMITH D.C. (1968a) Lichen physiology. IX. Carbohydrate movement from the *Trebouxia* symbiont of *Xanthoria aureola* to the fungus. *New Phytol.* **67**, 61–8.

RICHARDSON D.H.S. & SMITH D.C. (1968b) Lichen physiology. X. The isolated algal and fungal symbionts of *Xanthoria aureola. New Phytol.* **67**, 69–77.

RICHTER G. & PIRSON A. (1957) Enzyme von *Hydrodictyon* und ihre Beeinflussung durch Beleuchtungsperiodik. *Flora Jena*, **144**, 562–97.

RICKETTS T.R. (1966) On the chemical composition of some unicellular algae. *Phytochem.* **5**, 67–76.

RIS H. & SINGH R.N. (1961) Electron microscope studies on blue-green algae. *J. biophys. biochem. Cytol.* **9**, 63–80.

RUPPEL H-G. (1962) Untersuchungen über die Zusammensetzung von *Chlorella* bei Synchronisation im Licht-Dunkel-Weschel. *Flora Jena*, **152**, 113–38.

SANWAL G.G. & PREISS J. (1967) Biosynthesis of starch in *Chlorella pyrenoidosa*. II. Regulation of ATP- α-D-glucose 1-phosphate adenyl transferase (ADP-glucose pyrophosphorylase) by inorganic phosphate and 3-phosphoglycerate. *Archs. Biochem. Biophys.* **119**, 454–69.

SCHWEIGER R.G. (1967) Low-molecular weight compounds in *Macrocystis pyrifera* a marine alga. *Archs. Biochem. Biophys.* **118**, 383–7.

SERENKOV G.P. & ZLOCHEVSKAYA I.V. (1962) Chemical composition of two species of red algae. *Nauchn. Dokl. Vysshei Shkoly, Biol. Nauki*, No. 4, 151–6. *C.A.* **58**, 8236c.

SHAH H.N. & RAO A.V. (1969) Laminaran from Indian brown seaweeds. *Curr. Sci.* **38**, 413–14.

SHEELER P., MOORE J., CANTOR M. & GRANIK R. (1968) The stored polysaccharides of *Polytomella agilis. Life Sciences*, **7**, 1045–51.

SMILLIE R.M. (1968) Enzymology of *Euglena*. In *The biology of Euglena* II, ed. Buetow D.E., pp. 1–54. Academic Press, New York & London.

STEWART C.M., HIGGINS H.G. & AUSTIN S. (1961) Seasonal variation in alginic acid, mannitol, laminarin and fucoidin in the brown alga *Ecklonia radiata. Nature, Lond.* **192**, 1208.

SU J-C. & HASSID W.Z. (1962a) Carbohydrates and nucleotides in the red alga *Porphyra perforata*. I. Isolation and identification of carbohydrates. *Biochemistry, N.Y.* **1**, 468–74.

SU J-C. & HASSID W.Z. (1962b) Carbohydrates and nucleotides in the red alga *Porphyra perforata*. II. Separation and identification of nucleotides. *Biochemistry, N.Y.* **1**, 474–80.

TARCHEVSKII I.A. (1969) Products of the photosynthesis of thermophilic algae in the hot springs of Pauzhetka (Kamchatka). *Funkts. Osob. Khloroplastov*, pp. 105–8. *C.A.* **74**, 108297u.

TRENCH R.K. (1971) Physiology and biochemistry of zooxanthellae symbiotic with marine coelenterates. II. Liberation of fixed carbon-14 by zooxanthellae *in vitro. Proc. R. Soc.* B **177**, 237–50.

TSUSUÉ Y. & FUJITA Y. (1964) Mono- and oligo-saccharides in the blue-green alga, *Tolypothrix tenuis. J. gen. appl. Microbiol., Tokyo*, **10**, 283–94.

TSUSUÉ Y. & YAMAKAWA T. (1965) Chemical structure of oligosaccharides in the blue-green alga, *Tolypothrix tenuis. J. Biochem., Tokyo*, **58**, 587–94.

TURVEY J.R. (1961) Carbohydrates of *Porphyra umbilicalis. Colloq. Intern. Centre Nat. Rech. Sci. Paris*, No. 103, 29–37.

TURVEY J.R. & SIMPSON P.R. (1966) Polysaccharides from *Corallina officinalis. Proc. Int. Seaweed Symp.* **5**, 323–7.

VANDEN DRIESSCHE T. (1969) Influence of constant light on the inulin content of the chloroplasts in *Acetabularia mediterranea. Progr. Photosyn. Res., Proc. Int. Congr.* **1**, 450–7.

VANDEN DRIESSCHE T. & BONOTTO S. (1969) Nature du matériel accumulé par les chloroplastes d'*Acetabularia mediterranea. Arch. int. Physiol. Biochim.* **75**, 186–7.

VOGEL K. & BARBER A.A. (1968) Degradation of paramylon by *Euglena gracilis. J. Protozool.* **15**, 657–62.

WALLEN D.G. & GEEN G.H. (1971) Light quality in relation to growth, photosynthetic rates and carbon metabolism in two species of marine plankton algae. *Mar. Biol.* **10**, 34–43.

WALTER-LEVY L. & STRAUSS R. (1969) Sur les substances insolubilisées au cours de la dessiccation des Phéophycées. *C. r. hebd. Séanc. Paris*, **268**, 493–6.

WANKA P., JOPPEN M.M.J. & KUYPER CH.M.A. (1970) Starch degrading enzymes in synchronous cultures of *Chlorella. Z. Pfl. Physiol.* **62**, 146–57.

WEGMANN K. (1969) Pathways of carbon dioxide fixation in *Dunaliella. Progr. Photosyn. Res., Proc. Int. Congr.* **3**, 1559–64.

WEGMANN K. (1971) Osmotic regulation of photosynthetic glycerol production in *Dunaliella. Biochim. biophys. Acta*, **234**, 317–23.

WERZ G. & CLAUSS H. (1970) Über die chemische Natur der Reserve-Polysaccharide in *Acetabularia*-chloroplasten. *Planta*, **91**, 165–8.

WHYTE J.N.C. (1970) Extraction procedure for plants: extracts from the red alga *Rhodomela larix. Phytochem.* **9**, 1159–61.

WHYTE J.N.C. (1971) Polysaccharides of the red seaweed *Rhodymenia pertusa*. I. Water soluble glucan. *Carbohydr. Res.* **16**, 220–4.

WICKBERG B. (1957) Isolation of 2-L-amino-3-hydroxy-1-propane sulphonic acid from *Polysiphonia fastigiata. Acta chem. scand.* **11**, 506–11.

WÖBER G. & HOFFMANN-OSTENHOF O. (1969) Untersuchungen über die Biosynthese der Cyclite. XXII. Cyclite in *Chlorella fusca. Mschr. Chem.* **100**, 369–75.

WÖBER G. & HOFFMANN-OSTENHOF O. (1970a) Untersuchungen über die Biosynthese der Cyclite. XXVI. Bildung von Mytilit (C-methyl-*scyllo*-inosit) in *Chlorella fusca. Mschr. Chem.* **101**, 1861–3.

WÖBER G. & HOFFMANN-OSTENHOF O. (1970b) Biosynthesis of laminitol (D-4-C-methyl-myo-inositol) in *Porphyridium* species. *Eur. J. Biochem.* **17**, 393–6.

WÖBER G., RUIS H. & HOFFMANN-OSTENHOF O. (1971) Untersuchungen über die Biosynthese der Cyclite. XXVII. Ein Dehydrogenase-System, das die Epimerisierung von *myo*-Inosit zu D-*chiro*-Inosit in *Chlorella fusca* katalysieren. *Mschr. Chem.* **102**, 459–64.

YAMAGUCHI T., IKAWA T. & NISIZAWA K. (1966) Incorporation of radioactive carbon from $H^{14}CO_3$ into sugar constituents by a brown alga, *Eisenia bicyclis*, during photosynthesis and its fate in the dark. *Pl. Cell Physiol.*, Tokyo, **7**, 217–29.

YAMAGUCHI T., IKAWA T. & NISIZAWA K. (1969) Pathway of mannitol formation during photosynthesis in brown algae. *Pl. Cell Physiol.*, Tokyo, **10**, 425–40.

# CHAPTER 8

# FATTY ACIDS AND SAPONIFIABLE LIPIDS

## B. J. B. WOOD

Department of Applied Microbiology,
University of Strathclyde,
Glasgow, Scotland, U.K.

| | | | |
|---|---|---|---|
| **1** | **Introduction** 236 | 3.5 | Bacillariophyceae 250 |
| | | 3.6 | Phaeophyceae and Xantho- |
| **2** | **A note on lipid chemistry** 237 | | phyceae 250 |
| 2.1 | Fatty acids 237 | 3.7 | Chlorophyceae, Prasinophy- |
| 2.2 | Intact lipids 238 | | ceae and Euglenophyceae 251 |
| 2.3 | Triglycerides 238 | | |
| | (a) Lipids containing phosphate 239 | **4** | **Fatty acid metabolism** 251 |
| | (b) Other lipids of interest 240 | **5** | **Acyl lipids** 256 |
| | | **6** | **Lipids of nitrogen-fixing blue-green algae** 258 |
| **3** | **Fatty acid composition** 240 | | |
| 3.1 | Blue-green algae 241 | **7** | **Fatty acids, photosynthesis and phylogeny** 259 |
| 3.2 | Rhodophyceae 244 | | |
| 3.3 | Chrysophyceae, Haptophyceae and Cryptophyceae 245 | | |
| 3.4 | Dinophyceae 248 | **8** | **References** 261 |

## 1  INTRODUCTION

It is probably true that the major importance of algal lipids lies in their participation in biological membranes. Although references to 'fat' or 'oil' as a reserve material in algae are common in the earlier literature, it is not clear how far these claims would be substantiated by more modern analysis and the present survey is concerned mostly with the polar lipids, which occur in cell membranes.

Although much work has been carried out on algal lipids, it is clear that, despite the enormous number of species available in culture, only the surface of this topic has been scratched. Inevitably, the necessity to obtain sufficient amounts of material for analysis has directed attention towards the robust and easily-cultivated genera such as *Chlorella* and *Euglena*. To the worker concerned with algae it is pleasing to see his organisms find use as tractable model systems in the study of lipid biogenesis, although he cannot but wish that more attention be given to the more difficult and specialized algae, so that conclusions on algal lipid metabolism are not too narrowly based upon the 'weeds in the algologist's garden'.

There is also the feeling, as one peruses the literature, that many analyses are of cells of unknown physiological state and that often material from stationary growth phase has been used. There is a real need for more results derived from continuous and synchronized cultures and from batch cultures in the mid-logarithmic phase of growth, such cultures being produced using defined media under carefully controlled conditions.

The above remarks refer in the main to the micro-algae, but their general significance applies equally to the macro-algae. Here, indeed, the problems are even greater because of the impossibility of cultivating the majority of such algae under controlled conditions, let alone in axenic culture. One is therefore forced to rely upon analyses of material collected in the field, a problem particularly acute in groups such as the Rhodophyceae and Phaeophyceae with few representatives in laboratory cultivation.

This chapter does not seek to cover the field of algal lipids exhaustively, nor does it claim to include all the very latest results. Rather, it seeks to give a broad picture of the current state of the field. For the most recent news, the reader is directed to appropriate chapters in *Annual Reviews of Microbiology, Biochemistry* and *Plant Physiology* and to *Advances in Lipid Research*.

## 2 A NOTE ON LIPID CHEMISTRY

This chapter confines itself to the saponifiable lipids and to fatty acids, and thus excludes many materials which are extracted from tissues by lipid solvents such as chloroform and di-ethyl ether. The saponifiable lipids are compounds which can be hydrolysed into acids and alcohols on boiling with a solution of caustic soda or potash. The term is commonly further restricted to those compounds in which glycerol is the alcohol, and this will in general be the case here, although this must lead to exclusion of, for example, the waxes found in members of the Euglenophyceae. Steroids present a slightly anomalous case; they are classified with the non-saponifiable portion, in that they can be extracted by lipid solvents from the alkaline mixture after hydrolysis is complete, yet, in nature the sterols may occur esterified to fatty acids.

### 2.1 *Fatty acids*

Virtually all of the fatty acids found in algal lipids are straight chain molecules containing an even number of carbon atoms. This is a direct consequence of their biosynthesis from acetate by β-addition. The acids frequently possess one or more double bonds down the carbon chain, and acids are conveniently identified as to number of carbon atoms and double bonds by a double number, for example 18:0 (18 carbon atoms, no double bonds) stearic acid; 16:0 (16 carbons, no double bonds) palmitic acid; 20:4 would be one of the family of acids with 20 carbon atoms (deriving therefore from arachidic acid, 20:0) and 4 double bonds. Clearly, the introduction of one or more double bonds gives scope for considerable positional isomerism, but the choice is limited by the restrictions imposed by the mechanisms of lipid biogenesis. When it is necessary to describe the position of a double bond(s) this is indicated by $\Delta$ followed by a number denoting the distance of the double bond from the carboxyl end of the molecule. Thus 16:1 $\Delta$9 denotes palmitoleic acid with the double bond between the 9th and 10th carbon atoms, numbering from the carboxyl end. Similarly, the important plant acid α-linolenic acid has 3 double bonds situated between the 9th and 10th, 12th and 13th, and 15th and 16th pairs of carbon atoms, and so may be figured 18:3 $\Delta$9, 12, 15. The related γ-linolenic acid, is 18:3 $\Delta$6, 9, 12. Since chain-elongation in fatty acids normally occurs by the addition of acetate residues to the carboxyl end of the fatty acid, it will be obvious that such elongation will occasion the re-numbering of the double-bond positions. To overcome this, some workers prefer to number from the methyl end of the acid, which not only cuts out this re-numbering, but also makes relationships among groups of acids apparent at a glance. However, the bulk of the literature on fatty acids uses the older convention, and for the sake of simplicity and uniformity this is the one which will be followed here.

*Cis/trans* isomerism may occur around the double bond, but fortunately nearly all of the naturally occurring acids have the *cis* configuration. The one significant exception to this generalization, *trans*-hexadec-3-enoic acid (*trans*-16:1 $\Delta$3), is an important constituent of the phosphatidyl glycerol of photosynthetic eukaryotes.

## 2.2   *Intact lipids*

The occurrence of free fatty acids is probably unusual in living cells and reports of the presence of more than trace amounts are more likely to reflect the activities of lipases during extraction of lipids from cells. Usually, the acids are bound through ester links to co-enzyme A and the acyl carrier protein during degradation and synthesis, and to alcohols, notably glycerol, at all other times.

## 2.3   *Triglycerides*

These are the simplest glyceride esters, in which all three hydroxyl groups are esterified to fatty acids. These are usually regarded as storage materials rather than as structural membrane lipids. Occasionally, mono- and di-glycerides are reported, but their presence in intact algae is uncertain. The other major classes of glycerides have the middle hydroxyl and one terminal hydroxyl of the glycerol esterified to fatty acids; the remaining hydroxyl is attached to a more polar group, hence these compounds are referred to as the polar lipids. The following list is not exhaustive but comprises the classes of interest in the discussion of algal lipids. For details of structure, see, for example, Gurr and James (1971).

(a) *Lipids containing phosphate*

These may be formally regarded as derivatives of phosphatidic acid

$$
\begin{array}{c}
\qquad\qquad\underset{\displaystyle \|}{\overset{\displaystyle O}{}} \\
\qquad CH_2{-}O.\overset{\displaystyle O}{\overset{\displaystyle \|}{C}}{-}R^1 \\
\overset{\displaystyle O}{\overset{\displaystyle \|}{}}\qquad | \\
R^2{-}C.O.CH \\
\qquad CH_2.O{-}PO_3.X
\end{array}
$$

where $R^1$ and $R^2$ denote fatty acids, and are named for the residue represented by 'X', thus: phosphatidyl choline, phosphatidyl inositol etc. Table 8.1 lists the major compounds.

**Table 8.1.** Phospholipids

| Substituent group | | Abbreviation for lipid |
|---|---|---|
| Name | Structure | |
| Choline | $.O.CH_2.CH_2.N^+(CH_3)_3$ | P.C. (Lecithin) |
| Ethanolamine | $.OCH_2.CH_2.NH_2$ | P.E. (Cephalin) |
| Serine | $.OCH_2.CH.NH_2$ <br> $\qquad\quad |$ <br> $\qquad\ COOH$ | |
| Inositol | *myo*-inositol <br> | P.I. |
| Glycerol | $.OCH_2.CH(OH).CH_2OH$ | P.G. |
| Diphosphatidyl-Glycerol | -1-Glycerol-3-$PO_4$-Glycerol | Cardiolipin |

## (b) *Other lipids of interest*

*Plant sulpholipid.* This has the structure:

1,2 – diacyl – [6 – sulpho – α – D –
quinovopyranosyl – (1' → 3)] – *sn* – glycerol.

(6 – deoxy – D – glucose is normally called D – quinovose)

Other sulphur-containing lipids have been reported from certain algae, notably *Ochromonas,* but their structure is beyond the scope of the present work.

*Monogalactosyl diglyceride* (MGDG). This has the structure:

*Digalactosyl diglyceride* (DGDG). This has the structure:

Details of the chemistry and biochemistry of fatty acids and their glycerides are given by Gunstone (1967); the volume by Gurr and James (1971) provides a more recent introductory review, while that by Hitchcock and Nichols (1971) deals specifically with the biochemistry of plant lipids, including those of algae. Morris and Nichols (1966) and Nichols *et al.* (1966) have reviewed the chromatography of lipids.

## 3 FATTY ACID COMPOSITION

Tables 8.2 to 8.10 attempt to review the fatty acid composition of the various algal groups. While not exhaustive, they bring together a sufficient amount of information to permit the reader to appreciate the range of fatty acid composition exhibited by algae. Algae have not found use as sources of fatty acids because the quantities they contain are relatively small, but they could prove useful as sources of specific acids. For example, the important, powerful group of animal hormones called prostaglandins (see von Euler & Eliasson 1967) are biosynthesized from fatty acids of the arachidonic acid group. Some algae are comparatively rich in this acid and could provide a useful source of it.

The information contained in Tables 8.2 to 8.10 will be discussed, with particular reference to areas where the existing information seems to be inadequate or contradictory.

### 3.1  The blue-green algae

Table 8.2 is a compilation of information from various sources while Table 8.3 is drawn from the results of Kenyon (1972) and Kenyon *et al.* (1972), obtained from a large number of isolates from soil and fresh water. Some interesting differences exist between the results in the two tables. For example, the *Anabaena*-type algae examined by Kenyon *et al.* (1972) possess little or no linolenate, yet Nichols and Wood (1968a) found that both *A. cylindrica* and *A. flos-aquae* possess significant amounts of α-linolenate. The discovery that some blue-green algae possess 18-carbon acids with four double bonds, is also of interest. There is no obvious reason why blue-green algae should be confined to a maximum of 18 carbon atoms per fatty acid and there is a need for more work on the system responsible for their fatty acid biosynthesis. The biosynthesis of the 18:4 acid and its distribution among the various lipids of the algae also awaits elucidation.

The reason why the *Spirulina* species examined by Nichols and Wood (1968b) and by Pelloquin *et al.* (1970) contained γ-18:3, whereas the *Spirulina* isolated and identified by Kenyon *et al.* (1972) contained α-18:3 as the only linolenate species in each case, is unknown. The morphology of the algae is strikingly similar. If it was simply a matter of different *Spirulina* species containing a greater amount of one or the other acid, then there would be no problem about assigning them to the same genus, but such an absolute difference must raise a question as to the validity of the present classification.

In eukaryotic algae, the decrease in polyunsaturated fatty acid content when algae grow heterotrophically is well documented and correlates with the changes in etiolated leaves of higher plants. The situation in the blue-green algae is less clear. Holton *et al.* (1968) (see also Table 8.2) found a decrease in α-linolenic acid content from 12% to 0·4% in *Chlorogloea fritschii* when grown in the dark, as compared with the same organism grown autotrophically. This change was compensated for by an increase in oleic acid content. Kenyon *et al.* (1972) on the other hand report that when they examined the fatty acid content of two of their isolates which grow heterotrophically in the dark, they found no difference between light-grown and dark-grown cells.

**Table 8.2** Fatty acids of various Cyanophyceae (modified from Nichols 1973)

Fatty acids as % of total acid

| Fatty acid | Spirulina platensis FNH B*IV (1)** | Myxosarcina chroococcoides (1) | Trichodesmium erythraeum M (2) | Trichodesmium erythraeum M (2) | Nostoc muscorum Fr (3) | Chlorogloea fritschii F.H.IV(3)Fr | Chlorogloea fritschii D | Oscillatoria williamsii F.N.H.I. M (2) | Anabaena cylindrica F.H.I. Fr (1) | Anabaena flos-aquae F.H.I. Fr (1) | Mastigocladus laminosus Fr (1) | Hapalosiphon laminosus Fr (3) | Anacystis marina M (2) | Anacystis nidulans Fr (4) | Plectonema terebrans FNH III M (2) | Lyngbya lagerheimii FNH II M (2) |
|---|---|---|---|---|---|---|---|---|---|---|---|---|---|---|---|---|
| <16 carbon atoms | — | — | 52*** | 64*** | 5 | 1 | 1 | 6 | — | — | — | 2 | 24 | — | 3 | 4 |
| 16:0 | 43 | 38 | 17 | 11 | 32 | 41 | 39 | 36 | 46 | 40 | 38 | 54 | 32 | 47 | 35 | 40 |
| 16:1 | 10 | 9 | 4 | 7 | 15 | 16 | 19 | 24 | 6 | 6 | 42 | 24 | 36 | 39 | 13 | 15 |
| 16:2 | ⊢ | 1 | — | — | 6 | 1 | 1 | 14 | 6 | 4 | — | — | — | — | 5 | ⊢ |
| 17:1 | — | — | — | — | — | 1 | 1 | — | — | — | — | — | — | — | — | — |
| 18:0 | 3 | 4 | 3 | 3 | 2 | 2 | 2 | 2 | 4 | 1 | ⊢ | 3 | 2 | 1 | 2 | 2 |
| 18:1 | 5 | 7 | 3 | 7 | 7 | 14 | 26 | 11 | 6 | 5 | 17 | 18 | 4 | 10 | 2 | 31 |
| 18:2 | 12 | 9 | 4 | 1 | 10 | 13 | 13 | 4 | 24 | 36 | 2 | — | — | — | 11 | 7 |
| γ18:3 | 21 | — | — | — | — | — | — | — | — | — | — | — | — | — | — | — |
| α18:3 | ⊢ | 33 | 19 | 6 | 21 | 12 | 0·4 | — | 11 | 11 | — | — | — | — | 6 | — |

* Abbreviations: D, dark heterotrophic growth; Fr, fresh water; B, brackish water; M, marine; ⊢, less than 1%; – absent. The other abbreviations are intended to aid comparison with the classification employed by Kenyon (1972) and Kenyon *et al.* (1972) thus: F, filamentous; U, unicellular; H, heterocystous; NH, non-heterocystous; roman numerals refer to the sub-group.

** References: (1) Nichols & Wood 1968b; (2) Parker *et al.* 1967; (3) Holton *et al.* 1968; (4) Holton *et al.* 1964.

*** The two samples of *Trichodesmium* were derived from two natural blooms, and are therefore not pure cultures. They contained 27 and 50% decanoic acid (10:0) respectively.

**Table 8.3.** Fatty acids of various Cyanophyceae (selected data from Kenyon 1972 and Kenyon *et al.* 1972)

| Type, sub-group and Strain No. | Fatty acid as % of total acid | | | | | | | | | | |
|---|---|---|---|---|---|---|---|---|---|---|---|
| | 14:0 | 14:1 | 16:0 | 16:1 | 16:2 | 18:0 | 18:1* | 18:2 | 18:3 γ | 18:3 α | 18:4 |
| **Unicellular** | | | | | | | | | | | |
| *Synechococcus* types | | | | | | | | | | | |
| 1, 6708 | 14 | 2 | 29 | 40 | — | 2 | 6 | — | — | — | — |
| 4, 6301 | 1 | 6 | 32 | 46 | — | 1 | 4 | 2 | — | — | — |
| 5, 6312 | 1 | 1 | 29 | 61 | — | 1 | 4 | — | — | — | — |
| 3, 6715 (thermophile) | ⊦ | ⊦ | 31 | 29 | — | 2 | 27 | 1 | — | — | — |
| 2, 7001 | 7 | ⊦ | 14 | 16 | — | 3 | 16 | 17 | — | — | — |
| 6, 7003 | 1 | — | 29 | 11 | — | 2 | 18 | 18 | — | 10 | — |
| 9, 6801 | — | — | 41 | 2 | — | 1 | 9 | 21 | 16 | 3 | — |
| 9, 6903 | 18 | 1 | 11 | 8 | — | 1 | 20 | 26 | — | 7 | — |
| 7, 6605 | 6 | 1 | 20 | 2 | 7 | 2 | 3 | 43 | — | 4 | — |
| 8, 6910 | 1 | ⊦ | 29 | 30 | 21 | 1 | 5 | 5 | 3 | 1 | — |
| (Plus 14:2, 7% in Strain 6605, 21% in Strain 6910) | | | | | | | | | | | |
| *Microcystis* type | | | | | | | | | | | |
| 7005 | 1 | ⊦ | 40 | 3 | — | 2 | 6 | 15 | 27 | — | — |
| *Aphanocapsa* type | | | | | | | | | | | |
| 1, 6702 | 1 | — | 30 | 13 | — | 2 | 6 | 21 | 25 | — | — |
| 1, 6805 | 2 | 1 | 35 | 10 | — | 3 | 4 | 21 | 18 | 2 | — |
| 2, 6308 | 27 | 11 | 6 | 34 | — | 4 | 3 | 4 | — | — | — |
| 3, 6701 | 34 | 8 | 7 | 35 | — | 2 | 3 | 1 | — | — | — |
| 4, 6808 | 18 | 9 | 9 | 50 | — | 2 | 2 | 3 | — | — | — |
| 5, 6804 | 26 | 5 | 10 | 35 | — | 2 | 3 | 3 | — | — | — |
| *Gloeocapsa* type | | | | | | | | | | | |
| 6501 | 11 | 3 | 19 | 46 | — | 3 | 4 | 3 | — | — | — |
| *Chlorogloea* type | | | | | | | | | | | |
| 6712 | 26 | 3 | 9 | 27 | — | 7 | 3 | 2 | — | — | — |
| **Filamentous without heterocysts** | | | | | | | | | | | |
| I *Oscillatoria* type | | | | | | | | | | | |
| 1, ⎰6304 | ⊦ | ⊦ | 26 | 6 | — | 1 | 6 | 8 | — | 15 | 35 |
| ⎱6401 | 5 | — | 35 | 4 | — | 1 | 9 | 12 | — | 31 | 2 |
| 2, ⎰6407 | 1 | ⊦ | 12 | 3 | — | 4 | 15 | 30 | 5 | 5 | — |
| ⎱6506 | 1 | ⊦ | 22 | 6 | 15 | 1 | 6 | 29 | — | 15 | — |
| II *Lyngbya* type | | | | | | | | | | | |
| 1, 6703 | 1 | ⊦ | 28 | 9 | — | 4 | 11 | 21 | — | 15 | — |
| 2, 6409 | 1 | ⊦ | 18 | 9 | — | 3 | 21 | 32 | — | 1 | 2 |

**Table 8.3.**—continued

| Type, sub-group and Strain No. | 14:0 | 14:1 | 16:0 | 16:1 | 16:2 | 18:0 | 18:1* | 18:2 | 18:3 γ | 18:3 α | 18:4 |
|---|---|---|---|---|---|---|---|---|---|---|---|
| **III *Plectonema* type** | | | | | | | | | | | |
| 6306 | 1 | — | 21 | 30 | — | 2 | 16 | 25 | — | 1 | — |
| **IV *Spirulina* type** | | | | | | | | | | | |
| 1,　　6313 | 21 | 12 | 33 | 27 | — | 1 | 2 | — | — | — | — |
| 2,　　7106 | ⊦ | — | 33 | 2 | — | 4 | 7 | 12 | — | 27 | — |
| **Filamentous with heterocysts** | | | | | | | | | | | |
| **I *Anabaena* type** | | | | | | | | | | | |
| 1,　　6301 | ⊦ | 15 | 1 | 12 | — | — | 18 | 40 | — | ⊦ | — |
| 2,　　6309 | 1 | 5 | 1 | 7 | — | — | 25 | 26 | — | — | — |
| **II *Calothrix* type** | | | | | | | | | | | |
| 1,　　6303 | 1 | — | 34 | 6 | 46 | 1 | 4 | 14 | 2 | 7 | 17 |
| 2,　　7101 | 1 | 2 | 22 | 3 | — | 3 | 12 | 15 | 13 | 6 | 11 |
| **III *Microchaete* type** | | | | | | | | | | | |
| 1,　　6305 | 1 | ⊦ | 15 | 27 | — | 1 | 17 | 20 | 13 | ⊦ | 1 |
| **IV *Chlorogloeopsis* type** | | | | | | | | | | | |
| 6718 | 1 | ⊦ | 26 | 16 | — | 2 | 20 | 18 | — | 8 | — |

Table spanning header: Fatty acid as % of total acid

* 18:1 from the filamentous algae may also include 16:3
— indicates absent, ⊦ indicates less than 1%

The wide gap between even the simplest blue-green algae with only mono-unsaturated fatty acids (*Anacystis nidulans*, for example), and the photosynthetic bacteria, is illustrated by the fact that the former produce their unsaturated acids by the desaturation of saturated precursors, and so possess oleic acid (18:1 $\Delta$9), whereas the bacteria possess only the anaerobic pathway, unique to bacteria and so produce vaccenic acid (18:1 $\Delta$11) and do not contain poly-unsaturated fatty acids.

## 3.2　*Rhodophyceae*

The composition of the fatty acids of the red algae (Table 8.4) is more complex than that of the blue-green algae, and is remarkable for the high content of C20 and C22 acids with 4, 5 and even 6 double bonds. The classification of the anomalous hot-spring alga *Cyanidium caldarium* is problematical. I have followed Seckbach and Ikan (1972) in placing it with the red algae. However, the published results on its lipid composition (Adams *et al.* 1971, Allen *et al.* 1970, Kleinschmidt & McMahon 1970a,b) show a fatty acid composition quite different from that of most red algae, as it is totally deficient in C20 acids. This may indicate its primitive nature but alternatively could perhaps be a loss reflecting part of the adaptation necessary for life at 70°C. It is difficult to extract lipids from *Cyanidium*. Even prolonged exposure to the 2:1 chloroform/methanol mixture which is usually employed, does not seem to dissolve much chlorophyll from the cell, although the same mixture rapidly liberates practically all of the pigment from other algae. Harsher extraction may, however, reveal additional fatty acids.

**Table 8.4.** Fatty acids of certain Rhodophyceae

| | Fatty acid as % of total acid | | | | | | |
|---|---|---|---|---|---|---|---|
| Fatty acid | Porphyridium cruentum M* (1)** | Porphyridium aeruginosa Fr (3) | Plocamium coccineum M (2) | Ceramium rubrum M (2) | Rhodomela subfusca M (2) | Cyanidium caldarium Fr (4) 20°C | 55°C† |
| 14:0 | 1 | 8 | 10 | 6 | 4 | ⱶ | ⱶ |
| 16:0 | 23 | 30 | 27 | 39 | 29 | 18 | 53 |
| 16:1 | 2 | 7 | 6 | 7 | 5 | ⱶ | ⱶ |
| 16:2 | — | — | ⱶ | ⱶ | ⱶ | — | — |
| 16:3 | — | — | ⱶ | ⱶ | ⱶ | — | — |
| 16:4 | — | — | ⱶ | ⱶ | ⱶ | — | — |
| 18:0 | 2 | 11 | 1 | 1 | ⱶ | 7 | 5 |
| 18:1 | 3 | 28 | 6 | 14 | 15 | 34 | 21 |
| 18:2 | 16 | 8 | 3 | 1 | 1 | 11 | 21 |
| 18:3 | ⱶ | 3 | ⱶ | 2 | 1 | 30 | — |
| 18:4 | — | — | 1 | 1 | 1 | — | — |
| 20:3 | 2 | — | 7 | 1 | ⱶ | — | — |
| 20:4 | 36*** | — | 12 | 5 | 14 | — | — |
| 20:5 | 17 | — | 22 | 17 | 24 | — | — |
| 22:5 | — | — | — | ⱶ | — | — | — |
| 22:6 | — | — | ⱶ | 1 | 1 | — | — |

\* Abbreviations: M, marine; Fr, freshwater. ** References: (1) Nichols & Appleby 1969; (2) Klenk *et al.* 1963; (3) Wood, B.J.B. unpubl.; (4) Kleinschmidt & Mc-Mahon 1970a. *** Arachidonic acid (20:4 △ 5, 8, 11, 14). † The *Cyanidium* data are for cells grown at 20°C and 55°C.

*Porphyridium aeruginosa* is another simple eukaryote which may be blue-green in colour and shows differences from the majority of red algae. While the difficulty is in obtaining sufficient material for analysis and one must therefore be cautious in accepting the results, they do seem to have much in common with those reported for *Cyanidium*, especially the absence of C20 acids and the rather high levels of oleic and linoleic acids.

### 3.3   Chrysophyceae, Haptophyceae and Cryptophyceae

These algae are again notable for their high content of polyunsaturated acids containing four or more double bonds. It is of interest to note that the marine

I

**Table 8.5.** Fatty acids of certain Chrysophyceae and Haptophyceae

| | | Fatty acid as % of total acid | | | | | | | | | | |
|---|---|---|---|---|---|---|---|---|---|---|---|---|
| | | Chrysophyceae | | | | | Haptophyceae | | | | | |
| Fatty acid | Double-Bond Positions | Monochrysis lutheri M*(1)** | Pseudopedi- nella sp M(1) | Poteri- ochromonas stipitata FrHe(2) | Ochromonas danica FrHe(3) | Ochromonas malhamensis FrHe(3) | Prymnesium parvum M(1) | Isochrysis galbana M(1) | Dicrateria inornata M(1) | Coccolithus huxleyi M(1) | Cricosphaera carterae M(1) | Cricosphaera elongata M(1) |
| 14:0 | — | 9 | 5 | 5 | 13 | 13 | 6 | 6 | 1 | 6 | 9 | 7 |
| 16:0 | — | 10 | 8 | 18 | 12 | 12 | 16 | 16 | 16 | 17 | 9 | 9 |
| 16:1 | 9*** | 20 | 24 | 3 | 1 | 1 | 10 | 10 | 8 | 28 | 21 | 21 |
| 16:2 | 6,9 | 2 | 4 | — | — | — | + | + | + | 2 | 4 | 4 |
| 16:2 | 9,12 | 5 | 3 | — | — | — | 2 | 2 | 1 | 1 | 6 | 4 |
| 16:3 | 6,9,12 | 15 | 17 | 1 | — | — | 1 | 1 | 2 | 7 | 15 | 14 |
| 18:0 | — | + | — | 1 | 2 | 3 | — | — | + | + | + | — |
| 18:1 | 9*** | 6 | 3 | 7 | 14 | 9 | 25 | 25 | 17 | 10 | 3 | 2 |
| 18:2 | 9,12 | 2 | 2 | 18 | 20 | 26 | 18 | 18 | 5 | 2 | 3 | 2 |
| γ18:3 | 6,9,12 | — | — | 1 | 14 | 4 | + | + | + | — | + | + |
| α18:2 | 9,12,15 | 1 | — | 3 | 4 | 5 | 11 | 11 | 13 | — | 2 | — |
| 18:4 | 6,9,12,15 | 1 | — | — | 4 | 1 | + | + | 20 | 1 | 2 | 1 |
| 20:2 | 8,11 | — | — | 3 | 5 | — | + | + | + | + | — | 1 |
| 20:3 | 8,11,14 | 2 | 1 | — | 6 | 2 | 1 | 1 | 1 | + | — | + |
| 20:4 | 5,8,11,14 | + | 1 | 4 | — | 7 | 1 | 1 | + | 1 | 1 | + |
| 20:4 | 8,11,14,17 | + | 1 | † | — | — | 1 | 1 | + | 1 | 1 | — |
| 20:5 | 5,8,11,14,17 | 19 | 26 | — | — | — | 5 | 4 | 8 | 17 | 20 | 2 |
| 22:5 | 7,10,13,16,19 | 3 | 1 | — | 4 | 16 | — | 1 | 3 | + | 2 | 28 |
| 24:0 | — | + | + | — | — | — | + | + | — | + | 1 | + |

* Abbreviations: M, marine; Fr, freshwater; He, photoheterotrophic growth.   ** References: (1) Chuecas & Riley; 1969; (2) Erwin et al. 1964; (3) Nichols & Appleby 1969.   *** Other isomers may also be present.

† Other unidentified C-20 unsaturated fatty acids account for 27%. Traces of 16:4 △ 6,9,12, 15 were recorded in all marine species.

**Table 8.6.** Fatty acids of certain Cryptophyceae

| Fatty acid | Double-Bond Positions | Fatty acid as % of total acid | | | | | | | | | |
|---|---|---|---|---|---|---|---|---|---|---|---|
| | | Hemiselmis rufescens M*(1)** | Hemiselmis brunnescens M(1) | Cryptomonas appendiculata M(1) | Cryptomonas maculata M(1) | Cryptomonas sp. M(2) | Chroomonas sp. M(2) | Rhodomonas lens MHe(2) | Hemiselmis virescens M(2) | Cryptomonas ovata Fr(2) | Chilomonas paramecium Fr(2) |
| 14:0 | — | 1 | 1 | 2 | 4 | 6 | 2 | 18 | 8 | — | 18 |
| 16:0 | — | 21 | 12 | 16 | 15 | 4 | 16 | 13 | 23 | 10 | 18 |
| 16:1 | 9† | 10 | 3 | 12 | 6 | 2 | 2 | 5 | 5 | 4 | — |
| 16:1 | trans Δ3 | — | — | + | — | 3 | — | 1 | 2 | 1 | — |
| 16:2 | 6,9 | 1 | + | 1 | 3 | 2 } | 1 } | — | — | 2 } | — |
| 16:2 | 9,12 | 2 | 3 | 2 | 1 | | | — | — | | — |
| 16:3 | 6,9,12 | + | + | + | — | — | 1 | + | — | — | — |
| 16:4 | 6,9,12,15 | 1 | + | + | + | — | — | — | — | — | — |
| 18:0 | — | — | — | — | — | — | — | + | — | — | 1 |
| 18:1 | 9† | 2 | 2 | 2 | 4 | 5 | 5 | 10 | 7 | 2 | 9 |
| 18:2 | 9,12 | 1 | + | 4 | — | — | 3 | 2 | 3 | 5 | 12 |
| γ18:3 | 6,9,12 | 2 | + | — | — | 7 | — | — | — | — | — |
| α18:3 | 9,12,15 | 7 | 8 | 12 | 6 | — | 23 | 16 | 22 | 17 | 27 |
| 18:4 | 6,9,12,15 | 17 | 30 | 13 | 16 | 44 | 23 | 13 | 16 | 34 | — |
| 20:0 | — | 2 | + | 1 | — | — | — | — | — | — | — |
| 20:1 | 11 | 14 | 18 | 10 | 17 | — | — | — | — | — | — |
| 20:4 | 5,8,11,14 | — | + | 3 | 1 | 16 | — | — | — | — | — |
| 20:4 | 8,11,14,17 | 3 | + | 6 | 1 | — | — | — | — | — | — |
| 20:5 | 5,8,11,14,17 | 8 | 14 | 10 | 17 | — | 14 | 13 | 7 | 12 | 6 |
| 22:4 | 4,10,13,16 | 1 | — | — | — | — | 1 | 1 | 2 | 2 | — |
| 22:5 | 7,10,13,16,19 | 3 | 1 | 2 | 3 | — | — | — | — | — | — |
| 22:6 | 4,7,10,13,16,19 | — | — | — | — | 10 | 6 | 5 | 2 | 7 | 3 |

* Abbreviations: M, marine; He, photoheterotrophic growth; Fr, freshwater.   ** References: (1) Chuecas & Riley; 1969; (2) Beach et al. 1970.   † Other isomers may also be present.

members of the Chrysophyceae and Haptophyceae (Table 8.5) examined by Chuecas and Riley (1969) contain only small amounts of 20:4 acid, and that it is normally an acid with the structure 20:4 $\Delta$8, 11, 14, 17. The fresh water photoheterotrophs which they examined do not contain this acid, nor do they have the 20:5 acid which is so quantitatively important in some of the marine species, but they do contain significant amounts of arachidonic acid (which is only present in trace amounts in the marine algae examined). Most striking of all is the $\gamma$-linolenic acid content of *Ochromonas danica*. Even the related *O. malhamensis* has slightly more $\alpha$-linolenic acid than it has $\gamma$-linolenic acid, and in all members of the Chrysophyceae examined by Chuecas and Riley, $\gamma$-linolenic acid is present in only trace quantities.

Among the Cryptophyceae, too (Table 8.6), $\alpha$-linolenate is the more important acid. This group is characterized by comparatively high 18:4 contents, low levels of 20:4 acids and substantial amounts of 20:5 acid.

## 3.4 Dinophyceae

Few data are available (see Table 8.7) but members contain exceptional quantities of the very highly unsaturated 18:4, 20:5, and 22:6 acids. Neither of the

**Table 8.7.** Fatty acids of certain Dinophyceae

| | | Fatty acid content as % of total acid | | | | | |
|---|---|---|---|---|---|---|---|
| Acid | Double-Bond Positions | Proro-centrum micans (1)* | Peridinium trochoi-deum (1) | Gonyau-lax polyedra (2) | Gyro-dinium cohnii (3) | Amphi-dinium carteri (4) | Peri-dinium triquetrum (4) |
| 14:0 | — | 14 | 17 | 2 | 19 | 2 | 7 |
| 16:0 | — | 10 | 36 | 36 | 20 | 12 | 26 |
| 16:1 | 9 | 23 | 4 | 1 | 1 | 1 | 5 |
| 16:2 | 6, 9 | 4 | ⊦ | — | — | — | — |
| 16:2 | 9, 12 | 10 | 1 | 1 | — | 2 | — |
| 16:3 | 6, 9, 12 | 12 | 3 | — | — | — | — |
| 18:0 | — | — | — | ⊦ | — | 2 | 2 |
| 18:1 | 9 | 6 | 7 | 3 | 14 | 2 | 2 |
| 18:2 | 9, 12 | — | — | 2 | — | 1 | 9 |
| $\gamma$18:3 | 6, 9, 12 | 2 | ⊦ | ⊦ | — | — | — |
| $\alpha$18:3 | 9, 12, 15 | 6 | 3 | 3 | — | 3 | 3 |
| 18:4 | 6, 9, 12, 15 | — | 7 | 14 | — | 15 | 8 |
| 20:2 | 8, 11 | — | 2 | — | — | — | — |
| 20:4 | 8, 11, 14, 17 | — | 1 | — | — | — | — |
| 20:5 | 5, 8, 11, 14, 17 | 12 | 13 | 14 | — | 20 | 2 |
| 22:5 | 7, 10, 13, 16, 19 | — | 3 | ⊦ | ⊦ | — | 2 |
| 22:6 | ? | — | — | 23 | 30** | 22 | 16 |

*References: (1) Chuecas & Riley 1969; (2) Patton *et al.* 1966; (3) Harrington & Holz 1968; (4) Harrington *et al.* 1970.

All species are marine and were grown photoautotrophically.

** Unknowns with retention time greater than 22:6 = 5%

**Table 8.8.** Fatty acids of certain Bacillariophyceae

| Fatty acid | Double-Bond Positions | Skeletonema costatum M*(1)** | Ditylum brightwellii M (1) | Biddulphia sinensis M (3) | Chaetoceros septentrionale M (1) | Lauderia borealis M (1) | Asterionella japonica M (1) | Phaeodactylum tricornutum M (1) | Cylindrotheca fusiformis M (2) | Nitzschia angularis M (2) | N. thermalis M (2) | Cyclotella cryptica M (2) | Navicula pelliculosa Fr (2) |
|---|---|---|---|---|---|---|---|---|---|---|---|---|---|
| 14:0 | — | 6 | 8 | 11 | 6 | 7 | 6 | 9 | 3 | 6 | 2 | 5 | 3 |
| 16:0 | — | 11 | 13 | 14 | 11 | 12 | 11 | 11 | 9 | 25 | 30 | 24 | 16 |
| 16:1 | 9*** | 22 | 30 | 28 | 19 | 21 | 20 | 27 | 31 | 25 | 46 | 30 | 31 |
| 16:2 | 6,9 | 3 | 3 | + | 2 | 1 | 1 | 5 | 3 | 2 | 2 | 3 | 3 |
| 16:2 | 9,12 | 4 | 3 | 5 | 3 | 3 | 3 | 9 | [4] | [2] | [2] | [3] | [7] |
| 16:3 | 6,9,12 | 11 | 10 | 4 | 9 | 12 | 9 | 10 | 18 | 4 | 2 | 4 | 7 |
| 16:4 | 6,9,12,15 | 1 | + | 9 | 1 | 1 | 1 | 3 | + | 4 | | | |
| 18:0 | — | | + | 1 | + | | | 5 | + | 1 | 1 | 2 | 1 |
| 18:1 | 9*** | 2 | 6 | 2 | 3 | 2 | 3 | 5 | 6 | 3 | 2 | 3 | 2 |
| 18:2 | 9,12 | 2 | 2 | 1 | 1 | 1 | 1 | + | 4 | 2 | 2 | 2 | 1 |
| γ18:3 | 6,9,12 | | | + | + | | + | + | 3 | 2 | 1 | 1 | 1 |
| α18:3 | 9,12,15 | | 1 | 3 | | + | + | + | | | | | |
| 18:4 | 6,9,12,15 | 1 | + | — | 1 | — | 1 | — | — | — | — | — | — |
| 20:2 | 8,11 | + | | | | | | + | | | | | |
| 20:3 | 8,11,14 | | 2 | 2 | 1 | | 6 | + | | | | | |
| 20:4 | 5,8,11,14 | 2 | | + | + | | 4 | + | 4 | 9 | 4 | | |
| 20:4 | 8,11,14,17 | 2 | 1 | 1 | 1 | 1 | | 18 | | | | | |
| 20:5 | 5,8,11,14,17 | 30 | 11 | 15 | 21 | 30 | 20 | + | 14 | 17 | 7 | 21 | 26 |
| 22:1 | ? | + | + | — | 1 | 1 | 3 | + | | | | | |
| 22:2 | ? | | + | — | 6 | | | + | | | | | |
| 22:3 | ? | | 3 | 1 | 3 | 4 | 1 | + | | | | | |
| 22:5 | 4,7,10,13,16 | | | — | | | | + | | | | | |
| 22:5 | 7,10,13,16,19 | 2 | 1 | + | 4 | 1 | 3 | + | | | | | |

* Abbreviations: M, marine; Fr, freshwater. ** References: (1) Chuecas & Riley 1969; (2) Kates & Volcani 1966; (3) Klenk et al. 1964. *** Other isomers may be present. [ ], identity uncertain.

linolenic acids is present in quantity but the α isomer predominates over the γ form.

### 3.5 *Bacillariophyceae*

These algae have an odd but consistent pattern of distribution of fatty acids (Table 8.8). There is a considerable quantity of acids of the palmitic (C16) series, then nothing much else except for an acid with 20 carbon atoms and 5 double bonds. The very low levels of both linolenic acids are again noteworthy.

### 3.6 *Phaeophyceae and Xanthophyceae*

Data on the lipid composition of these two groups are rather sparse (Table 8.9) and only in *Monodus subterraneus* (Xanthophyceae) have the analyses been performed on axenic material cultured under controlled conditions. All other analyses were of plants collected in the field. Members of the Xanthophyceae resemble the diatoms in possessing considerable amounts of a 20:5 acid, with the balance consisting largely of palmitic and palmitoleic acids.

**Table 8.9.** Fatty acids of certain Phaeophyceae and Xanthophyceae

| | Fatty acid as % of total acid | | | | |
|---|---|---|---|---|---|
| | Phaeophyceae | | | Xanthophyceae | |
| Acid | *Fucus serratus* M* (1)** | *Fucus platycarpus* M (1) | *Fucus vesiculosus* M (1) | *Monodus subterraneus* Fr (2) | *Vaucheria* (? *sessilis*) Fr (3) |
| 14:0 | 10 | 12 | 11 | 2 | 12 |
| 16:0 | 26 | 24 | 26 | 24 | 15 |
| 16:1 | 2 | 2 | 2 | 24 | 32 |
| 16:2 | ⊦ | ⊦ | ⊦ | — | — |
| 16:3 | ⊦ | ⊦ | ⊦ | — | — |
| 16:4 | ⊦ | ⊦ | ⊦ | — | — |
| 18:0 | 1 | 1 | 1 | 1 | 1 |
| 18:1 | 19 | 16 | 17 | 9 | 13 |
| 18:2 | 9 | 8 | 7 | 4 | 2 |
| 18:3 | 6 | 7 | 8 | ⊦ | ⊦ |
| 18:4 | 6 | 7 | 6 | — | [2] |
| 20:0 | ⊦ | — | ⊦ | ⊦ | [1] |
| 20:2 | 1 | 1 | 1 | — | |
| 20:3 | ⊦ | 1 | 1 | 1 | 20:x*** |
| 20:4 | 10 | 11 | 10 | 5 | = 11% |
| 20:5 | 8 | 8 | 8 | 29 | |

* Abbreviations: M, marine; Fr, freshwater. ** References: (1) Klenk *et al.* 1963; (2) Nichols & Appleby 1969; (3) Wood B.J.B. unpubl. data, wild material. [ ] = identity uncertain. *** 20:x is unidentified 20-carbon unsaturated fatty acids.

The Phaeophyceae have rather less of the 20:5 acid, slightly more of the 20:4 acid, and oleic, linoleic and linolenic acids are more abundant. The general pattern is quite similar to that of the Cryptophyceae.

### 3.7   Chlorophyceae, Prasinophyceae and Euglenophyceae

Freshwater green algae in general, seem to contain few fatty acids with more than 3 double bonds or 18 carbon atoms (Table 8.10). Minor amounts of γ-linolenic acid occur sometimes, especially in the Volvocales but, as in higher plants, α-linolenate is the main 18:3 acid.

Substantial quantities of acids with more than 3 double bonds and more than 18 carbon atoms are found in the marine algae examined. It is not clear whether this correlation with the marine environment is fortuitous or whether it represents some form of adaptation necessary for survival in the sea. The composition of *Heteromastix rotunda* is remarkable in that it possesses 28% of a 20:5 acid and very little of the stearate family of acids. In this respect it resembles members of the Chrysophyceae rather than the Chlorophyceae.

## 4   FATTY ACID METABOLISM

The biosynthesis of fatty acids by algae is part of the fast-moving and exciting field of fatty acid biochemistry. So far as the general biochemistry of carbon-chain formation is concerned it does not appear to differ significantly in algae and other eukaryotes. The same holds true for the introduction of unsaturation into the carbon chain, even in the prokaryotic blue-green algae.

Harris *et al.* (1965) discussed differences in fatty acid synthesis in photosynthetic bacteria, algae and higher plants. The bacteria cannot desaturate preformed saturated acids (Wood *et al.* 1965) but use the anaerobic pathway described by Bloch *et al.* (1961). If provided with preformed acids, they apparently decompose the acids to 2-carbon units, then re-synthesize fatty acids *de novo*. Intact *Chlorella* on the other hand can desaturate monoenoic acids and in this resembles the intact plant leaf system described by James (1963) and can also directly desaturate stearate, unlike the plant leaf. These desaturations require molecular oxygen, unlike those in the photosynthetic bacteria (also see James *et al.* 1965).

The biosynthesis of unsaturated fatty acids by algae and protozoa was discussed by Erwin *et al.* (1964). They found that both the algae and the protozoa examined can progressively desaturate monoenoic acids, yielding di- and polyenoic acids, but that the organisms fall into two groups when examined for the ability to desaturate exogenously supplied saturated acids. The red alga *Porphyridium cruentum*, the amoeba *Hartmannella rhysodes*, the colourless euglenoid *Astasia longa* and several chrysomonads, can all effect direct desaturation, as can *Beggiatoa*. On the other hand, *Euglena gracilis*, *Chlamydomonas reinhardtii*, and

**Table 8.10.** Fatty acids of certain Chlorophyceae, Prasinophyceae and Euglenophyceae*

| Acid | Double bond positions | Chlorella vulgaris Fr.Au.** | Fr.He. | Fr.He.D.(1)*** | Chlorella pyrenoidosa Fr.Au.(2) | Scenedesmus obliquus Fr.Au.(2) | Scenedesmus quadricauda Fr.Au.(3) | Chlamydomonas reinhardtii Fr.Au.(7) | Chlamydomonas sp. M.Au.(4) | Chlamydomonas sp. M.Au.(4) | Dunaliella primolecta M.Au.(4) | Dunaliella tertiolecta M.Au.(4) |
|---|---|---|---|---|---|---|---|---|---|---|---|---|
| | | | | | Fatty acid as % of total acid | | | | | | | |
| 14:0 | — | 2 | 2 | t | — | 1 | — | 1 | 1 | 4 | 5 | 6 |
| 16:0 | — | 26 | 16 | 26 | 20 | 35 | 14 | 24 | 17 | 15 | 11 | 13 |
| 16:1 | 9 and/or 7 | 8 | 14 | 11 | 3 | 2 | 2 | 3 | 7 | 2 | 10 | 10 |
| 16:2 | 6,9 | 7 | 6 | 4 | — | t | 1 | — | t | 3 | — | — |
| 16:2 | 9,12 | | | | 7 | | | — | t | 4 | 8 | 3 |
| 16:3 | 6,9,12 | 2 | t | — | — | t | 2 | — | t | 7 | 7 | 5 |
| 16:4 | 6,9,12,15 | — | — | — | — | 15(Δ4,7,10,13) | (20) | — | t | t | 6 | 7 |
| 18:0 | — | 2 | 3 | 4 | — | — | — | 2 | t | — | t | — |
| 18:1 | 9 etc | 2 | 30 | 18 | 46 | 8 | 6 | 24 | 7 | 17 | 6 | 8 |
| 18:2 | 9,12 | 34 | 26 | 36 | 10 | 6 | 14 | 5 | 28 | 10 | 6 | 6 |
| γ18:3 | 6,9,12 | — | — | — | — | — | 1 | 6 | 1 | 3 | 2 | 1 |
| α18:3 | 9,12,15 | 20 | 4 | 1 | 12 | 30 | 34 | 31 | t | 28 | 10 | 8 |
| 18:4 | 6,9,12,15 | — | — | — | — | 2 | 4 | — | 19 | 1 | 7 | 8 |
| 20:0 | — | — | — | — | — | — | — | — | t | — | t | — |
| 20:1 | 11 etc | — | — | — | — | 1 | — | 2** | 16 | 1 | t | 1 |
| 20:2 | 8,11 | — | — | — | — | — | — | — | 1 | t | 1 | 2 |
| 20:3 | 8,11,14 | — | — | — | — | — | — | — | — | — | t | t |
| 20:4 | 5,8,11,14 | — | — | — | — | — | — | — | — | — | 1 | 4 |
| 20:4 | 8,11,14,17 | — | — | — | — | — | — | — | — | — | t | — |
| 20:5 | 5,8,11,14,17 | — | — | — | — | — | — | — | — | — | 2 | 10 |
| 22:0 | — | — | — | — | — | — | — | — | — | — | 10 | 1 |
| 22:5 | 4,7,10,13,16 | — | — | — | — | — | — | t | — | — | t | — |
| 22:5 | 7,10,13,16,19 | — | — | — | — | — | — | — | — | — | — | 6 |

Fatty acid as % of total acid

| Acid | Codium fragile M(2).Wild. | Enteromorpha compressa M(2).Wild. | Enteromorpha sp. Fr(3).Wild. | Spirogyra sp. Fr(3).Wild. | Heteromastix rotunda M.Au.(4). | Euglena gracilis Fr.He.(6) | Euglena gracilis z strain(6) Fr.He. | Euglena gracilis z strain(6) Fr.He.D. | Euglena gracilis Fr.Au.(7) | E. gracilis var. bacillaris Fr.He.(7) | Astasia longa Fr.He.(7) |
|---|---|---|---|---|---|---|---|---|---|---|---|
| 14:0 | 1 | 1 | 1 (2% 14:1) | 6 | 9 | 5 | 13 | 7 | 7 | 13 | 10 |
| 16:0 | 28 | 22 | 20 | 23 | 11 | 14 | 13 | 15 | 14 | 14 | 18 |
| 16:1 | 2 | +(2% 16:1Δ3) | 2 | 6 | 16 | 3(+2% trans-Δ3-16:1) | 5 | 6 | 6 | 4 | 1 |
| 16:2 | }1 | }1 | }3 | }1 | 1 | }8 | — | — | — | — | — |
| 16:2 | | | | | 2 | (5) | | | | | |
| 16:3 | 12 | 2 | 1 | 4 | 3 | | — | | | | — |
| 16:4 | — | 15 | 14(Δ4,7,10,13) | 8 | 1 | + | — | + | 16 | | — |
| 18:0 | 1 | — | + | 1 | — | 6 | — | 7 | — | — | — |
| 18:1 | 11 | 8 | 8 | 12 | 2 | 5 | 2 | 7 | 1 | 7 | 2 |
| 18:2 | 6 | 5 | 4 | 7 | 3 | 11 | 9 | 5 | 10 | 8 | 4 |
| γ18:3 | }27 | }26 | }18 | }17 | 4 | 15 | 3 | 21 | 4 | 3 | 1 |
| α18:3 | 2 | 9 | 16 | | 9 | | 1 | 1 | 32 | 16 | |
| 28:4 | + | — | | 4 | — | + | — | — | | | |
| 20:0 | — | — | — | — | — | | — | + | | | |
| 20:1 | — | — | + | + | — | (5) | — | | | | |
| 20:2 | — | 2 | + | + | — | + | — | | | | |
| 20:3 | — | | + | + | 1 | | — | | | | |
| 20:4 | 3 | }4 | }+ | }(1) | + | }(8) | — | | | | |
| 20:4 | 2 | 2 | (+20:6)3 | (+20:6)4 | 28 | | | | | | |
| 20:5 | 3 | 3 | | | | (9) | | | | | |
| 22:0 | — | — | — | — | 2 | — | — | | | | |
| 22:5 | }2 | }2 | }3 | 22:x =2% | 5 | }+ | C20+C22+C24 PUFA: 54 | C20+C22+C24 PUFA: 25 | C20 to C24 PUFA: 16 | C20 to C24 PUFA: 16 | C20 to C24 PUFA: 55 |
| 22:5 | | | | | | | | | | | |

* All genera listed in table are members of the Chlorophyceae apart from Heteromastix (Prasinophyceae) and Euglena and Astasia (Euglenophyceae). ** Abbreviations: Fr, freshwater; Au, photoautotrophic growth; He, photoheterotrophic growth; D, dark heterotrophic growth; M, marine. *** References: (1) Nichols 1965; (2) Klenk et al. 1963; (3) Shaw 1966; (4) Chuecas & Riley 1969; (5) Schlenk & Gellerman 1965; (6) Nichols & Appleby 1969; (7) Erwin & Bloch 1963.

the colourless *Polytoma uvella*, resemble higher plants in being unable to desaturate exogenously supplied radioactive stearic and palmitic acids. These observations, combined with the demonstration that the amoeba possesses only the γ-linolenic acid pathway of polyene synthesis, the green alga only the α-linolenic acid pathway, while the other organisms had both pathways, enabled them to propose a set of phylogenetic relationships, which, although now overtaken by more recent ideas on the relationships of chloroplasts and blue-green algae, display some interesting analogies with the views advanced by Lee (1972).

A review of the biosynthesis of polyunsaturated fatty acids (PUFA) has been presented by Nichols and Appleby (1969). They found that *Ochromonas danica* and *Porphyridium cruentum* synthesise arachidonic acid (20:4) by the pathway involving γ-linolenic acid, whereas *Euglena gracilis* employs the following route

$$18:2 \ (\Delta 9,12) \rightarrow 20:2 \ (\Delta 11,14) \rightarrow 20:3 \ (\Delta 8,11,14) \rightarrow 20:4 \ (\Delta 5,8,11,14)$$

and is incapable of converting γ-linolenic acid to arachidonic acid. They further conclude that in *Porphyridium cruentum* and in the xanthophycean alga *Monodus subterraneus* the site of synthesis of arachidonic acid appears to be the chloroplast, but that in *Ochromonas danica* and in *Euglena gracilis* little of this acid is accumulated in the photosynthetic apparatus. Because of difficulties in isolating reasonably intact chloroplasts, they reached these conclusions by the elegant method of first isolating, then analysing the fatty acids of, individual lipid species.

The route of *trans*-Δ3-hexadecenoic acid biosynthesis by *Chlorella vulgaris* was established by Nichols *et al.* (1965a). It arises by direct desaturation of palmitic acid in a reaction which requires light, and probably oxygen. They advance reasons for thinking that phosphatidyl glycerol, the only lipid in which this acid occurs in photosynthetic tissue, may be involved in the dehydrogenation reaction.

The question of the involvement of intact lipid species was also considered by Gurr and Brown (1970) who studied the composition of phosphatidyl choline species during the biosynthesis of linoleic acid by *Chlorella vulgaris*, and by Nichols (1968) who examined fatty acid metabolism in the chloroplast lipids of various algae. The latter author showed that blue-green algae and green algae incorporate radioactive acetate very efficiently into the fatty acids of their polar lipids and that there are marked differences in the rates of labelling in different lipids. Thus the fatty acids of digalactosyl diglyceride and sulpholipid are labelled more rapidly in blue-green algae than in *Chlorella*, while the reverse is true of phosphatidyl glycerol.

The ability of algae to desaturate exogenously supplied saturated acids was used by Morris *et al.* (1968) in a study of the stereochemistry of the reaction in *Chlorella*. They found high stereo-specificity, both in the conversion of stearic to oleic acid and of oleic to linoleic acid, the D9 and D10 hydrogen atoms of stearic acid being lost on desaturation, and the D12 and D13 hydrogen atoms being lost from oleic acid in the course of the further desaturation to linoleic acid.

Delo *et al.* (1971) and Ernst-Fonberg and Bloch (1971) examined the fatty acid synthetases of *Euglena gracilis* and demonstrated the existence of a system associated with the chloroplast. It will be of phylogenetic interest to compare these systems with those from blue-green algae.

There is little doubt that the pathway of fatty acid degradation in algae is the β-oxidation route (James *et al.* 1965) and the alternative α-oxidation route observed in higher plants (Hitchcock & James 1963, James *et al.* 1965) does not seem to be a route for complete degradation of fatty acids in algae. However, Kolattukudy (1970) has suggested that it is operative in the generation of the odd-carbon-number long-chain alcohols found in the waxes of *Euglena gracilis*. He has also examined, partially purified, and characterized, the two step reaction whereby an acid is reduced to the corresponding alcohol and the subsequent esterification to the wax. He showed that whereas added radioactive fatty acids are rapidly and efficiently converted into the alcohol and incorporated into wax, radioactive acetate, when supplied under similar conditions, is mainly synthesized into the fatty acids of the polar lipids. He proposed that this is because the coenzyme A derivatives of the fatty acids are needed for the reactions leading to the formation of alcohols and that the acetate goes to the formation of the fatty acid acyl carrier protein derivative which is then preferentially incorporated into the polar lipid. Wittels and Blum (1968) demonstrated in *Euglena* that cells grown with acetic acid as sole carbon source are capable (after a lag of about 50 hours) of growing on palmitic acid as sole carbon source. This is accompanied by a decrease in the level of free carnitine in the cells, but there is at the same time a three-fold increase in the level of the enzyme carnitine palmityl transferase, thus suggesting that carnitine derivatives are important in mobilizing fatty acids.

The effect of varying carbon dioxide concentrations on the fatty acid composition of heterotrophically grown cells, has been studied in *Chlorella fusca* by Dickson *et al.* (1969). They observed that autotrophically grown cells supplied with 1% carbon dioxide in air have 16:0, 16:3, 16:4, 18:1, 18:2 and 18:3 as the principal fatty acids. The rather high 16:4 acid content is noteworthy. When the cells are grown in the presence of glucose and air containing 1% carbon dioxide, the 16:4 acid disappears entirely, the linolenic acid content is reduced, and there are complementary increases in the levels of the other major acids. When the carbon dioxide concentration is raised step-wise to 30% of the gas phase supplied to a heterotrophic culture, the lipid content of the cells exhibits a 40% increase and 16:0, 18:1, 18:2 and 18:3 acids all increase. The authors discussed the biochemical interpretation of these findings. It would be of interest to know what effect such high levels of carbon dioxide have upon other *Chlorella* species, and on cells grown under autotrophic conditions.

A somewhat neglected aspect of lipid metabolism is the problem of what, if any, differences occur at different stages of the cell's life cycle. Reitz *et al.* (1967) have examined the fatty acid content of synchronously-grown *Chlorella* cells. They used two media, one giving fully autotrophic growth, the other containing a low (0·2%) concentration of glucose. Their test organism was the high temperature *Chlorella pyrenoidosa* strain 7–11–05. There was very little difference in lipid content between cells grown in the two media, perhaps due to the low glucose concentration employed, or perhaps because exposure to the glucose was for only a single generation, although the presence of the glucose had a very marked effect on the size and number of the cells. Even so, there were clear-cut changes in fatty acid composition over the first twelve hours of the growth cycle, the increase of 16:0 and 18:2 being especially marked. Further studies on fatty acid metabolism of synchronous cultures, using exogenously supplied radioactive compounds are required.

Graff *et al.* (1970) report on the preparation of fully deuterated fatty acids, using *Scenedesmus obliquus* grown in practically pure deuterium oxide. Since it was in no way relevant to their study, they did not perform parallel experiments with algae grown under otherwise identical conditions but using normal water instead of deuterium oxide, but the results of such experiments would be of interest to the algal physiologist. Comparisons between the fatty acid compositions which they report, and previously published reports on 'normal' *S. obliquus* show however that cells grown in deuterium oxide have a higher-than-usual level of 18:2 and a very low 18:3 content. Such comparisons, particularly if the individual lipids were also examined, could provide the membrane physiologist with useful information.

## 5 ACYL LIPIDS

Photosynthetic bacteria belonging to the Athiorhodaceae have only two lipids in common: phosphatidyl glycerol and phosphatidyl ethanolamine (Wood *et al.* 1965). The same study revealed that phosphatidyl choline, plant sulpholipid, cardiolipin and *O*-ornithyl phosphatidyl glycerol were all present in one or more of the five bacteria examined. Steiner *et al.* (1969) showed that *Chromatium* also contained phosphatidyl ethanolamine and phosphatidyl glycerol plus cardiolipin and lysophosphatidyl ethanolamine. None of these organisms contain the mono- and di-galactosyl diglycerides (MGDG and DGDG) which, as will soon become clear, appear to be ubiquitous among blue-green and eukaryotic algae.

The blue-green algae always contain four lipids: phosphatidyl glycerol, and the three glycolipids, MGDG, DGDG, and the plant sulpholipid (Nichols *et al.* 1965, Nichols & Wood 1968a,b, Kenyon *et al.* 1972). These lipids are also the four major acyl lipids of the chloroplast in all eukaryotic algae and higher plants. Occasionally, other classes of as yet unidentified lipids have been reported in blue-green algae (Nichols *et al.* 1965b, Allen *et al.* 1966, Nichols 1973). The lipid found in the heterocysts of nitrogen-fixing algae is discussed later (p. 258).

Eukaryotic algae resemble higher plants in having a greater variety of lipids, especially phospholipids. Kates (1970) has prepared a major review of the lipids in different plant species.

Certain algae, notably *Ochromonas danica*, possess a series of long chain diol disulphates, notably 1,14,S-docosanediol-1,14-disulphate and halogenated derivatives thereof (see Kates & Wassef 1970). The state of present knowledge of the plant sulphonoglycolipid is reviewed by Haines (1971). The metabolic role of the sulphur-containing lipids is not yet understood, although the ubiquitous nature of the plant sulpholipid seems to indicate that it at least has an important part to play, probably in photosynthesis. It is uncertain why the highest concentrations of plant sulpholipid apparently occur in marine red algae.

Studies by Alam *et al*. (1971) indicate that the brown alga *Fucus* may be unusual among eukaryotic algae so far examined in that it lacks phosphatidyl choline. It also seems to be rather rich in glycolipids, which constitute 30% of the total lipid. Ikawa *et al*. (1968) discuss the distribution of choline and inositol in algae and fungi.

With these exceptions, the intact saponifiable lipids of algae seem not to be unusual, and are best considered along with the lipids of higher plants. The reader is directed for further details of these to the book by Hitchcock and Nichols (1971) which gives an integrated account of plant lipid chemistry and biosynthesis.

The convenience with which unicellular algae may be handled under laboratory conditions has been turned to advantage in several studies. Thus Nichols (1965) has compared the fatty acid composition of individual lipids in light-grown *Chlorella vulgaris* with those in dark-grown cells. Particularly noteworthy are the sharp decreases in α-linolenic acid content of the four 'chloroplast' lipids of the etiolated cells when contrasted with those of the fully autotrophic cells: MGDG, from 45% to 5%; DGDG, from 37% to 4%; Sulpholipid, from 15% to 3%; PG, from 5% to a trace. In similar experiments comparing green and etiolated castor bean leaves, James and Nichols (1966) observed that the α-linolenic acid content of the four lipids was roughly halved when the leaves were grown in darkness. The *trans*-$\Delta^3$-hexadecenoic acid occurs exclusively in phosphatidyl glycerol, and in both the *Chlorella* and the castor bean, it virtually disappears from dark-grown plants.

Comparative fatty acid metabolism in the chloroplast lipids of the green alga *Chlorella vulgaris* and in the lipids of the blue-green algae *Anabaena cylindrica* and *Anacystis nidulans* was examined by Nichols (1968). In the blue-green algae, the sulpholipid and DGDG were the most rapidly labelled lipids, whereas in *Chlorella* it was the phosphatidyl glycerol. Different metabolic pathways may be involved, therefore, in the two groups of algae. In similar experiments comparing light-grown and dark-grown *Chlorella* cells, Nichols *et al*. (1967) showed that PG, MGDG, PC and neutral glyceride have a high turnover rate for certain fatty acids, whereas DGDG, PE, PI and the plant sulpholipid have a slow rate of fatty acid turnover. These results are similar whether the cells are grown autotrophically or heterotrophically in the dark, although the types of acid being synthesized are quite different, radioactivity being incorporated almost entirely into the saturated and monoenoic acids in the dark, whereas in light-grown cells, the polyunsaturated fatty acids are also rapidly labelled.

Goldberg and Ohad (1970), examining lipid and pigment changes during the synthesis of chloroplast membranes in a mutant of *Chlamydomonas reinhardtii*, found that when dark-grown cells are exposed to the light, the PG and glycolipids are the main lipids to be synthesized and this differs slightly from the results obtained with *Chlorella*. This may, of course, represent a purely transitory stage during adaptation to photosynthesis.

Constantopoulos and Bloch (1967) have examined the effect of light intensity on the lipid composition of *Euglena gracilis*. Chlorophyll and total lipid content both decrease as the light intensity increases but α-linolenic and 4,7,10,13 hexadecatetraenoic acid increase sharply. The increase in these acids is most marked in the chloroplast lipids, especially the MGDG. Arachidonic acid also increases, but γ-linolenate and linoleate decrease with increase of light intensity.

Manganese deficiency influences the biosynthesis of lipids in *Euglena* (Constantopoulos 1970). Deficiency of the metal strongly inhibits growth of the alga, but the chlorophyll content per cell is little affected. The galactosyl diglyceride content on the other hand may be reduced by as much as 40% in light-grown cells. Dark-grown cells also exhibit reduction in the concentration of galactosyl diglycerides when subjected to manganese deficiency. The fatty acids of manganese-deficient photoheterotrophic cells are rich in saturated acids, with an unusually high (45%) content of myristic acid. Despite this, the galactosyl diglycerides still contain a preponderance of poly-unsaturated fatty acids, testifying to the great importance of unsaturated acids in these lipids.

Matson *et al.* (1970) have studied the incorporation of galactose into the galactolipids of *Euglena gracilis* strain Z. They found that isolated chloroplasts from light-grown cells will incorporate galactose into galactolipids, but that extracts of dark-grown cells cannot do so.

## 6.  LIPIDS OF NITROGEN-FIXING BLUE-GREEN ALGAE

Nichols and Wood (1968a) found that a range of nitrogen-fixing algae possess a unique lipid class, a mixture of mono-hexoside derivatives of long chain polyhydroxy alcohols (Bryce *et al.* 1972). The lipid was present in all heterocyst-possessing blue-green algae examined, but absent even from non-heterocystous members of the same genus as the nitrogen-fixers. Further examination showed that, as expected, the lipid was confined to the heterocyst of the heterocystous nitrogen-fixing species examined (Walsby & Nichols 1969). It will be most interesting to examine the lipid composition of the nitrogen-fixing Chroococcacean species (Wyatt & Silvey 1969, Rippka *et al.* 1971). Rippka *et al.* (1971) have shown that their strain of *Gloeocapsa* exhibits no structural differences whether grown with $N_2$ or with nitrate as sole source of nitrogen.

Little work has been carried out on the lipids of aerobic nitrogen-fixing organisms other than Cyanophyceae, not least because of the difficulties imposed by, for example, the considerable quantities of poly-β-hydroxybutyric acid (poly-BHB) accumulated by *Azotobacter* species. Some unpublished work of a very preliminary nature by Rosalind Stewart and B. J. B. Wood indicates that these organisms do not possess a lipid chemically related to the lipid found in the blue-green algae. It is tempting to wonder if the poly-BHB accumulations may serve a role in the bacteria analogous to that thought to be filled by the fatty alcohol glycosides in the algae.

## 7. FATTY ACIDS, PHOTOSYNTHESIS AND PHYLOGENY

The range of fatty acids found in living systems and clear correlations between fatty acid compositions of related organisms have constantly excited speculation as to the phylogenetic significance of their distribution and biosynthetic routes (see, for example, Erwin *et al.* 1964, Wagner & Pohl 1966, Nichols & Appleby 1969, Nichols 1970).

Higher plants characteristically contain α-linolenic acid as a constituent of chloroplast lipids whereas at the other extreme the photosynthetic bacteria possess no polyunsaturated fatty acids. Since these latter are incapable of effecting the production of free oxygen during photosynthesis, there was some speculation that the presence of α-linolenic acid might be correlated in some way with the Hill reaction (Erwin & Bloch 1963). Support for this idea is supplied by the observation that when green algae such as *Chlorella* are grown in media containing organic compounds such as sugars, the content of α-linolenic acid decreases—most markedly in the dark. In a typical example, taken from Nichols (1965), the linolenic content of monogalactosyl diglyceride, digalactosyl diglyceride, and sulpholipid declined from 45%, 37% and 15% respectively in the light to 5%, 4% and 3% respectively in the dark, with compensatory increases in the 16:2, 18:1, and 18:2 acids.

Perhaps even more striking, is the change in *trans*-Δ³-hexadecenoic acid which is located exclusively in the phosphatidyl glycerol. This declined from 16% of the fatty acid of phosphatidyl glycerol in light-grown *Chlorella* to only a trace in the phosphatidyl glycerol of dark-grown cells.

Holton *et al.* (1964) studied the fatty acids of the blue-green alga *Anacystis nidulans* and found that the alga contained no poly-enoic acid, or di-enoic acid and thus resembled the photosynthetic bacteria, although of course the 18:1 acid present in the alga is oleic acid (Δ9) whereas that found (Wood *et al.* 1965) in the bacteria is vaccenic acid (Δ11). It is now clear that the blue-green algae encompass a range of fatty acid compositions, ranging from *Anacystis nidulans* at one extreme, to organisms such as *Myxosarcina chroococcoides* (possessing 33% of α-linolenic acid, Nichols and Wood 1968b) at the other. Stanier *et al.* (1971) have provided evidence of such a gradation even within the simple chroococcacean blue-green algae. Kenyon and Stanier (1970) discuss the relevance of the fatty acid composition of the blue-green algae to theories of the evolution of the chloroplast of eukaryotic algae from symbiotic blue-green algae (see also Fogg et al., 1973, and Carr & Whitton, 1973).

Nichols and Wood (1968a) have shown that in *Spirulina platensis* there is no α-linolenic acid, but that the alga possesses a considerable quantity (21·4%) of γ-linolenic acid, mainly concentrated in the mono- and di-galactosyl diglycerides. Pelloquin et al. (1970) note that *Spirulina geitleri* is even richer in γ-linolenic acid (22 to 26·6% of the total fatty acids) than is *S.platensis*, and the disparity might be even greater than their results suggest, since their extraction procedure— soxhlet extraction of the algal tissue with petroleum ether—would be only moderately effective at extracting the phospholipids in which the greater part of *S.platensis* γ-18:3 resides. They record 14 to 15·5% of γ-18:3 in the fatty acids of *S.platensis* against the 21·4% observed by Nichols and Wood (1968a). Stanier et al. (1971) have shown that Group IIA in their classification of the order Chroococcales possesses γ-18:3 and no α-18:3 acid. It is interesting to speculate upon the evolutionary significance of the presence of γ-linolenic acid in the two widely separated groups of Cyanophyceae—on the one hand the unicellular Chroococcales, and on the other, the filamentous *Spirulina* species. Did they develop the use of this acid independently, or is there an evolutionary link between the two groups?

With respect to the other fatty acid which Nichols et al. (1965) suggested had a connexion with the Hill reaction, namely trans-$\Delta^3$-hexadecenoic acid, the situation is, if anything, still more intriguing. They demonstrated its presence in the green tissue of a variety of higher plants and in green algae, and its absence both from the corresponding tissues when grown in the dark, and from photosynthetic bacteria when grown in darkness or in the light. Subsequently however, it has been shown to be absent from autotrophically grown blue-green algae. How, then, did this acid arise, and what bearing does its presence in all eukaryotic, light-grown, photosynthetic organisms so far examined, have on current speculations that modern chloroplasts are derived from once-independent blue-green algae? Clearly the acid is important to those organisms which possess it. It is unusual both in the position and in the trans-configuration of the double bond, and its synthesis must therefore require special enzyme systems. Nichols et al. (1965a) have shown that light is essential for its synthesis in *Chlorella vulgaris*. Its loss in darkness suggests its importance to photosynthesis and yet its absence from photosynthetic prokaryotes argues against this conclusion. It appears that a specific role for particular fatty acids in photosynthesis must be discounted, and, as suggested by Nichols et al. (1965b), the fatty acid containing lipids, sulphoquinovosyl diglyceride, phosphatidyl glycerol and the two galactosyl diglycerides should be looked on as materials providing a common link between photosynthetic cells, perhaps finding their role in the maintenance of structural configurations essential to the proper functioning of the photosynthetic electron transport chain.

Evolutionary relationships within the algae and related protista are among the most fundamental problems in our understanding of the early development of life; the current idea on the possibility of an origin of chloroplasts from blue-green algae providing a particularly stimulating area for discussion and argument. It is of interest to compare the views of Lee (1972), using reasoning from pigment composition and subcellular structure, with the evolutionary scheme put forward by Nichols (1970) employing lipid and fatty acid composition as his criteria. Lee proposes that cryptophycean cyanomes are the fundamental ancestors of all photosynthetic eukaryotes and suggests a divergence into three great groups: Chlorophyceae, Euglenophyceae and Prasinophyceae as constituting one group; the Cryptophyceae, together with the Phaeophyceae, Xanthophyceae, Chrysophyceae, Haptophyceae, and Dinophyceae constitute the second group; while the third group embraces the Rhodophyceae.

Nichols, using the alternative possession of either α-linolenic acid or of γ-linolenic acid as his major criterion, arrives at a very similar conclusion. The α-linolenic acid series comprises the Euglenophyceae and Chlorophyceae while the γ-linolenic acid series includes the Chrysophyceae, Haptophyceae, Rhodophyceae, Phaeophyceae, Bacillariophyceae and Xanthophyceae. There are of course differences in detail between the relationships suggested by their very different criteria, but the need to resolve these differences should stimulate much fascinating research.

Kenyon (1972) and Kenyon *et al.* (1972) with their demonstration of the occurrence of unicellular blue-green algae possessing both γ- and α-linolenic acids, and of the occurrence of an 18:4 acid in blue-green algae, provide further evidence for the credibility of a scheme in which the green algae (α-linolenic acid) and other (γ-linolenic acid) eukaryotic algae arose independently.

# 8  REFERENCES

ADAMS B.L., McMAHON V. & SECKBACH J. (1971) Fatty acids in the thermophilic alga *Cyanidium caldarium. Biochem. biophys. Res. Comm.* **42,** 359–65.

ALAM M., CHAKRAVARTI A. & IKAWA M. (1971) Lipid composition of the brown alga *Fucus vesiculosus. J. Phycol.* **7,** 267–8.

ALLEN C.F., GOOD P. & HOLTON R.W. (1970) Lipid composition of *Cyanidium. Pl. Physiol., Lancaster,* **46,** 748–51.

ALLEN C.F., HIRAYAMA O. & GOOD P. (1966) Lipid composition of photosynthetic systems. In Goodwin T.W. *The Biochemistry of Chloroplasts,* I, pp. 195–200. Academic Press, London & New York.

APPLEBY R.S., SAFFORD R. & NICHOLS B.W. (1971) The involvement of lecithin and monogalactosyl diglyceride in linoleate synthesis by green and blue-green algae. *Biochim. biophys. Acta,* **248,** 205–11.

BEACH D.H., HARRINGTON G.W. & HOLZ G.G. (1970) The polyunsaturated fatty acids of marine and fresh water cryptomonads. *J. Protozool.* **17,** 501–10.

BLOCH K., BARONOWSKY P., GOLDFINE H., LENNARZ W.J., LIGHT R., NORRIS A.T. & SCHEUERBRANDT G. (1961) Biosynthesis and metabolism of unsaturated fatty acids. *Federation Proc.* **20,** 921–7.

BRYCE T.A., WELTI D., WALSBY A.E. & NICHOLS B.W. (1972) Monohexoside derivatives of long chain polyhydroxy alcohols, a novel class of lipid specific to heterocystous algae. *Phytochemistry*, **11**, 295–302.

CARR N.G. & WHITTON B.A. (eds) (1973) *The Biology of Blue-green Algae*. Blackwell Scientific Publications, Oxford.

CHUECAS L. & RILEY J.P. (1969) Component fatty acids of the total lipids of some marine phytoplankton. *J. mar. biol. Ass. U.K.* **49**, 97–116.

CONSTANTOPOULOS G. (1970) Lipid metabolism of manganese-deficient algae. I. *Pl. Physiol., Lancaster*, **45**, 76–80.

CONSTANTOPOULOS G. & BLOCH K. (1967) Lipid composition of *Euglena gracilis*—The effect of light intensity. *J. biol. Chem.* **242**, 3538–42.

VAN DEENEN L.L.M. & HAVERKATE F. (1966) Chemical characterisation of phosphatidyl glycerol from photosynthetic tissues. In Goodwin T.W. *The Biochemistry of Chloroplasts*, I, pp. 195–200. Academic Press, London & New York.

DELO J., ERNST-FONBERG M.L. & BLOCH K. (1971) Fatty acid synthetases from *Euglena gracilis. Arch. Biochem. Biophys.* **143**, 384–91.

DICKSON L.G., GALLOWAY R.A. & PATTERSON G.W. (1969) Environmentally induced changes in the fatty acids of *Chlorella. Pl. physiol., Lancaster*, **44**, 1413–16.

DUNHAM J.E., HARRINGTON G.W. & HOLZ G.G. JR. (1966) Phytoplankton sources of the eicosapentaenoic and docosahexaenoic fatty acids characteristic of metazoa. *Biol. Bull.* **131**, 389.

ERNST-FONBERG M.L. & BLOCH K. (1971) A chloroplast-associated fatty acid synthetase system in *Euglena. Archs. Biochem. Biophys.* **143**, 392–400.

ERWIN J. & BLOCH K. (1963) Polyunsaturated fatty acids in some micro-organisms. *Biochem. Z.* **338**, 496–511.

ERWIN J., HULUNICKA D. & BLOCH K. (1964) Comparative aspects of unsaturated fatty acid synthesis. *Comp. Biochem. Physiol.* **12**, 191–207.

VON EULER U.S. & ELIASSON R. (1967) *Prostaglandins*. Academic Press, New York & London.

FOGG G.E., STEWART W.D.P., FAY P. & WALSBY A.E. (1973) *The Blue-green Algae*, Academic Press, London & New York.

GOLDBERG I. & OHAD I. (1970) Lipid and pigment changes during synthesis of chloroplast membranes in a mutant of *Chlamydomonas reinhardi* y.l. *J. Cell. Biol.* **44**, 563–571.

GOODWIN T.W. (ed.) (1966–7) *The biochemistry of chloroplasts* in 2 volumes, Proceedings of a NATO Advanced Study Institute held in Aberystwyth, in 1965. Academic Press, London & New York.

GRAFF, G., SZCZEPANIK P., KLEIN P.D., CHIPAULT J.R. & HOLMAN R.T. (1970) Identification and characterisation of fully deuterated fatty acids from *Scenedesmus obliquus* cultured in 99·7% deuterium oxide. *Lipids*, **5**, 786–92.

GUNSTONE F.D. (1967) *An Introduction to the Chemistry and Biochemistry of Fatty Acids and their Glycerides*. Chapman & Hall, London.

GURR M.I. & BROWN P. (1970) The composition of phosphatidyl choline species in *Chlorella vulgaris* during the formation of linoleic acid. *Eur. J. Biochem.* **17**, 19–22.

GURR M.I. & JAMES A.T. (1971) *Lipid Biochemistry: An Introduction*. Chapman & Hall, London.

HAIGH W.G., SAFFORD R. & JAMES A.T. (1969) Fatty acid composition and biosynthesis in ferns. *Biochim. biophys. Acta*, **176**, 647–50.

HAINES T.H. (1971) The plant sulpholipid. *Prog. in Chem. of Fats and Other Lipids*, **11**, 297–345.

HARRINGTON G.W., BEACH D.H., DUNHAM J.E. & HOLZ G.G. JR. (1970) The polyunsaturated fatty acids of marine dinoflagellates. *J. Protozool.* **17**, 213–19.

HARRINGTON G.W. & HOLZ G.G. JR. (1968) The monoenoic and docosahexaenoic fatty acids of a heterotrophic dinoflagellate. *Biochim. biophys. Acta*, **164**, 137–9.

HARRIS R.V., WOOD B.J.B. & JAMES A.T. (1965) Fatty acid biosynthesis in photosynthetic micro-organisms. *Biochem. J.* **94**, 22–3P.

HITCHCOCK C. & JAMES A.T. (1963) Breakdown of unsaturated fatty acids by plant tissue. *Biochem. J.* **89**, 22P.

HITCHCOCK C. & NICHOLS B.W. (1971) *Plant lipid biochemistry.* Academic Press, New York & London.

HOLTON R.W. & BLECKER H.H. (1970) Fatty acids of blue-green algae. In *Properties and Products of Algae*, ed. Zajic J.E. Plenum Press, New York.

HOLTON R.W., BLECKER H.H. & ONORE M. (1964) Effect of growth temperature on the fatty acid composition of a blue-green alga. *Phytochemistry*, **3**, 595–602.

HOLTON R.W., BLECKER H.H. & STEVENS T.S. (1968) Fatty acids in blue-green algae, possible relationship to phylogenetic position. *Science N.Y.*, **160**, 545–7.

IKAWA M., BOROWSKI P.T. & CHAKRAVARTI A. (1968) Choline and inositol distribution in algae and fungi. *App. Microbiol.* **16**, 620–3.

JAMES A.T. (1963) The biosynthesis of long-chain saturated and unsaturated fatty acids in isolated plant leaves. *Biochim. biophys. Acta*, **70**, 9–19.

JAMES A.T., HARRIS R.V., HITCHCOCK C., WOOD B.J.B. & NICHOLS B.W. (1965) Investigation of the biosynthesis and degradation of unsaturated fatty acids in higher plants and photosynthetic bacteria. *Fette-Seifen Anstrichmittel*, **67**, 393–6.

JAMES A.T. & NICHOLS B.W. (1966) Lipids of photosynthetic systems. *Nature, Lond.* **210**, 372–5.

KATES M. (1970) Plant phospholipids and glycolipids. *Adv. Lipid Res.* **8**, 225–67.

KATES M. & VOLCANI B.E. (1966) Lipid components of diatoms. *Biochim. biophys. Acta*, **116**, 264.

KATES M. & WASSEF M.K. (1970) Lipid chemistry. *A. Rev. Biochem.* **39**, 323–58.

KENYON C.N. (1972) The fatty acid composition of unicellular strains of blue-green algae. *J. Bact.* **109**, 827–34.

KENYON C.N., RIPPKA R. & STANIER R.Y. (1972) Fatty acid composition and physiological properties of some filamentous blue-green algae. *Arch. Mikrobiol.* **83**, 216–36.

KENYON C.N. & STANIER R.Y. (1970) Possible evolutionary significance of polyunsaturated fatty acids in blue-green algae. *Nature, Lond.* **227**, 1164–6.

KHOJA T. & WHITTON B.A. (1971) Heterotrophic growth of blue-green algae. *Arch. Mikrobiol.* **79**, 280–2.

KLEINSCHMIDT M.G. & MCMAHON V.A. (1970a) Effect of growth temperature on the lipid composition of *Cyanidium caldarium*. I. Class separation of lipids. *Pl. Physiol., Lancaster*, **46**, 286–9.

KLEINSCHMIDT M.G. & MCMAHON V.A. (1970b) Effect of growth temperature on the lipid composition of *Cyanidium caldarium*. II. Glycolipid and phospholipid components. *Pl. Physiol., Lancaster*, **46**, 290–3.

KLENK E., KNIPPRATH W., EBERHAGEN D. & KOOF H.D. (1963) Uber die Ungesättigten Fettsäuren der Fettstoffe von Süßwasser und Meeresalgen. *Z. Physiol. Chem.* **334**, 44.

KOLATTUKUDY P.E. (1970) Reduction of fatty acids to alcohols by cell-free preparations of *Euglena gracilis*. *Biochemistry*, **9**, 1095–102.

KORN E.D. (1964) The polyunsaturated 20-carbon and 22-carbon fatty acids of *Euglena*. *Biochem. biophys. Res. Comm.* **14**, 1–6.

LEE R.E. (1972) Origin of plastids and the phylogeny of algae. *Nature, Lond.* **237**, 44–5.

LEE R.F. & LOEBLICH A.R. (1971) Distribution of 21:6 hydrocarbon and its relationship to 22:6 fatty acid in algae. *Phytochemistry*, **10**, 593–602.

MATSON R.S., FEI M. & CHANG S.B. (1970) Comparative studies of the biosynthesis of galactolipids in *Euglena gracilis* Strain Z. *Pl. Physiol., Lancaster*, **45**, 531–2.

MORRIS L.J., HARRIS R.V., KELLY W. & JAMES A.T. (1968) The stereochemistry of desaturations of long-chain fatty acids in *Chlorella vulgaris*. *Biochem. J.* **109**, 673–8.

MORRIS L.J. & NICHOLS B.W. (1966) The chromatography of lipids, a literature review. In *Progress in Thin Layer Chromatography and Related Methods*, ed. Niederwieser A. Humphrey & Putati, Ann Arbor.

NICHOLS B.W. (1965) Light-induced changes in the lipids of *Chlorella vulgaris*. *Biochim. biophys. Acta*, **106**, 274–9.

NICHOLS B.W. (1968) Fatty acid metabolism in the chloroplast lipids of green and blue-green algae. *Lipids*, **3**, 354–60.

NICHOLS B.W. (1970) Comparative lipid biochemistry of photosynthetic organisms. In *Phytochemical Phylogeny*, ed. Harborne J. B. pp. 105–18. Academic Press, New York & London.

NICHOLS B.W. (1973) Lipid Metabolism. In *The biology of blue-green algae*, eds. Carr N. & Whitton B.A. pp. 144–61. Blackwell Scientific Publishers, Oxford.

NICHOLS B.W. & APPLEBY R.S. (1969) The distribution and biosynthesis of arachidonic acid in Algae. *Phytochem.* **8**, 1907–15.

NICHOLS B.W., HARRIS P. & JAMES A.T. (1965a) The biosynthesis of trans-$\triangle$³-hexadecenoic acid by *Chlorella vulgaris*. *Biochem. biophys. Res. Comm.* **21**, 473–9.

NICHOLS B.W., JAMES A.T. & BREUER J. (1967) Interrelationships between fatty acid biosynthesis and acyl lipid synthesis in *Chlorella vulgaris*. *Biochem. J.* **104**, 486–96.

NICHOLS B.W., MORRIS L.J. & JAMES A.T. (1966) Separation of lipids by chromatography. *Brit. Med. Bull.* **22**, 137–46.

NICHOLS B.W. & WOOD B.J.B. (1968a) New glycolipid specific to nitrogen-fixing blue-green algae. *Nature, Lond.* **217**, 767–8.

NICHOLS B.W. & WOOD B.J.B. (1968b) The occurrence and biosynthesis of gamma-linolenic acid in a blue-green alga *Spirulina platensis*. *Lipids*, **3**, 46–50.

NICHOLS B.W., WOOD B.J.B. & JAMES A.T. (1965b) The distribution of trans-$\Delta$³ hexadecenoic acid in plants and photosynthetic micro-organisms. *Biochem. J.* **95**, 6P.

PARKER P.L., VAN BAALEN C. & MAURER C. (1967) Fatty acids in eleven species of blue-green algae; geochemical significance. *Science N.Y.* **155**, 707–8.

PATTON S., FULLER G., LOEBLICH A.R. & BENSON A.A. (1966) Fatty acids of the 'red tide organism *Gonyaulax polyedra*'. *Biochim. biophys. Acta*, **116**, 577–9.

PELLOQUIN A., LAI R. & BUSSON F. (1970) Comparative study of lipids of *Spirulina platensis* Geitler and *S. geitleri* J. De Toni. *C.r. Acad. Sci. Paris*, **271**, 932–5.

REITZ R.C., HAMILTON J.G. & COLE F.E. (1967) Fatty acid concentration in synchronous cultures of *Chlorella pyrenoidosa* 7-11-05, grown in the presence and the absence of glucose. *Lipids*, **2**, 381–9.

RIPPKA R., NEILSON A., KUNISAWA R. & COHEN-BAZIRE G. (1971) Nitrogen fixation by unicellular blue-green algae. *Arch. Mikrobiol.* **76**, 341–8.

SCAGEL R.F., BANDONI R.J., ROUSE G.E., SCHOFIELD W.B., STEIN J.R. & TAYLOR T.M.C. (1969) *Plant diversity, an evolutionary approach*. Wadsworth Pub. Co. Inc., Belmont, Calif.

SCHLENK H. & GELLERMAN J.L. (1965) Arachidonic, 5, 11, 14, 17 eicosatetraenoic and related acids in plants—identification of unsaturated fatty acids. *J. Am. Oil Chem. Soc.* **42**, 504.

SCOTT R.P.W. (1966) Gas-lipid chromatography; recent development in apparatus and technique. *Brit. Med. Bull.* **22**, 131–6.

SECKBACH J. & IKAN R. (1972) Sterols and chloroplast structure of *Cyanidium caldarium*. *Pl. Physiol., Lancaster*, **49**, 457–9.

SHAW R. (1966) 'Polyunsaturated fatty acids of micro-organisms.' *Adv. Lipid Res.* **4**, 107–174.

STANIER R.Y., KUNISAWA R., MANDEL M. & COHEN-BAZIRE G. (1971) Purification and properties of unicellular blue-green algae (order Chroococcales). *Bact. Rev.* **32**, 171–205.

STEINER S., CONTI S.F. & LESTER R.L. (1969) Separation and identification of the polar lipids of *Chromatium* strain D. *J. Bact.* **98**, 10–15.

WAGNER H. & POHL P. (1966) A thesis: fatty acid biosynthesis and evolution in plants and animals. *Phytochem.* **5**, 903–20.

WALSBY A.E. & NICHOLS B.W. (1969) Lipid composition of heterocysts. *Nature, Lond.* **221**, 673–4.

WITTELS B. & BLUM J.J. (1968) Carnitine and carnitine palmityl transferase in *Euglena* grown with palmitate as sole carbon source. *Biochim. biophys. Acta,* **152**, 220–3.

WOLF F.T., CONIGLIO J.G. & BRIDGES R.B. (1966) The fatty acids of chloroplasts. In *The Biochemistry of Chloroplasts,* I, ed. Goodwin T.W. pp. 187–94. Academic Press, New York & London.

WOOD B.J.B., NICHOLS B.W. & JAMES A.T. (1965) The lipids and fatty acid metabolism of photosynthetic bacteria. *Biochim. biophys. Acta,* **106**, 261–73.

WYATT J.T. & SILVEY J.K.G. (1969) Nitrogen fixation by *Gloeocapsa. Science, N.Y.* **165**, 908–9.

# CHAPTER 9

# STEROLS

## T. W. GOODWIN

Department of Biochemistry,
University of Liverpool,
Liverpool, U.K.

1   Introduction   266

2   Chlorophyceae   269
2.1 4,4-Dimethyl- and 4-methyl sterols   269
2.2 Sterols   269

3   Rhodophyceae   270
3.1 4,4-Dimethyl- and 4-methyl sterols   270
3.2 Sterols   270

4   Xanthophyceae   274

5   Chrysophyceae and Haptophyceae   274

6   Phaeophyceae   274

7   Bacillariophyceae   276

8   Euglenophyceae   276
8.1 4,4-Dimethyl sterols   276
8.2 4-Methyl sterols   276
8.3 Sterols   276

9   Charophyceae   278

10  Cyanophyceae   278

11  Conclusions   278

12  References   279

## 1  INTRODUCTION

Triterpenes are $C_{30}$ compounds derived from squalene (1). Squalene is oxidized to squalene 1,2-oxide (2) which can undergo a number of primary cyclizations; in particular, in animals, lanosterol (3) is formed, which, after a number of further steps, involving *inter alia* loss of methyl groups at C-4 and C-14, is converted into the 27-C compound cholesterol (4). Lanosterol is also formed in fungi and is the precursor of fungal sterols. However, in higher plants and algae the first product of cyclization in sterol biosynthesis is not lanosterol (3) but cycloartenol (5). Apart from cholesterol, which is widely distributed in trace amounts in higher plants and algae, the phytosterols are characterized by 1-C or 2-C groups at C-24, by side-chain double bonds at C-22, C-24, (28) or C-25, and by nuclear double bonds at positions other than C-5. Intermediates in the conversion of cycloartenol into phytosterol may still retain one or both methyl

266

group at C-4. In this chapter these are grouped into 4,4-dimethylsterols and 4-methylsterols for convenience: the additional carbon atoms at C-24 arise by transmethylation and not via mevalonate. A full description of plant sterol structures and their relation to the problems of biosynthesis has recently appeared (Goad & Goodwin 1972).

**Table 9.1.** The trivial and systematic names of the most common algal sterols

| Trivial | Systematic |
| --- | --- |
| Cholesterol (4) | Cholest-5-en-3β-ol[a] |
| Lanosterol (3) | 4,4,14α-Trimethyl-5α-cholesta-8,24-dien-3β-ol |
| Cycloartenol (5) | 9β,19-*Cyclo*-4,14,14α-trimethyl-5a-cholest-24-en-3β-ol |
| Ergosterol (9) | 24R-Methylcholesta-5,7-22-trien-3β-ol[b] |
| Campesterol (24) | 24R-Methylcholest-5-en-3β-ol[b] |
| Poriferasterol (13) | 24R-Ethylcholesta-5,22-dien-3β-ol[b] |
| Sitosterol (20) | 24R-Ethylcholest-5-en-3β-ol[b] |
| 28-Isofucosterol (19) | 24(Z)-Ethylidene cholest-5-en-3β-ol[c] |
| Fucosterol (21) | 24(E)-Ethylidene cholest-5-en-3β-ol[c] |

(a) The prefix α- or β- indicates whether the substituent is above or below the plane of the paper; they are indicated thus α, ⅢⅢ β, ——— (see e.g. 4a on p. 268); they are not distinguished in the text.

(b) R- and S- is the convention used to ascribe absolute configuration at an asymmetric centre. It is based on a sequence rule applied to the substituents. It can lead to ambiguities; for example both poriferasterol (13) and sitosterol (20) have the assignment R at C-24 although ethyl groups are spatially distinct. This is the result of the change in sequence priority owing to the presence of a double bond at C-22. Similarly, ergosterol (9) and campesterol (10) also have the R assignment, although their stereochemistry at C-24 is the same. The old nomenclature α- for the stereochemistry in sitosterol (20) and β- for that in poriferasterol (13) has much in its favour for use outside strictly chemical papers.

(c) The prefixes Z and E indicate the geometrical isomerism around the double bond at C-28.

The nomenclature of sterols is complicated and often confusing, so in this chapter the simplest approach, consistent with accuracy, has been taken; all the structures are related to the parent hydrocarbon cholestane (4a). However, to help the more biologically inclined reader to deal with some of the more chemically orientated literature Table 9.1 has been drawn up giving the trivial and systematic names of some of the most common algal sterols. From Table 9.1 the structures of other sterols mentioned in the text can be easily worked out. The numbering of the sterol molecule is indicated for lanosterol (3). The missing numbers 28, 29 are applied to substituents at C-24, as in 24-ethylidene lophenol (8).

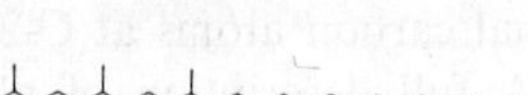

(1)

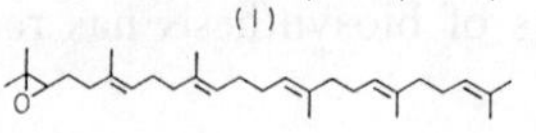

(2)

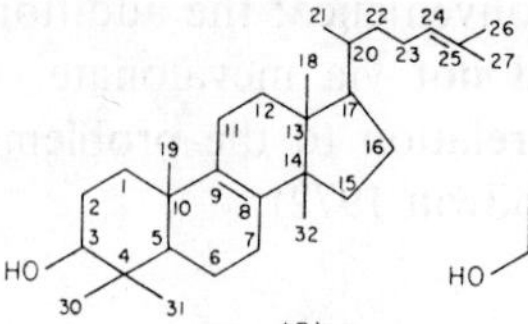


(3)

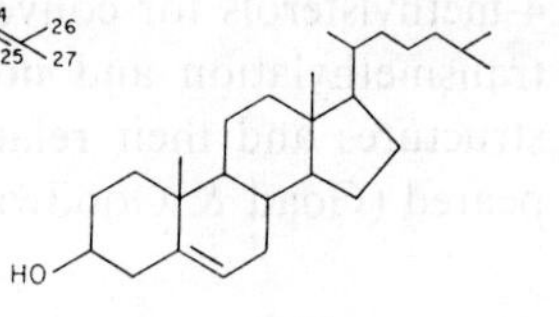

(4)

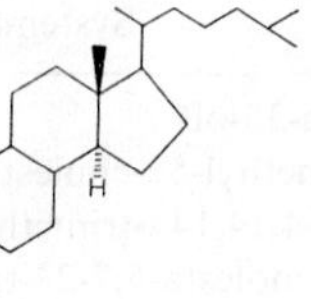

(4a)

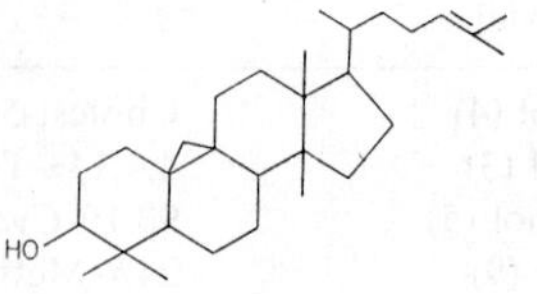

(5)

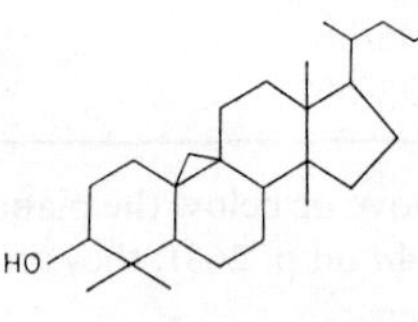

(6)

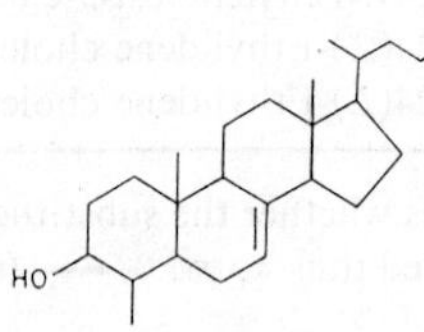

(7)

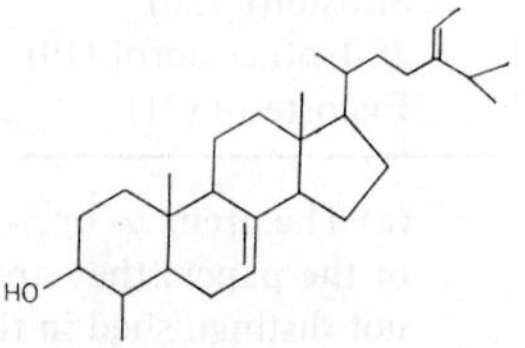

(8)

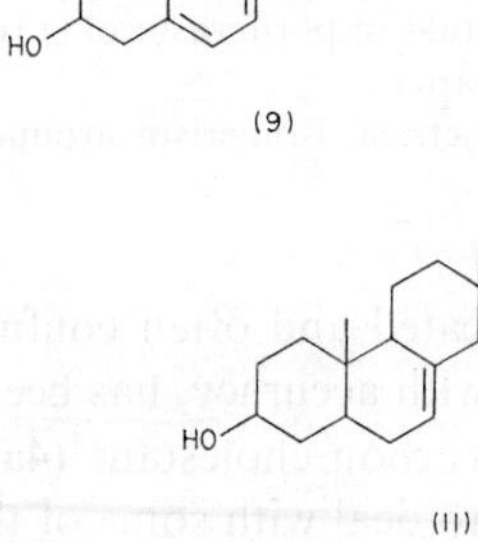

(9)

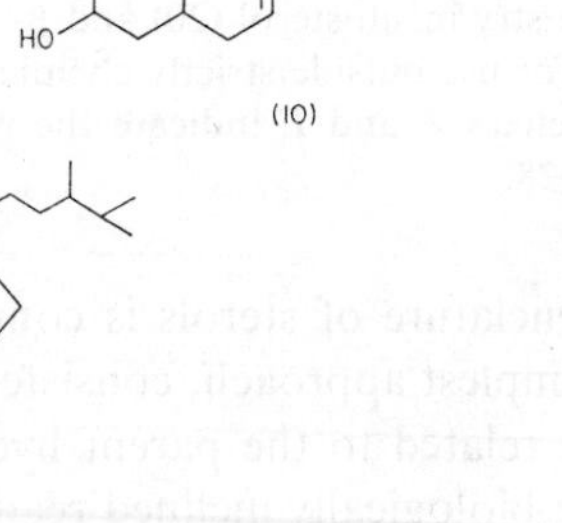

(10)

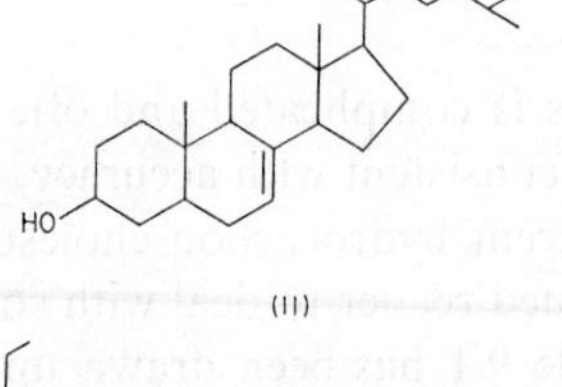

(11)

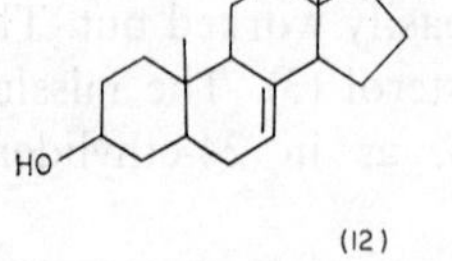

(12)

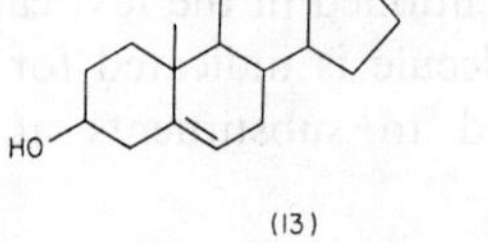

(13)

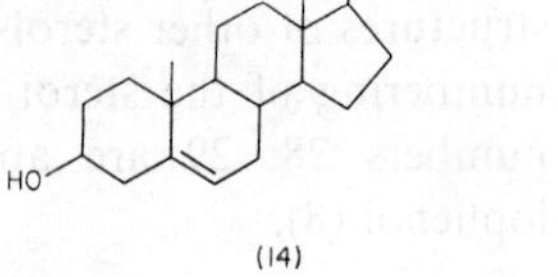

(14)

# 2 CHLOROPHYCEAE

## 2.1 4,4-*Dimethyl- and 4-Methyl sterols*

Cycloartenol (5) and 24-methylene cycloartenol (6) have been detected in *Ulva lactuca* (Gibbons 1968), and 24-methylene lophenol (7) and 24-ethylidene lophenol (8) in *Enteromorpha intestinalis* (Gibbons *et al.* 1968).

## 2.2 *Sterols*

A thorough study of a number of unicellular Chlorophyceae of the order Chlorococcales, all *Chlorella* spp., has resolved them into three main groups according to the nature of the double bonds in ring B (Table 9.2) (see for example,

**Table 9.2.** Classification of *Chlorella* spp. according to the major sterols which they accumulate (Patterson 1971).

| Species | Sterol type | | |
|---|:---:|:---:|:---:|
| | $\Delta^{5,7}$-sterols | $\Delta^{7}$-sterols | $\Delta^{5}$-sterols |
| C. candida | + | | |
| C. nocturna | + | | |
| C. prototothecoides v. communis | + | | |
| C. prototothecoides v. mannophila | + | | |
| C. simplex | + | | |
| C. sorokiniana | + | | |
| C. vannielii | + | | |
| C. ellipsoidea | | + | |
| C. saccharophila | | + | |
| C. emersonii | | | + |
| C. fusca | | | + |
| C. glucotropha | | | + |
| C. miniata | | | + |
| C. vulgaris | | | + |

Patterson 1971). The first group synthesizes mainly $\Delta^{5,7}$ sterols, particularly ergosterol (9); the second mainly $\Delta^{7}$ sterols, chondrillasterol (10) preponderating but with significant amounts of ergost-7-enol (11) and 22-dihydrochondrillasterol (12); and the third group synthesizes mainly $\Delta^{5}$ sterols, poriferasterol (13), ergost-5-enol (14) and clionasterol (15a). Recently it has been shown that *Chlorella ellipsoidea*, when grown in the presence of the sterol inhibitor AY-9944, accumulates 5α-ergosta-8,14-dien-3β-ol (15b) and 5α-stigmasta-8,14-dien-3β-ol (15c) (Dickson *et al.* 1972). *Trebouxia* sp., the algal component of the lichen *Cladonia impexa*, also synthesizes ergost-5-enol (14), clionasterol (15a) and poriferasterol (13) when cultured alone (J. Lenton unpubl.). *Scenedesmus obliquus* falls into the second group (Patterson 1971). *Oocystis polymorpha* also falls into the second group but in addition synthesizes cholesterol in small amounts, particularly in the sterol ester fraction (Orcutt & Richardson 1970). Thus all the sterols of members of the family Chlorococcaceae so far examined have the same absolute configuration at C-24 which is opposite to that in higher plants; compare for example poriferasterol (13) with stigmasterol (17). However, one member of the Chlorococcales, *Hydrodictyon reticulatum*, is said to make spinasterol (18) (Huneck 1969), which has the same configuration as the higher plant sterols, but the identification is not fully established. The order Ulotrichales can be clearly distinguished from the Chlorococcales. The major sterol of *Ulva lactuca* is 28-isofucosterol (19) (Gibbons *et al.* 1968) and not sitosterol (20) as originally thought (Heilbron *et al.* 1935). 28-Isofucosterol is the main sterol of *Enteromorpha intestinalis* (Gibbons *et al.* 1968) and *E. linza* (Tsuda & Sakai 1960). Fucosterol (21), not isofucosterol, is present in significant amounts (14·5% of the total sterols) in *Ulva pertusa* where cholesterol (4) is the main component (74%) and 24-methylene cholesterol (chalinasterol) (22) is also present in small amounts.

*Monostroma nitidum* is said to contain haliclonasterol (23), a C-20 isomer of campesterol (24) (Tsuda & Sakai 1960). An old report exists of a sterol glycoside in *Oedogonium* sp. (Heilbron *et al.* 1935).

According to Ikekawa *et al.* (1968) the major sterol of *Chaetomorpha crassa* (Cladophorales) is sitosterol (20), the epimer of clionasterol (15) and a widely distributed sterol in higher plants. It is accompanied by cholesterol (4), 24-methylene cholesterol (22) and two methyl sterols, one campesterol (24), with the same configuration as sitosterol at C-24, and one, brassicasterol (16), with the opposite configuration. However, the methods used were not adequate to allot configuration at C-24 with confidence. No other member of the Cladophorales has yet been examined.

## 3 RHODOPHYCEAE

### 3.1 *4,4-Dimethyl- and 4-Methyl sterols*

Cycloartenol (5) but not lanosterol (3) has been identified in *Rhodomela confervoides*, *Chondrus crispus* and *Rhodymenia palmata* (Alcaide *et al.* 1968) and in the unicellular *Porphyridium cruentum* grown in a chemically defined medium (Beastall *et al.* 1971). However, in spite of the absence of lanosterol, 24,25-dihydrolanosterol (25) is present in the last named (Beastall *et al.* 1971).

### 3.2 *Sterols*

Members from six orders of colonial Rhodophyceae have been examined and in almost all cases cholesterol, or its biosynthetic precursor desmosterol (26), is the main sterol component (Table 9.3). The relative amounts of the sterol can

(15a)                              (15b)                              (15c)

(16)                               (17)

(18)               (19)               (20)

(21)               (22)               (23)

(24)               (25)               (26)

(27)               (28)               (29)

**Table 9.3.** Major sterols in red algae

| Order/alga | Sterols | References |
|---|---|---|
| Porphyridiales | | |
| *Porphyridium cruentum* | 1, 5 | Beastall *et al.* (1971) |
| | | |
| Bangiales | | |
| *Porphyra purpurea* | 2 | Gibbons *et al.* (1967) |
| | | |
| Nemaliales | | |
| *Acanthopeltis japonica* | 3 | Tsuda *et al* (1958a) |
| *Gelidium amansii* | 3 | Tsuda *et al.* (1958a) |
| *Gelidium japonicum* | 3 | Tsuda *et al.* (1957) |
| *Gelidium subcostatum* | 2 | Tsuda *et al.* (1957) |
| *Pterocladia tenuis* | 3 | Tsuda *et al.* (1957) |
| | | |
| Cryptonemiales | | |
| *Corallina officinalis* | 3 | Gibbons *et al.* (1967) |
| *Cyrtymenia sparsa* | 3 | Tsuda *et al.* (1958b) |
| *Dilsea carnosa* | 3 | Gibbons *et al.* (1967) |
| *Gloiopeltis furcata* | 3 | Tsuda *et al.* (1958b) |
| *Grateloupia elliptica* | 3 | Tsuda *et al.* (1958b) |
| *Polyides caprinus* | 3 | Gibbons *et al.* (1967) |
| *Polyides rotundus* | 3 | Idler *et al.* (1968) |
| *Tichocarpus crinitus* | 3 | Tsuda *et al.* (1958b) |
| | | |
| Gigartinales | | |
| *Ahnfeltia stellata* | 3 | Gibbons *et al.* (1967) |
| *Chondrus crispus* | 3 | Saito & Idler (1966), Alcaide *et al.* (1968) |
| *Chondrus giganteus* | 3 | Tsuda *et al.* (1958b) |
| *Chondrus ocellatus* | 3 | Tsuda *et al.* (1958b) |
| *Furcellaria fastigiata* | 3 | Gibbons *et al.* (1967), Idler & Wiseman (1970) |
| *Gigartina stellata* | 3 | Gibbons *et al.* (1967) |
| *Gracilaria verrucosa* | 3 | Patterson (1971), Henriquez *et al.* (1972) |
| *Hypnea japonica* | 1 | Tsuda *et al.* (1960) |
| *Iridaea laminarioides* | 3 | Tsuda *et al.* (1960) |
| *Iridophycus cornucopiae* | 2 | Tsuda *et al.* (1958b) |
| *Phyllophora membranifolia* | 3 | Idler & Wiseman (1970) |
| *Plocamium vulgare* | 3 | Gibbons *et al.* (1967) |
| *Rhodoglossum pulcherrum* | 3 | Tsuda *et al.* (1957) |
| | | |
| Rhodymeniales | | |
| *Coeloseira pacifica* | 3 | Tsuda *et al.* (1958b) |
| *Halosaccion ramentaceum* | 2,3 | Idler *et al.* (1968) |
| *Rhodymenia palmata* | 2,3 | Gibbons *et al.* (1967), Idler *et al.* (1968) |

| Order/alga | Sterols | References |
|---|---|---|
| Ceramiales | | |
| *Ceramium rubrum* | 3 | Patterson (1971) |
| *Chondria dasyphylla* | 3 | Patterson (1971) |
| *Dasya pedicellata* | 3 | Patterson (1971) |
| *Grinnellia americana* | 2, 3 | Patterson (1971) |
| *Laurencia pinnatifida* | 3 | Gibbons *et al.* (1967) |
| *Polysiphonia lanosa* | | |
| *(fastigata)* | 3 | Gibbons *et al.* (1967) |
| *Polysiphonia nigrescens* | 3 | Gibbons *et al.* (1967) |
| *Polysiphonia subtillissima* | 3 | Patterson (1971) |
| *Ptilota serrata* | 3 | Idler & Wiseman (1970) |
| *Rhodomela confervoides* | 3 | Idler *et al.* (1968), Alcaide *et al.* (1968) |
| *Rhodomela larix* | 3 | Tsuda *et al.* (1958b) |
| *Rytiphlaea tinctoria* | 4 | Alcaide *et al.* (1969) |

Key: 1. Ergosterol; 2. Desmosterol; 3. Cholesterol; 4. Campesterol or ergost-5-enol.

vary with season, for example the desmosterol content of *Rhodymenia palmata* varied from 30·6% to 97·2% of the total sterol in samples collected between June and November in New Brunswick (Idler & Wiseman 1970). In *Hypnea japonica* 22-dehydrocholesterol (27) is the major component (Tsuda *et al.* 1960). In addition to these major components, trace components have been reported in *Rhodomela confervoides*, *Chondrus crispus* (Alcaide *et al.* 1968) and *Rhodymenia palmata* (Alcaide *et al.* 1968, Idler & Wiseman 1970). In the last named these include 22-dehydrocholesterol (27), brassicasterol (16), 22-dihydrobrassi-casterol (28), 24-methylene-cholesterol (22), sitosterol (20), fucosterol (21) and 28-isofucosterol (19).

An outstanding exception to this pattern is *Rytiphlaea tinctoria* which synthesizes $C_{28}$ sterol, either campesterol (24) or ergost-5-enol (14), as its major sterol (Alcaide *et al.* 1969). It also contains a mixture of α- and β-amyrins. Cholest-4-en-3-one (29) has been isolated from dried *Meristotheca papulosa* (Kanazawa & Yoshioka 1971).

The only unicellular red alga so far examined, *Porphyridium cruentum*, was said to contain no sterol (Aaronson & Baker 1961), but a reinvestigation using more sensitive techniques has revealed the presence of 22-dehydrocholesterol (27) as the major component and of trace amounts of cholesterol (4) and ergosterol (9) (Beastall *et al.* 1971). The individualistic thermophile *Cyanidium caldarium*, which has tentatively been placed close to the red algae (Ikan & Seckbach 1972), is often considered a transitional form between the blue-green and red algae (Klein & Cronquist 1967). Its sterol composition (Ikan & Seckbach 1972) is somewhat inexplicable on these grounds. Cholesterol (4) is only a minor component (0·3% of the total); the major components being ergosterol (9) (plus 5,6-dihydroergosterol) and sitosterol (20). There are also smaller amounts of campesterol (24), and traces of 7-dehydrositosterol. These identifications are based on an assumed configuration at C-24 (Seckbach & Ikan 1972). Up to now there have been no authenticated reports of ergosterol in red algae and campesterol and sitosterol have only been reported as minor components (see above). The pattern has some unimpressive similarities with what is known of sterols in blue-green algae (see p. 278).

## 4  XANTHOPHYCEAE

An old report indicates that sitosterol (20) is present in *Botryidium granulatum* and *Vaucheria hamata* (Heilbron 1942). This requires to be substantiated by modern techniques.

## 5  CHRYSOPHYCEAE AND HAPTOPHYCEAE

Early work on mixed cultures of *Apistonema carterae*, *Thallochrysis litoralis* and *Gloeochrysis maritima* indicated that fucosterol (21), the major sterol of the Phaeophyceae, also predominated in these algae (Heilbron, 1942). The only sterol present in detectable amounts in *Ochromonas malhamensis* is poriferasterol (13) (Williams *et al.* 1966, Gershengorn *et al.* 1968) and not its C-24 epimer, stigmasterol (17) as previously suggested (Bazzano 1965, Avivi *et al.* 1967). In the related *O. danica* sterols present in addition to poriferasterol (13) are brassicasterol (16), 22-dihydrobrassicasterol (28), clionasterol (15a) and, probably, 7-dehydroporiferasterol (30) (Gershengorn *et al.* 1968). The presence of ergosterol (9), reported earlier in *O. danica* (Stern *et al.* 1960, Aaronson & Baker 1961, Halevy *et al.* 1966), was confirmed, but stigmasterol (17) reported by Halevy *et al.* (1966) could not be detected. These authors also reported stigmasterol in *O. sociabilis*, but this has not been confirmed by W. Sach & L. J. Goad (unpublished observation) who identified poriferasterol as the main component. *Synura petersenii* is said to contain stigmasterol (17) and cholesterol (4) (Collins & Kalnins 1969).

## 6  PHAEOPHYCEAE

The pioneer work of Carter *et al.* (1939) revealed fucosterol (21) as the major sterol in this order of algae and this observation has been frequently confirmed over the years. The algae involved have been listed by Patterson (1971) and Goodwin (1973) and there is nothing to add here. The claim that sargasterol (31), the C-20 isomer of fucosterol is the major sterol of *Sargassum ringgoldianum* (Tsuda *et al.* 1958c) has not been confirmed in more recent investigations (Ikekawa *et al.* 1968). According to Knights (1970) the sterol saringosterol (32), recently reported in a number of Phaeophyceae (Ikekawa *et al.* 1966, 1968, Patterson 1968), may be an oxidative artefact arising from fucosterol. In addition to fucosterol traces of other sterols, such as 24-methylene cholesterol (22) and desmosterol (tentatively identified) (26), are often observed in *Laminaria faeroensis* and *L. digitata* (Patterson 1971); these tend to increase in old or milled samples (Knights 1970). 24-Oxocholesterol (33a) recently reported in *Pelvetia canaliculata* (Motzfeldt 1970) may also possibly be an artefact.

Recently two new sterols 3,5,24(28)-stigmastatrien-7-one (33b) and 5,24(28)-stigmasta-diene-3β,7α-diol (33c) have been reported in air-dried *Fucus evanescens* in addition to fucosterol (21) and saringosterol (32) (Ikekawa *et al.* 1972).

## 7 BACILLARIOPHYCEAE

*Navicula pelliculosa* is reported to synthesize chondrillasterol (10) (Low 1955), whilst *Cyclotella nana* and *Nitzschia closterium* (*Phaeodactylum tricornutum*) synthesize brassicasterol (16) (Kanazawa *et al.* 1971).

## 8 EUGLENOPHYCEAE

### 8.1 4,4-*Dimethyl sterols*

*Euglena gracilis* strain Z, when grown autotrophically in the light, contains cycloartenol (5), 24-methylene cycloartenol (6), 24-methylene lanosterol (34) and 24-methylene agnosterol (35) with 24-methylene cycloartenol predominating (80% of the total). Bleached cells produced heterotrophically in the dark contain the first two in about equal amounts but no 24-methylene lanosterol or -agnosterol. About 60% of the triterpenes in the bleached cells remain unidentified (Anding *et al.* 1971). It is interesting that the pentacyclic triterpene β-amyrin (36) is present in both green and white cultures. In the water-soluble sterol fraction (see below) no 4,4-dimethyl sterols were observed.

### 8.2 4-*methyl sterols*

The relative amounts of the 4-methyl sterols in green and bleached *Euglena gracilis* strain Z are given in Table 9.4. The 'undefined methylsterol' may be 4α,24-dimethyl-5α-cholesta-8(9)-en-3β-ol.

**Table 9.4.** 4α-Methyl sterols from green and bleached *Euglena gracilis* strain Z. (Anding *et al.* 1971).

| | Relative amount (% of total) | |
| Compound | Green Cells | White Cells* |
| --- | --- | --- |
| Obtusifoliol (37) | 30 | 35 |
| 24-Methylene lophenol (7) | 13 | 4 |
| 4α-Methyl zymosterol (38) | 7 | — |
| 'Undefined methyl sterol' | 40 | 13 |

\* about 50% of the fraction unidentified

### 8.3 *Sterols*

Ergosterol (9) was considered to be the major sterol of *Euglena gracilis* strain Z. (Stern *et al.* 1960, Avivi *et al.* 1967) until a recent reinvestigation indicated the components listed in Table 9.5. It will be seen that the major free sterols are $\Delta^7$ sterols and not $\Delta^{5,7}$ sterols, and that ergosterol was not detected (Brandt *et al.*

**Table 9.5.** Free and bound sterols in green and bleached *Euglena gracilis* strain Z (Brandt *et al.* 1970)

| Sterol | Free Green | Free Bleached | Esters Green | Esters Bleached | Water-soluble acid-treated Green | Water-soluble acid-treated Bleached | Water-soluble pyrogallol-treated Green | Water-soluble pyrogallol-treated Bleached |
|---|---|---|---|---|---|---|---|---|
| Cholest-7-enol (39) | Trace | Trace | | | | | | |
| Episterol (40) | — | 4 | | | | | | |
| Ergost-7-enol (11) | 33 | 30 | | | | | | |
| Chondrillast-7-enol (41) | 12 | 7 | | | | | | |
| Ergosta-5,22-dienol (42) | 18 | 6 | | | | | | |
| Chondrillasterol (12) | 35 | — | | | | | | |
| Cholesta-5,7-dienol (43) | Trace | Trace | — | — | — | — | — | — |
| Chondrillasta-5,7-dienol (44) | — | 5 | — | 8 | 2 | 6 | 2 | 3 |
| Cholesterol (4) | 1 | 12 | Trace | 16 | 52 | 42 | 54 | 56 |
| Chalinasterol (22) | — | — | — | 6 | — | — | — | — |
| 22-Dihydrobrassicasterol (28) | — | Trace | — | 3 | — | — | — | 1 |
| Clionasterol (15a) | — | 16 | — | 50 | 28 | 39 | 31 | 31 |
| Poriferasterol (13) | — | 3 | — | 6 | 4 | — | 5 | 3 |

Figures represent percentage of total sterols in fraction.

(39)

(40)

(41)

(42)

(43)

(44)

(45)

K

1970). Table 9.4 also shows that the sterols vary both quantitatively and qualitatively according to whether *Euglena* is grown autotrophically in the light (i.e. photosynthetically), or heterotrophically in the dark when it is colourless and contains no chloroplasts. Chondrillasterol (10) disappears in bleached cultures and its place is taken by chondrillasta-5,7-enol (44), clionasterol (15) and poriferasterol (13). The relative amount of cholesterol increases in dark cultures. Table 9.4 also shows that esterified sterols occur only in traces in green cells but are present in larger amounts in a characteristic distribution in the dark cells. Water-soluble sterols are found in both green and bleached cells with cholesterol (4) and clionasterol (15) as major components.

## 9   CHAROPHYCEAE

*Nitella opaca* contains sitosterol (20) and fucosterol (21) according to early work (Heilbron 1942) and probably also forms sterol glycosides (Heilbron *et al*. 1935). Recent work indicates the presence of clionasterol (15) and 28-isofucosterol (19) in *Nitella flexilis* and *Chara vulgaris* (Patterson 1972).

## 10   CYANOPHYCEAE

Blue-green algae were considered not to contain sterols (Levin & Bloch 1964) but more recently small amounts of cholesterol (4) and sitosterol (20) have been detected in *Anacystis nidulans*, *Fremyella diplosiphon* (Reitz & Hamilton 1968) and *Spirulina maxima* (Nadal 1971). In *Phormidium luridum* var. *olivaceae* 24-ethylcholest-7-enol (45) was the major sterol present and it was accompanied by $\Delta^{5-}$, $\Delta^{5,7}$ and $\Delta^{22-}$ ethyl derivatives of cholesterol; the stereochemistry around C-24 was not determined (de Souza & Nes 1968). In our laboratory sterols have been detected in two blue-green algae (L. J. Goad & N. G. Carr unpubl. observations). From the taxonomic point of view the determination of the configuration at C-24 of these is extremely important.

## 11   CONCLUSIONS

Although only relatively few algae have been examined thoroughly for sterols by modern methods, it is reasonable to conclude that these compounds are constant components of algae. The time is ripe for much careful survey work on pure cultures of the less common algae; a picture of some considerable phylogenetic significance could emerge.

# 12 REFERENCES

AARONSON S. & BAKER H. (1961) Lipid and sterol content of some protozoa. *J. Protozool.* **8**, 274–7.

ALCAIDE A., BARBIER M., POTIER P., MAGUEUR A.M. & TESTE J. (1969) Nouveaux resultats sur les sterols des algues rouges. *Phytochem.* **8**, 2301–3.

ALCAIDE A., DEVYS M. & BARBIER M. (1968) Remarques sur les sterols des algues rouges. *Phytochem.* **7**, 329–30.

ANDING C., BRANDT R.D. & OURISSON G. (1971) Sterol biosynthesis in *Euglena gracilis* Z. Sterol precursors in light-grown and dark-grown *Euglena gracilis* Z. *Eur. J. Biochem.* **24**, 259–63.

AVIVI L., IARON O. & HALEVY S. (1967) Sterols of some algae. *Comp. Biochem. Physiol.* **21**, 321–6.

BAZZANO G. (1965) Influence of vitamin $B_{12}$ and methionine on sterol synthesis in *Ochromonas malhamensis*. University Microfilm, 65–4100.

BEASTALL G.H., REES H.H. & GOODWIN T.W. (1971) Sterols in *Porphyridium cruentum*. *Tetrahedron Letters*, No. 52, 4935–8.

BRANDT R.D., PRYCE R.J., ANDING C. & OURISSON G. (1970) Sterol biosynthesis in *Euglena gracilis* Z. Comparative study of free and bound sterols in light and dark grown *Euglena gracilis* Z. *Eur. J. Biochem.* **17**, 344–9.

CARTER P.W., HEILBRON I.M. & LYTHGOE B. (1939) The lipochromes and sterols of the algal classes. *Proc. R. Soc.* **B128**, 82–108.

COLLINS R.P. & KALNINS K. (1969) Sterols produced by *Synura petersenii* (Crysophyta). *Comp. Biochem. Physiol.* **30**, 779–82.

DE SOUZA N.J. & NES W.R. (1968) Sterols: isolation from a blue-green alga. *Science, N.Y.* **162**, 363.

DICKSON L.G., PATTERSON G.W., COHEN C.F. & DUTKY S.R. (1972) Two novel sterols from inhibited *Chlorella ellipsoidea*. *Phytochem.* **11**, 3473–7.

GERSHENGORN M.C., SMITH A.R.H., GOULSTON G., GOAD L.J., GOODWIN T.W. & HAINES T.H. (1968) The sterols of *Ochromonas danica* and *Ochromonas malhamensis*. *Biochemistry, N.Y.* **7**, 1698–1706.

GIBBONS G.F. (1968) Aspects of sterol biosynthesis in plants and animals. Ph.D. Thesis, University of Liverpool.

GIBBONS G.F., GOAD L.J. & GOODWIN T.W. (1967) The sterols of some marine red algae. *Phytochem.* **6**, 677–83.

GIBBONS G.F., GOAD L.J. & GOODWIN T.W. (1968) The identification of 28-Isofucosterol in the marine green algae *Enteromorpha intestinalis* and *Ulva lactuca*. *Phytochem.* **7**, 983–8.

GOAD L.J. & GOODWIN T.W. (1972) Biosynthesis of plant sterols. *Progress in Phytochemistry*, **3**, 113–98.

GOODWIN T.W. (1973) Comparative biochemistry of sterols in eukaryotic micro-organisms. In *Lipids and Biomembranes of eukaryotic micro-organisms*, ed. Erwin J.A., pp. 1–40. Academic Press, New York & London.

HALEVY S., AVIVI L. & KATAN H. (1966) Sterols of soil amoebas and *Ochromonas danica*: Phylogenetic approach. *J. Protozool.* **13**, 480–3.

HEILBRON I.M. (1942) Some aspects of algal chemistry. *J. chem. Soc.* **79**, 78–89.

HEILBRON I.M., PARRY E.G. & PHIPERS R.F. (1935) The algae. II. Relationship between certain algal constituents. *Biochem. J.* **29**, 1376–81.

HENRIQUEZ P., TRUCCI R., SILVA M. & SAMMES P.G. (1972) Chlolesterol in *Iridaea laminaroides* and *Gracilaria verrucosa*. *Phytochem.* **11**, 1171.

HUNECK S. (1969) Spinasterin in *Hydrodictyon reticulatum*. *Phytochem.* **8**, 1313.

IDLER D.R., SAITO A. & WISEMAN P. (1968) Steroids in red algae (Rhodophyceae). *Steroids*, **11**, 465–73.

IDLER D.R. & WISEMAN P. (1970) Sterols in red algae (Rhodophyceae). Variation in the desmosterol content of dulse (*Rhodymenia palmata*). *Comp. Biochem. Physiol.* **35**, 679–87.

IKAN R. & SECKBACH J. (1972) Lipids of the thermophilic alga *Cyanidium caldarium*. *Phytochem.* **11**, 1077–82.

IKEKAWA N., MORISAKI N. & HIRAYA A.K. (1972) Two new sterols from *Fucus evanescens*. *Phytochem.* **11**, 2317–18.

IKEKAWA N., MORISAKI N., TSUDA K. & YOSHIDA T. (1968) Sterol composition of some green algae and brown algae. *Steroids*, **12**, 41–8.

IKEKAWA N., TSUDA K. & MORISAKI N. (1966) Saringosterol: A new sterol from brown algae. *Chem. Inds., Lond.* 1179–80.

KANAZAWA A. & YOSHIOKA M. (1971) Occurrence of Cholest-4-en-3-one in red alga *Meristotheca papulosa*. *Bull. Jap. Soc. scient. Fish.* **37**, 397.

KANAZAWA A., YOSHIOKA M. & TESHIMA S. (1971) The occurrence of Brassicasterol in diatoms *Cyclotella nana* and *Nitzschia closterium*. *Bull. Jap. Soc. scient. Fish.* **37**, 899–903.

KLEIN R.M. & CRONQUIST A. (1967) A consideration of the evolutionary and taxonomic significance of some biochemical, micromorphological, and physiological characters in the thallophytes. *Q. Rev. Biol.* **42**, 105–296.

KNIGHTS B.A. (1970) Sterols in *Ascophyllum nodosum*. *Phytochem.* **9**, 903–5.

LEVIN E.Y. & BLOCH K. (1964) Absence of sterols in blue-green algae. *Nature, Lond.* **202**, 90–1.

LOW E.M. (1955) Studies on some chemical constituents of diatoms. *J. mar. Res.* **14**, 199–204.

MOTZFELDT A.M. (1970) Isolation of 24-oxocholesterol from the marine brown alga *Pelvetia canaliculata* (Phaeophyceae). *Acta chem. scand.* **24**, 1846–7.

NADAL N.G.M. (1971) Sterols in *Spirulina maxima*. *Phytochem.* **10**, 2537–8.

ORCUTT D.M. & RICHARDSON B. (1970) Sterols of *Oocystis polymorpha*, a green alga. *Steroids* **16**, 429–46.

PATTERSON G.W. (1968) Sterols of *Laminaria*. *Comp. Biochem. Physiol.* **24**, 501–5.

PATTERSON G.W. (1971) The distribution of sterols in algae. *Lipids*, **6**, 120–7.

PATTERSON G.W. (1972) Sterols of *Nitella flexilis* and *Chara vulgaris*. *Phytochem.* **11**, 3481–3.

REITZ R.C. & HAMILTON J.G. (1968) The isolation and identification of two sterols from two species of blue-green algae. *Comp. Biochem. Physiol.* **25**, 401–16.

SAITO A. & IDLER D.R. (1966) Sterols in Irish moss (*Chondrus crispus*). *Can. J. Biochem.* **44**, 1195–9.

SECKBACH J. & IKAN R. (1972) Sterols and chloroplast structure of *Cyanidium caldarium*. *Pl. Physiol., Lancaster*, **49**, 457–9.

STERN A.I., SCHIFF J.A. & KLEIN H.P. (1960) Isolation of ergosterol from *Euglena gracilis*; distribution among mutant strains. *J. Protozool.* **7**, 52–5.

TSUDA K., AKAGI S. & KISHIDA Y. (1957) Discovery of cholesterol in some red algae. *Science, N.Y.* **126**, 927.

TSUDA K., AKAGI S. & KISHIDA Y. (1958a) Cholesterol in some red algae. *Pharm. Bull.* **6**, 101–4.

TSUDA K., AKAGI S., KISHIDA Y., HAYATSU R. & SAKAI K. (1958b) Sterols from ocean algae. *Pharm. Bull.* **6**, 724–7.

TSUDA K., HAYATSU R., KISHIDA Y. & AKAGI S. (1958c) Steroid studies. VI. Studies of the constitution of sargasterol. *J. Am. chem. Soc.* **80**, 921–5.

TSUDA K. & SAKAI K. (1960) Sterols from green ocean algae. *Chem. Pharm. Bull., Tokyo*, **8**, 554–8.

TSUDA K., SAKAI J., TANABE K. & KISHIDA Y. (1960) Isolation of 22-dehydrocholesterol from *Hypnea japonica*. *J. Am. chem. Soc.* **82**, 1442–3.

WILLIAMS B.L., GOODWIN T.W. & RYLEY J.F. (1966) The sterol content of some protozoa. *J. Protozool.* **13**, 227–30.

# CHAPTER 10

# NUCLEIC ACIDS AND THEIR METABOLISM

## BEVERLEY R. GREEN

Botany Department,
University of British Columbia,
Vancouver 8, Canada.

1   **Introduction**   281

2   **Blue-green algae**   282
2.1  DNA   282
2.2  RNA synthesis and degradation   284
2.3  Ribosomes   284
2.4  DNA-DNA and DNA-RNA hybridization   285

3   **Euglena**   285
3.1  DNA   286
3.2  Polysomes, ribosomes and RNA synthesis   288

4   **Chlamydomonas**   290
4.1  DNA   290
4.2  Ribosomes and RNA synthesis   292

5   **Chlorella**   294
5.1  DNA   294
5.2  Ribosomes and RNA synthesis   296

6   **Acetabularia**   297
6.1  DNA   298
6.2  Ribosomes and RNA synthesis   300

7   **Comparative aspects**   301

8   **Chloroplast autonomy and evolutionary origin**   304

9   **References**   304

## 1 INTRODUCTION

The term 'algae' includes organisms possessing a wide variety of cellular and intercellular levels of organization. As far as nucleic acids are concerned, only a few have been studied thoroughly. The writer of a review of this sort faces the problem of giving an adequate treatment of some fairly sophisticated information on a few species without neglecting the overall field. The prokaryotic Cyanophyceae are in a group by themselves, and some fairly valid generalizations can be made. However, the eukaryotic algae are classified into a number of distinct classes (Scagel *et al.* 1965) and are probably of polyphyletic origin (Raven 1970).

I have therefore chosen to concentrate on the 'big four' on which most bio-chemical work has been done: *Euglena*, *Chlamydomonas*, *Chlorella* and *Aceta-bularia*. In the last section, I will try to point out the limits within which this information can be generalized to other groups, especially the multicellular Rhodophyceae and Phaeophyceae, and to suggest profitable lines of investigation.

It is now generally accepted that the genetic information is coded in the nucleotide sequence of DNA, and transcribed into shorter segments of RNA which are used in the process of protein synthesis as either components of the ribosomes or as messenger RNA's. The details and controls thereof are the subject of this review.

# 2  BLUE-GREEN ALGAE (CYANOPHYCEAE)

Most of the molecular biology of the blue-green algae has been done within a comparative framework. They are either compared with the other large group of prokaryotes, the bacteria, or they are compared with chloroplasts in the light of Mereschkowsky's theory (1905) of the endosymbiotic origin of chloroplasts. For this reason, the biochemical information is mainly confined to a few easily-handled species.

## 2.1  *DNA*

DNA base composition has proven extremely useful as an exclusionary taxonomic criterion in the bacteria (Marmur *et al.* 1963, Mandel 1969) and other lower organisms (Storck & Alexopoulos 1970) where the generation time is short enough, and evolutionary history long enough, for considerable divergence to have occurred. Edelman *et al.* (1967) studied 21 filamentous blue-green algae (39 to 51% G+C) and five or six coccoid species (35 to 71% G+C), comparing their base compositions to those of chloroplasts, gliding bacteria and eukaryotic algae. Stanier *et al.* (1971) tackled the entire problem of the taxonomy of coccoid forms, using %G+C as one criterion. The two groups of data are plotted in Fig. 10.1. Some of the values have been confirmed by thermal denaturation (Craig *et al.* 1969). It is obvious that the Cyanophyceae are an extremely diverse group, with as large a range of base compositions as the bacteria. Due to the lack of morphological characteristics, DNA base composition will probably become extremely useful in the classification of these organisms. Whether it can be useful in discussing the evolutionary origin of chloroplasts will be considered later.

*Anacystis nidulans* is the only member of the Cyanophyceae in which DNA replication has been studied, probably because of its unusually low content of nucleases (Norton & Roth 1967a) and because cultures can be synchronized (Herdman *et al.* 1970, Asato & Folsome 1970). The actual results seem to depend on the growth conditions used, but it can be concluded that DNA replication (a) stops in the dark (Hayashi *et al.* 1969, Herdman *et al.* 1970, Asato & Folsome 1970); (b) occurs in the period just prior to cell division (Herdman *et al.* 1970), and (c) is followed by a period when no synthesis occurs (Herdman *et al.* 1970, Asato & Folsome 1970). These analyses were fortunately done using chemical estimations of DNA content rather than the incorporation of labelled precursors, as Piggott and Carr (1971) have shown that thymine, thymidine and almost all other nucleic acid precursors except uracil, are very poorly incorporated into intact cells. Herdman *et al.* (1970) used the increase in the reversion frequency of

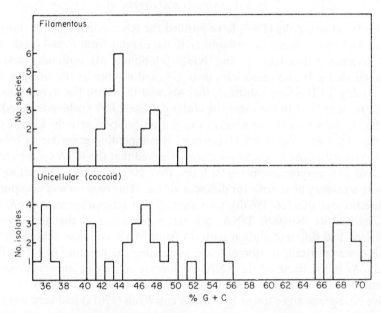

Fig. 10.1. DNA base composition distribution of blue-green algae. From Stanier *et al.* (1971) and Edelman *et al.* (1967).

a nitrate reductase mutant to determine the time of replication of that region of the DNA in the cell cycle. Asato and Folsome (1970) did the same thing with six different markers and were able to construct a temporal genetic map. Although the DNA polymerase has not been isolated, a light-requiring DNA photo-reactivating enzyme has been found in *A. nidulans* and *Plectonema boryanum* (Saito & Werbin 1970).

The genome size of those species studied is about the same as that of a bacterium. Synchronized cells of *A. nidulans* contain $3.7 \times 10^9$ daltons DNA per genome (Asato & Folsome 1970) compared to $2.5 \times 10^9$ daltons for *E. coli* (Cairns 1963). Somewhat higher values are found for exponentially growing, unsynchronized cultures (Craig *et al.* 1969) due to the presence of several genomes in various stages of replication. Kung *et al.* (1971) showed that DNA of a

*Lyngbya* sp. renatured at about the same rate as that of *E. coli* and *Micrococcus radiodurans*, and much faster than chloroplast or mitochondrial DNAs (see below). If blue-green algae were the ancestors of modern chloroplasts, they must have undergone a considerable reduction in genetic information.

There have been reports of genetic recombination (Bazin 1968) and transformation (Shestakov & Khyen 1970) in *Anacystis nidulans*, and it is to be hoped that more information on the molecular biology of 'the botanist's *E. coli*' will soon be available.

## 2.2   RNA synthesis and degradation

Von der Heim and Zillig (1967) have purified the RNA polymerase of *Anacystis nidulans* and shown that it is very similar to the enzyme from *E. coli* on the basis of its divalent cation requirement, $NH_4^+$ inhibition, pH optimum (8.4) and hexagonal shape. It also dissociates into $\alpha$, $\beta$, and $\sigma$ subunits (Herzfeld & Zillig 1971). Using T4 DNA as substrate, they showed that both the *A. nidulans* and *E. coli* enzymes bind to the same sites (about 1 site/2,000 nucleotide pairs), but that the *A. nidulans* enzyme would initiate transcription at only half of these (Von der Heim & Zillig 1969). Hybridization competition experiments between the RNAs synthesized by the two enzymes showed that the RNA synthesized by the *A. nidulans* enzyme competed with less than 50% efficiency, suggesting that initiation sites may be specific for different RNAs. This enzyme was also purified by Capesius and Richter (1967b), who showed that salmon sperm and *Chlorella* DNA, as well as *Anacystis* DNA, can act as primers, and that the RNAs so synthesized had different elution patterns from MAK columns. About 30% of the RNA was resistant to ribonuclease, suggesting the presence of DNA-RNA hybrids. As in the bacteria, the RNA polymerase activity in *Anacystis montana* is sensitive to the rifamycin antibiotics (Rodriguez-Lopez *et al.* 1970).

Five blue-green algae tested by Norton and Roth (1967a) had very low levels of ribonuclease activity compared to seaweeds. However, a ribonuclease was purified from *A. nidulans* (Norton & Roth 1967b). Capesius and Richter (1967a) purified a polynucleotide phosphorylase from the ribosomal fraction of the same organism.

## 2.3   Ribosomes

Ribosomes fall into two general classes on the basis of sedimentation coefficient: the '70s' prokaryotic type and the '80s' eukaryotic cytoplasmic type. The ribosomes of *Anabaena variabilis* (Craig & Carr 1968), *A. cylindrica* and *Anacystis nidulans* (Taylor & Storck 1964), and *Phormidium luridum* (Vasconcelos & Bogorad 1971) are of the prokaryotic type, as are their ribosomal RNAs. The molecular weights of rRNAs of several species are 0·55 to 0·56 and 1·06 to 1·07 million daltons compared to 0·56 and 1·07 million daltons for *E. coli* rRNAs (Loening 1968, Howland & Ramus 1971, Vasconcelos & Bogorad 1971).

In contrast, the cytoplasmic rRNAs of eukaryotic algae are 0·68 to 0·69 and 1·21 to 1·30 million daltons, while their chloroplast RNAs are of the prokaryotic type (Loening 1968). However, the ribosomal proteins of *Phormidium luridum* are distinctly different from those of *E. coli*, mung bean chloroplasts, and mung bean cytoplasmic ribosomes (Vasconcelos & Bogorad 1971).

Antibiotic sensitivity is another criterion for classifying ribosomes. The growth of *Anacystis nidulans* is inhibited at low concentrations by chloramphenicol, erythromycin, spiramycin, and lincomycin, which are bound by the ribosomes, but is inhibited very little by the same concentrations of cycloheximide, which does not bind to the ribosomes (Rodriguez-Lopez & Vasquez 1968).

### 2.4   *DNA-DNA and DNA-RNA hybridization*

Relationships between three species were investigated using DNA-DNA hybridization (Craig *et al.* 1969). *Nostoc muscorum* DNA bound to *Anabaena variabilis* DNA with 29% of the efficiency of *Anabaena* DNA; *Anacystis nidulans* DNA bound with 11% efficiency. Various bacterial DNAs gave 3 to 16%. Since *Anabaena* is in the same family of filamentous forms as *Nostoc*, while *Anacystis* is coccoid, the results are hardly startling. However, this technique should be very useful in straightening out the relationships among the various subgroups of the coccoid forms (see Stanier *et al.* 1971).

More interesting results concern possible homology between blue-green algal ribosomal RNA and *Euglena* chloroplast DNA (Piggott & Carr 1972). Their data showed that rRNAs from seven species of blue-green algae hybridized with filter-bound *Euglena* chloroplast DNA with 11 to 47% of the efficiency of Euglena rRNA. Two photosynthetic bacteria gave 4 to 6% and other bacteria were lower. Unfortunately, the authors did not determine the effect of salt concentration on specificity of hybridization, nor determine the thermal stability of the duplexes formed. Given the extreme conservatism of rRNA among eukaryotes (Sinclair & Brown 1971), one should be very careful about extrapolating data of this sort to conclusions about evolutionary relationships. The latter authors showed homology between *Xenopus* rRNA and pea DNA, and no modern biologist would conclude that these two organisms are on the same evolutionary line.

## 3   EUGLENA

The euglenoid algae are photosynthetic flagellates which have a characteristic morphology and pigment composition, reproduce asexually, and will grow either photoauxotrophically or heterotrophically under a wide range of physiological conditions. When grown heterotrophically in the dark, they contain proplastids which will differentiate into chloroplasts after a few hours in the light. This is the only algal group in which proplastids are found, and it is therefore a model system for studying the proplastid → chloroplast transition without the difficulties of working with higher plants. It is also possible to produce permanently bleached strains by treatment with streptomycin (Provasoli *et al.* 1941), ultraviolet light (Lyman *et al.* 1961) or heat (Pringsheim & Pringsheim 1952). A number of species of *Euglena* are available in axenic culture (Cook 1968), but all the work discussed here has been done on *E. gracilis*. Although there are differences between the Z and *bacillaris* strains (Buetow 1968b), there is no evidence for differences in nucleic acid properties or metabolism, so they will be treated interchangeably. Only the most recent work will be covered; for the older literature readers are referred to *The Biology of Euglena*, Vols I & II (Buetow 1968a).

### 3.1   DNA

Data on the physical properties and amounts of DNA in the various cell compartments is summarized in Table 10.1. Dark-grown cells and those which could

Table 10.1. The DNAs of *Euglena gracilis*

|  | Nucleus | Chloroplast | Mitochondrion | Reference |
|---|---|---|---|---|
| Buoyant density | 1·707 | 1·686 | 1·691 | (1–6) |
| Base composition calculated from buoyant density (%G+C) | 48 | 26 | 32 | |
| Thermal denaturation temperature ($T_m°C$) | 89–91 | 78–80 | — | (1) |
|  | 92 | 81·5 | — | (4) |
| Chemical analysis (%GC) | 50·6 | 27·3 | — | (1) |
|  | 53·2 | 23·6 | — | (2) |
| 5-methyl cytosine | 2·3% | undetectable | — | (1, 2) |
| Molecular weight (electron microscopy) | — | $9·2 \times 10^7$ daltons | — | (7) |
| Unique sequence length | — | $9 \times 10^7$ | — | (8) |
| DNA per organelle (daltons) | $138 \times 10^9$ | $45 - 72 \times 10^7$ | $9 - 22 \times 10^6$ | (4, 6) |

(1) Brawerman & Eisenstadt 1964a; (2) Ray & Hanawalt 1964; (3) Ray & Hanawalt 1965; (4) Edelman *et al.* 1964; (5) Edelman *et al.* 1965; (6) Edelman *et al.* 1966; (7) Manning & Richards 1972a; (8) Stutz 1970.

form at least a partial chloroplast contained chloroplast DNA but cells which were permanently bleached (incapable of forming chloroplasts under any conditions) did not contain it (Edelman *et al.* 1965, Ray & Hanawalt 1965). The colourless flagellate *Astasia longa*, which is thought to be derived from *Euglena*, has nuclear DNA of the same buoyant density as *E. gracilis* and a minor DNA at the density of its mitochondrial DNA (Kieras & Chiang 1971).

DNA isolated from chloroplasts without shearing is in the form of 44·6μm circles, which corresponds to a molecular weight of $9·2 \times 10^7$ daltons under the conditions used (Manning & Richards 1972a, Manning *et al.* 1971). This agrees very well with a unique sequence length of $9 \times 10^7$ daltons determined from renaturation rates (Stutz & Vandrey 1971, Stutz 1970). If estimates of the amount of DNA per chloroplast given in the table are correct, there would be 5 to 8 circular molecules per chloroplast.

DNA isolated from highly purified chloroplasts contains a minor band at $1 \cdot 701$ g cm$^{-3}$ in addition to the main chloroplast DNA band (Manning *et al.* 1971, Stutz & Vandrey 1971). This DNA is enriched in cistrons coding for chloroplast ribosomal DNA by the criteria of DNA/RNA hybridization (Stutz & Vandrey 1971). Since the DNA was sheared by this preparative technique to about $5 \times 10^6$ daltons, this does not conflict with the evidence that the DNA is circular but suggests that the cistrons for rRNA are close together in the circle and therefore end up in a fragment of distinct buoyant density. However, Brown and Haselkorn (1972) do not find this DNA in chloroplasts freed of nuclear contamination by flotation in a Renograffin gradient. It is clear that more work needs to be done on this DNA.

Manning and Richards (1972b), using the $^{15}$N-$^{14}$N density transfer technique, have clearly shown that both nuclear and chloroplast DNA replicate semi-conservatively. Their data suggest that the chloroplast DNA replicates faster than the nuclear DNA and 'turns over' to keep the amount per cell constant. However, the evidence is complicated by the presence of the $1 \cdot 701$ g cm$^{-3}$ satellite peak which is also replicating. Since the number of chloroplasts per cell depends greatly on culture conditions and the stage of the cell cycle (Davis & Epstein 1971), there are alternative interpretations for these data, as well as for the indirect evidence for 'metabolic DNA' (Gibor 1969). Isolated chloroplasts can incorporate radioactive deoxyribonucleoside triphosphates into DNA (Scott *et al.* 1968), and there is some other evidence that chloroplast and nuclear DNA replication are controlled independently (Davis & Epstein 1971, Cook 1966, Carell 1969).

An early report claimed that chloroplast DNA would hybridize with a certain fraction of nuclear DNA (Richards 1967), but competition controls were not done and the thermal stability of the duplexes was not tested. This work should be repeated using modern DNA renaturation techniques.

Nuclear DNA synthesis is strongly inhibited by cycloheximide; chloroplast and mitochondrial synthesis is somewhat inhibited (Richards *et al.* 1971). Chloramphenicol does not have a preferential effect on any type of DNA synthesis, up to 1 mg ml$^{-1}$. This is surprising in view of the specificity of chloramphenicol for chloroplast protein synthesis and of cycloheximide for cytoplasmic protein synthesis (Smillie *et al.* 1967). Richards *et al.* (1971) suggest that the proteins involved in organelle DNA synthesis are either very stable or are synthesized on cytoplasmic ribosomes. There is evidence for the latter explanation in *Chlamydomonas* (Surzycki 1969).

*E. gracilis* contains two deoxyribonucleases, one of which has a pH optimum of $9 \cdot 4$ and requires Ca$^{2+}$. It appears to be associated with the chloroplasts, as its activity increases during light-induction of chloroplast development (Carell *et al.* 1970). The other has an acid pH optimum and appears similar to DNase II.

### 3.2 *Polysomes, ribosomes and RNA synthesis*

Polysomes have been isolated from cytoplasmic, chloroplast and mitochondrial fractions (Rawson & Stutz 1969, Avadhani *et al.* 1971). The mitochondrial polysomes have been shown to incorporate labelled amino acids (Avadhani *et al.* 1971), as have polysomes from isolated chloroplasts (Harris & Eisenstadt 1971).

Problems with ribosomal instability and degradation of rRNA led to many conflicting reports on their size and sedimentation rates (for older references, see Schiff 1971). The cytoplasmic ribosomes require 0·1 to 0·2M NaCl for stability *in vitro* (Rawson *et al.* 1971, Krawiec & Eisenstadt 1970). $Zn^{2+}$ is required for their *in vivo* stability (Prask & Plocke 1971), which these authors suggest is due to a stabilizing effect on ribosomal conformation rather than inhibition of a nuclease. Heizmann (1970) showed that the instability of the large subunit rRNA was due to a detergent-activatable ribonuclease which formed part of the large cytoplasmic ribosomal subunit. This ribonuclease was inhibited by 0·25 M sodium thioglycolate or similar concentrations of sodium chloride or sodium acetate (Rawson *et al.* 1971). It therefore appears to be different from the ribonuclease isolated from whole cells (Fellig & Wiley 1960) which has optimum activity at 0·5M NaCl and pH 4·5. Chloroplast ribosomes require 0·05M KCl and 0·02M $Mg^{2+}$ for stability (Lee & Evans 1971).

The currently most probable values for the physical constants of ribosomes, subunits and their RNAs are given in Table 10.2. The cytoplasmic ribosomes and their RNAs are among the largest reported (Loening 1968). The formation of functional hybrid ribosomes between *Euglena* chloroplast 30s and *E. coli* 50s subunits supports the assignment of chloroplast ribosomes to the 'prokaryotic' type (Lee & Evans 1971).

Many studies on the synthesis of Euglena RNAs during light induction of chloroplast development were unfortunately done with methods which are now known to result in degradation (Zeldin & Schiff 1967, Schuit *et al.* 1970, Gnanam & Kahn 1967). More recent work has shown the following:

(a) There is a net increase in both cytoplasmic and chloroplast rRNAs (Heizmann 1970), but chloroplast RNA increases more—from 2 to 7% of the total RNA (Scott *et al.* 1971).

(b) Chloroplast rRNA incorporates $^3$H-orotic acid faster than cytoplasmic rRNA on transfer to light, but the difference becomes less as the chloroplasts complete their development (Munns *et al.* 1972).

(c) As measured by hybridization between chloroplast DNA and pulse-labelled RNA (Brown & Haselkorn 1971a), ribosomal RNA is the only chloroplast RNA which increases on transfer from dark to light. This technique is probably not sensitive enough to detect minor species of RNA (see (d) below).

(d) New species of isoleucyl-, phenylalanyl-, and glutamyl-tRNA are induced by light (Barnett *et al.* 1969). These are not found in the aplastidic mutant W₃BUL. The isoleucyl-tRNA was found only in the chloroplast, as was a light-induced synthetase specific for it (Reger *et al.* 1970). The phenylalanyl-tRNA was enriched in the chloroplast and could be acylated by only one of three constitutive synthetases. However, this synthetase was found in W₃BUL, and the authors suggest that in this case the enzyme is coded for by the nuclear DNA, synthesized on cytoplasmic ribosomes, and some of it transported to the chloroplast. The evidence suggests, but does not prove, that the light-inducible isoleucyl-tRNA and its specific synthetase are synthesized in the chloroplast (Reger *et al.* 1970).

(e) Pre-existing total RNA is methylated when cells are transferred from dark to light (Perl 1971).

**Table 10.2.** Characteristics of *Euglena gracilis* ribosomes

| | S rates of ribosomes | | | RNAs | |
| --- | --- | --- | --- | --- | --- |
| | Monomers | Subunits | S rate | Mol wt on P.A.G.E. ×10⁻⁶ | References |
| Cytoplasm | 88 | 67,46 | 26, 24,22 | | (1) |
| | 87 | 62,41 | 25,19 | | (2) |
| | 86 | 64,46 | 24,20 | | (3) |
| | 87 | — | 25,20 | 1·3, 0·85 | (4) |
| | — | — — | 25,19 | — | (5) |
| | — | – — | – | 1·3, 0·85 | (6) |
| Chloroplasts | 68 | 46,29 | 22,17 | – | (2) |
| | — | — | 23·5, 16·5 | | (7) |
| | 70 | 50,30 | | | (1) |
| Mitochondria | 72 | — | — | — | (8) |
| | — | — | 14,11 | — | (5) |

(1) Scott *et al.* 1970; (2) Rawson & Stutz 1969; (3) Rawson & Stutz 1968; (4) Rawson *et al.* 1971; (5) Krawiec & Eisenstadt 1970; (6) Loening 1968; (7) Scott & Smillie 1969; (8) Avadhani *et al.* 1971.

In dark-grown cells, only cytoplasmic rRNA incorporates ³²P (Brown & Haselkorn 1971b). The label appears to go first into a 3·5 × 10⁶ dalton precursor which gives rise to the 0·85 × 10⁶ dalton rRNA (small subunit) and to another large molecule which is cleaved to give the 1·35 × 10⁶ dalton rRNA (large subunit).

There have been a few reports of messenger RNA activity. Early preparations from chloroplasts may have consisted largely of degraded ribosomal RNA (Brawerman & Eisenstadt 1964b). Brown and Haselkorn (1971a,b) claimed that the same pulse-labelled RNAs were found in cells with and without functioning chloroplasts. They did not use any other criteria for defining 'messenger RNA'. Avadhani and Buetow (1972) isolated two rapidly-turning-over RNA fractions and a DNA-RNA fraction from an aplastidic mutant. Each one stimulated radio-active amino-acid incorporation in a cell-free *E. coli* system. In this case, radio-active peptides were isolated, separated by gel electrophoresis and isoelectric focusing, and shown to be different for the three different RNA fractions. The demonstration of a specific protein product is a much better definition of 'messenger' activity than rapid turnover of the RNA, as there is no *a priori* reason why messenger RNAs in eukaryotic cells should be turning over.

Ribosomal RNA is the only proven gene product of chloroplast DNA. Chloroplast rRNA is complementary to chloroplast DNA by hybridization (Scott & Smillie 1967). The 23s and 16s RNAs hybridize separately, indicating that the cistrons are distinct (Scott *et al.* 1971). When the chloroplast DNA is separated into 'heavy' and 'light' strands on alkaline denaturation, the former hybridizes with ten times more rRNA than the latter (Stutz & Rawson 1970), indicating that the RNA is transcribed off that strand. From the data given in the references, the number of rRNA cistrons per chloroplast can be calculated to be between 15 and 50 (Scott *et al.* 1971, Stutz & Vandrey 1971, Edelman *et al.* 1964).

Chloroplast rRNA synthesis is rifamycin-sensitive (Brown *et al.* 1970). Scott *et al.* (1971) reported hybridization of chloroplast rRNA with nuclear DNA, but they did not test the thermal stability of the duplexes, and were reluctant to draw any firm conclusions.


## 4   CHLAMYDOMONAS

*Chlamydomonas reinhardtii* is a model system for the study of nucleic acid synthesis and function in the various cell compartments. It is unicellular and grows reasonably quickly either photoautotrophically, heterotrophically, or mixotrophically. However, its big advantage is its sexuality. A large number of Mendelian and uniparentally inherited mutations have been isolated (Sager & Ramanis 1970a,b, Levine & Goodenough 1970) many of which are concerned with chloroplast function. As a result, the most sophisticated work on any algal species has been done on several closely-related strains of *C. reinhardtii*. No other species will be discussed.


### 4.1   *DNA*

The nuclear DNA of *C. reinhardtii* has a buoyant density of $1 \cdot 723$g cm$^{-3}$ (64% G+C) which conveniently separates it from the chloroplast DNA of $1 \cdot 692$g cm$^{-3}$ (33% G+C) (Chun *et al.* 1963, Sager & Ishida 1963, Leff *et al.* 1963, Chiang & Sueoka 1967a,b). There is also a minor DNA ($\gamma$ satellite) at $1 \cdot 715$g cm$^{-3}$ (56% G+C) which may or may not be the mitochondrial DNA (Chiang & Sueoka 1967a).

The nuclear DNA was shown to replicate semiconservatively in one of the early $^{15}$N-$^{14}$N transfer experiments (Sueoka 1960). In synchronous vegetative growth, there are two rounds of DNA synthesis, each followed by a nuclear division, then two successive cell divisions (Kates et al. 1968). However, if the culture is maintained under low light intensity, there are two rounds of DNA replication but only one cell division. During the next light-dark cycle there is a cell division without DNA synthesis. This implies that DNA synthesis and cell division are only loosely coupled and that the latter process is more light-dependent than the former (Kates et al. 1968). DNA synthesis under these conditions requires protein synthesis (Jones et al. 1968).

The $^{15}$N-$^{14}$N isotope transfer experiments, extended to the cytoplasmic DNAs (Chiang & Sueoka 1967a), showed that both chloroplast and γDNA went through two cycles of semiconservative replication, but during the light rather than the dark period. This effect was not due to the presence of a precursor pool (Chiang & Sueoka 1967b). This temporal separation implies at least some separation in the control of the two types of replication. The replication of chloroplast DNA has been shown to be independent of light (Chiang 1971).

If vegetative cells in the middle of the light period are transferred to nitrogen-free medium, DNA synthesis and cell division continue, but the four products of mitosis are competent gametes rather than vegetative cells (Kates et al. 1968). There is no net protein synthesis during this process, but there is protein turnover and resynthesis (Jones et al. 1968). Precursors for DNA synthesis come from the degradation of ribosomal RNA (Siersma & Chiang 1971). The gametes can then pair with gametes of opposite mating type to form zygospores (zygotes). After maturation of zygospores on agar plates for seven days (two in the light, five in the dark), a return to light causes the zygospores to germinate, releasing 4, 8, 16 or 32 zoospores depending on the strain.

$^{15}$N-$^{14}$N transfer experiments on a strain producing 8 zoospores (Sueoka et al. 1967) showed that there was no nuclear DNA synthesis during maturation and only one round of semiconservative replication during zygospore germination. The zoospores contained half the amount of DNA found in vegetative cells and gametes. After several generations of vegetative growth, the original amount of nuclear DNA was restored (Kates et al. 1968).

On the other hand, the chloroplast DNA had replicated once before the end of zygospore maturation, and by the end of germination there had been several more rounds of replication, as no $^{14}$N-$^{15}$N hybrid band was found in the gradient (Chiang & Sueoka 1967b). This is a further instance of lack of synchrony between nuclear and chloroplast DNA replication. The behaviour of the γ satellite DNA was not studied in the isotope transfer experiments because of the appearance of a very large band of unlabelled (i.e. both strands newly synthesized) DNA at 1·715g cm$^{-3}$, the density of unlabelled γ satellite DNA (Sueoka et al. 1967). This 'M band' DNA made up 50% of the total DNA during zygospore maturation but gradually disappeared during germination. It was not due to differential extraction or denaturation of other DNA components (Chiang & Sueoka 1967b). Its function is still unknown.

Using plus and minus gametes previously labelled with different isotopes, Chiang (1968) showed that both labels were conserved and were found in their usual place in preparative CsCl gradients throughout the process of sexual reproduction. In other words, there was conservation of cytoplasmic DNAs from gametes of both mating types. (The M DNA was present but did not contribute to the picture since it was synthesized after transfer to non-radioactive medium). This observation is in direct conflict with the uniparental inheritance of certain genetic markers which are believed to be located on extra-nuclear DNA (Sager & Ramanis 1968, 1970a,b). Under normal conditions, only the genes of the $mt^+$ parent are found in the progeny. It would therefore be expected that one or both cytoplasmic DNAs of the $mt^-$ parent would be lost before zygo-spore germination, or at least by the time a few cycles of vegetative reproduction had occurred. Some evidence for the non-replication of $mt^-$ chloroplast DNA was published in abstract form (Sager & Lane 1969). More recent data are given in Chapter 11, p. 314.

However, using a mixture of density and radioactive labelling, Chiang (1971) showed that the cytoplasmic DNAs were not only conserved but also underwent extensive recombination between the time of gamete fusion and the time of zoospore release. Since the two gamete chloroplasts fuse and undergo con-siderable morphological changes during this period (Bastia et al. 1969) the physical conditions for intra-organellar recombination are present. In the same experiment, the three drug resistance markers carried by the $mt^-$ parent virtually disappeared, even though the $mt^-$ parental DNA was fully conserved.

To further confuse the story, the pattern of chloroplast division during meiosis and zoospore formation could lead to Mendelian segregation of genes carried on chloroplast DNA (Chiang 1971, Levine & Goodenough 1970).

Although Sager and Ramanis (1968, 1970a,b) interpret some of their genetic evidence to indicate that there are one or two copies of each chloroplast gene per cell, both Sager's and Chiang's groups have shown by renaturation kinetics that there are at least 25 copies of the same genetic information per gamete chloroplast (Wells & Sager 1971, Bastia et al. 1971). The chloroplast DNA renatures with a kinetic complexity corresponding to a unique sequence length of $2 \times 10^8$ daltons, whereas there is $5 \times 10^9$ daltons of DNA per gamete chloro-plast (Chiang & Sueoka 1967a). As no electron microscopy has been done on the chloroplast DNA, it is not known whether all these copies are found in tandem on the same long molecule of DNA, or whether they exist as separate molecules, such as the 40μm circles found in Euglena (Richards et al. 1971). The existence of multiple copies of chloroplast genes would explain the behaviour of certain streptomycin-dependent mutants (Schimmer & Arnold 1970a,b).

## 4.2  Ribosomes and RNA synthesis

Chlamydomonas has 80s cytoplasmic ribosomes containing 25 and 18s RNAs, and 68s chloroplast ribosomes containing 23 and 16s RNAs (Sager & Hamilton

1967 Ohad *et al.* 1967, Hoober & Blobel 1969, Bourque *et al.* 1971). Both types of ribosome contain '5s' RNAs which can be separated from each other on MAK columns (Surzycki & Hastings 1969). The chloroplast ribosomes are particularly sensitive to low $Mg^{2+}$ concentrations (Sager & Hamilton 1967), undergoing dramatic changes in sedimentation rate before dissociation into subunits (Hoober & Blobel 1969, Bourque *et al.* 1971).

A number of mutants of both Mendelian and uniparental type make reduced amounts of 70s ribosomes, and instead, make 66s particles which contain both 16 and 23s RNAs (Gillham *et al.* 1970, Boynton *et al.* 1970, Bourque *et al.* 1971). In several cases, the defect appears to be due to a deficit of small ribosomal subunits, as 54s particles containing the 23s RNA were found in the gradient. They were not found in normal cells.

In synchronized cultures, the bulk of RNA synthesis occurs at a linear rate in the light period, but there is some RNA synthesis at a slower rate in the dark period (Jones *et al.* 1968). Net RNA synthesis proceeds at the same rate in vegetative cells and in those undergoing gametogenesis under conditions of complete nitrogen deprivation. However, if the cells are merely deprived of an essential amino acid, RNA synthesis starts to slow down after 6 hours. *Chlamydomonas* is thus analogous to a 'relaxed' strain of *E. coli* rather than a 'stringent' strain in which RNA synthesis stops as soon as the amino acid is removed (Stent & Brenner 1961). During this period, both tRNA and rRNA are synthesized (Jones *et al.* 1968). Surzycki and Hastings (1969) claim that only cytoplasmic RNA synthesis is relaxed, since the '5s' RNA of the chloroplast has a much lower specific activity than the '5s' cytoplasmic RNA in the experiments of Jones *et al.* (1968). Since the peaks in question are rather small and not well separated from the 4s RNA, this point should be reinvestigated.

Both cytoplasmic and chloroplast ribosomes are conserved during normal vegetative growth, i.e. there is no turnover of any rRNA. However, both RNAs are 90% degraded during gametogenesis (Siersma & Chiang 1971). The RNA degradation products are mainly used for the synthesis of DNA, but there is a small amount of new rRNA and tRNA synthesis. Gametes and vegetative cells have different arginyl-tRNA synthetases and different numbers of arginyl-tRNAs (Jones & Peng 1969).

Some very elegant studies have utilized the selective action of certain antibiotics. The drug rifampin (rifampicin) inhibits chloroplast DNA-dependent RNA polymerase completely, but does not affect the nuclear enzyme (Surzycki 1969). There is no synthesis of 16s and 23s RNAs, with a consequent loss of chloroplast ribosomes (Goodenough 1971). There is also a loss of the 5s RNA ascribed to the chloroplast (Surzycki *et al.* 1970). However, there is no cessation of chloroplast DNA synthesis, even when cells are kept in the drug for several generations. Chloroplast DNA synthesis is equally unaffected by spectinomycin, which specifically inhibits protein synthesis on chloroplast ribosomes (Surzycki *et al.* 1970). This indicates that the cistron coding for the chloroplast DNA polymerase is in the nucleus, and that the enzyme is synthesized on cytoplasmic ribosomes. Since cells can grow heterotrophically in the presence of rifampin, proteins coded by chloroplast DNA or synthesized on chloroplast ribosomes (these are not necessarily the same) are not essential for cell growth (Surzycki 1969). However, since cells grown in rifampin lose their ability to fix $CO_2$, some of these proteins are needed for photosynthetic functions. Since spectinomycin does not prevent the formation of chloroplast ribosomes (Surzycki *et al.* 1970), the chloroplast ribosomes are not involved in the synthesis of their own proteins.

These studies have been extended to show that some but not all of the components of the photosynthetic apparatus are coded for by chloroplast DNA, and that the synthesis of some of them appears to involve both cytoplasmic and chloroplast ribosomes (Armstrong *et al.* 1971, Eytan & Ohad 1970).

Experiments with rifampin and actinomycin D support a model in which 16s and 23s RNA cistrons are arranged in pairs on the chloroplast DNA, with the 16s cistron nearer the promoter (Surzycki & Rochaix 1971).

## 5  CHLORELLA

*Chlorella* has been studied largely because it can be synchronized easily with light-dark cycles, and is therefore a good subject for investigating the control of various events in the cell cycle. Unfortunately, it does not reproduce sexually, so its nucleic acid metabolism cannot be attacked with genetic tools.

### 5.1  *DNA*

The physical characteristics of the DNAs from the three species of *Chlorella* studied are given in Table 10.3. Several points deserve mention. First of all, the nuclear DNA base compositions of the three species are quite different, although all are high in G+C. Since *Chlorella* reproduces only asexually, each cell produces a clone which is genetically isolated from all others, which could lead to a fairly rapid divergence. Secondly, the low G+C satellite DNA is probably the chloroplast DNA, although it is clear from the published evidence that pure chloroplasts have not yet been isolated from any species. The satellite of density $1 \cdot 700$g $cm^{-3}$ in *C. pyrenoidosa* represents a very small proportion of the total DNA, consistent with its being mitochondrial DNA. The satellite of $1 \cdot 722$ to $1 \cdot 727$g $cm^{-3}$ found by two groups of workers (Rode & Bayen 1972, Gense-Vimon 1971) but not by a third (Wanka *et al.* 1970) is an anomaly.

When *C. pyrenoidosa* is synchronized by light-dark cycles, there is no nuclear DNA synthesis during the first nine hours of the light period. Only the two satellite DNAs incorporate labelled uracil (Wanka *et al.* 1970). This incorporation continues at a low level throughout the cell cycle, whereas nuclear DNA synthesis occurs only in the latter part of the light period, just prior to and during cell division. During the period of nuclear DNA synthesis, there is a

**Table 10.3.** DNAs of *Chlorella* spp.

| Species | Buoyant density (g cm⁻³) | | Tm (°C) | | Base composition | | Reference |
|---|---|---|---|---|---|---|---|
| | Nuclear | Satellite | Nuclear | Satellite | Nuclear | Satellite | |
| *C. pyrenoidosa* 211-8b (*C. fusca*) | 1·710 | 1·689, 1·700 | 89·1 | 79·9 | 48–51 | 26–28, 41 | (1) |
| | 1·710 | 1·687–1·689, 1·699–1·701, 1·723–1·727 | — | — | 51 | 28–30, 40–42, 65–68 | (2) |
| | 1·709, 1·710 | 1·694, 1·722, 1·724 | — | — | 50–51 | 35, 63–65 | (3) |
| *C. ellipsoidea* | 1·716 | 1·695 | — | — | 57 | 36 | (4) |
| | 1·717 | 1·692 | — | — | 58 | 33 | (5) |
| *C. protothecoides* | — | — | 95 | 81, 88 | 63 | 29, 46 | (6) |

(1) Wanka *et al.* 1970; (2) Rode & Bayen 1972; (3) Gense-Vimon 1971; (4) Chun *et al.* 1963; (5) Iwamura & Kuwashima 1969; (6) Oshio & Hase 1968.

doubling of the total DNA every 105 to 115 min (Wanka & Mulders 1967). The nuclei divide parallel to the increase in DNA content, but at a later time. Nuclear division is followed by cell division at a parallel rate.

Nuclear DNA synthesis is only indirectly controlled by light in *C. pyrenoidosa* (Wanka & Mulders 1967) and *C. protothecoides* (Sokawa & Hase 1968). It is not controlled through levels of enzymes involved in synthesis of DNA precursors (Wanka & Poels 1969, Shen & Schmidt 1966), as their activities parallel cell number and protein content. DNA polymerase increases almost continuously during the cell cycle, indicating that its level is not controlling DNA replication (Schönken & Wanka 1971). Deoxyribonuclease activity was the only enzyme activity correlated with DNA replication (Schönken *et al.* 1970).

Nuclear DNA synthesis is completely inhibited by $15 \times 10^{-6}$ M cycloheximide added any time during the period of synthesis (Wanka & Moors 1970, Wanka *et al.* 1972) implying that continuous protein synthesis is necessary for DNA synthesis. It appears that cycloheximide blocks DNA synthesis indirectly by interfering with some aspect of the regulatory process, as the activity of the DNA polymerase is not affected (Wanka & Moors 1970). Recent work indicates that a specific protein or proteins synthesized during the S phase is required for DNA replication (Wanka *et al.* 1972). Protein synthesis also appears to be involved in turning off DNA synthesis (Wanka & Geraedts 1972).

The satellite DNAs are almost completely insensitive to cycloheximide (Wanka & Moors 1970). This would suggest that the satellite DNA polymerases are not made on cytoplasmic ribosomes. If this is true, it is quite different from the case of *Chlamydomonas* (see above).

In *C. ellipsoidea*, Iwamura and Kuwashima (1969) claim that both the 1·717g cm⁻³ and 1·692g cm⁻³ DNAs are found in the chloroplast. Unfortunately, the chloroplasts were first sedimented at $20,000 \times g$, which would stick nuclear fragments and all other subcellular particles together, and the deoxyribonuclease-treated controls contain so little DNA that they are not convincing. These authors claim that the 1·692g cm⁻³ DNA is 'metabolic DNA', i.e. it is turning over (Iwamura 1966). However, this has not yet been independently confirmed.

### 5.2   *Ribosomes and RNA synthesis*

As in the other green algae, the cytoplasmic ribosomes are of the 80s type, dissociating into subunits of approximately 60 and 40s; the chloroplast ribosomes are of the 70s type, dissociating into 50 and 30s subunits (Rodriguez-Lopez & Vasquez 1968, Mihara & Hase 1969, Galling & Ssymank 1970). The cytoplasmic rRNAs have molecular weights of 1·3 and $0·69 \times 10^6$ daltons, compared to 1·1 and $0·56 \times 10^6$ daltons for the chloroplast (non-cytoplasmic) rRNAs (Loening 1968). In other words, *Chlorella* is similar to *Chlamydomonas* and higher plants but not to *Euglena*. Antibiotics which are specific for the 70s ribosomes in these organisms cause bleaching in *C. pyrenoidosa*, indicating that some of the proteins involved in photosynthesis are synthesized on organellar ribosomes (Rodriguez-Lopez & Vasquez 1968).

Enöckl (1968) showed that rRNA is synthesized only in the light period, whereas tRNA is synthesized throughout the cell cycle. For reasons which are not at all clear, rRNA and '5s' RNA synthesis are blocked by cycloheximide, but tRNA synthesis is not (Wanka & Schrauwen 1970).

A DNA-RNA complex isolated from *C. pyrenoidosa* (Richter & Senger 1965) was later shown to dissociate on gel filtration, giving an RNA which rechromatographed at the same position as the putative hybrid (Beiderbeck & Richter 1968). Unfortunately, this unique RNA has not been characterized further. It would be interesting to see if the same thing would happen to the DNA-RNA hybrids isolated from *Euglena* (Avadhani & Buetow 1972, Avadhani *et al.* 1970).

The fidelity of synchrony of *Chlorella* makes it a good subject for biochemical research. It is to be hoped that more strenuous efforts will be made to obtain pure chloroplast and mitochondrial fractions and publishable electron micrographs thereof.

## 6 ACETABULARIA

*Acetabularia* is unique. This large, uninucleate, single-celled, siphonous, green alga undergoes a complex morphogenesis at the end of its vegetative growth, producing an umbrella-like reproductive structure or cap. When the nucleus divides, the secondary nuclei migrate to the cap and form cysts, which are the resting-stage of the cell. Any time before the formation of the cap, the nucleus can be removed by simply snipping off the rhizoid. The enucleate cell will continue to grow for as long as five weeks, and under favourable conditions can even form a cap. Interspecific grafts and studies of enucleate cells in the 1930's led Hämmerling to postulate the existence of 'morphogenetic substances' which were elaborated in the nucleus and stored in the cytoplasm until they were needed to control the initiation and species-specific formation of the cap (see Hämmerling 1953). This was many years before messenger RNA was hypothesized. The ease of removal of the nucleus means that this was the first and only organism in which it was possible to show that the DNA associated with chloroplasts is not due to residual nuclear fragments (Gibor & Izawa 1963, Baltus & Brachet 1963).

It would appear that *Acetabularia* is an ideal experimental organism for tackling many problems. Unfortunately, this is not the case. Its life-cycle is 6 to 9 months. This makes it impossible to obtain the large quantities of material needed for careful biochemical studies. It also makes mutant hunts rather unattractive. Axenic cultures are difficult to obtain and impossible to maintain from one generation to the next. This limits the validity of radioisotope labelling experiments unless many controls are done.

In spite of all this, there is a current renewal of interest in *Acetabularia*, which is reflected in a recent symposium (Brachet & Bonotto 1970) and a book (Puiseux-Dao 1970). For a general review the reader should consult Vanden Driessche (1973), Schweiger (1969), Werz (1965) and Hämmerling (1953, 1963).

## 6.1   DNA

The first proof that DNA was associated with *Acetabularia* chloroplasts was obtained with direct fluorescence assay of partially purified chloroplasts from enucleate cells (Gibor & Izawa 1963, Baltus & Brachet 1963). They estimated $10^{-16}$ to $10^{-15}$g per chloroplast. However, there is some evidence that the chloroplasts contain a variable amount of DNA, and that 65 to 80% of them contain no detectable DNA at all (Woodcock & Bogorad 1970a). This was determined using acridine orange staining, radioautography of bound $^3$H-actinomycin DNA, electron microscopy of thin sections, and electron microscopy of DNA extruded from osmotically-shocked chloroplasts. Those that had DNA associated with them averaged between 200 and 400 μm, or 400 to 800 $\times$ $10^6$ daltons.

In another electron microscope study, 20 to 40% of the ruptured chloroplasts had DNA associated with them (Green & Burton 1970). However, some of those containing DNA appeared to have as much as a bacterium (Fig. 10.2). *Acetabularia* chloroplast DNA does not appear to be circular (Green & Burton 1970, Werz & Kellner 1968). DNA from detergent-lysed chloroplasts has an average length of 40μm (equivalent to 80 $\times$ $10^6$ daltons) but molecules ranging up to 200μm (400 $\times$ $10^6$ daltons) are also found (Green unpubl.). If the *Acetabularia* chloroplast has a genome size of about 200 $\times$ $10^6$ daltons, like *Euglena* and *Chlamydomonas*, and the number of copies of this genome are not distributed equally when the chloroplasts divide, the result would be a variable amount of DNA per chloroplast and none in some. The chloroplasts without DNA would be able to function photosynthetically using material synthesized before division, but they could not divide. There is some evidence for multiple amounts of DNA per chloroplast in higher plants (Herrmann & Kowallik 1970).

There is some debate about the base compositions of *Acetabularia* DNAs. The relevant data are given in Table 10.4. Berger and Schweiger (1969) and Gibor (1967) claimed that their cells were bacteria-free, and they used $^3$H-thymidine or $^{14}CO_2$ to label the DNA. However, only a very small amount of bacterial contamination can take up a lot of label, especially when the labelling period is 5 to 9 days. Green *et al.* (1967, 1970) and Heilporn and Limbosch (1971) used non-axenic cultures, but did not use any radioactive label. I have done very extensive controls of bacterial contamination in purified chloroplast fractions (Green unpubl.) and find at worst, one bacterium for 2,000 chloroplasts. The number is usually much less, especially with prior antibiotic treatment. If the average chloroplast contains one-tenth the amount of DNA of the average bacterium, the bacterial DNA would make up, at most, 0·5% of the total DNA, and would not be detected in the analytical ultracentrifuge. Even if the chloro-

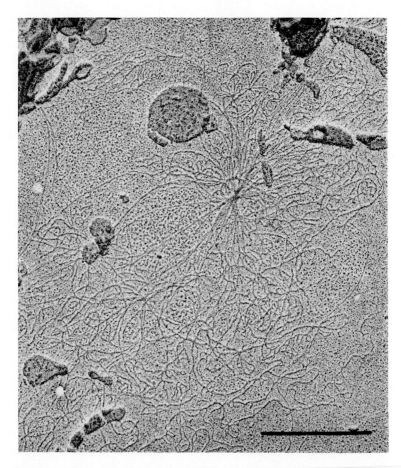

**Fig. 10.2.** DNA released from chloroplasts lysed by osmotic shock in a protein monolayer, and shadowed with platinum-palladium. The bar represents one micron. From Green & Burton (1970).

Table 10.4. DNAs of *Acetabularia mediterranea*

| | Chloroplast | | Mitochondrion | | Nucleus | | Unassigned | |
| --- | --- | --- | --- | --- | --- | --- | --- | --- |
| | Buoyant density (g cm⁻³) | %G+C | Buoyant density (g cm⁻³) | %G+C | Buoyant density (g cm⁻³) | %G+C | Buoyant density (g cm⁻³) | Reference |
| | 1·702–1·704 | 43–45 | 1·714–1·715 | 55–56 | 1·696 | 37 | — | (1–3) |
| | 1·695 | 36 | — | — | — | — | 1·717,1·724 | (4) |
| | 1·717 | 58 | 1·709 | 50 | 1·712 | 53 | — | (5) |

(1) Green et al. 1967, 1970; (2) Green unpubl.; (3) Heilporn & Limbosch 1971; (4) Gibor 1967, Granick & Gibor 1967; (5) Berger & Schweiger 1969.

plasts contained one-hundredth the DNA of a bacterium, their DNA would not constitute the major peak, and would probably be divided among a number of buoyant densities, as a number of different species are present. None of the common contaminants has DNA of $1.702$ to $1.704$g cm$^{-3}$ (Muir unpubl.).

It is hard to see how these big differences between the results of groups of workers could be due to anything except extraneous DNA, but it is possible that each laboratory's strain is from a different isolate and that the isolates have diverged during 15 to 20 years of domestication.

Chloroplasts in enucleate cells continue to multiply, although at a slower rate than controls. $^3$H-thymidine is incorporated into them in acid-insoluble, ribonuclease-insensitive form (Shephard 1965a). The total DNA of enucleate fragments more than doubles over a period of 15 days after enucleation (Heilporn-Pohl & Brachet 1966). These data appear to suggest that chloroplast DNA contains the information for its own DNA polymerase as well as all the rest of its replicative and photosynthetic machinery, and possibly even some information necessary for cap formation. However, the DNA polymerase could have been made before enucleation, and simply diluted somewhat as the chloroplasts divided (only once or twice under these experimental conditions). If a large fraction of the chloroplasts have no DNA anyway, the percentage with DNA would not change too much. Since there are long-lived 'morphogenetic substances' of nuclear origin concerned with cap formation, perhaps there are also some involved with formation of chloroplast components.

### 6.2   Ribosomes and RNA synthesis

Are Hämmerlings 'morphogenetic substances' really long-lived messenger RNAs or are they a host of zymogens waiting for the appropriate activators? Unfortunately, that question is not yet answered, although it has stimulated a lot of the recent interest in *Acetabularia* (Brachet & Bonotto 1970, Puiseaux-Dao 1970).

One of the major stumbling-blocks has been the isolation of intact ribosomes and undegraded RNA. Separate cytoplasmic and chloroplastic ribosomal peaks have not been isolated. The latest report gives them both a value of about 70s, with 50s and 30s subunits (Janowski *et al.* 1969). It is possible that the chloroplast ribosomes so outnumber the cytoplasmic ribosomes that the latter are not detectable, but partial dissociation is the more likely answer (cf. early work on *Euglena*). Most workers report only two rRNAs in whole cells, with S rates of 25 and 18s (Woodcock & Bogorad 1970b), 23 and 16s (Schweiger *et al.* 1967), 24 and 15s (Farber 1969), 25 and 16s (Janowski *et al.* 1968) and 24 and 16s (Dillard 1970). This state of affairs is probably due to the presence of at least two soluble ribonucleases (Schweiger 1966) and a ribosome-associated one (Dillard 1970). The kinetics of labelling indicate that both the large and small rRNA peaks are heterogeneous (Dillard & Schweiger 1967, 1968). Recently, Schweiger (1970) has claimed the separation of four rRNA peaks by gel electrophoresis, but they were not completely characterized. Clearly, better techniques will have to be found before the question of messenger RNA is tackled. It would be particularly interesting to know more about the stable 13 to 15s RNA reported by both Dillard (1970) and Janowski and Bonotto (1970).

On a more cheerful note, it is fairly well established that RNA is synthesized in isolated chloroplasts (Schweiger & Berger 1964, Berger 1967) and in chloroplasts of enucleate cells (Shephard 1965b), and that its synthesis is DNA-dependent.

A lot of the conclusions which have been drawn about the role of nucleic acids in *Acetabularia* morphogenesis are based on autoradiographic studies using various more-or-less specific inhibitors. Unfortunately, most of these studies were done with whole or enucleate cells, and the uptake of the drug and its precise action were not determined first. For the older references, readers should consult the papers of de Vitry (1965) and Shephard (1965a,b), and Puiseux-Dao's book (1970). One recent study deserves mention. A low concentration of rifampicin ($10\mu g$ $ml^{-1}$), which specifically inhibits chloroplast DNA-dependent RNA polymerase in *Chlamydomonas* (Surzycki 1969), strongly inhibits the incorporation of $^3$H-uridine into *Acetabularia* chloroplasts, but does not affect nuclear RNA synthesis (Brändle & Zetsche 1971). It has almost no effect on morphogenesis, but inhibits chlorophyll formation after a period of 6 to 12 days in the antibiotic. This points out the existence of a distinct chloroplast RNA polymerase, and suggests that chlorophyll synthesis is dependent on transcription of chloroplast RNA. However, the long lag time before inhibition of chlorophyll synthesis starts suggests that it is indirectly due to a lack of other proteins requiring chloroplast ribosomes for their synthesis. The regeneration of nucleate fragments is slowed down after a period of days (Puiseux-Dao *et al.* 1972), probably for the same reason. Rifampicin at 20 $\mu g$ $ml^{-1}$ has no effect on the circadian rhythm of photosynthesis even though total cell RNA synthesis is inhibited by 65 to 90% within 24 hours (Vanden Driessche *et al.* 1970).

In conclusion, although *Acetabularia* appears to be a model system for studying questions of chloroplast autonomy and cellular differentiation, there are formidable technical problems which have to be overcome before much progress can be made.

## 7  COMPARATIVE ASPECTS

The only area in which any attempt has been made to cover the algae as a whole is in their DNA base composition (Fig. 10.3). Data not cited in the text are from Mandel (1968) and Pakhomova *et al.* (1968), except for the red alga *Griffithsia* (Nasatir & Brooks 1966). The Chlorophyceae, as in other aspects of nucleic acids, are the most studied. The majority of species falls within the range of 50 to 65% GC. Some of those with lower % (G+C), such as *Nitella* and *Acetabularia*, are morphologically quite distinct from the rest of the division. The average base composition is quite different from the lower and narrow 35 to 45% GC range found for higher plants (Biswas & Sarkar 1970, Green 1971) and mosses (Green 1972). This does not mean that land plants were not descended from the Chlorophyceae but that (a) they were descended from a lower %(G+C) ancestor or (b) there has been considerable divergence in opposite directions in the aquatic and terrestrial environment. No generalities can be made about the other algal groups as yet, but in view of the usefulness of this parameter in the systematics of the fungi (Storck & Alexopoulos 1970), it would be profitable for workers familiar with the morphologically based systematics to undertake thorough surveys in the particular groups they are studying.

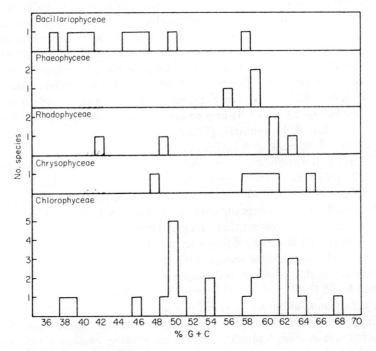

**Fig. 10.3.** DNA base composition distribution of some groups of eukaryotic algae. For references, see text.

Norton and Roth (1967a) found that several multicellular brown, red and green algae had considerable ribonuclease and deoxyribonuclease activity, mostly located in the blade. All algae investigated have DNA repair systems, including the colonial green alga, *Eudorina* (Kemp & Wentworth 1971).

Oddly enough, it appears that the diatom *Cylindrotheca fusiformis* requires silicon for DNA synthesis (Dailey & Volcani 1969). If this is indicative of coupling between cell wall formation and nuclear DNA synthesis, it should be looked into further.

It is quite understandable that a lot of algal biochemistry is concerned more with the chloroplast than with the alga as a whole. Two basic questions are being asked.

(a)  Is the chloroplast autonomous or semi-autonomous?

(b)  Did it evolve from a prokaryotic endosymbiont?

It is now accepted that all chloroplasts contain DNA (for references, see Woodcock & Bogorad 1971). Although this is based mainly on electron microscopy and autoradiography, DNA has been isolated from chloroplasts of a few algae other than those mentioned above. The chloroplast DNA of the red alga *Porphyra tenera* is 37% (G+C) (Ishida *et al.* 1969) and that of *Ochromonas danica*, a member of the Chrysophyceae, is 31% (G+C) (Schiff & Epstein 1965). The meagre data on algal chloroplast DNA base compositions is not adequate to support any theory of evolutionary origin.

*Euglena* appears to be the only alga with circular chloroplast DNA (Manning *et al.* 1971, Manning & Richards 1972a). However, the chloroplast genophore (DNA-containing area) in *Ochromonas danica* (Slankis & Gibbs 1968) and *Sphacelaria* (Bisalputra & Burton 1969) appears to be circular. Although Bisalputra and Burton (1970) claim that the DNA is in the form of 14μm circles attached to the inner membrane, their evidence is based on a limited number of molecules. Certainly the pattern of replication of this type of DNA needs to be investigated.

Chloroplast ribosomes from green algae and *Euglena* appear to be of the 70s prokaryotic type with RNA of about 16 and 23s, or approximately 0·56 and 1·07 × 10⁶ daltons (Loening 1968). The large chloroplast rRNAs of two red algae, *Porphyridium* and *Griffithsia*, appear to be slightly smaller (Howland & Ramus 1971). The cytoplasmic rRNAs do vary between divisions, with those of *Euglena* being heavier than those of members of the Chlorophyceae (Loening 1968), which are heavier than those of members of the Rhodophyceae (Howland & Ramus 1971). In spite of the conservatism of rRNAs as measured by DNA/RNA hybridization (Sinclair & Brown 1971), there has been considerable evolution in their size.

There are basically two types of biological research: the intensive 'take-it-apart-and-see-how-it-works' of the molecular biologist, and the comparative approach of the organismic biologist. There are many problems left for both. It would be profitable for some molecular biologists to study algae outside the Chlorophyceae. They might find some of these organisms more amenable to experimental manipulation as well as giving useful comparative data. However, in extending molecular studies into the comparative domain, it is important to ask meaningful questions which will truly illuminate the diversity of nature. Simply verifying the Central Dogma for another group of organisms is not enough. The important problems are those concerned with cellular and multicellular differentiation, and relationships among organelles and the rest of the cell.

## 8 CHLOROPLAST AUTONOMY AND EVOLUTIONARY ORIGIN

For an organelle to be genetically autonomous, its DNA should contain the information for the DNA and RNA polymerases, ribosomal RNA and ribosomal proteins, and all the proteins of the protein synthesizing and photosynthetic machinery. This is clearly not the case in *Chlamydomonas* (Armstrong *et al.* 1971, Surzycki 1969). It is probably not the case in *Acetabularia* (Apel & Schweiger 1972).

The symbiotic theory of chloroplast origin (Mereschkowsky 1905) has received recent support from the similarities between prokaryotic and chloroplast RNA and protein-synthesizing systems, and from studies of a number of modern-day cyanophycean symbionts (Raven 1970). The provocative book on the origin of eukaryotic cells by Margulis (1970), and an interesting symposium (Charles & Knight 1970) give recent evidence in support of this theory. However, it should be remembered that the modern chloroplast is completely integrated with the rest of its host cell, and appears to have much less genetic information than any known free-living prokaryote.

## 9 REFERENCES

APEL K. & SCHWEIGER H.G. (1972) Nuclear dependency of chloroplast proteins in *Acetabularia*. *Eur. J. Biochem.* **25**, 229–38.

ARMSTRONG J.J., SURZYCKI S.J., MOLL B. & LEVINE R.P. (1971) Genetic transcription and translation specifying chloroplast components in *Chlamydomonas reinhardi*. *Biochemistry, N.Y.* **10**, 692–701.

ASATO Y. & FOLSOME C.E. (1970) Temporal genetic mapping of the blue-green alga, *Anacystis nidulans*. *Genetics*, **65**, 407–19.

AVADHANI N.G. & BUETOW D.E. (1972) Studies on messenger RNA from *Euglena gracilis*. I. Techniques of saline fractionation and characterization of messenger RNAs from dividing cells. *Biochim. biophys. Acta*, **262**, 488–501.

AVADHANI N.G., LYNCH M.J. & BUETOW D.E. (1971) Protein synthesis on polysomes in mitochondria isolated from *Euglena gracilis*. *Expl. Cell Res.* **69**, 226–9.

AVADHANI N.G., ROE D.H. & BUETOW D.E. (1970) Isolation of a DNA-RNA complex from *Euglena gracilis*. *FEBS Letters*, **7**, 217–19.

BALTUS E. & BRACHET J. (1963) Presence of DNA in the chloroplasts of *Acetabularia mediterranea*. *Biochim. biophys. Acta*, **76**, 490–2.

BARNETT W.E., PENNINGTON C.J. & FAIRFIELD S.A. (1969) Induction of *Euglena* transfer RNA's by light. *Proc. natn. Acad. Sci. U.S.A.* **63**, 1261–8.

BASTIA D., CHIANG K.S. & SWIFT H. (1969) Chloroplast dedifferentiation and redifferentiation during zygote maturation and germination in *Chlamydomonas reinhardi*. *J. Cell Biol.* **43**, 11a.

BASTIA D., CHIANG K.S., SWIFT H. & SIERSMA P. (1971) Heterogeneity, complexity and repetition of the chloroplast DNA of *Chlamydomonas reinhardi*. *Proc. natn. Acad. Sci. U.S.A.* **68**, 1157–61.

BAZIN M.J. (1968) Sexuality in a blue-green alga: Genetic recombination in *Anacystis nidulans*. *Nature, Lond.* **218**, 282–3.

BEIDERBECK R. & RICHTER G. (1968) On the nature of DNA-associated RNA in *Chlorella* cells. *Biochim. biophys. Acta,* **157,** 658–60.

BERGER S. (1967) RNA synthesis in *Acetabularia.* II. RNA synthesis in isolated chloroplasts. *Protoplasma,* **64,** 11–25.

BERGER S. & SCHWEIGER H.G. (1969) Synthesis of chloroplast DNA in *Acetabularia. Physiol. Chem. Phys.* **1,** 280–92.

BISALPUTRA T. & BURTON H. (1969) Ultrastructure of chloroplast of a brown alga *Sphacelaria* sp. *J. Ultrastruct. Res.* **29,** 224–35.

BISALPUTRA T. & BURTON H. (1970) On the chloroplast DNA-membrane complex in *Sphacelaria* sp. *J. Microscopie,* **9,** 661–6.

BISWAS S.B. & SARKAR A.K. (1970) Deoxyribonucleic acid base composition of some angiosperms and its taxonomic significance. *Phytochem.* **9,** 2425–30.

BOURQUE D.P., BOYNTON J.E. & GILLHAM N.W. (1971) Studies on the structure and cellular location of various ribosome and rRNA species in the green alga *Chlamydomonas reinhardi. J. Cell Sci.* **8,** 153–83.

BOYNTON J.E., GILLHAM N.W. & BURKHOLDER B. (1970) Mutations altering chloroplast ribosome phenotype in *Chlamydomonas.* II. A new Mendelian mutation. *Proc. natn. Acad. Sci. U.S.A.* **67,** 1404–512.

BRACHET J. & BONOTTO S. (eds.) (1970) *The Biology of Acetabularia.* Academic Press, New York & London.

BRANDLE E. & ZETSCHE K. (1971) Die Wirkung von Rifampicin auf die RNA- und Proteinsynthese sowie die Morphogenese und den Chlorophyllgehalt kernhaltiger und kernloser *Acetabularia*-Zellen. *Planta,* **99,** 46–55.

BRAWERMAN G. & EISENSTADT J.M. (1964a) Deoxyribonucleic acid from the chloroplasts of *Euglena gracilis. Biochim. biophys. Acta,* **91,** 477–85.

BRAWERMAN G. & EISENSTADT J.M. (1964b) Template and ribosomal RNA's associated with the chloroplasts and cytoplasm of *Euglena gracilis. J. molec. Biol.* **10,** 403–17.

BROWN R.D., BASTIA D. & HASELKORN R. (1970) Effect of rifampicin on transcription in chloroplasts of *Euglena.* In *RNA Polymerase and Transcription,* ed. Silvestri L. pp. 309–28. North-Holland, Amsterdam.

BROWN R.D. & HASELKORN R. (1971a) Chloroplast RNA populations in dark-grown, light-grown and greening *Euglena gracilis. Proc. natn. Acad. Sci. U.S.A.* **68,** 2536–9.

BROWN R.D. & HASELKORN R. (1971b) Synthesis and maturation of cytoplasmic rRNA in *Euglena gracilis. J. molec. Biol.* **59,** 491–503.

BROWN R.D. & HASELKORN R. (1972) Isolation of *Euglena gracilis* chloroplasts uncontaminated by nuclear DNA. *Biochim. biophys. Acta,* **259,** 1–4.

BUETOW D.E. (ed.) (1968a) *The Biology of Euglena,* Vols. I and II. Academic Press, New York & London.

BUETOW D.E. (1968b) *Euglena*-cells for biological investigation. in *The Biology of Euglena,* ed. Buetow D.E. pp. 389–90. Academic Press, New York & London.

CAIRNS J. (1963) The chromosome of *E. coli. Cold Spring Harbor Symp. quant. Biol.* **28,** 43–6.

CAPESIUS I. & RICHTER G. (1967a) Enzyme der polynucleotid synthese in pflanzlichen organismen: I. Isolierung, reinigung und charakterisierung der polynucleotid-phosphorylase aus der blaualge *Anacystis. Z. Naturf.* **22b,** 204–15.

CAPESIUS I. & RICHTER G. (1967b) II. Isolierung, reinigung und charakterisierung der RNA-polymerase aus der blaualge *Anacystis. Z. Naturf.* **22b,** 876–85.

CHARLES H.P. & KNIGHT B.C.J.G. (eds.) (1970) *Organization and Control in Prokaryotic and Eukaryotic Cells. 20th Symp. Soc. Gen. Microbiol.* Cambridge University Press, London.

CARELL E.F. (1969) Studies on chloroplast development and replication in *Euglena.* I. Vitamin $B_{12}$ and chloroplast replication. *J. Cell. Biol.* **41,** 431–9.

CARELL E.F., EGAN J.E. & PRATT E.A. (1970) Studies on chloroplast development and replication in *Euglena.* II. Identification of two different deoxyribonucleases. *Archs. Bioch. Biophys.* **138,** 26–31.

CHIANG K.S. (1968) Physical conservation of parental cytoplasmic DNA through meiosis in *Chlamydomonas reinhardi*. *Proc. natn. Acad. Sci. U.S.A.* **60**, 194–200.

CHIANG K.S. (1971) Replication, transmission and recombination of cytoplasmic DNA's in *Chlamydomonas reinhardi*. In *Autonomy and Biogenesis of Mitochondria and Chloroplasts*, eds. Boardman N.K., Linnane A.W. & Smillie R.W. pp. 235–49. North-Holland, Amsterdam.

CHIANG K.S. & SUEOKA N. (1967a) Replication of chloroplast DNA in *Chlamydomonas reinhardi* during the vegetative cell cycle: Its mode and regulation. *Proc. natn. Acad. Sci. U.S.A.* **57**, 1506–13.

CHIANG K.S. & SUEOKA A. (1967b) Replication of chromosomal and cytoplasmic DNA during mitosis and meiosis in the eucaryote *Chlamydomonas reinhardi*. *J. Cell. Physiol.* **70**, Suppl. 1, 89.

CHUN E.H.L., VAUGHAN M.H. & RICH A. (1963) Isolation and characterization of DNA associated with chloroplast preparations. *J. molec. Biol.* **7**, 130–41.

COOK J.R. (1966) The synthesis of cytoplasmic DNA in synchronized *Euglena*. *J. Cell. Biol.* **29**, 369–73.

COOK J.R. (1968) The cultivation and growth of *Euglena*. In *The Biology of Euglena*, ed. Buetow D.E. pp. 244–314. Academic Press, New York & London.

CRAIG I.W. & CARR N.G. (1968) Ribosomes from the blue-green alga *Anabaena variabilis*. *Arch. Mikrobiol.* **62**, 167–77.

CRAIG I.W., LEACH C.K. & CARR N.G. (1969) Studies with DNA from blue-green algae. *Arch. Mikrobiol.* **65**, 218–27.

DAILEY W.M. & VOLCANI B.E. (1969) A silicon requirement for DNA synthesis in the diatom *Cylindrotheca fusiliformis*. *Expl. Cell Res.* **58**, 334–42.

DAVIS E.A. & EPSTEIN H.T. (1971) Some factors controlling stepwise variation of organelle number in *Euglena gracilis*. *Expl. Cell Res.* **65**, 273–80.

DILLARD W.L. (1970) RNA synthesis in *Acetabularia*. In *Biology of Acetabularia*, eds. Brachet J. & Bonotto S. pp. 13–15. Academic Press, New York & London.

DILLARD W.L. & SCHWEIGER H.G. (1967) Kinetics of RNA synthesis in *Acetabularia*. *Biochim. biophys. Acta*, **169**, 561–3.

DILLARD W.L. & SCHWEIGER H.G. (1968) RNA synthesis in *Acetabularia*. III. The kinetics of RNA synthesis in nucleate and enucleate cells. *Protoplasma*, **67**, 87–100.

EDELMAN M., COWAN C.A., EPSTEIN H.T. & SCHIFF J.A. (1964) Studies of chloroplast development in *Euglena*. VIII. Chloroplast-associated DNA. *Proc. natn. Acad. Sci. U.S.A.* **52**, 1214–19.

EDELMAN M., SCHIFF J.A. & EPSTEIN H.T. (1965) Studies of chloroplast development in *Euglena*. XII. Two types of satellite DNA. *J. molec. Biol.* **11**, 769–74.

EDELMAN M., SWINTON D., SCHIFF J.A., EPSTEIN H.T. & ZELDIN B. (1967) DNA of the blue-green algae (Cyanophyta). *Bact. Revs.* **31**, 315–31.

ENÖCKL F. (1968) Variation of tRNA content of *Chlorella pyrenoidosa* during synchronous development. *Z. Pfl. Physiol.* **58**, 241–7.

EYTAN G. & OHAD I. (1970) Cooperation between cytoplasmic and chloroplast ribosomes in synthesis of photosynthetic lamellar proteins during the greening process in a mutant of *Chlamydomonas reinhardi*. *J. biol. Chem.* **245**, 4297–307.

FARBER F.E. (1969) RNA metabolism in *Acetabularia mediterranea*. *Biochim. biophys. Acta*, **174**, 1–11.

FELLIG J. & WILEY C.E. (1960) Ribonuclease of *Euglena gracilis*. *Science, N.Y.* **132**, 1835–6.

GALLING G. & SSYMANK V. (1970) Bevorzugter einbau markerten uridins in die vorlaufer von chloroplasten-ribosomen in algenzellen. *Planta*, **94**, 203–7 (1970).

GENSE-VIMON M.T. (1971) Occurrence of the G + C rich satellite DNA in unicellular algae. *Biochem. biophys. Res. Commun.* **42**, 347–51.

GIBOR A. (1967) DNA synthesis in chloroplasts. In *Biochemistry of Chloroplasts*, Vol, II, ed. Goodwin T.W. pp. 321–8. Academic Press, New York & London.

GIBOR A. (1969) Effect of ultraviolet irradiation on DNA metabolism of *Euglena gracilis. J. Protozool.* **16**, 190–3.

GIBOR A. & IZAWA M. (1963) DNA content of chloroplasts of *Acetabularia. Proc. natn. Acad. Sci. U.S.A.* **50**, 1164–9.

GILLHAM N.W., BOYNTON J.E. & BURKHOLDER B. (1970) Mutations altering chloroplast ribosome phenotype in *Chlamydomonas.* I. Non-Mendelian mutations. *Proc. natn. Acad. Sci. U.S.A.* **67**, 1026–33.

GNANAM A. & KAHN J.S. (1967) Biochemical studies on the induction of chloroplast development in *Euglena.* I. Nucleic acid metabolism during induction. *Biochim. biophys. Acta,* **142**, 475–85.

GOODENOUGH U.W. (1971) Effects of inhibitors of RNA and protein synthesis on chloroplast structure and function in wild-type *Chlamydomonas reinhardi. J. Cell. Biol.* **50**, 35–60.

GRANICK S. & GIBOR A. (1967) DNA of chloroplasts, mitochondria and centrioles. *Prog. Nuc. Acid Res. Mol. Biol.* **6**, 143–86.

GREEN B.R. (1971) Isolation and base composition of DNA's of primitive land plants. I. Ferns and Fern-allies. *Biochim. biophys. Acta,* **254**, 402–6.

GREEN B.R. (1972) Isolation and base composition of DNA's of primitive land plants. II. Mosses. *Biochim. biophys. Acta,* **277**, 29–34.

GREEN B.R. & BURTON H. (1970) *Acetabularia* chloroplast DNA: Electron microscopic visualization. *Science, N.Y.* **168**, 981–2.

GREEN B.R., HEILPORN V., LIMBOSCH S., BOLOUKHERE M. & BRACHET J. (1967) The cytoplasmic DNA's of *Acetabularia mediterranea. Proc. natn. Acad. Sci. U.S.A.* **58**, 1351–8.

GREEN B.R., BURTON H., HEILPORN V. & LIMBOSCH S. (1970) The cytoplasmic DNA's of *Acetabularia mediterranea*: Their structure and biological properties. In *The Biology of Acetabularia,* eds. Brachet J. & Bonotto S. pp. 35–60. Academic Press, New York & London.

HAMMERLING J. (1953) Nucleocytoplasmic relationships in the development of *Acetabularia. Int. Rev. Cytol.* **2**, 475–98.

HAMMERLING J. (1963) Nucleocytoplasmic interactions in *Acetabularia* and other cells. *A. Rev. Pl. Physiol.* **14**, 65–92.

HARRIS E.H. & EISENSTADT J. (1971) Initiation of polysome formation in chloroplasts isolated from *Euglena gracilis. Biochim. biophys. Acta,* **232**, 167–70.

HAYASHI F., ISHIDA M. & KIKUCHI T. (1969) Macromolecular synthesis in a blue-green alga, *Anacystis nidulans,* in dark and light phases. *Ann. rep. Res. Reactor Inst. Kyoto Univ.* **2**, 56–66.

HEILPORN-POHL V. & BRACHET J. (1966) Net DNA synthesis in anucleate fragments of *Acetabularia mediterranea. Biochim. biophys. Acta,* **119**, 429–31.

HEILPORN V. & LIMBOSCH S. (1971) Recherches sur les acides desoxyribonucleiques d'*Acetabularia mediterranea. Eur. J. Biochem.* **22**, 573–9.

HEIM K. VON DER & ZILLIG W. (1967) Reinigung und Eigenschaften der DNA-abhängigen RNA-Polymerase aus *Anacystis nidulans. Hoppe-Seyler's Z. Physiol. Chem.* **348**, 902–12.

HEIM K. VON DER & ZILLIG W. (1969) Differential transcription of T-4 DNA by DNA-dependent RNA-polymerase of *E. coli* and *Anacystis nidulans. FEBS Letters,* **3**, 76–9.

HEIZMANN P. (1970) Propriétés des ribosomes et des RNA ribosomiques d'*Euglena gracilis. Biochim. biophys. Acta,* **224**, 144–54.

HERDMANN, M. FAULKNER B.M. & CARR N.G. (1970) Synchronous growth and genome replication in the blue-green alga *Anacystis nidulans. Arch. Mikrobiol.* **73**, 238–49.

HERRMANN R.G. & KOWALLIK K.V. (1970) Multiple amounts of DNA related to size of chloroplasts. II. Comparison of electron microscopic and autoradiographic data. *Protoplasma,* **69**, 365–72.

HERZFELD F. & ZILLIG W. (1971) Subunit composition of DNA-dependent RNA polymerase of *Anacystis nidulans. Eur. J. Bioch.* **24**, 242–8.

HOOBER J.K. & BLOBEL G. (1969) Characterization of the chloroplastic and cytoplasmic ribosomes of *Chlamydomonas reinhardi. J. molec. Biol.* **41**, 121–38.

HOWLAND G.P. & RAMUS J. (1971) Analysis of blue-green and red algal ribosomal RNA's by gel electrophoresis. *Arch. Mikrobiol.* **76**, 292–8.

ISHIDA M.R., KIKUCHI T., MATSUBARA T., HAYASHI F. & YOKOMURA E. (1969) Characterization of the satellite DNA from the cells of *Porphyra tenera. Ann. Rep. Res. Reactor Inst. Kyoto Univ.* **2**, 73–5.

IWAMURA T. (1966) Nucleic acids in chloroplasts and metabolic DNA. *Prog. Nucl. Acid. Res. Mol. Biol.* **5**, 133–55.

IWAMURA T. & KUWASHIMA S. (1969) Two DNA species in chloroplasts of *Chlorella. Biochim. biophys. Acta,* **174**, 330–9.

JANOWSKI M. & BONOTTO S. (1970) A stable RNA species in *Acetabularia mediterranea.* In *The Biology of Acetabularia,* eds. Brachet J. & Bonotto S. pp. 17–34. Academic Press, New York & London.

JANOWSKI M., BONOTTO S. & BRACHET J. (1968) Cinétique de l'incorporation d'uridine-H³ dans les RNA d'*Acetabularia mediterranea. Archs. int. Physiol. Biochim.* **76**, 934–5.

JANOWSKI M., BONOTTO S. & BOLOUKHERE M. (1969) Ribosomes of *Acetabularia mediterranea. Biochim. biophys. Acta,* **174**, 525–35.

JONES R.F. & PENG W. (1969) Changes in arginyl-tRNA and arginine synthetase during growth and gametic differentiation in *Chlamydomonas. Pl. Physiol. Lancaster,* **44**, Abstr. 38.

JONES R.F., KATES J.R. & KELLER S.J. (1968) Protein turnover and macromolecular synthesis during growth and gametic differentiation in *Chlamydomonas reinhardtii. Biochim. biophys. Acta,* **157**, 589–98.

KATES J.R., CHIANG K.S. & JONES R.F. (1968) Studies on DNA replication during synchronized vegetative growth and gametic differentiation in *Chlamydomonas reinhardtii. Expl. Cell Res.* **49**, 121–35.

KEMP C.L. & WENTWORTH J.W. (1971) Ultraviolet radiation studies on the colonial alga *Eudorina elegans. Can. J. Microbiol.* **17**, 1417–24.

KIERAS F.J. & CHIANG K.S. (1971) Characterization of DNA components from some colorless algae. *Expl. Cell Res.* **64**, 89–96.

KRAWIEC S. & EISENSTADT J.M. (1970) RNA's from mitochondria of bleached *Euglena gracilis* Z. II. Characterization of highly polymeric RNA's. *Biochim. biophys. Acta,* **217**, 132–41.

KUNG S.D., MOSCARELLO M.A., WILLIAMS J.P. & NADLER H. (1971) Renaturation of DNA from the blue-green alga *Lyngbya* sp. *Biochim. biophys. Acta,* **232**, 252–4.

LEE S.G. & EVANS W.R. (1971) Hybrid ribosome formation from *E. coli* and chloroplast ribosome subunits. *Science, N.Y.* **173**, 241–2.

LEFF J., MANDEL M., EPSTEIN H.T. & SCHIFF J.A. (1963) DNA satellites from cells of green and aplastidic algae. *Biochem. biophys. Res. Commun.* **13**, 125–30.

LEVINE R.P. & GOODENOUGH U.W. (1970) The genetics of photosynthesis and of the chloroplast in *Chlamydomonas reinhardi. Ann. Rev. Gen.* **4**, 397–408.

LOENING U.E. (1968) Molecular weights of ribosomal RNA in relation to evolution. *J. molec. Biol.* **38**, 355–65.

LYMAN H., EPSTEIN H.T. & SCHIFF J.A. (1961) Studies of chloroplast development in *Euglena.* Inactivation of green colony formation by ultraviolet light. *Biochim. biophys. Acta,* **50**, 301–9.

MANDEL M. (1968) In *Handbook of Biochemistry,* ed. Sober H.A. p. H27. Chemical Rubber Co., Cleveland.

MANDEL M. (1969) New approaches to bacterial taxonomy. *A. Rev. Microbiol.* **23**, 239–74.

MANNING J.E., WOLSTENHOLME D.R., RYAN R.S., HUNTER J.A. & RICHARDS O.C. (1971) Circular chloroplast DNA from *Euglena gracilis*. *Proc. natn. Acad. Sci. U.S.A.* **68**, 1169–73.

MANNING J.E. & RICHARDS O.C. (1972a). Isolation and molecular weight of circular chloroplast DNA from *Euglena gracilis*. *Biochim. biophys. Acta*, **259**, 285–96.

MANNING J.E. & RICHARDS O.C. (1972b) Synthesis and turnover of *Euglena gracilis* nuclear and chloroplast DNA. *Biochemistry, N.Y.* **11**, 2036–43.

MARGULIS L. (1970) *Origin of Eukaryotic Cells*. Yale Univ. Press, New Haven.

MARMUR J., FALKOW S. & MANDEL M. (1963) New approaches to bacterial taxonomy. *A. Rev. Microbiol.* **17**, 329–72.

MERESCHKOWSKY C. (1905) Über Natur und Ursprung der Chromatophoren in Pflanzenreiche. *Biol. Zbl.* **25**, 593–604.

MIHARA S. & HASE E. (1969) A note on ribosomes in cells of *Chlorella protothecoides*. *Pl. Cell Physiol., Tokyo*, **10**, 465–70.

MUNNS R., SCOTT N.S. & SMILLIE R.M. (1972) RNA synthesis during chloroplast development in *Euglena gracilis*. *Phytochem.* **11**, 45–52.

NASATIR M. & BROOKS A.E. (1966) DNA of the marine red alga *Griffithsia globulifera*. *J. Phycol.* **2**, 144–7.

NORTON J. & ROT J.S. (1967a) Some aspects of nuclease activity in *Anacystis*. *Comp. Biochem. F........l.* **23**, 361–71.

NORTON J. .....H J.S. (1967b) A ribonuclease specific for 2'-O-methylated RNA. *J. biol. Chem.......,* 2029–34.

OHAD I., SIEKEVITZ P. & PALADE G.E. (1967) Biogenesis of chloroplast membranes. I. Plastid dedifferentiation in a dark-grown algal mutant. *J. Cell. Biol.* **35**, 521–52.

OSHIO Y. & HASE E. (1968) Studies on nucleic acids in chloroplasts isolated from *Chlorella protothecoides*. *Pl. Cell Physiol., Tokyo*, **9**, 69–85.

PAKHOMOVA M.V., ZAJCEVA G.N. & BELOZERSKIJ A.N. (1968) Presence of 5-methylcytosine and 6-methylaminopurine in the DNA of some algae. *Dokl. Akad. Nauk. SSSR.* **182**, 712–14.

PERL M. (1971) Appearance of a methylated RNA on illumination of *Euglena gracilis* cells grown in the dark. *Biochem. J.* **125**, 401–5.

PIGGOTT G.H. & CARR N.G. (1971) Assimilation of nucleic acid precursors by intact cells and protoplasts of the blue-green alga *Anacystis nidulans*. *Arch. Microbiol.* **79**, 1–6.

PIGGOTT G.H. & CARR N.G. (1972) Homology between nucleic acids of blue-green algae and chloroplasts of *Euglena gracilis*. *Science, N.Y.* **175**, 1259–61.

PRASK J.A. & PLOCKE D.J. (1971) A role for zinc in the structural integrity of the cytoplasmic ribosomes of *Euglena gracilis*. *Pl. Physiol., Lancaster*, **48**, 150–5.

PRINGSHEIM F. & PRINGSHEIM O. (1952) Experimental elimination of chromotaphores and eye-spot in *Euglena gracilis*. *New Phytol.* **51**, 65–76.

PROVASOLI L., HUTNER S.H. & SCHATZ A. (1941) Streptomycin-induced chlorophyll-less races of *Euglena*. *Proc. Soc. exp. Biol. Med.* **69**, 279–82.

PUISEUX-DAO S. (1970) *Acetabularia and Cell Biology*. Springer, New York.

PUISEUX-DAO S., AKSIYOTE-BENBASSET J. & BONOTTO S. (1972) Effets biologiques de la rifampicine chez l'*Acetabularia mediterranea*. *C. r. hebd. Séance Acad. Sci. Paris*, **274**, 1678–81.

RAVEN P.H. (1970) A multiple origin for plastids and mitochondria. *Science*, **169**, 641–6.

RAWSON J.R. & STUTZ E. (1968) Characterization of *Euglena* cytoplasmic ribosomes and ribosomal RNA by zone velocity sedimentation in sucrose gradients. *J. molec. Biol.* **33**, 309–14.

RAWSON J.R. & STUTZ E. (1969) Isolation and characterization of *Euglena gracilis* cytoplasmic and chloroplast ribosomes and their rRNA components. *Biochim. biophys. Acta*, **190**, 368–80.

L

RAWSON J.R., CROUSE E.J. & STUTZ E. (1971) The integrity of the 25-S ribosomal RNA from *Euglena gracilis* 87-S ribosomes. *Biochim. biophys. Acta*, **246**, 507–16.

RAY D.S. & HANAWALT P.C. (1964) Properties of the satellite DNA associated with the chloroplasts of *Euglena gracilis*. *J. molec. Biol.* **9**, 812–24.

RAY D.S. & HANAWALT P.C. (1965) Satellite DNA components in *Euglena gracilis* cells which lack chloroplasts. *J. molec. Biol.* **11**, 760–8.

REGER B.J., FAIRFIELD S.A., EPLER J.L. & BARNETT W.E. (1970) Identification and origin of some chloroplast aminoacyl-tRNA synthetases and tRNAs. *Proc. natn. Acad. Sci. U.S.A.* **67**, 1207–13.

RICHARDS O.C. (1967) Hybridization of *Euglena gracilis* chloroplast and nuclear DNA. *Proc. natn. Acad. Sci. U.S.A.* **57**, 156–63.

RICHARDS O.C., RYAN R.S. & MANNING J.E. (1971) Effects of cycloheximide and chloramphenicol on DNA synthesis in *Euglena gracilis*. *Biochim. biophys. Acta*, **238**, 190–201.

RICHTER G. & SENGER H. (1965) Isolation of a DNA-RNA complex from *Chlorella* cells. *Biochim. biophys. Acta*, **95**, 362–4.

RODE A. & BAYEN M. (1972) Action de la 8-azaguanine sur la division cellulaire de *Chlorella pyrenoidosa*. I. Inhibition sélective de la synthèse de l'ADN nucléaire. *Planta*, **102**, 237–46.

RODRIGUEZ-LOPEZ M. & VASQUEZ D. (1968) Comparative studies on cytoplasmic ribosomes from algae. *Life Sci.* **7**, 327–36.

RODRIGUEZ-LOPEZ M., MUNOZ M.L. & VASQUEZ D. (1970) The effects of the rifamycin antibiotics on algae. *FEBS Letters*, **9**, 171–4.

SAGER R. & HAMILTON M.G. (1967) Cytoplasmic and chloroplast ribosomes of *Chlamydomonas*: Ultracentrifugal characterization. *Science, N.Y.* **157**, 709–11.

SAGER R. & ISHIDA M.R. (1963) Chloroplast DNA in *Chlamydomonas*. *Proc. natn. Acad. Sci. U.S.A.* **50**, 725–30.

SAGER R. & LANE D. (1969) Replication of chloroplast DNA in zygotes of *Chlamydomonas*. *Fed. Proc.* **28**, 347.

SAGER R. & RAMANIS Z. (1968) The pattern of segregation of cytoplasmic genes in *Chlamydomonas*. *Proc. natn. Acad. Sci. U.S.A.* **61**, 324–31.

SAGER R. & RAMANIS Z. (1970a) A genetic map of non-Mendelian genes in *Chlamydomonas*. *Proc. natn. Acad. Sci. U.S.A.* **65**, 593–600.

SAGER R. & RAMINIS Z. (1970b) Genetic studies of chloroplast DNA in *Chlamydomonas*. In *Control of Organelle Development*, ed. Miller P.L. *XXIV Symp. S.E.B.* pp. 401–17.

SAITO N. & WERBIN H. (1970) Purification of a blue-green algal DNA photoreactivating enzyme: An enzyme requiring light as a physical cofactor to perform its catalytic function. *Biochemistry, N.Y.* **9**, 2610–20.

SCAGEL R.F., BANDONI R.J., ROUSE G.E., SCHOFIELD W.B., STEIN J.R. & TAYLOR T.M.C. (1965) *An evolutionary survey of the plant kingdom*. Wadsworth, Belmont.

SCHIFF J.A. (1971) Developmental interactions among cellular compartments in *Euglena*. In *Autonomy and Biogenesis of Mitochondria and Chloroplasts*, eds. Boardman N.K., Linnane A.W. & Smillie R.W. pp. 98–118. North-Holland, Amsterdam.

SCHIFF J.A. & EPSTEIN H.T. (1965) The continuity of the chloroplast in *Euglena*. In *Reproduction: Molecular, Subcellular and Cellular*, ed. Locke M. pp. 131–89. Academic Press, New York & London.

SCHIMMER O. & ARNOLD C.G. (1970a) Untersuchungen uber reversions-und segregationsverhalten ein ausserkaryotischen gens von *Chlamydomonas reinhardii* zur bestimmung des erbtragers. *Mol. Gen. Gen.* **107**, 281–90.

SCHIMMER O. & ARNOLD C.G. (1970b) Uber die zahl der kopien eines ausserkaryotischen gens bei *Chlamydomonas reinhardii*. *Mol. Gen. Gen.* **107**, 366–71.

SCHÖNKEN O.J. & WANKA F. (1971) An investigation of DNA polymerase in synchronously growing *Chlorella* cells. *Biochim. biophys. Acta*, **232**, 83–93.

SCHÖNKEN O.J., WANKA F. & KUYPEI C.M.A. (1970) Periodic change of deoxyribonuclease activity in synchronous culture of *Chlorella*. *Biochem. biophys. Acta*, **224**, 74–9.

SCHUIT K.E., AVADHANI N.G. & BUETOW D.E. (1970) Analysis of the RNA of *Euglena gracilis* on polyacrylamide gels. *Arch. Mikrobiol.* **71**, 79–88.

SCHWEIGER H.G. (1966) Ribonuclease-activitat in *Acetabularia. Planta*, **68**, 247–57.

SCHWEIGER H.G. (1969) Cell biology of *Acetabularia. Current Topics Microb. Imm.* **50**, 1–36.

SCHWEIGER H.G. (1970) Synthesis of RNA in *Acetabularia*. In *Control of Organelle Development*, ed. Miller P.L. *XXIV Symp. S.E.B.* pp. 327–44.

SCHWEIGER H.G. & BERGER S. (1964) DNA-dependent RNA synthesis in chloroplasts of *Acetabularia. Biochim. biophys. Acta*, **87**, 533–5.

SCHWEIGER H.G., DILLARD W.L., GIBOR A. & BERGER S. (1967) RNA synthesis in *Acetabularia*. I. RNA synthesis in enucleate cells. *Protoplasma*, **64**, 1–12.

SCOTT N.S. & SMILLIE R.M. (1967) Evidence for the direction of chloroplast ribosomal RNA synthesis by chloroplast DNA. *Biochem. biophys. Res. Commun.* **28**, 598–603.

SCOTT N.S. & SMILLIE R.M. (1969) Ribosomal RNA in chloroplasts of *E. gracilis. Curr. Mod. Biol.* **2**, 339–42.

SCOTT N.S., SHAH V.C. & SMILLIE R.M. (1968) Synthesis of chloroplast DNA in isolated chloroplasts. *J. Cell. Biol.* **38**, 151–7.

SCOTT N.S., MUNNS R. & SMILLIE R.M. (1970) Chloroplast and cytoplasmic ribosomes in *Euglena gracilis. FEBS Letters*, **10**, 149–52.

SCOTT N.S., MUNNS R., GRAHAM D. & SMILLIE R.M. (1971) Origin and synthesis of chloroplast ribosomal RNA and photoregulation during chloroplast biogenesis. In *Autonomy and Biogenesis of Mitochondria and Chloroplasts*, eds. Boardman N.K., Linnane A.W. & Smillie R.M. pp. 383–92. North-Holland, Amsterdam.

SHEN S.R.C. & SCHMIDT R.R. (1966) Enzymic control of nucleic acid synthesis during synchronous growth of *Chlorella pyrenoidosa*. II. Deoxycytidine monophosphate deaminase. *Archs. Biochem. Biophys.* 13–20.

SHEPHARD D. (1965a) Chloroplast multiplication and growth in the unicellular alga *Acetabularia mediterranea. Expl. Cell Res.* **37**, 93–110.

SHEPHARD D. (1965b) An autoradiographic comparison of the effects of enucleation and actinomycin D on the incorporation of nucleic acid and protein precursors by *Acetabularia* chloroplasts. *Biochim. biophys. Acta*, **108**, 635–43.

SHESTAKOV S.B. & KHYEN N.T. (1970) Evidence for genetic transformation in the blue-green alga *Anacystis nidulans. Mol. Gen. Gen.* **107**, 372–5.

SIERSMA P.W. & CHIANG K.S. (1971) Conservation and degradation of cytoplasmic and chloroplast ribosomes in *Chlamydomonas reinhardtii. J. molec. Biol.* **58**, 167–85.

SINCLAIR J.H. & BROWN D.D. (1971) Retention of common nucleotide sequences in the ribosomal DNA of eukaryotes and some of their physical characteristics. *Biochemistry, N.Y.* **10**, 2761–9.

SLANKIS T. & GIBBS S.P. (1968) Localization of chloroplast DNA in a single peripheral ring-shaped nucleoid in *Ochromonas danica. J. Cell. Biol.* **39**, 126a.

SMILLIE R.M., GRAHAM D., DWYER M.R., GRIEVE A. & TOBIN N.F. (1967) Evidence for the synthesis *in vivo* of proteins of the Calvin cycle and of the photosynthetic electron transfer pathway on chloroplast ribosomes. *Biochem. biophys. Res. Commun.* **28**, 604–10.

SOKAWA Y. & HASE E. (1968) Suppressive effect of light on the formation of DNA and on the increase of deoxythymidine monophosphate kinase activity in *Chlorella protothecoides. Pl. Cell Physiol., Tokyo*, **9**, 461–6.

STANIER R.Y., KUNISAWA R., MANDEL M. & COHEN-BAZIRE G. (1971) Purification and properties of unicellular blue-green algae (Order Chroococcales). *Bact. Rev.* **35**, 171–205.

STENT G.S. & BRENNER S. (1961) A genetic locus for the regulation of ribonucleic acid synthesis. *Proc. natn. Acad. Sci. U.S.A.* **47**, 2005–14.

STORCK R. & ALEXOPOULOS C.J. (1970) Deoxyribonucleic acid of fungi. *Bact. Rev.* **34**, 126–54.

STUTZ E. (1970) The kinetic complexity of *Euglena gracilis* chloroplast DNA. *FEBS Letters*, **8**, 25–8.

STUTZ E. & RAWSON J.R. (1970) Separation and characterization of *Euglena gracilis* chloroplast single-stranded DNA. *Biochim. biophys. Acta*, **209**, 16–23.

STUTZ E. & VANDREY J.R. (1971) Ribosomal DNA satellite of *Euglena gracilis* chloroplast DNA. *FEBS Letters*, **17**, 277–80.

SUEOKA N. (1960) Mitotic replication of DNA in *Chlamydomonas reinhardi*. *Proc. natn. Acad. Sci. U.S.A.* **46**, 83–91.

SUEOKA N., CHIANG K.S. & KATES J.R. (1967) Deoxyribonucleic acid metabolism in meiosis of *Chlamydomonas reinhardi*. I. Isotopic transfer experiments with a strain producing eight zoospores. *J. molec. Biol.* **25**, 47–66.

SURZYCKI S.J. (1969) Genetic functions of the chloroplast of *Chlamydomonas reinhardi*: Effect of rifampin on chloroplast DNA-dependent RNA polymerase. *Proc. natn. Acad. Sci. U.S.A.* **63**, 1327–34.

SURZYCKI S.J. & HASTINGS P.J. (1969) Control of chloroplast RNA synthesis in *Chlamydomonas reinhardi*. *Nature, Lond.* **220**, 786–7.

SURZYCKI S.J. & ROCHAIX J.D. (1971) Transcriptional mapping of rRNA genes of the chloroplast and nucleus of *Chlamydomonas reinhardi*. *J. molec. Biol.* **62**, 89–109.

SURZYCKI S.J., GOODENOUGH U.W., LEVINE R.P. & ARMSTRONG J.J. (1970) Nuclear and chloroplast control of chloroplast structure and function in *Chlamydomonas reinhardi*. In *Control of Organelle Development*, ed. Miller P.L. *XXIV Symp. S.E.B.* pp. 13–37.

TAYLOR M.M. & STORCK R. (1964) Uniqueness of bacterial ribosomes. *Proc. natn. Acad. Sci. U.S.A.* **52**, 958–65.

VAN DEN DRIESSCHE T. (1973) The chloroplasts of *Acetabularia*: The controls of their multiplication and activities. *Subcellular Biochemistry*, **2**, 33–67.

VAN DEN DRIESSCHE T., BONOTTO S. & BRACHET J. (1970) Inability of rifampicin to inhibit circadian rhythmicity in *Acetabularia* despite inhibition of RNA synthesis. *Biochim. biophys. Acta*, **224**, 631–4.

VASCONCELOS A.C.L. & BOGORAD L. (1971) Proteins of cytoplasmic, chloroplast and mitochondrial ribosomes of some plants. *Biochim. biophys. Acta*, **228**, 492–502.

VITRY F. DE (1965) Incorporation de precursors-DNA-RNA-proteines. *Bull. Soc. Chim. Biol.* **47**, 1325–51.

WANKA F. & GERAEDTS (1972) Effect of temperature on the regulation of DNA synthesis in synchronous cultures of *Chlorella*. *Expl. Cell Res.* **71**, 188–92.

WANKA F. & MOORS J. (1970) Selective inhibition by cycloheximide of nuclear DNA synthesis in synchronous cultures of *Chlorella*. *Biochem. biophys. Res. Commun.* **41**, 81–90.

WANKA F., MOORS J. & KRIJZER F.N.C.M. (1972) Dissociation of nuclear DNA replication from concomitant protein synthesis in synchronous cultures of *Chlorella*. *Biochim. biophys. Acta*, **269**, 153–61.

WANKA F. & MULDERS P.F.M. (1967) The effect of light on DNA synthesis and related processes in synchronous cultures of *Chlorella*. *Arch. Mikrobiol.* **58**, 257–69.

WANKA F., PJOOSTEN H.F. & DE GRIP W.J. (1970) Composition and synthesis of DNA in synchronously growing cells of *Chlorella pyrenoidosa*. *Arch. Mikrobiol.* **75**, 25–36.

WANKA F. & POELS C.L.M. (1969) On the problem of an enzymatic regulation of DNA synthesis in *Chlorella*. *Eur. J. Biochem.* **9**, 478–82.

WANKA F. & SCHRAUWEN P.J.A. (1970) Selective inhibition by cycloheximide of rRNA synthesis in *Chlorella*. *Biochim. biophys. Acta*, **254**, 237–40.

WELLS R. & SAGER R. (1971) Denaturation and the renaturation kinetics of chloroplast DNA from *Chlamydomonas reinhardi*. *J. molec. Biol.* **58**, 611–22.

WERZ G. (1965) Determination and realization of morphogenesis in *Acetabularia*. *Brookhaven Symp. Biol.* 18185–201.

WERZ G. & KELLNER G. (1968) Molecular characteristics of chloroplast DNA in *Acetabularia* cells. *J. Ultrastruct. Res.* **24**, 109–15.

WOODCOCK C.F.L. & BOGORAD L. (1970a) Evidence for variation in the quantity of DNA among plastids of *Acetabularia*. *J. Cell. Biol.* **44**, 361–75.

WOODCOCK C.F.L. & BOGORAD L. (1970b) On the extraction and characterization of ribosomal RNA from *Acetabularia*. *Bioch. biophys. Acta*, **224**, 639–43.

WOODCOCK C.F.L. & BOGORAD L. (1971) Nucleic acids and information processing in chloroplasts. In *Structure and Function of Chloroplasts*, ed. Gibbs M. pp. 89–128. Springer, New York.

ZELDIN M.H. & SCHIFF J.A. (1967) RNA metabolism during light-induced chloroplast development in *Euglena*. *Pl. Physiol., Lancaster*, **42**, 922–32.

# CHAPTER 11

# NUCLEAR AND CYTOPLASMIC INHERITANCE IN GREEN ALGAE

## R. SAGER

Department of Biological Sciences,
Hunter College of the City University of New York,
695 Park Avenue, New York, U.S.A.

1    Introduction   314

2    Nuclear and cytoplasmic genes 315
2.1   Methods to distinguish between them   315
2.2   Functional differences   317

3    Nuclear gene inheritance   318
3.1   Mutagenesis of nuclear genes 319
3.2   Tetrad analysis, linkage and mapping   320
3.3   Phenotypes   323

3.4   Effects of radiation on recombination of nuclear genes in *Chlamydomonas*   324

4    Cytoplasmic inheritance   325
4.1   Mutagenesis   325
4.2   Phenotypes   328
4.3   Genetic analysis, linkage and mapping   329

5    Correlations of physical and genetic evidence   338

6    References   340

## 1 INTRODUCTION

The science of genetics has two principal facets: one looking inward to the properties of the genes themselves, with respect to mutation, segregation, recombination and linkage; and the other facet, looking outward to the functions of these genes in cell growth and development. Biochemists and physiologists have been traditionally more interested in the second aspect than in the first, but, as the history of phage and bacterial genetics has shown, the two facets are often experimentally intermixed, to the benefit of all concerned.

The aim of this chapter is to provide a brief summary of the present state of our knowledge about nuclear and cytoplasmic genetic systems in the algae, including both facets: properties of the genetic systems, and functions of individual genes so far as they are known. The discussion will be limited almost entirely to *Chlamydomonas*, since the most extensive genetic investigations of

314

algae have been carried out with this organism. The blue-green algae, which as prokaryotes, are genetically more closely related to the bacteria than to the eukaryotic green algae, will not be considered here (see however Fogg *et al.* 1973).

For what kinds of biochemical and physiological problems are mutant genes and genetic analysis of value? The use of mutations in establishing biosynthetic pathways was documented long ago with bacterial strains. Now, in addition to the study of biosynthetic pathways, mutants of bacteria and of eukaryotes are being utilized in the investigation of regulatory mechanisms, such as the control of DNA replication and cell division. Mutant strains are also providing valuable leads in the analysis of membrane transport, organelle biogenesis and neoplasia.

In the investigation of eukaryotic cells, a vital link between the bacteria and the multicellular organisms is provided by the eukaryotic microorganisms: algae, fungi, and yeasts. The algae provide the widest scope of all, because they contain chloroplasts and the photosynthetic apparatus as well as mitochondria and the other complex organelles of typical eukaryotic cells of multicellular forms. The biogenesis of cellular structures and organelles as well as their functions and inter-relations can be studied in a single organism, such as the sexual green alga, *Chlamydomonas*, which contains them all (Sager 1968, 1972, Sager & Ramanis 1970b,c).

## 2   NUCLEAR AND CYTOPLASMIC GENES

### 2.1   *Methods to distinguish between them*

Nuclear genes are identified by their Mendelian pattern of inheritance in crosses. If two parents differ in a pair of alleles carried on a particular chromosome, then the distribution of these alleles to progeny will be governed by the mechanics of chromosome distribution. More generally stated, Mendel's Laws of allelic segregation and reassortment are a reflection of the behaviour of nuclear chromosomes in meiosis.

Cytoplasmic genes were first discovered by virtue of their deviation from Mendel's Laws, and were referred to as non-Mendelian. Within the past few years, some cytoplasmic genes have been identified with the DNA of chloroplasts and mitochondria, and efforts are currently in progress to correlate the genetic segregation patterns with the replication and distribution of organelle DNA (Coen *et al.* 1970, Sager & Ramanis 1970b, Sager 1972). At present, cytoplasmic genes are distinguished from nuclear genes primarily by two criteria: (1) non-Mendelian inheritance, usually involving preferential transmission from one of the parents to all of the progeny (i.e. uniparental or maternal inheritance); and (2) frequent occurrence of allelic segregation and recombination in vegetative clones during mitotic cell division. Both of these properties are easy to recognize in suitable genetically marked stocks.

The difference in segregation pattern of nuclear and cytoplasmic genes is shown in Fig. 11.1, which is based on studies with *Chlamydomonas* (Sager 1954, 1955). Similar results have been reported with *Eudorina* (Mishra & Threlkeld 1968). Pairs of nuclear genes by which the parents differ show 2:2 segregation among the four meiotic products, or zoospores, recovered from each zygote. Cytoplasmic genes, on the other hand, show maternal inheritance, characterized by the disappearance of genes carried by the male (or mating type minus) parent. The molecular basis of maternal inheritance (Sager & Lane 1969, 1972), is analogous to systems of host modification and restriction of foreign DNAs in bacteria.

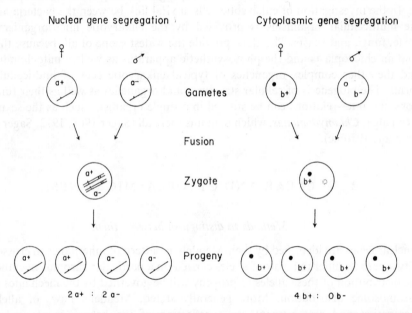

**Fig. 11.1** The segregation of chromosomal gene *a*, present as the *a*+ allele in the female parent and as the *a* − allele in the male parent, depends on the mechanics of chromosome duplication and segregation at meiosis. Each chromosome duplicates, and the homologous chromosomes from the two parents pair, giving rise to 4 strands for each chromosome of the haploid set. In meiosis the strands are distributed equally to the 4 progeny cells. In the same cells, the nonchromosomal genes, represented as *b*+ in the female and *b* − in the male, enter the zygote but the *b* − complement is generally not transmitted to the progeny.

In the simple haploid life-cycle of *Chlamydomonas* (Smith 1948, Smith & Regnery 1950), typical of the green algae, the only diploid cell in the life-cycle is the zygote which enlarges but does not divide mitotically. Exceptionally, a small percentage (less than 1%) of zygotes may divide mitotically, establishing themselves as diploid clones. These diploid clones are useful for studies of phenotypic dominance and interaction between alleles (i.e. complementation studies) of both nuclear and cytoplasmic genes (Ebersold 1963, 1967, Gillham 1963b, 1969).

## 2.2 *Functional differences*

The cytoplasmic genes so far identified have been associated with chloroplasts and mitochondria. No clear distinctions have yet been drawn between the functions of nuclear and cytoplasmic genes, primarily because the specific roles of cytoplasmic genes are not yet known.

Two principal classes of phenotypic effects have been associated with cytoplasmic gene mutations in microbial systems: (1) impairment or loss of organelle function: e.g. loss of respiratory activity by mitochondria or photosynthetic activity by chloroplasts; and (2) gain of resistance to antibiotics which are specific for the protein synthesizing systems of the organelles.

Mutations of the first type demonstrate that organelle genes are involved in the formation and function of the organelle, presumably by coding for particular key proteins. None of these gene products have yet been identified as such, primarily because they are membrane-bound components, and the technology of solubilizing and identifying membrane proteins is not yet adequate for the problem. It seems likely that when organelle-coded proteins are identified, they will turn out to be of particular importance in the regulation of organelle activity.

Mutations of the second type show that organelle genes are involved in the formation of the protein synthesizing apparatus of organelles. It is now well established that mitochondria and chloroplasts contain ribosomes and tRNA, which are different from those of the cell-sap, and it seems likely that some or all of the organelle amino acyl synthetases may also be unique. Protein synthesis in organelles is initiated with N-formyl-methionyl tRNA as in bacterial systems, and some if not all of the cofactors for initiation may be different in organelles from those of the cell-sap. Are all of these organelle-specific components coded by organelle genes? As yet there is no direct evidence of any protein coded by its DNA but considerable indirect evidence exists (see below, p. 328).

Considerable evidence that chloroplast DNA of algae is transcribed to produce the chloroplast RNA of the protein synthesizing apparatus has been reported from several laboratories using *Acetabularia* (Schweiger 1970, Shephard 1965), *Euglena* (Smillie *et al.* 1968, Stutz 1971) and *Chlamydomonas* (Surzycki *et al.* 1970, Surzycki & Rochaiz 1971). Chloroplast DNA has been shown to hybridize with plastid ribosomal RNA in *Euglena* (Stutz 1971), indicating that this RNA is transcribed from chloroplast DNA. Rifampicin, an antibiotic which blocks initiation of transcription by binding to the DNA-dependent RNA polymerase in some systems, has been shown to inhibit the formation of chloroplast ribosomal RNA in *Chlamydomonas* (Surzycki 1969) and in *Acetabularia* (Schweiger 1970). As yet, no similar studies have been reported with algal mitochondrial systems, which have proven technically difficult to purify. The mitochondrial origin of mitochondrial RNA has been demonstrated in yeast, fungi, and in several animal systems, leaving little doubt that similar results will be found in the algae (Sager 1972).

## 3   NUCLEAR GENE INHERITANCE

The life cycle of *Chlamydomonas reinhardtii* is shown in Fig. 11.2. This species is heterothallic and isogamous: mating is controlled by a pair of nuclear alleles, cells are either mating type plus (*mt+*) or mating type minus (*mt−*) and the two gametes contribute equal cell contents to the zygote (Sager 1955). Cells of the two mating types are morphologically indistinguishable except in early stages of mating when a special fertilization tubule is produced by cells of *mt−* (Friedmann *et al.* 1968).

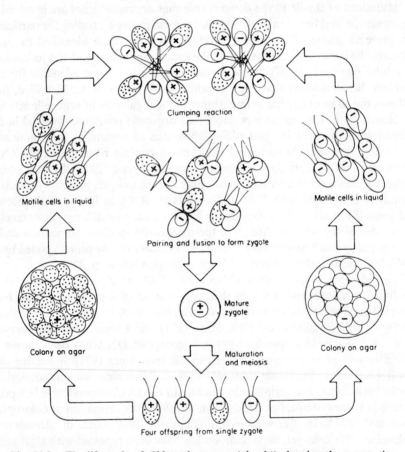

**Fig. 11.2.** The life cycle of *Chlamydomonas reinhardtii*, showing the segregation of mating type, denoted by + and −, and of an unlinked nuclear gene pair *a+/a−* denoted by the presence and absence of shading. Pairs of cells of opposite mating type fuse to form zygotes, which are the only diploid stage of the usual life cycle. After a period of maturation (several days) zygotes germinate with the release of 4 zoospores, the four products of meiosis (Sager 1972).

Sexuality in algae was previously reviewed by Coleman (1962). The genetic control of mating type has been analysed in *Eudorina* (Goldstein 1964, 1967, Mishra & Threlkeld 1968) and in *Cosmarium botrytis* (Starr 1954). Special mention should be made of the recent resurgence of interest in *Volvox* (Starr 1971), indicating that this organism is available for a new integrated genetic and biochemical investigation of differentiation.

Effects of the physiological state of the cell upon its ability to differentiate into a gamete and become capable of mating, have been studied in *Chlamydomonas* primarily in order to facilitate high efficiency mating for genetic analysis. The cells in exponentially growing cultures do not mate. It was established (Sager & Granick 1954) by depleting each of the components of the medium individually, that only the nitrogen source influences mating ability. Gametogenesis in *Chlamydomonas* is regulated by a balance between energy and nitrogen metabolism reminiscent of the carbon-nitrogen balance that regulates fruiting in higher plants.

Our discovery of the requirement for nitrogen starvation as an essential precondition for gametogenesis in *Chlamydomonas* has been of great value in subsequent studies of genetics and macromolecules in the sexual life cycle (Chiang 1971, Ebersold 1962, Ebersold *et al.* 1962, Gillham 1963b, 1965a,b, Kates & Jones 1964, Kates *et al.* 1968, Sager & Lane 1972, Sueoka *et al.* 1967), but the metabolic basis of this regulatory mechanism has not been further investigated.

The strains of this species now in common usage for genetic analysis, were isolated by G. M. Smith, who first established the sexual cycle using pure cultures in the laboratory (Smith 1948) and described the inheritance of mating types (Smith & Regnery 1950). Parallel studies were initiated with *Chlamydomonas eugametos* by Gowans (1960, 1963), McBride and Gowans (1967), and Nakamura and Gowans (1965, 1967).

### 3.1 *Mutagenesis of nuclear genes*

Mutant alleles of nuclear genes have been recovered by selection of pre-existing spontaneous mutants in untreated populations, and by mutagenic treatments. Early studies of mutagenesis and descriptions of mutants were reviewed about ten years ago (Levine & Ebersold 1960, Ebersold 1962) and should be consulted for references before 1960.

Most nuclear mutations so far described in *Chlamydomonas* were either of spontaneous origin or recovered after ultra-violet irradiation. Recently, N-methyl-N'-nitro-N-nitroso-guanidine (NTG) has been used as a mutagen for both nuclear and cytoplasmic genes (Gillham 1965b). Ethyl methane sulphonate has also been described as an effective mutagen for nuclear genes in *Chlamydomonas* (Loppes 1968, Ranneberg & Arnold 1968). Attempts to obtain mutants of *Chlamydomonas* with 2-amino-purine, an effective mutagen in bacteria, were unsuccessful (Sager & Ramanis unpubl.).

**Table 11.1.** Nuclear gene mutations in *Chlamydomonas reinhardtii*

| Mutant class | Mapped | Reference |
|---|---|---|
| *Auxotrophs* | | |
| arginine requiring | + | Ebersold 1956, Ebersole 1956 |
| nicotinamide requiring | + | Ebersole 1956, Nakamura & Gowans 1965 |
| *p*-amino benzoate requiring | + | Ebersole 1956 |
| thiamine requiring | + | Ebersold *et al.* 1962, Ebersole 1956, Ebersold 1962 |
| NH$_4$ requiring (unable to use NO$_3$) | + | Sager unpubl. |
| *Non-motile* | | |
| pf $-1$ through $-21$ | + | McVittie 1972, Randall *et al.* 1967, Starling 1969 |
| lf $-1$ and $-2$ | + | McVittie 1972 |
| *Abnormal cell division* | | |
| cyt $-1$ | + | Warr 1968, Warr & Durber 1971 |
| *Temperature sensitive* | − | Sager & Ramanis unpubl. |
| *Drug resistant* | | |
| low level streptomycin-1 | + | Sager 1954 |
| canavanine resistant | + | Levine & Goodenough 1970 |
| methionine sulphoximine resistant | + | Sager unpubl. |
| actidione (cycloheximide) resistant | + | Sager unpubl. |
| low level erythromycin resistant | − | Sager unpubl. |
| low level kanamycin resistant | − | Sager unpubl. |
| low level spectinomycin resistant | − | Sager unpubl. |
| *Acetate requiring* | | |
| Class A—pigment mutants | (some) | Chance & Sager 1957, Ebersole 1956, Goodenough *et al.* 1969, Goodenough & Levine 1969, Sager 1954 |
| Class B—non-photosynthetic | (some) | Chua & Levine 1969, Epel & Levine 1971, Garnier & Maroc 1971, Givan & Levine 1967, Goodenough *et al.* 1969, Goodenough & Levine 1969, 1970, Gorman & Levine 1966, Lavorel & Levine 1968, Levine & Gorman 1966, Levine & Paszewski 1970, Levine & Smillie 1963, Levine & Togasaki 1965, Smillie & Levine 1963, Togasaki & Levine 1970, Weaver 1969 |
| Class C—unstudied | (some) | Levine & Goodenough 1970, Sager unpubl. |

## 3.2 *Tetrad analysis, linkage and mapping*

The life cycle of *Chlamydomonas* is ideally suitable for genetic analysis, since all four products of meiosis are routinely recovered following zygote germination. The methods of tetrad analysis, originally developed with yeast and *Neurospora*, can be directly applied (Fincham & Day 1963). The four products are unordered, as in yeast, rather than ordered, as in *Neurospora*. The advantage of an ordered array is that one can distinguish allelic segregation occurring at the first meiotic division from that at the second division, and thereby determine the frequency of crossing-over between each gene and its centromere directly. However, the same result can be achieved with an unordered tetrad if one has a marker very closely linked to the centromere. In *Chlamydomonas*, the gene $y$–1 was found to segregate only at the first meiotic division (Sager 1955) and thereby it has served this function, converting unordered into the equivalent of ordered tetrads, in which centromere distances of all markers present can be determined directly.

The early genetic studies (Sager 1955, Ebersold & Levine 1958, Levine & Ebersold 1960) of nuclear gene segregation and linkage demonstrated that meiosis was conventional in *Chlamydomonas*, and no evidence of chromosomal aberrations was noted. Subsequently, extensive crosses between strains maintained in different laboratories (Sager 1962b) revealed neither incompatibility nor lethality, further supporting the view that this species possesses a very stable chromosomal apparatus. Ebersold (1962) identified six linkage groups. At present, 16 linkage groups have been described (Hastings *et al.* 1965) but the supporting genetic evidence has not yet been published. These linkage groups are shown in Fig. 11.3.

A problem arises concerning the mapping of $y$–1. This gene shows only first division segregation in crosses (Sager 1955), an indication of tight linkage with a centromere. In an attempt to map it, a strain carrying the $y$–1 allele was crossed with strains carrying centromere-linked markers for all the linkage groups then known, and no linkage was found (Sager 1962b). The susceptibility of $y$–1 to mutagenesis by streptomycin, which is a much more effective mutagen for cytoplasmic than for nuclear genes (see below, p. 325), led to the hypothesis that $y$–1 might be cytoplasmic despite its 2:2 pattern of inheritance. With the subsequent discovery of additional linkage groups, the mapping of $y$–1 should now be re-investigated.

Levine and Goodenough (1970) made the improbable proposal that eight of the 16 linkage groups shown in Fig. 11.3, are located not in the nucleus but in the chloroplast. This proposal contradicts all the genetic evidence of Mendelian inheritance, meiotic segregation and linkage relations exhibited by the markers in the 16 linkage groups, on the basis of which the linkage groups were identified. Furthermore, this proposal ignores all the published evidence about the genetic behaviour of cytoplasmic genes.

Specifically, Levine and Goodenough (1970) argued that those acetate-dependent genes known to be non-photosynthetic were located in only eight (actually nine) of the 16 linkage groups. However, as can be seen from Fig. 11.3, *ac*—markers are found on all but one of the 16 linkage groups. Finally, strong evidence against the Levine and Goodenough proposal comes from the work of McVittie and Davies (1971) who have demonstrated the presence of 16 chromosomes in *Chlamydomonas*, in a cytological study of meiosis. Thus, the

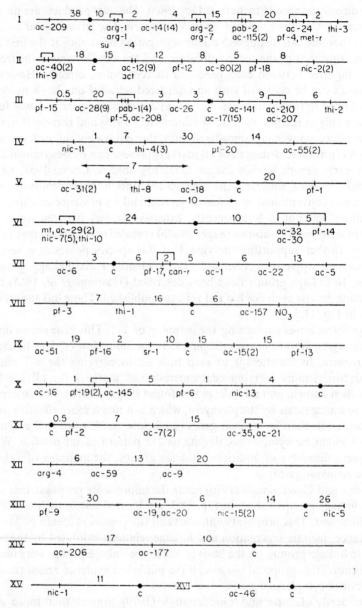

**Fig. 11.3.** The 16 nuclear linkage groups of *Chlamydomonas reinhardtii*. Figures in parenthesis after the name of a locus indicate the number of alleles known at that locus. Numbers above the line are map distances. The bracket above the group of markers indicates that their relative positions are uncertain or unknown. Abbreviations: *c*, centromere; *mt*, mating type; *arg*, arginine requiring; *su$^{arg-1}$*, suppressor of *arg*-1; *ac*, acetate requiring; *nic*, nicotinamide requiring; *pab*, p-aminobenzoate requiring; *thi*, thiamine requiring; *pf*, paralysed flagellae; *sr*, streptomycin resistant; *can-r*, canavanine resistant; *met-r*, methionine sulphoximine resistant; *act-r*, actidione resistant; NO$_3$, ability to grow on nitrate as sole nitrogen source (Levine & Goodenough 1970).

presence of 16 chromosomes correlates well with the genetic evidence of 16 linkage groups.

### 3.3 *Phenotypes*

The array of phenotypes found among mutant strains of *Chlamydomonas* is surprisingly restricted (Ebersold 1962, Levine & Ebersold 1960, Levine & Goodenough 1970, Sager 1955). Despite extensive efforts by several investigators, the only amino acid-requiring mutants found have been arginine-requiring, and the only other auxotrophic mutants blocked in biosynthetic pathways so far described have been adenine-requiring mutants (in *Chlamydomonas eugametos* but not in *C. reinhardtii*), vitamin-requiring mutants (para-amino benzoic acid-requiring, nicotinamide-requiring, and thiamine-requiring) and $NH_4^+$ requiring in the sense of having lost the ability to use $NO_3^-$ as a nitrogen source. The lack of auxotrophic mutants has been attributed to the relative impermeability of *Chlamydomonas* to amino acids, purines and pyrimidines. On this hypothesis, mutations to auxotrophy occur as in other organisms, but the mutants die while bathed in a medium containing an essential metabolite which they cannot transport into the cell.

The majority of mutants used in establishing the linkage groups of *Chlamydomonas* shown in Fig. 11.3 are either non-motile (*pf*) or acetate-requiring (*ac*−). Non-motile mutants, with apparently paralysed flagella, were first described by Lewin (1954) and subsequently have been under intensive genetic and biochemical investigation by Randall and co-workers (McVittie 1972, Randall 1969, Randall *et al.* 1964, 1967, Warr 1968, Warr & Durber 1971, Warr *et al.* 1966). Other nuclear gene mutations have been described with morphological effects on the eyespot (Hartshorne 1955); contractile vacuoles (Guillard 1960); cytokinesis (Warr 1968, Warr & Durber 1971); cell wall formation (Davies & Plaskitt 1971); stacking of chloroplast lamellae (Goodenough *et al.* 1969) and formation of chloroplast ribosomes (Goodenough & Levine 1970, Boynton *et al.* 1970). The acetate-requiring mutants have been studied most extensively by Levine and his collaborators and students, primarily as tools in the dissection of photosynthesis (Levine 1960, 1968, 1969) (see Chapter 14, p. 424). In addition to non-photosynthetic mutants, some acetate-requiring mutants have as yet no detectable lesions in the photosynthetic apparatus, but nonetheless cannot grow photosynthetically. Some non-photosynthetic mutants have reduced amounts of photosynthetic pigments, while others have a normal pigment content but are blocked elsewhere in photosynthesis (see Table 11.1 for references).

Other important classes of nuclear mutations are those resulting in resistance to antibiotics or antimetabolites. These include resistance to methionine sulphoximine and to cycloheximide (actidione) (Sager 1955, Sager & Ramanis 1963); canavanine-resistance (Levine & Goodenough 1970); low-level streptomycin resistance (Sager 1954, 1955); low-level resistance to erythromycin, spectinomycin, and kanamycin (Sager & Ramanis unpubl.) and a mutation amplifying the resistance level of strains carrying either nuclear or cytoplasmic mutations to streptomycin-resistance (Sager & Tsubo 1961). It is noteworthy that, in general, the nuclear gene mutations to drug-resistance are low-level. By analogy with bacteria, this low-level drug resistance may result from altered permeability, whereas the high level mutants to be discussed below, which are cytoplasmic, involve alterations at the ribosome level. Recently, a number of tempera-ture-sensitive mutants, growing at 25°C but not at 35°C have been isolated (Sager & Ramanis unpubl.) and a series of ultra-violet-sensitive mutants have been described (Rosen & Ebersold 1972). The known nuclear mutations, both mapped and unmapped, are listed in Table 11.1.

In the conventional life cycle of *Chlamydomonas* and related algae, the zygote is the only diploid stage. However, using two linked *arg* — mutations, Ebersold detected vegetative clones of diploids (Ebersold 1963, 1967), by screen-ing zygote colonies for arginine-independence. Diploids can be used for a number of purposes, among them the study of dominance relations as well as complementation of alleles. Ebersold found that *arg* — is recessive to *arg*+ and that the mating type allele *mt* — is dominant in the sense that all diploid clones are phenotypically *mt* — and will mate with *mt*+ strains to produce triploid zygotes. Gillham (1963b, 1969) used diploids produced by Ebersold's method to study the segregation of cytoplasmic genes.

### 3.4  *Effects of radiation on recombination of nuclear genes in* Chlamydomonas

*Chlamydomonas* has been used in a series of studies of the effects of non-lethal radiation on the frequency of recombination of selected pairs of linked genes (Davies & Lawrence 1967, Lawrence 1965a,b, 1967, 1968, 1971a,b, Lawrence & Davies 1967).

The ease with which the stages of meiosis can be synchronized in large populations of this alga makes it a useful organism for studies of this kind. It was shown that germinating zygotes have two periods of significant sensitivity to low-levels of $^{60}Co$ gamma radiation under conditions of almost no killing.

One period occurs 5·5 to 6·0 hours after the start of germination, a time that probably corresponds to a very early stage of meiosis (pre-leptotene). The second sensitive period occurs about one hour later, a time which is probably early pachytene. Irradiation during the first period decreases the frequency of re-combination and that during the second period increases it.

These effects on intergenic recombination were compared with effects of radiation on intragenic recombination between two alleles at the *ac*–14 locus (Lawrence 1967). Here the principal effect was a large increase in recombination frequency at times from 6·5 to 9 hours after germination. These results are similar to those found with higher plants and are interpreted as direct effects upon either DNA replication or repair.

## 4   CYTOPLASMIC INHERITANCE

By far the most extensive genetic information about any cytoplasmic system is based upon studies with *Chlamydomonas*, recently summarized in a book by Sager (1972), to which the interested reader is referred. Only the briefest account of these studies will be presented here.

Genetic analysis depends almost entirely on the availability of mutant alleles, whose transmission to progeny can be followed by comparison with the unmated, wild-type alleles. Consequently, the acquisition of mutant strains represents the first step in genetic analysis.

### 4.1   *Mutagenesis*

In *Chlamydomonas*, the most successful mutagen for cytoplasmic genes has been streptomycin, an antibiotic with little or no mutagenic action towards nuclear, bacterial or viral genes. Streptomycin is also mutagenic in *Euglena*, but in the absence of genetic analysis, the primary genetic target, i.e. nuclear or cytoplasmic DNA has not been determined in *Euglena*.

In our initial studies with *Chlamydomonas*, streptomycin was chosen as a selective agent, not as a mutagen, to isolate spontaneous nuclear mutations to drug-resistance for use as markers. This work led to the discovery of the first cytoplasmically inherited gene in *Chlamydomonas*, conferring a high level of streptomycin-resistance (Sager 1954). This finding led to further studies of streptomycin as a mutagen.

The cytoplasmic mutations to streptomycin-resistance selected by plating cells on streptomycin agar, were found to be induced by growth in the presence of the drug (Sager 1960). This finding was confirmed by Gillham and Levine (1962). In subsequent studies, streptomycin was shown to be mutagenic not only for streptomycin resistance but for other cytoplasmic genes as well (Sager 1962a).

Since that time, streptomycin has been the principal mutagen used in inducing cytoplasmic mutations in *Chlamydomonas* and all streptomycin-induced mutations so far mapped have been found to be linked within a single linkage group or cytoplasmic 'chromosome'. With the extensive evidence that this linkage group is located in chloroplast DNA, it now seems possible that streptomycin is primarily mutagenic for chloroplast DNA, thus explaining the failure to find mitochondrial mutations with this mutagen. The molecular mechanism of streptomycin mutagenesis is still unknown, but recent studies (Flechtner & Sager unpubl.) demonstrating an inhibitory effect of the drug on DNA replication, suggest that mutagenesis may be directly related to the inhibition of replication.

Another mutagenic treatment involves the withdrawal of streptomycin from a strain of streptomycin-dependent (*sd*) cells. These cells can grow for four to five doublings in the absence of the drug and then they die. Some clones survive,

**Table 11.2.** Cytoplasmic gene mutations in *Chlamydomonas*

| Gene | Origin[a] | Phenotype | Mapped[b] | Reference |
|---|---|---|---|---|
| *ac1* | SM induced | Requires acetate (*leaky*) | Yes | Sager & Ramanis 1963 |
| *ac2* | SM induced | Requires acetate (*stringent*) | Yes | Sager & Ramanis 1963 |
| *ac3* | SM induced | Requires acetate (*stringent*) | Yes | Sager & Ramanis 1971 |
| *ac4* | SM induced | Requires acetate (*leaky*) | Yes | Sager & Ramanis 1971 |
| *tm1* | SM induced | Cannot grow at 35°C | Yes | Sager & Ramanis in prep. |
| *tm2* | SM induced | Conditional: grows at 35°C only in the presence of streptomycin | No | Sager & Ramanis 1971 |
| Seven *tm* mutants | SM induced | Cannot grow at 35°C | No | Sager & Ramanis 1971 |
| *ti1* thru *ti5* | NG | Tiny colonies on all media | No | Sager & Ramanis 1971 |
| *ery1* | SM induced | Resistant to 50 μg ml$^{-1}$ erythromcyin | Yes | Sager & Ramanis in prep. |
| *kan1* | SM induced | Resistant to 100 μg ml$^{-1}$ kanamycin | No | Sager & Ramanis 1971 |
| *spc1* | SM induced | Resistant to 100 μg ml$^{-1}$ spectinomycin | Yes | Sager & Ramanis in prep. |
| *spi1* thru *5* | SM induced | Resistant to 100 μg ml$^{-1}$ spiramycin | Yes | Sager & Ramanis in prep. |
| *ole1* thru *3* | SM induced | Resistant to 50 μg ml$^{-1}$ oleandomycin | Yes | Sager & Ramanis in prep. |
| *car1* | SM induced | Resistant to 50 μg ml$^{-1}$ carbamycin | Yes | Sager & Ramanis in prep. |
| *cle1* | SM induced | Resistant to 50 μg ml$^{-1}$ cleosine | Yes | Sager & Ramanis in prep. |
| *ery3* | SM induced | Resistant to erythromycin, carbamycin, oleandomycin, spiramycin (same concentrations as single mutations above) | No | Sager & Ramanis 1971 |
| *ery11* | SM induced | Same as for *ery3* | Yes | Sager & Ramanis in prep. |
| *sm2* | SM induced | Resistant to 500 μg ml$^{-1}$ SM | Yes | Sager 1954 |
| *sm3* | SM induced | Resistant to 50 μg ml$^{-1}$ SM | Yes | Sager & Ramanis 1965 |
| *sm4* | SM induced | SM dependent | Yes | Sager & Ramanis 1965, 1970a |
| *sm5* | SM induced | Resistant to 500 μg ml$^{-1}$ SM; recombines with *sm2* | Yes | Sager & Ramanis 1971 |

| Gene | Origin[a] | Phenotype | Mapped[b] | Reference |
|---|---|---|---|---|
| D-371 and D-310 | Induced by growth of strain sm4 without SM | Resistant to 500 $\mu$g ml$^{-1}$ SM | No | Sager & Ramanis 1971 |
| Four D mutants | Induced by growth of strain sm4 without SM | Resistant to 500 $\mu$g ml$^{-1}$ SM; segregate like persistent hets (sd/sr) | No | Sager & Ramanis in prep. |
| Eleven D mutants | Induced by growth of strain sm4 without SM | Resistant to various low levels of SM: 20 $\mu$g ml$^{-1}$; 50 $\mu$g ml$^{-1}$; 100 $\mu$g ml$^{-1}$. Segregate like persistent hets (sd/low sr) | No | Sager & Ramanis 1971 |
| D-769 | Induced by growth of strain sm4 without SM | Conditional sd | Yes | Sager & Ramanis 1970a |
| Three D mutants | Induced by growth of strain sm4 without SM | Conditional sd; segregate like persistent hets (sd/cond. sd) | No | Sager & Ramanis 1971 |
| UV-16, UV-17 | UV induced in strain sm4 | Resistant to 500 $\mu$g ml$^{-1}$ SM | No | Sager & Ramanis 1971 |
| Four UV mutants | UV induced in strain sm4 | Resistant to 20 $\mu$g ml$^{-1}$ SM | No | Sager & Ramanis 1971 |
| Three UV mutants | UV induced in strain sm4 | Resistant to 20 $\mu$g ml$^{-1}$ SM; segregate like persistent hets (sd/low sr) | No | Sager & Ramanis 1971 |
| sr-2-1 sr-2-60 sr-2-280 sr-2-218 | Spontaneous mutations selected on SM | Resistant to SM 500 $\mu$g ml$^{-1}$ | No | Gillham 1969; Surzycki & Gillham 1971; Gillham 1969 |
| kan-1 | Spontaneous; selected on kanamycin | Resistant to kanamycin 50 $\mu$g ml$^{-1}$ | No | Gillham 1969 |
| ery-2-y | NG | Resistant to erythromycin 100 $\mu$g ml$^{-1}$ | No | Surzycki & Gillham 1971 |
| ery-3-6 | NG | Resistant to erythromycin 100 $\mu$g ml$^{-1}$ | No | Surzycki & Gillham 1971 |
| spr-1-27 | NG | Resistant to spectinomycin 100 $\mu$g ml$^{-1}$ | No | Surzycki & Gillham 1971 |
| sd-3-18 | NG | Dependent on at least 20 $\mu$g ml$^{-1}$ SM | No | Gillham 1969; Surzycki & Gillham, 1971 |
| nea-2-1 | NG | Resistant to neamine 1 mg ml$^{-1}$ | Yes[b] | Gillham 1969 |

[a] SM = streptomycin; NG = nitrosoguanidine    [b] Mapping references Sager & Ramanis 1970a and in prep.

and these turn out to be either streptomycin-resistant or -sensitive, but no longer phenotypically dependent. These new mutations are also cytoplasmic and genetically linked to the original *sd* gene. In addition to altered responses to streptomycin, some of these mutant strains carry additional cytoplasmic mutations, such as loss of photosynthetic ability, temperature sensitivity and poor growth on all media.

In contrast to streptomycin, NTG which is a potent mutagen in many systems, has been reported to induce both nuclear and cytoplasmic gene mutations in *Chlamydomonas* (Gillham 1965b). One drawback to the use of this mutagen for recovering cytoplasmic gene mutations is the preponderance of nuclear gene mutations that are simultaneously induced. Since NTG is particularly mutagenic towards replicating DNA, it should be possible to enrich for cytoplasmic mutations by treating synchronously growing cultures at the time when chloroplast and mitochondrial, but not nuclear, DNA is replicating.

The susceptibility of cytoplasmic genes to ultra-violet irradiation and to other mutagenic treatments has not been systematically investigated with *Chlamydomonas* or any other alga as yet.

### 4.2   *Phenotypes*

Four general phenotypic classes of cytoplasmic mutants have been described in *Chlamydomonas* to date: (1) acetate-requiring mutants; (2) mutants that are resistant to one or more antibiotics; (3) tiny colony mutants that are slow-growing on all supplemented media; and (4) temperature-sensitive mutants that grow better at one temperature than another, including both those that grow better at 25°C than at 35°C and the reverse. The mutations reported so far are listed in Table 11.2 (Sager 1972).

Biochemical characterization has been reported only for a few of the antibiotic-resistant mutants. Gillham *et al.* (1970) described apparent alterations in the chloroplast ribosomes of some mutant strains carrying cytoplasmic mutations to drug resistance and dependence, but not in others. They found a similar change in the chloroplast ribosomes of a strain carrying the nuclear gene mutations *cr*–1 (Boynton *et al.* 1970). Mets and Bogorad (1971) examined the erythromycin-binding capacity of chloroplast ribosomes from a number of erythromycin-resistant strains (three nuclear and one cytoplasmic). Only in the strain carrying a cytoplasmically-transmitted mutation did the chloroplast ribosomes show a decreased binding affinity for the drug.

The strongest evidence to date that chloroplast DNA codes for at least some chloroplast ribosomal proteins comes from studies of *in vitro* protein synthesizing ability of chloroplast ribosomes isolated from cytoplasmically inherited antibiotic-resistant mutants of *Chlamydomonas* (Schlanger *et al.* 1972). The chloroplast ribosomes of a carbomycin-resistant mutant were shown to be resistant to the drug in contrast to similar ribosomes from a wildtype strain. Subsequently, resistance to streptomycin and neamine have been localized to the 30S subunit, and resistance to carbamycin and cleosine to the 50S subunit. The likelihood that these ribosomal changes involve altered proteins rather than ribosomal RNA is strengthened by the fact that several different mutations are involved, separable by crossing-over, and that the different strains are not cross-resistant. (Schlanger *et al.* 1972.)

### 4.3   *Genetic analysis, linkage and mapping*

The pattern of inheritance of cytoplasmic streptomycin resistance, shown in Fig. 11.4, is typical of all cytoplasmic mutations so far described in *Chlamydomonas*. All the mutations which have been mapped lie in a single linkage group, or cytoplasmic chromosome, identified with chloroplast DNA (see below). No mitochondrial genes have yet been identified and their pattern of inheritance is unknown. The suggestion of Schimmer and Arnold (1970a-d) that the streptomycin dependent gene *sd3* may be mitochondrial is most unlikely, as will be discussed below.

Strict maternal inheritance of the sort shown in Fig. 11.4 essentially precludes the carrying out of genetic analysis, since all cytoplasmic genes are transmitted

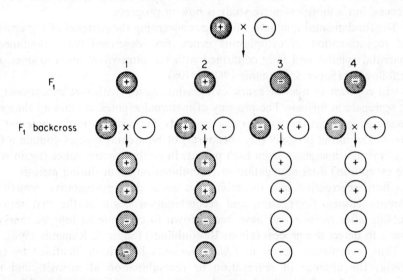

**Fig. 11.4.** Inheritance of cytoplasmic streptomycin resistance in *Chlamydomonas*. Plus and minus signs refer to mating type. In the initial cross, *sm−r mt+ yl+* × *sm−s mt− yl−*, all progeny were *sm−r*, though zoospores of each zygote segregated 2:2 for the nuclear gene pairs *mt+/mt−* and *yl+/yl−*. $F_1$ clones of *sm−r mt+* backcrossed to *sm−s mt+* produced all *sm−r* progeny (4:0 segregation) as in the initial cross, but $F_1$ clones of *sm−r mt−* backcrossed to *sm−s mt+* produced only *sm−s* progeny (0:4 segregation). Shaded cells indicate streptomycin resistant cells and unshaded cells indicate streptomycin sensitive cells (Sager 1954).

together from one parent to all progeny and none are transmitted from the other parent. Fortunately, spontaneous exceptions to the rule of maternal inheritance occur in *Chlamydomonas*, and were detected in my first study (Sager 1954) and confirmed by Gillham (1963a). From then until 1966, all genetic studies of cytoplasmic inheritance in *Chlamydomonas* were based upon the rare, spontaneously occurring, biparental zygotes which transmitted genes from both parents to the progeny (Sager 1960, Sager & Ramanis 1963, 1964, 1965, Gillham 1965a, 1969, Gillham & Fifer 1968).

The development of a systematic analysis of recombination and quantitative mapping procedures only became possible with the discovery of the ultra-violet effect on maternal inheritance (Sager & Ramanis 1967). Irradiation of the female parent ($mt+$) with a low dose of ultraviolet light, immediately before mating, was found to result in recovery of about 50% of all zygotes as biparental, in contrast to the 0·1% spontaneous biparentals found in unirradiated controls. This 500-fold enrichment of biparental zygotes occurred under conditions of little or no lethality, and no effects were seen upon the nuclear genome. Irradiation of the male parent ($mt -$) had no effect, and irradiation of newly-formed zygotes was highly lethal. Comparisons of recombination frequencies in spontaneous and ultra-violet induced biparental zygotes indicated little or no differences, but a more extensive study is now in progress.

The fundamental qualitative findings concerning the patterns of segregation and recombination of cytoplasmic genes, first described with spontaneous biparental zygotes and fully confirmed with the ultra-violet induced ones, are the following (Sager & Ramanis 1968, 1970a).

(1) In contrast to nuclear genes, cytoplasmic genes (with rare exceptions) do not segregate in meiosis. The progeny of maternal zygotes, contain all the cytoplasmic genes from the female ($mt+$) parents and none from the male ($mt -$) parent. The initial progeny (i.e. zoospores) of biparental zygotes contain a full set of cytoplasmic genes from each parent. In neither circumstance (again with rare exceptions) does segregation or recombination occur during meiosis.

(2) Both segregation and recombination occur during vegetative growth of zoospore clones. Segregation and recombination begin at the first mitotic doubling after meiosis and have been shown to continue so long as markers remain to detect the process (about 10 doublings) (Sager & Ramanis 1968).

Thus, cytoplasmic genes in *Chlamydomonas* have been identified by two criteria: the absence of segregation or recombination at meiosis, and the occurrence of both segregation and recombination in vegetative clones of zoospores just after meiosis.

The rest of this section will summarize present knowledge on segregation and recombination mechanisms, on mapping procedures, and on linkage of cytoplasmic genes in *Chlamydomonas*. The data have come from ultra-violet-induced biparental zygotes resulting from crosses in which the parents differed by at least three pairs of nuclear genes and four or more pairs of cytoplasmic genes. The nuclear markers were unlinked, and their independent reassortment in meiosis served to indicate whether all four products of meiosis were being recovered, and to distinguish the four zoospores from one another (see Sager & Ramanis 1970a, 1971, for further details).

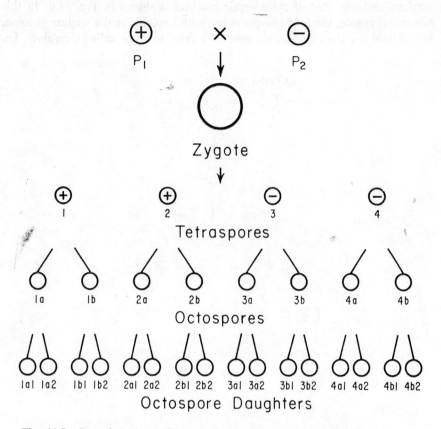

**Fig. 11.5.** Procedure for pedigree analysis. After germination, zoospores are allowed to undergo one mitotic doubling and then the eight cells (octospores) are transferred to a fresh petri plate and respread. After one further doubling each pair of octospore daughters is separated and allowed to form colonies. The sixteen colonies, derived from the first two doublings of each zoospore are then classified for all segregating markers (Sager & Ramanis 1970a.

Two procedures have been used for recovering progeny. Most studies have been based on pedigree analysis, as shown in Fig. 11.5. This method is too cumbersome to use in the examination of complete sets of progeny beyond the second or third doubling of zoospores. Subsequent events have been examined mainly by a liquid culture method, in which a population of zoospores is grown in liquid and samples are taken every few hours for classification of large numbers of random progeny by selective or replica plating. With pedigree analysis, one can score not only the progeny but also the events which have occurred, i.e. segregation or recombination, at each doubling. For mapping purposes both kinds of data are useful.

As an example, a cross between parents carrying three pairs of nuclear markers and one pair of cytoplasmic markers is shown in Fig. 11.6. In this biparental zygote, the four zoospores are each haploid for the nuclear genome, but diploid for the cytoplasmic gene. We refer to these cells as *cytohets*, i.e.

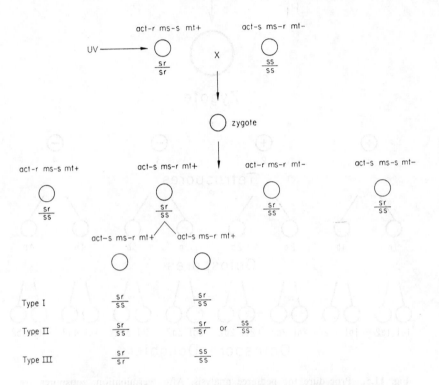

**Fig. 11.6.** Segregation patterns of nuclear and cytoplasmic markers. Cross showing nuclear and cytoplasmic gene segregation in exceptional zygotes. The female (mt+) parent differs from the male (mt −) by three pairs of unlinked nuclear genes (actidione resistance, methionine sulphoximine resistance, and mating type) and one pair of cytoplasmic genes (streptomycin resistance).

heterozygous for their cytoplasmic genome. At each mitotic doubling, a cytohet may behave in one of three possible ways, as shown in Fig. 11.6. These three alternative patterns of segregation are: Type I, in which both daughter cells are still cytohet; Type II, in which one is a cytohet while the other is a pure type like one or the other of the parents; and Type III, in which both daughters are pure types like the two parents. The diagram (Fig. 11.7) indicates how these segregation events may be visualized at the DNA level, for linear molecules, and Fig. 11.8 shows the equivalent events for circular molecules.

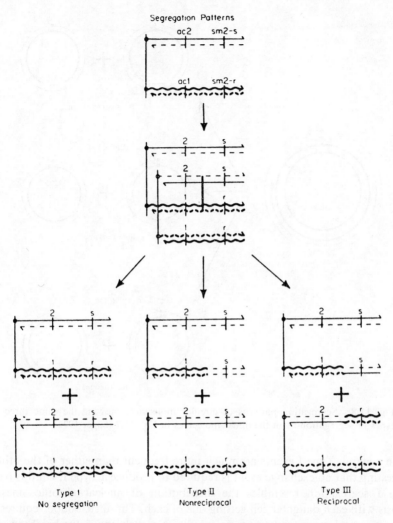

**Fig. 11.7.** Segregation patterns. The homologous DNAs from the two parents are shown as straight (*ac2* parent) and wavy (*ac1* parent) solid lines. The complementary replicated strands are shown as dashed lines. The homologous DNAs are bound to a hypothetical membrane by one of the complementary strands. The model predicts that the 'old' attachment points go to one daughter at cell division and the 'new' points to the other daughter, thus determining the regularity of distribution following semiconservative replication. Type II segregation is pictured as a double-stranded loss and replacement of a segment, and type III is shown as a double-stranded exchange. This model is consistent with the genetic data but otherwise speculative.

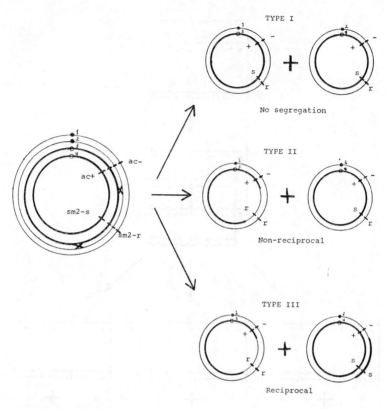

**Fig. 11.8.** Segregation patterns of circular molecules, showing the same three types of segregation as in the linear molecules in Fig. 11.7 (Sager 1972).

In general, Type I events are much more frequent than either of the others, suggesting that no exchange event is required to produce a Type I. On this basis, Type I segregation resembles the distribution of nuclear chromosomes at mitosis with each daughter cell getting one of each. This distribution requires an apparatus analogous to the mitotic spindle. No such apparatus has been seen associated with chloroplast or mitochondrial DNA, but the genetic evidence requires one. We therefore postulated the 'attachment point' (*ap*) shown in the figures, and found subsequently that *ap* could be mapped.

The mapping of *ap* is based on the observed frequencies of Type III segregations. In multifactor crosses involving several cytoplasmic genes, the frequency of Type II events was found to be about the same for each gene, but the Type III frequencies were different and characteristic for each gene. The same relative frequencies were seen in different crosses involving the same genes in different combinations. The differences in Type III frequencies provided an order on the basis of which the genes could be arranged, i.e. a map.

The same crosses also provided data on the frequencies of recombination between genes, on the basis of which a conventional map could be constructed. The arrangement and relative distances between genes, estimated from recombination fractions, were found to agree well with the arrangement based upon Type III segregation frequencies. Combining the data led to the composite map shown in Fig. 11.9, based on published data (Sager & Ramanis 1970a) and

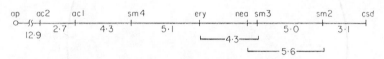

**Fig. 11.9.** Composite map based on published data (Sager & Ramanis 1970a). Symbols are: *ap*, attachment point; *ac2* and *ac1*, acetate-requiring; *sm4*, streptomycin dependence; *ery*, erythromycin resistance; *nea*, neamine resistance; *sm3*, low level streptomycin resistance; *sm2*, high level streptomycin resistance; *csd*, conditional streptomycin dependence.

discussed in several reviews (Sager & Ramanis 1970b,c, 1971). More recently, this set of genes has been found to lie on a segment of a larger linkage group, which appears from genetic data to be circular (Sager & Ramanis, in preparation). Our present view of the arrangement of genes in this linkage group is given in Fig. 11.10. Some difficulties in specifying order in the region 180° from the attachment point are still being investigated.

One problem in the analysis concerns the Type II segregations. These events appear similar to the gene conversions seen in yeast and *Neurospora*, where some kind of miscopying seems to occur, resulting in three strands carrying one of the parental alleles and only one strand carrying the other, giving rise to 3:1 ratios instead of the expected 2:2. These events are far more frequent in *Chlamydomonas* than in any system yet described, and methods to use these frequencies for mapping purposes are still being developed. The significance of high frequencies of Type II segregations is not clear, although it suggests that high frequencies of repair may be occurring in these chloroplast DNA molecules.

The liquid culture method provides another way to map genes based upon the frequency of Type III events. When populations of growing zoospores are scored for the number of segregated pure types and unsegregated cytohets for each of the markers in the cross, results of the type shown in Fig. 11.11 are found. The closer a gene lies to the attachment point, the less frequent will Type III segregation be. Although the liquid culture method actually detects a mixture of segregants from both Type II and Type III events, the differential rate depends on the Type III events alone, since the Type II occur with equal frequency around the map. In practice this method is quite sensitive, since large populations can easily be scored.

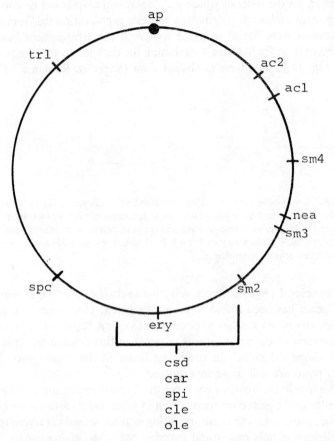

**Fig. 11.10.** Circular map incorporating data from several crosses. The qualitative evidence of gene order, relative distances, and circularity are well supported by the data but quantitative map distances need further clarification. The bracketed genes all map close to *ery* and no linear order has been established. Symbols: *ap*, attachment point; *ac2* and *ac1*, acetate requirement; *sm4*, streptomycin dependence; *nea*, neamine resistance; *sm3*, low level streptomycin resistance; *sm2*, high level streptomycin resistance; *ery* erythromycin resistance; *csd*, conditional streptomycin dependence; *car*, carbomycin resistance; *spi*, spiramycin resistance; *cle*, cleosine resistance; *ole*, oleandomycin resistance; *spc*, spectinomycin resistance; and *mt*, temperature sensitivity (Sager 1972).

A few comments may be made about the arrangement of genes on the map. Firstly, the arrangement of genes conferring antibiotic resistance is surprisingly similar to that found in *E. coli* and in *B. subtilis*. In both bacteria, the genes for spectinomycin and for streptomycin resistance, each coding for a 30S protein, are closely linked with a gene coding for a 50S protein, detected as erythromycin resistant by altered binding of the drug.

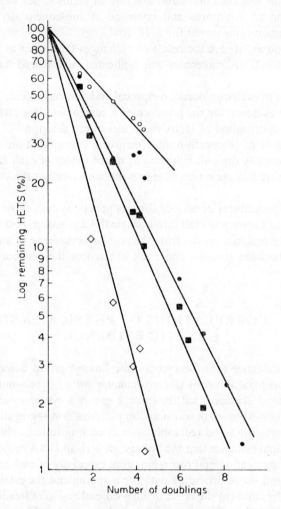

**Fig. 11.11.** Comparison of segregation rates in liquid culture of four genes in the progeny of a single cross. Symbols: ○, temperature sensitivity (*tml*); ■, streptomycin resistance (*sm2*); ●, spectinomycin resistance (*spc*); ◇, erythromycin resistance (*ery*). (Sager 1972).

The principal conclusions derived from the genetic analysis which has been briefly described above are the following:

(1)  Yields of up to 50% biparental zygotes, producing zoospore progeny which are cytohets can be obtained routinely by ultra-violet irradiation of the female (*mt*+) parent just before mating:

(2)  All cytoplasmic markers, in multiple marked crosses, show the same pattern of inheritance in individual zygotes: all maternal, all biparental, or all paternal.

(3)  Segregation and recombination are rare in meiosis, but begin at the first mitotic division of zoospores and continue at subsequent divisions with a constant probability (measured for 8–10 doublings).

(4)  Segregation events are the result of exchanges that occur at a 'four-strand stage' when the DNA molecules are replicated and paired but before cell division.

(5)  Two types of exchange occur: reciprocal and non-reciprocal. The reciprocal events provide evidence for the presence of a centromere-like attachment point regulating the distribution of DNA molecules at cell division.

(6)  The patterns of segregation and reciprocal recombination show that the genome is genetically diploid, behaving as if two copies of each DNA molecule carrying the entire linkage group were present in each cell of the young zoospore clones.

(7)  Mapping procedures developed for this genetic system show that the cytoplasmic genes so far mapped fall into a single linkage group, best described as a circle. Genetic circularity results from the regular occurrence of an even number of exchanges between strands, and need not reflect the physical state of the DNA.

## 5  CORRELATIONS OF PHYSICAL AND GENETIC EVIDENCE

The principal evidence that the cytoplasmic linkage group described above is located in chloroplast DNA of *Chlamydomonas* will now be summarized.

(1)  The principal evidence that the linkage group is cytoplasmic, not nuclear, is based on its non-Mendelian transmission patterns, non-segregation in meiosis, and regular segregation and recombination at each mitotic doubling.

(2)  The principal evidence that the linkage group is in DNA comes from (a) the regularities of segregation and recombination, especially of exchanges occurring at the four-strand stage, providing data for mapping and the establishment of a linear gene order; and (b) the effect of high dose ultra-violet irradiation of $mt+$ gametes leading to the production of paternal zygotes in which the cytoplasmic genes from the $mt+$ are lost unless rescued by photoreactivation.

(3)  Taken together, these findings correlate the cytoplasmic linkage groups with a cytoplasmic DNA. The question is which one? The two known candidates are chloroplast DNA and mitochondrial DNA. Of the two, only chloroplast DNA has yet been identified with certainty in *Chlamydomonas*, but it is assumed that mitochondrial DNA will also be identified when technical difficulties are resolved. Thus the cytoplasmic linkage group could be associated with either of the organelle DNAs. The following points summarize the reasons for choosing chloroplast rather than mitochondrial DNA as the likely carrier.

(4)  The genetic evidence of diploidy is consistent with the presence of one chloroplast per cell containing two copies of chloroplast DNA and would be

difficult to reconcile with the numerous mitochondria present, which presumably contain many copies of mitochondrial DNA.

(5)   The genetic evidence of diploidy is also consistent with the cytochemical evidence (Ris & Plaut 1962) of two Feulgen-positive bodies in the chloroplast.

(6)   Mutations of the cytoplasmic genes mapped in *Chlamydomonas* were obtained by streptomycin treatment, which is known to affect chloroplast development in *Euglena* and in higher plants, but not known to be an effective mutagen in any system except the chloroplast.

(7)   Studies of the different fates of chloroplast DNA from the two parents have shown that in maternal zygotes, only that from the female ($mt+$) parent is preserved and replicated, while that from the male ($mt-$) is degraded (Sager & Lane 1969, Sager & Lane 1972). This pattern of transmission parallels the maternal inheritance of cytoplasmic genes.

(8)   The phenotypic properties of cytoplasmic genes in this linkage group including loss of photosynthetic capacity and resistance to, and dependence on antibiotics that affect the functioning of chloroplast ribosomes, are consistent with the location of these genes in chloroplast DNA.

Questions have been raised about the designation of this linkage group as chloroplast rather than mitochondrial; they will now be briefly discussed.

(1)   Studies of chloroplast DNA (Sager & Ishida 1963) by reannealing kinetics have revealed the presence of a major genomic class present in multiple copies (Bastia *et al*. 1971, Wells & Sager 1971). Several explanations of this have been proposed including the possibility that the diploid linkage group is not located in chloroplast DNA. In our view, it is likely that chloroplast DNA contains redundant regions which are particularly prominent in reannealing experiments and mask the presence of slower fractions which may also be present.

(2)   In a series of papers, Schimmer and Arnold (1969, 1970a,b,c) have proposed that an *sd*3 marker (cytoplasmic streptomycin-dependence) gene that they have been studying may be mitochondrial because of its peculiar reversion rates to sensitivity and back to dependence. These reversion rates suggested the presence of many copies of the *sd*3 gene. However, extensive studies of *sd* strains in our laboratory (Sager & Ramanis, in preparation) have shown that growth of *sd* in the absence of streptomycin is itself a mutagenic treatment, and that streptomycin-sensitive strains recovered in experiments like those of Schimmer and Arnold are actually heterozygous diploids, i.e. cytohets, carrying one genetic copy of *sd* and one of either *ss* or *sr*. These clones, which we call 'persistent cytohets', segregate infrequently in mitosis, but when they do, they produce *sd* sub-clones, which are not revertants but segregants. The *sd* genes that behave in this way are linked in the cytoplasmic linkage group described above. Thus, the *sd* gene is a chloroplast, not a mitochondrial component.

# 6 REFERENCES

BASTIA D., CHIANG K.S., SWIFT H. & SIERSMA P. (1971) Heterogeneity, complexity and repetition of the chloroplast DNA of *Chlamydomonas reinhardtii*. *Proc. natn. Acad. Sci. U.S.A.* **68**, 1157–61.

BOYNTON J.E., GILLHAM N.W. & BURKHOLDER B. (1970) Mutations altering chloroplast ribosomes phenotype in *Chlamydomonas*. II. A new Mendelian mutation. *Proc. natn. Acad. Sci. U.S.A.* **67**, 1505–12.

CHANCE B. & SAGER R. (1957) Oxygen and light induced oxidations of cytochrome, flavoprotein, and pyridine nucleotide in a *Chlamydomonas* mutant. *Pl. Physiol., Lancaster*, **32**, 548–60.

CHIANG K.S. (1971) Replication, transmission and recombination of cytoplasmic DNA's in *Chlamydomonas reinhardi*. In *Autonomy and biogenesis of mitochondria and chloroplasts*. Boardman N.K., Linnane A.W. & Smillie R.M. eds. pp. 235–49. North-Holland, Publ. Amsterdam.

CHUA N.H. & LEVINE R.P. (1969) The photosynthetic electron transport chain of *Chlamydomonas reinhardi*. VIII. The 520nm light-induced absorbance change in the wild-type and mutant strains. *Pl. Physiol., Lancaster*, **44**, 1–6.

COEN D., DEUTSCH J., NETTER P., PETROCHILO E. & SLONIMSKI P. (1970) Mitochondrial genetics. I. Methodology and phenomenology. In *Control of Organelle Development*. Soc. Expt. Biol. Symp. 24, ed. Miller P.L. pp. 449–96, Cambridge Univ. Press, London.

COLEMAN A.W. (1962) Sexuality. In *Physiology and biochemistry of algae*, ed. Lewin R.A. pp. 711–29. Academic Press, New York & London.

DAVIES D.R. & LAWRENCE C.W. (1967) The mechanism of recombination in *Chlamydomonas reinhardi*. II. The influence of inhibitors of DNA synthesis on intergenic recombination. *Mutation Res.* **4**, 147–54.

DAVIES D.R. & PLASKITT A. (1971) Genetical and structural analyses of cell-wall formation in *Chlamydomonas reinhardi*. *Genet. Res.* **17**, 33–43.

EBERSOLD W.T. (1956) Crossing over in *Chlamydomonas reinhardi*. *Am. J. Bot.* **43**, 408–10.

EBERSOLD W.T. (1962) Biochemical genetics. In *Physiology and Biochemistry of Algae*, ed. Lewin R.A. pp. 731–9. Academic Press, New York & London.

EBERSOLD W.T. (1963) Heterozygous diploid strains of *Chlamydomonas reinhardi*. *Genetics, Princeton*, **48**, 888.

EBERSOLD, W.T. (1967) *Chlamydomonas reinhardi*: heterozygous diploid strains. *Science, N.Y.* **157**, 447–9.

EBERSOLD W.T. & LEVINE R.P. (1958) A genetic analysis of linkage group I of *Chlamydomonas reinhardi*. *Z. Vererb. Lehre*, **90**, 74–82.

EBERSOLD W.T., LEVINE R.P., LEVINE E.E. & OLMSTED M.A. (1962) Linkage maps in *Chlamydomonas reinhardi*. *Genetics, Princeton*, **47**, 531–43.

EBERSOLE R.A. (1956) Biochemical mutants of *Chlamydomonas reinhardi*. *Am. J. Bot.* **43**, 404–7.

EPEL B.L. & LEVINE R.P. (1971) Mutant strains of *Chlamydomonas reinhardi* with lesions on the oxidizing side of photosystem II. *Biochim. biophys. Acta*, **226**, 154–70.

FINCHAM J.R.S. & DAY P.R. (1963) *Fungal Genetics*. Blackwell Scientific Publications, Oxford.

FOGG G.E., STEWART W.D.P., FAY P. & WALSBY A.E. (1973) *The Blue-green Algae*. Academic Press, London & New York.

FRIEDMANN I., COLWIN A.L. & COLWIN L.H. (1968) Fine-structural aspects of fertilization in *Chlamydomonas reinhardi*. *J. Cell Science*, **3**, 115–28.

GARNIER J. & MAROC J. (1971) Studies of several carriers, particularly cytochromes b-559 and c-553 in three non-photosynthetic mutants of *Chlamydomonas reinhardi*. *Biochim. biophys. Acta*, **205**, 205–19.

GILLHAM N.W. (1963a) The nature of exceptions to the pattern of uniparental inheritance for high level streptomycin resistance in *Chlamydomonas reinhardi. Genetics,* **48,** 431–9.

GILLHAM N.W. (1963b) Transmission and segregation of a non-chromosomal factor controlling streptomycin resistance in diploid *Chlamydomonas. Nature, Lond.* **200,** 294.

GILLHAM N.W. (1965a) Linkage and recombination between non-chromosomal mutations in *Chlamydomonas reinhardi. Proc. natn. Acad. Sci. U.S.A.* **40,** 356–63.

GILLHAM N.W. (1965b) Induction of chromosomal and non-chromosomal mutation in *Chlamydomonas reinhardi* with N-methyl-N'-nitro-N-nitrosoguanidine. *Genetics, Princeton,* **52,** 529–37.

GILLHAM N.W. (1969) Uniparental inheritance in *Chlamydomonas reinhardi. Am. Nat.* **103,** 355–88.

GILLHAM N.W., BOYNTON J.E. & BURKHOLDER B. (1970) Mutations altering chloroplast ribosome phenotype in *Chlamydomonas.* I. Non-Mendelian mutations. *Proc. natn. Acad. Sci. U.S.A.* **67,** 1026–33.

GILLHAM N.W. & FIFER W. (1968) Recombination of non-chromosomal mutations: A three-point cross in the green alga *Chlamydomonas reinhardi. Science, N.Y.* **162,** 683–4.

GILLHAM N.W. & LEVINE R.P. (1962) Studies on the origin of streptomycin resistant mutants in *Chlamydomonas reinhardi. Genetics, Princeton,* **47,** 1463–74.

GIVAN A.L. & LEVINE R.P. (1967) The photosynthetic electron transport chain of *Chlamydomonas reinhardi.* VII. Photosynthetic phosphorylation by a mutant strain of *Chlamydomonas reinhardi* deficient in active P700. *Pl. Physiol. Lancaster,* **42,** 1264–8.

GOLDSTEIN M. (1964) Speciation and mating behavior in *Eudorina. J. Protozool.* **11,** 317–44.

GOLDSTEIN M. (1967) Colony differentiation in *Eudorina. Canad. J. Bot.* **45,** 1591–6.

GOODENOUGH U.W., ARMSTRONG J.J. & LEVINE R.P. (1969) Photosynthetic properties of *ac-31,* a mutant strain of *Chlamydomonas reinhardi* devoid of chloroplast membrane stacking. *Pl. Physiol., Lancaster,* **44,** 1001–12.

GOODENOUGH U.W. & LEVINE R.P. (1969) Chloroplast ultrastructure in mutant strains of *Chlamydomonas reinhardi* lacking components of the photosynthetic apparatus. *Pl. Physiol., Lancaster,* **44,** 990–1000.

GOODENOUGH U.W. & LEVINE R.P. (1970) Chloroplast structure and function in *ac-20,* a mutant strain of *Chlamydomonas reinhardi. J. Cell. Biol.* **44,** 547–62.

GORMAN D.S. & LEVINE R.P. (1966) Photosynthetic electron transport chain of *Chlamydomonas reinhardi.* VI. Electron transport in mutant strain lacking either cytochrome 553 or plastocyanin. *Pl. Physiol., Lancaster,* **41,** 1648–56.

GOWANS C.S. (1960) Some genetic investigations on *Chlamydomonas eugametos. Z. Vererb. Lehre,* **91,** 63–73.

GOWANS C.S. (1963) The conspecificity of *Chlamydomonas eugametos* and *Chlamydomonas moewusii:* an experimental approach. *Phycologia,* **3,** 37–44.

GUILLARD R.R.L. (1960) A mutant of *Chlamydomonas moewusii* lacking contractile vacuoles. *J. Protozool,* **7,** 262–8.

HARTSHORNE J.N. (1955) Multiple mutations in *Chlamydomonas reinhardi. Heredity,* **9,** 239–248.

HASTINGS P.J., LEVINE E.E., COSBEY E., HUDOCK M.O., GILLHAM N.W., SURZYCKI S.J. & LEVINE R.P. (1965) The linkage groups of *Chlamydomonas reinhardi. Microbiol. Genet. Bull.* **23,** 17–19.

KATES J.R. & JONES R.F. (1964) The control of gametic differentiation in liquid cultures of *Chlamydomonas. J. Cell Comp. Physiol.* **63,** 157–64.

KATES J.R., CHIANG K.S. & JONES R.F. (1968) Studies on DNA replication during synchronized vegetative growth and gametic differentiation in *Chlamydomonas reinhardtii. Expl. Cell Res.* **49,** 121–35.

LAVOREL J. & LEVINE R.P. (1968) Fluorescence properties of wild type *Chlamydomonas reinhardi* and three mutant strains having impaired photosynthesis. *Pl. Physiol., Lancaster,* **43,** 1049–55.

M

LAWRENCE C.W. (1965a) Influence of non-lethal doses of radiation on recombination in *Chlamydomonas reinhardi*. *Nature, Lond.* **206**, 789–91.

LAWRENCE C.W. (1965b) The effect of dose duration in the influence of irradiation on recombination in *Chlamydomonas*. *Mutation Res.* **2**, 487–93.

LAWRENCE C.W. (1967) Influence of non-lethal doses of radiation on allelic recombination in *Chlamydomonas reinhardi*. *Genet. Res.* **9**, 123–7.

LAWRENCE C.W. (1968) Radiation effect on genetic recombination in *Chlamydomonas reinhardi*. In *The Effect of Ionizing Radiations on Meiotic Systems*, I.A.E.A. Study Group Meeting, Vienna, p. 135–44.

LAWRENCE C.W. (1971a) Effect of gamma radiation and alpha particles on gene recombination in *Chlamydomonas reinhardi*. *Mutation Res.* **10**, 545–55.

LAWRENCE C.W. (1971b) Dose dependence for radiation-induced allelic recombination in *Chlamydomonas reinhardi*. *Mutation Res.* **10**, 557–66.

LAWRENCE C.W. & DAVIES D.R. (1967) The mechanism of recombination in *Chlamydomonas reinhardi*. I. The influence of inhibitors of protein synthesis on intergenic recombination. *Mutation Res.* **4**, 137–46.

LEVINE R.P. (1960) Genetic control of photosynthesis in *Chlamydomonas reinhardi*. *Proc. natn. Acad. Sci. U.S.A.* **46**, 972–8.

LEVINE R.P. (1968) Genetic dissection of photosynthesis. *Science, N.Y.* **162**, 768–71.

LEVINE R.P. (1969) The analysis of photosynthesis using mutant strains of algae and higher plants. *A. Rev. Pl. Physiol.* **20**, 523–40.

LEVINE R.P. & EBERSOLD W.T. (1960) The genetics and cytology of *Chlamydomonas*. *A. Rev. Microbiol.* **14**, 197–216.

LEVINE R.P. & GOODENOUGH U.W. (1970) Genetics of photosynthesis and of the chloroplast of *Chlamydomonas reinhardi*. *A. Rev. Genetics*, **4**, 397–408.

LEVINE R.P. & GORMAN D.S. (1966) Photosynthetic electron transport chain of *Chlamydomonas reinhardi*. III. Light-induced absorbance changes on chloroplast fragments of wild type and mutant strains. *Pl. Physiol., Lancaster*, **41**, 1293–300.

LEVINE R.P. & PASZEWSKI A. (1970) Chloroplast structure and function in *ac-20*, a mutant strain of *Chlamydomonas reinhardi*. II. Photosynthetic electron transport. *J. Cell. Biol.* **44**, 540–6.

LEVINE R.P. & SMILLIE R.M. (1963) Photosynthetic electron transport chain of *Chlamydomonas reinhardi*. I. Triphosphopyridine nucleotide-photoreduction in wild type and mutant strains. *J. biol. Chem.* **238**, 4052–7.

LEVINE R.P. & TOGASAKI R.K. (1965) A mutant strain of *Chlamydomonas reinhardi* lacking ribulose diphosphate carboxylase activity. *Proc. natn. Acad. Sci. U.S.A.* **53**, 987–90.

LEWIN R.A. (1954) Mutants of *Chlamydomonas moewusii* with impaired motility. *J. gen. Microbiol.* **11**, 358–63.

LOPPES R. (1968) Ethyl methanesulfonate: an effective mutagen in *Chlamydomonas reinhardi*. *Molec. Gen. Genetics*, **102**, 229–31.

McBRIDE J.C. & GOWANS C.S. (1967) Pyrithiamine resistance in *Chlamydomonas eugametos*. *Genetics*, **56**, 405–12.

McVITTIE A. (1972) Flagellum mutants of *Chlamydomonas reinhardii*. *J. gen. Microbiol.* **71**, 525–40.

McVITTIE A. & DAVIES D.R. (1971) The location of the linkage groups in *Chlamydomonas reinhardii*. *Molec. Gen. Genetics*, **112**, 225–8.

METS L.J. & BOGORAD L. (1971) Mendelian and uniparental alterations in erythromycin binding by plastid ribosomes. *Science, N.Y.* **174**, 707–9.

MISHRA N.C. & THRELKELD S.F.H. (1968) Genetic studies in *Eudorina*. *Genet. Res.* **11**, 21–31.

NAKAMURA K. & GOWANS C.S. (1965) Genetic control of nicotinic acid metabolism in *Chlamydomonas eugametos*. *Genetics, Princeton*, **51**, 931–45.

NAKAMURA K. & GOWANS C.S. (1967) Ionic remediability of a mutational transport defect in *Chlamydomonas*. *J. Bact.* **93**, 1185–7.

NOMURA M. (1970) Bacterial ribosome. *Bact. Rev.* **34**, 228–77.

RANDALL J. (1969) The flagellar apparatus as a model organelle for the study of growth and morphogenesis. *Proc. R. Soc., B,* **173**, 31–62.

RANDALL J., WARR J.R., HOPKINS J.M. & McVITTIE A. (1964) A single-gene mutation of *Chlamydomonas reinhardii* effecting motility: A genetic and electron microscope study. *Nature, Lond.* **203**, 912–14.

RANDALL J.T., CAVALIER-SMITH T., McVITTIE A., WARR J.R. & HOPKINS J.M. (1967) Developmental and control processes in the basal bodies and flagella of *Chlamydomonas reinhardii*. In *Control mechanisms in development processes. Symp. Soc. Devel. Biol.* **26**, 43–83.

RANNEBERG H. & ARNOLD C.G. (1968) Die wirkung von Athylmethansulfonat auf *Chlamydomonas reinhardi*. *Molec. Gen. Genetics*, **101**, 212–16.

RIS H. & PLAUT W. (1962) Ultrastructure of DNA containing areas in the chloroplast of *Chlamydomonas*. *J. Cell. Biol.* **13**, 383–91.

ROSEN H. & EBERSOLD W.T. (1972) Recombination in relation to ultraviolet sensitivity in *Chlamydomonas reinhardi*. *Genetics*, **71**, 247–53.

SAGER R. (1954) Mendelian and non-Mendelian inheritance of streptomycin resistance in *Chlamydomonas reinhardi*. *Proc. natn. Acad. U.S.A.* **40**, 356–63.

SAGER R. (1955) Inheritance in the green alga *Chlamydomonas reinhardi*. *Genetics, Princeton*, **40**, 476.

SAGER R. (1960) Genetic systems in *Chlamydomonas*. *Science, N.Y.* **132**, 1459–65.

SAGER R. (1962a) Streptomycin as a mutagen for non-chromosomal genes. *Proc. natn. Acad. Sci. U.S.A.* **48**, 2018–26.

SAGER R. (1962b) A non-mapable factor in *Chlamydomonas*. *Genetics*, **47**, 982.

SAGER R. (1968) Cytoplasmic genes and organelle formation. In *Formation and fate of cell organelles*. Annual Symp. of the International Society for Cell Biol., ed. Warren K.G. 317–34. Academic Press, New York & London.

SAGER R. (1972) *Cytoplasmic genes and organelles*. Academic Press, New York & London.

SAGER R. & GRANICK S. (1954) Nutritional control of sexuality in *Chlamydomonas reinhardi*. *J. gen. Physiol.* **37**, 729–42.

SAGER R. & ISHIDA M.R. (1963) Chloroplast DNA in *Chlamydomonas*. *Proc. natn. Acad. Sci. U.S.A.* **50**, 725–30.

SAGER R. & LANE D. (1969) Replication of chloroplast DNA in zygotes of *Chlamydomonas*. *Federation Proc.* **28**, 347.

SAGER R. & LANE D. (1972) Molecular basis of maternal inheritance. *Proc. natn. Acad. Sci. U.S.A.* **69**, 2410–3.

SAGER R. & RAMANIS Z. (1963) The particulate nature of non-chromosomal genes in *Chlamydomonas*. *Proc. natn. Acad. Sci. U.S.A.* **50**, 260–8.

SAGER R. & RAMANIS Z. (1964) Recombination of non-chromosomal genes in *Chlamydomonas*. *Genetics*, **50**, 282.

SAGER R. & RAMANIS Z. (1965) Recombination of non-chromosomal genes in *Chlamydomonas*. *Proc. natn. Acad. Sci. U.S.A.* **53**, 1053–60.

SAGER R. & RAMANIS Z. (1967) Biparental inheritance of non-chromosomal genes induced by ultraviolet irradiation. *Proc. natn. Acad. Sci. U.S.A.* **58**, 931–7.

SAGER R. & RAMANIS Z. (1968) The pattern of segregation of cytoplasmic genes in *Chlamydomonas*. *Proc. natn. Acad. Sci. U.S.A.* **61**, 324–31.

SAGER R. & RAMANIS Z. (1970a) A genetic map of non-Mendelian genes in *Chlamydomonas*. *Proc. natn. Acad. Sci. U.S.A.* **65**, 593–600.

SAGER R. & RAMANIS Z. (1970b) Genetic studies of chloroplast DNA in *Chlamydomonas*. In *Control of Organelle Development*, XXIV Symp. S.E.B. ed. Miller P.L. 401–17, Cambridge Univ. Press, London.

SAGER R. & RAMANIS Z. (1970c) Methods of genetic analysis of chloroplast DNA in *Chlamydomonas*. In *Autonomy and Biogenesis of Mitochondria and Chloroplasts*, eds. Boardman N.K., Linnane A.W. & Smillie R.M. pp. 250–9. North-Holland, Publ., Amsterdam.

SAGER R. & RAMANIS Z. (1971) Formal genetic analysis of organelle genetic systems. In *2nd L.J. Stadler Genetics Symp.*, 65. University of Missouri Press.

SAGER R. & TSUBO Y. (1961) Genetic analysis of streptomycin resistance and dependence in *Chlamydomonas*. *Z. Vererb. Lehre*, **92**, 430–8.

SCHIMMER O. & ARNOLD C.G. (1969) Untersuchungen zur Lokalisation eines ausserkaryotischen Gens bei *Chlamydomonas reinhardii*. *Arch. Mikrobiol.* **66**, 199–202.

SCHIMMER O. & ARNOLD C.G. (1970a) Untersuchungen uber Reversions und Segregationsverhalten eines ausserkaryotischen Gens von *Chlamydomonas reinhardii* zur Bestimmung des Erbtragers. *Molec. Gen. Genetics*, **107**, 281–90.

SCHIMMER O. & ARNOLD C.G. (1970b) Hin-und Racksegregation eines ausserkaryotischen Gens bei *Chlamydomonas reinhardii*. *Molec. Gen. Genetics*, **108**, 33–40.

SCHIMMER O. & ARNOLD C.G. (1970c) Uber die Zahl der Kopien eines ausserkaryotischen Gens bei *Chlamydomonas reinhardii*. *Molec. Gen. Genetics*, **107**, 366–71.

SCHIMMER O. & ARNOLD C.G. (1970d) Die Suppression der ausserkaryotisch bedingten streptomycin-abhaengigkeit bein *Chlamydomonas reinhardii*. *Arch. Mikrobiol.* **73**, 195–200.

SCHLANGER G., SAGER R. & RAMANIS Z. (1972) Mutation of a cytoplasmic gene in *Chlamydomonas* alters chloroplast ribosome function. *Proc. natn. Acad. Sci. U.S.A.* **69**, 3551–5.

SCHWEIGER H.G. (1970) Synthesis of RNA in *Acetabularia*. In *Control of Organelle Development*. XXIV Symp. S.E.B., ed. Miller P.L. 327–44. Cambridge University Press, London.

SHEPHARD D.C. (1965) An autoradiographic comparison of the effects of enucleation and actinomycin D on the incorporation of nucleic acid and protein precursors by *Acetabularia* chloroplasts. *Biochim. biophys. Acta*, **108**, 635–43.

SMILLIE R.M. & LEVINE R.P. (1963) Photosynthetic electron transport chain of *Chlamydomonas reinhardi*. II. Components of the triphosphopyridine nucleotide-photoreduction pathway in wild type and mutant strains. *J. biol. Chem.* **238**, 4058–62.

SMILLIE R.M., SCOTT N.S. & GRAHAM D. (1968) Biogenesis of chloroplasts: Role of chloroplast DNA and chloroplast ribosomes. In *Comparative Biochemistry and Biophysics of Photosynthesis*, eds. Shibata K., Takamiya A., Jagendorf A.T. & Fuller R.F. pp. 332–53. Univ. Tokyo Press, Tokyo.

SMITH G.M. (1948) Sexuality in *Chlamydomonas*. *Science, N.Y.* **108**, 680–1.

SMITH G.M. & REGNERY D.C. (1950) Inheritance of sexuality in *Chlamydomonas reinhardi*. *Proc. natn. Acad. Sci. U.S.A.* **36**, 246–8.

STARR R.C. (1954) Inheritance of mating type and a lethal factor in *Cosmarium botrytis* var. *subtumidum* Wittr. *Proc. natn. Acad. Sci. U.S.A.* **40**, 1060–3.

STARR R.C. (1971) Control of Differentiation in *Volvox*. *Symp. Soc. Devel. Biol.* **29**, 59.

STARLING D. (1969) Complementation tests on closely linked flagellar genes in *Chlamydomonas reinhardii*. *Genet. Res.* **14**, 343–7.

STUTZ E. (1971) Characterization of *Euglena gracilis* chloroplast single strand DNA. In *Autonomy and Biogenesis of Mitochondria and Chloroplasts*, eds. Boardman N.K., Linnane A.W. & Smillie R.M. p. 277–81. North-Holland, Publ., Amsterdam.

SUEOKA N., CHIANG K.S. & KATES J.R. (1967) Deoxyribonucleic acid replication in meiosis of *Chlamydomonas reinhardi*. *J. molec. Biol.* **25**, 47–66.

SURZYCKI S.J. (1969) Genetic functions of the chloroplast of *Chlamydomonas reinhardi*. Effect of rifampin on chloroplast DNA-dependent RNA polymerase. *Proc. natn. Acad. Sci. U.S.A.* **63**, 1327–34.

SURZYCKI S.J. & GILLHAM N.W. (1971) Organelle mutations and their expression in *Chlamydomonas reinhardi*. *Proc. natn. Acad. Sci. U.S.A.* **68**, 1301–6.

SURZYCKI S.J., GOODENOUGH U.W., LEVINE R.P. & ARMSTRONG J.J. (1970) Nuclear and chloroplast control of chloroplast structure and function in *Chlamydomonas reinhardi*.

In *Control of Organelle Development*, XXIV Symp. S.E.B. ed. Miller P.L. 13–38, Cambridge University Press, London.

SURZYCKI S.J. & ROCHAIZ J.D. (1971) Transcriptional mapping of r-RNA genes of the chloroplast and nucleus of *Chlamydomonas reinhardi*. *J. molec. Biol.* **62**, 89–109.

TOGASAKI R.K. & LEVINE R.P. (1970) Chloroplast structure and function in *ac-20*, a mutant strain of *Chlamydomonas reinhardi*. I. $CO_2$ fixation and ribulose-1,5-diphosphate carboxylase synthesis. *J. Cell. Biol.* **44**, 531–9.

WARR J.R. (1968) A mutant of *Chlamydomonas reinhardii* with abnormal cell division. *J. gen. Microbiol.* **52**, 243–51.

WARR J.R. & DURBER S. (1971) Studies on the expression of a mutant with abnormal cell division in *Chlamydomonas reinhardi*. *Expl. Cell Res.* **64**, 463–9.

WARR J.R., McVITTIE A., RANDALL J. & HOPKINS J.M. (1966) Genetic control of flagellar structure in *Chlamydomonas reinhardii*. *Genet. Res.* **7**, 335–51.

WEAVER E.C. (1969) Kinetic behavior of the paramagnetic resonance Signal I. II. Comparison of wild type and mutant (*ac-206*). *Chlamydomonas reinhardtii*. *Pl. Physiol., Lancaster,* **44**, 1538–41.

WELLS R. & SAGER R. (1971) Denaturation and renaturation kinetics of chloroplast DNA from *Chlamydomonas reinhardi*. *J. molec. Biol.* **58**, 611–22.

ZIMMERMAN R.A., GARVIN R.T. & GORINI L. (1971) Alteration of a 30S ribosomal protein accompanying the *ram* mutation in *Escherichia coli*. *Proc. natn. Acad. Sci. U.S.A.* **68**, 2263–7.

CHAPTER 12

# LIGHT ABSORPTION, EMISSION AND PHOTOSYNTHESIS

## GOVINDJEE AND BARBARA ZILINSKAS BRAUN

Department of Botany,
University of Illinois,
Urbana, Illinois 61801, U.S.A.

1    Introduction   346

2    Absorption Spectra   353
2.1  Pigments in solution   353
2.2  Chlorophylls   356
2.3  Carotenoids   361
2.4  Phycobilins   363

3    Separation of algal photosyn-
     thetic systems   365

4    Composition of the two pigment
     systems   369

5    Light-induced absorption chan-
     ges and photosynthesis   373
5.1  P680-P690   375
5.2  C550   376
5.3  P430   376

6    Adaptation   377

7    References   379

## 1  INTRODUCTION

Solar radiation provides the energy to maintain life on earth through photo-synthesis—the process by which green plants convert light energy into chemical energy. The first act of photosynthesis *is* light absorption. Out of all the electro-magnetic radiation falling on photosynthesizing plants (including algae) only the visible light (wavelength ($\lambda$) range, 400 to 720nm) is absorbed and used for photo-synthesis. This information is obtained by the measurements of the absorption spectra (absorbance as a function of wavelength of electromagnetic radiation) and of the action spectra of photosynthesis ($O_2$ evolution per incident quantum as a function of wavelength of light).

Algae have evolved various pigments for the purpose of light absorption; these can be classified into three major groups: (1) *chlorophylls* (Chl) that strongly absorb blue and red light (see Chapter 5, p. 161)—examples are Chl *a* (present in all algae) and Chl *b* (present in green algae); (2) *carotenoids* that absorb blue and green light—examples are β-carotene (present in all algae) and fucoxanthin (present in brown algae) (see Chapter 6, p. 176); (3) *phycobilins* that absorb green, yellow and orange light (see Chapter 6, p. 194). Examples are R-phycoerythrin (present in red algae) and C-phycocyanin (present in blue-green algae). These bulk pigments, called such as they are present in large quantities, provide the algae with *antennae* to capture the light energy.

Absorption spectra of algae provide us with the knowledge of the types and concentrations of pigments present in them. Such a knowledge is necessary for the complete understanding of photosynthesis, and it can also be used for taxonomic and phylogenetic purposes. Likewise, absorption spectra have furnished information regarding the adaptability of algae to varied environments. The photochemist has used absorption measurements in determining the natural lifetime of the excited state as well as the energy levels of the molecules under study. Absorption measurements are absolutely essential for calculations of quantum yields of photosynthesis (number of molecules of $O_2$ evolved or of an intermediate phototransformed into its oxidized or reduced form per absorbed quantum) and of Chl $a$ fluorescence (number of quanta emitted per quanta absorbed) as we need these measurements for the calculation of the number of absorbed quanta (Na) from the number of incident quanta (Ni) (Na = Ni × fractional absorption). The quantum yield of oxygen evolution ($\phi_{O_2}$) sets up the framework for the feasibility of various theories of photosynthesis; the quantum yield of a specific intermediate reaction allows us to judge whether it is in the main path of the process or not. Quantum yields of Chl $a$ fluorescence ($\phi_f$) and of $O_2$ evolution as a function of wavelength are necessary for calculations of the efficiency of excitation energy transfer from one pigment to another.

In the first approximation, absorption spectra can be used to predict photosynthetic spectral response as in the case of green algae, brown algae, diatoms and dinoflagellates (Emerson & Lewis 1943, Haxo & Blinks 1950, Tanada 1951, Haxo 1960). However, in these cases, the predictions are approximate because of the relative inefficiency of the carotenoids and the existence of the 'red drop' phenomena, that is, decline in the quantum yield of $O_2$ evolution ($\phi_{O_2}$) and of Chl $a$ fluorescence ($\phi_f$) in the red end of the spectrum (Emerson & Lewis 1943, Duysens 1952, Emerson *et al.* 1957, Govindjee 1960, Govindjee 1963a, Das & Govindjee 1967, Szalay *et al.* 1967, Das *et al.* 1968, R. Govindjee *et al.* 1968, Williams *et al.* 1969). In red and blue-green algae, absorption spectra cannot be used at all for these purposes as the red drop begins well before the decline of the main absorption band of Chl $a$ and is accompanied by a 'blue drop', that is, a drop in the yield in the blue end of the spectrum (Haxo & Blinks 1950, Duysens 1952, Brody & Emerson 1959a, Blinks 1960, Haxo 1960, Hoch & Kok 1961, Fork 1963, Papageorgiou & Govindjee 1967a).

*Two Light Reactions.* In photosynthesis, there are two primary photoreactions (I and II) and two pigment systems (photosystem I and photosystem II) that 'sensitize' them. This idea originated in the discovery of the Emerson enhancement effect (referred to by most authors as 'enhancement effect'; in earlier work it was called 'second Emerson effect' or simply 'Emerson effect'), that is, the rate of photosynthesis being greater when short wavelength light (absorbed mainly by Chl *b* or another accessory pigment) and far red light (absorbed mainly by Chl *a*) are used simultaneously than the sum of the two when they were used separately (Emerson *et al.* 1957, Emerson & Rabinowitch 1960, French *et al.* 1960, Govindjee & Rabinowitch 1960, Govindjee 1963a, Blinks 1963, Myers 1963, R. Govindjee *et al.* 1964, Myers 1971). It is widely accepted that two light reactions, arranged in series, are involved in this process (Hill & Bendall 1960, Kautsky *et al.* 1960, Duysens *et al.* 1961, Witt *et al.* 1961, see Chapter 13, p. 391). The first light reaction—arbitrarily called light reaction II—leads to the reduction of a cytochrome and oxidation of $H_2O$ to molecular $O_2$; the second light reaction (called light reaction I) results in reoxidation of the cytochrome and reduction of NADP. Along the electron transport pathway from $H_2O$ to NADP, a fraction of light energy is utilized to synthesize ATP from ADP and inorganic phosphate (Fig. 12.1). With sufficient NADPH and ATP available, enzymatic reduction of $CO_2$ to the carbohydrate level (Calvin-Benson cycle) becomes possible. Such a scheme and its details have been confirmed and elaborated and can be found in various reviews: Hoch and Kok 1961 (general), Rabinowitch

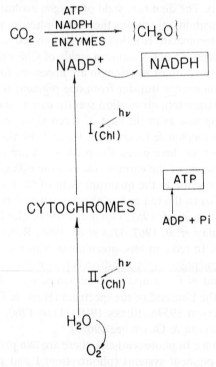

**Fig. 12.1.** Transfer of electrons (or hydrogen atoms) from water to carbon dioxide involving two light reactions. ADP, adenosine diphosphate; Pi, inorganic phosphate; ATP, adenosine triphosphate; $NADP^+$, nicotinamide adenine dinucleotide phosphate; NADPH, reduced $NADP^+$; I and II, light reactions; Chl, chlorophyll; $(CH_2O)$, carbohydrate moiety; h$\nu$, light quanta.

1963 (general), Smith and French 1963 (pigments); Duysens 1964 (biophysical aspects), Robinson 1964 (physico-chemical aspects); Vernon and Avron 1965 (biochemical aspects), Witt 1967 (fast absorbance changes), Avron and Neumann 1968 (photophosphorylation), Bendall and Hill 1968 (haem proteins), Hind and Olson 1968 (electron transport), Weaver 1968 (electron spin resonance studies), Fork and Amesz 1969 (energy transfer), Levine 1969 (mutants), Rabinowitch and Govindjee 1969 (general mechanisms), Boardman 1970 (physical separation of pigment systems), Cheniae 1970 ($O_2$ evolution), Packer et al. 1970 (ion movements and structural changes), Bishop 1971 (intermediates), Clayton 1971 (general, cooperation of pigment systems), Govindjee and Papageorgiou 1971 (chlorophyll fluorescence), Park and Sane 1971 (structure and function), Govindjee and Mohanty 1972 (photosynthesis of blue-green algae); see also Chapter 13, p. 391.

However, several other models have been proposed in the last ten years. Franck and Rosenberg (1964), Govindjee et al. (1967) (see Jackson & Volk 1970), Arnold and Azzi (1968) and Knaff and Arnon (1969) have proposed alternate models. References are given here so that the reader may consult them to keep an open mind toward future developments in this field. In this connexion, the experiment of Rurainski et al. (1971) is worth mentioning in which they could not find a relationship between NADP reduction and the energy trap of photosystem I in chloroplasts treated with $MgCl_2$—the implication being that NADP can be reduced by a system (photosystem II) independent of photosystem I!

*Two Pigment Systems.* It appears that most of the pigments are present in both photosystems but in different proportions; for example, photosystem II of the green algae has a lower ratio of Chl *a/b* than photosystem I, and those algae containing phycobilins usually have a larger proportion of these pigments in photosystem II. Another complexity is that as Chl *a* does not exist as a single homogeneous species *in vivo*, but rather in several spectroscopically distinguishable forms labelled Chl *a* 660, Chl *a* 670, Chl *a* 680, Chl *a* 690, Chl *a* 695, and Chl *a* 705, according to the location of the absorption maxima in the red end of the spectrum (see Butler & Hopkins 1970a,b, and French 1971). The chemical nature of these forms is not yet known. It has been suggested (e.g. by Seely 1971) that the existence of so many forms increases the efficiency of excitation energy transfer to the reaction centres; this may be the reason why algae evolved them.

Many experiments, especially the spectral measurements of separated photosystem II and photosystem I fractions from algae and higher plants, suggest rather strongly that a larger proportion of the long wavelength Chl *a* forms is found in photosystem I, while the reverse holds for photosystem II. The fact that the long wavelength form of Chl *a* is non-, or weakly, fluorescent at room temperature (being an aggregate form, it loses more energy by radiationless transitions as it has more modes of interaction than do its monomer molecules) makes the localization of these Chl *a* forms in one or the other photosystem amenable to fluorescence techniques that are often far more sensitive than absorption measurements. For instance, Brown (1969) has shown for algae of several groups a heavy, photosystem II enriched fraction that is significantly more fluorescent than the light, photosystem I fraction, and has correctly inferred from these and other data that the long wavelength Chl *a* forms are preponderant in the photosystem I fraction. Emission spectra are equally helpful. Thus certain Chl *a* forms at low temperatures (77°K or 4°K) fluoresce strongly at 710 to 740nm, while fluorescence at 685 and 695nm is due mainly to the short wavelength Chl *a* forms. There are some specialized pigment molecules (traps) to which the energy absorbed by the bulk pigments is transferred. These traps, present in small quantities (one per several hundred bulk Chl molecules), are simply Chl *a* molecules in specialized environments. The trap molecules, also called *reaction centres*, perform the primary photochemical reactions of photosynthesis transforming light energy, now in the form of 'excitons', into chemical energy. The primary oxidation-reduction reaction is:

$$\text{D.T.A.} \xrightarrow{hv} \text{D.T*.A} \rightarrow \text{D.T}^+.\text{A}^- \rightarrow \text{D}^+.\text{T.A}^-,$$

where D is the primary donor of electrons, A the primary acceptor of electrons *hv* the photon or exciton, T the energy trap, and T* the excited energy trap. Two types of reaction centres have been implicated: P700 (Kok 1957, 1959, 1961, Rumberg & Witt 1964, Witt 1967) and P680-P690 (Döring *et al.* 1967, 1968, 1969, Govindjee *et al.* 1970, Floyd *et al.* 1971), where P stands for pigment and the number following for the location of the long wavelength absorption band, in nanometres (nm). P680-690 has only been measured in chloroplasts from higher plants. However, its existence in algae has been inferred from a new fluorescence band at about 695nm appearing under conditions when photosynthesis is saturated or absent (Bergeron 1963, Brody & Brody 1963, Govindjee 1963a, Krey & Govindjee 1966, Cho & Govindjee 1970a, Govindjee & Briantais 1972, see also Satoh 1972). Most bulk pigments, except carotenoids (although there are conflicting reports), emit some light as fluorescence upon absorption of light, i.e. all the absorbed energy is not used for chemistry. In solution, extracted algal pigments have a high yield of fluorescence, but *in vivo* the yield is low as most energy is used for chemistry (Latimer *et al.* 1957, Weber 1960). The fluorescence yield of Chl *a* ($\phi_f$) in algae is related to the chemical reactions of photosynthesis by the following relationship:

$$\phi_f = \frac{k_f}{k_f + k_r + k_t + k_c[T]} \tag{1}$$

where $k_f$ is the rate constant of fluorescence (i.e. the number of transitions per sec. leading to fluorescence), $k_r$ for radiationless loss, $k_t$ for energy transfer from fluorescent to non- or weakly fluorescent Chl *a*, $k_c$ for chemical reactions, and $[T]$ is the concentration of the 'open' traps, i.e. traps that are in a state ready to perform chemistry. This relationship tells us that fluorescence yield can be used as a monitor of photochemistry in algae. (For background in the physical aspects of the interaction of light with matter, see Clayton 1965, 1971, and Rabinowitch & Govindjee 1969, and for a detailed discussion of Chl fluorescence studies in photosynthesis see, Rabinowitch 1951, 1956, Weber 1960, Duysens 1964, Butler 1966a,b, Goedheer 1966a, Govindjee et al. 1967, Hoch & Knox 1968, Fork & Amesz 1969, Govindjee & Papageorgiou 1971).

**Table 12.1.** The quantum efficiency* of each decay process of the excited first singlet state of Chl *a in vivo* for photosystem II (after Mar *et al.* 1972).

| Sample | $\phi_f$ | $\phi_h$ | $\phi_t$ |
|---|---|---|---|
| *Chlorella* | 0·04 | 0·30 | 0·66 |
| *Porphyridium* | 0·02 | 0·30 | 0·68 |
| *Anacystis* | 0·02 | 0·38 | 0·60 |

$$* \ \phi_f = \frac{k_f}{k_f + k_h + k_t}; \ \phi_h = \frac{k_h}{k_f + k_h + k_t}; \ \phi_t = \frac{k_t}{k_f + k_h + k_t},$$ where $k$'s refer to the

rate constants, and the subscripts $f$, $h$ and $t$ are for fluorescence, heat loss and energy transfer to weakly fluorescent photosystem I, and trapping respectively.

*Excitation Energy Migration.* Light energy absorbed by carotenoids, phycobilins and Chl *b* is transferred to Chl *a* leading to fluorescence of Chl *a*. This 'sensitized fluorescence' has been used in determining the efficiency of excitation energy transfer in photosynthesis from the various accessory pigments to Chl *a* (heterogenous energy transfer) (Dutton *et al.* 1943, Wassink & Kersten 1946, Duysens 1952, French & Young 1952, Tomita & Rabinowitch 1962, Ghosh & Govindjee 1966, Cho & Govindjee 1970b,c). Energy migration within molecules of the same kind (homogeneous energy transfer) is often demonstrated by the depolarization of fluorescence (Arnold & Meek 1956, Mar & Govindjee 1971), by the concentration quenching of fluorescence or by the observation that the quantum yield of the primary photochemical reaction of photosynthesis is close to 1·0. However, the quantum yield of $O_2$ evolution in algae is 0·12 as there are 8 primary photochemical reactions in the evolution of one molecule of $O_2$ when $CO_2$ is used as the oxidant (Emerson 1958, R. Govindjee *et al.*, 1968).

*Lifetimes of Excited States.* Strongly absorbing compounds have short radiative lifetimes while weakly absorbing compounds have long ones (see Rabinowitch 1957, Calvert & Pitts 1966). The radiative life of chlorophyll *a* in solution is about 15 nsec. (Brody & Rabinowitch 1957). It is difficult to calculate the exact radiative life for chlorophyll *a in vivo* because of the presence of different forms of Chl *a* as well as Chl *b* or other accessory pigments, and the absence of definite information regarding the extinction coefficient *in vivo*, but making certain assumptions we can calculate an order of magnitude of about 20 nsec. However, the measured lifetime of the excited state of Chl *a* in algae (*Chlorella, Porphyridium* and *Anacystis* sp.) is of the order of 0·5 to 2 nsec. (Brody & Rabinowitch 1957, Tomita & Rabinowitch 1962, Müller *et al.* 1965, Murty & Rabinowitch 1965, Nicholson & Fortoul 1967, Singhal & Rabinowitch 1969, Merkelo

Table 12.2. Spectral bands of chlorophylls (after Smith & Benitez 1955, also see Klein & Cronquist 1967).

| Pigment and Solvent | Absorption Maxima and Specific Extinction (Base 10) coefficients | | | | | | Fluorescence* |
|---|---|---|---|---|---|---|---|
| Chl a (Ethyl Ether) | 410 (85·2) | 430 (131·5) | 534 (4·22) | 578 (9·27) | 615 (16·3) | 662 (100·9) | 669, 723 |
| Chl b (Ethyl Ether) | 430 (62·7) | 455 (174·8) | 549 (7·07) | 595 (12·7) | | 644 (62) | 649, 708 |
| Chl c (Ethyl Ether) | (417) | 444 (277) | (545) | 578 (20·6) | | 626 (22) | 629, 690 |
| Chl d (Ethyl Ether) | 392 (58·4) | 447 (97·8) | 512 (1·98) | 549 (4·03) | 595 (9·47) | 643 (14·3) 688 (110·4) | 696, 752 |
| | UV-Blue | | Green-Yellow | | Orange-Red | Red or I.R. | |

\* Italic figures correspond to the location of the main peak, the others to the location of the vibrational satellite band.

*et al.* 1969, Müller *et al.* 1969, Briantais *et al.* 1972, Mar *et al.* 1972). The observed discrepancies between the calculated and experimentally measured $\phi_f$ are due to the presence of non- or weakly fluorescent chlorophyll *a* in algae (see Govindjee *et al.* 1967 for refs.). From measurements of the lifetime of the excited state of Chl *a* in several algae, in the presence and the absence of photosynthesis, Mar *et al.* (1972) calculated the quantum efficiencies of fluorescence, of heat loss and of transfer to weakly fluorescent photosystem I ($\phi_h$), and of energy trapping ($\phi_t$) to be 0·02–0·04, 0·30–0·38, and 0·60–0·68 respectively (Table 12·1). Mohanty (1972) has calculated the efficiency of energy transfer from photosystem II to photosystem I to be of the order of 0·10 from the increase in the fluorescence yield by the addition of $Mg^{2+}$ to 3-(3, 4-dichlorophenyl)-1, 1 di-methylurea-treated chloroplasts.

## 2   ABSORPTION SPECTRA

### 2.1   *Pigments in solution*

Pigments in intact cells can only be identified if we know the absorption spectra of known pigments in solutions. Thus, a brief discussion follows. No attempt is made here to cite and discuss all the work. Such data are reviewed elsewhere (Rabinowitch 1951, 1956, Rabinowitch & Govindjee 1969 (all pigments), O'hEocha 1965 (phycobilins), Goedheer 1966b (chlorophylls), Davies 1965 (carotenoids)). Recently, Singhal *et al.* (1968) have presented absorption spectra of chlorophyll and its derivatives at 77°K and have reviewed the past work in this area. The location of the absorption peaks of various chlorophylls are presented in Table 12.2 and Fig. 12.2.

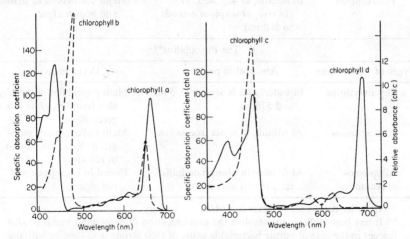

**Fig. 12.2.**   Absorption spectra of chlorophylls in diethyl ether. Curves on the left, ————, Chl *a*; — — — —, Chl *b*. Curves on the right, ————, Chl *d*; — — — —, Chl *c* (after Smith & Benitez 1955, redrawn from Allen *et al.* 1960).

**Table 12.3.** The photosynthetic pigments (after Rabinowitch & Govindjee 1969).

### A. The Chlorophylls*

| Type of Chlorophyll | Characteristic absorption peaks | | Occurrence |
| | In organic solvents, nm | In cells, nm | |
| --- | --- | --- | --- |
| Chl $a$ | 420, 662 | 435, 670–680 (several forms) | All algae |
| Chl $b$ | 455, 644 | 480, 650 (two forms?) | Green algae |
| Chl $c$ | 444, 626 | Red band at 645 | Diatoms and brown algae |
| Chl $d$ | 450, 690 | Red band at 740 | Reported in some red algae (?) |

### B. The Carotenoids***

| Types of Carotenoids | Characteristic absorption peaks, nm** | Occurrence |
| --- | --- | --- |

#### I. Carotenes

| | | |
| --- | --- | --- |
| α-carotene | In hexane, at 420, 440, 470 | In red algae and in siphonaceous green algae it is the major carotene |
| β-carotene | In hexane, at 425, 450, 480 (the 480nm band may be shifted to 500nm *in vivo*) | Main carotene of all other algae |

#### II. The Xanthophylls

| | | |
| --- | --- | --- |
| Lutein | In ethanol, at 425, 445, 475 | Major carotenoid of green algae and red algae |
| Fucoxanthin | In hexane, at 425, 450, 475 (*In vivo*, absorption extends to 580nm) | Major carotenoid of diatoms and brown algae |

### C. The Phycobilins***

| Types of Phycobilins | Absorption peaks | Occurrence |
| --- | --- | --- |
| Phycoerythrins | In water, and *in vivo*: 490, 546, and 576nm | Main phycobilin in red algae; also found in some blue-green algae |
| Phycocyanins | At 618nm, in water and *in vivo* | Main phycobilin of blue-green algae; also found in red algae |
| Allophyco-cyanin | At 654nm, in phosphate buffer (at pH 6·5) and *in vivo* | Found in blue-green and red algae |

* For fuller details see Chapter 5, p. 161.
** It has been difficult to establish the exact location of carotenoid bands *in vivo* (except in the case of purple bacteria) because of their strong over-lapping with the blue-violet bands of chlorophylls. The bands *in vivo* are estimated to be shifted by about 20–40nm to the long wavelength side from their position in solution.
*** For fuller details see Chapter 6, p. 176.

The main feature to remember is that the spectra of chlorophylls as well as of carotenoids are shifted upon extraction (Emerson & Lewis 1943). For example, the red peak of Chl *a* shifts from 675nm *in vivo* to 662nm in diethyl ether, whereas the blue peak shifts from 430nm to 435nm. In carotenoids, the shifts could be larger, e.g. one of the three peaks shifts from 470nm to 500nm. Fig. 12.3 shows

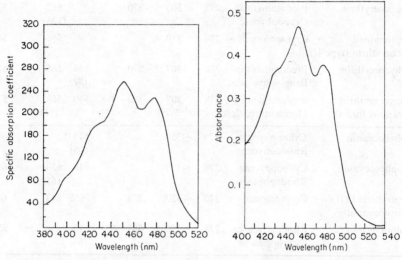

**Fig. 12.3.** Absorption spectra of β-carotene (left) and fucoxanthin (right) in hexane. Redrawn from Rabinowitch (1951); see the latter paper for references to original work.

absorption spectra of two well known carotenoids *in vitro*. Water soluble phycobilins, however, do not show any shifts. (See Table 12.3 for shifts observed in various pigments.) Fig. 12.4 shows absorption spectra of two phycoerythrins

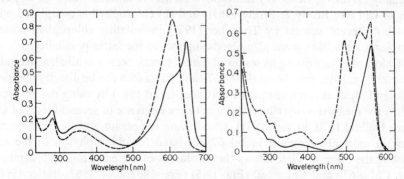

**Fig. 12.4.** Absorption spectra of aqueous solutions of phycoerythrins and phycocyanins (pH 6–7). Curves on the left, — — — —, C-phycocyanin; ————, allophycocyanin (both from *Nostoc muscorum*). Curves on the right, — — — —, R-phycoerythrin (*Ceramium rubrum*); ————, C-phycoerythrin (*Phormidium persicinum*) (redrawn from O'hEocha 1965).

**Table 12.4.** Physical properties of biliproteins (after O'hEocha 1965, also see Klein & Cronquist 1967, and references cited therein).

| Biliprotein | Organism | Absorption $\lambda_{max}$ (nm), and $E^{1\%}_{1cm}$ where available indicated by brackets | | | | | | Fluorescence ($\lambda_{max}$, nm) |
|---|---|---|---|---|---|---|---|---|
| C-phycoerythrin | Predominant in Cyanophyceae | 275 | 305 | ~370 | | | 562 (125) | 575 |
| Cryptomonad phycoerythrin (type II) | Cryptomonads | 274 | 310 | | | | 556 | 580 |
| B-phycoerythrin | Predominant in Bangiophycideae | 278 | 307 | ~370 | | 546 (82) | 565 | 578 |
| R-phycoerythrin (various forms) | Predominant in Florideophycideae | 278 | 307 | ~370 | 498 | 540 | 568 (81) | 578 |
| C-phycocyanin | Cyanophyceae Rhodophyceae | 278 | ~350 | | | | 615 (65) | 647 |
| Allophycocyanin | Cyanophyceae Rhodophyceae | 278 | ~350 | | | | 610 650 (65) | 660 |
| Cryptomonad (HV-) phycocyanin | Cryptomonads | 270 | ~350 | 583 | | | 625 643 | 660 |
| R-phycocyanin | Rhodophyceae Cyanophyceae | 278 | ~350 | 583 | 551 | 615 (66) | 565 637 | |

and two phycocyanins, and Table 12.4 lists characteristics of various phycobilins in several algae.

## 2.2 Chlorophylls

The physical state of chlorophyll in algae is not yet clear. There are two, not mutually exclusive, views: (a) chlorophyll forms are different states of aggregation of Chl (see Brody & Brody 1963); and (b) chlorophyll is complexed with proteins. Recent success by Thornber (1971) in isolating chlorophyll-protein complexes from blue-green algae lends support to the latter possibility.

Chlorophyll *a*, existing as several different forms, occurs in all algae (see also Chapter 5, p. 161). Two forms (Chl *a* 670 and Chl *a* 680) can be directly observed in several algae at room temperature (Cederstrand 1965) by using narrow band widths for measurements (for references to the existence of several forms of Chl *a*, see Halldal 1970, French 1971). Usual room temperature absorption spectra show in most algae only a broad 675nm band for chlorophyll *a* in the red. Cooling the algae to 77°K allows the resolution of the band into two components, Chl *a* 670 and Chl *a* 680 (Fig. 12.5) (Frei 1962, Kok 1963, Butler 1966a, Cho & Govindjee 1970b,c). Also a new band appears at 705 to 710nm which is thought to be different from P700 (Allen 1961 (in apple green mutant of *Chlorella pyrenoidosa*), Kok 1963 (in *Scenedesmus*)). Fig. 12.6 shows 77°K absorption spectra of *Euglena* chloroplasts with peaks at 672, 681, 693 and 706nm. Cho

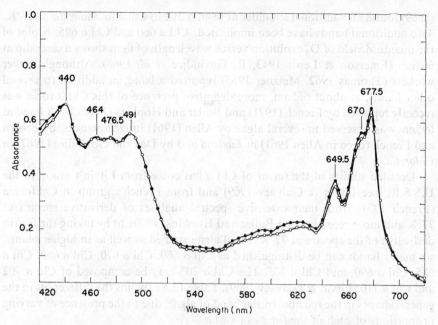

**Fig. 12.5.** Absorption spectra of *Chlorella pyrenoidosa* measured at 4°K (●—●) and 77°K (○—○). Note peaks at 440, 670 and 677·5nm due to Chl *a*, 476·5 and 649·5nm due to Chl *b*, and 464 and 491nm due to carotenoids (Cho & Govindjee 1970b).

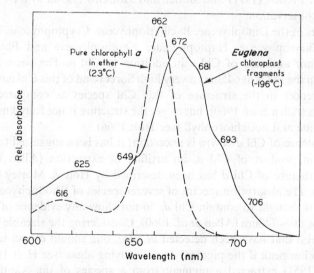

**Fig. 12.6.** Absorption spectra of *Euglena* at –196°C (77°K) and Chl *a* in ether at 23°C. The spectrum of *Euglena* shows bands at 672, 681, 693 and 706nm due to various forms of Chl *a*; the small band at 649nm is due to Chl *b*, and the band at 625nm is due to the vibrational bands of all Chl *a* forms (French 1971).

(1969) found an additional shoulder at 686nm due to Chl *a* in *Anacystis* at 77°K. Two additional bands have been implicated: Chl *a* 660 and Chl *a* 695. A plot of the quantum yield of $O_2$ evolution versus wavelength of light shows a clear dip at 660nm (Emerson & Lewis 1943, R. Govindjee *et al.* 1968). Although earlier workers (Thomas 1962, Metzner 1963) reported a band, in addition to several other bands, at about 665nm, more definitive presence of this Chl *a* form was recently reported by French (1971) and Butler and Hopkins (1970a,b). A band at 695nm was observed in several algae by Allen (1961) in *Ochromonas*, by Brown and French (cited in Allen 1961) in *Euglena* and by Das and Govindjee (1967) in *Chlorella*.

Detailed studies of the forms of Chl *a* has come from Litvin's group in the U.S.S.R. (see Litvin & Gulyaev 1969) and from French's group in California (French 1971) from their extensive spectral analyses of derivative spectra at 77°K and more recently from Butler and Hopkins (1970a,b) by taking the fourth derivative of the spectra at 77°K. In all algae studied as well as in higher plants, six major bands can be distinguished as Chl *a* 660, Chl *a* 670, Chl *a* 680, Chl *a* 685, Chl *a* 690, and Chl *a* 705. The Chl *a* 705 may be composed of Chl *a* 702 and Chl *a* 710 (Litvin & Gulyaev 1969). Table 12.5 explains the differences in the general shapes in the red absorption band of algae, due to the presence of varying proportions of each of the forms of Chl *a*.

Only the Chlorophyceae, Prasinophyceae and Euglenophyceae contain Chl *b*; absorption peaks appear at about 480 and 650nm (Fig. 12.5). Recently Thomas (1971) has shown that two forms of Chl *b* may exist, absorbing at 640 and 650nm. French (1971) and Butler and Hopkins (1970a,b) have both confirmed this observation.

Members of the Dinophyceae, Bacillariophyceae, Cryptophyceae, Rhaphidophyceae, Chrysophyceae, Haptophyceae, Xanthophyceae and Phaeophyceae contain minor amounts of Chl *c* in addition to Chl *a*. The weak absorption maximum in the red is at 645nm; a very high Soret band of this Chl has suggested some differences in the structure of this Chl species as compared to other chlorophylls (Allen *et al.* 1960), but as yet the structure is not fully known. It is a chlorophyllide and not chlorophyll (see Holt 1966).

The existence of Chl *d in vivo* is uncertain; it has been suggested to be simply an oxidation product of Chl *a*, an artifact of extraction (Allen 1966). The chemical structure of Chl *d* has been described by Holt & Morley (1959) and Holt (1961). The absorption spectra of several species of Rhodophyceae, the one group that is thought to contain Chl *d*, do not show any evidence of a pigment that absorbs at ∼740nm (Allen *et al.* 1960). Considering the sizeable concentrations of Chl *d* that have been detected *in vitro*, one should expect to see an *in vivo* absorption peak if the pigment exists in living algae (see Holt 1966).

Strain (1951) extracted a pigment from a species of the Xanthophyceae, *Tribonema bombycinum*, that absorbs maximally at 415nm and 654nm in methanol, and labelled it Chl *e*. Whether it exists *in vivo* is not certain.

Govindjee (1960, 1963b), Govindjee *et al.* (1961), and Gassner (1962), have

Table 12.5. Half band widths and proportions of chlorophyll a components in algae (after French et al. 1972).

| Curve No. and Material | Half band widths of Chl b forms | | Half band widths of Chl a forms; Proportions of Components* | | | | | |
| --- | --- | --- | --- | --- | --- | --- | --- | --- |
| | Chl b 640 | Chl b 650 | Chl a 662 | Chl a 670 | Chl a 677 | Chl a 683 | Chl a 691 | Chl a 704 |
| **Fraction 1 preparations** | | | | | | | | |
| *Very sharp spectra* | | | | | | | | |
| C34G, *Stichococcus* | 8·3 | 11·0 | 10·8; 24·7 | 8·9; 21·4 | 8·4; 30·7 | 11·3; 15·8 | 20·0; 7·4 | ...; 0 |
| C71C, *Scenedesmus* | 12·6 | 11·7 | 11·1; 21·9 | 9·8; 26·1 | 8·8; 27·4 | 9·9; 14·9 | 13·0; 8·0 | (14·2)**; 1·7 |
| *Typical green algal spectra* | | | | | | | | |
| C27D, *C. pyrenoidosa* | 12·1 | 12·4 | 11·6; 23·3 | 10·1; 24·4 | 9·8; 27·2 | 10·6; 16·2 | 13·1; 6·9 | (15·1); 2·0 |
| **Fraction 2 preparations** | | | | | | | | |
| *Very sharp spectra* | | | | | | | | |
| C35H, *Stichococcus* | (7·3) | 10·5 | 10·7; 27·3 | 8·9; 24·2 | 7·6; 34·6 | 8·8; 8·3 | 19·3; 5·6 | ...; 0 |
| C72A, *Scenedesmus* | (11·3) | 11·2 | 11·6; 24·6 | 10·4; 32·0 | 8·0; 30·6 | 8·2; 7·5 | 14·9; 5·3 | ...; 0 |
| *Typical green algal spectra* | | | | | | | | |
| C28E, *C. pyrenoidosa* | 11·3 | 11·2 | 12·0; 26·3 | 10·2; 26·3 | 9·3; 28·5 | 11·0; 12·4 | 17·1; 5·8 | ...; 0 |
| **Unfractionated preparations** | | | | | | | | |
| *Typical green algal spectra* | | | | | | | | |
| C76J, *C. vulgaris* | (7·0) | 11·8 | 10·9; 25·4 | 9·6; 29·6 | 8·4; 31·1 | 10·6; 7·3 | 18·8; 6·7 | ...; 0 |
| C61B, *Scenedesmus* mutant 8 | (13·1) | 12·1 | 11·3; 24·5 | 9·8; 27·3 | 8·6; 29·2 | 10·1; 11·8 | 18·8; 7·2 | ...; 0 |

* Proportions are given after the semicolon.   ** Parentheses indicate curves too small to be significant.

found an absorption band at 750nm (P750) in two species of blue-green algae (Fig. 12.7) which could be due to a long wavelength form of Chl *a* or to bacterio-pheophytin. However, Fischer and Metzner (1969) have shown that it is not a bacteriopheophytin, and it lacks chlorin characteristics. They suggest that P750 may be an open chain tetrapyrrole pigment, perhaps, of the bile pigment type.

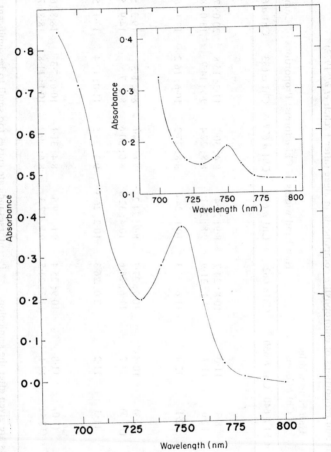

**Fig. 12.7.** Absorption spectrum of a thick suspension of *Anacystis nidulans* measured with an integrating dodecahedron spectrophotometer showing a peak at 750nm. Insert shows absorption spectrum of a similar sample measured with a Beckman DU spectrophotometer (Govindjee 1960).

Efficiency of energy transfer from Chl *b* to Chl *a* in algae, as measured by the excitation of Chl *a* fluorescence by Chl *b* is almost 100% both at room and liquid nitrogen temperature, and the efficiency of energy transfer from the various forms of Chl *a* to the reaction centre is also very high (Duysens 1952, Tomita & Rabinowitch 1962, Cho & Govindjee 1970b,c). Cho *et al.* (1966) and Cho and

Govindjee (1970b), however, found that energy transfer from the various forms of Chl $a$ to the reaction centre was temperature dependent being lower at 4°K than at 77°K. This can, perhaps, be taken to mean that the mechanism of energy transfer is by Förster's resonance 'slow' transfer.

### 2.3 Carotenoids

The variety of carotenoids present in algae is greater than in higher plants. Several excellent reviews on carotenoids are available: see Weedon (1965) for chemistry, Goodwin (1965a) for distribution, Goodwin (1965b) for biosynthesis, Burnett (1965) for function other than photosynthesis, Davies (1965) for chemical and spectral identification and this book (Chapter 6, p. 176). Several groups of algae have acquired common names that reflect their carotenoid content, e.g. the brown algae, Phaeophyceae, which contain several xanthophylls, most notably fucoxanthin. The carotenoids are usually yellow or orange in colour, but their colour is masked by the more predominant chlorophyll. β-Carotene, existing in fairly high concentration, is the only universal carotenoid of the algae. Chlorophyceae and Rhodophyceae, similar in this respect to the higher plants, also contain significant amounts of α-carotene, while only one species of diatom is known to have ε-carotene.

The types of xanthophylls are numerous in the algae. The similarity between higher plants and Chlorophyceae is again attested by the high lutein content of these algae. The Rhodophyceae and Cryptophyceae likewise contain considerable amounts of this pigment. The other xanthophylls of importance are myxoxanthin and myxoxanthophyll, characteristic only of the blue-green algae; peridinin, the primary xanthophyll of the Dinophyceae; and fucoxanthin, present primarily in members of the Phaeophyceae and Bacillariophyceae.

Often it is difficult to determine the precise absorption band of any one carotenoid in algae because in addition to there usually being more than one type of carotenoid present, there is also an overlap of absorption bands of the various chlorophylls present. In general, *in vivo* absorption bands for the carotenoids are to be found (apart from chlorophyll) between 400 and 540nm. Fig. 12.8 shows the calculated fractional absorption by carotenoids along with that by Chl $a$ and Chl $b$ in *Chlorella* (Govindjee 1960).

Fucoxanthin is unique in that its absorption *in vivo* is extended to 590nm and this pigment is very efficient, unlike most carotenoids, in transferring energy to Chl $a$. It is thought that fucoxanthin is probably complexed to some protein *in vivo*, as a wavelength shift of 40nm results when this pigment is extracted with organic solvents or when the algal cells are treated with detergents (Mann & Myers 1968). This shift can likewise be seen when cells are heated to 70°C for ten seconds (Goedheer 1970).

According to Goedheer (1970), β-carotene is the only carotenoid, with the exception of fucoxanthin, that can really be seen in the *in vivo* absorption spectra of algae. Making use of the fact that β-carotene can be extracted preferentially from algal cells and chloroplasts with light petroleum ether, Goedheer (1969a) showed that the bands seen at 465 and 495 nm at room temperature in blue-green and red algae are due primarily, if not entirely, to β-carotene; at 77°K, these bands are shifted to 471 and 504nm.

Cho and Govindjee (1970b) presented absorption spectra of *Chlorella* at 77°K and at 4°K which showed shoulders at 465 and 491nm due to carotenoids (Fig. 12.5). Goedheer (1969b), working with a Chl *b*-less species of the Chlorophyceae and applying the same extraction technique as described above, suggests that the bands at 460 and 485nm at room temperature (465 and 493nm at 77°K) are due to β-carotene. However, it must be pointed out that there is never complete disappearance of absorption from 460 to 500nm in any algae by extraction with petroleum ether, which can be explained by either incomplete extraction of the β-carotene or by the presence of additional minor carotenoids. The brown algae show shoulders at 495 and 545nm at 77°K for β-carotene and fucoxanthin respectively.

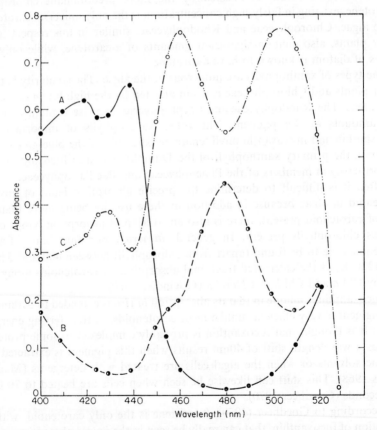

**Fig. 12.8.** Calculated fraction of total absorbed light, absorbed by Chl *a* (A), Chl *b* (B) and carotenoids (C) in a sample of *Chlorella pyrenoidosa* (Govindjee 1960).

From measurements of *in vitro* quenching of fluorescence of Chl *a* by various carotenoids, Teale (1958) concluded that β-carotene transfers energy to Chl *a* with 10% efficiency, lutein to Chl *a* with 60% efficiency and fucoxanthin to Chl *a* with 100% efficiency. In intact algae, energy transfer from the carotenoids to Chl *a* is 50% in green algae (Emerson & Lewis 1943, Duysens 1952, Cho & Govindjee 1970b), 10 to 15% in blue-green algae (Emerson & Lewis 1942, Duysens 1952, Papageorgiou & Govindjee 1967a, Cho & Govindjee 1970c), and 70 to 80% from fucoxanthin to Chl *a* (Tanada 1951, Duysens 1952). In view of the existence of two pigment systems (one strongly fluorescent photosystem II, and the other weakly fluorescent photosystem I) there is a need for reinvestigation of the energy transfer efficiencies in separated photosystem I and photosystem II particles. It is likely that carotenes transfer energy as efficiently as xanthophylls to photosystem I and photosystem II chlorophylls respectively.

### 2.4   *Phycobilins*

The water-soluble pigments, the phycobilins, are found in abundance only in the blue-green and red algae, although trace amounts of chemically different types of phycobilins have been found in isolated species of other algal groups (see Chapter

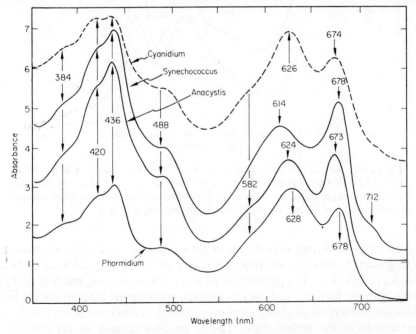

**Fig. 12.9.** Absorption spectra of several blue-green algae (solid curves) and of *Cyanidium* (dashed curve). The baselines for the different algae were shifted on the ordinate scale. The bands at 384, 420, 436, 673–678 and 712nm are due to Chl *a*; at 582 and 614–628nm due to phycocyanin, and at 490nm due to carotenoids (Govindjee & Mohanty 1972).

6, p. 194). Chemical and spectral characteristics of phycobilins have been extensively reviewed by O'hEocha (1960, 1962, 1965, 1971). Members of the Cyanophyceae contain largely the blue pigment, phycocyanin (see peaks around 620nm in Fig. 12.9), although several species may contain the red phycoerythrin in addition to, or in place of, phycocyanin; the reverse is true of members of the Rhodophyceae (see Hattori & Fujita (1959c). Both groups often also contain small amounts of allophycocyanin. (For chromoproteid pigments of crypto-monads, see Allen *et al.* 1959.) The existence of these pigments is very important

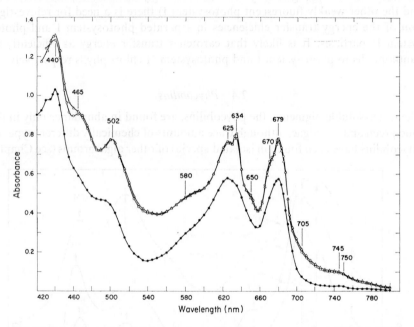

**Fig. 12.10.** Absorption spectra of *Anacystis nidulans* measured at 4°K (○), 77°K (Δ) and 295°K (●). Bands at 440, 670 and 679nm are due to Chl *a*; 580, 625 and 634nm due to phycocyanin; 650nm due to allophycocyanin; 465 and 502nm due to carotenoids; and 745nm due to P750 (Cho & Govindjee 1970c).

for the light-harvesting capabilities of these algae, as (excluding carotenoids) both groups in actuality contain only Chl *a* (if Chl *d* is considered to be an artefact of isolation). The phycobilins fill in, or at least narrow, much of the light energy gap left by Chl *a* and the carotenoids, allowing the algae to use the solar radiation much more efficiently in photosynthesis, in a manner much like that of fucoxanthin in the brown algae. This is possible because of the absorption characteristics of phycobilins: phycocyanins usually have a broad absorption band around 620nm, phycoerythrins around 545nm, and allophycocyanin around 650nm *in vivo* (Allen 1959, Brody & Emerson 1959b (*Porphyridium*); Thomas & Govindjee 1960 (*Porphyridium*), Ghosh & Govindjee 1966 (*Anacystis*),

Papageorgiou & Govindjee 1967a,b (*Anacystis*), Cho & Govindjee 1970c (*Anacystis*), Govindjee & Mohanty 1972 (*Cyanidium*)).

At room temperature, a phycocyanin peak can be seen in absorption spectra of blue-green and some red algae at about 625nm with shoulders at 580 and 635nm; at 4° and 77°K, the broad phycocyanin band becomes resolved into two maxima at 625 and 634nm, and a shoulder at 650nm due to allophycocyanin can be seen (Fig. 12.10). Absorption peaks of phycoerythrin can be seen *in vivo* and *in vitro* at 490, 546, and 576nm (Fig. 12.4) which may indicate why the red algae are so successful in sub-littoral marine habitats.

Excitation energy transfer from phycoerythrin to Chl *a* (via phycocyanin) and from phycocyanin to Chl *a* is very efficient (about 70 to 80%) (Emerson & Lewis 1942, French & Young 1952, Duysens 1952, Brody & Emerson 1959a, Brody & Brody 1959, Tomita & Rabinowitch 1962, Ghosh & Govindjee 1966)) but varies with experimental conditions. In fact, in blue-green and red algae a large proportion of the chlorophylls are in photosystem II, and thus one observes main peaks due to phycobilins in the action spectra of Chl $a_2$ fluorescence (for *Anacystis*, see Papageorgiou & Govindjee 1967a,b, Shimony *et al.* 1967, Cho & Govindjee 1970c; for *Chlorella*, see Cho & Govindjee 1970b). The energy transfer from phycocyanin to Chl *a* is temperature dependent and is about 20% lower at 4°K than at 77°K, perhaps, due to the operation of Förster's resonance 'slow' energy transfer mechanism. In *Cyanidium caldarium*, there is a relative increase in phycocyanin fluorescence (at 655nm) with respect to that in allophycocyanin (at 665nm) which allows the two bands at 77°K to be observed. This information may also be taken, with certain reservations to indicate that the energy transfer from phycocyanin to allophycocyanin is temperature dependent and is in agreement with 'slow' Förster type transfer (Mohanty *et al.* 1972).

# 3  SEPARATION OF ALGAL PHOTOSYNTHETIC SYSTEMS

The discovery of two pigment systems operating in photosynthesis has led to various attempts to physically fractionate the chloroplasts into photosystem II and photosystem I particles. This work has proceeded much more slowly in algae than in higher plants (see Boardman 1970), probably due to the relative ease of procuring ample higher plant material and breaking their cell walls. Nevertheless, fractionation procedures have been devised for several blue-green and green algae, two red algae, a diatom and a euglenoid.

Allen *et al.* (1963) were the first to attempt a separation of the photochemical systems. They subjected *Chlorella pyrenoidosa* to repeated freezing and grinding, followed by sonication and density gradient centrifugation and obtained two fractions: a heavy fraction with an absorption maximum at 680nm indicating that it was probably only broken cells, and a light fraction absorbing maximally at 672nm and containing also a large amount of Chl *b*. This light fraction is probably similar to the photosystem II particles isolated from higher plants, as they have been shown to contain most of the short wavelength form of Chl $a$ as well as Chl $b$.

Brown (1969), using the French press fractionation method of Michel and Michel-Wolwertz (1968), was able to obtain two fractions from three different species of the Chlorophyceae. It has been shown (Sane *et al.* 1970) that in higher plants this means of separation results in two particles, the lighter one, derived from the stroma lamellae, being photosystem I, and the heavier one being both photosystem I and photosystem II as it is the unseparated grana segment of the chloroplast membrane. Generally speaking, Brown found that the spectra of the green algae fractions resembled those obtained from higher plants. The light fraction contained more of the long wavelength Chl $a$ and less Chl $b$ compared to the heavy fraction (Fig. 12.11). In addition, the absorption spectra of the chloroplasts and the light fraction of *Scenedesmus obliquus* showed a band at 698nm, that was absent from the heavy fraction. These data suggest that there must be enough differentiation in green algae between stacked and unstacked regions in the chloroplast lamellae to allow the French pressure technique to be effective in fractionation. The electron micrographs of freeze-etched preparations from *Chlamydomonas* indicate that this is the case (Goodenough & Staehelin 1971), at least for this green alga. It is, however, not clear whether *Chlorella*, where most of the thylakoids appear stacked, has any unstacked thylakoids (Reger & Krauss 1970).

Treatment of *Euglena gracilis* cells with the French press and sodium deoxycholate followed by differential centrifugation results in the appearance of two fractions, i.e. a supernatant containing primarily Chl $a$ 670 and a sediment enriched in Chl $a$ 680 and Chl $a$ 695. If the latter fraction is indeed photosystem I, then it is not clear why a large proportion of Chl $b$ was found to be associated with it (Brown *et al.* 1965).

Brown (1969) also fractionated a red alga, *Porphyridium cruentum*, and a diatom, *Phaeodactylum tricornutum*, by the French pressure technique. (Neushul 1970, from a freeze-etching study of *Porphyridium*, suggests that both 'free' and 'stacked' thylakoids may be present in this alga.) There was no clear separation of the red alga into a long wavelength Chl $a$ enriched fraction and a short wavelength Chl $a$ fraction as witnessed by the absorption spectra, but the fluorescence emission spectra of the light fraction at 77°K showed a higher ratio of fluorescence from the long wavelength forms of Chl $a$ than from the short wavelength Chl $a$ forms, although the relative levels of Chl $a$ fluorescence at 685nm were drastically reduced in the light fraction. The effectiveness of fractionating *Phaeodactylum* is equally questionable. Breaking the cells alone resulted in a drastic change in the absorption spectrum with most of the long wavelength Chl $a$ forms being lost as well as some of the carotenoids. No visible separation in pigments could be achieved by a French press treatment although the relative fluorescence yield from the heavy fraction was several times greater than from the light fraction. This is an indication that there must be some

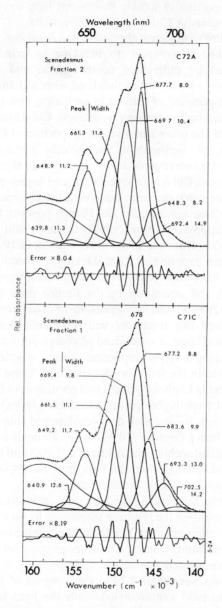

**Fig. 12.11.** Absorption spectra of fraction 2 (photosystems I + II) and fraction 1 (photosystem I) at $-196°C$ (77°K) from *Scenedesmus* chloroplasts, fitted by the sums of Gaussian components due to the different forms of Chl *a* and of Chl *b*. The error of fit is shown below each diagram with the designated magnification; peak locations and half widths are also indicated on the graph (French 1971).

difference in the proportion of weakly fluorescent long wavelength to strongly fluorescent short wavelength Chl *a* in the two fractions.

Ogawa *et al.* (1968) subjected the red alga, *Porphyra yezoensis*, and the diatom, *Phaeodactylum tricornutum*, to breaking by grinding or sonication respectively, followed by differential centrifugation and polyacrylamide-gel electrophoresis of their chloroplasts solubilized with sodium dodecyl sulphate. They found two 'components' of interest in each alga. One fraction (analogous to photosystem I) contained more long wavelength Chl *a* and carotenes (especially β-carotene), while the other had a larger proportion of Chl *a* 670 and more oxidized carotenoids, i.e. xanthophylls. Noteworthy is the fact that the component of *Phaeodactylum* corresponding to photosystem II is enriched in Chl *c*, in much the same way as Chl *b* is in green algae and higher plants.

Considerably more work has been done with the physical separation of the two photosystems in blue-green algae. The French pressure technique has been the least successful approach (Brown 1969); the net result has been only the removal of phycocyanin from the thylakoids. The difficulty is probably due to the fact that these algae have no thylakoid stacking, a necessity, at least in higher plants (where grana have both photosystem I and photosystem II, and stroma only photosystem I), for separation by the French press (Sane *et al.* 1970). Shimony *et al.* (1967) incubated a suspension of *Anacystis nidulans* in 0·6% digitonin and followed this treatment with differential centrifugation. Four fractions were obtained: one of solubilized phycocyanin, one of broken cells, a light fraction containing a large proportion of long wavelength Chl *a*, and a heavy fraction enriched in the short wavelength Chl *a*. These latter two fractions are probably comparable to photosystem I and photosystem II enriched particles respectively obtained from higher plants. Goedheer (1969a) prepared particles from *Synechococcus* cells treated with deoxycholate and subjected to differential centrifugation. The heavy fraction (10,000 × *g*) abounds in short wavelength Chl *a* and xanthophylls absorbing maximally at 525, 485 and 460nm (presumably myxoxanthophyll and lutein), while the light fraction (144,000 × *g*) contains mainly the long wavelength Chl *a* and β-carotene.

Ogawa *et al.* (1969), incubating sonicated *Anabaena variabilis* cells in 0·75% Triton-X-100 and separating the resulting suspension by sucrose density centrifugation, showed separation into two bands—a heavy, blue-coloured band and a light, orange-coloured band. Contrary to other photosystem separations, their heavy fragment corresponded chemically and spectrally to photosystem I (a preponderance of long wavelength Chl *a*, the presence of P700, and a higher ratio of β-carotene to total xanthophylls), while the light fraction (containing mainly short wavelength Chl *a* and a large amount of carotenoids, especially echinenone, zeaxanthin and myxoxanthophyll) looked more like photosystem II. These authors have suggested that unlike most other plants, Chl *a* of photosystem II of some blue-green algae such as *Anabaena* may be localized close to the outside of the thylakoid membrane such that it might be in contact with the phycobilisomes which are found on the exterior of the membrane (Gantt &

Conti 1966). However, unlike *Anabaena*, *Anacystis* which may by comparison also have 'exterior' phycobilisomes, shows the usual separation (Shimony *et al.* 1967). Obviously, more work on the structure-function relationships in algae is needed (see also Chapter 4, p. 124).

Lastly, in the way of particle preparations, Ogawa and Vernon (1969) treated lyophilized, carotenoid-extracted fragments from *Anabaena variabilis* with Triton-X-100 and then used sucrose-gradient centrifugation to isolate a fraction highly enriched in P700 (three P700 molecules per 100 bulk Chl *a* molecules). In addition, Thornber and Olson (1971) reported a Chl *a*-protein complex, isolated from sodium dodecyl sulphate treated lamellae of the blue-green alga *Phormidium luridum* lamellae, which contains one molecule of P700 per 60 to 90 Chl *a* molecules and has a red absorption band at 677nm at room temperature. This complex has a main emission band at 682nm and shoulders at 692, 720 and 740nm at room temperature, and two main emission bands at 681nm and 720nm with shoulders at 695nm and 730nm at 77°K (Mohanty *et al.* 1972). A similar Chl *a*-protein complex has been isolated from the red alga *Porphyridium cruentum*, although its red absorption maximum is at 671nm. Similar treatment of green algae and higher plants yields two Chl-protein complexes: one very similar to that of the blue-green algae and the other, containing equal amounts of Chl *a* and Chl *b* absorbing at 672 and 653nm respectively, which corresponds with the bulk pigments of photosystem II.

## 4   COMPOSITION OF THE TWO PIGMENT SYSTEMS

On the basis of action spectra of photosynthesis and fluorescence of the two pigment systems and spectral analyses of the physically separated photosystems (references and details given in the text; see Mohanty *et al.* 1972), it is commonly accepted that photosystem I and photosystem II of all plants, with the possible exception of blue-green and red algae, contain both Chl *a* 670 and Chl *a* 678 (and Chl *b* in Chlorophyceae, Euglenophyceae and higher plants), although the proportions of Chl *a* 670 (and Chl *b*) are somewhat greater in photosystem II. However, the long wavelength forms of Chl *a* (685–705) are mainly found in photosystem I (Fig. 12.12).

Cederstrand *et al.* (1966), using a precision integrating spectrophotometer, obtained absorption spectra of *Anacystis nidulans*, *Chlorella pyrenoidosa*, and *Porphyridium cruentum*. Comparing these spectra with those derived for photosystem I and photosystem II by Duysens (1963) from data of French *et al.* (1960) on action spectra of photosynthesis in the presence of strong background photosystem II or photosystem I light, Cederstrand *et al.* concluded that in *Chlorella* much of Chl *a* 668 and Chl *b* occurs in photosystem II, and a larger portion of Chl *a* 683 is found in photosystem I. However, in the phycobilin-containing algae, most of the Chl *a* 668 and Chl *a* 683 is associated with photosystem I; in fact, it is very difficult to see a peak at 670nm in the action spectrum of the

Emerson effect in blue-green and red algae, as there is a larger proportion of Chl $a$ in photosystem I than in photosystem II (Govindjee & Rabinowitch 1960, Jones & Myers 1964). Moreover, there is four fold more phycocyanin than Chl $a$ in photosystem II, and approximately an equal ratio of the two pigments in photosystem I. In addition, Cho and Govindjee (1970c), looking at the excitation spectra of *Anacystis nidulans* for fluorescence at 685 and 705nm (approximately photosystem II and photosystem I fluorescence respectively), showed a higher ratio of the excitation band of phycocyanin (637nm) to that of Chl $a$ (436nm) for fluorescence at 685nm than for 715nm fluorescence. Using a similar approach with *Chlorella pyrenoidosa*, Cho and Govindjee (1970a,b) confirmed that Chl $b$, Chl $a$ 670, and Chl $a$ 678 were present in both photosystems, photosystem II being enriched in Chl $b$ and photosystem I in Chl $a$ 685–705 (Fig. 12.12).

The curve analyses of the first and fourth derivative absorption spectra of photosystem I and photosystem II enriched fractions, from a number of algae and higher plants into various Chl $a$ components (Butler & Hopkins 1970a,b, French *et al.* 1971) show quantitatively the distinction of pigment forms existing in photosystem II and photosystem I (see Tables 12.5 and 12.6, and Fig. 12.11). Analyses of room and low temperature absorption spectra of fractions obtained from various green algae (French *et al.* 1971) indicate that the enrichment of Chl $a$ 670 in photosystem II fractions is 10 to 20%; moreover, there is 10% more Chl $a$ 678 in the photosystem II enriched fraction than in the photosystem I fraction. The latter, however, is enriched in the longer wavelength forms of Chl $a$ (i.e. Chl $a$ 684, Chl $a$ 693 and Chl $a$ 705). The placement of almost similar quantities of Chl $a$ 693 in the two photosystems raises some contradiction with fluorescence data which suggests that photosystem II contains much less of this Chl $a$ form than photosystem I (see Govindjee *et al.* 1967). As indicated in Table 12.6, this curve analysis assigns approximately 10% more Chl $b$ to photosystem II enriched particle fractions.

French's data on the placement of the various Chl $a$ forms and Chl $b$ in the two photosystems should be improved by changing the method used for separation of the photosystems because fractions prepared by the French pressure technique are actually not photosystem II and photosystem I fractions, but rather grana photosystem II *and* photosystem I and stroma lamellae (photosystem I alone) fractions (Sane *et al.* 1970). Curve analysis applied to absorption spectra obtained from a more purified photosystem II preparation rather than from grana would be valuable.*

* Recently R. A. Gasanor and C. S. French (personal communication, 1973) have analyzed absorption spectra of pigment systems I and II separated from grana. They have indeed shown the absence of Chl $a$ 693 in pigment system II. The fractions of different forms of Chl $a$ in pigment systems II and I from grana are: Chl $a$ 660 (25; 23), Chl $a$ 670 (30; 31), Chl $a$ 680 (38; 30), Chl $a$ 685 (7; 7), Chl $a$ 693 (0; 3), and Chl $a$ 705 (0; 6), the first and the second numbers (within parentheses) representing systems II and I, respectively. The fractions of different forms of Chl $b$ are: Chl $b$ 640 (26; 33) and Chl $b$ 650 (74; 67). These new data would slightly modify the scheme presented here in Fig. 12.12.

*Pigments beyond 720nm.* The possible role of pigments absorbing extreme red light ($>720$nm) in photosynthetic systems was first described by Rabinowitch *et al.* (1960) and Govindjee *et al.* (1960). They observed an inhibitory effect of $750 \pm 10$nm light on net $O_2$ evolution by *Porphyridium cruentum* at 700nm. The action spectrum of this effect in *Chlorella pyrenoidosa* and in *Porphyridium cruentum* had a maximum around 740 to 750nm. R. Govindjee (1961) discovered the same effect in the Hill reaction (quinone as oxidant) in *Chlorella* and in

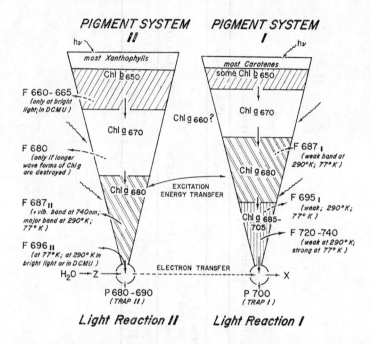

Fig. 12.12.   A working hypothesis for the approximate distribution of pigments in the two pigment systems in green algae. Chl *b*, chlorophyll *b*; Chl *a*, chlorophyll *a*; the numbers after Chl *a* indicate the approximate absorption maxima in the red end of the spectrum. (The position of chl *a* 660 is not yet clear to the authors.) The symbol F followed by numbers refers to the suggested fluorescence band, at appropriate wavelengths, in nm. The conditions under which these bands are observed are listed within parentheses. Z, primary electron donor of photosystem II; X, primary electron acceptor of photosystem I; P, pigment (trap). The photosystem I is located on the outer side of the thylakoid membrane and the photosystem II on the inner side such that excitation energy between them is possible. (We cannot, however, discount the possibility that the two systems may be side by side on the thylakoid membrane.) In diatoms and brown algae, Chl *c* replaces Chl *b*. In red and blue-green algae phycobilins replace Chl *b*. The proportion of phycobilins to Chl *a* is high in photosystem I, and low in photosystem II: a large portion of all Chl *a* forms is in photosystem I, and a large portion of phycobilins is in photosystem II. Phycobilins are located in phycobilisomes. The physical arrangement of pigment systems in these algae is not yet clear, and is under active investigation in some laboratories (modified after Govindjee *et al.* 1967).

**Table 12.6.** Approximate distribution of chlorophyll forms in two fractions, per cent (French et al. 1971).

| | Chlorophyll b | | | Chlorophyll a | | | | | |
| | Chlorophyll b as per cent of total | Chl b 640 | Chl b 650 | Chl a 662 | Chl a 670 | Chl a 677 | Chl a 683 | Chl a 692 | Chl a 705 |
|---|---|---|---|---|---|---|---|---|---|
| Fraction 1 | 20 | 14 | 86 | 22 | 23 | 29 | 17 | 7 | 3 |
| Fraction 2 | 23 | 14 | 86 | 25 | 27 | 33 | 8 | 5 | 0 |

pokeweed chloroplasts. However, using narrower band widths and a larger number of wavelengths, she found two peaks at 730 to 740 and 750 to 760nm both in the Hill reaction and in photosynthesis of algae. Govindjee et al. (1961) found minor absorption bands in *Chlorella* and *Porphyridium* and a major band at 750nm in *Anacystis*, but the latter alga did not show the inhibitory effect. The initial observations of these absorption bands in green and red algae may have been due to absorption by water or hydroxyl groups as even magnesium oxide suspensions showed similar bands (Govindjee & Cederstrand 1963). However, the fact remains that some absorbing species in algae must be responsible for the inhibitory effect observed by the above mentioned workers.

Lastly carotenoid distribution in the two pigment systems appears to follow a general rule that the oxidized carotenoids (xanthophylls) are mainly found in photosystem II, while the reduced carotenoids (carotenes, especially β-carotene) are identified with photosystem I.

# 5 LIGHT-INDUCED ABSORPTION CHANGES AND PHOTOSYNTHESIS

Relevant evidence for the two-step mechanism and the electron transport pathway of algae as derived from difference absorption spectroscopy may now be considered. (For information derived from mutants of algae, see Chapter 14, p. 424). What one observes in this type of study is differences between the absorption spectra of algal cells in weak measuring light, that is, with a very low rate of photosynthesis, and in the presence of strong light (when photosynthesis is going on at full speed). These difference absorption spectra reveal reversible changes in several components: active Chl $a$ of photosystem II (labelled as P680–P690, has not yet been observed in algae), C550 (a compound absorbing at 550nm has also not yet been seen in algae), plastoquinone (at 254nm), cytochrome $b_3$ (Cyt $b_{559}$), cytochrome $f$ (Cyt $f_{553}$), cytochrome $b_6$ (Cyt $b_{564}$), plastocyanin (PC, at 660nm), active Chl $a$ of photosystem I (P700), P430 (possibly 'X', the primary electron acceptor of photosystem I), NADP (at 340nm), and an absorbance change at 515–520nm. One scheme that includes these intermediates is shown in Fig. 12.13 (where Z is for 'primary' electron donor of photosystem II, Q is for 'primary' electron acceptor of photosystem I, Fd is for ferredoxin, FRS for ferredoxin reducing substance, and R for Fd-NADP reductase).

Levine and Gorman (1966) provided evidence for the existence and participation of Cyt $b_{559}$ in *Chlamydomonas*. Amesz (1964a) showed that in *Anacystis* 620nm light (absorbed by phycocyanin) reduces plastoquinone, and 680nm light (absorbed by Chl $a$) oxidizes it. Kouchkovsky and Fork (1964) showed in *Ulva* that plastocyanin was oxidized by photosystem I light and its reduction accelerated by photosystem II light. Duysens and Amesz (1962) were the first to show that in *Porphyridium* Cyt $f$ was oxidized by photosystem I light (red light, absorbed by Chl $a$) and was reduced by photosystem II light (green light,

N

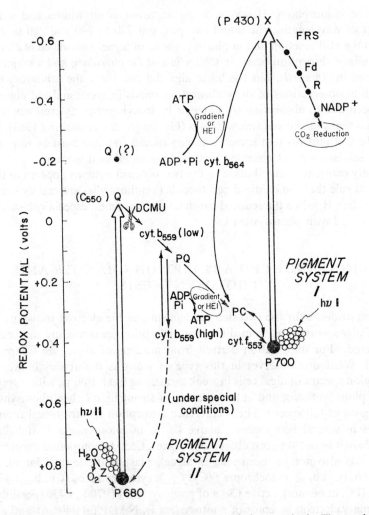

**Fig. 12.13.** A modified Hill and Bendall scheme of photosynthesis. The two bold vertical arrows represent the two light reactions; all others, dark reactions. Flow of electrons from $H_2O$ to $NADP^+$ is designated as 'non-cyclic' electron flow and from X to the intersystem intermediates as 'cyclic'. A similar cyclic flow of electrons involving only photosystem II shown as dashed line has also been suggested recently. Abbreviations used: Z, the primary donor of photosystem II; P690, the proposed trap of photosystem II; Q (C550), the proposed primary acceptor of photosystem II and the quencher of fluorescence; PQ, plastoquinone; Cyt $b_{559}$ (low and high potential forms) and Cyt $b_{564}$, cytochromes $b$; Cyt $f_{553}$, cytochrome $f$; PC, plastocyanin; P700, the trap for photosystem I; X (P430), the photosystem I electron acceptor; Fd, ferredoxin; R, Fd-NADP$^+$ reductase; NADP$^+$, nicotinamide adenine dinucleotide phosphate; ADP, ATP, adenosine di- and tri-phosphate; Pi, inorganic phosphate; HEI high energy intermediate; DCMU, 3-(3,4-dichlorophenyl)-1,1 dimethyl urea, blocks non-cyclic electron flow (see scissors). Note: the positions of PC and Cyt $f$ may have to be interchanged.

absorbed by phycoerythrin). Amesz and Duysens (1962) demonstrated the same phenomenon in *Anacystis*. This antagonistic effect of light of two different wavelengths is one of the best evidences for the operation of two light reactions in series as presented above. Kok and Gott (1960) (see also Kok & Hoch 1961) showed the crucial role of P700 in *Anacystis* where it was oxidized (bleached) by photosystem I light (Chl *a*) and reduced by photosystem II light (phycocyanin). Finally, Amesz (1964b) showed NADP reduction in intact cells of *Anacystis*. We do not discuss here the role of the complex 515nm change—a part of which may be an index of the membrane potential (Junge *et al.* 1969) and which was discovered by Duysens (1954) in *Chlorella* (see Rubinstein 1964 (*Chlorella*), Govindjee & R. Govindjee 1965 (*Chlorella*), Fork & Amesz 1967 (red and brown algae), Pratt & Bishop 1968 (*Scenedesmus*)). (See also reviews by Fork & Amesz 1969, Levine 1969 and this book Chapter 14, p. 424.)

Recent advances in the study of P690, C550 and P430 have been made mainly with higher plants. We shall briefly discuss the highlights of these studies due to the exciting nature of these discoveries; we suggest that these absorption changes, if they are due to important photosynthetic intermediates, should be looked for in algae.

### 5.1 *P680–P690*

From Witt's laboratory in Berlin came several papers on the light-induced absorption changes in chloroplast fragments of higher plants at 435nm and 682–690nm which they feel are due to the reaction centre chlorophyll of photosystem II (Döring *et al.* 1967, 1968, 1969). Using a repetitive flash technique, Döring and co-workers were able to distinguish this absorbance change that has a lifetime of $2 \times 10^{-4}$ seconds at room temperature, 100 fold shorter than that for P700. The concentration of P680–P690 is of the order of $(0.5–2.0) \times 10^{-3}$ of total chlorophyll. (This makes it doubtful that there is one P690 per photosynthetic unit of photosystem II; perhaps it is only a 'sensitizer' as suggested by Döring and coworkers). The light-induced absorbance changes at 680–690 and at 435nm are present in photosystem II enriched particles, but they cannot be seen in photosystem I fractions (Döring *et al.* 1968). This information, along with the sensitization of P680–690 by pigment system II and the same dependence on intensity as for $O_2$ evolution has placed it in photosystem II. Likewise, DCMU at appropriate concentrations (acting as a photosystem II inhibitor) blocks this absorbance change, although there is some difficulty in interpreting this information (Döring *et al.* 1969). Floyd *et al.* (1971) have demonstrated that this absorbance change is observable at 77°K which might suggest that in fact it is due to a primary quantum reaction; however to prove this point, one still needs to show that its yield is sufficiently high.

Butler (1972a) has questioned the validity of the assignment of this absorbance change to the photosystem II reaction centre, as it may be due simply to an increase in the fluorescence yield caused by actinic illumination leading to an apparent decrease in absorption. However, Döring (pers. comm.) has shown that the kinetics of changes at 435nm and 682nm are identical, and that this absorbance decrease is obtained when the photomultiplier is distant (1 metre) from the sample. Under these conditions, the contribution of fluorescence change is negligible. Also, Govindjee et al. (1970) have shown that the P680 change occurs in TRIS-washed chloroplasts, where no changes in fluorescence yield occur, and in 'wet' heptane-extracted chloroplasts, where the fluorescence yield changes are drastically reduced. It seems that the P680 change is a real absorption change, but more work is needed before this should be fully accepted. For example, a correlation between its concentration and fluorescence yield changes of photosystem II should be sought as Vredenberg and Duysens (1963) have made for photosynthetic bacteria. Floyd et al. (1971), Govindjee and Papageorgiou (1971) and Butler (1972a) considered the possibility that P680 may be an electron donor just as P700 is in photosystem I. Perhaps the available instruments and techniques do not monitor all the P680 change and/or the extinction coefficient of P680 is lower than that of P700!

### 5.2  C550

A second light-induced absorption change is due to a compound named C550. Discovered by Knaff and Arnon (1969), this absorption change that can be seen at 77°K or at room temperature (when all the cytochromes have been chemically oxidized and are not interfering in this spectral region) has been associated with Q, the 'unknown' primary electron acceptor of photosystem II (Erixon & Butler 1971a). Experiments with DCMU and photosystem II and photosystem I exciting light (Arnon et al. 1971) and with photosystem I and photosystem II particles or mutants (Erixon & Butler 1971a) have placed C550 in photosystem II. Erixon and Butler (1971b) found a good correlation between C550 and Q at low light intensities and at liquid nitrogen temperature, but difficulties arise in this correspondence either at room temperature or in far red light (Ben Hayyim & Malkin 1971). Butler (1972b) himself showed that at least one other 'component' (other than the possibility that C550 is indeed the photosystem II primary acceptor), the membrane potential, affects this light-induced absorption change. Whether Q is similarly affected is not yet known. Okayama and Butler (1972) showed that the photoreduction of C550, eliminated by hexane extraction, is restored by the addition of plastoquinone and β-carotene. This may suggest the possibility that C550 change is due to carotenes! This requires full investigation. In this connexion, we note that Erixon and Butler (1971a) earlier found an increase in absorbance at 543nm associated with the decrease at 550nm; this could imply a shift in the absorbance of a carotenoid type pigment.

### 5.3  P430

Finally, using flash kinetic spectrophotometry, Hiyama and Ke (1971a,b) have reported an absorption change at 430nm, observable in spinach and blue-green algal photosystem I particles (as well as spinach sub-chloroplast fractions). This light-induced absorption change, designated as P430, may be the long sought primary electron acceptor of photosystem I. Experiments show that P430 is bleached as fast as P700, the kinetics of the dark recovery of P430 are identical to those of the reduction of artificial electron acceptors that have redox potentials close to that of X (primary acceptor) and the quantum yield and the effective wavelengths for P430 'photoreduction'(?) are the same as for P700 photooxidation. The chemical component that is responsible for this light-induced absorbance change exists only in speculative terms now. However, light-induced EPR absorbance changes of chloroplasts and photosystem I fractions from higher plants have been associated with a bound ferredoxin that has also been suggested to be the primary electron acceptor of photosystem I (Malkin & Bearden 1971, Bearden & Malkin 1972). Lack of 1:1 correspondence, if proven, between the magnitude of EPR signals arising from P700 oxidation and from the reduction of this non-haem iron compound may, however, discount its role as a primary acceptor, but it is interesting to note that ferredoxin bound to a chlorophyll molecule might exhibit an absorption peak at 430nm (Yang & Blumberg 1972).

## 6  ADAPTATION

The section on absorption spectra did not mean to imply that the pigment composition of algae is static; actually the reverse is true—algae have considerable flexibility in responding to various environmental factors by altering their pigment composition and/or spectral response (see Halldal 1970).

Perhaps the best example of such adaptational strategy is that of the chlorophyll *a*- biliprotein system of the blue-green algae and red algae (see Rabinowitch 1951). The occurrence in Rhodophyceae of such a response to the environment enables the vertical distribution of these algae in nature. Yocum and Blinks (1958) have shown that the photosynthetic efficiency of Chl *a* in *Porphyridium cruentum* is very high in low intensity blue or red light, while in green light, absorbed by phycoerythrin, Chl *a* efficiency is low. Pigment ratio changes and alterations in photosynthetic spectral response have been observed in other members of Bangiophycidae, but the higher red algae, Florideophycidae, do not possess this adaptability. The pigment flexibility of the blue-green alga, *Tolypothrix tenuis*, which contains both phycoerythrin and phycocyanin, has been studied by Fujita and Hattori (Fujita & Hattori 1960, 1962, Hattori & Fujita 1959a,b). They find that blue or green light is the most favourable for phycoerythrin formation in the alga, while red light is the least favourable; the opposite is true for phycocyanin formation. However, their study in using only one intensity of light was not complete, as Brody (1958) and Brody and Emerson (1959b) found that for at least *Porphyridium cruentum* the light intensity determined in which way the chromatic adaptation is directed, with complementary chromatic adaptation being more probable at low light intensities than high.

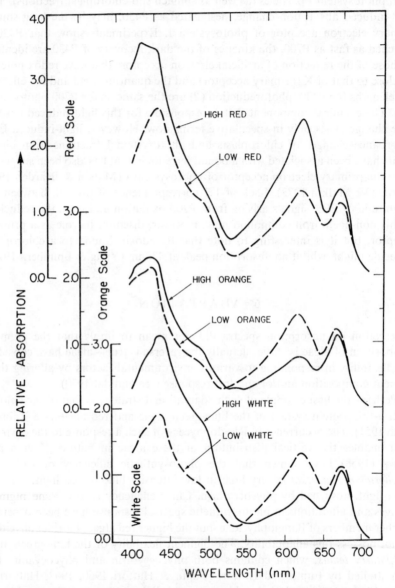

**Fig. 12.14.** Absorption spectra of different cultures of *Anacystis nidulans* grown in lights of different intensity and color. Spectra of cells grown in high light intensities are shown with solid lines and those in low light intensities with broken lines. The ordinate labelled 'white scale', 'orange scale', and 'red scale' are for cells grown in white, orange and red light, respectively (Ghosh & Govindjee 1966).

Ghosh and Govindjee (1966) showed the same in *Anacystis nidulans* where the pigment which best absorbed the light given during growth is decreased in high intensity light (Fig. 12.14); they also showed that when the ratio of Chl *a* to phycocyanin changed from the usual, the efficiency of energy transfer from phycocyanin to Chl $a_2$ decreased.

Diatoms, dinoflagellates, and brown algae (all with the Chl *a*-Chl *c*-carotenoid system) likewise show environmentally induced pigment changes. The Chl *c*/ Chl *a* ratio is greatest in dim light, but there is no drastic difference in the ratio of Chl *a* and Chl *c*/carotenoids with light of varying intensities (Brown & Richardson 1968). Most interesting is the large variability in the ratio of photosynthetically active/inactive carotenoids and action spectra indicate significant flexibility from the violet to green portions of the spectrum (Halldal 1970).

The green algae (Chl *a*-chl *b*-carotenoid system) show several types of variation in pigment composition and spectral response. First, Brown and Richardson (1968) demonstrated in three species of the Chlorophyceae that the Chl *b*/Chl *a* ratio increased in dim light. (It is well known that shade plants have a higher ratio of Chl *b*/Chl *a* than sun plants, see Rabinowitch 1945.) Moreover, the different forms of Chl *a* do vary in concentration and/or spectral response in extreme environmental conditions. The best example is the adaptation of *Ostreobium* to its environment within the coral *Favia pallida* (Halldal 1970). An analysis of the light that penetrates the coral and reaches the layer of algae (Shibata & Haxo 1969) indicates that only light above 680nm contributes significantly to the illumination of the algae; and, to be certain, the photosynthetic rates in the far-red light were very similar to those at 675nm and the absorption spectra showed an increase in the long wavelength forms of Chl *a*. This adaptation could be quickly reverted back to the more expected pattern for green algae on exposure to dim daylight. Oquist (1969) demonstrated a similar pigment flexibility in *Chlorella* exposed to far-red light during growth, although the extent of change was not as pronounced as in *Ostreobium*. Lastly, Halldal and French (1958) and Oquist (1969) showed that green algae grown in light of high intensity characteristically have a high amount of photosynthetically inactive carotenoids, as do Chl *c* containing algae. This can also occur under extreme environmental conditions such as high salt, low temperatures, and low $CO_2$ concentration (Halldal 1970).

## 7 REFERENCES

ALLEN M.B. (1959) Studies with *Cyanidium caldarium*, an anomalously pigmented chlorophyte. *Arch. Mikrobiol.* **32**, 270–7.

ALLEN M.B. (1961) Evidence for pigments absorbing at 705–10 mμ in photosynthetic organisms. In *Light and Life*, eds. McElroy W.D. & Glass B. pp. 479–82. Johns Hopkins Press, Baltimore.

ALLEN M.B. (1966) Distribution of the chlorophylls. In *The Chlorophylls*, eds. Vernon L.P. & Seely G.R. pp. 511–19. Academic Press, New York & London.

ALLEN M.B., DOUGHERTY E.C. & McLAUGHLIN J.J.A. (1959) Chromoprotein pigments of some cryptomonad flagellates. *Nature, Lond.* **184**, 1047–9.

ALLEN M.B., FRENCH C.S. & BROWN J.S. (1960) Native and extractable forms of chlorophyll in various algal groups. In *Comparative Biochemistry of Photoreactive Systems*, ed. Allen M.B. pp. 33–52. Academic Press, New York & London.

ALLEN M.B., MURCHIO J.C., JEFFREY S.W. & BENDIX S.A. (1963) Fractionation of the photosynthetic apparatus of *Chlorella pyrenoidosa*. In *Studies on Microalgae and Photosynthetic Bacteria*, pp. 407–12. Jap. Soc. Pl. cell Physiol. Univ. of Tokyo Press, Tokyo.

AMESZ J. (1964a) Spectrophotometric evidence for the participation of quinone in photosynthesis of intact blue-green algae. *Biochim. biophys. Acta*, **79**, 257–65.

AMESZ J. (1964b) Intracellular reactions of nicotinamide adenine dinucleotide in photosynthetic organisms. Ph.D. Thesis, University of Leiden, The Netherlands.

AMESZ J. & DUYSENS L.N.M. (1962) Action spectrum, kinetics and quantum requirement of phosphopyridine nucleotide reduction and cytochrome oxidation in blue-green alga *Anacystis nidulans*. *Biochim. biophys. Acta* **64**, 261–78.

ARNOLD W. & AZZI J.R. (1968) Chlorophyll energy levels and electron flow in photosynthesis. *Proc. natn. Acad. Sci., U.S.A.* **61**, 29–35.

ARNOLD W. & MEEK E.S. (1956) The polarization of fluorescence and energy transfer in grana. *Archs. Biochem. Biophys.* **60**, 82–90.

ARNON D.I., KNAFF D.B., McSWAIN B.D., CHAIN R.K. & TSUJIMOTO H.Y. (1971) Three light reactions and the two photosystems of plant photosynthesis. *Photochem. Photobiol.* **14**, 397–425.

AVRON M. & NEUMANN J. (1968) Photophosphorylation in chloroplasts. *A. Rev. Pl. Physiol.* **19**, 137–66.

BEARDEN A. & MALKIN R. (1972) The bound ferredoxin of chloroplasts: a role as the primary electron acceptor of PS I. *Biochem. biophys. Res. Commun.* **46**, 1299–305.

BENDALL D.S. & HILL R. (1968) Haem-proteins in photosynthesis. *A. Rev. Pl. Physiol.* **19**, 167–86.

BEN-HAYYIM G. & MALKIN S. (1971) The correlation of C550 change and the fluorescence induction in chloroplasts. Abstr. Proc. 2nd Intern. Conf. on Photosynthesis, Stresa, Italy. p. 1.

BERGERON J.A. (1963) Studies of the localization, physiochemical properties and action of phycocyanin in *Anacystis nidulans*. In *Photosynthetic Mechanisms of Green Plants*, eds. Kok B. & Jagendorf A.T. pp. 527–36. Nat. Acad. Sci.-Nat. Res. Council, Washington, D.C. **1145**,

BISHOP N. (1971) Photosynthesis: the electron transport system of green plants. *A. Rev. Biochem.* **40**, 197–226.

BLINKS L.R. (1960) Action spectra of chromatic transients and the Emerson effect in marine algae. *Proc. natn. Acad. Sci., U.S.A.* **46**, 327–33.

BLINKS L.R. (1963) Chromatic transients and enhancement recorded by the glass electrode. In *Photosynthetic Mechanisms of Green Plants*, eds. Kok B. & Jagendorf A.T. pp. 345–51. Nat. Acad. Sci.-Nat. Res. Council, Washington, D.C. **1145**.

BOARDMAN N.K. (1970) Physical separation of the photosynthetic photochemical systems. *A. Rev. Plant Physiol.* **21**, 115–40.

BRIANTAIS J.-M., MERKELO H. & GOVINDJEE (1972) Lifetime of the excited state *in vivo*. III. During fluorescence induction in *Chlorella pyrenoidosa*. *Photosynthetica*, **6**, 133–42.

BRODY M. (1958) I. The participation of chlorophyll and phycobilins in the photosynthesis of red algae. II. Observations on cellular structures of *Porphyridium cruentum*. Ph.D. Thesis. University of Illinois at Urbana-Champaign.

BRODY M. & EMERSON R. (1959a) The quantum yield of photosynthesis in *Porphyridium cruentum* and the role of chlorophyll *a* in the photosynthesis of red algae. *J. gen. Physiol.* **43**, 251–64.

BRODY M. & EMERSON R. (1959b) The effect of wavelength and intensity of light on the proportion of pigments in *Porphyridium cruentum*. *Am. J. Bot.* **46**, 433–40.

BRODY S.S. & BRODY M. (1959) Induced changes in the efficiency of energy transfer in *Porphyridium cruentum*. I. *Archs. Biochem. Biophys.* **82**, 161–78.

BRODY S.S. & BRODY M. (1963) Aggregated chlorophyll *in vivo*. In *Photosynthetic Mechanisms of Green Plants*, eds. Kok B. & Jagendorf A.T. pp. 455–78. Nat. Acad. Sci. Nat. Res. Council, Washington, D.C. **1145**.

BRODY S.S. & RABINOWITCH E. (1957) Excitation lifetimes of photosynthetic pigments *in vitro* and *in vivo*. *Science N.Y.* **125**, 555.

BROWN J. (1969) Absorption and fluorescence of chlorophyll *a* in particle fractions from different plants. *Biophys. J.* **9**, 1542–52.

BROWN J., BRIL C. & URBACH W. (1965) Fractionation of chlorophyll forms from *Euglena* and measurement of light-induced absorbance changes. *Pl. Physiol., Lancaster*, **40**, 1086–90.

BROWN T.E. & RICHARDSON F.T. (1968) The effect of growth environment on the physiology of algae: light intensity. *J. Phycol.* **4**, 38–54.

BURNETT J.H. (1965) Functions of carotenoids other than in photosynthesis. In *Chemistry and Biochemistry of Plant Pigments*, ed. Goodwin T.W. pp. 381–403. Academic Press, New York & London.

BUTLER W.L. (1966a) Spectral characteristics of chlorophyll in green plants. In *The Chlorophylls*, eds. Vernon L.P. & Seely G.R. pp. 343–79. Academic Press, New York & London.

BUTLER W.L. (1966b) Fluorescence yield in photosynthetic systems and its relation to electron transport. In *Current Topics in Bioenergetics*, Vol. I, ed. Sanadi D.R. pp. 49–73. Academic Press, New York & London.

BUTLER W.L. (1972a) On the relationship between P680 and C550. *Biophys. J.* **12**, 851–7.

BUTLER W.L. (1972b) The influence of membrane potential on measurements of C550 at room temperature. *FEBS Letters*, **20**, 333–8.

BUTLER W.L. & HOPKINS D.W. (1970a) Higher derivative analysis of complex absorption spectra. *Photochem. Photobiol.* **12**, 439–50.

BUTLER W.L. & HOPKINS D.W. (1970b) An analysis of fourth derivative spectra. *Photochem. Photobiol.* **12**, 451–6.

CALVERT J.G. & PITTS J.N. (1966) *Photochemistry*. John Wiley & Sons, Inc., New York & London.

CEDERSTRAND C.N. (1965) Spectrophotometric and spectrofluorometric characterization of the two pigment systems in photosynthesis. Ph.D. Thesis. University of Illinois at Urbana-Champaign.

CEDERSTRAND C.N., RABINOWITCH E. & GOVINDJEE (1966) Analysis of the red absorption band of chlorophyll *a in vivo*. *Biochim. biophys. Acta*, **126**, 1–12.

CHENIAE G. (1970) Photosystem II and O₂ evolution. *A. Rev. Pl. Physiol.* **21**, 467–98.

CHO F. (1969) Low temperature spectroscopy of algae. Ph.D. Thesis, University of Illinois at Urbana-Champaign.

CHO F. & GOVINDJEE (1970a) Fluorescence spectra of *Chlorella* in the 295–77°K range. *Biochim. biophys. Acta*, **205**, 371–8.

CHO F. & GOVINDJEE (1970b) Low temperature (4–77°K) spectroscopy of *Chlorella*; temperature dependence of energy transfer efficiency. *Biochim. biophys. Acta*, **216**, 139–150.

CHO F. & GOVINDJEE (1970c) Low temperature (4–77°K) spectroscopy of *Anacystis*; temperature dependence of energy transfer efficiency. *Biochim. biophys. Acta*, **216**, 151–61.

CHO F., SPENCER J. & GOVINDJEE (1966) Emission spectra of *Chlorella* at very low temperatures (–269°C to –196°C). *Biochim. biophys. Acta*, **126**, 174–6.

CLAYTON R.K. (1965) *Molecular Physics in Photosynthesis*. Balisdell Publishing Co., New York.

CLAYTON R.K. (1971) *Light and Living Matter*, vols. I and II. McGraw Hill Book Co., New York.

DAS M. & GOVINDJEE (1967) A long-wave absorbing form of chlorophyll *a* responsible for the 'red drop' in fluorescence at 298°K and the F723 band at 77°K. *Biochim. biophys. Acta*, **143**, 570–6.

DAS M., RABINOWITCH E. & SZALAY L. (1968) Red drop in the quantum yield of fluorescence of sonicated algae. *Biophys. J.* **8**, 1131–7.

DAVIES B.H. (1965) Analysis of carotenoid pigments. In *Chemistry and Biochemistry of Plant Pigments*, ed. Goodwin T.W. pp. 489–532. Academic Press, New York & London.

DÖRING G., BAILEY J.L., WEIKARD J. & WITT H.T. (1968) The action of chlorophyll *a*II in light reaction II of photosynthesis. *Naturwissenschaften*, **55**, 219–24

DÖRING G., RENGER G., VATER J. & WITT H.T. (1969) Properties of the photoactive chlorophyll *a*II in photosynthesis. *Z. Naturf.* **24b**, 1139–43.

DÖRING G., STIEHL H.H. & WITT H.T. (1967) A second chlorophyll reaction in the electron chain of photosynthesis—registration by the repetitive excitation techniques. *Z. Naturf.* **22b**, 639–44.

DUTTON H.J., MANNING W.M. & DUGGAR B.B. (1943) Chlorophyll fluorescence and energy transfer in the diatom *Nitzschia closterium*. *J. phys. Chem. Wash.* **47**, 308–13.

DUYSENS L.N.M. (1952) Transfer of excitation energy in photosynthesis. Ph.D. Thesis. University of Utrecht, The Netherlands.

DUYSENS L.N.M. (1954) Reversible changes in the absorbance spectrum of *Chlorella* upon illumination. *Science N.Y.* **120**, 353–4.

DUYSENS L.N.M. (1963) Studies of primary reactions and hydrogen or electron transport in photosynthesis by means of absorption and fluorescence difference spectrophotometry of intact cells. In *Photosynthetic Mechanisms of Green Plants*, eds. Kok B. & Jagendorf. A.T. pp. 1–17. Nat. Acad. Sci.-Nat. Res. Council, Washington, D.C. **1145**.

DUYSENS L.N.M. (1964) Photosynthesis. *Progress in Biophysics*, **14**, 1–100.

DUYSENS L.N.M. & AMESZ J. (1962) Function and identification of two photochemical systems in photosynthesis. *Biochim. biophys. Acta*, **64**, 243–60.

DUYSENS L.N.M., AMESZ J. & KAMP B.M. (1961) Two photochemical systems in photosynthesis. *Nature, Lond.* **190**, 510–11.

EMERSON R. (1958) The quantum yield of photosynthesis. *A. Rev. Pl. Physiol.* **9**, 1–24.

EMERSON R., CHALMERS R. & CEDERSTRAND C. (1957). Some factors influencing the long-wave limit of photosynthesis. *Proc. natn. Acad. Sci.*, *U.S.A.* **43**, 133–43.

EMERSON R. & LEWIS C.M. (1942) The photosynthetic efficiency of phycocyanin in *Chroococcus* and the problem of carotenoid participation in photosynthesis. *J. gen. Physiol.* **25**, 579–95.

EMERSON R. & LEWIS C.M. (1943) The dependence of the quantum yield of *Chlorella* photosynthesis on wavelength of light. *Am. J. Bot.* **30**, 165–78.

EMERSON R. & RABINOWITCH E. (1960) Red drop and role of auxiliary pigments in photosynthesis. *Pl. Physiol., Lancaster*, **35**, 477–85.

ERIXON K. & BUTLER W.L. (1971a) Light-induced absorption changes in chloroplasts at −196°C. *Photochem. Photobiol.* **14**, 427–33.

ERIXON K. & BUTLER W.L. (1971b) The relationships between Q, C550 and cytochrome b559 in photoreactions at −196°C in chloroplasts. *Biochim. biophys. Acta*, **234**, 381–9.

FISCHER K. & METZNER H. (1969) On chlorophyll and pigment P750 of *Anacystis nidulans*. In *Progress in Photosynthesis Research*, II, ed. Metzner H. pp. 547–51. Laupp Jr. H.. Tubingen.

FLOYD R.A., CHANCE B. & DEVAULT D. (1971) Low temperature photo-induced reactions in green leaves and chloroplasts. *Biochim. biophys. Acta*, **226**, 103–12.

FORK D.C. (1963) Observations on the function of chlorophyll *a* and accessory pigments in photosynthesis. In *Photosynthetic Mechanisms of Green Plants*, eds. Kok B. & Jagendorf A.T. pp. 352–61. Nat. Acad. Sci.-Nat. Res. Council, Washington, D.C. **1145**.

FORK D.C. & AMESZ J. (1967) Light-induced shifts in the absorption spectrum of carotenoids in red and brown algae. *Photochem. Photobiol.* **6**, 913–18.

FORK D.C. & AMESZ J. (1969) Action spectra and energy transfer in photosynthesis. *A. Rev, Pl. Physiol.* **20**, 305–28.

FRANCK J. & ROSENBERG J. (1964) A theory of light utilization in plant photosynthesis. *J. Theoret. Biol.* **7**, 276–301.

FREI Y.F. (1962) The derivative absorption spectra of chlorophyll in algae and leaves at low temperature. *Biochim. biophys. Acta*, **57**, 82–7.

FRENCH C.S. (1971) The distribution and action in photosynthesis of several forms of chlorophyll. *Proc. natn. Acad. Sci., U.S.A.* **68**, 2893–7.

FRENCH C.S., BROWN J.S. & LAWRENCE M.C. (1972) Four universal forms of chlorophyll *a*. *Pl. Physiol., Lancaster*, **49**, 421–9.

FRENCH C.S., BROWN J.S., WIESSNER W. & LAWRENCE M.C. (1971) Four common forms of chlorophyll *a*. *Carnegie Inst. Wash. Year Book*, **69**, 662–70.

FRENCH C.S., MYERS J. & McCLEOD G.C. (1960) Automatic recording of photosynthesis action spectra used to measure the Emerson enhancement effect. In *Comparative Biochemistry of Photoreactive Systems*, ed. Allen M.B. pp. 361–5. Academic Press, New York & London.

FRENCH C.S. & YOUNG V.M.K. (1952) The fluorescence spectra of red algae and the transfer of energy from phycoerythrin to phycocyanin and chlorophyll. *J. gen. Physiol.* **35**, 873–90.

FUJITA Y. & HATTORI A. (1960) Effect of chromatic light on phycobilin formation in a blue-green alga, *Tolypothrix tenuis*. *Pl. Cell Physiol., Tokyo*, **1**, 293–303.

FUJITA Y. & HATTORI A. (1962) Photochemical interconversion between precursors of phycobilin chromoproteins in *Tolypothrix tenuis*. *Pl. Cell Physiol., Tokyo*, **3**, 209–21.

GANTT E. & CONTI S.F. (1966) Granules associated with the chloroplast lamellae of *Porphyridium cruentum*. *J. Cell Biol.* **29**, 423–34.

GASSNER E.S. (1962) On the pigment absorbing at 750 mμ in some blue-green algae. *Pl. Physiol., Lancaster*, **37**, 637–9.

GHOSH A.K. & GOVINDJEE (1966) Transfer of excitation energy in *Anacystis nidulans* grown to obtain different pigment ratios. *Biophys. J.* **6**, 611–17.

GOEDHEER J.H.C. (1966a) Fluorescence polarization and other fluorescence properties of chloroplasts and cells in relation to molecular structure. In *Biochemistry of Chloroplasts*, Vol. I, ed. Goodwin T.W. pp. 75–82. Academic Press, New York & London.

GOEDHEER J.H.C. (1966b) Visible absorption and fluorescence of chlorophyll and its aggregates in solution. In *The Chlorophylls*, eds. Vernon L.P. & Seely G.R. pp. 147–84. Academic Press, New York & London.

GOEDHEER J.H.C. (1969a) Carotenoids in blue-green and red algae. In *Progress in Photosynthesis Research*, Vol. II, ed. Metzner H. pp. 811–17. Laupp Jr. H., Tubingen.

GOEDHEER J.H.C. (1969b) Energy transfer from carotenoids to chlorophyll in blue-green, red and green algae and greening bean leaves. *Biochim. biophys. Acta*, **172**, 252–65.

GOEDHEER J.H.C. (1970) On the pigment system of brown algae. *Photosynthetica*, **4**, 97–106.

GOODENOUGH V.W. & STAEHELIN L.A. (1971) Structural differentiation of stacked and unstacked chloroplast membranes: freeze-etch electron microscopy of wild-type and mutant strains of *Chlamydomonas*. *J. Cell Biol.* **48**, 594–619.

GOODWIN T.W. (1965a) Distribution of carotenoids. In *Chemistry and Biochemistry of Plant Pigments*, ed. Goodwin T.W. pp. 127–42. Academic Press, New York & London.

GOODWIN T.W. (1965b) The biosynthesis of carotenoids. In *Chemistry and Biochemistry of Plant Pigments*, ed. Goodwin T.W. pp. 143–73. Academic Press, New York & London.

GOVINDJEE (1960) Effect of combining two wavelengths of light on the photosynthesis of algae. Ph.D. Thesis. University of Illinois at Urbana-Champaign.

GOVINDJEE (1963a) Emerson enhancement effect and two light reactions in photosynthesis. In *Photosynthetic Mechanisms of Green Plants*, eds. Kok B. & Jagendorf A.T. pp. 318–34. Nat. Acad. Sci.-Nat. Res. Council, Washington, D.C. **1145**.

GOVINDJEE (1963b) Observation on P750 from *Anacystis nidulans*. *Naturwissenschaften*, **50**, 720–1.

GOVINDJEE & BRIANTAIS J.-M. (1972) Chlorophyll *b* fluorescence and an emission band at 700nm at room temperature in green algae. *FEBS Letters*, **19**, 278–80.

GOVINDJEE & CEDERSTRAND C.N. (1963) A letter to the editor. *Biophys. J.* **3**, 507–8.

GOVINDJEE, CEDERSTRAND C.N. & RABINOWITCH (1961) Existence of absorption bands at 730–740 and 750–760 millimicrons in algae of different divisions. *Science N.Y.* **134**, 391–2.

GOVINDJEE, DÖRING G. & GOVINDJEE R. (1970) The active chlorophyll $a_{II}$ in suspensions of lyophilized and Tris-washed chloroplasts. *Biochim. biophys. Acta*, **205**, 303–6.

GOVINDJEE & GOVINDJEE R. (1965) Action spectra for the appearance of difference absorption bands at 480 and 520 mμ in illuminated *Chlorella* cells and their possible significance to a two step mechanism of photosynthesis. *Photochem. Photobiol.* **4**, 793–801.

GOVINDJEE & MOHANTY P. (1972) Photochemical aspects of photosynthesis in blue-green algae. In *Biology and Taxonomy of Blue-Green Algae*, ed. Desikachary, T.V. pp. 171–96. U. Madras, Madras (India).

GOVINDJEE, MUNDAY J.C. & PAPAGEORGIOU G. (1967) Fluorescence studies with algae: changes with time and pre-illumination. In *Energy Conversion by the Photosynthetic Apparatus. Brookhaven Symposia in Biology*, **19**, 434–45.

GOVINDJEE & PAPAGEORGIOU G. (1971) Chlorophyll fluorescence and photosynthesis: fluorescence transients. In *Photophysiology*, ed. Giese A.C. pp. 1–46. Academic Press, New York & London.

GOVINDJEE, PAPAGEORGIOU G. & RABINOWITCH E. (1967) Chlorophyll fluorescence and photosynthesis. In *Fluorescence Theory, Instrumentation and Practice*, ed. Guilbualt G.G. pp. 511–64. Marcel Dekker, Inc., New York.

GOVINDJEE & RABINOWITCH E. (1960) Action spectrum of the second Emerson effect. *Biophys. J.* **1**, 73–89.

GOVINDJEE, RABINOWITCH E. & THOMAS J.B. (1960) Inhibition of photosynthesis in certain algae by extreme red light. *Biophys. J.* **1**, 91–7.

GOVINDJEE R. (1961) The action spectrum of the Hill reaction in whole algal cells and chloroplast suspensions (red drop, second Emerson effect and inhibition by extreme red light). Ph.D. Thesis. University of Illinois at Urbana-Champaign.

GOVINDJEE R., GOVINDJEE & HOCH G. (1964) Emerson enhancement effect in chloroplast reactions. *Pl. Physiol., Lancaster*, **39**, 10–14.

GOVINDJEE R., RABINOWITCH E. & GOVINDJEE (1968) Maximum quantum yield and action spectrum of photosynthesis and fluorescence in *Chlorella. Biochim. biophys. Acta*, **162**, 539–44.

HALLDAL P. (1970) The photosynthetic apparatus of microalgae and its adaptation to environmental factors. In *Photobiology of Microorganisms*, ed. Halldal P. pp. 17–55. John Wiley-Interscience, London, New York, Sydney, Toronto.

HALLDAL P. & FRENCH C.S. (1958) Algal growth in crossed gradients of light intensity and temperature. *Pl. Physiol., Lancaster*, **33**, 249–52.

HATTORI A. & FUJITA Y. (1959a) Formation of phycocyanin pigments in a blue-green alga, *Tolypothrix tenuis*, as induced by illumination with colored lights. *J. Biochem.* **46**, 521–4.

HATTORI A. & FUJITA Y. (1959b) Effect of pre-illumination on the formation of phycobilin pigments in a blue-green alga, *Tolypothrix tenuis. J. Biochem.* **46**, 1259–61.

HATTORI A. & FUJITA Y. (1959c) Spectroscopic studies on the phycobilin pigments obtained from blue-green and red algae. *J. Biochem.* **46**, 903–9.

HAXO F.T. (1960) The wavelength dependence of photosynthesis and the role of accessory pigments. In *Comparative Biochemistry of Photoreactive Systems*, ed. Allen M.B. pp. 339–60. Academic Press, New York & London.

HAXO F.T. & BLINKS L.R. (1950) Photosynthetic action spectra of marine algae. *J. gen. Physiol.* **33**, 389–422.

HILL R. & BENDALL F. (1960) Function of two cytochrome components in chloroplasts: a working hypothesis. *Nature, Lond.* **186**, 136–7.

HIND G. & OLSON J.M. (1968) Electron transport pathways in photosynthesis. *A. Rev. Pl. Physiol.* **19**, 249–82.

HIYAMA T. & KE B. (1971a) A new photosynthetic pigment, 'P430': its possible role as the primary electron acceptor of photosystem I. *Proc. natn. Acad. Sci., U.S.A.* **68**, 1010–13.

HIYAMA T. & KE B. (1971b) A further study of P430: a possible primary electron acceptor of photosystem I. *Archs. Biochem. Biophys.* **147**, 99–108.

HOCH G. & KOK B. (1961) Photosynthesis. *A. Rev. Pl. Physiol.* **12**, 155–94.

HOCH G. & KNOX R.S. (1968) Primary processes in photosynthesis. In *Photophysiology*, Vol. III, ed. Giese A.C. pp. 225–51. Academic Press, New York & London.

HOLT A.S. (1961) Further evidence of the relation between 2-desvinyl-2-formyl chlorophyll *a* and chlorophyll *d. Can. J. Bot.* **39**, 327–31.

HOLT A.S. (1966) Recently characterized chlorophylls. In *The Chlorophylls*, eds. Vernon L.P. & Seely G.R. pp. 111–18. Academic Press, New York & London.

HOLT A.S. & MORLEY H.V. (1959) A proposed structure for chlorophyll *d. Can. J. Chem.* **37**, 507–14.

JACKSON W.A. & VOLK R.J. (1970) Photorespiration. *A. Rev. Pl. Physiol.* **21**, 385–432.

JONES L.W. & MYERS J. (1964) Enhancement in the blue-green alga, *Anacystis nidulans*. *Pl. Physiol., Lancaster*, **39**, 938–46.

JUNGE W., SCHLIEPHAKE W.D. & WITT H.T. (1969) Experimental evidence for the chemiosmotic hypothesis. In *Progress in Photosynthesis Research*, Vol. III, ed. Metzner H. pp. 1383–91. Laupp Jr. H., Tübingen.

KAUTSKY H., APPEL W. & AMMAN H. (1960) Chlorophyll fluorescenz und kohlensaureassimilation XII. Mitteilung die fluorescenzcurve und die photochemie der pflanze. *Biochem. Z.* **332**, 277–92.

KLEIN R.M. & CRONQUIST A. (1967) A consideration of the evolutionary and taxonomic significance of some biochemical, micromorphological, and physiological characters in the thallophytes. *Q. Rev. Biol.* **42**, 105–296.

KNAFF D.B. & ARNON D.I. (1969) A concept of three light reactions in photosynthesis by green plants. *Proc. natn. Acad. Sci., U.S.A.* **64**, 715–22.

KOK B. (1957) Light-induced absorption changes in the photosynthetic organisms. *Acta Bot. Neerl.* **6**, 316–36.

KOK B. (1959) Light-induced absorption changes in photosynthetic organisms. II. A split-beam difference spectrophotometer. *Pl. Physiol., Lancaster*, **34**, 184–92.

KOK B. (1961) Partial purification and determination of oxidation reduction potential of the photosynthetic chlorophyll complex absorbing at 700 m$\mu$. *Biochim. biophys. Acta*, **48**, 527–33.

KOK B. (1963) Fluorescence studies. In *Photosynthetic Mechanisms of Green Plants*, eds. Kok B. & Jagendorf A.T. pp. 45–55. Nat. Acad. Sci.-Nat. Res. Council, Washington, D.C. **1145**.

KOK B. & GOTT W. (1960) Activation spectra of 700 m$\mu$ absorption change in photosynthesis. *Pl. Physiol., Lancaster*, **35**, 802–8.

KOK B. & HOCH G. (1961) Spectral change in photosynthesis. In *Light and Life*, eds. McElroy W.D. & Glass B. pp. 397–423. Johns Hopkins Press, Baltimore.

KOUCHKOVSKY Y.DE & FORK D.C. (1964) A possible functioning *in vivo* of plastocyanin in photosynthesis as revealed by a light-induced absorbance change. *Proc. natn. Acad. Sci., U.S.A.* **52**, 232–339.

KREY A. & GOVINDJEE (1966) Fluorescence studies on a red alga *Porphyridium cruentum*. *Biochim. biophys. Acta*, **120**, 1–18.

LATIMER P., BANNISTER T.T. & RABINOWITCH E. (1957) The absolute quantum yields of fluorescence of photosynthetically active pigments. In *Research in Photosynthesis*, eds. Gaffron H., Brown A.H., French C.S., Livingston R., Rabinowitch E., Strehler B.L. & Tolbert N.E. pp. 107–12. Interscience Publ., Inc., New York.

LEVINE R.P. (1969) Analysis of photosynthesis using mutant strains of algae and higher plants. *A. Rev. Pl. Physiol.* **20**, 523–40.

LEVINE R.P. & GORMAN D.S. (1966) Photosynthetic electron transport chain of *Chlamydomonas reinhardi*. III: light-induced absorbance changes in chloroplast fragments of the wild type and mutant strains. *Pl. Physiol., Lancaster*, **41**, 1293–300.

LITVIN F.F. & GULYAEV B.A. (1969) Resolution of the structure of the absorption spectrum of chlorophyll *a* and its bacterial analogs in the cell by measurement of the second derivative at 20° to −196°C. *Dokl. Biophysics*, **189**, 157–60.

MALKIN R. & BEARDEN A. (1971) Primary reactions of photosynthesis: photoreduction of a bound chloroplast ferredoxin at low temperature as detected by EPR spectroscopy. *Proc. natn. Acad. Sci., U.S.A.* **68**, 16–19.

MANN J.E. & MYERS J. (1968) On pigments, growth and photosynthesis of *Phaeodactylum tricornutum*. *J. Phycol.* **4**, 349–55.

MAR T. & GOVINDJEE (1971) Decrease in the degree of polarization of chlorophyll fluorescence upon the addition of DCMU to algae. Abstr. Proc. 2nd Intern. Conf. on Photosynthesis, Stresa, Italy, p. 135.

MAR T., GOVINDJEE, SINGHAL G.S. & MERKELO H. (1972) Lifetime of the excited state *in vivo* I. Chlorophyll *a* in algae at room and at liquid nitrogen temperature; rate constants of radiationless deactivation and trapping. *Biophys. J.* **12**, 797–808.

MERKELO H., HARTMAN S., MAR T., SINGHAL G.S. & GOVINDJEE (1969) Mode-locked lasers: measurements of very fast radiative decay in fluorescent systems. *Science N.Y.* **164**, 301–2.

METZNER H. (1963) Spektralphotometrische messungen an gefriergetrochneten Chlorellazellen. In *Studies on Microalgae and Photosynthetic Bacteria*, pp. 227–37. Jap. Soc. Pl. cell Physiol., University of Tokyo Press, Tokyo.

MICHEL J.-M. & MICHEL-WOLWERTZ M.-R. (1968) Fractionation of the photosynthetic apparatus from broken spinach chloroplasts by sucrose density-gradient centrifugation. *Carnegie Inst. Wash. Year Book*, **67**, 508–14.

MOHANTY P. (1972) Regulation of chlorophyll fluorescence during photosynthesis: a study of the factors affecting change in yield and emission of chlorophyll fluorescence in intact algal cells and isolated chloroplasts. Ph.D. Thesis. University of Illinois at Urbana-Champaign.

MOHANTY P., BRAUN B.Z., GOVINDJEE & THORNBER J.P. (1972) Chlorophyll fluorescence characteristics of system I chlorophyll *a*-protein complex and system II particles at room and liquid nitrogen temperature. *Pl. Cell Physiol., Tokyo*, **13**, 81–91.

MÜLLER A., LUMRY R. & KOKUBUN H. (1965) High performance phase fluorometer constructed from commercial subunits. *Rev. Sci. Instr.* **36**, 1214–26.

MÜLLER A., LUMRY R. & WALKER M.S. (1969) Light intensity dependence of the *in vivo* fluorescence lifetime of chlorophyll. *Photochem. Photobiol.* **9**, 113–26.

MURTY N.R. & RABINOWITCH E. (1965) Fluorescence decay studies of chlorophyll *a in vivo*. *Biophys. J.* **5**, 655–61.

MYERS J. (1963) Enhancement. In *Photosynthetic Mechanisms of Green Plants*, eds. Kok B. & Jagendorf A.T. pp. 301–17. Nat. Acad. Sci.-Nat. Res. Council, Washington, D.C. **1145**.

MYERS J. (1971) Enhancement studies in photosynthesis. *A. Rev. Pl. Physiol.* **22**, 289–312.

NEUSHUL M. (1970) A freeze-etching study of the red alga *Porphyridium*. *Am. J. Bot.* **57**, 1231–9.

NICHOLSON W.J. & FORTOUL J.I. (1967) Measurement of the fluorescent lifetimes of *Chlorella* and *Porphyridium* in weak light. *Biochim. biophys. Acta*, **143**, 577–82.

OGAWA T., KANAI R. & SHIBATA K. (1968) Distribution of carotenoids in the two photochemical systems of higher plants and algae. In *Comparative Biochemistry and Biophysics of Photosynthesis*, eds. Shibata K., Takamiya A., Jagendorf A.T. & Fuller R.C. pp. 22–35. University of Tokyo Press, Tokyo.

OGAWA T. & VERNON L.P. (1969) A fraction from *Anabaena variabilis* enriched in the reaction center chlorophyll P700. *Biochim. biophys. Acta*, **180**, 334–6.

OGAWA T., VERNON L.P. & MOLLENHAUER H.H. (1969) Properties and structure of fractions prepared from *Anabaena variabilis* by the action of Triton-X-100. *Biochim. biophys. Acta*, **172**, 216–29.

O'HEOCHA C. (1960) Chemical studies of phycoerythrins and phycocyanins. In *Comparative Biochemistry of Photoreactive Systems*, ed. Allen M.B. pp. 181–203. Academic Press, New York & London.

O'HEOCHA C. (1962) Phycobilins. In *Physiology and Biochemistry of Algae*, ed. Lewin R.A. pp. 421–35. Academic Press, New York & London.

O'HEOCHA C. (1965) Phycobilins. In *Chemistry and Biochemistry of Plant Pigments*, ed. Goodwin T.W. pp. 175–96. Academic Press, New York & London.

O'HEOCHA C. (1971) Pigments of the red algae. *Oceanogr. Mar. Biol. Ann. Rev.* **9**, 61–82.

OKAYAMA S. & BUTLER W.L. (1972) Extraction and reconstitution of photosystem II. *Pl. Physiol., Lancaster,* **49**, 769–74.

OQUIST G. (1969) Adaptations in pigment composition and photosynthesis by far-red radiation in *Chlorella pyrenoidosa. Physiologia Pl.* **22**, 516–28.

PACKER L., MURAKAMI S. & MEHARD C.W. (1970) Ion transport in chloroplasts and plant mitochondria. *A. Rev. Pl. Physiol.* **21**, 271–304.

PAPAGEORGIOU G. & GOVINDJEE (1967a) Oxygen evolution from lyophilized *Anacystis* with carbon dioxide as oxidant. *Biochim. biophys. Acta,* **131**, 173–8.

PAPAGEORGIOU G. & GOVINDJEE (1967b) Changes in intensity and spectral distribution of fluorescence: effect of light intensity on normal and DCMU-poisoned *Anacystis nidulans. Biophys. J.* **7**, 375–90.

PARK R.B. & SANE P.V. (1971) Distribution and function and structure in chloroplast lamellae. *A. Rev. Pl. Physiol.* **22**, 395–430.

PRATT L. & BISHOP N.I. (1968) The 520nm light-induced absorbance change in photosynthetic mutants of *Scenedesmus. Biochim. biophys. Acta,* **162**, 369–79.

RABINOWITCH E. (1945) *Photosynthesis and Related Processes*, Vol. I, 399–437, Interscience Pub., Inc., New York.

RABINOWITCH E. (1951) *Photosynthesis and Related Processes*, Vol. II, 603–828, Interscience Pub., Inc., New York.

RABINOWITCH E. (1956) *Photosynthesis and Related Processes*, Vol. III, 1375–432; 1793–885, Interscience Pub., Inc., New York.

RABINOWITCH E. (1957) Photosynthesis and energy transfer. *J. phys. Chem.* **61**, 870–7.

RABINOWITCH E. (1963) The mechanism of photosynthesis. In *Photosynthetic Mechanisms of Green Plants*, eds. Kok B. & Jagendorf A.T. pp. 112–21. Nat. Acad. Sci.-Nat. Res. Council, Washington, D.C. **1145**.

RABINOWITCH E. & GOVINDJEE (1969) *Photosynthesis*. John Wiley & Sons, Inc., New York.

RABINOWITCH E., GOVINDJEE & THOMAS J.B. (1960) Inhibition of photosynthesis in some algae by extreme-red light. *Science N.Y.* **132**, 422.

REGER B.J. & KRAUSS R.W. (1970) The photosynthetic response to a shift in the chlorophyll *a* to chlorophyll *b* ratio of *Chlorella. Pl. Physiol., Lancaster,* **46**, 568–75.

ROBINSON G.W. (1964) Quantum processes in photosynthesis. *A. Rev. phys. Chem.* **15**, 311–48.

RUBINSTEIN D. (1964) A study of absorption changes upon illumination of *Chlorella pyrenoidosa*. Ph.D. Thesis. University of Illinois at Urbana-Champaign.

RUMBERG B. & WITT H.T. (1964) Analyse der photosynthese mit blitzlicht I. die photooxydation von chlorophyll *a*: 430–703. *Z. Naturf.* **19b**, 693–707.

RURAINSKI H.J., RANDLES J. & HOCH G. (1971) The relationship between P700 and NADP reduction in chloroplasts. *FEBS Letters,* **13**, 98–100.

SANE P.V., GOODCHILD D.J. & PARK R.B. (1970) Characterization of chloroplast photosystems I and II separated by a non-detergent method. *Biochim. biophys. Acta,* **216**, 162–78.

SATOH K. (1972) Effect of urea and *o*-phenanthroline on F695 emission in chloroplasts. *Pl. Cell Physiol., Tokyo,* **13**, 23–34.

SEELY G.R. (1971) Facilitation of energy trapping by different forms of chlorophyll. Abstr. 2nd Intern. Conf. on Photosynthesis, Stresa, Italy, p. 70.

SHIBATA K. & HAXO F.T. (1969) Light transmission and spectral distribution through epi- and endozoic algal layers in the brain coral, *Favia. Biol. Bull.* **136**, 461–8.

SHIMONY C., SPENCER J. & GOVINDJEE (1967) Spectral characteristics of *Anacystis* particles. *Photosynthetica*, **1**, 113–25.

SINGHAL G.S. & RABINOWITCH E. (1969) Measurement of the fluorescence lifetime of chlorophyll *a in vivo. Biophys. J.* **9**, 586–91.

SINGHAL G.S., WILLIAMS W.P. & RABINOWITCH E. (1968) Fluorescence and absorption studies on chlorophyll *a in vitro* at 77°K. *J. phys. Chem.* **72**, 3941–51.

SMITH J.H.C. & BENITEZ A. (1955) Chlorophylls: analysis in plant materials. In *Modern Methods of Plant Analyses*, Vol. IV, eds. Paech K. & Tracey M.V. pp. 140–96. Springer-Verlag, Berlin.

SMITH J.H.C. & FRENCH C.S. (1963) The major and accessory pigments in photosynthesis. *A. Rev. Pl. Physiol.* **14**, 181–224.

STRAIN H. (1951) The pigments of algae. In *Manual of Phycology*, ed. Smith G.M. pp. 243–62. Chronica Botanica Co., Waltham, Mass.

SZALAY L., RABINOWITCH E., MURTY N.R. & GOVINDJEE (1967) Relationship between the absorption and emission spectra and the red drop in the action spectra of fluorescence *in vivo. Biophys. J.* **7**, 137–49.

TANADA T. (1951) The photosynthetic efficiency of carotenoid pigments in *Navicula minima Am. J. Bot.* **38**, 276–83.

TEALE F.W.J. (1958) Carotenoid sensitized fluorescence of chlorophyll *in vitro. Nature Lond.* **181**, 415–16.

THOMAS J.B. (1962) Structure of the red absorption band of chlorophyll *a* in *Aspidistra elatior. Biochim. biophys. Acta*, **59**, 202–10.

THOMAS J.B. (1971) The approximate red absorption band of chlorophyll *b* in *Ulva lactuca* at 77°K. *FEBS Letters*, **14**, 61–4.

THOMAS J.B. & GOVINDJEE (1960) Changes in quantum yield of photosynthesis in the red alga *Porphyridium cruentum* caused by stepwise reduction in the intensity of light preferentially absorbed by the phycobilins. *Biophys. J.* **1**, 63–72.

THORNBER J.P. (1971) Chlorophyll *a*-protein complex of blue-green algae. In *Methods in Enzymology*, Vol. XXIII, eds. Colowick S.P. & Kaplan N.O. pp. 682–7. Academic Press, New York & London.

THORNBER J.P. & OLSON J.M. (1971) Chlorophyll-proteins and reaction center preparations from photosynthetic bacteria, algae and higher plants. *Photochem. Photobiol.* **14**, 329–41.

TOMITA G. & RABINOWITCH E. (1962) Excitation energy transfer between pigments in photosynthetic cells. *Biophys. J.* **2**, 483–99.

VERNON L.P. & AVRON M. (1965) Photosynthesis. *A. Rev. Biochem.* **34**, 269–96.

VREDENBERG W.J. & DUYSENS L.N.M. (1963) Transfer of energy from bacteriochlorophyll to a reaction center during bacterial photosynthesis. *Nature, Lond.* **197**, 355–7.

WASSINK E.C. & KERSTEN J.A.H. (1946) Observations sûr le spectre d'absorption et sûr le role des carotenoids dans la photosynthèse des diatoms. *Enzymologia*, **12**, 3–32.

WEAVER E.C. (1968) EPR studies of free radicals in photosynthetic systems. *A. Rev. Pl. Physiol.* **19**, 283–94.

WEBER G. (1960) Fluorescence parameters and photosynthesis. In *Comparative Biochemistry of Photoreactive Systems*, ed. Allen M.B. pp. 395–411. Academic Press, New York & London.

WEEDON B.C.L. (1965) Chemistry of the carotenoids. In *Chemistry and Biochemistry of Plant Pigments*, ed. Goodwin T.W. pp. 75–125. Academic Press, New York & London.

WILLIAMS W.P., MURTY N.R. & RABINOWITCH E. (1969) The complexity of the fluorescence of *Chlorella pyrenoidosa*. *Photochem. Photobiol.* **9**, 455–69.

WITT H.T. (1967) On the analysis of photosynthesis by pulse techniques in the $10^{-1}$ to $10^{-8}$ second range. Nobel Symposium, Almqvist and Wiksell, Stockholm, **5**, 261–316.

WITT H.T., MÜLLER A. & RUMBERG B. (1961) Experimental evidence for the mechanism of photosynthesis. *Nature, Lond.* **191**, 194–5.

YANG C.S. & BLUMBERG W.E. (1972) Quantitative studies on the EPR signals of PS I and ferredoxin. *Biochem. biophys. Res. Commun.* **46**, 422–8.

YOCUM C.S. & BLINKS L.R. (1958) Light-induced efficiency and pigment alterations in red algae. *J. gen. Physiol.* **41**, 1113–17.

# CHAPTER 13

# PHOTOSYNTHETIC ELECTRON FLOW AND PHOTOPHOSPHORYLATION

## J. A. RAVEN

Department of Biological Sciences,
University of Dundee, Dundee, U.K.

1 Introduction 391

2 Criteria for involvement of intermediates in photosynthesis 392
2.1 Rate of reaction 392
2.2 Quantum yield 393
2.3 Mutants 393
2.4 Extraction/replacement 393
2.5 Amount present 393

3 Relation of the two photoreactions to cyclic and noncyclic-
photophosphorylation 393

4 Chemical nature of the two
photoreactions 398

5 Chemical nature of intermediates of non-cyclic electron transport 400
5.1 Reducing end of photosystem
I to NADP 400
5.2 Reducing end of photosystem
II to oxidizing end of photosystem I 401
5.3 Water to the oxidizing end of
photosystem II 402

6 Chemical nature of the intermediates of cyclic electron flow
402

7 Alternative schemes of electron
transport 403

8 Photosynthetic phosphorylation
404
8.1 General characteristics and
rates in vitro 404
8.2 Coupling and stoichiometry
405
8.3 Mechanisms and partial reactions 406
8.4 Location of coupling sites 407
8.5 ATP transport 408
8.6 Photophosphorylation in vivo
409
8.7 Regulation of cyclic and non-
cyclic photophosphorylation in
vivo 411

9 Summary 412

10 References 412

# 1  INTRODUCTION

This chapter attempts to outline the means by which the photosynthetic
apparatus of algae converts light energy absorbed by the photosynthetic pigments

(see Chapter 12, p. 346) into the biochemically useful forms of low-potential reductant and 'high-energy phosphate', which are used in $CO_2$ fixation (see Chapter 15, p. 434) and in other energy-requiring processes (Sections 5.1, 8.6). Recent reviews of photosynthesis are listed by Govindjee and Braun in Chapter 12; to this list may be added the excellent book by Gregory (1971).

Arnon (1959) describes the early experiments which distinguished the two major kinds of light-induced electron transport and phosphorylation involved in the photosynthesis of $O_2$-evolving organisms. One of these is cyclic photophosphorylation:

$$nADP + nP_i + 2 \text{ quanta} \rightarrow nATP + nH_2O \qquad (1)$$

in which a photoact transfers an electron against the free energy gradient. The energy released as this electron returns to the low energy state *via* a series of carriers is partly conserved as ATP. $n$ in equation (1) is probably 2.

By contrast, in non-cyclic photophosphorylation there is a net transfer of electrons from a donor (water) to an acceptor (NADP), equation (2). This electron transfer is again against the free energy gradient, and it is likely that two separate light reactions, using light harvested by two separate pigment systems, are involved in transferring each electron from water to NADP. This electron transport involves the use of 2 quanta per electron transferred (probably one in each photosystem), and is coupled to ATP synthesis; $n$ in equation (2) is probably 4, i.e. the $P/e_2$ is probably 2 (Section 8.2).

$$2NADP + nADP + nP_i + 2H_2O^* + 8 \text{ quanta}$$
$$\rightarrow 2NADPH_2 + O_2^* + nATP + nH_2O \qquad (2)$$

The details of these reactions will be dealt with below but before discussing these it is worth reviewing the criteria to be applied in deciding if a reaction qualifies as an obligate partial process of overall photosynthesis. This will be done, in illustration, for NADPH$_2$ and ATP in Section 2, and applied in later sections to the partial processes involved in their synthesis.

## 2  CRITERIA FOR INVOLVEMENT OF INTERMERDIATES IN PHOTOSYNTHESIS

### 2.1  *Rate of reaction*

Since algal photosynthesis can proceed at up to 200 $\mu$moles $CO_2$ fixed or $O_2$ evolved (mg chlorophyll)$^{-1}$ hour$^{-1}$, any partial process must proceed at a commensurate rate. If $CO_2$ fixation needs 2NADPH$_2$ and 3ATP per mole of $CO_2$, the rates of synthesis of these compounds must be at least 800 $\mu$eq (mg chlorophyll)$^{-1}$ hour$^{-1}$ and 600 $\mu$moles (mg chlorophyll)$^{-1}$ hour$^{-1}$ respectively (see Table 13.1).

## 2.2 Quantum yield

Photosynthesis has a quantum yield of 0·10 to 0·125 moles $CO_2$ fixed or $O_2$ evolved per Einstein absorbed. Table 13.1 shows that non-cyclic electron flow in algae has a high enough quantum yield to be an obligate process in photosynthesis; the data for ATP are sparse, but are not inconsistent with a role for ATP in photosynthesis. *In vivo* measurements of rates of reaction and quantum yield for photosynthetic intermediates are generally made using differential absorption spectrophotometry.

## 2.3 Mutants

Mutants of algae which lack photosynthesis, and also lack the ability to photoreduce NADP or synthesize ATP in photophosphorylation (Table 13.1) provide further evidence for the role of these compounds in photosynthesis (see Chapter 14, p. 424).

## 2.4 Extraction/replacement

Inhibition of the process by removal of the component under investigation, and restoration by its replacement, only applies to ATP and $NADPH_2$ in a trivial fashion, in that these, NADP and ADP must be supplied to thylakoid preparations *in vitro* in order to observe NADP reduction and ATP synthesis

## 2.5 Amount present

Flash yield experiments involve supplying a light flash so brief that only one turnover of the photochemical reaction centres can take place (Weiss *et al.* 1971). The $O_2$ evolved (or $CO_2$ fixed) indicates the concentration of reaction centres; on a chlorophyll basis, this involves one reaction centre of each photosystem per 400 to 600 chlorophylls (Myers & Graham 1971). Obligate electron transfer intermediates must be present at this concentration at least, unless they turn over more than once per reaction centre turnover. Both ATP (Holm-Hansen 1970, Strotmann & Berger 1969) and NADP (Table 13.2) are present in prokaryotic cells and eukaryotic chloroplasts at greater concentrations than this.

# 3 RELATION OF THE TWO PHOTOREACTIONS TO CYCLIC AND NON-CYCLIC PHOTOPHOSPHORYLATION

Oxygen-evolving photosynthesis involves the co-operation of two separate photochemical reactions, powered by excitation energy harvested by two separate pigment assemblies (Rabinowitch & Govindjee 1969, Myers 1971, Chapter 12, p. 346). Evidence for this is based largely on the Emerson enhancement effect (see Chapter 12, p. 347).

**Table 13.1.** Characteristics of overall photosynthesis compared with those of various possible partial reactions of photosynthesis

| Reaction | $E'_0$ range (volts) | Reaction centre concentration (reducing equivalents per 1000 moles chlorophyll) | Effects of light absorbed by pigment system one | Effects of light absorbed by pigment system two | Rate in saturating light, µeq (mg chl)⁻¹ h⁻¹, or µmoles ATP (mg chl)⁻¹ h⁻¹ | Inhibition of overall photosynthesis by treatments inhibiting the reaction | Equivalents transferred per Einstein absorbed by the relevant photosystem | Inhibition of overall photosynthesis by mutants lacking the reaction |
|---|---|---|---|---|---|---|---|---|
| Photosynthesis | +0·81-—0·32 | 1·6–2·5 (160, 223 224) | Both needed, enhancement (Chapter 12) | | ATP ≤600; electrons <800 (28, 87, 137, 185, 224) | | about 0·5 (135) | |
| Photoreaction one | +0·45 (P700)—<—0·52 (137, 153) [33, 139, 231] | 3 (3,164) 2 [131] | Needed (2,153) | Not needed (129, 153) | 300–2000 (29, 57, 58, 73, 98, 129, 186) | | about 1 (218) [203] | +(21, 145, 146) |
| Photoreaction two | >+0·81-+0·17—0·05 (5, 55, 137) | 2–3 (55) | Not needed | Needed (119, 153) | 450–700 (81, 84, 205) | DCMU: (79, 80) | about 1(1) [203] | +(31, 145, 146) |
| Non-cyclic electron flow | >+0·81-<—0·52 | 2 (65) | Both needed, enhancement (30, 89; ? 83) | | 380–968 (36, 56, 57, 58, 206) | DCMU: (79, 80) | about 0·5 (65, 135) | +(31, 145, 146) |
| Non-cyclic ATP synthesis in vitro | ,, | ,, | ,, | ,, | 180–400 (29, 37, 220) | ,, | ,, | +Mutant lacking active coupling factor common to both (186) |
| Cyclic photophosphorylation in vitro (PMS) | See photoreaction one | See photoreaction one | Needed (220) | Needed (220) | 370–750 (35, 62, 220) | | | |
| Cyclic photophosphorylation in vivo | ,, | ,, | Needed (174, 194, 211) | Not needed (166, 208, 216) | 30–100 (180, 209, 215) | None (126, 176, 195, 209) | about 0·5 (207, 211) | absent from mutant lacking system one (210) |

The figures in brackets refer to the references cited in the 'References' section. Those in curved brackets refer to work on algae; those in square brackets refer to work on higher plants where there is little or no relevant work on algae.

Table 13.2. Characteristics of the reactions of various potential intermediaries of non-cyclic electron transport in algae

| Component | E'₀, V. | Concentration (reducing equivalents per 1000 moles chlorophyll) | Effect of light absorbed by pigment system one; and half-time for redox reaction | Effect of light absorbed by pigment system two; and half-time for redox reaction | Adequacy of turnover in saturating light for photosynthesis at 800 μeq. (mg chl)⁻¹ h⁻¹ | Equivalents transformed per Einstein absorbed by the relevant photosystem | Inhibition of in vivo photosynthesis in mutants lacking the component | Inhibition of electron transport by removal of the component; restoration by replacement | Reaction at 77°K |
|---|---|---|---|---|---|---|---|---|---|
| NADP | $-0\cdot32$ | 8(1) 15—30 [99] | Reduction (2) | | (+) (32, 37, 87) | about 1 (2) | | | |
| Fd | $-0\cdot41$ (47) | 3–5 (81, 113) | Reduction [51] | | (+) (50) | | | + (36,42) | + |
| P430 | $<-0\cdot52$ (106) | as for P700 (106), [107] | Reduction; $t_{\frac{1}{2}} \leqslant$ 5μs (106) [107] | | + | about 1 [107] (106) | | | |
| P700 | $+0\cdot40$–$+0\cdot46$ (138, 213) | 3–10 (3, 164) | Oxidation; $t_{\frac{1}{2}} \approx$ 5μs (105,138,218) | Reduction; $t_{\frac{1}{2}} \approx$ 10ms (105,138, 218) | + | 0·2–1·0 (3,6, 106,107,218) | + (145,146,171) | | + (10,218) |
| Plastocyanin | $+0\cdot37$–$+0\cdot39$ (85,127) | 1–1·5 (85) 1–5 [170] | Oxidation; $t_{\frac{1}{2}} \approx$ 0·1s (141) [$t_{\frac{1}{2}} \approx$ 10μs 93?] | Reduction; $t_{\frac{1}{2}} \approx$ 0·1s (141) | (−) | | + (86) | | + (151) |
| Cytochrome f | $+0\cdot34$–$+0\cdot39$ (86,113,157,228) | 0·5–10 (3,29,113 147,165) | Oxidation; $t_{\frac{1}{2}} \approx$ 15μs (6,49,64,105, 162,167) | Reduction? $t_{\frac{1}{2}} \approx$ 10ms (6,49,162, 163,167) | + | 0·2–0·7 (2,3,6, 167,218) | + (86,171) | + (128) | — |
| Plastoquinone | $+0\cdot01$–$+0\cdot06$ (48) | Reactive: 6–8 (1, 183) Total: 10–100 (205) | Oxidation; $t_{\frac{1}{2}} \approx$ 10ms (1,6,183) | Reduction; $t_{\frac{1}{2}} \approx$ 100μs (1,6,183) [226] | + | about 1 (1,6) | (+) (146,191) | + (150) | |
| Cytochrome 559 | $+0\cdot32$ (114); 0·0–$+0\cdot37$ [24,219] | 3 (114) | Oxidation (300°K); $t_{\frac{1}{2}} \approx$ 15s (115,148, 149) | Reduction (300°K); $t_{\frac{1}{2}} \approx$ 15s (148,149) Oxidation (77°K); (26,68) | (−) | | (+) (146,191) | | + (26,68) |
| C550 | $+0\cdot025$–$+0\cdot05$ [69] | | | Reduction (77°K) (26,68) | | | (+) (146,191) | | + (26,68) |
| P690 | | | | Oxidation (bleaching) [61] | | | | | + [70] |

The figures in brackets refer to the references cited in the 'references' section.
Those in curved brackets refer to work on algae; those in square brackets refer to work on higher plants where there is little or no relevant work on algae.
+, = positive data available; —, negative data available.

The finding that enhancement could still occur even when the two light wavelengths were supplied sequentially rather than simultaneously showed that this interaction occurred at the level of stable chemical products rather than of excited states; the interval between the two lights which still permitted the demonstration of enhancement being much longer than the lifetime of excited chlorophyll.

Analysis of the wavelength dependence of enhancement allows description of the pigments associated with the two photosystems. One of the photosystems (photosystem I: Duysens 1963) contains a preponderance of long wavelength forms of chlorophyll *a*, while the other (photosystem II) contains most of the shorter wavelength forms of chlorophyll *a*, and most of the accessory pigments. While both of these systems are required for complete photosynthesis, it is clear (Section 8.6, Table 13.1) that photosystem I on its own can power cyclic photophosphorylation in algae *in vivo*. It is also agreed that $O_2$ evolution requires photosystem II. Most workers agree that both photosystems are involved in noncyclic photophosphorylation. However, Arnon and co-workers (Arnon *et al.* 1970, Arnon *et al.* 1971, Arnon 1971) suggest that the Emerson effect originates from a requirement for cyclic (photosystem I) and non-cyclic (photosystem II) photophosphorylations for complete photosynthesis. This hypothesis has been criticized from experiments on higher plants (Arntzen *et al.* 1972, Sun & Sauer 1972) and from algae (Raven 1970).

Most of the pertinent evidence from algae supports a requirement for both photosystems for non-cyclic photophosphorylation. The evidence comes from experiments on the Emerson effect on reactions in algae in which $CO_2$ is not the terminal electron acceptor. The Emerson effect has been found for the photoreduction of benzoquinone (Govindjee & Rabinowitch 1960) and the photoproduction of $H_2$ (Bishop & Gaffron 1963), both with $H_2O$ as electron donor. Since these reactions are stimulated by uncouplers (Gimmler & Avron 1971, Bishop & Gaffron 1963), enhancement cannot be due to ATP supply from cyclic photophosphorylation.

The generally accepted view of non-cyclic photophosphorylation which involves the co-operation of two photoacts is that based on the hypothesis of Hill and Bendall (1960), in which two photoreactions in series are involved in non-cyclic photophosphorylation. The energetic requirement for two photoacts in this process is discussed by Ross and Calvin (1967), and Knox (1969). This scheme accounts for the requirement for two quanta absorbed per electron moved in non-cyclic photophosphorylation.

The Hill and Bendall hypothesis describes electron transport from water ($E'_0 = +0.81v$) to NADP ($E'_0 = -0.32v$), i.e. over an energy span of some $1.13v$. The two light reactions each cover a range of $0.8ev$, and they overlap by about $0.4v$. The two light reactions cover the range to $+0.81v$ to $0.0v$; (based on the observed photoreduction of a *b*-type cytochrome at about $0.0v$) and $+0.4$ to $-0.32v$ (based on the observed photo-oxidation of cytochrome *f*, $E'_0$ about $+0.4v$). The two photoacts are connected by a downhill electron transport chain which reduces cytochrome $f$ and oxidizes cytochrome $b$; by analogy with the respiratory chain, this might contain a phosphorylation site (Section 8.4, below).

This scheme does not necessitate two separate pigment systems to harvest the light energy for the two light reactions. However, the above-mentioned association of $O_2$ evolution with photosystem II suggested that this photosystem might be the $H_2O$—cytochrome $b$ part of this scheme, while photosystem I would be left for the cytochrome $f$—NADP part. Further evidence for this came from effects of light predominantly absorbed by one or other of the photosystems on the redox state of intermediates whose redox potential suggested that they were located between the photosystems; thus, far-red light oxidized cytochrome $f$, while red light reduced it (e.g. Duysens et al. 1961, see Table 13.2). While it is likely that the two photoreactions of algae and higher plants correspond, in terms of redox span covered, to the two light reactions proposed by Hill and Bendall, it is important to realize that this does not necessarily prove that they interact in non-cyclic photophosphorylation in the manner suggested by this scheme (Hoch & Owens 1963, Govindjee et al. 1966, Hoch & Randles 1971, Section 7 below).

The evidence as to the identity of the two photoreactions, in terms of the redox span over which they operate, comes from experiments on the action spectrum of light-dependent redox reactions in subcellular particles; photo-reaction I is driven by light absorbed by pigment system I, and photoreaction II is driven by light absorbed by pigment system II. The Hill-Bendall scheme suggests that redox compounds which are more oxidizing than $0·0v$ can be reduced by photosystem II, while redox compounds which are more reducing than $+0·4v$ can be oxidized by photosystem I. This provides the rationale for separating the two photosystems by the use of redox reagents (Losada et al.1961).

Ludlow and Park (1969) have measured the action spectra for various redox reactions in algal cells treated with formaldehyde; this fixes the cells and allows penetration of redox reagents which otherwise would not enter the cell, and thus overcomes the difficulties experienced with extracting chloroplasts from many algae (Graham & Smillie 1971, Gimmler & Avron 1971, Wallach et al. 1972). The $0·81$ to $0·0v$ segment was tested using DCPIP as an electron acceptor with a redox potential more oxidizing than $0·0v$. The action spectrum for this reaction (DCPIP reduction coupled to $O_2$ evolution) in Chlorella, Porphyridium and Anacystis was that of photosystem II. Since no enhancement experiments were carried out, it is not possible to rule out the possibility that both photosystems were being used, since DCPIP (and similar high-potential oxidants, such as benzoquinone and ferricyanide), can be reduced by either photosystem I or photosystem II (Kok et al. 1966, Fujita & Myers 1965, Givan & Levine 1969). Despite this ambiguity, it is clear that all reactions in algae which involve $O_2$ evolution (i.e. $H_2O$ as electron donor) have an action spectrum which implicates photosystem II, and have a 'red drop' in quantum efficiency (see e.g. Govindjee & Rabinowitch 1960, Bishop & Gaffron 1963, Joliot 1965).

The $+0·4$ to $—0·32v$ segment of the pathway was tested using reduced

DCPIP as electron donor (less oxidizing than $+0.4v$), and methyl viologen (which has a more negative potential than NADP) as electron acceptor. Reactions of photosystem II were abolished by the presence of DCMU (Gingras & Lemasson 1965). This photoreaction had an action spectrum which resembled that of photosystem I, and showed a 'red rise' in quantum efficiency. Using intact algal cells, Amesz and Duysens (1962) showed that the action spectrum for NADP photoreduction and cytochrome $f$ photo-oxidation was that of photosystem I. More complete data for the action spectra and quantum requirements of the two photosystems are available for higher plant chloroplasts (Sun & Sauer 1971). Indirect evidence (Table 13.1) suggests that each photoreaction in algae proceeds with a quantum requirement of 1 quantum absorbed by the pigment system per electron used in the photoreaction, and that the two kinds of reaction centres occur in equal concentrations.

More detailed analysis of the photoreactions involves a closer specification of the redox spans over which they operate (Table 13.1). Taking first photosystem one, algal preparations can photoreduce bipyridilum compounds, but no systematic attempt to find the maximum reducing potential generated by algae (as opposed to higher plants) has been carried out. As regards the oxidizing end of photosystem I, it is assumed that the oxidant generated is P700—(see Section 4 and Table 13.2), and the redox value is that of this compound.

Photosystem II is more difficult. Various estimates for algae and higher plants for the reducing end, based on various assumptions, range from $+0.17$ to around $-0.05v$ (Table 13.1). The oxidizing end of photosystem I must be able to extract electrons from water, i.e. must be more oxidizing than $H_2O$ at $+0.81v$; however, all known artificial donors to photosystem II are *less* oxidizing than this (e.g. Katoh & San Pietro 1968, Bishop & Wang 1971, Ogawa *et al.* 1970, Mohanty *et al.* 1971, Epel & Levine 1971).

## 4  CHEMICAL NATURE OF THE TWO PHOTOREACTIONS

Implication of a chemical component in photosynthesis involves establishing its chemical identity and properties, and application of the criteria mentioned in Section 2 for showing involvement as an obligate member of the photosynthetic reaction.

The position of a chemical component in the reaction sequence is identified by consideration of (a) its redox potential; (b) the effect of light absorbed by the two photosystems (e.g. a compound reduced by photosystem II and oxidized by photosystem I lies between the two photosystems); (c) consideration of the activities of systems lacking the component, either by mutation or by extraction; and (d) analysis of the kinetics of oxidation or reduction of the component after a light flash (see Witt 1971, Levine 1969b). Identification of a component as participating in a photoreaction requires the use of two further criteria. One criterion is that a 'pure' photochemical reaction is temperature-insensitive, and so can occur at very low temperatures (e.g. liquid nitrogen temperature, 77°K). However, it appears that certain redox reagents which are not part of the primary photochemical event, but which are related to it by chemical processes with very low activation energies can also show this temperature independence (electron tunnelling, Chance 1967).

A further point is that photochemical events must, by definition, occur within the lifetime of the relevant excited state of chlorophyll; for the first excited singlet this is about 10nsec (Clayton 1966), so the primary photochemical event must occur within this time. Observation of photoreactions occuring within this time span can only be by optical means (Witt 1971), and even so they are used at the limits of their resolution.

The best investigated example of a primary reaction in photosynthesis is that involving P700, a form of chlorophyll *a* which is involved at the oxidizing end of photosystem I. The evidence that P700 is an obligate intermediate of photosynthesis is that it is oxidized by photosystem one and reduced by photosystem II, and the rapidity and temperature-insensitivity of its oxidation, is surveyed in Table 13.2. Its mode of action is probably as follows. Excitation energy from pigment system I is channelled into P700, which acts as an energy trap since it absorbs at longer wavelengths than the bulk chlorophyll *a*. This produces the excited state of P700; of the 1·8eV of energy available in this state, some 1·0eV is stabilized as the oxidized form of P700 (+0·45v) and the reduced form of the acceptor (about —0·6v: Table 13.1).

The nature of the electron acceptor for photosystem I is still unclear. Recent evidence (see Chapter 12, p. 346) supports the view that a chemically undefined component (P430) is the reaction partner of P700 in higher plants and blue-green algae. Chemical data on possible primary electron acceptors for photoact I in algae are reviewed by Trebst (1970), Fujita and Myers (1971), Bearden and Malkin (1972), and Yang and Blumberg (1972).

The situation with regard to photoreaction II is even less clear. Doring *et al.* (1967) detected a pigment P690 in higher plant chloroplasts which may have the same relationship to photosystem II as does P700 to photosystem I (Witt 1971, see Chapter 12, p. 375). There are however (see Table 13.2) several gaps in our knowledge to be filled before P690 can be accepted as an obligate member of photoreaction II or as a participant in the primary photoact.

The reducing end of photosystem II has usually been characterized in terms of $Q$, a quencher of fluorescence, or $E$, an electron acceptor whose presence has been deduced from kinetic studies of $O_2$ evolution in algae and chloroplasts (Kok & Cheniae 1966, Cheniae 1970). The work of Knaff and Arnon (1969) recently extended to green (but possibly not blue-green) algae (Bendall & Sofrova 1971), (Table 13.2) suggests that a pigment variously called C550 and P546, and not yet chemically characterized, may be the primary electron acceptor of photosystem II. Thus the data on the identity of the primary donors and acceptors of algal photosynthesis are mainly based on absorption changes whose chemical basis is unclear.

Physical separation of the two photosystems has been achieved in higher plants (Boardman 1970, Park & Sane 1971) but there has been less success with algae (see Section 8.7, and Chapter 12, p. 365).

# 5 CHEMICAL NATURE OF THE INTERMEDIATES OF NON-CYCLIC ELECTRON TRANSPORT

Compared with the primary reactants in the photoreactions, the possible chemical intermediates of the non-photochemical portions of non-cyclic electron transport are much better characterized: ferredoxin, cytochrome $f$, plastocyanin and plastoquinone have all been obtained in pure form from algal sources (Buchanan & Arnon 1971, Levine 1969b, Bendall & Hill 1968, Barr & Crane 1971, Yakushiji 1971, Malkin 1969). The components believed to be involved in these reactions are shown in Fig. 12.13 (p. 374). Table 13.2 gives documentation for many of these components.

## 5.1 *Reducing end of photosystem I to NADP*

Ferredoxin has been implicated in electron transport by a variety of data (Table 13.2). This kind of ferredoxin, also found in higher plants, is ferredoxin $b$ in the terminology of Buchanan and Arnon (1971). A bound form of ferredoxin, distinct from the chemically characterized form which is relatively easily extracted, has been proposed as a possible primary acceptor in photosystem I, although the evidence is equivocal.

In iron-deficient algae (*Anacystis, Anabaena, Chlorella*), ferredoxin is partly replaced by the flavoprotein flavodoxin (phytoflavin) (Smillie & Entsch 1971, Zumft & Spiller 1971). This replaces ferredoxin $b$ in many reactions, and differs from most flavoproteins in acting as a donor and an acceptor of a single electron (Bothe *et al.* 1971).

The photoreduction of NADP by reduced ferredoxin requires ferredoxin-NADP oxidoreductase which occurs in many algae (see e.g. Boger 1970, Bothe & Berzborn 1970, Powls *et al.* 1969). Its essentiality for electron transport to NADP in *Anacystis* has been shown by extraction and replacement studies, although attempts to confirm this by immunological reactions were, unlike corresponding experiments on higher plant chloroplasts, unsuccessful (Bothe & Berzborn 1970). An absorbance change in a pale green mutant of *Chlamydomonas* has been attributed to photoreduction of this flavoprotein, or its complex with ferredoxin, by photosystem I (Hiyami *et al.* 1970).

NADP is commonly accepted as the terminal electron acceptor in non-cyclic electron flow, and as the reductant for the photosynthetic carbon reduction cycle (Table 13.2, see Chapter 15, p. 436). However, this has been challenged (Ben-Amotz & Avron 1972b). Evidence obtained with antibodies to ferredoxin and ferredoxin-NADP oxidoreductase in higher plants (Berzborn *et al.* 1966), suggests that the NADPH$_2$ generated is in the stroma rather than inside the thylakoids, contrary to the suggestion of Weier *et al.* (1966). This means that photoproduced NADPH$_2$ is available to soluble dehydrogenases of the chloroplast stroma. The use of photosynthetically generated reducing power for reductive synthesis outside the chloroplasts may involve 'shuttles' to overcome the barrier to direct permeation of pyridine-adenine dinucleotides presented by the outer chloroplast membranes (Hall 1972, Heber & Krause 1971). Evidence for the use of photoproduced reductant (NADPH$_2$, or ferredoxin) for processes other than CO$_2$ fixation in algal cells is discussed by Raven (1969b).

### 5.2 *Reducing end of photosystem II to oxidizing end of photosystem I*

The three best characterized possible intermediates in this segment of non-cyclic electron flow are cytochrome *f*, plastocyanin and plastoquinone.

The redox potential of cytochrome *f* associates it with the oxidizing end of photosystem I; other characteristics of its reactions agree with this (Table 13.2). However, the quantum yield for its photo-oxidation can be low, and it is possible that not all of the electron flux of non-cyclic electron flow pass through this component. It is also present in very small amounts in *Anabaena variabilis*, unless the alga is grown in the presence of diphenylamine (Ogawa & Vernon 1971).

The cuproprotein (plastocyanin) has a similar redox potential to that of cytochrome *f* and has certain physical properties in common with algal (but not higher plant) cytochrome *f* which make these two components interchangeable in a number of reaction systems. This weakens evidence for the specific role of either of these components based on extraction-replacement studies (e.g. Lee *et al.* 1969). In any case, mutant and extraction-replacement studies cannot show that a component has a biochemical rather than a purely structural role. The observed slow *in vivo* redox reactions of plastocyanin (Table 13.2) suggest that not all of the electron flux can be carried by plastocyanin.

One explanation for the poor quantum yield of photo-oxidation of cytochrome *f*, or for the very slow oxidation of plastocyanin, is that reduction by a cyclic electron flow could be occurring, thereby obscuring the photo-oxidation. However, in some cases the measured efficiency of NADP reduction in parallel experiments rules this out (Amesz & Duysens 1962). It is uncertain whether cytochrome *f* and plastocyanin act in parallel or, if in series, cytochrome *f* is nearer photosystem II (Levine 1968) or photosystem I (cf. Hind & Olsen 1969).

The involvement of plastoquinone close to photosystem II is well established (Table 13.2), although the precise chemical identity of the compound(s) involved in the ultra-violet absorption shifts is not clear (Wallwork & Crane 1970). Plastoquinone probably corresponds to the large pool of 'secondary reductant' of photosystem II revealed by kinetic experiments (Cheniae 1970) and the transfer of electrons from plastoquinone to the electron acceptors associated with the oxidizing end of photosystem I, is probably the rate-determining step in overall non-cyclic electron transport (Witt 1971, Amesz *et al.* 1972). The role of rubi-medin (Henninger & Crane 1966) in algal non-cyclic electron transport appears not to have been investigated.

The action of *b*-type cytochromes in non-cyclic electron transfer is difficult to evaluate. Cytochrome$_{559}$ (Table 13.2) can be photoreduced by photosystem II and photo-oxidized by photosystem I in certain algae at room temperature, although this is too slow a reaction to carry all the electron flow between the photosystems (Table 13.2). Cytochrome$_{559}$ exists in at least two forms with high and low redox potentials (Table 13.2). Levine (Chapter 14, p. 427) discusses evidence that it is the high-potential form which participates in these reactions, and which is absent in certain mutants which lack photosystem II activity. However, these mutants also lack C550, and part of their plastoquinone content (cf. Schweitz 1972), and the role of cytochrome$_{559}$ is difficult to define (see below, and Chapter 14, p. 427). Cytochrome $b_{563}$ is generally assigned to the cyclic electron transport pathway (see Section 6 below).

### 5.3   *Water to the oxidizing end of photosystem II*

The low-temperature oxidation of cytochrome$_{559}$ by photoreaction II in algae (Table 13.2) suggests that this cytochrome (or at least its high-potential form) is involved in reactions on the oxidizing side of photosystem II. However, no data are available for its involvement at normal temperatures, and its relatively low potential suggests that not all of the electrons flowing from water to photo-system II pass through this component (Bendall & Sofrova 1971). Studies of higher plants (Hiller & Boardman 1971, Ke *et al.* 1972, Knaff 1972) cast doubt on its role in the non-cyclic electron transport chain.

Manganese in organically bound form may be an intermediate close to the $O_2$-evolving step, and also possibly at another site closer to the oxidizing end of photosystem II in algae (Kok & Cheniae 1966, Cheniae 1970). The involvement of $Cl^-$ (or other anions of strong acids) in photosystem II of algae is unclear (Satoh *et al.* 1970, Bishop & Senger 1971). Inorganic carbon may also be a cata-lyst of photosystem II (see Chapter 15, p. 434).

## 6   CHEMICAL NATURE OF THE INTERMEDIATES
## OF CYCLIC ELECTRON FLOW

The existence and rate of cyclic electron flow is generally inferred from the existence of ATP synthesis dependent on photosystem I which does not result in net production or consumption of redox compounds. Section 8 and Table 13.2 present data on the process *in vivo* and *in vitro*. *In vitro* cyclic electron trans-port in algal chloroplasts, in contrast to certain higher plant chloroplasts, requires the addition of some exogenous redox cofactor, such as PMS or ferre-doxin (see below). The *in vitro* process dependent on added PMS is less like the *in vivo* reaction; thus there is evidence that both plastocyanin and cytochrome *f* are involved in cyclic electron flow *in vivo*, but not in the PMS-catalysed process *in vivo* or *in vitro* (Lee *et al.* 1969, Powls *et al.* 1969, Gorman & Levine 1966c Levine 1969b, Teichler-Zallen & Hoch 1967).

Cytochrome $b_{563}$ ($b_6$) occurs in algae at a concentration of about 3 to 6 moles per 1,000 chlorophyll (Ikegami *et al.* 1968, Levine & Armstrong 1972). It is both oxidized and reduced by photosystem I in *Chlamydomonas* and *Porphyridium* cells (but not in cell fragments) (Levine 1969a, Hiyami *et al.* 1970, Amesz *et al.* 1972b,c), and may be involved in cyclic electron flow pathway. It is also involved in PMS-catalysed cyclic photophosphorylation (Levine 1969a). Extraction-replacement and inhibitor studies indicate that plastoquinone is required for the PMS-catalysed reaction in algae and higher plants, and for the ferredoxin-catalysed reaction in higher plants and for *in vivo* cyclic electron flow in algae (Lightbody & Krogmann 1966, Bosshardt & Arnon 1970, Bohme *et al.* 1971, Urbach & Kaiser 1972). However, endogenous cyclic electron flow in higher plant chloroplasts is not inhibited by a plastoquinone antagonist (Forti & Rosa 1972).

A contradiction between experiments on endogenous cyclic electron flow in chloroplasts and that catalysed by added ferredoxin occurs with regard to the role of ferredoxin-NADP oxidoreductase; antibodies to this component inhibit endogenous (Forti & Rosa 1972) but not ferredoxin-induced (Bothe & Berzborn 1970) cyclic electron flow. Hiyami *et al.* (1970) concluded that this component was involved in a different cyclic electron transport pathway to that involving cytochrome $b_{563}$.

The role of ferredoxin is also unclear. The ferredoxin-catalysed process *in vitro* generally resembles the *in vivo* process more than do other *in vitro* cyclic photophosphorylations (Avron & Neumann 1968, Raven 1969b), although Antimycin A is relatively ineffective in inhibiting ferredoxin-catalysed cyclic electron flow in *Anacystis* (Bothe 1969). *Phormidium* preparations which retain enough ferredoxin to support high rates of non-cyclic photophosphorylation and NADP reduction have no cyclic photophosphorylation (Biggins 1967b). This agrees with the requirement for larger amounts of ferredoxin needed for cyclic than for non-cyclic electron flow *in vitro* (Buchanan & Arnon 1970, Bothe 1969, Boger 1970), and with the finding that the ferredoxin antagonist DSPD inhibits cyclic photophosphorylation *in vivo* to a greater extent than non-cyclic photophosphorylation (see Raven 1969b).

Evidence on the electron transport pathway of cyclic electron flow is thus very incomplete and almost entirely qualitative (but see Teichler-Zallen & Hoch (1967)).

# 7 ALTERNATIVE SCHEMES OF ELECTRON TRANSPORT

Most of the experimental evidence on electron transport in algal photosynthesis is most readily explained by the series formulation of non-cyclic electron flow (see Fig. 12.13, p. 374). However, a number of findings are not readily explained by this scheme. The most popular alternative scheme is that of Hoch (Hoch & Owens 1963, Govindjee *et al.* 1966, Hoch & Randles 1971) which involves both photosystem I and photosystem II in non-cyclic electron flow, but the interaction is at the level of a hypothetical 'high energy intermediate' rather than of redox intermediates. This intermediate, produced by photosystem I, can be used *inter alia* to supply the energy needed to bridge the gap between the reducing end of photosystem II and the more reducing products ($NADPH_2$, reduced ferredoxin) which are the products of non-cyclic electron flow. Such a scheme can explain the antagonistic effects of light absorbed by the two photosystems on the redox state of intermediates with $E'_0$ in the range $0 \cdot 0$ to $+0 \cdot 45V$, and on the fluoresence yield of photosystem II (Govindjee *et al.* 1966).

The specific data which the scheme may explain more convincingly than does the series formulation are:

(a) The absence of enhancement in some cultures of *Chlorella* which have a high rate of dark respiration (Govindjee *et al.* 1966) (the high energy intermediate could be produced from respiration rather than photosystem I). However, the absence of enhancement is not necessarily inconsistent with the operation of two photosystems (Myers 1971).

(b) The inability of the series formulation of non-cyclic photophosphorylation, together with cyclic photophosphorylation, to make enough ATP and $NADPH_2$ from 8 to 10 quanta to provide for the photosynthesis and growth corresponding to the fixation of one mole of $CO_2$ (Hoch & Owens 1963). The problems here are less serious if a $P/e_2$ of 2 is assumed for non-cyclic photophosphorylation (Section 8.2). In any case, a new mechanism for electron transfer cannot alter the measured stoichiometry of phosphorylation to electrons transferred.

(c) The absence of a stoichiometric relationship between the operation of photoreaction I and overall non-cyclic electron transport. Hoch and Randles (1971) find that photoreaction I proceeds more or less independently of photoreaction II. However, in algae (cf. Rurainski *et al.* 1971, Arntzen *et al.* 1972), this discrepancy usually involves an excess of photoreaction I, which could be explained in terms of cyclic electron transport superimposed on the series-type non-cyclic electron flow.

Thus there are several results which are as easily explained on the alternate hypothesis as by the series formulation supported in this review. It is therefore as well to regard the series formulation as a currently useful peg to hang most of the data on rather than as an immutable scheme.

# 8  PHOTOSYNTHETIC PHOSPHORYLATION

## 8.1  *General characteristics and rates* in vitro

Boger (1971) and Graham and Smillie (1971) review the methods available for extracting particles from algae which retain the capacity for light-dependent

ATP synthesis. Such extracts have been prepared from Cyanophyceae (e.g. Petrack & Lipmann 1961, Bothe 1969, Biggins 1967b, Lee et al. 1969), Euglenophyceae (Chang & Kahn 1970, Schneyour & Avron 1970), Xanthophyceae (Boger 1969), Haptophyceae (Jeffrey et al. 1966) and Chlorophyceae (e.g. Powls et al. 1969, Wallach et al. 1972, Ben-Amotz & Avron 1972a, Smith & West 1969).

On a chlorophyll basis, these preparations can synthesize ATP at a rate which is adequate to support the in vivo rate of $CO_2$ fixation (Table 13.1, Section 2 above). While the highest rates have been obtained with such 'non-physiological' cofactors as PMS and ferricyanide, it is likely that these agents are only involved in establishing a high rate of electron transport through the coupling site(s).

The first light-dependent synthesis of ATP in algal extracts were of the cyclic type (Petrack & Lipmann 1961, Powls et al. 1969); no photophosphorylation coupled to non-cyclic electron transport was observed. Later experiments showed that non-cyclic photophosphorylation could also occur. Cyclic photophosphorylation dependent on PMS, Vitamin K + FMN, and ferredoxin has now been established, as has non-cyclic photophosphorylation with water as electron donor and NADP, ferricyanide or viologen dyes as electron acceptors (e.g. Boger 1970, Bothe 1969, Wallach et al. 1972). The greater fragility of the non-cyclic compared with the cyclic process is consistent with the occurrence of separate phosphorylation sites for cyclic and for non-cyclic electron flow (Lee et al. 1969).

## 8.2 *Coupling and stoichiometry*

In general, it is not possible to measure the rate of electron transport during cyclic photophosphorylation. In non-cyclic photophosphorylation, simultaneous measurements of electron transport and phosphorylation have led to the concept of coupling between the two processes analogous to that developed for oxidative phosphorylation. Thus the two processes are inhibited in parallel by treatments which interfere with electron transport. Phosphorylation is completely inhibited, and electron transport is partially inhibited, if the ADP and $P_i$ necessary for ATP synthesis are removed, or if the coupling factor responsible for the final enzymic reaction of phosphorylation is inhibited. Finally, uncoupling agents, which dissipate the high energy intermediate inhibit phosphorylation and stimulate electron flow and also overcome the inhibition of electron transport by treatments which reduce the activity of the coupling factor (see Avron 1971). In algal preparations, this 'photosynthetic control' has also been found (see e.g. Smith & West 1969, Boger 1970). Photosynthetic control is generally correlated with high values for the $P/e_2$ ratio, i.e. the number of moles of ATP made per pair of electrons passing along the electron transport chain (cf. Hall 1972). In algae this ratio is generally less than 1 (Boger 1971), although it is greater than 1 in *Chara corallina* (Smith & West 1969), and may be greater than 1 in *Anabaena* (Lee et al. 1969). Indirect evidence for more than one phosphorylation site in non-cyclic photophosphorylation has also been presented (Lee et al. 1969,

o

Kahn 1970). On the assumption that each coupling site has a $P/e_2$ of 1, this means that the real $P/e_2$ is 2.

The interpretation of $P/e_2$ ratios in non-cyclic photophosphorylation *in vitro* is complicated by the possible occurrence of a simultaneous cyclic photophosphorylation (e.g. Brandon 1969). Nevertheless this and other complicating factors are inadequate to invalidate claims for a $P/e_2$ ratio greater than 1 (Hall *et al.* 1971).

The direct measurement of the $P/e_2$ ratio for cyclic photophosphorylation is not possible. However, from the quantum requirement of *in vivo* processes dependent on cyclic photophosphorylation, and assuming that each quantum absorbed by pigment system I brings about the transfer of one electron, it is possible to calculate a $P/e_2$ ratio. For anaerobic glucose photoassimilation in *Chlorella* (Tanner *et al.* 1968), and taking into account the ATP used for uptake (Tanner 1969, Decker & Tanner 1972) as well as the metabolism of glucose, a $P/e_2$ ratio of 2 seems reasonable.

Thus there is evidence for a $P/e_2$ ratio of more than 1 for both cyclic and non-cyclic photophosphorylation in algae.

### 8.3   *Mechanisms and partial reactions*

There are two main hypotheses regarding the coupling of electron transport to phosphorylation: the chemical and the chemi-osmotic mechanisms, and their relative merits are discussed by Mitchell (1966), Greville (1969) and Schwartz (1971). Experiments with algae generally confirm the findings for higher plants, but do not contribute any crucial points to the argument.

Inhibitors of algal photophosphorylations and higher plant photophosphorylations generally exert similar effects (e.g. Kahn 1968, Chang & Kahn 1972, Smith & West 1969, Boger 1970). Algal chloroplasts have the protein 'coupling factor' which is believed to act as the terminal step in the reactions of ATP synthesis. This has been shown to be involved in photophosphorylation by inhibitor (Kahn 1968) and mutant (Sato *et al.* 1971) studies, but not, by antibody techniques (see Avron & Neumann 1968). This coupling factor appears to be involved at both the cyclic site(s) and the non-cyclic site(s) of photophosphorylation suggesting that the difference between the coupling sites for cyclic and non-cyclic photophosphorylation must reside in a reaction prior to that catalyzed by the coupling factor. This is borne out by *in vivo* experiments in which the sensitivity of an ATP-requiring process to energy transfer inhibitors, which inhibit the activity of the coupling factor, was the same for cyclic and non-cyclic electron flow (Raven 1971c, but see Jeschke 1972, for an alternative interpretation of the activity of these inhibitors).

Algae also possess the thylakoid $H^+$ pump, which is thought to be an obligate intermediate in ATP synthesis (chemi-osmotic hypothesis), or to be closely related kinetically and energetically to an as yet uncharacterized intermediate (chemical hypothesis). Such an inwardly-directed $H^+$ pump has been demonstrated in subcellular preparations of *Chlamydomonas* (Sato *et al.* 1971, Wallach *et al.* 1971) and *Euglena* (Kahn 1971). *Anabaena* exhibits light-dependent $H^+$ extrusion from whole cells (Mitchell 1967). The polarity of the $H^+$ pump in these two situations is analogous if it is assumed that the thylakoids of chloroplasts are derived from invaginations of the inner limiting chloroplast membrane which has the same polarity as a whole prokaryote cell. The work on $H^+$ fluxes in algal photosynthesis is qualitative in that no attempt has been made to determine quantitative relations between electron transport, $H^+$ transport and phosphorylation (Schwartz 1971).

It has been suggested (Ben-Amotz & Ginzberg 1969, Schuldiner & Ohad 1969) that the light-induced pH increase in the medium bathing cells of *Dunaliella* and *Chlamydomonas* reflects $H^+$ uptake by the thylakoids. However, subsequent work (Atkins & Graham 1971, Neumann & Levine 1971) suggests that much of the observed pH change is the result of photosynthetic carbon dioxide uptake.

## 8.4 *Location of coupling sites*

The location of coupling sites in the electron transport chain of algae is unclear. It is assumed on the chemical hypothesis of coupling that electron transport in the energetically downhill direction (i.e. on the scheme shown in Fig. 12.13 for non-cyclic photophosphorylation, P430 $\rightarrow$ NADP; C550 $\rightarrow$ P700; and $H_2O$ $\rightarrow$ P690) releases sufficient energy to bring about the synthesis of ATP. Of these three sites, the segment between the two photosystems, which is analogous to coupling site II of oxidative phosphorylation, has received most attention. Controversy still surrounds the adequacy of the energy span in this region to support a phosphorylation site with a $P/e_2$ of 1. Storey (1970) has proposed a scheme for oxidative phosphorylation, which could be modified to include photophosphorylation, in which energy from two distinct redox spans can interact to provide the energy for ATP synthesis.

On the chemi-osmotic hypothesis, coupling sites are less easy to define, and may involve the light reactions themselves as well as regions of 'downhill' electron transport (e.g. Mitchell 1966, Kok *et al.* 1966, Witt 1971). Suitable arrangement of carriers within the membrane would involve uptake of $H^+$ from the stroma when photosystem II reduces PQ, and photosystem I reduces NADP, and $H^+$ release into the thylakoids when $H_2O$ donates electrons to photosystem II, and $PQH_2$ reduces plastocyanin or cytochrome $f$.

Two general approaches have been used in attempting to define coupling sites. In one of these the effect of uncouplers on the redox state of electron transport intermediates has been studied; uncouplers should increase electron flux through the coupling site relative to that through the rest of the chain, and hence lead to reduction of intermediates on the oxidizing side and oxidation of those on the reducing side (Avron & Neumann 1968, Avron 1971, Böhme & Cramer 1972). *In vivo* studies in algae support the existence of a coupling site on the photosystem II side of P700 and cytochrome $f$ (e.g. Amesz & Vredenberg

1966, Nishimura 1968); further localization of this site with respect to photo-system II has not been attempted for algae (Cheniae 1970, Wraight & Crofts 1971, Böhme & Cramer 1971).

The other approach is to isolate portions of the electron transport chain and determine if coupled phosphorylation is associated with them. Experiments with a mutant of *Chlamydomonas* (Levine 1969b) support the concept of a coupling site on the photosystem II side of P700. Controversy still surrounds the hypo-thesis of a non-cyclic photophosphorylation site associated with electron supply to photosystem I using an artificial donor such as $DCPIPH_2$ (Arntzen *et al.* 1971). In algae ATP synthesis accompanies this non-cyclic electron flow (Levine 1963, Duane *et al.* 1965, Gee *et al.* 1969). One possibility is that the coupling is at a normal non-cyclic site (Losada *et al.* 1961). However, since this photo-phosphorylation is observed in preparations which have no ATP synthesis associated with non-cyclic electron flow from water to NADP (e.g. Duane *et al.* 1965), but do possess cyclic photophosphorylation catalysed by PMS, it is possible that the coupling site involved is the cyclic site. This could be due either to cyclic electron transport, or to the cyclic site operating as part of a non-cyclic sequence (Arnon *et al.* 1965, Avron & Neumann 1968). The non-cyclic hypothesis is supported by the finding that electron transport from $DCPIPH_2$ to NADP in *Porphyridium* is uncoupler-stimulated (Gimmler & Avron 1971).

Thus the role of photosystem I in non-cyclic photophosphorylation is uncertain; current evidence does not support the idea of a site on the reducing side of photosystem I.

In cyclic photophosphorylation, no direct evidence on phosphorylation sites is available; the *in vivo* evidence (see Section 8.2) suggests a $P/e_2$ of 2. On energetic grounds, it is likely that one site is between the reducing end of photo-system I and cytochrome $b_6$, while the other is between cytochrome $b_6$ and P700. A $P/e_2$ ratio of two demands that both flavoprotein and plastoquinone are involved in cyclic electron transport if the two redox loops demanded by the chemi-osmotic hypothesis (Mitchell 1966) are to be provided.

## 8.5   *ATP transport*

Recent evidence (see Raven 1968) suggests that the terminal steps of photo-phosphorylation occur at the outer surface of the thylakoid membrane, and that ATP is released into the stroma phase of the chloroplast (or the cytosol of prokaryotes). Since this is the location of the ATP-consuming reactions of the photosynthetic carbon reduction cycle, no membrane transport step is required for the use of ATP made in photophosphorylation for processes located in the stroma (contrast Weier *et al.* 1966).

There is evidence (see Raven 1972) that photophosphorylation can also support processes occurring outside the chloroplasts of eukaryotes. Evidence from algae (Strotmann & Berger 1969, Forsee & Kahn 1972) supports the conclusion of Hall (1972) that isolated chloroplasts which retain the *in vivo* rate of $CO_2$ fixation have very low rates of transfer of adenylates across the outer chloroplast membranes. The apparent contradiction between the *in vivo* and the *in vitro* evidence may be resolved by the hypothesis that 'shuttles' catalyse the transfer of 'high energy phosphate' across the chloroplast outer membrane (Heber & Krause 1971, Smith 1971).

## 8.6 *Photophosphorylation* in vivo

Evidence as to the occurence, rates and types of photophosphorylation *in vivo* is discussed by Kandler (1960), Simonis (1960), Avron and Neumann (1968), Halldal (1970), Raven (1972) and Simonis and Urbach (1973).

Direct evidence for the occurrence of photophosphorylation *in vivo* comes from measurements of transients in the concentrations of phosphate compounds in intact cells during dark-light-dark transitions (e.g. Wintermans 1958), or by comparison of the rates of incorporation of supplied $^{32}P$ into organic compounds in the light and the dark steady states (Kanai & Simonis 1968, Urbach & Gimmler 1970). The former measurements are complicated by the non-steady-state conditions, while the latter are difficult to interpret quantitatively (Rowan 1966). Despite these difficulties, important conclusions can be drawn from the rates of $^{32}P$ incorporation *in vivo* under various conditions.

Indirect evidence comes from measurements of the rates of ATP-requiring processes under various conditions in the light and the dark (Halldal 1970). With both types of experiment, a light-stimulation of the process investigated under conditions in which oxidative phosphorylation is reduced or prevented is taken as evidence for photophosphorylation. Most of the available evidence suggests that oxidative phosphorylation is inhibited or unchanged when green tissues are illuminated (Raven 1972), so most light-stimulated ATP turnover in green tissues may be attributed to photophosphorylation.

In the absence of $CO_2$ it is believed that most of the photophosphorylation is of the cyclic type; this is based on insensitivity to removal of $O_2$ and other potential non-cyclic electron acceptors, and the absence of requirement for photosystem II, as shown by the action spectrum and the insensitivity to DCMU (Table 13.1). This type of photophosphorylation is also characterized by inhibition by a number of inhibitors which have little or no effect on $CO_2$ fixation. This suggests that cyclic photophosphorylation is not needed for $CO_2$ fixation (Table 13.1).

The quantitative role of cyclic photophosphorylation is difficult to assess. Experiments in which the initial rate of increase in the ATP level in intact cells is measured under conditions which exclude oxidative phosphorylation or open-chain phosphorylations, yield rates of 5 (*Ankistrodesmus*, Urbach & Kaiser 1972) to 80 (*Anacystis*, Bornefeld *et al.* 1972, Owens & Krauss 1972) μmoles ATP (mg chl)$^{-1}$ h$^{-1}$. These rates are likely to be minimal estimates (Simonis & Urbach 1973). Values within this range may be estimated from comparisons of the rates of various processes supported by cyclic photophosphorylation in comparison with the rates supported by oxidative phosphorylation. The rate of ATP synthesis by cyclic photophosphorylation is given by the relationship $ATP_c = ATP_0 (r_c/r_0)$, where $ATP_c$ is the rate of ATP synthesis by oxidative phosphorylation, $ATP_0$ is the rate of oxidative phosphorylation (computed from the rate of oxygen uptake in dark respiration, assuming a $P/e_2$ of 3; Kandler 1958), $r_c$ is the rate of ATP-consuming reactions supported by cyclic photophosphorylation, and $r_0$ is the rate of ATP-consuming reactions supported by oxidative phosphorylation (Tanner et al. 1969, Raven 1972, cf. Urbach & Gimmler 1970, Halldal 1970).

Non-cyclic photophosphorylation occurs in vivo when such electron acceptors as $CO_2$, $NO_3^-$ and $SO_4^{2-}$ are present, since the incorporation of $^{32}P_i$ into organic combination is stimulated by the presence of these compounds in the light (Urbach & Gimmler 1970, Ullrich 1971, Simonis & Urbach 1973). This view is supported by the finding that variations in light intensity, and the presence of inhibitors such as DCMU, have parallel effects on $CO_2$ fixation and $CO_2$-dependent $^{32}P$ incorporation (Simonis 1967, Urbach & Gimmler 1970, Gimmler et al. 1971). However, such a parallelism could come about either from a dependence of ATP synthesis on non-cyclic electron flow, or from a dependence of ATP consumption on non-cyclic electron flow, with ATP coming from elsewhere, or both ATP synthesis and consumption being related to non-cyclic electron flow. The latter assumption is the one in best agreement with current knowledge, but available evidence does not strictly rule out either of the other two.

It is difficult to make quantitative estimates of the rate of non-cyclic photophosphorylation in vivo by more direct means than assuming that 3 ATP are produced and consumed per $CO_2$ fixed. Measurements of the initial rates of ATP increase in algal cells under conditions in which non-cyclic photophosphorylation can occur, yield rates of 15 (Ankistrodesmus, Urbach & Kaiser 1972) to 150 to 200 (Anacystis, Bornefeld et al. 1972, Owens & Krauss 1972) μmoles ATP (mg chl)$^{-1}$ h$^{-1}$. These rates are probably not complicated by oxidative phosphorylation (Urbach & Kaiser 1972). The response to DCMU suggests that a variable fraction may be due to cyclic photophosphorylation, although it is possible that cyclic photophosphorylation is only important when non-cyclic is inhibited (Urbach & Kaiser 1972, Bornfeld et al. 1972). As with measurements of cyclic photophosphorylation, these are likely to be minimal estimates. Quantitative estimates from the $^{32}P$-incorporation data of Urbach and Gimmler (1970) are difficult to make (Simonis & Urbach 1973). Thus it is likely that the rate of non-cyclic photophosphorylation in vivo is adequate to support $CO_2$ fixation with an $ATP/CO_2$ ratio of 3, although more work is needed (Simonis & Urbach 1973).

There is some evidence (Raven 1971a) that non-cyclic photophosphorylation can also supply ATP to other ATP-requiring reactions in vivo. The relative contributions of non-cyclic, cyclic and oxidative phosphorylation to ATP-requiring processes during the normal growth of an algal cell are not precisely known (Raven 1971a,b). The contribution of pseudocyclic photophosphoryla-tion (i.e. non-cyclic photophosphorylation in which $O_2$ is the terminal electron acceptor) *in vivo* is unknown, but probably small (Raven 1969b, 1970, Ullrich 1971, Urbach & Gimmler 1970). The overall importance to algal growth of ATP supplied from photophosphorylation to processes other than $CO_2$ fixation is discussed by Hoch and Owens (1963) and by Raven (1971b, 1972). The contribution is probably large (greater than the ATP supply from oxidative phosphorylation) in many cases.

### 8.7  *Regulation of cyclic and non-cyclic photophosphorylation* in vivo

The regulation of the relative rates of cyclic and non-cyclic photophosphoryla-tion *in vivo* is discussed by Raven (1971a) and Gimmler *et al.* (1971). One hypo-thesis is that all cyclic electron transfer is carried out by photosystem I units also involved in non-cyclic electron flow; the other is that all cyclic electron transfer is carried out by photosystem I units distinct from the photosystem I units involved in non-cyclic electron flow.

The first hypothesis requires a switch mechanism which regulates the cyclic and non-cyclic activities of photosystem I (see Raven 1971a). This hypothesis agrees well with experiments which suggest that such a switch of excitation energy absorbed by photosystem I can occur (Raven 1971a), and also with data on the 'over-reduction' of the cyclic electron transport chain by photosystem II, and its 'over-oxidation' by electron acceptors for non-cyclic electron flow (Raven 1971a). The absence of photosystem I units involved only in cyclic electron flow also agrees with observations of quantum yields of 1 $O_2$ from 8 Einsteins (Kok 1960a). However, in the only case in which a direct comparison between the quantum requirements of cyclic photophosphorylation and photosynthesis was made (Tanner *et al.* 1968), the results did not rule out the existence of such independent units.

The second hypothesis could explain the absence of interactions between cyclic electron transport and conditions. The existence of separate photosystem I units, isolated from non-cyclic electron flow, is suggested by experiments on dis-ruption of algal chloroplasts by French press treatment (Brown 1971). In higher plants, it appears that the grana thylakoids have both photosystems, while the stroma thylakoids only have photosystem I (Park & Sane 1971, but see Hall & Evans 1972, Smillie *et al.* 1972).

The separation of cyclic and non-cyclic photophosphorylation reactions in different thylakoids can also explain the differential effects of certain uncoupling agents (Table 13.1) on cyclic and on non-cyclic photophosphorylation *in vivo*. If the high energy state which these agents dissipate is (chemi-osmotic hypothesis), or equilibrates with (chemical hypothesis), an ionic electrochemical potential gradient across the thylakoid membrane, then such differential effects require different properties of the membranes involved in the two kinds of phosphoryla-tion (cf. Witt 1971, Arntzen *et al.* 1971).

Thus, as concluded by Raven (1971a), elements of both hypotheses are required to account for all the observations. Whatever the mode of regulation, it appears that the activity of cyclic, relative to non-cyclic, photophosphorylation varies during the cell cycle of *Scenedesmus* (Senger 1970) and *Ankistrodesmus* (Gimmler *et al.* 1971). An extreme case of variation occurs in two types of algal cells which only have photosystem I and cyclic photophosphorylation. These are green algae such as *Chlamydobotrys* (*Pyrobotrys*) grown on acetate (Chapter 19, p. 546), and the heterocysts of Cyanophyceae (Chapter 20, p. 569). The existence of algal cells without cyclic photophosphorylation is as yet unproved, although none of the energy-requiring processes so far tested in *Chara corallina* has given evidence for the existence of cyclic electron transport or photophosphorylation. (F. A. Smith & J. A. Raven, in press.)

## 9  SUMMARY

Non-cyclic electron transport and coupled photophosphorylation in algae uses 8 quanta to reduce 2NADP and produce 4ATP. Enhancement and other studies suggest that two separate light reactions, each with its own distinct light-harvesting system, are involved, and that these two photoreactions occur in series (Fig. 12.13, p. 374). One of these photosystems also drives cyclic photophosphorylation, which produces ATP as its sole product.

The nature of the primary photochemical reactions, and of the dark reactions linking them, are considered (Tables 13.1 & 13.2). Photophosphorylation *in vitro* in algae is similar to that in higher plants.

## 10  REFERENCES

1  AMESZ J. (1964) Spectrophotometric evidence for the participation of a quinone in photosynthesis of intact blue-green algae. *Biochim. biophys. Acta*, **79**, 257–65.
2  AMESZ J. & DUYSENS L.N.M. (1962) Action spectrum, kinetics and quantum requirement of phosphopyridine nucleotide reduction and cytochrome oxidation in the blue-green alga *Anacystis nidulans. Biochim. biophys. Acta*, **64**, 261–78.
3  AMESZ J. & FORK D.C. (1967) The function of P700 and cytochrome f in the photosynthetic reaction centre of system one in red algae. *Photochem. Photobiol.* **6**, 903–12.
4  AMESZ J. & VREDENBERG W.J. (1966) Reaction kinetics of photosynthetic intermediates in intact algae. In *Biochemistry of Chloroplasts*, vol. II, ed. Goodwin T.W. pp. 593–600. Academic Press, London & New York.
5  AMESZ J., VAN DER BOS P. & DIRKS M.P. (1970) Oxidation reduction potentials of photosynthetic intermediates. *Biochim. biophys. Acta*, **197**, 324–7.
6  AMESZ J., VISSER J.W.H., VAN DEN ERGH G.J. & DIRKS M.P. (1972a) Reaction kinetics of intermediates of the photosynthetic chain between the two photosystems. *Biochim. biophys. Acta* **256**, 370–80.
6a  AMESZ J., VISSER J.W.H., VAN DEN ERGH G.J. & PULLES M.P.J. (1972b) Components

of cyclic and non-cyclic electron transport in *Porphyridium aerugineum. Physiol. Veg.* **10**, 319–28.

6b AMESZ J., PULLES M.P.J., VISSER J.W.H. & SUBBING F.A. (1972c) Reactions of b-cytochromes in the red alga *Porphyridium cruentum. Biochem. biophys. Acta*, **275**, 42–52.

7 ARNON D.I. (1959) Conversion of light energy into chemical energy in photosynthesis. *Nature, Lond.* **184**, 10–21.

8 ARNON D.I. (1971) The light reactions of photosynthesis. *Proc. natn. Acad. Sci., U.S.A.* **68**, 2883–92.

9 ARNON D.I., TSUJIMOTO H.Y. & McSWAIN B.D. (1965) Photosynthetic phosphorylation and electron transport. *Nature, Lond.* **207**, 1367–72.

10 ARNON D.I., CHAIN R.K., McSWAIN B.D., TSUJIMOTO H.Y. & KNAFF D.B. (1970) Evidence from chloroplast fragments for 3 photosynthetic light reactions. *Proc. natn. Acad. Sci., U.S.A.* **67**, 1404–9.

11 ARNON D.I., KNAFF D.B., McSWAIN B.D., CHAIN R.K. & TSUJIMOTO H.Y. (1971) Three light reactions and two photosystems in plant photosynthesis. *Photochem. Photobiol.* **14**, 397–425.

12 ARNTZEN C.J., NEUMANN J. & DILLEY R.A. (1971a) Inhibition of electron transport in chloroplasts by a quinone analogue: evidence for two sites of DPIPH₂ oxidation. *Bioenergetics*, **2**, 73–83.

13 ARNTZEN D.R., DILLEY R.A. & NEUMANN J. (1971b) Localization of photophosphorylation and proton transport activities in various regions of the chloroplast lamellae. *Biochim. biophys. Acta*, **245**, 409–24.

14 ARNTZEN C.J., DILLEY R.A., PETERS G.A. & SHAW E.R. (1972) Photochemical activity and structural studies of photosystems derived from chloroplast grana and stroma lamellae. *Biochim. biophys. Acta*, **256**, 85–107.

15 ATKINS C.A. & GRAHAM D. (1972) Light-induced pH changes by cells of *Chlamydomonas reinhardi*: dependence on CO₂ uptake. *Biochim. biophys. Acta*, **226**, 481–5.

16 AVRON M. (1971) Biochemistry of photophosphorylation. In *Structure and Function of Chloroplasts*, ed. Gibbs M. pp. 149–68. Springer-Verlag, Berlin, Heidelberg and New York.

17 AVRON M. & NEUMANN J. (1968) Photophosphorylation in chloroplasts. *A. Rev. Pl. Physiol.* **19**, 137–66.

18 BARR R. & CRANE F.L. (1971) Quinones in algae and higher plants. In *Methods in Enzymology*, XXIIIA, ed. san Pietro A. pp. 372–408. Academic Press, New York & London.

19 BEARDEN A.J. & MALKIN R. (1972) The bound ferredoxin of chloroplasts: a role as the primary electron acceptor of photosystem I. *Biochem. biophys. Res. Commun.* **46**, 1299–305.

20 BEINERT H., KOK B. & HOCH G. (1962) The light-induced electron paramagnetic resonance signal of photocatalyst P700. *Biochem. biophys. Res. Commun.* **7**, 209–12.

21 BEN-AMOTZ A. & AVRON M. (1972a) Photosynthetic activities of the halophilic alga *Dunaliella parva. Pl. Physiol., Lancaster*, **49**, 240–3.

22 BEN-AMOTZ A. & AVRON M. (1972b) Is nicotinamide adenine dinucleotide phosphate an obligatory intermediate in photosynthesis? *Pl. Physiol., Lancaster*, **49**, 244–8.

23 BEN-AMOTZ A. & GINZBURG B.Z. (1969) Light-induced proton uptake in whole cells of *Dunaliella parva. Biochim. biophys. Acta*, **183**, 144–54.

24 BENDALL D.S. (1968) Oxidation-reduction potentials of cytochromes in chloroplasts from higher plants. *Biochem. J.* **109**, 46P–47P.

25 BENDALL D.S. & HILL R. (1968) Haem-proteins in photosynthesis. *A. Rev. Pl. Physiol.* **19**, 167–86.

26 BENDALL D.S. & SOFROVA D. (1971) Reactions at 77°K in photosystem II of green plants. *Biochim. biophys. Acta*, **234**, 371–80.

27 BERZBORN R., MENKE W., TREBST A. & PISTORIUS E. (1966) Über die Hemmung photosynthetischer Reaktionen isolierter Chloroplasten durch Chloroplasten-Antiköper. *Z. Naturf.* **21B**, 1057–9.

28 BIGGINS J. (1967a) Preparation of metabolically active protoplasts from the blue-green alga, *Phormidium luridum. Pl. Physiol., Lancaster*, **42**, 1442–6.

29 BIGGINS J. (1967b) Photosynthetic reactions of lysed protoplasts and particle preparations from the blue-green alga, *Phormidium luridum. Pl. Physiol., Lancaster*, **42**, 1447–56.

30 BISHOP N.I. & GAFFRON H. (1963) On the interrelation of the mechanisms for oxygen and hydrogen evolution in adapted algae. In *Photosynthetic mechanisms of green plants*, eds. Kok B. & Jagendorf A.T. pp. 441–51. N.A.S.-N.R.C. Washington, D.C. **1145**.

31 BISHOP N.I. & SENGER H. (1971) Preparation and photosynthetic properties of synchronous cultures of *Scenedesmus*. In *Methods in Enzymology*, XXIIIA, ed. san Pietro A. pp. 53–66. Academic Press, New York & London.

32 BISHOP N.I. & WONG J. (1971) Observations on photosystem II mutants of *Scenedesmus*: pigments and proteinacious components of the chloroplasts. *Biochim. biophys. Acta*, **234**, 433–45.

33 BLACK C.C. (1966) Chloroplast reactions with dipyridyl salts. *Biochim. biophys. Acta*, **120**, 332–40.

34 BOARDMAN N.K. (1970) Physical separation of the photosynthetic photochemical systems. *A. Rev. Pl. Physiol.* **21**, 115–40.

35 BOGER P. (1969) Photophosphorylierung mit chloroplasten aus *Bumilleriopsis filiformis* Vischer. *Z. PflPhysiol.* **61**, 85–97.

36 BOGER P. (1970) Ferredoxin-catalysed reactions in a cell-free system of the alga *Bumilleriopsis filiformis* Vischer. *Z. PflPhysiol.* **61**, 447–61.

37 BOGER P. (1971) Algal preparations with photophosphorylation activity. In *Methods in Enzymology*, XXIIIA, ed. san Pietro A. pp. 242–8. Academic Press, New York & London.

38 BÖHME H. & CRAMER W.A. (1972) Localisation of a site of energy coupling between plastoquinone and cytochrome f in the electron-transport chain of spinach chloroplasts. *Biochemistry, N.Y.* **11**, 1155–60.

39 BÖHME H., REIMER S. & TREBST A. (1971) The effect of dibromothymoquinone, an antagonist of plastoquinone, on non-cyclic electron flow systems in isolated chloroplasts. *Z. Naturf.* **26B**, 341–51.

40 BORNEFELD T., DOMANSKI J. & SIMONIS W. (1972) Influence of light conditions, gassing and inhibitors on photophosphorylation and ATP-level to *Anacystis nidulans*. In *Proc. 2nd International Congress on Photosynthetic Research*, eds. Forti G., Avron M. & Melandri A. pp. 1379–86. Dr. W. Junk, N.V., The Hague.

41 BOSSHARDT H. & ARNON D.I. (1970) Effect of plastoquinone and vitamin K derivatives on photophosphorylation and electron transfer. Funkns. Biokhim. Kletochryl. Strukt. 97–103. (*Chem. Abstr.* **74**, 7292 SF (1971)).

42 BOTHE H. (1969) Ferredoxin als Kofactor der cyclischen Photophosphorylierung in einem zellfrien System aus das Blaualge *Anacystis nidulans. Z. Naturf.* **24B**, 1574–82.

43 BOTHE H. & BERZBORN R.J. (1970) Wirkung von Antikorpen gegen die Ferredoxin-. NADP-reduktase aus Spinat auf Photosynthetischen Reaktionen in einem zellfrein System aus das Blaualge *Anacystis nidulans. Z. Naturf.* **24B**, 529–34.

44 BOTHE H., HEMMEREICH P. & SUND H. (1971) Some properties of phytoflavin isolated from the blue-green alga *Anacystis nidulans*. In *Flavins and Flavoproteins*, ed. Kamen M. pp. 211–37. University Park, Baltimore.

45 BRANDON P.C. (1969) Cyclic photophosphorylation in the presence of NADP and ferricyanide. *Abstr. Int. Bot. Congress*, XI, Seattle, Abstract 21.

46 BROWN J.S. (1971) Photochemical properties of chloroplast particle preparations. *Yearbook of the Carnegie Institution of Washington*, **70**, 499–504.

47 BUCHANAN B.B. & ARNON D.I. (1971) Ferredoxins from photosynthetic bacteria, algae and higher plants. In *Advances in Enzymology*, XXIIIA, ed. san Pietro A. pp. 413–40. Academic Press, New York & London.

48 CARRIER J-M. (1966) Oxidation-reduction potentials of plastoquinones. In *Biochemistry of Chloroplasts*, vol. II, ed. Goodwin T.W. pp. 551–7. Academic Press, New York & London.

49 CHANCE B. (1967) The reactivity of haemoproteins and cytochromes. *Biochem. J.* **103**, 1–18.

50 CHANCE B. & SAN PIETRO A. (1963) On the light-induced bleaching of photosynthetic pyridine nucleotide reductase in the presence of chloroplasts. *Proc. natn. Acad. Sci., U.S.A.* **49**, 633–8.

51 CHANCE B., SAN PIETRO A., AVRON M. & HILDRETH W.W. (1965) The role of spinach ferredoxin (photosynthetic pyridine nucleotide reductase) in photosynthetic electron transfer. In *Non-Haem Iron Proteins*, ed. san Pietro A. pp. 225–36. Antioch Press, Yellow Springs, Ohio.

52 CHANG I.C. & KAHN J.S. (1970) Factors affecting the rate of photophosphorylation by isolated chloroplasts of *Euglena gracilis*. *J. Protozool.* **17**, 556–63.

53 CHANG I.C. & KAHN J.S. (1972) Light-activation of $Mg^{2+}$-dependent ATPase in isolated *Euglena* chloroplasts. *Pl. Physiol., Lancaster*, **49**, 299–302.

54 CHANG S.B. & VEDVICK T.S. (1968) A study of plastoquinones in the photochemical reactions in the chloroplasts of *Euglena gracilis* strain Z. *Pl. Physiol., Lancaster*, **43**, 1661–5.

55 CHENIAE G.M. (1970) Photosystem II and oxygen evolution. *A. Rev. Pl. Physiol.* **21**, 467–98.

56 CHENIAE G.M. & MARTIN I.F. (1966) Studies on the function of manganese in photosynthesis. *Brookhaven Symposia in Biology*, **19**, 406–17.

57 CHUA N-H. (1972) Photooxidation of 3,3'-diaminobenzidine by blue-green algae and *Chlamydomonas reinhardii*. *Biochim. biophys. Acta*, **267**, 179–89.

58 CHUA N-H. & LEVINE R.P. (1969) The photosynthetic electron transport chain of *Chlamydomonas reinhardii*. *Pl. Physiol., Lancaster*, **44**, 1–6.

59 CLAYTON R.K. (1966) *Molecular physics in photosynthesis*. Blaisdell Publishing Company, New York, Toronto, London.

60 DECKER M. & TANNER W. (1972) Respiratory increase and active hexose uptake of *Chlorella vulgaris*. *Biochim. biophys. Acta*, **266**, 661–9.

61 DÖRING G., STIEHL H.H. & WITT H.T. (1967) A second chlorophyll reaction in the electron transport chain of photosynthesis—registration by the repetitive excitation technique. *Z. Naturf.* **22B**, 639–44.

62 DUANE W.C., HOHL M.S. & KROGMANN D.W. (1965) Photophosphorylation in cell free preparations of a blue-green alga. *Biochim. biophys. Acta*, **109**, 108–16.

63 DUYSENS L.N.M. (1963) Studies on primary reactions and hydrogen or electron transport in photosynthesis by means of absorption and fluorescence difference spectrophotometry of intact cells. In *Photosynthetic Mechanisms of green plants*, eds. Kok B. & Jagendorf A.T. pp. 1–17. N.A.S.-N.R.C., Washington, D.C. **1145**.

64 DUYSENS L.N.M., AMESZ J. & KAMP B.H. (1961) Two photochemical systems in photosynthesis. *Nature, Lond.* **190**, 510–11.

65 EHRMENTRAUT H. & RABINOWITCH E. (1952) Kinetics of the Hill reaction. *Archs. Biochem. Biophys.* **38**, 67–84.

66 EMERSON R., CHALMERS R.V. & CEDERSTRAND C. (1957) Dependence of yield of photosynthesis in long wave red on wavelength and intensity of supplementary light. *Science, N.Y.* **125**, 746.

67 EPEL B.L. & LEVINE R.P. (1971) Mutant studies of *Chlamydomonas reinhardi* with

lesions on the oxidising side of photosystem II. *Biochim. biophys. Acta,* **226,** 154–60

68  ERIXON K. & BUTLER W.L. (1971a) Light-induced absorbance changes in chloroplasts at −196°C. *Photochem. Photobiol.* **14,** 427–33.

69  ERIXON K. & BUTLER W.L. (1971b) The relationship between C550 and cytochrome b559 in photoreactions at −196°C in chloroplasts. *Biochim. biophys. Acta,* **234,** 381–9.

70  FLOYD R.A., CHANCE B. & DE VAULT D. (1971) Low temperature photoinduced reactions in green leaves and chloroplasts. *Biochim. biophys. Acta,* **226,** 103–12.

71  FORSEE W.J. & KAHN J.S. (1972) Carbon dioxide fixation by isolated chloroplasts of *Euglena gracilis.* II. Inhibition of carbon dioxide fixation by AMP. *Archs. Biochem. Biophys.* **150,** 302–9.

72  FORTI G. & ROSA L. (1972) Cyclic photophosphorylation. In *Proc. 2nd International Congress on Photosynthetic Research,* eds. Forti G., Avron M. & Melandri A. pp. 1261–70. Dr. W. Junk, N.V., The Hague.

73  FUJITA Y. & MURANO F. (1968) Occurrence of back flow of electrons against the action of photosystem I in isolated lamellar fragments. In *Comparative Biochemistry and Biophysics of Photosynthesis,* eds. Shibata K., Takamiya A., Jagendorf A.T. & Fuller R.C. pp. 161–9. University of Tokyo Press, Tokyo/University Park Press, State College, Pennsylvania.

74  FUJITA Y. & MYERS J. (1965) The 2,6 dichlorophenol indophenol-Hill reaction by a cells-free preparation of *Anabaena cylindrica. Archs. Biochem. Biophys.* **112,** 506–11.

75  FUJITA Y. & MYERS J. (1971) Cytochrome reducing substance. In *Methods in Enzymology,* XXIIIA, ed. san Pietro A. pp. 613–18. Academic Press, New York & London.

76  GEE R., SALTMAN P. & WEAVER E.C. (1969) Studies on three photosynthetic mutants of *Scenedesmus. Biochim. biophys. Acta,* **189,** 106–15.

77  GIMMLER H. & AVRON M. (1971) On the mechanism of benzoquinone penetration and photoreduction by whole cells. *Z. Naturf.* **26B,** 585–8.

78  GIMMLER H., NEIMANI S., EILMAN I. & URBACH W. (1971) Photophosphorylation and photosynthetic $^{14}CO_2$-fixation *in vivo.* Comparison of cyclic and non-cyclic photophosphorylation with photosynthetic $^{14}CO_2$-fixation during the synchronous life cycle of *Ankistrodesmus braunii. Z. PflPhysiol.* **64,** 358–66.

79  GINGRAS G., LEMASSON C. & FORK D.C. (1963) A study of the mode of action of CMU on photosynthesis. *Biochim. biophys. Acta,* **69,** 438–40.

80  GINGRAS G. & LEMASSON C. (1965) A study of the mode of action of CMU on photosynthesis. *Biochim. biophys. Acta,* **109,** 67–78.

81  GIVAN A.L. & LEVINE R.P. (1967) The photosynthetic electron transport chain of *Chlamydomonas reinhardi.* VII. Photosynthetic phosphorylation by a mutant strain of *C. reinhardi* deficient in active P700. *Pl. Physiol., Lancaster,* **42,** 1264–8.

82  GIVAN A.L. & LEVINE R.P. (1969) The photosynthetic electron transport chain of a mutant strain of *Chlamydomonas reinhardi* lacking P700. *Biochim. biophys. Acta,* **189,** 404–10.

83  GORDON S.A. (1963) Observations on enhancement and inhibition by light of NADP reduction in preparations of *Laurencia obtusa* (Huds.) Lam. *Pl. Physiol., Lancaster,* **38,** 153–6.

84  GORMAN D.S. & LEVINE R.P. (1965) Cytochrome f and plastocyanin: their sequence in the electron transport chain in *Chlamydomonas reinhardi. Proc. natn. Acad. Sci., U.S.A.* **54,** 1665–9.

85  GORMAN D.S. & LEVINE R.P. (1966a) Photosynthetic electron transport chain of *Chlamydomonas reinhardi.* IV. Extraction and properties of plastocyanin. *Pl. Physiol., Lancaster,* **41,** 1637–42.

86  GORMAN D.S. & LEVINE R.P. (1966b) Photosynthetic electron transport chain of

*Chlamydomonas reinhardi*. V. Purification and properties of cytochrome 553 and ferredoxin. *Pl. Physiol., Lancaster,* **41**, 1643–7.

87 GORMAN D.S. & LEVINE R.P. (1966c) Photosynthetic electron transport chain of *Chlamydomonas reinhardi*. VI. Electron transport in mutant strains lacking either cytochrome 553 or plastocyanin. *Pl. Physiol., Lancaster,* **41**, 1648–56.

88 GOVINDJEE, MOHANTY J.C. Jr. & PAPAGEORGIOU G. (1966) Fluorescence studies with algae: changes with time and preillumination. *Brookhaven Symposia in Biology,* **19**, 434–45.

89 GOVINDJEE R. & RABINOWITCH E. (1960) Studies on the second Emerson effect in the Hill reaction in algal cells. *Biophys. J.* **1**, 377–88.

90 GRAHAM D. & SMILLIE R.M. (1971) Chloroplasts (and lamellae): algal preparations. In *Methods of Enzymology,* XXIIIA, ed. San Pietro A. pp. 228–42. Academic Press, New York & London.

91 GREGORY R.P.F. (1971) *Biochemistry of Photosynthesis.* Wiley-Interscience, London, New York, Sidney & Toronto.

92 GREVILLE G.D. (1969) A scrutiny of Mitchell's chemiosmotic hypothesis of respiratory chain and photosynthetic phosphorylation. In *Current Topics in Bioenergetics,* vol. 3, ed. Sanadi D.R. pp. 1–78. Academic Press, New York and London.

93 HAEHNEL W., DORING G. & WITT H.T. (1971) On the reaction between chlorophyll $a_1$ and its primary electron donors in photosynthesis. *Z. Naturf.* **26B**, 1171–4.

94 HALL D.O. (1972) Nomenclature for isolated chloroplasts. *Nature New Biology,* **235**, 125–6.

95 HALL D.O. & EVANS M.C.W. (1972) Photosynthetic phosphorylation in chloroplasts. *Sub-cell Biochem.* **1**, 197–206.

96 HALL D.O., REEVES S.G. & BALTSHEFFSKY H. (1971) Photosynthetic control in isolated chloroplasts with endogenous and artificial electron acceptors. *Biochem. biophys. Res. Commun.* **43**, 359–66.

97 HALLDAL P. (ed.) (1970) *Photobiology of Microorganisms.* Wiley-Interscience, London, New York, Sydney & Toronto.

98 HATTORI A. & UESUGI I. (1968) Ferredoxin-dependent photoreduction of nitrate and nitrite by subcellular preparations of *Anabaena cylclindrica*. In *Comparative Biochemistry and Biophysics of Photosynthesis,* eds. Shibata K., Takamiya A., Jagendorf A.T. & Fuller R.C. pp. 201–5. Univeristy of Tokyo Press, Tokyo/University Park Press, State College, Pennsylvania.

99 HEBER U.W. & SANTARIUS K.A. (1966) Compartmentation and reduction of pyridine nucleotides in relation to photosynthesis. *Biochim. biophys. Acta,* **109**, 390–408.

100 HEBER U. & KRAUSE G.H. (1971) Transfer of carbon, phosphate energy and reducing equivalents across the chloroplast envelope. In *Photosynthesis and Photorespiration,* eds. Hatch M.D., Osmond C.B. & Slatyer R.O. pp. 218–25. Wiley-Interscience, London, Sydney, New York & Toronto.

101 HENNINGER M.D. & CRANE F.L. (1966) Electron transport in chloroplasts: a new redox protein, rubimedin. *Biochem. biophys. Res. Commun.* **24**, 386–90.

102 HILL R. & BENDALL F. (1960) Function of two cytochrome components in chloroplasts: a working hypothesis. *Nature, Lond.* **186**, 136–7.

103 HILLER R.G. & BOARDMAN N.K. (1971) Light driven redox changes of cytochrome f and the development of photosystems I and II during greening of bean leaves. *Biochim. biophys. Acta,* **253**, 449–58.

104 HIND G. & OLSEN J.M. (1968) Electron transport pathways in photosynthesis. *A. Rev. Pl. Physiol.* **19**, 249–82.

105 HIYAMI T. & KE B. (1971a) Laser-induced reactions of P700 and cytochrome f in a blue-green alga, *Plectonema boryanum. Biochim. biophys. Acta,* **226**, 320–7.

106 HIYAMI T. & KE B. (1971b) A new photosynthetic pigment, 'P430': its possible role as

the primary electron acceptor of photosystem I. *Proc. natn. Acad. Sci., U.S.A.* **68**, 1010–13.

107  HIYAMI T. & KE B. (1971c) A further study of P430: a possible primary electron acceptor of photosystem I. *Archs. Biochem. Biophys.* **147**, 99–108.

108  HIYAMI T. & KE B. (1972) Difference spectra and extinction coefficients of P700. *Biochim. biophys. Acta*, **267**, 160–71.

109  HIYAMI T., NISHIMURA M. & CHANCE B. (1970) Energy and electron transfer systems of *Chlamydomonas reinhardi*. II. Two cyclic pathways of photosynthetic electron transfer in the pale green mutant. *Pl. Physiol., Lancaster*, **46**, 163–8.

110  HOCH G. & OWENS O.v.H. (1963) Photoreactions and respiration. In *Photosynthetic mechanisms of green plants*, eds. Kok B. & Jagendorf A.T. pp. 409–20. N.A.S.-N.R.C., Washington, D.C. **1145**.

111  HOCH G.E. & RANDLES J. (1971) On the interaction between photosystems I and II in algal cells. *Photochem. Photobiol.* **14**, 435–49.

112  HOLM-HANSEN O. (1970) ATP levels in algal cells as influenced by environmental conditions. *Pl. Cell Physiol., Tokyo*, **11**, 689–700.

113  HOLTON R.W. & MYERS J. (1963) Cytochromes of a blue-green alga: Extraction of a C-type with a strongly negative redox potential. *Science, N.Y.* **142**, 234–5.

114  IKEGAMI I., KATOH S. & TAKAMIYA A. (1968) Nature of haem moiety and oxidation-reduction potential of cytochrome 558 in *Euglena* chloroplasts. *Biochim. biophys. Acta*, **162**, 604–7.

115  IKEGAMI I., KATOH S. & TAKAMIYA Y. (1970) Light-induced changes of b-type cytochromes in the electron transport chain of *Euglena* chloroplasts. *Pl. Cell Physiol., Tokyo*, **11**, 777–91.

116  JEFFREY S.W., ULRICH J. & ALLEN M.B. (1966) Some photochemical properties of chloroplast preparations from the chrysophyte *Hymenomonas* sp. *Biochim. biophys. Acta*, **112**, 35–44.

117  JESCHKE W.D. (1972) The effect of the inhibitor of photophosphorylation Dio-9 and the uncoupler Atebrin on the light-dependent $Cl^-$ influx of *Elodea densa*; direct inhibition of membrane transport? *Z. PflPhysiol.* **66**, 397–408.

118  JESCHKE W.D., GIMMLER H. & SIMONIS W. (1967) Incorporation of $^{32}P$ and $^{14}C$ into photosynthetic products of *Ankistrodesmus braunii* as affected by X-rays. *Pl. Physiol., Lancaster*, **42**, 380–6.

119  JOLIOT P. (1965) Cinetiques des reactions liees a l'emission d'oxygene photosynthetique. *Biochim. biophys. Acta*, **102**, 116–34.

120  KAHN J.S. (1968) Chlorotri-n-butyl tin, an inhibitor of photophosphorylation in isolated chloroplasts. *Biochim. biophys. Acta*, **153**, 203–10.

121  KAHN J.S. (1970) Absence of a common intermediate pool among individual enzyme chains in the energy conserving pathway in chloroplasts of *Euglena gracilis*. *Biochem. J.* **116**, 55–60.

122  KAHN J.S. (1971) Evidence for a two-directional hydrogen ion transport in chloroplasts of *Euglena gracilis*. *Biochim. biophys. Acta*, **245**, 144–50.

123  KANAI R. & SIMONIS W. (1968) Einbau von $^{32}P$ in verscheidne Phosphatfractionem, besonders Polyphosphate, bei einzelligen Grunalgen (*Ankistrodesmus braunii*). *Arch. Mikrobiol.* **62**, 56–71.

124  KANDLER O. (1958) The effect of 2-4,dinitrophenol on respiration, oxidative assimilation and photosynthesis in *Chlorella*. *Physiologia Pl.* **11**, 675–84.

125  KANDLER O. (1960) Energy transfer through phosphorylation mechanisms in photosynthesis. *A. Rev. Pl. Physiol.* **11**, 37–54.

126  KANDLER O. & TANNER W. (1966) Die Photassimilation von Glucose als Indikator für die Lichtphosphorylierung *in vivo*. *Ber. dt. bot. Ges.* **79**, 48–57.

127  KATOH S. (1960) A new copper protein from *Chlorella ellipsoidea*. *Nature, Lond.* **186**, 533–4.

128  KATOH S. & SAN PIETRO A. (1967) The role of c-type cytochrome in the Hill reaction with *Euglena* chloroplasts. *Archs. Biochem. Biophys.* **119**, 488–96.

129  KATOH S. & SAN PIETRO A. (1968) Photoreactions of chloroplasts: NADP photoreduction by *Euglena* chloroplasts. In *Comparative Biochemistry and Biophysics of photosynthesis*, eds. Shibata K., Takamiya A., Jagendorf A.T. & Fuller R.C. pp. 148–60. University of Tokyo Press, Tokyo/University Park Press, State College, Pennsylvania.

130  KE B., VERNON L.P. & CHANEY T.H. (1972) Photoreduction of cytochrome b559 in a photosystem II subchloroplast particle. *Biochim. biophys. Acta*, **256**, 345–57.

131  KELLY J. & SAUER K. (1968) Functional photosynthetic unit sizes for each of the light reactions in spinach chloroplasts. *Biochemistry, N.Y.* **7**, 882–90.

132  KNAFF D.B. (1972) The effect of a plastiquinone antagonist on the oxidation-reduction reactions of chloroplast cytochrome b559. *FEBS Letters*, **23**, 142–4.

133  KNAFF D.B. & ARNON D.I. (1969) Spectral evidence for a new photoreactive component of the oxygen-evolving system in photosynthesis. *Proc. natn. Acad Sci., U.S.A.* **63**, 963–9.

134  KNOX R.S. (1969) Thermodynamics and the primary processes of photosynthesis. *Biophys. J.* **9**, 1351–62.

135  KOK B. (1960) The efficiency of photosynthesis. *Handb. PflPhysiol.* vol. 1, ed. Ruhland H. pp. 566–633. Springer-Verlag, Berlin, Heidelberg & New York.

136  KOK B. & CHENIAE G.M. (1966) Kinetics and intermediates of the oxygen evolution steps in photosynthesis. In *Current Topics in Bioenergetics*, vol. 1, ed. Sanddi D.R. pp. 1–47. Academic Press, New York and London.

137  KOK B. & DATKO E.A. (1965) Reducing power generated in the second photoact of photosynthesis. *Pl. Physiol., Lancaster*, **40**, 1171–7.

138  KOK B. & HOCH G. (1961) Spectral changes in photosynthesis. In *A Symposium on Light and Life*, eds. McElroy W.D. & Glass B. pp. 397–423. Academic Press, New York & London.

139  KOK B., RURAINSKI H.J. & OWENS O.v.H. (1965) The reducing power generated in photoact one of photosynthesis. *Biochim. biophys. Acta*, **109**, 347–56.

140  KOK B., MALKIN S., OWENS O. & FORBUSH B. (1966) Observations on the reducing side of the $O_2$-evolving photoact. *Brookhaven Symposia in Biology*, **19**, 446–59.

141  KOUCHKOVSKY Y.DE & FORK D.C. (1964) A possible functioning *in vivo* of plastocyanin in photosynthesis as revealed by a light-induced absorbance change. *Proc. natn. Acad. Sci., U.S.A.* **52**, 232–9.

142  LEE S.S., YOUNG A.M. & KROGMANN D.W. (1969) Site-specific inactivation of the photophosphorylation reactions of *Anabaena variabilis*. *Biochim. biophys. Acta*, **180**, 130–6.

143  LEVINE R.P. (1963) The electron transport system of photosynthesis deduced from experiments with mutants of *Chlamydomonas reinhardi*. In *Photosynthetic Mechanisms of Green Plants*, eds. Kok B. & Jagendorf A.T. pp. 158–73. N.A.S.-N.R.C. Washington, D.C. **1145**.

144  LEVINE R.P. (1969a) A light-induced absorbance change at 564nm in wild-type and mutant strains of *Chlamydomonas reinhardi*. In *Progress in Photosynthesis Research*, vol. II, ed. Metzner H. pp. 971–7. International Biological Union, Tubingen.

145  LEVINE R.P. (1969b) The analysis of photosynthesis using mutant strains of algae and higher plants. *A. Rev. Pl. Physiol.* **20**, 523–40.

146  LEVINE R.P. (1971) Preparation and properties of mutant strains of *Chlamydomonas reinhardi*. In *Methods in Enzymology*, XXIIIA, ed. san Pietro A. pp. 119–29. Academic Press, New York & London.

147  LEVINE R.P. & ARMSTRONG J. (1972) The site of synthesis of two chloroplast cytochromes in *Chlamydomonas reinhardi*. *Pl. Physiol., Lancaster*, **49**, 661–2.

148  LEVINE R.P. & GORMAN D.S. (1966) Photosynthetic electron transport chain of

*Chlamydomonas reinhardi.* III. Light-induced absorbance changes in chloroplast fragments of the wild-type and mutant strains. *Pl. Physiol., Lancaster,* **41,** 1293–300.

149 LEVINE R.P., GORMAN D.S., AVRON M. & BUTLER W.L. (1966) Light-induced absorbance changes in wild-type and mutant strains of *Chlamydomonas reinhardi. Brookhaven Symposia in Biology,* **19,** 143–8.

150 LIGHTBODY J.J. & KROGMANN D.W. (1966) The role of plastoquinone in the photosynthetic reactions of *Anabaena variabilis. Biochim. biophys. Acta,* **120,** 57–64.

151 LIGHTBODY J.J. & KROGMANN D.W. (1967) Isolation and properties of plastocyanin from *Anabaena variabilis. Biochim. biophys. Acta,* **131,** 508–15.

152 LOSADA M., WHATLEY F.R. & ARNON D.I. (1961) Separation of two light reactions in non-cyclic photophosphorylation of plants. *Nature, Lond.* **190,** 606–10.

153 LUDLOW C.J. & PARK R.B. (1969) Action spectra of photosystems I and II in formaldehyde fixed algae. *Pl. Physiol., Lancaster,* **44,** 540–3.

154 MALKIN S. (1969) On the equilibrium between the reaction centres of the two photosystems in photosynthesis. The effect of independent electron-transport chains. *Biophys. J.* **9,** 489–99.

155 MITCHELL P. (1966) *Chemiosmotic coupling in oxidative and photosynthetic phosphorylation.* Research Report 6611, Glynn Research Ltd., Bodmin, Cornwall.

156 MITCHELL P. (1967) Proton-translocation phosphorylation in mitochondria, chloroplasts and bacteria: natural fuel cells and solar cells. *Fedn. Proc. Fedn. Am. Socs. Exptl. Biol.* **26,** 1370–9.

157 MITSUI A. (1971) *Euglena* cytochromes. In *Methods in Enzymology,* XXIIIA, ed. san Pietro A. pp. 368–71. Academic Press, New York & London.

158 MOHANTY P., MAR T. & GOVINDJEE (1971) Action of hydroxylamine in the red alga *Porphyridium cruentum. Biochim. biophys. Acta,* **253,** 213–21.

159 MYERS J. (1971) Enhancement studies in photosynthesis. *A. Rev. Pl. Physiol.* **22,** 289–312.

160 MYERS J. & GRAHAM J-R. (1971) The photosynthetic unit in *Chlorella* measured by repetitive short flashes. *Pl. Physiol., Lancaster,* **48,** 282–6.

161 NEUMANN J. & LEVINE R.P. (1971) Reversible pH changes in cells of *Chlamydomonas reinhardi* resulting from $CO_2$ fixation in the light and its evolution in the dark. *Pl. Physiol., Lancaster,* **47,** 700–4.

162 NISHIMURA M. & TAKAMIYA A. (1966) Energy- and electron-transfer systems in algal photosynthesis. I. Actions of two photochemical systems in oxidation-reduction reactions of cytochrome in *Porphyra. Biochim. biophys. Acta,* **120,** 45–56.

163 NISHIMURA M. (1968) Energy- and electron-transfer reactions in algal photosynthesis. II. Oxidation-reduction reactions of two cytochromes and interactions of two photochemical systems in red algae. *Biochim. biophys. Acta,* **153,** 838–47.

164 OGAWA T. & VERNON L.P. (1970) Properties of partially purified photosynthetic reaction centres from *Scenedesmus* mutant 6E and *Anabaena variabilis* grown with diphenylamine. *Biochim. biophys. Acta,* **197,** 292–301.

165 OGAWA T. & VERNON L.P. (1971) Increased content of cytochromes 554 and 562 in *Anabaena variabilis* cells grown in the presence of diphenylamine. *Biochim. biophys. Acta,* **197,** 302–7.

166 OLSEN J.M. & SMILLIE R.M. (1963) Light-driven cytochrome reactions in *Anacystis* and *Euglena.* In *Photosynthetic Mechanisms in Green Plants,* eds. Kok B. & Jagendorf A.T. pp. 56–65. N.A.S.-N.R.C. Washington, D.C. **1145.**

167 OWENS O. & KRAUSS R.W. (1972) Kinetics of photophosphorylation in cells of *Anacystis nidulans. Pl. Physiol., Lancaster,* **50,** 52s.

168 PARK R.B. & SANE P.V. (1971) Distribution of function and structure in chloroplast lamellae. *A. Rev. Pl. Physiol.* **22,** 395–430.

169 PETRACK B. & LIPMANN F. (1961) Photophosphorylation and photohydrolysis in cell-

free preparations of blue-green algae. In *A Symposium on Light and Life*, eds. McElroy W.D. & Glass B. pp. 621–30. Academic Press, New York & London.

170 PLESNICAR M. & BENDALL D.S. (1970) The plastocyanin content of chloroplasts from some higher plants estimated by a sensitive enzyme assay. *Biochim. biophys. Acta*, **216**, 192–9.

171 POWLS R., WONG J. & BISHOP N.I. (1969) Electron transfer components of wild-type and photosynthetic mutant strains of *Scenedemus obliquus* D₃. *Biochim. biophys. Acta*, **180**, 490–9.

172 RABINOWITCH E. & GOVINDJEE (1969) *Photosynthesis*. John Wiley & Sons, Inc., New York.

173 RAVEN J.A. (1968) The action of phlorizin on photosynthesis and light-stimulated ion transport in *Hydrodictyon africanum*. *J. exp. Bot.* **19**, 712–23.

174 RAVEN J.A. (1969a) Action spectra for photosynthesis and light-stimulated ion transport in *Hydrodictyon africanum*. *New Phytol.* **68**, 45–62.

175 RAVEN J.A. (1969b) Effects of inhibitors on photosynthesis and active influxes of K and Cl in *Hydrodictyon africanum*. *New Phytol.* **68**, 1089–113.

176 RAVEN J.A. (1970) The role of cyclic and pseudocyclic photophosphorylation in photosynthetic $^{14}CO_2$ fixation in *Hydrodictyon africanum*. *J. exp. Bot.* **22**, 1–16.

177 RAVEN J.A. (1971a) Cyclic and non-cyclic photophosphorylation as energy sources for active K influx in *Hydrodictyon africanum*. *J. exp. Bot.* **22**, 420–33.

178 RAVEN J.A. (1971b) Energy metabolism in green cells. *Trans. Proc. bot. Soc. Edinb.* **41**, 219–25.

179 RAVEN J.A. (1971c) Inhibitor effects on photosynthesis, respiration and active ion transport in *Hydrodictyon africanum*. *J. membr. Biol.* **6**, 89–107.

180 RAVEN J.A. (1972) Endogenous inorganic carbon sources in plant photosynthesis. I. Occurrence of the dark respiratory pathways in illuminated green cells. *New Phytol.* **71**, 227–47.

181 ROSS R.T. & CALVIN M. (1967) Thermodynamics of light emission and free energy storage in photosynthesis. *Biophys. J.* **7**, 595–614.

182 ROWAN K.S. (1966) Phosphorus metabolism in plants. *Int. Rev. Cytol.* **19**, 301–90.

183 RUMBERG B., SCHMIDT-MENDE P. & WITT H.T. (1964) Different demonstrations of the coupling of two light reactions in photosynthesis. *Nature, Lond.* **201**, 466–8.

184 RURAINSKI H.J., RANDLES J. & HOCH G.E. (1971) The relationship between P700 and NADP reduction in chloroplasts. *FEBS Letters*, **13**, 98–100.

185 RUSSELL G.K. & GIBBS M. (1966) Regulation of photosynthetic capacity in *Chlamydomonas reinhardi*. *Pl. Physiol., Lancaster*, **41**, 885–90.

186 SATO V.L., LEVINE R.P. & NEUMANN J. (1971) Photosynthetic phosphorylation in *Chlamydomonas reinhardi*. Effects of a mutation altering an ATP-synthesising enzyme. *Biochim. biophys. Acta*, **253**, 437–48.

187 SATOH K., KATOH S. & TAKAMIYA A. (1970) Effects of chloride ions on Hill reactions in *Euglena* chloroplasts. *Pl. Cell. Physiol., Tokyo*, **11**, 453–66.

188 SCHNEYOUR A. & AVRON M. (1970) High biological activity in chloroplasts from *Euglena gracilis* prepared with a new gas pressure device. *FEBS Letters*, **8**, 164–6.

189 SCHULDINER S. & OHAD I. (1969) Biogenesis of chloroplast membranes. II. Light-dependent induction of proton pump activity in whole cells and its correlation with cytochrome f photo-oxidation during greening of a *Chlamydomonas reinhardii* mutant (y-1). *Biochim. biophys. Acta*, **180**, 165–77.

190 SCHWARTZ M. (1971) The relation of ion transport to phosphorylation. *A Rev. Pl. Physiol.* **22**, 469–84.

191 SCHWEITZ F.D. (1972) Structure and function in photosynthetic membranes. *Diss. Abstr.* **32**, 5038B.

192 SENGER H. (1970) Quantenausbete und Unterscheidliches Verhalten der beiden Photo-

systems des Photosyntheseapparates wahrend des Entwicklungseblaufes von *Scenedesmus obliquus* in Synchronkulturen. *Planta*, **92**, 327–46.

193  SIMONIS W. (1960) Photosynthese und Lichtahangige Phosphorylierung. *Handb. PflPhysiol.* vol. 1, ed. Ruhland H. pp. 966–1013. Springer-Verlag, Berlin, Heidelberg & New York.

194  SIMONIS W. (1964) Untersuchungen zur Photosynthese-Phosphorylierung an intakter Algenzelle (*Ankistrodesmus braunii*). *Ber. dt. Bot. Ges.* **77**, (5)–(11).

195  SIMONIS W. (1967) Zyclische und nichtzyclische Photophosphorylierung *in vivo*. *Ber. dt. Bot. Ges.* **80**, 395–402.

196  SIMONIS W. & URBACH W. (1973) Photosynthetic phosphorylation *in vivo*. *A. Rev. Pl. Physiol.* **24**, 89–114.

197  SMILLIE R.M. & ENTSCH B. (1971) Phytoflavin. In *Methods in Enzymology*, XXIIIA, ed. San Pietro A. pp. 504–14. Academic Press, New York & London.

198  SMILLIE R.M., ANDERSEN K.S., TOBIN N.F., ENTSCH B. & BISHOP D.G. (1972) Nicotinamide adenine dinucleotide phosphate photoreduction from water by agranal chloroplasts isolated from bundle sheath cells of maize. *Pl. Physiol., Lancaster*, **49**, 471–5.

199  SMITH F.A. (1971) Transport of solutes during $C_4$-photosynthesis. In *Photosynthesis and Photorespiration*, eds. Hatch M.D., Osmond C.B. & Slatyer R.O. pp. 302–6. Wiley-Interscience, London, New York, Sydney & Toronto.

200  SMITH F.A. & WEST K.R. (1969) A comparison of the effects of metabolic inhibitors on chloride uptake and photosynthesis in *Chara corallina*. *Austr. J. biol. Sci.* **22**, 351–64.

201  STOREY B.T. (1970) Chemical hypothesis for energy conservation in the mitochondrial respiratory chain. *J. Theoret. Biol.* **28**, 233–59.

202  STROTMANN H. & BERGER S. (1969) Adenine nucleotide translocation across the membrane of isolated *Acetabularia* chloroplasts. *Biochem. biophys. Res. Commun.* **35**, 20–6.

203  SUN A.S.K. & SAUER K. (1971) Pigment systems and electron transport in chloroplasts. I. Quantum requirements for the two light reactions in spinach chloroplasts. *Biochim. biophys. Acta*, **234**, 399–414.

204  SUN A.S.K. & SAUER K. (1972) Pigment systems and electron transport in chloroplasts. II. Emerson enhancement in broken spinach chloroplasts. *Biochim. biophys. Acta*, **256**, 409–27.

205  SUN E., BARR R. & CRANE F.L. (1968) Comparative studies of plastoquinones. IV. Plastoquinones in algae. *Pl. Physiol., Lancaster*, **43**, 1935–40.

206  SUSOR W.A. & KROGMANN D.W. (1964) Hill activity in cell-free preparations of a blue-green alga. *Biochim. biophys. Acta*, **88**, 11–19.

207  TANNER W. (1969) Light-driven uptake of 3-0-Methylglucose via an inducible hexose uptake system in *Chlorella*. *Biochem. biophys. Res. Commun.* **36**, 278–83.

208  TANNER W., DACHSEL E. & KANDLER O. (1965) Effects of DCMU and Antimycin A on photoassimilation of glucose in *Chlorella*. *Pl. Physiol., Lancaster*, **40**, 1151–6.

209  TANNER W., LOFFLER M. & KANDLER O. (1969) Cyclic photophosphorylation *in vivo* and its relation to photosynthetic $CO_2$-fixation. *Pl. Physiol., Lancaster*, **44**, 422–9.

210  TANNER W., ZINECKER U. & KANDLER O. (1967) Die Anaeroben Photoassimilation von Glucose bei Photosynthese-Mutanten von *Scenedesmus*. *Z. Naturf.* **22B**, 358–9.

211  TANNER W., LOOS E., KLOB W. & KANDLER O. (1968) The quantum requirement for light dependent anaerobic glucose assimilation by *Chlorella vulgaris*. *Z. PflPhysiol.* **59**, 301–3.

212  TEICHLER-ZALLEN D. & HOCH G. (1967) Cyclic electron flow in algae. *Archs. Biochem. Biophys.* **120**, 227–30.

213  THORNBER J.P. & OLSEN J.M. (1971) Chlorophyll-proteins and reaction centre prepar-

ations from photosynthetic bacteria, algae and higher plants. *Photochem. Photobiol.* **14**, 329–41.

214 TREBST A. (1970) Neue Reaktionen und Substanzen im photosynthetischen Elektronen-transport. *Ber. dt. bot. Ges.* **83**, 373–98.

215 ULLRICH W. (1971) Nitratabhangige nichtcyclische Photophosphorylierung bei *Ankistrodesmus braunii* in Abwesenheit von $CO_2$ und $O_2$. *Planta*, **100**, 18–30.

216 URBACH W. & GIMMLER H. (1970) Photophosphorylierung und Photosynthetische $^{14}CO_2$-Fixierung *in vivo*. I. Vergleich von cyclischer und nichtcyclischer Photophosphorylierung mit der photosynthetischen $^{14}CO_2$-Fixierung. *Z. PflPhysiol.* **62**, 276–86.

217 URBACH W. & KAISER W. (1972) Changes of ATP levels in green algae and intact chloroplasts during photosynthetic reactions. In *Proc. 2nd International Congress on Photosynthetic Research*, eds. Forti G., Avron H. & Melandri A. pp. 1401–11. Dr. W. Junk N.V., The Hague.

218 VREDENBERG W.J. & DUYSENS L.N.M. (1965) Light-induced changes in absorbancy and fluorescence of chlorophyllous pigments associated with the pigment systems I and II in blue-green algae. *Biochim. biophys. Acta*, **94**, 355–70.

219 WADA K. & ARNON D.I. (1971) Three forms of cytochrome b559 and their relation to the photosynthetic activity of chloroplasts. *Proc. natn. Acad. Sci., U.S.A.* **68**, 3064–8.

220 WALLACH D., BAR-NUN S. & OHAD I. (1972) Biogenesis of chloroplast membranes. XI. Development of photophosphorylation and proton pump activities in greening *Chlamydomonas reinhardi* y-1 as measured with an open-cell preparation. *Biochim. biophys. Acta*, **267**, 125–37.

221 WALLWORK J.C. & CRANE F.L. (1970) *The nature, distribution and biosynthesis of prenyl phytoquinones and related compounds*. In *Progress in Phytochemistry*, II, eds. Reinhold L. & Linschitz Y. pp. 267–342. Interscience, London.

222 WEIER T.E., STOCKING C.R. & SHUMWAY L.K. (1966) The photosynthetic apparatus in chloroplasts of higher plants. *Brookhaven Symposia in Biology*, **19**, 353–74.

223 WEISS C.Jr., SOLNIT K.T. & VON GUTFIELD R.J. (1971) Flash activation kinetics and photosynthetic unit size for $O_2$ evolution using 3-nsec flashes. *Biochim. biophys. Acta*, **253**, 298–301.

224 WILD A. & EGLE K. (1968) The size of the photosynthetic unit and its variability. I. Calculation of the size of the photosynthetic unit. *Beitr. biol. Pflanzen*, **45**, 213–41.

225 WINTERMANS J.F.G.M. (1958) Some observations on phosphate metabolism in the induction phase of photosynthesis in *Chlorella. Acta bot. neerl.* **7**, 489–502.

226 WITT H.T. (1971) Coupling of quanta, electrons, fields, ions and phosphorylation in the functional membrane of photosynthesis. Results by the pulse spectroscopic methods. *Q. Rev. Biophys.* **4**, 365–477.

227 WRAIGHT C.A. & CROFTS A.R. (1971) Delayed fluorescence and the high-energy state of chloroplasts. *Eur. J. Biochem.* **19**, 386–97.

228 YAKUSHIJI E. (1971) Cytochromes: algae. In *Methods in Enzymology*, XXIIIA, ed. san Pietro A. pp. 364–8. Academic Press, New York & London.

229 YANG C.S. & BLUMBERG W.E. (1972) Quantitative studies on the EPR signal of photosystem I and ferredoxin. *Biochem. biophys. Res. Commun.* **46**, 422–8.

230 ZUMFT W.G. & SPILLER M. (1971) Characterization of a flavodoxin from the green alga *Chlorella. Biochem. biophys. Res. Commun.* **45**, 112–18.

231 ZWEIG G. & AVRON M. (1965) On the oxidation-reduction potential of the photo-produced reductant of isolated chloroplasts. *Biochim. biophys. Res. Commun.* **19**, 397–400.

# CHAPTER 14

# MUTANT STUDIES ON PHOTOSYNTHETIC ELECTRON TRANSPORT

R. P. LEVINE

The Biological Laboratories,
Harvard University,
Cambridge, Massachusetts 02138, U.S.A.

1   Introduction 424

2   Mutations affecting photosystem II 425

3   Mutations affecting the oxidizing side of photosystem II 426

4   A mutation affecting photosynthetic phosphorylation 427

5   A structure-function syndrome 429

6   Mutations affecting chloroplast membrane structure 430

7   References 431

## 1 INTRODUCTION

The previous chapter dealt in general with photosynthetic electron transport, and in particular with photosynthetic phosphorylation in algae. In this chapter information contributing to the elucidation of the electron transport chain, obtained from studies on algal mutants, will be presented. The use of these organisms illustrates a unique way in which algae contribute to our basic understanding of the mechanisms involved.

Mutant strains of algae, particularly *Chlamydomonas reinhardi* and *Scenedesmus obliquus*, have proved to be invaluable for delineating the nature and function of components of the photosynthetic electron transport chain (see reviews by Bishop 1964, Levine 1969, Levine & Goodenough 1970). The strains cannot carry out photosynthesis and thus require a source of reduced carbon for growth, and analyses of broken cell preparations reveal that the chloroplast membranes do not possess an intact photosynthetic electron transport chain. The normal chain comprises a series of electron donor/acceptor molecules that lie in a pathway between water, the primary electron donor, and NADP, the terminal electron acceptor (Fig. 14.1, and also Fig. 12.13), and if a component in the chain is missing or inactive as a consequence of mutation, then electron flow will proceed to that point and stop. The mutant strains of *C. reinhardi* and *S. obliquus* are, however, often able to carry out partial reactions *in vitro* when artificial electron donors and/or acceptors are provided to accept or donate electrons in specific regions of the chain. By comparing the various mutant strains of *C. reinhardi* with regard to the components they lack and the partial reactions they are able to perform, the sequence of many of the components in the chain has been elucidated (Fig. 14.1). The pathway deduced from the study of the mutant strains is in agreement with the generally accepted Hill-Bendall scheme (Hill & Bendall 1960) (see also Chapter 13, p. 391).

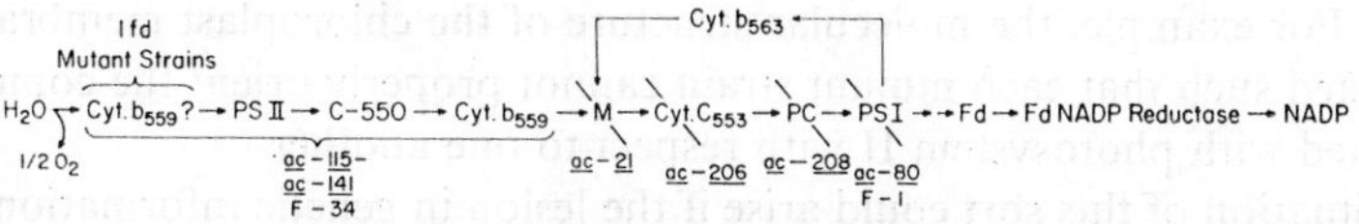

**Fig. 14.1.** The photosynthetic electron transport chain of *C. reinhardi*. Several components of the chain are affected by gene mutations; consequently electron transport from water to NADP is blocked. Electron transport in the *lfd* strains is blocked on the oxidizing side of photosystem II. These strains have half of the wild-type level of a high potential cytochrome $b_{559}$. The strains *ac-115*, *ac-141*, and F-34 have no photosystem II activity, and they lack both C-550 and a high potential cytochrome $b_{559}$. P700 is affected in *ac-80* and F-1 and the remaining strains lack the components indicated. PC = plastocyanin, M = unidentified component, Fd = ferredoxin.

In this chapter the properties of certain of the mutant strains of *C. reinhardi* will be considered in detail. It will be shown that some of the mutations have pleiotropic effects and that an understanding of the pleiotropy provides further understanding of the nature of the electron transport chain.

## 2 MUTATIONS AFFECTING PHOTOSYSTEM II

The nature of photosystem II is by far the most baffling part of the photosynthetic process. Recently, a detailed analysis (Epel & Butler 1971) of the properties of two classes of mutant strains of *C. reinhardi* has helped to diminish the complexity surrounding photosystem II by providing information about the relationship of some of the components closely associated with it.

Among the earliest mutant strains of *C. reinhardi* to be described (Levine & Smillie 1963) are two that cannot transfer electrons from photosystem II to any of a number of artificial electron acceptors. It is now apparent that there are at least three genetically distinct groups of such mutant strains (Levine 1969, Levine & Goodenough 1970) and that they have similar and complex phenotypes. Each strain lacks C-550 (Epel & Butler 1971), the primary electron acceptor of photosystem II (Bendall & Sofrova 1971), thus explaining the high invariant fluorescence yield that is characteristic of these mutant strains (Epel & Butler

1971, Lavorel & Levine 1968, Epel & Levine 1971). Each strain also lacks a high potential, ascorbate-reducible cytochrome $b_{559}$, which in the wild-type strain can be reduced by photosystem II and oxidized by photosystem I (Epel & Butler 1971, Levine & Gorman 1966). At least three of the strains (ac-115, ac-141, and F-34) are characterized by altered chloroplast membrane conformation (Goodenough & Levine 1969), and this altered conformation is correlated with changes in the amounts and molecular weights of some of the chloroplast membrane polypeptides (Levine et al. 1972).

Since each of the mutant strains differs from the wild-type strain by a single gene mutation (Levine 1969), the complex nature of their phenotypes suggests that the primary genetic lesion in each case is of an exceedingly fundamental nature. For example, the molecular structure of the chloroplast membrane may be affected such that each mutant strain cannot properly orient the components associated with photosystem II with respect to one another.

A situation of this sort could arise if the lesion in genetic information results in altered membrane structure into which membrane-bound components such as C-550 and cytochrome $b_{559}$ cannot conform. Alternatively, the loss of the capacity to form C-550 could itself result in altered membrane structure so that another component associated with photosystem II (in this case cytochrome $b_{559}$) would not find its 'fit' in the membrane. There is no evidence to support either of these alternatives, but it is interesting to note that although these strains do not demonstrate the light-induced redox changes of cytochrome $b_{559}$ that are characteristic of the wild-type strain and although they do not exhibit the presence of a high potential ascorbate reducible cytochrome $b_{559}$, they do possess a cytochrome $b_{559}$ (Epel & Butler 1971) that is reducible with dithionite and thus of low potential. It is tempting to speculate that in each instance the mutational events have not affected cytochrome synthesis but rather the structure of photosystem II and that the cytochrome in its altered environment has an altered redox potential. This would mean that although the cytochrome is present and photosystem I is unaffected by the altered structural circumstances, the cytochrome cannot be oxidized by the light that sensitizes photosystem I.

In this regard, it is of interest that the amount of low potential cytochrome $b_{559}$ that can be detected in normal chloroplasts depends on how the chloroplasts have been treated. For example, exposure of chloroplasts to high concentrations of TRIS buffer (Erixon et al. 1972), to heat (Wada & Arnon 1971), to extraction with acetone (Hind & Nakatini 1970), or to lipases (Okayama et al. 1971, Epel et al. unpubl.) all enhance the amount of low potential cytochrome $b_{559}$ that is detected.

## 3  MUTATIONS AFFECTING THE OXIDIZING SIDE OF PHOTOSYSTEM II

In contrast to the mutations described above, a group of mutant strains exists that possesses photosystem II activity provided that an artificial electron donor is substituted for the natural donor, water. With such a donor these mutant strains operate a normal electron transport chain that is capable of reducing NADP (Epel & Levine 1971). This observation indicates that photosystem II is operable in these mutant strains, and that they possess lesions on the oxidizing, water-splitting side of photosystem II. In this respect they resemble TRIS-washed chloroplasts in which electron transport to NADP is lost but is restorable upon the addition of an electron donor to photosystem II (Yamashita & Butler 1968).

A spectroscopic analysis of these mutant strains (Epel *et al.* 1972) has revealed that they lack *half* of the high potential, ascorbate-reducible cytochrome $b_{559}$ possessed by the wild-type strain. Other known components do not appear to be affected and the fine structural appearance of the chloroplast membranes is similar to wild type, at least in the single mutant strain examined to date (U.W. Goodenough unpubl). Thus, in contrast to the mutant strains described in the preceding section, only one component appears to be affected.

It thus appears at this point that cytochrome $b_{559}$ may act on both the oxidizing and the reducing sides of photosystem II. Perhaps the situation is relatively simple, with half of the cytochrome acting on each side. However, the genetics of the situation does not help to provide a simple picture. The mutant strains having lesions on the oxidizing side of photosystem II might be explained by assuming that their cytochrome deficiency is the consequence of either a structural alteration in the chloroplast membranes or a loss in the capacity to synthesize the cytochrome unique to the oxidizing side of photosystem II. Whether the cytochrome $b_{559}$ postulated to act on the oxidizing side is chemically identical to the cytochrome acting on the reducing side and whether the genetic specification of these cytochromes is identical is unknown at this time.

# 4 A MUTATION AFFECTING PHOTOSYNTHETIC PHOSPHORYLATION

It should be clear from the foregoing sections that it may be difficult to distinguish the structural from the functional attributes of chloroplast membranes. Indeed, such distinctions may not prove to be very meaningful in view of the intimacy between structure and function that characterizes biological membranes in general. This point of view is reinforced by studies of F-54, a mutant strain of *C. reinhardi* that is incapable of both cyclic and non-cyclic photosynthetic phosphorylation (Sato *et al.* 1971). As in the preceding examples, however, the properties of the mutant strain provide information regarding the mechanism of photosynthesis even though the mutant phenotype is complex.

Cells of F-54 carry a single gene mutation. They possess the capacity for photosynthetic electron transport but not photosynthetic phosphorylation. The general effect of the mutation has been localized more precisely by studies of light-induced pH changes or proton movement. These changes allow one to examine early reactions leading to ATP synthesis that are considered to be closely related to the generation of a high-energy intermediate (Jagendorf & Neumann 1965, Jagendorf & Hind 1963, 1965, Neumann & Jagendorf 1964, Jagendorf & Uribe 1966, McCarty & Racker 1966). Two distinct light-induced pH changes can be measured using chloroplast fragments of *C. reinhardi*. One is associated with cyclic electron flow around photosystem I and is mediated by phenazine methosulphate (PMS); the other is associated with non-cyclic electron flow from photosystem II, in which *p*-benzoquinone is an electron acceptor. Since both types of pH change are catalysed by the mutant strain at rates and extents of proton movement that are comparable to those observed in the wild type, it appears that the mutation affects the terminal stages of phosphorylation, past the proposed site at which proton movement takes place.

The affected step in F-54 appears to centre on the $Ca^{2+}$-dependent chloroplast ATPase which, in intact chloroplasts, is involved in ATP synthesis (McCarty & Racker 1966). The enzyme is present and active in the mutant strain but, in contrast to the enzyme of the wild-type strain and of higher plant chloroplasts, it is not latent, meaning that hydrolytic activity does not require a heat activation step at 58°C. The mutant and wild-type enzyme activities are, however, strongly similar with respect to pH optimum, dependence on $Ca^{2+}$ with an optimum at 7·5mM, oligomycin insensitivity, and inhibition by ADP. These observations suggest that the non-latent mutant enzyme is the analogue of the latent wild-type enzyme. Furthermore, evidence has been obtained (Sato *et al.* 1971) to show that the non-latent enzyme does not bring about the rapid hydrolysis of any of the ATP that might be synthesized by the chloroplast. Thus, it would seem that the inability of F-54 to synthesize ATP and the loss of latency of the ATPase are both functional expressions of a structural change in the ATP synthesizing enzyme.

Genetic analysis has shown that the mutation behaves in a Mendelian fashion as in a single gene alteration (E. Cosbey unpubl.), so that the alteration would be expected to affect a single enzyme. Furthermore, since the coupling enzymes involved at all sites of ATP synthesis are affected in F-54, it appears that the enzymes are either identical or similar enough to be equally affected by the same genetic lesion. Hence, if there is a common coupling enzyme for all sites of ATP synthesis, then the substrates involved in the transphosphorylation sequence of ADP to ATP are identical at all sites. In addition, if the final reaction in the sequence is represented as ADP + X P $\rightleftharpoons$ ATP, it is probable that the high energy intermediate, X~P, is the same at each coupling site. Accordingly, this leads to the conclusion that X is not a component of the electron transport chain. Therefore, the properties of F-54 lend strong positive evidence to a mechanism of phosphorylation (Boyer 1968) in which a common high energy phosphorylated intermediate is generated via non-phosphorylated, site-specified intermediates.

To these details of the phosphorylation process that are revealed by F-54 must be added the observations regarding its membrane structure. Electron microscopy studies of the chloroplast membranes from wild type and F-54 show that the mutant strain lacks the 10 nm particles on the surface of the chloroplast membranes believed to be associated with both coupling factor and ATPase activities (Howell & Moudrianakis 1967). The relationship between the altered $Ca^{2+}$-ATPase and the structural change on the surface of the chloroplast membranes remains to be clarified.

## 5 A STRUCTURE-FUNCTION SYNDROME

The manifold effects that a given mutation can have on chloroplast structure and function are perhaps best illustrated by the ac-20 locus where a single gene mutation in C. reinhardi has led to a morphological and functional syndrome (Goodenough & Levine 1970, 1971, Togasaki & Levine 1970, Levine & Paszewski 1970, Levine & Armstrong 1972). The most fundamental aspect of the mutation known so far appears to be its effect on the formation of chloroplast ribosomes. The low level of chloroplast ribosomes that characterizes the cells of ac-20 is in turn responsible for abnormally arranged chloroplast membranes, a rudimentary pyrenoid, an inactive photosystem II because of a deficiency of cyt $b_{559}$ and C-550, a cyt $c_{553}$ deficiency, and a deficiency of the carbon cycle enzyme ribulose-1,5-diphosphate carboxylase. The syndrome of effects in the ac-20 case is relatively simple to understand in terms of a set of features of the photosynthetic apparatus that depend on translational steps that occur on chloroplast ribosomes.

When cells of ac-20 are cultured mixotrophically (in the light in acetate-supplemented medium), they possess levels of chloroplast ribosomes that are 5 to 10% of those of wild type. On the other hand, when cells of ac-20 are cultured phototrophically in minimal medium, the number of chloroplast ribosomes increases to about 40% of the wild-type level. This increase requires at least 12 hours if the transferred cells are maintained in minimal medium in the dark; however, when the transfer is carried out in the light, the increase takes place within a few hours (Goodenough & Levine 1970).

The in vivo photosynthetic capacity of ac-20 cultured in acetate-supplemented medium is, at best, 4% of wild type, whereas for phototrophically growing cells it is at least 25%. Cells transferred from acetate-supplemented to minimal medium increase their photosynthetic capacity and reform a normal pyrenoid and normally-organized chloroplast membranes only after the chloroplast ribosome increase has occurred, and then only if the cells are placed in the light. For example, the number of chloroplast ribosomes can be increased in the dark, as described above, but there is no concomitant increase in photosynthetic capacity; when such cells are placed in the light, photosynthetic rates rise rapidly and reach a steady state within a few hours. If, however, cells are placed in the light immediately after transfer, there is a lag in the onset of increased photosynthetic capacity which corresponds in duration to the time taken for the number of chloroplast ribosomes to increase (Goodenough & Levine 1970).

Clearly, the photosynthetic capacity of these cells seems to be correlated with the level of chloroplast ribosomes present, and it has been proposed (Good-enough & Levine 1970, 1971) that the photosynthetic capacity is dependent upon translational steps that take place on these ribosomes.

Cycloheximide and spectinomycin inhibit respectively translation on cyto-plasmic (Ellis 1970) and chloroplast (Burton 1972) ribosomes. When mixo-trophically-grown cells of *ac-20* are transferred to phototrophic growth condi-tions and allowed to attain the phototrophic level of chloroplast ribosomes and then exposed to either spectinomycin or cycloheximide, it is found (Goodenough & Levine 1971) that the recovery of a normal chloroplast and pyrenoid as well as the recovery of photosynthetic capacity is inhibited by spectinomycin but not by cycloheximide.

The properties of *ac-20* show that the mutation of a single gene can have profound effects on chloroplast structure and function and that except in a case such as described here it may be difficult to identify the nature of the primary genetic effect.

## 6  MUTATIONS AFFECTING CHLOROPLAST MEMBRANE STRUCTURE

Various particles that are associated with chloroplast membranes have been observed in electron micrographs of shadowed, negatively-stained, and freeze-etched preparations. These particles have at various times been said to represent some sort of photosynthetic unit and have been termed quantosomes (Park & Biggins 1964, Park & Branton 1966, Park & Pfeifhofer 1969). Several studies (Howell & Moudrianakis 1967) have indicated that the particles seen on the surface of negatively-stained membranes correspond to the $Ca^{2+}$-ATPase. Thus the internal particles have remained as possible structural representatives of a photosynthetic unit.

Most biological membranes when viewed by freeze-etch electron microscopy exhibit within their interior, particles with diameters of 10 to 12nm (Branton 1969). Chloroplast membranes are unique in possessing closely packed arrays of large particles having a mean diameter of 17·5nm (Park & Branton 1966, Park & Pfeifhofer 1969, Arntzen *et al.* 1969, Goodenough & Staehelin 1971), and it has been suggested that these particles represent either an entire photo-synthetic unit (Park & Biggins 1964) or, more recently, photosystem II (Arntzen *et al.* 1969). Studies of two mutant strains of *C. reinhardi*, *ac-5* and *ac-31*, have shown (Goodenough & Staehelin 1971) that membranes without the closely packed arrays of large particles have normal photosynthetic electron transport chain activity. On the other hand, the presence of this morphology can be cor-related with the occurrence of membrane fusion (stacking, grana formation) between chloroplast membranes (Goodenough & Staehelin 1971).

With mutant strains such as *ac-5*, *ac-31*, and others of *C. reinhardi* in which chloroplast membrane structure and/or function is altered, one can clearly begin to probe the nature of the chloroplast membrane in molecular terms. Current research in this laboratory has begun such a study by comparing the membrane polypeptides and lipids from the wild-type and the mutant strains. For example, the chloroplast of *ac-5* does not possess stacked membranes when cells are grown under mixotrophic conditions nor does the chloroplast have normal levels of at least three major chloroplast membrane polypeptides. However, all three polypeptides are present in near-normal amounts in the membranes when cells of this mutant strain are grown phototrophically, and such membranes are capable of stacking. It is through this kind of analysis that some of the functional roles of membrane components may be understood.

# 7 REFERENCES

ARNTZEN C.J., DILLEY R.A. & CRANE F.L. (1969) A comparison of chloroplast membrane surfaces visualized by freeze-etch and negative staining techniques; and ultrastructural characterization of membrane fractions obtained from digitonin-treated spinach chloroplasts. *J. Cell Biol.* **43**, 16–31.

BENDALL D.S. & SOFROVA D. (1971) Reactions at 77°K in photosystem II of green plants. *Biochim. biophys. Acta*, **234**, 371–80.

BISHOP N.I. (1964) Mutations of unicellular green algae and their application to studies on the mechanism of photosynthesis. *Rec. chem. Prog.* **25**, 181–95.

BOYER P. (1968) Oxidative phosphorylation. In *Biological Oxidations*, ed. Singer T.P. pp. 193–235. John Wiley, New York.

BRANTON D. (1969) Membrane structure. *A. Rev. Pl. Physiol.* **20**, 209–38.

BURTON W.G. (1972) Dihydrospectinomycin binding to chloroplast ribosomes from antibiotic-sensitive and resistant strains of *Chlamydomonas reinhardi*. *Biochim. biophys. Acta*, **272**, 305–11.

ELLIS R.J. (1970) Further similarities between chloroplast and bacterial ribosomes. *Planta*, **91**, 329–35.

EPEL B.L. & BUTLER W.L. (1971) A spectroscopic analysis of a high fluorescent mutant of *Chlamydomonas reinhardi*. *Biophys. J.* **12**, 922–9.

EPEL B.L., BUTLER W.L. & LEVINE R.P. (1972) A spectroscopic analysis of low fluorescent mutants of *Chlamydomonas reinhardi* blocked in their water-splitting oxygen evolving apparatus. *Biochim. biophys. Acta*, **275**, 395–400.

EPEL B.L. & LEVINE R.P. (1971) Mutant strains of *Chlamydomonas reinharai* with lesions on the oxidizing side of photosystem II. *Biochim. biophys. Acta*, **226**, 154–70.

ERIXON K., LOZIER R. & BUTLER W.L. (1972) The relationship between Q, C-550, and cytochrome $b_{559}$ in photoreactions at —196° in chloroplasts. *Biochim. biophys. Acta*, **234**, 387–9.

GOODENOUGH U.W. & LEVINE R.P. (1969) Chloroplast ultrastructure in mutant strains of *Chlamydomonas reinhardi* lacking components of the photosynthetic apparatus. *Pl. Physiol., Lancaster*, **44**, 990–1000.

GOODENOUGH U.W. & LEVINE R.P. (1970) Chloroplast structure and function in *ac-20*, a mutant strain of *Chlamydomonas reinhardi*: Chloroplast ribosomes and membrane organization. *J. Cell Biol.* **44**, 547–62.

GOODENOUGH U.W. & LEVINE R.P. (1971) The effects of inhibitors of RNA and protein synthesis on the recovery of chloroplast ribosomes, membrane organization, and photosynthetic electron transport in the *ac-20* strain of *Chlamydomonas reinhardi*. *J. Cell Biol.* **50**, 50–62.

432 CHAPTER 14

GOODENOUGH U.W. & STAEHELIN L.A. (1971) Structural differentiation of stacked and un-stacked chloroplast membranes. *J. Cell Biol.* **48**, 594–619.

HILL R. & BENDALL F. (1960) Function of the two cytochrome components in chloroplasts: A working hypothesis. *Nature, Lond.* **186**, 136–7.

HIND G. & NAKATANI H.Y. (1970) Determination of the concentration and the redox potential of chloroplast cytochrome 559. *Biochim. biophys. Acta*, **216**, 223–5.

HOWELL S.H. & MOUDRIANAKIS E.N. (1967) Function of the 'quantosome' in photosynthesis: Structure and properties of membrane-bound particle active in the dark reactions of photophosphorylation. *Proc. natn. Acad. Sci., U.S.A.* **58**, 1261–8.

JAGENDORF A.T. & HIND G. (1963) Studies on the mechanism of photophosphorylation. In *Photosynthetic Mechanisms of Green Plants*, ed. Kok. B. & Jagendorf A.T. pp. 599–609. N.A.S.-N.R.C. Washington DC, **1145**.

JAGENDORF A.T. & HIND G. (1965) Relation between electron flow, phosphorylation, and a high energy state of chloroplasts. *Biochem. biophys. Res. Commun.* **18**, 702–9.

JAGENDORF A.T. & NEUMANN J. (1965) Effect of uncouplers on the light-induced pH rise with spinach chloroplasts. *J. Biol. Chem.* **240**, 3210–14.

JAGENDORF A.T. & URIBE E. (1966) ATP formation caused by acid-base transition of spinach chloroplasts. *Proc. natn. Acad. Sci., U.S.A.* **55**, 170–7.

LAVOREL J. & LEVINE R.P. (1968) Fluorescence properties of wild-type *Chlamydomonas reinhardi* and three mutant strains having impaired photosynthesis. *Pl. Physiol., Lancaster*, **43**, 1049–55.

LEVINE R.P. (1969) The analysis of photosynthesis using mutant strains of algae and higher plants. *A. Rev. Pl. Physiol.* **20**, 523–40.

LEVINE R.P. & ARMSTRONG J. (1972) The site of synthesis of two chloroplast cytochromes in *Chlamydomonas reinhardi*. *Pl. Physiol., Lancaster*, **49**, 661–2.

LEVINE R.P., BURTON W.G. & DURAM H.A. (1972) Membrane polypeptides associated with photochemical systems. *Nature, Lond.* **237**, 176–7.

LEVINE R.P. & GOODENOUGH U.W. (1970) The genetics of photosynthesis and of the chloroplast in *Chlamydomonas reinhardi*. *A. Rev. Genetics*, **4**, 397–408.

LEVINE R.P. & GORMAN D.S. (1966) Photosynthetic electron transport chain of *Chlamydomonas reinhardi*. III. Light-induced absorbance changes in chloroplast fragments of the wild type and mutant strains. *Pl. Physiol., Lancaster*, **41**, 1293–300.

LEVINE R.P. & PASZEWSKI A. (1970) Chloroplast structure and function in *ac-20*, a mutant strain of *Chlamydomonas reinhardi*: $CO_2$ photosynthetic electron transport. *J. Cell Biol.* **44**, 540–6.

LEVINE R.P. & SMILLIE R.M. (1963) The photosynthetic electron transport chain of *Chlamydomonas reinhardi*. I. Triphosphopyridine nucleotide photoreduction in wild-type and mutant strains. *J. biol. Chem.* **238**, 4052–7.

MCCARTY R.E. & RACKER E. (1966) Effect of a coupling factor and its antiserum on photophosphorylation and hydrogen ion transport. *Brookhaven Symposium*, **19**, 202–14.

NEUMANN J. & JAGENDORF A.T. (1964) Light-induced pH changes related to phosphorylation by chloroplasts. *Archs. Biochem. Biophys.* **107**, 109–19.

OKAYAMA S., EPEL B.L., ERIXON K., LOZIER R. & BUTLER W.L. (1971) The effects of lipase on spinach and *Chlamydomonas* chloroplasts. *Biochim. biophys. Acta*, **253**, 476–84.

PARK R.B. & BIGGINS J. (1964) Quantosome: Size and composition. *Science, N.Y.* **144**, 1009–111.

PARK R.B. & BRANTON D. (1966) Freeze-etching of chloroplasts from glutaraldehyde-fixed leaves. *Brookhaven Symposium*, **19**, 341–51.

PARK R.B. & PFEIFHOFER A.O. (1969) Ultrastructural observations on deep-etched thylakoids. *J. Cell. Sci.* **5**, 299–311.

SATO V.L., LEVINE R.P. & NEUMANN J. (1971) Photosynthetic phosphorylation in *Chlamy-*

*domonas reinhardi*: Effects of a mutation altering an ATP-synthesizing enzyme. *Biochim. biophys. Acta*, **253**, 437–48.

TOGASAKI R. & LEVINE R.P. (1970) Chloroplast structure and function of *ac-20*, a mutant strain of *Chlamydomonas reinhardi*: $CO_2$ fixation and ribulose-1,5-diphosphate carboxylase synthesis. *J. Cell. Biol.* **44**, 531–9.

WADA K. & ARNON D.I. (1971) Three forms of cytochrome $b_{559}$ and their relation to the photosynthetic activity of chloroplasts. *Proc. natn. Acad. Sci.*, *U.S.A.* **68**, 3064–8.

YAMASHITA T. & BUTLER W.L. (1968) Photoreduction and photophosphorylation with TRIS-washed chloroplasts. *Pl. Physiol.*, *Lancaster*, **43**, 1978–86.

# CHAPTER 15

# CARBON DIOXIDE FIXATION

## J. A. RAVEN

Department of Biological Sciences,
University of Dundee, Dundee, U.K.

1 Introduction 434

2 Sources of inorganic carbon for algae 435

3 Autotrophic $CO_2$ fixation 435
3.1 Introduction 435
3.2 Sequence of labelling after feeding $^{14}CO_2$ 437
3.3 Intramolecular labelling pattern in products 438
3.4 Activities of enzymes of the photosynthetic carbon reduction cycle 438
3.5 Location of the photosynthetic carbon reduction cycle in the algal cell 439
3.6 Energetics 439
3.7 Regulation of the rate of autotrophic $CO_2$ fixation 440

3.8 Regulation of the products of the photosynthetic carbon reduction cycle 441

4 Heterotrophic $CO_2$ fixation 444
4.1 Introduction 444
4.2 β-carboxylation 444
4.3 Carbamoyl phosphate synthesis 446
4.4 Other dark $CO_2$ fixation reactions 447

5 Catalytic roles of $CO_2$ 448

6 Summary 448

7 References 449

## 1 INTRODUCTION

Carbon dioxide is the source of cell carbon during the characteristic photo-autotrophic growth of algae. In photo-autotrophic growth, light energy is converted into the chemical energy of ATP and $NADPH_2$, most of which is used to convert $CO_2$ into reduced carbon compounds (see Chapter 13, p. 391). In a number of algae (obligate photo-autotrophs: see Chapter 19, p. 549), $CO_2$ is the only carbon compound which can support growth. In several algae in which the 'normal' mode of nutrition is photoheterotrophic, good growth on $CO_2$ as the sole carbon source can be observed under suitable conditions (Wiessner 1969). Even when the main source of cell carbon is not $CO_2$ (heterotrophic growth),

$CO_2$ fixation reactions are still important as the 'dark' fixation reactions essential for the operation of many biosynthetic pathways. In both autotrophic and heterotrophic growth, $CO_2$ can also play a catalytic role.

Since the review by Holm-Hansen (1962), a number of review articles have appeared which deal with $CO_2$ fixation in algae, either exclusively or in a wider context (see e.g. Bassham 1964, 1971, Gibbs *et al.* 1970, Levedahl 1968, Raven 1970a).

## 2 SOURCES OF INORGANIC CARBON

Most algae are submerged aquatic organisms, and so are exposed to $H_2CO_3$ and its ions, $HCO_3^-$ and $CO_3^{2-}$, as well as the unhydrated $CO_2$ available to terrestial plants. Water in equilibrium with air at 15°C contains about 10 μM $CO_2$ dissolved in it; the amount of $HCO_3^-$ and $CO_3^{2-}$ increases with pH.

All algae seem to be able to take up free $CO_2$, which readily diffuses across cell membranes. Not all algae can take up $HCO_3^-$ (Guyomarch 1970, Joliffe & Tregunna 1970, Raven 1970a). Uptake of $HCO_3^-$ at rates sufficient to support net photosynthesis requires active transport of $HCO_3^-$ (Raven 1970a, see Chapter 25, p. 676). It is possible that the rate at which inorganic carbon enters the algal cell can limit the rate of photosynthesis under natural conditions (Raven 1970a, Schindler 1971).

Inorganic carbon is also produced inside algal cells by decarboxylation reactions in the light as well as in the dark. Some of this is used in carboxylation reactions before it can escape to the medium (reassimilation, Raven 1972a,b).

Many carboxylases use unhydrated $CO_2$; these include the ribulose diphosphate carboxylase of the photosynthetic carbon reduction cycle (see below). The biotin-dependent carboxylases and probably phosphoenolpyruvate carboxylase (see 4 below) on the other hand use $HCO_3^-$ (see Raven 1970a, Cooper & Wood 1971). Thus when $HCO_3^-$ is the form of inorganic carbon which enters the algal cell, it must be converted to $CO_2$ before it can be used in photosynthesis. This dehydration is slow if uncatalysed, and photosynthetic cells contain the enzyme carbonic anhydrase which catalyses the reversible hydration of $CO_2$. There is evidence that this enzyme is specifically involved in the use of $HCO^-_3$ in photosynthesis (Raven 1970a, Loeblich 1970). Carbonic anhydrase may also be involved in inorganic carbon transport in the cytoplasm, and in preparatory reactions of inorganic carbon prior to the action of ribulose diphosphate carboxylase (Raven 1970a, Graham *et al.* 1971, see also Section 5 below).

## 3 AUTOTROPHIC $CO_2$ FIXATION

### 3.1 *Introduction*

Most of the available evidence suggests that the pathway of autotrophic $CO_2$ fixation in algae is the photosynthetic carbon reduction cycle, or Calvin cycle

(Fig. 15.1). Much of the experimental evidence for this pathway was obtained by Calvin and his co-workers using unicellular green algae. The advantages of these organisms for tracer kinetic experiments on photosynthesis are discussed by Holm-Hansen (1962).

Other pathways of $CO_2$ fixation, such as the four carbon acid cycle which apparently acts as a $CO_2$-concentrating mechanism for the photosynthetic carbon reduction cycle in certain angiosperms (Hatch & Slack 1970), and the reductive tricarboxylic acid cycle found in certain photosynthetic bacteria (Buchanan & Arnon 1970) have not, as yet, been found in algae.

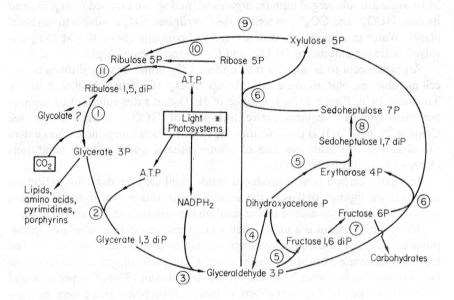

**Fig. 15.1.** The photosynthetic carbon reduction cycle. The enzymes of the cycle, referred to by numbers, are: (1) Ribulose diphosphate carboxylase (regulated enzyme), (2) 3-phosphoglycerate kinase, (3) 3-phosphoglyceraldehyde dehydrogenase, NADP-linked (regulated enzyme), (4) Phosphotriose isomerase, (5) Fructose diphosphate aldolase, (6) Transketolase, (7) Fructose-1,6-diphosphate-1-phosphatase (regulated enzyme), (8) Sedoheptulose-1,7-diphosphate-1-phosphatase (regulated enzyme), (9) Phosphopentose epimerase, (10) Phosphoribose isomerase, (11) Phosphoribulokinase (regulated enzyme).   * = catalytic role of $CO_2$.

The evidence for the occurrence of the cycle in a variety of algae is discussed below. Norris *et al.* (1955) showed that, in fairly long-term (5 min) experiments, a range of algae incorporated label from $^{14}CO_2$ into many of the intermediates shown in Fig. 15.1. This is consistent with the operation of the cycle, but does not rule out the alternatives mentioned in the last paragraph. Similarly, the occurrence of the enzymes of the cycle is not conclusive evidence that the cycle occurs *in vivo*, although the correlations between enzyme activities and photosynthetic rate strongly support involvement of the cycle in autotrophic $CO_2$ fixation (see below). The experiments *in vivo* with $^{14}CO_2$, showing the time sequence of labelling and the intramolecular distribution of label afford the most direct evidence for the widespread occurrence of the cycle in algae.

## 3.2 Sequence of labelling after feeding $^{14}CO_2$

The first stable product of $^{14}CO_2$ fixation in the photosynthetic carbon reduction cycle (Fig. 15.1) is 3-phosphoglycerate. The occurrence of the cycle is inferred if 3-phosphoglycerate is the major product of short-term (a few seconds) $^{14}CO_2$ fixation, and the fraction of the total label present in 3-phosphoglycerate decreases with increasing time of labelling (Bassham 1964). In order to see if there are any intermediates prior to 3-phosphoglycerate, the fixation time should be kept as short as possible.

Experiments with algae which suggest that 3-phosphoglycerate is the initial product of $CO_2$ fixation are discussed by Raven (1970a). Later experiments at similar short times (less than eight seconds; Codd & Merrett 1971a, Kluge quoted by Luttge *et al.* 1971, Pelroy & Bassham 1972) or at rather longer times (15 seconds; Coombs & Volcani 1968, Patil & Joshi 1970, Kremer & Willenbrink 1972; or 10 seconds, Wegman 1969, Gingras 1966; or 35 seconds, Bidwell *et al.* 1970) establish the photosynthetic carbon reduction cycle as the main $CO_2$ fixation pathway under conditions allowing photo-autotrophic growth.

Longer-term experiments are consistent with the photosynthetic carbon reduction cycle being the major $CO_2$ fixation pathway in both red and blue light during a 30 second exposure (Ogasawara & Miyachi 1970a) and in $CO_2$ fixation with $H_2$ as electron donor in the dark (oxyhydrogen reaction) and in the light (photoreduction, with or without a functional photosystem II, see Chapter 16, p. 456, Gingras 1966).

The photosynthetic carbon reduction cycle is quantitatively less important in $CO_2$ fixation under certain conditions in the light which do not allow photosynthetic growth. Thus at low light intensities which do not allow net photosynthesis, four carbon dicarboxylic acids and citrulline may be the major products of $^{14}CO_2$ fixation in *Chlorella* (Ogasawara & Miyachi 1970a). Transfer of *Chlorella* grown in high $CO_2$ to a low $CO_2$ level (Graham & Whittingham 1968) involves an adaptation period in which the rate of $CO_2$ fixation is low and the first product of photosynthesis, judged from a fairly long (30 seconds) exposure to $^{14}CO_2$, is malate. After this lag, which is correlated with synthesis of carbonic anhydrase (Graham *et al.* 1971), the photosynthetic carbon reduction cycle is re-established. There is no evidence that the carbon fixed into malate in this lag period is on the pathway to carbohydrate, as is found in the four carbon acid pathway found in certain angiosperms (Hatch & Slack 1970). In both of these sets of experiments the fairly long exposure time (30 seconds) makes the conclusions drawn somewhat equivocal.

P

### 3.3   *Intramolecular labelling pattern in products*

Degradation of various sugar phosphates labelled after short-term photo-synthesis in $^{14}CO_2$, and determination of the specific activity of $^{14}C$ in their various carbon atoms, was an important technique in establishing the nature of the cycle which regenerated the $CO_2$ acceptor in the photosynthetic carbon reduction cycle (Holm-Hansen 1962, Bassham 1964). It has been used in implicating the photosynthetic carbon reduction cycle in photosynthesis in *Anacystis* (Kindel & Gibbs 1963) and in *Eisenia* (Yamaguchi *et al.* 1969), and in photoreduction and the oxyhydrogen reaction in *Scenedesmus* (Russell & Gibbs 1968, see Chapter 16, p. 456).

These experiments showed that the label in hexose derived from storage carbohydrate (starch or mannitol) was mainly in $C_3$ and $C_4$, with little in the outer pairs of carbon atoms. It must be noted, however, that this technique does not distinguish the ordinary photosynthetic carbon reduction cycle from the variant with the four carbon acid cycle added to it, and, unless each carbon of the hexose is analysed (as was done by Kindel & Gibbs 1963, Russell & Gibbs 1968, but not by Yamaguchi *et al.* 1969), confusion with the reductive tricarboxylic acid cycle followed by gluconeogenesis from phosphoenolpyruvate could occur (see Buchanan & Arnon 1970, and Section 4.4 below).

### 3.4   *Activities of enzymes of the photosynthetic carbon reduction cycle*

Jacobi (1962) reviews early attempts to find enzymes of the photosynthetic carbon reduction cycle in algae; in many of the algae tested, no activity of one or more enzymes of the cycle (see Fig. 15.1) was found. Although negative results with enzymes are not conclusive, the state of research reviewed by Jacobi (1962) did suggest that the cycle might not be functional in the form shown in Fig. 15.1.

Subsequent work showed that all of the enzymes of the cycle shown in Fig. 15.1 are present in *Chlorella*, *Euglena* and *Tolypothrix* (Petterkoffsky & Racker 1961, Smillie 1968, Latzko & Gibbs 1969). Less complete data for other genera suggests that these enzymes are found in all algal classes, although not every attempt to demonstrate a particular enzyme in a particular alga has been successful (e.g. Ziegler & Ziegler 1967, Yamaguchi *et al.* 1969, Hellebust *et al.* 1967, Antia 1967, Kawashima & Wildman 1970, Moll & Levine 1970, Tan *et al.* 1971).

Thus in qualitative terms the enzyme data support the presence of the photosynthetic carbon reduction cycle in algae. Quantitatively, it is often (Petterkoffsky & Racker 1961, Latzko & Gibbs 1969) but not invariably (Hellebust *et al.* 1967) found that the *in vitro* activity of certain enzymes of the cycle, assayed under optimal conditions, is inadequate to account for the observed rate of photosynthesis *in vivo*. Most commonly it is ribulose diphosphate carboxylase and the two phosphatases (fructose-1,6-diphosphate-1-phosphatase and sedoheptulose-1,7-diphosphate-1-phosphatase) which appear to be inadequate. Bassham (1971) points out that these enzymes catalyse irreversible reactions, whose control is essential for regulation of the photosynthetic carbon reduction cycle (see 3.7 below). Perhaps the low activity *in vitro* reflects the loss of some *in vivo* activation mechanism.

Other enzymic evidence is consistent with the photosynthetic carbon reduction cycle being the major pathway of autotrophic $CO_2$ fixation in algae. One line of evidence is the parallelism of changes in rates of photosynthesis and of activity of certain enzymes of the cycle when environmental conditions are changed (see e.g. Russell & Gibbs 1966, Huth 1967, Latzko & Gibbs 1969). A further correlation is found in the mutant F-60 of *Chlamydomonas reinhardtii*. This mutant lacks the ability to fix $CO_2$ in photosynthesis, and the only functional deficiency detected (NADPH$_2$ and ATP synthesis, and eight enzymes of the carbon reduction cycle were tested) was phosphoribulokinase.

### 3.5 *Location of the photosynthetic carbon reduction cycle in the algal cell*

Two main lines of evidence have been used in determining where in the eukaryotic algal cell the photosynthetic carbon reduction cycle occurs. One sort of evidence comes from the metabolic capabilities of isolated chloroplasts. Jeffrey *et al.* (1966) found that isolated chloroplasts of *Hymenomonas* required the addition of a soluble cell extract before they would photosynthesize (fix $CO_2$) at an appreciable fraction of the *in vivo* rate. Later workers, using siphonaceous green algae and *Euglena* have obtained rates of $CO_2$ fixation in isolated chloroplasts in the range 50 to 100% of the *in vivo* rate, expressed on a chlorophyll basis (Bidwell *et al.* 1970, Giles & Sanofis 1972, Forsee & Kahn 1972). This suggests that the carbon reduction cycle is confined to the chloroplasts of eukaryotic algae, and is supported by the observations of Trench and Smith (1970) on the ability of *Codium* chloroplasts to fix $CO_2$ after ingestion by molluscs.

Another line of evidence comes from determinations of the intracellular distribution of the enzymes of the carbon reduction cycle. Most of these data (Smillie 1968, Cremona 1968, Vasconcelos *et al.* 1971) but not all of them (Guerrini *et al.* 1971) suggest that all the enzymes required by the cycle are found in chloroplasts, and that the enzymes specific to the cycle are only found in the chloroplasts.

### 3.6 *Energetics*

As shown in Fig. 15.1, the photosynthetic carbon reduction cycle needs three moles of ATP and two moles of NADPH$_2$ for the reduction of one mole of $CO_2$ to the level of carbohydrate. The synthesis of a wider range of cell materials would probably increase both the ATP and the NADPH$_2$ requirement (Bassham 1971, Raven 1971c).

The low rate of phosphate turnover in *Chlorella* during photosynthesis, and the effects of certain inhibitors (Kandler & Liesenkotter 1961) led to the suggestion that less than 3ATP are needed per $CO_2$ reduced (Kandler 1960). However, no further evidence for this view has been produced (Raven 1970b) and more recent experiments on algae suggest that the turnover of phosphate is adequate to support $CO_2$ fixation with the stoichiometry of Fig. 15.1 (Urbach & Gimmler 1970).

Thus the high-energy phosphate turnover is probably adequate for the photosynthetic carbon fixation cycle. Some evidence suggests that this might be used as a phosphorylated high-energy intermediate of photophosphorylation rather than as ATP itself (see Raven 1971a).

There is evidence that both the ATP and the $NADPH_2$ required for the photosynthetic carbon reduction cycle come from non-cyclic photophosphorylation, without the need for extra supplies from cyclic or pseudocyclic photophosphorylation, or from oxidative phosphorylation (Simonis 1967, Tanner *et al.* 1969, Raven 1970b, 1971b). This implies that the $P/e_2^-$ ratio in non-cyclic photophosphorylation in algae is greater than $1 \cdot 0$. This, in fact, has been shown in *Chara* (Smith & West 1969).

Different energy sources are presumably used during $CO_2$ fixation in hydrogenase-containing algae in the presence of $H_2$ (see Kessler, Chapter 16, p. 458, and 3.2 and 3.3 above, for evidence that the photosynthetic carbon reduction cycle is then involved in $CO_2$ fixation). Photoreduction may involve the generation of $NADPH_2$ by a direct interaction of $H_2$ via hydrogenase with ferredoxin; or photosystem I, or both photosystems, may be involved. Tanner *et al.* (1969) have presented evidence that cyclic photophosphorylation may be an important source of ATP for photoreduction of $CO_2$ in *Ankistrodesmus*, even though their experimental conditions did not rule out the participation of photosystem II in electron transfer from $H_2$ to NADP, i.e. non-cyclic photophosphorylation could have occurred.

The oxyhydrogen reaction presumably uses $NADPH_2$ generated from $H_2$ interacting with ferredoxin via hydrogenase. The ATP for this reaction probably comes from oxidative phosphorylation (Gingras *et al.* 1963). Raven (1972a) reviews evidence that extraplastidic ATP could enter the chloroplast at a sufficient rate to support the $CO_2$ fixation observed during the oxyhydrogen reaction.

Evidence discussed below (4.3) suggests that, in some algae, the photosynthetic carbon reduction cycle can occur slowly in the dark even in the absence of $H_2$; here both $NADPH_2$ and ATP must be supplied from respiratory sources.

### 3.7 Regulation of the rate of autotrophic $CO_2$ fixation

The rate at which an algal cell fixes $CO_2$ in photosynthesis is conditioned by a number of external and internal factors and the interaction between them. Internal factors include the algal species (Brown & Tregunna 1967, McAllister *et al.* 1964); the stage in the cell cycle for rapidly dividing cells (Tamiya 1966, Pirson & Lorenzen 1966), and circadian rhythms for slowly dividing cells (Cumming & Wagner 1968, Hellebust *et al.* 1967).

External factors controlling the rate of $CO_2$ fixation include the inorganic carbon supply and pH (e.g. Brown & Tregunna 1967, Raven 1970a); the intensity and wavelength of light (see e.g. Paasche 1966, Raven 1969); the oxygen level (Turner & Brittain 1962), and the organic and inorganic nutrient supply (Russell & Gibbs 1966, Kanazawa *et al.* 1970, 1970a,b).

The way in which these factors can influence the rate at which $CO_2$ is fixed include: first, the control of the *amount* of enzyme present (Smillie 1968, Russell & Gibbs 1966); second, the control of the *activity* of the enzyme that is present (Preiss & Kosuge 1970, Bassham 1971); and third, control of the supply of substrates $CO_2$, ATP and $NADPH_2$ (Bassham 1971, Raven 1970a).

The best understood case of a factor which influences the rate of $CO_2$ fixation is that of light. In many algae the enzymes of the photosynthetic carbon reduction cycle, and of photoproduction of ATP and $NADPH_2$, are absent or reduced in amount if the cells are grown heterotrophically in the dark (Latzko & Gibbs 1969). This limits the rate of $CO_2$ fixation under otherwise optimal conditions, until the necessary enzymes have been synthesized (see 3.1 above).

Light also alters the activity of pre-existing photosynthetic carbon reduction cycle enzymes, both by activation (increasing of the rate of enzyme activity in the presence of a fixed amount of substrate) and by energising the synthesis of the substrates ATP and $NADPH_2$. Light activation of enzymes occurs at those steps which consume ATP and $NADPH_2$, *independently* of the action of light in supplying those substrates; and at irreversible steps (Bassham 1971, Raven 1970a, Preiss & Kosuge 1970). In this way the rate of the photosynthetic carbon reduction cycle can be kept in step with the rate at which $NADPH_2$ and ATP are supplied to it. Light activation of enzymes occurs *via* changes in the cation distribution between stroma and thylakoids, the redox state of NADP and ferredoxin, and possibly the levels of ATP, ADP, AMP and inorganic phosphate.

Thus it is possible to account for regulation of the rate of the photosynthetic carbon reduction cycle in relation to cofactor supply at various light intensities. When the cofactors are being supplied by dark processes (photosynthetic carbon reduction cycle powered by the oxyhydrogen reaction (see 3.2 above), or by dark respiration (see 4.3 below)), it must be assumed that the suppression of cofactor use in the cycle normally found in the dark is incomplete.

Much less is known of the ways in which factors other than light alter the rate of $CO_2$ fixation by activation/inhibition mechanisms (Bassham 1971, Kanazawa *et al.* 1970, 1970a,b).

3.8 *Regulation of the products of the photosynthetic carbon reduction cycle*

Figs. 15.1 and 15.2 show some of the major metabolic pathways leading out of the photosynthetic carbon reduction cycle. The kinetic experiments of Kanazawa *et al.* (1970a) suggest that the products shown in Fig. 15.2 arise mainly from 3-phosphoglycerate produced in the photosynthetic carbon reduction cycle rather than from 3-phosphoglycerate derived from carbon withdrawn from the cycle as hexose. Thus the two main points at which carbon is withdrawn from the cycle are hexose and as 3-phosphoglycerate (for the role of glycolate, see Chapter 17, p. 474). Factors which influence the main point at which carbon is removed from the cycle include the species of alga, the stage in the division cycle, the supply of nutrients, the intensity and wavelength of light, the concentration of $O_2$ and $CO_2$, and the osmotic pressure of the medium.

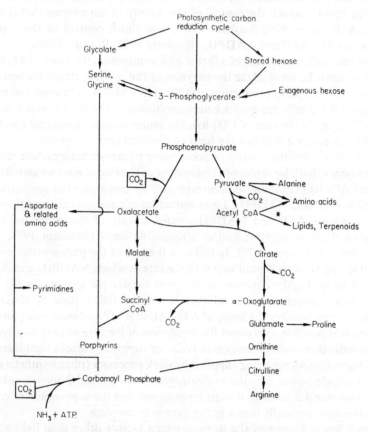

**Fig. 15.2.** Relation of $CO_2$ fixation by the phosphoenolpyruvate carboxylase carbamoyl phosphate synthetase reactions to the synthesis of essential metabolites.
* = catalytic role of $CO_2$.

With respect to the differences between species, a major difference between the composition of algae arises from the storage of lipid rather than carbohydrate as the main carbon and energy reserve. This has a major influence on the point at which carbon is removed from the cycle. Hexose phosphates are used in the synthesis of storage carbohydrates, while 3-phosphoglycerate is used for lipid synthesis. Bassham (1971) discusses the regulation of fructose-1,6-diphosphate-1-phosphatase in relation to carbon loss from the cycle as 3-phosphoglycerate (prior to phosphatase action) or as hexose monophosphate (after phosphatase action). Such regulation is probably important in the case of other factors which alter the main exit point from the cycle (see below).

The composition of algae varies with the stage in the division cycle (Tamiya 1966, Pirson & Lorenzen 1966). These changes are paralleled by changes in the major exit points from the photosynthetic carbon reduction cycle at different stages in the synchronous life cycle of *Chlorella* (Ahmed & Ries 1969, Kanazawa *et al.* 1970, 1970a,b) and *Euglena* (Codd & Merrett 1971a,b).

The nutrient which has been investigated most in relation to its effect on the carbon pathway is inorganic nitrogen. References, and recent work on *Chlorella*, may be found in Kanazawa *et al.* (1970), Kanazawa *et al.* (1970b) and Hiller (1970). The presence of an inorganic nitrogen source increases the carbon flux from the photosynthetic carbon reduction cycle into the tricarboxylic acid cycle and its nitrogenous derivatives, and to a smaller extent into lipids, at the expense of carbon flux into reserve carbohydrate. Thus the presence of a nitrogen source favours the 3-phosphoglycerate exit point.

Light intensity and quality have important effects on the carbon flux out of the photosynthetic carbon reduction cycle. Cook (1963) showed that, in *Euglena*, synchronous growth at light intensities higher than those needed to saturate synthetic rates of other cell components lead to additional synthesis of the reserve polysacharide, paramylon.

Kowallik (1970) reviews the evidence that blue light favours protein and RNA synthesis, while red light favours carbohydrate formation, in algae. This effect is not dependent on photosynthesis, since it is found when photosynthesis is completely blocked (e.g. by DCMU, 3'(3,4 dichlorophenyl) 1',1' dimethyl urea) and biosynthesis is occurring at the expense of an exogenous reduced carbon source. However, when photosynthesis can occur, it is clear that light wavelength does influence the exit point from the cycle (e.g. Ogasawara & Miyachi 1970a,b, Kowallik 1970). Thus the light wavelength effect does not act solely on the metabolic fate of hexose, but can influence whether carbon is withdrawn from the cycle as 3-phosphoglycerate or as hexose. As is discussed below (4.2) blue light also stimulates β-carboxylation, which is required for the biosynthesis of nitrogen-containing compounds.

The blue light effect on photosynthetic products is widespread in algae (Hauschild *et al.* 1962, Ahmed & Ries 1969, Wallen & Geen 1971a,b). Its apparent absence from the red alga *Rhodosorus marinus* (Champigny 1959) may be related to pretreatment effects (Hauschild *et al.* 1962). Oxygen and $CO_2$ levels, as well as light wavelength and intensity, have large effects on the synthesis and metabolism of glycolate in photosynthesizing algae (see Chapter 17, p. 474).

A high osmotic pressure in the medium can increase the fraction of photosynthate appearing as soluble carbohydrate and this may be a way of increasing the internal osmotic pressure (Wetherell 1963, Kauss 1968, Wegman 1971).

Possible control mechanisms which relate to the exit points from the cycle to endogenous and exogenous factors are discussed by Bassham (1971).

# 4  HETEROTROPHIC CO$_2$ FIXATION

## 4.1  Introduction

As in other organisms, 'dark' CO$_2$ fixation reactions are important in both the heterotrophic and the autotrophic growth of algae. This is attested to by the frequent inhibition of dark growth by removal of respiratory CO$_2$, and the stimulation of growth by added CO$_2$ (Meffert 1960, Cramer & Myers 1952, van Dreal & Padilla 1964, Tremmell & Levedahl 1966). However, in a number of other cases, retention of respiratory CO$_2$ is apparently adequate to supply any CO$_2$ requirement (Myers 1951, Nuhrenberg et al. 1968).

Some reactions of heterotrophic growth (also important in autotrophic growth) which are believed to require CO$_2$, either by net fixation or catalytically, are shown in Fig. 15.2. Quantitatively the most important is β-carboxylation, which is required as an anaplerotic pathway related to the role of the tricarboxylic acid cycle in supplying carbon skeletons for biosynthesis (Wood & Utter 1965, Kornberg 1966). It is also involved in regulation of intracellular pH. Another important reaction is the synthesis of carbamoyl phosphate, required in arginine and pyrimidine synthesis. Levedahl (1968) tabulates the products of heterotrophic $^{14}$CO$_2$ fixation in a number of algae over fairly long incubation times (30 to 60 minutes). This shows that the products are generally those predicted from the pathways of Fig. 15.2.

The individual dark fixation reactions will now be considered.

## 4.2  β-carboxylation

In this reaction a four carbon dicarboxylic acid is produced from CO$_2$ and a three carbon precursor (pyruvate or phosphoenolpyruvate). For various reasons (Walker 1962, Raven 1970a), it is believed that the reaction in plants is catalysed by phosphoenolpyruvate carboxylase:

phosphoenolpyruvate $+$ HCO$^-_3$ → oxaloacetate $+$ inorganic phosphate.

The reason that short-term dark fixation leads to labelling of malate and aspartate, rather than oxaloacetate, from $^{14}$CO$_2$ is probably that oxaloacetate is unstable under the normal killing conditions used in such experiments.

In most algae the major products of short-term (2·5 to 30 seconds) $^{14}$CO$_2$ fixation in the dark are the four carbon dicarboxylic acids malate and aspartate (Moses et al. 1959, Kauss & Kandler 1962, Ogasawara & Miyachi 1970a, Codd & Merrett 1971c, Grünberg & Galloway 1971, Akagawa et al. 1972). These four carbon acids are also early products of $^{14}$CO$_2$ fixation in the light. Thus it is possible that they arise from β-carboxylation, i.e. are closely related to the primary carboxylation product (oxaloacetate).

This is confirmed when the molecule is degraded and the radioactivity in individual carbon atoms is determined. This shows that most of the label is in

$C_4$, i.e. the carbon atom labelled in $\beta$-carboxylation, in both light and dark (Benson & Bassham 1948, Calvin *et al*. 1951, Ferrari & Passera 1964) short term experiments. Light generally enhances the labelling of these acids (see Raven 1970a). Some of the increased label is due to labelling of the carboxylation substrate (phosphophenolpyruvate) from photosynthetically produced 3-phosphoglycerate (Fig. 15.2), but there is also a genuine increase in the rate of $\beta$-carboxylation, i.e. labelling of $C_4$ (Bassham & Kirk 1960).

For a variety of reasons (Walker 1962, Raven 1970a) it is believed that phosphoenolpyruvate carboxylase is the enzyme which catalyses $\beta$-carboxylation in plants. The presence of this enzyme in amounts adequate to support the observed *in vivo* rate of $\beta$-carboxylation has been reported for a number of algae (Vasilenok *et al*. 1969, Ohman & Plhak 1969, Chen & Jones 1970, Codd & Merrett 1971c, Miyachi & Hogetsu 1970). Algae also contain two other enzymes which, under certain circumstances, might catalyse $\beta$-carboxylation: the malic enzyme (Ammon & Friedrich 1967, Heinrich & Cook 1967) and phosphoenolpyruvate carboxykinase (Ohman 1969, Hase 1971). A new form of phosphoenolpyruvate carboxylase, characterized by a low pH optimum, sensitivity to cyanide, and insensitivity to thermal inactivation, has recently been found in algae; the relationship of this enzyme to the one previously described in algae is not clear (Pan & Waygood 1971).

As regards the function and regulation of $\beta$-carboxylation, its main function appears to be in the net synthesis of four carbon acids and their derivatives. If intermediates of the tricarboxylic acid cycle are withdrawn for biosynthetic purposes, four carbon acids as well as acetyl CoA must be supplied to the cycle. The alternative to $\beta$-carboxylation in supplying four carbon acids is the glyoxylate cycle, which converts two moles of a two carbon compound (acetate) into malate (see Chapter 19, p. 545). Since the glyoxylate cycle is largely confined to algae growing on two carbon compounds, it cannot be a universal means of net synthesis of four carbon acids in algae. However, when the glyoxylate cycle does occur, it is of interest that both the rate of dark fixation of $CO_2$ (Heinrich & Cook 1967) and the activity of phosphoenolpyruvate carboxylase (Ohman & Plhak 1969, Stabenau 1971) are lower than in cultures lacking the glyoxylate cycle.

In quantitative terms, the measured rate of $\beta$-carboxylation (Bassham & Kirk 1960) in algae lacking the glyoxylate cycle is adequate to supply the requirements of the cell for four carbon acids and their derivatives during growth (see discussion by Raven 1970a). There is a lot of other data which correlates the rates of $\beta$-carboxylation with the biosynthetic requirement for four carbon acids and their derivatives. Much of this stems from treatments which alter the rate of synthesis of derivatives of $\beta$-carboxylation, such as varying the wavelength of light supplied, or varying the nitrogen supply (most of the products of four carbon acid metabolism contain nitrogen, see Fig. 15.2).

As was mentioned in 3.8 above, blue light enhances the synthesis of tricarboxylic acid cycle derivatives compared with the rate in red light, under both photosynthetic and non-photosynthetic (+DCMU) conditions (see, Kowallik 1970, Raven 1970a, 1972a). Blue light treatments also increase the rate of β-carboxylation *in vivo* both in the presence and the absence of CMU (3' (*p*-chlorophenyl) 1', 1' dimethylurea), and the activity of phosphoenolpyruvate carboxylase in extracts of cells treated with blue light (Ogasawara & Miyachi 1970a,b).

Similarly, the addition of a nitrogen source to cells lacking nitrogen stimulates β-carboxylation (see Syrett 1962, Hiller 1964), and the presence of $CO_2$ stimulates nitrogen assimilation in the dark (Meffert 1960). This effect of nitrogen supply, and the effect of blue light, both show a correlation between the rate of β-carboxylation and the rate of synthesis of nitrogen-containing derivatives of the tricarboxylic acid cycle. The rate of β-carboxylation also varies with the stage in the algal life cycle, and here again there is a relationship between the rate of $CO_2$ fixation and the nature of cell syntheses (Galloway *et al.* 1967, Codd & Merrett 1971c).

Thus the rate of β-carboxylation appears to be regulated by the cell's requirements for four carbon acids and their derivatives. In addition to providing monomers for protein and nucleic acid synthesis, products of β-carboxylation are also important in regulating intracellular pH. β-carboxylation yields a fairly strong acid (malate and its derivatives) from neutral molecules (for example, carbohydrates) and the weak acid, $H_2CO_3$. In many plants this synthesis occurs in response to cytoplasmic alkalinity such as occurs during influx of $K^+$ in exchange for endogenous $H^+$ (Shieh & Barber 1971), or when the $OH^-$ generated during $NO_3^-$ reduction is not excreted (see Raven & Smith 1973). In algae it is likely that the first of these processes occurs. During $K^+$ and $H^+$ exchange in *Chlorella*, a synthesis of organic acids and amino acids occurs (Schaedle & Hernandez 1969) and many of these syntheses involve β-carboxylation.

In the case of $NO_3^-$ reduction, however, it is likely that the $OH^-$ produced is excreted. Syrett (1956) found that the ratio of dark $CO_2$ fixation to nitrogen assimilated in *Chlorella* was the same for $NO_3^-$ as for $NH_4^+$, that is, no carboxylation associated with an internal production of an organic acid to neutralize the $OH^-$ was observed. Similarly, the $OH^-$ generated inside algal cells during photosynthesis with $HCO_3^-$ all appear to be excreted.

Thus β-carboxylation in algae provides carbon skeletons for the synthesis of monomers for protein and nucleic acid synthesis, and also makes acids involved in pH regulation.

### 4.3 *Carbamoyl phosphate synthesis*

Carbamoyl phosphate synthesis is an essential part of the biosynthetic pathway to arginine and pyrimidines in most organisms (Fig. 15.2). Its synthesis involves the reaction

$$CO_2 + NH_3 + ATP \rightarrow NH_2 \cdot COO\textcircled{P} + ADP.$$

In some cases the ammonia is derived from the amide group of glutamine (Jones 1970).

Evidence for $CO_2$ fixation in intact algae via the carbamoyl phosphate synthetase reaction is indirect, in that carbamoyl phosphate has not been demonstrated as a product of $^{14}CO_2$ fixation. However, its presumed derivatives citrulline and arginine are common products of both the light and dark fixation of $CO_2$ in algae (Linko et al. 1957, Ferrari et al. 1963a,b, Hiller 1964, Ogasawara & Miyachi 1970a).

In short-term (90 seconds) experiments in the light (Ferrari et al. 1963a,b) and in longer term (15 minutes) experiments in the dark (Hiller 1964), the $^{14}C$ label from $^{14}CO_2$ is mainly in $C_6$ of citrulline and arginine. This is the carbon atom derived from $CO_2$ via carbamoyl phosphate in other organisms. Extracts of algae contain the enzyme carbamoyl phosphate synthetase (Holm-Hansen 1962, Hiller 1964, Vasilenok et al. 1969) and also the enzymes involved in the use of carbamoyl phosphate in the synthesis of arginine and pyrimidines (see e.g. Cole & Schmidt 1964, Strijkert & Sussenbach 1969, Holden & Morris 1970). The relation of algal carbamoyl phosphate synthetase to the enzyme(s) from other prokaryotes and eukaryotes is not clear (Jones 1970, Tramell & Campbell 1971).

Quantitatively, the labelling of citrulline during photosynthesis in Chlorella (Ogasawara & Miyachi 1970a) is similar to the rate predicted for the synthesis of the known arginine content of these cells (Raven 1970a). Citrulline labelling in Chlorella is relatively higher at low than at high light intensities (Ogasawara & Miyachi 1970a). The relative insensitivity to CMU suggests that the light effect could involve the supply of ATP from cyclic photophosphorylation. The rate of citrulline labelling in Cyanophyceae is generally greater than the rate of net synthesis of citrulline and arginine. The significance of this apparent turnover is not clear (Holm-Hansen 1962).

### 4.4 Other dark $CO_2$ fixation reactions

Sugar phosphates are products of dark fixation in most algae, although the rate at which they become labelled is very variable (Lynch & Calvin 1953, Wegman 1969, Grünberg & Galloway 1971, Moses et al. 1959). It is possible that the cases of rapid labelling of the sugar phosphates could be due to incomplete inactivation of the photosynthetic carbon reduction cycle in the dark (see 3.7 above). However, alternative explanations cannot be ruled out. One possibility is the reductive reversal of the decarboxylation reaction of the oxidative pentose phosphate pathway (Holm-Hansen 1962). Another possibility is the formation of malate labelled in $C_4$ from dark fixation, randomization of the label into $C_1$ by the activity of fumarase, and conversion of atoms 1,2 and 3 of malate into hexose by gluconeogenesis. Investigation of the labelling pattern in the hexose produced would distinguish the reversal of the oxidative pentose phosphate pathway (which labels $C_1$) from the other two possibilities (which label $C_3$ and $C_4$).

No evidence appears to be available as to the occurrence in algae of the carboxylation reaction used in the synthesis of purines (Flaks & Lukens 1963). It is also likely that the carboxylation of propionyl CoA to form methylmalonyl CoA is involved in the metabolism of propionate by *Ochromonas malhamensis* (Arnstein & White 1962).

## 5  CATALYTIC ROLES OF $CO_2$

In many organisms the synthesis of fatty acids requires a catalytic action of $CO_2$. Carbon dioxide is added to acetyl CoA to form malonyl CoA, which is then added to the growing fatty acid chain and the $CO_2$ is then released (Stumpf 1969). However, this is not always the mechanism (Seubert 1970) and the evidence for *Euglena* is contradictory as to the involvement of $CO_2$ in fatty acid synthesis (Cheniae 1964, Delo *et al.* 1971, Ernst-Forberg & Bloch 1971). Vasilenok *et al.* (1969) have shown that the enzyme acetyl CoA carboxylase, which brings about the carboxylation involved in fatty acid synthesis, is present in *Anacystis*. Any $CO_2$ requirement for fatty acid synthesis in *Chlorella* must saturate at low $CO_2$ levels, since Paschinger (1969) found that glucose could be converted to lipid even when the cells were illuminated and a trap for $CO_2$ was present.

Another catalytic role for $CO_2$ is in urea assimilation by *Chlorella*, which appears to proceed via a biotin-dependent carboxylation of urea (Thompson & Muenster 1971). Biotin is also involved in the carboxylation of propionyl CoA (see 4.4 above) and of acetyl CoA.

Carbon dioxide is a catalyst of photosystem II of photosynthesis and of photophosphorylation in higher plants (Avron & Neumann 1968). In algae only the effect on photosystem II has so far been demonstrated (see e.g. Heise & Gaffron 1963, Vennesland 1966). Suggestions that the effect of $CO_2$ concentration on the rate of photosynthesis is predominantly related to the $CO_2$ effect on photosystem II (see e.g. Warburg & Krippahl 1958, Bunt 1969) are not supported by the majority of relevant data (see e.g. Heise & Gaffron 1963, Raven 1970b). Thus the widely held view that the effects of $CO_2$ concentration on the rate of $CO_2$ fixation predominantly reflects limitations imposed directly on the rate of carboxylation is probably correct.

## 6  SUMMARY

The major carbon source for photosynthesis is unhydrated $CO_2$, which enters the cell by passive diffusion. At high pH values some algae are able to actively transport $HCO_3^-$ into their cells. Carbonic anhydrase then yields $CO_2$, which is fixed by carboxydismutase, and $OH^-$, which is excreted from the cell. Some reassimilation of endogenous (respiratory) $CO_2$ also occurs.

Autotrophic (photosynthetic) $CO_2$ fixation in algae occurs via the photosynthetic carbon reduction cycle (Fig. 15.1). The main carboxylation reaction is that catalysed by ribulose diphosphate carboxylase. The evidence for this comes from experiments on kinetics of $^{14}CO_2$ incorporation, intramolecular labelling patterns, and the presence of the requisite enzymes.

In eukaryotic algae the photosynthetic carbon reduction cycle occurs in the plastids. The ATP and $NADPH_2$ required for autotrophic $CO_2$ fixation usually come from non-cyclic photophosphorylation. The rate of autotrophic $CO_2$ fixation, and its products, are regulated by a variety of factors (e.g. light intensity and wavelength, nutrient supply, $O_2$ and $CO_2$ levels).

Heterotrophic (dark) $CO_2$ fixation is essential for both autotrophic and heterotrophic growth of algae (Fig. 15.2). β-carboxylation, catalysed by phosphoenolpyruvate carboxylase is an important anaplerotic pathway related to the biosynthetic functions of the tricarboxylic acid cycle, as well as to pH regulation within the cell. Carbamoyl phosphate synthesis, using $CO_2$ as carbon source, is important in the biosynthesis of arginine and pyrimidines. Carbon dioxide also has a number of catalytic roles in algae, e.g. in lipid synthesis, and in electron transport and phosphorylation in photosynthesis.

## 7  REFERENCES

AHMED A.M.M. & RIES E. (1969) The pattern of $^{14}CO_2$ fixation in different phases of the life cycle and under different wavelengths in *Chlorella pyrenoidosa*. In *Progress in Photosynthetic Research III*, ed. Metzner H. pp. 1662–8. International Biological Union, Tubingen.

AKAGAWA H., IKAWA T. & NISIZAWA K. (1972) $^{14}CO_2$ fixation in marine algae with special reference to the dark-fixation in brown algae. *Botanica mar.* **15**, 126–32.

AMMON R. & FRIEDRICH G. (1967) Enzyme behaviour in *Euglena gracilis*. *Acta biol. med. germ.* **19**, 669–72.

ANTIA N.J. (1967) Comparative studies on aldolase activity in marine planktonic algae, and their evolutionary significance. *J. Phycol.* **3**, 81–5.

ARNSTEIN H.R.V. & WHITE A.M. (1962) The function of vitamin $B_{12}$ in the metabolism of propionate by the protozoan *Ochromonas malhamensis*. *Biochem. J.* **83**, 264–70.

AVRON M. & NEUMANN J. (1968) Photophosphorylation in chloroplasts. *A. Rev. Pl. Physiol.* **19**, 137–66.

BASSHAM, J.A. (1964) Kinetic studies of the photosynthetic carbon reduction cycle. *A. Rev. Pl. Physiol.* **15**, 101–20.

BASSHAM J.A. (1971) The control of photosynthetic carbon metabolism. *Science, N.Y.* **172**, 526–34.

BASSHAM, J.A. & KIRK M.R. (1960) Dynamics of the synthesis of carbon compounds. I. Carboxylation reactions. *Biochim. biophys. Acta*, **43**, 447–64.

BENSON A.A. & BASSHAM J.A. (1948) Chemical degradation of isotopic succinic and malic acids. *J. Am. Chem. Soc.* **70**, 3939–40.

BIDWELL R.G.S., LEVIN W.B. & SHEPPARD D.C. (1970) Intermediates of photosynthesis in *Acetabularia mediterranea* chloroplasts. *Pl. Physiol., Lancaster*, **45**, 70–5.

BROWN D.L. & TREGUNNA E.B. (1967) Inhibition of respiration during photosynthesis by some algae. *Can. J. Bot.* **45**, 1135–43.

BUCHANAN B.B. & ARNON D.I. (1970) Ferredoxins: chemistry and function in photosynthesis, nitrogen fixation and fermentative metabolism. *Adv. Enzymol.* **33**, 119–76.

BUNT J.S. (1969) The $CO_2$ compensation point, Hill activity and photorespiration. *Biochem. biophys. Res. Commun.* **35**, 748–53.

CALVIN M., BASSHAM J.A., BENSON A.A., LYNCH V.H., OUELLET C., SCHOU L., STEPKA W. & TOLBERT N.E. (1951) Carbon dioxide assimilation in plants. *V Symp. S.E.B.* 284–305.

CHAMPIGNY M.L. (1959) Sur l'influence de l'intensite des differentes radiations lumineuses dans la photosynthese de *Rhodosorus marinus* Geitler (Rhodophyceae). La repartition du $^{14}C$ assimile entre les substances synthesisees en presence de $^{14}CO_3HNa$; influence de $NO_3Na$ et de l'uree. *Rev. de Cytol. et de Biol. Vegetales*, **21**, 1–43.

CHEN J.M. & JONES F.T. (1970) Multiple forms of phosphoenolpyruvate carboxylase from *Chlamydomonas reinhardii*. *Biochim. biophys. Acta*, **214**, 318–25.

CHENIAE G.M. (1964) Fatty acid synthesis by extracts of *Euglena* IV. *Archs. Biochem. Biophys.* **106**, 163–9.

CODD G.A. & MERRETT M.J. (1971a) Photosynthetic products of division synchronised cultures of *Euglena*. *Pl. Physiol., Lancaster*, **47**, 635–9.

CODD G.A. & MERRETT M.J. (1971b) The regulation of glycolate metabolism in division synchronised cultures of *Euglena*. *Pl. Physiol., Lancaster*, **47**, 640–3.

CODD G.A. & MERRETT M.J. (1971c) Phosphopyruvate carboxylase activity and $CO_2$ fixation via $C_4$ acids over the division cycle in synchronised *Euglena* cultures. *Planta*, **100**, 124–30.

COLE F.E. & SCHMIDT R.R. (1964) Control of aspartate transcarbamylase activity during synchronous growth of *Chlorella pyrenoidosa*. *Biochim. biophys. Acta*, **90**, 616–18.

COOK J.R. (1963) Adaptations in growth and division in *Euglena* affected by energy supply. *J. Protozool.* **10**, 436–44.

COOMBS J. & VOLCANI B.E. (1968) Studies on the biochemistry and fine structure of silica shell formation in diatoms. Silicon-induced metabolic transients in *Navicula pelliculosa* (Bréb.) Hilse. *Planta*, **80**, 264–79.

COOPER T.G. & WOOD H.G. (1971) The carboxylation of phosphoenolpyruvate and pyruvate. II. The active species of '$CO_2$' utilized by PEP carboxylase and pyruvate carboxylase. *J. biol. Chem.* **246**, 5488–90.

CRAMER M. & MYERS J. (1952) Growth and photosynthetic characteristics of *Euglena gracilis*. *Arch. Mikrobiol.* **17**, 384–402.

CREMONA T. (1968) Control mechanisms at the level of FDP aldolases in algae. *G. Bot. ital.* **102**, 253–9.

CUMMING B.G. & WAGNER E. (1968) Rhythmic processes in plants. *A. Rev. Pl. Physiol.* **19**, 381–416.

DELO J., ERNST-FORBERG H.L. & BLOCH K. (1971) Fatty acid synthetases from *Euglena gracilis*. *Archs. Biochem. Biophys.* **143**, 384–91.

ERNST-FORBERG H.L. & BLOCH K. (1971) A chloroplast-associated fatty acid synthetase system in *Euglena*. *Archs. Biochem. Biophys.* **143**, 392–400.

FERRARI G. & PASSERA C. (1964) Ricerche sul meccanismo di sintesi degli amminoacidi nei vegetali. Nota IV. Genesi dell'acido aspartico in *Chlorella vulgaris*. *Riv. Sci. Rend B.* **4**, 631–43.

FERRARI G., PASSERA C. & CULTRERA R. (1963a) Ricerche sul meccanismo di sintesi amminoacidi nei vegetali. Nota I. Il cammino del $^{14}C$ nella formazione degli amminoacidi in *Chlorella vulgaris*. *Riv. Sci. Rend. B.* **3**, 181–8.

FERRARI G., PASSERA C. & BENEDETTI E. (1963b) Ricerche sul meccanismo di sintesi degli amminoacidi nei vegetali. Nota II. Rapida assimilazione in *Chlorella vulgaris* di carbonio di $NaHCO_3$ nel gruppe guanidilico dell arginina. *Riv. Sci. Rend. B.* **3**, 189–92.

FLAKS J.G. & LUKENS L.N. (1963) 5-aminoimidazole ribotide carboxylase. In *Methods in Enzymology*, VI, eds. Colowick S.P. & Kaplan N.O. pp. 79–82. Academic Press, New York & London.

FORSEE W.T. & KAHN J.S. (1972) Carbon dioxide fixation by isolated chloroplasts of *Euglena gracilis*. I. Isolation of functionally intact chloroplasts and their characterisation. *Archs. Biochem. Biophys.* 150, 296–301.

GALLOWAY R.A., SOEDER C.J., CARTER J.C. & KAUSCH B. (1967) Some aspects of dark fixation of $CO_2$ by synchronous *Chlorella*. *Pl. Physiol., Lancaster*, 42, 1053–8.

GIBBS M., LATZKO M., HARVEY M.J., PLAUT A. & SHAIN Y. (1970) Photosynthesis in the algae. *Ann. N.Y. Acad. Sci.* 175, 541–54.

GILES V.L. & SANOFIS V. (1972) Chloroplast survival and division *in vivo*. *Nature New Biol.* 236, 56–7.

GINGRAS G. (1966) Étude comparative, chez quelques algues, de la photosynthese et de la photoreduction realisée en presence de l'hydrogene. *Physiol. Veg.* 4, 1–65.

GINGRAS G., GOLDSBY R.A. & CALVIN M. (1963) Carbon dioxide metabolism in hydrogen-adapted *Scenedesmus*. *Archs. Biochem. Biophys.* 100, 178–84.

GRAHAM D. & WHITTINGHAM C.P. (1968) The path of carbon in photosynthesis in *Chlorella pyrenoidosa* at high and low carbon concentrations. *Z. Pfl. Physiol.* 58, 418–27.

GRAHAM D.A., ATKINS C.A., REED M.L., PATTERSON B.P. & SMILLIE R.M. (1971) Carbonic anhydrase, photosynthesis and light-induced pH changes. In *Photosynthesis and photorespiration*, eds. Hatch M.D., Osmond C.B. & Slatyer R.O. pp. 267–74. Wiley-Interscience, London, New York, Sydney & Toronto.

GRUNBERG M. & GALLOWAY R.A. (1971) Heterotrophic carbon dioxide fixation in *Chlorella* at high carbon dioxide concentration. *J. Phycol.* 7, *suppl.* 5.

GUERRINI A.M., CREMONA T. & PREDDIE E.C. (1971) The aldolases of *Chlamydomonas reinhardii*. *Archs. Biochem. Biophys.* 146, 249–55.

GUYOMARCH C. (1970) Influence des pH alcalins et des ions bicarbonates sur la photosynthese de *Chlorella vulgaris*. *Beij. Bull. Soc. scient. Bretagne*, 45, 113–26.

HASE E. (1971) Studies on the metabolism of nucleic acid and protein associated with the processes of de- and re-generation of chloroplasts in *Chlorella protothecoides*. In *Autonomy and biogenesis of mitochondria and chloroplasts*, eds. Boardman N.K., Linnane A.W. & Smillie R.M. pp. 434–46. North-Holland, Amsterdam.

HATCH M.D. & SLACK C.R. (1970) The $C_4$-dicarboxylic acid pathway of photosynthesis. In *Progress in Phytochemistry*, vol. II, eds. Reinhold L. & Linschitz Y. pp. 35–106. Interscience Publishers, London.

HAUSCHILD A.H.W., NELSON C.D. & KROTKOV G. (1962). The effect of light quality on the products of photosynthesis in green and blue-green algae and in photosynthetic bacteria. *Can. J. Bot.* 40, 1519–30.

HEINRICH B. & COOK J.R. (1967) Studies on the respiratory physiology of *Euglena gracilis* cultured on acetate or glucose. *J. Protozool.* 14, 548–53.

HEISE J.J. & GAFFRON H. (1963) Catalytic effects of $CO_2$ in $CO_2$ assimilating cells. *Pl. Cell Physiol., Tokyo*, 4, 1–11.

HELLEBUST J.A., TERBORGH J. & MCLEOD G.C. (1967) The photosynthetic rhythm of *Acetabularia crenulata*. II. Measurements of photoassimilation of carbon dioxide and the activities of enzymes of the reductive pentose cycle. *Biol. Bull.* 133, 670–8.

HILLER R.G. (1964) Synthesis of citrulline and arginine in *Chlorella pyrenoidosa*. *J. exp. Bot.* 15, 15–20.

HILLER R.G. (1970) Transients in the photosynthetic carbon reduction cycle produced by iodoacetic acid and ammonium ions. *J. exp. Bot.* 21, 628–38.

HOLDEN J. & MORRIS I. (1970) Regulation of arginine biosynthesis in *Chlamydomonas reinhardii*: studies *in vivo* and of ornithine transcarbamylase and arginosuccinate lyase levels. *Arch. Mikrobiol.* 74, 58–68.

HOLM-HANSEN O. (1962) Assimilation of carbon dioxide. In *Physiology and Biochemistry of algae*, ed. Lewin R.A. pp. 25–45. Academic Press, New York & London.

HUTH W. (1967) Dependence of enzymes in green algae on the carbon supply. *Flora, Jena*, 158, 58–87.

JACOBI G. (1962) Enzyme systems. In *Physiology and Biochemistry of algae*, ed. Lewin R.A. pp. 125–40. Academic Press, New York & London.

JEFFREY S.W., ULLRICH J. & ALLEN M.B. (1966) Some photochemical characteristics of chloroplast preparations from the chrysomonad *Hymenomonas* sp. *Biochim. biophys. Acta*, **112**, 35–44.

JOLIFFE E.A. & TREGUNNA E.B. (1970) Studies on $HCO_3^-$ ion uptake during photosynthesis in benthic marine algae. *Phycologia*, **9**, 292–303.

JONES M.E. (1970) Regulation of pyrimidine and arginine biosynthesis in mammals. *Adv. Enz. Regn.* **9**, 19–49.

KANAZAWA T., KIRK M.R. & BASSHAM J.A. (1970) Regulatory effects of ammonia on carbon metabolism in photosynthesising *Chlorella pyrenoidosa*. *Biochim. biophys. Acta*, **205**, 401–8.

KANAZAWA T., KANAZAWA K., KIRK M.R. & BASSHAM J.A. (1970a) Regulation of photosynthetic carbon metabolism in synchronously growing *Chlorella pyrenoidosa*. *Pl. Cell Physiol., Tokyo*, **11**, 149–60.

KANAZAWA T., KANAZAWA K., KIRK M.R. & BASSHAM J.A. (1970b) Differences in nitrate reduction in 'light' and 'dark' stages of synchronously growing *Chlorella pyrenoidosa* and resultant metabolic changes. *Pl. Cell. Physiol., Tokyo*, **11**, 445–52.

KANDLER O. (1960) Energy transfer through phosphorylation mechanisms in photosynthesis. *A. Rev. Pl Physiol.* **11**, 37–54.

KANDLER O. & LIESENKOTTER I. (1961) The effect of monoiodo-acetic acid, arsenate and dinitrophenol on the path of carbon in photosynthesis. *Proc. 5th int. Biochem. Congr., Moscow*, **6**, 326–39.

KAUSS H. (1968) α-Galactosyglyzeride und osmo-regulation in Rotalgen. *Z. PflPhysiol.* **58**, 428–33.

KAUSS H. & KANDLER O. (1962) Die Kohlensaureassimilation von *Ochromonas malhamensis* bei Thiamin- und Biotinmangel. I. Die heterotrophe Kohlensaureassimilation. *Arch. Mikrobiol.* **42**, 204–18.

KAWASHIMA N. & WILDMAN S.G. (1970) Fraction one protein. *A. Rev. Pl. Physiol.* **21**, 325–58.

KINDEL, P. & GIBBS M. (1963) Distribution of carbon-14 in polysaccharide after photosynthesis in carbon dioxide labelled with carbon-14 by *Anacystis nidulans*. *Nature, Lond.* **200**, 260–1.

KORNBERG H.L. (1966) Anaplerotic sequences and their role in metabolism. In *Essays in Biochemistry*, vol. 2, eds. Campbell P.N. & Greville G.D. pp. 1–31. Academic Press, New York & London.

KOWALLIK W. (1970) Light effects on carbohydrate and protein metabolism in algae. In *Photobiology of micro-organisms*, ed. Halldal P. pp. 333–79. Wiley-Interscience, London, New York, Sydney & Toronto.

KREMER B.P. & WILLENBRINK J. (1972) $CO_2$-Fixierung und Stofftransport in benthischen marinen algen. I. Zur Kinetik der $^{14}CO_2$-Assimilation bei *Laminaria saccharina*. *Planta*, **103**, 55–64.

LATZKO E. & GIBBS M. (1969) Enzyme activities of the carbon reduction cycle in some photosynthetic organisms. *Pl. Physiol., Lancaster*, **44**, 295–300.

LEVEDAHL B.H. (1968) Heterotrophic $CO_2$ fixation in *Euglena*. In *The Biology of Euglena*, II, ed. Buetow D.E. pp. 85–96, Academic Press, New York & London.

LINKO P., HOLM-HANSEN P., BASSHAM J.A. & CALVIN M. (1957) Formation of radioactive citrulline during photosynthetic $C^{14}O_2$ fixation by blue-green algae. *J. exp. Bot.* **8**, 147–56.

LOEBLICH L.A. (1970) Growth limitation of *Dunaliella salina* by $CO_2$ at high salinity. *J. Phycol.* **6**, *suppl.* 9.

LÜTTGE U., BALL E. & VON WILLERT K. (1971) A comparative study of the coupling of ion uptake to light reactions in leaves of higher plant species having the $C_3$- and $C_4$-pathway of photosynthesis. *Z. PflPhysiol.* **65**, 336–50.

LYNCH V.H. & CALVIN M. (1953) $CO_2$ fixation by *Euglena*. *Ann. N.Y. Acad. Sci.* **56**, 890–900.

MCALLISTER C.D., SHAH N. & STRICKLAND J.D.H. (1964) Marine phytoplankton photosynthesis as a function of light intensity: a comparison of methods. *J. Fish. Res. Bd. Can.* **21**, 159–81.

MEFFERT M-E. (1960) Uber den Einfluss von Kohlendioxyd auf die Stickstoff-assimilation von *Scenedesmus obliquus* im Licht-Dunkel-Wechsel. *Arch. Mikrobiol.* **37**, 49–56.

MIYACHI S. & HOGETSU D. (1970) Effects of preillumination with light of different wavelengths on subsequent dark $CO_2$ fixation in *Chlorella* cells. *Can. J. Bot.* **48**, 1203–7.

MOLL B. & LEVINE R.P. (1970) Characterisation of a photosynthetic mutant strain of *Chlamydomonas reinhardi* deficient in phosphoribulokinase activity. *Pl. Physiol., Lancaster*, **46**, 576–80.

MOSES V., HOLM-HANSEN O. & CALVIN M. (1959) Non-photosynthetic fixation of carbon dioxide by micro-organisms. *J. Bact.* **77**, 70–8.

MYERS J. (1951) Physiology of algae. *A. Rev. Microbiol.* **5**, 157–80.

NÜHRENBERG B., LESEMANN D. & PIRSON A. (1968) Zur Frage eines anaeroben Wachstums von einzelligen Grünalgen. *Planta*, **79**, 162–80.

OGASAWARA N. & MIYACHI S. (1970a) Regulation of $CO_2$-fixation in *Chlorella* by light of varied wavelengths and intensities. *Pl. Cell Physiol., Tokyo*, **11**, 1–14.

OGASAWARA N. & MIYACHI S. (1970b) Effects of DSPD and near far-red light on $^{14}CO_2$ fixation in *Chlorella* cells. *Pl. Cell Physiol., Tokyo*, **11**, 411–16.

OHMAN E. (1969) Die regulation der pyruvate-kinase in *Euglena gracilis*. *Arch. Mikrobiol.* **67**, 273–92.

OHMAN E. & PLHAK F. (1969) Reinigung und Eigenschaften von Phosphoenolpyruvate-carboxylase aus *Euglena gracilis*. *Eur. J. Biochem.* **10**, 43–55.

PAASCHE E. (1966) Action spectrum of coccolith formation. *Physiologia Pl.* **19**, 770–9.

PAN D. & WAYGOOD E.R. (1971) A fundamental thermostable cyanide-sensitive phosphoenolpyruvate acid carboxylase in photosynthetic and other organisms. *Can. J. Bot.* **49**, 631–43.

PASCHINGER H. (1969) Photochemical oxygen evolution by *Chlorella fusca* in glucose. *Arch. Mikrobiol.* **67**, 243–50.

PATIL B.A. & JOSHI G.V. (1970) Photosynthetic studies in *Ulva lactuca*. *Botanica mar.* **13**, 111–15.

PELROY R.A. & BASSHAM J.A. (1972) Photosynthetic and dark carbon metabolism in unicellular blue-green algae. *Arch. Mikrobiol.* **86**, 25–38.

PETTERKOFFSKY A. & RACKER E. (1961) The reductive pentose phosphate cycle. III. Enzyme activities in cell-free extracts of photosynthetic organisms. *Pl. Physiol., Lancaster*, **36**, 409–13.

PIRSON A. & LORENZEN H. (1966) Synchronized dividing algae. *A. Rev. Pl. Physiol.* **17**, 439–58.

PREISS J. & KOSUGE T. (1970) Regulation of enzyme activity in photosynthetic systems. *A. Rev. Pl. Physiol.* **21**, 433–66.

RAVEN J.A. (1969) Action spectra for photosynthesis and light-stimulated ion transport processes in *Hydrodictyon africanum*. *New Phytol.* **68**, 45–62.

RAVEN J.A. (1970a) Exogenous inorganic carbon sources in plant photosynthesis. *Biol. Rev.* **45**, 167–221.

RAVEN J.A. (1970b) The role of cyclic and pseudocyclic photophosphorylation in photosynthetic $^{14}CO_2$ fixation in *Hydrodictyon africanum*. *J. exp. Bot.* **21**, 1–16.

RAVEN J.A. (1971a) Inhibitor effects on photosynthesis, respiration and active ion transport in *Hydrodictyon africanum*. *J. membrane Biol.* **6**, 89–107.

RAVEN J.A. (1971b) Cyclic and non-cyclic photophosphorylation as energy sources for active K influx in *Hydrodictyon africanum*. *J. exp. Bot.* **22**, 420–33.

454 CHAPTER 15

RAVEN J.A. (1971c) Energy metabolism in green cells. *Trans. Bot. Soc. Edinb.* **41**, 219–25.

RAVEN J.A. (1972a) Endogenous inorganic carbon sources in plant photosynthesis. I. Occurrence of the dark respiratory pathways in illuminated green cells. *New Phytol.* **71**, 227–47.

RAVEN J.A. (1972b) Endogenous inorganic carbon sources in plant photosynthesis. II. Comparison of total $CO_2$ production with measured $CO_2$ evolution in the light. *New Phytol.* **71**, 995–1014.

RAVEN J.A. & SMITH F.A. (1973) The regulation of intracellular pH as a fundamental biological process. In *Ion Transport in Plants*, ed. Anderson W.P. pp. 271–8. Academic Press, New York & London.

RUSSELL G.K. & GIBBS M. (1966) Regulation of photosynthetic capacity in *Chlamydomonas mundana*. *Pl. Physiol., Lancaster*, **41**, 885–90.

RUSSELL G.K. & GIBBS M. (1968) Evidence for the participation of the reductive pentose phosphate cycle in photoreduction and the oxy-hydrogen reaction. *Pl. Physiol., Lancaster*, **43**, 649–52.

SCHAEDLE M. & HERNANDEZ R. (1969) Cation stimulated anion synthesis in *Chlorella pyrenoidosa*. *Abstracts XI International Botanical Congress*, 191.

SCHINDLER D.W. (1971) Carbon, nitrogen and phosphorus and the eutrophication of freshwater lakes. *J. Phycol.* **7**, 321–9.

SEUBERT W. (1970) Mitochondrial Fettsaure-Synthese und pathologische Ketonkorperbildung. *Naturwissenschaften*, **57**, 443–9.

SHIEH Y.J. & BARBER J. (1971) Intracellular sodium and potassium concentrations and net cation movements in *Chlorella pyrenoidosa*. *Biochim. biophys. Acta*, **233**, 594–603.

SIMONIS W. (1967) Zycklische und nichtzycklische Photophosphorylierung *in vivo*. *Ber. dt. bot. Ges.* **80**, 395–402.

SMILLIE R.M. (1968) Enzymology of *Euglena*. In *The Biology of Euglena*, II, ed. Buetow D.E. pp. 1–54. Academic Press, New York & London.

SMITH F.A. & WEST J.R. (1969) A comparison of the effects of metabolic inhibitors on chloride uptake and photosynthesis in *Chara corallina*. *Austr. J. biol. Sci.* **22**, 351–64.

STABENAU H. (1971) Die regulation des Photosyntheseapparates bei *Chlorogonium elongatum* Dangeard unter dem Einfluss von Licht und Acetat. *Biochem. Physiol. Pfl.* **162**, 371–85.

STRIJKERT P.J. & SUSSENBACH J.S. (1969) Arginine metabolism in *Chlamydomonas reinhardi*. Evidence for a specific regulatory mechanism of the biosynthesis. *Eur. J. Biochem.* **8**, 408–12.

STUMPF P.K. (1969) Metabolism of fatty acids. *A. Rev. Biochem.* **38**, 159–212.

SYRETT P.J. (1956) The assimilation of ammonia and nitrate by nitrogen-starved cells of *Chlorella vulgaris*. IV. The dark fixation of carbon dioxide. *Physiologia Pl.* **9**, 165–71.

SYRETT P.J. (1962) Nitrogen assimilation. In *Physiology and Biochemistry of algae*, ed. Lewin R.A. pp. 171–88. Academic Press, New York & London.

TAMIYA H. (1966) Synchronous cultures of algae. *A. Rev. Pl. Physiol.* **17**, 1–26.

TAN C.K., BADOUR S.S. & WAYGOOD E.R. (1971) Photosynthesis of the unicellular green alga *Gloeomonas* in synchronous cultures. *Pl. Physiol., Lancaster*, **47**, 9s.

TANNER W., LOFFLER W. & KANDLER O. (1969) Cyclic photophosphorylation *in vivo* and its relation to photosynthetic $CO_2$ fixation. *Pl. Physiol., Lancaster*, **44**, 422–8.

THOMPSON J.F. & MUENSTER A-M.E. (1971) Separation of *Chlorella* ATP:urea amido-lyase into two components. *Biochem. biophys. Res. Commun.* **43**, 1049–55.

TRAMELL P.R. & CAMPBELL J.W. (1971) Carbamyl phosphate synthesis in invertebrates. *Comp. Biochem. Physiol.* **40B**, 395–406.

TREMMELL R.D. & LEVEDAHL B.H. (1966) Effect of $CO_2$ on the growth of a bleached *Euglena*. *J. Cell Physiol.* **67**, 361–5.

TRENCH R.K. & SMITH D.C. (1970) Synthesis of pigments in symbiotic chloroplasts. *Nature Lond.* **227**, 196–7.

TURNER J.S. & BRITTAIN E.G. (1962) Oxygen as a factor in photosynthesis. *Biol. Rev.* **37**, 130–70.

URBACH W. & GIMMLER H. (1970) Photophosphorylation und photosynthetische $^{14}CO_2^-$ fixierung *in vivo*. I. Vergleich von cyclischer und non-cyclischer Photophosphorylierung mit der $^{14}CO_2$-fixierung. *Z. PflPhysiol.* **62**, 276–86.

VAN DREAL P.A. & PADILLA G.M. (1964) $CO_2$ and cytokinesis in temperature-synchronised *Astasia longa*. *Biochim. biophys. Acta*, **93**, 668–70.

VASCONCELOS A., POLLOCK M., MENDIOLA L.R., HOFFMANN H.P., BROWN D.H. & PRICE C.A. (1971) Isolation of intact chloroplasts from *Euglena gracilis* by zonal centrifugation. *Pl. Physiol.*, *Lancaster*, **47**, 217–21.

VASILENOK L.I., VOLKOVA N.V. & PERKOVSKAYA E.K. (1969) Carbonic acid fixation by *Anacystis nidulans*. *Gidrobiol. Zh. A.N.U.S.S.R.* **5**, 69–70 (*Chemical Abstracts*, **71**, 19719d).

VENNESLAND B. (1966) Involvement of $CO_2$ in the Hill reaction. *Fed. Proc.* **25**, 893–8.

WALKER D.A. (1962) Pyruvate carboxylation and plant metabolism. *Biol. Rev.* **37**, 215–56.

WALLEN D.G. & GEEN G.H. (1971a) Light quality in relation to growth, photosynthetic rates and carbon metabolism in two species of marine plankton algae. *Mar. Biol.* **10**, 34–43.

WALLEN D.G. & GEEN G.H. (1971b) The nature of the photosynthate in natural phytoplankton populations in relation to light quality. *Mar. Biol.* **10**, 157–68.

WARBURG O. & KRIPPAHL G. (1958) Hill-Reaktionen. *Z. Naturf.* **13B**, 509–14.

WEGMAN K. (1969) On the pathways of $CO_2$ fixation in *Dunaliella*. In *Progress in Photosynthetic Research*, vol. III, ed. Metzner H. pp. 1559–64. International Biological Union, Tubingen.

WEGMAN K. (1971) Osmotic regulation of photosynthetic glycerol production in *Dunaliella*. *Biochim. biophys. Acta*, **234**, 317–23.

WETHERELL D.F. (1963) Osmotic equilibration and growth of *Scenedesmus obliquus* in saline media. *Physiologia Pl.* **16**, 82–91.

WIESSNER W. (1969) Effect of autotrophic or photo-heterotrophic growth conditions on *in vivo* absorption of visible light by green algae. *Photosynthetica*, **3**, 225–32.

WOOD H.G. & UTTER M.F. (1965) The role of $CO_2$ fixation in metabolism. In *Essays in Biochemistry*, vol. I, eds. Campbell P.N. & Greville G.D. pp. 1–27. Academic Press, London & New York.

YAMAGUCHI T., IKAWA T. & NISAZAWA K. (1969) Pathway of mannitol formation during photosynthesis in brown algae. *Pl. Cell Physiol.*, *Tokyo*, **10**, 425–40.

ZIEGLER, H. & ZIEGLER I. (1967) Die Lichtinduzierte aktivitat-steigerung der $NADP^+$-abhangigen glycerinaldehyd-3-phosphat dehydrogenase. V. Das verhalten von Meeresalgen. *Planta*, **72**, 162–9.

CHAPTER 16

# HYDROGENASE, PHOTOREDUCTION AND ANAEROBIC GROWTH

E. KESSLER

Botanisches Institut der Universität,
852 Erlangen, Germany

1   Introduction   456

2   Hydrogenase and hydrogen met-
    abolism   458
2.1 Reactions involving hydrogen-
    ase   458
    (a) Photoreduction   458
    (b) Dark reactions with $H_2$
    461
    (c) Production of $H_2$   463

2.2 Properties   of   hydrogenase
    464
2.3 Taxonomic   and   phylogenetic
    significance   466

3   Anaerobic growth   467

4   References   468

## 1 INTRODUCTION

Only a limited number of organisms are able to metabolize molecular hydrogen. The responsible enzyme, hydrogenase, is present in certain autotrophic, heterotrophic, anaerobic and aerobic bacteria, and in members of most major groups of algae. Table 16.1 shows that hydrogenase activity has been detected in about 50% of the algae examined so far. Depending upon the presence of suitable acceptors or donors, the enzyme can act in the consumption or evolution of $H_2$:

$$H_2 + 2R \rightleftharpoons 2RH.$$

Apart from its intrinsic interest, the study of hydrogen metabolism has provided important contributions to our knowledge of such fundamental processes as photosynthesis and nitrate reduction, and, more recently, has served to clarify the taxonomy of the widely used algae of the genus *Chlorella*.

Various aspects of hydrogen metabolism in algae have been reviewed by Gaffron (1944, 1960), Kessler (1960, 1962), Spruit (1962) and Bishop (1966).

**Table 16.1.** Algae with and without hydrogenase.

| Hydrogenase present |
| --- |

Cyanophyceae
*Anabaena cylindrica* (Hattori 1963)
*Nostoc muscorum* (Ward 1970a)
*Synechococcus elongatus* (Frenkel *et al.* 1950)
*Synechocystis* sp. (Frenkel *et al.* 1950)

Euglenophyceae
*Euglena gracilis* (Hartman & Krasna 1963)
*Euglena* sp. (Krasna & Rittenberg 1954)

Chlorophyceae
*Ankistrodesmus* spp. (Gaffron 1940, Kessler & Czygan 1967)
*Chlamydomonas debaryana* (Healey 1970a)
*Chlamydomonas dysosmos* (Healey 1970a)
*Chlamydomonas eugametos* (Abeles 1964)
*Chlamydomonas intermedia* (Ward 1970a)
*Chlamydomonas moewusii* (Frenkel 1949)
*Chlamydomonas reinhardtii* (Hartmann & Krasna 1963)
*Chlorella fusca* (Kessler 1967)
*Chlorella homosphaera* (Kessler & Zweier 1971)
*Chlorella kessleri* (Kessler 1967)
*Chlorella vulgaris* f. *tertia* (Kessler 1967)
*Chlorococcum vacuolatum* (Kessler unpubl.)
*Coelastrum* spp. (Kessler & Maifarth 1960, Sodomková 1969)
*Crucigenia apiculata* (Kessler unpubl.)
*Rhaphidium* spp. (Gaffron 1940, Kessler & Czygan 1967)
*Scenedesmus* sp. (Gaffron 1940, Kessler & Czygan 1967)
*Selenastrum gracile* (Kessler & Maifarth 1960)
*Ulva faciata* (Ward 1970a)
*Ulva lactuca* (Frenkel & Rieger 1951)

Phaeophyceae
*Ascophyllum nodosum* (Frenkel & Rieger 1951)

Rhodophyceae
*Porphyra umbilicalis* (Frenkel & Rieger 1951)
*Porphyridium cruentum* (Frenkel & Rieger 1951)

| No hydrogenase |
| --- |

Cyanophyceae
*Anacystis nidulans* (Richter 1961)
*Cylindrospermum* sp. (Frenkel & Rieger 1951)
*Gloeocapsa* sp. (Ward 1970a)
*Oscillatoria* sp. (Gaffron, unpubl.)

Chlorophyceae
*Bracteacoccus* spp. (Kessler unpubl.)
*Caulerpa* sp. (Ward 1970a)

**Table 16.1.**—continued

| No hydrogenase |
| --- |

Chlorophyceae – *contd.*
*Chlamydomonas pseudogloe* (Ward 1970a)
*Chlorella luteoviridis* (Kessler 1967)
*Chlorella minutissima* (Kessler 1967)
*Chlorella protothecoides* (Kessler & Zweier 1971)
*Chlorella saccharophila* (Kessler 1967)
*Chlorella vulgaris* (Kessler 1967)
*Chlorella zofingiensis* (Kessler 1967)
*Chlorococcum* spp. (Kessler, unpubl.)
*Coccomyxa* sp. (Gaffron 1940)
*Dictyosphaeria* sp. (Ward 1970a)
*Hormidium flaccidum* (Kessler & Maifarth 1960)
*Keratococcus bicaudatus* (Kessler & Czygan 1967)
*Oocystis marssonii* (Kessler, unpubl.)
*Protosiphon botryoides* (Kessler, unpubl.)
*Stichococcus bacillaris* (Kessler & Maifarth 1960)

Bacillariophyceae
*Nitzschia* sp. (Gaffron, unpubl.)

Phaeophyceae
*Sargassum* sp. (Ward 1970a)
*Turbinaria* sp. (Ward 1970a)

Rhodophyceae
*Hypnea* sp. (Ward 1970a)

## 2 HYDROGENASE AND HYDROGEN METABOLISM

### 2.1 *Reactions involving hydrogenase*

(a) *Photoreduction*

When algae containing hydrogenase are illuminated after several hours of anaerobic incubation under $H_2$, they take up $H_2$ and $CO_2$ and do not evolve $O_2$. This photosynthetic reaction, discovered and named 'photoreduction' by Gaffron (1940), can be described by the equation

$$CO_2 + 2H_2 \xrightarrow{\text{light}} (CH_2O) + H_2O.$$

It is observed only at low light intensities; above 500 to 1,000 lux its rate decreases and a reversion to normal photosynthesis with evolution of $O_2$ occurs. The reduction of $CO_2$ follows the same path, with the same intermediates, in photoreduction and in normal photosynthesis (Badin & Calvin 1950, Gingras et al. 1963, Gingras 1966a, Russell & Gibbs 1968).

The occurrence in the same algae of photosynthetic reactions with and without $O_2$ evolution provides an opportunity for studying this still least accessible partial reaction of photosynthesis. Work with specific inhibitors has led to the discovery of three groups of poisons interacting, respectively, with $O_2$ evolution, $CO_2$ reduction, and the activation of $H_2$.

Inhibitors of the first group turned out to be most important for the understanding of the mechanism of photosynthesis. Poisons which specifically inhibit the photosynthetic evolution of $O_2$ from water, acting at, or close to, photosystem II, are potent inhibitors of normal photosynthesis. Photoreduction with $H_2$, by contrast, is more or less insensitive and, at the same time, is stabilized against the reversion to photosynthetic $O_2$ evolution which is normally observed at higher light intensities. High concentrations of most of these poisons, however, also produce a 50% inhibition of photoreduction. Inhibitors of this type are hydroxylamine (Gaffron 1942a), phthiocol, menadione and other derivatives of vitamin K (Gaffron 1945a), o-phenanthroline (Gaffron 1945b), hydrogen sulphide (Frenkel et al. 1949), phenylurethane (Frenkel, cf. Kessler 1960), and herbicides of the 3' (3,4-dichlorophenyl) 1',1' dimethyl urea (DCMU) (Bishop 1958) and amino triazine type (Bishop 1962b, Rau & Grimme 1971). The same kind of inhibition is brought about by manganese deficiency. From this it has been concluded that manganese is specifically involved in photosynthetic $O_2$ evolution (Kessler 1957a). In addition, Bishop (1962a, 1964a) and Pratt and Bishop (1968) obtained X-ray mutants of Scenedesmus ('mutant 11' and others) which are incapable of normal photosynthesis but are still able to photoreduce under hydrogen-adapted conditions. These mutants appear to have lost the $O_2$-evolving photosystem II, whereas their photosystem I is still intact.

Inhibitors acting equally on photoreduction and photosynthesis, and which thus are assumed to affect $CO_2$ reduction, are 2,4-dinitrophenol (Gaffron 1942a), iodoacetamide (Gaffron 1945b), and phosphate deficiency (Kessler 1957a).

With copper deficiency (Bishop 1964b) and chloride deficiency (Grimme & Kessler 1970, Grimme 1972), i.e. deficiencies of ions supposed to act at various sites of the electron transport system of photosynthesis and photoreduction, the inhibition of photoreduction is more pronounced than that of photosynthesis.

Cyanide, carbon monoxide (Gaffron 1942a), chloretone (Gaffron 1945b), and iron deficiency (Kessler 1957a) affect photoreduction much more strongly than they affect photosynthesis. They seem to inhibit primarily the hydrogenase system.

For an explanation of photoreduction, three different mechanisms have been suggested.

(i) The possibility that 'photoreduction' is simply a combination of photosynthetic $O_2$ evolution with the hydrogenase-mediated oxyhydrogen reaction has been considered (Gaffron 1940). Evidence in support of such a mechanism is the finding that fully adapted, photoreducing algae can evolve small amounts of $O_2$ (Horwitz & Allen 1957). On the other hand, there are conditions where this simple explanation cannot apply and a 'true photoreduction' must exist. This is the case, for example, when photosynthetic $O_2$ evolution has been completely inhibited by means of sufficiently high concentrations of hydroxylamine, $o$-phenanthroline, DCMU and related inhibitors (Gaffron 1942a, 1945a,b, Bishop 1958). Mutants lacking photosystem II and the ability to evolve $O_2$ (Bishop 1962a, 1964a) are another example.

(ii) Gaffron (1944) originally suggested that in photoreduction the $H_2$ activated by hydrogenase serves to reduce a photosynthetic precursor of $O_2$ ('YOH' or a peroxide 'Z$(OH)_2$') with the formation of $H_2O$ instead of $O_2$. This would imply only a minor difference in the mechanisms of photosynthesis and photoreduction, with both photosystems and most of the photosynthetic electron transport system involved also in photoreduction. The equality of the quantum yields (Rieke 1949, Bishop 1967) and the kinetics (Rieke & Gaffron 1943) of photosynthesis and photoreduction have been considered as evidence in favour of this mechanism. It is further supported by the observation that cyclic photophosphorylation, a process involving only one of the light reactions of photosynthesis, has a quantum requirement of six (Tanner et al. 1968), compared to about 10 for photosynthesis and photoreduction.

(iii) The most recent and most widely accepted concept (Arnon et al. 1961, Stanier 1961) is that photoreduction involves photosystem I only, which produces ATP by means of cyclic phosphorylation, whereas the hydrogen activated by hydrogenase serves for $CO_2$ reduction via ferredoxin and NADP.

Unanimous agreement concerning the mechanism of photoreduction has not been reached so far. There is no doubt that, under certain conditions, the reaction can proceed with photosystem I only, i.e. in mutants lacking photosystem II, in normal algae when photosystem II has been completely inhibited, or when the organisms are illuminated with far-red light ($\lambda = 705$ to 720nm) (Bishop & Gaffron 1962, Ying et al. 1963). In addition, the lack of a dichromatic enhancement effect (Bishop & Gaffron 1963, Gingras 1966a) and the action spectrum of photoreduction (Gingras 1966b, Bishop 1967) support this conclusion. On the other hand, under normal conditions, when both photosystems are active and receive light of suitable wavelengths, there seems little reason to believe that photosystem II does not participate in the reaction. Apart from the fact that photoreduction of normal cells is usually accompanied by the evolution of small amounts of oxygen (Horwitz & Allen 1957), several recent observations provide further evidence in support of this interpretation. Thus Kessler's (1968b) work on fluorescence of aerobic and hydrogen-adapted, normal and manganese-deficient algae indicated at least some activity of photosystem II during photoreduction. Stuart and Kaltwasser (1970) were able to show that photoreduction is far less effective in far-red light than a typical photosystem I reaction (i.e. the photoproduction of hydrogen). In addition, Grimme's (1972) work on the effect of chloride deficiency and dibromothymoquinone on photoreduction indicates that a non-cyclic electron flow including to some extent also photosystem II is involved in this reaction. It seems likely that hydrogen activated by hydrogenase can enter the photosynthetic electron transport chain not only at the site of ferredoxin, but also at other points at, or close to, photosystem II (Gaffron & Bishop 1963, Gingras & Lavorel 1965, Kessler 1968b, 1970, Gibbs *et al.* 1970, Grimme 1972).

Photosynthetic bacteria are also able to use in photoreduction other hydrogen donors such as $H_2S$ and organic compounds (e.g. isopropanol), in addition to molecular hydrogen. A photoreduction with a stoichiometric uptake of $H_2S$ has been reported to occur in various algae (Nakamura 1938, Frenkel *et al.* 1949), but could not be confirmed by Knobloch (1966). Similarly, Hellmann and Kessler (unpubl.) were unable to substantiate the occurrence in algae of a photo-reduction with isopropanol, which had been reported by Vishniac and Reazin (1957).

## (b) *Dark reactions with $H_2$*

After anaerobic incubation for some time, algae containing hydrogenase can utilize molecular hydrogen to reduce a number of acceptors in the dark (Table 16.2). In the absence of added compounds, there is a weak uptake of $H_2$, which seems to be due to a reduction of cellular substances (Gaffron 1940).

**Table 16.2.** Substances reduced and not reduced with molecular hydrogen in the dark.

| *reduced* | *not reduced* |
|---|---|
| benzyl viologen (Abeles 1964) | acetate (Kessler & Maifarth 1960) |
| chloramphenicol (Czygan 1964) | chromate (Kessler & Maifarth 1960) |
| dichlorophenol indophenol (Kessler & Mai-farth 1960) | formate (Kessler & Maifarth 1960) |
| | fumarate (Kessler & Maifarth 1960) |
| dinitrophenol (Ahmed & Morris 1968) | glycine (Kessler & Maifarth 1960) |
| dithionite (Kessler & Maifarth 1960) | lactate (Kessler & Maifarth 1960) |
| FAD (Lee & Stiller 1967) | malate (Kessler & Maifarth 1960) |
| ferredoxin (Fujita & Myers 1965) | neutral red (Fujita *et al.* 1964) |
| ferricyanide (Kessler & Maifarth 1960) | phenosafranine (Fujita *et al.* 1964) |
| FMN (Lee & Stiller 1967) | sulphate (Kessler & Maifarth 1960) |
| hydroxylamine (Kessler 1957b) | tetrathionate (Kessler & Maifarth 1960) |
| methyl viologen (Abeles 1964) | thiosulphate (Kessler & Maifarth 1960) |
| methylene blue (Kessler 1957b) | |
| nitrate (Kessler 1957b) | |
| nitrite (Kessler 1957b) | |
| nitrophenol (Ahmed & Morris 1968) | |
| NAD(P) (Abeles 1964) | |
| oxygen (Gaffron 1942b) | |
| phenazine methosulphate (Fujita *et al.* 1964) | |
| pyrosulphite (Kessler & Maifarth 1960) | |
| pyruvate (Kessler & Maifarth 1960) | |
| quinone (Frenkel 1949, Kessler & Maifarth 1960) | |
| sulphite (Kessler & Maifarth 1960) | |
| toluidine blue (Fujita *et al.* 1964) | |
| triphenyl tetrazolium chloride (Abeles 1964) | |

When a small amount of $O_2$ (about 0·1 to 0·2%) is introduced after activation of the hydrogenase system, it is reduced in an oxyhydrogen reaction. In the presence of $CO_2$ there is a concomitant chemosynthetic reduction of $CO_2$ to carbohydrate (Gaffron 1942b). The theoretical quotient $H_2:O_2 = 2$ is only obtained in the presence of $CO_2$; otherwise a quotient close to 1 is observed. The coupled reduction of $CO_2$ proceeds with the consumption of 2 $H_2$. Under optimum conditions, 1 mole of $CO_2$ is reduced per 2 moles of $O_2$ taken up:

$$2O_2 + 4H_2 \rightarrow 4H_2O$$
$$CO_2 + 2H_2 \rightarrow (CH_2O) + H_2O.$$

Low concentrations of dinitrophenol inhibit primarily the coupled reduction of $CO_2$ and do not impair the oxyhydrogen reaction itself (Gaffron 1942b). On the whole, this process has certain properties in common with respiration and others with photosynthesis (Horwitz 1957). The intermediates of chemosynthetic $CO_2$ reduction are the same as in photosynthesis and photoreduction (Badin & Calvin 1950, Russell & Gibbs 1968).

In addition to $O_2$, inorganic nitrogen compounds are also of interest as possible physiological acceptors for $H_2$ activated by hydrogenase. The reduction of nitrite is by far the fastest reaction of this kind (Kessler 1957b, Damaschke & Lübke 1958b, Stiller 1966). It can be described by the equation

$$HNO_2 + 3H_2 \rightarrow NH_3 + 2H_2O.$$

The rate of this process, like that of the oxyhydrogen reaction, is increased in the presence of $CO_2$, and only under these conditions can the theoretical value of 3 $H_2$ for every nitrite reduced be observed; without $CO_2$, only 2·6 $H_2$ are consumed. This reaction, however, is not coupled with a chemosynthetic reduction of $CO_2$. In most algae it is a dark process, but in *Anabaena cylindrica* it proceeds only in the light (Hattori 1963). Thus $H_2$ activated by hydrogenase, in addition to hydrogen donors produced by respiration or photosynthesis, can be used for the reduction of nitrite and, to a lesser extent, nitrate. The superiority of nitrite as an acceptor for $H_2$ might be due to the role of ferredoxin as an electron carrier for hydrogenase and nitrite reductase, whereas nitrate reductase requires NADH.

Likewise, some inorganic sulphur compounds can be reduced. Sulphite is a reasonably good acceptor for hydrogen, but there is no reaction with sulphate (Kessler & Maifarth 1960).

Other substances that can be reduced in the dark with molecular hydrogen activated by hydrogenase include some compounds commonly used as Hill reagents in the study of photosynthesis, i.e. quinone, dichlorophenol indophenol and ferricyanide (Kessler & Maifarth 1960), and some redox dyes such as methylene blue.

Of special interest is the fact that cell-free hydrogenase preparations are able to reduce also physiological electron carriers like ferredoxin (Fujita & Myers 1965), NAD(P) (Abeles 1964, Fujita & Myers 1965) and flavines (Lee & Stiller 1967).

(c) *Production of $H_2$*

Depending upon the presence of suitable cellular or added hydrogen donors, the hydrogenase system can also evolve molecular hydrogen under anaerobic conditions. Such reactions occur predominantly under atmospheres of nitrogen, argon or helium. Their metabolic function seems to be to enable the organisms to dispose of excessive amounts of reducing power in the absence of $O_2$.

In the dark, a rather slow production by fermentation of $H_2$ and $CO_2$ is found. The rate of this reaction can be increased by an addition of glucose, pyruvate or lactate (Gaffron & Rubin 1942, Damaschke 1957, Damaschke & Lübke 1958a, Syrett & Wong 1963).

Illumination in the absence of $CO_2$ leads to a much stronger 'photoproduction' of $H_2$ and $CO_2$ (Gaffron & Rubin 1942). This reaction, which is likewise accelerated by glucose, is increased in the presence of dinitrophenol and carbonyl cyanide m-chlorophenyl hydrazone (CCCP) (Kaltwasser *et al.* 1969, Healey 1970a). These uncouplers of phosphorylation, by contrast, inhibit the evolution of $H_2$ in the dark.

The origin of this $H_2$ and the mechanism of its production have been the subject of some controversy. Two possibilities, evolution from cellular organic substances or from the water-splitting reaction of photosynthesis, have been envisaged. Some earlier investigators found a simultaneous production of $H_2$ and small amounts of $O_2$ which were assumed to originate from the photosynthetic splitting of water (Frenkel 1952, Spruit 1958, Bishop & Gaffron 1963). On the other hand, the photoproduction of $H_2$ has been shown to be insensitive against inhibitors of photosynthetic $O_2$ evolution, i.e. hydroxylamine, phenylurethane, o-phenanthroline (Damaschke & Lübke 1958c) and DCMU (Kaltwasser *et al.* 1969, Stuart & Kaltwasser 1970, Healey 1970b, Stuart & Gaffron 1972b), and to occur also in mutants unable to evolve $O_2$ (Kaltwasser *et al.* 1969, Stuart & Kaltwasser 1970). This indicates that organic substances rather than water are the ultimate source of the $H_2$ produced in the light, and that the process involves photosystem I only (cf. also Arnon *et al.* 1961, Stanier 1961). Further support for this interpretation comes from the observation that the photoproduction of $H_2$ proceeds at a high rate in far-red light, and that it does not show an Emerson enhancement effect (Stuart & Kaltwasser 1970, Healey 1970b). When photosynthesis, photoreduction, Hill reaction, light-dependent uptake of glucose and cyclic photophosphorylation have been inhibited by means of salicylaldoxime or a heat treatment, the photoproduction of $H_2$ continues as the only photochemical activity of the algae (Stuart 1971). These results indicate that this process does not depend on cyclic phosphorylation. It seems to be due to a noncyclic electron flow from organic substances through photosystem I to hydrogenase. Kinetic studies have shown that a fast initial phase of photoproduction from a pool of hydrogen donors, limited by the light reaction, can be distinguished from a slow, long-lasting second phase which is limited by the flow of electrons from fermentation. Upon addition of glucose, 0·5 mole $H_2$ is evolved per mole of glucose (Stuart & Gaffron 1971, 1972a). In spite of these results, however, the possibility has to be considered that the photoproduction of $H_2$ can follow different pathways in different organisms and under different experimental conditions (Healey 1970a). Thus photosystem II can contribute up to 50% of the electrons for this process (Stuart & Gaffron 1972c).

## 2.2  *Properties of hydrogenase*

Hydrogenase activity or, more correctly, hydrogen metabolism, can be measured by a variety of methods. For manometric determinations, the test reactions have been photoreduction (Gaffron 1940, Frenkel & Rieger 1951), the reduction of nitrite (Kessler 1957b, Kessler & Maifarth 1960, Kessler *et al.* 1963, Chiba & Sasaki 1963, Stiller 1966, Yanagi & Sasa 1966, Langner 1968, Sodomková 1969) or methylene blue (Kessler 1957b, Chiba & Sasaki 1963, Ward 1970a, Oesterheld 1971), the photoproduction of $H_2$ (Healey 1970a) and the evolution of $H_2$ from reduced methyl viologen (Ward 1970a). The exchange reaction between $H_2$ and $D_2O$ or HTO has been used by Krasna and Rittenberg (1954), Hartman and Krasna (1963), Abeles (1964) and Anand and Krasna (1965), and the ortho-para conversion of hydrogen by Krasna and Rittenberg (1954). Various electro-chemical methods have been applied by Damaschke (1957), Damaschke and Lübke (1958c), Spruit (1958) and Wang *et al.* (1971), and mass spectrometry by Stuart *et al.* (1972) and Stuart and Gaffron (1972a). It is important to note that other enzymes, in addition to hydrogenase, are involved in most of these reactions (e.g. photosynthetic enzymes and electron carriers in photoreduction and photoproduction of $H_2$, and nitrite reductase in nitrite reduction). The exchange reaction, however, seems to be a process catalysed by hydrogenase only.

The most striking property of hydrogenase is its sensitivity against $O_2$, which seems to be much more pronounced in algae than in certain aerobic bacteria like, for example, the chemosynthetic hydrogen bacteria. Thus, in algae, several hours of anaerobic incubation are required for the enzyme to become fully active. This process has been termed 'adaptation' by Gaffron (1940). However, it seems to involve a reductive activation of the aerobically inactive, constitutive enzyme rather than an adaptive formation of hydrogenase. This follows from the observation that hydrogenase activity appears as quickly under atmospheres of nitrogen or noble gases, i.e. in the absence of its substrate, as it does under $H_2$ (Gaffron & Rubin 1942, Ward 1970a). In addition, Gaffron (1942a) has shown that dinitrophenol, an inhibitor of the formation of adaptive enzymes, does not prevent the appearance of hydrogenase activity. The occurrence in some algae of hydrogen metabolism after only a few minutes of anaerobic incubation (Frenkel & Rieger 1951, Frenkel & Lewin 1954, Gingras *et al.* 1963) points in the same direction. On the other hand, Oesterheld (1971) has found that inhibitors of protein synthesis like actinomycin, puromycin and gentamycin produce a partial inhibition of the appearance of hydrogenase activity under anaerobic conditions.

Thus 'adaptation' might consist of both reductive activation of existing enzyme and adaptive formation of more hydrogenase.

The time course of hydrogenase activation has been measured by Kessler (1957b, 1968a), Kessler and Maifarth (1960), Kessler *et al.* (1963), Ward (1970a) and Wang *et al.* (1971). It shows a rise to an optimum which is reached after 2 to 30 or more hours and followed by a more or less pronounced decline of activity. Some reports that an addition of carrot extract, glucose or iron + EDTA produces a full activity of hydrogenase immediately after anaerobic conditions have been established (Chiba & Sasaki 1963, Stiller & Lee 1964a,b, Stiller 1966, Sasaki 1966), could not be confirmed by Kessler (1966, 1968a) and Oesterheld (1971).

Activation of hydrogenase can be prevented by hydroxylamine, phthiocol, o-phenanthroline and hydrogen sulphide; if added after adaptation, however, these inhibitors do not interfere with, and even stabilize hydrogen metabolism against inactivation by high light intensities (Gaffron 1942a, 1945a,b, Frenkel *et al.* 1949). By contrast, cyanide, carbon monoxide and chloretone (Gaffron 1942a, 1945b), like iron deficiency (Kessler 1957a), seem to be rather specific inhibitors of hydrogenase.

Hydrogenase activity can be increased and stabilized by means of an improved supply of iron + EDTA (Sasaki 1966, Yanagi & Sasa 1966, Kessler 1968a). The effect of glucose, on the other hand, is variable and ranges from inhibition to increased hydrogen metabolism under different experimental conditions (Stiller & Lee 1964a,b, Sasaki 1966, Yanagi & Sasa 1966, Kessler 1966, Oesterheld 1971). Under nitrogen deficiency, there remains some hydrogenase activity, but nitrite can no longer serve as an acceptor for $H_2$ and has to be replaced by methylene blue (Oesterheld 1971).

Hydrogenase activity in a strain of *Chlorella fusca* has been shown to exhibit seasonal variations under constant experimental and growth conditions (Kessler & Langner 1962, Langner 1968).

Inactivation or 'deadaptation' of hydrogenase can be brought about by rather low concentrations of $O_2$ which are only slightly higher than those required for the oxyhydrogen reaction (Gaffron 1940, 1942b, Horwitz 1957). Likewise, illumination with higher light intensities produces an inactivation of hydrogenase and a 'reversion' of photoreduction to normal photosynthesis (Gaffron 1940). This effect is due to increasing photosynthetic $O_2$ evolution even in adapted algae (Horwitz & Allen 1957, Stuart & Gaffron 1972a). Gaffron (1942a) observed that in the presence of certain concentrations of hydroxylamine, which inhibit photosynthetic $O_2$ evolution completely, there is still an inactivation of photoreduction by higher light intensities. This indicates also that precursors of photosynthetic $O_2$ can inactivate the hydrogenase system.

In spite of its high sensitivity to $O_2$, there seems to be some slight residual activity of hydrogenase, or a system closely associated with it, even under normal, aerobic conditions. This conclusion was derived from the observation that algae without hydrogenase, like higher plants, lose most of their chlorophyll under manganese-deficient conditions, whereas algae with hydrogenase do not become chlorotic (Kessler 1968c). The induction of fluorescence of normal algae with and without hydrogenase also supports this assumption (Kessler 1970).

Our knowledge concerning the chemical nature of algal hydrogenase and its mechanism of action is still very scanty. Work with whole cells has indicated that the algal enzyme, like that of bacteria, might contain iron (Kessler 1957a, 1968a, Sasaki 1966, Yanagi & Sasa 1966). Algal hydrogenase, like that of bacteria, seems to be closely linked to ferredoxin (Fujita & Myers 1965) and to require active SH-groups (Hartman & Krasna 1964). A heterolytic splitting of $H_2$ is supposed to take place during activation by hydrogenase, with the formation of an enzyme hydride ($EH^-$) (Krasna & Rittenberg 1954, Hartman & Krasna 1963):

$$E + H_2 \rightleftharpoons EH^- + H^+.$$

Cell-free preparations with hydrogenase activity have been obtained from species of *Anabaena* (Fujita *et al.* 1964, Fujita & Myers 1965), *Chlamydomonas* (Abeles 1964, Ward 1970a,b), *Chlorella* (Lee & Stiller 1967) and *Scenedesmus* (Ward 1970a). These usually particulate systems reduce a variety of acceptors (cf. Table 16.2), evolve $H_2$ from reduced methyl viologen and catalyze the exchange reaction. Ward (1970a,b) was able to solubilize the hydrogenase (molecular weight about 60,000), to show the presence of isoenzymes, and to obtain an activation of hydrogenase even in cell-free extracts of *Chlamydomonas*.

### 2.3 *Taxonomic and phylogenetic significance*

Hydrogenase shows a random distribution among the bacteria, blue-green algae and all major groups of eukaryotic algae (cf. Table 16.1). Thus its range extends beyond that of nitrogenase which seems to be restricted to prokaryotic organisms. Like nitrogenase, and in contrast to other biochemical characters such as the presence of chlorophyll *b* or phycobilins, it cannot be used for the delineation of higher systematic groups. On the genus, species and subspecies level, however, hydrogenase has proved to be a valuable taxonomic character. Thus members of the genera *Scenedesmus* and *Ankistrodesmus* (including *Rhaphidium*) all seem to contain hydrogenase, and *Keratococcus* can be distinguished by its lack of hydrogenase activity from the closely related *Ankistrodesmus* (Kessler & Czygan 1967). The *Chlorella* species, taxonomically ill-defined for a long time, could be separated on the basis of the presence or absence of hydrogenase (and other physiological and biochemical criteria) (Kessler *et al.* 1963, Kessler 1967, Kessler & Zweier 1971), and the biochemical characterization of the species was confirmed and supported by morphological and cytological criteria (Fott & Nováková 1969). Hydrogenase might also turn out to be important as a character in other taxonomically difficult genera such as *Chlamydomonas* (Ward 1970a) and *Chlorococcum* (Kessler unpubl.).

The special importance of hydrogenase in algal taxonomy stems from the assumption that it is very likely to be a truly primitive, phylogenetically significant characteristic (Kessler 1962, 1967). There seems to be general agreement that the primaeval atmosphere of the earth was rich in $H_2$ and free of $O_2$. Under such conditions, hydrogenase was probably a very important enzyme, enabling the early organisms to utilize for various biochemical processes an ubiquitous source of reducing power. Thus photoreduction with $H_2$ most likely preceded normal photosynthesis (cf. Gaffron 1962). The present algae, however, are aerobic organisms usually living in an environment with a high concentration of $O_2$ which tends to keep the hydrogenase in a largely inactive state. Therefore it can be regarded as a phylogenetic relic, and it seems very unlikely that such an enzyme could appear anew under aerobic conditions, after the transformation due to photosynthetic $O_2$ evolution of the primaeval reducing atmosphere.

For a long time, the persistence into the geological present of an aerobically inactive and seemingly useless enzyme like algal hydrogenase was considered rather enigmatic (Kessler 1962). However, the enzyme has been shown to be strikingly resistant towards mutation (Kessler & Czygan 1966, Pratt & Bishop 1968). In addition, its slight residual aerobic activity might render it less useless than it was previously assumed to be (Kessler 1968c, 1970).

## 3  ANAEROBIC GROWTH

Occasionally algae are found at locations free of $O_2$ and, sometimes, rich in $H_2S$. However, it has been doubted that they are able to grow under these conditions, and algal fermentation has been supposed to be insufficient for the maintenance of growth (Gibbs 1962, Gibbs et al. 1970).

Scenedesmus and Chlamydomonas do not grow with the oxyhydrogen reaction (Gaffron 1944, Lewin 1950), and Nührenberg et al. (1968) were unable to obtain growth of Chlorella with photoreduction. Frenkel and Rieger (1951), however, found an increase of dry weight and pigments with Porphyridium, and Ochromonas malhamensis has been reported to grow anaerobically in the presence of isopropanol with a process allegedly resembling photoreduction (Vishniac & Reazin 1957).

Concerning 'anaerobic growth' of algae, one has to distinguish between truly anaerobic conditions, i.e. in the dark or with photosynthetic $O_2$ evolution inhibited in the light, and less strict anaerobiosis with at least some photosynthetic $O_2$ production in the light.

Anaerobic, heterotrophic growth in the dark has been observed with a strain of Nostoc from the coralloid roots of cycads (Hoare et al. 1971) and with several green algae from waste stabilization ponds (Wiedeman & Bold 1965).

Mixotrophic or autotrophic anaerobic growth in the light has been reported for Anabaena (Stewart & Pearson 1970), Plectonema (Stewart & Lex 1970), Nostoc (Hoare et al. 1971), Ochromonas malhamensis (Vishniac & Reazin 1957), Cyanidium caldarium (Seckbach et al. 1970), several green flagellates (Pringsheim

& Wiessner 1960), various green algae (Wiedeman & Bold 1965), some *Chlorella* strains (Enebo 1967, Battley & Haditirto 1968, Orcutt *et al.* 1970) and *Pseudochlorococcum* (Archibald 1970). The necessity of photosynthetic $O_2$ evolution for the growth of *Anabaena* and *Chlorella*, however, has been stressed by Stewart and Pearson (1970) and Nührenberg *et al.* (1968).

## 4 REFERENCES

ABELES F.B. (1964) Cell-free hydrogenase from *Chlamydomonas*. *Pl. Physiol., Lancaster*, **39**, 169–76.

AHMED J. & MORRIS I. (1968) The effects of 2,4-dinitrophenol and other uncoupling agents on the assimilation of nitrate and nitrite by *Chlorella*. *Biochim. biophys. Acta*, **162**, 32–8.

ANAND S.R. & KRASNA A.I. (1965) Catalysis of the $H_2$-HTO exchange by hydrogenase. A new assay for hydrogenase. *Biochemistry, N.Y.* **4**, 2747–53.

ARCHIBALD P.A. (1970) *Pseudochlorococcum*, a new chlorococcalean genus. *J. Phycol.* **6**, 127–32.

ARNON D.I., LOSADA M., NOZAKI M. & TAGAWA K. (1961) Photoproduction of hydrogen, photofixation of nitrogen and a unified concept of photosynthesis. *Nature, Lond.* **190**, 601–6.

BADIN E.J. & CALVIN M. (1950) The path of carbon in photosynthesis. IX. Photosynthesis, photoreduction and the hydrogen-oxygen-carbon dioxide dark reaction. *J. Am. chem. Soc.* **72**, 5266–70.

BATTLEY E.H. & HADITIRTO S. (1968) Studies on the anaerobic growth of a biotype of *Chlorella vulgaris. Antonie van Leeuwenhoek*, **34**, 234–8.

BISHOP N.I. (1958) The influence of the herbicide, DCMU, on the oxygen-evolving system of photosynthesis. *Biochim. biophys. Acta*, **27**, 205–6.

BISHOP N.I. (1962a) Separation of the oxygen-evolving system of photosynthesis from the photochemistry in a mutant of *Scenedesmus. Nature, Lond.* **195**, 55–7.

BISHOP N.I. (1962b) Inhibition of the oxygen-evolving system of photosynthesis by aminotriazines. *Biochim. biophys. Acta*, **57**, 186–9.

BISHOP N.I. (1964a) Mutations of unicellular green algae and their application to studies on the mechanism of photosynthesis. *Rec. chem. Progr.* **25**, 181–95.

BISHOP N.I. (1964b) Site of action of copper in photosynthesis. *Nature, Lond.* **204**, 401–2.

BISHOP N.I. (1966) Partial reactions of photosynthesis and photoreduction. *A. Rev. Pl. Physiol.* **17**, 185–208.

BISHOP N.I. (1967) Comparison of the action spectra and quantum requirements for photosynthesis and photoreduction of *Scenedesmus. Photochem. Photobiol.* **6**, 621–8.

BISHOP N.I. & GAFFRON H. (1962) Photoreduction at λ 705 mμ in adapted algae. *Biochem. biophys. Res. Commun.* **8**, 471–6.

BISHOP N.I. & GAFFRON H. (1963) On the interrelation of the mechanisms for oxygen and hydrogen evolution in adapted algae. In: *Photosynthetic Mechanisms in Green Plants.* eds. Kok B. & Jagendorf A.T. pp. 441–51. N.A.S.-N.R.C., Washington, D.C. **1145**.

CHIBA Y. & SASAKI H. (1963) Hydrogenase activity in *Scenedesmus* $D_3$ cultured in a medium containing carrot extract. *Pl. Cell Physiol., Tokyo*, **4**, 41–7.

CZYGAN F.C. (1964) Untersuchungen über den Abbau von Chloramphenicol durch Grünalgen. *Naturwissenschaften*, **51**, 541–2.

DAMASCHKE K. (1957) Die Wasserstoffgärung von *Chlorella* im Dunkeln nach Anaerobiose unter Stickstoff. *Z. Naturf.* **12b**, 441–3.

DAMASCHKE K. & LÜBKE M. (1958a) Hemmung der Wasserstoffgärung von *Chlorella pyrenoidosa* durch Gifte. *Z. Naturf.* **13b,** 54–5.

DAMASCHKE K. & LÜBKE M. (1958b) Über die Fähigkeit der *Chlorella pyrenoidosa* zur anaeroben Nitritreduktion. *Z. Naturf.* **13b,** 134–5.

DAMASCHKE K. & LÜBKE M. (1958c) Die Wirkung verschiedener Gifte auf die Entstehung von Wasserstoff bei Belichtung anaerob inkubierter *Chlorella pyrenoidosa. Z. Naturf.* **13b,** 172–82.

ENEBO L. (1967) A methane-consuming green alga. *Acta chem. scand.* **21,** 625–32.

FOTT B. & NOVÁKOVÁ M. (1969) A monograph of the genus *Chlorella.* The fresh water species. In *Studies in Phycology,* ed. Fott B. pp. 10–74. Academia, Prague.

FRENKEL A. (1949) A study of the hydrogenase systems of green and blue-green algae. *Biol. Bull.* **97,** 261–2.

FRENKEL A.W. (1952) Hydrogen evolution by the flagellate green alga, *Chlamydomonas moewusii. Archs. Biochem. Biophys.* **38,** 219–30.

FRENKEL A., GAFFRON H. & BATTLEY E.H. (1949) Photosynthesis and photoreduction by a species of blue-green alga. *Biol. Bull.* **97,** 269.

FRENKEL A., GAFFRON H. & BATTLEY E.H. (1950) Photosynthesis and photoreduction by the blue-green alga, *Synechococcus elongatus,* Näg. *Biol. Bull.* **99,** 157–62.

FRENKEL A.W. & LEWIN R.A. (1954) Photoreduction by *Chlamydomonas. Am. J. Bot.* **41,** 586–9.

FRENKEL A.W. & RIEGER C. (1951) Photoreduction in algae. *Nature, Lond.* **167,** 1030.

FUJITA Y. & MYERS J. (1965) Hydrogenase and NADP-reduction reactions by a cell-free preparation of *Anabaena cylindrica. Archs. Biochem. Biophys.* **111,** 619–25.

FUJITA Y., OHAMA H. & HATTORI A. (1964) Hydrogenase activity of cell-free preparation obtained from the blue-green alga, *Anabaena cylindrica. Pl. Cell Physiol., Tokyo,* **5,** 305–14.

GAFFRON H. (1940) Carbon dioxide reduction with molecular hydrogen in green algae. *Am. J. Bot.* **27,** 273–83.

GAFFRON H. (1942a) The effect of specific poisons upon the photoreduction with hydrogen in green algae. *J. gen. Physiol.* **26,** 195–217.

GAFFRON H. (1942b) Reduction of carbon dioxide coupled with the oxyhydrogen reaction in algae. *J. gen. Physiol.* **26,** 241–67.

GAFFRON H. (1944) Photosynthesis, photoreduction and dark reduction of carbon dioxide in certain algae. *Biol. Rev. Cambridge Phil. Soc.* **19,** 1–20.

GAFFRON H. (1945a) Some effects of derivatives of vitamin K on the metabolism of unicellular algae. *J. gen. Physiol.* **28,** 259–68.

GAFFRON H. (1945b) o-Phenanthroline and derivatives of vitamin K as stabilizers of photoreduction in *Scenedesmus. J. gen. Physiol.* **28,** 269–85.

GAFFRON H. (1960) Energy storage: Photosynthesis. In *Plant Physiology,* vol. IB, ed. Steward F.C. pp. 3–277. Academic Press, New York & London.

GAFFRON H. (1962) On dating stages in photochemical evolution. In *Horizons in Biochemistry,* eds. Kasha M. & Pullman B. pp. 59–89. Academic Press, New York & London.

GAFFRON H. & BISHOP N.I. (1963) The photolysis of water in living cells and the role of the two-pigment system. La Photosynthèse, *Coll. Int. Centre Nat. Rech. Sci. No.* **119,** 229–41.

GAFFRON H. & RUBIN J. (1942) Fermentative and photochemical production of hydrogen in algae. *J. gen. Physiol.* **26,** 219–40.

GIBBS M. (1962) Fermentation. In *Physiology and Biochemistry of Algae,* ed. Lewin R.A. pp. 91–7. Academic Press, New York & London.

GIBBS M., LATZKO E., HARVEY M.J., PLAUT Z. & SHAIN Y. (1970) Photosynthesis in the algae. *Ann. N.Y. Acad. Sci.* **175,** 541–54.

GINGRAS G. (1966a) Étude comparative, chez quelques algues, de la photosynthèse et de la photoréduction réalisée en présence d'hydrogène. *Physiol. végét.* **4,** 1–65

Q

GINGRAS G. (1966b) Pigments sensibilisateurs de la photoréduction. *Currents in Photosynthesis*, eds. Thomas J.B. & Goedheer J.C. pp. 187–95. Donker Publishers, Rotterdam.

GINGRAS G., GOLDSBY R.A. & CALVIN M. (1963) Carbon dioxide metabolism in hydrogen-adapted *Scenedesmus*. *Archs. Biochem. Biophys.* 100, 178–84.

GINGRAS G. & LAVOREL J. (1965) Propriétés de fluorescence de *Scenedesmus* adapté à l'hydrogène. *Physiol. végét.* 3, 109–20.

GRIMME L.H. (1972) Evidence for the participation of the $O_2$ evolving system and a non-cyclic electron transport in $CO_2$ photoreduction of $H_2$-adapted *Chlorella fusca*. *Proc. 2nd Internat. Congress on Photosynthesis Research*, Stresa, pp. 2011–19.

GRIMME L.H. & KESSLER E. (1970) Chloride effect on photosynthesis and photoreduction in *Chlorella*. *Naturwissenschaften*, 57, 133.

HARTMAN H. & KRASNA A.I. (1963) Studies on the 'adaptation' of hydrogenase in *Scenedesmus*. *J. biol. Chem.* 238, 749–57.

HARTMAN H. & KRASNA A.I. (1964) Properties of the hydrogenase of *Scenedesmus*. *Biochim. biophys. Acta*, 92, 52–8.

HATTORI A. (1963) Effect of hydrogen on nitrite reduction by *Anabaena cylindrica*. *Studies on Microalgae and Photosynthetic Bacteria. Jap. Soc. Pl. Cell Physiol.* pp. 485–92. Tokyo University Press.

HEALEY F.P. (1970a) Hydrogen evolution by several algae. *Planta*, 91, 220–6.

HEALEY F.P. (1970b) The mechanism of hydrogen evolution by *Chlamydomonas moewusii*. *Pl. Physiol., Lancaster*, 45, 153–9.

HOARE D.S., INGRAM L.O., THURSTON E.L. & WALKUP R. (1971) Dark heterotrophic growth of an endophytic blue-green alga. *Arch. Mikrobiol.* 78, 310–21.

HORWITZ L. (1957) Observations on the oxyhydrogen reaction in *Scenedesmus* and its relation to respiration and photosynthesis. *Archs. Biochem. Biophys.* 66, 23–44.

HORWITZ L. & ALLEN F.L. (1957) Oxygen evolution and photoreduction in adapted *Scenedesmus*. *Archs. Biochem. Biophys.* 66, 45–63.

KALTWASSER H., STUART T.S. & GAFFRON H. (1969) Light-dependent hydrogen evolution by *Scenedesmus*. *Planta*, 89, 309–22.

KESSLER E. (1957a) Stoffwechselphysiologische Untersuchungen an Hydrogenase enthaltenden Grünalgen. I. Über die Rolle des Mangans bei Photoreduktion und Photosynthese. *Planta*, 49, 435–54.

KESSLER E. (1957b) Stoffwechselphysiologische Untersuchungen an Hydrogenase enthaltenden Grünalgen. II. Dunkel-Reduktion von Nitrat und Nitrit mit molekularem Wasserstoff. *Arch. Mikrobiol.* 27, 166–81.

KESSLER E. (1960) Biochemische Variabilität der Photosynthese: Photoreduktion und verwandte Photosynthesetypen. *Handb. d. Pflanzenphysiol.* 5/1, 951–65.

KESSLER E. (1962) Hydrogenase und $H_2$-Stoffwechsel bei Algen. *Vortr. Gesamtgebiet Bot.*, N.F. 1, 92–101.

KESSLER E. (1966) The effect of glucose on hydrogenase activity in *Chlorella*. *Biochim. biophys. Acta*, 112, 173–5.

KESSLER E. (1967) Physiologische und biochemische Beiträge zur Taxonomie der Gattung *Chlorella*. III. Merkmale von 8 autotrophen Arten. *Arch. Mikrobiol.* 55, 346–57.

KESSLER E. (1968a) Iron supply and hydrogenase activity in green algae. *Arch. Mikrobiol.* 61, 77–80.

KESSLER E. (1968b) Effect of hydrogen adaptation on fluorescence in normal and manganese-deficient algae. *Planta*, 81, 264–73.

KESSLER E. (1968c) Effect of manganese deficiency on growth and chlorophyll content of algae with and without hydrogenase. *Arch. Mikrobiol.* 63, 7–10.

KESSLER E. (1970) Photosynthesis, photooxidation of chlorophyll and fluorescence of normal and manganese-deficient *Chlorella* with and without hydrogenase. *Planta*, 92, 222–34.

KESSLER E. & CZYGAN F.-C. (1966) Physiologische und biochemische Beiträge zur Taxonomie der Gattung *Chlorella*. II. Untersuchungen an Mutanten. *Arch. Mikrobiol.* **54**, 37–45.

KESSLER E. & CZYGAN F.-C. (1967) Physiologische und biochemische Beiträge zur Taxonomie der Gattungen *Ankistrodesmus* und *Scenedesmus*. I. Hydrogenase, Sekundär-Carotinoide und Gelatine-Verflüssigung. *Arch. Mikrobiol.* **55**, 320–6.

KESSLER E. & LANGNER W. (1962) Jahresperiodische Aktivitätsschwankungen bei einer *Chlorella*. *Naturwissenschaften*, **49**, 331–2.

KESSLER E., LANGNER W., LUDEWIG I. & WIECHMANN H. (1963) Bildung von Sekundär-Carotinoiden bei Stickstoffmangel und Hydrogenase-Aktivität als taxonomische Merkmale in der Gattung *Chlorella*. *Studies on Microalgae and Photosynthetic Bacteria.* *Jap. Soc. Pl. Cell Physiol.* pp. 7–20. Tokyo University Press.

KESSLER E. & MAIFARTH H. (1960) Vorkommen und Leistungsfähigkeit von Hydrogenase bei einigen Grünalgen. *Arch. Mikrobiol.* **37**, 215–25.

KESSLER E. & ZWEIER I. (1971) Physiologische und biochemische Beiträge zur Taxonomie der Gattung *Chlorella*. V. Die auxotrophen und mesotrophen Arten. *Arch. Mikrobiol.* **79**, 44–8.

KNOBLOCH K. (1966) Photosynthetische Sulfid-Oxydation grüner Pflanzen. I. *Planta*, **70**, 73–86.

KRASNA A.I. & RITTENBERG D. (1954) The mechanism of action of the enzyme hydrogenase. *J. Am. chem. Soc.* **76**, 3015–20.

LANGNER W. (1968) Untersuchungen zur Jahresperiodik bei Grünalgen. *Beitr. Biol. Pfl.* **45**, 1–38.

LEE J.K.H. & STILLER M. (1967) Hydrogenase activity in cell-free preparation of *Chlorella*. *Biochim. biophys. Acta*, **132**, 503–5.

LEWIN J.C. (1950) Obligate autotrophy in *Chlamydomonas moewusii* Gerloff. *Science, N.Y.* **112**, 652–3.

NAKAMURA H. (1938) Über die Kohlensäureassimilation bei niederen Algen in Anwesenheit des Schwefelwasserstoffs. *Acta phytochim., Tokyo*, **10**, 271–81.

NÜHRENBERG B., LESEMANN D. & PIRSON A. (1968) Zur Frage eines anaeroben Wachstums von einzelligen Grünalgen. *Planta*, **79**, 162–80.

OESTERHELD H. (1971) Das Verhalten von Nitratreduktase, Nitritreduktase, Hydrogenase und anderen Enzymen von *Ankistrodesmus braunii* bei Stickstoffmangel. *Arch. Mikrobiol.* **79**, 25–43.

ORCUTT D.M., RICHARDSON B. & HOLDEN R.D. (1970) Effects of hypobaric and hyperbaric helium atmospheres on the growth of *Chlorella sorokiniana*. *Appl. Microbiol.* **19**, 182–183.

PRATT L.H. & BISHOP N.I. (1968) Chloroplast reactions of photosynthetic mutants of *Scenedesmus obliquus*. *Biochim. biophys. Acta*, **153**, 664–74.

PRINGSHEIM E.G. & WIESSNER W. (1960) Photo-assimilation of acetate by green organisms. *Nature, Lond.* **188**, 919–21.

RAU I. & GRIMME L.H. (1971) Zum Einfluß verschieden substituierter s-Triazine auf Stoffwechselreaktionen der Grünalge *Ankistrodesmus braunii*. *Z. Naturf.* **26b**, 919–21.

RICHTER G. (1961) Die Auswirkungen von Mangan-Mangel auf Wachstum und Photosynthese bei der Blaualge *Anacystis nidulans*. *Planta*, **57**, 202–14.

RIEKE F.F. (1949) Quantum efficiencies for photosynthesis and photoreduction in green plants. In *Photosynthesis in Plants*, eds. Franck J. & Loomis W.E. pp. 251–72. Iowa.

RIEKE F.F. & GAFFRON H. (1943) Flash saturation and reaction periods in photosynthesis. *J. Phys. Chem.* **47**, 299–308.

RUSSELL G.K. & GIBBS M. (1968) Evidence for the participation of the reductive pentose phosphate cycle in photoreduction and the oxyhydrogen reaction. *Pl. Physiol., Lancaster*, **43**, 649–52.

SASAKI H. (1966) Effects of culture conditions on hydrogenase activity of *Scenedesmus* $D_3$. *Pl. Cell Physiol., Tokyo*, **7**, 231–41.

SECKBACH J., BAKER F.A. & SHUGARMAN P.M. (1970) Algae thrive under pure $CO_2$. *Nature, Lond.* **227**, 744–5.

SODOMKOVÁ M. (1969) Physiologische und biochemische Charakterisierung einiger *Coelastrum*-Arten. *Arch. Protistenkunde,* **111**, 223–7.

SPRUIT C.J.P. (1958) Simultaneous photoproduction of hydrogen and oxygen by *Chlorella. Mededel. Landbouwhogeschool Wageningen,* **58**, 1–17.

SPRUIT C.J.P. (1962) Photoreduction and anaerobiosis. In *Physiology and Biochemistry of Algae,* ed. Lewin R.A. pp. 47–60. Academic Press, New York & London.

STANIER R.Y. (1961) Photosynthetic mechanisms in bacteria and plants: Development of a unitary concept. *Bact. Rev.* **25**, 1–17.

STEWART W.D.P. & LEX M. (1970) Nitrogenase activity in the blue-green alga *Plectonema boryanum* strain 594. *Arch. Mikrobiol.* **73**, 250–60.

STEWART W.D.P. & PEARSON H.W. (1970) Effects of aerobic and anaerobic conditions on growth and metabolism of blue-green algae. *Proc. R. Soc. B.* **175**, 293–311.

STILLER M. (1966) Hydrogenase mediated nitrite reduction in *Chlorella. Pl. Physiol., Lancaster,* **41**, 348–52.

STILLER M. & LEE J.K.H. (1964a) Hydrogenase-mediated reactions in *Chlorella. Pl. Physiol., Lancaster,* **39**, xv.

STILLER M. & LEE J.K.H. (1964b) Hydrogenase activity in *Chlorella. Biochim. biophys. Acta,* **93**, 174–6.

STUART T.S. (1971) Hydrogen production by photosystem I of *Scenedesmus:* Effect of heat and salicylaldoxime on electron transport and photophosphorylation. *Planta,* **96**, 81–92.

STUART T.S. & GAFFRON H. (1971) The kinetics of hydrogen photoproduction by adapted *Scenedesmus. Planta,* **100**, 228–43.

STUART T.S. & GAFFRON H. (1972a) The gas exchange of hydrogen-adapted algae as followed by mass spectrometry. *Pl. Physiol., Lancaster,* **50**, 136–40.

STUART T.S. & GAFFRON H. (1972b) The mechanism of hydrogen photoproduction by several algae. I. The effect of inhibitors of photophosphorylation. *Planta,* **106**, 91–100.

STUART T.S. & GAFFRON H. (1972c) The mechanism of hydrogen photoproduction by several algae. II. The contribution of photosystem II. *Planta,* **106**, 101–12.

STUART T.S., HEROLD E.W. & GAFFRON H. (1972) A simple combination mass spectrometer inlet and oxygen electrode chamber for sampling gases dissolved in liquids. *Anal. Biochem.* **46**, 91–100.

STUART T.S. & KALTWASSER H. (1970) Photoproduction of hydrogen by photosystem I of *Scenedesmus. Planta,* **91**, 302–13.

SYRETT P.J. & WONG H.A. (1963) The fermentation of glucose by *Chlorella vulgaris. Biochem. J.* **89**, 308–15.

TANNER W., LOOS E., KLOB W. & KANDLER O. (1968) The quantum requirement for light-dependent anaerobic glucose assimilation by *Chlorella vulgaris. Z. PflPhysiol.* **59**, 301–3.

VISHNIAC W. & REAZIN G.H. (1957) Photoreduction in *Ochromonas malhamensis.* In *Research in Photosynthesis,* eds. Gaffron H., Brown A.H., French C.S., Livingston R., Rabinowitch E., Strehler B.L. & Tolbert N.E. pp. 239–42. Interscience Publ., Inc., New York.

WANG R., HEALEY F.P. & MYERS J. (1971) Amperometric measurement of hydrogen evolution in *Chlamydomonas. Pl. Physiol., Lancaster,* **48**, 108–10.

WARD M.A. (1970a) Whole cell and cell-free hydrogenases of algae. *Phytochemistry,* **9**, 259–66.

WARD M.A. (1970b) Adaptation of hydrogenase in cell-free preparations from *Chlamydomonas. Phytochemistry,* **9**, 267–74.

WIEDEMAN V.E. & BOLD H.C. (1965) Heterotrophic growth of selected waste-stabilization pond algae. *J. Phycol.* **1**, 66–9.

YANAGI S. & SASA T. (1966) Changes in hydrogenase activity during the synchronous growth of *Scenedesmus obliquus* D$_3$. *Pl. Cell Physiol., Tokyo*, **7**, 593–8.

YING H.-C., LI T.-Y. & SHEN Y.-K. (1963) Comparative studies on the mechanism of photosynthesis. I. Photoreduction by *Scenedesmus* sp. under light of different wavelengths. *Acta biochim. biophys. sin.* **3**, 497–501.

CHAPTER 17

# PHOTORESPIRATION

## N. E. TOLBERT

Department of Biochemistry,
Michigan State University,
East Lansing, Michigan 48823, U.S.A.

**1    Introduction**   474

**2    Glycolate pathway**   476
2.1   Magnitude   477
2.2   $^{14}$C-labelling experiments on glycolate formation and metabolism   477
2.3   Glycolate biosynthesis   478
2.4   P-glycolate phosphatase and P-glycerate phosphatase   480
2.5   Glycolate oxidation to glyoxylate   480
2.6   Catalase   481
2.7   Glycolate-glyoxylate shuttle   482
2.8   Conversion of glyoxylate to glycine and serine   483
2.9   Interconversions between phosphoglycerate and serine   483

**3    Enzyme location and microbodies**   484
3.1   Peroxisomes   484
3.2   Glyoxysomes   485

**4    Excretion during photosynthesis**   486
4.1   General   486
4.2   Glycolate assimilation   488

**5    Function**   488

**6    Physiological experiments**   489
6.1   $O_2$ uptake during photorespiration   490
6.2   $CO_2$ release during photorespiration   491
6.3   Effect of $O_2$ and $CO_2$ concentration   492
6.4   $CO_2/O_2$ quotient   493
6.5   Synchronous cultures   494
6.6   pH effects   494
6.7   Light intensity and quality   495
6.8   Inhibitors   496
6.9   Phylogeny   497

**7    References**   498

## 1   INTRODUCTION

Photorespiration is defined as a light dependent $O_2$ uptake and $CO_2$ release that occurs in photosynthetic tissues, and it has been extensively investigated with higher plants (Hatch *et al.* 1971, Jackson & Volk 1970, Tolbert 1971, Zelitch 1964, 1971, Richardson 1974). In the present summary of this phenomenon in algae, it is concluded that photorespiration is also present and important in this group of plants. Photorespiration can be explained metabolically by glycolate biosynthesis in the chloroplasts followed by its metabolism, which in leaves occurs in peroxisomes and mitochondria. The total sequence, as shown in Fig. 17.1, has also been referred to as the glycolate pathway (Tolbert 1963, 1971).

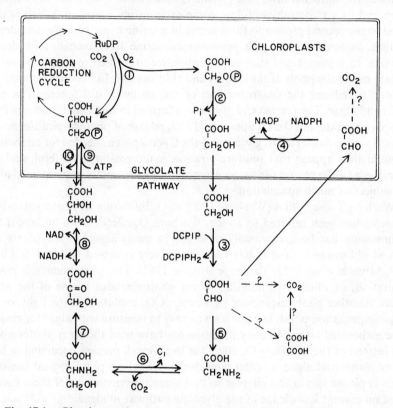

**Fig. 17.1.** Glycolate pathway in algae: 1. Ribulose diphosphate oxygenase or carboxylase; 2. Phosphoglycolate phosphatase; 3. Glycolate:DCPIP oxidoreductase or glycolate dehydrogenase; 4. NADPH glyoxylate reductase; 5. Glutamate-glyoxylate aminotransferase; 6. Serine hydroxymethyl transferase and glycine oxidase; 7. Serine-glyoxylate aminotransferase or glutamate-hydroxypyruvate aminotransferase; 8. NAD hydroxypyruvate reductase or glycerate dehydrogenase; 9. Glycerate kinase; 10. 3-Phosphoglycerate phosphatase.

These reactions occurring in unicellular green algae are similar to those in higher plants with a few modifications, particularly in the enzymatic oxidation and fate of glycolate. Photorespiration differs from mitochondrial respiration in that it does not occur in the dark, does not conserve energy as ATP, and does not utilize substrates of the tricarboxylic acid cycle.

In the aerobic glycolate pathway of higher plants, $O_2$ uptake occurs during glycolate formation in the light, during its oxidation by glycolate oxidase, and during the oxidation of glycine. Carbon dioxide release occurs during glycolate metabolism in which two glycolates are oxidized to two glyoxylates by glycolate oxidase, converted to two glycines by specific transaminases in the peroxisomes, and then to one $CO_2$ and one serine (Fig. 17.1). In the latter reaction, which occurs in the mitochondria, one glycine is oxidized to $CO_2$ from the carboxyl group and to a $C_1$-tetrahydrofolate complex from the $\alpha$ carbon, which is then added to the second glycine to form serine in a serine hydroxymethyltransferase reaction. Subsequently, in the peroxisomes, serine is converted to hydroxy-pyruvate, to glycerate and then to phosphoglycerate which re-enters the carbo-hydrate metabolic pools of the cytosol and chloroplasts. In the dark or when the $O_2$ level is reduced the anaerobic part of this pathway still functions in both plants and algae. Then serine and glycine are formed in significant amounts from phosphoglycerate, but the $O_2$ uptake and $CO_2$ release of photorespiration do not occur. Serine is converted to glycine and the $C_1$ complex as needed for anabolism. It would thus appear that photorespiration is unnecessary, wasteful, and un-desirable, and the reason for its occurrence in large amounts in plants and algae is a subject of much speculation.

Warburg's description (Warburg 1920) of $O_2$ inhibition of photosynthesis by *Chlorella* has been referred to as the 'Warburg $O_2$ effect'. Since his report this phenomenon has been repeatedly observed in many algae, higher plants and isolated chloroplasts, and it has been extensively reviewed (Ellyard & Gibbs 1969, Miyachi *et al.* 1955, Turner & Brittain 1962). This phenomenon is photo-respiration, in which $O_2$ inhibition of net photosynthesis is one of the prime effects. Another photorespiratory parameter, $CO_2$ evolution in the light, or the $CO_2$ compensation point, has not been as easy to measure with algae for reasons to be elaborated later. Recently investigators have used the term photorespira-tion instead of the Warburg $O_2$ effect for this overall process occurring in both higher plants and algae. Further consideration of these physiological measure-ments is placed last in this chapter so that some interpretations of them may be based on current knowledge of the glycolate pathway in algae.

The work reviewed has been carried out mainly with strains of *Chlorella vulgaris* or *C. pyrenoidosa*, *Chlamydomonas reinhardtii*, *Scenedesmus* spp., *Ankistrodesmus braunii*, *Euglena* spp., and two blue-green algae, *Anabaena cylindrica* and *Anacystis nidulans*. Quantitative differences have not been well documented because of the many physiological variables, but qualitatively, glycolate metabolism and photorespiration seem similar in all photosynthetic algae.

## 2  GLYCOLATE PATHWAY

The glycolate pathway in algae (Fig. 17.1) is mainly substantiated by [14]C label-ling of its compounds with [14]$CO_2$ or added substrates and by investigation of the enzymes in algae (Bruin *et al.* 1970, Hess & Tolbert 1967a, Lord & Merret 1970a, Murray *et al.* 1971, Tolbert 1963, Tolbert *et al.* 1971), as well as in other plants.

## 2.1 *Magnitude*

Values for the magnitude of glycolate production during photosynthesis in algae can vary, depending upon many parameters, from a few percent to nearly 100% of the total $CO_2$ fixation (Bassham & Kirk 1962, Benson & Calvin 1950, Coombs & Whittingham 1966b, Hess *et al.* 1967, Warburg & Krippahl 1960). High $O_2$, low $CO_2$, high light intensity, and high pH of the medium all favour increased glycolate production (see *Physiological Experiments*, below). Since glycolate is not the first product of $CO_2$ fixation, measurements of its pool sizes with $^{14}C$ involve experiments lasting longer than a minute. However, since glycolate is also excreted and becomes an accumulating end-product of photosynthesis, unusually high values for total $^{14}C$ in glycolate can be obtained with algae relative to other intermediary and metabolically active products. Warburg and Krippahl (1960) found that 92% of the fixed $CO_2$ appeared as glycolate during photosynthesis for 10 minutes by *Chlorella* in red light. Many other investigators report for algal photosynthesis that glycolate, glycine, and serine may represent 10 to 30% of the fixed $^{14}C$ after 2 to 10 minutes of $^{14}CO_2$ fixation under conditions for which glycolate excretion was not significant. Smallest pools of these products were generally observed in blue-green algae, but more recent data indicate that these algae assimilate and metabolize glycolate rapidly. Certainly the compounds of the glycolate pathway always represent major carbon products of photosynthesis. These studies did not indicate how much more of the carbon had flowed through the glycolate pathway back to the photosynthetic carbon cycle. Glycolate, mainly excreted into the medium, becomes the major product (over 50% of the total $^{14}C$) of photosynthesis by algae at high $O_2$ and very low $CO_2$ levels (Hess *et al.* 1967) or in media over pH 8·5 (Orth *et al.* 1966).

## 2.2 $^{14}C$-*Labelling experiments on glycolate formation and metabolism*

Most of the early photosynthetic investigations with $^{14}CO_2$ were done with algae and included key experiments on glycolate formation and metabolism (Benson & Calvin 1950, Calvin *et al.* 1951a,b, Schou *et al.* 1950, Wilson & Calvin 1955). Uniformly labelled glycolate and glycine are formed from $^{14}CO_2$ in short periods of 10 to 30 seconds (Bruin *et al.* 1970, Calvin *et al.* 1951b, Hess & Tolbert 1967a, Planondan & Bassham 1966, Schou *et al.* 1950, Wilson & Calvin 1955). From even shorter $^{14}CO_2$ experiments of 1 to 10 seconds on *Chlorella* or *Chlamydomonas* it is apparent that the $\alpha$ carbon of glycolate is labelled faster than the carboxyl carbon (Bruin *et al.* 1970, Hess & Tolbert 1967a). A similar labelling pattern is observed in carbons 1 and 2 of RuDP and in carbons 2 and 3 of 3-phosphoglyceric acid. These data are consistent with the formation of glycolate from the first 2 carbon atoms of RuDP. In blue light glycolate is even more predominantly $\alpha$-labelled (Cayle & Emerson 1957).

Schou *et al.* (1950) first fed glycolate-2-$^{14}C$ to *Scenedesmus* and found that it was converted to glycine-2-$^{14}C$ and to serine and to 3-phosphoglycerate labelled *in* carbons 2 and 3. Likewise *Scenedesmus* converted glycolate-1-$^{14}$C to glycine-1-$^{14}$C and serine-1-$^{14}$C (Bruin *et al.* 1970). Similar results from the metabolism of specifically labelled glycolate-$^{14}$C by leaves of higher plants (Bowes *et al.* 1971), *Euglena* (Murray *et al.* 1970, 1971), and *Chlorella* (Lord & Merrett 1970a) provide convincing evidence for the glycolate pathway in photosynthetic tissue.

During $^{14}CO_2$ fixation by $C_3$-plants only uniformly labelled glycine and serine are formed rapidly from uniformly labelled glycolate at time periods so short that 3-phosphoglycerate is predominantly (70 to 90%) carboxyl labelled (Rabson *et al.* 1962). Thus most of the serine arises by the aerobic glycolate pathway. In *Chlorella*, *Scenedesmus* and *Chlamydomonas* the serine pool becomes $^{14}$C labelled before glycine and the initial serine is carboxyl labelled. However, the serine and glycine pools become uniformly labelled before 3-phosphoglycerate. These data are consistent with the flow of carbon from the photosynthetic carbon cycle into glycine and serine by both routes shown in Fig. 17.1, namely from phospho-glycolate and phosphoglycerate. In algae the formation of uniformly labelled serine from glycolate is slower than in higher plants, apparently because of lower activity of the enzymes catalysing glycolate oxidation and glycine conversion to serine. Thus carbon flow from carboxyl-labelled 3-phosphoglycerate to glycerate, hydroxypyruvate and carboxyl labelled serine is detectable (Bruin *et al.* 1970, Hess & Tolbert 1967a), and anaerobically or in the dark only this route could function for essential serine and glycine synthesis in both algae and plants.

## 2.3 *Glycolate biosynthesis*

Glycolate is formed in chloroplasts of leaves during photosynthesis, and it is also assumed to occur in algal chloroplasts. In the absence of this substrate formation in the dark or in a DCMU inhibited photosystem, photorespiration does not occur. Biochemical details of glycolate biosynthesis have remained obscure, but current research favours the hypothesis for its formation from RuDP. Homogeneous preparations of RuDP carboxylase from leaves catalyse the oxidation of RuDP to phosphoglycolate and 3-phosphoglycerate (Andrews *et al.* 1971, 1973, Bowes *et al.* 1971, Bowes & Ogren 1972, Lorimer *et al.* 1973). This activity of the enzyme may be considered as a RuDP oxygenase. The phosphoglycolate comes from the top two carbon atoms of RuDP, and 3-phosphoglycerate from the lower 3 carbons. For the oxidation reaction the $K_m$ (RuDP) is about $10^{-4}$ M, the pH optimum is between 9 and 9·3, and $Mg^{2+}$ is required. Half maximal activity for the oxidation reaction, the operational $K_m$, is achieved in water equilibrated with about 68% $O_2$, so the oxidase activity is not saturated by 100% $O_2$, but it is significant at 20% $O_2$. The $K_i$ for $O_2$ on the carboxylase activity has been reported to be 0·8mM. The $O_2$ concentration inside a photosynthesizing cell may be very high. During the reaction one atom of oxygen from $^{18}O_2$ is incorporated into the carboxyl group of phosphoglycolate, but not into 3-phosphoglycerate, and $H_2^{18}O$ labels the carboxyl group of 3-phosphoglycerate (Lorimer *et al.* 1973). *In vivo* experiments with $^{18}O_2$ confirm this method of synthesis and the reactions of the glycolate pathway, since $^{18}O_2$ in the carboxyl group of phosphoglycolate is retained in the carboxyl group of glycine as well as serine (Andrews *et al.* 1971). *In vivo* 3-phosphoglycerate of the photosynthetic carbon cycle is not labelled by $^{18}O_2$. This mechanism for glycolate biosynthesis is consistent with all previous $^{14}C$ labelling data and with the fact that *Chlamydomonas* mutants without RuDP carboxylase or Ru-5-P kinase do not excrete glycolate when grown photoheterotrophically (Togasaki & Levine 1970). The mechanism explains the requirement of high $O_2$ for glycolate synthesis, and why it is maximal in low $CO_2$, which restricts the carboxylation reaction. The action of RuDP oxygenase and RuDP carboxylase results in a competition between $O_2$ and $CO_2$ for the common substrate RuDP. The role of light in the reaction is indirect through the photosynthetic generation of RuDP. There are no known photochemical reactions in the immediate steps in glycolate biosynthesis or metabolism.

It has often been proposed that glycolate may be formed by a peroxidation of a thiamine pyrophosphate-glycolaldehyde (TPP-$C_2$) complex generated by transketolase during the reactions of the photosynthetic carbon cycle (see, e.g. Bassham 1971, Bradbeer & Anderson 1967, Coombs & Whittingham 1966a, Ellyard & Gibbs 1969, Gibbs 1969, Shain & Gibbs 1971, Wilson & Calvin 1955). There is certainly no doubt that any strong oxidant such as ferricyanide or $H_2O_2$ will oxidize TPP-$C_2$ to glycolate *in vitro*. Furthermore $H_2O_2$ generation from the oxidation of the electron acceptors from photosystem I such as reduced ferredoxin, does take place *in vitro*. Whether these reactions occur *in vivo* has not been proven. By these reactions, equal $^{14}C$ distribution in glycolate from the top two carbon atoms of a sugar phosphate is possible. However, the process should not incorporate $^{18}O_2$ into the carboxyl group of the products of the glycolate pathway, as does occur *in vivo* (Andrews *et al.* 1971), and it does not account for the presence of phosphoglycolate phosphatase in the chloroplast. An unidentified $CO_2$ plus $CO_2$ condensation has also been proposed for glycolate formation (Zelitch 1965). However, the kinetics for $^{14}C$ labelling of glycolate with time clearly show that it is formed after the labelling of 3-phosphoglycerate of the photosynthetic carbon cycle in algae (Orth *et al.* 1966), and in tobacco leaves with or without an inhibitor of glycolate oxidase (Hess & Tolbert 1966a).

Glycolate formation in the chloroplast from several metabolic routes producing glyoxylate may also account for part of the glycolate pool. All algae so far examined contain a NADPH-glyoxylate reductase (Gruber *et al.* unpubl.) which would be equivalent to the chloroplast glyoxylate reductase (Tolbert *et al.* 1970). The functioning of the glycolate-glyoxylate shuttle will be discussed in detail in a later section. Likewise the formation of glyoxylate and then glycolate during photoheterotrophic growth on $C_4$ (malate) and $C_2$ (acetate) substrates is elaborated in the section on glyoxysomal metabolism.

## 2.4   *P-glycolate phosphatase and P-glycerate phosphatase*

A specific and very active P-glycolate phosphatase (Anderson & Tolbert 1966, Richardson & Tolbert 1961) is present in the chloroplasts of higher plants (Randall *et al.* 1971, Thompson & Whittingham 1967, Yu *et al.* 1964), and large amounts of this phosphatase activity have been found in all algae examined (Bruin *et al.* 1970, Codd & Merrett 1970, 1971b, Gruber *et al.* unpubl., Randall *et al.* 1971). The enzyme has been partially purified and characterized from *Chlorella* (unpubl.), and the pH optimum at 6·3 and $K_m$ at about $2 \times 10^{-3}$ are similar to those of the enzyme in leaves. If *Chlorella* cells are first suspended in 1 M NaCl and then in water, most of this phosphatase is solubilized by the osmotic shock. A nearly constant level of the enzyme is present in synchronized *Euglena* at all stages of development (Codd & Merrett 1971b). The formation of phosphoglycolate-$^{14}$C during $CO_2$ fixation has been reported (Bassham & Kirk 1962, Calvin *et al.* 1951b, Lorimer *et al.* 1973, Orth *et al.* 1966, Zelitch 1965). *Ankistrodesmus braunii* (Naeg. Collins, No. 245) forms a larger pool of phosphoglycolate-$^{14}$C than any other algae that we have studied (unpubl.)). To detect phosphoglycolate-$^{14}$C it is necessary to kill the algae by strong acid and not by boiling ethanol or methanol, which do not inactivate phosphoglycolate phosphatase (Ullrich 1963). Phosphoglycolate is a potent inhibitor of triose-phosphate isomerase with a $K_i$ of $2·2 \times 10^{-6}$ (Wolfenden 1970). The accumulation of phosphoglycolate in the chloroplasts would severely inhibit the photosynthetic carbon cycle, and thus this very active and specific phosphatase is essential.

Algae also contain phosphoglycerate phosphatase activity (Codd & Merrett 1971b, Randall *et al.* 1971) and its properties and specificity as far as they have been studied (D. D. Randall & N. E. Tolbert, unpubl.), are similar to the enzyme in leaves (Randall & Tolbert 1971a,b). Action of this phosphatase would initiate the conversion of 3-phosphoglycerate to serine and glycine and is analogous to phosphoglycolate phosphatase initiation of glycine and serine formation from glycolate. So far no unique inhibitors for either phosphatase have been found for tests of *in vivo* function.

## 2.5   *Glycolate oxidation to glyoxylate*

Glycolate oxidase of peroxisomes from leaves has been extensively studied (Tolbert 1963,1971). Failure to detect this flavin oxidase in *Chlorella, Scenedesmus* and *Chlamydomonas* (Hess & Tolbert 1967a) was consistent with a tendency for algae to excrete glycolate during photosynthesis (see below). However, from the labelling pattern after fixation of $^{14}CO_2$, it was evident that a part of the glycolate was also converted to glycine and serine by algae by reactions similar to the glycolate pathway. Subsequently the oxidation of glycolate by algal extracts was reported. The total activity of this enzyme in algae ranges around 3 to 13 nmoles mg protein$^{-1}$ minute$^{-1}$ as compared to leaf glycolate oxidase activity of 50 to 100 nmoles mg protein$^{-1}$ minute$^{-1}$. The enzyme which oxidizes glycolate in green unicellular algae, in most cases, does not couple to $O_2$ and does not produce $H_2O_2$. It has been called glycolate:DCPIP oxidoreductase or glycolate dehydrogenase to distinguish it from glycolate:oxygen oxidoreductase or glycolate oxidase. Properties of the enzyme in cell homogenates consistent with a dehydrogenase have been reported for various strains of *Chlorella pyrenoidosa* (Bruin *et al.* 1970, Codd *et al.* 1969, Lord & Merrett 1968, 1971a, Nelson & Tolbert 1970, Zelitch & Day 1968b), for *Chlamydomonas reinhardtii* (Cooksey 1971, Nelson & Tolbert 1970), for *Scenedesmus* (Nelson *et al.* 1969a), for *Euglena* (Codd & Merrett 1970, Graves *et al.* 1971a, Lord & Merrett 1971a, Nelson & Tolbert 1970) and for blue-green algae (Grodzinski & Colman 1970). An exception has been reported by Schmid and Kowallik (Codd *et al.* 1971, Kowallik 1971, Kowallik & Schmid 1971, Schmid & Schwarze 1969), who observed $O_2$ uptake during the oxidation of glycolate by homogenates from a wild type *Chlorella pyrenoidosa* and its yellow mutant. The rate was small relative to glycolate oxidation by leaf extracts and to DCPIP reduction by algal extracts; furthermore $H_2O_2$ production was not shown. This claim has not yet been confirmed by others.

Only limited purification of the algal glycolate dehydrogenase has been accomplished because of its lability and low activity. Its pH optimum at 8·3 to 8·7 and inhibition by sulphonates (Lord & Merrett 1968, Nelson & Tolbert 1970) are similar to the properties of glycolate oxidase. The algal enzyme differs from the plant glycolate oxidase in that it does not couple to $O_2$, it oxidizes D-lactate rather than L-lactate, and it is inhibited by $10^{-3}$ M cyanide (Nelson & Tolbert 1970). Although the algal glycolate dehydrogenase has some properties similar to the D-lactate dehydrogenase in *Euglena* and *Chlorella*, the D-lactate dehydrogenase and glycolate dehydrogenase of *Chlorella* are different proteins (Gruber *et al.* unpubl.). The natural electron acceptor for the algal glycolate dehydrogenase has not been elucidated except that quinones such as DCPIP (Nelson & Tolbert 1970) and others (Codd *et al.* 1972) will function *in vitro*. The yellow *Chlorella* mutant, with high rates of respiration is of particular interest, because it is reported to contain an active $O_2$-linked glycolate oxidase (Codd *et al.* 1972, Kowallik 1971, Kowallik & Schmid 1971, Schmid 1970) typical of higher plants. This yellow mutant has high amounts of glycolate oxidase and high rates of photorespiration. Further interesting experiments with the glycolate oxidizing system in the mutant can be expected from these analogies. Kowallik (1971) has stated that the consumption of exogenous glycolate by their yellow *Chlorella* mutant is not $O_2$ dependent, which is inconsistent with a typical plant glycolate oxidase.

### 2.6 *Catalase*

Excess catalase in leaf peroxisomes is characteristic of glycolate oxidation by this flavin oxidase system. Nelson and Tolbert (1969, 1970) pointed out, after examining only five unicellular green algae, that the glycolate dehydrogenase did not produce $H_2O_2$ and that catalase was about 10% of that in the higher plant. In fact *Euglena* with glycolate dehydrogenase has no catalase. On the other hand the yellow *Chlorella* mutant with glycolate oxidase has high levels of catalase (Codd *et al.* 1972, Kowallik 1971). The significance of relatively low (20μmoles mg protein$^{-1}$ minute$^{-1}$) levels of catalase in green algae is not clear. In leaves catalase activity is about 300 to 500 μmoles mg protein$^{-1}$ minute$^{-1}$. The relatively low algal catalase is still an active system compared to other algal enzymes. Because of assay difficulties catalase activity in algae has not been put on a quantitative basis. Other proteins, such as peroxidase and cytochrome oxidase, which contain haem prosthetic groups, exhibit catalase activity *in vitro* so that low levels of catalase are often meaningless.

A hypothesis for further work is that the glycolate oxidase and catalase system of leaf peroxisomes represent the major difference in the glycolate pathway between vascular plants and unicellular green algae, which contain a glycolate dehydrogenase and little or no catalase. If true, then $O_2$ uptake during algal photorespiration should be less than in leaves, but $O_2$ uptake would still occur during glycolate formation and glycine oxidation.

An interesting chapter on catalase and microbodies is yet to be elucidated with the anaerobic photosynthetic bacterium, *Rhodopseudomonas spheroides*, which contains as much as 25% of its total protein as catalase (see Clayton 1960, one of 15 papers on this by Clayton).

## 2.7   *Glycolate-glyoxylate shuttle*

The concept of this shuttle is based on two enzymatic reactions at two different cellular locations. Glyoxylate reductase in the chloroplasts catalyses the reduction of glyoxylate to glycolate (reaction 4, Fig. 17.1). Glycolate is then oxidized elsewhere by glycolate oxidase or glycolate dehydrogenase (reaction 3, Fig. 17.1). This shuttle potentially could consume excess photosynthetic reducing power (NADPH) with the cycling of relatively small pools of the two acids (Tolbert *et al.* 1970, Tolbert 1971). It might function to alter the ratio of NADPH/ATP production, or to dispose of excess reduction power. If the unknown cofactors in the glycolate dehydrogenase reaction were coupled to energy-conserving reactions, the shuttle could be utilized for moving reducing equivalents from the chloroplasts. Algae contain an NADPH-glyoxylate reductase (Gruber *et al.*, unpubl.) which seems similar to that in chloroplasts of spinach leaves in that it is specific for NADPH and glyoxylate. The NADH-hydroxypyruvate reductase of algae is like that in plants (Bruin *et al.* 1970, Tolbert *et al.* 1970) and probably does not function as a glyoxylate reductase because of a $K_m$ (glyoxylate) of $2 \times 10^{-2}$ M. Little experimental evidence for a glycolate-glyoxylate shuttle as a terminal respiratory mechanism in algae has yet been published. At the stage of cell division, when *Scenedesmus* will assimilate glycolate and glycolate production is minimal, $O_2$ evolution during photosynthesis is stimulated one- to threefold by small amounts of exogenous glycolate or glyoxylate (Nelson *et al.* 1969a).

In this case the terminal glycolate-glyoxylate shuttle would remove NADPH and permit more electron flow so that $O_2$ evolution would be stimulated. The light dependent uptake of glucose by *Chlorella* is blocked by hydroxysulphonate inhibitors of glycolate oxidation, and the results perhaps indicate a requirement for the glycolate-glyoxylate shuttle to reoxidize NADPH so that sufficient ATP from photophosphorylation is available for glucose uptake (Butt & Peel 1963).

## 2.8 *Conversion of glyoxylate to glycine and serine*

Enzymes for these reactions have been detected in algal extracts but have not been extensively studied. A glutamate-glyoxylate aminotransferase was noted in *Chlamydomonas*, *Chlorella* and *Euglena* (Bruin *et al.* 1970, Codd & Merrett 1971b), but the serine-glyoxylate aminotransferase of leaf peroxisomes has not been studied in algae. The conversion of two glycines to serine, $CO_2$, and ammonia or a partial reaction of one glycine to $CO_2$, and perhaps ammonia and a methylene tetrahydrofolate complex has not been studied in algae but is under investigation in plants (Bird *et al.* 1972, Kisaki & Tolbert 1969, 1970, Kisaki *et al.* 1971a,b). The subcellular location for the interconversion between glycine and serine is not known for algae, but in higher plants it is in the mitochondria. Two reports of serine hydroxymethyl transferase activity involved in the conversion of glycine to serine and $CO_2$ indicated that this activity in algae may be lower than in leaves (Bruin *et al.* 1970, Lord & Merrett 1971a). These results are consistent with a slower rate of interconversion of glycine and serine in algae as indicated by the $^{14}$C labelling data. The mitochondrial oxidation of glycine to $CO_2$ and ammonia or conversion to serine and ammonia is accompanied by NAD reduction and this mitochondrial respiration would be accompanied by ATP synthesis and oxygen uptake.

## 2.9 *Interconversions between phosphoglycerate and serine*

Aminotransferase activity for conversion between serine and hydroxypyruvate has not been studied in algae. Hydroxypyruvate reductase or D-glycerate dehydrogenase in homogenates of *Chlorella*, *Scenedesmus* and *Chlamydomonas* is an active (40nmoles mg protein$^{-1}$ minute$^{-1}$) enzyme (Bruin *et al.* 1970, Goulding & Merrett 1967, Lord & Merrett 1970a), which is easy to assay and relatively stable. In partially purified preparations its properties appear similar to the NAD-hydroxypyruvate reductase of leaf peroxisomes. No reports of glycerate kinase for 3-phosphoglyceric acid formation were noted for algae. However, phosphoglycerate phosphatase activity is present in unicellular green algae and the ratio of activity for phosphoglycolate phosphatase and phosphoglycerate phosphatase is about 4:1, as in $C_3$ plants (Randall *et al.* 1971).

# 3 ENZYME LOCATION AND MICROBODIES

The chloroplasts of higher plants contain enzymes involved in glycolate bio-synthesis; the enzymes for glycolate conversion to glycine and for glycerate conversion to serine are in leaf peroxisomes; and mitochondria contain the system for glycine interconversion (Tolbert 1971). Little is known concerning the subcellular distribution of the reactions of photorespiration in algae. In plants the light dependent development of these enzymes has been associated with organelle development. Results with greening *Euglena gracilis* have similarly shown that phosphoglycolate phosphatase and glycolate dehydrogenase develop during greening by *de novo* synthesis (Codd & Merrett 1970, Murray *et al.* 1970, 1971).

## 3.1 *Peroxisomes*

Since part of glycolate metabolism occurs in leaf peroxisomes and accounts for a portion of photorespiration (Tolbert 1971), a current question is the nature of a similar organelle in algae. The initial hypothesis was that green algae would not contain peroxisomes, since there is no algal peroxide generating glycolate oxidase and thus no need for such an organelle with catalase (Nelson & Tolbert 1970). This theory seemed consistent with the observed absence of peroxisomes in previously published electron micrographs of algae. According to this hypothesis the yellow *Chlorella* mutants with glycolate oxidase and catalase should have peroxisomes. Indeed organelles similar in appearance to micro-bodies have been observed in this mutant (Gergis 1971), although no *in situ* cytological stain for catalase was reported. However, the green wild type *Chlorella* of this mutant also contained visible microbodies.

The subcellular location of glycolate dehydrogenase appears to be in a large and stable macromolecular complex, perhaps similar to peroxisomes. Part of the enzyme activity in algal homogenates from the French pressure cell can be pelleted by high speed centrifugation (Codd *et al.* 1972, Lord & Merrett 1971a, Nelson & Tolbert 1970, Zelitch & Day 1968b). A few small microbodies have been detected in *Euglena* grown phototrophically (Graves *et al.* 1971a,b), and as mentioned above in a green strain of *Chlorella* (Gergis 1971). The glycolate dehydrogenase and glutamate-glyoxylate aminotransferase of *Euglena* has also been found in sucrose gradients at a density characteristic of microbodies (Graves *et al.* 1971a). Likewise from *Chlorella* partial isolation of a fraction enriched in glycolate oxidase and catalase and appearing to be similar to peroxi-somes has now been reported (Codd *et al.* 1972). Ultrastructural evidence has also been presented for microbodies in *Micrasterias* (Kiermayer 1970), *Polyto-mella caeca* and *Chlorogonium* (Gerhardt 1971, Gerhardt & Berger 1971), although these may be glyoxysomes.

Microbodies from *Chlorella*, *Euglena*, *Polytomella* and *Chlorogonium* do not show a catalase reaction or only a small amount of activity either when isolated on sucrose gradients or when stained *in situ*. The near absence of catalase is consistent with oxidation of glycolate in algae by a dehydrogenase rather than a flavin peroxide-generating oxidase. Thus the criterion of catalase as an essential component for microbodies may not be valid for algae. Microbodies have also been seen but not isolated from a brown alga (Bouck 1965), which has a high $CO_2$ compensation point (Brown & Tregunna 1967) like a $C_3$-plant. It is probable that with improved techniques and experience, organelles similar to microbodies, but often without much catalase, will be recognized and characterized in algae.

The reports that isolated chloroplasts without microbodies from *Acetabularia mediterranea* carried out complete photorespiration, including glycolate biosynthesis and metabolism (Bidwell *et al.* 1969, 1970) have been qualified by the fact that these isolates were membrane bound droplets of cytoplasm and chloroplasts. This alga contains glycolate dehydrogenase (Nelson & Tolbert 1970).

## 3.2 *Glyoxysomes*

The general term, microbodies, designates both peroxisomes with the glycolate pathway and glyoxysomes with the glyoxylate pathway (Tolbert 1971). Organisms which can grow photoheterotrophically or heterotrophically on $C_4$ compounds, such as malate, or $C_2$ compounds, such as acetate, develop the glyoxylate cycle for production of oxaloacetate, necessary for phosphoenolpyruvate synthesis during gluconeogenesis. In most such tissue, glyoxysomes, containing the glyoxylate cycle have now been recognized. Some data are now available for species of *Euglena* (Cook 1970, Graves *et al.* 1971a,b, Haigh & Beevers 1964a), *Polytomella* (Gerhardt 1971, Haigh & Beevers 1964b), *Chlorella* (Goulding & Merrett 1966, Haigh & Beevers 1964a, Harrop & Kornberg 1966), *Chlamydomonas* (Giraud & Czaninsky 1971, Haigh & Beevers 1964a, Harrop & Kornberg 1966), *Chlorogonium* (Haigh & Beevers 1964a), *Gonium* (Haigh & Beevers 1964a) and *Gloeomonas* (Foo *et al.* 1971). The characteristic enzymes of the glyoxylate cycle are malate synthetase and isocitrate lyase. The presence of glyoxysomes in algae grown on $C_2$ or $C_4$ compounds has been suggested by ultrastructural studies, by assays for malate synthetase and isocitrate lyase, by their development upon addition of the inducing substrate, and in a few cases by isolating the glyoxysomes on sucrose gradients. Since this metabolic sequence does not generate $H_2O_2$, glyoxysomes also need not contain catalase. In the unique case of glyoxysomes from fatty seedlings, $H_2O_2$ is generated during acetyl CoA formation and glyoxysomes from these tissues do contain catalase, but this situation has not so far been reported in an alga.

The glyoxylate cycle is functionally and metabolically different from glycolate formation and metabolism in photorespiration. Both the glyoxylate cycle and glycolate metabolism occur in microbodies of similar appearance but different enzymatic composition. Both generate glyoxylate and the glyoxylate cycle may form glycolate by reduction of glyoxylate. Evidence for this source of glycolate as significant or related to photorespiration has yet to be obtained *in vivo* by gas exchange measurements. Likewise the role of some isocitrate lyase in algae grown photoautotrophically with $CO_2$ is not known (Foo *et al.* 1971, Wiessner 1968).

In several extensive investigations, *Chlorella*, as well as other algae, has been reported to convert acetate-$^{14}$C to glycolate-$^{14}$C *in vivo* (Goulding & Merrett 1967, Goulding *et al.* 1969, Merrett & Goulding 1967, Wiessner 1967). Since the direct oxidation of acetate to glycolate seems unlikely, the glyoxylate cycle may account for part of these observations (Wiessner 1967). Acetate-$^{14}$C incorporated into the glyoxylate cycle would first label isocitrate which would be cleaved to succinate and glyoxylate. The latter could be reduced to glycolate by a reductase. In some experiments with *Chlorella* the distribution and specific activity of $^{14}$C and $^3$H in glycolate from labelled acetate appeared to suggest a direct conversion. Re-examination of these experiments however is necessary, since trace contamination of labelled glycolate in the acetate substrate would appear as the initial product of acetate metabolism. No explanation was offered for the rapid initial appearance of glycolate-$^{14}$C during $^{14}CO_2$ fixation by *Rhodospirillum rubrum* grown on malate (Anderson & Fuller 1967). Since this did not occur in autotrophically grown cells, the probable induction of the glyoxylate cycle by growth on malate may relate to the formation of labelled glycolate.

## 4 EXCRETION DURING PHOTOSYNTHESIS

### 4.1 *General*

After characterization of glycolate among the initial $^{14}CO_2$ fixation products, it was shown in 1956 that much of the glycolate produced by laboratory cultures of *Chlorella* and *Chlamydomonas* was excreted (Lewin 1957, Tolbert & Zill 1956). These observations have been extended to many unicellular algae in nature (see Chapter 30, p. 842). From our experience, glycolate is the major photosynthetic product excreted by algae, but by no means the only compound. However, the magnitude and significance of glycolate excretion remains unclear, and it may not occur in significant amounts in nature. In this chapter, glycolate excretion is considered as part of the phenomenon of photorespiration. Isolated plant chloroplasts produce and excrete glycolate (Gibbs 1969, Kearney & Tolbert 1962, Shain & Gibbs 1971), but similar investigations with intact algal chloroplasts have not been done. An analogy is that the higher plant chloroplasts and algae make and excrete glycolate into the cytoplasm or medium, respectively. Photorespiratory measurement of the $CO_2$ evolution in algae is reduced by the amount of glycolate excretion.

Excreted glycolate in the media has been measured by the Calkins colourimetric test with acidified 2,7-dihydroxynaphthalene (Calkins 1943), after rapid removal of the algae or chloroplast by filtration on Celite or a Millipore filter or by centrifugation (Kearney & Tolbert 1962). The color development is not specific and, in particular, nitrate and aldehyde or compounds oxidizable to aldehydes by the procedure will interfere. Often it is necessary to partially purify the glycolate by anion exchange resin before the colorimetric assay (Zelitch 1965). Estimations of glycolate excretion have also been based on the amount of $^{14}C$ excreted into the media after $^{14}CO_2$ fixation, and in this procedure it is necessary to check by chromatography what percent of the excreted $^{14}C$ is glycolate (Tolbert & Zill 1956).

Many factors affect the observed magnitude of glycolate excretion, such as the rate of its formation, adaptiveness of the algae to utilize the glycolate before excretion, rate of reutilization by other plankton, $CO_2$ and $O_2$ concentration, pH of the media, variability among algae, and the stage of algal development. Nearly any value can be obtained by changing the many variables discussed in this chapter that affect glycolate synthesis. After laboratory cultures are grown on 1% or more $CO_2$, glycolate excretion during the test period with low levels of $CO_2$ may be a significant 5 to 20% of the total $^{14}CO_2$ fixation. High $CO_2$ reduces glycolate synthesis and may repress glycolate dehydrogenase (Nelson & Tolbert 1969). Then during a $^{14}CO_2$ test at low $CO_2$ to conserve the tracer, glycolate will be formed more rapidly than it can be metabolized, and the excess excreted. Alternatively, growth on low $CO_2$ (aeration by air) to derepress glycolate dehydrogenase will reduce glycolate excretion. When the pH of the medium is 9 or above, excreted glycolate is the major photosynthetic product (Orth et al. 1966), and this pH is reached quite rapidly during bicarbonate uptake by unbuffered cultures of algae. This situation is consistent with enzyme studies on glycolate biosynthesis; at pH 9 RuDP oxidation to glycolate is maximal, and RuDP carboxylation to phosphoglyceric acid is inhibited (Andrews et al. 1973).

The amount of glycolate excretion by synchronized algal cultures parallels their photosynthetic rates, being maximal for young growing cells, low for mature cells, and lowest for dividing cells (Chang & Tolbert 1970, Gimmler et al. 1969, Codd & Merrett 1971b, Murray et al. 1971). For example, at the end of the light phase Ankistrodesmus excreted 35% of the total $^{14}C$ as glycolate, but dividing cells excreted very little. These results are substantiated by large changes during the life cycle in enzyme activities associated with glycolate metabolism (Codd & Merrett 1971b).

The mechanism of glycolate excretion is unknown. Tolbert and Zill (1956) showed that glycolate excretion by Chlorella required the presence of some $CO_2$ and proposed that this implicated an anion exchange which would be particularly important at pH 9 for bicarbonate. Phosphate esters and phosphatase are often involved in active membrane transport systems, and although phosphoglycolate phosphatase is active in algae and chloroplasts, no definitive experiments have been reported to test its role in glycolate excretion. Chang and Tolbert (1970) have also speculated that Ankistrodesmus might excrete glycolate as the unstable diglycolide lactone in an analogous way to the excretion of the stable isocitrate lactone by the same alga.

## 4.2  *Glycolate assimilation*

Although green algae and chloroplasts rapidly excrete glycolate, they do not assimilate it or phosphoglycolate as readily (Hess & Tolbert 1967a). For many experiments in which glycolate-$^{14}$C was given to algae to establish its assimilation, either the total amount of glycolate-$^{14}$C uptake was a very small part of the total added or the amount of uptake was not reported. To measure glycolate stimulation of respiration by *Chlorella vulgaris* the algae had to be ruptured by a French press (Kowallik & Schmid 1971). Glycolic acid is taken up as the acid by most algae at pH 3·8 or lower (Tolbert 1963), which is the $pK_a$ of this strong acid. Light or glucose as an additional energy source is necessary for sustained glycolate uptake and growth of *Euglena* on this acid (Murray *et al.* 1970). One strain of *Chlorella pyrenoidosa* has been found to grow photoheterotrophically on glycolate (Lord & Merrett 1971b), and this strain does not excrete glycolate (Goulding *et al.* 1969). Synchronized *Chlamydomonas* (Nelson *et al.* 1969a) and *Euglena* (Codd & Merrett 1971b) will excrete glycolate during their log phase of growth, but will not assimilate it. In the dark stage of cell division, when they do not form it, they will assimilate glycolate at neutral pH. Nevertheless, from tests on many algae (Droop & McGill 1966) it is unlikely that much of the glycolate excreted in natural waters is reutilized by the algae.

Blue-green algae readily assimilate and utilize glycolate (Lex *et al.* 1972, Miller *et al.* 1971) and *Scenedesmus* (Kowallik & Schmid 1971, Schou *et al.* 1950), *Chlorella* (Kowallik & Schmid 1971, Lord & Merrett 1970a), *Ankistrodesmus braunii* (Kowallik & Schmid, 1971), and *Euglena* (Murray *et al.* 1970) are reported to oxidize exogenous glycolate to some extent. Other investigators have not observed significant glycolate uptake and metabolism by various algae.

## 5  FUNCTION

Various hypotheses for photorespiration and glycolate biosynthesis and metabolism in higher plants (Tolbert 1971, Zelitch 1971) are equally applicable to algae. As in most biological processes, multiple and interrelated functions seem to exist, and only three hypotheses are mentioned here.

(a) Most of the glycine and serine biosynthesis in algae arises from the glycolate pathway. These two amino acids are essential for synthesis of protein, $C_1$-compounds, porphyrins and certain other amino acids. However, algae can also form serine and glycine from 3-phosphoglycerate without photorespiration. Thus glycine and serine are useful byproducts from photorespiration, but do not constitute an essential function.

(b) Phosphoglycolate and then glycolate formation may be an unavoidable aerobic reaction of photosynthesis that occurs in the presence of high $O_2$. The competition between $CO_2$ and $O_2$ for RuDP carboxylation or oxidation may be so inherent in the chemistry of RuDP and the enzyme that phosphoglycolate formation has not been eliminated during the course of evolution. Subsequent glycolate conversion to glycine, serine and glycerate, represents a gluconeogenic pathway for use of part of this carbon. Excreted glycolate may be used in a similar manner by other plankton. Thus glycolate biosynthesis may be unnecessary but unavoidable, and its aerobic metabolism salvages part of the carbon for the cell. As more is learned about glycolate biosynthesis, however, this hypothesis may be greatly modified.

(c) Peroxisomal metabolism in higher plants (Tolbert 1971) appears as if it were a mechanism to regulate cellular growth by disposing of the excess carbon by wasteful respiration. Glycolate production and excretion by algae is one example of this hypothesis. During the aerobic formation and metabolism of phosphoglycolate in the cell considerable energy and $CO_2$ are lost. The glycolate-glyoxylate shuttle, as currently envisaged, serves to dispose of excess reducing capacity (NADPH) of the chloroplasts. This may serve to keep the cofactors in the electron transport chains within some oxidized/reduced ratio. The glycolate-glyoxylate cycle by disposing of excess NADPH may permit further ATP synthesis by photophosphorylation, without invoking a cyclic phosphorylation process.

# 6 PHYSIOLOGICAL EXPERIMENTS

Many observations with algae relating to photorespiration have not previously been unified, but in this decade organizational hypotheses are feasible. Professor Gaffron has indicated that a decent hypothesis should last for at least two years. In this section physiological experiments are grouped together, and the results related to the concepts of photorespiration and the glycolate pathway. The quantitative magnitude of photorespiration as measured by gas exchange rates or as glycolate biosynthesis has not been well established in algae, perhaps because it is so variable. That it exists and is not mitochondrial respiration seems certain.

Glycolate, glycine and serine formation during $^{14}CO_2$ fixation has been observed for all photosynthetic organisms with photosystem II, or which evolve $O_2$. The formation of glycolate to some extent in all these algae is consistent with the ubiquitous presence of RuDP carboxylase activity and the probable activity of this protein as an oxygenase to catalyse P-glycolate synthesis. Thus the carboxylase from *Chlorella* appears similar to the enzyme from higher plants, although to date glycolate biosynthesis by enzymes isolated from algae has not been studied.

Most past experiments with algae in the laboratory have been run under conditions which favour photosynthesis and inhibit photorespiration. Growth of certain algae on 1% or more $CO_2$ represses glycolate dehydrogenase, and measurement of $^{14}CO_2$ fixation products in high concentrations of $CO_2$ favours RuDP carboxylase action over RuDP oxygenase activity. Gaseous exchange measurements in the presence of excess $CO_2$ or low $O_2$, as has been the case with $^{18}O_2$ experiments, cannot be expected to measure much photorespiration. Measurements at low light intensities or at the $CO_2$ compensation point are often made, but then glycolate biosynthesis and metabolism are also minimal. Assessment of photorespiration in algae under physiologically favourable and normal conditions is needed, as well as under conditions which increase photorespiration.

Oxygen uptake in the glycolate pathway may occur in three different reactions each in different subcellular locations; namely, during glycolate biosynthesis in the chloroplast, during glycolate oxidation to glyoxylate in the peroxisomal system, and during glycine oxidation and conversion to serine in the mitochondria. The first two reactions require saturating $O_2$ concentration for maximal rates, whereas the mitochondrial conversion of glycine to serine occurs at low oxygen pressures (Bird *et al.* 1972). In unicellular algae with glycolate dehydrogenase the second reaction is modified, so less $O_2$ uptake would occur. A total of 7 atoms of oxygen is required for the formation of one serine by the glycolate pathway in algae and plants with glycolate oxidase, four atoms for the formation of two glycolates, two atoms for the oxidation of the glycolate, and one atom for the conversion of two glycines to one serine. In this glycolate pathway only one mole of $CO_2$ would be evolved, providing all the serine was used for synthetic processes. This low $CO_2$ production has led Zelitch (1971), to propose that part of the glyoxylate must be oxidized to $CO_2$. However, serine conversion back to glycine and further oxidation of glycine again to serine sets up an equilibrium for terminal oxidation of all the glycine or serine and evolution of much larger amounts of $CO_2$. An area of further ambiguity in nomenclature for photorespiration is that the glycine-serine interconversion of the glycolate pathway occurs in the mitochondria and is really a mitochondrial type of respiration, although not involving reactions of the citric acid cycle.

## 6.1   $O_2$ uptake during photorespiration

In earlier algal experiments designed to measure changes in respiration rate in light and darkness, confusing or negative results were obtained primarily because the investigators used one or more conditions unfavourable for photorespiration, such as high $CO_2$, low $O_2$ or low light intensity. Direct measurements of algal respiration in light or darkness by rates of $O_2$ uptake or $^{18}O_2$ and $^{16}O_2$ exchange were first made by Brown and associates (Brown & Weis 1959, Good & Brown 1961). They did not observe significant alterations in respiration rates between light and darkness, unless high light intensities were used, which would have favoured glycolate formation. Later Hoch *et al.* (1963), using *Anacystis* and *Scenedesmus* and improved equipment, observed an increased rate of $^{18}O_2$ uptake in the light compared to the dark rate. However, their use of only low $O_2$ concentrations to conserve on the tracer doomed the observations to minimal rates of photorespiration. Low rates of photorespiration as $^{18}O_2$ uptake were also observed with *Chlorella*, *Scenedesmus*, and *Chlamydomonas* (Bunt 1969,

Bunt & Heeb 1970), but again the values were probably not maximal. Most recently, by taking advantage of the accumulating knowledge on photorespiration very significant rates of $O_2$ uptake during photorespiration by *Anabaena* have been obtained and the rates were linear with increasing $O_2$ up to 0·23 atmosphere (Lex *et al.* 1972). Using a paramagnetic gas analyser for $O_2$ very substantial reduction in net $O_2$ evolution has also been observed during photosynthesis by algae aerated with gas phases of air or higher $O_2$ concentration (Fock & Egle 1966, Fock *et al.* 1968, 1971). Reduction of the apparent value for net $O_2$ evolution can be interpreted as due to a stimulation of photorespiratory $O_2$ uptake. These results are consistent with the concept of photorespiration in algae, and its acceleration by a high concentration of $O_2$.

### 6.2   *$CO_2$ release during photorespiration*

Due to $CO_2$ fixation in the light, only indirect measurements of $CO_2$ evolution during photorespiration have been devised. Photorespiration was first observed in plants as a rapid release of $CO_2$ for a short dark period immediately after illumination. This has not always been observed for algae (Brown & Weis 1959), but in more recent investigations it has been reported to occur to a limited extent in some algae (Döhler & Braun 1971, Fock & Egle 1966, Schaub & Egle 1965). In fact opposite exchanges have been observed; a $CO_2$ burst at the start of illumination (Brown & Whittingham 1955) and a gulp of $CO_2$ in a post illumination period (Hiller & Whittingham 1959), i.e. more reminiscent of crassulacian acid metabolism than photorespiration. However, the $CO_2$ burst and gulp in changing light and dark phases do not respond to $O_2$ partial pressure as would be expected of photorespiration (Bunt 1970).

The glycolate pathway in algae should release $CO_2$ during the conversion of two glycines to one serine and one $CO_2$. Other possible sources of photorespiratory $CO_2$, which have not been explored with algae, involve the oxidation of glyoxylate to $CO_2$ or to oxalate, which in turn may be oxidized to $CO_2$. The glycolate pathway in algae seems less capable of releasing as much $CO_2$ as in higher plants. In brief test periods favourable for photorespiration, the overproduction of glycolate would not lead to a major increase in $CO_2$ production, but rather to glycolate excretion. Both glycolate dehydrogenase and serine hydroxymethyl transferase in algae may be present in low limiting activity so that this metabolic potential for rapid $CO_2$ release is low. In addition $CO_2$ may not diffuse from algal cells as rapidly as from mesophyll cells surrounding the stomatal cavity in leaves.

Assays for photorespiration in plants involve the loss of $CO_2$ into $CO_2$-free air as measured either by $^{14}CO_2$ (Fock *et al.* 1971, Goldsworthy 1966, Schaub & Egle 1965, Zelitch & Day 1968a) or by an infra-red gas analyser (Brown & Tregunna 1967, Downton & Tregunna 1968, Fock & Egle 1966, Fock *et al.* 1968, 1971). Assays by infra red-analysis indicate that some algae lose little $CO_2$ and others much during photosynthesis in the light. The $^{14}C$ assay showed a high level of photorespiration in *Chlamydomonas* but only a low level in *Chlorella*. $CO_2$ assays underestimate photorespiration by the amount of $CO_2$ refixation, which can be large, and which has not been estimated in algae. Assays involving the conversion of glycolate-$^{14}C$ to $^{14}CO_2$ by algae (Brown & Tregunna 1967, Zelitch & Day 1968a) are also not reliable, since another limiting factor is a slow rate of glycolate uptake by most algae. $CO_2$ assays are even more unreliable for photorespiration in algae than in plants, because in $CO_2$-free air the algae excrete large amounts of glycolate rather than oxidize it to $CO_2$ (Hess *et al.* 1967). Glycolate excretion is indicative of photorespiration in algae, although then only $O_2$ uptake would occur during its biosynthesis. In $CO_2$- free air, glycolate excretion plus $CO_2$ release from cellular metabolism would be a better approximation of photorespiration than either alone.

In plants the $CO_2$ compensation point is a useful indication of the magnitude of photorespiration. At that $CO_2$ concentration the rate of photosynthetic $CO_2$ fixation equals the rate of $CO_2$ release mainly by photorespiration. Many unicellular green algae have a low $CO_2$ compensation (Bunt 1969, Downton & Tregunna 1968, Zelitch & Day 1968a). Algae such as *Chlorella*, *Scenedesmus* and *Fucus* have compensation points near zero, whereas *Nitella* with glycolate oxidase, and the brown algae *Polyneura* and *Sargassum* have compensation points of over 25 ppm $CO_2$ at 25°C, similar to $C_3$-plants (Brown & Tregunna 1967, Downton & Tregunna 1968). A low value, as in $C_4$-plants, does not mean that photorespiration is absent, but only that $CO_2$ loss from photorespiration is not great. For both algae and $C_4$-plants repression of net $O_2$ evolution by $O_2$ still occurs. The $CO_2$ compensation point has been used to differentiate between $C_3$- and $C_4$-plants by measuring survival in air without $CO_2$. In the $C_3$-plant $CO_2$ loss from glycolate oxidation in the absence of sufficient $CO_2$ for photosynthesis soon kills the plant. Equivalent phenomena may occur in algae. In some the production and excretion of glycolate in $CO_2$-free air will occur (Hess *et al.* 1967). In other algae, such as *Nitella*, which do not excrete glycolate (unpublished), the glycolate may be metabolized in the cell and the compensation point will be high. Extensive formation of malate and aspartate during $^{14}CO_2$ fixation by some algae is also indicative of a $C_4$-dicarboxylic acid cycle, which efficiently refixes the $CO_2$ from glycolate oxidation. More data are needed for evaluation of these alternatives.

### 6.3   *Effect of $O_2$ and $CO_2$ concentration*

Dark mitochondrial respiration in leaves is saturated by about 2% $O_2$ whereas photorespiration increases with $O_2$ concentration and is not saturated even by 100% $O_2$. Therefore, increasing partial pressure of $O_2$ and decreasing $CO_2$ concentration greatly favour photorespiration or glycolate formation. Many past observations with algae (Bassham & Kirk 1962, Bidwell *et al.* 1969, Brown & Tregunna 1967, Calvin *et al.* 1951a, Coombs & Whittingham 1966a,b, Downton & Tregunna 1968, Marker & Whittingham 1966, Pritchard *et al.* 1962, Tolbert &

Zill 1956, Whittingham & Pritchard 1963, Whittingham *et al.* 1967) indicate that glycolate formation from $^{14}CO_2$ was greatest in 100% $O_2$, and these results can be explained as a competition between $CO_2$ and $O_2$ for the carboxylation or oxidation of RuDP. The pool size of glycolate in algae during steady state photosynthesis at 1 to 5% $CO_2$ is small but increases immensely when the $CO_2$ level is lowered to 0·003% or air (Coombs & Whittingham 1966b, Lord & Merrett 1971a, Pritchard *et al.* 1962, Wilson & Calvin 1955). High $CO_2$ prevents RuDP oxidation and stimulates all aspects of photosynthesis ($O_2$ evolution and $CO_2$ fixation), whereas high $O_2$ stimulates photorespiration which reduces net photosynthesis. The resultant reduction in algal growth in high $O_2$ and/or low $CO_2$ has been explored as the Warburg effect and more recently as photorespiration (see Introduction), but this information has not yet been used to evaluate algal growth in nature. Photorespiration in algae may be limited by the carrying capacity of water for dissolved $O_2$, but, on the other hand, saturated $O_2$ conditions may exist in the cell during photolysis of water and photorespiration may be as limiting on growth as in $C_3$-plants.

In addition to the rapid and reversible effect of $CO_2$ levels on glycolate biosynthesis, the $CO_2$ concentration may have longer-term growth effects upon the level of enzymic activities, particularly carbonic acid anhydrase (Nelson *et al.* 1969b, Reed & Graham 1968). Continuous growth on high levels of $CO_2$ (1% or more) represses glycolate dehydrogenase. Growth on very low levels of $CO_2$ (air) increases the activity of glycolate dehydrogenase in *Euglena* about 16 fold (Codd *et al.* 1969, Lord & Merrett 1971a), about 3 fold for *Chlamydomonas* (Cooksey 1971, Nelson & Tolbert 1969), but not for *Chlorella* (Codd *et al.* 1969, Codd & Merrett 1971b), which does not excrete glycolate in 5% $CO_2$. Nitrogen deficient *Chlamydomonas* cells lose glycolate dehydrogenase activity (Cooksey 1971). This alga synthesizes the enzyme when transferred from an acetate-containing to an inorganic medium (Lord & Merrett 1970b). It appears as if glycolate dehydrogenase levels are influenced in some algae by a slow feedback inhibition from the carbon and nitrogen source available to the cell.

### 6.4    $CO_2/O_2$ quotient

Total $CO_2$ release and $O_2$ uptake during algal photorespiration might be expected to be less pronounced than in plants. $O_2$ uptake occurs during glycolate formation, during glycolate oxidation if glycolate oxidase is present, and during glycine oxidation or conversion to serine. In most unicellular green algae that part of $O_2$ uptake due to glycolate oxidation is absent, because the glycolate is either excreted or oxidized slowly by glycolate dehydrogenase which does not couple to oxygen. Whereas $O_2$ uptake remains significant due to glycolate formation, $CO_2$ release may be less, if the glycolate is excreted or not metabolized further than glycine. Therefore the hypothetical $CO_2/O_2$ quotient of photorespiration would be expected to vary, but to be less than 1. The measurable $CO_2/O_2$ ratio during photosynthesis, which is about 1 in an atmosphere of low

$O_2$ concentration is profoundly effected by higher $O_2$ concentrations (Fock *et al.* 1971). With 44% $O_2$ in the gas phase the photosynthetic $CO_2/O_2$ ratio reaches 1·6 for *Chlorella* due to decreased net $O_2$ evolution from increased photo-respiratory $O_2$ uptake. Thus at high $O_2$ concentrations recent photosynthate and stored reserves are converted to oxidized compounds (Fock *et al.* 1971, Hess *et al.* 1967), which according to experiments on [18]O incorporation are products of the glycolate pathway (Andrews *et al.* 1971). Certainly the observed $CO_2/O_2$ quotients in the light reflect gas exchanges from both photosynthesis and photo-respiration and values will vary depending upon the concentration of $O_2$ and $CO_2$ and a competition between the two processes.

## 6.5 *Synchronous cultures*

During periods of high photosynthetic ability but inhibited growth, as at the end of the light phase or dark phase, overproduction of glycolate and its excretion seem to occur. This manifestation of photorespiration fulfils its postulated function as a system for disposing of excess energy. Glycolate synthesis and excretion reached a maximum of 35% of the total [14]C fixed by synchronized *Ankistrodesmus* during or at the end of the light growing phase and were nearly absent in dividing cells in the dark phase, if placed back in the light (Chang & Tolbert 1970, Gimmler *et al.* 1969). The amount of the glycolate-[14]C pool in *Euglena* also varied widely during a life cycle, being maximal at the end of the dark phase and beginning of the light phase and minimal at the beginning of the dark phase (Codd & Merrett 1971a,b). When the glycolate pool was minimal, maximum [14]C appeared in glycine and serine, as if the capacity to metabolize glycolate varied. In fact glycolate dehydrogenase activity in the dark phase was maximal when the glycolate-[14]C pool was minimal. Although no similar studies on gas exchange measurements of photorespiration with synchronized algal cultures have been made, it is predictable that the magnitude of $O_2$ uptake would correspond with glycolate formation. Since the activity of RuDP carboxylase (Codd & Merrett 1971a) and phosphoglycolate phosphatase (Codd & Merrett 1971b) does not appear to change significantly during the life cycle, the change in glycolate pools may more nearly represent the algal ability or need to metabolize it. In fact only during the dark phase of cell division when little glycolate is produced can *Chlamydomonas* and *Euglena* assimilate exogenous glycolate (Codd & Merrett 1971b, Nelson *et al.* 1969a).

## 6.6 *pH effects*

Glycolate formation and excretion increase with the pH of the medium reaching maximal rates at pH values between 8 and 9 (Dodd & Bidwell 1971, Orth *et al.* 1966). The pH optimum of RuDP oxygenase is around 9·3 whereas the pH optimum of RuDP carboxylase is about 7·8 and the activity almost nil at pH 9

(Andrews *et al.* 1973). Maximum $CO_2$ photosynthesis is around 7·6 to 7·8, the maximum for RuDP carboxylase. At pH above 8·3 the carboxylase activity drops rapidly while the RuDP oxygenase activity increases. Therefore, glycolate synthesis would predominate at higher pH, as is observed. The pH of an algal medium rapidly increases in the light during bicarbonate uptake (Atkins & Graham 1971, Orth *et al.* 1966), so that the physiological pH of the medium favours phosphoglycolate formation.

### 6.7 *Light intensity and quality*

It is established that photorespiration and glycolate production in plants and algae are light dependent and increase with light intensity (Jackson & Volk 1970, Pederson *et al.* 1966, Tolbert 1963, Whittingham & Pritchard 1963). This effect may be explained by increased photosynthetic production of RuDP, the precursor of phosphoglycolate. A direct photo-involvement in glycolate formation or oxidation does not occur, except perhaps for some blue light effects. The action spectra for photorespiration in plants is that of photosynthesis, but it has not been reported in detail for algae.

During photosynthesis, *Chlorella* and *Anacystis* form and excrete large amounts of glycolate in red or white light but in blue light they accumulate little and excrete none (Becker *et al.* 1968, Döhler & Braun 1971, Lord *et al.* 1970). This is consistent with the report by Warburg and Krippahl (1960) that in red light *Chlorella* converted 92% of the photosynthate into glycolate. The reason for this phenomenon remains unexplained. It has been proposed that in blue light little glycolate is formed (Lord *et al.* 1970), or alternatively glycolate, glycine, and serine may be rapidly used in blue light for accelerated protein synthesis. The latter possibility is supported by the fact that in blue light glycine and serine are of higher specific activity (Cayle & Emerson 1957), as if glycolate were more rapidly being utilized for protein synthesis. When *Chlorella* cells were adapted by continuous growth in monochromatic light, they produced mainly glycolate in blue light and insignificant amounts in red light (Hess & Tolbert 1967b). Under these conditions growth and thus glycolate utilization was restricted even in blue light.

Kowallik, Schmid and Gaffron have described an enhanced respiration by small amounts of blue light for algae and particularly for a *Chlorella* strain and its chlorophyll-free mutant (Kowallik & Gaffron 1967, Kowallik 1967, 1971, Schmid & Schwarze 1969). That this widespread phenomenon may be related to photorespiration and glycolate metabolism was suggested by their studies on its action spectrum, which could be that of flavo-protein, and by *in vitro* stimulation by blue light of various flavin enzymes, including glycolate oxidase (Schmid 1970). However, other observations indicate that blue light stimulation of respiration is different from photorespiration. It occurs in a non-photosynthetic mutant of *Chlorella* or in DCMU-poisoned algae and in both instances glycolate synthesis should not occur. Only a very small amount of blue light saturates the enhanced respiration, so that stimulation of glycolate biosynthesis by high light intensity should not be involved. There is also no $O_2$ stimulation of the blue-light enhanced respiration of the yellow *Chlorella* mutant.

## 6.8 *Inhibitors*

Manganese deficiency in *Chlorella* results in a preferential inhibition of glycolate formation during $CO_2$ fixation (Hess & Tolbert 1967a, Tanner *et al.* 1960) as well as $O_2$ evolution during electron transport. These results seem consistent with the utilization of $O_2$ in glycolate formation. Earlier interpretations of these data implicating glycolate biosynthesis with a Mn-protein or directly linked to photosystem II are probably superseded by the $O_2$ effect.

Sulphonates, particularly α-hydroxypyridine methane sulphonate, have been used to inhibit glycolate metabolism, resulting in glycolate accumulation during photosynthesis (Zelitch 1971). The sulphonate inhibits algal glycolate dehydrogenase *in vitro*. The sulphonates have been used with algae to force glycolate excretion or to alter algal respiration. Care must be exercised in its use. How much of it is taken up by the alga has generally not been determined, and it is rapidly photoxidized in solution with $O_2$ uptake, which has generally not been considered (Bunt & Heeb 1970). Evidence for enhanced glycolate accumulation or excretion in the presence of the sulphonate has been reported for *Chlamydomonas* (Nelson & Tolbert 1969), *Chlorella* (Lord & Merrett 1969), and *Ankistrodesmus* (Gimmler *et al.* 1969, Urbach & Gimmler 1968). It inhibited photoheterotrophic growth of a strain of *Chlorella* on glycolate (Lord & Merrett 1971b). Sulphonate enhanced $^{14}CO_2$ fixation rates by *Ankistrodesmus*, *Chlorella*, and *Scenedesmus* by keeping the $CO_2$ fixation rate from dropping between pH 8 and 9 (Gimmler *et al.* 1969, Hess & Tolbert 1966b). With *Chlorella* and *Scenedesmus* (Hess & Tolbert 1966b), RuDP accumulated in the presence of the sulphonate at media pH above 8. An exact interpretation is not yet possible for these pronounced results.

Isonicotinyl hydrazide primarily inhibits the conversion of glycine to serine, for in its presence much glycolate and glycine accumulate and glycolate is secreted during photosynthesis (Clayton 1960, Coombs & Whittingham 1966b, Lord *et al.* 1972, Miflin & Whittingham 1966, Pritchard *et al.* 1962, Whittingham & Pritchard 1963, Whittingham *et al.* 1967). Isonicotinyl hydrazide is an inhibitor of aminotransferase, but it does not prevent glycine formation from glyoxylate in the glycolate pathway.

Disalicylidenepropanediamine has been reported to block the photoreduction of ferredoxin-dependent reactions (Trebst & Burba 1967) or the reduction of ferredoxin (Ben-Amotz & Avron 1972) and to stimulate glycolate formation and excretion (Gimmler *et al.* 1969, Urbach & Gimmler 1968). In some cases it does not inhibit photosynthetic $CO_2$ fixation at selected low concentrations but the formation of glycolate is favoured over that of sugars (Gimmler *et al.* 1968, 1969). These results would be consistent with the formation of $H_2O_2$ by aerobic oxidation of the electron acceptor from photosystem I, and the formation of glycolate by a peroxidation reaction.

## 6.9 *Phylogeny*

A working hypothesis has been that perhaps photorespiration and peroxisomes are primitive features which originated when $O_2$ first appeared in the environment and before mitochondria were present (Tolbert 1971). By this hypothesis photorespiration is not now so essential, and in fact it is a wasteful and undesirable type of respiration. A more efficient biological system should be anaerobic glycolysis and mitochondrial respiration, as does exist in some tissues without peroxisomes (i.e. muscle). Current and future research, particularly with algae, should elaborate on this hypothesis. The absence of peroxisomes, glycolate oxidase and most of the catalase in nearly all unicellular green algae suggest that catalase rich peroxisomes might be of more recent origin. Most of these algae have a glycolate dehydrogenase reaction instead. During evolution, the higher plants, perhaps upon exposure to an atmosphere of more $O_2$, seem to have developed the peroxisomal organelle.

Already several arguments suggest that the photorespiration in plants with photosystem II must in some way be so obligatory that it has not disappeared during evolution. Data summarized in this chapter show that photorespiration is very substantial and ubiquitous in photosynthetic algae, including blue-green algae. Both $C_4$-plants, which may represent a more recent evolutionary plant development, and algae have little loss of $CO_2$ during photorespiration (Hatch *et al.* 1971, Jackson & Volk 1970, Tolbert 1971), but photorespiration in both is a significant metabolic process. The $C_4$-plant efficiently refixes the $CO_2$ from photorespiration, and the algae may also do that, as well as excrete the glycolate. Thus, neither the algae or the $C_4$-plant seems to be without photorespiration. The reason for photorespiration will become more apparent as more is learned about glycolate biosynthesis. RuDP carboxylase is the primary $CO_2$ fixation process, yet it is also a RuDP oxygenase responsible for P-glycolate formation. The inevitable competition between $CO_2$ and $O_2$ for RuDP carboxylation and RuDP oxidation suggests that photorespiration is unavoidable in photosynthetic systems that evolve oxygen. Is there something so essential about glycolate biosynthesis that through evolution it has not been eliminated? By flushing higher plants or algae with low levels of $CO_2$ this wasteful photorespiration can be reduced and growth is faster. Thus by manipulating the atmosphere, man can increase some plant growth rates over those in nature. It almost seems as if glycolate formation and photorespiration are inevitable during photosynthesis and cannot be avoided, or else nature would have done it long ago.

# 498 CHAPTER 17

## 7 REFERENCES

ANDERSON D. & TOLBERT N.E. (1966) Phosphoglycolate phosphatase. In *Methods in Enzymol.* IX, ed. Wood W.A. pp. 646–50. Academic Press, New York & London.

ANDERSON L. & FULLER R.C. (1967) The rapid appearance of glycolate during photosynthesis in *Rhodospirillum rubrum. Biochim. biophys. Acta*, **131**, 198–201.

ANDREWS T.J., LORIMER G.H. & TOLBERT N.E. (1971) Incorporation of molecular oxygen into glycine and serine during photorespiration in spinach leaves. *Biochemistry, N.Y.* **10**, 4777–82.

ANDREWS T.J., LORIMER G.H. & TOLBERT N.E. (1973) Ribulose diphosphate oxygenase. I. Synthesis of phosphoglycolate by fraction-1 protein of leaves. *Biochemistry, N.Y.* **12**, 11–18.

ATKINS C.A. & GRAHAM P. (1971) Light-induced pH changes by cells of *Chlamydomonas reinhardii.* Dependence on $CO_2$ uptake. *Biochim. biophys. Acta*, **226**, 481–5.

BASSHAM J.A. (1971) The control of photosynthetic carbon metabolism. *Science, N.Y.* **172**, 526–34.

BASSHAM J.A. & KIRK M. (1962) The effect of oxygen on the reduction of $CO_2$ to glycolic acid and other products during photosynthesis by *Chlorella. Biochem. biophys. Res. Commun.* **9**, 376–80.

BECKER J.D., DÖHLER G. & EGLE K. (1968) Die Wirkung monochromatischen Lichts auf die extrazelluläre Glykolsäure-Ausscheidung bei der Photosynthese von *Chlorella. Z. PflPhysiol.* **58**, 212–21

BEN-AMOTZ A. & AVRON M. (1972) Is nicotinamide adenine dinucleotide phosphate an obligatory intermediate in photosynthesis? *Pl. Physiol., Lancaster*, **49**, 244–8.

BENSON A.A. & CALVIN M. (1950) The path of carbon in photosynthesis VII. Respiration and photosynthesis. *J. exp. Bot.* **1**, 63–8.

BIDWELL R.G.S., LEVIN W.B. & SHEPHARD D.C. (1969) Photosynthesis, photorespiration and respiration of chloroplasts from *Acetabularia mediterrania. Pl. Physiol., Lancaster*, **44**, 946–54.

BIDWELL R.G.S., LEVIN W.B. & SHEPHARD D.C. (1970) Intermediates of photosynthesis in *Acetabularia mediterrania* chloroplasts. *Pl. Physiol., Lancaster*, **45**, 70–5.

BIRD I.F., CORNELIUS M.J., KUP A.J. & WHITTINGHAM C.P. (1972) Oxidation and phosphorylation associated with the conversion of glycine to serine. *Phytochem.* **11**, 1587–94.

BOUCK G.B. (1965) Fine structure and organelle associations in brown algae. *J. Cell Biol.* **26**, 523–7.

BOWES G., OGREN W.L. & HAGEMAN R.H. (1971) Phosphoglycolate production catalyzed by ribulose diphosphate carboxylase. *Biochem. biophys. Res. Commun.* **45**, 716–22.

BOWES G. & OGREN W.L. (1972) Oxygen inhibition and other properties of soybean ribulose 1,5-diphosphate carboxylase. *J. biol. Chem.* **247**, 2171–6.

BRADBEER J.W. & ANDERSON C.M.A. (1967) Glycolate formation in chloroplast preparations. In *Biochemistry of Chloroplasts*, vol. II, ed. Goodwin T.W. pp. 175–9. Academic Press, New York & London.

BROWN A.H. & WHITTINGHAM C.P. (1955) Identification of the carbon dioxide burst in *Chlorella* using the recording mass spectrometer. *Pl. Physiol., Lancaster*, **30**, 231–7.

BROWN A.H. & WEIS D. (1959) Relation between respiration and photosynthesis in the green alga, *Ankistrodesmus braunii. Pl. Physiol., Lancaster*, **34**, 177–364.

BROWN D.L. & TREGUNNA E.B. (1967) Inhibition of respiration during photosynthesis by some algae. *Can. J. Bot.* **45**, 1135–43.

BRUIN W.J., NELSON E.B. & TOLBERT N.E. (1970) Glycolate pathway in green algae. *Pl. Physiol., Lancaster*, **46**, 386–91.

BUNT J.S. (1969) The $CO_2$ compensation point, Hill activity and photorespiration. *Biochem. biophys. Res. Commun.* **35**, 748–53.

BUNT J.S. (1970) Some observations on the carbon dioxide burst in *Chlorella* and *Chlamydomonas*. *Pl. Physiol., Lancaster*, **45**, 139–42.

BUNT J.S. & HEEB M.A. (1970) Consumption of $O_2$ in the light by *Chlorella pyrenoidosa* and *Chlamydomonas reinhardtii*. *Biochim. biophys. Acta*, **226**, 354–9.

BUTT V.S. & PEEL M. (1963) The participation of glycolate oxidase in glucose uptake by illuminated *Chlorella* suspensions. *Biochem. J.* **88**, 31 P.

CALKINS V.P. (1943) Micro determination of glycolic and oxalic acids. *Ind. Engng Chem. analyt. Edn.* **15**, 762–3.

CALVIN M., BASSHAM J.A., BENSON A.A., LYNCH V.H., OUELLET C., SCHOU L., STEPKA W. & TOLBERT N.E. (1951a) Carbon dioxide assimilation in plants. *V Symp. S.E.B.* 284–305.

CALVIN M., BASSHAM J.A., BENSON A.A., KAWAGUCHI S., LYNCH V.H., STEPKA W. & TOLBERT N.E. (1951b) O caminho do carbono na fotosynthesis XIV. *Selecta Chim.* **10**, 143–79.

CAYLE T. & EMERSON R. (1957) Effect of wavelength on the distribution of carbon-14 in the early products of photosynthesis. *Nature, Lond.* **179**, 89–90.

CHANG W.-H. & TOLBERT N.E. (1970) Excretion of glycolate, mesotartrate, and isocitrate lactone by synchronized cultures of *Ankistrodesmus braunii*. *Pl. Physiol., Lancaster*, **46**, 377–85.

CLAYTON R.K. (1960) Kinetics of $H_2O_2$ destruction in *Rhodopseudomonas spheroides*: roles of catalase and other enzymes. *Biochem. biophys. Acta*, **40**, 165–7.

CODD G.A., LORD J.M. & MERRETT M.J. (1969) The glycolate oxidizing enzyme of algae. *FEBS Letters*, **5**, 341–2.

CODD G.A. & MERRETT M.J. (1970) Enzymes of the glycolate pathway in relation to greening in *Euglena gracilis*. *Planta*, **95**, 127–32.

CODD G.A. & MERRETT M.J. (1971a) Photosynthetic products of division synchronized cultures of *Euglena*. *Pl. Physiol., Lancaster*, **47**, 635–9.

CODD G.A. & MERRETT M.J. (1971b) The regulation of glycolate metabolism in division synchronized cultures of *Euglena*. *Pl. Physiol., Lancaster*, **47**, 640–3.

CODD G.A., SCHMID G.H. & KOWALLIK W. (1972) Enzymic evidence for peroxisomes in a mutant of *Chlorella vulgaris*. *Arch. Mikrobiol.* **81**, 264–72.

COOK J.R. (1970) Properties of partially purified malate synthase from *Euglena gracilis*. *J. Protozool.* **17**, 232–5.

COOKSEY K.E. (1971) Glycolate:dichlorophenolindophenol oxidoreductase in *Chlamydomonas reinhardtii*. *Pl. Physiol., Lancaster*, **48**, 267–9.

COOMBS J. & WHITTINGHAM C.P. (1966a) The mechanism of inhibition of photosynthesis by high partial pressures of $O_2$ in *Chlorella*. *Proc. R. Soc., Ser. B.* **164**, 511–20.

COOMBS J. & WHITTINGHAM C.P. (1966b) The effect of high partial pressures of oxygen on photosynthesis in *Chlorella* I. The effect on end products of photosynthesis. *Phytochem.* **5**, 643–51.

DODD W.A. & BIDWELL R.G.S. (1971) The effect of pH on the products of photosynthesis in $^{14}CO_2$ by chloroplast preparations from *Acetabularia mediterranea*. *Pl. Physiol., Lancaster*, **47**, 779–83.

DÖHLER G. & BRAUN F. (1971) Untersuchung der Beziehung zwischen extra-cellulärer Glykolsäure-Ausscheidung und der photosynthetischen $CO_2$-Aufnahme bei der Blau alge *Anacystis nidulans*. *Planta*, **98**, 357–61.

DOWNTON W.J.S. & TREGUNNA E.B. (1968) Photorespiration and glycolate metabolism: A re-examination and correlation of some previous studies. *Pl. Physiol., Lancaster*, **43**, 923–9.

DROOP M.R. & McGILL S. (1966) The carbon nutrition of some algae: the inability to utilize glycolic acid for growth. *J. mar. biol. Ass. U.K.* **46**, 679–84.

ELLYARD P.W. & GIBBS M. (1969) Inhibition of photosynthesis by oxygen in isolated spinach chloroplasts. *Pl. Physiol., Lancaster*, **44**, 1115–21.

FOCK H. & EGLE K. (1966) Über die 'Lichtatmung' bie grünen pflanzen. *Beitr. Biol. Pfl.* **42**, 213–39.

FOCK H., SCHAUB H. & HILGENBERG W. (1968) Uber den Sauerstaff und Kohlendioxide-gaswechsel von *Chlorella* und *Conocepholum* während der Lichtphase. *Z. PflPhysiol.* **60**, 56–63.

FOCK H., CANVIN D.T. & GRANT B.R. (1971) Effects of oxygen and carbon dioxide on photosynthetic $O_2$ evolution and $CO_2$ uptake in sunflower and *Chlorella*. *Photosynthetica*, **5**, 389–94.

FOO S.K., BADOUR S.S. & WAYGOOD E.R. (1971) Regulation of isocitrate lyase from autotrophic and photoheterotrophic cultures of *Gloeomonas* sp. *Can. J. Bot.* **49**, 1647–53.

GERGIS M.S. (1971) The presence of microbodies in three strains of *Chlorella*. *Planta*, **101**, 180–4.

GERHARDT B. (1971) Zur lokalisation von enzymen der microbodies in *Polytomella caeca*. *Arch. Mikrobiol.* **80**, 205–18.

GERHARDT B. & BERGER C. (1971) Microbodies and Diaminobenzidin-Reaktion in den acetate-Flagellaten *Polytomella caeca* and *Chlorogonium elongatum*. *Planta*, **100**, 155–66.

GIBBS M. (1969) Photorespiration, Warburg effect and glycolate. *Ann. N.Y. Acad. Sci.* **168**, 356–68.

GIMMLER H., URBACH W., JESCHKE W.D. & SIMONIS W. (1968) Die unterschiedliche Wirkung von Disalicylidenpropandiamin auf die cyclische und nichtcyclische Photophosphorylierung *in vivo* sowie auf die $^{14}C$-Markierung einzelner Photosyntheseprodukte. *Z. PflPhysiol.* **58**, 353–64.

GIMMLER H., ULLRICH W., DOMANSKI-KADEN J. & URBACH W. (1969) Excretion of glycolate during synchronous culture of *Ankistrodesmus braunii* in the presence of disalicylidenepropanediamine or hydroxypyridinemethanesulfonate. *Pl. Cell Physiol., Tokyo*, **10**, 103–12.

GIRAUD G. & CZANINSKI Y. (1971) Localisation ultrastructurale d'activites oxydasiques chez le *Chlamydomonas reinhardi*. *C. r. hebd. Séanc. Acad. Sc. Paris*, **273**, 2500–3.

GOLDSWORTHY A. (1966) Experiments on the origin of $CO_2$ released by tobacco leaf segments in the light. *Phytochem.* **5**, 1013–19.

GOOD N.E. & BROWN A.H. (1961) The contribution of endogenous oxygen to the respiration of photosynthesizing *Chlorella* cells. *Biochim. biophys. Acta*, **50**, 544–54.

GOULDING K.H. & MERRETT M.J. (1966) The photometabolism of acetate by *Chlorella. J. exp. Bot.* **17**, 678–89.

GOULDING K.H., LORD J.M. & MERRETT M.J. (1969) Glycollate formation during the photorespiration of acetate by *Chlorella. J. exp. Bot.* **20**, 34–45.

GOULDING K.H. & MERRETT M.J. (1967) The role of glycollic acid in the photoassimilation of acetate by *Chlorella pyrenoidosa. J. exp. Bot.* **18**, 620–30.

GRAVES L.B., TRELEASE R.N. & BECKER W.M. (1971a) Particulate nature of glycolate dehydrogenase in *Euglena*: possible location in microbodies. *Biochem. biophys. Res. Commun.* **44**, 280–6.

GRAVES L.B., HANZELY L. & TRELEASE R.N. (1971b) The occurrence and fine structure characterization of microbodies in *Euglena gracilis. Protoplasma*, **72**, 141–52.

GRODZINSKI B. & COLMAN B. (1970) Glycolic acid oxidase activity in cell-free preparations of blue-green algae. *Pl. Physiol., Lancaster*, **45**, 735–7.

GRUBER P., FREDERICK S. & TOLBERT N.E. unpublished.

HAIGH W.G. & BEEVERS H. (1964a) The occurrence and assay of isocitrate lyase in algae. *Archs. Biochem. Biophys.* **107**, 147–51.

HAIGH W.G. & BEEVERS H. (1964b) The glyoxylate cycle in *Polytomella caeca. Archs. Biochem. Biophys.* **107**, 152–7.

HARROP L.C. & KORNBERG H.L. (1966) The role of isocitrate lyase in the metabolism of algae. *Proc. R. Soc. B.* **166**, 11–29.

HATCH M.D., OSMOND C.B. & SLATYER R.O., eds. (1971) *Photosynthesis and Photorespiration.* Wiley-Interscience, London, New York, Sydney & Toronto.

HESS J.L. & TOLBERT N.E. (1966a) Glycolate, glycine, serine and glycerate formation during photosynthesis by tobacco leaves. *J. biol. Chem.* **241,** 5707–11.

HESS J.L. & TOLBERT N.E. (1966b) The effect of hydroxymethylsulfonates on $^{14}CO_2$ photosynthesis by algae. *J. biol. Chem.* **241,** 5712–5.

HESS J.L. & TOLBERT N.E. (1967a) Glycolate pathway in algae. *Pl. Physiol. Lancaster,* **42,** 371–9.

HESS J.L. & TOLBERT N.E. (1967b) Changes in Chlorophyll a/b ratio and products of $^{14}CO_2$ fixation by algae grown in blue or red light. *Pl. Physiol., Lancaster,* **42,** 1123–30.

HESS J.L., TOLBERT N.E. & PIKE L.M. (1967) Glycolate biosynthesis by *Scenedesmus* and *Chlorella* in the presence or absence of $NaHCO_3$. *Planta,* **74,** 278–85.

HILLER R.G. & WHITTINGHAM C.P. (1959) Further studies on the carbon dioxide burst in algae. *Pl. Physiol., Lancaster,* **34,** 219–22.

HOCH G., OWENS O. & KOK B. (1963) Photosynthesis and respiration. *Archs. Biochem. Biophys.* **101,** 171–80.

JACKSON W.A. & VOLK R.J. (1970) Photorespiration. *A. Rev. Pl. Physiol.* **21,** 385–432.

KEARNEY P.C. & TOLBERT N.E. (1962) Appearance of glycolate and related products of photosynthesis outside of chloroplasts. *Archs. Biochem. Biophys.* **98,** 164–71.

KIERMAYER O. (1970) Electron microscopic studies on the problem of cytomorphogenesis in *Micrasterias denticulata* Bréb. *Protoplasma,* **69,** 97–132.

KISAKI T. & TOLBERT N.E. (1969) Glycolate and glyoxylate metabolism by isolated peroxisomes or chloroplasts. *Pl. Physiol., Lancaster,* **44,** 242–50.

KISAKI T. & TOLBERT N.E. (1970) Glycine as a substrate for photorespiration. *Pl. Cell Physiol., Tokyo,* **11,** 247–58.

KISAKI T., IMAI A. & TOLBERT N.E. (1971a) Intracellular localization of enzymes related to photorespiration in green leaves. *Pl. Cell Physiol., Tokyo,* **12,** 267–73.

KISAKI T., YOSHIDA N. & IMAI A. (1971b) Glycine decarboxylase and serine formation in spinach leaf mitochondria with reference to photorespiration. *Pl. Cell Physiol., Tokyo,* **12,** 275–88.

KOWALLIK W. (1967) Action spectrum for an enhancement of endogenous respiration by light in *Chlorella. Pl. Physiol., Lancaster,* **42** 672–6.

KOWALLIK W. (1971) Light-stimulated respiratory gas exchange in algae and its relation to photorespiration. In *Photosynthesis and Photorespiration,* eds. Hatch M.D., Osmond C.B. & Slatyer R.O. pp. 514–22. Wiley-Interscience, London, New York, Sydney & Toronto.

KOWALLIK W. & GAFFRON H. (1967) Enhancement of respiration and fermentation in algae by blue light. *Nature, Lond.* **215,** 1038–40.

KOWALLIK W. & SCHMID G.H. (1971) On the oxidation of glycolate by unicellular green algae. *Planta,* **96,** 224–37.

LEX M., SILVESTER W.B. & STEWART W.D.P. (1972) Photorespiration and nitrogenase activity in the blue-green alga, *Anabaena cylindrica. Proc. R. Soc. B.* **180,** 87–102.

LEWIN R.A. (1957) Excretion of glycolic acid by *Chlamydomonas. Bull. Japan Soc. Phycol.* **5,** 74–5.

LORD J.M. & MERRETT M.J. (1968) Glycolate oxidase in *Chlorella pyrenoidosa. Biochim. biophys. Acta,* **159,** 543–4.

LORD J.M. & MERRETT M.J. (1969) The effect of hydroxymethanesulfonate on photosynthesis in *Chlorella pyrenoidosa. J. exp. Biol.* **20,** 743–50.

LORD J.M. & MERRETT M.J. (1970a) The pathway of glycolate utilization in *Chlorella pyrenoidosa. Biochem. J.* **117,** 929–37.

LORD J.M. & MERRETT M.J. (1970b) The regulation of glycolate oxidoreductase with photosynthetic capacity in *Chlamydomonas mundana. Biochem. J.* **119,** 125–7.

LORD J.M. & MERRETT M.J. (1971a) The intracellular localization of glycolate oxidoreductase in *Euglena gracilis. Biochem. J.* **124**, 275–81.

LORD J.M. & MERRETT M.J. (1971b) The growth of *Chlorella pyrenoidosa* on glycolate. *J. exp. Bot.* **22**, 60–9.

LORD J.M., CODD G.A. & MERRETT M.J. (1970) The effect of light quality on glycolate formation and excretion in algae. *Pl. physiol., Lancaster,* **46**, 855–6.

LORIMER G.H., ANDREWS T.J. & TOLBERT N.E. (1973) Ribulose diphosphate oxygenase II. Further proof of reaction products and mechanism of action. *Biochemistry,* **12**, 18–23.

MARKER A.F.H. & WHITTINGHAM C.P. (1966) The photoassimilation of glucose in *Chlorella* with reference to the role of glycollic acid. *Proc. R. Soc. B.* **165**, 473–85.

MERRETT M.J. & GOULDING K. (1967) Short-term products of $^{14}$C-acetate assimilation by *Chlorella pyrenoidosa* in the light. *J. exp. Bot.* **18**, 128–39.

MIFLIN B.J. & WHITTINGHAM C.P. (1966) The effect of inhibitors on the path of carbon in photosynthesis by *Chlorella* at low partial pressures of $CO_2$. *Ann. Bot.* **30**, 339–47.

MILLER A.G., CHANG K.H. & COLMAN B. (1971) The uptake and oxidation of glycolic acid by blue-green algae. *J. Phycol.* **7**, 97–100.

MIYACHI S., IZAWA S. & TAMIYA H. (1955) Effect of oxygen on the capacity of carbon dioxide fixation by green algae. *J. Biochem., Tokyo,* **42**, 221–44.

MURRAY D.R., GIOVANELLI J. & SMILLIE R.M. (1970) Photometabolism of glycolate, glycine and serine by *Euglena gracilis. J. Protozool.* **17**, 99–104.

MURRAY D.R., GIOVANELLI J. & SMILLIE R.M. (1971) Photometabolism of glycolate by *Euglena gracilis. Austr. J. biol. Sci.* **24**, 23–33.

NELSON E.B. & TOLBERT N.E. (1969) The regulation of glycolate metabolism in *Chlamydomonas reinhardtii. Biochim. biophys. Acta,* **184**, 263–70.

NELSON E.B. & TOLBERT N.E. (1970) Glycolate dehydrogenase in green algae. *Archs. Biochem. Biophys.* **141**, 102–10.

NELSON E.B., TOLBERT N.E. & HESS J.L. (1969a) Glycolate stimulation of oxygen evolution during photosynthesis. *Pl. Physiol., Lancaster,* **44**, 55–9.

NELSON E.B., CENEDELLA A. & TOLBERT N.E. (1969b) Carbonic anhydrase levels in *Chlamydomonas. Phytochem.* **8**, 2305–6.

NORRIS L., NORRIS R.E. & CALVIN M. (1955) A survey of the rates and products of short term photosynthesis in plants of nine phyla. *J. exp. Bot.* **6**, 64–70.

ORTH G.M., TOLBERT N.E. & JIMENEZ E. (1966) Rate of glycolate formation during photosynthesis at high pH. *Pl. Physiol., Lancaster,* **41**, 143–7.

PEDERSON T.A., KIRK M. & BASSHAM J.A. (1966) Light-dark transients in levels of intermediate compounds during photosynthesis in air-adapted *Chlorella. Physiologia Pl.* **19**, 219–31.

PLANONDAN J.E. & BASSHAM J.A. (1966) Glycolic acid labeling during photosynthesis with $^{14}CO_2$ and tritiated water. *Pl. Physiol., Lancaster,* **41**, 1272–5.

PRITCHARD G.G., GRIFFIN W.J. & WHITTINGHAM C.P. (1962) The effect of carbon dioxide concentration, light intensity and isonicotinyl hydrazide on the photosynthetic production of glycolic acid by *Chlorella. J. exp. Bot.* **13**, 176–84.

RABSON R., TOLBERT N.E. & KEARNEY P.C. (1962) Formation of serine and glyceric acid by the glycolate pathway. *Archs. Biochem. Biophys.* **98**, 154–63.

RANDALL D.D. & TOLBERT N.E. (1971a) 3-Phosphoglycerate phosphatase in plants. I. Isolation and characterization from sugar cane leaves. *J. biol. Chem.* **246**, 5510–17.

RANDALL D.D. & TOLBERT N.E. (1971b) Two phosphatases associated with photosynthesis and the glycolate pathway. In *Photosynthesis and photorespiration,* eds. Hatch M.D., Osmond C.B. & Slatyer R.O. pp. 259–66. Wiley-Interscience, London, New York, Sydney & Toronto.

RANDALL D.D., TOLBERT N.E. & GREMEL D. (1971) 3-Phosphoglycerate phosphatase in plants. II. Distribution, physiological considerations, and comparison with P-glycolate phosphatase. *Pl. Physiol., Lancaster,* **48**, 480–7.

REED M.L. & GRAHAM D. (1968) Control of photosynthetic carbon dioxide fixation during an induction phase in *Chlorella*. *Pl. Physiol., Lancaster*, **43**, 5–29.

RICHARDSON K.E. & TOLBERT N.E. (1961) Phosphoglycolic acid phosphatase. *J. biol. Chem.* **236**, 1285–90.

RICHARDSON M. (1974) Microbodies (glyoxysomes and peroxisomes) in plants. *Sci. Progr.* **61**, 41–61.

SCHAUB H. & EGLE K. (1965) Uber die $CO_2$-Kompensationslage bei der Photosynthese von Algensedimenten. *Beitr. Biol. Pfl.* **41**, 5–10.

SCHOU L., BENSON A.A., BASSHAM J.A. & CALVIN M. (1950) The path of carbon in photosynthesis. XI. The role of glycolic acid. *Physiologia Pl.* **3**, 487–95.

SCHMID G.H. (1970) The effect of blue light on some flavin enzymes. *Z. phys. Chem.* **351**, 575–8.

SCHMID G.H. & SCHWARZE P. (1969) Blue light enhanced respiration in a colorless *Chlorella* mutant. *Hoppe-Seyler's Z. phys. Chem.* **350**, 1513–20.

SHAIN Y. & GIBBS M. (1971) Formation of glycolate by a reconstituted spinach chloroplast preparation. *Pl. Physiol., Lancaster*, **48**, 325–30.

TANNER H.A., BROWN T.E., EYSTER H.C. & TREHARNE R.W. (1960) A manganese dependent photosynthesis process. *Biochem. biophys. Res. Commun*, **3**, 205–10.

THOMPSON C.M. & WHITTINGHAM C.P. (1967) Intracellular localization of phosphoglycolate phosphatase and glyoxylate reductase. *Biochim. biophys. Acta*, **143**, 642–4.

TOGASAKI R.K. & LEVINE R.P. (1970) Chloroplast structure and function in ac-20, a mutant strain of *Chlamydomonas reinhardi* I. Carbon dioxide and ribulose-1,5-diphosphate carboxylase synthesis. *J. Cell Biol.* **44**, 531–9.

TOLBERT N.E. & ZILL L.P. (1956) Excretion of glycolic acid by algae during photosynthesis. *J. biol. Chem.* **222**, 895–906.

TOLBERT N.E. (1963) Glycolate pathway. In *Photosynthetic mechanism in green plants*, pp. 648–62. N.A.S./N.R.C., Washington D.C. **1145**.

TOLBERT N.E. (1971) Microbodies—peroxisomes and glyoxysomes. *A. Rev. Pl. Physiol.* **22**, 45–74.

TOLBERT N.E., YAMAZAKI R.K. & OESER A. (1970) Localization and properties of hydroxypyruvate and glyoxylate reductase in spinach leaf particles. *J. biol. Chem.* **245**, 5129–36.

TOLBERT N.E., NELSON E.B. & BRUIN W.J. (1971) Glycolate pathway in algae. In *Photosynthesis and photorespiration*, eds. Hatch M.D., Osmond C.B. & Slatyer R.O. pp. 506–13. Wiley-Interscience, London, New York, Sydney & Toronto.

TURNER J.S. & BRITTAIN E.G. (1962) Oxygen as a factor in photosynthesis. *Biol. Rev.* **37**, 130–70.

TREBST A. & BURBA M. (1967) Uber die Hemmung photosynthetischer Reaktionen in isolierten Chloroplasten und *Chlorella* durch Disalicylidenpropandiamin. *Z. PflPhysiol.* **57**, 419–33.

ULLRICH W. (1963) Phosphatase action on phosphoglycolic, 3-phosphoglyceric and phosphoenol pyruvic acids in spinach chloroplast fragments in the presence and absence of high concentrations of methanol. *Biochim. biophys. Acta*, **71**, 589–94.

URBACH W. & GIMMLER H. (1968) Stimulation of glycolate excretion of algae by disalicylidenepropanediamine and hydroxypyridinemethanesulfonate. *Z. Naturf.* **23**, 1282–3.

WARBURG O. (1920) Über die Geschwindiglkit der photochemischen Kohlensaurezersetzung in lebenden Zellen. *Biochem. Z.* **103**, 188–217.

WARBURG O. & KRIPPAHL G. (1960) Glycolsaurebildung in *Chlorella*. *Z. Naturf.* **15b**, 197–200.

WHITTINGHAM C.P. & PRITCHARD G.G. (1963) The production of glycollate during photosynthesis in *Chlorella*. *Proc. R. Soc. B.* **157**, 366–82.

WHITTINGHAM C.P., COOMBS J. & MARKER A.F.H. (1967) The role of glycollate in photosynthetic carbon fixation. In *Biochemistry of Chloroplasts*, Vol. II, ed. Goodwin T.W. pp. 155–73. Academic Press, New York & London.

WIESSNER W. (1967) The problem of glycolate formation from acetate in green algae. *Arch. Mikrobiol.* **58**, 366–9.

WIESSNER W. (1968) Studies on the connection between nutrition and enzyme activity in several green algae. *Planta*, **79**, 92–8.

WILSON A.T. & CALVIN M. (1955) The photosynthetic cycle. $CO_2$ dependent transients. *J. Am. chem. Soc.* **77**, 5948–57.

WOLFENDEN R. (1970) Binding of substrate and transition state analogs to triosephosphate isomerase. *Biochemistry*, **9**, 3404–5.

YU Y.L., TOLBERT N.E. & ORTH G.M. (1964) Isolation and distribution of phosphoglycolate phosphatase. *Pl. Physiol., Lancaster*, **39**, 643–7.

ZELITCH I. (1964) Organic acids and respiration in photosynthetic tissues. *A. Rev. Pl. Physiol.* **15**, 121–42.

ZELITCH I. (1965) The relation of glycolic acid synthesis to the primary photosynthetic carboxylation reaction in leaves. *J. biol. Chem.* **240**, 1869–76.

ZELITCH I. (1971) *Photosynthesis, photorespiration, and plant productivity*. Academic Press, New York & London.

ZELITCH I. & DAY P.R. (1968a) Variation in photorespiration. The effect of genetic differences in photorespiration on net photosynthesis in tobacco. *Pl. Physiol., Lancaster*, **43**, 1838–44.

ZELITCH I. & DAY P.R. (1968b) Glycolate oxidase activity in algae. *Pl. Physiol., Lancaster*, **43**, 289–91.

# CHAPTER 18

# DARK RESPIRATION

## D. LLOYD

Department of Microbiology,
University College, Cardiff, U.K.

1   Introduction   505

2   Utilization of exogenous carbon sources: pathways of carbon metabolism involved in respiration   506

3   Effects of environmental changes on respiration and the respiratory apparatus 508
3.1   Oxygen tension   508
3.2   Carbon sources   509
3.3   Growth with inhibitors   511
3.4   pH and temperature   511
3.5   Light   511
3.6   Changes in respiration during the cell cycle   512
3.7   Starvation and differentiation   513

4   Fractionation of extracts   514
4.1   Methodology   514
4.2   Properties of isolated 'particulate' and mitochondrial fractions; electron transport and oxidative phosphorylation   515
    (a) *Prototheca zopfii*   516
    (b) *Euglena gracilis*   517
    (c) *Astasia longa*   518
    (d) *Polytoma* and *Polytomella*   519
    (e) Blue-green algae   521
4.3   Organelles other than mitochondria   522

5   Conclusion   523

6   References   524

## 1   INTRODUCTION

When dark respiration, or classical respiration, as distinct from photorespiration, was last reviewed by Gibbs in 1962, it was evident that despite the extent of literature then available, a number of important unsolved problems existed which clearly indicated the paths which research would take. The past decade has seen some of these paths pursued vigorously and others neglected; it is hoped that this review will highlight some of the outstanding gaps in our understanding of algal respiration. Technological advances over the last ten years have provided more powerful tools for overcoming some of the methodological barriers which are particularly great in studies of algae, and the emphasis throughout this review will be on the availability of methods for extending work on algal respiratory metabolism both at the whole cell and molecular level.

In 1962, as pointed out by Gibbs, there was no evidence for the *in vivo* operation of the pentose phosphate cycle or measure of the relative importance of pathways alternative to those of glycolysis. The tricarboxylic acid enzymes had been demonstrated in extracts of several algae, but evidence for the role of the cycle at the whole cell level was lacking. The glyoxylate cycle had not been demonstrated in any alga. Studies on electron transport pathways suggested that these 'differ somewhat from those in animal cells'. No subcellular fractionation studies or separation of sub-cellular components had been attempted, and no cell-free system of oxidative phosphorylation had been obtained from any alga. These are the major topics covered by the present review and represent major areas of activity in recent years.

## 2  UTILIZATION OF EXOGENOUS CARBON SOURCES— PATHWAYS OF CARBON METABOLISM INVOLVED IN RESPIRATION

Before a pathway may be considered to be conclusively established as a functional mechanism involved in cellular oxidation or growth, the following criteria must be satisfied:

(a) The enzymes of the pathway are present in cell-free extracts. Where there is no alternative metabolic route the specific activity of the rate-limiting enzyme of the sequence should be commensurate with $O_2$ uptake rates or growth rates.

(b) The key metabolic intermediates are present (demonstrated by using metabolic blocks produced by an inhibitor or in a genetically defective strain, or by radioisotopic analysis of intracellular pools).

(c) The intermediates become labelled after presentation of the isotopically-labelled substrate in the sequence expected of the postulated pathway.

(d) The position of labelled carbon atoms within the labelled intermediates or products fits the postulated reaction sequence.

Bearing in mind these criteria it is evident that our knowledge of algal carbon metabolism is still far from complete, but it is now clear that energy production involves glycolysis, the pentose-phosphate cycle, the tricarboxylic acid (TCA) cycle and the glyoxylate cycle. The review of Gibbs (1962) has presented the evidence then available for the operation of these pathways, and a more recent review by Danforth (1967) has presented a detailed account of the recent work on phytoflagellates (including species of *Ochromonas*, *Euglena*, *Astasia*, *Chilomonas*, *Polytomella*, *Polytoma* and *Chlamydomonas*). No attempt will be made to repeat information contained in these two reviews, but some recent developments will be outlined.

The Embden-Meyerhoff pathway of glycolysis is the basic fermentation pathway involved in glucose utilization by all the algal species investigated (see e.g. Ciferri 1962). However, detailed analyses of stoichiometry of lactate production indicate major differences between various strains of *Chlorella* (Firdousi &

Syrett 1970). Changes in the net flux of carbon entering the TCA cycle by way either of glycolysis or the oxidative pentose-phosphate cycle have been studied in *Chlorella pyrenoidosa* during the transition from photosynthesis to dark metabolism (Kanazawa *et al.* 1972). In the absence of a nitrogen source, this change triggers the oxidative pentose-phosphate cycle by activation of glucose-6-phosphate dehydrogenase and inactivation of key steps of the Calvin cycle. A transient, but not permanent activation of pyruvate kinase, and a decreased rate of fatty acid synthesis, increase the flow of carbon into the TCA cycle. In the presence of $NH_4^+$, the reductive amination of 2-oxoglutarate also occurs.

The key enzymes of the pentose phosphate cycle (glucose-6-phosphate and 6-phosphogluconate dehydrogenases, transketolase and transaldolase) have been detected in extracts from *Prototheca zopfii* (Ciferri 1962) and from *Euglena gracilis* var. *bacillaris* (Hurlbert & Rittenberg 1962). Few attempts have been made to assess the relative importance of glycolysis or the pentose phosphate cycle as alternative routes of glucose utilization (Casselton & Syrett 1962), although some recent estimates have been made. Preferential production of $^{14}CO_2$ from cell suspensions supplied with $(1-^{14}C)$-glucose compared with that from identical suspensions oxidizing $(6-^{14}C)$-glucose may be taken as evidence for the operation of the pentose phosphate pathway. Provision of $NH_4^+$ to nitrogen-starved *Prototheca* led to a decrease in the $C_6:C_1$ ratio from 0·53 to 0·25 (Casselton & Stacey 1969). It was concluded on this basis that NADPH used in the reductive amination of 2-oxoglutarate is provided by way of the pentose phosphate cycle. Evidence for the simultaneous operation of glycolysis and the pentose phosphate cycle has been obtained under both autotrophic and heterotrophic conditions in *Chlorella ellipsoidea* (Kandler *et al.* 1961); measurement of the $C_6:C_1$ ratio indicates that inhibition of the TCA cycle by monofluoroacetate leads to a stimulation of the pentose phosphate cycle (Oaks 1962). Changes in this ratio have also been correlated with changes in specific activities of the pentose phosphate cycle enzymes of *Chlorella* (Devlin & Galloway 1968); this pathway was less active in autotrophically-grown cells than after heterotrophic growth on glucose. Several other investigations have shown changes in glucose-6-phosphate dehydrogenase activities under different growth conditions in *Euglena gracilis* (Huth 1967, Ohmann *et al.* 1969) and in *Chlorella pyrenoidosa* (Huth 1967). During the cell cycle in synchronous cultures of *Chlorella* (Berger 1966), the lowest activity of this enzyme coincides with the middle of the light period, which is also the time at which the respiratory quotient reached a minimum (Reid *et al.* 1963); this finding is consistent with the production of a reductant for biosynthesis by way of the pentose phosphate cycle. This cycle also plays a dominant role in glucose utilization by blue-green algae (Wildon & Rees 1965, Cheung & Gibbs 1966, Pearce & Carr 1969).

The inability of 'acetate flagellates' to utilize glucose arises from the lack of a transport mechanism rather than the absence of key enzymes (Barry 1962, Chapman *et al.* 1965, see also Chapter 19, p. 540). Adaptation of acetate-grown *Polytomella agilis* to propionate or butyrate has been studied by Cantor and

James (1965). Pathways of oxidation of higher fatty acids in these organisms have received little attention. The mechanism of butyryl-CoA formation has been studied in *Polytoma obtusum* (Chapman *et al.* 1965). Oxidation of propionate may be by way of the vitamin $B_{12}$-requiring methyl-malonyl CoA pathway, as in *Ochromonas malhamensis* (Arnstein & White 1962), or by a β-oxidation pathway which involves decarboxylation of malonic semialdehyde (Callely & Lloyd 1964b, Lloyd & Callely 1965).

The TCA and glyoxylate cycles operate in growing cultures of *Prototheca zopfii* when either acetate or propionate are present as sole carbon sources; thus the distribution of radioactivity from $[C^{14}]$-labelled growth substrates prior to the establishment of steady-state conditions indicates that both mechanisms are involved in carbon assimilation (Lloyd & Callely 1965). A similar result was obtained with *Polytomella caeca* growing on propionate (Lloyd *et al.* 1968). The adaptive formation of glyoxylate cycle enzymes necessary for growth on acetate in *Chlorella vulgaris* (Syrett *et al.* 1964) involves the synthesis of new protein; isocitrate lyase can account for 0·6 to 1·2% of the cellular dry weight (John & Syrett 1965). Control of the glyoxylate cycle in *Chlorella* has been further studied (Harrop & Kornberg 1963, 1966, John & Syrett 1968, John *et al.* 1970). Enzymes of the glyoxylate cycle have been demonstrated in extracts of *Euglena* (Reeves *et al.* 1962). The repressive effect of growth with glucose on isocitrate lyase activity is not marked (Heinrich & Cook 1967), but levels of activity of malate synthetase show great variation during the growth of batch cultures (Cook & Heinrich 1968). Isocitrate lyase activity has also been demonstrated in cell-free extracts from acetate-grown *Chlamydomonas dysosmos*, *Chlorogonium elongatum*, *Gonium quadratum* and *Polytomella agilis* (Haigh & Beevers 1964a). Both the key enzymes of the glyoxylate cycle have been detected in *Polytomella caeca* after growth with acetate (Haigh & Beevers 1964b).

An incomplete TCA cycle and a glyoxylate cycle have been demonstrated in *Anabaena variabilis* (Pearce *et al.* 1969) and it is of interest that inclusion of acetate in the growth medium did not significantly alter enzyme activities measured in extracts.

## 3 EFFECTS OF ENVIRONMENTAL CHANGES ON RESPIRATION AND THE RESPIRATORY APPARATUS

### 3.1 *Oxygen tension*

The high affinity of cytochrome $c$ oxidase for $O_2$ (apparent $K_m = 0·02\mu M$, Chance 1965) allows a zero-order reaction for the classical CO-sensitive mitochondrial oxidase down to this low ('critical') $O_2$ tension. However, the CO-insensitive respiration of *Chlorella pyrenoidosa* also proceeds normally at a partial pressure of $O_2$ as low as 5mm mercury (Tang & French 1933), suggesting that terminal oxidation is not flavoprotein mediated. This unidentified oxidase is not inhibited by 1mM KCN, 1mM 8-hydroxyquinoline or 0·1mM thiocyanate and has an apparent $K_m$ for $O_2$ of $6·7\mu M$ (Sargent & Taylor 1972). A report that decreased $O_2$ tension results in decreased respiration in algae of fast-flowing aerated waters (e.g. *Hydrurus* and *Batrachospermum* are more affected than algae more characteristic of standing waters such as *Cladophora*) suggests that the respiratory chains of these algae may have unusual characteristics, especially as the reduction in respiration rate occurs as soon as the $O_2$ content falls below 100% air saturation (Gessner & Pannier 1958).

High partial pressures of $O_2$ can immediately inhibit respiration and growth of *Astasia longa* (Begin-Heick & Blum 1967) and the effect is reversible. Exposure of organisms to a 95:5 ratio of $O_2:CO_2$ increased the ergosterol content of the cells but not the coenzyme $Q_9$ content of the mitochondrial fraction. Rhodoquinone in the mitochondrial fraction is increased, whereas succinate dehydrogenases, linked either to phenazine methosulphate or mammalian cytochrome $c$, and cytochrome $c$ oxidase are all decreased. It was suggested that rhodoquinone plays a part in electron transport of this organism. Similar inhibition of respiration under conditions of high $O_2$ tension is observed in *Euglena gracilis* var. *bacillaris* strain SMLI (streptomycin-bleached), but in this case, phosphate deprivation produces a sensitizing effect on the mitochondrial system. The opposite effect has also been observed, e.g. the respiration of *Pithopora* increased up to 300% saturation then decreased, and unicellular planktonic species *Asterionella formosa* and *Fragilaria crotonensis* show the same phenomenon (Gessner & Pannier 1958). High concentrations of $O_2$ inhibit the respiration, photosynthesis, nitrogen fixation and growth of *Anabaena flos-aquae* (Stewart & Pearson 1970). Thus the level of dissolved oxygen in the environment may affect markedly the physiology and ecological distribution of blue-green algae.

## 3.2 Carbon sources

Normal cells of *Chlorella protothecoides* are bleached when incubated aerobically, either in the light or dark in a glucose-medium without added nitrogen source (Matsuka & Hase 1965). Chloroplast degeneration does not however occur under an atmosphere of $N_2$ in the dark. Inhibitor studies suggested that growth and the ability to carry out oxidative phosphorylation are both necessary before 'glucose-bleaching' occurs in the dark and that an $O_2$ requiring step is necessary for chlorophyll degradation.

The respiratory system of many algae can be modified extensively by changes of energy sources which result in the induced synthesis of the enzymes of new energy-yielding pathways. Thus adaptation of acetate-grown *Polytoma uvella* to oxidation of and growth on butyrate requires a nitrogen source (Cirillo 1956, 1957). Non-proliferating suspensions of *Prototheca zopfii*, grown on acetate, oxidize propionate only after a lag of 20 to 30 min, and the adaptation involves induction of the enzymes for β-oxidation of the new energy source (Lloyd & Venables 1967). These enzymes are partly located in mitochondria, but difficulties of fractionation necessitated a more easily-disrupted organism for further studies. It was possible to demonstrate that adaptation of *Polytomella caeca* to propionate also involves the formation of β-oxidation enzymes. These enzymes are synthesized extra-mitochondrially, and are apparently integrated with the inner mitochondrial membranes of existing mitochondria. A progressive increase in the efficiency of energy conservation as adaptation proceeds was confirmed by the observation that the ability for oxidative phosphorylation coupled to β-oxidation in mitochondria isolated at various stages in the process only reaches a maximum some 18h after exposure to the new energy source. Phosphorylation is then as tightly linked to the oxidation of malonic semialdehyde or

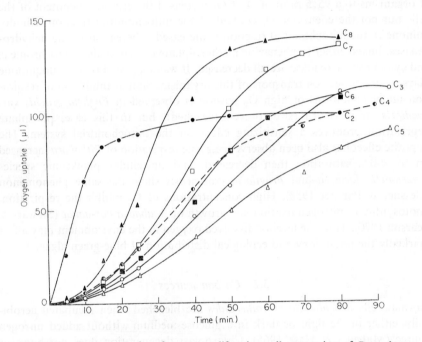

**Fig. 18.1.** Oxygen uptake of a non-proliferating cell suspension of *Prototheca zopfii* (grown with acetate as sole carbon source (Callely & Lloyd 1964a)), incubated at 30°C with straight-chain fatty acids ($C_2$—$C_8$). Fatty acids were present in the Warburg flasks in 20 mM-K phosphate buffer at pH 7·2. Concentrations were as follows: butyrate and valerate 2 mM, acetate and propionate 1·6 mM, hexoate and heptoate 1·5 mM, octoate 1 mM. The flasks contained 700 μmoles KOH in the centre wells and 4·5 mg dry wt. of cells in a final volume of 2·9 ml.
(Data of Lloyd & Venables, unpubl.)

β-hydroxypropionate as is the case in mitochondria from organisms grown with propionate (Lloyd *et al.* 1968). This is shown by the dependence of oxidation rates on the presence of ADP. Acetate-grown *Prototheca zopfii* is particularly versatile in its adaptation to higher fatty acid oxidation (Fig. 18.1); it seems likely that inducible mitochondrial β-oxidations are also involved here.

### 3.3 *Growth with inhibitors*

Modifications of the respiratory components of algal mitochondria result from growth in the presence of inhibitors of energy metabolism. Thus *Euglena gracilis* (bleached strain) grown in the presence of antimycin A produces a novel mitochondrial succinoxidase (Fig. 18.3a) insensitive to antimycin A or 2·7mM KCN. (Sharpless & Butow 1970a). The activity of the induced pathway is AMP stimulated, but is non-phosphorylating and produces malate, fumarate and water (not hydrogen peroxide). Oxidation of NADH through this pathway is coupled to phosphorylation. Changes in activity of the alternate oxidase are reflected in the redox states of components of the normal electron transport chains, particularly those of the *b*-cytochromes, indicating that the by-pass is a functional one. The induction of the new pathway is accompanied by marked morphological changes with mitochondria fusing to form a giant sheath-like structure in the periphery of the cell (Calvayrac & Butow 1971, Calvayrac *et al.* 1971). A similar modification of mitochondrial ultrastructure can be induced by changing the carbon source from L-glutamate + DL-malate to DL-lactate (Calvayrac 1970). Enhanced respiratory activity accompanies derepression of the glyoxylate cycle in the presence of acetate or ethanol, as shown by the formation of an active malate synthetase (Heinrich & Cook 1967). Sharpless and Butow (1970a,b) suggest that the alternate succinoxidase may play a physiological role in the control of the oxidative steps of the glyoxylate cycle. Key enzymes of this cycle are associated with the peroxisomes of *Euglena* (Graves *et al.* 1972, Brody & White 1972).

Antibiotics, such as chloramphenicol, specifically inhibit mitochondrial protein synthesis, and when included in the growth medium decrease growth and respiration in *Polytomella caeca* (Lloyd *et al.* 1970). No morphological changes in the mitochondria are detectable under the electron microscope, but succinoxidase, NADH oxidase and cytochrome oxidase activities are all decreased. No major cytochrome dislocations are detectable and the lesion responsible for decreased cellular respiration occurs at the level of the flavoprotein dehydrogenases.

### 3.4 *pH and temperature*

The effect of pH on cellular respiration of algae is primarily on the availability of external substrates. Thus oxidation of organic acids by acetate flagellates proceeds most quickly at an acid pH, as the organisms are more permeable to the undissociated acids than to the charged anions (Hutner & Provasoli 1951) (see Chapter 19, p. 532). Respiration rates are more temperature-dependent than are rates of photosynthesis (Tang & French 1933).

### 3.5 *Light*

Interactions between photosynthesis and respiration were discussed by Gibbs (1962). A hypothesis for the interaction of chloroplasts and mitochondria was developed to explain observations on the pale green mutant of *Chlamydomonas reinhardtii* (Hiyama *et al.* 1969). Thus the light-induced oxidation of cyto-chrome $b_{563}$ is eliminated by its $O_2$-induced oxidation, whilst $O_2$-induced oxidation is greatly diminished under illumination. Antimycin A diminished the $O_2$ induced oxidation of cytochrome $b_{563}$ but did not affect the light-induced oxidation. Thus it was proposed that mitochondrial reducing equivalents either pass to $O_2$ down the electron transport chain, or proceed by way of cytochrome $b_{563}$ to $O_2$ or to cytochrome *f*. The possibility that the apparent derepression of respiratory $O_2$ uptake on activation of photosystem I in *Chlamydomonas reinhardtii* (Kok effect) also involves diversion of a respiratory reductant has been discussed by Healey and Myers (1971). The processes of photosynthesis and respiration may also share some components in blue-green algae (Bisalputra *et al.* 1969), and this may explain the dependence of respiration rates on the presence or absence of light (Jones & Myers 1963, Padan *et al.* 1971, Lex *et al.* 1972). The latter workers have studied photorespiration by blue-green algae in detail (see also Chapter 17, p. 474).

Blue light enhances the endogenous respiration of a starved chlorophyll-free mutant of *Chlorella vulgaris* with saturation, which was achieved at low intensi-ties (500 ergs $cm^{-2}$ $sec^{-1}$), giving a three-fold stimulation of oxygen uptake (Kowallik 1967). The action spectra reveal two maxima (at 450nm and 375nm) which implicate either a flavin or a carotenoid (Kowallik 1968), but further experiments to determine which were not conclusive. Pickett (1966) has sug-gested that a similar effect on *Chlorella pyrenoidosa* may involve the cyanide-sensitive portion of dark respiration.

The inhibitory effects of high light intensities ($2 \cdot 9 \times 10^4$ ergs $cm^{-2}$ $sec^{-1}$) on the growth of *Prototheca zopfii* (Epel & Krauss 1966) have been traced to the photodestruction of cytochrome $c_{551}$, $b_{559}$ and $a_3$ (Epel & Butler 1969, 1970). Oxygen is required during irradiation for cytochrome loss and cyanide protects cytochrome $a_3$ from damage.

### 3.6  *Changes in respiration during the cell-cycle*

Studies with synchronized *Chlorella* cultures indicate that rates of endogenous respiration and of $O_2$ uptake with added respiratory substrates vary through the cell-cycle. The literature is difficult to summarize because the extent of changes of the respiratory quotient observed depend on the experimental conditions, and many parameters must be considered (Talbert & Sorokin 1971). Experiments with *Astasia longa* (Wilson & James 1966) indicate that the respiration rate is intimately geared to growth and cell division, although the metabolic bases for cyclic increases and decreases remain to be investigated. A sharp doubling of respiratory quotient in synchronized cultures of *Polytomella agilis* (Cantor & Klotz 1971), may reflect a doubling of effective mitochondrial mass per cell or simply indicate increased respiratory activity associated with increased bio-synthetic energy requirements. Marked changes in mitochondrial morphology occur during the cell cycle in synchronized cultures of *Euglena gracilis* (Calvayrac *et al.* 1972). The reversible formation of giant mitochondria by the fusion of many smaller mitochondria has been reported to occur at defined stages in the

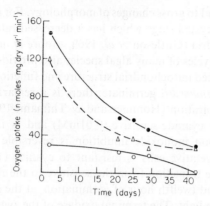

**Fig. 18.2.** The decline of respiration rates of *Prototheca zopfii* during starvation. Organisms were grown on a minimal medium containing acetate as sole carbon source (Callely & Lloyd 1964a). After harvesting and washing under aseptic conditions, organisms were resuspended at a cell density of $10^6$ organisms ml$^{-1}$ in 20 mM K-phosphate buffer (pH 7·2) and force aerated at 30°C for 40 days. Respiration rates were measured polarographically.

    (●) Respiration rates in the presence of 10 mM Na acetate
    (○) Endogenous respiration rates
    (△) Respiration rates with acetate (endogenous substracted)

      (Data of Lloyd & Venables, unpubl.)

cell cycle of *Chlamydomonas reinhardtii*, and this coincides with a decreased $O_2$ uptake of cell suspensions (Osafune *et al.* 1972). However, changes in the highly branched mitochondria of this organism cannot be easily deduced without serial sectioning (Arnold *et al.* 1972). Cell selection techniques, rather than induction synchrony, will often be the methods of choice for further work, as these are less likely to cause major metabolic disturbances as artefacts of experimental procedure.

### 3.7 Starvation and differentiation

Withdrawal of carbon sources leads to a succession of ultrastructural changes and ultimately to increased autophagy of *Euglena* mitochondria (Brandes *et al.* 1964, Leedale & Buetow 1970). Respiratory activity of organisms declines during starvation. For example in *Prototheca zopfii* both the ability to respire added acetate and the endogenous respiration decrease, but with little loss of cell viability, over a period of forty days (Fig. 18.2).

Replacement of silicon in a medium deficient in this element, which is essential for the growth of the diatom *Navicula pelliculosa*, gave a transient over-shoot in the rate of $O_2$ uptake (Coombs *et al.* 1967). Studies on effects of nutrient-limitation offer many exciting possibilities for future experimentation on the limits of structure and function of the respiratory apparatus under conditions of environmental stress.

Starvation may lead to gross changes of morphology. For instance *Polytomella agilis* produces an encysted stage which has a decreased rate of respiration and degenerate mitochondria (Gittleson *et al.* 1969, Moore *et al.* 1970).

The complex life-cycles of many algal species also provide interesting systems for the study of modified mitochondrial structure and function. When zygospores of *Chlamydomonas reinhardtii* germinate, there is a remarkable change in the characteristics of respiration (Hommersand & Thimann 1965). The $O_2$ uptake of the zygospores is cyanide-sensitive (0·01mM) and is inhibited by carbon monoxide. This carbon monoxide inhibition is reversible by blue light. The respiration of the vegetative cells is resistant to cyanide (1·0mM) and carbon monoxide (99%). The transition from the sensitive to the resistant state occurs between the eighth and twelfth hour of germination, at the same time as photosynthesis begins in the light. The terminal oxidase of the zygospores has a higher affinity for $O_2$ than that present in vegetative cells, and these observations suggest that a classical inhibitor-sensitive electron transport chain present in the zygospores is not the major route of electron transport during the greatly increased respiratory activity of the vegetative organisms.

## 4  FRACTIONATION OF EXTRACTS

### 4.1  *Methodology*

A fraction highly enriched in a particular subcellular component should display the physiological functions assumed to be associated with that organelle *in vivo*, and has to obey several minimum criteria which, in the case of mito-chondria are: (a) a high ratio of pyridine nucleotide to cytochrome $c$; (b) respiratory control, i.e. a low respiration rate in the absence of added ADP (state 4, Chance & Williams 1956) and a high respiration rate (state 3), produced immediately on adding ADP; (c) low endogenous respiration in the absence of added substrates; (d) integrity of mitochondrial membranes and a low degree of contamination by other organelles as revealed by examination in the electron microscope. This necessitates a sacrifice of yields for high 'purity'.

To characterize different components of the extract by their differing distributions (for instance through a density gradient) using 'marker' enzymes unique to a particular organelle the following parameters can be investigated: size distribution (determined from sedimentation coefficients), density, and proportion of the total enzyme complement of the homogenate present in any particular organelle. This approach proves invaluable for study of enzymes located in organelles which represent a small proportion of the total extract protein as in cases where preparative techniques involve manipulation of small amounts of materials. It is also essential to calculate recoveries of enzymes (based on the number of enzyme units detected in the whole homogenate), i.e. to construct a balance sheet (de Duve 1971).

Both approaches assume that a high proportion of the organelles and other membranous structures will be preserved without loss of enzymes or physiological function during the preparation of the original homogenate. Thus perhaps the most crucial step is the process of cell disruption, the aim being to effect efficient breakage with little loss of information content (i.e. subcellular organization). This is particularly difficult in the cases of algae with walls of high tensile strength; hence the emphasis in most fractionation studies on the more fragile phyto-flagellates. Some methods developed primarily for breakage of bacteria (Hughes et al. 1971) are also effective when applied to algae. Where it is necessary to minimize damage to intracellular structures enzymic methods are often best. Trypsin has been used to digest the pellicle of Euglena gracilis (Price & Bourke 1966). In our experience the walls of various species of Chlorella and also Prototheca zopfii are highly resistant to attack, for instance by snail digestive enzymes, but a method of overcoming these difficulties in yeasts may also be applicable to algae (Poole & Lloyd 1973). Mutants of Chlamydomonas rein-hardtii which possess defective cell walls have recently been isolated by Hyams and Davies (1972). Certain blue-green algae can be lysed by lysozyme (Crespi et al. 1962). Another approach is to isolate cell-walls, then use enrichment cultures containing these as sole carbon source to isolate organisms which produce suitable wall-degrading enzymes.

The crude homogenate thus obtained may then be fractionated by differential centrifugation, density gradient centrifugation or by electrophoresis. In a study of enzyme distributions in Ochromonas malhamensis, mitochondria, peroxi-somes and acid hydrolase-containing organelles were characterized on the basis of size and density (Lui et al. 1968) and successful fractionation of homogenates has been achieved with Polytomella caeca (Cooper et al. 1969, Cartledge et al. 1971, Cooper & Lloyd 1972) and Euglena gracilis (Graves et al. 1972).

### 4.2 Properties of isolated 'particulate' and mitochondrial fractions; electron transport and oxidative phosphorylation

The process of terminal oxidation in algae is evidently quite different from that in mammalian cells. Emerson (1927) observed that in Chlorella respiration is resistant to CO and stimulated by cyanide; as much as 60% of the $O_2$ uptake of this organism proceeds uninhibited at high concentrations of cyanide (Syrett 1951). Addition of cupric ions followed by fluoride inhibits $O_2$ consumption (Hassal 1967, 1969). An early report that a pale green mutant of Chlamydo-monas has no detectable cytochrome a (Chance & Sager 1957) has been checked, and it is evident that the very low content of this cytochrome (detected as the pyridine haemochromogen) suggests the possibility of a novel pathway of mitochondrial electron transport (Hiyama *et al.* 1969). Methods of preparation of mitochondria from protozoa and algae have been reviewed by Buetow (1970).

## (a) *Prototheca zopfii*

Hand-grinding of *Prototheca zopfii* with glass beads gives extracts containing active mitochondria. The mitochondrial fraction has a low endogenous respiration and the rates of substrate oxidations are ADP-dependent, i.e. they carry out 'tightly coupled' oxidative phosphorylation (Lloyd 1965). Other methods of cell disruption, including shaking with glass beads or breakage in a Hughes Press produce preparations of uncoupled mitochondria (Lloyd 1966a). Cytochromes $b$, $c$, $a$ and $a_3$ and coenzyme $Q_7$ are present in both types of mitochondria and mammalian cytochrome $c$ can serve as an electron acceptor and donor. Respiratory control ratios as high as 5·5 were found for 2-oxoglutarate, and $P/O$ ($P/e_2$) ratios of 1·6 for succinate, 2·0 for malate, between 2 and 3 for pyruvate, and about 3 for 2-oxoglutarate suggest that energy conservation in some algae is as efficient as in mammalian mitochondria.

Difference spectra of whole cells and particulate fractions (sedimenting at $2 \times 10^5 g$ min$^{-1}$) at 77°K confirm the presence of cytochromes $b$, $c$, $a$ and $a_3$, and notable features of the spectra include the presence of excess cytochrome $b$ and a split Soret band at 440nm and 446nm ($a + a_3$) (Webster & Hackett 1965). Ratios of cytochromes $a + a_3$:$b$:$c$ reducible in these particles by NADH were 1:0·5:0·8, and by dithionite 1:1·9:1·5. Further resolution of the respiratory pigments of *Prototheca zopfii* has recently been reported by Epel and Butler (1970) using whole cells and particulate preparations. Seven different cytochromes were identified, $c_{549}$ (soluble), $c_{551}$ (membrane bound), $b_{555}$, $b_{559}$, $b_{564}$ and cytochromes $a$ and $a_3$. The endogenous respiration of *Prototheca zopfii* is insensitive to high ratios of $CO:O_2$ (19:1), and is stimulated 48% by cyanide (10μM), whereas respiration with glucose is sensitive to these inhibitors (Webster & Hackett 1965). Thus KCN (0·1mM) gave 45% inhibition, and a 19:1 ratio of $CO:O_2$ gave 33% inhibition of glucose supported respiration. Antimycin A (5·5μM) gave 95% inhibition of $O_2$ uptake by whole cells in the presence of glucose. Acetate oxidation by intact cells was 90% inhibited by KCN (0·1mM) and, assuming no effect on the endogenous rate, 75% inhibited by a 19:1 ratio of $CO:O_2$ (Lloyd 1966b). The succinoxidase activity of the isolated mitochondria was sensitive to inhibition by azide, cyanide, antimycin A, amytal and rotenone, and a number of compounds related to rotenone inhibited NADH oxidase in disrupted mitochondria. The concentrations of inhibitors reducing activities to 50% were similar to those required with mammalian mitochondria. A residual activity insensitive to antimycin A, which acted at a site between cytochromes $b$ and $c$, was noted.

Cyanide reacted with cytochrome $a_3$, and azide reacted primarily either with the oxidized form of cytochrome $a_3$ or the reduced form of cytochrome $a$ (Epel & Butler 1970). Cytochrome oxidase inhibitors were completely effective against the respiration of glycerol-grown cultures and cyanide and azide led to a two- and five-fold stimulation of the endogenous respiration of starved cells. Azide also stimulated the respiration of cyanide-inhibited cells. An alternative pathway of electron transport to $O_2$ via the $b$-type cytochromes was proposed in which the rate-limiting step in this bypass was the reaction of cytochrome $b$ with $O_2$. Azide also reacts with cytochrome $b_{564}$ to give a complex which was distinguished in difference spectra by its non-reducibility by dithionite. It was proposed that cyanide or azide may react with cytochrome $b_{564}$ to give complexes which react with $O_2$ at an increased rate, hence the stimulation by these compounds of the endogenous respiration.

### (b) Euglena gracilis

Oxidative phosphorylation was first demonstrated in algal mitochondria by Buetow and Buchanan (1964, 1965) who showed that after disruption of streptomycin-bleached *Euglena gracilis* by grinding with beads, it was possible to isolate mitochondria which would oxidize NADH, succinate, L-malate, L-glutamate, 2-oxoglutarate, lactate, malate, and malate + pyruvate and catalyse loosely-coupled oxidative phosphorylation. Respiratory control was not observed by these preparations. $P/O$ ratios for NADH were about 0·4, for lactate and succinate 0·9 to 1·0, and between 1·6 and 2·0 for NAD-linked substrates. β-hydroxybutryrate was not oxidized, and isocitrate and citrate were oxidized very slowly (0·1 to 0·2 μatoms $O_2$ (mg mitochondrial protein)$^{-1}$ 30min$^{-1}$). The results suggested that these mitochondria possess one phosphorylation site less than mammalian mitochondria but more recently it has been shown that this is not the case. Cyanide, amytal, rotenone and antimycin A were all inhibitory; 2, 4-dinitrophenol inhibited both oxidation and coupled phosphorylation with malate as substrate.

Difference spectra of whole cells and of particulate preparations rich in mitochondria indicate the presence of cytochromes $b_{561}$ and $a_{609}$, reducible by succinate, oxidized by $O_2$, and reacting with a soluble cytochrome $c_{556}$ (Raison & Smillie 1969). Measurement of steady state reduction levels suggested that these three pigments were part of an electron transport chain with a sequence $b \rightarrow c \rightarrow a$. Although respiration was only 60% inhibited by cyanide (1mM), oxidation of the $c$- and $a$-type cytochromes was completely abolished at this inhibitor concentration. As the alternative electron transport pathway was also insensitive to antimycin A inhibition, it was suggested that cyanide-insensitive respiration is mediated by an autooxidizable cytochrome $b$.

Mitochondria isolated from a permanently bleached *Euglena gracilis* strain Z, which had been grown on glutamate + malate as carbon source, were shown by specific assays to possess all three phosphorylation sites characteristic of mammalian mitochondria (Sharpless & Butow 1970a). An interesting observation was that both NADH and D- and L-lactate were oxidized by an antimycin A-insensitive route involving one less than the normal number of coupling sites. $P/O$ ratios for NADH oxidase fell from over 2 to just below 2 when antimycin A was added; corresponding decreases from 2·1 to 1·1 and 2·0 to 1·3 were observed for D- and L-lactate. The antimycin A insensitive respiration was inhibited by cyanide (2mM). Cytochromes detected were $a_{607}$, with a $\gamma$-band at 453nm, a second cytochrome $a$ with an ill-defined absorption at 593nm and a $\gamma$-band at 444nm, $b_{568}$, $b_{561}$, $b_{558}$, $c_{555}$ which was soluble, and $c_{551}$. Carbon monoxide reacted both with cytochrome $a_{593}$ and more slowly with a $b$-type cytochrome which also reacted with cyanide.

Of the four cytochromes detected in light-grown wild type *Euglena gracilis* strain *bacillaris* (cytochromes $c_{552}$, $b_{561}$, $a_{605}$ and $c_{556}$), only cytochrome $a_{605}$ and cytochrome $c_{556}$ were detected in dark-grown wild type, in the Albino ($W_3$ and $W_8$) mutants which lack chloroplasts completely, or in the Yellow ($Y_3$) mutant which has only a rudimentary plastid (Perini *et al.* 1964). Cytochrome $c_{552}$ was enriched in a plastid fraction from normal green cells whereas a six-fold enrichment of cytochrome $c_{556}$ was obtained in a 'small-particle' fraction from the Albino mutant together with cytochrome $a_{605}$, diaphorase and succinate dehydrogenase. The $a$-type cytochrome content of this fraction was 0·55μmol (g protein)$^{-1}$. Purified reduced cytochrome $c_{556}$ could be reoxidized by $O_2$ in the presence of yeast mitochondria at one tenth of the rate for reduced mammalian cytochrome $c$, but three times more rapidly than reduced cytochrome $c_{552}$. The respiration of sonicated cell suspensions of normal cells of *Euglena* $W_3$ with succinate was 95% inhibited by cyanide (1·0mM) but only 15% inhibited by a 19:1 ratio of $CO:O_2$. Cytochrome $a_{605}$ and cytochrome $c_{556}$ were reducible in whole cells by succinate or acetate, and the $\alpha$-band of cytochrome $a_{605}$ disappeared more rapidly on aeration than that of cytochrome $c_{556}$. The addition of cyanide (0·5mM) led to the disappearance of the 605nm band but did not affect the 556nm band. Perini *et al.* (1964) suggested on the basis of these results that although cytochrome $a_{605}$ and cytochrome $c_{556}$ are mitochondrially located, the function of the $a$-type cytochrome is not clear in view of the CO-insensitive respiration. The autooxidizable cytochrome $c_{556}$ may provide an alternative terminal oxidation pathway.

### (c) *Astasia longa*

$P/O$ ratios of between 0·43 and 0·67 were measured for succinoxidase of mitochondria isolated from *Astasia longa* (Buetow & Buchanan 1969). Particulate preparations also oxidized NADH, L-malate and 2-oxoglutarate (Kahn & Blum 1967). Mitochondria isolated from *Astasia longa* did not reoxidize reduced mammalian cytochrome $c$ and showed a complex spectrum (when reduced with dithionite) in the $\alpha$-regions of cytochrome $b$ and $c$, with a predominant absorption band at 554nm (Webster & Hackett 1965) and shoulders at 563nm, 557nm,

552nm and 547nm. The Soret bands of $a$-type cytochromes were at 448nm and 440nm, and CO-reacting haemoproteins other than cytochrome oxidase were revealed in difference spectra but these have not been further investigated. Further studies on the respiratory change of this organism would be worthwhile.

## (d) *Polytoma* and *Polytomella*

The colourless flagellates *Polytoma uvella* and *Polytomella agilis* both have cytochromes $b$, $c$ and $a + a_3$ (Webster & Hackett 1965). The NADH-oxidase of *Polytoma uvella* mitochondria is sensitive to KCN, antimycin A, rotenone, diphenylamine and amytal. Difference spectra of acetone-extracted whole cells show the presence of a dithionite-reducible CO-reacting haemoprotein with absorption maxima at 570nm, 540nm and 414nm in addition to cytochrome oxidase (trough at 445nm, shoulder at 430nm); the 414nm Soret band was also found in CO-spectra of isolated mitochondria. This CO-binding pigment was not a peroxidase, but may have been a denatured cytochrome $b$.

Mitochondria isolated from *Polytomella caeca* (Lloyd & Chance 1968) contain cytochromes $b$, $c_1$, $c$, $a$ and $a_3$. Electron transport is sensitive to the classical inhibitors; succinoxidase is inhibited by 94% in the presence of KCN (10mM) and to a similar extent by 4 nmoles of antimycin A per mg protein. In intact mitochondria the oxidation of endogenously generated NADH with citrate as substrate was more sensitive to inhibition by rotenone or Piericidin A than that of exogenous NADH. This difference was abolished in disrupted mitochondrial preparations and the site of antimycin A inhibition was located between cytochromes $b$ and $c$. Several different flavoproteins were distinguished spectrophotometrically and fluorimetrically by their characteristic fluorescent yields, and by differing kinetic behaviour, and the distinct species involved in the oxidation of endogenous NADH, endogenously generated NADH and succinate have been characterized. A large amount of enzymically non-reducible 'flavoprotein' with a low fluorescent yield detected on dithionite addition may be contributed to by non-haem iron which also has an absorption at 475 to 410 nm. A 'g = 1·94' signal in electron paramagnetic resonance spectra was produced on reduction of mitochondrial components with succinate, ethanol, or in damaged mitochondria by NADH. When anaerobic mitochondrial suspensions with succinate as substrate were rapidly mixed with air-saturated buffer in a regenerative flow apparatus, cytochromes $a_3$, $a$, $b$ and flavoprotein were reoxidized with half-times of 35, 40, 120 and 200ms respectively while the reoxidation of cytochrome $c$ was anomalously slow (600ms). This observation warrants further study, although there is no doubt that cytochrome $c$ is a functional intermediate in the respiratory chain of the flagellate. The steady state levels of reduction of components in the absence of added ADP (state 4, Chance & Williams 1956) were as follows $a_3$ (+a) 0%, $a$ (+$a_3$) 5%, $c$ 21%, and $b$ 48%, suggesting that this series represents a functional sequence (Fig. 18.3b) (Lloyd *et al.* 1970). Flavoprotein reduction is the rate limiting step in electron transport

from succinate to $O_2$ as shown by the kinetics of reduction of the respiratory
chain components during the transition to anaerobiosis (Lloyd *et al.* 1970).

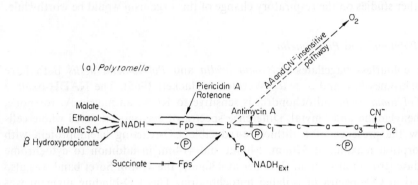

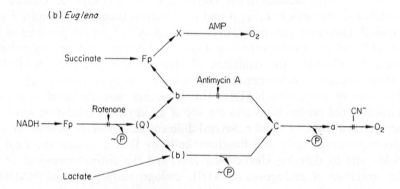

**Fig. 18.3.** Pathways of electron transport in algae.

(a) Based on results with mitochondria from *Polytomella caeca* (Lloyd & Chance
1968, Lloyd, Evans & Venables 1968). $F_pD$ and $F_{ps}$ are two NADH dehydrogenases, and succinate dehydrogenase flavoproteins respectively. NADH Ext represents NADH added to mitochondria. Phosphorylation sites are indicated by $\sim ℗$.

(b) Scheme of electron transport proposed for *Euglena* mitochondria (Sharpless &
Butow 1970b). The alternative pathway stimulated by AMP is inducible. (For a
schematic diagram of the mammalian respiratory chain see Chance *et al.* 1969.)

Mitochondria from *Polytomella caeca* carry out oxidative phosphorylation
and exhibit respiratory control (Lloyd *et al.* 1968), 2-oxoglutarate oxidation
gives *P/O* ratios in the range 3·1 to 3·4, while succinoxidase usually shows ratios
between 1·6 and 1·8, and NADH-linked substrates give 2 to 2·9. The mitochondria appear intact in the electron microscope with clearly defined outer and
inner membranes. The integrity of the NADH-impermeable barrier (the inner

membrane) is indicated by latency of tricarboxylic acid cycle enzymes of the matrix (D. A. Evans & D. Lloyd unpubl.), and by the clear differences between responses to externally added and endogenously generated NADH (Lloyd & Chance 1968).

### (e) *Blue-green algae*

In blue-green algae, as in other prokaryotes, there are no discrete respiratory organelles; possible reduction sites of tetra-nitro blue tetrazolium (or tellurite) linked to succinate oxidation in the dark were demonstrated only in the photosynthetic lamellae of vegetative cells of *Nostoc sphaericum* (Bisalputra *et al.* 1969). Whole cells of the colourless filamentous prokaryotes *Saprospira grandis*, *Leucothrix mucor* and *Vitreoscilla* sp. were examined at 77°K spectrophotometrically and showed four peaks of $b$ and $c$-type cytochromes at around 560nm. No $\alpha$-band of cytochrome $a$ was detected (Webster & Hackett 1966). The NADH oxidase activity of extracts was particulate and resembled that of bacteria in its response to inhibitors; particles from *Saprospira grandis* were insensitive to inhibition by antimycin A or rotenone but were 56% inhibited by HOQNO (7·7µM) or KCN (1·0mM). Different results were obtained with the particulate NADH oxidase from *Vitreoscilla* which showed 10% inhibition by antimycin A (2·2µM) and 68% inhibition by rotenone (60µM). Although this preparation was 48% inhibited by KCN (1mM) it was completely insensitive to azide (1mM). Diphenylamine (0·33mM) and amytal (1mM) gave inhibitions of 17% and 53% respectively. Although cytochrome $c$ oxidase activity, using reduced mammalian cytochrome $c$ as substrate, was not detectable in homogenates of any of the three organisms, a light-reversible inhibition of respiration by CO was demonstrated and CO-difference spectra suggested that this terminal oxidase was cytochrome $o$ (absorption maxima at 415 to 416, 535 to 536, and 570nm in dithionite + CO *minus* dithionite difference spectra). The photochemical action spectrum for the light-reversal of CO inhibition confirmed this suggestion. Stimulation of the endogenous respiration of *Leucothrix* and *Vitreoscilla* by cyanide was noted.

Webster and Frenkel (1952) reported that the endogenous respiration of *Anabaena* is inhibited by cyanide, azide and CO. Horton (1968) reported that particles isolated from *Anacystis nidulans* and *Anabaena variabilis* possess a cyanide-sensitive NADH oxidoreductase but no cytochrome $c$ oxidoreductase, succinate oxidoreductase or NADH-cytochrome $c$ oxidoreductase. Absence of NADH oxidoreductase activity was reported in extracts of *Anacystis nidulans*, *Coccochloris peniocystis*, and *Gloeocapsa alpicola* (Smith *et al.* 1967). The slow rate of endogenous respiration of *Anacystis nidulans* and *Phormidium luridum* was not increased by the addition of respiratory substrates (Biggins 1969), but was inhibited by low concentrations of cyanide or high partial pressures of CO, and was stimulated by the uncouplers DNP and FCCP. Measurements of endogenous NADPH/NADP$^+$ and adenine nucleotide pools indicated that oxidative phosphorylation was coupled to oxidation of NADPH. Membrane fragments oxidized succinate, malate, ferrocytochrome $c$ (mammalian), ascorbate-TMPD and reduced DCPIP, but did not oxidize NADH or NADPH in a cyanide-sensitive system. A dark ATP-$P_i$ exchange was inhibited by phlorizin, atebrin, and uncouplers, but only at very high concentrations. These results (Biggins 1969), and those of Leach and Carr (1969, 1970) invalidate the hypothesis of Smith *et al.* (1967) that the basis for obligate autotrophy is the inability of such organisms to carry out oxidative phosphorylation. The development of a proton gradient during oxygenation of a suspension of intact cells of *Anabaena variabilis* in the dark, may also be evidence for the process of oxidative phosphorylation (Scholes *et al.* 1969).

The first demonstration of oxidative phosphorylation in a cell-free system from a blue-green alga was that by Leach and Carr (1969, 1970) with extracts of *Anabaena variabilis* which couple oxidation of NADPH in the dark to ATP synthesis. NADPH oxidation (2·0nmol (mg protein)$^{-1}$ min$^{-1}$) was 50% inhibited by cyanide (20mM) or rotenone (0·1mM), while antimycin A, amytal, and azide were less inhibitory. Oxidation of NADPH linked to menadione or mammalian cytochrome $c$ reduction was detected, as was succinate dehydrogenase and cytochrome $c$ oxidase. The rate of phosphorylation coupled to NADPH oxidation was 0·4nmol (mg protein)$^{-1}$ min$^{-1}$. This was strictly dependent on $O_2$ and was inhibited 25% by FCCP (5$\mu$M). NADPH could be replaced by NADP + isocitrate. NADH oxidase was also present (1·0nmol (mg protein)$^{-1}$ min$^{-1}$) and was by way of a separate pathway to the level of cytochrome $c$, but the non-additive activities of the two pyridine nucleotide oxidases suggest that at least parts of the respiratory chains involved are common to both. Oxidative phosphorylation was also detected with NADH as electron donor and succinate gave very low rates of phosphorylation. A slow azide and copper-chelator-insensitive, $O_2$-dependent oxidation of reduced mammalian cytochrome $c$ has also been found in photosynthetic lamellae isolated from *Anabaena variabilis* (Krogman & Tang 1971).

### 4.3   *Organelles other than mitochondria*

Extra-mitochondrial oxidation of NADH and NADPH can occur by way of the microsomal electron transport chain. Membrane fragments from *Polytomella caeca* which have cytochromes $b$, *P*-450 and *P*-416 and which show cytochrome $c$ reductase activities linked to pyridine nucleotide oxidation have been characterized (Cooper & Lloyd 1972). The presence of peroxisomes (microbodies) has recently been reported in *Chlorella* (Gergis 1971). The first reports of isolation of algal peroxisomes were those of *Polytomella caeca* (Cooper *et al.* 1969, Gerhardt & Berger 1971, Cooper & Lloyd 1972). The only enzyme definitely located in peroxisomes was catalase; none of the flavoprotein oxidases or glyoxylate cycle enzymes associated with several types of protozoal peroxisomes (Muller 1969) was detected in organelles from *Polytomella*. It remains to be seen whether the glyoxylate cycle of other algae or the enzymes of glycolate oxidation (Lord & Merrett 1971, Kowallik & Schmid 1971) are peroxisomal (Graves *et al.* 1971) (see also Chapter 17, p. 484), as has been recently demonstrated to be the case in *Euglena gracilis* (Graves *et al.* 1972). The bioluminescent organelles (scintillons) of the dinoflagellate, *Gonyaulax polyedra* have been separated from homogenates, and are thought to represent evolutionary relics of an ancestral $O_2$-scavenging system with a very high affinity for $O_2$ (DeSa & Hastings 1968).

## 5 CONCLUSION

It is evident from our understanding of the biochemistry of algae that many novel respiratory mechanisms may await discovery. The quantitative importance in algae of pathways of electron transport alternative to the main phosphorylating respiratory chain is not yet clear, neither is the extent to which modification of these pathways can be achieved in various environmental situations. It is not yet established that these ancillary redox components are mitochondrial but the increasing efficiency of fractionational methods should enable precise subcellular distributions to be determined.

Recent advances in studies of higher plant respiration also suggest that intensive investigations of certain aspects of algal respiration would be rewarding. For instance one of the CO-reacting haemoproteins formerly attributed to plant mitochondria has been shown on careful fractionation to be a peroxidase bound to non-mitochondrial membranes which still contaminate fractions enriched in mitochondria (Plesnicar *et al.* 1967). Cytochrome oxidase is the only CO-binding pigment of higher plant mitochondria (Chance *et al.* 1968). Recent work has also enabled a critical reassessment of hypothetical mechanisms of cyanide-insensitive respiration in higher-plant mitochondria (Bendall & Bonner 1971) and suggests that autooxidizable flavoproteins or cytochrome $b$, do not play an essential role, as was formerly believed. Inhibition of the alternative terminal oxidase by thiocyanate, $\alpha,\alpha'$-dipyridyl and 8-hydroxyquinoline suggest the active participation of a non-haem iron protein with a high affinity for $O_2$. Further support for this suggestion is provided by the demonstration that specific inhibition of the cyanide-resistant respiration by hydroxamic acids (e.g. 30 mM *m*-chlorobenzhydroxamic acid gives 50% inhibition in mung bean mitochondria without discernible effect on the cytochrome oxidase pathway or energy coupling) also enhances a 'g = 2' signal in electron paramagnetic spectra (Schonbaum *et al.* 1971).

# 6 REFERENCES

ARNOLD C.G., SCHIMMER O., SCHÖTZ F. & BATHELT H. (1972) Die mitochondrien von *Chlamydomonas reinhardii*. *Arch. Mikrobiol.* **81**, 50–67.

ARNSTEIN H.R.V. & WHITE A.M. (1962) The function of vitamin $B_{12}$ in the metabolism of propionate by the protozoan *Ochromonas malhamenis*. *Biochem. J.* **83**, 264–70.

BARRY S.C. (1962) Utilization of glucose by *Astasia longa*. *J. Protozool.* **9**, 395–400.

BEGIN-HEICK N. & BLUM J.J. (1967) Oxygen toxicity in *Astasia*. *Biochem. J.* **105**, 813–19.

BENDALL D.S. & BONNER JR. W.D. (1971) Cyanide-insensitive respiration in plant mitochondria. *Pl. Physiol., Lancaster*, **47**, 236–65.

BERGER C. (1966) Aktivitätsänderungen einiger enzyme synchronisierter *Chlorella*-zellen im licht-dunkel-wechsel. I Veränderungen während des entwicklungszyklus. *Flora Jena, Abt. A.*, **157**, 211–32.

BIGGINS J. (1969) Respiration in blue-green algae. *J. Bact.* **99**, 570–5.

BISALPUTRA T., BROWN D.L. & WEIER T.E. (1969) Possible respiratory sites in a blue-green alga *Nostoc sphaericum* as demonstrated by potassium tellurite and tetranitro-blue tetrazolium reduction. *J. Ultrastruct. Res.* **27**, 182–97.

BRANDES D., BUETOW D.E., BERTINI I.F. & MALKOFF D.B. (1964) Role of lysosomes in cellular lytic processes. I. Effect of carbon starvation in *Euglena gracilis*. *Exp. Mol. Path.* **3**, 583–609.

BRODY M. & WHITE J.E. (1972) Environmental factors controlling enzymatic activity in microbodies and mitochondria of *Euglena gracilis*. *FEBS Letters*, **23**, 149–52.

BUETOW D.E. (1970) Preparation of mitochondria from protozoa and algae. In *Methods in Cell Physiology*, IV, ed. Prescott D.M. pp. 83–115. Academic Press, New York & London.

BUETOW D.E. & BUCHANAN P.J. (1964) Isolation of mitochondria from *Euglena gracilis*. *Expl. Cell Res.* **36**, 204–7.

BUETOW D.E. & BUCHANAN P.J. (1965) Oxidative phosphorylation in mitochondria isolated from *Euglena gracilis*. *Biochim. biophys. Acta*, **96**, 9–17.

BUETOW D.E. & BUCHANAN P.J. (1969) Isolation of phosphorylating mitochondria from *Astasia longa*. *Life Sci.* **8**, 1099–102.

CALLELY A.G. & LLOYD D. (1964a) The metabolism of acetate in the colourless alga, *Prototheca zopfii*. *Biochem. J.* **90**, 483–9.

CALLELY A.G. & LLOYD D. (1964b) Studies on the metabolism of propionate in the colourless alga *Prototheca zopfii*. *Biochem. J.* **92**, 338–45.

CALVAYRAC R. (1970) Relation entre les substrates, la respiration et la structure mitochondriale chez *Euglena gracilis* (Z). *Arch. Mikrobiol.* **73**, 308–14.

CALVAYRAC R. & BUTOW R.A. (1971) Action de l'antimycine A sur la respiration et la structure des mitochondries d'*Euglena gracilis* Z. *Arch. Mikrobiol.* **80**, 62–9.

CALVAYRAC R., BUTOW R.A. & LEFORTTRAN M. (1972) Cyclic replication of DNA and changes in mitochondrial morphology during the cell cycle of *Euglena gracilis*. *Expl. Cell Res.* **71**, 422–7.

CALVAYRAC R., VANLENTE F. & BUTOW R.A. (1971) *Euglena gracilis*: formation of giant mitochondria. *Science, N.Y.* **173**, 252–4.

CANTOR M.H. & JAMES T.W. (1965) Fatty acid adaptation in *Polytomella agilis*. *J. cell comp. Physiol.* **65**, 285–90.

CANTOR M.H. & KLOTZ J. (1971) Synchronous growth of *Polytomella agilis*. *Experientia*, **27**, 801–3.

CARTLEDGE T.G., COOPER R.A. & LLOYD D. (1971) Subcellular fractionation of eukaryotic microorganisms. In *Separations with zonal rotors*, ed. Reid E. pp. V-4.1-V-4.16. Wolfson Bioanalytical Centre, University of Surrey, Guildford.

CASSELTON P.J. & SYRETT P.J. (1962) The oxidation of C$^{14}$-labelled glucose by *Chlorella vulgaris. Ann. Bot. N.S.* **26**, 71–82.

CASSELTON P.J. & STACEY J.L. (1969) Observations on the nitrogen metabolism of *Prototheca* Krüger. *New Phytol.* **68**, 731–49.

CHANCE B. (1965) Reaction of oxygen with the respiratory chain in cells and tissues. *J. gen. Physiol.* **49**, 163–95.

CHANCE B., BONNER W.D. & STOREY B.T. (1968) Electron transport in respiration. *A. Rev. Pl. Physiol.* **19**, 295–320.

CHANCE B. & SAGER R. (1957) Oxygen and light-induced oxidation of cytochrome, flavoprotein, and pyridine nucleotide in a *Chlamydomonas* mutant. *Pl. Physiol., Lancaster,* **32**, 548–61.

CHANCE B. & WILLIAMS G.R. (1956) The respiratory chain and oxidative phosphorylation. *Adv. Enzymol.* **17**, 65–134.

CHANCE B., AZZI A., LEE I.Y., LEE C.P. & MELA L. (1969) The nature of the respiratory chain; localization of energy conservation sites, the high energy store, electron transferlinked conformation changes, and the closedness of submitochondrial vesicles. *FEBS Symposium,* **17**, 233–73.

CHAPMAN L.F., CIRILLO V.P. & JAHN T.L. (1965) Permeability to sugars and fatty acids in *Polytoma obtusum. J. Protozool.* **12**, 47–51.

CHEUNG W.Y. & GIBBS (1966) Dark and photometabolism of sugars by a blue-green alga: *Tolypothrix tenuis. Pl. Physiol., Lancaster,* **41**, 731–7.

CIFERRI O. (1962) Carbohydrate metabolism of *Prototheca zopfii.* I. Enzymes of the glycolytic and hexose monophosphate pathways. *Enzymologia,* **24**, 283–97.

CIRILLO V.P. (1956) Induced enzyme synthesis in the phytoflagellate *Polytoma uvella. J. Protozool.* **3**, 69–74.

CIRILLO V.P. (1957) Long term adaptation to fatty acids by the phytoflagellate *Polytoma uvella. J. Protozool.* **4**, 60–2.

COOK J.R. & HEINRICH B. (1968) Unbalanced respiratory growth of *Euglena. J. gen. Microbiol.* **53**, 237–51.

COOMBS J., SPANIS C. & VOLCANI B.E. (1967) Studies on the biochemistry of silica shell formation in diatoms. Photosynthesis and respiration in silicon-starvation synchrony of *Navicula pelliculosa. Pl. Physiol., Lancaster,* **42**, 1607–11.

COOPER R.A. & LLOYD D. (1972) Subcellular fractions of the colourless alga, *Polytomella caeca* by differential and zonal centrifugation. *J. gen. Microbiol.* **72**, 59–70.

COOPER R.A., JONES M., VENABLES S.E. & LLOYD D. (1969) Subcellular fractionation of *Polytomella caeca* by zonal centrifugation. *Biochem. J.* **114**, 65P.

CRESPI H.L., MANDEVILLE S.E. & KATZ J.J. (1962) The action of lysozyme on several bluegreen algae. *Biochem. biophys. Res. Commun.* **9**, 569–73.

DANFORTH W.F. (1967) Respiratory metabolism. In *Research in Protozoology,* vol. 1, ed. Chen T.T. pp. 205–306. Pergamon Press, Oxford, London & New York.

DEDUVE C. (1971) Tissue fractionation, past and present. *J. Cell Biol.* **50**, 20d–55d.

DESA R. & HASTINGS J.W. (1968) The characterisation of scintillons. Bioluminescent particles from the marine dinoflagellate, *Gonyaulax polyedra. J. gen. Physiol.* **51**, 105–22.

DEVLIN R.M. & GALLOWAY R.A. (1968) Oxidative enzymes and pathways of hexose and triose metabolism in *Chlorella. Physiologia Pl.* **21**, 11–25.

EMERSON R. (1927) Effect of certain respiratory inhibitors on *Chlorella. J. gen. Physiol.* **10**, 469–79.

EPEL B.L. & BUTLER W.L. (1969) Cytochrome $a_3$: destruction by light. *Science, N.Y.* **166**, 621–2.

EPEL B.L. & BUTLER W.L. (1970) Inhibition of *Prototheca* respiration by light. *Pl. Physiol., Lancaster,* **45**, 728–34.

EPEL B.L. & KRAUSS R.W. (1966) The inhibitory effect of light on the growth of *Prototheca zopfii. Biochim. biophys. Acta,* **120**, 73–83.

FIRDOUSI B. & SYRETT P.J. (1970) Fermentation of glucose by *Chlorella. Arch. Mikrobiol.* **72**, 344–52.

GERGIS M.S. (1971) The presence of microbodies in three strains of *Chlorella. Planta*, **101**, 180–4.

GERHARDT B. & BERGER C. (1971) Microbodies und diamino-benzidin-reaktion in den acetat-flagellaten *Polytomella caeca* and *Chlorogonium elongatum. Planta*, **100**, 155–166.

GESSNER F. & PANNIER F. (1958) The oxygen consumption of water plants with different oxygen tensions. *Limnol. Oceanogr.* **3**, 478–90.

GIBBS M. (1962) Respiration. In *Physiology and Biochemistry of Algae*, ed. Lewin R.A. pp. 61–90. Academic Press, London & New York.

GITTLESON S.M., ALPER R.E. & CONTI S.F. (1969) Ultrastructure of trophic and encysted *Polytomella agilis. Life Sci.* **8**, 591–9.

GRAVES L.B., TRELEASE R.N. & BECKER W.M. (1971) Particulate nature of glycollate dehydrogenase in *Euglena gracilis* var. *bacillaris* (SM L-1). *Pl. Physiol., Lancaster*, **47**, (*Suppl.*) A 170.

GRAVES L.B., TRELEASE R.N., GRILL A. & BECKER W.B. (1972) Localization of glyoxylate cycle enzymes in glyoxysomes in *Euglena. J. Protozool.* **19**, 527–32.

HAIGH W.G. & BEEVERS H. (1964a) Occurrence and assay of isocitrate lyase in algae. *Archs. Biochem. Biophys.* **107**, 147–51.

HAIGH W.G. & BEEVERS H. (1964b) The glyoxylate cycle in *Polytomella caeca. Archs. Biochem. Biophys.* **107**, 152–7.

HARROP L.C. & KORNBERG H.L. (1963) Enzymes of the glyoxylate cycle in *Chlorella vulgaris. Biochem. J.* **88**, 42P.

HARROP L.C. & KORNBERG H.L. (1966) The role of isocitrate lyase in the metabolism of the algae. *Proc. R. Soc. B.* **166**, 11–29.

HASSAL K.A. (1967) Inhibition of respiration of *Chlorella vulgaris* by simultaneous addition of cupric and fluoride ions. *Nature, Lond.* **215**, 521.

HASSAL K.A. (1969) An asymmetric respiratory response occurring when fluoride and copper ions are applied jointly to *Chlorella vulgaris. Physiologia Pl.* **22**, 304–11.

HEINRICH B. & COOK J.R. (1967) Studies on the respiratory physiology of *Euglena gracilis* cultured on acetate or glucose. *J. Protozool.* **14**, 548–53.

HEALEY F.P. & MYERS J. (1971) The Kok effect in *Chlamydomonas reinhardi. Pl. Physiol., Lancaster*, **47**, 373–9.

HIYAMA T., NISHIMURA M. & CHANCE B. (1969) Energy and electron transfer system of *Chlamydomonas reinhardi.* Photosynthetic and respiratory cytochrome systems of the pale green mutant. *Pl. Physiol., Lancaster*, **44**, 527–34.

HOMMERSAND M.H. & THIMANN K.V. (1965) Terminal respiration of vegetative cells and zygospores in *Chlamydomonas reinhardi. Pl. Physiol., Lancaster*, **40**, 1220–7.

HORTON A.A. (1968) NADH oxidase in blue-green algae. *Biochem. biophys. Res. Commun.* **32**, 839–45.

HUGHES D.E., WIMPENNY J.W.T. & LLOYD D. (1971) The disintegration of microorganisms. In *Methods in Microbiology*, vol. 5B, eds. Norris J.R. & Ribbons D.W. pp. 1–55. Academic Press, London & New York.

HURLBERT R.E. & RITTENBERG S.C. (1962) Glucose metabolism of *Euglena gracilis bacillaris.* Growth and enzymatic studies. *J. Protozool.* **9**, 170–82.

HUTH W. (1967) Enzyme in grunen Einzellen in Abhangigkeit von der Kohlenstoffversorgung. *Flora Jena*, **158**, 57–87.

HUTNER S.H. & PROVASOLI L. (1951) eds. The phytoflagellates. In *Biochemistry and Physiology of the Protozoa*, vol. 1, ed. Lwoff A. pp. 27–128. Academic Press, New York & London.

HYAMS J. & DAVIES D.R. (1972) The induction and characterisation of cell-wall mutants of *Chlamydomonas reinhardi. Mutation Res.* **14**, 381–9.

JOHN P.C.L. & SYRETT P.J. (1965) The development of isocitrate lyase in *Chlorella vulgaris* demonstrated by acrylamide gel electrophoresis. *Biochem. J.* **95**, 49P.

JOHN P.C.L. & SYRETT P.J. (1968) The inhibition by intermediary metabolites of isocitrate lyase from *Chlorella pyrenoidosa*. *Biochem. J.* **110**, 481–4.

JOHN P.C.L., THURSTON C.F. & SYRETT P.J. (1970) Disappearance of isocitrate lyase enzyme from cells of *Chlorella pyrenoidosa*. *Biochem. J.* **119**, 913–19.

JONES L.W. & MYERS J. (1963) A common link between photosynthesis and respiration in a blue-green alga. *Nature, Lond.* **199**, 670–2.

KAHN V. & BLUM J.J. (1967) The rate of mitochondrial protein synthesis during synchronized division of *Astasia*. *Biochemistry, N.Y.* **6**, 817–26.

KANAZAWA T., KANAZAWA K., KIRK M.R. & BASSHAM J.A. (1972) Regulatory effects of ammonia on carbon metabolism in *Chlorella pyrenoidosa* during photosynthesis and respiration. *Biochim. biophys. Acta*, **256**, 656–69.

KANDLER O., LIESENKÖTTER I. & OAKS B.A. (1961) Die Wirkung von Monojodessigsäure auf Atmung und Photosynthese von *Chlorella*. *Z. Naturf.* **16b**, 50–61.

KOWALLIK W. (1967) Action spectrum for an enhancement of endogenous respiration by light in *Chlorella*. *Pl. Physiol., Lancaster*, **42**, 672–6.

KOWALLIK W. (1968) On the effect of KI on light-enhanced endogenous respiration of algae. *Planta*, **79**, 122–7.

KOWALLIK W. & SCHMID S.J. (1971) On glycollate oxidation in unicellular green algae. *Planta*, **96**, 224–37.

KROGMANN D.W. & TANG F. (1971) Cytochrome oxidase in subcellular preparations from a blue-green alga. *Pl. Physiol., Lancaster*, **47**, (*Suppl.*) A 157.

LEACH C.K. & CARR N.G. (1969) Oxidative phosphorylation in an extract of *Anabaena variabilis*. *Biochem. J.* **112**, 125–6.

LEACH C.K. & CARR N.G. (1970) Electron transport and oxidative phosphorylation in the blue-green alga, *Anabaena variabilis*. *J. gen. Microbiol.* **64**, 55–70.

LEEDALE G.F. & BUETOW D.E. (1970) Observations on the mitochondrial reticulum in *Euglena gracilis*. *Cytobiologie*, **1**, 195–202.

LEX M., SILVESTER W.B. & STEWART W.D.P. (1972) Photorespiration and nitrogenase activity in the blue-green alga, *Anabaena cylindrica*. *Proc. R. Soc. B.* **180**, 87–102.

LLOYD D. (1965) Respiratory control in mitochondria isolated from the colourless alga, *Prototheca zopfii*. *Biochim. biophys. Acta*, **110**, 425–6.

LLOYD D. (1966a) Isolation of mitochondria from the colourless alga, *Prototheca zopfii*. *J. Expl. Cell Res*, **45**, 120–32.

LLOYD D. (1966b) Inhibition of electron transport in *Prototheca zopfii*. *Phytochem.* **5**, 527–30.

LLOYD D. & CALLELY A.G. (1965) The assimilation of acetate and propionate by *Prototheca zopfii*. *Biochem. J.* **97**, 176–9.

LLOYD D. & CHANCE B. (1968) Electron transport in mitochondria isolated from the flagellate, *Polytomella caeca*. *Biochem. J.* **107**, 829–37.

LLOYD D. & VENABLES S.E. (1967) The regulation of propionate oxidation in *Prototheca zopfii*. *Biochem. J.* **104**, 639–46.

LLOYD D., EVANS D.A. & VENABLES S.E. (1968) Propionate assimilation in the flagellate, *Polytomella caeca*. *Biochem. J.* **109**, 897–907.

LLOYD D., EVANS D.A. & VENABLES S.E. (1970) The effects of chloramphenicol on growth and mitochondrial function of the flagellate *Polytomella caeca*. *J. gen. Microbiol.* **61**, 33–41.

LORD J.M. & MERRETT M.J. (1971) The intercellular localization of glycollate oxidoreductase in *Euglena gracilis*. *Pl. Physiol., Lancaster*, **47** (*Suppl.*) A 169.

LUI N.S.T., ROELS O.A., TROUT M.E. & ANDERSON O.R. (1968). Subcellular distribution of enzymes in *Ochromonas malhamensis*. *J. Protozool.* **15**, 536–42.

MATSUKA M. & HASE E. (1965) Metabolism of glucose in the process of 'glucose bleaching' of *Chlorella protothecoides*. *Pl. Cell Physiol., Tokyo*, **6**, 721–41.

MOORE J., CANTOR M.H., SHEELER P. & KAHN W. (1970) The ultrastructure of *Polytomella agilis*. *J. Protozool.* **17**, 671–6.

MULLER M. (1969) Peroxisomes of protozoa. *Ann. N.Y. Acad. Sci.* **168**, 292–301.

OAKS A. (1962) Influence of glucose and light on pyruvate metabolism by starved cells of *Chlorella elipsoida*. *Pl. Physiol., Lancaster*, **37**, 316–22.

OHMANN E., RINDT K.P. & BORRISS R. (1969) Glucose-6-phosphat-dehydrogenase in autotrophen mikroorganismen I. Die regulation der synthese der glucose-6-phosphat-dehydrogenase in *Euglena gracilis* und *Rhodopseudomonas spheroides* in abhängigkeit von der kulturbed-ingungen. *Z. allg. Mikrobiol.* **9**, 557–64.

OSAFUNE T., MIHARA S., HASE E. & OHKURO I. (1972) Electron microscopic studies on the vegetative cellular life cycle of *Chlamydomonas reinhardi* Dangeard in synchronous culture I. Some characteristics of changes in subcellular structures during the cell cycle, especially in formation of giant mitochondria. *Pl. Cell Physiol., Tokyo*, **13**, 211–27.

PADAN E., BILHA R. & SHILO M. (1971) Endogenous dark respiration of the blue-green alga, *Plectonema boryanum*. *J. Bact.* **106**, 45–50.

PEARCE J. & CARR N.G. (1969) The incorporation and metabolism of glucose by *Anabaena variabilis*. *J. gen. Microbiol.* **54**, 451–62.

PEARCE J., LEACH C.K. & CARR N.G. (1969) The incomplete tricarboxylic acid cycle in the blue-green alga *Anabaena variabilis*. *J. gen. Microbiol.* **55**, 371–8.

PERINI F., SCHIFF J.A. & KAMEN M.D. (1964) Iron-containing proteins in *Euglena*. II. Functional localization. *Biochim. biophys. Acta*, **88**, 91–8.

PICKETT J.M. (1966) The action spectra for blue-light stimulated oxygen uptake in *Chlorella*. Carnegie Institute Yearbook, Washington, **7B**, 197–203.

PLESNICAR M., BONNER JR. W.D. & STOREY B.T. (1967) Peroxidase associated with higher plant mitochondria. *Pl. Physiol., Lancaster*, **42**, 366–70.

POOLE R.K. & LLOYD D. (1973) Effect of 2-deoxyglucose on growth and cell walls of *Schizosaccharomyces pombe* 972h⁻. *Arch. Mikrobiol.* **88**, 257–72.

PRICE D.A. & BOURKE M.E. (1966) 'Sphaeroplasts' prepared from *Euglena gracilis* by proteolysis. *J. Protozool.* **13**, 474–6.

RAISON J.K. & SMILLIE R.M. (1969) Respiratory cytochromes of *Euglena gracilis*. *Biochim. biophys. Acta*, **180**, 500–8.

REEVES H.C., KADIS S. & AJL S. (1962) Enzymes of the glyoxylate by-pass in *Euglena gracilis*. *Biochim. biophys. Acta*, **57**, 403–4.

REID A., SOEDER C.J. & MULLER I. (1963) Respiration in synchronous cultures of *Chlorella*. I. Changes in respiratory gaseous exchange during the course of the developmental cycle. *Arch. Mikrobiol.* **45**, 343–58.

SARGENT D.F. & TAYLOR C.P.S. (1972) Terminal oxidases of *Chlorella pyrenoidosa*. *Pl. Physiol., Lancaster*, **49**, 775–9.

SCHOLES P., MITCHELL P. & MOYLE J. (1969) The polarity of proton translocation in some photosynthetic microorganisms. *Eur. J. Biochem.* **8**, 450–4.

SCHONBAUM G.R., BONNER JR. W.D., STOREY B.T. & BAHR J. (1971) Specific inhibition of the CN-insensitive respiratory pathway in plant mitochondria by hydroxamic acids. *Pl. Physiol., Lancaster*, **47**, 124–8.

SHARPLESS T.K. & BUTOW R.A. (1970a) Phosphorylation sites, cytochrome complement and alternate pathways of coupled electron transport in *Euglena gracilis* mitochondria. *J. biol. Chem.* **245**, 50–7.

SHARPLESS T.K. & BUTOW R.A. (1970b) An inducible alternate terminal oxidase in *Euglena gracilis* mitochondria. *J. biol. Chem.* **245**, 58–70.

SMITH A.J., LONDON J. & STANIER R.Y. (1967) Biochemical basis of obligate autotrophy in blue-green algae and thiobaccilli. *J. Bact.* **94**, 972–83.

STEWART W.D.P. & PEARSON H.W. (1970) Effects of aerobic and anaerobic conditions on growth and metabolism of blue-green algae. *Proc. R. Soc. B*, **175**, 292–311.

SYRETT P.J. (1951) The effect of cyanide on the respiration and the oxidative assimilation of glucose by *Chlorella vulgaris. Ann. Bot.* **15**, 473–82.

SYRETT P.J., BOCKS S.M. & MERRETT M.J. (1964) The assimilation of acetate by *Chlorella vulgaris. J. Expl. Bot.* **15**, 35–47.

TALBERT D.M. & SOROKIN C. (1971) Respiration in cell development. *Arch. Mikrobiol.* **78**, 281–94.

TANG I.S. & FRENCH C.S. (1933) The rate of oxygen consumption of *Chlorella pyrenoidosa*, as a function of temperature and of oxygen tension. *Chin. J. Physiol.* **1**, 353–78.

WEBSTER G.C. & FRENKEL A.W. (1952) Some respiratory characteristics of the blue-green alga, *Anabaena. Pl. Physiol., Lancaster*, **28**, 63–9.

WEBSTER D.A. & HACKETT D.P. (1965) Respiratory chain of colourless algae. I. *Chlorophyta* and *Euglenophyta. Pl. Physiol., Lancaster*, **40**, 1091–100.

WEBSTER D.A. & HACKETT D.P. (1966) Respiratory chain of colourless algae. II. Cyanophyta. *Pl. Physiol., Lancaster*, **41**, 599–605.

WILDON D.C. & AP REES T. (1965) Metabolism of glucose-C[14] by *Anabaena cylindrica. Pl. Physiol., Lancaster*, **40**, 332–5.

WILSON D.W. & JAMES T.W. (1966) Energetics and the synchronized cell cycle. In *Cell Synchrony*, eds. Cameron I.L. & Padilla G.M. pp. 236–55. Academic Press, New York & London.

# CHAPTER 19

# HETEROTROPHY OF CARBON

## M. R. DROOP

Scottish Marine Biological Association,
Oban, Scotland, U.K.

1   Occurrence of heterotrophic algae   530

2   Acetate and sugar algae   532

3   Phagotrophic algae   538

4   Acid tolerance and permeability 538

5   The glucose block   540

6   Growth kinetics   541

7   Oxidative assimilation   544

8   Phototrophic assimilation   546

9   Obligate phototrophy   549

10   Apochlorosis   551

11   References   552

## 1   OCCURRENCE OF HETEROTROPHIC ALGAE

Heterotrophy is defined as the utilization of organic compounds for growth. In this chapter I shall discuss organic carbon as a major nutrient, confining myself, to use the jargon of Lwoff's nutritional *schema* (Lwoff 1951), to chemo- and photo-organotrophy, that is, to the oxidative and photosynthetic assimilation of organic carbon.

Species lacking photosynthetic pigments occur in most phyla of algae. These are necessarily heterotrophic, as none are known to be able to derive their energy from the oxidation of inorganic compounds. Some even depend on particulate food and many others utilize organic carbon whilst retaining their photosynthetic apparatus and potential. Indeed few physiologists would disagree with Myers (1962) that 'for most algae carbon may also be provided as organic solutes such as glucose or acetate in the medium rather than carbon dioxide in the gas phase'. Nevertheless, the statement is misleading, because it only strictly applies to the most commonly used laboratory algae to which Myers was no doubt referring. Nothing is known about the vast majority of algae in this respect, while most apparently strict photoautotrophic strains cultivated have been slow in finding their way to the experimental bench, for the simple reason that they seem to be generally less tolerant of laboratory conditions than the fast growing potentially heterotrophic types. (An exception to this is the marine *Phaeodactylum tricornutum*, which has been much used in experimental work, often under the name *Nitzschia closterium* forma *minutissima*.) Notwithstanding, the impression obtained from the algal literature is likely to be biased in favour of heterotrophy.

Few Rhodophyceae or Phaeophyceae have been cultivated and none show any ability to grow in the dark on organic carbon. Pelagic diatoms and coloured Dinophyceae have similarly thus far exhibited no chemotrophic ability (Provasoli & McLaughlin 1963), while the versatility of *Ochromonas malhamensis* is exceptional among Chrysophyceae. Khoja and Whitton (1971) list 17 heterotrophic Cyanophyceae out of a collection of 24, and many littoral pennate diatoms have been shown to be facultative chemotrophs (Lewin & Lewin 1960), although the writer found only two out of two dozen diatom isolates from off sand grains of a Scottish sea loch to be able to grow in the dark. Few pigmented Euglenophyceae have been studied and one should not judge the phylum by the performance of the various strains of *Euglena gracilis*. In the Chlorophyceae chemotrophy has been reported more often from the Chlorococcales than Volvocales. Indeed, the first algae of this group to be cultivated were able to grow in the dark (Beijerinck 1890, Krüger 1894). Nevertheless, all *Chlorococcum* spp. are reported to be obligate phototrophs (Parker *et al.* 1961), as are many *Chlorella* strains (Kessler 1967, Shihira & Krauss 1965) and, in the writer's experience, all marine *Nannochloris* species thus far isolated. Among all groups of the Scottish Marine Biological Association's collection, isolated in the main from supralittoral rock pools, there are 24 facultative chemotrophs and 91 apparently obligate phototrophs.

On the other hand, obligate phototrophy can never be proved except with respect to specified conditions. No matter how many substrates one tests for dark growth, and Aldrich (1962) tested over 100 on *Gymnodinium breve*, it is possible that the vital one was missed out or that other factors, such as the nitrogen source or the concentration of substrate, were unfavourable. After a lifetime of experience, though admittedly with algae in the main from eutrophic habitats, Pringsheim (1963, p. 296) wrote 'Chlorophyllführende Algen und Flagellaten *können* strikt photoautotroph sein und organische Nährstoffe ablehnen; aber ansehen kann man es ihnen meist nicht, und es ist nicht die Regel'. As if to point to this, the toxic *Prymnesium parvum*, a member of the Haptophyceae long thought to be obligately phototrophic, has been shown to assimilate and grow in the dark on glycerol, provided the concentration of this substrate is over 0·25M (Rahat & Jahn 1965, Rahat & Spira 1967), and the same has been reported of the marine cryptomonad *Chroomonas salina* (Antia *et al.* 1969). These algae must be regarded physiologically as facultative chemotrophs and yet the concentration of glycerol needed for dark growth is so large as never to be encountered in the natural habitat and these organisms remain, ecologically speaking, obligate phototrophs.

The aphorism that autotrophs manufacture organic matter while heterotrophs merely shuffle it around naturally oversimplifies but does make its point.
Perhaps the fact that there are heterotrophic algae is of no great consequence
from the broad standpoint of ecology: as primary producers algae are dominant
in the aquatic environment, whereas as heterotrophs they are in competition
with the rest of the microbial world for whom numbers, small size, metabolic
speed and versatility are overriding advantages. Nevertheless, there are situations, such as sewage oxidation ponds and similar habitats, where the ability of
algae to metabolize organic carbon is significant (Eppley & MaciasR 1963).
Wherever crowding puts $CO_2$ and light at a premium heterotrophs will naturally
be favoured over other algae, but being able to dispense with light when necessary may be an important survival factor, even in open waters. From time
to time there have been reports of apparently healthy and even multiplying
populations of pelagic phytoplankton in places where light was excluded, for
example under the ice in arctic Sweden (Rodhe 1955) or below the photic zone
in the Mediterranean (Bernard 1963). They pose the awkward question as to
the origin, in such impoverished habitats, of the organic carbon upon which the
algae are presumably living.

## 2  ACETATE AND SUGAR ALGAE

It is customary to distinguish between acetate-utilizing and sugar-utilizing algae.
This distinction was made first by Pringsheim (1921, 1937) to contrast the
nutrition of certain colourless flagellates, such as *Polytoma*, with that of the
chlorococcoid algae already known to grow in the dark on glucose. Because it
was found later that many of these so called 'acetate flagellates' could utilize
other simple acids such as pyruvic acid and lactic acid, Lwoff (1932) proposed
the term *oxytroph* for the acetate flagellate and *haplotroph* for the sugar-utilizing
organism. However, Lwoff (1944) also equated oxytroph with the ability to
manufacture glucoside polymers in the absence of photosynthesis, usually
coupled with the inability to oxidize glucose directly, and haplotroph with the
non-involvement of starch-like polymers coupled with the ability to oxidize
glucose and other carbon compounds. It is clear from this that oxytroph does
not have quite the same connotation as acetate organism and leaves the status
of all the sugar-utilizing Chlorococcales and pennate diatoms in doubt, since the
reserve materials in these two groups are respectively starch and β-linked glucans
such as chrysolaminaran.

In his *Evolution Physiologique* Lwoff (1944) envisaged a progression from
green algae to protozoa which was characterized first by the loss of chlorophyll
(*leucophytes*) then of the remainder of the plastidic apparatus (*protozoa*), the
transition from oxytrophy to haplotrophy occurring with the latter. Although
remnants of the chromatophore can be identified in such leucophytes as *Polytoma
uvella* (Lang & Cook 1962) and *Prototheca ciferrii* (Menke & Fricke 1962), the
latter oxidize hexose sugars directly and moreover the reserve product is prob-

ably a branched polysaccharide like glycogen (Barker & Bourne 1955). Indeed the distinction between animal glycogens and the various plant polysaccharides has proved not to be nearly as clear as Lwoff supposed (Barker & Bourne 1955, Meeuse 1962). Neither is the retention of the plastidic apparatus necessarily a requirement of oxytrophy. For example, the dinoflagellate, *Oxyrrhis marina*, manufactures true starch when encysted and is limited in its carbon assimilation to acetate and ethanol, and can therefore be considered as an oxytroph although it has no trace of a plastidic apparatus. Moreover it can be phagotrophic and considered as a protozoan (Droop 1959, Droop & Pennock 1971, Dodge & Crawford 1971). Another view (Gibbs 1970) distinguishes between algae which have their polysaccharide synthesizing enzymes associated with the plastids (e.g. some Chlorophyceae and Prasinophyceae) and algae which do not, irrespective of the nature of the polysaccharide. In the latter group, the fate of the plastids would not be expected to have any direct bearing on polysaccharide synthesis.

It is now realized that most chemotrophic algae can use acetate. Hutner and Provasoli (1951) stated 'It is difficult to find a protist which while dissimilating glucose and showing during the process a simultaneous uptake of $O_2$ cannot also utilize acetate'. Many examples do, however, exist among algae, for example some species of *Chlorella* (Shihira & Krauss 1965), *Tribonema minus* (Belcher & Miller 1960) and several littoral diatoms (Lewin & Lewin 1960). The range of substrates utilized varies greatly from species to species, both among acetate and sugar algae, and closely related strains may differ markedly in their versatility and preferences, so much so that categorical boundaries become blurred. The Mainx and Vischer strains of *Euglena gracilis* for instance, apparently use only acetic acid and butyric acid, while the Pringsheim strain of this species uses a great range of straight-chain fatty acids including lauric acid and myristic acid, several Krebs' cycle acids, and straight chain monohydric alcohols (Cramer & Myers 1952), while the substrates available to Provasoli's *E. gracilis* var. *urophora* include the lower branched acids and alcohols as well (Provasoli 1938). Sugars and sugar alcohols are not used by any of the above *Euglena* strains. A further widening of the range of substrates to include some carbohydrates is found in *E. gracilis* var. *bacillaris* and Pringsheim's *E. gracilis* var. *saccharophila* (Pringsheim 1955). The former, according to Cramer and Myers (1952) and Hurlbert and Rittenberg (1962), can also use the carbon skeleton of some amino acids as sole source of carbon. Generally, the availability of fatty acids and alcohols for heterotrophic growth decreases with chain length, while acids and alcohols with an even number of carbon atoms are more often used than those with an odd number. The latter tend to be more toxic. A particularly good example of this pattern is provided by the colourless member of the Cryptophyceae, *Chilomonas paramecium* studied by Cosgrove and Swanson (1952), who suggested that the enzymes responsible for the oxidation of fatty acids to $C_2$ fragments would only work if the reaction could go to completion. *Polytomella caeca*. on the other hand has not nearly the range of *Chilomonas* but

**Table 19.1.** Some algae exhibiting dark growth on acetate but not on glucose.

| | | |
|---|---|---|
| *Amphora coffaeiformis*[1,2]* | *Chlamydomonas subglobosa*[25] | *Haematococcus pluvialis*[16,17] |
| *Astrephomene gubernaculifera*[15] | *C. spreta*[11] | *Nitzschia curvilineata*[1,2] |
| *Astasia longa*[3,4]* | *Chlorellidium tetrabotrys*[29] | *N. putrida*[26]*,** |
| *Beggiatoa* spp[30,31]* | *Chlorogonium elongatum*[7,12] | *N. leucosigma*[26]*,** |
| *Brachiomonas submarina*[5] | *C. euchlorum*[7,13,14] | *N. alba*[26]*,** |
| *B. submarina* var *pulsifera*[5] | *Cylindrotheca fusiformis*[32] | *Oxyrrhis marina*[19]* |
| *Chilomonas paramecium*[6,7,23]* | *Diplostauron elegans*[27] | *Polytoma ocellatum*[7,14]** |
| *Chlamydomonas dysosmos*[8,9] | *Euglena gracilis* (Vischer, Mainx and | *P. obtusum*[7,16] |
| *C. dorsocentralis*[25] | Pringsheim strains)[7,18] | *P. uvella*[7]* |
| *C. monoica*[25] | *E. gracilis* var *urophora*[7] | *Polytomella caeca*[7,14,20,21,22]* |
| *C. pulchra*[25] | *Furcilla stigmatophora*[24] | *Stephanosphaera pluvialis*[15] |
| *C. reinhardtii*[10] | *Gonium quadratum*[15] | *Volvulina steinii*[15] |
| | *G. octonarium*[15] | |

* Lacks chromatophores. ** Lactate preferred to acetate. References: 1, Lewin J.C. 1963; 2, Lewin & Lewin 1960; 3, Barry 1962; 4, Buetow & Padilla 1963; 5, Droop & McGill 1966; 6, Lwoff & Dusi 1934; 7, Provasoli 1938; 8, Lewin R.A. 1954; 9, Haigh & Beevers 1962; 10, Sager & Granick 1953; 11, Droop unpubl.; 12, Loefer 1935; 13, Lwoff & Dusi 1935; 14, Pringsheim 1937; 15, Pringsheim & Pringsheim 1959; 16, Lwoff 1932; 17, Droop 1961; 18, Cramer & Myers 1952; 19, Droop 1959; 20, Pringsheim 1935; 21, Lwoff 1935; 22, Wise 1959; 23, Cosgrove & Swanson 1952; 24, Belcher 1967; 25, Lucksch 1933; 26, Lewin & Lewin 1967; 27, Lynn & Starr 1970; 28, Little et al. 1951; 29, Belcher & Miller 1960; 30, Faust & Wolfe 1961; 31, Pringsheim 1967a; 32, Lewin & Hellebust 1970.

uses the 3 and 5 carbon straight chain acids and alcohols, which are toxic to *Chilomonas*. At the other end of the scale very many, especially photosynthetic, acetate organisms utilize only acetate. The most often cited examples of these are *Chlorogonium euchlorum* and *Hyalogonium klebsii* (Provasoli 1938). Any further reduction in permitted substrates results, as Hutner and Provasoli (1951) pointed out, in obligate phototrophy.

The sugar algae show equal diversity in the range of substrates that they can utilize. Glucose is available to the great majority and galactose and fructose, are nearly as widely used, but disaccharides are less generally available, and of the polyhydric alcohols only glycerol is frequently utilized. Mannitol is available to *Ochromonas malhamensis* (Pringsheim 1952). Sugar algae usually cannot use the carbon of amino acids, although Belcher and Miller (1960) found that glycine supported dark growth of two members of the Xanthophyceae. *Prototheca zopfii* utilizes only hexoses and glycerol, but otherwise is a versatile acetate organism. The range of acids utilized by sugar algae is probably greater than indicated in Tables 19.1–5, which summarize information on the ability of certain algae to utilize organic substrates, because data are wanting in many instances.

**Table 19.2.** Some algae exhibiting dark growth on glucose but not on acetate.

| | | |
|---|---|---|
| *Amphora coffaeiformis* [1,2] | *Chlorogloeopsis* type[19]** | *Ochromonas malhamensis*[10] |
| *Anabaena* type, 3 strains[19]** | *Cyclotella* sp.[1] | *Oocystis naegelii*[4] |
| *Ankistrodesmus falcatus*[16] | *Dictyochloris* spp.[3] | *Oscillatoria* spp.[6] |
| *Bracteacoccus engadiensis*[3] | *Lyngbya* sp.[6] | *Phormidium foveolarum*[6] |
| *B. minor*[3] | *Lyngbya* type, 4 strains[19]** | *Plectonema notatum*[6] |
| *Calothrix* type, 2 strains[19]** | *Melosira nummuloides*[7] | *Scenedesmus costulatus*[13] |
| *Chlorella candida*[17] | *Navicula incerta*[1,2] | *S. obtusiusculus*[4] |
| *C. ellipsoidea*[17] | *N. pelliculosa*[1,2] | *Scenedesmus* sp.[12] |
| *C. emersonii*[17] | *Neochloris* spp.[4] | *Spongiococcum* sp.[4] |
| *C. emersonii* var *globosa*[17] | *Nitzschia angularis*[1,2] | *Spongiochloris* sp.[4] |
| *C. fusca*[17] | *N. curvilineata*[1,2] | *Stichococcus bacillaris*[16] |
| *C. fusca* var *vacuolata*[17] | *N. filiformis*[1,2] | *S. diplosphaera*[4] |
| *C. miniata*[17] | *N. frustulum*[1,2] | *Tolypothrix tenuis*[14] |
| *C. pringsheimii*[17] | *N. laevis*[1,2] | *Tribonema minus*[15] |
| *C. rubescens*[4] | *N. marginata*[1,2] | *Tetraselmis chui*[18] |
| *C. saccharophila*[17] | *N. punctata*[1,2] | *T. subcordiformis*[18] |
| *C. simplex*[17] | *N. thermalis*[1,2] | *T. tetrathele*[18] |
| *C. sorokiniana*[17] | *Nostoc muscorum*[8] | *Tetraselmis* spp.[18] |
| *C. vannielii*[17] | *N. punctiforme*[9] | |
| *Chlorogloea fritschii*[5]* | *Ochromonas danica*[11] | |

\* Prefers sucrose to glucose. \*\* No data on acetate. References: 1, Lewin J.C. 1963; 2, Lewin & Lewin 1960; 3, Parker *et al.* 1961; 4, Chodat & Schopfer 1960; 5, Fay 1965; 6, Allen 1952; 7, Droop unpubl.; 8, Allison *et al.* 1937; 9, Harder 1917; 10, Pringsheim 1952; 11, Aaronson & Baker 1959; 12, Samejima & Myers 1958; 13, Bristol Roach 1927, 1928; 14, Kiyohara *et al.* 1960; 15, Belcher & Miller 1960; 16, Algeus 1946; 17, Shihira & Krauss 1965; 18, Turner 1970a; 19, Kenyon *et al* 1972.

**Table 19.3.** Some algae exhibiting dark growth on both glucose and acetate

| | | |
|---|---|---|
| *Amphora coffaeiformis*[1,2] | *Chlorella protothecoides*[5,20] | *Nitzschia frustulum*[1,2**] |
| *Astasia longa* (mutant)[3*] | *C. pyrenoidosa*[5,20] | *N. laevis*[1,2**] |
| *Bracteacoccus cinnibarinus*[4] | *C. regularis*[20] | *N. leucosigma*[19*] |
| *B. terrestris*[4] | *C. vulgaris*[6,7,17,20] | *N. putrida*[14*] |
| *Botrydiopsis intercedens*[5] | *Cyanidium caldarium*[21] | *N. tenuissima*[1,2] |
| *Bumilleriopsis brevis*[5] | *Euglena gracilis* var *bacillaris*[9,10] | *Prototheca zopfii*[15,16] |
| *Chlamydomonas pseudogloe*[18] | *E. gracilis* var *saccharophila*[11] | *Scenedesmus obliquus*[17] |
| *C. pseudococcum*[18] | *Gyrodinium cohnii*[12,13*] | *S. quadricauda*[6] |
| *Chlorella ellipsoidea*[8,20] | *Neochloris alveolaris*[4] | *S. dimorphus*[6] |
| *C. mutabilis*[20] | *Nitzschia alba*[19*] | *Tribonema aequale*[5] |
| *C. nocturna*[20] | *N. closterium*[1,2] | |

\* Lacks chromatophores. \*\* Prefers lactate to acetate. References: 1, Lewin J.C. 1963; 2, Lewin & Lewin 1960; 3, Barry 1962; 4, Parker *et al.* 1961; 5, Belcher & Miller 1960; 6, Algeus 1946; 7, Neish 1951; 8, Samejima & Myers 1958; 9, Cramer & Myers 1952; 10, Cook & Heinrich 1965; 11, Pringsheim 1955; 12, Provasoli & Gold 1962; 13, Ishida & Katoda 1965; 14, Pringsheim 1967b; 15, Barker 1935; 16, Anderson 1945; 17, Mineeva 1961; 18, Lucksch 1933; 19, Lewin & Lewin 1967; 20, Shihira & Krauss 1965; 21, Allen 1952.

**Table 19.4.** Some examples of dark growth with amino acids as sole carbon source.

| | |
|---|---|
| *Bumilleriopsis brevis* | glycine[1] |
| *Chlamydomonas pseudogloe* | |
| *C. monoica* | |
| *C. dorsocentralis* | asparagine[7*] |
| *C. pulchra* | |
| *Chlorellidium tetrabotrys* | glycine[1] |
| *Chlorogloea fritschii* | glycine, glutamine[2] |
| *Euglena gracilis* var *bacillaris* | alanine, aspartate, asparagine, glutamine[3,4] |
| *Haematococcus pluvialis* | asparagine[5*] |
| *Nitzschia alba* | glutamate[6] |
| *N. leucosigma* | glutamate[6] |

\* slow but repeatable growth. References: 1, Belcher & Miller 1960; 2, Fay 1965; 3, Cramer & Myers 1952; 4, Hurlbert & Rittenberg 1962; 5, Lwoff & Lwoff 1929; 6, Lewin & Lewin 1967; 8, Lucksch 1933.

*Achnanthes brevipes*[1,2]
*Agmenellum quadruplicatum*[28]
*Amphipleura rutilans*[1,2]
*Amphiprora paludosa*[1,2]
*Amphora coffaeiformis*[1,2]
*A. lineolata*[1,2]
*Anabaena flos-aquae*[7]
*A. cylindrica*[9]
*A. variabilis*[8,9]
*Anabaena* type, 5 strains[32]
*Anacystis marina*[28]
*A. nidulans*[8]
*Balticola buetschlii*[3]
*B. droebakensis*[3]
*Calothrix parietina*[9]
*Calothrix* type, 2 strains[32]
*Carteria mediterranea*[24]
*Chaetoceros ceratosporum*[1]
*C. pelagicus*[1]
*C. pseudocrinatus*[1]
*Chlamydobotrys stellata*[10]
*Chlamydomonas eugametos*[11]
*C. humicola*[24]
*C. moewusii*[12]
*C. mundana*[13]
*C. pulsatilla* (7 strains)[3,4]
*Chlorella autotrophica*[26]
*C. infusionum*[26]
*Chlorella infusionum auxenophila*[26]

*Chlorella luteoviridis* (6 strains)[15]*
*C. photophila*[26]
*C.*[2] *saccharophila* 7 strains[15]*
*C. variabilis*[25]
*C. vulgaris* (17 strains)[15,16]*
*C. pyrenoidosa* (9 strains)[15]*
*C. zofingiensis* (3 strains)[15]*
*Chlorella* spp.[15,17]
*Chlorococcum* (8 spp.)[14]
*Coccochloris asteromphalus*[1]
*Cricosphaera elongata*[3]
*Chroococcus turgidus*[9]
*Cyclotella caspia*[1]
*C. nana*[1]
*Dunaliella salina*[27]
*D. primolecta*[3]
*D. viridis*[27]
*Eudorina elegans*[22]
*Euglena anabaenae*[18]
*E. deses*[18]
*E. klebsii*[18]
*E. pisciformis*[18]
*E. stellata*[18]
*E. viridis*[25]
*Gloeocapsa polydermatica*[9]
*Gonyaulax polyedra*[19]
*Gonium multicoccum*[22]
*G. pectorale*[22]

*Gonium sacculiferum*[22]
*G. sociale*[22]
*Gymnodinium breve*[20]
*G. foliacium*[3]
*Haematococcus pluvialis* (Naumann strain)[31]
*Hemiselmis virescens*[3]
*Isochrysis galbana*[3]
*Lyngbya aestuarii*[21]
*L. lagerheimii*[28]
*Lyngbya* type[32]
*Mallomonas epithalattia*[3]
*Melosira nummuloides*[1]
*M. lineata*[3]
*Microchaete* type, 2 strains[32]
*Microglena arenicola*[3]
*Microcoleus chthonoplastes*[28]
*Monochrysis lutheri*[3]
*Monodus subterraneus*[21]
*Nannochloris oculata* (4 strains)[3,5]
*Navicula meniscus*[1,2]
*Nitzschia frustulum*[1,2]
*N. hybridaeiformis*[1,2]
*N. closterium*[2]
*N. obtusa*[1,2]
*N. ovalis*[1,2]
*Nostoc muscorum*[8,9]
*Oscillatoria amphibia*[28]
*O. princeps*[9]

*Oscillatoria subtilissima*[28]
*Oscillatoria* type, 2 strains[32]
*Pandorina morum*[22]
*Pascherella tetras*[22]
*Phaeodactylum tricornutum* (2 strains)[3]
*Phormidium luridum*[9]
*Plectonema* type, 2 strains[32]
*Pleodorina californica*[22]
*P. illinoiensis*[22]
*Polyedriella helvetica*[21]
*Porphyridium cruentum*[3]
*Prorocentrum micans*[3,6]
*Prymnesium parvum*[3]
*Radiosphaera dissecta*[14]
*Rhodella maculata*[29]
*Scenedesmus acuminatus*[23]
*S. acutiformis*[23]
*Skeletonema costatum*[1,3]
*Spirulina* type[32]
*Stauroneis amphoroides*[1,2]
*Stephanoptera gracilis*[27]
*Synechococcus cedrorum*[9]
*Synedra affinis*[1,2]
*Tetraselmis carteriformis*[30]
*T. tetrathele*[30]
*Thalassiosira fluviatilis*[1]
*Waerniella lucifuga*[3]

* Kessler does not state whether or not the strains are *obligate* phototrophs. References: 1, Lewin J.C. 1963; 2, Lewin & Lewin 1960; 3, Droop unpubl.; 4, Droop & McGill 1966; 5, Droop 1966; 6, Barker 1935; 7, Hoare et al. 1967; 8, Kratz & Myers 1955; 9, Allen 1952; 10, Pringsheim & Wiessner 1960; 11, Wetherell 1958; 12, Lewin J.C. 1950; 13, Eppley et al.1963; 14, Parker et al. 1961; 15, Kessler 1967; 16, Finkle et al. 1950; 17, Fogg & Belcher 1961; 18, Dusi 1933; 19, W.H. Thomas personal communication; 20, Aldrich 1962; 21, Belcher & Miller 1960; 22, Pringsheim & Pringsheim 1959; 23, Algeus 1946; 24, Lucksch 1933; 25, Dusi 1944; 26, Shihira & Krauss 1965; 27, Gibor 1956; 28, Van Baalen 1962; 29, Turner 1970b; 30, Turner 1970a; 31, McLachlan & Craigie 1965; 32, Kenyon et al. 1972.

## 3 PHAGOTROPHIC ALGAE

A function of phagotrophy, that is, the ingestion of particulate food, is to allow a high rate of heterotrophic assimilation under conditions of low net permeability. Phagotrophy becomes relatively more important the larger the organism, because of the relatively reduced absorbing surface, and is the mode of nutrition adopted by most animals. Among algae phagotrophy is common only in colourless members of the Euglenophyceae, Dinophyceae and Chrysophyceae, and only three genera have been worked on to any extent, namely *Peranema*, *Oxyrrhis* and *Ochromonas*.

With the possible exception of *Peranema*, whose complex auxotrophic requirements still cloud the issue (Storm & Hutner 1953, Allen *et al.* 1966) none of these algae depend completely on phagotrophy for their gross carbon requirements. Both *Ochromonas malhamensis* and *O. danica*, which incidentally are photosynthetic, grow in a completely defined monophasic medium (Aaronson & Baker 1959), while with *Oxyrrhis marina* phagotrophy can be limited to the two insoluble micronutrients, ubiquinone and ergosterol (Droop & Pennock 1971). Nevertheless phagotrophy on living food permits faster growth of *O. marina* (Droop 1959) and heavier growth of *Ochromonas* (Aaronson & Baker 1959), either of which may or, more probably, may not have been connected with availability of the major carbon source. On the other hand, no source of nonliving particulate carbon can replace acetate or ethanol for *Oxyrrhis* in an otherwise complete medium. Thus, olive oil and starch although ingested failed to support growth in the absence of acetate (Droop unpubl.). This is curious in an acetate organism which stores starch at certain stages of its life history. Pringsheim (1952) showed that *O. malhamensis* excreted an amylase when grown with starch and a lipase when grown on a mixture of olive oil and egg lecithin. Both substrates supported growth in the absence of glucose.

## 4 ACID TOLERANCE AND PERMEABILITY

Hutner and Provasoli (1951) drew attention to the fact that many acetate organisms, in contrast to most algae, show extreme tolerance both of acid conditions and of high concentrations of penetrating fatty acids and alcohols. They suggested that these organisms, the most extreme examples of acetate flagellates, were characterized by their relative impermeability to the substrates. Passive penetration of living cell membranes by organic molecules is easier the smaller the molecule and the greater the fat solubility. With the fatty acids, whose undissociated molecules are fat-soluble, the fat solubility of the dissociated ion, or rather the sodium salt thereof, begins to increase steeply as the number of carbon atoms increases beyond five, and the surface activity increases accordingly. In the undissociated acid the main consideration is molecular size, so that the lower acids penetrate most easily. A progressive decrease in permeability of the cell would thus progressively limit the size of the molecule able to penetrate. Acetic acid would be the last to be excluded before complete impermeability to organic solutes imposed obligate phototrophy.

Ionization in the lower acids of course decreases penetration so that a weaker acid should penetrate more easily than a stronger at a given pH. The dissociation constants of fatty acids up to $C_9$ are all of the order $1.5 \times 10^{-5}$ and in media more alkaline than pH 6 the concentration of undissociated acid decreases by a factor of ten for each unit of pH. Acid toxicity has often been shown to increase with decreasing pH. For example, Provasoli (1938) was able to train *Polytomella caeca* in media containing 0·0015M acetate to a limit of pH 3·5, whereas Lwoff (1941) reduced the permissible pH to 2·2 when the acetate concentration was halved. Similarly, Samejima and Myers (1958) found that acetate became toxic to *Chlorella pyrenoidosa* at a concentration of 0·004M at pH 4·5, while the threshold was 0·12M at pH 6·7. Erickson *et al.* (1955) also stressed the importance of undissociated acid in acetate toxicity in *Chlorella*. A corollary is that the optimum pH for acid utilization decreases with the strength of the acid (Holz 1954, Wise 1959). Most di- and tricarboxylic acids are bulkier, more hydrophilic and stronger acids than the equivalent fatty acids and should penetrate therefore less easily, and one might expect them to be rather poor substrates. *Nannochloris oculata*, an obligate phototroph, will not tolerate $3 \times 10^{-5}$M acetate at pH 7·0 but is indifferent to many dicarboxylic acids at concentrations up to at least 0·01M at this pH (Droop 1966).

The penetrating power of carbohydrates through living membranes is poor; their molecules are relatively large and insoluble in lipids and it is probable that carbohydrate utilization always involves active transport as in bacteria (Cohen & Monod 1957). Taylor's experiments on sugar uptake by *Scenedesmus quadricauda* (Taylor 1960) showed uptake to be ATP dependent and Tanner and co-workers have confirmed that constitutive and inducible active glucose transport occurs in various Chlorococcales (Tanner 1969, Tanner & Kandler 1967, Tanner *et al.* 1970). The utilization of anions at high pH is also likely to be an active process, for example, the phototrophic utilization of acetate by the marine *Brachiomonas submarina* at pH 9 (Droop & McGill 1966).

The relative impermeability of acetate organisms may indeed fit them for the highly polluted situation favoured by many of them. However, the permeability hypothesis does not explain why many obligate phototrophs cannot utilize fatty acids. If toxicity is mainly a question of permeability one should find a pH range in which the acid does not penetrate so fast as to be harmful and yet penetrates fast enough for the requirements of growth. Possibly the use of acetate by *Haematococcus pluvialis* or *Brachiomonas submarina*, which can be grown only with difficulty below pH 7, are instances of this, but for other organisms, for example *Nannochloris oculata*, fatty acids are toxic, even at pH 8, and therefore presumably penetrate, but cannot be used for growth. Erickson *et al.* (1955) found that in contrast to the toxicity of undissociated acetic acid for *Chlorella*, the synthesis of carbohydrate reserves in this organism depends on the concentration of the anion in the medium, which again points to an uptake mechanism obeying the laws of enzyme kinetics. Wise (1961) reached the same conclusion studying uptake by *Polytomella caeca*. Similarly Abraham and Bachhawat (1962) inferred that an acetate kinase system was active in acetate assimilation by *Euglena gracilis*, while a succinate transport system in this species was postulated by Levedahl (1965) on the evidence of specific non-penetrating inhibitors. It is likely that in most cases where there is utilization of an organic source of carbon it will be possible to demonstrate an active, or at least mediated, transport mechanism at work. Without active transport the substrate concentration would need to be very high, as, for instance, in glycerol utilization during dark growth of *Prymnesium parvum*.

The intolerance which most algae show to even a moderate excess of hydrogen ions cannot always be due to the penetration of lower fatty acids, for it also occurs in exclusively mineral media and sometimes is due to heavy metal toxicity. The concentration of free heavy metal ions tends to be higher the lower the pH. On the other hand the metal binding capability of many organic anions, especially the so-called chelating agents, can greatly reduce the heavy metal effect, though to lower the pH is still to increase the relative free metal concentration.

Acid tolerance in acetate organisms is as likely to be a question of the ability to dispose of an internal excess of hydrogen ions as one of impermeability, for the utilization of fatty acids at low pH implies that entry of the acid is fast enough for the requirements of growth, while on the other hand, the metabolizing of an undissociated acid involves an accretion of hydrogen ions to the cell sap. It seems that acetate organisms are not necessarily less permeable than other organisms. Votta *et al.* (1971), for example, who measured internal pH during succinic acid utilization by *Euglena*, found no lowering of the internal pH. On the contrary the internal pH rose as the medium became more alkaline. These authors also showed growth to be dependent on the undissociated acid.

## 5  THE GLUCOSE BLOCK

The probable involvement of permease systems in sugar and anion uptake suggests that the so-called glucose block is merely a question of specificity.

The pH effect on glucose utilization observed by Cramer and Myers (1952) with *Euglena gracilis* var. *bacillaris* could be due to the requirements of such an enzyme. Furthermore, the involvement of an enzyme system, though not necessarily a permease, is indicated by the reported need for a training period when *E. gracilis* is first transferred to a glucose medium. On the other hand, the demand associated with glucose utilization by this organism for a high carbon dioxide concentration (Cramer & Myers 1952) or meat extract (Pringsheim 1955) or various Krebs' cycle acids (Baker *et al.* 1955) is warning against too facile an explanation.

An alternative hypothesis for the glucose block, namely that acetate flagellates lack sugar phosphorylating enzymes, notably hexokinase, was proposed by Lwoff, who demonstrated the presence of a starch phosphorylase and glucose-1-phosphate in *Polytomella caeca*. This alga stores starch but is unable to utilize glucose and the phosphorylating enzymes for glucose, hexokinase and phosphoglucomutase are absent (Lwoff *et al.* 1949, 1950). The absence of either enzyme could explain the inability to utilize glucose, while possessing a glycogen-based economy. Hexosephosphates should support growth if they penetrate. Hexokinases occur in other algae, for instance *Euglena gracilis* (Belskey & Schulz 1962) where, as in *P. caeca*, they are probably located in a mitochondrial fraction, and in *Astasia longa*, which does not utilize glucose although a mutant does so (Barry 1962). Chapman *et al.* (1965) also found evidence of glucokinase in cell-free extracts of *Polytoma obtusum*. These are clear indications of a permeability barrier, since whole cells of these species do not utilize glucose.

The glucose block, of course, may not have the same cause in every case, or, it may be (see Darnforth 1962) that the kinase and permeases are more intimately linked than their different locations seem to imply. Indeed Taylor (1960) suggested that a hexokinase system was situated on the outside of the plasma membrane.

## 6  GROWTH KINETICS

Table 19.6 lists specific growth rates of some algae under heterotrophic and autotrophic conditions. It establishes that the maxima for phototrophic growth, presumably when light intensities are near saturating, are usually considerably higher than the chemotrophic maxima. Bristol Roach (1928) found with *Scenedesmus costulatus* that at high light intensities the heterotrophic contribution to the total was negligible, whereas at low light intensities the heterotrophic and autotrophic components were additive, being 0·35, 0·48 and 0·87, respectively, in the dark with glucose and in the light with and without glucose.

Assuming the Monod model for growth of micro-organisms (Monod 1942):

$$\mu = \mu_{max}\left(\frac{s}{K_s + s}\right),$$

in which $s$ is the substrate concentration, $\mu$ the specific growth rate and $K_s$ the saturation constant, $K_s$, is a good measure of the net demand on the environment with respect to any substrate, while $\mu_{max}$ measures the organism's potential in the circumstance. The concentrations of organic substrate are shown in parenthesis in Table 19.6; they range between 0·6 and 10 g $1^{-1}$ and are probably saturating. The growth rates indicated can therefore be taken as equivalent to $\mu_{max}$ of the Monod equation. Monod (*loc. cit.*) records growth saturation

**Table 19.6.** Comparison of some specific growth rates recorded for autotrophic and heterotrophic growth, expressed in natural units per day, i.e., $\log_e 2 \div$ doubling time in days.

| Species | Illumination | Temperature °C | Substrate | Specific growth rate | Reference |
|---|---|---|---|---|---|
| Bumilleriopsis brevis | Light<br>Dark<br>Dark | 20–<br>25 | $CO_2$<br>Acetate (0·06%)<br>Glucose (0·2%) | 2·03<br>0·51<br>0·35 | Belcher & Miller 1960* |
| Chlamydomonas reinhardtii | Light<br>Dark | 25 | $CO_2$, acetate (0·2%)<br>Acetate (0·2%) | 2·37<br>0·92 | Sager & Granick 1953 |
| Chlorella pyrenoidosa | Light<br>Dark<br>Dark | 25 | $CO_2$<br>Acetate<br>Glucose | 1·96<br>0·40<br>0·92 | J.N. Phillips (in Myers 1953) |
| Chlorella vulgaris | Light<br>Dark | 23 | $CO_2$, glucose (1·0%)<br>Glucose (1·0%) | 1·67<br>1·10 | Pearsall & Bengry 1940<br>Pearsall & Loose 1937 |
| Chlorogloea fritschii | Light<br>Dark | 30 | $CO_2$<br>Sucrose (0·36%) | 0·13<br>0·07 | Fay 1965* |
| Cylindrotheca fusiformis | Light<br>Dark | 20 | $CO_2$<br>Lactate (0·5%) | 1·32<br>0·50 | Lewin & Hellebust 1970 |
| Cyclotella sp | Light<br>Dark | 20 | $CO_2$<br>Glucose (9·5%) | 0·92<br>0·55 | Lewin 1963 |
| Euglena gracilis (Vischer) | Light<br>Dark<br>Dark | 25 | $CO_2$<br>Acetate (0·5%)<br>Butyrate (0·5%) | 1·40<br>0·58<br>0·58 | Cramer & Myers 1952 |
| Euglena gracilis var bacillaris | Light<br>Dark<br>Dark | 25 | $CO_2$<br>Acetate (0·5%)<br>Butyrate (0·5%) | 0·97<br>0·92<br>0·97 | Cramer & Myers 1952 |
| Haematococcus pluvialis | Light<br>Dark | | $CO_2$, acetate (0·05%)<br>Acetate (0·05%) | 0·89<br>0·34 | Droop 1955 a, b* |
| Nitzschia closterium | Light<br>Dark | 20 | $CO_2$<br>Lactate (0·2%) | 1·38<br>0·21 | Lewin 1963 |
| Nitzschia marginata | Light<br>Dark | 20 | $CO_2$<br>Glucose (0·5%) | 1·38<br>1·38 | Lewin 1963 |
| _Scenedesmus obliquus_ | Light | 24·5 | $CO_2$, glucose (1·0%) | 1·08 | Bristol Roach 1928 |

| | | | | | |
|---|---|---|---|---|---|
| Astasia longa | 25 | Dark<br>Dark | Ethanol (0·03–0·6%)<br>Acetate (0·03–0·06%) | 1·66<br>1·66 | Buetow & Padilla 1963 |
| Astasia longa (mutant) | 30 | Dark<br>Dark | Pyruvate (0·2%)<br>Glucose (0·2%) | 0·97<br>0·36 | Barry 1962 |
| Chilomonas paramecium | 25 | Dark<br>Dark | Acetate (0·2%)<br>Ethanol (0·2%)<br>n-Butyrate (0·26%) | 1·61<br>1·62<br>1·40 | Cosgrove & Swanson 1952 |
| Gyrodinium cohnii | 27 | Dark<br>Dark | Glucose (0·6%)<br>Glucose (0·6%),<br>Acetate (0·2%) | 1·13<br>1·36 | Ishida & Katoda 1965 |
| Oxyrrhis marina | 22 | Dark<br>Dark | Live yeast<br>Acetate (0·2%) | 1·38<br>0·53 | Droop 1959* |
| Anabaena variabilis | 30 | Light<br>Light | $CO_2$<br>$CO_2$, acetate (0·03%) | 2·09<br>2·51 | Hoare et al. 1967 |
| Anacystis nidulans | 30 | Light<br>Light | $CO_2$<br>$CO_2$, acetate (0·03%) | 2·09<br>2·44 | Hoare et al. 1967 |
| Brachiomonas submarina | | Light<br>Light | $CO_2$<br>$CO_2$, acetate (0·1%) | 0·46<br>0·92 | Droop & McGill 1966 |
| Brachiomonas submarina var pulsifera | | Light<br>Light | $CO_2$<br>$CO_2$, acetate (0·1%) | 0·57<br>0·72 | Droop & McGill 1966 |
| Chlamydomonas mundana | 29 | Light<br>Light | $CO_2$<br>$CO_2$, acetate (0·34%) | 2·01<br>3·30 | Eppley & MaciasR 1962 |
| Chlamydomonas mundana var astigmata | 32 | Light<br>Light | $CO_2$<br>$CO_2$, acetate (0·34%) | 2·00<br>5·50 | Eppley & MaciasR 1962 |
| Chlamydomonas pulsatilla (Finnish strain) | | Light<br>Light | $CO_2$<br>$CO_2$, acetate (0·1%) | 0·27<br>1·21 | Droop & McGill 1966 |
| Chlamydomonas pulsatilla (Ailsa Craig strain) | | Light<br>Light | $CO_2$<br>$CO_2$, acetate (0·1%) | 0·29<br>0·37 | Droop & McGill 1966 |
| Chlorogloea fritschii | 30 | Light<br>Light | $CO_2$<br>$CO_2$, sucrose (0·36%) | 0·13<br>0·19 | Fay 1965* |
| Nostoc muscorum | 30 | Light<br>Light | $CO_2$<br>$CO_2$, acetate (0·03%) | 1·61<br>1·29 | Hoare et al. 1967 |

\* Computed from the published growth curves.

constants between 2 and 20 mg $1^{-1}$ for bacteria on glucose, mannitol, glycerol and lactose. The equivalent constants for algae may be somewhat higher. Growth curves of *Navicula pelliculosa* suggest a saturation constant of 75 mg $1^{-1}$ for glucose (Lewin 1953) and those of *Cylindrotheca fusiformis* a constant of 57 mg $1^{-1}$ for lactate (Lewin & Hellebust 1970). The rate/concentration curve for acetate and encystment in *Haematococcus pluvialis* in the dark gives a saturation constant of 190 mg $1^{-1}$ (Droop 1955a). Low permeability of *Prymnesium* to glycerol is reflected in a saturation constant of some 30 g $1^{-1}$ for this substrate (Rahat & Jahn 1965).

## 7  OXIDATIVE ASSIMILATION

Energy required for chemotrophic assimilation must be supplied by oxidation of part of the substrate. The processes involved conform in the main to the general pattern found in other heterotrophic microorganisms, but algae are, in general, obligate aerobes: they appear to be dependent on aerobic pathways for the terminal oxidations, possibly on account of their lack of dehydrogenases for lactic acid and ethanol and consequent inability to re-oxidise $NADH_2$ anaerobically (Jacobi 1957, Richter 1957, Gibbs 1962).

The oxidative assimilation of glucose presents no difficulty, in principle, since phosphorylation results in compounds directly available for storage, cell syntheses, and respiration according to the needs of the cell. Whereas the equivalent of a single phosphate bond is required per mole of glucose assimilated to hexosephosphate, some 30 are generated by the aerobic oxidation of a mole of glucose. In other words, only three per cent of the substrate is theoretically needed to provide for the chemical energy change when the substrate assimilated is glucose. The assimilation of acetate and other fatty acids, on the other hand, is uncertain although solved in principle by Kornberg (see Kornberg & Elsden 1961). Acetylization of coenzyme A would be the obvious starting point for acetate assimilation were it not for the fact that the tricarboxylic acid cycle only provides for respiration since it oxidises acetate completely.

The observation (Anderson 1945) that withholding thiamine from *Prototheca zopfii* represses respiration of glucose but not acetate has been quoted as evidence that pyruvate decarboxylations are not involved in acetate assimilation, thiamine pyrophosphate being the coenzyme of pyruvic acid carboxylase (Griffiths 1965). Strictly, however, it only shows that *respiration* of acetate does not involve this particular decarboxylation, as indeed it would not if the terminal mechanism is the tricarboxylic acid cycle. The carboxylation of acetate to pyruvate does not take place (Bidwell & Ghosh 1963), whereas the existence of acetate flagellates with obligate requirements for thiamine, e.g. *Oxyrrhis marina* (Droop 1959), indicates that pyruvate decarboxylation is obligatory, presumably for the terminal oxidation of endogenous reserves, whether the external carbon source is glucose or acetate.

The mechanism for fatty acid assimilation proposed by Kornberg is the so-called glyoxylate shunt, the net effect of which is a yield of a molecule of succinate from two of active acetate, the succinate being available for cell syntheses. The key enzymes associated with reactions in the glyoxylate cycle, isocitratase, catalysing the breakdown of isocitrate to glyoxylate and succinate, and malate synthetase, catalysing the synthesis of malate from glyoxylate and active acetate, have been identified in a number of algae (Haigh & Beevers 1962, Reeves *et al.* 1962, Callely & Lloyd 1964, Syrett *et al.* 1963, Wiessner & Kuhl 1962, Cook & Carver 1966, Goulding & Merrett 1967). Goulding and Merrett (1967) showed further that the glyoxylate cycle, but not some tricarboxylic acid cycle, enzymes were induced in *Chlorella* on acetate in the dark. Light and glucose both suppress the formation of isocitratase and the enzymes induced are sufficiently active to account for the rates of dark growth on acetate. Tracer studies by Syrett *et al.* (1964) and Haigh and Beevers (1962) with *Polytomella caeca* show that when acetate is being assimilated most of the malate recovered could not have been derived from succinate in the tricarboxylic acid cycle, which argues strongly for the operation of the glyoxylate shunt. On the other hand the reported lack of isocitrate lyase in certain *Euglena* strains during acetate assimilation (Ross & Jahn 1971) and its suppression in *Chlorella* (Syrett 1966) is a reminder of the possibility of alternative routes for fatty acid assimilation.

**Table 19.7.** Excess high energy bonds per mole of hexosephosphate synthesized, assuming one mole in four of substrate respired in the process of assimilation.

| Substrate | Excess high energy bonds |
|---|---|
| Glucose | 8 |
| Acetate | 8 |
| Pyruvate | 2 |
| Malate | 2 |
| Succinate | 2 |
| Oxaloacetate | −4 |

In the synthesis of hexosephosphate from succinate, oxaloacetate is decarboxylated and phosphorylated, two moles of succinate going to make one of hexosephosphate. Thus, in assimilation by the glyoxylate cycle four moles of acetate are required per mole of hexosephosphate synthesized, one of the four being respired in the process. This is not quite as inefficient as at first appears, because about half the energy is available for other purposes. Table 19.7 shows the number of excess high energy bonds per mole of hexosephosphate synthesized, assuming for argument the respiration of one mole in four of substrate. Seen in this light acetate is comparable to glucose and superior to the other substrates along the assimilatory pathway. On the other hand, one should only *need* to respire one in 30 moles when the substrate assimilated is glucose and two in seven when it is pyruvate, against one in four for the 2- and 4-carbon compounds. In practice of course very much more substrate is respired during growth than these calculations suggest due to such other energy demands on the cells as nitrogen assimilation and transport (see Griffiths 1965). In his review Payne (1970) quoted yield values for bacteria growing aerobically on glucose of 72 to 90g dry wt per mole of substrate, which gives an economic coefficient (moles incorporated per mole taken up) of 0·4 to 0·5 whereas when the substrate is acetate the coefficient works out at only 0·3 to 0·4.

An idea of the overall efficiency, including the assimilatory process can be obtained from the ratio of the heat of combustion of the cells to that of the substrate producing them, or alternatively from the ratio of $CO_2$ excreted in respiration to substrate utilized. Such measurements tend to confirm the superiority of glucose over acetate. For example, Samejima and Myers (1958) found that with cells of *Chlorella pyrenoidosa* growing on ammonia the combustion ratio is 0·58 on glucose and 0·17 on acetate. Essentially the same conclusions were reached regarding glucose from calculations based on $CO_2$ excretion (Cramer & Myers 1948). On the other hand, with non-growing cells of *Chilomonas*, when little synthesis of cell materials other than carbohydrate could be expected, combustion ratios of 0·8 and 0·3 respectively have been recorded (Hutchins *et al.* 1948, Blum *et al.* 1951). Not surprisingly *Chlorella pyrenoidosa* and *C. ellipsoidea*, grow faster in the dark on glucose than on acetate (Samejima & Myers 1958). Cramer and Myers (1948) noted that the respiratory quotient in *Chlorella* is sensitive to the degree of reduction of the nitrogen source and this has been amply confirmed since. It is obviously not possible to study the energetics of carbon assimilation in isolation in a growing cell.

When the major source of energy is light and oxidative energy is not needed for assimilation the respiratory level can be much lower. Expressing Samejima and Myers' figure quoted above for *Chlorella* in the dark on glucose and ammonia as moles respired per mole incorporated we get 0·72, which contrasts with one of 0·3 quoted for autotrophic cells (Raven 1971). Gaffron as early as 1939 observed respiratory quotients of between 1·2 and 2·0 for *Chlorella* and *Scenedesmus* grown in the light on glucose, compared with 1·0 for autotrophically grown cells.

## 8 PHOTOTROPHIC ASSIMILATION

Carrying out the cyclic process of photosynthesis should present no difficulty in algae and photoassimilation or organic carbon should always be possible providing permeability allows and the glyoxylate shunt operates. This though is being wise after the event; the requirement for acetate on the part of certain obligate phototrophs was certainly puzzling to the uninitiated (Droop 1961, Darnforth 1962). Moreover, it is only in the last decade that it has been recognized that this type of photosynthesis is by no means uncommon in algae (Table 19.8). It is most easily observed in species that neither assimilate organic carbon oxidatively nor $CO_2$ phototrophically, *Chlamydobotrys* (*Pyrobotrys*) of Pringsheim and Wiessner (1960) being the classical example. This colonial green alga had a near obligate requirement for acetate and yet would not grow in the dark. A similar state of affairs exists in the *pulsatilla* group of *Chlamydomonas* and, though not quite so extreme, in *Brachiomonas submarina* (Droop 1961, Droop & McGill 1966). Pringsheim & Wiessner, however, showed the light dependent acetate uptake to be unaccompanied by oxygen evolution and $CO_2$ uptake. The similarity to bacterial photosynthesis is apparent, but the fact that the quantum efficiency of acetate photoassimilation in *Chlamydobotrys* is not only maintained as in bacteria but increases as wavelengths move into the far red shows there to be a significant distinction (Wiessner 1965, 1966).

The inability to utilize $CO_2$ is not complete in any of the algae studied, but varies from one to another. In *Chlamydobotrys* the aldolase activity associated with photoassimilation of $CO_2$ is adaptive upon withdrawal of acetate (Wiessner 1963). Wiessner (1969) has also shown that the change from photoheterotrophic to autotrophic growth is accompanied by characteristic changes in the chlorophyll absorption spectra.

Wiessner and Gaffron (1964) found that under anaerobic conditions, when no oxidative energy is available, the cells became dependent on photosystem II for reductant and no acetate uptake occurred in the presence of DCMU (3' (3,4-dichlorophenyl), 1', 1' dimethyl urea), which inhibits photosystem II. DCMU was without effect when $O_2$ was present. The requirement for photosystem II suggests that assimilation of acetate is $O_2$ dependent (presumably to effect the re-oxidation of $NADH_2$) irrespective of whether the ATP is generated photosynthetically or oxidatively. However, Wiessner and Gaffron believe the anaerobic requirement in *Chlamydobotrys* to be specifically for $CO_2$ *per se*; they suppose reoxidation of $NADH_2$ depends on the oxidative power of $CO_2$, which of necessity must act via the photosynthetically generated $NADPH_2$. The parallel between oxidative and photosynthetic assimilation of acetate is emphasized by the fact that in several species, e.g. *Chlamydobotrys stellata* (Wiessner & Kuhl 1962), *Chlamydomonas mundana* (Eppley, *et al.* 1963) and *Chlorella* (Syrett 1966), photoassimilation is associated with high activity of the glyoxylate shunt enzyme, isocitrate lyase. The converse effect, namely the suppression of the enzymes of the carbon reduction cycle, has been observed during photoheterotrophic growth of *Chlamydomonas mundana*, *Euglena gracilis*, *Chlorella pyrenoidosa* and *Tolypothrix tenuis* (Russell & Gibbs 1966, Huth 1967, Latzko & Gibbs 1969).

Photoassimilation by *Chlorella pyrenoidosa*, *Euglena gracilis* and *Anacystis nidulans* appears to differ from that found in *Chlamydobotrys* and *Chlamydomonas*, and, surprisingly, to be dependent on non-cyclic phosphorylation (Goulding & Merrett 1966, Cook 1967, Hoare *et al.* 1967). Several things suggested this. First, the inhibitor desaspidine, which specifically inhibits cyclic

**Table 19.8.** Some algae able to photoassimilate organic compounds.

| Species | Code | Species | Code | Code |
|---|---|---|---|---|
| Anabaena type 3 strains | G*36 | Gonium sacculiferum | G10 | G10 |
| Anabaena flos-aquae | G1 | Lyngbya type 4 strains | G36 | A11 |
| A. variabilis | A1,2,24 G3 | Microchaete type 2 strains | G36 | A10 |
| Anacystis nidulans | A*4,23,24 | Navicula pelliculosa | glycerol16 | A10 |
| Ankistrodesmus braunii | G32 | Nitella translucens | G12 | A30 |
| Brachiomonas submarina | A5,6 | Nostoc muscorum | A1,2 | A27 G14,31 |
| Calothrix type 2 strains | G36 | Ochromonas malhamensis | G17 | G28 |
| Chlamydobotrys stellata | A7 | O. danica | G18 | G13 |
| Chlorella spp. | A8 | Plectonema type 2 strains | G36 | G36 |
| Chlamydomonas mundana | A22 | Scenedesmus acuminatus | G19 | A29 |
| C. dysosmos | A5,6 | S. acutiformis | G19 | A21 |
| C. pulsatilla (7 strains) | A9 | S. obliquus | G14 | A25,26,34 |
| C. reinhardtii | G10 | S. quadricauda | G13 | A15 |
| C. dorsocentralis | G10 | Tolypothrix tenuis | G20 | A35 |
| C. humicola | G10 | | | |

* G, glucose; A, acetate. References: 1, Hoare et al. 1967; 2, Allison et al. 1937; 3, Pearce & Carr 1969; 4, Hoare & Moore 1965; 5, Droop 1961; 6, Droop & McGill 1966; 7, Pringsheim & Wiessner 1960; 8, Eppley et al. 1963; 9, Stross 1960; 10, Lucksch 1933; 11, Pimenova & Kondratieva 1965; 12, Smith 1967; 13, Taylor 1960; 14, Dvorakova-Hladka 1966; 15, Belcher 1967; 16, Lewin J.C. 1953; 17, Vishniac & Reazin 1957; 18, Aaronson & Baker 1959; 19, Algeus 1946; 20, Cheung & Gibbs 1966; 21, Lynn & Starr 1970; 22, Lewin R.A. 1954; 23, Van Baalen 1965; 24, Pearce & Carr 1967a; 25, Cook 1967; 26, Fischer & Wiessner 1968; 27, Goulding & Merrett 1966; 28, Kandler 1954; 29, Wiessner 1968; 30, Oaks 1962; 31, Myers 1947; 32, Bishop 1961; 33, Fogg & Belcher 1961; 34, Lynch & Calvin 1953; 35, McLachlin & Craigie 1965; 36, Kenyon et al. 1972.

photophosphorylation, had no effect on assimilation of acetate by *Chlorella*. Secondly, although light increased the incorporation of $^{14}C$ acetate some three-fold over the dark rate and, as in *Chlamydobotrys*, $CO_2$ was not necessary in aerobic conditions, the addition of DCMU switched the metabolism to an oxidative pathway, as shown by the pattern of labelled products. In *Anacystis*, which is an obligate phototroph, the effect of DCMU was to suppress acetate assimilation. A preponderance of fats, which is to be expected in the products of non-cyclic photoassimilation because of the additional reducing power generated, has been observed in *C. pyrenoidosa* photosynthesizing acetate (Cook 1967, see also Wiessner 1970). It also appears that whereas in *Chlamydobotrys* glyoxylate is one of the first compounds to become labelled when $^{14}C$ acetate is photo-assimilated, in *Chlorella* the labelling is found in glycolic acid and succinic acid (Goulding & Merrett 1967, Merrett & Goulding 1967).

It is, however, possible that acetate is metabolized in the light by *Chlorella* by pathways other than the glyoxylate cycle, for Syrett (1966) showed that isocitrate lyase formation in this alga could be suppressed by a combination of light and $CO_2$ but, significantly, the suppression was inhibited by DCMU. On the other hand, it was not stated whether acetate continued to be taken up during the suppression. In this connexion, Wiessner (1968) suggested that acetate carbon might be assimilated indirectly by *Chlorella*, that is by the respiration of the acetate and photoassimilation of the resultant carbon dioxide. Goulding and Merrett (1966), however, had previously dismissed this idea on the grounds that the pattern of compounds labelled when $^{14}C$ bicarbonate was assimilated in the presence of unlabelled acetate differed from that obtained with $^{14}C$ acetate assimilation.

Photoassimilation of glucose and various other organic compounds by a number of algae has been reported. It appears that in this cyclic photophos-phorylation is the relevant ATP generator for glucose as for acetate assimilation. The reader is referred to Wiessner (1970) for a detailed account of the photo-metabolism of organic substrates by microorganisms.

# 9  OBLIGATE PHOTOTROPHY

There seems not to be any general biochemical reason why algae having a glycolytic based economy should not metabolize organic carbon in the dark. The truth is likely to be various and sometimes trivial. Two obvious explanations, namely impermeability and acid toxicity, have already been discussed. They probably account satisfactorily for obligate phototrophy in many algae especially in those, like the phytoplankton, accustomed to oligotrophic conditions.

Instances are known of organic acids being respired while not being able to support dark growth; acetate, succinate, pyruvate in *Chlamydomonas moewusii* (Lewin 1950), and acetate, succinate, propionate, fumarate in *Monodus* (Miller & Fogg 1958). Lewin (1954) reasoned that to be able to respire an acid without being able to assimilate it implies a lack of coupling between terminal oxidation of the substrate and phosphorylation. Lewin was studying an obligately phototrophic mutant of *C. dysosmos* which respired acetate, whereas the wild type also assimilated the acid, except in the presence of dinitrophenol when it behaved just like the mutant. The occurrence of photophosphorylation makes Lewin's hypothesis entirely feasible.

It is true that in the absence of a glyoxylate pathway, or the equivalent, fatty acids would be expected to be respired completely by the TCA cycle without being assimilated. This hypothesis, however, like the permeability hypothesis, cannot account for the absence of dark growth in some of the algae discussed in the previous section which assimilate acetate photosynthetically, whereas Lewin's hypothesis can. Even though organic acids are being assimilated by the same pathway as in oxidative assimilation, the reducing power, $NADH_2$, being oxidative in origin in cyclic photosynthesis, uncoupling of the phosphorylations here would leave photosynthesis the sole source of ATP.

A dearth of ATP could affect any energy requiring process, including those involving active transport. Thus one might have a situation where membranes were permeable in the light but impermeable in the dark. A certain amount of evidence points to there being gradations in the dependence on light; in other words, competition for ATP or reductant as the case may be, can be more or less severe in the dark. For example, some algae fail to reduce nitrate in the dark though they do so perfectly well in the light, the reduction of $NO_3$ requiring a comparatively large amount of energy. Thus, Lwoff and Lwoff (1929) reported that their strain of *Haematococcus pluvialis* required ammonia for dark growth whereas most strains of this species, including that of Lucksch (1933), can be cultivated in complete darkness with $NO_3$ as nitrogen source. *Brachiomonas submarina* is another alga seeming to require ammonia only for dark growth. The inability to reduce $NO_3$ is common among apochlorotic algae (see Pringsheim 1963) and, interestingly, among algae dependent on cyclic photophosphorylation (e.g. all strains of the *Chlamydomonas pulsatilla* group (Droop 1961)).

The absence of a complete Krebs' cycle, as has been reported for blue-green algae, could account for a failure to assimilate fatty acids oxidatively and could therefore be a cause of obligate phototrophy if coupled with an inability to assimilate carbohydrate (Smith *et al.* 1967, Pearce & Carr 1967b, Pearce *et al.* 1969, Hoare *et al.* 1967, 1969, Stanier *et al.* 1971). It is perhaps significant that while acetate can be photoassimilated by a number of blue-green algae dark growth among them is recorded only in the three species also able to photoassimilate glucose. Cheung and Gibbs (1966), Pearce and Carr (1969) and Pelroy *et al.* (1972) believe that the main pathway of glucose dissimilation is the pentose phosphate cycle, at the expense of which growth in the dark presumably takes place.

## 10 APOCHLOROSIS

That very many colourless algae are distinguished from coloured forms solely by the absence of chromatophores and chlorophyll suggests that the loss of the photosynthetic apparatus is a common hazard that can sometimes be survived. Pringsheim (1963) lists a great many; here we need only mention the more familiar (with the coloured counterparts in parenthesis): *Prototheca* (*Chlorella*), *Polytoma* (*Chlamydomonas*), *Polytomella* (*Dunaliella*), *Astasia longa* (*Euglena gracilis*) and the colourless species of *Nitzschia* and *Gyrodinium*.

The spontaneous and permanent disappearance of chromatophores has been observed on several occasions in *Euglena*, though it has never been clear how it came about (Pringsheim 1948), whether by the failure of the chromatophores, which are autonomous, to reproduce with the cells, or by destruction of the chromatophores, or by their failure simply to manufacture chlorophyll. Only in the last instance will the cells be left with the colourless remnants of the chromatophore, i.e. the leucoplasts. Colourless algae are found both with (Lang & Cook 1962, Menke & Fricke 1962), and without (Dodge & Crawford 1971), leucoplasts. Leucoplasts are not found in experimentally produced apochlorotic strains.

Light *per se* may be a factor with such algae as require light for chlorophyll maintenance, e.g. *Euglena gracilis*, *Ochromonas malhamensis*, and *Brachiomonas submarina*. The chromatophores of dark grown *E. gracilis* degenerate into colourless vesicles, the so-called 'proplastids', which continue to reproduce with the cells and which on restoration of light redevelop into normal chloroplasts (Epstein & Schiff 1961). It has been suggested that it is damage to the proplastids that is the cause of permanent loss of chlorophyll. According to Provasoli *et al.* (1951) the effect of chlorophyll bleaching agents like streptomycin or heat is to cause the chloroplasts to degenerate into proplastid-like forms, which may or may not be destroyed according to the severity of the treatment. De Deken-Grenson and Messin (1958) and Schiff and Epstein (1965) confirm that it is the disorganization of the chloroplast *per se* (in its ability to reproduce) rather than the mere destruction of chlorophyll synthesis, though the latter certainly occurs.

However, by no means all facultative heterotrophs bleach when grown in the dark. For example Bristol Roach (1926) observed cultures of *Scenedesmus costulatus* to green on glucose in the dark, while according to Droop (1955b) *Haematococcus pluvialis* keeps its colour indefinitely growing in the dark on acetate. Furthermore, the writer has observed that the dark death of such obligate phototrophs as *Chlamydomonas pulsatilla* or *Balticola droebakensis*, or indeed *H. pluvialis* in the absence of substrate, is not accompanied by loss of chlorophyll. It is perhaps significant that permanent loss of chlorophyll has only been observed to occur in a species which normally bleaches reversibly when grown in the dark. It would also be interesting to know how generally the ability to reduce $NO_3$ in the dark is correlated with that of dark synthesis of chlorophyll and *vice versa*. The five facultative heterotrophs mentioned in this and the previous paragraph show this correlation, but *Tolypothrix*, which keeps its colour will not reduce nitrate in the dark (Kiyohara *et al.* 1960).

It is well known that shortage of nitrogen usually results in greatly diminished cell chlorophyll content (e.g. Hase *et al.* 1957, Fogg 1959). More recent work with *Chlorella prototheecoides* (Matsuka *et al.* 1969a,b) has shown that bleaching during nitrogen starvation is greatly enhanced when organic carbon, either acetate or glucose, is assimilated. Although these authors believe that the main effect is due in some way to the enhanced lipid formation under the circumstances, whatever the carbon source, Ochiai and Hase (1970) showed that glucose specifically inhibits both the synthesis of the precursor δ-aminolevulinic acid and that of chlorophyll from it. According to Ochiai and Hase these two steps are both light-dependent, presumably indirectly as a source of energy or reducing power, since otherwise one would need to postulate an entirely different mechanism for algae that maintain their chlorophyll during prolonged dark growth.

## 11 REFERENCES

AARONSON S. & BAKER H. (1959) A comparative biochemical study of two species of *Ochromonas*. *J. Protozool.* **6**, 282–4.

ABRAHAM A. & BACHHAWAT B.K. (1962) Studies on acetocoenzyme A kinase from *Euglena gracilis*. *Biochim. biophys. Acta*, **62**, 376–84.

ALDRICH D.V. (1962) Photoautotrophy in *Gymnodinium breve* Davis. *Science, N.Y.* **137**, 988–90.

ALGEUS S. (1946) Untersuchungen über die Ernhährungsphysiologie der Chlorophyceen. *Bot. Notiser* 1946, 129–278.

ALLEN M.B. (1952) The cultivation of Myxophyceae. *Arch. Mikrobiol.* **17**, 34–53.

ALLEN J.R., LEE J.J., HUTNER S.H. & STORM J. (1966) Prolonged culture of the voracious flagellate *Peranema* in antioxidant-containing media. *J. Protozool.* **13**, 103–8.

ALLISON F.E.; HOOVER F.R. & MORRIS H.J. (1937) Physiological studies with the nitrogen fixing blue-green alga *Nostoc muscorum*. *Bot. Gaz.* **98**, 433–63.

ANDERSON E.H. (1945) Nature of the growth factor for the colorless alga *Prototheca zopfii*. *J. gen. Physiol.* **28**, 287–327.

ANTIA N.J., CHENG J.Y. & TAYLOR F.J.R. (1969) The heterotrophic growth of a marine photosynthetic cryptomonad (*Chroomonas salina*). *Proc. int. Seaweed Symp.* **6**, 17–29.

BAKER H., HUTNER S.H. & SOBOTKA H. (1955) Nutritional factors in thermophily, a comparative study of bacilli and *Euglena*. *Ann. N.Y. Acad. Sci.* **62**, 349–76.

BARKER H.A. (1935) The metabolism of the colorless alga. *Prototheca zopfii. J. cell comp. Physiol.* **7**, 73–93.

BARKER S.A. & BOURNE E.J. (1955) Composition and synthesis of the starch of *Polytomella caeca*. In *Biochemistry and Physiology of Protozoa*, vol. 2, eds. Hutner S.H. & Lwoff A. pp. 45–56. Academic Press, New York & London.

BARRY S.-N.C. (1962) Utilization of glucose by *Astasia longa. J. Protozool.* **9**, 395–400.

BEIJERINCK M.W. (1890) Culturversuche mit Zoochlorellen, Lichenengonidien und anderen niederen Algen. *Bot. Ztg.* **48**, 727–39, 743–54, 759–86.

BELCHER J.H. (1967) Notes on the carbon and nitrogen requirements of *Furcilla stigmatophora. Arch. Mikrobiol.* **58**, 181–5.

BELCHER J.H. & MILLER J.D.A. (1960) Studies on the growth of Xanthophyceae in pure culture. IV. Nutritional types amongst the Xanthophyceae. *Arch. Mikrobiol.* **36**, 219–28.

BELSKY M.M. & SCHULTZ J. (1962) Partial characterization of hexokinase from *Euglena gracilis. J. Protozool.* **9**, 195–200.

BERNARD F. (1963) Density of flagellates and Myxophyceae in the heterotrophic layers related to environment. In *Symposium on Marine Microbiology*, ed. Oppenheimer C. pp. 215–28. Charles C. Thomas, Springfield, Illinois.

BIDWELL R.G.S. & GHOSH N.R. (1963) Photosynthesis and metabolism of marine algae. V. Respiration and metabolism of $^{14}C$ labelled glucose and organic acids supplied to *Fucus vesiculosus. Can. J. Bot.* **41**, 155–63.

BISHOP N.I. (1961) The photometabolism of glucose by a hydrogen-adapted alga. *Biochim. biophys. Acta*, **51**, 323–32.

BLUM J.J., PODOLSKY B. & HUTCHINS J.O. (1951) Heat production in *Chilomonas. J. cell. comp. physiol.* **37**, 403–26.

BRISTOL ROACH B.M. (1926) On the relation of certain soil algae to some soluble organic compounds. *Ann. Bot.* **40**, 149–201.

BRISTOL ROACH B.M. (1927) On the carbon nutrition of some algae isolated from soil. *Ann. Bot.* **41**, 509–17.

BRISTOL ROACH B.M. (1928) On the influence of light and of glucose on the growth of a soil alga. *Ann. Bot.* **42**, 317–45.

BUETOW D.E. & PADILLA G.M. (1963) Growth of *Astasia longa* in ethanol. I. Effects of ethanol on generation time, population density and biochemical profile. *J. Protozool.* **10**, 121–3.

CALLELY, A.G. & LLOYD D. (1964) The metabolism of acetate in the colourless alga *Prototheca zopfii. Biochem. J.* **90**, 483–9.

CHAPMAN L.F., CIRILLO V.P. & JAHN T.L. (1965) Permeability to sugars and fatty acids in *Polytoma obtusum. J. Protozool.* **12**, 47–51.

CHEUNG W.Y. & GIBBS M. (1966) Dark and photometabolism of sugars by a blue-green alga *Tolypothrix tenuis. Pl. Physiol., Lancaster*, **41**, 731–7.

CHODAT F. & SCHOPFER J.F. (1960) Éffects de synergie des conditions simultanées de carbo-autotrophie et de carbohétérotrophie observées dans la croissance de quelques algues. *Schweiz. Z. Hydrol.* **22**, 103–10.

COHEN G.N. & MONOD J. (1957) Bacterial permeases. *Bact. Rev.* **21**, 109–94.

COOK J.R. (1967) Photoassimilation of acetate by an obligately phototrophic strain of *Euglena gracilis. J. Protozool.* **14**, 382–4.

COOK J.R. & CARVER M. (1966) Partial repression of the glyoxylate bypass in *Euglena gracilis. Pl. Cell Physiol., Toyko*, **7**, 377–83.

COOK J.R. & HEINRICH B. (1965) Glucose vs acetate metabolism in *Euglena. J. Protozool.* **12**, 581–3.

COSGROVE W.B. & SWANSON B.K. (1952) Growth of *Chilomonas paramoecium* in simple organic media. *Physiol. Zool.* **25**, 287–92.

CRAMER M. & MYERS J. (1948) Nitrate reduction and assimilation in *Chlorella. J. gen. Physiol.* **32**, 93–102.

CRAMER M. & MYERS J. (1952) Growth and photosynthetic characteristics of *Euglena gracilis. Arch. Mikrobiol.* **17**, 384–402.

DARNFORTH W.F. (1962) Substrate assimilation and heterotrophy. In *Physiology and Biochemistry of Algae*, ed. Lewin R.A. pp. 99–123. Academic Press, New York & London.

DE DEKEN-GRENSON M. & MESSIN S. (1958) La continuité génétique des chloroplastes chez les Euglènes. I. Mécanisme de l'apparition des lignées blanches dans les cultures tractées par la streptomycine. *Biochim. biophys. Acta*, **27**, 145–55.

DODGE J.D. & CRAWFORD R.M. (1971) Fine structure of the dinoflagellate *Oxyrrhis marina*. I. The general structure of the cell. *Protistologica*, **7**, 295–304.

DROOP M.R. (1955a) Some factors governing encystment in *Haematococcus pluvialis*. *Arch. Mikrobiol.* **21**, 267–72.

DROOP M.R. (1955b) Carotogenesis in *Haematococcus pluvialis*. *Nature, Lond.* **175**, 42.

DROOP M.R. (1959) Water-soluble factors in the nutrition of *Oxyrrhis marina*. *J. mar. biol. Ass. U.K.* **38**, 605–20.

DROOP M.R. (1961) *Haematococcus pluvialis* and its allies. III. Organic nutrition. *Revue algol.* **4**, 247–59.

DROOP M.R. (1966) Organic acids and bases and the lag phase in *Nannochloris oculata*. *J. mar. biol. Ass. U.K.* **46**, 673–8.

DROOP M.R. & MCGILL S. (1966) The carbon nutrition of some algae; the inability to utilize glycollic acid for growth. *J. mar. biol. Ass. U.K.* **46**, 679–84.

DROOP M.R. & PENNOCK J.F. (1971) Terpenoid quinones and steroids in the nutrition of *Oxyrrhis marina*. *J. mar. biol. Ass. U.K.* **51**, 455–70.

DUSI H. (1933) Recherches sur la nutrition de quelques Euglènes. II. *Euglena stellata, klebsii, anabaenae, deses* et *pisciformis. Annls. Inst. Pasteur, Paris*, **50**, 840–89.

DUSI H. (1944) Le pouvoir de synthèse d'*Euglena viridis. Annls. Inst. Pasteur, Paris*, **70**, 311–12.

DVORÁKOVA-HLADKÁ J. (1966) Utilization of organic substrates during mixotrophic and heterotrophic cultivation of algae. *Biologia Pl.* **8**, 354–61.

EPPLEY R.W., GEE R. & SALTMAN P. (1963) Photometabolism of acetate by *Chlamydomonas mundana. Physiologia Pl.* **16**, 777–92.

EPPLEY R.W. & MACIASR F.M. (1962) Rapid growth of a sewage lagoon *Chlamydomonas* with acetate. *Physiol. Pl.* **15**, 72–9.

EPPLEY R.W. & MACIASR F.M. (1963) Role of the alga *Chlamydomonas mundana* in anaerobic waste stabilization lagoons. *Limnol. Oceanogr.* **8**, 411–16.

EPSTEIN H.T. & SCHIFF J.A. (1961) Studies on chloroplast development in *Euglena*. IV. Electron and fluorescence microscopy of the proplastid and its development into a mature chloroplast. *J. Protozool.* **8**, 427–82.

ERICKSON L.C., WEDDING R.F. & BRANNERMAN B.L. (1955) Influence of pH on 2,4-dichlorophenoxy-acetic acid activity in *Chlorella. Pl. Physiol., Lancaster*, **30**, 69–74.

FAUST L. & WOLFE R.S. (1961) Enrichment and cultivation of *Beggiatoa alba. J. Bact.* **81**, 99–110.

FAY P. (1965) Heterotrophy and nitrogen fixation in *Chlorogloea fritschii. J. gen. Microbiol.* **39**, 11–20.

FINKLE B.J., APPLEMAN D. & FLEISCHER F.K. (1950) Growth of *Chlorella vulgaris* in the dark. *Science, N.Y.* **111**, 309.

FISCHER E. & WIESSNER W. (1968) Acetate utilization in *Euglena gracilis. Pl. Physiol., Lancaster*, **43**, S31.

FOGG G.E. (1959) Nitrogen nutrition and metabolic patterns in algae. *Symp. Soc. expl. Biol.* **13**, 106–25.

FOGG G.E. & BELCHER J.H. (1961) Physiological studies on a planktonic μ alga. *Verh. int. verein theor. angew. Limnol.* **14**, 893–6.

GAFFRON H. (1939) Über Anomalien des Atmungsquotienten von Algen aus Zuckerkulturen. *Biol. Zbl.* **59**, 288–302.

GIBBS M. (1962) Fermentation. In *Physiology and Biochemistry of Algae*, ed. Lewin R.A. pp. 91–7, Academic Press, New York & London.

GIBBS S.P. (1970) The comparative ultrastructure of the algal chloroplast. *Ann. N.Y. Acad. Sci.* **175**, 454–73.

GIBOR A. (1956) The culture of brine algae. *Biol. Bull. mar. biol. Lab., Woods Hole*, **111**, 223–9.

GOULDING K.H. & MERRETT M.J. (1966) The photometabolism of acetate by *Chlorella pyrenoidosa*. *J. exp. Bot.* **17**, 678–89.

GOULDING K.H. & MERRETT M.J. (1967) The photometabolism of acetate by *Pyrobotrys stellata*. *J. gen. Microbiol.* **48**, 127–37.

GRIFFITHS D.J. (1965) The oxidative assimilation of organic substrates by algae. *Sci. Prog.* **53**, 553–66.

HAIGH W.G. & BEEVERS H. (1962) Acetate metabolism in algae. *Pl. Physiol., Lancaster*, **37**, (suppl.) 60.

HARDER R. (1917) Ernährungsphysiologische Untersuchungen an Cyanophyceen, hauptsächlich dem endophytischen *Nostoc punctiforme*. *Z. Bot.* **9**, 145–242.

HASE E., MORIMURA Y. & TAMIYA H. (1957) Some data on the growth physiology of *Chlorella* studied by the technique of continuous culture. *Archs. Biochem. Biophys.* **69**, 149–65.

HOARE D.S., HOARE S.L. & MOORE R.B. (1967) The photoassimilation of organic compounds by autotrophic blue-green algae. *J. gen. Microbiol.* **49**, 351–70

HOARE D.S., HOARE S.L. & SMITH A.J. (1969) Assimilation of organic compounds by blue-green algae and photosynthetic bacteria. *Prog. Photosynth. Res.* **3**, 1570–3.

HOARE D.S. & MOORE R.B. (1965) Photoassimilation of organic compounds by autotrophic blue-green algae. *Biochim. biophys. Acta*, **109**, 622–5.

HOLZ G.G. (1954) The oxidative metabolism of a cryptomonad flagellate *Chilomonas paramoecium*. *J. Protozool.* **1**, 114–20.

HURLBERT R.E. & RITTENBERG S.C. (1962) Glucose metabolism of *Euglena gracilis* var *bacillaris*: growth and enzymatic studies. *J. Protozool.* **9**, 170–82.

HUTCHINS J.O., PODOLSKY B. & MORALES M.F. (1948) Studies on the kinetics and energetics of carbon and nitrogen metabolism of *Chilomonas paramoecium*. *J. cell. comp. Physiol.* **32**, 117–41.

HUTH W. (1967) Enzyme in grünen Einzellern in Abhängigkeit von der Kohlenstoffversorgung. *Flora, Jena*, **158**, 58–87.

HUTNER S.H. & PROVASOLI L. (1951) The phytoflagellates. In *Biochemistry and Physiology of Protozoa*, vol. 1, ed. Lwoff A. pp. 27–128. Academic Press, New York & London.

ISHIDA Y. & KATODA H. (1965) Metabolism of acetate in *Gyrodinium cohnii*. *Mem. Res. Inst. Fd. Sci. Kyoto Univ.* **26**, 10–17.

JACOBI G. (1957) Vergleichende enzymatische Untersuchungen an marinen Grün—und Rotalgen. *Kieler Meeresforsch.* **13**, 212–9.

KANDLER O. (1954) Über die Beziehungen zwischen Phosphathanstalt und Photosynthese II. Gesteigerter Glucoseeinban im Licht als Indikator einer lichtabhängigen Phosphorylierung. *Z. Naturf.* **96**, 625–44.

KENYON C.N., RIPPKA R. & STANIER R.Y. (1972) Fatty acid composition and physiological properties of some filamentous blue-green algae. *Arch. Mikrobiol.* **83**, 216–36.

KESSLER E. (1967) Physiologische und biochemische Beiträge zur Taxonomie der Gattung *Chlorella*. III. Merkmale von 8 autotrophen Arten. *Arch. Mikrobiol.* **55**, 346–57.

KHOJA T. & WHITTON B.A. (1971) Heterotrophic growth of blue-green algae. *Arch. Mikrobiol.* **79**, 280–2.

KIYOHARA T., FUJITA Y., HATTORI A. & WATANABE A. (1960) Heterotrophic culture of a blue-green alga *Tolypothrix tenuis*. *J. gen. appl. Microbiol., Tokyo*, **6**, 176–82.

KORNBERG H.L. & ELSDEN S.R. (1961) The metabolism of 2-carbon compounds by microorganisms. *Adv. Enzymol.* **23**, 401–70.

KRATZ J. & MYERS J. (1955) Nutrition and growth of several blue-green algae. *Am. J. Bot.* **42**, 282–7.

KRÜGER W. (1894) Beiträge zur Kenntnis der Organismen des Saftflussers der Laubbäume. *Beitr. Physiol. Morph. niederer Organismen*, **4**, 69.

LANG N.J. & COOK P.W. (1962) Ultrastructure and pigment analysis of *Polytoma*. *Am. J. Bot.* **49**, 672.

LATZKO E. & GIBBS M. (1969) Enzyme activities of the carbon reduction cycle in some photosynthetic organisms. *Pl. Physiol., Lancaster*, **44**, 295–300.

LEVEDAHL B.H. (1965) Succinate transport in *Euglena. Expl. Cell Res.* **39**, 233–41.

LEWIN J.C. (1950) Obligate autotrophy in *Chlamydomonas moewusii* Gerloff. *Science, N.Y.* **112**, 652–3.

LEWIN J.C. (1953) Heterotrophy in diatoms. *J. gen. Microbiol.* **9**, 305–13.

LEWIN J.C. (1963) Heterotrophy in marine diatoms. In *Symposium on Marine Microbiology*, ed. Oppenheimer C.H. pp. 227–35. Charles C. Thomas, Springfield, Illinois.

LEWIN J.C. & HELLEBUST J.A. (1970) Heterotrophic nutrition of the marine pennate diatom *Cylindrotheca fusiformis. Can. J. Microbiol.* **11**, 1123–9.

LEWIN J.C. & LEWIN R.A. (1960) Auxotrophy and heterotrophy in marine littoral diatoms. *Can. J. Microbiol.* **6**, 127–34.

LEWIN J.C. & LEWIN R.A. (1967) Culture and nutrition of some apochlorotic diatoms. *J. gen. Microbiol.* **46**, 361–7.

LEWIN R.A. (1954) Utilization of acetate by wild type and mutant *Chlamydomonas dysosmos. J. gen. Microbiol.* **11**, 459–71.

LITTLE P.A., OLSON J.J. & WILLIAMS J.W. (1951) Growth studies on *Polytomella agilis. Proc. Soc. exp. Biol. Med.* **78**, 510–13.

LOEFER J.B. (1935) Effects of certain carbohydrates and organic acids on the growth of *Chlorogonium* and *Chilomonas. Arch. Protistenk.* **84**, 456.

LUCKSCH I. (1933) Ernährungsphysiologische Untersuchungen an Chlamydomonadeen. *Beih. bot. Zbl.* **A50**, 64–94.

LWOFF A. (1932) *Recherches Biochimiques sur la Nutrition des Protozoaires.* Masson, Paris.

LWOFF A. (1935) La nutrition azotée et carbonée de *Polytomella agilis. C.r. Séanc. Soc. Biol.* **119**, 974–6.

LWOFF A. (1941) Limites de concentration en ions H et OH compatibles avec le développment *in vitro* du flagellé *Polytomella caeca. Annls. Inst. Pasteur, Paris*, **66**, 407–16.

LWOFF A. (1944) *L'Évolution Physiologique.* Hermann, Paris.

LWOFF A. (1951) Introduction to biochemistry of protozoa. In *Biochemistry and Physiology of Protozoa*, vol. 1, ed. Lwoff A. pp. 1–26. Academic Press, New York and London.

LWOFF A. & DUSI H. (1934) L'oxitrophy et la nutrition des flagellés leucophytes. *Annls. Inst. Pasteur, Paris*, **53**, 641.

LWOFF A. & DUSI H. (1935) La nutrition azotée et carbonée de *Chlorogonium euchlorum* à l'obscurité; l'acide acétique envisagé comme produit de l'assimilation chlorophyllienne. *C. r. Séanc. Soc. Biol.* **119**, 1260–3.

LWOFF A., IONESCO H. & GUTMANN A. (1949) Metabolism de l'amidon chez un flagellé sans chlorophyll incapable d'utilizer le glucose. *C. r. hebd. Séanc. Acad. Sci., Paris*, **228**, 342–4.

LWOFF A., IONESCO H. & GUTMANN A. (1950) Synthèse et utilization de l'amidon chez un flagellé sans chlorophylle incapable d'utilizer les sucres. *Biochim. biophys. Acta*, **4**, 270–5.

LWOFF A. & LWOFF M. (1929) Le pouvoir de synthèse de *Chlamydomonas agloeformis* et d'*Haematococcus pluvialis* en culture pure à l'obscurité. *C. r. Séanc. Soc. Biol.* **102**, 569.

LYNCH V.H. & CALVIN M. (1953) $CO_2$ fixation by *Euglena. Ann. N.Y. Acad. Sci.* **56**, 890–900.

LYNN R.I. & STARR R.C. (1970) The biology of the acetate flagellate *Diplostauron elegans* Skuja. *Arch. Protistenk.* **112**, 283–302.

MATSUKA M., MIYACHI S. & HASE E. (1969a) Further studies on the metabolism of glucose in the process of glucose bleaching of *Chlorella protothecoides. Pl. Cell Physiol., Tokyo*, **10**, 503–12.

MATSUKA M., MIYACHI S. & HASE E. (1969b) Acetate metabolism in the process of 'acetate bleaching' of *Chlorella protothecoides. Pl. Cell Physiol., Tokyo*, **10**, 513–26.

McLachlan J. & Craigie J.S. (1965) Effects of carboxylic acids on growth and photosynthesis of *Haematococcus pluvialis*. *Can. J. Bot.* **43**, 1449–56.

Meeuse B.J.D. (1962) Storage products. In *Physiology and Biochemistry of Algae*, ed. Lewin R.A. pp. 289–314. Academic Press, New York & London.

Menke W. & Fricke B. (1962) Einige Beobachtungen an *Prototheca ciferrii*. *Portugalia Acta biol.* **A6**, 243–63.

Merrett M.J. & Goulding K.H. (1967) Short term products of $^{14}C$ acetate assimilation by *Chlorella pyrenoidosa* in the light. *J. exp. Bot.* **18**, 128–39.

Miller J.D.A. & Fogg G.E. (1958) Studies on the growth of Xanthophyceae in pure culture II. The relations of *Monodus subterraneus* to organic substances. *Arch. Mikrobiol.* **30**, 1–16.

Mineeva L.A. (1961) The use of various organic compounds by *Chlorella vulgaris* and *Scenedesmus obliquus* cultures. *Mikrobiologiya*, **30**, 586–92.

Monod J. (1942) *Recherches sur la Croissance des Cultures Bactériennes*. Hermann, Paris.

Myers J. (1947) Oxidative assimilation in relation to photosynthesis in *Chlorella*. *J. gen. Physiol.* **30**, 217–27.

Myers J. (1953) Growth characteristics of algae in relation to the problems of mass culture. In *Algal Culture from Laboratory to Pilot Plant*, ed. Burlew J.S. pp. 37–54. Carnegie Inst., Washington.

Myers J. (1963) Laboratory cultures. In *Physiology and Biochemistry of Algae*, ed. Lewin R.A. pp. 603–15. Academic Press, New York & London.

Neish A.C. (1951) Carbohydrate nutrition of *Chlorella vulgaris*. *Can. J. Bot.* **29**, 68–78.

Oaks A. (1962) Influence of glucose and light on pyruvate metabolism of starved cells of *Chlorella ellipsoidea*. *Pl. Physiol., Lancaster*, **37**, 316–22.

Ochiai S. & Hase E. (1970) Studies on the chlorophyll formation in *Chlorella protothecoides*. I. Enhancing effects of light and added δ-aminolevulinic acid and suppressive effect of glucose on chlorophyll formation. *Pl. Cell Physiol., Tokyo*, **11**, 663–73.

Parker B.C., Bold H.C. & Deason T.R. (1961) Facultative heterotrophy in some chlorococcacean algae. *Science, N.Y.* **133**, 761–3.

Payne W.J. (1970) Energy yields and growths of heterotrophs. *A. Rev. Microbiol.* **24**, 17–52.

Pearce J. & Carr N.G. (1967a) The metabolism of acetate by the blue-green algae, *Anabaena variabilis* and *Anacystis nidulans*. *J. gen. Microbiol.* **49**, 301–13.

Pearce J. & Carr N.G. (1967b) An incomplete tricarboxylic acid cycle in the blue-green alga *Anabaena variabilis*. *Biochem. J.* **105**, 45P.

Pearce J. & Carr N.G. (1969) The incorporation and metabolism of glucose by *Anabaena variabilis*. *J. gen. Microbiol.* **54**, 451–62.

Pearce J., Leach C.K. & Carr N.G. (1969) The incomplete tricarboxylic acid cycle in the blue-green alga *Anabaena variabilis*. *J. gen. Microbiol.* **55**, 371–8.

Pearsall W.H. & Bengry R.P. (1940) The growth of *Chlorella* in darkness and in glucose solution. *Ann. Bot.* **4**, 365–77.

Pearsall W.H. & Loose L. (1937) The growth of *Chlorella vulgaris* in pure culture. *Proc. R. Soc. B*, **121**, 451–501.

Pelroy R.A., Rippka R. & Stanier R.Y. (1972) Metabolism of glucose by unicellular blue-green algae. *Arch. Mikrobiol.*, **87**, 303–22.

Pimenova M.N. & Kondratieva T.F. (1965) A contribution to the use of acetate by *Chlamydomonas globosa*. *Mikrobiologiya*, **34**, 230–5.

Pringsheim E.G. (1921) Zur Physiologie von *Polytoma uvella*. *Ber. dt. bot. Ges.* **38**, 8–9.

Pringsheim E.G. (1935) Über Azetatflagellaten. *Naturwissenschaften*, **23**, 110–14.

Pringsheim E.G. (1937) Beiträge zur Physiologie saprotropher Algen und Flagellaten. *Planta*, **27**, 69–72.

Pringsheim E.G. (1948) The loss of chromatophores in *Euglena gracilis*. *New Phytol.* **47**, 52–87.

Pringsheim E.G. (1952) On the nutrition of *Ochromonas*. *Q. Jl. Microsc. Sci.* **93**, 71–96.

PRINGSHEIM E.G. (1955) Kleine Mitteilungen über Flagellaten und Algen. II. *Euglena gracilis* var *saccharophila* n. var und einer vereinfachten Nährlösung zur Vitamin $B_{12}$ Bestimmung. *Arch. Mikrobiol.* **21**, 414–19.

PRINGSHEIM E.G. (1963) *Farblose Algen.* Gustav Fischer, Stuttgart.

PRINGSHEIM E.G. (1967a) Die Mixotrophie von *Beggiatoa. Arch. Mikrobiol.* **59**, 247–54.

PRINGSHEIM E.G. (1967b) Zur Physiologie der farblosen Diatomee *Nitzschia putrida. Arch. Mikrobiol.* **56**, 60–7.

PRINGSHEIM E.G. & PRINGSHEIM O. (1959) Die Ernährung koloniebildender Volvocales. *Biol. Zbl.* **78**, 937–71.

PRINGSHEIM E.G. & WIESSNER W. (1960) Photoassimilation of acetate by green organisms. *Nature, Lond.* **188**, 919–21.

PROVASOLI L. (1938) Studi sulla nutrizione dei protozoi. *Boll. Lab. Zool. agr. Bachic. R. Ist. sup. agr. Milano,* **9**, 1–124.

PROVASOLI L. & GOLD K. (1962) Nutrition of the American strain of *Gyrodinium cohnii. Arch. Mikrobiol.* **42**, 196–203.

PROVASOLI L., HUTNER S.H. & PINTNER I.J. (1951) Destruction of chloroplasts by streptomycin. *Cold Spring Harb. Symp. quant. Biol.* **16**, 113–20.

PROVASOLI L. & McLAUGHLIN J.J.A. (1963) Limited heterotrophy of some photosynthetic dinoflagellates. In *Symposium on Marine Microbiology,* ed. Oppenheimer C.H. pp. 105–13. Charles C. Thomas, Springfield, Illinois.

RAHAT M. & JAHN T.L. (1965) Growth of *Prymnesium parvum* in the dark: a note on ichthiotoxin formation. *J. Protozool.* **12**, 266–80.

RAHAT M. & SPIRA Z. (1967) Specificity of glycerol for dark growth of *Prymnesium parvum. J. Protozool.* **14**, 45–8.

RAVEN J.A. (1971) Energy metabolism in green cells. *Trans. bot. Soc. Edin.* **41**, 219–25.

REEVES H.C., KADIS S. & AJL S. (1962) Enzymes of the glyoxylate bypass in *Euglena gracilis. Biochim. biophys. Acta,* **57**, 403–4.

RICHTER G. (1957) Nachweis und quantitative Bestimmung einiger Enzyme des Kohlenhydrat-Stoffwechsels in Grünalgen. *Z. Naturf.* **126**, 662–3.

RODHE W. (1955) Can plankton production proceed during winter darkness in arctic lakes? *Verh. Int. Verein. theor. angew. Limnol.* **12**, 117–22.

ROSS M.R. & JAHN T.L. (1971) Mechanism for acetate metabolism in *Euglena gracilis. J. Protozool.* **18**, (Suppl.) 20.

RUSSELL G.H. & GIBBS M. (1966) Regulation of photosynthetic capacity in *Chlamydomonas mundana. Pl. Physiol., Lancaster,* **41**, 885–90.

SAGER R. & GRANICK S. (1953) Nutritional studies with *Chlamydomonas reinhardi. Ann. N.Y. Acad. Sci.* **56**, 831–8.

SAMEJIMA H. & MYERS (1958) On the heterotrophic growth of *Chlorella pyrenoidosa. J. gen. Microbiol.* **18**, 107–17.

SCHIFF J.A. & EPSTEIN H.T. (1965) The continuity of the chloroplast in *Euglena.* In *Reproduction: Molecular and Cellular,* ed. Locke M. pp. 131–89. Academic Press, New York & London.

SHIHIRA I. & KRAUSS R.W. (1965) *Chlorella: Physiology and taxonomy of 41 isolates.* Univ. Maryland Press, College Park, U.S.A.

SMITH F.A. (1967) Links between glucose uptake and metabolism in *Nitella translucens. J. exp. Bot.* **18**, 348–58.

SMITH A.J., LONDON J. & STANIER R.Y. (1967) Biochemical basis of obligate autotrophy in blue-green algae and thiobacilli. *J. Bact.* **94**, 972–83.

STANIER R.Y., KUNISAWA R., MANDEL M. & COHEN-BAZIRE G. (1971) Purification and properties of unicellular blue-green algae (order Chlorococcales). *Bact. Rev.* **35**, 171–205.

STORM J. & HUTNER S.H. (1953) Nutrition of *Peranema. Ann. N.Y. Acad. Sci.* **56**, 901–9.

STROSS R.G. (1960) Growth response of *Chlamydomonas* and *Haematococcus* to the volatile fatty acids. *Can. J. Microbiol.* **6**, 611–17.

SYRETT P.J. (1966) The kinetics of isocitrate lyase formation in *Chlorella*: Evidence for the promotion of enzyme synthesis by photophosphorylation. *J. exp. Bot.* **17**, 641–54.

SYRETT P.J., MERRETT M.J. & BOCKS S.M. (1963) Enzymes of the glyoxylate cycle in *Chlorella vulgaris. J. exp. Bot.* **14**, 249–64.

SYRETT P.J., BOCKS S.M. & MERRETT M.J. (1964) The assimilation of acetate by *Chlorella vulgaris. J. exp. Bot.* **15**, 35–47.

TANNER W. (1969) Light-driven active uptake of 3-O.methylglucose via an inducible hexose uptake system of *Chlorella. Biochem. biophys. Res. Commun.* **36**, 278–83.

TANNER W., GRÜNES R. & KANDLER O. (1970) Spezifität und Turnover des induzierbaren Hexose-Aufnahmesystems von *Chlorella. Z. Pfl. Physiol.* **62**, 376–86.

TANNER W. & KANDLER O. (1967) Die Abhängigkeit der Adaptation der Glucose-Aufnahme von der oxydativen und der photosynthetischen Phosphorylierung bei *Chlorella vulgaris. Z. Pfl. Physiol.* **58**, 24–32.

TAYLOR F.J. (1960) Absorption of glucose by *Scenedesmus quadricauda*. II. The nature of the absorptive process. *Proc. R. Soc. B.* **151**, 483–96.

TURNER M.F. (1970a) In *Scottish Marine Biological Association, Report of the Council for 1969/70*, p. 13.

TURNER M.F. (1970b) A note on the nutrition of *Rhodella. Br. Phycol. J.* **5**, 15–18.

VAN BAALEN C. (1962) Studies on marine blue-green algae. *Botanica mar.* **4**, 128–39.

VAN BAALEN C. (1965) The photooxidation of citric acid by *Anacystis nidulans. Pl. Physiol., Lancaster*, **40**, 368–71.

VISHNIAC W. & REAZIN G.H. (1957) Photo-reduction in *Ochromonas malhamensis*. In *Research in Photosynthesis*, ed. Gaffron H. pp. 239–42. Interscience, New York.

VOTTA J.J., JAHN T.L. & LEVEDAHL B.H. (1971) The mechanism of onset of the stationary phase in *Euglena gracilis* grown with 10 mM succinate: intracellular pH values. *J. Protozool.* **18**, 166–70.

WETHERELL D.F. (1958) Obligate phototrophy in *Chlamydomonas eugametos. Physiologia Pl.* **11**, 260–74.

WIESSNER W. (1963) Stoffwechselleistung und Enzymaktivität bei *Chlamydobotrys* (Volvocales). *Arch. Mikrobiol.* **45**, 33–45.

WIESSNER W. (1965) Quantum requirement for acetate assimilation and its significance for quantum measurements in photosynthesis. *Nature, Lond.* **205**, 56–7.

WIESSNER W. (1966) Vergleichende Studien zum Quantenbedarf der Photoassimilation von Essigsäure durch photoheterotrophe Purpurbakterien und Grünalgen. *Ber. dt. bot. Ges.* **79**, 58–62.

WIESSNER W. (1968) Enzymaktivität und Kohlenstoffassimilation bei Grünalgen unterschiedlichen ernährungsphysiologischen Typs. *Planta*, **79**, 92–8.

WIESSNER W. (1969) Effect of autotrophic or photoheterotrophic growth conditions on *in vivo* absorption of visible light by green algae. *Photosynthesis*, **3**, 225–32.

WIESSNER W. (1970) Photometabolism of organic substrates. In *Photobiology of Microorganisms*, ed. Halldal P. pp. 95–133. Wiley-Interscience, London & New York.

WIESSNER W. & GAFFRON H. (1964) Role of photosynthesis in the light-induced assimilation of acetate by *Chlamydobotrys. Nature, Lond.* **201**, 725–6.

WIESSNER W. & KUHL A. (1962) Die Bedeutung des Glyoxylsäurezyklus für die Photoassimilation von Acetat bei phototrophen Algen. *Vortr. GesGeb. Bot. dt. bot. Ges.* **N.F.1**, 102–8.

WISE D.L. (1959) Carbon nutrition and metabolism of *Polytomella caeca. J. Protozool.* **6**, 19–23.

WISE D.L. (1961) Absorption of acid nutrients by an acetate flagellate. *J. Protozool.* **8**, (Suppl.) 8–9.

# CHAPTER 20

# NITROGEN FIXATION

## G. E. FOGG

Marine Sciences Laboratory,
University College of North Wales,
Menai Bridge, Anglesey, U.K.

1   Introduction   560

2   Biochemistry of the fixation
    process   561

3   The heterocyst as the site of
    nitrogen fixation   564

4   Nitrogen fixation in non-hetero-
    cystous species   567

5   Relationships between nitrogen
    fixation and photosynthesis
    569

6   Physiological ecology of nitro-
    gen fixation   570

6.1   Light   571
6.2   Combined nitrogen   571
6.3   Mineral nutrition   572
6.4   Salinity   572
6.5   Herbicides   573
6.6   Temperature   573
6.7   Transfer of fixed nitrogen to
      other organisms   573
6.8   Economic importance   574

7   Nitrogen fixation in symbioses
    involving blue-green algae   575

8   Nitrogen fixation by algae other
    than Cyanophyceae   576

9   References   576

## 1  INTRODUCTION

Definite proof that a blue-green alga can assimilate or 'fix' molecular nitrogen ($N_2$) was not published until 1928 (Drewes 1928) although the suspicion that certain members of this group possessed the property had been current for forty years before this. By about 1955 it had become evident that many species of blue-green algae are nitrogen-fixing, but they were comparatively little studied until around 1967. Progress in the biochemical investigation of nitrogen fixation accelerated about this time following the preparation of cell-free extracts in which the process took place actively, but two circumstances in particular drew attention to the blue-green algae. First, newly devised methods for the estimation of nitrogen fixation in the field began to show that these organisms are more active in this respect than had hitherto been thought. Second, the suggestion that the peculiar type of differentiated cell known as the heterocyst was the site of nitrogen fixation in blue-green algae provoked controversy and stimulated new lines of research. This chapter will be largely concerned with these more recent developments, of which accounts have also been given by Stewart (1971a, 1973) and Fogg et al. (1973). Information on general and historical aspects may be sought in the earlier reviews by Fogg and Wolfe (1954), Fogg (1956, 1962), Fogg and Stewart (1965), Laporte and Pourroit (1967) and Stewart (1966, 1969, 1970a).

## 2  BIOCHEMISTRY OF THE FIXATION PROCESS

Our knowledge of the biochemistry of nitrogen fixation has been reviewed by Burris (1969), Postgate (1971) and Hardy et al. (1971). In vitro, the electrons needed for the reduction of $N_2$ or alternative substrates such as acetylene or cyanide may be supplied by dithionite but, in vivo, ferredoxin reduced by hydrogenase or some other source of reductant acts as the electron donor, with the additional energy needed for the reduction being supplied as ATP. Evolution of hydrogen is another alternative to $N_2$ reduction, the two processes competing for electrons. Extracted nitrogenase may be fractionated into two proteins, neither of which shows enzymic activity by itself although on recombination they yield the active enzyme, but there is no definite evidence permitting one to assign electron-activating or substrate-reducing functions to one or the other. It is thought that $N_2$ is reduced stepwise via enzyme-bound diimide and hydrazine at a site composed of iron and molybdenum atoms linked by a sulphur atom, but evidence for this is so far of an indirect nature.

Nitrogenases isolated from various sources show minor chemical and physical differences but there is no reasonable doubt that the mechanism of nitrogen fixation is basically the same in all organisms, blue-green algae included. A cell-free extract fixing nitrogen was first reported from a blue-green alga by Schneider et al. (1960) using Mastigocladus laminosus. Cox et al. (1964) obtained an extract from Anabaena cylindrica in which the activity was associated with a fraction containing photosynthetic lamellae sedimenting after 20 min. at 35,000 × g. In Mastigocladus extracts the activity remained in the supernatant after centrifuging at 45,000 × g for 45 min. Stewart et al. (1969) found that activity in disrupted preparations of A. cylindrica was not associated with photosynthetic lamellae but with a fraction containing heterocysts. Smith and Evans (1970) obtained extracts from the same alga in which the activity remained in the supernatant after 10 min at 40,000 × g. This was confirmed by Haystead et al. (1970) with extracts both of A. cylindrica and Plectonema boryanum. Sedimentation of activity in crude extracts is ascribed to adsorption of the enzyme on particulate matter by Haystead and Stewart (1972), who found that the partially purified enzyme remained in suspension after centrifuging at 144,000 × g for 3 hours. The characteristics of the nitrogenase thus isolated are generally similar to those of nitrogenases isolated from other kinds of organism. Fay and Cox

(1967) observed that it is oxygen-sensitive and this has been confirmed by Haystead *et al.* (1970), Smith and Evans (1970) and Bothe (1970). ATP, $Mg^{++}$, and dithionite as an electron donor, are required for nitrogenase activity (Bothe 1970, Haystead *et al.* 1970, Smith & Evans 1970). Like that from *Clostridium*, but not to the same extent, the enzyme is cold-labile (Haystead *et al.* 1970), being 60% inactivated by 12 hours at 0°C in comparison with controls held at room temperature. The most active preparation reported to date has an acetylene reduction rate of 50 to 100 n moles (mg protein)$^{-1}$ min$^{-1}$ (Haystead & Stewart 1972). With this preparation evidence was obtained that *Anabaena* nitrogenase is a metallo-protein containing iron and reduced thiol groups. Smith *et al.* (1971b) fractionated the partially purified enzyme into two components probably corresponding to fractions 1 and 2 of heterotrophic bacteria. Their fraction 1 combined with fraction 2 of *Chloropseudomonas ethylicum* to give a nitrogenase rather more active than that obtained by recombining the two *Anabaena* fractions.

The presence of a hydrogenase in *Anabaena cylindrica* was demonstrated by Fujita *et al.* (1964), and Haystead *et al.* (1970) have demonstrated ATP-dependent hydrogen evolution by a cell-free extract of the same alga. Haystead and Stewart (1972) showed that nitrogenase from *A. cylindrica* is able to accept electrons from *Clostridium kluyveri* hydrogenase via ferredoxin and suggested that under some conditions *Anabaena* hydrogenase could act as a source of electrons for nitrogen fixation.

In extracts of *Clostridium pasteurianum* nitrogen fixation is supported by pyruvate, which yields both electrons and ATP via the phosphoroclastic reaction carried out by this organism. Cox (1966) found that pyruvate is also particularly efficacious in supporting fixation by *Anabaena cylindrica*. Cox and Fay (1967, 1969) showed that decarboxylation of pyruvate is coupled to nitrogen fixation in cell-free extracts of this alga and that the ratio of the rates of these two processes is 3:1, which is in accord with the theory that pyruvate is acting as the hydrogen donor. Other workers could not demonstrate this pyruvate-supported fixation (Bothe 1970, Smith & Evans 1970, Haystead & Stewart 1972) but it has been confirmed by Smith *et al.* (1971a). Leach and Carr (1971), working with a non-nitrogen fixing strain of *Anabaena variabilis*, could not demonstrate the presence in it of a pyruvate dehydrogenase but detected a pyruvate: ferredoxin oxido-reductase. The latter enzyme has recently been demonstrated in the nitrogen-fixing *Anabaena cylindrica* (Bothe & Falkenberg, 1973) and may mediate the supply of electrons to nitrogenase. Glucose-6-phosphate may also act as an electron donor for *Anabaena* nitrogenase in a system containing glucose-6-phosphate dehydrogenase, NADP, ferredoxin and ferredoxin-NADP reductase (Bothe 1970). Nitrogen fixation can occur in the dark in blue-green algae (Fay 1965, Cox 1966) and, even in the light, is sensitive to inhibitors of respiration such as chlorpromazine and cyanide (Cox 1966). It therefore seems likely that respiratory processes such as those described above may act as electron donors for nitrogen fixation in the intact alga.

The possible importance of photosynthetically generated reducing power has been shown by several workers. Bothe (1970) found that ferredoxin from either spinach or *Anabaena*, or phytoflavin from iron-deficient *Anacystis* could transfer electrons from illuminated chloroplasts to *Anabaena* nitrogenase. This was confirmed by Smith and Evans (1971) and Smith *et al.* (1971a) who also showed that a particulate fraction from *A. cylindrica* could donate electrons to nitrogenase via ferredoxin when 2,6-dichloro-phenolindophenol (DCPIP) and ascorbate were supplied in the light. However, the finding of Cobb and Myers (1964), Cox and Fay (1969) and Bothe (1970) that nitrogen fixation in intact cells of *A. cylindrica* is scarcely affected by inhibitors of photosystem II, raises doubts as to whether photoreduction occurs *in vivo*. Lyne and Stewart (1973) were unable to find an Emerson enhancement effect with acetylene reduction by *Anabaena cylindrica* under conditions in which enhancement of photosynthetic carbon dioxide fixation was clearly observed. This, again, is consistent with the idea that electron flow from water mediated by photosystems I and II does not directly provide the main source of reductant for nitrogenase in *A. cylindrica*.

Since nitrogen fixation may take place in the dark, it is evident that ATP can be supplied by oxidative phosphorylation. However, in nitrogen-starved cells of *Anabaena cylindrica*, nitrogen fixation is strongly light dependent although not inhibited by CMU at concentrations which reduce the fixation of carbon dioxide by over 90% (Cox & Fay 1969). The action spectrum for light stimulated nitrogen fixation in this alga corresponds closely to the action spectrum of photosystem I (Fay 1970). These facts strongly suggest that ATP is being supplied by photo-phosphorylation under the conditions of these experiments, but direct proof of this has not yet been obtained.

Like nitrogenases from other sources, the enzyme from blue-green algae is capable of reducing substrates other than $N_2$. Stewart *et al.* (1967, 1968) demonstrated acetylene reduction by intact cells of various blue-green algae and found an acetylene reduction: nitrogen fixation ratio of nearly the theoretical value, 3:1. Acetylene and cyanide reduction have been shown to occur in cell-free extracts of both *Anabaena* and *Plectonema* (Haystead *et al.* 1970).

Ammonia has been long recognized as a key intermediate and probably the first stable product in nitrogen fixation. Magee and Burris (1954) found it to be the first substance, together with glutamic acid, to become labelled with $^{15}N$ when nitrogen-fixing blue-green algae were provided with $N_2$ enriched with this isotope. Similar results were obtained with *Westiellopsis prolifica* by Fogg and Pattnaik (1966). Production of ammonia during nitrogen fixation in cell-free extracts of *Anabaena cylindrica* has been demonstrated by Bothe (1970). Little advance has been made until recently in our knowledge of the entry of ammonia into organic combination in nitrogen-fixing algae. It is generally assumed that the common path of entry by reductive amination of α-ketoglutarate to give glutamate is the one which operates. However recent work has shown the presence of alanine dehydrogenase (Scott & Fay 1972, Haystead *et al.* 1973; Neilson & Doudoroff 1973), and high levels of glutamine synthetase (Dharmawardene *et al.* 1973). A variety of other ammonia-incorporating enzymes have also been reported in *A. cylindrica* (Haystead *et al.* 1973). These include glutaminase, aspartate dehydrogenase, carbamoyl phosphate synthetase and various amino-transferases. The idea that citrulline is another substance through which ammonia enters organic combination (Fogg & Than Tun 1960) was based on the finding that this happens in alder root nodules. It is also supported by the finding of carbamoyl-phosphate synthetase in *A. cylindrica* and by the finding that citrulline is an early and abundant product of photosynthesis in *Nostoc* (Linko *et al.* 1957).

### 3 THE HETEROCYST AS THE SITE OF NITROGEN FIXATION

Heterocysts, empty-looking cells with thick refractive walls, are a conspicuous feature of many blue-green algae. As the distribution amongst members of the group of the capacity to fix $N_2$ became better known it became obvious that most of the species possessing this capacity have this kind of differentiated cell, whereas those unable to fix nitrogen do not (Fogg 1956). This led to the idea that heterocysts might be the nitrogen-fixing organs of blue-green algae, but an attempt to demonstrate nitrogen fixation in a suspension of isolated heterocysts, which on the evidence of their respiratory activity were judged to be healthy, was unsuccessful (Fay & Walsby 1966). Nevertheless the weight of indirect evidence in favour of heterocysts being the site of nitrogen fixation was so great that Fay *et al.* (1968) felt justified in advancing the hypothesis in print. Although it is now known that some non-heterocystous blue-green algae are able to fix $N_2$ under certain circumstances (see p. 567) subsequent investigations have confirmed the essential correctness of this hypothesis. The principal evidence is as follows—

(1)  Ammonia, the key intermediate in nitrogen fixation, suppresses the formation of nitrogenase and also inhibits the formation of heterocysts (Fogg 1949, Stewart *et al.* 1968, Ogawa & Carr 1969, Jewell & Kulasooriya 1970). If filaments grown in the presence of an ammonium salt are deprived of this source of combined nitrogen, nitrogenase activity and numbers of mature heterocysts increase in a parallel fashion (Kulasooriya *et al.* 1972).

(2)  The pattern of development of heterocysts is consistent with the idea that they produce ammonia. In non-polar filaments such as those of *Anabaena*, new heterocysts arise midway between two mature heterocysts at the point where the concentration of ammonia, inhibitory to heterocyst differentiation, would be expected to be minimal. In polar filaments, such as those of *Gloeotrichia*, with a heterocyst at one end there is a gradient of growth, spores being developed next to the heterocyst and the distal cells becoming vacuolated and ceasing to divide. In the presence of a supplied ammonium salt *Gloeotrichia* tends to develop non-polar filaments (Fay *et al.* 1968). Van Gorkom and Donze (1971) have provided an elegant demonstration that heterocysts are a source of combined nitrogen by using the phycocyanin content of cells as an index of their nitrogen content. In

nitrogen-starved filaments of *Anabaena cylindrica* a gradient of phycocyanin, with highest concentrations next to the heterocysts and lowest in intermediate positions, was found to develop when $N_2$ was supplied. If combined nitrogen is supplied the distribution of phycocyanin along the filament is uniform.

(3) Heterocysts are strongly reducing and in the presence of triphenyl-tetrazolium chloride (TTC) accumulate crystals of its reduction product, formazan, more rapidly than do vegetative cells. If filaments of a blue-green alga are left in a TTC solution for a sufficient time so that most of the heterocysts but few of the vegetative cells contain formazan crystals, it is found that nitrogenase activity is largely suppressed, although photosynthetic activity is scarcely affected (Stewart *et al*. 1969). That is to say, a substance which reacts differentially with heterocysts and vegetative cells has a parallel effect on nitrogen fixation.

(4) Direct evidence of nitrogenase activity in heterocysts has been provided by Stewart *et al*. (1969), who found that heterocysts of *Anabaena cylindrica* isolated by sonic disruption of vegetative cells were capable of reducing acetylene in the presence of ATP and dithionite under dark anaerobic conditions. The rate was low, but the finding has been confirmed by Wolk and Wojciuch (1971a), who reported that similarly isolated heterocysts reduced acetylene in the light, without added cofactors, at rates up to 30% of those shown by the intact aerobically grown filaments from which they were derived. The negative results of Fay and Walsby (1966) may be explained by the circumstances that cofactors were not supplied and the experiment was conducted under aerobic conditions. Fay and Lang (1971) subsequently found by examination with the electron microscope of heterocysts isolated in various ways that considerable damage to the ultra-structure resulted when sonic disruption was used and the only acceptable method, among those which they tested, was disruption of vegetative cells with lysozyme. Examination of the nitrogen fixing activity of heterocysts isolated by this less drastic method is desirable, but meanwhile it appears from the results of Wolk and Wojciuch (1971a) that heterocysts have greater nitrogenase activity than the vegetative cells with which they were associated. Whether vegetative cells in aerobically grown nitrogen-fixing material completely lack nitrogenase cannot yet be stated with certainty (Wolk & Wojciuch 1971b). An experiment of Stewart *et al*. (1969), however, indicates that this is so. Filaments of *A. cylindrica* were progressively disrupted by sonication so that heterocysts became increasingly detached as more of the vegetative cells were disrupted. The nitrogenase activity was measured at intervals during this process and was found to decrease to zero at a point at which the percentage of attached heterocysts became zero but when short chains of up to 10 intact vegetative cells still remained. It appears from this that vegetative cells lack a capacity to fix nitrogen but that fixation in heterocysts normally occurs only when they are attached to vegetative cells.

Thus, there seems good evidence that the heterocysts of aerobically grown *Anabaena cylindrica*, at least, are the sites of nitrogen fixation. The objections that have been raised against the hypothesis can mostly be discounted. Allsopp (1968) urged that heterocysts have important morphological functions. This may well be true but it does not preclude them from also being organs for nitrogen fixation. Smith and Evans (1971) found nitrogenase in the supernatant from preparations in which vegetative cells but not heterocysts had been disrupted. Their argument that the nitrogenase must have come from the vegetative cells is not valid in view of Fay and Lang's (1971) demonstration that the method of sonic disruption which they used seriously damages heterocysts so that the nitrogenase which they contain could leak out.

A similar objection applies to the finding of Ohmori and Hattori (1971) that after short exposure to $^{15}N_2$ heterocysts had no greater concentration of the tracer than vegetative cells, since soluble products of fixation may have been lost from damaged heterocysts. An experiment by the same authors in which nitrogen fixation apparently occurred in short, actively growing filaments without heterocysts is, however, not easily reconciled with the hypothesis that heterocysts are the site of nitrogen fixation. The finding of Kurtz and La Rue (1971) that filaments with few heterocysts may fix nitrogen at high rates whereas others with abundant heterocysts may have low nitrogenase activity, does not take into account the possibility that the nitrogenase activity of the individual heterocyst, may vary (see p. 567). Wolk (1970) suggested that all the evidence for nitrogen fixation in heterocysts could be explained on the assumption that vegetative cells contain nitrogenase but that its activity depends on a specific product of heterocysts. In the absence of any evidence as to the nature of this substance this seems to be multiplying hypotheses unnecessarily.

The biochemical and structural features of heterocysts seem well suited to their nitrogen-fixing function. It may be that the thick, several-layered wall (Lang & Fay 1971) plays some part in excluding oxygen which might inactivate nitrogenase but this remains speculative. There is no doubt, however, that the heterocyst is the site of intense reducing activity. Fay and Walsby (1966) found with heterocysts isolated by the French press and therefore damaged, that the rate of respiration on a dry weight basis was almost twice that of vegetative cells. The powerful reducing activity of heterocysts with respect to TTC has been mentioned above. They are also capable of reducing the silver salt in photographic emulsions (Stewart *et al.* 1969). Such reducing conditions would favour nitrogen fixation. On the other hand, heterocysts lack a complete photosynthetic system which, by producing oxygen, might counteract the reducing activity. Fay and Walsby (1966) found that isolated heterocysts were incapable of fixing $^{14}C$-labelled carbon dioxide, whereas filaments which had also been passed through the French press, though at a lower pressure, did so actively. Heterocysts in their preparations were undoubtedly damaged but, if intact filaments are exposed for short periods to $^{14}C$-carbon dioxide in the light, it can be shown by radio-autography that little or no fixation occurs in the heterocysts as compared with vegetative cells (Wolk 1968, Stewart *et al.* 1969). Since they do not fix carbon dioxide, heterocysts are unlikely to produce oxygen. This is in line with the finding that heterocysts lack photosystem II, which is responsible for release of oxygen. Although they contain chlorophyll *a* to the extent of about 77% of that in vegetative cells, heterocysts of *Anabaena cylindrica* have little phycocyanin and appear to lack myxoxanthophyll, both of which pigments are especially associated with photosystem II (Fay 1969, Wolk & Simon 1969). This has been confirmed on single cells of an *Anabaena* sp. using a microspectrophotometric technique (Thomas 1970). Donze *et al.* (1972) showed that heterocysts of *A. cylindrica* contain a high concentration of the photosystem I reaction centre, P700, have a low yield of chlorophyll *a* fluorescence, and lack Hill-reaction activity, all these facts being consistent with the idea that heterocysts contain photosystem I only.

Scott and Fay (1972) have demonstrated, using $^{32}$P as tracer, that photophosphorylation takes place in isolated heterocysts. It is inhibited by $O_2$ and its rate is at least twice that of oxidative phosphorylation.

Traces of molybdenum are essential for nitrogen fixation and if this element is deficient heterocyst frequency in *Anabaena cylindrica* may double, a response to the nitrogen deficiency which develops in the filaments, but nitrogenase activity falls to a fifth of that in control material with an ample molybdenum supply (Fay unpubl.). This shows clearly that the heterocyst structure and its nitrogenase content are independent variables. Jewell and Kulasooriya (1970) found that in cultures of various blue-green algae in which the rate of nitrogen fixation varied more than seventy-fold the variation per heterocyst was, at the most, nine-fold. With *Anabaena flos-aquae* in continuous culture, Bone (1971a) found that the ratio of nitrogenase activity to heterocyst numbers varied ninety-fold depending on the source of nitrogen.

The ultrastructure of heterocysts was studied in detail by Lang and Fay (1971). In the course of differentiation there is an extensive elaboration of membranes, involving dense layering and, frequently, coiling, in addition to reorganization of thylakoids. This is no doubt related to the loss of photosystem II and the development of the nitrogenase system, but the function of the various structures remains to be determined. Nitrogenase activity is not manifest until heterocysts are fully differentiated (Kulasooriya *et al.* 1972).

## 4 NITROGEN FIXATION IN NON-HETEROCYSTOUS SPECIES

Reports of nitrogen fixation by non-heterocystous blue-green algae have often been equivocal and most species have failed to grow in media, lacking combined nitrogen, which supported the growth of heterocystous species (Fogg & Wolfe 1954). Nevertheless, Wyatt and Silvey (1969) found a *Gloeocapsa* strain which was able to grow in axenic culture in the absence of combined nitrogen and showed by the acetylene reduction technique that it is able to fix nitrogen. These results were confirmed by Rippka *et al.* (1971) with what appears to be an independent isolate of the same species. Vigorous nitrogen fixation was observed to take place only over a short period in aerobic batch culture in the light. Other species of unicellular blue-green algae show no evidence of nitrogen fixation under aerobic or micro-aerophilic conditions.

Various filamentous non-heterocystous species have been reported as growing in media free from combined nitrogen, but reports of nitrogen fixation have been most persistent in respect of the marine plankton genus *Trichodesmium*, which is characteristic of nutrient-poor waters. There is much evidence from tracer experiments with $^{15}N_2$ on natural populations obtained from widely separated sea areas which shows that nitrogen fixation is associated with *Trichodesmium* blooms (Dugdale *et al.* 1961, Dugdale *et al.* 1964, Goering *et al.* 1966, Dugdale & Goering 1967). Since it is light dependent, this property presumably appertains to the alga rather than to associated bacteria. Observed rates of fixation have been as high as $2\mu g$ N $1^{-1}$ $h^{-1}$ and up to 10 times greater than the rate of removal of ammonium nitrogen from the water by the alga. Bunt *et al.* (1970) obtained high acetylene reduction rates by unialgal and almost bacteria-free *Trichodesmium* collected off the coast of Florida. Nevertheless, convincing evidence of the ability of *Trichodesmium* to fix $N_2$ in pure culture is so far lacking. Ramamurthy and Krishnamurthy (1968) compared increases in cell-nitrogen in cultures of *Trichodesmium* in 'nitrogen-free' medium and similar media containing nitrate or ammonium salts. However, the observation that greater increase took place in the 'nitrogen-free' medium cannot be accepted as evidence of nitrogen fixation since the concentrations at which the other nitrogen sources were used were so high as probably to be inhibitory.

Convincing evidence for nitrogen fixation was obtained with *Plectonema boryanum* by Stewart and Lex (1970) using acetylene reduction, incorporation of $^{15}N_2$, and growth in a medium free from combined nitrogen, as criteria. Fixation only occurred in the absence of added oxygen and was completely inhibited in air. This has been confirmed in other non-heterocystous filamentous algae by Kenyon *et al.* (1972) and by Singh (1972). Cell-free extracts capable of fixing nitrogen have been obtained from *Plectonema* by Haystead *et al.* (1970).

These findings suggest that, while unicellular algae may fix $N_2$ aerobically, non-heterocystous filamentous species are only able to fix nitrogen vigorously under micro-aerophilic conditions, and are in line with the results of Stewart and Pearson (1970) showing that nitrogen fixation by *Anabaena flos-aquae*, which has heterocysts, is greater the lower the oxygen concentration. In view of this, it will be necessary to re-examine under micro-aerophilic conditions the capacity for nitrogen fixation of non-heterocystous algae such as *Oscillatoria* and *Phormidium*, some species of which have been reported as nitrogen fixers on inadequate evidence (see, for example, Copeland 1932). Stewart (1971a) has, in fact, reported acetylene reduction by *Lyngbya* and *Phormidium* spp. in the absence of oxygen. However, nitrogen fixation by *Trichodesmium* does not seem to fit into this picture since these forms are usually found under fully aerobic conditions. The difficulty might be resolved if it were found that *Trichodesmium* underwent its main phase of growth and nitrogen fixation towards the bottom of the photic zone in water of low oxygen content, only rising to the surface, oxygenated, layers under certain circumstances. Suitable situations for this exist over broad areas of tropical seas where the oxygen content between 150 and 500m depth approaches zero (Vaccaro 1965). Another possibility is that sufficiently reduced oxygen concentrations for nitrogen fixation may occur in the middle of the bundles of filaments of *Trichodesmium*. It may be significant that *Aphanizomemon flos-aquae*, which fixes nitrogen but often has very few heterocysts (see Horne & Fogg 1970), also occurs in bundles.

## 5 RELATIONSHIPS BETWEEN NITROGEN FIXATION AND PHOTOSYNTHESIS

The recent work described above has largely resolved the uncertainty about the relationship of nitrogen fixation and photosynthesis. A close connexion between these two processes was postulated by Fogg and Than-Tun (1960), who found that their rates were closely correlated under a wide range of growth conditions. Rate of nitrogen fixation by nitrogen-starved *Anabaena cylindrica* increases with light intensity, becoming saturated at about 6,000 lx (Cox & Fay 1969). In the photosynthetic bacteria, nitrogenase activity is dependent on photosynthesis for hydrogen donor and ATP (Schick 1971). On the other hand, nitrogen fixation in *Chlorogloea fritschii* (Fay 1965) and *Anabaenopsis circularis* (Watanabe & Yamamoto 1967) can take place during growth in the dark and even in the obligate phototroph, *Anabaena cylindrica*, nitrogen fixation is decreased to a greater extent by inhibitors specific for respiration than by those for photosynthesis (Cox 1966). Finally the finding that nitrogenase in cell-free extracts is inactivated by oxygen (Fay & Cox 1967, Haystead *et al.* 1970) and that nitrogenase activity in intact algae is greater under microaerophilic conditions (Stewart & Pearson 1970) would seem to preclude any close association of this enzyme with the photosynthetic mechanism.

These apparently conflicting facts are largely reconciled if it is accepted that in heterocystous algae fixation is confined to heterocysts. The absence of competition in the short term between photosynthetic fixation of carbon dioxide and light stimulated acetylene reduction, which has been demonstrated by Lyne and Stewart (1973), argues for some compartmentalization with separate pools of reductant and ATP for the two processes. The absence of photosystem II and the retention of photosystem I in heterocysts means that $O_2$ is not released in the vicinity of nitrogenase yet the availability of ATP produced by photophosphorylation could account for the close dependence of nitrogen fixation on light intensity without precluding the possibility that it could also use ATP produced by oxidative phosphorylation. This would leave a requirement for hydrogen donor and carbon skeletons which would have to be provided by full photosynthesis by adjacent vegetative cells. By autoradiography with $^{14}C$ as a tracer, Wolk (1968) demonstrated that products of photosynthesis in *Anabaena cylindrica* pass via an intrafilamentous route from vegetative cells to heterocysts in the dark. This transfer might well be dependent on respiration, thus accounting for the sensitivity of nitrogen fixation towards inhibitors specific for respiration. In algae under micro-aerophilic conditions, nitrogen fixation could occur in vegetative cells and be dependent on photosynthesis for ATP, hydrogen donor, and carbon skeleton, as it is in the photosynthetic bacteria, provided that diffusion and respiration were sufficient to prevent $O_2$ accumulating in inactivating concentrations in the vicinity of the nitrogenase.

This picture is complicated by the occurrence of photorespiration. Lex *et al.* (1972) found that light-stimulated $O_2$ uptake in *A. cylindrica* may take place at a rate up to twenty times that of dark respiration. The rate of $O_2$ uptake in the light increases linearly with $pO_2$ whereas dark respiration becomes saturated at a $pO_2$ of about 0·05 atm. The rate of photorespiration approaches that of true photosynthesis at the carbon dioxide compensation point but becomes zero at carbon dioxide concentrations of about 0·02 atm. Conditions which stimulate photorespiration, i.e. high $pO_2$ or low $pCO_2$, inhibit nitrogenase activity. In short-term studies, DCMU inhibits nitrogen fixation as measured by acetylene reduction, under conditions which favour photorespiration, but has little effect when photorespiration is inhibited. These results suggest that photorespiration and nitrogenase activity compete indirectly for reducing power and that, in addition to inactivating nitrogenase, $O_2$ inhibits nitrogen fixation via stimulation of photorespiration. As Lex *et al.* (1972) point out, this hypothesis can account for many of the anomalous features that have been observed by other workers in studies of nitrogen fixation under various conditions of light intensity and supply of $CO_2$ and $O_2$.

## 6  PHYSIOLOGICAL ECOLOGY OF ALGAL NITROGEN FIXATION

The adaptation of the $^{15}N_2$ (Neess *et al.* 1962, Stewart 1967a, Fogg & Horne 1967) and acetylene reduction techniques (Stewart *et al.* 1967, 1968, Hardy *et al.* 1968) for use in the field has resulted in a great deal of information about the performance of nitrogen-fixing blue-green algae in their natural habitats. It should be noted however, that when field samples are fixed and kept some time before determination of ethylene, the acetylene reduction method is liable to considerable error if not used critically. In various habitats, fixation has been found to be correlated with the presence of heterocystous algae (Stewart 1965, Ogawa & Carr 1969, Horne & Fogg 1970, Granhall & Lundgren 1971), and heterocyst counts may enable a rapid semi-quantitative assessment of the nitrogen-fixing capacity of a natural algal sample to be made (Jewell & Kulasooriya 1970). In view of the work described above on non-heterocystous species under micro-aerophilic conditions it is likely that this method, as well as the $^{15}N$ and acetylene reduction techniques, which so far seem mainly to have been used under fully aerobic conditions, underestimates the importance of blue-green algae as nitrogen fixers. Nevertheless, as tables of nitrogen-fixing activity for various habitats compiled by Stewart (1969, 1971b) and Fogg (1971a) show, they make appreciable contributions to the nitrogen budget of certain habitats. A full discussion of the ecological aspects of this would be out of place here but certain physiological aspects may be touched on. General accounts of the ecology of nitrogen fixation by blue-green algae are given by Stewart (1969, 1970a) and by Fogg *et al.* (1973) and more specialized reviews relating to tropical soils by Singh (1961) and Watanabe and Yamamoto (1971), to temperate soils by Shtina (1969) and Henriksson (1971), to freshwaters by Stewart (1968) and Fogg (1971a), and to the sea by Stewart (1971b).

## 6.1 *Light*

In the field, nitrogen fixation by blue-green algae is usually light-dependent. This is so in freshwater (Dugdale & Dugdale 1962, Goering & Neess 1964, Horne & Fogg 1970), the rate showing much the same relationship as photosynthesis itself to the light intensity as it varies with depth, with inhibition at the surface in full sunlight, a maximum some way below the surface and light limitation below this (Horne & Fogg 1970, Horne & Viner 1971). Low, but statistically significant rates of fixation have been recorded below the photic zone (Dugdale & Dugdale 1962, Goering & Neess 1964, Horne & Fogg 1970, Horne & Viner 1971) but there is the possibility that this may be carried out by photosynthetic bacteria. Dark fixation was found by Dugdale and Dugdale (1962) to be greater if the samples had been previously exposed to light, but it was not increased by addition of glucose as a substrate (Goering & Neess 1964). Fixation in a sand dune slack was also light dependent (Stewart 1965), but such situations do not lend themselves as readily as aquatic habitats to precise investigations of the role of light.

## 6.2 *Combined nitrogen*

In laboratory cultures, heterocyst formation and nitrogen fixation are suppressed in the presence of a readily available source of combined nitrogen such as an ammonium salt or nitrate. However, the situation is not as simple as it once appeared. Combined nitrogen inhibits synthesis of nitrogenase rather than the activity of the existing enzyme complex (Stewart *et al.* 1968). Suppression of heterocyst formation by ammonium salts is usually complete but by nitrate may be only partial. Ogawa and Carr (1969), indeed, found heterocysts to be still present in *Anabaena variabilis* grown in the presence of 28 mg $1^{-1}$ $NO_3$-N. With an *Anabaena* sp., Thomas and David (1971) found heterocysts to be totally inhibited when nitrate was supplied in batch culture in which ammonium accumulated in the medium, but not when it was supplied in continuous culture at high dilution rates. It is not surprising, therefore, that although nitrogen fixation in lakes is generally confined to periods when the concentration of combined inorganic nitrogen in the water is low, there is no close inverse correlation between rate of nitrogen fixation and this concentration (Dugdale & Dugdale 1962, Horne & Fogg 1970). Goering and Neess (1964) found that addition of ammonia to plankton samples had no consistent effect on nitrogen fixation, sometimes even increasing its rate, and Horne (unpubl.) found appreciable nitrogen fixation in Lake Windermere, England, even when nitrate was at its high winter level. It should also be remembered that both in soil and water, $N_2$ is present in relatively high concentration and diffuses more rapidly than either nitrate or ammonium ions (Fogg 1971a). It seems a general rule that, in freshwater, rates of nitrogen fixation are positively correlated with concentrations of dissolved organic nitrogen (Dugdale & Dugdale 1962, Goering & Neess 1964, Horne & Fogg 1970, Fogg 1971a). This is probably a reflection of the general growth promoting effect, perhaps dependent on chelation of inorganic ions, which dissolved organic matter has for blue-green algae in general.

### 6.3 Mineral nutrition

A general summary of the mineral requirements of blue-green algae is given by Fogg et al. (1973). Okuda and Yamaguchi (1956) concluded that growth of nitrogen fixing algae in paddy fields was most usually limited by low pH and deficiency of phosphorus. Molybdenum is also likely to be a limiting factor (Okuda et al. 1962). Addition of lime, superphosphate and sodium molybdate has been reported by Indian workers to enhance nitrogen fixation by algae as judged by yield of rice or nitrogen content of the grain and straw (Subrahmanyan & Sahay 1964, Sankaram et al. 1967). Addition of phosphate to natural seawater increased nitrogen fixation by blue-green algae grown in it (Stewart 1964). Phosphorus-starved cells of *Anabaena flos-aquae* rapidly increase their capacity to reduce acetylene when phosphate is supplied, and this has been proposed as a basis for bioassay of available phosphorus in aquatic ecosystems (Stewart et al. 1970, Stewart & Alexander 1971).

### 6.4 Salinity

Marine blue-green algae fix nitrogen as actively as their freshwater counterparts. The marine species *Calothrix scopulorum* and *Nostoc entophytum* grow well in media of salinities ranging from 1 ‰ to 40 ‰, the growth of the former being totally inhibited at 106 ‰ and of the latter at 60 ‰ (Stewart 1964). Nitrogen fixation by *C. scopulorum* was at a maximum at 5 ‰, and was reduced to about half this at 45 ‰ (Jones & Stewart 1969a). A *Calothrix* sp. isolated from a rice field showed good growth at 1·6 ‰ but was severely inhibited at 3·4 ‰ (El-Nawawy et al. 1968).

### 6.5 *Herbicides*

Lundkvist (1970) has reported that 2,4-dichlorophenoxyacetic acid and 2-methyl-4-chlorophenoxyacetic acid inhibited nitrogen fixation by *Nostoc punctiforme*, *N. muscorum* and *Cylindrospermum* sp. when applied at concentrations used in weed-killing (approximately 0·01M). Nitrogen fixation was, however, stimulated at 1·0M $\times$ $10^{-4}$ to $10^{-5}$ concentrations.

### 6.6 *Temperature* ⌣

Active nitrogen fixation by blue-green algae has been observed, both at nearly freezing point, in the Antarctic (Fogg & Stewart 1968, Horne 1972) and in hot springs (Stewart 1970b). *Mastigocladus laminosus* and *Calothrix* sp. are the most abundant nitrogen fixing species in Yellowstone Park hot spring streams (the former having an optimum of 42·5°C and an upper limit of 54°C for fixation) and evidently contribute substantially to the productivity of the water. *Nostoc commune*, in the Antarctic, had a temperature coefficient ($Q_{10}$) for fixation as high as 6 and accomplishes most fixation when its microhabitat reaches 10°C or above. Fixation is prevented by desiccation (Fogg & Stewart 1968). Goering and Neess (1964) found a $Q_{10}$ of about 3 for fixation by planktonic algae in a Wisconsin lake, and Horne and Fogg (1970) obtained about the same value for English lakes although, in this case, the results were obtained *in situ*, and were barely statistically significant.

### 6.7 *Transfer of fixed nitrogen to other organisms*

That blue-green algae liberate relatively large quantities of combined nitrogen, mainly in the form of polypeptides with lesser amounts of free amino acids, during the course of healthy growth has been confirmed by a number of workers (Stewart 1963, Fay *et al.* 1964, Fogg & Pattnaik 1966, Rzhanova 1967, Jones & Stewart 1969a). Stewart (1967b) demonstrated by using $^{15}N_2$ as a tracer that higher plants in a sand-dune slack region take up nitrogen fixed by blue-green algae in the soil. Appreciable transfer of combined nitrogen occurred after one week but it was not clear whether this was *via* autolysis. However, Jones and Stewart (1969b) found in both long-term growth experiments and short-term tracer experiments that a proportion of the extracellular combined nitrogen liberated by *Calothrix scopulorum* can be taken up by a variety of algae, fungi and bacteria. Uptake was by both active assimilation and passive processes, the relative importance of these varying with the test organism and the nature of the extracellular product. Adsorption of extracellular products onto suspended inorganic particles also occurred, a process which might facilitate their utilization by bacteria. The decomposition of the nitrogen-fixing species, *Tolypothrix tenuis*, *Calothrix brevissima*, *Nostoc muscorum* and *Anabaenopsis circularis*, by various bacteria has been investigated by Watanabe and Kiyohara (1960). A particular strain of *Bacillus subtilis* was found to give the most active liberation of ammonia, releasing about 40% of the total cell nitrogen in this form in 10 days at 30°C. Other soluble nitrogenous products of decomposition were scarcely evident.

### 6.8 *Economic importance*

The importance of nitrogen fixation by blue-green algae in contributing to the fertility of rice-fields is now generally recognized (Singh 1961, Fogg 1971b, Watanabe *et al.* 1971). Within the limits of accuracy of the counting methods used, blue-green algae were found to be as numerous as photosynthetic bacteria and heterotrophic nitrogen-fixing bacteria in paddy-fields in south-east Asia (Kobayashi *et al.* 1967). Inoculation of paddy-fields with nitrogen-fixing blue-green algae has been found to result in substantial increases in rice-yields. Thus in various districts of Japan, inoculation with *Tolypothrix tenuis* grown in mass culture produced increases in rice yield mounting after three years to a mode of about 10% and a maximum of 55% (Watanabe *et al.* 1971). In the United Arab Republic, inoculation with about 250 (g dry wt) ha$^{-1}$ of the same species resulted in a 19·5% increase in rice yield as compared with 16·6% produced by a dressing of 25 kg ha$^{-1}$ of ammonium sulphate (Aboul-Fadl *et al.* 1967). Rather similar results were reported from the Indian Central Rice Research Institute (Subrahmanyan *et al.* 1964). Partial sterilization of the soil by burning the rice straw before inoculation and fertilization with lime, superphosphate and molybdenum enhanced the effect of the algae, giving yields as much as 185% of the control in field experiments (Subrahmanyan *et al.* 1965a,b). Such increased yields are obtained with the improved rice varieties as well as traditional strains. A simple means of cultivating algal inoculum, operable by peasant farmers, has been devised (Venkataraman 1971, 1972).

It is now realized that the importance of nitrogen fixing blue-green algae in temperate soils was discounted prematurely (Henriksson 1971), and that perhaps much of the gain in soil nitrogen which takes place when soil lies fallow may be attributed to these organisms. In freshwater, too, there may be a considerable contribution to productivity. Horne and Viner (1971) estimate that perhaps 33% of the annual nitrogen turnover in Lake George, Uganda, which supports an important fishery, is met by nitrogen fixation by blue-green algae. On the other hand, it may aggravate eutrophication problems: measurements made by the acetylene technique over a period of five months in Lake Erken, Sweden, showed that nitrogen fixation by blue-green algae in this moderately eutrophic lake increases the annual loading with combined nitrogen by 40% (Granhall & Lundgren 1971).

## 7 NITROGEN FIXATION IN SYMBIOSES INVOLVING BLUE-GREEN ALGAE

A general account of the great variety of symbiotic associations formed between blue-green algae and other plants or animals has been given by Fogg *et al.* (1973). In some of these, the blue-green alga is of a proven nitrogen-fixing species. In many instances there is direct evidence of nitrogen fixation by the intact symbiotic system, e.g. by the lichens *Collema* spp. (Bond & Scott 1955, Rogers *et al.* 1966, Henriksson & Simu 1971, Fogg & Stewart 1968, Stewart 1971a), *Leptogium lichenoides* (Bond & Scott 1955), *Peltigera* spp. (Scott 1956, Watanabe & Yamamoto 1971), *Stereocaulon,* sp. (Fogg & Stewart 1968) and *Lichina* spp. (Stewart 1971b); by liverworts and blue-green algal symbionts such as *Blasia*, and *Cavicularia* (Bond & Scott 1955, Pankow & Martens 1964, Watanabe & Kiyohara 1963, Watanabe & Yamamoto 1971), and by cycads with coralloid roots containing blue-green algae (Watanabe & Kiyohara 1963, Stewart *et al.* 1968, Watanabe & Yamamoto 1971); and by water fern, *Azolla* (Saubert 1949). Lichens which have phycobionts other than blue-green algae do not fix nitrogen (Scott 1956, Stewart 1971b), except at marginal rates (Rogers *et al.* 1966) which may be attributable to the presence of bacteria. In lichens in which the principal phycobiont is a green alga, nitrogen fixation only occurs if cephalodia containing blue-green algae are present (Fogg & Stewart 1968, Millbank & Kershaw 1969).

Physiological studies show that the metabolism of nitrogen-fixing blue-green algae is greatly modified by symbiosis. In *Peltigera aphthosa*, the *Nostoc* phycobiont shows a notably rapid rate of fixation and almost complete transfer of fixed nitrogen to the mycobiont (Kershaw & Millbank 1970). This is paralleled by a high rate of transfer of carbohydrate from phycobiont to mycobiont in *P. polydactyla* (Drew & Smith 1967). The loss of carbohydrate from the *Nostoc* cells was considerably reduced once they were isolated from the lichen (Drew & Smith 1967). Millbank (1972) estimated the *Nostoc* content of *Peltigera canina* as 2·7% of the total thallus nitrogen. The nitrogenase activity of the symbiotic *Nostoc* on this basis appeared to be two or three times that of the free-living alga. Examination by light microscopy often gives the impression that heterocysts are rare in the symbiotic state and the suggestion has been made that under micro-aerophilic conditions produced by the mycobiont the vegetative cells of the alga develop nitrogenase activity (Millbank 1972). However, it has been shown by electron microscopy that the heterocyst frequency of the *Nostoc* in a sample of *Peltigera canina* was 3·27% (Griffiths *et al.* 1972). Nitrogen fixation ceases when lichens are desiccated (Fogg & Stewart 1968) but even after 30 weeks storage in the dry condition *Collema tuniforme* and *Peltigera rufescens* showed almost complete recovery of nitrogen-fixing ability after soaking in water for 1 day (Henriksson & Simu 1971).

# 8 NITROGEN FIXATION BY ALGAE OTHER THAN CYANOPHYCEAE

In view of the failure of Millbank (1969) to confirm nitrogen fixation in various eukaryote organisms which had been reported to possess the property, it seems likely that it is confined to the Prokaryota and therefore to be found only in the Cyanophyceae among the algal groups. In accordance with this view, Stewart (1971b), who examined a number of green, brown and red algae by the acetylene reduction technique, found no evidence for nitrogen fixation by any of them, although *Calothrix* sp. under similar conditions reduced acetylene vigorously.

# 9 REFERENCES

ABOUL-FADL M., TAHA E.M., HAMISSA M.R., EL-NAWAWY A.S. & SHOUKRY A. (1967) The effect of the nitrogen fixing blue-green alga, *Tolypothrix tenuis* on the yield of paddy. *J. Microbiol. U.A.R.*, **2**, 241–9.

ALLSOPP A. (1968) Germination of hormocysts of *Scytonema javanicum* and the function of blue-green algal heterocysts. *Nature, Lond.* **220**, 810.

BOND G. & SCOTT G.D. (1955) An examination of some symbiotic systems for fixation of nitrogen. *Ann. Bot. N.S.* **19**, 67–77.

BONE D.H. (1971a) Nitrogenase activity and nitrogen assimilation in *Anabaena flos-aquae* growing in continuous culture. *Arch. Mikrobiol.* **80**, 234–41.

BONE D.H. (1971b) Kinetics of synthesis of nitrogenase in batch and continuous culture of *Anabaena flos-aquae*. *Arch. Mikrobiol.* **80**, 242–51.

BOTHE H. (1970) Photosynthetische Stickstoffixierung mit einem zellfreien Extrakt aus der blaualge *Anabaena cylindrica*. *Ber. Dtsch. Bot. Ges.* **88**, 421–82.

BOTHE H. & FALKENBERG R. (1973) Demonstration and possible role of a ferredoxin-dependent pyruvate decarboxylation in the nitrogen-fixing blue-green alga *Anabaena cylindrica*. *Plant Sci. Lett.* **1**, 151–6.

BUNT J.S., COOKSEY K.E., HEEB M.A., LEE C.C. & TAYLOR B.F. (1970) Assay of algal nitrogen fixation in the marine subtropics by acetylene reduction. *Nature, Lond.* **227**, 1163–4.

BURRIS R.H. (1969) Progress in the biochemistry of nitrogen fixation. *Proc. R. Soc. B.* **172**, 339–54.

COBB H.D. & MYERS J. (1964) Comparative studies on nitrogen fixation and photosynthesis in *Anabaena cylindrica*. *Am. J. Bot.* **51**, 753–62.

COPELAND J.J. (1932) Nitrogen fixation by Myxophyceae. *Am. J. Bot.* **19**, 844.

COX R.M. (1966) Physiological studies on nitrogen fixation in the blue-green alga *Anabaena cylindrica*. *Arch. Mikrobiol.* **53**, 263–76.

COX R.M. & FAY P. (1967) Nitrogen fixation and pyruvate metabolism in cell-free preparations of *Anabaena cylindrica*. *Arch. Mikrobiol.* **58**, 357–65.

COX R.M. & FAY P. (1969) Special aspects of nitrogen fixation by blue-green algae. *Proc. R. Soc. B.* **172**, 357–66.

COX R.M., FAY P. & FOGG G.E. (1964) Nitrogen fixation and photosynthesis in a subcellular fraction of the blue-green alga, *Anabaena cylindrica*. *Biochim. biophys. Acta*, **88**, 208–10.

DHARMAWARDENE M.W.N., HAYSTEAD A. & STEWART W.D.P. (1973) Glutamine synthetase of the nitrogen-fixing alga *Anabaena cylindrica*. *Arch. Mikrobiol.* **90**, 281–5.

DONZE M., HAVEMAN J. & SCHIERECK P. (1972) Absence of photosystem II in heterocysts of the blue-green alga *Anabaena*. *Biochim. biophys. Acta*, **256**, 157–61.

DREW E.A. & SMITH D.C. (1967) Studies in the physiology of lichens. VIII. Movement of glucose from alga to fungus during photosynthesis in the thallus of *Peltigera polydactyla*. *New Phytol.* **66**, 389–400.

DREWES K. (1928) Über die Assimilation des Luftstickstoffs durch Blaualgen. *Zentr. Bakteriol. Parasitenk Abt. II.* **76**, 88–101.

DUGDALE R.C. & DUGDALE V.A. (1962) Nitrogen metabolism in Lakes. II. Role of nitrogen fixation in Sanctuary Lake, Pennsylvania. *Limnol Oceanogr.* **7**, 170–7.

DUGDALE R.C. & GOERING J.J. (1967) Uptake of new and regenerated forms of nitrogen in primary productivity. *Limnol. Oceanogr.* **12**, 196–206.

DUGDALE R.C., GOERING J.J. & RYTHER J.H. (1964) High nitrogen fixation rates in the Sargasso Sea and the Arabian Sea. *Limnol. Oceanogr.* **9**, 507–10.

DUGDALE R.C., MENZEL D.W. & RYTHER J.W. (1961) Nitrogen fixation in the Sargasso Sea. *Deep-Sea Res.* **7**, 298–300.

EL-NAWAWY A.S., IBRAHIM A.N. & ABOUL-FADL M. (1968) Nitrogen fixation by *Calothrix* sp. as influenced by certain sodium salts and nitrogenous compounds. *Acta Agron. Acad. Sci. Hungar.* **17**, 323–7.

FAY P. (1965) Heterotrophy and nitrogen fixation in *Chlorogloea fritschii*. *J. gen. Microbiol.* **39**, 11–20.

FAY P. (1969) Cell differentiation and pigment composition in *Anabaena cylindrica*. *Arch. Mikrobiol.* **67**, 62–70.

FAY P. (1970) Photostimulation of nitrogen fixation in *Anabaena cylindrica*. *Biochim. biophys. Acta* **216**, 353–6.

FAY P. & COX R.M. (1967) Oxygen inhibition of nitrogen fixation in cell-free preparations of blue-green algae. *Biochim. biophys. Acta* **143**, 562–9.

FAY P., KUMAR H.D. & FOGG G.E. (1964) Cellular factors affecting nitrogen fixation in the blue-green alga *Chlorogloea fritschii*. *J. gen. Microbiol.* **35**, 351–60.

FAY P. & LANG N.J. (1971) The heterocysts of blue-green algae. I. Ultrastructural integrity after isolation. *Proc. R. Soc. B.* **178**, 185–92.

FAY P., STEWART W.D.P., WALSBY A.E. & FOGG G.E. (1968) Is the heterocyst the site of nitrogen fixation in blue-green algae? *Nature, Lond.* **220**, 810–12.

FAY P. & WALSBY A.E. (1966) Metabolic activities of isolated heterocysts of the blue-green alga *Anabaena cylindrica*. *Nature, Lond.* **209**, 94–5.

FOGG G.E. (1949) Growth and heterocyst production in *Anabaena cylindrica* Lemm. II. In relation to carbon and nitrogen metabolism. *Ann. Bot. N.S.* **13**, 241–59.

FOGG G.E. (1956) Nitrogen fixation by photosynthetic organisms. *A. Rev. Pl. Physiol.* **7**, 51–70.

FOGG G.E. (1962) Nitrogen fixation. In *Physiology and Biochemistry of Algae*, ed. Lewin R.A. pp. 161–70. Academic Press, New York & London.

FOGG G.E. (1971a) Nitrogen fixation in lakes. *Plant & Soil*, Special Vol. 1971, 393–401.

FOGG G.E. (1971b) Blue-green algae in rice cultivation. *Proc. Third Int. Conference on the Global Impacts of Applied Microbiology*. University of Bombay, pp. 46–52.

FOGG G.E. & HORNE A.J. (1967) The determination of nitrogen fixation in aquatic environments. In *Chemical Environment in the Aquatic Habitat*, ed. Golterman H.L. & Clymo R.S. pp. 115–20. Noord-Hollandsche Uitgerers Maatschappij, Amsterdam.

FOGG G.E. & PATTNAIK H. (1966) The release of extracellular nitrogenous products by *Westiellopsis prolifica* Janet. *Phykos.* **5**, 58–67.

FOGG G.E. & STEWART W.D.P. (1965) Nitrogen fixation in blue-green algae. *Sci. Prog.* **53**, 191–201.

FOGG G.E. & STEWART W.D.P. (1968) *In situ* determinations of biological nitrogen fixation in Antarctica. *Br. Antarct. Survey Bull.* **15**, 39–46.

FOGG G.E., STEWART W.D.P., FAY P. & WALSBY A.E. (1973) *The Blue-green Algae*. Academic Press, London & New York.

FOGG G.E. & THAN-TUN (1960) Interrelations of photosynthesis and assimilation of elementary nitrogen in a blue-green alga. *Proc. R. Soc. B.* **153**, 111–27.

Fogg G.E. & Wolfe M. (1954) Nitrogen metabolism in blue-green algae. *Symp. Soc. gen. Microbiol.* **4**, 99–125.

Fujita Y., Ohama H. & Hattori A. (1964) Hydrogenase activity of cell-free preparation obtained from the blue-green alga, *Anabaena cylindrica*. *Pl. Cell Physiol., Tokyo*, **5**, 305–14.

Goering J.J., Dugdale R.C. & Menzel D.W. (1966) Estimates of *in situ* rates of nitrogen uptake by *Trichodesmium* sp. in the tropical Atlantic Ocean. *Limnol. Oceanogr.* **11**, 614–20.

Goering J.J. & Neess J.C. (1964) Nitrogen fixation in two Wisconsin lakes. *Limnol. Oceanogr.* **9**, 535–9.

Granhall U. & Lundgren A. (1971) Nitrogen fixation in Lake Erken. *Limnol. Oceanogr.* **16**, 711–19.

Griffiths H.B., Greenwood A.D. & Millbank J.W. (1972) The frequency of heterocysts in the *Nostoc* phycobiont of the lichen *Peltigera canina* Willd. *New Phytol.* **71**, 11–13.

Hardy R.W.F., Burns R.C. & Parshall G.W. (1971) The biochemistry of N₂ fixation. *Advances in Chemistry Series*, No. **100**, 219–47.

Hardy R.W.F., Holsten R.D., Jackson E.K. & Burns R.C. (1968) The acetylene-ethylene assay for N₂ fixation: Laboratory and field evaluation. *Pl. Physiol., Lancaster*, **43**, 1185–207.

Haystead A., Dharmawardene M.W.N. & Stewart W.D.P. (1973) Ammonia assimilation in a nitrogen-fixing blue-green alga. *Plant Sci. Letts.* **1**, 439–45.

Haystead A., Robinson R. & Stewart W.D.P. (1970) Nitrogenase activity in extracts of heterocystous and non-heterocystous blue-green algae. *Arch. Mikrobiol.* **74**, 235–43.

Haystead A. & Stewart W.D.P. (1972) Characteristics of the nitrogenase system of the blue-green alga *Anabaena cylindrica*. *Arch. Mikrobiol.* **82**, 325–36.

Henriksson E. (1971) Algal nitrogen fixation in temperate regions. *Plant & Soil*, Special Vol. 1971, 415–19.

Henriksson E. & Simu B. (1971) Nitrogen fixation by lichens. *Oikos*, **22**, 119–21.

Horne A.J. (1972) The ecology of nitrogen fixation on Signy Island, South Orkney Islands. *Brit. Antarct. Survey Bull.* **27**, 1–18.

Horne A.J. & Fogg G.E. (1970) Nitrogen fixation in some English lakes. *Proc. R. Soc. B.* **175**, 351–66.

Horne A.J. & Viner A.B. (1971) Nitrogen fixation and its significance in tropical Lake George, Uganda. *Nature, Lond.* **232**, 417–18.

Jewell W.J. & Kulasooriya S.A. (1970) The relation of acetylene reduction to heterocyst frequency in blue-green algae. *J. exp. Bot.* **21**, 874–80.

Jones K. & Stewart W.D.P. (1969a) Nitrogen turnover in marine and brackish habitats. III. The production of extracellular nitrogen by *Calothrix scopulorum*. *J. mar. biol. Ass. U.K.* **49**, 475–88.

Jones K. & Stewart W.D.P. (1969b) Nitrogen turnover in marine and brackish habitats. IV. Uptake of the extracellular products of the nitrogen-fixing alga *Calothrix scopulorum*. *J. mar. biol. Ass. U.K.* **49**, 701–16.

Kenyon C.N., Rippka R. & Stanier R.Y. (1972) Fatty acid composition and physiological properties of some filamentous blue-green algae. *Archiv für Mikrobiologie*, **83**, 216–36.

Kershaw K.A. & Millbank J.W. (1970) Nitrogen metabolism in lichens. II. The partition of cephalodial-fixed nitrogen between the mycobiont and phycobionts in *Peltigera aphthosa*. *New Phytol.* **69**, 75–9.

Kobayashi M., Takahashi E. & Kawaguchi K. (1967) Distribution of nitrogen-fixing microorganisms in paddy soils of southeast Asia. *Soil Sci.* **104**, 113–18.

Kulasooriya S.A., Lang N.J. & Fay P. (1972) The heterocysts of blue-green algae. III. Differentiation and nitrogenase activity. *Proc. R. Soc. B.* **181**, 199–209.

Kurtz W.G.W. & La Rue T.A. (1971) Nitrogenase in *Anabaena flos-aquae* filaments lacking heterocysts. *Naturwissenschaften*, **58**, 417.

LANG N.J. & FAY P. (1971) The heterocysts of blue-green algae. II. Details of ultrastructure. *Proc. R. Soc. B.* **178**, 193–203.

LAPORTE G.S. & POURROIT R. (1967) Fixation de l'azote atmosphérique par les algues Cyanophycées. *Rev. Ecol. Biol. Sol.* **4**, 81–112.

LEACH C.K. & CARR N.G. (1971) Pyruvate: ferredoxin oxidoreductase and its activation by ATP in the blue-green alga *Anabaena variabilis. Biochim. biophys. Acta,* **245**, 165–74.

LEAF G., GARDNER I.C. & BOND G. (1958) Observations on the composition and metabolism of the nitrogen-fixing root nodules of *Alnus. J. exp. Bot.* **9**, 320–31.

LEX M., SILVESTER W.B. & STEWART W.D.P. (1972) Photorespiration and nitrogenase activity in the blue-green alga *Anabaena cylindrica. Proc. R. Soc. B.* **180**, 87–102.

LINKO P., HOLM-HANSEN O., BASSHAM J.A. & CALVIN M. (1957) Formation of radioactive citrulline during photosynthetic $C^{14}O_2$-fixation by blue-green algae. *J. exp. Bot.* **8**, 147–56.

LUNDKVIST I. (1970) Effect of two herbicides on nitrogen fixation by blue-green algae. *Svensk bot. Tidskr.* **64**, 460–1.

LYNE R.L. & STEWART W.D.P. (1973) Emerson enhancement of carbon fixation but not of acetylene reduction (nitrogenase activity) in *Anabaena cylindrica. Planta, Berlin,* **109**, 27–38.

MAGEE W.E. & BURRIS R.H. (1954) Fixation of $N_2$ and utilization of combined nitrogen by *Nostoc muscorum. Am. J. Bot.* **41**, 777–82.

MILLBANK J.W. (1969) Nitrogen fixation in moulds and yeasts—a reappraisal. *Arch. Mikrobiol.* **68**, 32–9.

MILLBANK J.W. (1972) Nitrogen metabolism in lichens. IV. The nitrogenase activity of the *Nostoc* phycobiont in *Peltigera canina. New Phytol.* **71**, 1–10.

MILLBANK J.W. & KERSHAW K.A. (1969) Nitrogen metabolism in lichens. I. Nitrogen fixation in the cephalodia of *Peltigera aphthosa. New Phytol.* **68**, 721–9.

NEESS J.C., DUGDALE R.C., DUGDALE V.A. & GOERING J.J. (1962) Nitrogen metabolism in lakes. I. Measurement of nitrogen fixation with $N^{15}$. *Limnol. Oceanogr.* **7**, 163–9.

NEILSON A.H. & DOUDOROFF M. (1973) Ammonia assimilation in blue-green algae. *Arch. Mikrobiol.* **89**, 15–22.

OGAWA R.E. & CARR J.F. (1969) The influence of nitrogen on heterocyst production in blue-green algae. *Limnol. Oceanogr.* **14**, 342–51.

OHMORI M. & HATTORI A. (1971) Nitrogen fixation and heterocysts in the blue-green alga *Anabaena cylindrica. Pl. Cell Physiol., Tokyo,* **12**, 961–7.

OKUDA A. & YAMAGUCHI M. (1956) Nitrogen-fixing microorganisms in paddy soils. II. Distribution of blue-green algae in paddy soils and the relationship between the growth of them and soil properties. *Soil & Plant Food, Tokyo,* **2**, 4–7.

OKUDA A., YAMAGUCHI M. & NIOH I. (1962) Nitrogen-fixing microorganisms in paddy soils. X. Effect of molybdenum on the growth and the nitrogen assimilation of *Tolypothrix tenuis. Soil Sci. & Plant Nutrition,* **8**, 35–9.

PANKOW H. & MARTENS B. (1964) Ueber *Nostoc sphaericum* Vauch. *Arch. Mikrobiol.* **48**, 203–12.

POSTGATE J. (1971) Relevant aspects of the physiological chemistry of nitrogen fixation. *Symp. Soc. gen. Microbiol.* **21**, 287–307.

RAMAMURTHY V.D. & KRISHNAMURTHY S. (1968) Nitrogen fixation by the blue-green alga *Trichodesmium erythraeum* (Ehr.). *Current Sci.* **37**, 21–2.

RIPPKA R., NEILSON A., KUNISAWA R. & COHEN-BAZIRE G. (1971) Nitrogen fixation by unicellular blue-green algae. *Arch. Mikrobiol.* **76**, 341–8.

ROGERS R.W., LANGE R.T. & NICHOLAS D.J.D. (1966) Nitrogen fixation by lichens of arid soil crusts. *Nature, Lond.* **209**, 96–7.

RZHANOVA G.N. (1967) Extracellular nitrogen-containing compounds of two nitrogen-fixing species of blue-green algae. *Mikrobiologiya,* **36**, 639–45 (*Microbiology,* **36**, 536–40).

SANKARAM A. (1971) *Work done on blue-green algae in relation to agriculture.* Indian Council of Agricultural Research Technical Bulletin (Agric.), No. 27.

SANKARAM A., MUDHOLKAR N.J. & SAHAY M.N. (1967) Inoculation of blue-green algae on the yield of rice under field conditions. *Indian J. Microbiol.* **7**, 57–62.

SAUBERT G.G.P. (1949) Provisional communication on the fixation of elementary nitrogen by a floating fern. *Ann. bot. Gdn. Buitenz.* **51**, 177–97.

SCHICK H.-J. (1971) Substrate and light-dependent fixation of molecular nitrogen in *Rhodospirillum rubrum. Arch. Mikrobiol.* **75**, 89–101.

SCHNEIDER K.C., BRADBEER C., SINGH R.N., WANG L.C., WILSON P.W. & BURRIS R.H. (1960) Nitrogen fixation by cell-free preparations from microorganisms. *Proc. natn. Acad. Sci. U.S.A.* **46**, 726–33.

SCOTT G.D. (1956) Further investigations of some lichens for fixation of nitrogen. *New Phytol.* **55**, 111–16.

SCOTT W.E. & FAY P. (1972) Phosphorylation and amination in heterocysts of *Anabaena variabilis. Brit. Phycol. Bull.* **7**, 283–4.

SHTINA E.A. (1969) Some regularities in the distribution of blue-green algae in soils. In *Biology of the Cyanophyta,* eds. Fedorova V.D. & Telichenko M.M. pp. 21–45. Moscow University Press (in Russian with English summary).

SINGH R.N. (1961) *Role of blue-green algae in nitrogen economy of Indian agriculture.* Indian Council of Agricultural Research, New Delhi.

SINGH R.N. (1972) *Physiology and Biochemistry of Nitrogen Fixation by Blue-Green Algae (Final Technical Report),* 1967–1972 Dep. Bot. Banaras Hindu University, Varanasi-5, India.

SMITH R.V. & EVANS M.C.W. (1970) Soluble nitrogenase from vegetative cells of the blue-green alga *Anabaena cylindrica. Nature, Lond.* **225**, 226–8.

SMITH R.V. & EVANS M.C.W. (1971) Nitrogenase activity in cell-free extracts of the blue-green alga, *Anabaena cylindrica. J. Bact.* **105**, 913–17.

SMITH R.V., NOY R.J. & EVANS M.C.W. (1971a) Physiological electron donor systems to the nitrogenase of the blue-green alga *Anabaena cylindrica. Biochim. Biophys. Acta,* **253**, 104–9.

SMITH R.V., TELFER A. & EVANS M.C.W. (1971b) Complementary functioning of nitrogenase components from a blue-green alga and a photosynthetic bacterium. *J. Bact.* **107**, 574–5.

STEWART W.D.P. (1963) Liberation of extracellular nitrogen by two nitrogen-fixing blue-green algae. *Nature, Lond.* **200**, 1020–1.

STEWART W.D.P. (1964) Nitrogen fixation by Myxophyceae from marine environments. *J. gen. Microbiol.* **36**, 415–22.

STEWART W.D.P. (1965) Nitrogen turnover in marine and brackish habitats. I. Nitrogen fixation. *Ann. Bot. N.S.* **29**, 229–39.

STEWART W.D.P. (1966) *Nitrogen Fixation in Plants.* Athlone Press of the University of London.

STEWART W.D.P. (1967a) Nitrogen turnover in marine and brackish habitats. II. Use of [15]N in measuring nitrogen fixation in the field. *Ann. Bot. N.S.* **31**, 385–407.

STEWART W.D.P. (1967b) Transfer of biologically fixed nitrogen in a sand dune slack region. *Nature, Lond.* **214**, 603–4.

STEWART W.D.P. (1968) Nitrogen input into aquatic ecosystems. In *Algae, Man, and the Environment,* ed. Jackson D.F. pp. 53–72. Syracuse University Press.

STEWART W.D.P. (1969) Biological and ecological aspects of nitrogen fixation by free-living micro-organisms. *Proc. Soc. B.* **172**, 367–88.

STEWART W.D.P. (1970a) Algal fixation of atmospheric nitrogen. *Plant & Soil,* **32**, 555–588.

STEWART W.D.P. (1970b) Nitrogen fixation by blue-green algae in Yellowstone thermal areas. *Phycologia,* **9**, 261–8.

STEWART W.D.P. (1971a) Physiological studies on nitrogen-fixing blue-green algae. *Plant & Soil*, Special Vol. (1971), 377–91.

STEWART W.D.P. (1971b) Nitrogen fixation in the sea. In *Fertility of the Sea*, ed. Costlow J.D. pp. 537–64. Gordon & Breach Science Publishers, London.

STEWART W.D.P. (1973) Nitrogen fixation. In *The Biology of Blue-green Algae*, eds. Carr N. & Whitton B.A. pp. 260–78. Blackwell Scientific Publications, Oxford.

STEWART W.D.P. & ALEXANDER G. (1971) Phosphorus availability and nitrogenase activity in aquatic blue-green algae. *Freshwat. Biol.* **1**, 389–404.

STEWART W.D.P., FITZGERALD G.P. & BURRIS R.H. (1967) *In situ* studies on $N_2$ fixation using the acetylene reduction technique. *Proc. natn. Acad. Sci. U.S.A.* **58**, 2071–8.

STEWART W.D.P., FITZGERALD G.P. & BURRIS R.H. (1968) Acetylene reduction by nitrogen-fixing blue-green algae. *Arch. Mikrobiol.* **62**, 336–48.

STEWART W.D.P., FITZGERALD G.P. & BURRIS R.H. (1970) Acetylene reduction assay for determination of phosphorus availability in Wisconsin lakes. *Proc. natn. Acad. Sci. U.S.A.* **66**, 1104–11.

STEWART W.D.P., HAYSTEAD A. & PEARSON H.W. (1969) Nitrogenase activity in heterocysts of filamentous blue-green algae. *Nature, Lond.* **224**, 226–8.

STEWART W.D.P. & LEX M. (1970) Nitrogenase activity in the blue-green alga *Plectonema boryanum* Strain 594. *Arch. Mikrobiol.* **73**, 250–60.

STEWART W.D.P. & PEARSON H.W. (1970) Effects of aerobic and anaerobic conditions on growth and metabolism of blue-green algae. *Proc. R. Soc. B.* **175**, 293–311.

SUBRAHMANYAN R. & SAHAY M.N. (1964) Observations on nitrogen-fixation by some blue-green algae and remarks on its potentialities in rice culture. *Proc. Indian Acad. Sci. B.* **60**, 145–54.

SUBRAHMANYAN R., RELWANI L.L. & MANNA G.B. (1964) Role of blue-green algae and different methods of partial soil sterilization on rice yield. *Proc. Indian Acad. Sci. B.* **60**, 293–7.

SUBRAHMANYAN R., RELWANI L.L. & MANNA G.B. (1965a) Fertility build-up of rice field soils by blue-green algae. *Proc. Indian Acad. Sci. B.* **62**, 252–72.

SUBRAHMANYAN R., RELWANI L.L. & MANNA G.B. (1965b) Nitrogen enrichment of rice soils by blue-green algae and its effect on the yield of paddy. *Proc. Indian Acad. Sci. A.* **35**, 382–6.

THOMAS J. (1970) Absence of the pigments of photosystem II of photosynthesis in heterocysts of a blue-green alga. *Nature, Lond.* **228**, 181–3.

THOMAS J. & DAVID K.A.V. (1971) Studies on the physiology of heterocyst production in the nitrogen-fixing blue-green alga *Anabaena* sp. L-31 in continuous culture. *J. gen. Microbiol.* **66**, 127–31.

VACCARO R.F. (1965) Inorganic nitrogen in sea water. In *Chemical Oceanography*, eds. Riley J.P. & Skirrow G. pp. 365–408. Academic Press, London & New York.

VAN GORKOM H.J. & DONZE M. (1971) Localization of nitrogen fixation in *Anabaena*. *Nature, Lond.* **234**, 231–2.

VENKATARAMAN G.S. (1971) Role of blue-green algae in the tropical agriculture. *Proc. Third Int. Conference on the Global Impacts of Applied Microbiology*, University of Bombay, p. 72.

VENKATARAMAN G.S. (1972) *Algal biofertilisers and rice cultivation.* Today & Tomorrow's Publ., New Delhi.

WATANABE A. & KIYOHARA T. (1960) Decomposition of blue-green algae as effected by the action of soil bacteria. *J. gen. appl. Microbiol., Tokyo*, **5**, 175–9.

WATANABE A. & KIYOHARA T. (1963) Symbiotic blue-green algae of lichens, liverworts and cycads. In *Studies on Microalgae and Photosynthetic Bacteria*, Jap. Soc. Pl. Physiol., Tokyo; pp. 189–96.

WATANABE A., SHIROTA M., ENDO H. & YAMAMOTO Y. (1971) An observation of the practical applications of nitrogen-fixing blue-green algae for rice cultivation. *Proc. Third Int.*

*Conference on the Global Impacts of Applied Microbiology*, University of Bombay, pp. 53–64.

WATANABE A. & YAMAMOTO Y. (1967) Heterotrophic nitrogen fixation by the blue-green alga *Anabaenopsis circularis*. *Nature, Lond.* **214**, 788.

WATANABE A. & YAMAMOTO Y. (1971) Algal nitrogen fixation in the tropics. *Plant & Soil,* Special Vol. 1971, 403–13.

WOLK C.P. (1968) Movement of carbon from vegetative cells to heterocysts in *Anabaena cylindrica. J. Bact.* **96**, 2138–43.

WOLK C.P. (1970) Aspects of the development of a blue-green alga. *Ann. N.Y. Acad. Sci.* **175**, 641–7.

WOLK C.P. & SIMON R.D. (1969) Pigments and lipids of heterocysts. *Planta,* **86**, 92–7.

WOLK C.P. & WOJCIUCH E. (1971a) Photoreduction of acetylene by heterocysts. *Planta,* **97**, 126–34.

WOLK C.P. & WOJCIUCH E. (1971b) Biphasic time course of solubilization of nitrogenase during cavitation of aerobically grown *Anabaena cylindrica. J. Phycol.* **7**, 339–44.

WYATT J.T. & SILVEY J.K.G. (1969) Nitrogen fixation by *Gloeocapsa. Science, N.Y.* **165**, 908–9.

CHAPTER 21

# NITROGEN ASSIMILATION AND PROTEIN SYNTHESIS

I. MORRIS

Department of Botany and Microbiology,
University College London,
Gower Street, London, W.C.1, U.K.

1 Introduction 583

2 Metabolism of inorganic N-compounds 584
2.1 Preferential utilization of various inorganic N-compounds 584
2.2 Enzymology of nitrate reduction 584
   (a) Nitrate reductase 585
   (b) Nitrite reductase 587
2.3 Regulation of nitrate assimilation 588
   (a) Nitrate as an inducer 589
   (b) Nature of the feedback inhibitor 589
   (c) Macromolecular events accompanying induction of nitrate-reducing enzymes 590
   (d) Disappearance of nitrate reductase 590
2.4 The effect of light on nitrate assimilation 591
2.5 General energetic considerations of nitrate assimilation. 592

3 Metabolism of organic N-compounds 593
3.1 Primary assimilation of ammonium-N into amino acids and amides 593
3.2 Utilization of organic N-compounds as N-sources for growth 595
3.3 Metabolism of specific organic N-compounds 596
   (a) Urea 596
   (b) Synthesis and utilization of arginine and related amino acids 597
   (c) Metabolism of other amino acids 599

4 Protein synthesis 599
4.1 Introduction 599
4.2 Effects of inhibitors of algal protein synthesis 600
4.3 The physical characteristics of algal ribosomes 601
4.4 Cell-free protein synthesis 602

5 References 602

## 1 INTRODUCTION

This chapter is concerned with three aspects of the physiology and biochemistry of nitrogen-containing compounds in algae. The first is the metabolism of inorganic N-compounds; the second the metabolism of soluble organic N-compounds; and the third the synthesis of polypeptides and proteins. The work of the past ten years is emphasized. For earlier work, the reviews of Syrett (1962), Fowden (1962) and Kessler (1964) are recommended.

## 2 METABOLISM OF INORGANIC N-COMPOUNDS

### 2.1 *Preferential utilization of various inorganic N-compounds*

The two most common sources of nitrogen used for growth of algae are nitrate and ammonium ions. Nitrite can also be used, but its toxicity at higher concentrations makes it less convenient than either nitrate or ammonium-N. For most algae ammonium-N is utilized preferentially (Syrett 1962); that is, when an alga is supplied with both ammonium-N and nitrate-N, nitrate is not utilized until all ammonium ions have disappeared. This preferential utilization of ammonium-N is now known to be related to the regulation of nitrate assimilation and will be discussed below (Section 2.3). However, there are some interesting exceptions to this general preferential assimilation of ammonium-N. For example, in a study of 38 freshwater 'chlamydomonad algae', Cain (1965) noted that *Chlamydomonas gloeopara*, *C. microsphaerella*, *C. peterfi*, *Gloeocystis gigas* and *G. maxima* differed from the others in not showing preferential utilization of ammonium-N. Stross (1963) has extended the earlier observations of Proctor (1957) on the unusual preferential utilization of nitrate (when supplied with both nitrate-N and ammonium-N) by *Haematococcus* spp. Stross demonstrated that the preferential utilization of nitrate could be altered; preferential utilization of ammonium-N being favoured by using cells from the exponential phase of growth and maintaining the pH of the medium at more acid levels.

Although, when both ammonium-N and nitrate-N are supplied together, the former is generally used preferentially, the growth rate is usually the same on ammonium-N as it is on nitrate (Syrett 1962). Apparent improved growth of some algae on nitrate (e.g. Cain 1965, Dushkova 1970) might result from the highly acid conditions which can result from the removal of ammonium-N from the medium. Such an explanation does not appear to be responsible for the observations of Paasche (1971). He noted that the cell growth of *Dunaliella tertiolecta* was 30% faster when ammonium was the N-source than with nitrate. Paasche (1971) suggests that a higher content of photosynthetic enzymes is responsible for this improved growth on ammonium-N.

### 2.2 *Enzymology of nitrate reduction*

It is generally accepted that nitrate-N is reduced to the ammonium level before being incorporated into organic compounds (see Syrett 1962, for earlier references dealing with the physiological evidence for this view). Czygan (1965) has proposed an alternative scheme in which oximes are formed and then reduced to amino acids. That is, in Czygan's scheme N is incorporated into organic compounds before being reduced to the reduction level of ammonium. This proposal of Czygan was based on the ability of six strains of green algae to utilize hydroxylamine, oximes and γ-glutamyl hydroxamic acid as sources of nitrogen, and on the enzymatic reduction of certain oximes. However, no further work appears to have been carried out to establish that oximes participate in the 'normal' process of nitrate assimilation of algae.

The oxidation/reduction state of the N-atom in nitrate is +5 and in ammonium is −3. For many years it has been assumed that three intermediates occur between nitrate and ammonium, thus allowing four reductive steps, each involving the addition of a pair of electrons. The most generally accepted sequence (Nicholas 1959, Syrett 1962, Kessler 1964) is:

$$\overset{+5}{NO_3^-} \rightarrow \overset{+3}{NO_2^-} \rightarrow \overset{+1}{N_2O_2^{2-}} \rightarrow \overset{-1}{NH_2OH} \rightarrow \overset{-3}{NH_4^+}$$

However, more recent work with higher plants (Beevers & Hageman 1969, Hewitt 1970) and algae (Hattori & Myers 1966, 1967, Zumft et al. 1969, Aparicio et al. 1971) suggests that only two enzymes catalyze the entire reduction of nitrate to ammonium-N. The first is *nitrate reductase* (NAD(P)H: nitrate oxidoreductase) which catalyses the reduction of nitrate to nitrite. The second is *nitrite reductase* (NAD(P)H: nitrite oxidoreductase) which catalyses the reduction of nitrite to ammonium.

## (a) *Nitrate reductase*

Schafer et al. (1961) first reported that the activity of nitrate reductase in extracts of *Chlorella pyrenoidosa* required NADH as electron donor and was not active when NADPH was used. FAD stimulated the activity. These workers also observed that mutants (induced by ultra-violet irradiation) unable to grow with nitrate as N source, but able to utilize nitrite-N, ammonium-N and some organic compounds, did not contain this enzyme activity. Czygan (1963) reported the activity of a soluble nitrate reductase in cell-free extracts of *Ankistrodesmus braunii*; this activity required NADH and FAD. Nitrate reductase activity in *A. braunii* appeared to show seasonal changes (Kessler & Czygan 1963).

Osajima and Yamafuji (1964) measured nitrate reductase activity in extracts of *Chlorella* and *Scenedesmus* spp. using NADH as an electron donor. The activity was stimulated by FAD but not by FMN. Syrett and Morris (1963) reported that the enzymatic reduction of nitrate to nitrite by extracts from *Chlorella vulgaris* (later, the name was changed to *C. pyrenoidosa*, Syrett 1966; and most recently changed to *C. fusca* var. *vacuolata*, P. J. Syrett pers. comm.) required NADH as electron donor, and was unable to function with NADPH. These early studies of nitrate reductase activity were made with crude extracts. Although Trebst and Burba (1967) partially purified the enzyme from *Chlorella pyrenoidosa*, they used artificial electron donors (methyl viologen and dithionite) and did not investigate the characteristics of the enzyme further.

More precise information on the nature of nitrate reductase from *Chlorella fusca* Shihiva et Krauss (= *pyrenoidosa*) came from work by Losada and his colleagues (Zumft *et al.* 1969, 1970, Aparicio *et al.* 1971. Cardenas *et al.* 1971, 1972, Vega *et al.* 1971). After preliminary ammonium sulphate precipitation, nitrate reductase was purified by the same techniques as previously used for the spinach enzyme (Paneque *et al.* 1965). This involved treatment with calcium phosphate gel, ammonium sulphate precipitation and chromatography on hydroxylapatite columns. From gel filtration on agarose a highly purified *Chlorella* nitrate reductase was isolated and estimated to have a molecular weight of about 500,000 (Zumft *et al.* 1969).

This purified preparation of nitrate reductase was unable to accept directly electrons from reduced ferredoxin or from NADPH. The enzyme could function however with NADH, reduced flavin nucleotides ($FNH_2$) or with artificial electron donors such as reduced viologen dyes. With NADH as the electron donor, two enzyme activities were required for the reduction of nitrate: (1) a NADH-specific diaphorase and (2) nitrate reductase proper. When $FNH_2$ or reduced viologens were used as electron donors, only the second enzyme was required. In later papers from Losada's laboratory the involvements of FAD (Zumft *et al.* 1970) and molybdenum (Aparicio *et al.* 1971, Cardenas *et al.* 1971, Vega *et al.* 1971) were established.

Slightly different results have been obtained with extracts from *Anabaena cylindrica* Lemm. (Fogg strain) (Hattori & Myers 1967). Firstly, activity of the enzyme appeared in a particulate fraction. Secondly, the nitrate reductase could accept electrons from reduced ferredoxin, artificial donors such as methyl viologen, benzyl viologen and diquat, or NADH. But the system could not use NADPH directly. Also, although reduced flavin nucleotides ($FADH_2$ and $FMNH_2$) could serve as electron donors, their affinity towards the enzyme was very low. The essential difference between these results of Hattori and Myers (1967) and those obtained with higher plants (Ramirez *et al.* 1966, Del Campo *et al.* 1965) is in the way in which ferredoxin acts. In higher plants reduced ferredoxin acts as an electron donor for the reduction of nitrate to nitrite only when $NAD^+$ mediates the transfer of electrons. However, the system from *A. cylindrica* does not require $NAD^+$ to be present when reduced ferredoxin is functioning as the electron donor. Moreover, NADPH can function as a donor if ferredoxin (or an artificial dye) is present but not in its absence. Thus, the conclusion Hattori and Myers (1967) reached was that nitrate reductase from *Anabaena* could accept electrons directly from reduced ferredoxin. Possibly, these differences between the *Anabaena* and *Chlorella* (and higher plant) systems will become clearer when the enzyme from *Anabaena* is purified.

Although nitrate reductase activity has been detected in extracts from other algae, crude extracts have been used and have yielded little information on the precise nature of the enzyme and its mechanism of reaction. Eppley and his colleagues (Eppley & Coatsworth 1968, Eppley *et al*. 1969) measured the enzyme in extracts of the marine planktonic diatom *Ditylum brightwellii*. It required NADH as electron donor, was inactive with NADPH, was unaffected by the addition of FAD, but was stimulated by magnesium sulphate. Enzyme activity also required the reduced sulphur compounds glutathione or dithiothreitol, and phosphate. Similar properties were noted for the nitrate reductase extracted from *Coccolithus huxleyi*, *Cyclotella nana*, *Dunaliella tertiolecta* and *Gonyaulax polyedra* (Eppley *et al*. 1969), except that the enzyme from *D. tertiolecta* was also active with NADPH. Using NADH as electron donor, Eppley *et al*. (1969, 1970) were also able to measure nitrate reductase in some natural phytoplankton populations. Thacker and Syrett (1972b) have measured nitrate reductase activity in extracts from *Chlamydomonas reinhardtii* using reduced benzyl viologen as electron donor.

## (b) *Nitrite reductase*

This enzyme catalyses the reduction of nitrite ($NO_2^-$) to ammonium ($NH_4^+$). Although this involves the net addition of three pairs of electrons, and might therefore be expected to occur in three steps, a single enzyme appears to catalyse the entire reduction sequence without the accumulation of any free intermediates.

Czygan (1963) first measured the activity of nitrite reductase in extracts of *Ankistrodesmus braunii*. Activity was particulate and utilized NADH, NADPH, FAD or FMN as electron donors. Czygan (1963) also suggested that there were two nitrite reductases; one, a dissimilatory one found on large particles; the other, an assimilatory one requiring 'high-energy phosphate', and localized on smaller particles. Unfortunately these interesting preliminary results do not appear to have been followed up by studies on purified preparations. Osajima and Yamafugi (1964) also reported the activity of nitrite reductase in extracts from *Chlorella* and *Scenedesmus* spp. Their conditions were essentially the same as those used for the assay of nitrate reductase. Also, these workers reported the presence of hyponitrite and hydroxylamine reductases, but did not take any of the precautions Cresswell *et al*. (1965) suggest should be taken when assaying these apparent activities. Hence, it is uncertain as to whether the activities reported by Osajima and Yamafugi (1964) are caused by distinct enzymes participating in the reduction of nitrate to ammonium.

Hattori and Myers (1966) partially purified a nitrite reducing activity from *Anabaena cylindrica*. The enzyme appeared in the supernatant fraction and the partial purification was achieved by chromatography on Sephadex G-75. NADPH was active as an electron donor but the transfer was mediated by ferredoxin. NADH was inactive. Ferredoxin could be replaced by methyl viologen, benzyl viologen or diquat. The stoichiometry of the reduction showed a 1:1 ratio between nitrite loss and ammonium production. Activity of nitrite reductase was inhibited by cyanide, 2,4-dinitrophenol and arsenate.

Purified nitrite reductase from *Chlorella fusca* (= *pyrenoidosa*) is similar to that from *Anabaena*. Zumft *et al*. (1969) purified the *Chlorella* nitrite reductase by ammonium sulphate precipitation and chromatography on DEAE-cellulose and Sephadex G-100. From its movement through Sephadex G-100 the molecular weight of nitrite reductase was estimated to be about 63,000. Reduction could be achieved by reduced ferredoxin or reduced viologens but not with NADH, NADPH, FAD (+ dithionite) or FMN (+ dithionite). NADPH could function as the electron donor if ferredoxin and NADP reductase (catalysing the reduction of ferredoxin by NADPH) were also present.

Unlike nitrate reductase, the enzyme catalysing the reduction of nitrite to ammonium in *Chlorella* does not require molybdenum (Cárdenas *et al*. 1971); instead, iron appears to be a component (Aparicio *et al*. 1971). The iron content of the growth medium specifically increases the capacity for nitrite reduction in *Ankistrodesmus braunii* and *Chlorella fusca* (Kessler & Czygan 1968) and the level of nitrite reductase in *C. fusca* (Cárdenas *et al*. 1972). In both of these studies the iron supply had no effect on nitrate reducing activity.

This nitrate reducing system in *Chlorella* is therefore very similar to that in higher plants (Losada *et al*. 1965, Paneque *et al*. 1965, Ramirez *et al*. 1966, Beevers & Hageman 1969, Hewitt 1970). That is, it consists of a flavomolybdo-protein which catalyses the reduction of nitrate to nitrite with reduced flavin nucleotide as the natural electron donor and uses NADH only when a NADH-specific diaphorase is also present, and an iron protein ferredoxin-nitrite reductase which utilizes ferredoxin as the electron donor and reduces nitrite to ammonium. It will be of interest to see how general this picture is when the purified enzymes from other algae are examined. Of particular interest will be the possibility that nitrate reductase from *Anabaena* is different in that it uses reduced ferredoxin as electron donor.

Since ferredoxin is usually found only in the photosynthetic tissues of plants (for example, see Beevers & Hageman 1969, Hewitt 1970), it is interesting to establish the nature of the cofactor used for nitrite reduction in the dark. Using a colourless mutant of *Chlorella*, Guerrero *et al*. (1971) reported that the nitrate-reducing system of this dark-grown mutant was the same as that operating in autotrophically-grown organisms. That is, the dark-grown mutant contained NADH-nitrate reductase, ferredoxin-nitrite reductase and NADP reductase (the last named catalysing the reduction of ferredoxin by NADPH).

## 2.3  *Regulation of nitrate assimilation*

The preferential utilization of ammonium-N when algae are supplied with both ammonium and nitrate is now known to be related to the control of nitrate assimilation. In general terms, assimilation of ammonium causes the feedback inhibition and repression of the enzymes responsible for nitrate reduction. This has been observed for several species of *Chlorella* (Syrett & Morris 1963, Morris & Syrett 1963b, Knutsen 1965, Losada *et al*. 1970, Smith & Thompson 1971),

*Ankistrodesmus braunii* (Czygan 1963), several marine planktonic algae (Eppley *et al.* 1969), *Anabaena cylindrica* (Ohmori & Hattori 1970) and *Chlamydomonas reinhardtii* (Thacker & Syrett 1972a,b).

Several interesting aspects of metabolic control have emerged from these studies.

## (a) *Nitrate as an inducer*

A high activity of nitrate reductase in algae grown with nitrate and a low activity in ammonium-grown algae does not necessarily mean that nitrate is an inducer. The control might merely operate through repression by ammonium. That the level of nitrate reductase was higher when *Chlorella fusca* var. *vacuolata* was grown on ammonium nitrate than when grown on ammonium alone was interpreted by Morris and Syrett (1963a) as evidence for the inducer role of nitrate. However, the studies of Losada *et al.* (1970) with *Chlorella fusca*, and those of Thacker and Syrett (1972a,b) with *Chlamydomonas reinhardtii* revealed no difference between ammonium nitrate and ammonium alone, and so argue against an inducer role of nitrate. Ohmori and Hattori (1970) suggest that in *Anabaena cylindrica* nitrate *per se* does not induce the formation of nitrite reductase but that it needs to be reduced to nitrite, which then induces synthesis of nitrite reductase.

The nitrate reductase activity of *Chlorella* increases when the alga is transferred from ammonium-containing medium to N-free medium, although the magnitude of the increase is less than when nitrate is present (Morris & Syrett 1963a). This increase in the absence of nitrate is usually interpreted as indicating derepression following the removal of ammonium-N. However, Kessler and Oesterheld (1970) and Oesterheld (1971) have suggested that N-deficiency of *Ankistrodesmus braunii* is accompanied by oxidation of organic N-compounds leading to intracellular formation of nitrite and nitrate which then induces formation of the nitrate-reducing enzymes.

## (b) *Nature of the feedback inhibitor*

Although addition of ammonium to algal cultures assimilating nitrate causes immediate and complete inhibition of nitrate assimilation, ammonium does not inhibit nitrate reductase activity in cell-free extracts (Syrett & Morris 1963, Smith & Thompson 1971). Moreover, under conditions when ammonium is not assimilated, it does not inhibit nitrate assimilation by the intact cells of *Chlorella fusca* var. *vacuolata* (Syrett & Morris 1963). A similar result has been obtained with *Chlamydomonas reinhardtii* (Thacker & Syrett 1972a). These observations have been interpreted as indicating that some product of ammonium assimilation, instead of ammonium *per se*, is the feedback inhibitor of nitrate reductase activity. However, attempts to inhibit nitrate reductase with various amino acids and amides have not yet identified likely inhibitors (Syrett & Morris 1963,

Smith & Thompson 1971). Morris and Syrett (1963b) observed that cyanate was a potent inhibitor of nitrate reductase from *Chlorella* but, although cyanate is a breakdown product of carbamoyl phosphate, the physiological significance, if any, of this cyanate inhibition remains unknown. Smith and Thompson (1971) observed repression of nitrate reductase by eight different nitrogenous compounds, none of which inhibited nitrate reductase activity, and so concluded 'that nitrate reduction in *Chlorella* is controlled by repression of enzyme synthesis and not by feedback inhibition'. However, it is difficult to reconcile such a conclusion with the observed short-term ammonium inhibition of nitrate assimilation.

### (c) *Macromolecular events accompanying induction of nitrate-reducing enzymes*

The increase in activity of nitrate and nitrite reductases when *Chlorella* is incubated with medium containing nitrate as sole N-source (Losada *et al.* 1970, Smith & Thompson 1971) is assumed to be due to the *de novo* synthesis of protein. This conclusion is based on inhibition studies with cycloheximide (Losada *et al.* 1970, Smith & Thompson 1971), actinomycin D and puromycin (Smith & Thompson 1971). One other aspect of the molecular events associated with synthesis of the nitrate-reducing enzymes is the investigation by Knutsen (1965). This worker observed that the inducibility of nitrite reductase in synchronized cultures of *Chlorella pyrenoidosa* was confined to a nine-hour period beginning about five hours after exposure of autospores to favourable photosynthetic conditions, and proposed that this was determined by a variability in the accessibility of the nitrite reductase gene for transcription.

### (d) *Disappearance of nitrate reductase*

In addition to its induction by nitrate, there have been several studies of the disappearance of nitrate reductase. Morris and Syrett (1965) observed a disappearance of nitrate reductase activity when *Chlorella* was transferred from nitrate medium to N-free or ammonium-containing medium. This was a disappearance relative to the general protein content and was interpreted as indicating that, under these conditions at least, the rate of nitrate reductase turnover is greater than that of general cellular protein. Losada *et al.* (1970) observed a similar disappearance of all nitrate-reducing enzymes ($FNH_2$-nitrate reductase, NADH diaphorase and nitrite reductase) when *Chlorella fusca* was transferred from nitrate medium to ammonium-containing medium. However, this long-term disappearance of these enzymes was preceded by a more rapid deactivation of nitrate reductase (but not of the diaphorase or the nitrite reductase) on ammonium medium.

Thacker and Syrett (1972b) have observed the disappearance of nitrate reductase from *Chlamydomonas reinhardtii* cultures when photosynthesis is prevented by (a) darkening cultures, (b) removing carbon dioxide, or (c) adding

DCMU. This disappearance can be prevented under conditions which allow nitrate reduction. Possibly, in addition to the synthesis of nitrate reductase when required for algal growth, there is a controlled removal of the enzyme under conditions when it is no longer required.

### 2.4    The effect of light on nitrate assimilation

Most algae assimilate nitrate more rapidly in the light than in darkness. However, there is no general view of the mechanism of this light stimulation of nitrate assimilation (see Syrett 1962, Kessler 1964). A direct photochemical reduction of nitrate and nitrite has been observed with isolated chloroplasts and chloroplast fragments (e.g. Losada *et al.* 1965, Joy & Hageman 1966). That is, with these isolated systems light reduces cofactors such as flavoproteins, ferredoxins and pyridine nucleotides which then become used as electron donors for nitrate and nitrite reduction. The question is whether this is the only role for light in the stimulation of nitrate assimilation in whole algal cells. Essentially, this question has been approached by investigating the requirements for carbon dioxide in the light stimulation of nitrate assimilation. Kessler (1955) observed nitrite reduction, but little reduction of nitrate, when *Ankistrodesmus braunii* was illuminated under anaerobic conditions in the absence of $CO_2$. Hattori (1962) also observed a light-dependent reduction of nitrate, nitrite and hydroxylamine by *Anabaena cylindrica*. At low light intensity, $CO_2$ did not affect the rate of light-dependent nitrate and nitrite reduction; rather nitrite inhibited the rate of $CO_2$ fixation. At high light intensities, however, the rates of nitrate and nitrite reduction were stimulated by the presence of $CO_2$. Also, Morris and Ahmed (1969) observed a light-dependent assimilation of nitrate and nitrite when *Ankistrodesmus braunii* and *Chlorella fusca* var. *vacuolata* (then called *C. pyrenoidosa*) were incubated in the absence of carbon dioxide under both aerobic and anaerobic conditions. The presence of $CO_2$ did not affect nitrate and nitrite assimilation by *Chlorella* but did have a stimulatory effect on *Ankistrodesmus*.

Thus, in some algae, it is possible to observe light-dependent nitrate and nitrite assimilation under conditions when the direct photochemical reduction is emphasized. However, two additional observations of Morris and Ahmed (1969) make it difficult to draw a firm conclusion on the role of light in nitrate assimilation. First, there are not the 'expected' ratios of $O_2:NO_3$ and $O_2:NO_2$ (see Syrett 1962, for a discussion of these ratios). Secondly, DCMU has a more pronounced inhibitory effect on $O_2$ evolution than on the accompanying assimilation of nitrate and nitrite. Thus, merely measuring nitrate and nitrite assimilation when algae are illuminated in the absence of carbon dioxide does not necessarily provide definitive evidence for the 'direct' role of light.

For most algae, it appears that light only stimulates nitrate and nitrite assimilation in the presence of carbon dioxide (Myers 1949, Davis 1953, Kessler 1955, Strotmann 1966, 1967, Grant 1967, 1968, Grant & Turner 1969, Thacker & Syrett 1972a). Grant and Turner (1969) studied several different algae and concluded that only in *Chlorella pyrenoidosa* was the light stimulation of nitrate and nitrite assimilation accompanied by the evolution of 'extra' oxygen and thus equated with direct photoreduction to ammonium. In the other algae Grant and Turner (1969) 'concluded that in light the rate-limiting step in nitrate assimilation is the formation of cell nitrogen rather than nitrate reduction or photosynthetic electron flow. Further it is suggested that only carbon skeletons formed directly from $CO_2$ can be used in this assimilation of nitrogen in light, and that the whole process goes on within the algal photosynthetic apparatus'. This idea suggests that light not only stimulates nitrate assimilation by producing suitable electron donors for the nitrate assimilation but, in most algae, the important requirement is for photosynthetic $CO_2$ assimilation so that suitable carbon skeletons can be made available for the formation of organic N-compounds.

Thacker and Syrett (1972a), on the other hand, have suggested that the light stimulation of both nitrate and ammonium assimilation by *Chlamydomonas reinhardtii* cannot be explained solely by providing suitable organic carbon. In addition they suggest a requirement for ATP produced by cyclic photophosphorylation (and thus insensitive to DCMU); a suggestion also made by Morris and Ahmed (1969).

## 2.5  *General energetic considerations of nitrate assimilation*

In addition to the above-mentioned observations of Morris and Ahmed (1969) and Thacker and Syrett (1972a,b) on the light-stimulation of nitrate assimilation, other indirect evidence suggests that the process of nitrate assimilation requires ATP. Kessler (1955, 1959) originally observed that nitrite reduction by *Ankistrodesmus braunii* was inhibited by the uncoupling agent 2,4-dinitrophenol (2,4-DNP) but that the conversion of nitrate to nitrite was insensitive. Later, Ahmed and Morris (1967, 1968) showed that nitrate assimilation by this alga and by *Chlorella fusca* var. *vacuolata* was partially inhibited by 2,4-DNP and that this inhibition could be overcome by increasing the concentration of nitrate. These workers therefore suggest that inhibition of nitrate assimilation is caused by inhibition of the active uptake of nitrate, whereas inhibition of nitrite assimilation is caused by inhibition of the reduction process.

Kessler and Czygan (1963) observed an inhibition by 2,4-DNP of nitrite reductase activity in cell-free extracts from *Ankistrodesmus braunii*, and that the uncoupling agent had no effect on nitrate reductase activity. Also, Kessler and Bücker (1960) observed that arsenate also inhibited nitrite reductase activity, and that 2,4-DNP inhibition could be reversed by adding ATP. 2,4-DNP inhibition of nitrite assimilation (Hattori 1962) and of nitrite reductase activity in extracts (Hattori & Myers 1966) has also been observed in *Anabaena cylindrica*.

It has been suggested that the effect of 2,4-DNP on nitrite metabolism is due to the reduction of the nitro-group in DNP competing with nitrite for available electron donors (Del Campo *et al.* 1965, Ahmed & Morris 1967). However, Kessler *et al.* (1970) have pointed out that such an explanation would be difficult to apply to uncouplers such as arsenate carbonyl cyanide, *m*-chlorophenyl-hydrazone (CCCP) and pentachlorophenol (PCP) which lack the reducible nitro-group. Also, no rapid rates of 2,4-DNP reduction have been measured with algae (Ahmed & Morris 1967, Kessler *et al.* 1970). If the reduction of nitrite to ammonium by algae does require ATP, the mechanism of this reaction remains unknown and should provide an interesting challenge for the future.

# 3   METABOLISM OF ORGANIC N-COMPOUNDS

## 3.1   *Primary assimilation of ammonium-N into amino acids and amides*

Algae appear to resemble other plants and microorganisms in having three main reactions whereby inorganic nitrogen, in the form of ammonium, is assimilated into organic compounds (e.g. Sims *et al.* 1968). The first is the reductive amination of certain keto acids with the formation of certain amino-acids. The second is the further amination of amino-acids to form amides. The third is the reaction between $CO_2$, ATP and ammonium to form carbamoyl phosphate; that is, the reaction catalysed by the enzyme carbamoyl phosphate synthetase (ATP:carbamate phosphotransferase).

Two main enzymes catalysing reductive aminations of keto acids have been detected in extracts of algae. One is glutamic dehydrogenase (L-glutamate:NAD(P)oxidoreductase) catalysing the reductive amination of α-ketoglutarate to glutamate (Fowden 1962, Kates & Jones 1964a, Belmont & Miller 1965, Kretovich *et al.* 1970). A second is alanine dehydrogenase (L-alanine:NAD oxidoreductase) catalysing the conversion of pyruvate to alanine (Kates & Jones 1964a, Belmont & Miller 1965, Romanov *et al.* 1965, Shatilov *et al.* 1968, Kretovich *et al.* 1970). A third reaction has been suggested by Jacobi (1957a) to operate in extracts of *Ulva lactuca*, this being a reaction analogous to glutamic dehydrogenase but yielding aspartate.

Although the activities of these various aminating enzymes can be detected in algal extracts, their significance in the primary assimilation of ammonium into amino acids by intact cells is unclear. The most critical experiments relating to this problem appear to be those which follow the kinetics of labelling when [15]N-ammonium is supplied. From such experiments with yeasts (Sims & Folkes 1964) and algae (Bassham & Kirk 1964), the data suggest that glutamic dehydrogenase is the most important route whereby inorganic nitrogen is assimilated into amino acids; the other keto acids being converted to the appropriate amino acids by transamination, instead of by direct reductive amination. The original [14]C-kinetic data of Smith *et al.* (1961) suggested that alanine dehydrogenase was also an important reaction for ammonium incorporation into amino acids in *Chlorella pyrenoidosa*. Also Kates and Jones (1964b) interpreted the failure of fluoroacetate to inhibit the photosynthetic incorporation of [14]C-carbon dioxide into alanine as evidence of the importance of alanine dehydrogenase. But the later data with [15]N-ammonium (Bassham & Kirk 1964) labelling indicated the prime importance of glutamic dehydrogenase. Transaminases have been measured in a variety of green, red and brown alga (see e.g. Jacobi 1957b, Nakamura *et al.* 1968). A specific δ-aminolevulinic acid transaminase catalysing the transamination of γ, δ-dioxovaleric acid with L-α alanine, L-glutamic acid or L-phenylalanine has been detected in extracts of *Chlorella vulgaris* (Gassman *et al* 1968).

One important aspect of the assimilation of ammonium into amino acids has been the regulation of this process (e.g. Sims *et al.* 1968). Kretovich *et al.* (1970) observed that the NADP-glutamate dehydrogenase activity increased in *Chlorella pyrenoidosa* growing on ammonium, and decreased when nitrate was used as the N-source. This change in N-source had no effect on the activities of NAD-glutamate dehydrogenase and had little effect on alanine dehydrogenase. On the basis of these results Kretovich *et al.* (1970) concluded that the NADP-linked enzyme was the assimilatory one, and its synthesis was induced by ammonium-N. However, these workers point out the difficulty of explaining the decreased activity of this 'biosynthetic' enzyme on nitrate, since nitrate is readily assimilated. The enhanced activity of NAD-glutamate dehydrogenase on N-starvation (Kretovich *et al.* 1970) might suggest a 'degradative' role for this enzyme.

During the past decade, few investigations appear to have been made into the further amination of amino acids to yield amides. Syrett (1962) mentioned the work of Reisner *et al.* (1960) which showed a rapid increase in glutamine concentration (but a much slower increase in asparagine) when N-deficient cells of *Chlorella vulgaris* were supplied with ammonium-N. Also, Fowden (1962) mentioned the earlier work of Loomis (1959) on the activity of glutamine synthetase in extracts of *Chlorella* sp. and *Scenedesmus* sp. Based on the kinetics of labelling with $^{15}$N-ammonium, Bassham and Kirk (1964) concluded that 'the role of glutamine in ammonia incorporation appears to be minor' in *Chlorella pyrenoidosa*. However, these authors recognized that their 'data did not preclude the possibility that there is a small actively-turning pool of glutamine which saturates very quickly and which accounts for substantial amounts of $NH_4^+$ incorporation'. Dharmawardene *et al.* (1973) and Haystead *et al.* (1973) reported on glutamine synthetase and other $NH_4^+$ incorporating enzymes in *Anabaena cylindrica*.

The third method of incorporating ammonium into organic N-compounds is via the activity of carbamoyl phosphate synthetase. Carbamoyl phosphate is a precursor of citrulline and early indirect evidence for the presence of the synthetase came from the rapid incorporation of radioactivity into citrulline when *Nostoc muscorum* was supplied with $^{14}$C-carbon dioxide (Linko *et al.* 1957). Incorporation was as rapid in the dark as in the light, and the radioactivity was largely confined to the ureido carbon atom of citrulline. Holm-Hansen and Brown (1963) later measured the activity of carbamoyl phosphate synthetase in extracts of *Nostoc muscorum*. The high labelling of arginine when *Chlamydomonas reinhardtii* assimilated $^{14}CO_2$ in the dark affords indirect evidence for the activity of the carbamoyl phosphate synthetase (Holden & Morris 1970).

### 3.2    *Utilization of organic N-compounds as N-sources for growth*

Although most algae are able to synthesize all the organic N-compounds required for growth from the inorganic N-source, some are also able to use certain organic N-compounds as sole sources of nitrogen (see Syrett 1962, for a review of earlier work). Birdsey and Lynch (1962) observed that eight species of the Chlorophyceae were able to use urea and five able to use uric acid and xanthine as N-sources. In the same study it was found that *Anacystis nidulans* and *Synechococcus cedrorum* (Cyanophyceae), *Porphyridium cruentum* (Rhodophyceae) and *Euglena gracilis* (Euglenophyceae) failed to utilize these three compounds. *A. nidulans* decomposed uric acid to allantoin; none of the algae were able to utilize allantoin. Stacey and Casselton (1966) observed that the colourless alga *Prototheca zopfii* utilized adenine and ammonium, but not nitrate. Bollard (1966) measured the growth of *Chlorella vulgaris* on a wide range of organic N-compounds. Growth was estimated qualitatively and the tests made under only one set of experimental conditions. However, from this study the main general points were these: (a) of the twenty protein-amino acids, all except cysteine, histidine, hydroxy-proline, lysine, methionine, phenylalanine and tryptophan supported good growth; (b) D-glutamine, D-serine and D-threonine supported slight growth, but D-aspartic acid, D-glutamic acid and D-asparagine supported none; (c) several dipeptides (e.g. glycyl-glycine, glycyl-serine, etc.) supported growth, as did several amides; (d) amines (except putrescine) and amino alcohols did not allow growth; (e) other amino acids and their derivatives which could be used for growth included ornithine, citrulline, $\alpha,\varepsilon$-diaminopimelic acid, etc. A similar wide use of 23 amino acids has been reported by den Dooren de Jong (1967) for *Chlorella vulgaris*. Later den Dooren de Jong (1969) observed that this wide utilization of amino acids could be seen in two other strains of *Chlorella vulgaris*, but that two other strains and four strains of *Ankistrodesmus* were much more restricted; one strain of *Ankistrodesmus* being unable to utilize any of the amino acids. In the study of den Dooren de Jong (1969) L-amino acids were found to support better growth than the D-forms (except that three strains of *Ankistrodesmus* preferred D-serine to L-serine). Few of the amino acids supported growth in the dark, only L-leucine producing good growth (den Dooren de Jong 1967, 1969).

The observations of Bollard (1966) and den Dooren de Jong (1967, 1969) showed that glutamic acid supported the growth of *Chlorella*. This conflicts with other observations (Lynch & Gillmor 1966) which suggest that, probably because of its negative charge, glutamic acid is not taken up by *Chlorella pyrenoidosa*. Possibly, strain differences between the algae are responsible for this apparent conflict.

### 3.3  *Metabolism of specific organic N-compounds*

(a) *Urea*

Urea can be used as N-source for the growth of a wide range of algae (Syrett 1962). The metabolic problem posed by this utilization comes from the fact that, in some algae, urea is utilized in the apparent absence of urease (Syrett 1962, Baker & Thompson 1962). Such observations led to the conclusion that there exists an enzymatic process, distinct from that catalysed by urease, converting urea into intermediates of known pathways of N-metabolism. Roon and Levenberg (1968) described a new enzyme from a urea-grown yeast and from urea-grown *Chlorella ellipsoidea* and *C. pyrenoidosa*. This enzyme catalysed the ATP-dependent cleavage of urea to carbon dioxide and ammonia:

$$\text{Urea} + \text{ATP} \xrightarrow{\text{(Mg}^{2+}, \text{K}^+)} CO_2 + 2NH_3 + ADP + Pi$$

The enzyme was given the trivial name of ATP:urea amido-lyase (ADP). Hodson and Thompson (1969) also measured the ATP-dependent breakdown of urea in extracts of *Chlorella* and later Thompson and Muenster (1971) separated the enzyme into two components. Enzyme 1 is urea:$CO_2$ ligase (ADP) catalysing this reaction:

$$
\begin{array}{c}
NH_2 \\
| \\
C = O + HCO_3^- + ATP \\
| \\
NH_2
\end{array}
\xrightarrow[\text{Biotin}]{Mg^{2+}}
\begin{array}{c}
COO^- \\
| \\
NH \\
| \\
C = O + ADP + Pi \\
| \\
NH_2
\end{array}
$$

The allophanate produced by this reaction is then used as the substrate for the second enzyme (allophanate amidohydrolase):

$$
\begin{array}{c}
COO^- \\
| \\
NH \\
| \\
C = O \\
| \\
NH_2
\end{array}
+ 2H_2O + OH^- \longrightarrow 2HCO_3^- + 2NH_3
$$

In *Chlorella fusca* var. *vacuolata* urea amido-lyase activity is induced by urea, repressed by ammonium, and appears when ammonium-grown cells are N-starved (Adams 1971). Leftley and Syrett (1973) studied the distribution of urease and urea amidolyase in a number of different algae. Interestingly, the amido-lyase was found in crude cell-free extracts of ten species of the Chlorophyceae, but urease (and not the amido lyase) was detected in extracts from members of the Prasinophyceae, Xanthophyceae, Haptophyceae and Bacillariophyceae.

Urea is secreted into the medium when *Ochromonas malhamensis* is grown on complete nutrient medium, and Lui and Roels (1970) suggest that it is formed from uric acid via allantoin and allantoic acid as intermediates, and also formed from arginine through the activity of arginase (L-arginine ureohydrolase). In this organism, too, urease activity increases when urea is the sole N-source and when the cells are N-starved.

### (b) *Synthesis and utilization of arginine and related amino acids*

During the past decade there have been several studies of the mechanism by which algae synthesize and degrade arginine. The pathway of arginine bio-synthesis in *Chlamydomonas reinhardtii* (see Fig. 21.1) appears to resemble that in yeast (Hudock 1962, Staub & Dénes 1966, Farago & Dénes 1967, Sussenbach & Strijkert 1969a).

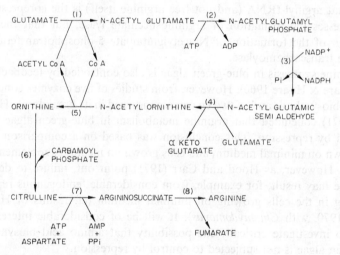

Fig. 21.1. Proposed pathway of arginine biosynthesis in *Chlamydomonas*.

This series of reactions is catalysed by the following enzymes:

(1) Acetylglutamate synthetase (Acetyl CoA:L-glutamate N-acetyltransferase)
(2) Acetylglutamate kinase (ATP:N-acetylglutamate-5-phosphotransferase)
(3) Acetylglutamylphosphate reductase (N-Acetyl L-glutamate semialde-hyde:NADP oxidoreductase (phosphorylating)
(4) Acetylornithine-transaminase (-N-Acetyl L-ornithine:2 oxoglutarate aminotransferase)
(5) Acetylornithine-glutamate transacetylase (N-Acetylornithine: L-glutamate N-acetyltransferase)

(6)   Ornithine   transcarbamoylase   (carbamoylphosphate : L-ornithine   carbamoyl transferase)

(7)   Argininosuccinate synthetase (L-citrulline : L-aspartate ligase (AMP))

(8)   Argininosuccinate lyase (L-argininosuccinate arginine lyase)

A similar pathway of arginine biosynthesis appears to operate in blue-green algae (Holm-Hansen & Brown 1963, Hood *et al*. 1969).

Much of the work on arginine biosynthesis in algae has been concerned with the regulation of this process. In *Chlamydomonas* the biosynthesis of arginine appears to be regulated by both feedback inhibition and repression of certain enzymes in the biosynthetic sequence. Farago and Dénes (1967) demonstrated that the second enzyme (N-acetylglutamate-5-phosphotransferase) was allosteric and was inhibited by arginine. The details of repression are less clear. Hudock (1962) observed repression of argininosuccinate lyase when *Chlamydomonas* was grown on arginine. However, this was not confirmed by Strijkert and Sussenbach (1969), Sussenbach and Strijkert (1969a) and Holden and Morris (1970). From studies with two arginine-requiring mutants of *Chlamydomonas*, Strijkert and Sussenbach (1969) and Sussenbach and Strijkert (1969a) have suggested that arginyl-tRNA (and not free arginine itself) is the corepressor, that this represses the formation of argininosuccinate lyase, but that there is no repression of the formation of N-acetylglutamate 5-phosphotransferase or of ornithine transcarbamoylase.

Arginine synthesis in blue-green algae is also controlled by feedback inhibition (Hoare & Hoare 1966). However, from studies of five enzymes concerned in arginine biosynthesis and of two enzymes concerned in its breakdown, Hood and Carr (1971) concluded that arginine metabolism in blue-green algae was not regulated by repression. This conclusion was based on a comparison between cells grown on minimal medium and cells grown on medium supplemented with arginine. However, as Hood and Carr (1971) point out, failure to detect any difference may result, for example, from considerable 'endogenous repression' occurring in the cells growing on minimal medium (cf. results of Holden & Morris 1970, with *Chlamydomonas*). It will be of considerable interest in the future to investigate critically the possibility that amino acid biosynthesis in blue-green algae is not subjected to control by repression.

In addition to its synthesis, there has been interest in the degradation and utilization of arginine as an N-source. Baker and Thompson (1962) reported that *Chlorella* converts arginine to citrulline and later, Shafer and Thompson (1968) partially purified the enzyme catalysing this reaction, arginine desmidase = arginine deiminase (L-arginine iminohydrolase).

$$\text{arginine} + \text{H}_2\text{O} \longrightarrow \text{citrulline} + \text{NH}_4{}^+$$

Sussenbach and Strijkert (1969b) have characterized arginine deiminase in extracts from *Chlamydomonas reinhardtii*. Later these same workers (Sussenbach & Strijkert 1970) suggested that the further degradation of citrulline in extracts of *Chlamydomonas* occurred by some hitherto unknown reaction (i.e. not a catabolic ornithine transcarbamoylase and not citrulline hydrolase). Originally, it was suggested that the formation of arginine deiminase (Sussenbach & Strijkert 1969b) and the citrulline degrading enzyme (Sussenbach & Strijkert 1970) was induced by arginine, but later Strijkert *et al.* (1971) suggested that the important regulatory phenomenon was ammonium inhibition of arginine uptake and ammonium repression of arginine deiminase formation. These authors suggested that 'neither the uptake system nor arginine deiminase seem to be inducible by arginine'.

### (c) *Metabolism of other amino acids*

Little appears to be known of the metabolism of most amino acids in algae. There has been some interest in the synthesis of lysine, but this has mainly been concerned with the phylogenetic significance of the fact that certain groups of organisms utilize a pathway which has $\alpha$, $\varepsilon$-diaminopimelic acid as a precursor and others have $\alpha$-aminoadipic acid as the precursor to lysine (Vogel 1963, 1964, Lé John 1971). A similar phylogenetic emphasis has been placed on the fact that algae resemble bacteria, liverworts, and some lichens in that the metabolism of D-methionine is qualitatively the same as that of L-methionine (Pokorny *et al.* 1970). Another study of methionine metabolism (Hochberg & Rahat 1971) has revealed that *Ochromonas danica* incorporates methionine into protein and partially de-methylates some to cysteine.

## 4 PROTEIN SYNTHESIS

### 4.1 *Introduction*

The review by Fowden (1962) contains a discussion of the N-compounds found within algal cells; this includes free amino acids, peptides and the amino acid composition of algal protein. Although the composition of algal protein has continued to be of interest, particularly to those concerned with its nutritive value, there has been little significant change in the ideas of the general nature of algal protein. For this reason, the present section does not include a discussion of the composition of algae, and emphasizes, instead, the biochemical processes leading to the synthesis of protein in algae.

The mechanism of protein synthesis in microbial and animal systems has been the subject of intensive investigations during the past decade (e.g. Lengyel & Söll 1969, Lucas-Lenard & Lipmann 1971). The studies of protein synthesis in algae have been more restricted. Essentially, these have been concerned with three aspects of the problem: the use of selected inhibitors to emphasize the contrast between chloroplast and cytoplasmic protein synthesis; the physical characteristics of algal ribosomes, again placing emphasis on the difference between those of the chloroplast and those of the cytoplasm; and a few studies of a cell-free protein-synthesizing system.

## 4.2   *Effects of inhibitors on algal protein synthesis*

Two inhibitors have been most widely used, chloramphenicol and cycloheximide. The usefulness of these compounds arises from the fact that chloramphenicol, at least at low concentrations, inhibits the growth of bacteria (McLean *et al.* 1949) and blue-green algae, but does not inhibit growth of green algae (except at much higher concentrations) (Galloway & Krauss 1959, Morris 1966a). Moreover, inhibition of bacterial growth results from specific inhibition of protein synthesis (Brock 1961). Cycloheximide, on the other hand, inhibits protein synthesis in yeasts (e.g. Kerridge 1958), mammalian cells (e.g. Godchaux *et al.* 1967) and algae (Kirk & Allen 1965, Morris 1966b, 1967, Kirk 1968, Smillie *et al.* 1968) but does not inhibit bacterial protein synthesis (e.g. Ennis & Lubin 1964).

Thus, the combined use of chloramphenicol and cycloheximide has afforded a means of contrasting protein synthesis in the prokaryotic blue-green algae with that in other (eukaryotic) algae, and of emphasizing, within the latter group of algae, the difference between chloroplast and cytoplasmic protein synthesis.

Chloramphenicol, at concentrations 10 to 100 times those necessary to inhibit bacterial growth, inhibits the growth of many eukaryotic algae, including several strains of *Chlorella* (Galloway & Krauss 1959, Tamiya *et al.* 1962, Czygan 1964, Aoki *et al.* 1965, Morris 1966a), *Chlamydomonas reinhardtii* (Hudock *et al.* 1964), *Scenedesmus quadricauda* (Taylor 1965), *Ankistrodesmus braunii* (Czygan 1964), *Coelastrum proboscideum* (Czygan 1964) and several strains of *Euglena gracilis* (Pogo & Pogo 1965, Linnane & Stewart 1967, Aaronson *et al.* 1967, Ben-Shaul & Markus 1969, Smillie *et al.* 1971). This inhibition appears to result from inhibition of protein synthesis by the chloroplast ribosomes and not from any effect on cytoplasmic ribosomes. Evidence for this comes from differential effects on photoautotrophic and heterotrophic growth (Smillie *et al.* 1963, Smillie *et al.* 1971), from the fact that 'greening' of dark-grown *Euglena* can be inhibited under conditions when growth is unaffected (Smillie *et al.* 1971) and from inhibition of amino acid incorporation into protein by isolated plastids from *Euglena* by chloramphenicol concentrations which do not inhibit protein synthesis by cytoplasmic ribosomes (Eisenstadt & Brawerman 1964, Gnanum & Kahn 1967). These differences between chloroplast and cytoplasmic protein synthesis are reinforced by complementary studies with cycloheximide (Kirk & Allen 1965, Kirk 1968, Hoober & Siekevitz 1968, Smillie *et al.* 1968, 1971, Ben-Shaul & Ophir 1970). In addition to revealing the differences between chloroplast and cytoplasmic protein synthesis, and the essential similarity of the former with protein synthesis in prokaryotic cells, these inhibitor studies have been used to understand the relative importance of chloroplast and cytoplasmic ribosomes on the synthesis of chloroplast proteins (e.g. Scott *et al.* 1971, Smillie *et al.* 1971).

### 4.3    *The physical characteristics of algal ribosomes*

More precise information on the nature of protein synthesis in algae has come from physical studies of the ribosomes which are the sites of such synthesis. The most widely studied property of ribosomes is the rate at which they sediment when centrifuged in various gradients. This property is indicated by a sedimentation coefficient—the 's' value. On the basis of this coefficient, ribosomes can be classified into two types, 70s ribosomes found in prokaryotic cells and 80s found in the cytoplasm of eukaryotic cells (e.g. Vazquez 1964, Vazquez & Monro 1967). Furthermore, the 70s ribosomes of bacteria contain 23s and 16s ribosomal RNA (Taylor & Storck 1964) whilst the 80s ribosomes from plants contain 25s and 16s or 18s RNA and those from animals contain 28s and 18s RNA (Stutz & Nöll 1967, Click & Tint 1967).

With this background information, the studies of algal ribosomes have been orientated around two main questions; the first a survey of a wide range of algae in an attempt to understand their evolution, and the second a study of the contrast between chloroplast and cytoplasmic ribosomes. Rodriguez-Lopez and Vazquez (1968) observed that the ribosomes from *Anacystis montana* resembled bacterial ribosomes in being 70s, in having an RNA/protein ratio of 63/37 and in being able to bind chloramphenicol and spiramycin III, but were not so readily dissociated in buffer containing $Mg^{2+}$ and $NH_4^+$. These authors also observed that the 80s ribosomes of *Chlamydomonas reinhardtii* have an RNA/protein ratio resembling that of 70s ribosomes and their ionic requirement for stability appears to be intermediate between the two types of ribosomes. Apart from being more readily dissociated, ribosomes of *Chlorella pyrenoidosa* resembled those of 'higher organisms' (Rodriguez-Lopez & Vazquez 1968). In a more detailed study, Loening (1968) reported that bacteria, actinomycetes, blue-green algae and higher plant chloroplasts all have 23s and 16s ribosomal RNA with molecular weights of 1·1 million and 0·56 million, and that the 25s and 18s RNA of higher plants, algae, fungi and some protozoa had molecular weights of 1·3 million and 0·7 million confirming the sharp distinction between prokaryotic and eukaryotic organisms. Loening (1968) found no ribosomal RNA intermediate in weight between the bacterial type and that of plants.

Most of the work on chloroplast ribosomes has been undertaken with higher plants (e.g. Boardman *et al.* 1965, Eisenstadt 1967), but some studies with algae have been made (Brawerman 1963, Eisenstadt & Brawerman 1964, Brawerman 1966, Sager & Hamilton 1967, Craig & Carr 1968). These various studies confirm the general observation that the 70s of the chloroplasts of eukaryotic algae resemble those of blue-green algae, bacteria and higher plant chloroplasts and differ from the cytoplasmic ribosomes of eukaryotic cells.

## 4.4 Cell-free protein synthesis

Apart from the studies of amino-acid incorporation into proteins in isolated chloroplasts and chloroplast fragments mentioned in the previous section, few studies have been made on protein synthesis by cell-free extracts from algae. Galling (1966) prepared such a system from *Chlorella pyrenoidosa* and observed that activity of the system depended on GTP (with an energy generating system of phospho(enol)pyruvate + pyruvate kinase) and, to a lesser extent, on CTP and UTP. By using a 'pH-5-enzyme preparation' the polypeptide synthesis could be stimulated by adding nucleic acids. Later, Galling (1967a) observed stimulation of this protein synthesis by DNA from various organisms. Galling (1967b) also compared the cell-free protein synthesizing system of *Chlorella* with that from *Anacystis* and later (Galling 1968) observed that the system from *Chlorella* was stimulated by a low molecular weight RNA sedimenting between the soluble RNA and the smaller ribosomal RNA component. Galling (1968) concluded that the stimulating RNA was not identical with transfer-RNA. Erben (1967) has also characterized the cell-free synthesis of polypeptide from *Chlorella pyrenoidosa* growing synchronously.

## 5  REFERENCES

AARONSON S., ELLENBOGEN B.B., YELLEN L.K. & HUTNER S.H. (1967) *In vivo* differentiation of *Euglena* cytoplasmic and chloroplast protein synthesis with chloramphenicol and DL ethionine. *Biochem. biophys. Res. Commun.* **27**, 535–8.

ADAMS A.A. (1971) The regulation of nitrogen assimilation in *Chlorella*. Ph.D. Thesis. University College, Swansea.

AHMED J. & MORRIS I. (1967) Inhibition of nitrate and nitrite reduction by 2,4-dinitrophenol in *Ankistrodesmus*. *Arch. Mikrobiol.* **56**, 219–24.

AHMED J. & MORRIS I. (1968) The effects of 2,4-dinitrophenol and other uncoupling agents on the assimilation of nitrate and nitrite by *Chlorella*. *Biochim. biophys. Acta*, **162**, 32–8.

AOKI S., MATSUBARA J.K. & HASE E. (1965) De- and re-generation of chloroplasts in the cells of *Chlorella protothecoides*. IV. Effects of 5-fluorouracil, dihydrostreptomycin, chloramphenicol and acridine orange on the processes of greening and division of 'glucose-bleached' algal cells. *Pl. Cell Physiol., Tokyo*, **6**, 475–97.

APARICHIO P.J., CARDENAS J., ZUMFT W.G., VEGA J.M., HERRERA J., PANEQUE A. & LOSADA M. (1971) Molybdenum and iron as constituents of the enzymes of the nitrate-reducing system from *Chlorella*. *Phytochemistry*, **10**, 1487–95.

BAKER J.E. & THOMPSON J.F. (1962) Metabolism of urea and ornithine cycle intermediates by nitrogen-starved cells of *Chlorella vulgaris*. *Pl. Physiol., Lancaster*, **37**, 618–24.

BASSHAM J.A. & KIRK M. (1964) Photosynthesis of amino acids. *Biochim. biophys. Acta*, **90**, 553–62.

BEEVERS L. & HAGEMAN R.H. (1969) Nitrate reduction in higher plants. *A. Rev. Pl. Physiol.* **20**, 495–522.

BELMONT L. & MILLER J.D.A. (1965) The utilization of glutamine by algae. *J. exp. Bot.* **16**, 318–24.

BEN-SHAUL Y. & MARKUS Y. (1969) Effects of chloramphenicol on growth, size distribution, chlorophyll synthesis and ultrastructure of *Euglena gracilis*. *J. Cell Sci.* **4**, 627–44.

BEN-SHAUL Y. & OPHIR I. (1970) Structural and developmental aspects of cycloheximide effects on the chloroplast of *Euglena gracilis*. *Can. J. Bot.* **48**, 929–34.

BIRDSEY E.C. & LYNCH V.H. (1962) Utilization of nitrogen compounds by unicellular algae. *Science, N.Y.* **137**, 763–4.

BOARDMAN N.K., FRANCKI R.I.B. & WILDMAN S.G. (1965) Protein synthesis by cell-free extracts from tobacco leaves. II. Association of activity with chloroplast ribosomes. *Biochemistry, N.Y.* **4**, 872–6.

BOLLARD E.G. (1966) A comparative study of the ability of organic nitrogenous compounds to serve as sole sources of nitrogen for the growth of plants. *Plant & Soil*, **25**, 153–66.

BRAWERMAN G. (1963) The isolation of specific species of ribosomes associated with chloroplast development in *Euglena gracilis*. *Biochim. biophys. Acta*, **72**, 317–31.

BRAWERMAN G. (1966) Nucleic acids associated with the chloroplasts of *Euglena gracilis*. In *Biochemistry of Chloroplasts*, vol. I, ed. Goodwin T.W. pp. 301–17. Academic Press, New York & London.

BROCK T.D. (1961) Chloramphenicol. *Bact. Rev.* **25**, 32–48.

CAIN J. (1965) Nitrogen utilization in 38 freshwater chlamydomonad algae. *Can. J. Bot.* **43**, 1367–78.

CÁRDENAS J., RIVAS J., PANEQUE A. & LOSADA M. (1971) Molybdenum and the nitrate-reducing system from *Chlorella*. *Arch. Mikrobiol.* **79**, 367–76.

CÁRDENAS J., RIVAS J., PANEQUE A. & LOSADA M. (1972) Effect of iron supply on the activities of the nitrate-reducing system from *Chlorella*. *Arch. Mikrobiol.* **81**, 260–3.

CLICK R.E. & TINT B.L. (1967) Comparative sedimentation rates of plant, bacterial and animal ribosomal RNA. *J. mol. Biol.* **25**, 111–22.

CRAIG I.W. & CARR N.G. (1968) Ribosomes from the blue-green alga *Anabaena variabilis*. *Arch. Mikrobiol.* **62**, 167–77.

CRESSWELL C.F., HAGEMAN R.H., HEWITT E.J. & HUCKLEBY D.P. (1965) The reduction of nitrate, nitrite and hydroxylamine to ammonia by enzymes from *Cucurbita pepo* L. in the presence of reduced benzyl viologen as electron donor. *Biochem. J.* **94**, 40–53.

CZYGAN F.-C. (1963) Untersuchungen über die Nitratreduktion der Grünalge *Ankistrodesmus braunii in vivo* und *in vitro*. *Planta*, **60**, 225–42.

CZYGAN F.-C. (1964) Untersuchungen über die wirkung von Chloramphenicol auf den auf- und abbau der Pigmente einiger Grünalgen. *Arch. Mikrobiol.* **47**, 251–9.

CZYGAN F.-C. (1965) Zur frage der zwischenprodukte der Nitratreduktion bei Grünalgen. *Planta*, **64**, 301–11.

DAVIS E.A. (1953) Nitrate reduction by *Chlorella*. *Pl. Physiol., Lancaster*, **28**, 539–44.

DEL CAMPO F.F., PANEQUE A., RAMIREZ J.M. & LOSADA M. (1965) Nitrate reduction with molecular hydrogen in a reconstituted enzymatic system. *Nature, Lond.* **205**, 387–8.

DEN DOOREN DE JONG L.E. (1967) Dark and light metabolism of amino acids in *Chlorella vulgaris*. *Antonie van Leeuwenhoek J. Microbiol. & Serol.* **33**, 166–70.

DEN DOOREN DE JONG L.E. (1969) Light and dark metabolism of L- and D-amino acids in 5 strains of *Chlorella vulgaris* and 4 strains of the genus *Ankistrodesmus*. *Antonie van Leeuwenhoek J. Microbiol. & Serol.* **35**, 107–12.

DHARMAWARDENE M.W.N., HAYSTEAD A. & STEWART W.D.P. (1973) Glutamine synthetase of the nitrogen-fixing alga *Anabaena cylindrica*. *Arch. Mikrobiol.* **90**, 281–95.

DUSHKOVA P.I. (1970) Use of inorganic and organic nitrogen sources for differentiation of strains of unicellular algae. *Nauch. Tp. Vissh. Pedagog. Inst. Plordiv Mat. Fiz. Khim. Biol.* **8**, 131–9.

EISENSTADT J.M. (1967) Protein synthesis in chloroplasts and chloroplast ribosomes. In *Biochemistry of Chloroplasts*, vol. II, ed. Goodwin T.W. pp. 341–9. Academic Press, New York & London.

EISENSTADT J.M. & BRAWERMAN G. (1964) The protein synthesizing systems from the cytoplasm and chloroplasts of *Euglena gracilis*. *J. Mol. Biol.* **10**, 392–402.

ENNIS M.L. & LUBIN M. (1964) Cycloheximide aspects of inhibition of protein synthesis in mammalian cells. *Science, N.Y.* **146**, 1474–6.

EPPLEY R.W. & COATSWORTH J.L. (1968) Nitrate and nitrite uptake by *Ditylum brightwellii* —kinetics and mechanisms. *J. Phycol.* **4**, 151–6.

EPPLEY R.W., COATSWORTH J.L. & SOLÓRZANO L. (1969) Studies of nitrate reductase in marine phytoplankton. *Limnol. Oceanog.* **14**, 194–205.

EPPLEY R.W., PACKARD T.T. & MACISAAC J.J. (1970) Nitrate reductase in Peru current phytoplankton. *Marine Biology*, **6**, 195–9.

ERBEN K. (1967) Zur quantitativen Erfassung der Proteinsynthese-Kapazität in zellfreien Fraktionen von *Chlorella*. *Z. PflPhysiol.* **57**, 329–38.

FARAGO A. & DÉNES G. (1967) Mechanism of arginine biosynthesis in *Chlamydomonas reinhardii*. *Biochim. biophys. Acta*, **136**, 6–18.

FOWDEN L. (1962) Amino acids and proteins. In *Physiology and Biochemistry of Algae*, ed. Lewin R.A. pp. 189–209. Academic Press, New York & London.

GALLING G. (1966) Über ein zellfreies System der Proteinsynthese aus *Chlorella*. *Z. Naturf.* **216**, 993–5.

GALLING G. (1967a) Über die Stimulienung des Aminosäure-Einbaus *in vitro* durch DNS in zellfreien Systemen aus *Chlorella pyrenoidosa*. *Z. Naturf.* **22b**, 348–51.

GALLING G. (1967b) Stimulierung des Aminosäure-Einbaus durch verschiedene Nucleibsäure-Präparationen in zellfreien Systemen aus *Chlorella* und *Anacystis*. *Z. Naturf.* **22b**, 687–8.

GALLING G. (1968) Stimulierung des Aminosäure-Einbaus *in vitro* durch niedermolekulare RNA in *Chlorella*. *Planta*, **79**, 44–9.

GALLOWAY R.A. & KRAUSS R.W. (1959) The differential action of chemical agents, especially polymyxin B on certain algae, bacteria and fungi. *Am. J. Bot.* **46**, 40–9.

GASSMANN M., PLUSCEE J. & BOGORAD L. (1968) δ-Aminolevulinic acid transaminase in *Chlorella vulgaris*. *Pl. Physiol., Lancaster*, **43**, 1411–14.

GNANUM A. & KAHN J.S. (1967) Biochemical studies on the induction of chloroplast development in *Euglena gracilis*. III. Ribosome metabolism associated with chloroplast development. *Biochim. biophys. Acta*, **142**, 493–9.

GODCHAUX W., ADAMSON S.D. & HERBERT E. (1967) Effects of cycloheximide on polyribosome function in reticulocytes. *J. mol. Biol.* **27**, 57–72.

GRANT B.R. (1967) The action of light on nitrate and nitrite assimilation by the marine chlorophyte, *Dunaliella tertiolecta*. *J. gen. Microbiol.* **48**, 379–89.

GRANT B.R. (1968) Effect of carbon dioxide concentration and buffer system on nitrate and nitrite assimilation in *Dunaliella tertiolecta*. *J. gen. Microbiol.* **54**, 444–55.

GRANT B.R. & TURNER I.M. (1969) Light-stimulated nitrate and nitrite assimilation in several species of algae. *Comp. Biochem. Physiol.* **29**, 995–1004.

GUERRERO M.G., RIVAS J., PANEQUE A. & LOSADA M. (1971) Mechanism of nitrate and nitrite reduction in *Chlorella* cells grown in the dark. *Biochem. biophys. Res. Commun.* **45**, 82–9.

HATTORI A. (1962) Light-induced reduction of nitrate, nitrite and hydroxylamine in a blue-green alga, *Anabaena cylindrica*. *Pl. Cell Physiol., Tokyo*, **3**, 355–69.

HATTORI A. & MYERS J. (1966) Reduction of nitrate and nitrite by subcellular preparations of *Anabaena cylindrica*. I. Reduction of nitrite to ammonia. *Pl. Physiol., Lancaster*, **41**, 1031–6.

HATTORI A. & MYERS J. (1967) Reduction of nitrate and nitrite by subcellular preparations of *Anabaena cylindrica*. II. Reduction of nitrate to nitrite. *Pl. Cell Physiol., Tokyo*, **8**, 327–37.

HAYSTEAD A., DHARMAWARDENE M.W.N. & STEWART W.D.P. (1973) Ammonia assimilation in a nitrogen-fixing blue-green alga. *Plant Sci. Letts.* **1**, 439–45.

HEWITT E.J. (1970) Physiological and biochemical factors controlling the assimilation of inorganic nitrogen supplies by plants. In *Nitrogen Nutrition of the Plant*, ed. Kirby E.A. pp. 78–103. University of Leeds.

HOARE D.S. & HOARE S.L. (1966) Feedback regulation of arginine biosynthesis in blue-green algae and photosynthetic bacteria. *J. Bacteriol.* **92**, 375–9.

HOCHBERG A. & RAHAT M. (1971) Ethionine and methionine metabolism by the chrysomonad flagellate *Ochromonas danica*. *J. Protozool.* **18**, 487–90.

HODSON R.C. & THOMPSON J.F. (1969) Metabolism of urea by *Chlorella vulgaris*. *Pl. Physiol., Lancaster*, **44**, 691–6.

HOLDEN J. & MORRIS I. (1970) Regulation of arginine biosynthesis in *Chlamydomonas reinhardii*: studies *in vivo* and of ornithine transcarbamoylase and argininosuccinate lyase activities. *Arch. Mikrobiol.* **74**, 58–68.

HOLM-HANSEN O. & BROWN G.W. (1963) Ornithine cycle enzymes in the blue-green alga *Nostoc muscorum*. *Pl. Cell Physiol., Tokyo*, **4**, 299–306.

HOOBER J.K. & SIEKEVITZ P. (1968) Effects of chloramphenicol and cycloheximide on chloroplast membrane formation in *Chlamydomonas reinhardi* Y-1. *J. Cell. Biol.* **39**, 62a.

HOOD W. & CARR N.G. (1971) Apparent lack of control by repression of arginine metabolism in blue-green algae. *J. Bact.* **107**, 365–7.

HOOD W., LEAVER A.C. & CARR N.G. (1969) Extracellular nitrogen and the control of arginine biosynthesis in *Anabaena variabilis*. *Biochem. J.* **114**, 12–13P.

HUDOCK G.A. (1962) The pathway of arginine biosynthesis in *Chlamydomonas reinhardi*. *Biochem. biophys. Res. Commun.* **9**, 551–5.

HUDOCK G.A., MCLEOD G.C., MORAVKOVA-KIELY J. & LEVINE R.P. (1964) The relation of oxygen evolution to chlorophyll and protein synthesis in a mutant strain of *Chlamydomonas reinhardi*. *Plant Physiol., Lancaster*, **39**, 898–903.

JACOBI G. (1957a) Enzymes in the amino acid metabolism of *Ulva lactuca*. *Naturwissenschaften*, **44**, 265–6.

JACOBI G. (1957b) Enzymes in the amino acid metabolism of *Ulva lactuca*. Transaminases and amino acid dehydrogenases. *Planta*, **49**, 561–7.

JOY K.W. & HAGEMAN R.H. (1966) The purification and properties of nitrite reductase from higher plants, and its dependence on ferredoxin. *Biochem. J.* **100**, 263–73.

KATES J.R. & JONES R.J. (1964a) Variation in alanine dehydrogenase and glutamate dehydrogenase during the synchronous development of *Chlamydomonas*. *Biochim. biophys. Acta*, **86**, 438–47.

KATES J.R. & JONES R.J. (1964b) Fluoroacetate inhibition of amino acids during photosynthesis by *Chlamydomonas reinhardii*. *Science, N.Y.* **143**, 145–6.

KERRIDGE D. (1958) The effects of actidione and other antifungal agents on nucleic acid and protein synthesis in *Saccharomyces carlsbergensis*. *J. gen. Microbiol.* **19**, 497–506.

KESSLER E. (1955) Über die Wirkung von 2,4-dinitrophenol aus Nitratreduktion und Atmung von Grünalgen. *Planta*, **45**, 94–105.

KESSLER E. (1959) Reduction of nitrate by green algae. *Symp. Soc. exp. Biol.* **13**, 87–105.

KESSLER E. (1964) Nitrate assimilation by plants. *A. Rev. Pl. Physiol.* **15**, 57–71.

KESSLER E. & BÜCKER W. (1960) Über die Wurkung von Arsenat auf Nitratreduktion, atmung und Photosynthese von Grünalgen. *Planta*, **55**, 512–24.

KESSLER E. & CZYGAN F.-C. (1963) Seasonal changes in the nitrate-reducing activity of a green alga. *Experientia*, **19**, 89–92.

KESSLER E. & CZYGAN F.-C. (1968) The effect of iron supply on the activity of nitrate and nitrite reduction in green algae. *Arch. Mikrobiol.* **60**, 282–4.

KESSLER E. & OESTERHELD H. (1970) Nitrification and induction of nitrate reductase in nitrogen-deficient algae. *Nature, Lond.* **228**, 287–8.

KESSLER E., HOFMANN A. & ZUMFT W.G. (1970) Inhibition of nitrite assimilation by uncouplers of phosphorylation. *Arch. Mikrobiol.* **72**, 23–6.

KIRK J.T.O. (1968) Studies on the dependence of chlorophyll synthesis on protein synthesis in *Euglena gracilis*, together with a nomogram for determination of chlorophyll concentration. *Planta*, **78**, 200–7.

KIRK J.T.O. & ALLEN R.L. (1965) Dependence of chloroplast pigment synthesis on protein synthesis: effect of actidione. *Biochem. biophys. Res. Commun.* **21**, 523–30.

KNUTSEN G. (1965) Induction of nitrite reductase in synchronized cultures of *Chlorella pyrenoidosa. Biochim. biophys. Acta*, **103**, 495–502.

KRETOVICH W.L., EVSTIGNEEVA Z.G. & TOMOVA N.G. (1970) Effect of nitrogen source on glutamate dehydrogenase and alanine dehydrogenase of *Chlorella. Can. J. Bot.* **48**, 1179–83.

LEFTLEY J.W. & SYRETT P.J. (1973) Urease and ATP: urea amido lyase in unicellular algae. *J. gen. Microbiol.* **77**, 109–15.

LÉ JOHN H.B. (1971) Enzyme regulation, lysine pathways and cell wall structures as indicators of major lines of evolution in fungi. *Nature, Lond.* **231**, 164–8.

LENGYEL P. & SÖLL D. (1969) Mechanism of protein synthesis. *Bact. Rev.* **33**, 264–301.

LINKO P., HOLM-HANSEN O., BASSHAM J.A. & CALVIN M. (1957) Formation of radioactive citrulline during photosynthetic $^{14}CO_2$ by blue-green algae. *J. exp. Bot.* **8**, 147–56.

LINNANE A.W. & STEWART P.R. (1967) The inhibition of chlorophyll formation in *Euglena* by antibiotics which inhibit bacterial and mitochondrial protein synthesis. *Biochem. biophys. Res. Commun.* **27**, 511–16.

LOENING U.E. (1968) Molecular weights of ribosomal RNA in relation to evolution. *J. mol. Biol.* **38**, 355–65.

LOOMIS W.D. (1959) Amide metabolism in higher plants. II. Distribution of glutamyl transferase and glutamine synthetase activity. *Pl. Physiol., Lancaster*, **34**, 541–6.

LOSADA M., RAMIREZ J.M., PANEQUE A. & DEL CAMPO F.F. (1965) Light and dark reduction of nitrate in a reconstituted chloroplast system. *Biochim. biophys. Acta*, **109**, 86–96.

LOSADA M., PANEQUE A., APARICHIO P.J., VEGA J.M., CÁRDENAS J. & HERRERA J. (1970) Inactivation and repression by ammonium of the nitrate-reducing system in *Chlorella. Biochem. biophys. Res. Commun.* **38**, 1009–15.

LUCAS-LENARD J. & LIPMANN F. (1971) Protein biosynthesis. *Ann. Rev. Biochem.* **40**, 409–48.

LUI N.S.T. & ROELS O.A. (1970) Nitrogen metabolism of aquatic organisms. I. The assimilation and formation of urea in *Ochromonas malhamensis. Archs. Biochem. Biophys.* **139**, 269–77.

LYNCH V.H. & GILLMOR G.G. (1966) Utilization of glutamine and glutamic acid by *Chlorella pyrenoidosa. Biochem. biophys. Acta*, **115**, 253–9.

MCLEAN JR. I.W., SCHWAB J.L., HILLEGAS A.B. & SCHLINGMAN A.S. (1949) Susceptibility of microorganisms to chloramphenicol (Chloromycetin). *J. clin. Invest.* **28**, 953–63.

MILLBANK J.W. (1953) Demonstration of transaminase systems in the alga *Chlorella. Nature, Lond.* **171**, 476.

MORRIS I. (1966a) Some effects of chloramphenicol on the metabolism of *Chlorella*. I. The effect on protein, polysaccharide and nucleic acid synthesis. *Arch. Mikrobiol.* **54**, 160–8.

MORRIS I. (1966b) Inhibition of protein synthesis by cycloheximide (actidione) in *Chlorella. Nature, Lond.* **211**, 1190–2.

MORRIS I. (1967) The effect of cycloheximide (actidione) on protein and nucleic acid synthesis by *Chlorella. J. exp. Bot.* **18**, 54–64.

MORRIS I. & AHMED J. (1969) The effect of light on nitrate and nitrite assimilation by *Chlorella* and *Ankistrodesmus. Physiologia Pl.* **22**, 1166–74.

MORRIS I. & SYRETT P.J. (1963a) The development of nitrate reductase in *Chlorella* and its repression by ammonium. *Arch. Mikrobiol.* **47**, 32–41.

MORRIS I. & SYRETT P.J. (1963b) Cyanate inhibition of nitrate reductase. *Biochim. biophys. Acta*, **77**, 649–50.

MORRIS I. & SYRETT P.J. (1965) The effect of nitrogen starvation on the activity of nitrate reductase and other enzymes in *Chlorella*. *J. gen. Microbiol.* **38**, 21–8.

MYERS J. (1949) The pattern of photosynthesis in *Chlorella*. In *Photosynthesis in Plants*, eds. Franck J. & Loomis W.E. pp. 349–64. Iowa.

NAKAMURA S., ASHINO K. & YAMAMOTO A. (1968) Distribution of transaminases in marine algae. *Shokubutsugaku Zasshi*, **81**, 74–8.

NICHOLAS D.J.D. (1959) Metallo-enzymes in nitrate assimilation of plants with special reference to microorganisms. *Symp. Soc. exp. Biol.* **13**, 1–13.

OESTERHELD H. (1971) Das Verhalten von Nitratreductase, Nitritreductase, Hydrogenase und andeven Enzymen von *Ankistrodesmus braunii* bei Stickstoffmangel. *Arch. Mikrobiol.* **79**, 25–43.

OHMORI K. & HATTORI A. (1970) Induction of nitrate and nitrite reductases in *Anabaena cylindrica*. *Plant & Cell Physiol.* **11**, 873–8.

OSAJIMA Y. & YAMAFUJI K. (1964) Reduction of nitrate to ammonia by enzymes isolated from green algae. *Enzymologia*, **27**, 129–40.

PAASCHE E. (1971) Effect of ammonia and nitrate on growth, photosynthesis and ribulose-diphosphate carboxylase content of *Dunaliella tertiolecta*. *Physiologia Pl.* **25**, 294–299.

PANEQUE A., DEL CAMPO F.F., RAMIREZ J.M. & LOSADA M. (1965) Flavin nucleotide nitrate reductase from spinach. *Biochim. biophys. Acta*, **109**, 79–85.

PANEQUE A., RAMIREZ J.M., DEL CAMPO F.F. & LOSADA M. (1964) Light and dark reduction of nitrite in a reconstituted enzymic system. *J. biol. Chem.* **239**, 1737–41.

POGO B.G.T. & POGO A.O. (1965) Inhibition by chloramphenicol of chlorophyll and protein synthesis and growth in *Euglena gracilis*. *J. Protozool.* **12**, 96–100.

POKORNY M., MARČENKO E. & KEGLEVIC D. (1970) Comparative studies of L- and D-methionine metabolism in lower and higher plants. *Phytochemistry*, **9**, 2175–88.

PROCTOR V.W. (1957) Preferential assimilation of nitrate ion by *Haematococcus pluvialis*. *Am. J. Bot.* **44**, 141–3.

RAMIREZ J.M., DEL CAMPO F.F., PANEQUE A. & LOSADA M. (1966) Ferredoxin-nitrite reductase from spinach. *Biochim. biophys. Acta*, **118**, 58–71.

REISNER G.S., GERING R.K. & THOMPSON J.F. (1960) The metabolism of nitrate and ammonia by *Chlorella*. *Pl. Physiol., Lancaster*, **35**, 48–52.

ROMANOV V.I., EVSTIGNEEVA Z.G. & KRETOVICH W.L. (1965) About amino acid dehydrogenases of *Chlorella*. *Prikl. Biokhim. Mikrobiol.* **1**, 495.

RODRIGUEZ-LOPEZ M. & VAZQUEZ D. (1968) Comparative studies on cytoplasmic ribosomes from algae. *Life Sciences*, **7**, 327–36.

ROON R.J. & LEVENBERG B. (1968) An adenosine triphosphate-dependent, Avidin-sensitive enzymatic cleavage of urea in yeast and green algae. *J. biol. Chem.* **243**, 5213–15.

SAGER R. & HAMILTON M.G. (1967) Cytoplasmic and chloroplast ribosomes of *Chlamydomonas*: ultracentrifugal characterization. *Science, N.Y.* **157**, 709–11.

SCOTT N.S., GRAHAM R.M.D. & SMILLIE R.M. (1971) Origin and synthesis of chloroplast ribosomal R.N.A. and photoregulation during chloroplast biogenesis. In *Autonomy and Biogenesis of Mitochondria and Chloroplasts*, eds. Boardman N.K., Linnane A.W. & Smillie R.M. pp. 383–92. North-Holland.

SHAFER J. & THOMPSON J.F. (1968) Arginine desimidase in *Chlorella*. *Phytochemistry*, **7**, 391–9.

SHAFER J., BAKER J.E. & THOMPSON J.F. (1961) A *Chlorella* mutant lacking nitrate reductase. *Am. J. Bot.* **48**, 896–9.

SHATILOV V.R., EVSTIGNEEVA Z.G. & KRETOVICH W.L. (1968) Alanine dehydrogenase of *Chlorella*. *Dokl. Akad. Nauk.* **178**, 482.

SIMS A.P. & FOLKES B.F. (1964) A kinetic study of the assimilation of $^{15}$N-ammonia and the synthesis of amino acids in an experimentally growing culture of *Candida utilis*. *Proc. R. Soc. B.* **159**, 479–502.

SIMS A.P., FOLKES B.F. & BUSSEY A.H. (1968) Mechanisms involved in the regulation of nitrogen assimilation in microorganisms and plants. In *Recent Aspects of Nitrogen Metabolism in Plants*, eds. Hewitt E.J. & Cutting C.V. pp. 91–114. Academic Press, New York & London.

SMILLIE R.M., EVANS W.R. & LYMAN H. (1963) Metabolic events during the formation of a photosynthetic from a nonphotosynthetic cell. *Brookhaven Symp. Biol.* **16**, 89–108.

SMILLIE R.M., SCOTT N.S. & GRAHAM D. (1968) Biogenesis of chloroplasts: roles of chloroplast DNA and chloroplast ribosomes. In *Comparative Biochemistry and Biophysics of Photosynthesis*, eds. Shibata K., Takamiya A., Jagendorf A.T. & Fuller R.C. pp. 332–53. Univ. Tokyo Press.

SMILLIE R.M., BISHOP D.G., GIBBONS G.C., GRAHAM D., GRIEVE A.M., RAISON J.K. & REGER B.J. (1971) Determination of the sites of synthesis of proteins and lipids of the chloroplast using chloramphenicol and cycloheximide. In *Autonomy and Biogenesis of Mitochondria and Chloroplasts*, eds. Boardman N.K., Linnane A.W. & Smillie R.M. pp. 422–33. North-Holland.

SMITH D.C., BASSHAM J.A. & KIRK M. (1961) Dynamics of the photosynthesis of carbon compounds. II. Amino acid synthesis. *Biochim. biophys. Acta*, **48**, 299–313.

SMITH F.W. & THOMPSON J.F. (1971) Regulation of nitrate reductase in *Chlorella vulgaris*. *Pl. Physiol., Lancaster*, **48**, 224–7.

STACEY J.L. & CASSELTON P.J. (1966) Utilization of adenine but not nitrate as nitrogen source by *Prototheca zopfii*. *Nature, Lond.* **211**, 862.

STAUB M. & DÉNES G. (1966) Mechanism of arginine biosynthesis in *Chlamydomonas reinhardii*. I. Purification and properties of ornithine acetyltransferase. *Biochim. biophys. Acta*, **128**, 82–91.

STRIJKERT P.J. & SUSSENBACH J.S. (1969) Arginine metabolism in *Chlamydomonas reinhardi*. Evidence for a specific regulatory mechanism of the biosynthesis. *Eur. J. Biochem.* **8**, 408–12.

STRIJKERT P.J., LOPPES R. & SUSSENBACH J.S. (1971) Arginine metabolism in *Chlamydomonas reinhardi*. Regulation of uptake and breakdown. *FEBS Letters*, **14**, 329–32.

STROSS R.G. (1963) Nitrate preference in *Haematococcus* as controlled by strain, age of inoculum and pH of the medium. *Can. J. Microbiol.* **9**, 33–40.

STROTMANN H. (1966) Kinetik der photosynthetischen Nitritreduktion *in vivo*. *Ber. dt. bot. Ges.* **79**, 118–20.

STROTMANN H. (1967) Untersuchungen zur lichtablängigen Nitritreduktion von *Chlorella*. *Planta*, **77**, 32–48.

STUTZ E. & NÖLL (1967) Characterization of cytoplasmic and chloroplast polysomes in plants: evidence for 3 classes of ribosomal RNA in nature. *Proc. natn. Acad. Sci. U.S.A.* **57**, 774–81.

SUSSENBACH J.S. & STRIJKERT P.J. (1969a) Arginine metabolism in *Chlamydomonas reinhardi*. On the regulation of the arginine biosynthesis. *Eur. J. Biochem.* **8**, 403–7.

SUSSENBACH J.S. & STRIJKERT P.J. (1969b) Arginine metabolism in *Chlamydomonas reinhardi*. Arginine deiminase: the first enzyme of the catabolic pathway. *FEBS Letters*, **3**, 166–8.

SUSSENBACH J.S. & STRIJKERT P.J. (1970) Arginine metabolism in *Chlamydomonas reinhardi*. A new type of citrulline degradation. *FEBS Letters*, **7**, 274–6.

SYRETT P.J. (1962) Nitrogen assimilation. In *Physiology and Biochemistry of Algae*, ed. Lewin R.A. pp. 171–88. Academic Press, New York & London.

SYRETT P.J. (1966) The kinetics of isocitrate lyase formation in *Chlorella*: evidence for the promotion of enzyme synthesis by photophosphorylation. *J. exp. Bot.* **17**, 641–54.

SYRETT P.J. & MORRIS I. (1963) The inhibition of nitrate assimilation by ammonium in *Chlorella*. *Biochim. biophys. Acta*, **67**, 566–75.

TAMIYA H., MORIMURA Y. & YOKOTO M. (1962) Effects of various antimetabolites upon the life-cycle of *Chlorella*. *Arch. Mikrobiol.* **42**, 4–16.

TAYLOR F.J. (1965) Chloramphenicol inhibition of the growth of green algae. *Nature, Lond.* **207**, 783.

TAYLOR M.M. & STORCK R. (1964) Uniqueness of bacterial ribosomes. *Proc. natn. Acad. Sci. U.S.A.* **52**, 958–64.

THACKER A. & SYRETT P.J. (1972a) The assimilation of nitrate and ammonium by *Chlamydomonas reinhardi. New Phytol.* **71**, 423–33.

THACKER A. & SYRETT P.J. (1972b) Disappearance of nitrate reductase activity from *Chlamydomonas reinhardi. New Phytol.* **71**, 435–41.

THOMPSON J.F. & MUENSTER A.E. (1971) Separation of the *Chlorella* ATP: urea amidolyase into two components. *Biochem. biophys. Res. Commun.* **43**, 1049–55.

TREBST A. & BURBA M. (1967) Über die Lemmung photosynthetischer Reaktionen in isolierten Chloroplasten und in *Chlorella* durch disalicyliden propandiamin. *Z. PflPhysiol.* **57**, 419–33.

VAZQUEZ D. (1964) Uptake and binding of chloramphenicol by sensitive and resistant organisms. *Nature, Lond.* **203**, 257–8.

VAZQUEZ D. & MONRO R.E. (1967) Effects of some inhibitors of protein synthesis on the binding of aminoacyl tRNA to ribosomal subunits. *Biochim. biophys. Acta,* **142**, 155–73.

VEGA J.M., HERRERA J., APARICHIO P.J., PANEQUE A. & LOSADA M. (1971) Role of molybdenum in nitrate reduction by *Chlorella. Plant Physiol.* **48**, 294–9.

VOGEL H.J. (1963) Lysine pathways as biochemical fossils. *Proc. Int. Congr. Biochem. Moscow 1961,* vol. 3, p. 341.

VOGEL H.J. (1964) Distribution of lysine pathways among fungi: evolutionary implications. *Am. Naturalist,* **98**, 435–46.

ZUMFT W.G., PANEQUE A., APARICHIO P.J. & LOSADA M. (1969) Mechanism of nitrate reduction in *Chlorella. Biochem. biophys. Res. Commun.* **36**, 980–6.

ZUMFT W.G., APARICHIO P.J., PANEQUE A. & LOSADA M. (1970) Structural and functional role of FAD in the NADH-nitrate-reducing system from *Chlorella. FEBS Letters,* **9**, 157–60.

# CHAPTER 22

# INORGANIC NUTRIENTS

## J. C. O'KELLEY

Department of Biology,
University of Alabama,
University, Alabama, U.S.A.

| | | | |
|---|---|---|---|
| **1** | **Introduction** 610 | 3.4 | Zinc 620 |
| | | 3.5 | Molybdenum 620 |
| | | 3.6 | Chlorine 621 |
| **2** | **Macronutrient elements** 611 | | |
| 2.1 | Sulphur 611 | **4** | **Elements required by some** |
| 2.2 | Potassium 613 | | **algae** 621 |
| 2.3 | Calcium 614 | 4.1 | Cobalt 621 |
| 2.4 | Magnesium 615 | 4.2 | Boron 622 |
| | | 4.3 | Silicon 623 |
| **3** | **Micronutrient elements consid-** | 4.4 | Sodium 624 |
| | **ered essential to all algae** 616 | 4.5 | Vanadium 624 |
| 3.1 | Iron 616 | 4.6 | Iodine 625 |
| 3.2 | Manganese 617 | | |
| 3.3 | Copper 619 | **5** | **References** 625 |

## 1 INTRODUCTION

Reviews dealing with aspects of algal nutrition have been published at intervals (Krauss 1958, Provasoli 1958, Lewin 1962, Gerloff 1963, Nicholas 1963, Hunter & Provasoli 1964, Healey 1973). The recent well-distributed review by O'Kelley (1968) dealt specifically with mineral nutrition and considered two questions in particular; namely, (a) what elements are known to be essential to algae, and (b) what are their respective metabolic roles?

Essentiality of a mineral element may be proposed whenever Arnon's classical criteria are satisfied (Arnon 1953). For the algae, experimental techniques should involve optimal or near optimal growth rates when the element in question is provided (Allen & Arnon 1955a, Bowen *et al.* 1965), and the purification of the growth medium should be such that the alga dies or ceases growth completely without the element. An element may also be considered essential if it is possible in *in vitro* experiments to demonstrate that it has a non-replaceable role in a fundamental cellular process; for example, participation as a catalyst in a step in photosynthesis. It should be recognized that growth stimulation by an element added to the nutrient medium, in itself does not prove this element to be essential.

Compared to vascular plants, the algae as a group are physiologically, as well as morphologically, very heterogeneous. This heterogeneity makes generalizations about their nutrition difficult. Whereas higher plants are believed to have the same elemental requirements (excluding, perhaps, those with nodulating bacteria), there appear to be differences in the elemental requirements of different algal species. In addition, substituting a chemically related element for an 'essential' one, that is, one that is otherwise essential, appears possible in some algae.

There is presently good evidence that the following inorganic elements (in addition to the organic elements C, H and O) are required by one or more algal species: N, P, K, Mg, Ca, S, Fe, Cu, Mn, Zn, Mo, Na, Co, V, Si, Cl, B, I. Of these, N, P, Mg, Fe, Cu, Mn, Zn and Mo are considered to be required by all algae and not replaceable even in part by other elements; however, to some extent at least S, K and Ca may be replaced. Cobalt is known to be required as vitamin $B_{12}$ by algal flagellates (Hutner $et$ $al.$ 1949), or in the inorganic form by some blue-green algae (Holm-Hansen $et$ $al.$ 1954). Other blue-green algae have been demonstrated to require Na (Allen & Arnon 1955b). Vanadium has been shown to be an essential micro-nutrient for the growth of $Scenedesmus$ $obliquus$ (Arnon & Wessel 1953). The essentiality of boron for diatoms appears to be well established (Lewin 1965, 1966a,b). While no demonstration has been provided that Cl is required in the medium of a growing alga, the chloride role in photosynthesis indicates its essentiality (Arnon 1955). Silicon is known to be needed for diatom growth (Lewin & Guillard 1963), and there is good evidence that at least one marine algal species requires iodine (Fries 1966).

A considerable part of the current study of mineral element nutrition in the algae is concerned with ion uptake. Since this is considered in Chapter 25 (p. 676) it will not be considered extensively here. Much of the remainder of the current interest in mineral nutrition centres around the metabolic functions, or roles, of the elements. Metabolic relations of nitrogen and phosphorus are considered in Chapters 20 and 21 (p. 560 and p. 583) and Chapter 23 (p. 636) respectively. With the exception of these elements and the specific processes, calcification and silicification (see Chapter 24, p. 655), metabolic functions of mineral elements are the major concern of this article. These are considered on an element-by-element basis even though it is recognized that each mineral element functions in the presence of others and is affected by them.

## 2 MACRONUTRIENT ELEMENTS

### 2.1 Sulphur

Since most algae can supply all of their sulphur requirement by reducing sulphate, the most abundant form of sulphur in nature, few studies have been made involving the use by algae of other sources of sulphur. *Chlorella vulgaris* will absorb methionine rapidly for a brief time, and more slowly for a prolonged period (Wedding & Block 1960). Streptomycin-bleached *Euglena* can use methionine, cysteine or homocysteine as a sole sulphur-source (Buetow 1965). Of twenty strains and seven species of *Chlorella* studied, four used D- or L-methionine and one used the L-form preferentially (Shrift & Sproul 1963a). Sulpholipid also can be utilized, but it probably is deacylated on the cell surface before the sulphur is absorbed (Miyachi & Miyachi 1966). Species of *Anabaena* and *Anacystis* can use sulphite (Prakash & Kumar 1971).

While a large part of the sulphur in most algae is incorporated into protein, sulphur may also exist in considerable quantity in algae as a component of other materials. Some of these materials appear to be unique to the algae. *Chlorella* and other green as well as brown algae produce sulpholipids (Benson & Shibuya 1962, Kennedy & Collier 1963). Fifty per cent of the sulphur of *Ochromonas danica* may exist as sulpholipid (Haines 1964), and sulpholipid may also be abundant in *Ochromonas malhamensis, Chlorella pyrenoidosa* and *Chlamydomonas* sp. (Haines 1965, Haines & Block 1962). Taurine and derivatives (sulphur at the sulphite level of reduction) occur in the red algae (Lindberg 1955). Sulphonium compounds have been isolated from various algae (Schiff 1962b). Sulphate may tend to be excluded by some algae, as *Valonia* and *Halicystis*, but in others it may accumulate; the marine alga *Desmarestia* contains much sulphate as sulphuric acid in the vacuolar sap (Blinks 1951). Polysaccharides esterified with sulphuric acid are common in several groups of algae (O'Colla 1962). Metabolically active forms of sulphur including S-adenosyl methionine (Schiff 1959) and adenosine-3'-phosphate-5'-phosphosulphate (Goldberg & Delbrück 1959) have also been found in algae.

A cell-free sulphate-reducing system from a strain of *Chlorella pyrenoidosa* shows maximal activity when fortified with ATP, an ATP-generating system, NADP, an NADP-reducing system and $MgCl_2$ (Schiff & Levinthal 1968). Thiosulphate is its major product (Levinthal & Schiff 1963, 1965, 1968); S-adenosyl methionine (Schiff 1959), adenosine-3'-phosphate-5'-phosphosulphate (Hodson *et al.* 1968a) and related compounds (Schiff 1962a, 1964) are also formed. Thiosulphate is readily utilized by this *Chlorella* strain; the molecule appears to undergo a dismutation, with the SH-sulphur preferentially incorporated in a reduced state (Hodson *et al.* 1967, 1968b). A sulphite-reducing system has been demonstrated in *Euglena* and species of green, red and blue-green algae but not brown algae (Saito *et al.* 1969). In *Porphyra tenera* the sulphite reduction can be coupled to reduced ferredoxin (Saito *et al.* 1970), but a purified sulphite-reductase from *Porphyra yezoensis* could only use reduced methyl viologen and not NADPH, NADH or reduced spinach ferredoxin (Saito & Tamura 1971).

Many studies on algae indicate a special requirement for sulphur in cell division. Some of the earliest of these involved using selenium as a sulphur analogue in *Chlorella vulgaris* (Shrift 1954a,b). Selenium also influences the morphological growth pattern of, and inhibits sulphate absorption in, *Chlorella vulgaris*, as well as phosphate uptake in *Scenedesmus* (Kylin 1966). In *Chlorella*, methionine partially reverses the selenium inhibition. Also in *Chlorella*, when selenomethionine is supplied, cell enlargement continues but division is temporarily arrested and giant cells form. In time, these giant cells adapt to selenomethionine to the extent that cell division resumes. If the abnormal large cells are supplied [35]S-sulphate and selenomethionine together, they form [35]S-protein which initially lacks any [35]S-methionine; after resumption of cell division [35]S-methionine is again found in the proteins being formed (Shrift 1959). The induced resistance to selenomethionine is believed to follow repression of a methionine-absorbing system, perhaps of a 'permease' type (Shrift & Sproul 1963b). It appears that sulphur in the form of methionine is necessary in this species of *Chlorella* for protein synthesis that leads to cell division. Selenium, probably as Se-methionine, temporarily prevents this synthesis by preventing the use of methionine and consequent formation of protein required for division.

Cytoplasmic cleavage leading to zoospore production in *Protosiphon botryoides* was blocked by removing sulphate or nitrate from an otherwise sufficient medium into which multi-nucleate sacs were placed (O'Kelley & Deason 1962). In synchronized *Chlorella ellipsoidea* cultures cell development in sulphur-deficient medium is arrested at the early ripening stage. Both nuclear and cytoplasmic division require sulphur (Hase *et al.* 1959). Sulphate is assimilated most actively in the ripening phases of cell development. Just before and after nuclear division sulphur-containing peptide-nucleotide components accumulate, and other sulphur-containing materials appear prior to and during autospore formation (Hase 1962). Similar sulphur-containing nucleotides are found in synchronized *Euglena* cells (Cook & Hess 1964).

## 2.2 *Potassium*

Evans and Sorger (1966) reviewed the role of univalent cations in plant metabolism and concluded that the potassium requirement could be based entirely upon use of this ion as the activator of enzymes. Not much has been written about the effects of potassium deficiency in the algae, but Pirson and Badour (1961) found that green algae pass through an early stage involving a high carbohydrate level and undergo changes in protein level and turnover in later stages of deficiency.

Sodium may replace potassium, at least in part (Emerson & Lewis 1942, Allen 1952). The replacement of potassium by rubidium has been reported for a number of algal species (Pirson 1939, Kellner 1955, Osretkar & Krauss 1965, Baum & O'Kelley 1966). In *Ankistrodesmus braunii* replacement of potassium by rubidium was considered to be partial at first and then complete following mutative adaptation (Kellner 1955). *Chlamydomonas reinhardtii* continued to grow in the rubidium replacement but lost motility (Baum & O'Kelley 1966). In *Chlorella pyrenoidosa* the rate of growth in the rubidium-containing medium was only 20% of that with potassium; cells that had been grown in the rubidium medium did not immediately respond to potassium when the latter element was re-supplied; the major compositional difference was a diminished porphyrin content in the rubidium-substituted cells (Osretkar & Krauss 1965). Whether the replacement can be complete in any of these species is unknown since the purest rubidium salts available contain traces of potassium.

Potassium uptake is considered in Chapter 25, p. 676. Rubidium absorption by algae, which appears to be by a similar mechanism or mechanisms, has been studied by Cohen (1962a,b), Schaedl and Jacobson (1966), Brenner and Maynard (1966) and by Springer-Lederer and Rosenfeld (1968).

### 2.3   *Calcium*

Partly because of the general macronutrient requirement for calcium in higher plants, it has been accepted that algae also require calcium, at least in the absence of strontium. The quantity required, however, appears to vary greatly between species. For at least one *Chlorella* strain calcium is a micro-nutrient (Walker 1953); other algal species, including *Protosiphon* (O'Kelley & Herndon 1961), respond to macronutrient levels. The calcium level for optimal growth of *Gonium pectorale* is lower by a factor of 100 than the level for optimal 16-cell colony formation (Groves & Kostir 1961); *Gonium* cells failed to adhere and produce colonies at the low levels of Ca that supported cell division optimally. The magnitude of calcium in the medium may be related to the quantity of pectic substances produced in the cell walls of algae that have the capacity to synthesize such materials, and the level of calcium required may be related to whether the species does or does not produce pectic substances (O'Kelley & Deason 1962). In some algae, such as *Chara*, excess calcium is inhibitory; in *Chara* a calcium level of 20mg $1^{-1}$ greatly reduces the rate of photosynthesis (Wetzel & McGregor 1968).

Some effects of calcium ions can also be demonstrated using magnesium. For example, either ion will stimulate sheath formation in *Oscillatoria limosa* (Foerster 1964) and either will promote flagella agglutination in heterothallic *Chlamydomonas* spp. (Wiese & Jones 1963). The thickness of the cell covering of a number of species of blue-green algae is both calcium- and magnesium-dependent (Sirenko 1967). Calcium will cause endoplasmic drops (cell fragments) of *Nitella* to form a surface membrane, doing so more effectively than magnesium (Shimizu 1965).

Early reports that strontium could substitute for calcium in algae were based on low growth rates. However, in carefully controlled experiments, strontium was found to be as effective in supporting growth of a *Chlorella* strain as was calcium (Walker 1953). The response of other algae to this substitution varies. Strontium inhibits calcium utilization in *Coccomyxa* (Walker 1956). In *Pandorina morum*, strontium stimulates growth only if a micronutrient level of calcium is supplied (Nichols 1960). A strontium replacement for calcium in *Protosiphon* produced cultures lacking free-swimming zoospores (O'Kelley & Herndon 1961). Accompanying this strontium replacement there are differences in cell wall composition (Denton *et al.* 1969); similar wall composition differences occur when strontium is supplied to *Chlorococcum echinozygotum* (Gilbert & O'Kelley 1964). In other algal species a strontium-replacement results in an increase in overall cell size in cultures (Kylin & Das 1967, Moss *et al.* 1971). Studies of the strontium-replacement phenomenon, particularly those involving *Protosiphon*, point to a multiple calcium role in the life of algal cells. Some functions of calcium appear to be carried out equally well by strontium, while others seem not to be mediated by the latter element. In *Protosiphon*, for example, digestion of sac walls leading to the release of zoospores specifically requires calcium.

## 2.4 *Magnesium*

Since nearly all algae possess chlorophyll and all are expected to carry out molecular phosphate transfers, magnesium is without doubt needed universally by algal species. *Chlorella* cells deprived of magnesium become chlorotic, enlarged and extensively vacuolated (Retovsky & Klasterska 1961). Magnesium-deficient algae can exhibit a number of metabolic disturbances. For example, nitrogen metabolism can be disturbed and there can be a temporary accumulation of carbohydrate material (Pirson & Badour 1961); an abnormally high quantity of labile phosphate may be produced (Badour 1961). Net synthesis of RNA may stop immediately following magnesium withdrawal from a culture while protein synthesis remains unaffected for several hours; soluble nitrogen compounds including uracil, orotic acid and hypoxanthine may accumulate, along with carbohydrates (Galling 1963).

The internal workings of magnesium have not been studied much in algae specifically. Magnesium markedly stimulated the Hill reaction in cell-free preparations from *Anabaena variabilis* (Susor & Krogmann 1964). Magnesium functions effectively in *in vitro* protein synthesis involving preparations of other kinds of organisms (Bonner 1965) because of a magnesium role in sub-unit association. However, in algae, other divalent cations may have this function; a recent paper suggests that zinc, rather than magnesium, has this role in *Euglena* (Prask & Plocke 1971). An S-adenosylmethionine-magnesium-protoporphyrin methyltransferase in *Euglena gracilis*, strain Z, is in part located on chloroplast lamellae, but in dark-grown cells may also exist as soluble protein in part (Ebbon & Tait 1969). *Euglena gracilis* cells possess a magnesium-dependent ATPase and also an unspecific calcium-dependent ATPase that can be partially activated by magnesium (Chang & Kahn 1966). A cell-free system from cells of this species requires magnesium for incorporating amino acids into protein (Eisenstadt & Brawerman 1964), and there is an allosteric effect of magnesium upon the ribulose-1,5-diphosphate carboxylase of *Chlorella ellipsoidea* (Sugiyama *et al.* 1969).

# 3  MICRONUTRIENT ELEMENTS CONSIDERED ESSENTIAL TO ALL ALGAE

## 3.1  *Iron*

The iron requirement in biological oxidation-reduction applies to algae as well as to other living organisms. *Ankistrodesmus braunii* required a minimum of 3 mg $1^{-1}$ FeSO$_4$.7H$_2$O for growth which increased as the iron salt was increased up to 70 mg $1^{-1}$; higher concentrations of 80 to 280 mg $1^{-1}$ were toxic (Palamar-Mordvyntseva 1968). The critical concentration of iron for heterotrophic growth of *Chlorella pyrenoidosa* has been shown to be $1 \times 10^{-9}$M, while for autotrophic growth it is $1 \cdot 8 \times 10^{-5}$ M (Eyster 1962); electron spin resonance studies indicate an interaction of part of the iron in this organism with manganese (Treharne & Eyster 1962). The effects of iron and manganese on productivity and on photosynthesis and respiration rates have recently been studied in several species of *Microcystis* (Velichko 1968). Droop (1961) has indicated the importance of redox potentials of marine media in relation to iron; these are believed to have important effects upon the uptake of iron by diatoms and other marine algae. Chelated iron as a factor in algal media has been investigated by MaciasR (1964, 1965) and more recently by Davies (1970). The formation of full-sized 16-cell colonies of *Gonium pectorale* is also a function of the content and availability of iron in the medium (Groves & Kostir 1961).

Ribonucleic acid and chlorophyll synthesis have been studied in *Chlorella vulgaris* during recovery from iron deficiency (Van Noort 1964). The iron level of the medium affects chlorophyll production, as well as growth, in both wild-type *Chlorella vulgaris* and its C-10 mutant. There appears to be a synergistic effect of iron and light upon chlorophyll production in the mutant, or light can 'substitute' for iron in circumventing a genetic lesion of C-10 (Bryan & Bogorad 1963). *Euglena gracilis* cells grown under low light and then transferred to high light in non-nutrient buffer synthesize chlorophyll in proportion to the iron content of the cells (Karali 1963, Price & Carell 1964). An iron requirement exists for the conversion of coproporphyrinogen to protoporphyrin in *Euglena gracilis* and also at another, unidentified, step in the synthesis of chlorophyll (Karali & Price 1963, Carell & Price 1965). In *Anacystis nidulans* both the 679nm peak of chlorophyll *a* in the absorption spectrum and a corresponding peak in the photosynthesis action spectrum is shifted when the organism becomes iron-deficient (Öquist 1971).

The proteins of *Euglena gracilis* var. *bacillaris* include two that resemble cytochrome *c* and a third characterized as cytochrome *b*; an *a*-type cytochrome is also believed to exist (Perini *et al.* 1964a); one *c*-type and the $b_6$ cytochromes are located in plastids; the other *c*-type and the *a*-type cytochromes appear to be involved in respiration (Perini *et al.* 1964b). Significant changes in the *b*-type cytochromes of *Euglena* are induced by light (Ikegami *et al.* 1970). A *c*-type cytochrome has been extracted from *Anacystis nidulans* (Holton & Myers 1963).

The respiratory chains in *Prototheca zopfii* and *Polytomella* have been reported to contain $b$, $c$, and $a$-$a_3$ cytochromes; *Astasia* also was reported to possess $a$, $b$, and $c$-type cytochromes, but with special characteristics (Webster & Hackett 1965). Colourless Cyanophyceae contain similar cytochromes, but show no cytochrome oxidase activity (Webster & Hackett 1966). More recently, seven respiratory cytochromes have been identified in *Prototheca*; some of these are soluble, but others are membrane-bound (Epel & Butler 1970).

The spectral properties of ferredoxin from blue-green algae have been studied by Smillie (1965), Susor and Krogmann (1966), San Pietro (1967), Bothe (1969) and Yamanaka *et al.* (1969). The ferredoxin-chlorophyll ratio in *Euglena* was earlier reported to be 1:400 (Perini *et al.* 1964a); more recent estimates include a range from 1:145 to 1:200 in *Anacystis* (Holton & Myers 1967), a range from 1:266 to 1:956 in *Euglena* (Boger & San Pietro 1967) depending upon the age of the cells, and a ratio of 1:50 ferredoxin/chlorophyll $a$ in *Nostoc* (Mitsui & Arnon 1971). Purified ferredoxin from *Scenedesmus* (Matsubara 1968), from *Euglena* (Mitsui 1970), and from *Nostoc* (Mitsui & Arnon 1971) exists as needle-shaped crystals. Mössbauer spectra of *Euglena* ferredoxin indicate that in the reduced state a single electron is shared by both iron atoms in a two-iron centre of the molecule (Johnson *et al.* 1968). Iron has been reported to be involved in nitrate reduction by *Chlorella* (Trubachev 1968), and a ferredoxin-dependent photoreduction of nitrate and nitrite has been demonstrated in subcellular preparations of *Anabaena cylindrica* (Hattori & Uesugi 1968). *Nostoc* ferredoxin has been shown to catalyse the photoreduction of NADP (Shin & Arnon 1965) and to be involved in the production and consumption of hydrogen gas (Tagawa & Arnon 1962). Hydrogenase activity of cell-free preparations of *Anabaena cylindrica* reduced NADP through the *Anabaena* PPNR (Fujita *et al.* 1964). In *Scenedesmus* the level of iron is directly related to the extent of hydrogenase activity (Sasaki 1966). Iron has been demonstrated to be essential to hydrogenase development in *Scenedesmus* (Yanagi & Sasa 1966). An enzyme has been reported from *Chlorella* that contains one per cent iron, is red in colour, and is linked to non-autoxidizable flavoprotein (Gewitz & Völker 1962), and iron containing acids believed to be associated with coenzyme A have been isolated from species of several genera of green algae (Boichenko & Zarin 1965).

## 3.2   *Manganese*

The initial report of a manganese requirement for algae was made by Hopkins (1930). Like higher plants, some algae become chlorotic when made manganese-deficient. Algae without hydrogenase, including *Chlorella vulgaris* and *C. saccharophila*, particularly show chlorosis if deprived of manganese, but the chlorophyll content of hydrogenase-containing algae is more stable (Kessler 1968). In photoautotrophic *Euglena gracilis*, growth was strongly dependent upon manganese while chlorophyll formation was not; also, the galactosylglyceride content was about 40% lower in manganese-deficient autotrophic cells than in normal autotrophic cells. However, when dark-grown *Euglena* cultures were transformed to light sufficient for greening of cells, both chlorophyll formation and the galactosyldiglyceride content were affected by the manganese supply (Constantopoulos 1970). Other lipid differences were observed in manganese-deficient *Euglena*. Manganese-deficient *Chlorella* cells were inhibited by high oxygen concentrations that had no effect on healthy cells (Brown 1961). Glycolate synthesis in *Chlorella* depends upon manganese (Tanner et al. 1960). In one study 30% of the $^{14}CO_2$ fixed in 10 minutes by manganese-sufficient *Chlorella pyrenoidosa* cells was in glycolate, while in deficient cells only two per cent was fixed in this acid and much more appeared in glycine and serine (Hess & Tolbert 1967).

The most studied effect of manganese on algal metabolism is the role of manganese in the $O_2$-evolving system of photosynthesis (Vernon 1962). Mixotrophically grown *Chlamydomonas* cells were unable, when manganese-deficient, to carry out photosynthetic reactions involving photosystem II; in contrast, the electron transport reactions of photosystem I did not appear to be dependent upon manganese (Teichler-Zallen 1969). A photoreactivation by manganese of the $O_2$-evolution in the Hill reaction in manganese-deficient *Anacystis* required electron flow through photosystem II (Cheniae 4 Martin 1969). In both *Scenedesmus* and *Ankistrodesmus* photosystem I-mediated phosphorylation was much more sensitive to manganese deficiency than was photosystem I photoreduction of NADP (Homann 1967). *Scenedesmus* that was made manganese-deficient lost the capacity to evolve $O_2$ before there was any gross alteration in chloroplast lamellar structure, pigmentation or respiratory capacity (Cheniae & Martin 1968). Studies of changes with light in the intensity of manganese electron spin resonance (ESR), originally suspected to be related to oxygen evolution in photosynthesis, are now believed to relate to an increased capacity of algal cells to take up manganese when illuminated (Teichler-Zallen & Levine 1967). Warburg and Krippahl (1967) have proposed that in *Chlorella*, manganese is also involved catalytically in $CO_2$ utilization in photosynthesis.

Manganese exists in algae in other material than protein. Complexes of manganese and galactosyldiglyceride exist in both green and blue-green algae (Udel'nova et al. 1968). The amount of manganese that is not bound to the photosynthetic apparatus can be measured by electron paramagnetic resonance (EPR) (Donnat & Briantais 1967). This element can be inhibitory or toxic in excess amounts. Manganese in the anionic form was toxic to *Microcystis* at a level of 2 mg $1^{-1}$ (Velichko 1968). A special case of toxicity occurred with *Anacystis nidulans* in medium containing manganese and citrate; upon being autoclaved or irradiated at 365nm the medium produced toxic levels of $H_2O_2$ (Marler & Van Baalen 1965). Manganese was found to be a negative fertility factor in the centric diatom *Ditylum brightwellii* in that auxospores and sperms were formed preferentially in manganese-free medium (Steele 1965).

### 3.3 Copper

Since the early paper describing a relationship of copper to algal growth (Guseva 1940) it has come to be accepted that all algae have a micronutrient copper requirement. One metabolic basis for this requirement is a probable role in photosynthesis for plastocyanin, which has been found in *Chlorella ellipsoidea* (Katoh 1960) and in other algae (Katoh *et al.* 1961). In *Anabaena variabilis* copper exists in protein of the photosynthetic apparatus; copper chelators inhibit the Hill reaction, but not plastocyanin reactions, suggesting more than one form of active copper in this organism (Lightbody & Krogmamn 1967). Copper is also implicated in photoreduction by *Scenedesmus obliquus* strain $D_3$; presumably it is essential here for photosystem I activity (Bishop 1964).

Most algae are extremely sensitive to excess copper; thus copper salts are frequently used as algicides. The conditions under which copper toxicity occurs and its physiological effects have been studied recently. A toxicity to *Enteromorpha* by sodium dimethyl dithiocarbamate and tetramethylthiuram disulphide has been attributed to the formation of toxic copper complexes (Lindahl 1962). In the experimental culture of *Porphyra tenera* the optimal copper level was 100 μg $1^{-1}$ (Shimo & Nakatani 1967); ethylenediamine-tetraacetic acid reduced inhibitory effects of additional copper. In *Chlorella pyrenoidosa* copper toxicity was reduced by increasing the potassium level within the medium (Nielsen *et al.* 1969). One effect of copper toxicity is a loss of cell potassium, but more potassium is lost than can be accounted for on the basis of exchange with copper (McBrien & Hassell 1965). In the same species, the inhibition by copper of both photosynthesis and growth was found to be influenced by such factors as whether or not cells were dividing, cell concentration, light intensity, and the presence of other ions including those of iron and hydrogen (Kamp-Nielsen 1969). Synchronized *Chlorella* cells at the 'ripened' stage were prevented from dividing in darkness by cupric ions, especially at pH 6·3 (Kanazawa & Kanazawa 1969). Other *Chlorella* cells were not necessarily killed at a concentration at which no growth would occur, and copper toxicity was greater at pH 8 than at pH 5 (Nielsen & Kamp-Nielsen 1970). In *Nitzschia palea*, in contrast to *Chlorella ellipsoidea*, copper inhibited photosynthesis more than it did growth (Nielsen & Wium-Anderson 1971). In *Chlorella vulgaris* copper absorbed under anaerobic conditions inhibited respiration, photosynthesis and growth more severely than did copper absorbed aerobically; under anaerobic conditions copper appears to bind to sites, possibly SH groups, that are not available with aerobiosis (McBrien & Hassall 1967). Copper and fluoride applied together inhibit respiration in particular (Hassall 1967, 1969). In spite of the sensitivity of algae to copper, it has been reported that no relation exists between the copper concentration of sea water and plankton growth (Godoy & Barth 1967).

## 3.4  Zinc

The earliest demonstration of a zinc requirement in algae was in *Stichococcus bacillaris* (Eilers 1926). The essentiality of zinc has been demonstrated in many other species since then and it is now assumed to be universally required by algae.

*Euglena gracilis*, which has a zinc-dependent lactic dehydrogenase (Price 1962), may be particularly useful in investigating the biochemical consequences of zinc-deficiency (Price & Vallee 1962). Much of the study of zinc metabolism in algae has involved *Euglena*; the growth rate of *Euglena* has been shown to be a linear function of its zinc content (Price & Quigley 1966), but the oxidation of heterotrophic substrates is less clearly affected by zinc (Price & Millar 1962). In zinc-deficient photosynthesizing *Euglena* the RNA content was found to be low, the amino acid content high, and the DNA content double that of zinc-sufficient cells (Wacker 1962). Altman *et al.* (1968) have shown a parallel increase in the zinc-content of *Chlorella* cells and their messenger-RNA content. Recently, the disappearance of cytoplasmic ribosomes has been demonstrated to occur in zinc-deficient *Euglena*; their reappearance requires an addition of zinc which may play a role in preserving the structure of ribosomal components (Prask & Plocke 1971).

Much of the interest in zinc and algae is related to uptake mechanisms. At the time of the most recent review of mineral nutrition in the algae (O'Kelley 1968), [65]Zn accumulation factors up to 1,200 had been reported for marine algae. However, it had also been noted that killed as well as living cells accumulate [65]Zn, and that [65]Zn can be removed rapidly from some algal cells after having been accumulated. Concentration factors for [65]Zn in Columbia River plankton have been shown to range from 300 to as high as 19,000 (Cushing 1967). Harvey and Patrick (1967) determined similar high concentration factors in freshwater algae, and also found that [65]Zn was rapidly desorbed in inactive medium. Cushing and Watson (1968) reported that killed plankton accumulated more [65]Zn than did living material. A study involving *Laminaria digitata* (Bryan 1969), however, indicated gradual uptake of inactive zinc or [65]Zn by this organism, and little tendency for the zinc to be lost by exchange. Matzku and Broda (1970), after studying zinc uptake in asynchronous *Chlorella fusca*, concluded that some [65]Zn taken up can be removed by exchange with non-labelled zinc, while another zinc fraction is more tightly bound. The exchangeable zinc is believed to be in the free space while the other is probably within the protoplast. Chemical evidence for more than one site of zinc uptake has also been obtained from recent studies involving *Chlorella pyrenoidosa* (Schuster & Broda 1970).

## 3.5  Molybdenum

Bortels (1940) showed a requirement for molybdenum in nitrogen fixation in blue-green algae and since then molybdenum has been demonstrated as essential for *Chlorella* (Walker 1953, Loneragan & Arnon 1954), and involved specifically in the assimilation of nitrate by *Scenedesmus* (Arnon *et al.* 1955, Arnon & Ichioka 1955). Molybdenum is required for the nitrate-reducing activity of the nicotinamide adenine dinucleotide nitrate-reducing complex that has been isolated from *Chlorella fusca* (Vega *et al.* 1971). Molybdenum-deficient *Oocystis* cells have been shown to have a low total nitrogen content (Rao 1963), and the nitrate reduction coupled with hydrogenase in *Anabaena* can be stimulated by 0·025 micromolar levels of molybdenum (Fujita *et al.* 1964). Nitrogen fixation accompanied by oxygen evolution in *Anabaena oscillarioides* and *Hapalosiphon fontinalis* f. *globosus*, was stimulated by molybdenum and manganese (Darkanbaev & Sachkova 1970). The addition of trace quantities of ammonium molybdate increased the dry weight of the marine algae *Ulva lactuca*, *Dictyota dichotoma* and *Pterocladia capillacea* (Nasr & Bekheet 1970). Molybdenum has been shown to be a limiting factor in nature for the growth of algae, in Castle Lake, California (Goldman 1967).

### 3.6 *Chlorine*

All photosynthesizing algae are believed to require chloride, on the basis of photosynthesis studies indicating such a requirement for the Hill reaction, for ATP-formation and for FMN-catalysed phosphorylation (Vernon 1962). Whole cell nutritional experiments corroborate this only in an inconclusive way. Eyster (1962) demonstrated that *Chlorella* growth could be doubled by adding $3·4 \times 10^{-6}$ M NaCl to medium otherwise suitable for growth but lacking chloride.

In contrast to chloride ions, free chlorine at 0·2 parts per million acts as a long term algistat while bromine is algicidal (Kott *et al.* 1966). In field experiments the combined use of chlorine and bromine proved more toxic to algae than either halogen alone (Kott 1969).

## 4 ELEMENTS REQUIRED BY SOME ALGAE

### 4.1 *Cobalt*

Hutner and colleagues (Hutner *et al.* 1949) were the first to demonstrate an algal requirement for cobalt, as vitamin $B_{12}$ required by *Euglena*; a cobalt requirement by blue-green algal species that could be satisfied using the inorganic form of this element was demonstrated a few years later (Holm-Hansen *et al.* 1954). Recently, cobalt has been shown to cause an increase in both the dry weight and the nitrogen content of the marine algae *Ulva*, *Dictyota* and *Pterocladia* (Nasr & Bekheet 1970).

A number of marine algae either require vitamin $B_{12}$ or have their growth stimulated by its addition to the culture medium. These include the marine diatom *Skeletonema costatum* (Droop 1962), and species of the red algal genera

*Goniotrichum* (Fries 1959), *Nemalion* (Fries 1961), *Polysiphonia* (Fries 1964), *Bangia* (Provasoli 1964), *Antithamnion* (Tatewaki & Provasoli 1964), *Asterocytis* (Fries & Pettersson 1968) and the *Conchocelis* stage of *Porphyra tenera* (Iwasaki 1965). Among the brown algae, *Pylaiella* (*Pilayella*) *littoralis* zoospores have a vitamin $B_{12}$ requirement (Pedersen 1969b); species of *Ectocarpus* (Boalch 1961) and *Litosiphon* (Pedersen 1969b) are stimulated to grow by additions of vitamin $B_{12}$ to the medium. Most of the other divisions (or phyla) of algae include a few species known to require vitamin $B_{12}$ (Droop 1962).

*Euglena gracilis*, widely used as a vitamin $B_{12}$ assay organism, responds not only to cyanocobalamin but also to pseudovitamin $B_{12}$ and to factors A, G and H (Droop 1962). Other organisms that might be more useful in assays include *Chlamydomonas pallens* which responds to cyanocobalamin and factor I but not to pseudovitamin $B_{12}$ or to factor B (Pringsheim 1962), and *Ochromonas* which is reported to respond to the same variants as do human beings (Coats & Ford 1955). In *Antithamnion* analogues containing benzimidazole and factor $Z_1$ and $Z_3$, presumed to have no nucleotide, may replace the vitamin (Tatewaki & Provasoli 1964).

In *Ochromonas malhamensis*, vitamin $B_{12}$ stimulated the incorporation of formate- and formaldehyde-$^{14}C$ cobalamin-donated methyl groups for choline synthesis (Lust & Daniel 1964). Propionate metabolism in *Ochromonas* is also regulated in some manner by vitamin $B_{12}$, since in its absence there is a nearly complete block of propionate oxidation. One analogue stimulated propionate oxidation but inhibited growth, while a second stimulated growth but increased the oxidation of propionate only slightly (Arnstein & White 1962). In the same organism, there is evidence of alternate biosynthetic pathways to thymine from 5-methyl cytosine; one in the presence of vitamin $B_{12}$, a second in its absence (Letendre & Daniel 1966). In *E. gracilis* var. *bacillaris* a vitamin $B_{12}$ deficiency is accompanied by decreased concentrations of RNA, DNA, and protein. The addition of $B_{12}$ stimulates the incorporation of $^{14}C$-formate into the above compounds, as well as into the thymine of DNA, and into the serine and methionine of proteins (Venkataraman *et al.* 1965). These and other effects have been interpreted to indicate that, in DNA synthesis, vitamin $B_{12}$ has a significant role in the reduction of ribotides to deoxyribotides, and in the synthesis of thymine, through a vitamin $B_{12}$-dependent conversion of glutamate to beta-methyl aspartate. In *Prymnesium parvum*, methionine counteracted some inhibitory effects of $B_{12}$ analogues substituted at the benzimidazole part of the molecule (Rahat & Reich 1963), indicating that either methionine or the benzimidazole moiety is required for the transfer of methyl groups. Other aspects of vitamin $B_{12}$ requirements are considered later (Chapter 27, p. 742).

## 4.2 Boron

Hercinger (1940) was first to report a boron requirement for algae, but this was followed by experiments indicating that algae had no such requirement. Lewin

(1965, 1966a) provided firm evidence that boron is required by the marine pennate diatom *Cylindrotheca fusiformis*. Germanium, a possible replacement element for boron, was inactive (Neales 1967) in the nutrition of this organism. Other marine and also freshwater diatoms require boron (Lewin 1966b). Prior reports of boron requirements in other algal species than diatoms, and boron stimulation of growth, have been criticized on the basis of techniques used (Gerloff 1968). Bowen *et al*. (1965), paying special attention to *Chlorella*, used techniques involving near-optimal or optimal growth rates and could not demonstrate a requirement in four species tested. Gerloff (1968) failed to demonstrate a requirement for boron in *Chlorella pyrenoidosa* or in two other species of green algae. Dear and Aronoff (1968) concluded that neither cell number nor dry weight were limited by a boron concentration of 0·5 μg $1^{-1}$ (5 × $10^{-8}$ M). McBride *et al*. (1971) also found that boron concentrations, in culture media, from 0·001 to 10·0 mg $1^{-1}$ did not affect growth of a strain of *Chlorella vulgaris*.

A stimulation of blue-green algal growth has been reported for *Nostoc muscorum* (Eyster 1952, 1959), and for this organism and two other nitrogen-fixing blue-green algae by Gerloff (1968); boron, however, had no effect upon the growth of the blue-green alga *Microcystis aeruginosa*, which does not fix nitrogen.

Boron accumulation in freshwater algal species has been observed in cultures (Gerloff 1968) and also in nature (Boyd 1970). In *Chlorella vulgaris*, however, no accumulation was observed and no significant relation between external and internal boron levels appeared to exist (McBride *et al*. 1971).

In *Ulva* and in *Dictyota* boron is reported to stimulate reproductive growth specifically (Nasr & Bekheet 1970). Whatever the role of boron may be in plant nutrition, it is hardly likely that it is involved in any part of physiological processes, such as photosynthesis or DNA production, which are common to all green plants. Rather, the clue should be sought in the commonality, perhaps, between the metabolism of *Cylindrotheca* and higher plants that is not evident in *Chlorella* sp. or in fungi which also do not appear to require boron (Gerloff 1968).

### 4.3 *Silicon*

Silicon occurs abundantly in nature and is a constituent of many plants. It exists, among the algae, particularly in the cell walls of diatoms and some chrysophycean flagellates (Lewin & Reimann 1969). A silicon requirement in diatoms, for valve formation, was proposed first by Bachrach and Lefevre (1929). It is now generally accepted that diatom cells do not divide unless a supply of silicon, as silicic acid, is available (Jørgensen 1952, 1957, Lewin 1955). In cells of *Cyclotella cryptica* deprived of silicate, the synthesis of protein, DNA, RNA, chlorophyll, xanthophyll and lipid, as well as photosynthesis, are impaired (Werner 1966). *Navicula pelliculosa* and *Cylindrotheca fusiformis* are reported to require silicon for mitosis, and the latter species is reported to have a specific requirement for silicon in the net synthesis of DNA (Darley 1969). It is hard to understand how this represents a primary effect of the silicon deficiency, however, unless silicon is required by all living organisms. Silicon utilization by diatoms requires a supply of reduced sulphur (Lewin 1954). Germanium, the next higher analogue of silicon, interferes with utilization of the latter element (Lewin 1966c, Werner 1966, 1967). Silicon uptake, including the physiological and morphological aspects of silicification, is considered in detail in Chapter 24 (p. 655).

## 4.4  *Sodium*

Since sodium and potassium have similar chemical properties, the early experiments dealing with sodium and algal growth were concerned with its possible replacement for potassium. Benecke (1898) described an *Oscillatoria* sp. that grew when potassium in the medium was replaced by sodium. Emerson and Lewis (1942) grew a *Chroococcus* sp. in the presence of sodium and 1mg $l^{-1}$ of potassium; however, the *Chroococcus* grew very poorly in the absence of sodium. Allen (1952) found 23 strains of blue-green algae that would grow in sodium salt media lacking added potassium. Kratz and Myers (1955) found that logarithmic growth of three blue-green algae, *Anabaena variabilis*, *Nostoc muscorum* and *Anacystis nidulans* ceased in the absence of sodium and was re-established upon its addition. Allen and Arnon (1955b) determined that 5 mg $l^{-1}$ of sodium is required for optimal growth of *Anabaena cylindrica*, and that the sodium could not be replaced completely by potassium, nor potassium by sodium. Similarly, *Anabaena flos-aquae* has requirements for both potassium and sodium (Bostwick *et al.* 1968).

The apparently complete replacements of potassium by sodium may have been only partial; non-replaceable potassium in micro quantities may have existed as contaminant. In the other situations, where sodium appears to be needed in addition to potassium, a metabolic basis for the sodium requirement has not been established. Droop (1958) suggested a non-osmotic sodium requirement in euryhaline algae, and McLachlan (1960) reported such a requirement in *Dunaliella tertiolecta*. By definition, this would be a nutritional requirement, but the nutritional role for sodium was not identified. Nitrite accumulates in sodium-deficient cultures of *Anabaena cylindrica*, and nitrate reductase is markedly increased within the *Anabaena* cells. Largely on this basis, it has been proposed that sodium is required in nitrogen-fixing blue-green algae for the transformation of molecular nitrogen into ammonia (Brownell & Nicholas 1967).

## 4.5  *Vanadium*

Although earlier papers suggested an essential role for vanadium in algae, the first conclusive evidence was that of Arnon and Wessel (1953) for *Scenedesmus obliquus*. In *Chlorella* vanadium is reported to stimulate $CO_2$ uptake in photosynthesis at low light intensities, possibly by serving as a catalyst for $CO_2$ reduction (Warburg *et al.* 1955). Vanadium has also been reported to accelerate photosynthesis in both species under high light intensity (Eyster 1962). Experimental difficulties involved in demonstrating the necessity for vanadium result from an affinity of vanadium for iron and a resulting common contamination of iron compounds with traces of vanadium. Recently a study of the distribution of vanadium in algae has shown that it varies from as little as $0 \cdot 3$ to as much as $10 \cdot 6$ mg $l^{-1}$ in marine species (Yamamoto *et al.* 1970).

### 4.6 *Iodine*

*Polysiphonia urceolata* appears to have an absolute requirement for iodine (Fries 1966), and Iwasaki (1967) has shown a growth stimulation of the *Conchocelis* stage of *Porphyra tenera* upon the addition of iodine to the medium. *Ectocarpus fasciculatus*, when bacteria-free, also would not grow in the absence of iodine (Pedersen 1969a). Other marine algae, including *Ceramium*, *Ulva* and *Polysiphonia* concentrate iodine from sea water or from artificial media (Roche & Andre 1962), as do also some freshwater organisms such as *Callitriche aquatica* and *Chara globularis* (Roche & Andre 1966). The uptake of iodine is believed to be by active transport, since ouabaine inhibits it (Andre 1965). Internally, the iodine may be incorporated into mono- and di-iodotyrosine, although no metabolic role for these amino acids has been demonstrated in algae (Shaw 1962). In some edible marine algae the major chemical forms of iodine appear to be low molecular weight organic compounds (Meguro *et al.* 1967).

Iodine ions inhibit both positive and negative phototaxis of *Mougeotia* chloroplasts (Schonbohm 1966). At least one blue-green alga, *Synechococcus salina*, is tolerant of iodide as evidenced by its being found growing abundantly in waters from a crude-oil well containing $15 \cdot 1$ mg $l^{-1}$ (Nindak 1968).

## 5 REFERENCES

ALLEN M.B. (1952) The cultivation of Myxophyceae. *Arch. Mikrobiol.* **17**, 34–53.
ALLEN M.B. & ARNON D.I. (1955a) Studies on nitrogen-fixing blue-green algae. I. Growth and nitrogen fixation by *Anabaena cylindrica* Lemm. *Pl. Physiol., Lancaster*, **30**, 366–72.
ALLEN M.B. & ARNON D.I. (1955b) Studies on nitrogen-fixing blue-green algae. II. The sodium requirements of *Anabaena cylindrica*. *Physiologia Pl.* **8**, 653–60.
ALTMAN H., FETTER F. & KAINDL K. (1968) Untersuchungen über den Einfluss von Zn-Ionen auf die m-RNA-Synthese in *Chlorella* zellen. *Z. Naturf.* **23b**, 395–6.
ANDRE S. (1965) Sur la biochimie comparee du transport des iodures. Action de divers inhibiteurs sur la fixation des iodures par un Bryozoaire (*Bugula neritina* L.) et par deux Algues marines (*Laminaria flexicaulis* et *Enteromorpha* sp.). *C.r. Séanc. Soc. Biol.* Paris, **159**, 2327–32.
ARNON D.I. (1953) Growth and function as criteria in determining the essential nature of inorganic nutrients. In *Mineral Nutrition of Plants*, ed. Truog E. pp. 313–41. The University of Wisconsin Press, Madison.
ARNON D.I. (1955) The chloroplast as a complete photosynthetic unit. *Science, N.Y.* **122**, 9–16.
ARNON D.I. & ICHIOKA P.S. (1955) Molybdenum in relation to nitrogen metabolism II.

Assimilation of ammonia and urea without molybdenum by *Scenedesmus. Physiologia Pl.* **8**, 552–60.

ARNON D.I., ICHIOKA P.S., WESSEL G., FUJIWARA A. & WOOLLEY J.T. (1955) Molybdenum in relation to nitrogen metabolism. I. Assimilation of nitrate nitrogen by *Scenedesmus. Physiologia Pl.* **8**, 538–51.

ARNON D.I. & WESSEL G. (1953) Vanadium as an essential element for green plants. *Nature, Lond.* **172**, 1039–40.

ARNSTEIN H.R.V. & WHITE A.M. (1962) The function of vitamin $B_{12}$ in the metabolism of propionate by the protozoan *Ochromonas malhamensis. Biochem. J.* **83**, 264–70.

BACHRACH E. & LEFEVRE M. (1929) Contribution a l'etude du role de la silice chez les etres vivants. Observations sur la biologie des Diatomées. *J. Physiol. Path. Gen.* **27**, 241–9.

BADOUR S.S.A. (1961) Kennzeichnung von Mineralsalzmangelzustaenden bei Gruenalgen mit analytisch-chemischen Methodik II. Phosphat fractionen bei Kaliummangel im Vergleich mit Magnesium- und Manganmangel. *Flora Jena,* **151**, 99–119.

BAUM L.S. & O'KELLEY J.C. (1966) Rubidium substitution for potassium in several species of fresh water algae. *Pl. Physiol., Lancaster,* **41** (*Suppl.*), xxxiv.

BENECKE W. (1898) Über Culturbedingungen einiger Algen. *Bot. Zeit.* **56**, 83–96.

BENSON A.A. & SHIBUYA I. (1962) Surfactant lipids. In *Physiology and Biochemistry of Algae,* ed. Lewin R.A. pp. 371–84. Academic Press, New York & London.

BISHOP N.I. (1964) Site of action of copper in photosynthesis. *Nature, Lond.* **204**, 401–2.

BLINKS L.R. (1951) Physiology and biochemistry of algae. In *Manual of Phycology,* ed. Smith G.M. pp. 263–91. Chronica Botanica Company, Waltham. Mass.

BOALCH G.T. (1961) Studies on *Ectocarpus* in culture. II. Growth and nutrition of a bacteria-free culture. *J. mar. biol. Ass. U.K.* **41**, 287–304.

BOGER P. & SAN PIETRO A. (1967) Ferredoxin and cytochrome f in *Euglena gracilis. Z. PflPhysiol.* **58**, 70–5.

BOICHENKO E.A. & ZARIN V.E. (1965) Svedinenie zheleza a kofermentom A v rasteniyakh. *Dokl. Akad. Nauk. S.S.S.R.* **165**, 1423–6.

BONNER J. (1965) Ribosomes. In *Plant Biochemistry,* eds. Bonner J. & Varner J.E. pp. 21–37. Academic Press, New York & London.

BORTELS H. (1940) Über die Bedeutung des Molybdäns für stickstoffbindende Nostocaceen. *Arch. Mikrobiol.* **11**, 155–86.

BOSTWICK C.D., BROWN L.R. & TISCHER R.G. (1968) Some observations on the sodium and potassium interactions in the blue-green alga *Anabaena flos-aquae* A-37. *Physiologia Pl.* **21**, 466–9.

BOTHE H. (1969) The role of phytoflavin in photosynthetic reactions. In *Progress in Photosynthesis Research,* eds. Metzner H. H. Laupp Jr, Tübingen, **3**, 1483–91.

BOWEN J.E., GAUCH H.G., KRAUSS R.W. & GALLOWAY R.A. (1965) The non-essentiality of boron for *Chlorella. J. Phycol.* **1**, 151–4.

BOYD C.E. (1970) Boron accumulation by native algae. *Amer. Midland Natur,* **84**, 565–67.

BRENNER M.L. & MAYNARD D.N. (1966) Study of rubidium accumulation in *Euglena gracilis. Pl. Physiol., Lancaster,* **41**, 1285–8.

BROWN T.E. (1961) Physiological aspects of the growth of normal and manganese-deficient *Chlorella* in the presence of various atmospheres. *Pl. Physiol., Lancaster,* **36** (*Suppl.*), iii.

BROWNELL P.F. & NICHOLAS D.J.D. (1967) Some effects of sodium on nitrate assimilation and $N_2$ fixation in *Anabaena cylindrica. Pl. Physiol., Lancaster,* **42**, 915–21.

BRYAN G.W. (1969) The absorption of zinc and other metals by the brown seaweed *Laminaria digitata. J. mar. biol. Ass. U.K.* **49**, 225–43.

BRYAN G.W. & BOGORAD L. (1963) Protochlorophyllide and chlorophyll formation in response to iron nutrition in a *Chlorella* mutant. In: *Studies on Microalgae and Photosynthetic Bacteria* pp. 399–405. Japanese Society of Plant Physiology, University of Tokyo Press, Tokyo, Japan.

BUETOW D.E. (1965) Growth, survival and biochemical alteration of *Euglena gracilis* in medium limited in sulfur. *J. cell. comp. Physiol.* **66**, 235–42.

CARELL E.F. & PRICE C.A. (1965) Porphyrins and the iron requirement for chlorophyll formation in *Euglena*. *Pl. Physiol., Lancaster*, **40**, 1–7.

CHANG F.C. & KAHN J.S. (1966) Isolation of a possible coupling factor for photophosphorylation from chloroplasts of *Euglena gracilis*. *Archs. Biochem. Biophys.* **117**, 282–8.

CHENIAE G.M. & MARTIN I.F. (1968) Site of manganese function in photosynthesis. *Biochim. biophys. Acta*, **153**, 819–37.

CHENIAE G.M. & MARTIN I.F. (1969) Photoreactivation of manganese catalyst in photosynthetic oxygen evolution. *Pl. Physiol., Lancaster*, **44**, 351–60.

COATS M.E. & FORD J.E. (1955) Methods of measurement of vitamin $B_{12}$. *Symp. Biochem. Soc.* **13**, 36–51.

COHEN D. (1962a) Specific binding of rubidium in *Chlorella*. *J. Gen. Physiol.* **45**, 959–77.

COHEN D. (1962b) The chemistry of sites binding rubidium in *Chlorella*. *J. gen. Physiol.* **45**, 979–87.

CONSTANTOPOULOS G. (1970) Lipid metabolism of manganese-deficient algae. I. Effect of manganese deficiency on the greening and the lipid composition of *Euglena gracilis Z. Pl. Physiol., Lancaster*, **45**, 76–80.

COOK J.R. & HESS M. (1964) Sulfur-containing nucleotides associated with cell division in synchronized *Euglena gracilis*. *Biochim. biophys. Acta*, **80**, 148–51.

CUSHING C.E. (1967) Concentration and transport of $^{32}P$ and $^{65}Zn$ by Columbia River plankton. *Limnol. Oceanogr.* **12**, 330–2.

CUSHING C.E. & WATSON D.E. (1968) Accumulation of $^{32}P$ and $^{65}Zn$ by living and killed plankton. *Oikos*, **19**, 143–5.

DARKANBAEV T.B. & SACHKOVA O.P. (1970) Vliyanie Mo i Mn na fiksatsiya azota i fotosintez nekotorykh sinezelnykh vodoroslei. *Izv. Akad. Nauk. Kaz. U.S.S.R. Ser. Biol.* **1**, 26–9.

DARLEY W.M. (1969) Silicon requirements for growth and macromolecular synthesis in synchronized cultures of the diatoms, *Navicula pelliculosa* (Brebisson) Hilse and *Cylindrotheca fusiformis* Reimann and Lewin. (Ph.D. Dissertation, University of California, San Diego), 1–148.

DAVIES A.G. (1970) Iron, chelation and the growth of marine phytoplankton. I. Growth kinetics and chlorophyll production in cultures of the euryhaline flagellate *Dunaliella tertiolecta* under iron-limiting conditions. *J. mar. biol. Assoc. U.K.* **50**, 65–86.

DEAR J.M. & ARONOFF S. (1968) The non-essentiality of boron for *Scenedesmus*. *Pl. Physiol., Lancaster*, **43**, 997–8.

DENTON T.E., MESHAD M., HOLADAY J.W. & O'KELLEY J.C. (1969) Ca and Sr influence on carbohydrate synthesis and composition in *Protosiphon*. *Pl. Cell Physiol., Tokyo*, **10**, 711–14.

DONNAT P. & BRIANTAIS J-M. (1967) Contenu et etat du manganese dans l'appareil photosynthetique. *C.r. hebd. Séanc. Acad. Sci., Ser. D., Paris*, **264**, 2903–8.

DROOP M.R. (1958) Optimum relative and actual ionic concentrations for growth of some euryhaline algae. *Verhandl. intern. Ver. Limnol.* **13**, 722–30.

DROOP M.R. (1961) Some chemical considerations in the design of synthetic culture media for marine algae. *Botanica. Mar.* **2**, 231–46.

DROOP M.R. (1962) Organic micronutrients. In *Physiology and Biochemistry of Algae*, ed. Lewin R.A. pp. 141–59. Academic Press, New York & London.

EBBON J.G. & TAIT G.H. (1969) Studies on S-adenosyl-methionine-magnesium protoporphyrin methyltransferase in *Euglena gracilis* strain Z. *Biochem. J.* **111**, 573–82.

EILERS H. (1926) Zur Kenntnis der Ernährungsphysiologie von *Stichococcus bacillaris* Näg. *Rec. trav. botan. Neerl.* **23**, 362–95.

EISENSTADT J. & BRAWERMAN G. (1964) Characteristics of a cell-free system from *Euglena gracilis* for the incorporation of amino acids into protein. *Biochim. biophys. Acta*, **80**, 463–72.

EMERSON R. & LEWIS C.M. (1942) The photosynthetic efficiency of phycocyanin in *Chroococcus*, and the problem of carotenoid participation in photosynthesis. *J. gen. Physiol.* **25**, 579–95.

EPEL B.L. & BUTLER W.L. (1970) The cytochromes of *Prototheca zopfii*. *Pl. Physiol., Lancaster*, **45**, 723–7.

EVANS H.J. & SORGER G.J. (1966) Role of mineral elements with emphasis on the univalent cations. *A. Rev. Pl. Physiol.* **17**, 47–76.

EYSTER C. (1952) Necessity of boron for *Nostoc muscorum*. *Nature, Lond.* **170**, 755.

EYSTER C. (1959) Mineral requirements of *Nostoc muscorum* for nitrogen fixation. *Proc. IX Int. Bot. Cong., Montreal*, **2**, 109.

EYSTER C. (1962) Requirements and functions of micronutrients by green plants with respect to photosynthesis. In *Biologistics for Space Systems Symposium. U.S. Air Force Tech. Doc. Rept.* AMRL-TDR-62-116, 199–209.

FOERSTER J.W. (1964) The use of calcium and magnesium ions to stimulate sheath formation in *Oscillatoria limosa* (Roth) C. A. Agardh. *Trans. Am. Microsc. Soc.* **83**, 420–7.

FRIES L. (1959) *Goniotrichum elegans*: a marine red alga requiring vitamin $B_{12}$. *Nature, Lond.* **183**, 558–9.

FRIES L. (1961) Vitamin requirements of *Nemalion multifidum*. *Experientia*, **17**, 75.

FRIES L. (1964) *Polysiphonia urceolata* in axenic culture. *Nature, Lond.* **202**, 110.

FRIES L. (1966) Influence of iodine and bromine on growth of some red algae in axenic culture. *Physiologia Pl.* **19**, 800–8.

FRIES L. & PETTERSSON H. (1968) Physiology of *Asterocytis*. *Br. Phycol. Bull.* **3**, 417–22.

FUJITA Y., OHAMA H. & HATTORI A. (1964) Hydrogenase activity of cell-free preparation obtained from the blue-green alga *Anabaena cylindrica*. *Pl. Cell Physiol., Tokyo*, **5**, 305–14.

GALLING G. (1963) Analyse des Magnesium-Mangels bei synchronisierten *Chlorellen*. *Arch. Mikrobiol.* **46**, 150–84.

GERLOFF G.C. (1963) Comparative mineral nutrition of plants. *A. Rev. Pl. Physiol.* **14**, 107–23.

GERLOFF G.C. (1968) The comparative boron nutrition of several green and blue-green algae. *Physiologia Pl.* **21**, 369–77.

GEWITZ H.-S. & VÖLKER W. (1962) The respiratory enzymes of *Chlorella*. *Hoppe-Seyler's Zeit. Physiol. Chem.* **330**, 124–31.

GILBERT W.A. & O'KELLEY J.C. (1964) The effects of replacement of calcium by strontium on the reproduction of *Chlorococcum echinozygotum*. *Am. J. Bot.* **51**, 866–9.

GODOY O.T. & BARTH R. (1967) Concentracao de cobre na aquae sua influencia sobre o planeton. *Notas Tec. Inst. Pesqui Mar. Rio de Janeiro*, **1**, 1–16.

GOLDBERG I.H. & DELBRÜCK A. (1959) Transfer of sulfate from 3′-phosphoadenosine-5′-phosphosulfate to lipids, mucopolysaccharides and aminoalkyl phenols. *Fed. Proc.* **18**, 235.

GOLDMAN C.R. (1967) Molybdenum as an essential micronutrient and useful water-mass marker in Castle Lake, California. In *Chemical Environment in the Aquatic Habitat*. Proceedings of an International Biological Programme Symposium, 10–16 Oct. 1966, Amsterdam and Nieuwersluis, Neth. pp. 229–36. North Holland Publishing Co., Amsterdam.

GROVES JR.L.B. & KOSTIR W.J. (1961) Some factors affecting the formation of colonies in *Gonium pectorale*. *Ohio J. Science*, **61**, 321–31.

GUSEVA K.A. (1940) Dyeystvye myedei na vodoroslei. *Mikrobiologiya*, **9**, 480–99.

HAINES T.H. (1964) A new sulfolipid in microbes (*Ochromonas danica*). *Diss. Abstr.* **24**, 2203.

HAINES T.H. (1965) A microbial sulfolipid. I. Isolation and physiological studies. *J. Protozool.* **12**, 655–9.

HAINES T.H. & BLOCK R.J. (1962) Sulfur metabolism in algae. I. Synthesis of metabolically inert chloroform-soluble sulfate esters by two chrysomonads and *Chlorella pyrenoidosa*. *J. Protozool.* **9**, 33–8.

HARVEY R.S. & PATRICK R. (1967) Concentration of $Cs^{137}$, $Zn^{65}$, and $Sr^{89}$ by fresh water algae. *Biotechnol. Bioeng.* **9**, 449–56.

HASE E. (1962) Cell division. In *Physiology and Biochemistry of Algae*, ed. Lewin R.A. pp. 617–24. Academic Press, New York & London.

HASE E., OTSUKA H., MIHARA S. & TAMIYA H. (1959) Role of sulfur in the cell division of *Chlorella*, studied by the technique of synchronous culture. *Biochim. biophys. Acta*, **35**, 180–9.

HASSALL K.A. (1967) Inhibition of respiration of *Chlorella vulgaris* by simultaneous application of cupric and fluoride ions. *Nature, Lond.* **215**, 521.

HASSALL K.A. (1969) An asymmetric respiratory response occurring when fluoride and copper ions are applied jointly to *Chlorella vulgaris*. *Physiologia Pl.* **22**, 304–11.

HATTORI A. & UESUGI I. (1968) Ferredoxin-dependent photoreduction of nitrate and nitrite by subcellular preparations of *Anabaena cylindrica*. In *Comparative Biochemistry and Biophysics of Photosynthesis*, 201–5. Univ. Tokyo Press, Tokyo, Univ. Park Press. Pa.

HEALEY F.P. (1973) Inorganic nutrient uptake and deficiency in algae. *Crit. Rev. Microbiol.* **3**, 69–113.

HERCINGER F. (1940) Beiträge zum Wirkungskreislauf des Bors. *Bodenk. u. Pflanzenernähr.* **16**, 141–68.

HESS J.L. & TOLBERT N.E. (1967) Glycolate pathway in algae. *Pl. Physiol., Lancaster*, **42**, 371–9.

HODSON R.C., SCHIFF J.A. & SCARSELLA A.J. (1967) Metabolism of thiosulfate by *Chlorella pyrenoidosa*. *Pl. Physiol., Lancaster*, **42**, (*Suppl.*) xxxvi.

HODSON R.C., SCHIFF J.A., SCARSELLA A.J. & LEVINTHAL M. (1968a) Studies of sulfate utilization by algae. 6. Adenosine-3′-phosphate-5′-phosphosulfate formation from sulfate by cell-free extracts of *Chlorella*. *Pl. Physiol., Lancaster*, **43**, 563–9.

HODSON R.C., SCHIFF J.A. & SCARSELLA A.J. (1968b) Studies of sulfate utilization by algae. 7. *In vivo* metabolism of thiosulfate by *Chlorella*. *Pl. Physiol., Lancaster*, **43**, 570–7.

HOLM-HANSEN O., GERLOFF G.C. & SKOOG F. (1954) Cobalt as an essential element for blue-green algae. *Physiologia Pl.* **7**, 665–75.

HOLTON R.W. & MYERS J. (1963) Cytochromes of a blue-green alga: extraction of a c-type with a strongly negative redox potential. *Science, N.Y.* **142**, 234–5.

HOLTON R.W. & MYERS J. (1967) Water-soluble cytochromes from a blue-green alga. II. Physicochemical properties and quantitative relationships of cytochrome c (549, 552 and 554, *Anacystis nidulans*). *Biochim. biophys. Acta*, **131**, 375–84.

HOMANN P.H. (1967) Studies on the manganese of the chloroplast. *Pl. Physiol., Lancaster*, **42**, 997–1007.

HOPKINS E.F. (1930) The necessity and function of manganese in the growth of *Chlorella* sp. *Science, N.Y.* **72**, 609–10.

HUTNER S.H. & PROVASOLI L. (1964) Nutrition of algae. *A. Rev. Pl. Physiol.* **15**, 37–56.

HUTNER S.H., PROVASOLI L., STOCKSTAD E.L.R., HOFFMAN C.E., BELT M., FRANKLIN A.L. & JUKES J.H. (1949) Assay of antipernicious anemia factor with *Euglena*. *Proc. Soc. expl. Biol. Med.* **70**, 117–20.

IKEGAMI I., KATOH S. & TAKAMIYA A. (1970) Light-induced changes of b-type cytochromes in the electron transport chain of *Euglena* chloroplasts. *Pl. Cell Physiol., Tokyo*, **11**, 777–91.

IWASAKI H. (1965) Nutritional studies of the edible seaweed *Porphyra tenera*. I. The influence of different $B_{12}$ analogues, plant hormones, purines and pyrimidines on the growth of *Conchocelis*. *Pl. Cell Physiol., Tokyo*, **6**, 325–36.

IAWSAKI H. (1967) Nutritional studies of the edible seaweed *Porphyra tenera*. II. Nutrition of *Conchocelis*. *J. Phycol.* **3**, 30–4.

X

JOHNSON C.E., ELSTNER E., GIBSON J.F., BENFIELD E., EVANS M.C.W. & HALL D.O. (1968) Mössbauer effect in the ferredoxin of *Euglena*. *Nature, Lond.* **220**, 1291–3.

JØRGENSEN E.G. (1952) Effects of different silicon concentrations on the growth of diatoms. *Physiologia Pl.* **5**, 161–70.

JØRGENSEN E.G. (1957) Diatom periodicity and silicon assimilation. Experimental and ecological investigations. *Dansk. bot. Ark.* **18**, 6–54.

KAMP-NIELSEN L. (1969) The influence of copper on the photosynthesis and growth of *Chlorella pyrenoidosa*. *Dansk Tidsskr. Farm.* **43**, 249–54.

KANAZAWA T. & KANAZAWA K. (1969) Specific inhibitory effect of copper on cellular division in *Chlorella*. *Pl. Cell Physiol., Tokyo*, **10**, 495–502.

KARALI E.F. (1963) Iron, growth and chlorophyll synthesis in *Euglena*. *Diss. Abstr.* **24**, 1675.

KARALI E.F. & PRICE C.A. (1963) Iron, porphyrins and chlorophyll. *Nature, Lond.* **198**, 708.

KATOH S. (1960) A new copper protein from *Chlorella ellipsoidea*. *Nature, Lond.* **186**, 533–4.

KATOH S., SUGI I., SHIRATORI I. & TAKAMIYA I. (1961) Distribution of plastocyanin in plants, with special reference to its localization in chloroplasts. *Archs. Biochem. Biophys.* **94**, 136–41.

KELLNER K. (1955) Die Adaptation von *Ankistrodesmus braunii* an Rubidium und Kupfer. *Biol. Zentr.* **74**, 662–91.

KENNEDY G.Y. & COLLIER R. (1963) The sulpholipids of plants. II. The isolation of sulpholipid from green and brown algae. *J. mar. biol. Assoc., U.K.* **43**, 613–19.

KESSLER E. (1968) Effect of manganese deficiency on growth and chlorophyll content of algae with and without hydrogenase. *Arch. Mikrobiol.* **63**, 7–10.

KOTT Y. (1969) Effect of halogens on algae: III. Field experiment. *Water Res.* **3**, 265–71.

KOTT Y., HERSHKOVITZ G., SHEMTOB A. & SLESS J.B. (1966) Algicidal effect of bromine and chlorine on *Chlorella pyrenoidosa*. *Appl. Microbiol.* **14**, 8–11.

KRATZ W.A. & MYERS J. (1955) Nutrition and growth of several blue-green algae. *Am. J. Bot.* **42**, 282–7.

KRAUSS R.W. (1958) Physiology of the fresh-water algae. *A. Rev. Pl. Physiol.* **9**, 207–44.

KYLIN A. (1966) The influence of photosynthetic factors and metabolic inhibitors on the uptake of phosphate in P-deficient *Scenedesmus*. *Physiologia Pl.* **19**, 644–9.

KYLIN A. & DAS G. (1967) Calcium and strontium as micronutrient and morphogenetic factors for *Scenedesmus*. *Phycologia*, **6**, 201–10.

LETENDRE C.H. & DANIEL L.J. (1966) Some aspects of thymine biosynthesis in *Ochromonas malhamensis*. *Archs. Biochem. Biophys.* **113**, 182–8.

LEVINTHAL M. & SCHIFF J.A. (1963) Formation of thiosulfate by a cell-free sulfate reducing system from *Chlorella*. *Pl. Physiol., Lancaster*, **38**, (*Suppl.*) xli.

LEVINTHAL M. & SCHIFF J.A. (1965) Intermediates in thiosulfate formation by cell-free sulfate-reducing system from *Chlorella*. *Pl. Physiol., Lancaster*, **40**, (*Suppl.*) xvi.

LEVINTHAL M. & SCHIFF J.A. (1968) Studies of sulfate utilization by algae. 5. Identification of thiosulfate as a major acid-volatile product formed by a cell-free sulfate-reducing system from *Chlorella*. *Pl. Physiol., Lancaster*, **43**, 555–62.

LEWIN J.C. (1954) Silicon metabolism in diatoms. I. Evidence for the role of reduced sulfur compounds in silicon utilization. *J. gen. Physiol.* **37**, 589–99.

LEWIN J.C. (1955) Silicon metabolism in diatoms. II. Sources of silicon for growth of *Navicula pelliculosa*. *Pl. Physiol., Lancaster*, **30**, 129–34.

LEWIN J.C. (1965) The boron requirement of a marine diatom. *Naturwissenschaften*, **52**, 70.

LEWIN J.C. (1966a) Physiological studies of the boron requirement of the diatom, *Cylindrica fusiformis* Reimann and Lewin. *J. exp. Bot.* **17**, 473–9.

LEWIN J.C. (1966b) Boron as a growth requirement for diatoms. *J. Phycol.* **2**, 160–3.

LEWIN J.C. (1966c) Silicon metabolism in diatoms. V. Germanium dioxide, a specific inhibitor of diatom growth. *Phycologia*, **6**, 1–12.

LEWIN J.C. & GUILLARD R.R.L. (1963) Diatoms. *A. Rev. Microbiol.* **17**, 373–414.

LEWIN J.C. & REIMANN B.E.F. (1969) Silicon and plant growth. *A. Rev. Pl. Physiol.* **20**, 289–304.

LEWIN R.A. ed. (1962) *Physiology and Biochemistry of Algae*. Academic Press, New York & London.

LIGHTBODY J.J. & KROGMANN D.W. (1967) Isolation and properties of plastocyanin from *Anabaena variabilis. Biochim. biophys. Acta*, **131**, 508–15.

LINDAHL P.E.B. (1962) The inhibition of growth of *Enteromorpha linza* (L.) J. Ag. by sodium dimethyl dithiocarbamate and tetramethylthiuram disulphide. *Physiologia Pl.* **15**, 607–22.

LINDBERG B. (1955) Methylated taurines and choline sulfate in red algae. *Acta chem. scand.* **9**, 1323–6.

LONERAGAN J.F. & ARNON D.I. (1954) Molybdenum in the growth and metabolism of *Chlorella. Nature, Lond.* **174**, 459.

LUST G. & DANIEL L.J. (1964) Vitamin $B_{12}$ and the synthesis of the methyl group of choline in *Ochromonas malhamensis. Archs. Biochem. Biophys.* **108**, 414–19.

MACIASR F.M. (1964) Growth of *Chlamydomonas mundana* in the presence of different synthetic chelators. *Appl. Microbiol.* **12**, 391–4.

MACIASR F.M. (1965) Effect of pH of the medium on the availability of chelated iron for *Chlamydomonas mundana. J. Protozool.* **12**, 500–4.

MARLER J.E. & VAN BAALEN C. (1965) Role of $H_2O_2$ in single-cell growth of the blue-green alga *Anacystis nidulans. J. Phycol.* **1**, 180–5.

MATSKU S. & BRODA E. (1970) Die Zinkaufnahme in das Innere von *Chlorella. Planta*, **92**, 29–40.

MATSUBARA H. (1968) Purification and some properties of *Scenedesmus* ferredoxin. *J. biol. Chem.* **243**, 370–5.

MCBRIDE L., CHORNEY W. & SKOK J. (1971) Growth of *Chlorella* in relation to boron supply. *Bot. Gaz.* **132**, 10–13.

MCBRIEN D.C.H. & HASSALL K.A. (1965) Loss of cell potassium by *Chlorella vulgaris* after contact with toxic amounts of copper sulphate. *Physiologia Pl.* **18**, 1059–65.

MCBRIEN D.C.H. & HASSALL K.A. (1967) The effect of toxic doses of copper upon the respiration, photosynthesis and growth of *Chlorella vulgaris. Physiologia Pl.* **20**, 113–17.

MCLACHLAN J. (1960) The culture of *Dunaliella tertiolecta* Butcher—a euryhaline organism. *Can. J. Microbiol.* **6**, 363–79.

MEGURO H., ABE T., OGASAWARA T. & TUZIMURA K. (1967) Analytical studies of iodine in food substances. I. Chemical form of iodine in edible marine algae. *Agr. Biol. Chem.* **31**, 999–1002.

MITSUI A. (1970) Crystallization and some properties of algal ferredoxins. *Pl. Physiol., Lancaster*, **46** (*Suppl.*) xxxix.

MITSUI A. & ARNON D.I. (1971) Crystalline ferredoxin from a blue-green alga, *Nostoc* sp. *Physiologia Pl.* **25**, 135–40.

MIYACHI S. & MIYACHI S. (1966) Sulfolipid metabolism in *Chlorella. Pl. Physiol., Lancaster*, **41**, 479–86.

MOSS S.W., THOMAS J.P. & O'KELLEY J.C. (1971) Strontium substitution for calcium and algal cell size. *Physiologia Pl.* **25**, 184–7.

NASR A.H. & BEKHEET I.A. (1970) Effect of certain trace elements and soil extract on some marine algae. *Hydrobiologia*, **36**, 53–60.

NEALES T.F. (1967) The boron nutrition of the diatom *Cylindrotheca fusiformis*, grown on agar, and the biological activity of some substituted phenylboric acids. *Aust. J. biol. Sci.* **20**, 67–76.

NICHOLAS D.J.D. (1963) Inorganic nutrient nutrition of microorganisms. In *Plant Physiology*, vol. III, ed. Steward F.C. pp. 363–447. Academic Press, New York & London.

NICHOLS H.W. (1960) *Strontium replacement for calcium in the growth of four volvocalean algae* (M.S. Thesis, University of Alabama, Library, University, Ala.), pp. 1–40.

NIELSEN E.S. & KAMP-NIELSEN L. (1970) Influence of deleterious concentrations of copper on the growth of *Chlorella pyrenoidosa*. *Physiologia Pl.* **23**, 828–40.

NIELSEN E.S. & WIUM-ANDERSON S. (1971) The influence of Cu on photosynthesis and growth in diatoms. *Physiologia Pl.* **24**, 480–4.

NIELSEN E.S., KAMP-NIELSEN L. & WIUM-ANDERSON S. (1969) The effect of deleterious concentrations of copper on the photosynthesis of *Chlorella pyrenoidosa*. *Physiologia Pl.* **22**, 1121–3.

NINDAK F. (1968) Mass development of the chroococcalean blue-green alga *Synechocystis salina* Wisl. in iodine and bromine rich waters of a crude oil-well. *Biologia* (*Bratislova*), **23**, 841–4.

O'COLLA P.S. (1962) Mucilages. In *Physiology and Biochemistry of Algae*, ed. Lewin R.A. pp. 337–56. Academic Press, New York & London.

O'KELLEY J.C. (1968) Mineral nutrition of algae. *A. Rev. Pl. Physiol.* **19**, 89–112.

O'KELLEY J.C. & DEASON T.R. (1962) Effect of nitrogen, sulfur and other factors on zoospore production by *Protosiphon botryoides*. *Am. J. Bot.* **49**, 771–7.

O'KELLEY J.C. & HERNDON W.R. (1961) Alkaline earth elements and zoospore release and development in *Protosiphon botryoides*. *Am. J. Bot.* **48**, 796–802.

ÖQUIST G. (1971) Changes in pigment composition and photosynthesis induced by iron-deficiency in the blue-green alga *Anacystis nidulans*. *Physiologia Pl.* **24**, 188–91.

OSRETKAR A. & KRAUSS R.W. (1965) Growth and metabolism of *Chlorella pyrenoidosa* Chick during substitution of Rb for K. *J. Phycol.* **1**, 23–33.

PALAMAR-MORDVYNTSEVA H.M. (1968) Potreba ankistrodesma brauna (*Ankistrodesmus braunii* Brunnth.) v zalizi. *Ukr. Bot. Zh.* **25**, 21–8.

PEDERSEN M. (1969a) The demand for iodine and bromine of three marine brown algae grown in bacteria-free cultures. *Physiologia Pl.* **22**, 680–5.

PEDERSEN M. (1969b) Marine brown algae requiring vitamin $B_{12}$. *Physiologia Pl.* **22**, 977–83.

PERINI F., KAMEN M.D. & SCHIFF J.A. (1964a) Iron-containing proteins in *Euglena*. I. Detection and characterization. *Biochim. biophys. Acta* **88**, 74–90.

PERINI F., SCHIFF J.A. & KAMEN M.D. (1964b) Iron-containing proteins in *Euglena*. II. Functional localization. *Biochim. biophys. Acta*, **88**, 91–8.

PIRSON A. (1939) Über die Wirkungen alkalionen auf Wachstum und Stoffwechsel von *Chlorella*. *Planta*, **29**, 231–61.

PIRSON A. & BADOUR S.S.A. (1961) Kennzeichnung von Mineralsalzmangelzustanden bei Gruenalgen mit analytischenchemischen Methodik. I. Kohlenhydratspiegel, organischen Stickstoff und Chlorophyll bei Kalimangel im Vergleich mit Magnesium- und Manganmangel. *Flora Jena* **150**, 243–58.

PRAKASH G. & KUMAR H.D. (1971) Studies of sulphur-selenium antagonism in blue-green algae. I. Sulfur nutrition. *Arch. Mikrobiol.* **77**, 196–202.

PRASK J.A. & PLOCKE D.J. (1971) A role for zinc in the structural integrity of the cytoplasmic ribosomes of *Euglena gracilis*. *Pl. Physiol.*, *Lancaster*, **48**, 150–5.

PRICE C.A. (1962) A zinc-dependent lactate dehydrogenase in *Euglena gracilis*. *Biochem. J.* **82**, 61–6.

PRICE C.A. & CARELL E.F. (1964) Control by iron of chlorophyll formation and growth in *Euglena gracilis*. *Pl. Physiol.*, *Lancaster*, **39**, 862–8.

PRICE C.A. & MILLAR E. (1962) Zinc, growth and respiration in *Euglena*. *Pl. Physiol.*, *Lancaster*, **37**, 423–7.

PRICE C.A. & QUIGLEY J.W. (1966) A method for determining quantitative zinc requirements for growth. *Soil Sci.* **101**, 11–16.

PRICE C.A. & VALLEE B.L. (1962) *Euglena gracilis*, a test organism for the study of zinc. *Pl. Physiol.*, *Lancaster*, **37**, 428–33.

PRINGSHEIM E.G. (1962) *Chlamydomonas pallens*, a new organism proposed for assays of vitamin $B_{12}$. *Nature, Lond.* **195**, 604.

PROVASOLI L. (1958) Nutrition and ecology of protozoa and algae. *A. Rev. Microbiol.* **12**, 279–308.

PROVASOLI L. (1964) Growing marine seaweeds. *Proc. 4th Intern. Seaweed Symp.* 1961, eds. Devirville A.D. & Feldman J. 9–17. Pergamon Press, Oxford, London & New York.

RAHAT M. & REICH K. (1963) The $B_{12}$ vitamins and methionine in the metabolism of *Prymnesium parvum. J. Gen. Microbiol.* **31**, 203–9.

RAO K.V.M. (1963) The effect of molybdenum on the growth of *Oocystis marssonii* Lemm. *Indian J. Pl. Physiol.* **6**, 142–9.

RETOVSKY R. & KLASTERSKA I. (1961) Study of the growth and development of *Chlorella* populations in the culture as a whole. V. The influence of $MgSO_4$ on autospore formation. *Folia Microbiol.* **6**, 115–26.

ROCHE J. & ANDRE S. (1962) Etude radioautographique de la fixation de l'iode ($^{131}$I) par des algues marines. *C.r. Séanc. Soc. Biol.* **156**, 1968–71.

ROCHE J. & ANDRE S. (1966) Concentration de l'iode par des vegetaux d'eau douce et transport actif des iodures. *C.r. Séanc. Soc. Biol.* **160**, 1800–5.

SAITO E. & TAMURA G. (1971) Studies on the sulfite reducing system of algae. Part II. Purification and properties of the reduced methyl viologen-linked sulfite reductase from a red alga, *Porphyra yezoensis. Agr. Biol. Chem.* **35**, 491–500.

SAITO E., TAMURA G. & SHINANO S. (1969) Studies on the sulfite reducing system of algae. Part I. Sulfite reduction by algal extracts using reduced methylviologen as a hydrogen donor. *Agr. Biol. Chem.* **33**, 860–7.

SAITO E., WAKASA K., OKUMA M. & TAMURA G. (1970) Studies on the sulfite reducing system of algae. Part III. Sulfite reduction by algal extract coupling to the reduced ferredoxin. *Bull. Ass. Nat. Sci., Senshy Univ.* **3**, 45–50.

SAN PIETRO A. (1967) Plant and algal ferredoxins. *8th Internat. Cong. Biochem., Tokyo.* Abstr. Col. XIII-3, 559.

SASAKI H. (1966) Effects of culture conditions on hydrogenase activity of *Scenedesmus* $D_3$. *Pl. Cell Physiol., Tokyo*, **7**, 231–41.

SCHAEDL M. & JACOBSON L. (1966) Ion absorption and retention by *Chlorella pyrenoidosa*. II. Permeability of the cell to sodium and rubidium. *Pl. Physiol., Lancaster*, **41**, 248–54.

SCHIFF J.A. (1959) Studies on sulfate utilization by *Chlorella pyrenoidosa* using sulfate-$S^{35}$; the occurrence of S-adenosyl methionine. *Pl. Physiol., Lancaster*, **34**, 73–80.

SCHIFF J.A. (1962a) Sulfate reduction by cell-free extracts of *Chlorella* and by intact cells. *Pl. Physiol., Lancaster*, **37**, *(Suppl.)* xlii.

SCHIFF J.A. (1962b) Sulfur. In *Physiology and Biochemistry of Algae*, ed. Lewin R.A. pp. 239–66. Academic Press, New York & London.

SCHIFF J.A. (1964) Studies on sulfate utilization by algae. II. Further identification of reduced compounds formed from sulfate by *Chlorella. Pl. Physiol., Lancaster*, **39**, 176–9.

SCHIFF J.A. & LEVINTHAL M. (1968) Studies of sulfate utilization by algae. 4. Properties of a cell-free sulfate-reducing system from *Chlorella. Pl. Physiol., Lancaster*, **43**, 547–54.

SCHONBOHM E. (1966) Die Hemmung der positiven und negativen Phototaxis des Mougeotia-Chloroplasten durch Iodid-Ionen. *Z. Pfl Physiol.* **56**, 366–74.

SCHUSTER I. & BRODA E. (1970) Die Bindung von Zink durch Zellwande von *Chlorella. Mschr. Chem.* **101**, 285–95.

SHAW T.I. (1962) Halogens. In *Physiology and Biochemistry of Algae*, ed. Lewin R.A. p.p 247–53. Academic Press, New York & London.

SHIMIZU A. (1965) Formation of surface membrane of the endoplasmic drop isolated *in vitro* from the *Nitella* internode. *Sci. Rept. (Osaka Univ.)* **14**, 21–5.

SHIMO S. & NAKATANI S. (1967) The influence of various concentrations of copper and mercury ions on the growth of *Porphyra tenera* and its tolerance to them. *J. agr. Lab. (Chiba)*, **9**, 109–25.

SHIN M. & ARNON D.I. (1965) Enzyme mechanisms of pyridine nucleotide reduction in chloroplasts. *J. biol. Chem.* **240**, 1405–11.

SHRIFT A. (1954a) Sulfur-selenium antagonism. I. Antimetabolite action of selenate on the growth of *Chlorella vulgaris. Am. J. Bot.* **41**, 223–30.

SHRIFT A. (1954b) Sulfur-selenium antagonism. II. Antimetabolite action of selenomethionine on the growth of *Chlorella vulgaris. Am. J. Bot.* **41**, 345–52.

SHRIFT A. (1959) Nitrogen and sulfur changes associated with growth uncoupled from cell division in *Chlorella vulgaris. Pl. Physiol., Lancaster,* **34**, 505–12.

SHRIFT A. & SPROUL M. (1963a) Sulfur nutrition and the taxonomy of *Chlorella. Phycologia,* **3**, 85–100.

SHRIFT A. & SPROUL M. (1963b) Nature of the stable adaptation induced by selenomethionine in *Chlorella vulgaris. Biochim. biophys. Acta,* **71**, 332–44.

SIRENKO L.A. (1967) Fizioloho-biokhimichne osoblyvosti metabolizmusyn o-zelenykh, vodorostei v svitli novitnikh danykh. I. Tsytoloxichni osoblyvosti. *Ukr. Bot. Zh.* **24**, 3–18.

SMILLIE R.M. (1965) Isolation of phytoflavin, a flavoprotein with chloroplast ferredoxin activity. *Pl. Physiol., Lancaster,* **40**, 1124–8.

SPRINGER-LEDERER H. & ROSENFELD D.L. (1968) Energy sources for the absorption of rubidium by *Chlorella. Physiologia Pl.* 435–44.

STEELE R.L. (1965) Induction of sexuality in two centric diatoms. *Bioscience,* **15**, 298.

SUGIYAMA T., MATSUMOTO C. & AKAZAWA T. (1969) Structure and function of chloroplast proteins. VII. Ribulose-1,5-diphosphate carboxylase of *Chlorella ellipsoidea. Archs. Biochem. Biophys.* **129**, 597–602.

SUSOR W.A. & KROGMANN D.W. (1964) Hill activity in cell-free preparations of a blue-green alga. *Biochim. biophys. Acta,* **88**, 11–19.

SUSOR W.A. & KROGMANN D.W. (1966) Triphosphopyridine nucleotide photoreduction with cell-free preparations of *Anabaena variabilis. Biochim. biophys. Acta,* **120**, 67–72.

TAGAWA K. & ARNON D.I. (1962) Ferredoxins as electron carriers in photosynthesis and in the biological production and consumption of hydrogen gas. *Nature, Lond.* **195**, 537–43.

TANNER H.A., BROWN T.E., EYSTER C. & TREHARNE R.W. (1960) A manganese-dependent photosynthetic process. *Biochem. biophys. Res. Commun.* **3**, 205–10.

TATEWAKI M. & PROVASOLI L. (1964) Vitamin requirements of three species of *Antithamnion. Botanica Mar.* **6**, 193–203.

TEICHLER-ZALLEN D. (1969) The effect of manganese on chloroplast structure and photosynthetic ability of *Chlamydomonas reinhardi. Pl. Physiol., Lancaster,* **44**, 701–10.

TEICHLER-ZALLEN D. & LEVINE R.P. (1967) An electron spin resonance study of manganese in wild-type and mutant strains of *Chlamydomonas reinhardi. Pl. Physiol., Lancaster,* **42**, 1643–7.

TREHARNE R.W. & EYSTER H.C. (1962) Electron spin resonance study of manganese and iron in *Chlorella pyrenoidosa. Biochim. biophys. Res. Commun.* **8**, 477–80.

TRUBACHEV I.N. (1968) Ob uchasti askorbinovoi kisloty, perekisi vodoroda i zheleza v. vosstanovlenii nitratov klorelloi. *Fiziol. Rast.* **15**, 658–64.

UDEL'NOVA T.M., KONDRAT'EVA E.N. & BOICHENKO E.A. (1968) Soderzhanie zheleza i margantsa u razlichnykh fotosinteziruyushchikh mikroorganismov. *Mikrobiologiya,* **37**, 197–200.

VAN NOORT D. (1964) Ribonucleic acid and chlorophyll synthesis in *Chlorella vulgaris* during recovery from iron deficiency. *Diss. Abstr.* **25**, 1545.

VEGA J.M., HERRERA J., APARICIO P.J., PANEQUE A. & LOSADA M. (1971) Role of molybdenum in nitrate reduction by *Chlorella. Pl. Physiol., Lancaster,* **48**, 294–9.

VELICHKO I.M. (1968) The role of iron and manganese in the vital activities of blue-green algae of the genus *Microcystis. Mikrol. Lem. Selskokhoz Med. Respub. Mezhvedom SB.* **4**, 11–17. (Translated from Ref. Zh. Biol. 1969, No. 10V62.)

VENKATARAMAN S., NETRAWALI M.S. & SREENWASON A. (1965) The role of vitamin $B_{12}$ in the metabolism of *Euglena gracilis* var. *bacillaris. Biochem. J.* **96**, 552–6.

VERNON L.P. (1962) Mechanism of oxygen evolution in photosynthesis. In *Biologistics for Space Systems*. (U.S. Air Force Doc. Rept. AMRL-TDR-62-116), 131–98.

WACKER W.E. (1962) Nucleic acids and metals. III. Changes in nucleic acid, protein, and metal content as a consequence of zinc deficiency in *Euglena gracilis. Biochemistry, N.Y.* **1**, 859–65.

WALKER J.B. (1953) Inorganic micronutrient requirements of *Chlorella*. I. Requirements for calcium (or strontium), copper and molybdenum. *Archs. Biochem. Biophys.* **46**, 1–11.

WALKER J.B. (1956) Strontium inhibition of calcium utilization by a green alga. *Archs. Biochem. Biophys.* **60**, 264–5.

WARBURG O. & KRIPPAHL G. (1967) Photolyt und Mangan in *Chlorella. Biochem. Z.* **346**, 429–33.

WARBURG O., KRIPPAHL G. & BUCHHOLZ W. (1955) Wirkung von Vanadium auf die Photosynthese. *Z. Naturf.* **10b**, 422.

WEBSTER D.A. & HACKETT D.P. (1965) The respiratory chain of colorless algae. I. Chlorophyta and Euglenophyta. *Pl. Physiol., Lancaster*, **40**, 1091–1100.

WEBSTER D.A. & HACKETT D.P. (1966) The respiratory chain of colorless algae. II. Cyanophyta. *Pl. Physiol., Lancaster*, **41**, 599–605.

WEDDING R.T. & BLOCK M.K. (1960) Uptake and metabolism of sulfate by *Chlorella*. II. Sulfate accumulation and active sulfate. *Pl. Physiol., Lancaster*, **35**, 72–80.

WERNER D. (1966) Die Kieselsäure im Stoffwechsel von *Cyclotella cryptica* Reimann, Lewin und Guillard. *Arch. Mikrobiol.* **55**, 278–308.

WERNER D. (1967) Untersuchungen uber die Rolle der Kieselsäure in der Entwicklung hoherer Pflanzen. *Planta*, **76**, 25–36.

WETZEL R.G. & McGREGOR D.L. (1968) Axenic cultures and nutritional studies of aquatic macrophytes. *Am. Midland Natur.* **80**, 52–63.

WIESE L. & JONES R.F. (1963) Studies on gamete copulation in heterothallic chlamydomonads. *J. Cell. comp. Physiol.* **61**, 265–74.

YAMANAKA T., TAKENAMI S., WADA K. & OKUNUKI K. (1969) Purification and some properties of ferredoxin from the blue-green alga, *Anacystis nidulans. Biochem. biophys. Acta*, **180**, 196–8.

YAMAMOTO T., FUJITA T. & ISHIBUSHI M. (1970) Chemical studies on the seaweeds. 25. Vanadium and titanium contents in seaweeds. *Rec. Oceanogr. Works, Japan*, **10**, 125–35.

YANAGI S. & SASA T. (1966) Changes in hydrogenase activity during the synchronous growth of *Scenedesmus obliquus* D₃. *Pl. Cell Physiol., Tokyo*, **7**, 593–8.

# CHAPTER 23

# PHOSPHORUS

## A. KUHL

Pflanzenphysiologisches Institut der Universität,
34 Göttingen, Untere Karspüle 2, Germany.

1    **Introduction**   636

2    **The study of phosphate metabolism: a note on methodology** 637

3    **Algal metabolism under normal or limited phosphorus supply** 639
   3.1   Optimum phosphorus levels for growth   639
   3.2   Phosphorus levels within algal cells   639
   3.3   Phosphorus uptake by phosphorus-sufficient and phosphorus-limited cultures   639
   3.4   Effect of phosphorus on phosphatase activity   641
   3.5   Luxury uptake of phosphorus   642

3.6   Utilization of phosphorus sources other than phosphate   643

4    **Phosphate metabolism and the generation of 'high-energy' phosphate** 644

5    **Inorganic condensed phosphates (polyphosphates)** 645
   5.1   General properties of algal polyphosphates   645
   5.2   Synthesis and metabolic functions of polyphosphates in algae   647

6    **References** 649

## 1 INTRODUCTION

Phosphorus plays a significant role in most cellular processes, especially those involved in generating and transforming metabolic energy. It is thus indispensable for the growth and reproduction of living organisms and its metabolic pathways have been extensively studied. Particular attention has been paid to the role of phosphorylated compounds in the conversion of light energy to biological energy in green plants, an area of research where studies with algae have contributed significantly to our understanding of the processes involved.

The literature dealing with phosphorus metabolism of algae has accumulated to such an extent that a condensation of the whole within the scope of this article is impossible. I shall therefore take the liberty of restricting myself to facts which in my opinion may raise problems for future research in this field. For a survey of the earlier literature the reader is referred to reviews by Myers (1951), Ketchum (1954), Krauss (1958) Provasoli (1958), Kuhl (1962a, 1968) and O'Kelley (1968). General problems of phosphorus metabolism in plants have been extensively compiled by Rowan (1966) and Hagemann (1969). The monograph by Matheja and Degens (1971) on the 'Structural molecular biology of phosphates' is recommended to those who are interested in phosphorus metabolism at the molecular level. The metabolism of phosphate uptake by algae (see Smith 1966, Wiessner 1970) and the role of phosphorus as an ecological factor in natural waters (see Edmondson 1970, Lange 1970, Ryther & Dunstan 1971, Healey 1973) are not considered in detail in this article.

## 2 THE STUDY OF PHOSPHATE METABOLISM: A NOTE ON METHODOLOGY

Investigations of the participation of phosphate or phosphorylated organic compounds in algal metabolism can be carried out using a variety of analytical and physiological methods, but they are of two basic types.

The first method depends on extracting and analysing the various phosphorylated compounds present in algae, because changing patterns of these reflect changes in phosphate turnover and its relationship to general cell metabolism. The sensitivity of this method can be increased if radioactive phosphorus is also used.

The distribution of the total cellular phosphates extracted from algae by the method of Pirson and Kuhl (1958) is shown in Fig. 23.1. Phosphorylated compounds can be separated into three main fractions by successive extractions with non-polar solvents and cold acids. For the acid extraction trichloroacetic acid (TCA) is used most commonly. After the separation of the phosphatides by a non-polar solvent, ice-cold TCA is used to extract the acid-soluble fraction: inorganic orthophosphate, inorganic polyphosphate (mostly of low condensation grade), phosphorylated carbohydrates, and nucleotides. Acid insoluble compounds are: ribonucleic acid (RNA) and deoxyribonucleic acid (DNA), high molecular weight inorganic polyphosphates and phosphoproteins. For further characterization the phosphates of these two main fractions are treated with 1 N hydrochloric acid at 100°C for seven minutes. This procedure by Lohmann (1931) totally splits all inorganic polyphosphates to orthophosphate thus allowing their rapid determination. More recently extraction methods coupled with ion-exchange-, paper-, and thin-layer chromatography, have enabled the separation and identification of single phosphate compounds to be carried out. These latter methods are dealt with by Bieleski and Young (1963), Lee et al. (1965), and Rowan (1966).

The second approach is to study the inter-relationships between phosphorus availability and the basic physiological reactions such as growth, reproduction, photosynthesis or respiration. This is useful in many studies but a disadvantage is that only the overall effects of phosphorus on metabolism are measured, and the direct sites and modes responsible for the action of phosphates remain obscure.

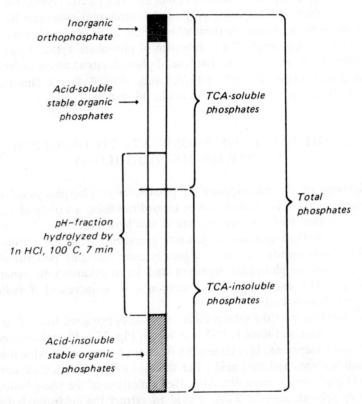

Fig. 23.1. Diagrammatic representation of conventional phosphate fractions obtained from algae (Pirson & Kuhl 1958).

A third method which, although specific, is of considerable interest in relation to phosphorus metabolism in algae is the measurement of *in vivo* ATP levels in algal cells. Techniques using $^{32}$P incorporation and subsequent chromatography can be used but the most satisfactory and sensitive is based on the firefly luciferin-luciferase technique (McElroy & Strehler 1949, Strehler 1968). It has been used to measure ATP levels in Euglenophyceae, Chlorophyceae, Bacillariophyceae, Chrysophyceae, Haptophyceae, Dinophyceae, Rhodophyceae and Cyanophyceae by Holm-Hansen (1970) and other workers have used it with *Chlorella* (St John 1970) and blue-green algae (Batterton & Van Baalen 1968, Stewart & Alexander 1971). During exponential growth ATP levels remain uniform at about 0·35% of the cellular organic carbon, and decrease under conditions of phosphorus, or nitrogen-deficiency (Holm-Hansen 1970). The levels are similar in the light and dark, and ATP determination may be a highly sensitive technique of measuring algal biomass (see Strickland 1971).

## 3 ALGAL METABOLISM UNDER NORMAL OR LIMITED PHOSPHORUS SUPPLY

### 3.1 *Optimum phosphorus levels for growth*

The phosphorus requirements for optimal algal growth differ considerably from species to species even if no other external factor is limiting. Chu (1943) studied the optimal phosphate concentrations for the growth of diatoms and green algae under defined laboratory conditions. Concentrations below 50 μg $1^{-1}$ of phosphorus were limiting, those about 20 mg $1^{-1}$ were inhibitory and 100 to 2,000 μg $1^{-1}$ were optimum. Rodhe (1948) later differentiated three main groups of freshwater algae according to their ability to tolerate phosphate within the ranges below, around, and above 20 μg $1^{-1}$ of phosphorus. Most planktonic algae fall into the groups with low or medium phosphorus tolerance. Many workers have found low optima for growth on phosphorus. Forsberg (1964, 1965a,b) observed best growth of *Chara* at levels below 20 μg $1^{-1}$ of phosphorus and Stewart and Alexander (1971) observed saturation of nitrogenase activity in *Anabaena* at 20 μg $1^{-1}$. Blum (1966) observed saturation of phosphate uptake by *Euglena* at 3·1 mg $1^{-1}$ of phosphorus. However Batterton and Van Baalen (1968) find limiting growth of blue-green algae at levels below 3·7 mg $1^{-1}$ of phosphorus.

### 3.2 *Phosphorus levels within algal cells*

The phosphorus levels within algal cells may fluctuate widely depending on whether the algae are growing under phosphorus-limited conditions or not. Even at very low levels in the environment the algae can accumulate and store phosphorus as Mackereth (1953) noted with *Asterionella formosa*. He showed that as little as 1 μg $1^{-1}$ of phosphorus was sufficient to provide optimum growth in his experiments. Phosphorus levels reported in algal cells are $6 \times 10^{-8}$ to $4 \times 10^{-6}$μg P cell$^{-1}$ for *Asterionella formosa* (Lund 1950); $1 \times 10^{-7}$ to $1·5 \times 10^{-6}$μg P cell$^{-1}$ for *Chlorella* (Al Kholy 1956), and $9·2 \times 10^{-7}$ to $8·7 \times 10^{-6}$μg P cell$^{-1}$ for *Scenedesmus* (Franzen 1932).

### 3.3 *Phosphorus uptake by phosphorus-sufficient and phosphorus-limited cultures*

The basic metabolic reactions which are responsible for the close connexion between growth and phosphate supply have been extensively studied, particular attention having been paid to the metabolism of algae growing under phosphorus-limited or phosphorus-sufficient conditions.

Ketchum (1939) studied phosphorus deficiency, and recovery from phosphorus deficiency in *Nitzschia closterium* (*Phaeodactylum tricornutum*) and

*Chlorella pyrenoidosa.* He took the quantity of phosphate absorbed in the dark by phosphorus-deficient cells as a measure of the deficiency and found it directly proportional to the number of cells present. Scott (1945) studied the mineral composition of phosphate-deficient *Chlorella* cells during restoration of phosphate and found its absorption to be independent of the concentration either in the light or in the dark. Other early studies include those of Emerson *et al.* (1944), Gest and Kamen (1948), Wassink *et al.* (1951), Wintermans (1955), Simonis and Kating (1956), Jacobi (1959), Balavskaya and Weber (1959), Badour (1959) and Pirson and Kuhl (1958). More recently evidence of active uptake has been obtained by workers such as Smith (1966) using *Nitella translucens*, Blum (1966) using *Euglena*, and Kylin (1966), and Kylin and Tillberg (1967a,b) using *Scenedesmus.* Smith (1966) found that CCCP ($5 \times 10^{-6}$M), which uncouples respiratory and photosynthetic ATP synthesis *in vitro* inhibits phosphate uptake, but that DCMU ($1 \times 10^{-6}$M) and light filters which inhibit photosystem II activity have little or no effect. De-oxygenation had little effect on uptake in the light. These results together indicate that the active process depends on cyclic photophosphorylation rather than oxidative phosphorylation. Kylin (1966) found that anaerobiosis had more effect in the dark than in the light. Blum (1966) showed that uptake was saturated at low phosphate concentrations and that arsenate was a competitive inhibitor of phosphate uptake. Other studies using inhibitors include those of Simonis and Gimmler (1969) and Ullrich and Simonis (1969). As emphasized by Wiessner (1970), the light-dependent phosphate uptake would result from phosphate utilization in photophosphorylation itself, as well as in other processes.

Pirson *et al.* (1952) found that in phosphate-deficient *Ankistrodesmus* cultures, dry weight production, cell division, photosynthetic oxygen production, and chlorophyll synthesis are inhibited under conditions of phosphorus starvation, but on adding phosphate, recovery resulted and photosynthesis ($O_2$-production) soon reached the rate of normally grown cells. Only in the cases of severe and longer lasting deficiency could the photosynthetic rate not be restored to normal. Thus, lack of phosphorus, at least during the early stages of phosphorus-deficiency, has no effect on the photosynthetic apparatus. Since these studies, rather similar results have been obtained in studies on a variety of metabolic processes.

The endogenous respiration of phosphorus-deficient cells is higher than in phosphorus-sufficient cells, but is significantly increased further by the addition of phosphate. This may be due to the fact that phosphorus-deficient cells have a higher carbohydrate content (Kuhl 1968). In heterotrophically grown *Chlorella*, the metabolic changes resulting from phosphate deficiency are, in general, the same as in autotrophic cultures. Under conditions of phosphorus-deficiency, the respiration of glucose decreases while the percentage glucose incorporated in the course of oxidative assimilation is higher and an accumulation of fat is reported (Daniel 1956, Bergmann 1955). In *Ankistrodesmus*, too, the storage of fat has been observed. Prolonged phosphorus-deficiency results in the synthesis of astaxanthin, which is absent from normal cells, to such an extent that the cultures appear orange in colour. Such cells restart normal growth on the addition of phosphate and the astaxanthin disappears (Dersch 1960).

The rate of ion uptake may also be influenced by the availability of phosphorus. Kylin (1964a) demonstrated that practically no sodium is taken up by phosphorus-deficient cells, but that uptake occurs when phosphate is added. A contrary effect is observed on the incorporation of sulphate-sulphur from the external medium into the following sulphur fractions of *Scenedesmus*: soluble reduced sulphur, lipid-sulphur, DNA-sulphur, and protein-sulphur. Phosphorus-deficient cells used in these experiments contained 0·2 to 0·3 mg P (g fresh weight)$^{-1}$ compared to 3 to 5 mg g$^{-1}$ for normal cells. Although there are differences in the incorporation into different fractions (DNA-sulphur and protein-sulphur) the total uptake of sulphate is inhibited by the presence of phosphate in the medium (Kylin 1964b,c). Kylin (1967a,b) extended these observations to other ions. The uptake of chloride in the light by phosphorus-starved cells of *Scenedesmus* is inhibited, just as the addition of phosphate (0·5 to 5·0mM) inhibits uptake of sulphate. A similar situation occurs in the dark except that there is an additional stimulatory effect on adding low phosphate levels, which develops mainly during the first and after the second hour of uptake. Light is inhibitory to chloride uptake at low phosphate levels but becomes stimulatory if higher levels of phosphate are added. The situations for the uptake of different cations (Rb$^+$, Cs$^+$, Ca$^{2+}$, and Sr$^{2+}$) and the influence of phosphate on these processes is more complicated.

The effect of phosphorus-deficiency and phosphorus-availability on nitrogenase activity of various blue-green algae has been reported by Stewart *et al.* (1970) and Stewart and Alexander (1971). They found that phosphorus-deficient cultures showed only low levels of acetylene reduction but that on the addition of phosphorus there was a rapid stimulation in nitrogenase activity, often within 15 to 30 min and a levelling off in the response after one to two hours. These authors proposed that the response could be used as a sensitive bioassay for detecting available phosphorus in aquatic ecosystems. They also noted slower responses of dark respiration and $^{14}CO_2$ uptake when phosphorus was supplied to phosphorus-deficient cells.

An increase in endogenous ATP levels on adding phosphorus to phosphorus-deficient cells has been noted by Batterton and Van Baalen (1968) and Stewart and Alexander (1971). This increase occurred both in the light and in the dark indicating that there was a stimulation both of oxidative phosphorylation and photophosphorylation.

### 3.4   *Effect of phosphorus on phosphatase activity*

The repressive action of phosphate (2mM) on the synthesis of acid phosphatase has been shown in *Euglena gracilis* by Price (1962). In marine algae, especially

Chrysophyceae and Bacillariophyceae phosphate-repressible alkaline phosphatases have been found by Kuenzler and Perras (1965). The synthesis of these
enzymes begins when the algae become phosphorus-deficient and stops if
phosphate is added to the medium. These enzymes appear to be firmly bound
near the cell surface and may enable phosphorus-deficient algae to obtain
phosphorus from phosphorylated compounds present in natural sea water. The
induction of the alkaline phosphatase under phosphorus-limited conditions
seems to be a characteristic phenomenon which allows this enzyme assay to be
used as a measure of phosphorus-limited growth. Blum (1965) distinguished
between two acid phosphatases in *Euglena*. One was a constitutive enzyme which
was insensitive to *p*-fluorophenylalanine (PFA), which inhibits the synthesis of
messenger RNA; the other was PFA sensitive. The enzymes are differentially
sensitive to pH and this fact can be used to assay them separately. Fluoride and
tartrate both inhibited phosphatase activity in *Euglena*, as in other organisms.
Sommer and Blum (1965) showed further that the inducible phosphatase is
located in special helicle regions in the pellicle while the constitutive enzyme is
located intracellularly. Algae which are phosphorus-limited have as much as
25 times more alkaline phosphatase activity than algae with surplus available
phosphorus (Fitzgerald & Nelson 1966). These data have been confirmed by
Bone (1971) who found that the activity of alkaline phosphatase in *Anabaena*
varied 20-fold with the lowest values being found in cells containing excess
phosphate. Similar regulatory mechanisms exist under natural conditions
(Reichardt & Overbeck 1968). Fitzgerald (1969) showed that the addition of
high levels of phosphorus in cells showing phosphatase activity did not inhibit
activity immediately and suggested that inhibition by phosphorus depended on a
dilution out of the enzyme rather than on end-product repression.

### 3.5   *Luxury uptake of phosphorus*

Phosphorus-deficient algae of different species commonly possess the ability to
incorporate phosphate extremely rapidly if they are provided with this compound. In most cases the incorporated phosphate exceeds by far the actual need
of the cells as was noted by early workers such as Ketchum (1939), Scott (1945),
Mackereth (1953) and Kuenzler and Ketchum (1962), so that a surplus amount
must be stored inside the cell. In various algae this occurs almost exclusively as
high molecular inorganic polyphosphates (Kuhl 1968), and Blum (1966) has
shown, using $^{32}$P-labelled phosphate, that within one minute of uptake at 25°C
over 95% of this was converted to organic compounds in *Euglena*. Smith (1966),
using *Nitella translucens*, found that after 30 min exposure to phosphate, about
75% of the uptake was into the cytoplasm, and thereafter the levels in the
vacuole rose rapidly. Stewart and Alexander (1971) studying various blue-green
algae noted by electron microscopy that new polyphosphate bodies formed
within one to two hours of re-adding phosphate to phosphorus-deficient cells.
There is some evidence (Kylin & Tillberg 1967a) that polyphosphate synthesis is

stimulated by conditions which inhibit ATP synthesis, but on the other hand
Simonis (1967) found that light preferentially stimulates uptake into organic compounds. The accumulation of polyphosphates is of considerable ecological
importance, because this surplus amount enables the algae to grow even in
waters where phosphorus is limiting. The amount of this surplus phosphorus in
algae can be separated from essential phosphorus compounds and determined
after extracting for 60 minutes with boiling water (Fitzgerald & Nelson 1966,
Fitzgerald 1968). Much of it is apparently released as orthophosphate after
killing the algae, indicating that it is highly labile (Fitzgerald & Faust 1967).
Closely related with the problems of 'excess phosphate' in algae are those of the
phosphorus requirements of different algae for optimal growth rates under a
given set of external conditions. Newer and improved techniques which allow
continuous cultures of axenic algae under steady state conditions using chemo-
stats or turbidostats have revealed many interesting relationships between
phosphorus supply and other factors influencing algal growth (see Eichhorn
1969, Carpenter 1970, Müller 1970, Azad & Borchardt 1970, Bone 1971).

### 3.6   *Utilization of phosphorus sources other than phosphate*

The growth of many algal species can proceed when orthophosphates are
substituted by other sources of phosphorus. *Chlorella* can utilize inorganic
polyphosphates (up to a chain length of 55 phosphate units) at the same rate as
potassium phosphate is utilized (Galloway & Krauss 1963). On the other hand
Overbeck (1962a,b) showed that *Scenedesmus quadricauda* was unable to use
pyrophosphate or some organic phosphorus compounds as phosphate source.
The utilization of glucose-6-phosphate as a phosphate source by marine algae
seems to be possible through the action of phosphatase present at the cell surface
which hydrolyses the ester so that the phosphate groups become available
(Kuenzler 1965). Marine algae are also reported to assimilate phosphorus from
adenosine monophosphate and α-glycerophosphate at the same rate as from
glucose-6-phosphate (Kuenzler & Perras 1965). Stewart and Alexander (1971)
observed that blue-green algae could utilize phosphorus present in phosphorus-
based detergents.

The use of synchronous cultures of unicellular algae has opened up the
possibility of studying phosphate metabolism during the course of the life cycle
of these organisms. Representative of a great number of such investigations are
those of Soeder *et al.* (1967) and Soeder (1970). They confirm that *Chlorella fusca*
tolerates a relatively high phosphate concentration and that the growth rate is
not influenced by a variation of the phosphate concentration between $0.0295^{-1}$
g $1^{-1}$ and $2.95$ g $1^{-1}$. The phosphate content changes markedly in the course of the
life cycle. On a dry weight basis the rate of phosphate uptake reaches a maximum
during the first hours of the light period, while it is close to zero during the first
hour of the dark period. In the course of one light:one dark cycle (16 h light:8 h

dark) one cell consumes, on average, about $1.5 \times 10^{-5} \mu g$ P and divides into 20 daughter cells. If the cellular phosphorus falls below about $5 \times 10^{-8} \mu g$ phosphate-phosphorus, cell division ceases. O'Kelley (1968) and Azad and Borchardt (1970) should be consulted for further information.

## 4 PHOSPHATE METABOLISM AND THE GENERATION OF 'HIGH-ENERGY' PHOSPHATE

There are three major processes by which algae as well as other green plants can incorporate orthophosphate into organic 'high-energy' compounds: substrate phosphorylation, oxidative phosphorylation, and photophosphorylation. Thus:

$$\text{ADP} + \text{orthophosphate} \xrightarrow{\text{energy}} \text{ATP}$$

$$\left.\begin{array}{l} \text{photo-} \\ \text{oxidative-} \\ \text{substrate-} \end{array}\right\} \text{phosphorylation}$$

By these energy-generating systems adenosine triphosphate (ATP) as the main 'high-energy' product is formed by transferring orthophosphate to adenosine diphosphate.

Substrate phosphorylation is coupled directly to the oxidation of the respiratory substrate while the phosphorylation reactions of oxidative phosphorylation are linked to the electron transport system of the mitochondria. In plants a third mechanism exists by which the energy-requiring (endergonic) transfer of orthophosphate to ATP is mediated. By this mechanism of photophosphorylation light energy is converted into the 'energy-rich' phosphate bonds of ATP (see Chapter 13, p. 391), a compound occupying a central position in the metabolism of algae as in other living organisms (Holm-Hansen 1970). By the hydrolysis of its β- or γ-phosphate bond it provides energy which enables cell synthesis or breakdown to occur. It is certain that all energy-requiring reactions of plant metabolism such as photosynthetic $CO_2$-fixation, uptake and transport of ions, activation of amino acids for protein synthesis, formation of nucleic acids, and polyphosphate synthesis can be supported by the 'energy-rich' phosphate bond of ATP generated by this special light reaction.

In a series of investigations with intact *Ankistrodesmus* cells the presence of photophosphorylation could clearly be demonstrated and many conditions for its *in vivo* action have been determined (Simonis 1967, Urbach & Gimmler 1970a,b, Ullrich et al. 1970, Gimmler et al. 1971, Ullrich 1971, 1972). To deal with all the facts of phosphorylation reactions in detail would be far beyond the scope of this chapter. For this the reader is referred to summaries by Kuhl (1962a, 1968); Rowan (1966); Gibbs (1967); Hagemann (1969); Cheniae (1970) and Bishop (1971) (see also Chapter 13, p. 391).

# 5  INORGANIC CONDENSED PHOSPHATES (POLYPHOSPHATES)

As mentioned above condensed inorganic phosphates are normal constituents of algal cells and play an important role in the general metabolism of these organisms, as they do in all microorganisms. The natural distribution of these chemical compounds seems to be limited to microorganisms, although there are a few exceptions (Miyachi 1961, Tewari & Singh 1964). Polyphosphates were first discovered in algae by Sommer and Booth (1938) using *Chlorella* and since then increasing attention has been paid to them by algologists and plant physiologists. General reviews of the progress made are available (Schmidt 1951, Wiame 1958, Kuhl 1960, 1962a, Langen 1965, Harold 1966, Bieleski 1973). Here I shall concentrate only on the basic facts and more recent approaches.

## 5.1  *General properties of algal polyphosphates*

The naturally occurring condensed phosphates are classified by their molecular structures into two different groups (Thilo 1959). Cyclic condensed phosphates (closed rings of $PO_4$-units) are called *metaphosphates* while compounds with unbranched chains of $PO_4$-groups linked together by oxygen bridges are named *polyphosphates* (see Fig. 23.2). The inorganic condensed phosphates isolated from microorganisms are polymers which in most cases belong to the latter type but metaphosphates (up to hexaphosphates) have been identified in extracts from *Anacystis nidulans* (Niemeyer & Richter 1969) and *Acetabularia* sp. (Niemeyer & Richter 1972).

Polyphosphates are acid-labile substances which are completely split to orthophosphate units within seven minutes by 1 N HCl at 100°C. This is the reason why they are often labelled as 'seven-minute-phosphates'. Using histochemical techniques distinct polyphosphate granules can be localized as normal cell constituents of many microorganisms and their size and number often reflect the physiological state of the cell (Stich 1955). Their disappearance under conditions of phosphorus deficiency proves that they can actively participate in metabolism and that they are not only deposits of excess phosphorus. In many electron microscopic studies, especially with blue-green algae, polyphosphate granules (polyphosphate bodies) have been thoroughly examined (Jensen 1968, Lang 1968, Stewart & Alexander 1971). They appear as electron dense particles which are in general closely associated with the neoplasm and polyhedral bodies. Sometimes they are enclosed in what appears to be a membrane. The fine structure of developing polyphosphate bodies in *Plectonema boryanum* is shown particularly well in the electron micrographs of Jensen (1969). In cells of the isolated lichen phycobiont *Trebouxia erici* polyphosphate granules have also been demonstrated by electron microscopy (Fischer 1971) and in most algal cells their size varies between 30 and 500 nm in diameter. For light miscroscope studies on algal polyphosphate bodies see Fuhs (1968, 1973).

Different methods can be used to extract polyphosphates from algal cells. The smaller part of these compounds is extracted easily by treatment of the cell with ice cold dilute trichloroacetic acid (TCA) while the remainder (acid insoluble polyphosphates) can be extracted with hot TCA, perchloric acid or alkaline or neutral salt solutions of different concentrations and temperatures.

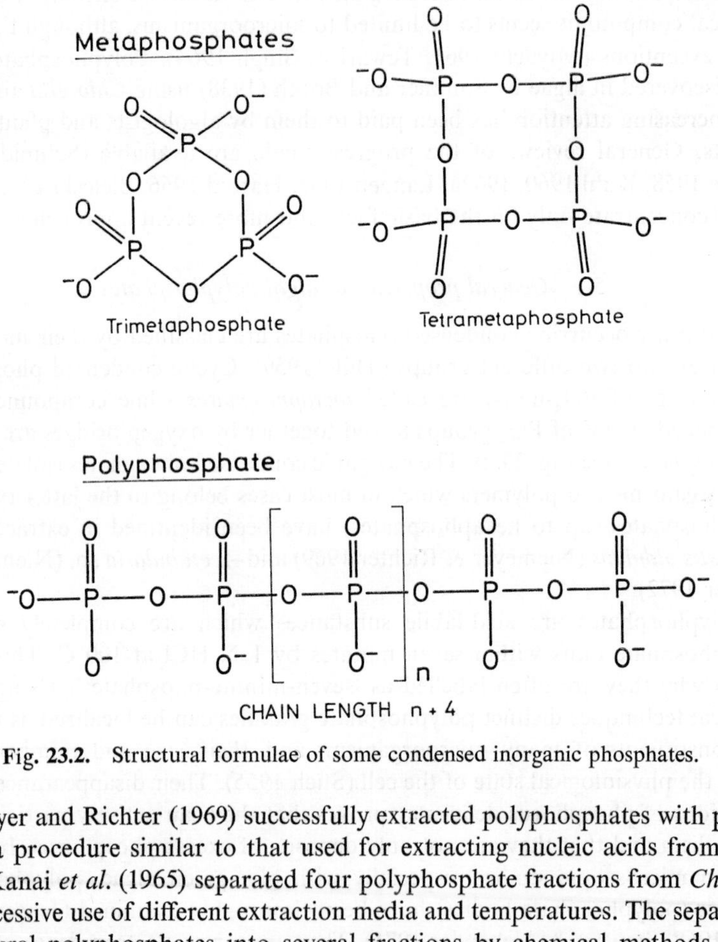

## Metaphosphates

Trimetaphosphate                  Tetrametaphosphate

## Polyphosphate

CHAIN LENGTH  n + 4

Fig. 23.2.   Structural formulae of some condensed inorganic phosphates.

Niemeyer and Richter (1969) successfully extracted polyphosphates with phenol using a procedure similar to that used for extracting nucleic acids from plant cells. Kanai et al. (1965) separated four polyphosphate fractions from *Chlorella* by successive use of different extraction media and temperatures. The separation of natural polyphosphates into several fractions by chemical methods seems necessary since it is known that single fractions are differently involved in metabolic processes and may be located in different subcellular structures in algal cells.

In most cases a mixture of polyphosphates with different chain lengths is extracted from the cells, but most show a high degree of polymerization. Although there are no data for algal polyphosphates it seems reasonable to assume that they will be of the same order as for other microorganisms, that is, between 3 and 500 phosphate groups per molecule (Harold 1966). Polyphosphates may also be separated using paper-, ion-exchange-, or thin-layer chromatography (Hettler 1958, Aoki & Miyachi 1964, Rössel 1963, Niemeyer & Richter 1969).

Correll and Tolbert (1962, 1964) examined the physical and chemical properties of isolated ribonucleic acid-polyphosphate complexes from *Chlorella* and *Anabaena*. The properties of the RNA-polyphosphate indicate that in the union of these two substances weak bonding and covalent linkages may be involved. Separation of the RNA-polyphosphate of *Chlorella* by DEAE-cellulose chromatography into different fractions showed that these fractions changed in properties during different stages of synchronous growth of this alga. The results of different hydrolytic treatments on isolated RNA-polyphosphate from synchronous *Chlorella* cells and the analysis of various hydrolysates by ion-exchange chromatography indicate that in *Chlorella*, polyphosphate is not a simple anhydride chain (Correll 1965). In this connexion it is interesting that Correll (1966) finds that at least part of the polyphosphate isolated from *Chlorella* contains both phosphoanhydride and imidodiphosphate linkages. This result conflicts with the view on polyphosphate structure accepted by most authors and requires confirmation.

## 5.2 *Synthesis and metabolic functions of polyphosphates in algae*

The first careful and detailed investigation to elucidate the conditions of polyphosphate synthesis in algae was carried out by Wintermans (1955). He found that cells of *Chlorella vulgaris* were able to synthesize polyphosphates from external orthophosphate in the light. The polyphosphate thus synthesized was mainly insoluble in TCA. The conversion of external orthophosphate to intracellular polyphosphate was light saturated at a much lower intensity than was photosynthesis and the rate of polyphosphate synthesis was higher in the absence than in the presence of $CO_2$, and was optimum at about pH 4. Nitrate had no effect, glucose was somewhat inhibitory and oxygen was not required. These results and those obtained by the action of different metabolic inhibitors led Wintermans to conclude that energy-rich phosphate groups generated by the photochemical reactions of photosynthesis are transformed to polyphosphate when the energy requiring reactions of photosynthesis are saturated, or are limited, for example by the lack of $CO_2$.

The influence of light on polyphosphate synthesis in *Chlorella* was clearly demonstrated by Kuhl (1962b). Immediately after the addition of phosphate to an algal suspension precultivated for four days in continuous light in a medium with a very low phosphate concentration an increase of acid-insoluble polyphosphate could be observed in the light together with the synthesis of carbohydrates, while the polyphosphate synthesis in the dark practically ceased two or three hours after the addition of phosphate. The amount of polyphosphate synthesized within four hours after the addition of phosphate was twice as high in the light as in the dark. The stimulation of polyphosphate synthesis by light could be suppressed by inhibitors of photosynthesis, thus indicating the participation of photosynthetic phosphorylation. The energy for polyphosphate synthesis in the dark is provided by oxidative phosphorylation, as shown using inhibitors of cell respiration. The phosphorus uptake/$O_2$ uptake ratio for the transformation of orthophosphate into polyphosphate in the dark is 0·75.

Since Wintermans' investigations interest in the inter-relationships between photosynthesis and phosphorus metabolism has increased steadily, and the general conclusion is that in algae inorganic polyphosphates are synthesized in the light at the expense of ATP which is generated by photophosphorylation reactions.

Van Rensen (1969) has demonstrated that polyphosphate formation in *Scenedesmus* in $N_2$ results from cyclic photophosphorylation. Ullrich and Simonis (1969) came to a similar conclusion on investigating the effects of red and far-red light (683nm and 712nm) on the rate of $^{32}P$ incorporation and on the distribution pattern of various phosphate fractions in *Ankistrodesmus* under $N_2$. The use of $N_2$ is to avoid competition between polyphosphate synthesis and $CO_2$-fixation for energy-rich phosphate bonds generated by the light reactions. Their data enable the conclusion to be reached that far-red light produces conditions for cyclic photophosphorylation only and that polyphosphate formation in the light is favoured by this form of photophosphorylation.

Further investigations by Ullrich (1970a,b, 1972), however, have demonstrated that polyphosphate formation in the light is influenced by a number of other metabolic conditions as well, including the presence or absence of $O_2$ or $CO_2$ and the pH value of the medium. The actions and interactions of these on polyphosphate formation seem complex and require further investigation.

There have been many discussions about the role of polyphosphates in the metabolism of microorganisms. Hoffman *et al.* (1952) postulated that they function by storing energy and phosphate under conditions where phosphate and energy-rich ATP are available and in excess of metabolic needs. This seems possible because the P–O–P bonds of polyphosphates are so-called' energy-rich' phosphate bonds (Meyerhof *et al.* 1953) and enzymes which can reversibly transfer phosphate from ATP to polyphosphate have been discovered in some heterotrophic microorganisms. In algae, however, the existence of such enzymes has not yet been demonstrated although it is clear from many experiments that polyphosphates are always accumulated when the energy generated by photophosphorylation cannot be used for other energy-requiring reactions. There is also no direct proof that the energy stored in polyphosphate can be used by algal cells if no other energy is available. From the recent literature it seems more likely that the metabolic role of inorganic polyphosphate is mainly to supply the cells with phosphorus for special synthetic reactions, e.g. synthesis of nucleic acids or those closely connected with cell division. Markarova and Baslavskaya (1969) found that in *Ankistrodesmus* total polyphosphates and nucleic acids were divided without loss between the daughter cells formed from maternal cells in the dark. Polyphosphates may also be used as a phosphorus source at certain developmental stages during the life-cycle of unicellular algae (see Kuhl 1968). Lysek and Simonis (1968) showed, for example, that *Ankistrodesmus* cells use polyphosphates for the phosphorylation of absorbed glucose by the action of an inorganic polyphosphate-glucose-phosphotransferase, an enzyme which has been demonstrated previously in bacteria (Harold 1966).

In general there is agreement that inorganic polyphosphates in algae function as deposits of phosphorus which can be easily used. They enable algae to continue growth for a period under conditions of phosphorus depletion and therefore under natural conditions may be of considerable ecological importance.

## 6 REFERENCES

AL KHOLY (1956) On the assimilation of phosphorus in *Chlorella pyrenoidosa*. *Physiologia Pl.* **9**, 135–43.

AOKI S. & MIYACHI S. (1964) Chromatographic analyses of acid-soluble polyphosphates in *Chlorella* cells. *Pl. Cell Physiol., Tokyo*, **5**, 241–50.

AZAD H.S. & BORCHARDT J.A. (1970) Variations in phosphorus uptake by algae. *Environmental Science and Technology*, **4**, 737–43.

BADOUR A.S.S. (1959) Analytisch-chemische Untersuchung des Kaliummangels bei *Chlorella* im Vergleich mit andersen Mangelzuständen. Dissertation, University of Göttingen, Germany.

BALAVSKAYA S.S. & WEBER H. (1959) The effect of light upon transformations of phosphates in plants. *Dokl. Acad. Nauk. S.S.S.R.* **124**, 227–30 (in Russ.).

BATTERTON J.C. & VAN BAALEN C. (1968) Phosphorus deficiency and phosphate uptake in the blue-green alga *Anacystis nidulans*. *Can. J. Microbiol.* **14**, 341–8.

BERGMANN L. (1955) Stoffwechsel und Mineralsalzernährung einzelliger Grünalgen. II. Vergleichende Untersuchungen über den Einfluss mineralischer Faktoren bei heterotropher und mixotropher Ernährung. *Flora, Jena*, **142**, 493–539.

BIELESKI R.L. (1973) Phosphate pools, phosphate transport, and phosphate availability. *A. Rev. Pl. Physiol.* **24**, 225–52.

BIELESKI R.L. & YOUNG R.E. (1963) Extraction and separation of phosphate esters from plant tissues. *Analyt. Biochem.* **6**, 54–68.

BISHOP N.I. (1971) Photosynthesis: The electron transport system of green plants. *A. Rev. Biochem.* **40**, 197–226.

BLUM J.J. (1965) Observations on the acid phosphatases of *Euglena gracilis*. *J. Cell Biol.* **24**, 223–34.

BLUM J.J. (1966) Phosphate uptake by phosphate-starved *Euglena*. *J. gen. Physiol.* **49**, 1125–37.

BONE D.H. (1971) Relationship between phosphates and alkaline phosphatase of *Anabaena flos-aquae* in continuous culture. *Arch. Mikrobiol.* **80**, 147–53.

CARPENTER E.J. (1970) Phosphorus requirements of two planktonic diatoms in steady state culture. *J. Phycol.* **6**, 28–30.

CHENIAE G.M. (1970) Photosystem II and $O_2$ evolution. *A. Rev. Pl. Physiol.* **21**, 467–98.

CHU S.P. (1943) The influence of the mineral composition of the medium on the growth of planktonic algae. II. The influence of the concentration of inorganic nitrogen and phosphate phosphorus. *J. Ecol.* **31**, 109–48.

CORRELL D.I. & TOLBERT N.E. (1962) Ribonucleic acid-polyphosphate from algae. I. Isolation and physiology. *Pl. Physiol., Lancaster*, **37**, 627–36.

CORRELL D.L. & TOLBERT N.E. (1964) Ribonucleic acid-polyphosphate from algae. II. Physical and chemical properties of the isolated complexes. *Pl. Cell Physiol., Tokyo*, **5**, 171–91.

CORRELL D.L. (1965) Ribonucleic acid-polyphosphate from algae. III. Hydrolysis studies. *Pl. Cell Physiol., Tokyo*, **6**, 661–9.

CORRELL D.L. (1966) Imidonitrogen in *Chlorella* 'Polyphosphate'. *Science, N.Y.* **51**, 819–821.

DANIEL A.L. (1956) Stoffwechsel und Mineralsalzernährung einzelliger Grünalgen. III. Atmung und oxydative Assimilation von *Chlorella*. *Flora*, **143**, 31–66.

DERSCH G. (1960) Mineralsalzmangel und Sekundärcarotinoide in Grünalgen. *Flora, Jena*, **149**, 566–603.

EDMONDSON W.T. (1970) Phosphorus, nitrogen and algae in Lake Washington after diversion of sewage. *Science, N.Y.* **169**, 690–1.

EICHHORN M. (1969) Zur Stoffproduktion kontinuierlicher Kulturen von *Scenedesmus obliquus* (Turp.) Kützing in Dauerlicht bei Phosphat- und Nitrat Limitation. *Flora, Jena Abt. A.* **159**, 494–506.

EMERSON R.L., STAUFFER J.F. & UMBREIT W.D. (1944) Relationship between phosphorylation and photosynthesis in *Chlorella*. *Am. J. Bot.*, **31**, 107–20.

FISCHER K.A. (1971) Polyphosphate in a chlorococcalean alga. *Phycologia*, **10**, 177–82.

FITZGERALD G.P. (1968) Detection of limiting or surplus nutrients in algae. *Final report to National Institutes of Health* (1961–1968), pp. 1–46. University of Wisconsin, Madison.

FITZGERALD G.P. (1969) Some factors in the competition or antagonism among bacteria, algae and aquatic weeds. *J. Phycol.* **5**, 351–9.

FITZGERALD G.P. & FAUST S.I. (1967) Effect of water sample preservation methods on the release of phosphorus from algae. *Limnol. Oceanogr.* **12**, 332–4.

FITZGERALD G.P. & NELSON T.C. (1966) Extractive and enzymatic analysis for limiting or surplus phosphorus in algae. *J. Phycol.* **2**, 32–47.

FORSBERG C. (1964) Phosphorus, a maximum factor in the growth of Characeae. *Nature, Lond.* **201**, 517–18.

FORSBERG C. (1965a) Environmental conditions of Swedish charophytes. *Symb. Bot. Ups.* **18**, 1–67.

FORSBERG C. (1965b) Sterile germination of oospores of *Chara* and seeds of *Najas marina*. *Physiologia Pl.* **18**, 128–37.

FRANZEN W.A.W. (1932) Ein Versuch der physiologisches Erforschung der Productionsfähigkeit des Moskauflosswassers. *Microbiologiya*, **1**, 122–30.

FUHS G.W. (1968) Cytology of blue-green algae: light microscopic aspects. In *Algae, Man and the Environment*, ed. Jackson D.F. pp. 213–33. Syracuse University Press, Syracuse.

FUHS G.W. (1973) Cytochemical examination of blue-green algae. In: *The Biology of blue-green algae*, ed. N.G. Carr & B.A. Whitton, pp. 117–143. Blackwell, Oxford.

GALLOWAY R.A. & KRAUSS R.W. (1963) Utilization of phosphorus sources by *Chlorella*. In *Studies on microalgae and photosynthetic bacteria. Jap. Soc. Pl. Cell Physiol., Tokyo*. University Press, Tokyo, 569–75.

GEST H. & KAMEN M.D. (1948) Studies on the phosphorus metabolism of green algae and purple bacteria in relation to photosynthesis. *J. biol. Chem.* **176**, 299–318.

GIBBS M. (1967) Photosynthesis. *A. Rev. Biochem.* **36**, 757–84.

GIMMLER H., NEIMANIS S., EILMANN I. & URBACH W. (1971) Photophosphorylation and photosynthetic $^{14}CO_2$-fixation. II. Comparison of cyclic and non-cyclic photophosphorylation with photosynthetic $^{14}CO_2$-fixation during the synchronous life cycle of *Ankistrodesmus braunii*. *Z. PflPhysiol.* **64**, 358–66.

HAGEMANN R.H. (1969) Phosphorus metabolism in plants. *Hort. Science*, **4**, 311–14.

HAROLD F.M. (1966) Inorganic polyphosphates in biology: structure, metabolism, and function. *Bact. Rev.* **30**, 772–94.

HEALEY F.P. (1973). Inorganic nutrient uptake and deficiency in algae. *Crit. Rev. Microbiol.* **3**, 69–113.

HETTLER H. (1958) Zur Papierchromatographie der Phosphorverbindungen. I. Anorganische Phosphorverbindungen. *Z. Chromatog.* **1**, 389–410.

Hoffmann-Ostenhof O. & Weigert W. (1952) Über die mögliche Funktion des polymeren Metaphosphats als Speicher energiereichen Phosphats in der Hefe. *Naturwissenschaften,* **39,** 303–4.

Holm-Hansen O. (1970) ATP levels in algal cells as influenced by environmental conditions. *Pl. Cell Physiol., Tokyo,* **11,** 689–700.

Jacobi G. (1959) Uber den Zusammenhang von Glykolsaure und lichtabhängigen Phosphorylierung. *Planta,* **53,** 402–11.

Jensen T.E. (1968) Electron microscopy of polyphosphate bodies in a blue-green alga, Nostoc pruniforme. *Arch. Mikrobiol.* **62,** 144–52.

Jensen T.E. (1969) Fine structure of developing polyphosphate bodies in a blue-green alga, Plectonema boryanum. *Arch. Mikrobiol.* **67,** 328–38.

Kanai R., Aoki S. & Miyachi S. (1965) Quantitative separation of inorganic polyphosphates in Chlorella cells. *Pl. Cell Physiol., Tokyo,* **6,** 467–73.

Ketchum B.H. (1939) The development and restoration of deficiencies in the phosphorus and nitrogen composition of unicellular plants. *J. cell. comp. Physiol.* **13,** 373–81.

Ketchum B.H. (1954) Mineral nutrition of phytoplankton. *A. Rev. Pl. Physiol.* **5,** 55–74.

Krauss R.W. (1958) Physiology of the freshwater algae. *A. Rev. Pl. Physiol.* **9,** 207–44.

Kuenzler E.J. (1965) Glucose-6-phosphate utilization by marine algae. *J. Phycol.* **1,** 156–64.

Kuenzler E.J. & Ketchum B.H. (1962) Rate of phosphorus uptake by Phaeodactylum tricornutum. *Biol. Bull. mar. biol. Lab., Woods Hole,* **123,** 134–45.

Kuenzler E.J. & Perras J.P. (1965) Phosphatases of marine algae. *Biol. Bull. mar. biol. Lab., Woods Hole,* **128,** 271–84.

Kuhl A. (1960) Die Biologie der kondensierten anorganischen Phosphate. In *Ergebnisse der Biologie,* ed. Autrum H., vol. 23. pp. 144–86. Springer Verlag, Berlin.

Kuhl A. (1962a) Inorganic phosphorus uptake and metabolism. In *Physiology and Biochemistry of Algae,* ed. Lewin R.A. pp. 211–29. Academic Press, New York.

Kuhl A. (1962b) Zur Physiologie der Speicherung kondensierter anorganischer Phosphate in Chlorella. In *Vorträge aus dem Gesamtgebiet der Botanik,* **1,** 157–66. Fischer Verlag, Stuttgart.

Kuhl A. (1968) Phosphate metabolism of green algae. In *Algae, Man and the Environment,* ed. Jackson D.F. pp. 37–52. Syracuse University Press, Syracuse.

Kylin A. (1964a) An outpump balancing phosphate-dependent sodium uptake in Scenedesmus. *Biochem. biophys. Res. Commun.* **16,** 497–500.

Kylin A. (1964b) The influence of phosphate nutrition on growth and sulphur metabolism of Scenedesmus. *Physiologia Pl.* **17,** 384–402.

Kylin A. (1964c) Sulphate uptake and metabolism in Scenedesmus as influenced by phosphate, carbon dioxide and light. *Physiologia Pl.* **17,** 422–33.

Kylin A. (1966) The influence of photosynthetic factors and metabolic inhibitors on the uptake of phosphate in P-deficient Scenedesmus. *Physiologia Pl.* **19,** 644–9.

Kylin A. (1967a) Ion transport in P-deficient Scenedesmus upon readditions of phosphate in light and darkness. I. Uptake and loss of Cl⁻, and measurements of oxygen consumption. *Z. PflPhysiol.* **56,** 70–80.

Kylin A. (1967b) Ion transport in P-deficient Scenedesmus upon readditions of phosphate in light and darkness. II. Uptake of Rb⁺, Cs⁺, Ca²⁺, and Sr²⁺. *Z. PflPhysiol.* **56,** 81–90.

Kylin A. & Tillberg J.E. (1967a) The relation between total photophosphorylation, level of ATP, and O₂ evolution in Scenedesmus as studied with DCMU and Antimycin A. *Z. PflPhysiol.* **58,** 165–74.

Kylin A. & Tillberg J.E. (1967b) Action sites of the inhibitor-B complex from potato and of phloridzin in the light-induced energy transfer in Scenedesmus. *Z. PflPhysiol.* **57,** 72.

Lang N. (1968) The fine structure of blue-green algae. *A. Rev. Microbiol.* **22,** 15–46.

Lange W. (1970) Cyanophyta-bacteria systems: effects of added carbon compounds or phosphate on algal growth at low nutrient concentrations. *J. Phycol.* **6,** 230–4.

LANGEN P. (1965) Vorkommen und Bedeutung von Polyphosphaten in Organismen. *Biol. Rdsch.* **2,** 145–52.

LEE F.G., CLESCERI N.L. & FITZGERALD G.P. (1965) Studies on the analysis of phosphates in algae cultures. *Int. J. Air Wat. Pollut.* **75,** 715–22.

LOHMANN K. (1931) Darstellung der Adenylpyrophosphorsäure aus Muskulatur. *Biochem. Z.* **233,** 460–9.

LUND J.W.G. (1950) Studies on *Asterionella formosa.* Hass II. Nutrient depletion and the spring maximum. Part I. Observations on Windermere, Esthwaite Water and Blelham Tarn. Part II. Discussion. *J. Ecol.* **38,** 1–14, 15–35.

LYSEK G. & SIMONIS W. (1968) Substrataufnahme und Phosphatstoffwechsel bei *Ankistrodesmus braunii.* I. Beteiligung der Polyphosphate an der Aufnahme von Glucose und 2-Desoxyglucose im Dunkeln und im Licht. *Planta,* **79,** 133–45.

MACKERETH F.J. (1953) Phosphorus utilization by *Asterionella formosa.* Hass. *J. exp. Bot.* **4,** 296–313.

MCELROY W.D. & STREHLER B.L. (1949) Factors influencing the response of the bioluminescent reaction to adenosine triphosphate. *Archs. Biochem.* **22,** 420–33.

MARKAROVA E.M. & BASLAVSKAYA S.S. (1969) Content and accumulation of polyphosphates in cells of *Scenedesmus obliquus* (Turpin) Kützing of various age. *Fiziol. Rastenij.* **16,** 702–7.

MATHEJA J. & DEGENS E.T. (1971) Structural molecular biology of phosphates. In *Fortschritte der Evolutionsforschung,* vol. 5, eds. Heberer G. & Schwanitz F. pp. 1–180. Fischer Verlag, Stuttgart.

MEYERHOF O., SHATAS R. & KAPLAN A. (1953) Heat of hydrolysis of trimetaphosphate. *Biochim. biophys. Acta,* **12,** 121–7.

MIYACHI S. (1961) Inorganic polyphosphate in spinach leaves. *J. Biochem., Tokyo,* **50,** 367–71.

MÜLLER H. (1970) Das Wachstum von *Nitzschia actinastroides* (Lemm.) v. Goor im Chemostaten bei limitierender Phosphatkonzentration. *Ber. dt. bot. Ges.* **83,** 537–44.

MYERS J. (1951) Physiology of the algae. *A. Rev. Microbiol.* **5,** 157–80.

NIEMEYER R. & RICHTER G. (1969) Schnellmarkierte Polyphosphate und Metaphosphate bei der Blaualge *Anacystis nidulans. Arch. Mikrobiol.* **69,** 54–9.

NIEMEYER R. & RICHTER G. (1972) Rapidly labelled polyphosphates in *Acetabularia.* In *Biology and Radiobiology of Anucleate systems II: Plant Cells,* ed. Bonotto S. pp. 225–36, Academic Press, New York & London.

O'KELLEY J. (1968) Mineral nutrition of algae. *A. Rev. Pl. Physiol.* **19,** 89–112.

OVERBECK J. (1962a) Untersuchungen zum Phosphathaushalt von Grünalgen. III. Das Verhalten der Zellfraktionen von *Scenedesmus quadricauda* (Turp.) Bréb. im Tageszyklus unter verschiedenen Belichtungsbedingungen und bei verschiedenen Phosphatverbindungen. *Arch. Mikrobiol.* **41,** 11–26.

OVERBECK J. (1962b) Untersuchungen zum Phosphathaushalt von Grünalgen. II. Die Verwertung von Pyrophosphat und organisch gebundenen Phosphaten und ihre Beziehung zu den Phosphatasen von *Scenedesmus quadricauda* (Turp.) Bréb. *Arch. Hydrobiol.* **58,** 281–308.

PIRSON A. & KUHL A. (1958) Über den Phosphathaushalt von Hydrodicyton I. *Arch. Mikrobiol.* **30,** 211–25.

PIRSON A., TICHY C. & WILHELMI G. (1952) Stoffwechsel und Mineralsalzernährung einzelliger Grünalgen. I. Vergleichende Untersuchungen an Mangelkulturen von *Ankistrodesmus. Planta,* **40,** 199–253.

PRICE C.A. (1962) Repression of acid phosphatase synthesis in *Euglena gracilis. Science, N.Y.* **135,** 46.

PROVASOLI L. (1958) Nutrition and ecology of protozoa and algae. *A. Rev. Microbiol.* **12,** 279–308.

REICHARDT W. & OVERBECK J. (1968) Zur enzymatischen Regulation der Phosphatmono-esterhydrolyse durch Cyanophyceenplankton. *Ber. dt. bot. Ges.* **81**, 391–6.

RENSEN J.J.S. VAN (1969) Polyphosphate formation in *Scenedesmus* in relation to photo-synthesis. In *Progress in photosynthetic research*, vol. III, ed. Metzner H. pp. 1769–76. Tübingen.

RODHE W. (1948) Environmental requirements of freshwater plankton algae. *Symb. Bot. Ups.* **10**, 1–149.

RÖSSEL T. (1963) Die chromatographische Analyse von Phosphaten. II. Die Dünnschicht-chromatographie der kondensierten Phosphate. *Z. analyt. Chem.* **197**, 333–47.

ROWAN K.S. (1966) Phosphorus metabolism in plants. *Intern. Rev. Cytol.* **19**, 301–90.

RYTHER J.H. & DUNSTAN W.H. (1971) Nitrogen, phosphorus and eutrophication in the coastal marine environment. *Science, N.Y.* **171**, 1008–31.

ST. JOHN J.B. (1970) Determination of ATP in *Chlorella* with the luceriferin-luciferase enzyme system. *Analyt. Biochem.* **37**, 409–16.

SCOTT G.T. (1945) The mineral composition of phosphate deficient cells of *Chlorella pyrenoidosa* during the restoration of phosphate. *J. cell. comp. Physiol.* **26**, 35–42.

SCHMIDT G. (1951) The biochemistry of inorganic pyrophosphates and metaphosphates. In *Phosphorus metabolism*, vol. I. eds. McElroy W.D. & Glass B. pp. 443–75. Johns Hopkins Press, Baltimore.

SIMONIS W. (1967) Zyklische und nichtzyklische Photophosphorylierung *in vivo*. *Ber. dt. bot. Ges*, **20**, 395–402.

SIMONIS W. & KATING H. (1956) Untersuchungen zur lichtabhängigen Phosphorylterung. I. Die Beeinflussung der lichtabhängigen Phosphorylierung von Algen durch Glucone-gaben. *Z. Naturf.* **116**, 115–72.

SIMONIS W. & GIMMLER H. (1969) Effect of inhibitors of photophosphorylation on light-induced dark incorporation of $^{32}$P into *Ankistrodesmus braunii*. *Progr. Phot. Res.* **3**, 1155–61.

SMITH F.A. (1966) Active phosphate uptake by *Nitella translucens*. *Biochim. biophys. Acta*, **126**, 94–9.

SOEDER C.J., SCHULZE G. & THIELE D. (1967) Einfluβ verschiedener Kulturbedingungen auf das Wachstum in Synchronkulturen von *Chlorella fusca* Sh. et Kr. *Arch. Hydrobiol.* (*Suppl.*) **33**, Falkau-Arbeiten 6, 127–71.

SOEDER C.J. (1970) Zum Phoshat-Haushalt von *Chlorella fusca* Sh. et Kr. *Arch. Hydrobiol. Suppl.* **38**, Falkau-Arbeiten 7, 1–17.

SOMMER A.L. & BOOTH T.E. (1938) Meta- and pyro-phosphate within the algal cell. *Pl. Physiol., Lancaster*, **13**, 199–205.

SOMMER J.R. & BLUM J.J. (1965) Cytochemical localization of acid phosphatases in *Euglena gracilis*. *J. Cell Biol.* **24**, 235–51.

STEWART W.D.P., FITZGERALD G.P. & BURRIS R.H. (1970) Acetylene reduction assay for determination of phosphorus availability in Wisconsin lakes. *Proc. natn. Acad. Sci., U.S.A.* **66**, 1104–11.

STEWART W.D.P. & ALEXANDER G. (1971) Phosphorus availability and nitrogenase activity in aquatic blue-green algae. *Freshwater Biol.* **1**, 389–404.

STICH H. (1955) Synthese und Abbau der Polyphosphate von *Acetabularia* nach autoradio-graphischen Untersuchungen des $^{32}$P-Stoffwechsels. *Z. Naturf.* **10b**, 282–4.

STREHLER B.L. (1968) Bioluminescence assay: Principles and practice. In *Methods of Bio-chemical Analysis*, **16**, ed. Glick D. pp. 99–181. Wiley-Interscience, New York.

STRICKLAND J.D.H. (1971) Microbial activity in aquatic environments. *Symp. Soc. Gen. Microbiol.* **21**, 231–54.

TÈWARI K.K. & SINGH M. (1964) Acid soluble and acid insoluble inorganic polyphosphates in *Cuscuta reflexa*. *Phytochemistry*, **3**, 341–7.

THILO E. (1959) Die kondensierten Phosphate. *Naturwissenschaften*, **46**, 367–73.

ULLRICH W.R. & SIMONIS W. (1969) Die Bildung von Polyphosphaten bei *Ankistrodesmus braunii* durch Photophosphorylierung in Rotlicht von 683 und 712nm unter Stickstoffatmosphäre. *Planta*, **84**, 358–67.

ULLRICH W.R., EBERIUS C.I. & SIMONIS W. (1970) Der Einfluß von Natrium- und Kalium-Ionen auf die Photophosphorylierung bei *Ankistrodesmus braunii*. *Planta*, **92**, 358–73.

ULLRICH W.R. (1970a) Die Wirkung von $O_2$ und $CO_2$ auf die $^{32}$P-Markierung der Polyphosphate von *Ankistrodesmus braunii* bei der Photosynthese. *Ber. dt. bot. Ges.* **83**, 435–7.

ULLRICH W.R. (1970b) Zur Wirkung von Sauerstoff auf die $^{32}$P-Markierung von Polyphosphaten und organischen Phosphaten bei *Ankistrodesmus* im Licht. *Planta*, **90**, 272–290.

ULLRICH W.R. (1971) Nitratabhängige nichtzyklische Photophosphorylierung bei *Ankistrodesmus braunii* in Abwesenheit von $CO_2$ und $O_2$. *Planta*, **100**, 18–30.

ULLRICH W.R. (1972) Der Einfluß von $CO_2$ und pH auf die $^{32}$P-Markierung von Polyphosphaten und organischen Phosphaten bei *Ankistrodesmus braunii* im Licht. *Planta*, **102**, 37–54.

URBACH W. & GIMMLER H. (1970a) Photophosphorylierung und photosynthetische $^{14}CO_2$-Fixierung *in vivo*. I. Vergleich von zyklischer und nichtzyklischer Photophosphorylierung mit der photosynthetischen $^{14}CO_2$-Fixierung. *Z. PflPhysiol.* **62**, 276–86.

URBACH W. & GIMMLER H. (1970b) Das Verhalten der zyklischen und nicht-zyklischen Photophosphorylierung während des Entwicklungszyklus von *Ankistrodesmus braunii*. *Ber. dt. bot. Ges.* **83**, 439–42.

WASSINK E.G., WINTERMANS J.F.G.M. & TJLA J.E. (1951) Phosphate exchanges in *Chlorella* in relation to conditions for photosynthesis. *K. Nederl. Akad. Wetenschap. Proc.* **54**, 41–52.

WIAME J.M. (1958) Accumulation de l'acide phosphorique (Phytine, Polyphosphates). In *Handb. PflPhysiol.* vol. 9, ed. Ruhland W. pp. 136–48. Springer-Verlag, Berlin.

WIESSNER W. (1970) Light effects on ion fluxes in microalgae. In *Photobiology of Microorganisms*, ed. Halldal P. pp. 135–62. Wiley-Interscience, London, New York, Sydney & Toronto.

WINTERMANS J.F.G.M. (1955) Polyphosphate formation in *Chlorella* in relation to photosynthesis. *Mededel. Landbouwhogeschool Wageningen*, **55**, 69–126.

# CHAPTER 24

# SILICIFICATION AND CALCIFICATION

## W. M. DARLEY

Department of Botany,
University of Georgia,
Athens, Georgia 30601, U.S.A.

1    **Introduction**  655

2    **Silicification**  656
2.1  Algae that deposit silica  656
2.2  The diatom cell wall  657
    (a) Structure  657
    (b) Formation  657
    (c) Chemical composition  658
2.3  Physiology of silica deposition  658
    (a) Silicic acid uptake  658
    (b) Initiation of silica deposition  659
2.4  Silicon as an essential nutrient  660
2.5  Germanium: A competitive inhibitor of silicon metabolism  661

3    **Calcification**  662
3.1  Algae that deposit calcium carbonate  662
3.2  Calcification in coccolithophorids  662
    (a) Ultrastructure and biochemistry  662
    (b) Physiology  665
3.3  Calcification in macroscopic algae  668
3.4  Mechanisms of calcification  668

4    **General Discussion**  669

5    **References**  670

## 1  INTRODUCTION

Mineralization in the algae is most frequently in the form of siliceous or calcareous deposits associated in some form with the cell wall or skeletal structures. While silicification is an intracellular process resulting in one mineral form, calcification may take place both intracellularly or extracellularly and results in at least two mineral forms. Although the physiological and biochemical aspects of mineralization are poorly known, it is clear that the process is not an accidental, haphazard event but rather a species specific, meticulously ordered, and carefully controlled process. Morphological aspects of mineralization have been worked out in some detail in a number of organisms and provide a framework for an understanding of the molecular events of silicification and calcification.

## 2 SILICIFICATION

### 2.1 *Algae that deposit silica*

Although silicon is the second most abundant element in the earth's crust, only a few groups of algae have utilized this element for skeletal structures. In addition to the well known silicified frustule enclosing diatom cells, siliceous structures are found in the silicoflagellates and in certain Chrysophyceae, Chlorophyceae, Phaeophyceae and Xanthophyceae. In virtually all cases examined silicon is found as hydrated amorphous silica, $SiO_2 . nH_2O$, frequently referred to as opal (Lewin & Reimann 1969). Recent reviews of silicification and diatom biology include Lewin (1962a), Lewin and Guillard (1963), Lewin and Reimann (1969), Darley (1969b) and Werner (1969a).

Parker (1969) found significant amounts of silica in all the Phaeophyceae and Hydrodictyaceae (Chlorophyceae) *pro parte* examined. In the brown algae the cell wall was suggested as the primary location of the silica. The siliceous component of the cell wall of *Pediastrum* has been more thoroughly investigated (Gawlik & Millington 1969, Millington & Gawlik 1967, 1970). Silica is localized in the thin outer wall which is contoured into a reticulate pattern of triangles, reflecting the uneven thickening of the non-siliceous inner wall. The first indication of the wall pattern is in the extracellular deposition of plaques of outer wall material at the corners of a hexagonal pattern.

The morphogenesis of the silicified body scales of the chrysophyte, *Synura petersenii*, has recently been revealed by Schnepf and Deichgräber (1969). These scales are formed in vesicles which lie adjacent to the periplastidal endoplasmic reticulum. The shape of the scale vesicle and hence the shape of the scale is determined by a complicated process of membrane morphogenesis. The margins of the vesicle fold over while the central part of the vesicle is distended by an outgrowth of the outer periplastidal membrane to form a hollow cylinder. The deposition of silica in this pre-shaped mould completes scale formation, after which the scale is released through the plasmalemma to form part of the cell covering. Unfortunately, there is no work on the physiology of silica deposition in chrysophytes to complement this excellent study.

Silicoflagellates are a poorly known group of marine unicellular algae possessing an 'internal', tubular siliceous skeleton variously described as star- or crown-shaped. The only ultrastructural studies of contemporary silicoflagellates (Van Valkenburg 1971a,b) show the surface details of the skeleton and the unusual protoplast. The structure of the skeleton suggests that it is formed sequentially beginning with a series of dichotomous branches which later fuse to complete the rings and arches of the skeleton. No evidence was presented for the presence of organic material surrounding the silica, although it is unlikely that there is no enclosing membrane of some kind, at least during the initial silicification process.

## 2.2   The diatom cell wall
(a) *Structure*

The cell wall of all diatoms which have been examined is composed of a silica shell which is surrounded by an organic membrane or casing during its formation as well as in the mature condition (see Helmcke & Krieger 1962–66, for morphological details of the siliceous frustules). Isolated cell walls of *Navicula pelliculosa* retain a replica of the inner silica shell after its removal by hydrofluoric acid vapours (Reimann *et al.* 1966). The cell wall of *Cylindrotheca fusiformis* is unusual in that the silicified portions are reduced leaving the organic casing as the only cell wall component over a large portion of the cell (Reimann *et al.* 1965). The fusiform cells of *Phaeodactylum tricornutum*, an atypical diatom used in many physiological experiments, lack any organized siliceous structures, yet contain as much silicon per cell as do the partially silicified oval cells of this species (Lewin *et al.* 1958). The organic wall of the fusiform cells is composed of three layers, the outer one of which has a corrugated surface composed of ridges, this being the only reported case of substructure in the organic casing of the diatom cell wall (Reimann & Volcani 1968).

Since diatom taxonomy is based almost exclusively on wall morphology, reports of polymorphism in clonal cultures raise questions not only concerning the validity of the use of skeletal structures for taxonomic purposes, but also concerning the more general aspects of the factors controlling the specific patterns found on the frustules. Valve patterns characteristic of several species were found during the life cycle of two *Coscinodiscus* clones (Holmes 1967, Holmes & Reimann 1966). Vegetative division of *Nitzschia alba* often results in both *Nitzschia* and *Hantzschia* daughter cells (Lauritis *et al.* 1967, Geitler 1969, but see Round 1970). Badour (1968) described internal 'lateral' shells formed by *Cyclotella cryptica* under experimental conditions which inhibited growth. Variations in valve morphology have been related to salinity changes in both pennate diatoms (Stoermer 1967) and centric diatoms (Schultz 1971). Polyploid clones of *Navicula pelliculosa* produce enlarged valves which are otherwise normal (Coombs *et al.* 1968b).

(b) *Formation*

Ultrastructural studies of cell wall morphogenesis in diatoms have been carried out on *Amphipleura pellucida* (Stoermer *et al.* 1965), *Cylindrotheca fusiformis* (Reimann 1964), *Gomphonema parvulum* (Drum & Pankratz 1964), *Navicula pelliculosa* (Reimann *et al.* 1966, Coombs *et al.* 1968a) and *Nitzschia alba* (Lauritis *et al.* 1968). Wall formation normally is initiated at the completion of mitosis while the two daughter protoplasts are still enclosed by the parent frustule. The first evidence of new wall formation is the appearance of the silica deposition vesicle (Drum & Pankratz 1964) in the central region of the cell immediately inside the plasmalemma. The membrane of this vesicle has been termed the silicalemma (Reimann *et al.* 1966). Formation of the silica deposition vesicle proceeds laterally, presumably being formed from Golgi-derived vesicles (Stoermer *et al.* 1965, Coombs *et al.* 1968a) and is followed closely by advancing silica deposition. The silicalemma adheres to the surface of the growing shell rather than forming a pre-shaped mould as in *Synura*. At the completion of silica deposition, a new plasmalemma, the origin of which is not clear, appears underneath the new cell wall. The silicalemma, possibly including additional material such as the old plasmalemma, continues to surround and adhere to the mature cell wall. The two daughter cells separate when the new valves are complete. Round (1971) has studied the formation of girdle bands in *Stephanodiscus* during cell division.

## (c) *Chemical composition*

Information on the chemistry of the inorganic fraction of the wall has been reviewed by Lewin and Reimann (1969). The organic cell wall layer, which has been extensively studied by Volcani and his group, is a complex assortment of sugars, lipids, amino acids and uronic acids (Coombs & Volcani 1968b). Lipids account for 1 to 13% of the wall organic matter and differ significantly from cellular lipids (Kates & Volcani 1968). Nakajima and Volcani (1969, 1970) have discovered new derivatives of hydroxyproline and lysine in diatom cell walls. Labelling studies during and following cell wall formation in *Navicula pelliculosa* (Coombs & Volcani 1968b) suggest that proteinaceous compounds are added for the most part during cell wall formation while carbohydrates are continually being added to mature walls.

## 2.3  *Physiology of silica deposition*

### (a) *Silicic acid uptake*

The form of silicon utilized by diatoms (Lewin 1955a), and presumably other siliceous algae, is silicic acid, $H_4SiO_4$, the un-ionized form of which predominates in the pH range found in most biological systems (Siever & Scott 1963). Early investigations by Lewin (1954, 1955b) on *Navicula pelliculosa* demonstrated that silicic acid uptake only occurs in living cells grown aerobically. The process is temperature dependent, is linked to energy yielding processes and in some way requires the presence of divalent sulphur (see Lewin 1962a, for a more complete discussion). More recently Lewin and Chen (1968) found that silicic acid uptake in the colourless marine diatom *Nitzschia alba* was impaired by washing, but that amino acids, notably glutamine, aspartic acid, and glutamic acid, reversed this inhibition. In this regard it is interesting that Werner (1968) found that the aspartic acid and glutamic acid pools in *Cyclotella cryptica* were decreased significantly during 3 hours of silicic acid starvation.

Several investigations by Volcani's group (Coombs *et al.* 1967b, c, d) have provided additional information on the energy requirements associated with silicic acid uptake. Thus the transients in the concentration of nucleoside triphosphates and in $O_2$ exchange rates when silicic acid is re-introduced to silicon starved cultures are consistent with higher energy requirements associated with silicic acid uptake. Hemmingsen (1971) has demonstrated the presence of a membrane bound, mono-silicic acid stimulated ATPase in *Nitzschia alba*. The activity of the enzyme increases in proportion to the silicic acid concentration up to 0·2mM. This enzyme is not associated with the mitochondrial fraction and is stimulated by low concentrations of germanic acid (see below), further suggesting that ATPase is involved with the uptake of silicic acid.

Synchronized cultures of diatoms, produced either by silicon starvation or by light-dark cycles (see Darley & Volcani 1971) have proved to be useful tools in the study of diatom physiology and biochemistry. Light-dark synchronized cultures of *Navicula pelliculosa* (Darley 1969a), *Cylindrotheca fusiformis* (Coombs *et al.* 1967b, Darley & Volcani 1969) and *Ditylum brightwellii* (Eppley *et al.* 1967) take up silicic acid from the growth medium only during the cell wall formation period of the cell cycle; silicic acid is not taken up and stored in the cell for use later during cell wall formation. Studies with $^{31}Si$ (Coombs & Volcani 1968b) suggest that once it is taken up by the cell, silicic acid is rapidly incorporated into the growing shell. Exponential cells of *Cyclotella cryptica* when placed in the dark continued to take up silicic acid for five hours following the cessation of cell division (Werner 1966, Werner & Pirson 1967). Although the form of this excess silicic acid in the cell is not yet known with certainty, it is clear that it is not used for future cell wall formation since no increase in cell number occurred when these cells were placed in silicon-free medium in the light (Werner 1966). Under these conditions (silicon-free medium in the light), however, these cells did show an enhanced biosynthetic capacity compared to exponential cells subjected to silicon starvation. This phenomenon was ascribed to the excess silicon accumulated in the dark (Werner & Pirson 1967), but may be explained on the basis of different cell 'ages' at the beginning of the experiment (see Darley 1969b).

### (b) *Initiation of silica deposition*

When Geitler (1963) pointed out that cell wall formation in diatoms is always preceded by mitosis and usually cytokinensis as well, he was indirectly raising a question concerning the molecular event which triggers wall formation. Although there is general agreement about the necessity of mitosis preceding wall formation under natural conditions (von Stosch & Kowallik 1969, Geitler 1970), this observation provides little information about the physiological and biochemical interdependence of the two processes. This problem has been approached experimentally using a number of inhibitors on the centric diatom *Cyclotella cryptica* (Badour 1968, Oey & Schnepf 1970). Inhibiting cytokinesis or mitosis (but not cell growth) by calcium deficiency or colchicine respectively, results in the formation of two internal lateral shells which lie against the girdle bands in the mid-region of the cell. From these experiments Badour (1968) concluded that wall formation is dependent on plasmatic growth. Oey and Schnepf (1970) measured DNA content per nucleus after treatment with colchicine and inhibitors of nucleic acid synthesis. They found that lateral shell formation depended on prior DNA synthesis. Thus, although in a morphological sense, mitosis always precedes wall formation, the actual initiating event apparently occurs during or slightly after the DNA synthetic phase of the cell cycle. The formation of lateral shells might be restricted to centric diatoms since colchicine treatment of *Navicula pelliculosa* (Coombs *et al.* 1968a) caused many aberrations in cell division and wall formation, but did not result in the formation of any structures comparable to lateral shells.

### 2.4   *Silicon as an essential nutrient*

All diatoms examined to date have an absolute requirement for silicon and are the only algae in which a silicon requirement has been demonstrated (although it is likely to be found in other siliceous algae when looked for). In addition to a metabolic requirement for silicon in at least some diatoms (see below), it seems that the actual presence of the siliceous cell wall is also required for normal growth. There are various reports of diatoms lacking silica shells (see Lewin 1962a), but these naked cells have not yet been maintained in culture nor have they been induced to regenerate their cell walls under defined conditions. Apelt (1969) has described naked, actively dividing cells of the genus *Licmophora* as symbionts of the flatworm *Convoluta convoluta*. Attempts to culture these protoplasts outside the host or induce shell formation were unsuccessful. A deficiency of calcium and/or magnesium results in protoplast formation by *Nitzschia alba* cells (Hemmingsen 1971); no attempt was made to induce shell regeneration by these protoplasts. Werner (1969b) reports that borosilicates act as growth factors for diatoms. These carbon-free compounds are required for the growth of some diatoms and stimulate the growth of others. This requirement is in addition to the requirement for silicic acid.

The direct effect of silicon starvation on the diatom division cycle is to block cell development either before mitosis or during cell wall formation, depending on the organism and the experimental system. Indirect effects of silicon starvation are a general decline in active metabolism caused by the arresting of further cell development. Lipid synthesis continues undiminished, however, in both *Navicula pelliculosa* (Coombs *et al.* 1967a) and *Cyclotella cryptica* (Werner 1966, 1968). Silicon starvation has therefore been used to synchronize cell division in *Navicula pelliculosa* cultures (Lewin *et al.* 1966, Busby & Lewin 1967, Darley & Volcani 1971), an excellent system for studying silicic acid uptake, cell wall formation and the cell metabolism associated with these processes (Coombs *et al.* 1967a,c,d, Healey *et al.* 1967, Coombs & Volcani 1968a,b). Silicon starvation of exponential cells blocks cell development at the cell wall formation stage of the cell cycle; active cell metabolism gradually decreases during the period of silicon starvation. The re-introduction of silicic acid results in an immediate stimulation of energy metabolism primarily directed toward silicic acid uptake and cell wall formation. Following the separation of daughter cells, energy metabolism is redirected toward organic biosynthesis for the start of a new cell cycle.

Silicon starvation of young light-dark synchronized *Cylindrotheca fusiformis* cells, on the other hand, inhibited DNA net synthesis, and hence mitosis, while other cell components, including RNA, increased by at least 75% (Darley & Volcani 1969). Following this report Sullivan (1971) provided evidence that silicic acid regulates DNA synthesis by stimulating the activity of DNA polymerase at some post-transcriptional level. Re-introduction of silicic acid to silicon starved cells resulted in a rapid doubling in DNA polymerase specific activity and a doubling of cellular DNA, followed by mitosis, cytokinesis, cell wall formation and cell separation. Additional data suggest that this silicon requirement for DNA synthesis may be a general occurrence among diatoms. If young light-dark synchronized cells of *Navicula pelliculosa* are deprived of silicic acid, a significant number show a block prior to mitosis (Darley 1969a). Stationary phase (presumably silicic acid limited) cells of *Cyclotella cryptica* were arrested prior to the DNA synthetic phase of the cell cycle (Oey & Schnepf 1970).

Werner (1966) studied the effect of silicon deprivation on *Cyclotella cryptica* although he did not correlate his biochemical data with the division cycle. He noted the same general decline in cell metabolism during which the biosynthesis of certain cell components was suppressed before others. His results are comparable to those for *Navicula pelliculosa* and *Cylindrotheca fusiformis* if one assumes that his experimental material consisted of young cells at the beginning of the period of silicon deprivation (see Darley 1969b).

## 2.5 *Germanium: a competitive inhibitor of silicon metabolism*

Germanium, an element which has chemical properties similar to those of silicon (Glockling 1969), is a potent inhibitor of growth in diatoms, although usually not affecting the growth of other algae (Lewin 1966, Werner 1966, 1967, Darley 1969a, Darley & Volcani 1969). Data showing that the inhibition is a function of the germanium/silicon molar ratio (a ratio of 0·1–0·2 is usually inhibitory) suggest that germanic acid acts as a competitive inhibitor of silicic acid metabolism. The effect of germanium on several silicon requiring processes has been investigated. In *Navicula pelliculosa*, the only effect seems to be the inhibition of silicic acid uptake involved in cell wall formation, even at high germanium/silicon molar ratios (Darley 1969a). In *Cylindrotheca fusiformis*, however, the effect depends on the germanium/silicon molar ratio. Low ratios inhibit wall formation and mitosis but not DNA synthesis, while higher ratios inhibit overall growth including DNA synthesis (Darley & Volcani 1969). Regarding the silicon requirement for DNA synthesis in this organism, Sullivan (1971) found that germanic acid did not substitute for silicic acid *in vivo*, although it slightly

Y

stimulated *in vitro* DNA polymerase activity. In *Cyclotella cryptica* (Werner 1966, 1967) germanic acid inhibits silicon uptake, chlorophyll synthesis and perhaps the synthesis of other plastid components, although it does not inhibit carbohydrate synthesis, as did silicon starvation. Thus in some cases germanic acid mimics the effect of silicon starvation while in others, its action appears independent of silicon requirements.

## 3 CALCIFICATION

### 3.1 *Algae that deposit calcium carbonate*

Calcium deposition by algae occurs primarily in tropical marine waters and is almost exclusively in the form of carbonate. Macroscopic calcareous algae are a principal agent in atoll reef formation and unicellular forms are an important component of marine phytoplankton. Lewin (1962b) summarized the principal groups of calcareous algae and their habitats, as well as important aspects of the mineralogy of the deposit. Calcification in unicellular organisms and plants has also been the subject of recent reviews (Arnott & Pautard 1970, Pautard 1970). There are two naturally occurring forms of crystalline calcium carbonate: calcite, deposited by certain members of the Cyanophyceae, Haptophyceae, Cryptonemiales (Rhodophyceae) and Charophyceae, and aragonite, deposited by certain members of the Chlorophyceae, Phaeophyceae and Nemaliales (Rhodophyceae). No alga is known to deposit a mixture of calcite and aragonite (Stark *et al.* 1969). *Tydemania* has recently been added to the genera of siphonous algae known to deposit aragonite (McConnell & Colinvaux 1967). Several dinoflagellates of the order Peridiniales have been found to produce calcareous resting cysts. Using cultured material, Wall *et al.* (1970) have shown that the cyst of *Peridinium trochoideum* is composed of more than 50 calcite spines which form between the inner and outer cyst membrane. Crystals found in the terminal vacuoles of *Closterium* and certain other desmids are thought to be composed of calcium sulphate but deserve further study (see Arnott & Pautard 1970, Pickett-Heaps & Fowke 1970).

### 3.2 *Calcification in coccolithophorids*

(a) *Ultrastructure and biochemistry*

The generally unicellular algae of the class Haptophyceae are characterized in part by the possession of organic surface scales with distinctive surface patterns. The coccolithophorid members of this class are further distinguished by the presence of calcite on the margin of certain of these organic scales. The calcified scale, or coccolith, and its morphogenesis have been the subject of a number of ultrastructural studies since Paasche (1968a) reviewed the subject. A marine species, *Hymenomonas* (syns. *Cricosphaera, Syracosphaera*) *carterae*, has been extensively investigated (Manton & Leedale 1969, Pienaar 1969a,b, 1971, Outka & Williams 1971); a fresh water member of this genus, *H. roseola*, has also been studied (Manton & Peterfi 1969). In addition, ultrastructural investigations of the stages in the life cycle of '*Apistonema-Syracosphaera*' *sensu* von Stosch (Leadbeater 1970) and *Pleurochrysis scherffelii* (Leadbeater 1971) suggest that these isolates also belong to the genus *Hymenomonas*. There has been a report on the development of the relatively large coccoliths of *Coccolithus pelagicus* (Manton & Leedale 1969) and a preliminary ultrastructural study of both coccolith-forming and naked cells of *Coccolithus huxleyi* (Klaveness & Paasche 1971).

The descriptive coccolith terminology which has developed to deal with the wide diversity of living and fossil coccolith forms is beyond the scope of this article (see McIntyre & Bé 1967, Black 1963, 1968). However, some details of coccolith morphology revealed in conjunction with the studies mentioned above, will prove useful in the discussion of coccolith morphogenesis. The coccolith of *Hymenomonas carterae* consists of a subtending oval, organic plate or scale the proximal face of which is patterned with ridges radiating from a longitudinal suture (Outka & Williams 1971, Manton & Leedale 1969, Pienaar 1969a, Leadbeater 1970, 1971). The distal face (inside the ring of calcite) appears amorphous in shadowed preparations (Manton & Leedale 1969, Leadbeater 1971) and has a rim to which the calcitic elements are attached. The structure and arrangement of the calcareous elements of the *H. carterae* coccolith are clearly shown by Outka and Williams (1971) in partially decalcified material. An outer ring of 13–16 anvil-like (Braarud *et al.* 1952) elements alternate with an inner ring of a similar number of smaller anvil elements. Each element is thought to be composed of one to a few calcium carbonate crystals as previously reported for the coccolith of *Coccolithus huxleyi* (Watabe 1967). Both the smaller and larger elements have a groove matching the rim of the margin of the subtending scale.

The calcite elements are surrounded by an organic matrix, which may be demonstrated by decalcifying the coccoliths (Manton & Leedale 1969, Outka & Williams 1971) and was first noticed by Braarud *et al.* (1952) in coccoliths of *Coccolithus huxleyi*. Some strains of *H. carterae* lose the ability to produce calcified structures in culture, although these strains continue to produce structurally recognizable coccoliths consisting of a base plate and organic matrix (Manton & Peterfi 1969, Manton & Leedale 1969, Leadbeater 1971). On a structural and developmental (see below) basis, therefore, the term coccolith has been applied to the complete structure, rather than solely to the calcified component (Outka & Williams 1971).

Between the layer of coccoliths and the cell membrane, there are one or more layers of smaller, thin, circular to elliptical, unmineralized scales with a proximal surface of radiating ridges and a distal surface of concentric lines. These scales are found on *Hymenomonas carterae* (Manton & Leedale 1969, Pienaar 1969a), *H. roseola* (Manton & Peterfi 1969) and *Coccolithus pelagicus* (Manton & Leedale 1969), although the coccolith-bearing cells of *Coccolithus huxleyi* appear to lack them (Klaveness & Paasche 1971). They are found as components of the laminated cell wall of the *Apistonema* and *Pleurochrysis* phases and on the coccolith-free swarmers of these stages (Brown *et al.* 1970, Leadbeater 1970, 1971). A coccolith-free form of *Coccolithus huxleyi*, which may represent a stage of the normal life cycle, has organic scales, the structure of which is not yet described (Klaveness & Paasche 1971).

It is now firmly established that the coccoliths and unmineralized scales of *Coccolithus pelagicus* (Manton & Leedale 1969), *Hymenomonas carterae* (Outka & Williams 1971) and *H. roseola* (Manton & Peterfi 1969) are derived from the cisternae of the Golgi body. A similar origin had previously been demonstrated for the unmineralized scales of the related genera, *Chrysochromulina* (Manton 1967a,b) and *Prymnesium* (Manton 1966). Brown *et al.* (1970) proposed an elaborate scheme for the Golgi directed formation of the organic scales of *Pleurochrysis scherffelii.* An 'intracellular coccolith precursor' proposed by Pienaar (1969b) as the site of coccolith formation is now thought to be an auto-phagic vacuole (Pienaar 1971).

The morphogenesis of the coccolith in *H. carterae* is described in the most detail by Outka and Williams (1971) who confirm and extend earlier work on this organism. The strongly polarized Golgi body is located anteriorly in the cell near the flagella. The organic base plate (scale) first appears in cisternae near the central region of the Golgi body while small ($\sim$28nm) densely staining particles called coccolithosomes accumulate in the marginal pockets of these Golgi cisternae. The coccolithosomes are then released into the base plate vesicle where they are found at the plate margin and become closely associated with the developing organic matrix, presumably contributing to it. Similar particles have been seen in *H. carterae* by others (Manton & Leedale 1969, Pienaar 1971). Leadbeater (1971) also suggested that they are involved in the formation of the organic matrix. Throughout this process, the only constant association of the cisternal membrane with the developing coccolith is at the outer border of the base plate. When complete, the mature coccolith is liberated to the cell exterior, presumably by fusion of its Golgi vesicle with the cell membrane. As many as four developing coccoliths may be present in a cell at one time. A columnar deposit on the cell membrane may function to maintain coccolith position on the cell exterior (Manton & Leedale 1969, Outka & Williams 1971).

The coccolith of *Coccolithus pelagicus* is formed in a similar manner (Manton & Leedale 1969) except that coccolithosomes were not observed and it is not clear that calcification occurs in an area previously delimited by the organic matrix. The large size of coccolith necessitates some rearrangement of the structures involved. The cisterna in which the coccolith develops is in the form of a 'T'; the stem is still associated with the Golgi stack while the cross piece contains the coccolith with its future distal surface facing the stack. Only one coccolith is formed in a cell at a time. Manton and Leedale (1963) found no evidence of intracellular calcification in *Crystallolithus hyalinus* (the motile form of *Coccolithus pelagicus*), although physiological evidence (Paasche 1969) suggests that calcification is intracellular. The situation in *Coccolithus huxleyi* has not yet been thoroughly investigated but work to date implicates a membranous 'reticular body' in coccolith formation (Wilbur & Watabe 1963, Klaveness & Paasche 1971).

Chemical analysis of coccoliths or unmineralized scales show that carbohydrates are the major component. Galactose and ribose were the principal sugars detected in the unmineralized scales of *Chrysochromulina chiton* by Green and Jennings (1967). Untreated scales from the cell wall of *Pleurochrysis* also contained large amounts of galactose and ribose in addition to 3 to 9% protein (Brown *et al.* 1970). Mild alkali treatment of *Pleurochrysis* scales removed these sugars and proteins, and left only cellulose microfibrils which were identified as the radial and concentric lines on the scales. Isenberg *et al.* (1966) identified hydroxyproline in the protein fraction of a distilled water extract of cell-free coccoliths of *Hymenomonas*.

Taking the liberty of generalizing from the limited and/or incomplete studies available, the following interpretations may be postulated, bearing in mind that further investigations are required. Both organic and calcified scales of haptophycean algae are composed of two layers of material (Manton & Leedale 1969). The homologous basic structure is a cellulose network of radially and concentrically arranged microfibrils which constitutes the proximal layer of the scale (Brown *et al.* 1970). The outer layer of the scale, presumably including the organic matrix of coccoliths is composed of acidic polysaccharides (containing galactose and ribose) and small amounts of protein, probably containing hydroxyproline. The variation in scale morphology, including mineralization in coccolithophorids, would be a function of this noncellulosic outer layer (Brown *et al.* 1970, Franke & Brown 1971), which Manton (1967b) has suggested may be under independent genetical control in *Chrysochromulina*.

## (b) *Physiology*

Studies on the physiology of calcification in coccolithophorids are limited to *Coccolithus huxleyi*, *C. pelagicus* and *Hymenomonas* sp. Paasche (1968a) and Wilbur and Watabe (1967) have reviewed all but the most recent work in this area. Most of the studies have centred on the question of the relationship between photosynthetic carbon fixation and the precipitation of carbonate in coccolith formation. The initial demonstration that light stimulates coccolith formation (Paasche 1963) and that the carbonate is taken into the cell as bicarbonate (Paasche 1964), suggested that photosynthetic carbon dioxide assimilation could promote the deposition of carbonate by shifting the equilibrium

$$2HCO_3^- \rightleftharpoons CO_2 + CO_3^{2-} + H_2O$$

to the right. The summation of experimental results to date suggests, however, that the stimulatory effect of light on calcification is not directly connected to the assimilation of $CO_2$, but is more likely the result of a general stimulation of cellular metabolism.

Experimental approaches to this question have included a comparison of photosynthetic carbon fixation and coccolith formation in the presence of photosynthetic inhibitors (Crenshaw 1964, Paasche 1965), inhibitors of carbonic anhydrase (Isenberg *et al*. 1963, Paasche 1964, Crenshaw 1964) and respiratory inhibitors (Paasche 1964). Conflicting results from some of these studies have not been completely resolved, although the differences are thought to be less fundamental than the data would seem to indicate (Paasche 1968a). Calcium levels low enough to inhibit coccolith formation did not reduce photosynthesis in short term experiments (Paasche 1964). Comparative studies of naked and coccolith forming clones of *C. huxleyi* with respect to growth rates and the efficiency of photosynthesis (Steemann Nielsen 1966) suggested that the lower growth rates observed in naked clones were caused by their inability to use bicarbonate for coccolith formation. In a recent detailed comparison of naked and coccolith forming cells, Paasche and Klaveness (1970) found that the lower growth rate of the naked clone was due to a decrease in the photosynthetic rate, presumably caused by a similar decrease in chlorophyll *a* content. These workers found no positive indication that the two cell types differed in their ability to use bicarbonate. An additional close interaction between photosynthesis and coccolith formation was suggested by the approximately one-to-one ratio of photosynthetic and coccolith carbon fixed by *C. huxleyi* (Paasche 1964). This relation is likely to be largely coincidental in view of the results obtained with *C. pelagicus* (Paasche 1969). In the non-motile form, coccolith carbon may exceed photosynthetic carbon by several fold while in the motile form coccolith carbon amounts to less than 2% of the photosynthetic carbon.

More direct approaches to the role of light in coccolith formation have also been used. The action spectrum of coccolith formation in *C. huxleyi* (Paasche 1966b) exhibits peaks in the red and blue corresponding to regions of maximum photosynthesis, although blue light had an additional stimulatory effect not seen in photosynthesis. In another approach, Paasche (1966a) transferred autotrophically grown, decalcified cells of *C. huxleyi* to darkness in an inorganic medium and observed that coccolith formation continued at one seventh to one tenth the light saturated rate but no cell division occurred under these conditions. In a similar type of experiment without the decalcifying treatment, Blankley (1971) found that cell division and coccolith formation continued at the full light saturated rate for 1·0 and 0·44 doublings respectively. Differences in these two sets of data were attributed to the different techniques for measuring coccolith production and to the deleterious effect of the decalcification procedure used by Paasche (Blankley 1971), although this decalcifying treatment did not affect short term photosynthesis (Paasche 1964). In cultures of *Hymenomonas carterae* coccolith formation continues at the light saturated rate for 0·53 doublings before ceasing (Blankley 1971).

More definitive evidence for the light-independence of coccolith formation has been obtained using heterotrophic cultures. Although many earlier attempts to grow coccolithophorids under heterotrophic conditions had failed, Blankley (1971) found that all coccolithophorids tested could grow and produce coccoliths through many subcultures during heterotrophic growth on 0·5M glycerol. Even though such a high glycerol concentration is of limited ecological significance, it does provide a useful system for the study of coccolith formation. In *H. carterae* the relative rates of growth and coccolith formation were preserved under light and dark growth, even though the generation time under heterotrophic conditions was 3 to 4 times longer than that of the autotrophic cultures (Blankley 1971). Dark grown coccoliths were shown to have a normal crystallographic structure. Thus although much remains to be learned about the action of light on coccolith production, present evidence suggests that light stimulates coccolith formation indirectly through an overall stimulation of the cell's energy metabolism, perhaps through cyclic photophosphorylation (Paasche 1964, 1965, Blankley 1971) rather than through a separate light-dependent mechanism of coccolith formation (Paasche 1966a).

In other physiological studies Paasche (1968b) showed that coccolith production per unit cell volume is relatively insensitive to temperature (see Watabe & Wilbur 1966). External coccoliths may be dissolved by lowering the pH of the medium or by lowering the ion product of calcium and carbonate below the solubility product of calcite (Paasche 1964, Crenshaw 1964, Swift & Taylor 1966), although coccoliths apparently are still synthesized under these conditions (Paasche 1964). Thus limited inorganic carbon in the medium might account for the report by Isenberg *et al.* (1965) that log phase cultures of *Hymenomonas carterae* did not produce coccoliths (Blankley 1971). Decalcified cells of *Hymenomonas* sp. and *Coccolithus huxleyi* will regain a complete coccolith cover in 15 to 40 hours respectively, independent of cell division (Crenshaw 1964, Paasche 1966a). Paasche (1967) found that light-dark synchronized cells of *C. huxleyi* had complete coverings throughout the cell cycle but the rate of calcification during the cell cycle does not appear to have been measured.

The possible function of coccoliths is interesting in that the presence of coccoliths in *C. huxleyi* was shown to increase sinking rates more than four-fold (Eppley *et al.* 1964). There is no evidence to support Steemann Nielsen's suggestion that coccolith production increased the total inorganic carbon available to the cells (see Paasche & Klaveness 1970, Blankley 1971). Experimental evidence has similarly not supported an early suggestion (Braarud *et al.* 1952) that coccoliths function as a protective light screen (see Paasche & Klaveness 1970, Blankley 1971). Based on the observation that coccolith forming strains of *C. huxleyi* and *H. carterae* have lower saturating light intensities for growth than naked strains of these species, Blankley (1971) has suggested that coccoliths function to scatter light, thus increasing its path length through the water column and increasing the proportion of light absorbed by the cells.

### 3.3    Calcification in macroscopic algae

In contrast to the amount of study on the ultrastructure of coccolithophorids, relatively little effort has been expended on the large calcareous algae, undoubtedly due to difficulties encountered growing them in laboratory culture and obtaining adequate fixation of these heavily mineralized plants. Several calcareous green algae have been successfully grown in laboratory culture (Colinvaux *et al.* 1965) and the fine structure of one of these, *Halimeda*, has been investigated (Wilbur *et al.* 1969). Crystals of aragonite form extracellularly in the interutricular spaces of the thallus. The solution in these spaces is not in free communication with the surrounding sea water due to the continuous outer surface formed by the peripheral utricles. Calcification first appears on a fibrous material which is attached to the wall and perhaps acts as a nucleation site for crystal formation; no organic matrix surrounds the growing or mature crystals. The first instance of intracellular formation of aragonite crystals has been reported in a related green alga, *Penicillus dumetosus*. The internal crystals differed in form and size from the extracellularly deposited aragonite crystals (Perkins *et al.* 1972).

The recent application of freeze-etching and scanning electron microscopy has overcome some of the problems encountered with thin-sectioning techniques in the heavily encrusted coralline red algae (Bailey & Bisalputra 1970). Many vesicles produced as a result of Golgi activity and possibly from the degradation of endoplasmic reticulum contribute to the calcareous region of the wall as well as to an inner amorphous wall layer. The vesicle contents appear amorphous or slightly granular; their chemical composition is not known.

Physiological studies have demonstrated that calcification is usually stimulated by light (Stark *et al.* 1969, Ikemori 1970, Okazaki *et al.* 1970); although in a survey of thirty-two species Goreau (1963) found seven species with calcification rates which were higher in the dark than in the light. Uptake of $^{45}$Ca by *Halimeda* showed a diurnal rhythm and was stimulated by light both during the day and night (Stark *et al.* 1969). Calcification is associated with the development of chloroplasts in young filaments (Wilbur *et al.* 1969) and with the diurnal movement of chloroplasts to the peripheral areas of the plant (Stark *et al.* 1969). Inhibitors of carbonic anhydrase and photosynthesis markedly reduced the incorporation of $^{45}$Ca in several calcareous algae (Ikemori 1970). Differential washout rates for calcium adsorbed in the light or in the dark suggest that calcification in *Halimeda* may occur by a two-step mechanism (Stark *et al.* 1969).

### 3.4    Mechanisms of calcification

Although it is clear that light generally stimulates algal calcification, the mechanism by which this occurs remains obscure. It now seems unlikely that calcification is a simple precipitation reaction due to the photosynthetic utilization of carbon dioxide. The suggestion by Paasche (1965) that light provides an additional energy source through photophosphorylation is probable. In this respect it is interesting to note that light also stimulates calcification in corals containing symbiotic algae. Pearse and Muscatine (1971) have demonstrated that the zooxanthellae act indirectly through the production of organic products which are then translocated to the areas with the highest rates of calcification where they may function as a supplementary energy source or possibly as part of the organic matrix. Morphological studies alone suggest that the mechanism of calcification may vary among the different groups of algae. Thus the highly ordered intracellular, matrix enclosed, calcite precipitation of coccolith formation (Outka & Williams 1971) may differ in several respects from the more random extracellular growth of aragonite crystals in *Halimeda* (Wilbur *et al.* 1969). Pearse (1972) has recently studied the effect of light on the rate and stability of $^{45}$Ca-labelling in a coralline member of the Rhodophyceae, *Bossiella orbigniana*.

## 4 GENERAL DISCUSSION

Silicification in diatoms and calcification in coccolithophorids are the best known examples of mineralization in algae. In both cases mineralization occurs intracellularly within a vesicle shown to be of Golgi derivation, at least in the coccolithophorids. The coccolith develops within a second pre-shaped organic matrix contained in the Golgi vesicle, while in diatoms the vesicle itself acts as the organic matrix, presumably directly responsible for the polymerization process, and only reflecting the final form of the shell upon completion. In each case the organic layer immediately responsible for the deposition continues to envelop the mature structure, probably preventing its solubilization (Lewin 1961). In addition, in both cases this layer has been found to contain hydroxyproline, an important component of collagen, the organic matrix associated with bone calcification. Calcium, of course, is required for many cellular processes besides calcification, while until recently it was assumed that silicification was relatively independent of other metabolic processes. Now it seems that in diatoms silicon is required for DNA synthesis and that DNA synthesis, in turn, is a prerequisite for silicification. Coccolithophorids can exist quite well without their coccolith covering, while diatoms have not yet been cultured without their siliceous cell wall (the cases of the fusiform cell of *Phaeodactylum* and the symbiotic cell of *Licmophora* deserve further study). In addition, coccolith formation is not tied as closely to the division cycle as is silicification. Thus while silicon is not as thoroughly integrated with general cell metabolism as is calcium, silicification would appear to be more of an absolute requirement for diatoms than calcification is for coccolithophorids.

Perhaps the best known aspect of the physiology and biochemistry of both silicification and calcification in algae is the general requirement for metabolic energy including the demonstration of a silicic acid stimulated, membrane bound

ATPase in a diatom. Beyond this, the role of the deposition vesicle (or the significance of its absence in extracellular mineralization), the precipitation or polymerization reaction itself and the mechanism by which the cell controls the specific form of the deposit await further study.

# 5 REFERENCES

APELT G. (1969) Die Symbiose zwischen dem acoelen Turbellar *Convoluta convoluta* und Diatomeen der Gattung *Licmophora. Mar. Biol.* **3**, 165–87.

ARNOTT H.J. & PAUTARD F.G.E. (1970) Calcification in plants. In *Biological Calcification*, ed. Schraer, H. pp. 375–446. Appleton-Century-Crofts, New York.

BADOUR S.S. (1968) Experimental separation of cell division and silica shell formation in *Cyclotella cryptica. Arch. Mikrobiol.* **62**, 17–33.

BAILEY A. & BISALPUTRA T. (1970) A preliminary account of the application of thin-sectioning, freeze-etching, and scanning electron microscopy to the study of coralline algae. *Phycologia*, **9**, 83–101.

BLACK M. (1963) The fine structure of the mineral parts of Coccolithophoridae. *Proc. Linn. Soc. Lond.* **174**, 41–6.

BLACK M. (1968) Taxonomic problems in the study of coccoliths. *Paleontology*, **11**, 793–813.

BLANKLEY W.F. (1971) Auxotrophic and heterotrophic growth and calcification in coccolithophorids. (Doctoral thesis, University of California, San Diego, California.)

BRAARUD T., GAARDER K.R., MARKLAKI J. & NORDLI E. (1952) Coccolithophorids studied in the electron miscroscope. Observations on *Coccolithus huxleyi* and *Syracosphaera carterae. Nytt Mag. Botan.* **1**, 129–33.

BROWN R.M. JR., FRANKE W.W., KLEINIG H., FALK H. & SITTE P. (1970) Scale formation in Chrysophycean algae I. Cellulosic and noncellulosic wall components made by the Golgi apparatus. *J. Cell Biol.* **45**, 246–71.

BUSBY W.F. & LEWIN J.C. (1967) Silicate uptake and silica shell formation by synchronously dividing cells of the diatom *Navicula pelliculosa* (Bréb) Hilse. *J. Phycol.* **3**, 127–31.

COLINVAUX L.H., WILBUR K.M. & WATABE N. (1965) Tropical marine algae: growth in laboratory culture. *J. Phycol.* **1**, 69–78.

COOMBS J., DARLEY W.M., HOLM-HANSEN O. & VOLCANI B.E. (1967a) Studies on the biochemistry and fine structure of silica shell formation in diatoms. Chemical composition of *Navicula pelliculosa* during silicon-starvation synchrony. *Pl. Physiol., Lancaster*, **42**, 1601–6.

COOMBS J., HALICKI P.J., HOLM-HANSEN O. & VOLCANI B.E. (1967b) Studies on the biochemistry and fine structure of silica shell formation in diatoms. Changes in concentration of nucleoside triphosphates during synchronized division of *Cylindrotheca fusiformis* Reimann and Lewin. *Expl. Cell Res.* **47**, 302–14.

COOMBS J., HALICKI P.J., HOLM-HANSEN O. & VOLCANI B.E. (1967c) Studies on the biochemistry and fine structure of silica shell formation in diatoms. II. Changes in concentration of nucleoside triphosphates in silicon-starvation synchrony of *Navicula pelliculosa* (Bréb) Hilse. *Expl. Cell Res.* **47**, 315–28.

COOMBS J., SPANIS C. & VOLCANI B.E. (1967d) Studies on the biochemistry and fine structure of silica shell formation in diatoms. Photosynthesis and respiration in silicon-starvation synchrony of *Navicula pelliculosa. Pl. Physiol., Lancaster*, **42**, 1607–11.

COOMBS J., LAURITIS J.A., DARLEY W.M. & VOLCANI B.E. (1968a) Studies on the biochemistry and fine structure of silica shell formation in diatoms. V. Effects of colchicine on wall formation in *Navicula pelliculosa* (Bréb) Hilse. *Z. PflPhysiol.* **59**, 124–52.

COOMBS J., LAURITIS J.A., DARLEY W.M. & VOLCANI B.E. (1968b) Studies on the biochemistry and fine structure of silica shell formation in diatoms. VI. Fine structure of colchicine-induced polyploids of *Navicula pelliculosa* (Bréb) Hilse. *Z. PflPhysiol.* **59**, 274–84.

COOMBS J. & VOLCANI B.E. (1968a) Studies on the biochemistry and fine structure of silica shell formation in diatoms. Silicon-induced metabolic transients in *Navicula pelliculosa* (Bréb) Hilse. *Planta*, **80**, 264–79.

COOMBS J. & VOLCANI B.E. (1968b) Studies on the biochemistry and fine structure of silica shell formation in diatoms. Chemical changes in the wall of *Navicula pelliculosa* during its formation. *Planta*, **82**, 280–92.

CRENSHAW M.A. (1964) Coccolith formation by two marine coccolithophorids, *Coccolithus huxleyi* and *Hymenomonas* sp. (Doctoral thesis, Duke University, Durham, N.C.).

DARLEY W.M. (1969a) Silicon requirements for growth and macromolecular synthesis in synchronized cultures of the diatoms, *Navicula pelliculosa* (Bréb) Hilse and *Cylindrotheca fusiformis* Reimann and Lewin. (Doctoral thesis, University of California, San Diego, California.)

DARLEY W.M. (1969b) Silicon and the division cycle of the diatoms *Navicula pelliculosa* and *Cylindrotheca fusiformis*. *Proc. N. Amer. Paleontological Conv.*, Sept. 1969. pp. 994–1009.

DARLEY W.M. & VOLCANI B.E. (1969) A silicon requirement for deoxyribonucleic acid synthesis in the diatom *Cylindrotheca fusiformis* Reimann and Lewin. *Expl. Cell Res.* **58**, 334–42.

DARLEY W.M. & VOLCANI B.E. (1971) Synchronized cultures: Diatoms. In *Methods in Enzymology*, XXIII A, ed. San Pietro A., 85–96. Academic Press, New York & London.

DRUM R.W. & PANKRATZ H.S. (1964) Post mitotic fine structure of *Gomphonema parvulum*. *J. Ultrastruct. Res.* **10**, 217–23.

EPPLEY R.W., HOLMES R.W. & PAASCHE E. (1967) Periodicity in cell division and physiological behavior of *Ditylum brightwellii*, a marine planktonic diatom, during growth in light-dark cycles. *Arch. Mikrobiol.* **56**, 305–23.

EPPLEY R.W., HOLMES R.W. & STRICKLAND J.D.H. (1964) Sinking rates of marine phytoplankton measured with a fluorometer. *J. expl. mar. Biol. Ecol.* **1**, 191–208.

FRANKE W.W. & BROWN R.M. JR. (1971) Scale formation in Chrysophycean algae. III. Negatively stained scales of the coccolithophorid *Hymenomonas*. *Arch. Mikrobiol.* **77**, 12–19.

GAWLIK S.R. & MILLINGTON W.F. (1969) Pattern formation and the fine structure of the developing cell wall in colonies of *Pediastrum boryanum*. *Am. J. Bot.* **56**, 1084–93.

GEITLER L. (1963) Alle Schalenbildungen der Diatomeen treten als Folge von Zell- oder Kernteilungen auf. *Ber. dt. bot. Ges.* **75**, 393–6.

GEITLER L. (1969) Die Lage der Raphen in den Zellen von *Nitzschia*-Arten. *Ber. dt. bot. Ges.* **81**, 411–13.

GEITLER L. (1970) Die Entstehung der Innerschalen von *Amphiprora paludosa* unter acytokinetischer Mitose. *Österr. Bot. Z.* **118**, 591–6.

GLOCKLING F. (1969) *The Chemistry of Germanium*, 234 pp. Academic Press, New York & London.

GOREAU T.F. (1963) Calcium carbonate deposition by coralline algae and corals in relation to their roles as reef-builders. *Ann. N.Y. Acad. Sci.* **109**, 127–67.

GREEN J.C. & JENNINGS D.H. (1967) A physical and chemical investigation of the scales produced by the Golgi apparatus within and found on the surface of the cells of *Chrysochromulina chiton* Parke et Manton. *J. exp. Bot.* **18**, 359–70.

HEALEY F.P., COOMBS J. & VOLCANI B.E. (1967) Changes in pigment content of the diatom *Navicula pelliculosa* (Bréb) Hilse in silicon-starvation synchrony. *Arch. Mikrobiol.* **59**, 131–42.

HELMCKE J.G. & KRIEGER W. (1962–1966) *Diatomeenschalen im Elektronenmikroskipischen Bild*. J. Cramer, Weinheim,

HEMMINGSEN B.B. (1971) A mono-silicic acid stimulated adenosinetriphosphatase from protoplasts of the apochlorotic diatom, *Nitzschia alba*. (Doctoral thesis, University of California, San Diego, California.)

HOLMES R.W. (1967) Auxospore formation in two marine clones of the diatom genus *Coscinodiscus*. *Am. J. Bot.* **54**, 163–8.

HOLMES R.W. & REIMANN B.E.F. (1966) Variation in valve morphology during the life cycle of the marine diatom *Coscinodiscus concinnus* W. Smith. *Nature, Lond.* **209**, 217–18.

IKEMORI M. (1970) Relation of calcium uptake to photosynthetic activity as a factor controlling calcification in marine algae. *Bot. Mag., Tokyo*, **83**, 152–62. (In Japanese.)

ISENBERG H.D., DOUGLAS S.D., LAVINE L.S., SPICER S.S. & WEISSFELLNER H. (1966) A protozoan model of hard tissue formation. *Ann. N.Y. Acad. Sci.* **136**, 155–90.

ISENBERG H.D., LAVINE L.S. & WEISSFELLNER H. (1963) The suppression of mineralization in a coccolithophorid by an inhibitor of carbonic anhydrase. *J. Protozool.* **10**, 477–9.

ISENBERG H.D., LAVINE L.S., WEISSFELLNER H. & SPOTNITZ A. (1965) The influence of age and heterotrophic nutrition on Ca deposition in a marine coccolithophorid protozoan. *Trans. N.Y. Acad. Sci. Ser.* 2, **27**, 530–45.

KATES M. & VOLCANI B.E. (1968) Studies on the biochemistry and fine structure of silica shell formation in diatoms. Lipid components of the cell walls. *Z. PflPhysiol.* **60**, 19–29.

KLAVENESS D. & PAASCHE E. (1971) Two different *Coccolithus huxleyi* cell types incapable of coccolith formation. *Arch. Mikrobiol.* **75**, 382–5.

LAURITIS J.A., COOMBS J. & VOLCANI B.E. (1968) Studies on the biochemistry and fine structure of silica shell formation in diatoms. IV. Fine structure of the apochlorotic diatom *Nitzschia alba* Lewin and Lewin. *Arch. Mikrobiol.* **62**, 1–16.

LAURITIS J.A., HEMMINGSEN B.B. & VOLCANI B.E. (1967) Propagation of *Hantzschia* sp. Grunow daughter cells by *Nitzschia alba* Lewin and Lewin. *J. Phycol.* **3**, 236–7.

LEADBEATER B.S.C. (1970) Preliminary observations on differences of scale morphology at various stages in the life cycle of '*Apistonema-Syracosphaera*' *sensu* von Stosch. *Br. Phycol. Bull.* **5**, 57–69.

LEADBEATER B.S.C. (1971) Observations on the life history of the haptophycean alga *Pleurochrysis scherffelii* with special reference to the microanatomy of the different types of motile cells. *Ann. Bot.* **35**, 429–39.

LEWIN J.C. (1954) Silicon metabolism in diatoms. I. Evidence for the role of reduced sulfur compounds in silicon utilization. *J. gen. Physiol.* **37**, 589–99.

LEWIN J.C. (1955a) Silicon metabolism in diatoms. II. Sources of silicon for growth of *Navicula pelliculosa*. *Pl. Physiol.* **30**, 129–34.

LEWIN J.C. (1955b) Silicon metabolism in diatoms. III. Respiration and silicon uptake in *Navicula pelliculosa*. *J. gen. Physiol.* **39**, 1–10.

LEWIN J.C. (1961) The dissolution of silica from diatom walls. *Geochim. Cosmochim. Acta*, **21**, 182–98.

LEWIN J.C. (1962a) Silicification. In *The physiology and biochemistry of algae*, ed. Lewin R.A. pp. 445–55. Academic Press, New York & London.

LEWIN J.C. (1962b) Calcification. In *The physiology and biochemistry of algae*, ed. Lewin R.A. pp. 457–65. Academic Press, New York & London.

LEWIN J.C. (1966) Silicon metabolism in diatoms. V. Germanium dioxide, a specific inhibitor of diatom growth. *Phycologia* **6**, 1–12.

LEWIN J.C. & CHEN C-H. (1968) Silicon metabolism in diatoms. VI. Silicic acid uptake by a colorless marine diatom, *Nitzschia alba* Lewin and Lewin. *J. Phycol.* **4**, 161–6.

LEWIN J.C. & GUILLARD R.R.L. (1963) Diatoms. *A. Rev. Microbiol.* **17**, 373–414.

LEWIN J.C., LEWIN R.A. & PHILPOTT D.E. (1958) Observations on *Phaeodactylum tricornutum*. *J. gen. Microbiol.* **18**, 418–26.

LEWIN J.C. & REIMANN B.E. (1969) Silicon and plant growth. *A. Rev. Pl. Physiol.* **20**, 289–304.

LEWIN J.C., REIMANN B.E., BUSBY W.F. & VOLCANI B.E. (1966) Silica shell formation in synchronously dividing diatoms. In *Cell synchrony-Studies in biosynthetic regulation*, eds. Cameron I.L. & Padilla G.M. pp. 169–88. Academic Press, New York & London.

MANTON I. (1966) Observations on scale production in *Prymnesium parvum. J. Cell Sci.* **1**, 375–80.

MANTON I. (1967a) Further observations on the fine structure of *Chrysochromulina chiton* with special reference to the haptonema, 'peculiar' Golgi structure and scale production. *J. Cell Sci.* **2**, 265–72.

MANTON I. (1967b) Further observations on scale formation in *Chrysochromulina chiton. J. Cell Sci.* **2**, 411–18.

MANTON I. & LEEDALE G.F. (1963) Observations on the microanatomy of *Crystallolithus hyalinus* Gaarder and Markali. *Arch. Mikrobiol.* **47**, 115–36.

MANTON I. & LEEDALE G.F. (1969) Observations on the microanatomy of *Coccolithus pelagicus* and *Cricosphaera carterae*, with special reference to the origin and nature of coccoliths and scales. *J. mar. Biol. Ass. U.K.* **49**, 1–16.

MANTON I. & PETERFI L.S. (1969) Observations on the fine structure of coccoliths, scales, and the protoplast of a freshwater coccolithophorid, *Hymenomonas roseola* Stein, with supplementary observations on the protoplast of *Cricosphaera carterae. Proc. R. Soc. B.* **172**, 1–15.

MCCONNELL D. & COLINVAUX L.H. (1967) Aragonite in *Halimeda* and *Tydemania* (order Siphonales). *J. Phycol.* **3**, 198–200.

MCINTYRE A. & BÉ A.W.H. (1967) Modern coccolithophoridae of the Atlantic Ocean. I. Placoliths and cystoliths. *Deep Sea Res.* **14**, 561–97.

MILLINGTON W.F. & GAWLIK S.R. (1967) Silica in the wall of *Pediastrum. Nature, Lond.* **216**, 68.

MILLINGTON W.F. & GAWLIK S.R. (1970) Ultrastructure and initiation of wall pattern in *Pediastrum boryanum. Am. J. Bot.* **57**, 552–61.

NAKAJIMA T. & VOLCANI B.E. (1969) 3,4-dihydroxy-proline: a new amino acid in diatom cell walls. *Science, N.Y.* **164**, 1400–1.

NAKAJIMA T. & VOLCANI B.E. (1970) ε-N-trimethyl-L-δ-hydroxylysine phosphate and its nonphosphorylated compound in diatom cell walls. *Biochem. biophys. Res. Commun.* **39**, 28–33.

OEY J.L. & SCHNEPF E. (1970) Über die Auslösung der Valvenbildung bei der Diatomee *Cyclotella cryptica. Arch. Mikrobiol.* **71**, 199–213.

OKAZAKI M., IKAWA T., FURUYA K., NISIZAWA K. & MIWA T. (1970) Studies on calcium carbonate deposition of a calcareous red alga *Serraticardia maxima. Bot. Mag., Tokyo,* **83**, 193–201. (In Japanese.)

OUTKA D.E. & WILLIAMS D.C. (1971) Sequential coccolith morphogenesis in *Hymenomonas carterae. J. Protozool.* **18**, 285–97.

PAASCHE E. (1963) The adaptation of the carbon-14 method for the measurement of coccolith production in *Coccolithus huxleyi. Physiologia Pl.* **16**, 186–200.

PAASCHE E. (1964) A tracer study of the inorganic carbon uptake during coccolith formation and photosynthesis in the coccolithophorid *Coccolithus huxleyi. Physiologia Pl. Suppl.* **3**, 1–82.

PAASCHE E. (1965) The effect of 3-(p-chlorophenyl)-1, 1-dimethylurea (CMU) on photosynthesis and light-dependent coccolith formation in *Coccolithus huxleyi. Physiologia Pl.* **18**, 138–45.

PAASCHE E. (1966a) Adjustment to light and dark rates of coccolith formation. *Physiologia Pl.* **19**, 271–8.

PAASCHE E. (1966b) Action spectrum of coccolith formation. *Physiologia Pl.* **19**, 770–9

PAASCHE E. (1967) Marine plankton algae grown with light-dark cycles. I. *Coccolithus huxleyi. Physiologia Pl.* **20**, 946–56.

PAASCHE E. (1968a) Biology and physiology of coccolithophorids. *A. Rev. Microbiol.* **22**, 71–86.

PAASCHE E. (1968b) The effect of temperature, light intensity and photoperiod on coccolith formation. *Limnol. Oceanogr.* **13**, 178–81.

PAASCHE E. (1969) Light-dependent coccolith formation in the two forms of *Coccolithus pelagicus. Arch. Mikrobiol.* **67**, 199–208.

PAASCHE E. & KLAVENESS D. (1970) A physiological comparison of coccolith-forming and naked cells of *Coccolithus huxleyi. Arch. Mikrobiol.* **73**, 143–52.

PARKER B.C. (1969) Occurrence of silica in brown and green algae. *Can. J. Bot.* **47**, 537–40.

PAUTARD F.G.E. (1970) Calcification in unicellular organisms. In *Biological calcification*, ed. Schraer H. pp. 105–201. Appleton-Century-Crofts, New York.

PEARSE V.B. (1972) Radioisotopic study of calcification in the articulated coralline alga *Bossiella orbigniana. J. Phycol.* **8**, 88–97.

PEARSE V.B. & MUSCATINE L. (1971) Role of symbiotic algae (zooxanthellae) in coral calcification. *Biol. Bull. mar. biol. Lab., Woods Hole*, **141**, 350–63.

PERKINS R.D., MCKENZIE M.D. & BLACKWELDER P.L. (1972) Aragonite crystals with Codiacean algae: distinctive morphology and sedimentary implications. *Science, N.Y.* **175**, 624–6.

PICKETT-HEAPS J.D. & FOWKE L.C. (1970) Mitosis, cytokinesis and cell elongation in the desmid, *Closterium littorale. J. Phycol.* **6**, 189–215.

PIENAAR R.N. (1969a) The fine structure of *Cricosphaera carterae*. I. External morphology. *J. Cell Sci.* **4**, 561–7.

PIENAAR R.N. (1969b) The fine structure of *Hymenomonas (Cricosphaera) carterae*. II. Observations on scale and coccolith production. *J. Phycol.* **5**, 321–31.

PIENAAR R.N. (1971) Coccolith production in *Hymenomonas carterae. Protoplasma*, **73**, 217–24.

REIMANN B.E.F. (1964) Deposition of silica inside a diatom cell. *Expl. Cell Res.* **34**, 605–8.

REIMANN B.E.F., LEWIN J.C. & VOLCANI B.E. (1965) Studies on the biochemistry and fine structure of silica shell formation in diatoms. I. The structure of the cell wall of *Cylindrotheca fusiformis* Reimann and Lewin. *J. Cell Biol.* **24**, 39–55.

REIMANN B.E.F., LEWIN J.C. & VOLCANI B.E. (1966) Studies on the biochemistry and fine structure of silica shell formation in diatoms. II. The structure of the cell wall of *Navicula pelliculosa* (Bréb) Hilse. *J. Phycol.* **2**, 74–84.

REIMANN B.E.F. & VOLCANI B.E. (1968) Studies on the biochemistry and fine structure of silica shell formation in diatoms. III. The structure of the cell wall of *Phaeodactylum tricornutum* Bohlin. *J. Ultrastruct. Res.* **21**, 182–93.

ROUND F.E. (1970) The genus *Hantzschia* with particular reference to *H. virgata* v. *intermedia* (Grun.) comb. nov. *Ann. Bot.* **34**, 75–91.

ROUND F.E. (1971) Observations on girdle bands during cell division in the diatom *Stephanodiscus. Br. Phycol. J.* **6**, 135–43.

SCHNEPF E. & DEICHGRÄBER G. (1969) Über die Feinstruktur von *Synura petersenii* unter besonderer Berücksichtigung der Morphogenese ihrer Kieselschuppen. *Protoplasma*, **68**, 85–106.

SCHULTZ M.E. (1971) Salinity-related polymorphism in the brackish-water diatom *Cyclotella cryptica. Can. J. Bot.* **49**, 1285–9.

SIEVER R. & SCOTT R.A. (1963) Organic geochemistry of silica. In *Organic geochemistry*, ed. Breger I.A. pp. 579–95. Macmillan Co. New York.

STARK L.M., ALMODOVAR L. & KRAUSS R.W. (1969) Factors affecting the rate of calcification in *Halimeda opuntia* (L.) Lamouroux and *Halimeda discoidea* Decaisne. *J. Phycol.* **5**, 305–12.

STEEMANN NIELSEN E. (1966) The uptake of free $CO_2$ and $HCO_3^-$ during photosynthesis of plankton algae with special reference to the coccolithophorid *Coccolithus huxleyi*. *Physiologia Pl.* **19**, 232–40.

STOERMER E.F. (1967) Polymorphism in *Mastogloia*. *J. Phycol.* **3**, 73–7.

STOERMER E.F., PANKRATZ H.S. & BOWEN C.C. (1965) Fine structure of the diatom *Amphipleura pellucida*. II. Cytoplasmic fine structure and frustule formation. *Am. J. Bot.* **52**, 1067–78.

SULLIVAN C.W. (1971) A silicic acid requirement for DNA polymerase, thymidylate kinase and DNA synthesis in the marine diatom *Cylindrotheca fusiformis*. (Doctoral thesis, University of California, San Diego, California.)

SWIFT E., 5TH & TAYLOR W.R. (1966) The effect of pH on the division rate of the coccolithophorid *Cricosphaera elongata*. *J. Phycol.* **2**, 121–5.

VAN VALKENBURG S.D. (1971a) Observations on the fine structure of *Dictyocha fibula* Ehrenberg. I. The skeleton. *J. Phycol.* **7**, 113–18.

VAN VALKENBURG S.D. (1971b) Observations on the fine structure of *Dictyocha fibula* Ehrenberg. II. The protoplast. *J. Phycol.* **7**, 118–32.

VON STOSCH H.A. & KOWALLIK K. (1969) Der von L. Geitler aufgestellte Satz über die Notwendigkeit einer Mitose für jede Schalenbildung von Diatomeen. Beobachtungen über die Reichweite und Überlegungen zu seiner zellmechanischen Bedeutung. *Österr. Bot. Z.* **116**, 454–74.

WALL D., GUILLARD R.R.L., DALE B., SWIFT E. & WATABE N. (1970) Calcitic resting cysts in *Peridinium trochoideum* (Stein) Lemmermann, an autotrophic marine dinoflagellate. *Phycologia*, **9**, 151–6.

WATABE N. (1967) Crystallographic analysis of the coccolith of *Coccolithus huxleyi*. *Calc. Tiss. Res.* **1**, 114–21.

WATABE N. & WILBUR K.M. (1966) Effects of temperature on growth and calcification and coccolith form in *Coccolithus huxleyi* (Coccolithineae). *Limnol. Oceanogr.* **11**, 567–75.

WERNER D. (1966) Die Kieselsäure im Stoffwechsel von *Cyclotella cryptica* Reimann, Lewin and Guillard. *Arch. Mikrobiol.* **55**, 278–308.

WERNER D. (1967) Hemmung der Chlorophyllsynthese und der NADP$^+$-abhängigen Glycerinaldehyd-3-Phosphat-Dehydrogenase durch Germaniumsäure bei *Cyclotella cryptica*. *Arch. Mikrobiol.* **57**, 51–60.

WERNER D. (1968) Stoffwechselregulation durch den Zellwandbaustein Kieselsäure: Poolgrössenänderungen von α-Ketoglutarsäure, Aminosäuren und Nucleosidphosphaten. *Z. Naturf.* **23b**, 268–72.

WERNER D. (1969a) Beiträge zur Physiologie und Biochemie der Kieselsäure. *Ber. Deutsch. Bot. Ges.* **81**, 425–9.

WERNER D. (1969b) Silicoborate als erste nicht C-haltige Wachstumsfaktoren. *Arch. Mikrobiol.* **65**, 258–74.

WERNER D. & PIRSON A. (1967) Über reversible Speicherung von Kieselsäure in *Cyclotella cryptica*. *Arch. Mikrobiol.* **57**, 43–50.

WILBUR K.M., COLINVAUX L.H. & WATABE N. (1969) Electron microscope study of calcification in the alga *Halimeda* (order Siphonales). *Phycologia*, **8**, 27–35.

WILBUR K.M. & WATABE N. (1963) Experimental studies on calcification in molluscs and the alga *Coccolithus buxleyi*. *Ann. N.Y. Acad. Sci.* **109**, 82–112.

WILBUR K.M. & WATABE N. (1967) Mechanisms of calcium carbonate deposition in coccolithophorids and molluscs. *Stud. Trop. Oceanogr. Miami*, **5**, 133–54.

# CHAPTER 25

# ION UPTAKE

## E. A. C. MACROBBIE

Botany School,
University of Cambridge,
Cambridge, U.K.

1    **Introduction**  676
1.1  General introduction  676
1.2  Types of ion movement: definitions  677
1.3  Experimental procedure  682

2    **Giant algal cells**  683
2.1  Fresh and brackish water species  683
   (a) Flux measurements: K, Na and Cl  683
   (b) Active fluxes: energy coupling  687
   (c) Other ion pumps  688
   (d) Non-active fluxes: K, Na and Cl  691
   (e) Electrical properties  692
   (f) Vacuolar fluxes  693

2.2  Marine species  697
   (a) *Griffithsia*  697
   (b) *Acetabularia*  701
   (c) *Chaetomorpha darwinii*  703
   (d) *Valonia ventricosa*  704
   (e) *Halicystis*  705
   (f) *Valoniopsis*  706

3    **Small-celled algae**  706

4    **References**  708

## 1 INTRODUCTION

### 1.1 *General introduction*

The part that algal work has played in the development of ion transport studies in plant cells is made clear by the fact that it is possible to write a chapter under this title in a book on the algae, with only restricted reference to other plant groups, whereas the converse, a chapter with only restricted reference to the algae and emphasis on other plant groups, would omit most of the definitive experimental work. The relevance of algal work to the ionic relations of higher plant cells is sometimes questioned, but the fact remains that, up to the present, work in which it has proved experimentally feasible to study the mechanism of ion transport processes by a series of well-defined specific questions, has largely been confined to the algae. This is only in part due to the existence of giant cells in various algal groups; it arises also from the fact that experimental conditions for labelling most algae with radioactive ions present less disturbance to the natural state than is involved in comparable experiments with most higher plant tissues. Thus, although the predominant role played by the algal work of Osterhout, Blinks and others in the early days of the subject (1920–40) was followed by a period of reduced interest in algae, the development of current ideas of ion transport, made possible by modern experimental techniques, has again been largely based on algal work.

This chapter aims first to set out the theoretical framework to which most basic studies of the mechanism of ion transport are related, presenting the specific questions to which answers are required. It then aims to review the experimental work in which answers to these specific questions have been sought, and to draw what conclusions we can about the nature of specific ion transport processes in specific algal cells. The discussion is restricted to those systems in which the problem can be reduced to that of movement of a given ion across a specified membrane under measurable driving forces, or can be said to approach this ideal. Because of this restriction the emphasis will necessarily be on giant algal cells, whether freshwater or marine, since the experimental difficulties in extending this approach to cells of ordinary size are considerable.

Previous general reviews of ion transport in plant cells, with, of necessity, considerable emphasis on algal cell work, are given by Dainty (1962, 1969), Schilde (1968a), Lüttge (1969), Pitman (1970), MacRobbie (1970a, 1971a), and Bentrup (1971); a thorough review of the ionic relations of marine algae is provided by Gutknecht and Dainty (1968). The theoretical background and its application to a wide range of plant and animal cells is treated in a recent book (Hope 1971).

## 1.2 *Type of ion movement: definitions*

In practice, the identification of specific ion fluxes as passive diffusion, exchange diffusion, or active transport, is done by a process of elimination. We start by establishing whether the flux behaviour has the characteristics we expect of a simple diffusion process, and the terms 'exchange diffusion' and 'active transport' arise from various experimentally observed discrepancies from the expected behaviour.

The driving force for passive diffusion of an ion is the difference of electrochemical potential ($\Delta\bar{\mu}$) between two points along the path; that is, the force has two components, one arising from the difference in concentration $\Delta C$ (or more correctly activity), and the other from the difference in electrical potential, $\Delta E$. The simplest assumption is that this is the only force acting on the ions, and that the net flux ($J$) of an ion of valency $z$ is therefore given by the equation (Teorell 1953):

$$J = \frac{uC\,(-d\bar{\mu})}{dx} \tag{1a}$$

where $u$ is the ion mobility.

Since the electrochemical potential has the form $\bar{\mu} = \mu_0 + RT \ln C + zFE$ equation (1) reduces to the form:

$$J = \frac{-u.RT.dC}{dx} - \frac{u.C.zF.dE}{dx} \tag{1b}$$

This equation assumes that the ions move independently through the membrane, unaffected by any other forces, including frictional forces generated by interactions between moving components in the membrane. To use this equation to describe the net ion flux across a cell membrane, with external and internal concentrations $C_0$ and $C_i$ respectively, and a membrane potential $E_i$ (potential of the inside relative to the outside), it must be integrated across the membrane. The simplest integration makes a further assumption that the electric potential gradient is linear within the membrane, and we arrive at the Goldman equation (Goldman 1943), in the form:

$$J = \frac{-P.\, zFE_i/RT}{1 - \exp zFE_i/RT} \left[ C_0 - C_i \exp \frac{zFE_i}{RT} \right] \tag{2}$$

where $P$, the ion permeability, is equal to $uRT.K/a$, with $K$ the partition coefficient for the ion in the membrane $C_0{}^m/C_0$, and $a$ equal to the membrane thickness.

This equation separates into a term depending on $C_0$, and one depending on $C_i$, taken to be the influx and efflux. Thus it is predicted that the fluxes should be:

$$\text{influx} = P.\, \frac{zFE_i/RT}{1 - \exp zFE_i/RT} \cdot C_0 \tag{3a}$$

$$\text{efflux} = \frac{P.\, (zFE_i/RT) \exp zFE_i/RT.\, C_i}{1 - \exp zFE_i/RT} \tag{3b}$$

$$\text{flux ratio} = \frac{\text{influx}}{\text{efflux}} = \frac{C_0 \exp (-zFE_i/RT)}{C_i} \tag{3c}$$

The last of these equations, first introduced by Ussing (1949) and Teorell (1949), provides the commonest test of whether a given ion is moving by simple passive diffusion only, independent of any other driving force. We see that it demands measurements of two-way ion fluxes across the membrane, of ion concentrations on each side of the membrane, and of the membrane potential. Non-independence and interaction of ions in the membrane may change the magnitude of the logarithm of the flux ratio but not its sign. If in fact we find a flux ratio in which the sign is wrong, then we must suppose that we have omitted the main driving force for the movement of this ion from equation (1). If there is an obvious flow component across the membrane we may then argue that the ion flux is coupled to that flow (for example, of water), and we may test this hypothesis by experimental manipulations of the flow. Irreversible thermodynamics allows for the coupling of the flow of one substance to that of another, the entrainment of two flows by non-zero cross coefficients in the Onsager transport equations, and we must therefore look for such instances. If we find no such obvious flow to which the ion flux could be coupled, then we identify a process of 'active transport', and we define such processes by this type of discrepancy.

In spite of the absence of a visible coupled flow of another substance across the membrane we envisage entrainment of two flows within the membrane. Since the ion flux is against the electrochemical gradient and involves an apparent increase in Gibb's free energy, the entraining flow must involve a greater decrease in free energy; it must therefore be in some way linked to an energy-yielding process of metabolism, and it must involve vectorial chemical reactions within the membrane, supplying the necessary asymmetry to the flow. It has been suggested that a useful definition of active transport is provided by the link with metabolism, the existence of a non-zero coupling coefficient between the ion flux and a metabolic reaction flow $J_r$ (Kedem 1961). This is in some instances a useful interim measure, but there are difficulties in making proper experimental tests in these terms (MacRobbie 1970a). Further, the indirectness of this coupling diverts attention from events in the membrane, where we are required to establish interactions between the ion and specific membrane components and reactants, and to consider their direct coupling coefficients—in principle even if not yet in practice.

Thus we start by considering the characteristics of passive, independent, diffusion of the ion and we predict behaviour of the type given by equations (1)–(3). We may test this by looking at the flux ratio, or by looking at the flux changes resulting either from changes in $C_0$, with possible consequent changes in $E_i$, or from changes in $E_i$, by voltage clamping.

If the results of one or more of these three tests suggest that movement by passive diffusion is an inadequate explanation, then we must extend the hypothesis. We may assume that, although $\Delta\bar{\mu}$ is still the most important driving force, there is interaction of ions moving in the membrane, and equation (1) is not an adequate flow equation; this might arise, for example, by the presence of long pores in the membrane. We may still have only passive diffusion, but the permeability is now a function of all the interacting constituents, and will change when any of them is changed; it is no longer a constant of the particular ion and the membrane, but involves also cross coefficients and entraining flows. We are however still considering only passive driving forces ($\Delta\bar{\mu}$ for each of the interacting ions, or of water), and movement as free ion. A consequence of this extension is that we no longer have any equations to test the hypothesis; equations (1)–(3) are no longer valid but we have as yet no substitute. We may look for qualitative dependence of the ion fluxes on $C_0$, $C_i$ and $E$, but this is as far as we can go.

There are two other ways in which the Goldman analysis is commonly used in the analysis of electrical changes in the membrane. The membrane potential arises from asymmetric distribution of ions across the membrane, and the restriction of the consequent ion movements by the condition of zero current flow across the membrane; the very large potentials set up by quite small charge transfers across the membrane ensure that the membrane potential will be set, under any set of conditions, to its value for zero net charge transfer. The asymmetric ion distributions may arise from active transport processes, or from the presence of metabolically-produced impermeant charged molecules inside the cell, giving a Donnan distribution. In the simplest analysis, of independent movement of ions in the membrane, the Goldman potential equation may be derived from equations (1)–(3), and the condition of zero net current. If we consider only monovalent ions (since cell membranes are much less permeable to divalent ions) the Goldman potential equation may be written, in terms of $Na^+$, $K^+$, $Cl^-$, and other charge-carrying species $X^+$ and $Y^-$:

$$E_i = \frac{RT}{F} \ln \frac{P_K K_o + P_{Na} Na_o + P_{Cl} Cl_i + P_X X^+_o + P_Y Y^-_i}{P_K K_i + P_{Na} Na_i + P_{Cl} Cl_o + P_X X^+_i + P_Y Y^-_o} \tag{4}$$

Since internal concentrations in many cells change only slowly after changes in external concentration, the usual experimental test consists of changing external concentrations and trying to fit the resultant potential changes to some combination of values of relative permeabilities $P_{Na}/P_K$ ($\alpha$), $P_{Cl}/P_K$ ($\beta$) etc. If there is an electrogenic (charge-carrying) pump for one or more ions, then the charge balance equation may be modified, by setting the net current carried by passive ion movements equal and opposite to the current carried by the pump (Briggs 1962, Moreton 1969). The potential equation thus derived is complex, but if the electrogenic pump contributes a major part of the membrane potential the equation may be simplified (Hansen & Gradmann 1971) to:

$$\frac{EF}{RT} = \frac{A}{P_K \bar{C}_o} \tag{5}$$

where A is the rate of electrogenic transport, and $\bar{C}_o = K_o + \alpha Na_o + \beta Cl_i + \ldots$, with other terms if necessary, as they appear in the top line of equation (4). This equation assumes a constant rate of active transport; if $A$ is a function of $E$ then a more complex solution is required. If we simply assume that the electrogenic transport is limited by the membrane potential, and that the carrier complex is removed at the receiving side of the membrane, then we may write an equation (3a) or (3b) for the pump flux in equation (5), and we may simply equate the active current with the net passive fluxes as before. For electrogenic anion influx (or $Cl^-$ say) or cation efflux (of $H^+$ say) again on the assumption that the electrogenic pump makes a major contribution to the membrane potential, we finish with an equation of the form:

$$\frac{EF}{RT} = \frac{\ln P_X \cdot C_X}{P_K \bar{C}_o} \tag{6}$$

Where $P_X$ is the membrane permeability to the charged complex, and $C_X$ is the concentration of the complex at the side of the membrane where it forms.

This equation implies that the membrane potential in the presence of the electrogenic pump is displaced from the membrane potential in its absence (given by equation (4)), by a constant amount; its dependence on $K_0$, for example, will have the same form as in the absence of an electrogenic pump. Such behaviour is observed in *Acetabularia* (see later). In any event we recognize the contribution of an electrogenic pump to the membrane potential by immediate changes in potential and resistance on inhibition of the active transport system, in a time too short for appreciable changes of internal concentrations.

Since membrane conductance arises from the movement of all the charge-carrying species in the membrane under an electrical driving force, we may also hope to get some information on ion movements from resistance measurements. In the interpretation of either potential or resistance measurements the same problem arises, that we must consider *all* the charge-carrying species, and we may be in difficulty in identifying or measuring them all. Controversy centres on the number and nature of ion species to be considered, the question of whether sodium and potassium are the only important charge-carrying species, or whether we must also include hydrogen ion and/or electrogenic ion pumps. The other difficulty is that if we have good reason to believe that ions interact in the membrane, even when moving passively, then we should not be using the Goldman equations, and values of $P_K$, $P_{Na}$ etc. derived from equations (3) and (4) have little value.

Electrical measurements do provide useful qualitative information, and are particularly valuable in that rapid transients can be measured, whereas flux measurements are slow and insensitive. The aim must be to identify the particular ion or ions responsible for such transients, but it is sometimes difficult to provide clear-cut identifications.

In many instances however it is clear from the experimental results that further extensions of the original postulates are necessary, in that particular ion fluxes may change markedly as a result of only very small changes in $C_0$ or $C_i$ or $E_i$, and hence of $\Delta\bar{\mu}$, or that net ion fluxes may be observed against the electrochemical gradient. These deviations are taken to reflect chemical interactions of the ion with components of the membrane, and movement of the ion bound or associated with some other chemical constituent, with a new driving force determined primarily by the association. The chemical association may have no vectorial component, and take place equally at both sides of the membrane; we then have carrier-mediated diffusion and the ion flux will show characteristics of the chemical reaction. It may be strongly temperature-dependent and it may show Michaelis-Menten type kinetics, implying the interaction of the ion with a membrane constituent present in very small amounts. If the carrier can only cross the membrane in association with the ion then we have a process of exchange diffusion, a 1:1 ion exchange process which does not achieve any net transfer across the membrane; this is recognized by the very strong dependence of the ion flux on the presence of the ion in the solution to which it is being delivered, usually by efflux measurements.

If on the other hand there is asymmetry in the chemical reaction, by the supply of a substrate from one side of the membrane only, then the process will achieve a net flux which is not governed by the electrochemical activity gradient of the ion concerned, and we have a process of active transport. This will only be recognized as such when the asymmetry imposed by the chemistry is against the electrochemical activity gradient of the ion. The process may involve direct chemical binding to a carrier molecule, and movement of a carrier complex, Na X etc., or it could involve a looser association. For example if we had charged pores in the membrane in which $K^+$ was more strongly adsorbed by negative charges on the wall than was $Na^+$, and we had a flow of water maintained through the pore as a result of a chemical reaction at its inner end, we would achieve some separation of $Na^+$ and $K^+$ across the membrane; the outward driving force on $Na^+$, predominantly in the centre of the pore, would be enhanced, whereas $K^+$ would be much less affected, and indeed the streaming potential would provide an inward driving force on $K^+$.

The process of asymmetric, chemically-mediated ion movement is primary active transport, but there may be secondary consequences of such asymmetry. There may be other ion fluxes coupled to the primary active transport, which may be referred to as 'secondary' active transport. The coupling may involve frictional interactions with the species involved in the primary transport, or alterations in the driving forces across the membrane for other ions by a contribution to the electrical gradient. If an ion-carrier complex, or the return form of the carrier, is charged then there is a net charge transfer involved in the operation of the transport process, and the pump is electrogenic; the net ion fluxes of the secondary movements must balance the primary rate of charge transfer. The best examples of such electrogenic pumps are the vectorial redox systems postulated to exist in membranes of mitochondria, chloroplasts and prokaryotic cells by Mitchell (1961, 1962, 1966) in his theory of chemi-osmotic coupling, but such mechanisms may well be of wider distribution. In this instance the primary movement is of hydrogen ion, by the reduction of a hydrogen carrier at one side of the membrane and its oxidation by an electron carrier at the opposite side of the membrane; as a consequence of this primary charge transfer an electric potential will be set up, with consequent effects on any charged species which is capable of moving through the membrane.

### 1.3  *Experimental procedure*

The study of ion transport processes in a given cell follows certain well-defined stages, in which answers are sought to certain specific questions, with the final aim of describing and understanding the ion traffic across the cell membranes, including the intracellular boundaries. The first stage is descriptive, the measurement of the two-way ion fluxes across the membrane and the passive driving forces, preferably under some range of experimental conditions; the aim is to identify processes of active transport, or of chemically-mediated transport, of specific ions. Study of these processes then follows, particularly of the active transport processes which need a link with metabolism. The aim is to identify the specific energy-yielding sequence to which the flux is linked, and the means of coupling of flux and exergonic reaction sequence, the agent by which energy is transferred to the pump. The final questions concern the mechanism of the transfer process across the membrane, but in most instances we are some way off a detailed description of molecular events in the membrane.

Since these measurements demand access to the cytoplasm as a separate phase (and indeed strictly demand access to *each* membrane-bound cytoplasmic phase), most experimental work aimed at the mechanisms of ion transport has been based on the giant algal cells. This chapter will deal with such work, on freshwater species and on marine species, before considering the inevitably less well-defined studies on more normal cells.

## 2 GIANT ALGAL CELLS

### 2.1 *Fresh and brackish water species*

#### (a) *Flux measurements: K, Na and Cl*

Ion concentrations and potentials in a number of species have been measured, and extensive flux measurements are available on four species—*Nitella translucens*, *Chara corallina* (formerly *australis*), *Tolypella intricata* and *Hydrodictyon africanum*. Other species in which tracer flux measurements have been made include *Nitella clavata*, *Nitellopsis obtusa*, and *Chara globularis*.

Table 25.1 gives figures for ion concentrations and potentials, and Table 25.2 lists the fluxes for which an active component is presumed, from evidence of the maintenance of a non-equilibrium distribution of ions in the face of measurable fluxes. Since cytoplasmic concentrations and potentials are not available for all species the information on localization of the pumps is incomplete; in the table a negative sign is given where measurements show a flux which could be purely passive, and blanks are left where we have no information. Mechanisms for active extrusion of sodium and active uptake of chloride are common to all cells for which we have figures; in some cells there is also active influx of potassium but in others potassium is close to its equilibrium value. The high K/Na and high Cl concentration are both developed in the cytoplasm, and the plasmalemma is therefore the primary site of the active fluxes by which the cellular ion concentrations are regulated. It is not, however, the only site of active ion transport, and there is evidence of transport of both anions and cations from cytoplasm to vacuole in several species. It is also clear that the cytoplasm is a heterogeneous phase, and that concentrations of both anions and cations in the chloroplasts are much higher than those in the flowing cytoplasm. However, in the absence of any figure for the potential across the chloroplast membrane(s) we cannot identify primary active transport of either anions or cations. Thus we

**Table 25.1.** Ion concentrations and potentials found in various algae

|  | Concentration (mM) | | |
|---|---|---|---|
|  | K | Na | Cl |
| *Nitella translucens* $E_{co} - 140mV$, $E_{vc} + 18mV$ (3) |  |  |  |
| outside | 0·1 | 1·0 | 1·3 |
| vacuole (1,2,3) | 75 | 65 | 150–170 |
| flowing cytoplasm F (3) | 119 | 14·0 | 65 |
| F (4) |  |  | 87 |
| cytoplasm including chloroplasts C+F (2) | 150 | 55 | 240 |
| *Nitella flexilis* $E_{co} - 170mV$, $E_{vc} + 15mV$ (5) |  |  |  |
| outside | 0·1 | 0·2 | 1·3 |
| vacuole (5) | 80 | 28 | 136 |
| flowing cytoplasm F | 125 | 5 | 36 |
| chloroplast layer C | 110 | 26 | 136 |
| *Tolypella intricata* $E_{vo} - 140mV$ (6) |  |  |  |
| $- 120$ mV (7) |  |  |  |
| outside | 0·4 | 1·0 | 1·4 |
| vacuole (6) | 90–110 | 3–10 | 120 |
| (7) | 110–119 | 8–39 | 110–136 |
| flowing cytoplasm (7) | 87–97 | 4–22 | 23–31 |
| chloroplasts (7) | 340 | 36 | 340 |
| *Chara corallina* $E_{vo} - 159mV$(8) |  |  |  |
| $E_{co} - 170mV$, $E_{vc} + 18mV$(9) |  |  |  |
| outside | 0·1 | 1·0 | 1·6 |
| vacuole (8) | 66 | 54 | 100–150 |
| (9) | 48* |  | 84* |
| flowing cytoplasm (9) | 115* |  |  |
| vacuole (10) |  |  | 100* |
| flowing cytoplasm (10) |  |  | 10* |
| cytoplasm C+F (11) |  |  | 130 |
| *Hydrodictyon africanum* $E_{co} - 116mV$, $E_{vc} + 26mV$ (12) |  |  |  |
| outside | 0·1 | 1·0 | 1·3 |
| vacuole (12) | 40 | 17 | 38 |
| cytoplasm (12) | 93 | 51 | 58 |
| *Lamprothamnium succinctum* $E_{vo} - 100mV$ (13) |  |  |  |
| outside | 6 | 289 | 337 |
| vacuole (13) | 250 | 136 | 373 |
| cytoplasm C+F (13) | 137 | 47 | 86 |

Table 25.1 (cont.)

| | Concentration (mM) | | |
| | K | Na | Cl |
|---|---|---|---|
| *Nitellopsis obtusa* $E_{vo}-120mV$ (14) | | | |
| $E_{co}-141mV$, $E_{vc}+19mV$ (15) | | | |
| outside | 0·65 | 30 | 35 |
| vacuole (14) | 113 | 54 | 206 |
| outside | 1·0 | 24·5 | 32 |
| vacuole (15) | 95 | 60 | 174 |
| | | | |
| *Nitella clavata* $E_{vo}-106$ to $-120mV$ (16) | | | |
| outside | 0·1 | 3 | 3·1 |
| vacuole (16) | 75–83 | 34–36 | 120–124 |

(1) MacRobbie 1962, (2) MacRobbie 1964, (3) Spanswick & Williams 1964, (4) Hope, Simpson & Walker 1966, (5) Kishimoto & Tazawa 1965a, (6) Smith 1968a, (7) Larkum 1968, (8) Hope & Walker 1960, (9) Vorobiev 1967, (10) Coster 1966, (11) Coster & Hope 1968, (12) Raven 1967a, (13) Kishimoto & Tazawa 1965b, (14) MacRobbie & Dainty 1958, (15) Findlay 1970, (16) Barr & Broyer 1964.
* These values are determined by the use of intracellular electrodes specific for the ion in question and therefore represent activities rather than concentrations.

**Table 25.2.** Active fluxes in various algae

| Species | Fluxes presumed to have active components | | | | |
| | $M_{oc}{}^{K}$ | $M_{co}{}^{Na}$ | $M_{cv}{}^{Na}$ | $M_{oc}{}^{Cl}$ | $M_{cv}{}^{Cl}$ |
|---|---|---|---|---|---|
| *Nitella translucens* | + | + | + | + | + |
| *N. flexilis* | − | + | + | | + |
| *N. clavata** | + | + | | + | |
| *Tolypella intricata** | − | + | + | + | + |
| *Chara corallina* | − | + | | + | + |
| *Lamprothamnium succinctum** | | + | | + | |
| *Nitellopsis obtusa** | − | + | | + | |
| *Hydrodictyon africanum* | + | + | | + | |

$M_{oc}$ = flux from outside to cytoplasm; $M_{co}$ = flux from cytoplasm to outside; $M_{cv}$ = flux from cytoplasm to vacuole; $E_{co}$ = potential of cytoplasm with respect to outside; $E_{vc}$ = potential of vacuole with respect to cytoplasm.
* In these species only $E_{vo}$ has been measured and not the separate potentials at plasmalemma and tonoplast, but it has been assumed that the potential distribution is similar to that in other species. In *Tolypella* the inferred active fluxes at the tonoplast depend on this assumption.

have evidence for active transport at each of the three membranes for which the figures allow conclusions to be drawn, and we must expect the list to grow as observations are extended to the membranes of other organelles.

Most work has been done on plasmalemma fluxes, and the effects of environmental or metabolic factors on them, with the aim of classifying individual fluxes in the terms discussed earlier. In the usual pond water the plasmalemma fluxes are typically about 1 pmole cm$^{-2}$ sec$^{-1}$ for K, 0·5 pmole cm$^{-2}$ sec$^{-1}$ for Na, and 1–2 pmole cm$^{-2}$ sec$^{-1}$ for Cl. The pattern of pumps, and their metabolic dependence, seems to be very similar in *Nitella translucens* (MacRobbie 1962, 1964, 1965, 1966a, Smith 1967), and in *Hydrodictyon africanum* (Raven 1967a,b, 1968a), whereas *Chara corallina* (Coster & Hope 1968, Smith & West 1969, Findlay *et al.* 1969a) differs in some respects. In *Nitella* and in *Hydrodictyon* the following partition of the plasmalemma fluxes is suggested:

1. A cation exchange pump responsible for an active component of K influx, and an active component of Na efflux which depends on the presence of potassium in the solution. These amount in light to about 0·8 pmole cm$^{-2}$ sec$^{-1}$ for K and 0·5 pmole cm$^{-2}$ sec$^{-1}$ for Na in *Hydrodictyon*, and about three-quarters of this in *Nitella*. This pump is light- and temperature-sensitive, sensitive to ouabain (a specific inhibitor of coupled K/Na transport in animal cells), and sensitive to uncouplers. However, recent work (Raven 1971a) suggests that the ouabain-sensitive component of K influx is not present in all coenobia of *Hydrodictyon*; in about a quarter of experiments the inhibition of the K influx by ouabain was less than 25%, compared with the 25 to 80% inhibition found in the other three-quarters of experiments. Hence the active component of K influx is not a universal attribute of *Hydrodictyon*, and its magnitude seems to depend in some unspecified way on the culture conditions.

2. A (chloride + cations) system. This is responsible for the influx of chloride and for associated influxes of cations, although the chloride-stimulated fluxes of monovalent cations account for only 30 to 70% of the stimulation of chloride influx produced by the presence of monovalent cations in the solution. This transport system is responsible for the net salt uptake during growth; it is light- and temperature-sensitive but insensitive to low concentrations of uncouplers. No significant effect of ouabain is seen in short influx periods (MacRobbie 1962, Raven 1967a, 1971a) but Janacek and Rybova (1966) found that prolonged treatment of *Hydrodictyon* with ouabain produced large net influxes of potassium and chloride and considerable swelling. In three days the contents of potassium and chloride rose by 175 and 157 mmole (kg dry weight)$^{-1}$ respectively, although their concentrations remained unchanged; by contrast the contents of sodium and calcium remained unchanged, and hence their concentrations must have fallen. This effect is obviously distinct from immediate effects on the transport mechanisms, but may be related to the increased influx of potassium seen in some *Nitella* cells after prolonged ouabain treatment (MacRobbie 1962); the effect has, as yet, no explanation.

3. Residual components of cation influxes and effluxes; chloride efflux. These are the cation components which are insensitive to ouabain, and independent of external chloride.

In *Chara corallina* there is no evidence for an active component of potassium influx, and hence it is not surprising that this flux is not sensitive to ouabain;

however, sodium efflux, which has an active component, is also insensitive to ouabain (Findlay *et al.* 1969a). In some respects the chloride transport in *Chara* is similar to that in *Nitella* and *Hydrodictyon*, in that it is light-stimulated, and usually linked to monovalent cations (Findlay *et al.* 1969a); however, it differs in that it shows sensitivity to uncouplers (Coster & Hope 1968, Smith & West 1969).

## (b) *Active fluxes: energy coupling*

The cation transport system is sensitive to uncouplers, but can be supported in far-red light when only cyclic photophosphorylation is possible (MacRobbie 1965, 1966a, Raven 1967b). It shows an increased quantum efficiency in the far-red (red rise), suggesting it can be driven by photosystem I alone (Raven 1967b); this is confirmed by comparison of its action spectrum with that of $CO_2$ fixation (Raven 1969a). Hence the Na/K exchange pump may have ATP as its energy source which may be supplied by non-cyclic or cyclic photophosphorylation, or by oxidative phosphorylation. This pump may be ouabain-sensitive membrane-bound ATPase.

The contributions of the two forms of photophosphorylation to the cation transport in *Hydrodictyon* are assessed by Raven (1971b) in a detailed study of the effects of $CO_2$ on K influx under conditions when either non-cyclic or cyclic phosphorylation should be inhibited. He finds no effect of $CO_2$ on K influx in anaerobic light conditions, suggesting that either form of electron flow is capable of supporting K influx. However, in the presence of antimycin, thought to be a selective inhibitor of cyclic photophosphorylation, K influx is stimulated by $CO_2$, i.e. by the provision of an electron acceptor for non-cyclic electron flow. This contrasts with the inhibitory effect of $CO_2$ on K influx in normal conditions but in low light, an inhibition which can be partially reversed by inhibitors of non-cyclic electron flow (Raven 1968b) and the consequent transition to cyclic flow; thus in low light there may be competition between K influx and $CO_2$ fixation for the available energy. Either cyclic or non-cyclic flow can provide ATP for cation transport, depending on the experimental conditions; in $CO_2$-free nitrogen only cyclic photophosphorylation is available, whereas in nitrogen plus $CO_2$ both cyclic and non-cyclic photophosphorylation probably contribute (since K influx is more sensitive to inhibitors of cyclic flow than is photosynthesis). The light-saturated rate of K influx is independent of the ATP source.

The insensitivity to uncouplers, and the fact that conditions in which only cyclic photophosphorylation is possible (in far-red light or in the presence of DCMU) do not support chloride influx, suggest that in *Nitella* and *Hydrodictyon* ATP is not the energy source for this transport system (MacRobbie 1965, 1966a, Raven 1967b). Raven's demonstration that chloride influx shows a red drop in its quantum efficiency, and has the same action spectrum as photosynthesis (1969a), in marked contrast to the results on K influx, suggests that chloride influx is independent of ATP.

The distinction between the energy sources for K and Cl influxes also emerges clearly from studies of the effects of inhibitors on ion fluxes and $CO_2$ fixation (Raven 1968b,c, 1969b, 1971c, Raven et al. 1969). Uncouplers (proton-translocating uncouplers or their equivalent, such as CCCP, DNP and desaspidin; but also quinacrine and arsenate), ethionine (as an ATP trap), and energy-transfer inhibitors (Dio-9, DCCD, phlorizin), all inhibit chloride influx much less than they inhibit $CO_2$ fixation or K influx. Hence there is strong evidence against ATP as the energy source for chloride transport in the light, and some other product or consequence of non-cyclic electron flow must be sought. In the dark it appears also that mitochondrial electron transport is required, or a product of it other than ATP (Raven 1969b, 1971c).

Chloride transport in Chara differs markedly from that in Nitella and Hydrodictyon in that it is sensitive to the uncoupler CCCP, and to the energy-transfer inhibitors, phlorizin and Dio-9 (Coster & Hope 1968, Smith & West 1969). We have no information on the relative efficiencies of far-red light and of shorter wavelength light, but this inhibition evidence suggests that chloride influx is dependent on ATP. In Chara foetida Penth and Weigl (1971) measured both chloride influx and ATP levels in light and dark, and in the presence of CCCP. They found no effect of light on the ATP level of the cell as a whole, although the chloride influx in the dark was only 50% of that in light. In light CCCP ($10^{-5}$ M) neither reduced the ATP level or the chloride influx, whereas in the dark CCCP reduced the ATP level by 30 to 50% and the chloride influx by a factor of 13. Again they argue for ATP as the energy source for chloride influx.

In Tolypella intricata chloride influx shows intermediate sensitivity to uncouplers, and Smith (1968a) suggested that the effect might be indirect; if CCCP reduced cation permeability then its effect on chloride influx might be similar to that of removing monovalent cations. The sensitivity of the chloride influx in Hydrodictyon to proton-translocating uncouplers was significantly increased by stirring (Raven 1971c), an effect which is not yet understood. However, the effect of stirring did not extend to other uncouplers (quinacrine, arsenate), or to energy-transfer inhibitors, or to ethionine, and hence the evidence still suggests that in Hydrodictyon the chloride influx is independent of ATP.

(c) *Other ion pumps*

Smith (1966) has given evidence of the active uptake of phosphate in Nitella translucens, by a mechanism distinct from that of chloride uptake. Thus the phosphate transport is much less light-sensitive than the fluxes so far discussed, and seems to derive its energy from ATP (the phosphate influx in the dark is about 60 to 70% of the value in light, compared with only about 5 to 10% for the fluxes so far discussed; the light-dependent phosphate influx can be supported in far-red light, is insensitive to low concentrations of DCMU, but sensitive to uncouplers).

Sulphate influx has been studied in a number of Characean species by Robinson (1969a,b), who finds a transient stimulation of the flux on transfer from light to dark. He argues that in the dark at least, the energy supply is ATP from oxidative phosphorylation. By contrast, Penth and Weigl (1971) find the sulphate influx in *Chara foetida* is 20% higher in light than in dark; they find marked inhibition by CCCP in the dark, but little effect in light.

The existence of active uptake of bicarbonate, and its metabolic connexions, have been studied by Raven (1968d) in *Hydrodictyon*, and by Smith (1968b) in *Nitella translucens*. The bicarbonate entry process, like that of chloride, appears to be powered in some way other than by the provision of ATP, and seems to be capable of fairly high rates. Measured rates of bicarbonate entry are up to 9 to 10 pmole $cm^{-2}$ $sec^{-1}$, compared with maximum rates of chloride transport of about 5 pmole $cm^{-2}$ $sec^{-1}$, or more usually 2 to 3 pmole $cm^{-2}$ $sec^{-1}$; since the rate of fixation of $HCO_3^-$ is likely to underestimate the rate of entry of $HCO_3^-$, because of back leakage of $CO_2$, the true maximum rate of bicarbonate transport is likely to be about 5 to 10 times that of chloride.

Probably the most active ion transport process in the cell, but that least accessible to experiment, is the hydrogen ion extrusion postulated by Kitasato (1968). It is suggested that the discrepancy between the measured electrical conductance of the membrane of *Nitella clavata* and that calculated from the fluxes of sodium, potassium and chloride implies that the membrane is permeable to hydrogen ions. This seems to be confirmed by the observation that considerable current flows can be induced by pH changes in the solution bathing a cell voltage-clamped at the potassium equilibrium potential. This explanation was disputed by Walker and Hope (1969), but hydrogen ion permeability seems to be accepted by Hope and Richards (1971). If the cell is permeable to $H^+$, then there must be a considerable net passive influx of $H^+$, since the membrane potential is about 100mV more negative than $E_H$, the $H^+$ equilibrium potential. If therefore the internal pH is not to fall steadily to very acid levels, there must be a balancing active proton extrusion. Kitasato estimated a hydrogen ion transport of about 40 pmole $cm^{-2}$ $sec^{-1}$ and, even if the permeability is not as high under all conditions as he suggests, the efflux may still be comparable with the maximum rates of other transport processes. Spear *et al.* (1969) in *Nitella clavata*, and Smith (1970) in *Chara corallina*, have observed net acid secretion from parts of the cell, at rates of the order of 5 to 20 pmole $cm^{-2}$ $sec^{-1}$. However, regions of acid secretion are separated by regions of base secretion, although the reasons for compartmentation on such a gross scale are still obscure. Smith (1970) and Lucas (1971) found the alkaline bands in *Chara corallina* to be very intense in the presence of bicarbonate in the solution, and suggest that the base-secretion is associated with the uptake and fixation of bicarbonate; Lucas showed that the base secretion appears to be localized in specific spots on the cell, and the bands of alkalinity spread outwards from such highly localized regions. The base secreting spots are responsible for the development of banded deposits of $CaCO_3$ on Characean cells (Arens 1939), which arise when the alkalinity displaces the

$CO_2/HCO_3^-/CO_3^{2-}$ equilibrium enough to bring the value of $[Ca^{2+}]$ $[CO_3^{2-}]$ above its solubility product.

Two proposals have been put forward in which the hydrogen ion extrusion is the primary process of active transport, of which the active chloride influx is a secondary consequence. The first arose from the observations of Spear *et al.* (1969), that chloride influx in *Nitella clavata* is largely confined to the acid regions of the cell; they propose that the extruded $H^+$ is released within the membrane, in a region to which $Cl^-$ has access, and from which molecular HCl can partition into the lipids of the membrane and move preferentially inward. This puts very considerable restrictions on access to this site, and to account for net chloride uptake from $0.2$ mM external chloride to $160$ mM internal chloride a pH of about $2.5$ is required at the site. Whether these requirements are feasible remains to be established. In the form proposed, the pump is electrogenic, since the fluxes between the outside solution and the site of acid secretion consist of $H^+$ outwards and $Cl^-$ inwards; we might therefore expect to find net cation influxes, which would appear as chloride-linked cation fluxes.

The second proposal (Smith 1970, 1972) that the primary process is a Mitchell-type charge transfer across the plasmalemma, by which a pH gradient is set up, with the cytoplasm alkaline and the medium made acid; since this is an electrogenic hydrogen ion extrusion there will also be an electrical gradient set up, with the cytoplasm negative. It is then proposed that a $Cl^-/OH^-$ exchange is responsible for chloride uptake, and that cations move inwards along the electrical gradient. This model arose from observations that the chloride influx is very sensitive to the external pH, falling as external pH rises, and that chloride influx can be stimulated by the presence of ammonium ions in the solution (or by a number of other weak bases); Smith argues that such bases act by their ability to penetrate the cell and increase the internal pH. He has also shown (Smith 1972) that, although a high pH in the labelling solution inhibits chloride influx, pretreatment at high pH can put the cells into a state in which their ability to take up chloride in the dark is higher than it is after pretreatment at low pH. Again he argues that this arises from an increase in the cytoplasmic pH by pretreatment at high pH. His results are a clear indication of the importance of pH in the control of the transport process, and they are consistent with his hypothesis. But before we accept the hypothesis some further evidence seems to be required, for example by measurements of internal pH. Smith proposes a passive $Cl^-/OH^-$ exchange, with no further energy input at this stage, and argues that a pH gradient of 2 units would be adequate; since chloride uptake can occur from external solutions up to pH 9, and occurs readily from pH $7.5$, very high values of cytoplasm pH are required to make the mechanism possible in this form. The other necessary evidence, so far lacking, is the demonstration that it is a pH gradient across the membrane which is important, rather than the absolute values of pH inside or outside the membrane, and that the time course of the decay of the pretreatment effects (not yet measured) bears some relation to a decay of a pH differential. Direct measurement of cytoplasmic pH, and of the relations between the $H^+$ efflux and $Cl^-$ influx, may throw light on the subject.

### (d) *Non-active fluxes: K, Na and Cl*

The existence of chloride-linked cation fluxes has already been mentioned, and we must now consider their nature, and their relation to the residual 'passive' fluxes. We must also consider whether the so-called passive fluxes obey the Goldman equations. The nature of the link between the fluxes of chloride and cations is not yet established beyond doubt, but the evidence favours chemical rather than electrogenic coupling. The changes in membrane potential observed on substituting sulphate for chloride in the solution are small, and may even vary in sign (Spanswick *et al.* 1967, Findlay *et al.* 1969a, Lefebvre & Gillet 1970). Calculations on the basis of the Goldman equations suggest that the observed potential changes are too small to account for the changes in cation influxes. However, measurements of fluxes in voltage-clamped cells show that ion fluxes vary much more steeply with membrane potential than the Goldman equation predicts (Walker & Hope 1969). But even if the observed potential dependence is used as a basis for calculation, a hyperpolarization of some 20mV would be required to achieve a net cation influx of 1 pmole $cm^{-2}$ $sec^{-1}$ in *Nitella*, part by increased influx and part by decreased efflux; such a hyperpolarization on adding chloride is not observed. Hence the evidence is against electrogenic coupling of chloride and cations (unless local hyperpolarizations are suggested, with current circulation in the membrane), and favours instead chemical coupling, by the transport of salt. Since the chloride-linked cation fluxes are smaller than the chloride fluxes we must suppose that a part of the chloride influx is accompanied by some other cation (perhaps $H^+$), or is part of an anion exchange system. The marked dependence of chloride efflux on external chloride in the solution has led to the suggestion that exchange diffusion may contribute to chloride fluxes in *Chara corallina* (Hope *et al.* 1966, Findlay *et al.* 1969a), but to a relatively small extent (0·1 to 0·4 pmole $cm^{-2}$ $sec^{-1}$).

The fact that light has such marked effects on the cation influxes, but only a small effect on the resistance (Hope 1965, Hogg *et al.* 1969), is also consistent with the suggestion of chemical coupling between chloride and cations. The residual cation influxes (in sulphate solutions, in the presence of ouabain) amount to only 0·1–0·2 pmole $cm^{-2}$ $sec^{-1}$ for either potassium or sodium, in both *Hydrodictyon* and *Nitella translucens* (Raven 1968a, MacRobbie 1962, Smith 1967), and on this argument these should be considered as 'passive' fluxes. These fluxes are light-dependent, with a requirement for photosystem II (Raven 1968a, 1969a, Smith 1967), and the membrane appears to be less permeable to cations in the dark than in the light. By contrast the chloride efflux may increase in the dark in *Nitella translucens* and in *Chara corallina*, an effect attributed to an increased chloride permeability (Hope *et al.* 1966). But this effect is not common to all species—in *Hydrodictyon* and in *Nitella clavata* (Raven 1967b,

Spear *et al.* 1969) the chloride efflux decreased somewhat in the dark. The reason for the light-dependence of what are allegedly passive fluxes is unclear.

Before leaving the subject of the nature of the non-active fluxes we should notice that the argument that a large fraction of the cation influxes are not simple passive fluxes, but are chemically linked to the chloride transport, is difficult to reconcile with the results of Walker and Hope (1969) on the effects of potential on the cation fluxes in voltage-clamped cells. They found that both influx and efflux of potassium are very sensitive to potential, roughly equally so, and their results would therefore suggest that the cation influxes do in fact represent 'passive' influxes, dependent on concentration gradients and on the electrical gradient. Hence there is still some doubt on the nature of the non-active fluxes, and on the driving forces responsible. It is however clear that even the fluxes which are to be considered as passive do not obey the Goldman equations.

### (e) *Electrical properties*

Work on the electrical properties of Characean cells has been aimed at establishing the particular ion movements which control the membrane potential and resistance, and has shown that these vary considerably in different conditions. In the complete absence of calcium the membrane potential is sensitive to $K_o$, and can be represented by a Goldman equation with terms in sodium and potassium only (Hope & Walker 1961, Spanswick *et al.* 1967); in this state the ion fluxes are much higher than in solutions containing calcium (Smith 1967). However, in the presence of calcium the membrane potential is insensitive to changes in K/Na outside, and cannot be fitted to any form of Goldman equation containing terms for the major ions (Spanswick *et al.* 1967). Kitasato (1968) found that the membrane potential was very sensitive to external pH and argued that hydrogen ions contributed most of the membrane conductance; the potential would then be largely controlled by the electrogenic extrusion of hydrogen ions by an active mechanism, and their passive inward diffusion. Walker and Hope (1969) produced alternative explanations of some of Kitasato's results, but not of the key observation of pH-dependent current flow in a cell clamped at the potassium equilibrium potential. It is clear that in some conditions the membrane potential may be a good deal more negative than the Nernst potentials for any of the ions concerned, including probable values for the hydrogen ion (Hope 1965, Spanswick *et al.* 1967, Kitasato 1968, Hope & Richards 1971), and some contribution from one or more electrogenic ion pumps seems to be required.

The effect of light on the membrane potential has also been much studied. The direction and time course of the potential changes produced by light/dark changes are variable, depending in a complex way on the composition of the solution and the pretreatment of the cell. Thus hyperpolarization in light was observed by Nagai and Tazawa (1962) in *Nitella* and by Metlicka and Rybova (1967) in *Hydrodictyon*, but depolarization in light by Andrianov *et al.* (1965, 1968), Findlay *et al.* (1969a), Hogg *et al.* (1969) and Vredenberg (1969). The time course in *Chara braunii* was studied by Nishizaki (1968), who found that the response depended on the pretreatment; in light-adapted cells the response to a light-on signal was an initial very rapid hyperpolarization and increase in resistance, followed by a rapid depolarization and decrease in resistance. However, after long dark pretreatment the response to light-on was a small depolarization, followed by slow large hyperpolarization, followed by a second fall in potential. It is likely that the response depends on the extent to which the membrane is sensitive to potassium and to hydrogen ion, and on the activity of any electrogenic pumps, e.g. of $H^+$. Hence the complexity is perhaps to be expected if the various contributory effects oppose each other.

Two recent studies suggest such opposing effects, and perhaps point to methods of clarifying the situation. Hansen (1971) measured the response of the membrane potential to light whose intensity was sinusoidally modulated at varying frequencies, determining the frequency-dependence of the light-induced potential change, i.e. of the amplitude of the depolarization produced by increasing light intensity, and the phase responses to increasing and decreasing light. The author concludes that at least three light-dependent processes are involved in the effects on membrane potential, and further discussion of their identification with particular membrane processes is promised (Hansen & Gradmann, in press). The other work which may throw some light on the ion movements responsible is that of Vredenberg (1969, 1970a,b, 1971), who showed that light-induced changes in the fluorescence yield of chlorophyll *a* in *Nitella* are correlated with changes in the membrane potential. He suggests that ion movements across the chloroplast membranes (including both $H^+$ and $K^+$) are responsible for changes in the ionic environment both in the chloroplast (thereby affecting the fluorescence yield), and at the plasmalemma inner surface (thereby affecting the membrane potential). The predicted light-induced increase in $K^+$ and fall in $H^+$ outside the chloroplast would be expected to have opposite effects on the membrane potential. Vredenberg (1970a) also measured the quantum requirement for the light-induced current in voltage-clamped cells; he found values of 5 to 15 quanta per equivalent, but argues that this may be an underestimate of the true value.

## (f) *Vacuolar fluxes*

The kinetics of the appearance in the vacuole of tracer from the outside solution can be used to give information on the equilibration of ions between cytoplasm and vacuole, or of ions within different phases of the cytoplasm (if these have sufficiently different exchange constants). We may test the kinetic behaviour observed against that predicted for various models of the cell, starting with the simplest possible, that of two phases, cytoplasm and vacuole, in series with one another. If this proves inadequate then we add extra compartments as necessary to fit the results, and then try to identify the kinetic compartments with structural phases in the cell. On the simple two-phase model, the cytoplasmic specific

z

activity during uptake will rise initially on a (1—exp) time course, characterized by a single rate constant $k$, determined by the ratio of the fluxes out of the cytoplasm (to the vacuole and to the outside) to the cytoplasmic content. The amount of activity in the vacuole should show an initial lag, the rate of transfer to the vacuole rising slowly as the cytoplasm fills; the percentage of total activity which is in the vacuole should rise with time, linearly from the origin. In the first use of this method (MacRobbie 1966b) it appeared that for chloride uptake times of 15 min upwards, vacuolar fractions of 20 to 60%, the cell of *Nitella translucens* did behave as cytoplasm and vacuole in series, and the simple model appeared to be adequate. The rate constant $k$ was found to be proportional to influx. (Notice that this means that if cell A has twice the chloride influx of cell B, then after uptake for equal times the amount of tracer in the vacuole of cell A is four times that in the vacuole of cell B; as the influx increases a correspondingly higher fraction of the total entry is transferred to the vacuole.) The method was also used by Coster and Hope (1968) in *Chara corallina*, but without doing a time course; they found the same relationship between $k$ and influx to the cell. However, in subsequent experiments, looking at the early part of the time course, it was found that the behaviour was more complex (MacRobbie 1969). At shorter times of uptake much more activity was found in the vacuole than expected; there appeared to be no initial lag in the appearance of tracer chloride in the vacuole (i.e. the lag was less than 1 min), and the percentage of total activity which was in the vacuole was linear with time, but did not go through the origin. Thus the cytoplasm must be heterogeneous with respect to its ion exchange, and two distinct compartments in the cytoplasm are involved in the transfer of chloride to the vacuole—an initial rapid component in which the incoming tracer mixes with only a very small fraction of the cytoplasmic chloride, and a second cytoplasmic compartment filling with a (1—exp) time course. In *Nitella* the path of rapid transfer involves chloride and sodium, and may also involve a small and variable amount of potassium, but in *Tolypella* potassium has a significant component of rapid transfer; these are the cations which have chloride-linked components in their influxes. The shortness of the initial lag allows an upper limit to be set on the content of the fast phase, and this turns out to be very small indeed; under low flux conditions the fast phase must contain less than 3 nmole cm$^{-2}$ chloride, whereas the figures in Table 25.1 suggest that even the flowing cytoplasm has a chloride content of 33 to 44 nmole cm$^{-2}$. Hence the fast compartment can involve only a very small membrane-bound phase within the cytoplasm, and we have to identify this with some structural phase.

The most curious feature of the results is the very close link, in both phases of the exchange, between entry of chloride to the cell and transfer to the vacuole. This was established first by comparing mean values in groups of cells under different experimental conditions, when it was found (MacRobbie 1969) that the fast component was a constant fraction of the total influx, and also that the rate constant for exchange in the slow cytoplasmic compartment was proportional to influx. The essential point is that the level of activity in the cytoplasm does not specify the rate of transfer to the vacuole directly, but only as a fraction of the total influx. In these results the percentage in the vacuole ($P$) was related to the total entry ($Q_T^*$) by the relation, $P = 0.13 + 0.08 \, Q_T^*$, i.e. the slope of its time dependence is proportional to influx. If we consider instead the instantaneous rate of transfer to the vacuole ($M_V$), as a fraction of the total influx ($M_T$) we have the relation, $M_V/M_T = 0.13 + 0.16 \, Q_T^*$. Thus influx and vacuolar transfer are so intimately linked that we can predict only their ratio at any given level of activity in the cell. Hence it appears that the distribution of tracer between cytoplasm and vacuole is controlled by the influx, and we need a model of the system which will account for this.

The fast component of vacuolar transfer was later examined (MacRobbie 1970b), in the individual cells in a single experiment, rather than in the means of groups of cells under different conditions. It was found that values of the ratio $M_{ov}/M_T$ were not normally distributed, but fell into groups whose means were in the ratios of $1:2:3$, although the mean influxes in the groups were not significantly different. The essential feature is that the distribution of values of $M_{ov}/M_T$ was not Gaussian but was peaked. The existence of quantization was questioned by Findlay et al. (1971b), who examined the distribution of values of $M_{ov}/M_T$ in Chara, after 10 min uptake periods at 24°C; they report that they found no evidence of quantization, although their distribution of values of vacuolar percentage was not Gaussian but bimodal, and they conclude explicitly that there may be unexpected features in the distribution. In fact, since their influxes were very high, the total chloride entry in their experiments might have been expected to bring their values into the period of the slow phase of exchange, and therefore their conditions were not suited for detecting any abnormalities in the fast component, should such exist. We must await more detailed measurements on Chara, at lower total entry and with a time course, before we can draw definite conclusions.

The kinetics of vacuolar transfer were further examined (MacRobbie 1971b), in experiments using both $^{36}$Cl and $^{82}$Br designed to measure both fast and slow components in the same cell. In spite of very different influxes of bromide and chloride, the vacuolar percentages for the two ions were equal for cells labelled for equal times with bromide and chloride, whether the loading was for a short or a longer time. Hence both slow and fast components are equal for the two ions. The fast component again appeared to be quantized; the contribution of the slow component to the vacuolar transfer increased with time, with a slope which was proportional to the influx to the cell, but was not quantized. For chloride the relation between the instantaneous rate of transfer to the vacuole as a fraction of the total influx, and the tracer content of the cell, was established as linear with a slope of $0.25$ per nmole cm$^{-2}$ of tracer chloride content. It was argued that the unexpected equality of vacuolar fractions for bromide and chloride must reflect some inherent property of the exchange process, and that the relation should therefore be written in terms of halide, rather than simply chloride; the slope then becomes $0.19$ per nmole cm$^{-2}$ of tracer halide in the cell, very similar to the previously quoted value. The surprising thing is that this slope is so steep, and it must mean that even the slow compartment of the cytoplasm can represent only a part of the cytoplasm with much higher specific activity than average; the main bulk of cytoplasmic chloride must exchange at an even slower rate (because of restricted transfer across chloroplast membranes but probably also because of restricted diffusion from the region just inside the plasmalemma).

The present situation is that we have complex kinetics for chloride transfer from cytoplasm to vacuole, with a number of curious features, which have to be related to exchange with structural compartments in the cytoplasm. In view of the smallness of the fast compartment, and of the link between vacuolar transfer and influx, it was suggested that salt entry took place by a process involving the formation of salt-filled vesicles at the plasmalemma and their movement through the cytoplasm with some exchange of ions on the way, to discharge either directly at the tonoplast, or by fusion with the endoplasmic reticulum from which new vacuoles subsequently formed. The equality of vacuolar percentages of bromide and chloride, in spite of very different influxes, is entirely consistent with the hypothesis that the slow component reflects entry with pinocytotic entry vesicles, whose total area is directly proportional to the influx, but it is difficult to explain in terms which do not involve a common entry process of salt, involving both chloride and bromide. It should be stressed however that indiscriminate uptake of all small molecules in the outside solution is not involved, and some form of selective adsorption on the membrane must precede the membrane movements. It is suggested that the result is the uptake of chloride and cations (K$^+$ and Na$^+$, with perhaps also some H$^+$), in the form of salt-filled vesicles.

The existence of a quantized discharge to the vacuole suggested that entry to a cytoplasmic phase, such as the endoplasmic reticulum, must precede this discharge to the vacuole, but the subsequent finding that only the fast component is quantized, and the slow component is not, suggesting the two are in parallel, makes interpretation difficult. The kinetics of the slow component give rise to the hypothesis of transfer as salt-filled vesicles, but it is then difficult to relate this to the quantized fast component. The alternative explanation of the fast component, as transfer to the vacuole during one or more action potentials on cutting the cell, was discarded because of results in bicarbonate-containing solutions. It is possible that this argument should be re-examined under further conditions, but for the present the relation between fast and slow components remains obscure. What is clearly established, however, is that the kinetics of vacuolar transfer place severe restrictions on the type of process likely to be involved, and seem impossible to explain in terms of fixed compartments.

Active transport by means of membrane flow and vesiculation was suggested by Bennett (1956) and discussed by Sutcliffe (1962). Hall (1970) suggested that vesicles derived from the plasmalemma by pinocytosis can be identified in the cells of maize root tips, and that these increase in number in conditions in which cells are accumulating salt. Steward and Mott (1970) suggest that ion accumulation is only one facet of a more general process, of internal secretion of solutes in vacuoles, of the creation of new internal solution by phase separation. The results on *Nitella* are certainly difficult to explain in terms of static membrane systems, and can more readily be explained in terms of the formation of new vacuole, rather than as the transfer of individual ions to the existing vacuole. It is clear that we do not understand the process, but the results suggest that a shift in outlook, away from that previously held by membrane physiologists, may be required.

## 2.2 *Marine species*

A number of marine algae with giant cells have been studied, with varying degrees of completeness. It is difficult to summarize this work as the cells studied differ markedly with respect to both active and passive fluxes, and also in respect of their electrical properties, and we may therefore consider each species separately. Table 25.3 summarizes concentration, potential and flux measurements in the algae studied, and gives the active fluxes inferred from such measurements. The values quoted for tonoplast fluxes are derived from flux analysis, and depend on the assumption that the cell may be considered as two phases, cytoplasm and vacuole, in series. Where comparison influx and efflux kinetics have been made (*Chaetomorpha* and *Griffithsia*) there is some discrepancy (see MacRobbie 1971a) and hence a more complex model may be required. If the transfer of ions to the vacuole does involve formation and transfer of small vacuoles in the cytoplasm, then the figures derived from the simple analysis are invalid.

### (a) *Griffithsia*

Measurements of both fluxes and electrical properties have been made in a number of species of *Griffithsia* (Rhodophyceae) (Findlay *et al.* 1969b, 1970, Lilley & Hope 1971a,b, Lilley 1971). The properties of the four species studied (*G. pulvinata, G. monile, G. teges, G. flabelliformis*) seem rather similar.

Ion concentrations, potentials and resistances are shown in Table 25.3, with the inferred active fluxes, the familiar pattern of active influx of chloride and potassium, and active efflux of sodium. The plasmalemma potential and resistance are very sensitive to external potassium, and $E_{co}$ can be fitted to a Goldman equation in terms of sodium and potassium only (equation 4), from which estimates of $\alpha$ ($P_{Na}/P_K$) and cytoplasmic concentration ($K_c + \alpha Na_c$) can be made (Findlay *et al.* 1969b). The figures obtained were $\alpha = 0.004$, which is extremely low compared with other cells, and ($K_c + \alpha Na_c$) = 300–350 mM. (The concentration figure assumes that the activity coefficient in cytoplasm is close to that in sea water, 0.68.) Since $\alpha$ is so small, the concentration figure must be close to $K_c$. The measured cytoplasmic activity of potassium in *Griffithsia* is

**Table 25.3.** Marine algae: concentrations, potentials and fluxes

| | Concentrations mM, fluxes pmole cm$^{-2}$ sec$^{-1}$ | | |
|---|---|---|---|
| | K | Na | Cl |
| Seawater concentration | 10–13 | 470–510 | 520–600 |
| *Griffithsia* spp. (1,2) | | | |
| $E_{co}$ −85 mV $E_{vc}$ 20–35 mV | | | |
| $R_P$ 0·2        $R_T$ 5 $k$ '$\Omega$cm$^2$ | | | |
| vacuolar concentration | 550 | 50 | 620 |
| plasmalemma flux $M_1$ | 50–380 | 12–17 | 10–40 |
| tonoplast flux $M_2$ | (3–4)$M_1$ | 5–12 | $\gg M_1$ |
| active fluxes | $M_{cv}$ | $M_{co}$ | $M_{oc}$ |
| *Acetabularia mediterranea* (3,4,5) | | | |
| $E_{co}$−170 mV $E_{vc} \sim 0$ | | | |
| $R_P$ 0·1 $k$ $\Omega$ cm$^2$ $R_T$ very low | | | |
| vacuolar concentration | 355 | 65 | 480 |
| cytoplasmic concentration | 400 | 57 | |
| plasmalemma flux | 11–40 | 11–50 | 200–790 |
| active fluxes | uncertain; results equivocal | $M_{co}$ | $M_{oc}$ |
| *Chaetomorpha darwinii* (6,7) | | | |
| $E_{co}$ −72 mV $E_{vc}$ (a)+75 mV or (b)+46 mV | | | |
| (a) $R_P$ 0·5        $R_T$ 4·9 $k$ $\Omega$ cm$^2$ | | | |
| or (b) $R_P$ 0·75        $R_T$ 7·1 $k$ $\Omega$ cm$^2$ | | | |
| vacuolar concentration | 530 | 56 | 620 |
| cytoplasmic concentration | 400–450 | 50 | 140 |
| plasmalemma flux | 100–200 | 100 | 200 |
| tonoplast flux | 200 | 5–10 | 50–250 |
| active fluxes | $M_{oc}$, $M_{cv}$ | $M_{co}$, $M_{cv}$ | |
| *Valonia ventricosa* (8,9) | | | |
| $E_{co}$−71 mV $E_{vc}$+88 mV | | | |
| $R$ 9$k$ $\Omega$ cm$^2$ | | | |
| vacuolar concentration | 625 | 44 | 643 |
| cytoplasmic concentration | 434 | 40 | 138 |
| plasmalemma flux | | | |
| tonoplast flux | | | |
| active fluxes | $M_{oc}$, $M_{cv}$ | $M_{co}$, $M_{cv}$ | $M_{oc}$. |
| *Halicystis ovalis* (10) | | | |
| $E_{co}$ −80 mV $E_{vc} \sim 0$ | | | |
| vacuolar concentration | 337 | 257 | 543 |
| active fluxes | | $M_{co}$ | $M_{cv}$ |
| *Valoniopsis* sp. (11) | | | |
| $E_{vo}$ −5 to −15 mV | | | |
| $R$ 5 $k$ $\Omega$ cm$^2$ | | | |
| vacuolar concentration | 50 | 650 | 700 |
| plasmalemma flux | 50–100 | 30–600 | 130 |
| tonoplast flux | 1–5 | 20–50 | 80 |

(1) Findlay *et al.* 1969b, (2) Findlay *et al.* 1970, (3) Saddler 1970a, (4) Saddler 1970b, (5) Saddler 1970c, (6) Dodd *et al.* 1966, (7) Findlay *et al.* 1971a, (8) Gutknecht 1966, (9) Gutknecht 1967, (10) Blount & Levedahl 1960, (11) Hope 1971.

only 153 mM (Vorobiev 1967), corresponding to a concentration of only 225 mM if the activity coefficient is 0·68. Findlay *et al.* (1969b) suggest that the activity coefficient may be much lower in cytoplasm, nearer the figure of 0·43 required to reconcile the two estimates. The other possibility is that *Griffithsia* resembles *Acetabularia* in that its potential can be fitted by adding an extra constant term to the Goldman equation; in *Acetabularia* the term is large ($-85$ mV), but in *Griffithsia* only a small correction is required ($-9$ mV). We have seen previously that such a term may arise from the operation of an electrogenic pump whose flux is limited by the opposing membrane potential it sets up.

The allocation of the active potassium flux to plasmalemma or tonoplast or both demands a knowledge of the cytoplasmic concentration. For potassium to be distributed passively across the tonoplast, with a vacuolar concentration of 550 mM and $E_{vc} = 20\text{--}35$ mV, a cytoplasmic potassium concentration of 1·2–2·2 M would be required, which seems excessively high. Findlay *et al.* therefore suggest active transfer of potassium from cytoplasm to vacuole, with passive movement at the plasmalemma. The potassium influx at the plasmalemma was unaffected by dark, or by various metabolic inhibitors, and its behaviour was consistent with that expected of a passive flux. However estimates of cytoplasmic content of potassium from the kinetics of uptake or washing-out of tracer potassium in the cell give much higher figures (Findlay *et al.* 1970). The cytoplasmic content of potassium estimated by separation of sap and cytoplasm after short times of uptake was estimated as 0·24 to 0·76 $\mu$mole cm$^{-2}$, whereas the figure derived from analysis of efflux curves was even higher, 1·8 $\mu$mole cm$^{-2}$; in a phase 5 $\mu$m thick these correspond to concentrations of 0·5 to 1·5 M and 3·6 M respectively. The discrepancy suggests that the assumptions of the flux analysis, that cytoplasm and vacuole are simply in series, may be in error, and the earlier figures quoted are likely to be nearer the correct value for the concentration in the cytoplasm outside the chloroplasts.

Figures for the fluxes of both plasmalemma and tonoplast are given by Findlay *et al.* (1970), calculated on the assumption that the two-phase series model for cytoplasm and vacuole is valid. The initial rate of influx to the cell gives $M_{oc}$, whereas the influx measured over long uptake times, or calculated from the rate constant for exchange in the slow phase in loading or efflux, is $M_{cv}.M_{oc}/(M_{cv} + M_{co})$. Measurements were made of sodium, potassium and chloride fluxes, using this method of analysis. Values of 17 and 12 pmole cm$^{-2}$ sec$^{-1}$ are quoted for the sodium flux at the plasmalemma in two experiments, with 5 and 12 pmole cm$^{-2}$ sec$^{-1}$ for the corresponding tonoplast flux. Chloride fluxes at the plasmalemma of 10 to 40 pmole cm$^{-2}$ sec$^{-1}$ were found, with much higher tonoplast fluxes. Potassium fluxes were a good deal higher, with values of 50 to 380 pmole cm$^{-2}$ sec$^{-1}$ at the plasmalemma, and calculated tonoplast fluxes about 3 to 4 times those at the plasmalemma. Potassium fluxes at both membranes rose steeply with increasing external potassium. The high tonoplast fluxes are surprising in view of the very high resistance of this membrane compared with the plasmalemma, and the fact that the tonoplast resistance increases sharply as $K_0$ increases, an unexpected effect for which we have no explanation as yet. The authors suggest that a vesicular transport of potassium to the vacuole in *Griffithsia* might explain both the discrepancies in the flux analysis, and the lack of correlation between the behaviour of the fluxes and that of the resistance; alternatively it was suggested that most of the flux of potassium at the tonoplast was exchange diffusion, but this suggestion removes only one of the anomalies.

The metabolic dependence of the chloride influx in *Griffithsia* has been studied by Lilley and Hope (1971a,b), who also measured $O_2$ consumption and ATP levels. The figures for chloride influx, $O_2$ and $CO_2$ fluxes are as follows: chloride influx, 20 to 70 pmole $cm^{-2}$ $sec^{-1}$ in light, 5 to 15 pmole $cm^{-2}$ $sec^{-1}$ in the dark; apparent $O_2$ output in light, 45 to 80 pmole $cm^{-2}$ $sec^{-1}$, $O_2$ consumption in the dark, 5 to 20 pmole $cm^{-2}$ $sec^{-1}$; $CO_2$ fixation in the light, 20 to 50 pmole $cm^{-2}$ $sec^{-1}$, but less than 3% of this rate in the dark. The apparent oxygen output is the difference between output in light and consumption in the dark, and may therefore be an underestimate of the light-driven electron flow; this makes the imbalance between $O_2$ and $CO_2$ even more marked. In the light there is little difference in chloride influx in air, $CO_2$-free nitrogen, or $CO_2$-free $O_2$, but the dark influx is stimulated in $O_2$ and inhibited in nitrogen. The light influx is as sensitive to DCMU as is the apparent oxygen evolution, hence cyclic photophosphorylation seems ruled out as the energy source. In the light there is a parallel inhibition of Cl influx and $CO_2$ fixation by CCCP in the range 3 to 10 μM, but, curiously, at 1 μM CCCP there is stimulation of chloride influx by about 10% and inhibition of $CO_2$ fixation by about 20%; in the dark CCCP in the range 1 to 10 μM stimulates oxygen consumption and, even more markedly, chloride influx. The authors suggest that the chloride influx in light is related to non-cyclic phosphorylation, and that in the dark there may be an exchange diffusion system linked somehow to mitochondrial electron flow. In the absence of any information on the metabolic dependence of chloride efflux this must remain speculative. The measurements of ATP levels (Lilley & Hope 1971b) do not help a great deal, since the level of ATP does not change in light, dark or in the presence of CCCP in the light; it is however greatly reduced by CCCP in the dark. Since the measurements are of total ATP in the cell they may not reflect changing levels of ATP at the transport site, but the authors conclude that the light stimulation of the chloride influx is unlikely to be achieved by an increase in ATP as an energy source. Lilley (1971) suggests that the level of cytoplasmic glyceraldehyde-3-phosphate may affect the activity of the chloride transport.

Two other aspects of the electrical properties deserve comment. Throm (1970) observed a small depolarization of the vacuolar potential in light, in *Griffithsia setacea*, by an amount that was dependent on the external potassium level. However, at high pH (9·5–10) depolarization was replaced by hyperpolarization, and at intermediate pH values (8–9) transients were observed, hyperpolarization followed by depolarization. Lilley (1971) showed that changes at the tonoplast were responsible for the light-dark potential effects, and that the plasmalemma potential was unchanged; the main resistance changes were also associated with the tonoplast.

The second effect is the observation of punch-through at the tonoplast, i.e. of a sudden very large increase in current with increasing applied potential which produces a reversible breakdown of the membrane and a sudden increase in permeability (Findlay et al. 1969b). It is interesting that in *Griffithsia* this occurs at positive potentials, when the tonoplast potential is increased to 80 to 90 mV, whereas the phenomenon is observed at the plasmalemma of Characean cells at large negative potentials (Coster 1965, Williams & Bradley 1968). If the explanation of the effect provided by Coster (1965) is valid, in terms of a membrane containing two fixed charge regions of opposite signs, then we might perhaps expect that the applied polarity should differ in the two instances; in each case punch-through occurs when the cytoplasmic side of the membrane is driven negative to some threshold value. However, this is contradicted by the observation of punch-through at the tonoplast of *Chara* at $E_{vc}$ $-30$mV, i.e. when the cytoplasm is driven more positive with respect to the vacuole (Coster 1969), and some other explanation must be sought.

## (b) *Acetabularia*

Saddler (1970a,b,c, 1971) has made a study of both fluxes and electrical properties of *Acetabularia mediterranea*, and electrical work has also been carried out on *Acetabularia crenulata* (Schilde 1966, 1968b, Gradmann 1970, Gradmann & Bentrup 1970, Hansen & Gradmann 1971).

The electrochemical data in Table 25.3 (Saddler 1970a) suggest that there is active transport of sodium outwards and of chloride inwards at the plasmalemma; this is supported by the sensitivity of sodium efflux to light and temperature (although not to CCCP), and chloride influx to light, temperature and CCCP (Saddler 1970b). However there are problems over the interpretation of potassium fluxes in *Acetabularia*. The vacuolar potential of the cell is about $-170$ mV, which is a good deal more negative than the potassium equilibrium potential, $E_K = -93$ mV, and hence we would expect potassium efflux to be active, whereas potassium influx is downhill and could be passive. But flux measurements show that potassium influx is inhibited at low temperature and by CCCP, although it is not light sensitive, and potassium efflux is stimulated in the cold and insensitive to light or CCCP; hence the metabolic dependence of influx and efflux is the reverse of that predicted from electrochemical potential measurements. Also the kinetics of the efflux of tracer potassium defy explanation (Saddler 1970b). Wash-out curves of whole cells show three components, with half-times for exchange of $<10$ min, $1$–$2$ h and $>8$ h respectively, with tracer contents after loading (incomplete labelling) of 11 to 28%, 15 to 31% and 53 to 67% respectively. Since the ratio of potassium contents of vacuole:cytoplasm is at least 10:1, we must identify the slowest component with the vacuole. But wash-out curves of non-vacuolate cell fragments show components with half-times similar to the fast and slow components above, with relative tracer contents of 30 to 45% and 55 to 70% respectively, and are lacking in the medium component. No simple model can be offered in explanation, and the author suggests that there are two independent pathways for $K^+$ efflux at the plasmalemma.

The most interesting aspect of the results on *Acetabularia* is the contribution of an electrogenic pump to the membrane potential (Gradmann & Bentrup 1970, Saddler 1970c), and Saddler's demonstration that this is an electrogenic chloride pump. The membrane potential of both species behaves as if it were the sum of an electrogenic component and a Goldman-type diffusion potential. At low temperature the potential in *A. mediterranea* fits a Goldman equation with $\alpha = 0\cdot02$–$0\cdot06$ and $(K_c + \alpha Na_c) = 290$–$350$ mM; at normal temperatures it can be fitted by a Goldman term with similar values of $\alpha$ and $(K_c + \alpha Na_c)$, added to a constant electrogenic term of about $-85$ mV. Under conditions when chloride influx is inhibited (low temperature, darkness, CCCP or low external chloride) the membrane is depolarized from the normal $-170$ mV to a new level of about $-90$ mV. The chloride fluxes at the plasmalemma, 200–790 pmole $cm^{-2}$ $sec^{-1}$, are very high compared with either sodium or potassium fluxes, each of which are about 10–40 pmole $cm^{-2}$ $sec^{-1}$. If we calculate the electrogenic flux required to hyperpolarize the membrane to the observed level, using equations (3a) and (6) and Saddler's values of $\alpha = 0\cdot04$, $(K_c + \alpha Na_c) = 320$ mM, we find it should be about 17 times the sodium influx; thus we expect to find chloride influxes of 170 to 670 pmole $cm^{-2}$ $sec^{-1}$, as indeed we do. In *A. crenulata* the electrical effects are very similar, and the conclusion is the same, that the potential is in part controlled by potassium diffusion and in part by an electrogenic component.

The chloride fluxes in *A. mediterranea* are so high that in low chloride solutions the chloride content of cells will fall to a new low level in a few hours; the cell then reaches a new state of reduced chloride fluxes and content, with very little turgor. In the course of this the membrane repolarizes to $-140$ to $-170$ mV and the resistance increases from its normal low level of $0\cdot1$ $k\Omega$ $cm^2$, in sea water and initially in low chloride solutions, to a new level of $1\cdot5$ $k\Omega$ $cm^2$ (Saddler 1970c, 1971). In the approach to the new state the net chloride efflux can only be balanced by $H^+$ efflux, as the $K^+$ and $Na^+$ fluxes are too small. Since even at pH 1 the vacuolar reservoir of $H^+$ is not adequate to balance a chloride flux of 500 pmole $cm^{-2}$ $sec^{-1}$ for long, Saddler suggests that oxalate synthesis is induced in the cytoplasm, with active transfer of oxalate to the vacuole to achieve the required charge balance, and $H^+$ loss at the plasmalemma; oxalate is in fact present in the vacuole at a concentration of 110 mM (Saddler 1970a).

The repolarization of the cell potential described above, with associated increase in resistance, is one of a number of curious electrical effects in *Acetabularia* described by Saddler (1971) and by Gradmann (1970). These include spontaneous depolarizing spikes, similar to but slower than Characean action potentials, and a hyperpolarizing response produced by applying current to the cell in the depolarized state. It is suggested that the electrical effects observed are associated with changes in the permeability of the cell to chloride. As in other cells the membrane shows punch-through, the rapid decrease in membrane differential resistance with applied bias; in *Acetabularia mediterranea* this occurs at hyperpolarizing potentials, when the potential reaches about $-230$ mV (Saddler 1971).

The effects of light on the membrane potential of *A. crenulata* have been studied by Schilde (1966, 1968a,b) and by Gradmann (1970); Schilde found an initial very rapid depolarization of a few mV whose action spectrum had a peak at 540 nm, and a slow, somewhat larger hyperpolarization with an action spectrum like that of photosynthesis. Hansen and Gradmann (1971) studied the effect of modulated light intensity (sinusoidal or square wave modulation) on the membrane potential, as a function of the frequency of modulation. The hyperpolarization induced by increasing light intensity has a maximum at a frequency of $\frac{1}{3}$ cycle per minute, and the phase angles between maximum light and maximum membrane potential (most negative), and minimum light and minimum potential also vary with the frequency of modulation. The authors consider the type of membrane processes required to produce the observed responses, and conclude that either a feedback system, or two light-dependent ion fluxes with opposing effects on potential, might be involved. They suggest that a feedback loop might be created by the existence of ion fluxes at the plasmalemma driven by products of light reactions in the chloroplasts, if the light reactions were themselves sensitive to the ionic concentrations in these organelles (cf. Vredenberg 1969, on light-induced potential changes in *Nitella*).

### (c) *Chaetomorpha darwinii*

Electrical and flux measurements have been made by Dodd *et al.* (1966) and by Findlay *et al.* (1971a). The concentration and potential measurements in Table 25.3 demand active influx of potassium and efflux of sodium at the plasmalemma, and also of both ions from cytoplasm to vacuole. The cytoplasmic chloride content (from flux analysis) is 100 nmole $cm^{-2}$, which corresponds to a mean concentration of 140 mM; Findlay *et al.* suggest that if this were distributed as 250 mM in the chloroplasts and 35 mM outside the chloroplasts, which they argue is similar to the distribution in *Chara australis*, then chloride movement could be passive at both plasmalemma and tonoplast. While this is true it does not provide evidence that this distribution is the correct one, and the question of active transport of chloride is not settled finally by these measurements. Findlay *et al.* in fact suggest that most of the chloride fluxes at either membrane are exchange diffusion, to remove two discrepancies in the observations—the first between $P_{Cl}/P_K$ calculated from flux measurements and the much lower value required to fit the Goldman potential equation, and the second between the measured electrical conductance of the tonoplast and the much higher conductance calculated from the flux measurements (the reverse of the usual form of anomaly from such comparisons). Exchange diffusion is one way of producing such a discrepancy, but neutral salt transport would provide an alternative explanation. Findlay *et al.* partition the various plasmalemma and tonoplast fluxes into active, passive and exchange diffusion components—on the basis that the Goldman equations provide satisfactory descriptions of the passive fluxes, and that the assumptions of the flux analysis are valid and hence the method yields correct values for tonoplast fluxes. They suggest that at the plasmalemma the active and passive components of potassium influx are 115 and 85 pmole $cm^{-2}$ $sec^{-1}$ respectively, whereas the chloride fluxes are made up of only 30 pmole $cm^{-2}$ $sec^{-1}$ passive transport and 170 pmole $cm^{-2}$ $sec^{-1}$ exchange diffusion. At the tonoplast they suggest that about 160 pmole $cm^{-2}$ $sec^{-1}$ of the potassium flux is exchange diffusion, with the remaining 40 pmole $cm^{-2}$ $sec^{-1}$ wholly passive in the direction vacuole to cytoplasm, but largely active in the opposite direction (37 pmole $cm^{-2}$ $sec^{-1}$), with only a small passive component (3 pmole $cm^{-2}$ $sec^{-1}$). For chloride at the tonoplast they again suggest a large component of exchange diffusion. The other possibility is of course that the assumptions of the flux analysis are invalid, that cytoplasm and vacuole are not simply in series, and that the tonoplast fluxes are in error; nor can we be sure, in view of results with other cells, that the Goldman equations are adequate. For potassium there is poor agreement between the kinetics of cytoplasmic exchange measured from short-term influx and from washing-out, which must throw doubt on the assumptions of the flux analysis. The possibility that the kinetics may be complicated by restricted diffusion across the cytoplasm and by the formation and discharge of small vacuoles into the main vacuole, leading to overestimates of tonoplast fluxes, was discussed by Dodd *et al.* (1966). Hence the partition of fluxes into such categories may be uncertain.

The potential $E_{co}$ is sensitive to both $K_o$ and $Cl_o$, hence the need for terms in all three ions K, Na and Cl, in the Goldman potential equation. Findlay *et al.* found that $E_{co}$ could be fitted to a Goldman equation with $P_{Na}/P_K = 0.01$–$0.001$, and $P_{Cl}/P_K = 0.05$–$0.1$. As in *Griffithsia* they found the curious effect that the tonoplast resistance increases as external potassium increases, for which there is no obvious explanation. In the usual state of the cell $E_{vo}$ is slightly positive, as the cytoplasm is negative to both outside and vacuole, but the tonoplast potential is slightly larger than the plasmalemma potential. However there is another state of the cell in which $E_{vo}$ is $-25$ to $-40$ mV, because the tonoplast potential is less than the plasmalemma potential by this amount. The authors suggest that an increase in $P_{Cl}$ at the tonoplast is responsible for the change from the positive state to the negative state.

### (d) *Valonia ventricosa*

*Valonia* provided one of the classic subjects of study of the ionic relations of algal cells (see for example the review of early work by Blinks (1955)), and it has also been studied more recently by Gutknecht (1966, 1967, 1968) and by Aikman and Dainty (1966).

In *Valonia ventricosa* there is evidence for active sodium efflux and potassium influx at the plasmalemma, but once again the evidence on chloride is not clear-cut. In the vacuole the chloride concentration is close to equilibrium with the outside, but the cytoplasmic concentration of 138 mM is well above that for passive distribution at either membrane; however, Gutknecht (1966) points out that this figure may reflect chloride accumulation in the chloroplasts, and the bulk cytoplasmic concentration may not be higher than that for passive distribution. On the other hand Gutknecht's demonstration (1967) of a large net influx of chloride in short-circuited cells seems to point to the existence of active transport of chloride inwards. In *Valonia* short-circuited and perfused with artificial sap inside and out, he found net influxes of 2 pmole Na $cm^{-2}$ $sec^{-1}$, 80 pmole K $cm^{-2}$ $sec^{-1}$ and 50 pmole Cl $cm^{-2}$ $sec^{-1}$; these summed fluxes however accounted for only about half the inward positive current. Both the potential and the short-circuit current (equivalent to an influx of 30 to 100 pmole $cm^{-2}$ $sec^{-1}$ of monovalent cation) were strongly light-dependent, suggesting the operation of a light-dependent electrogenic pump. The very marked correlation between net potassium flux and short-circuit current might suggest an electrogenic potassium influx, but we then need a neutral chloride transport system to account for the discrepancy in current. A very large water transport was associated with the ion fluxes, equivalent to the transfer of a salt solution isotonic with the cell sap, much larger than that predicted from the osmotic driving force of the increasing salt concentration and the hydraulic conductivity of the protoplast. Gutknecht therefore argues for coupling between water and ion fluxes similar to that found in animal epithelia. The other marked property of *Valonia* is the very strong inhibitory effect of hydrostatic pressure on potassium influx (Gutknecht 1968); this may be related to the old observation of very large net influx of salt in impaled cells of *Valonia macrophysa* (or of *Halicystis osterhoutii*) whose turgor is reduced to zero (Jacques 1939a,b).

## (e) *Halicystis*

*Halicystis ovalis* and *H. osterhoutii* are also classic species in algal ion work, and more recent flux measurements have been made on *H. ovalis* by Blount (1958) and Blount and Levedahl (1960). Again the technique of short-circuiting was used to identify processes of active transport. Under short-circuit conditions, with sea water inside and out, a net influx of chloride (300 pmole $cm^{-2}$ $sec^{-1}$) and a net efflux of sodium (60 pmole $cm^{-2}$ $sec^{-1}$) were found; it was argued that these two ions together would account for all the current. However, the two ions were measured in separate experiments in which the short circuit currents were very different, equivalent to 460 and 120 pmole $cm^{-2}$ $sec^{-1}$ respectively, and therefore measurements of potassium fluxes in *Halicystis* seem also to be required. Blount and Levedahl allocated the sodium pump to the plasmalemma, and the chloride pump to the tonoplast (mainly to allow osmotic equilibrium between cytoplasm and vacuole by pumping some ion into the vacuole). There are very rapid effects of light on current and potential in this species also, and the existence of an electrogenic pump seems to be indicated. The differential sensitivity of the sodium transport and the chloride transport to dinitrophenol suggest that the chloride pump may not be ATP-dependent.

### (f) *Valoniopsis*

*Valoniopsis* is another giant algal cell which has been studied (Findlay *et al.* quoted by Hope 1971, Findlay pers. comm.). The potassium concentration in the vacuole of only 50 mM is much lower in this species than in any already discussed, although the K/Na ratio in the cytoplasm is probably more normal. Also in *Valoniopsis* the tonoplast fluxes are much lower relative to the plasma-lemma fluxes than in the other cells for which figures are available.

## 3  SMALL-CELLED ALGAE

There is a wide literature on ion uptake in a range of more normal algae, but there are very few species in which the system can be experimentally well-defined, in that potential measurements can be made and vacuole and cytoplasm can be separated. Gutknecht and Dainty (1968) review a number of examples, but in most instances, values must be assumed for quantities which cannot be measured before any discussion of mechanism is possible.

However, there is one recent series of investigations in which enough information is available for a discussion of mechanisms to be useful, the study of *Chlorella pyrenoidosa* by Barber (1968a-d, 1969, Shieh & Barber 1971). The non-vacuolate cells are 5 μm or larger in size, and measurements of electrical potential were made, using electrode tips of less than 0·5 μm. The electrochemical potential measurements show the existence of active influxes of potassium and chloride and active efflux of sodium ($E_{co}$ is $-40$ mV; the internal concentrations are 1 mM Na, 114 mM K and 1·3 mM Cl, maintained in an external solution of 1 mM Na, 6·5 mM K and 1 mM Cl). Measurements of fluxes under various conditions support this. Sodium efflux is light sensitive, and sensitive to CCCP, but not to DCMU in low concentration; it is also inhibited to some extent by the removal of external potassium. Potassium influx is light-sensitive, temperature-sensitive, and inhibited by CCCP but not by ouabain; it is suggested that about 25% of the potassium influx is passive. The fluxes are much lower than in the vacuolate algae studied, being only about 0·14 pmole cm$^{-2}$ sec$^{-1}$ for sodium, and about 1 pmole cm$^{-2}$ sec$^{-1}$ for potassium from the fairly high external concentration of 6·5 mM. There is very little efflux of potassium from the cells in the absence of potassium outside, and Barber suggests that there is very little passive diffusion across the membrane, that exchange fluxes of one sort or another must be invoked. He argues for the existence of K$^+$/K$^+$ exchange diffusion, but also for K$^+$/H$^+$ exchange and K$^+$/Na$^+$ exchange driven by ATP. Cells maintained a constant internal potassium concentration of 160 to 170 mM when grown at external levels of 1 to 100 mM K, in the presence of 7 to 4 mM Na, but were unable to maintain this level of potassium at external levels below 1 mM. However, when 3 mM potassium was supplied to potassium-depleted cells, the cells gained K and lost Na by a temperature-sensitive light-dependent process. The initial rates of K/Na exchange were very high, about 20 pmole $cm^{-2}$ $sec^{-1}$, and the exchange was inhibited by CCCP and by DCCD. A small fraction of the $K^+$ uptake was balanced by $H^+$ extrusion rather than by $Na^+$ extrusion, and subsequently the cells reached a steady state of $K^+/K^+$ exchange. Barber suggests that the ionic relations of *Chlorella* may be more akin to those of microorganisms studied than to current views of giant algal cells, in that ions may cross the *Chlorella* membrane almost exclusively by exchange mechanisms. Rothstein (1964) has argued that low ion permeabilities may be a prerequisite for microorganisms growing in dilute media. Thus significant passive leakage of ions to the medium, by diffusion rather than by exchange, may be something that can be tolerated only by cells growing in an environment with an adequate supply of salt, and whose surface/volume ratios are favourable, i.e. not too large.

There are a number of other species discussed by Gutknecht and Dainty (1968), but our information is in general incomplete; it does however suggest that the pattern of ion pumps discussed is a general one, of sodium efflux, potassium influx and chloride influx. One system studied recently deserves final comment, as being of more general importance outside the field of ion transport work—namely the developing *Fucus* egg. A polarity can be imposed on the *Fucus* egg by light, or by external gradients of potential or concentration, thereby determining the point from which the rhizoid grows out; it is suggested that internal currents are generated which may be concerned with the development of morphogenetic polarity within the cell (Jaffe 1966, 1968, Bentrup *et al.* 1967) (see also Chapter 28, p. 799). The membrane potential in the unfertilized egg is very low ($-19$ mV), but in the 12 hours after fertilization this increases to $-78$ mV, with a transient depolarization of about 13 mV at about 9 hours. The potential change arises from the development of potassium selectivity in the membrane ($P_{Na}/P_K$ falls from about 0·1 to 0·001), and appears to be a potassium diffusion potential (Bentrup 1970). Robinson *et al.* (1970) observed a considerable increase in intracellular potassium over this time course, comparable with that inferred from the potential measurements. Once a polarity is established in the zygote there is a measurable electrical gradient across the zygote, of about 270 nvolt per egg (Jaffe 1966) and Jaffe (1969) suggests that this arises from differences in the membrane properties at the two poles, with the membrane depolarized at the rhizoid pole. He suggests that such a local depolarization is the primary event, which gives rise to a current flow through the egg, thereby polarizing the cell contents. Hence membrane properties and their resultant effects on intracellular concentrations may be important as regulators in morphogenesis and in the establishment of intracellular polarities.

# 4 REFERENCES

AIKMAN D.P. & DAINTY J. (1966) Ionic relations of *Valonia ventricosa*. In *Some Contemporary Studies in Marine Science*, ed. Barnes H. pp. 37–43. Allen & Unwin, London.

ANDRIANOV V.K., KURELLA G.A. & LITVIN F.F. (1965) Changes of resting potential of alga *Nitella* cells under light effect and the connection of this effect with photosynthesis. *Biophysics*, **10**, 588–91.

ANDRIANOV V.K., KURELLA G.A. & LITVIN F.F. (1968) Influence of light on the electrical activity of *Nitella* cells. *Abh. dt. Akad. Wiss. Berl.*, **4A**, 187–96.

ARENS K. (1939) Physiologische Multipolarität der Zelle von *Nitella* während der Photosynthese. *Protoplasma*, **33**, 295–300.

BARBER J. (1968a) Measurement of the membrane potential and evidence for active transport of ions in *Chlorella pyrenoidosa*. *Biochim. biophys. Acta*, **140**, 618–25.

BARBER J. (1968b) Sodium efflux from *Chlorella pyrenoidosa*. *Biochim. biophys. Acta*, **150**, 730–3.

BARBER J. (1968c) The influx of potassium into *Chlorella pyrenoidosa*. *Biochim. biophys. Acta*, **163**, 141–9.

BARBER J. (1968d) The efflux of potassium from *Chlorella pyrenoidosa*. *Biochim. biophys. Acta*, **163**, 531–8.

BARBER J. (1969) Light-induced net uptake of sodium and chloride by *Chlorella pyrenoidosa*. *Archs. Biochem. Biophys.* **130**, 389–92.

BARR C.E. & BROYER T.C. (1964) Effect of light on sodium influx, membrane potential and protoplasmic streaming in *Nitella*. *Pl. Physiol., Lancaster*, **39**, 48–52.

BENNETT H.S. (1956) The concept of membrane flow and membrane vesiculation as mechanisms for active transport and ion pumping. *J. biophys. biochem. Cytol.* **2**, 94–103.

BENTRUP F.W. (1971) Zellphysiologie: Elektrophysiologie der Zelle. *Fortschr. Botan.* **33**, 51–61.

BENTRUP F.W. (1970) Elektrophysiologische Untersuchungen am Ei von *Fucus serratus*: Das Membranpotential. *Planta*, **94**, 319–32.

BENTRUP F.W., SANDAN T. & JAFFE L.F. (1967) Induction of polarity in *Fucus* eggs by potassium ion gradients. *Protoplasma*, **64**, 254–66.

BLINKS L.R. (1955) Some electrical properties of large plant cells. In *Electrochemistry in Biology and Medicine*, ed. Shedlovsky T. pp. 187–212. Wiley, New York.

BLOUNT R.W. (1958) A quantitative analysis of active ion transport in the single-celled alga *Halicystis ovalis*. Ph.D. Thesis, University of California, L.A.

BLOUNT R.W. & LEVEDAHL B.H. (1960) Active sodium and chloride transport in the single-celled marine alga *Halicystis ovalis*. *Acta physiol. scand.* **49**, 1–9.

BRIGGS G.E. (1962) Membrane potential differences in *Chara australis*. *Proc. R. Soc. B.*, **156**, 573–7.

COSTER H.G.L. (1965) A quantitative analysis of the voltage-current relationships of fixed charge membranes and the associated property of 'punch-through'. *Biophys. J.* **5**, 669–86.

COSTER H.G.L. (1966) Chloride in cells of *Chara australis*. *Aust. J. biol. Sci.* **19**, 545–54.

COSTER H.G.L. (1969) The role of pH in the punch-through effect in the electrical characteristics of *Chara australis*. *Aust. J. biol. Sci.* **22**, 365–74.

COSTER H.G.L. & HOPE A.B. (1968) Ionic relations of *Chara australis*. XI. Chloride fluxes. *Aust. J. biol. Sci.* **21**, 243–54.

DAINTY J. (1962) Ion transport and electrical potentials in plant cells. *A. Rev. Pl. Physiol.* **13**, 379–402.

DAINTY J. (1969) The ionic relations of plants. In *Physiology of Plant Growth and Development*, ed. Wilkins M.B. pp. 453–85. McGraw-Hill, London.

DODD W.A., PITMAN M.G. & WEST K.R. (1966) Sodium and potassium transport in the marine alga *Chaetomorpha darwinii*. *Aust. J. biol. Sci.* **19**, 341–54.

FINDLAY G.P. (1970) Membrane electrical behaviour in *Nitellopsis obtusa*. *Aust. J. biol. Sci.* **23**, 1033–45.

FINDLAY G.P., HOPE A.B., PITMAN M.G., SMITH F.A. & WALKER N.A. (1969a) Ion fluxes in cells of *Chara corallina*. *Biochim. biophys. Acta*, **183**, 565–76.

FINDLAY G.P., HOPE A.B. & WILLIAMS E.J. (1969b) Ionic relations of marine algae. I. *Griffithsia*: membrane electrical properties. *Aust. J. biol. Sci.* **22**, 1163–78.

FINDLAY G.P., HOPE A.B. & WILLIAMS E.J. (1970) Ionic relations of marine algae. II. *Griffithsia*: ionic fluxes. *Aust. J. biol. Sci.* **23**, 323–38.

FINDLAY G.P., HOPE A.B., PITMAN M.G., SMITH F.A. & WALKER N.A. (1971a) Ionic relations of marine algae. III. *Chaetomorpha*: membrane electrical properties and chloride fluxes. *Aust. J. biol. Sci.* **24**, 731–45.

FINDLAY G.P., HOPE A.B. & WALKER N.A. (1971b) Quantization of a flux ratio in Charophytes? *Biochim. biophys. Acta*, **233**, 155–62.

GOLDMAN D.E. (1943) Potential, impedance and rectification in membranes. *J. gen. Physiol.* **27**, 37–60.

GRADMANN D. (1970) Einfluß von Licht, Temperatur und Außenmedium auf das elektrische Verhalten von *Acetabularia crenulata*. *Planta*, **93**, 323–53.

GRADMANN D. & BENTRUP F.W. (1970) Light-induced membrane potential changes and rectification in *Acetabularia*. *Naturwissenschaften*, **57**, 46–7.

GUTKNECHT J. (1966) Sodium, potassium and chloride transport and membrane potentials in *Valonia ventricosa*. *Biol. Bull. mar. biol. Lab.*, *Woods Hole*, **130**, 331–344.

GUTKNECHT J. (1967) Ion fluxes and short-circuit current in internally perfused cells of *Valonia ventricosa*. *J. gen. Physiol.* **50**, 1821–34.

GUTKNECHT J. (1968) Salt transport in *Valonia*: inhibition of potassium uptake by small hydrostatic pressures. *Science, N.Y.* **160**, 68–70.

GUTKNECHT J. & DAINTY J. (1968) Ionic relations of marine algae. *Oceanogr. mar. Biol.* **6**, 163–200.

HALL J.L. (1970) Pinocytotic vesicles and ion transport in plant cells. *Nature, Lond.* **226**, 1253–4.

HANSEN U.P. (1971) The frequency response of the action of light on the membrane potential of *Nitella*. *Biophysik*, **7**, 223–7.

HANSEN U.P. & GRADMANN D. (1971) The action of sinusoidally modulated light on the membrane potential of *Acetabularia*. *Pl. Cell Physiol.*, *Tokyo*, **12**, 335–48.

HANSEN U.P. & GRADMANN D. (in press).

HOGG J., WILLIAMS E.J. & JOHNSTON R.J. (1969) Light intensity and the membrane parameters of *Nitella translucens*. *Biochim. biophys. Acta*, **173**, 564–6.

HOPE A.B. (1965) Ionic relations of cells of *Chara australis*. X. Effects of bicarbonate ions on electrical properties. *Aust. J. biol. Sci.* **18**, 789–801.

HOPE A.B. (1971) *Ion Transport and Membranes*. Butterworths, London.

HOPE A.B. & RICHARDS J.L. (1971) Proton fluxes and permeability in Characean cells. In *Proceedings of the First European Biophysics Congress*, eds. Broda E., Locker A. & Lederer. H., Vienna, Springer-Verlag der Wiener Medizinischen Akademie.

HOPE A.B., SIMPSON A. & WALKER N.A. (1966) The efflux of chloride from cells of *Nitella* and *Chara*. *Aust. J. biol. Sci.* **19**, 355–62.

HOPE A.B. & WALKER N.A. (1960) Ionic relations of cells of *Chara australis*. III. Vacuolar fluxes of sodium. *Aust. J. biol. Sci.* **13**, 277–91.

HOPE A.B. & WALKER N.A. (1961) Ionic relations of cells of *Chara australis*. IV. Membrane potential differences and resistances. *Aust. J. biol. Sci.* **14**, 26–44.

JACQUES A.G. (1939a) The kinetics of penetration. XV. The restriction of the cellulose cell wall. *J. gen. Physiol.* **22**, 147–63.

JACQUES A.C. (1939b) The kinetics of penetration. XIX. Entrance of electrolytes and of water into impaled *Halicystis*. *J. gen. Physiol.* **22**, 757–73.

JAFFE L.F. (1966) Electric currents through the developing *Fucus* egg. *Proc. natl. Acad. Sci. U.S.A.* **56**, 1102–9.

JAFFE L.F. (1968) Localization in the developing *Fucus* egg and the general role of localizing currents. *Adv. Morphogenesis*, **7**, 295–329.

JAFFE L.F. (1969) On the centripetal course of development, the *Fucus* egg, and self-electrophoresis. *Developm. Biol. Suppl.* **3**, 83–111.

JANACEK K. & RYBOVA R. (1966) The effect of ouabain on the alga *Hydrodictyon reticulatum*. *Cytologia*, **31**, 199–202.

KEDEM O. (1961) Criteria for active transport. In *Membrane Transport and Metabolism*, eds. Kleinzeller A. & Kotyk A. pp. 87–93. Academic Press, London & New York.

KISHIMOTO U. & TAZAWA M. (1965a) Ionic composition of the cytoplasm of *Nitella flexilis*. *Pl. Cell Physiol., Tokyo*, **6**, 507–18.

KISHIMOTO U. & TAZAWA M. (1965b) Ionic composition and electric response of *Lamprothamnium succinctum*. *Pl. Cell Physiol., Tokyo*, **6**, 529–36.

KITASATO H. (1968) The influence of $H^+$ on the membrane potential and ion fluxes of *Nitella*. *J. gen. Physiol.* **52**, 60–87.

LARKUM A.W.D. (1968) Ionic relations of chloroplasts *in vivo*. *Nature, Lond.* **218**, 447–9.

LEFEBVRE J. & GILLET C. (1970) Variations du potentiel électrochimique des chlorures chez *Nitella* en présence de benzène sulphonate. *Experientia*, **26**, 482–3.

LEFEBVRE J. & GILLET C. (1971) Influence de la teneur relative en potassium et en sodium de milieu extérieur sur la différence de potentiel électrochimique des chlorures chez *Nitella*. *Bull. Soc. Roy. Bot. Belgique*, **104**, 137–49.

LILLEY R.M. (1971) Chloride active transport and photosynthesis in *Griffithsia*. Ph.D. Thesis, Flinders University.

LILLEY R.M. & HOPE A.B. (1971a) Chloride transport and photosynthesis in cells of *Griffithsia*. *Biochim. biophys. Acta*, **226**, 161–71.

LILLEY R.M. & HOPE A.B. (1971b) Adenine nucleotide levels in cells of the marine alga *Griffithsia*. *Aust. J. biol. Sci.* **24**, 1351–4.

LUCAS W.J. (1971) Some aspects of the pH-banding phenomenon along the cell wall of *Chara corallina*. B.Sc. Hon. Thesis, University of Adelaide.

LÜTTGE U. (1969) Aktiver Transport (Kurzstreckentransport bei Pflanzen). In *Protoplasmatologia VIII*, **7b**. Springer, Wien-New York.

MACROBBIE E.A.C. (1962) Ionic relations of *Nitella translucens*. *J. gen. Physiol.* **45**, 861–78.

MACROBBIE E.A.C. (1964) Factors affecting the fluxes of potassium and chloride ions in *Nitella translucens*. *J. gen. Physiol.* **47**, 859–77.

MACROBBIE E.A.C. (1965) The nature of the coupling between light energy and active ion transport in *Nitella translucens*. *Biochim. biophys. Acta*, **94**, 64–73.

MACROBBIE E.A.C. (1966a) Metabolic effects on ion fluxes in *Nitella translucens*. I. Active fluxes. *Aust. J. biol. Sci.* **19**, 363–70.

MACROBBIE E.A.C. (1966b) Metabolic effects on ion fluxes in *Nitella translucens*. II. Tonoplast fluxes. *Aust. J. biol. Sci.* **19**, 371–83.

MACROBBIE E.A.C. (1969) Ion fluxes to the vacuole of *Nitella translucens*. *J. exp. Bot.* **20**, 236–56.

MACROBBIE E.A.C. (1970a) The active transport of ions in plant cells. *Q. Rev. Biophys.* **3**, 251–94.

MACROBBIE E.A.C. (1970b) Quantized fluxes of chloride to the vacuole of *Nitella translucens*. *J. exp. Bot.* **21**, 335–44.

MACROBBIE E.A.C. (1971a) Fluxes and compartmentation in plant cells. *A. Rev. Pl. Physiol.* **22**, 75–96.

MACROBBIE E.A.C. (1971b) Vacuolar fluxes of chloride and bromide in *Nitella translucens*. *J. exp. Bot.* **22**, 487–502.

MACROBBIE E.A.C. & DAINTY J. (1958) Ion transport in *Nitellopsis obtusa. J. gen. Physiol.* **42**, 335–53.

METLIČKA R. & RYBOVA R. (1967) Oscillations of the trans-membrane potential difference in the alga *Hydrodictyon reticulatum. Biochim. biophys. Acta*, **135**, 563–5.

MITCHELL P. (1961) Coupling of phosphorylation to electron and hydrogen transfer by a chemi-osmotic type of mechanism. *Nature, Lond.* **191**, 144–8.

MITCHELL P. (1962) Molecule, group and electron translocation through natural membranes. *Biochem. Soc. Symp.* **22**, 142–69.

MITCHELL P. (1966) Chemi-osmotic coupling in oxidative and photosynthetic phosphorylation. *Biol. Rev.* **41**, 445–502.

MORETON R.B. (1969) An investigation of the electrogenic sodium pump in snail neurones, using the constant field theory. *J. exp. Biol.* **51**, 181–201.

NAGAI R. & TAZAWA M. (1962) Changes in resting potential and ion asorption by light in a single plant cell. *Pl. Cell Physiol., Tokyo*, **3**, 323–39.

NISHIZAKI Y. (1968) Light-induced changes of bioelectric potential in *Chara. Pl. Cell Physiol., Tokyo*, **9**, 377–87.

PENTH B. & WEIGL J. (1971) Anionen-Influx, ATP Spiegel und $CO_2$-Fixierung in *Limnophila gratioloides* und *Chara foetida. Planta, Berlin*, **96**, 212–23.

PITMAN M.G. (1970) Ion transport in plant cells. In *Intestinal Absorption of Metal Ions, Trace Elements and Radionuclides*, eds. Skoryna S.C. & Waldron-Edward D. pp. 115–33. Pergamon Press, Oxford, New York & London.

RAVEN J.A. (1967a) Ion transport in *Hydrodictyon africanum. J. gen. Physiol.* **50**, 1607–25.

RAVEN J.A. (1967b) Light-stimulation of active ion transport in *Hydrodictyon africanum. J. gen. Physiol.* **50**, 1627–40.

RAVEN J.A. (1968a) The linkage of light-stimulated Cl influx to K and Na influxes in *Hydrodictyon africanum. J. exp. Bot.* **19**, 233–53.

RAVEN J.A. (1968b) Photosynthesis and light-stimulated ion transport in *Hydrodictyon africanum. Abh. dt. Akad. Wiss., Berl.*, **4A**, 145–51.

RAVEN J.A. (1968c) The action of phlorizin on photosynthesis and light-stimulated ion transport in *Hydrodictyon africanum. J. exp. Bot.* **19**, 712–23.

RAVEN J.A. (1968d) The mechanism of photosynthetic use of bicarbonate by *Hydrodictyon africanum. J. exp. Bot.* **19**, 193–206.

RAVEN J.A. (1969a) Action spectra for photosynthesis and light-stimulated ion transport processes in *Hydrodictyon africanum. New Phytol.* **68**, 45–62.

RAVEN J.A. (1969b) Effects of inhibitors on photosynthesis and the active influxes of K and Cl in *Hydrodictyon africanum. New Phytol.* **68**, 1089–113.

RAVEN J.A. (1971a) Ouabain-insensitive K influx in *Hydrodictyon africanum. Planta*, **97**, 28–38.

RAVEN J.A. (1971b) Cyclic and non-cyclic phosphorylation as energy sources for active K influx in *Hydrodictyon africanum. J. exp. Bot.* **22**, 420–33.

RAVEN J.A. (1971c) Inhibitor effects on photosynthesis, respiration and active ion transport in *Hydrodictyon africanum. J. Membrane Biol.* **6**, 89–107.

RAVEN J.A., MACROBBIE E.A.C. & NEUMANN J. (1969) The effect of Dio-9 on photosynthesis and ion transport in *Nitella, Tolypella* and *Hydrodictyon. J. exp. Bot.* **20**, 221–35.

ROBINSON J.B. (1969a) Sulphate influx in Characean cells. I. General characteristics. *J. exp. Bot.* **20**, 201–11.

ROBINSON J.B. (1969b) Sulphate influx in Characean cells. II. Links with light and metabolism in *Chara australis. J. exp. Bot.* **20**, 212–20.

ROBINSON K.R., JOAQUIN J.C., WEISENSEEL M.H. & JAFFE L.F. (1970) Potassium ion movements in the developing egg of a Fucoid alga. *Abstr. Biophys. Soc.* **10**, 227a.

ROTHSTEIN A. (1964) Membrane function and physiological activity of microorganisms. In *Cellular Functions of Membrane Transport*, ed. Hoffman J.F. Prentice-Hall Inc., Englewood Cliffs, New Jersey.

SADDLER H.D.W. (1970a) The ionic relations of *Acetabularia mediterranea. J. exp. Bot.* **21**, 345–59.

SADDLER H.D.W. (1970b) Fluxes of sodium and potassium in *Acetabularia mediterranea. J. exp. Bot.* **21**, 605–16.

SADDLER H.D.W. (1970c) The membrane potential of *Acetabularia mediterranea. J. gen. Physiol.* **55**, 802–21.

SADDLER H.D.W. (1971) Spontaneous and induced changes in the membrane potential and resistance of *Acetabularia mediterranea. J. Membrane Biol.* **5**, 250–60.

SCHILDE C. (1966) Zur Wirkung des Lichts auf das Ruhepotential der grünen Pflanzenzelle. *Planta*, **71**, 184–8.

SCHILDE C. (1968a) Zellphysiologie, Elektrophysiologie der Zelle. *Fortschr. Botan.* **30**, 44–56.

SCHILDE C. (1968b) Schnelle photoelektrische Effekte der Alge *Acetabularia. Z. Naturf.* **23b**, 1369–76.

SCHULTZ S.G., EPSTEIN W. & SOLOMON A.K. (1963) Cation transport in *Escherichia coli.* IV. Kinetics of net K uptake. *J. gen. Physiol.* **47**, 329–46.

SHIEH Y.J. & BARBER J. (1971) Intracellular sodium and potassium concentrations and net cation movements in *Chlorella pyrenoidosa. Biochim. biophys. Acta*, **233**, 594–603.

SMITH F.A. (1966) Active phosphate uptake by *Nitella translucens. Biochim. biophys. Acta*, **126**, 94–9.

SMITH F.A. (1967) The control of Na uptake into *Nitella translucens. J. exp. Bot.* **18**, 716–731.

SMITH F.A. (1968a) Metabolic effects on ion fluxes in *Tolypella intricata. J. exp. Bot.* **19**, 442–51.

SMITH F.A. (1968b) Rates of photosynthesis in Characean cells. II. Photosynthetic $^{14}CO_2$ fixation and $^{14}$C-bicarbonate uptake by Characean cells. *J. exp. Bot.* **19**, 207–17.

SMITH F.A. (1970) The mechanism of chloride transport in Characean cells. *New Phytol.* **69**, 903–17.

SMITH F.A. (1972) Stimulation of chloride transport in *Chara* by external pH changes. *New Phytol.* **71**, 595–601.

SMITH F.A. & WEST K.R. (1969) A comparison of the effects of metabolic inhibitors on chloride uptake and photosynthesis in *Chara corallina. Aust. J. biol. Sci.* **22**, 351–63.

SPANSWICK R.M. & WILLIAMS E.J. (1964) Electric potentials and Na, K and Cl concentrations in the vacuole and cytoplasm of *Nitella translucens. J. exp. Bot.* **15**, 193–200.

SPANSWICK R.M., STOLAREK J. & WILLIAMS E.J. (1967) The membrane potential of *Nitella translucens. J. exp. Bot.* **18**, 1–16.

SPEAR D.G., BARR J.K. & BARR C.E. (1969) Localization of hydrogen ion and chloride ion fluxes in *Nitella. J. gen. Physiol.* **54**, 397–414.

STEWARD F.C. & MOTT R.L. (1970) Cells, solutes, and growth: salt accumulation in plants re-examined. *Int. Rev. Cytol.* **28**, 275–370.

SUTCLIFFE J.F. (1962) *Mineral Salts Absorption in Plants.* Pergamon Press, Oxford, London, New York & Paris.

TEORELL T. (1949) Membrane electrophoresis in relation to bio-electrical polarisation effects. *Arch. Sci. physiol.* **3**, 205–19.

TEORELL T. (1953) Transport processes and electrical phenomena in ionic membranes. *Progr. Biophys. biophys. Chem.* **3**, 305–69.

THROM G. (1970) Die lichtabhängige Änderung des Membranpotentials bei *Griffithsia setacea. Z. PflPhysiol.* **63**, 162–80.

USSING H.H. (1949) The distinction by means of tracers between active transport and diffusion. *Acta physiol. scand.* **19**, 43–56.

VOROBIEV L.N. (1967) Potassium ion activity in the cytoplasm and the vacuole of cells of *Chara* and *Griffithsia. Nature, Lond.* **216**, 1325–7.

VREDENBERG W.J. (1969) Light-induced changes in membrane potential of algal cells associated with photosynthetic electron transport. *Biochem. biophys. Res. Commun.* **37**, 785–92.

VREDENBERG W.J. (1970a) Application of the voltage-clamp technique for measuring the quantum efficiency of light-induced potential changes in *Nitella translucens. Biochim. biophys. Acta*, **216**, 431–4.

VREDENBERG W.J. (1970b) Chlorophyll *a* fluorescence induction and changes in the electric potential of the cellular membranes of green plants. *Biochim. biophys. Acta*, **223**, 230–9.

VREDENBERG W.J. (1971) Changes in membrane potential associated with cyclic and non-cyclic electron transport in photochemical system I in *Nitella translucens. Biochem. biophys. Res. Comm.* **42**, 111–18.

WALKER N.A. & HOPE A.B. (1969) Membrane fluxes and electrical conductance in Chara-cean cells. *Aust. J. biol. Sci.* **22**, 1179–95.

WILLIAMS E.J. & BRADLEY J. (1968) Steady-state membrane hyperpolarisation by large applied currents in *Nitella translucens. Biophys. J.* **8**, 145–7.

# CHAPTER 26

# PHYSICO-CHEMICAL FACTORS AFFECTING METABOLISM AND GROWTH RATE

## C. SOEDER and E. STENGEL

Kohlenstoffbiologische Forschungsstation, e.V.,
Dortmund, Germany.

1   Introduction   714

2   **Light**   715
2.1 Light and algal growth   715
2.2 Light and heterotrophic growth 716
2.3 Light adaptations   716
2.4 Inhibition by light   717
2.5 Effect of light on various metabolic processes   718
2.6 Light and the chemical composition of algae   720

3   **Temperature**   721
3.1 Temperature requirements and temperature tolerance   721
3.2 Temperature and growth rate 723
3.3 Temperature and cell growth 723
3.4 Adaptation to temperature 724
3.5 Effect of temperature on various metabolic processes   724

4   **Hydrogen ion concentration** 726
4.1 General   726
4.2 pH requirements and pH tolerance   726
4.3 Effect of pH on various metabolic processes   727

5   **Osmotic effects**   727
5.1 General   727
5.2 Adaptation to changes of osmotic pressure   728
5.3 Effect of osmotic pressure and salinity on various metabolic processes   728

6   **Mechanical factors**   729
6.1 Turbulence   729
6.2 Hydrostatic pressure   730

7   **References**   730

## 1   INTRODUCTION

This chapter is concerned with the effects of various physico-chemical factors on algal growth and physiology. It is important to appreciate at the outset that the effects of external factors show complex interactions and that an optimum level of one factor under certain conditions may be sub-optimum under other conditions. A good example of this is the finding that the optimum temperature for growth may vary depending on the light intensity employed. Responses also vary depending on the physiological state of the experimental material and whether it is unadapted, adapting, or adapted. The physiological ecology of algae is not dealt with in detail here but is considered in Lewin (1962), Lund (1965) and Allen (1969). Brock (1969) has considered the growth of algae under extreme environmental conditions.

## 2 LIGHT

### 2.1 *Light and algal growth*

The relationships between light intensity and the rates of photosynthesis and photoautotrophic growth show a rectangular hyperbolic function with an inhibition of growth occurring at supersaturating light intensities. A similar situation also applies to algal suspensions growing in a light-and-dark cycle (Sorokin & Krauss 1962). The shape of light-photosynthesis curves and of light-growth curves is markedly affected by temperature (Sorokin & Krauss 1962, Setlik *et al.* 1969, Smayda 1969) and by other factors such as salinity (McCombie 1960) and nutrient level (Maddux & Jones 1964). Various mathematical models have been designed to describe light-photosynthesis curves and their integrals. The most widely accepted model is probably that of Shelef *et al.* (1969) which considers light intensity, light quality and quantum conversion efficiency. Interactions of light intensity and suspension density are analysed by Belyanin and Kovrov (1968) and Govindjee *et al.* (1968). Simmer (1969) and Trukhin (1970) deal with important quantitative aspects of light energy conversion (see also Kok 1965).

The light dependency of photosynthesis is important for calculating primary production rates in pelagic ecosystems. A mathematical model for the integration of pelagic photosynthesis in space and time, and which includes light inhibition effects, has been developed by Vollenweider (1965, 1969). *In situ* measurements show that the decrease of primary production rates with depth is usually a function of green light intensity, i.e. of the most penetrating light component (Rodhe 1965). Light quality at various depths in natural waters is considered by Talling (1971).

The relativity of light-photosynthesis curves has been demonstrated by Setlik (1968). He found that the curve obtained by a stepwise increase in light intensity was quite different from that obtained by gradually decreasing the light intensity from saturating levels. In addition, the characteristics of the light photosynthesis curves and of photosynthetic rates obtained under standard conditions vary with the developmental stage in green algae (Metzner & Lorenzen 1960, Sorokin 1960, Senger 1969). There are also variations in quantum yield and in the degree of Emerson enhancement depending on the stage of the cellular life cycle of the algae used (Senger 1970).

Extracellular release of photosynthetic carbon products also varies with light intensity and seems to reach its maximum at light intensities which inhibit photosynthesis slightly (Watt 1969). At even higher light intensities there may be an inhibition of extracellular photosynthate release (Fogg *et al.* 1965). The release of extracellular carbon is also high at very low light intensities (Watt 1966, 1969, Watt & Fogg 1966). Light quality may also affect the release of extracellular carbon. For example, photosynthetic excretion of glycolate by *Chlorella* is stimulated by red light and inhibited by blue light (Becker *et al.* 1968, Lord *et al.* 1970).

## 2.2   *Light and heterotrophic growth*

In the presence of suitable organic substrates growth and metabolism of photosynthetic algae are often enhanced by light of appropriate intensity (see e.g. Senger 1962). This light stimulation, which is apparently lacking in *Euglena* (Cook 1965) is usually attributable to light-driven transport and phosphorylation via photosystem I and has been studied in the presence (Lysek & Simonis 1968) and absence (see e.g. Tanner 1969) of $CO_2$. The linkage of photoheterotrophy to photosystem I has been demonstrated in certain members of the Volvocales (Wiessner 1969) and the quantum requirement of *Chlorella* for anaerobic glucose uptake was determined by Tanner *et al.* (1968). Very dim light stimulates heterotrophic growth in *Chromulina* sp. (Pintner & Provasoli 1968) and certain *Chlorella* strains cannot grow heterotrophically but are able to utilize organic substrate in light well below the compensation point of photosynthesis (Karlander & Krauss 1966). The action spectrum of this light effect is characterized by a major peak at 425nm and a smaller one around 575nm. A porphyrin precursor of cytochromes may be the photoreceptor molecule. The action spectrum obtained by Karlander and Krauss (1966) differs clearly from the action spectrum for the induction of cell division and nucleic acid synthesis in glucose-mixotrophic *Chlorella fusca* (Senger & Bishop 1966, Senger & Schoser 1966). Blue light directs the metabolism of exogenous glucose mainly towards the synthesis of protein while carbohydrates are preferentially synthesized from glucose in red light (Laudenbach & Pirson 1969). Many blue-green algae appear to be obligate photoautotrophs (see Holm-Hansen 1968, Lange 1970) but nevertheless some may grow heterotrophically (Khoja & Whitton 1971) and a light-enhanced uptake of sugars and acetate (Hoare *et al.* 1967) and amino acids (Kuz'menko 1968) has been demonstrated in some species (see also Fogg *et al.* 1973).

## 2.3   *Light adaptations*

A number of freshwater and marine planktonic algae have been investigated with respect to the kinetics of their light-photosynthesis curves (cf. Steemann

Nielsen & Jørgensen 1968), after long-term precultivation at different light intensities and subsequent to changes in light intensity (Jørgensen 1969). Two types of adaptive reaction can be distinguished. The most usual is the 'Chlorella type' which is characterized by an inverse relationship between the light intensity to which the algae are exposed and their chlorophyll a content (see also Beale & Appleman 1971). That is, the light adaptation of this type is mainly accomplished by changes in pigment concentration. Algae belonging to the 'Cyclotella type' on the other hand show an inverse correlation between the activities and/or concentrations of photosynthetic enzymes and light intensity.

The time required by Cyclotella (Jørgensen 1964), and Chlorella vulgaris (Steemann Nielsen et al. 1962) to adapt to a new light intensity is less than 30 hours but the blue-green alga Oscillatoria rubescens takes 7 to 14 days to adapt to a light intensity change from 400 to 1,400 lx and sometimes, in fact, there is no evidence of adaptation. This fits in with the ecological finding (Meffert 1971) that planktonic species of Oscillatoria are usually found in the deeper layers of the euphotic zone and are easily injured by intense illumination (Baker et al. 1969, Zimmermann 1969, Meffert 1971). The preference of Cryptomonas sp. for greater depths (Ichimura et al. 1968) may also be a reflection of its preference for low light intensities. Response to light could be a factor regulating the daily rhythm of vertical migration of phytoplankton (see e.g. Berman & Rodhe 1971, Talling 1971).

Siphonaceous green algae living in the skeleton of reef corals below the anthozoa layer provide an interesting example of adaptation both to very low light intensity (less than 150 lx) and to light quality. Photosynthesis by these algae is saturated at 10 lx and photoinhibition begins at 500 lx (Frankzisket 1968). Oxygen evolution at 720nm equals the rate at 680nm which points to an unusually high content of chlorophyll $a_{720}$ (Halldal 1968). Chlorella can also adapt to unusual light qualities (Öquist 1969). If grown in continuous far-red light, the action spectrum extends up to 740 nm and also shows increased activity in blue light. On the other hand certain algae show extreme light hardiness. Examples are Zygogonium (Lynn & Brock 1969) and certain Chlorella strains which in dense turbulent suspensions tolerate light intensities as high as 30,000 lx, the latter being equivalent to about 30 times full sunlight (Matthern et al. 1969). Light adaptations of natural populations of benthic diatoms and phytoplankton have been considered by Patrick (1968), and by Javornicky (1966, 1970) respectively.

### 2.4  Inhibition by light

Impairment of photosynthesis and photoautotrophic growth by supra-optimal intensities of visible light is well known from laboratory (see e.g. Sorokin & Krauss 1962) and field studies (Goldman et al. 1963, Lund 1965, Talling 1971). The so-called surface inhibition observed in pelagic communities seems to be due to damage by ultraviolet radiation (Steemann Nielsen 1964, Elster & Motsch 1966, Findenegg 1966).

The effects of ultraviolet light on algae are summarized by Halldal (1970). The short wavelength action spectrum for inhibition of regreening in *Euglena* (maximal effect at 260nm) points to a direct action upon nucleic acids (Lyman *et al*. 1960). Photoreactivation, as accomplished by exposure of ultraviolet-treated cells to visible light, is optimal at about 350nm (Lyman *et al*. 1960). The site for both ultraviolet damage and its photoreactivation is chloroplast DNA (Shneyour *et al*. 1969). Related studies have been carried out on *Chlorella* (Gilet & Terrier 1969), *Anacystis* (Asato & Folsome 1969), and *Chlamydomonas* (see e.g. Davies *et al*. 1969). In all, a reversion of the mutagenic effect of ultraviolet light has been noted on exposure of the cells to visible light.

High light intensities may also inhibit respiration of actively photosynthesizing cells (see, e.g. Brown & Tregunna 1967). Such photoinhibition is greatest in blue light, and the size of the effect depends in *Chlorella* on respiratory activity prior to illumination (Ried 1969a). The action spectrum corresponds closely to that of photosystem I (Ried 1970). This, together with the sensitivity to various inhibitors, indicates that cyclic photophosphorylation reduces the rate of respiration via an ADP drain from the mitochondria. The inhibition of colourless algae such as *Prototheca* (Epel & Butler 1970) by high light intensities is also due to an inhibition of respiration. Blue light leads to a destruction of cytochromes, especially of cytochrome $a_3$ (Epel & Butler 1969).

Destruction of chlorophyll by intense light is favoured by $O_2$ and involves a stimulation of chlorophyllase activity (Ziegler & Schanderl 1969). Margulis (1968) has studied the reversibility of light bleaching in *Euglena*. In experiments with two heterotrophic mutants of *Chlorella*, Gross and Dugger (1969) found that light, tolerable to a yellow-in-the-dark mutant, severely inhibited the non-pigmented mutant. Blue light was most inhibitory and green light was ineffective. During the regreening of *Chlorella protot500ides* an inhibition by light of DNA synthesis and of subsequent cell division has been observed with blue light again being most active (Sokawa 1967a,b Sokawa & Hase 1968). *In vitro* inhibition of algae by white light depends to some extent upon the type of light source. For instance, fluorescent light of moderate intensity inhibits growth and division in green *Euglena* although the same intensity of incandescent light has no adverse effect (Cook 1968). Both qualities of white light are, however, inhibitory to bleached cultures of the same strain. Sensitivity of *Chlorella* to strong light or to a sudden increase of light intensity varies with cellular age in synchronous cultures (Sorokin 1960, Pirson *et al*. 1959). The role of blue light in the most susceptible stage is discussed by Pirson and Ruppel (1962). Close to the upper threshold of temperature tolerance *Chlorella* can be grown only in a light-and-dark cycle and not in continuous light (Lorenzen 1963).

## 2.5 *Effect of light on various metabolic processes*

Ion transport is often light dependent and is closely interrelated with changes in membrane potentials. By inserting electrodes directly into individual *Chlorella* cells, Barber (1968a) was able to show that the equilibrium potentials for $K^+$, $Na^+$, and $Cl^-$ are all light dependent. Throm (1971) concluded from studies on the light-induced changes of membrane potential between the medium and the vacuole in *Griffithsia*, that ion uptake was linked to non-cyclic electron transport via photosystem II. Light-dependent changes of the resting potential and of the membrane potential have also been studied in *Acetabularia* by Gradmann (1970) (see also Bentrup 1971, and Chapter 25, p. 676).

Concomitant with the light-stimulated uptake of $K^+$ or $Na^+$ in *Hydrodictyon africanum* there is a stimulation of $Cl^-$ uptake. The action spectrum for the responsible Na-K-Cl pump coincides with photosystem II (Raven 1969a). By contrast, the chloride-independent $K^+$ influx is linked to photosystem I (Raven 1969b) but the active $K^+$ transport, involving the ATPase system of the plasmalemma, is more sensitive to the inhibitor Dio-9 than is photosynthesis (Raven *et al.* 1969). The light-stimulated portions of the alkali-ion effluxes share several important characteristics with the respective influxes (Barber 1968a, Raven 1968a). The light-dependent uptake of $HCO_3^-$ at high pH may depend on a specific, ATP-independent $HCO_3^-$ pump coupled to photosystem I activity (Raven 1968b).

Sulphate uptake and transport in several charophytes has been studied by Robinson (1969a,b). There was an enhancement by light of sulphate transport across the plasmalemma and tonoplast. The transient stimulation of sulphate influx subsequent to darkening resembles the effect of adding DCMU. Although not directly $CO_2$-dependent (Robinson 1969a,b), the stimulation of sulphate influx by light is enhanced by $CO_2$ and accompanied by a preferential incorporation of sulphur into lipids (Kylin 1966).

Nitrate and nitrite uptake is light dependent and is linked to photosystem II activity in chloroplasts (see e.g. Grant 1967, Morris & Ahmed 1968). Two types of interaction between nitrite uptake rate and $CO_2$ supply can be distinguished (Strotmann 1967a). At constant quantum flow nitrite uptake in blue light (465 nm) is much greater than uptake in red light (647nm) or white light (Strotmann 1967b). This blue light effect consists of a relative stimulation of respiration-linked reduction of nitrite, and the quantum requirement is lower than that for photosynthetic nitrite reduction via photosystem II (Strotmann & Ried 1969). In the presence of nitrate a transient excretion of nitrite occurs upon illumination (Strotmann & Ried 1969).

The stimulation of algal respiration by blue light has been investigated by numerous authors (see e.g. Kowallik 1967, Kowallik & Gaffron 1966, Ried 1967, 1969a). Sargent and Taylor (1971) consider that there are two different blue light effects. Firstly, there is a stimulation of normal endogenous respiration (see Ried 1970). Secondly, a stimulation occurs via the FMN-dependent amino acid oxidase system (Schmid & Schwarze 1969). The action spectra for blue light-stimulated respiration (Kowallik & Kowallik 1969, Ried 1969a) has not yet solved the question of whether carotenoids or flavoproteins are the photoreceptor molecules in this reaction.

Nitrogen fixation by blue-green algae and its light dependency has been studied by various workers (see Chapter 20, p. 560). The action spectrum for this photometabolic activity as measured by the nitrogenase-dependent reduction of acetylene to ethylene indicates a primary involvement of photosystem I (Fay 1970, Lyne & Stewart 1973).

Coccolith formation in *Coccolithus* sp. is also light-dependent (Paasche 1968, 1969). The non-motile form of this flagellate may deposit several times more carbon into the calcium carbonate of its coccoliths than into photosynthetic products. Another light-dependent process is the recovery of *Anacystis* from manganese deficiency (Gerhardt & Wiessner 1967).

The influence of light on the toxic effects of various poisons on algae are of practical as well as physiological significance. For example, the threshold concentrations of several antibiotics which inhibit *Anacystis* are distinctly lower in the light than in the dark, although the toxicity of copper and mercury is reduced by illumination (Whitton 1968). In dark-grown but not in light-grown cells of *Euglena*, chloramphenicol produced permanently bleached cells and it has been suggested that this is due to an inhibition of proplastid replication in the dark (Miyoshi & Tsubo 1969).

### 2.6 *Light and the chemical composition of algae*

During adaptation of autotrophic *Euglena* to a higher light intensity a decrease of chlorophyll and lipid content is observed. At the same time the relative concentrations of unsaturated fatty acids and of monogalactosyl-glycerides increase markedly in the chloroplasts (Constantopoulos & Bloch 1967). In apochlorotic *Euglena* light stimulates carotenogenesis, the optimal effect occurring in blue light (Dolphin 1970). In *Chlorococcum wimmeri* the synthesis of red carotenoids (astaxanthin and its esters) is a function of light intensity (Brown *et al.* 1967).

The induction by light of chlorophyll synthesis in glucose-bleached *Chlorella prototothecoides* (Sokawa 1967a) parallels the kinetics of plastoquinone biosynthesis (Oku *et al.* 1968). Upon illumination of dark-grown *Chlamydomonas* cells, there is a transformation of protochlorophyll into chlorophyll. The process is stimulated by a preillumination of only 10 seconds and is slowed down by high light intensities (Matsuda *et al.* 1971). In regreening *Euglena* marked changes in the action spectrum (Kirk & Reade 1970) may be indicative of shifts in the proportions of the various pigments.

The enhancement of protein synthesis by blue light and the increases in the rate of carbohydrate synthesis by red light have been studied in various algae (e.g. Hauschild *et al.* 1962, 1964, Kowallik 1962 and Trukhin 1968 for *Chlorella*, and Peterfi *et al.* 1969 for *Scenedesmus*). Despite the frequent response of green algae to either blue or red light, one has to differentiate between *Chlorella* which tolerates long-term cultivation in red light (Öquist 1969) and *Acetabularia* where extended exposure to red light leads to degeneration (Clauss 1968, 1970).

Starvation of *Chlorella* in red flashing light leads to a decrease of protein content, while blue flashing light keeps the protein/carbohydrate ratio at a constant level (Pickett 1971). The inhibition of glycolate excretion by blue light has been related to the blue light effect on protein synthesis (Becker *et al.* 1968).

A detailed comparison of the properties of synchronous *Chlorella* cells grown in either red or blue light has been carried out by Kowallik (1962, 1963, 1965). The cultures were adjusted to give the same yield of dry matter per unit culture volume in one light-and-dark cycle. During the first cycles after transfer to monochromatic light, red-light cells attain a higher carbohydrate content, a lower content in protein and RNA, and the average cell divides into a greater number of daughter cells (14) than do blue-light cells (10 to 12). The reduction of autospore number by blue light is due to a partial inhibition of DNA synthesis. These initial discrepancies between red-light cells and blue-light cells disappear, however, after prolonged cultivation in monochromatic light. In the case of the blue-light cells this could perhaps be due to the selection of a specific mutant during the experimental period (Kowallik 1963).

An action spectrum for the stimulation of protein synthesis by blue light has been obtained by growing *Chlorella* on glucose in monochromatic light. One broad maximum between 450 and 490nm was found (Kowallik 1965) and this agrees fairly well with the action spectra for nucleic acid synthesis and for the induction of cell division in the same strain (Senger & Bishop 1966, Senger & Schoser 1966).

# 3  TEMPERATURE

## 3.1  *Temperature requirements and temperature tolerance*

Temperature is an important factor determining the general geographical distribution of certain algae. An impressive example is the relationship in benthic algae between heat hardiness and geographical latitude (see Biebl 1970). Algal life at extreme temperatures has been reviewed by Brock (1969) with special reference to the heat tolerance of thermophilic forms, the upper limit of which seems to be 74°C. The monograph by Kol (1968) deals with algae inhabiting the surfaces of snow and ice and the physiological characteristics of cryophilic algae are dealt with by Fogg and Horne (1968). The vertical zonation of benthic marine algae in the littoral zone, once thought to be due solely to heat hardiness on the basis of cytological examination of the tissue after temperature shock does not in fact appear to be the whole answer as studies on the restoration kinetics of photosynthesis by the algae show clearly (Ried 1969b). The tropical diatom, *Skeletonema tropicum*, ceases to grow if the temperature drops below 13°C (Hulburt & Guillard 1968). Another alga confined for physiological reasons to the warmer seas is *Caulerpa* (Kajimura 1968). Upper temperature limits for *Fragilaria capucina* and *F. crotonensis* are 26 and 30°C respectively, whereas

*Nitzschia actinastroides* isolated from the same lake shows almost optimal growth at 34°C (Müller 1970).

Thermophilic algae can be cultured at comparatively low temperatures. For example *Cyanidium* which has an upper temperature limit of 55 to 60°C (Doemel & Brock 1970) grows well at 15°C (Becker 1972). Cryophilic algae, on the other hand, seem to be much more stenothermic. For example, the temperature optimum of *Koliella tatrae* is 4°C and the maximum is 10°C (Hindak & Komarek 1968). Lange and Metzner (1965) have measured active photosynthesis below 0°C in lichens, and *Nostoc* sp. and *Prasiola crispa* growing on Antarctic soil can photosynthesize down to −5 and −20°C respectively (Becker 1972). Algae from liquid Antarctic and Arctic freshwaters, however, are much more sensitive to low temperatures (Biebl 1969a, see also Fogg & Stewart 1968), while the Antarctic marine diatom, *Fragilaria sublinearis* (optimum 7°C) shows good photosynthetic activity at −2°C (Bunt 1965, Bunt *et al.* 1966).

There may be significantly different temperature optima for different species of the same genus. For example, species of the genus *Chlorella* differ markedly in their temperature optima which may be at 25°C, 30°C or 39°C (Sorokin 1959, Lorenzen 1963). A stenothermic *Chlorella* strain growing only between 31 and 35°C has been described by Spektorova (1967) and the existence of clones of *Rhizosolenia fragilissima* with differing temperature requirements has been investigated by Ignatiades and Smayda (1970).

The artificial character of laboratory studies using constant temperatures should always be appreciated. For example, a strain of *Scenedesmus* with a constant-temperature maximum of 34°C (Hegewald unpubl.) grows without inhibition under outdoor conditions in the tropics at noon peaks of 45°C (Payer unpubl.).

Prolonged incubation in the dark at 0°C can be tolerated by various algae (Müntz 1965a,b, Müller-Stoll & Müntz 1971). Stock cultures of green algae remain viable for up to six years if stored in the refrigerator (Odoevskaya *et al.* 1969). The success of lyophilization of blue-green and green algae depends on pretreatment and storage conditions (Holm-Hansen 1963, 1967, 1968). Lyophilized *Nostoc muscorum* shows no loss of viability after five years of storage at +25°C, whereas all green algae which were tested died off gradually under the same conditions (Holm-Hansen 1967). Frozen thalli of *Porphyra* survive six months at −12°C (Taniguchi 1969). Temperature hardiness of *Porphyridium* has been studied by Rieth (1967).

Resting stages of many algae possess a greater tolerance to extreme temperatures than do the vegetative stages. For example, desiccated cysts of freshwater members of the Prasinophyceae remain viable after exposure to 100°C for one hour (Belcher 1970). A similar heat hardiness is reported for zygotes of *Furcilla* although the upper temperature limit of the vegetative flagellate does not exceed 35°C (Belcher 1967).

Heat hardiness is based, in part at least, on the heat stability of enzymes (Brock 1969). Thus, benzylviologen-stimulated nitrate reductase is stable at temperatures up to 60°C (Rigano 1971). The molecular basis for cold hardiness in algae is not yet clear. It may be, from considering data for other groups of organisms (for higher plants, see Heber 1969), that the synthesis of substances such as free sugars, oligosaccharides etc., prevents the formation of ice crystals in the cell and thus decreases the freezing point of the cytoplasm.

### 3.2 Temperature and growth rate

The frequent interaction of light intensity and temperature on photoautotrophic growth of algae has already been emphasized (see p. 715). Three dimensional plots by Setlik *et al.* (1969) of photosynthetic rate versus temperature and light intensity resemble in principle the curves for the growth of microalgae in crossed gradients of the two latter factors (Halldal & French 1958). Provided that only temperature is limiting, a more or less exponential increase of yield with temperature occurs in synchronous *Chlorella* cells between 20 and 30°C, the $Q_{10}$ being about 6 (Soeder *et al.* 1967). With increasing optical density of the suspension, a marked decrease of $Q_{10}$ occurs and the temperature optimum also decreases somewhat (unpubl. data of Hegewald for *Scenedesmus*). The interactions of temperature with light supply in *Chlorella* have been studied and mathematically analysed by Slobodskoi *et al.* (1969). Temperature has little effect on photosynthesis and growth in light-limited systems and it is not surprising, therefore, that the temperature profile of standing waters is seldom considered in current models of vertical profiles of pelagic photosynthesis (see e.g. Rodhe 1965, Vollenweider 1965). In adapted cultures of the marine diatom, *Skeletonema costatum*, growth is only 50% better at 20°C than at 7°C (Steemann-Nielsen & Jørgensen 1968, Jørgensen 1968).

### 3.3 Temperature and cell growth

The differential effect of temperature on growth and cell division has been studied particularly in *Chlorella* (Sorokin & Krauss 1962, Lorenzen 1963, Semenenko *et al.* 1967), in *Astasia* (Wilson & James 1966) and in bleached *Euglena*. In general, the processes involving DNA synthesis and subsequent cell division are more sensitive to sub-optimal temperatures than is the growth phase of the cellular life cycle. A synchronization of algae by periodic temperature changes can be accomplished using this feature in *Anacystis* (Venkataraman *et al.* 1969), *Astasia* (James 1962) and *Euglena* (Terry & Edmunds 1969). At a particular period of the cellular life cycle of *Chlorella fusca*, growth is impaired by decreasing the temperature from 30 to 4°C for two hours (Pirson *et al.* 1959). The resultant bleaching of the alga, which is affected by light intensity and $O_2$ level, follows an inhibition of photosynthesis (Lorenzen 1963). A cooling from 30°C to either 1–2°C, or 7°C, and even freezing at −20°C for one hour are, by comparison, tolerated well. Bleaching of *Chlorella fusca* can also be induced by heating to 46°C for 15 minutes (Lorenzen 1963) and the most heat-sensitive stage is different from the stage which is most susceptible to cold shock.

On transfer of *Chlorella fusca* cells with a temperature optimum of about 33°C, and a temperature maximum of 35°C, to 39°C in the light, there is a stimulation of photosynthesis and dry matter production which continues for at least 14 hours, although the capacity for cell division is irreversibly lost. The resulting giant cells attain an abnormally high protein content and the activity of glutamate-oxaloacetate transaminase is higher than in normal cells (Tischner pers. comm., see also Semenenko *et al.* 1967).

### 3.4  *Adaptation to temperature*

Algae can, within certain limits, adapt to changes in temperature. In *Cyanidium* the optimum temperature of photosynthesis coincides with the precultivation temperature irrespective of whether the latter is 15°C or 39°C (Becker 1972). Luknitskaya (1963) has reported that 'heat hardening' comparable to that of higher plants exists in several green microalgae (see also Schölm 1968, Lyntova *et al.* 1968) but not necessarily in *Chlamydomonas* (Luknitskaya 1967).

Jørgensen (1968) studied the mechanism of temperature adaptation after precultivation by transferring *Skeletonema costatum* from 20°C to 7°C and *vice versa*. From the kinetics of light-photosynthesis curves prior to, during, and after adaptation to the new temperature, it appears that this alga adapts to lower temperatures by synthesizing more enzyme protein. Likewise, the response to temperature increase is thought to result in a reduction of enzyme content per cell. This conclusion agrees with observed changes in total protein, but there is, as yet, no evidence that these proteins are enzymes.

The time course of adaptation to supra-optimal temperatures has been studied by Hegewald (unpubl.) using *Scenedesmus*. Following a temperature increase of only 1°C two to three days are required before vigorously growing cultures can be regarded as being fully adapted. Synchronous cultures of *Chlorella fusca* growing in a light-and-dark cycle respond to temperature changes by transitory shifts of the developmental rhythm (Soeder 1966). This adaptation phenomenon disappears after one or two days. Slow growing microalgae may require even longer periods of adaptation comparable to those reported for benthic marine algae. The recovery of the latter from a moderate heat shock may take over a week in some species (Ried 1969b). Temperature adaptation at the community level has been analysed by Patrick (1968) in complex diatom populations.

### 3.5  *Effect of temperature on various metabolic processes*

The temperature dependence of membrane potentials has been studied in *Chara* and *Nitella*. Within the physiological temperature range the membrane potential increases with temperature. According to Hogg *et al.* (1969) a differential action of temperature on the passive permeabilities of $Na^+$ and $K^+$ may be responsible for the observed change of about 1·6 mV per 1°C change in temperature. By contrast, Barber (1968b) and Gradmann (1970) emphasize the importance of active $K^+$ transport which is sensitive to temperature and to various inhibitors. Stepanskii and Yaglova (1968) conclude that conformational changes of cytoplasmic proteins are also involved in temperature effects of membrane potential (Gradmann 1970).

The $Q_{10}$ of active uptake of sulphate ions by *Chlorella* was found to be $3 \cdot 1$ to $4 \cdot 4$ (Vallee & Jeanjean 1968). Another strongly temperature-dependent phenomenon related to ion transport is the formation of coccoliths in *Coccolithus* (Paasche 1968). Iodine uptake and accumulation in marine algae is also positively related to temperature (Parekh *et al.* 1969).

Respiration is usually temperature dependent and there is a linear increase in $O_2$ uptake with temperature; that noted in *Euglena* by Buetow (1962) is fairly typical. Temperature-respiration curves may, however, vary. For example, the temperature response of marine planktonic diatoms may vary greatly depending on the temperatures at which they usually grow. Thus the respiration rate of the slightly warm-stenothermic *Thalassiosira fluviatilis* increases more with increased temperature, the lower the precultivation temperature has been. The opposite effect of raising the temperature is found in the moderately cold-stenothermic *Rhizosolenia setigera* (Ryther & Guillard 1962). If *Chlorella* is kept at 0°C in the dark and then brought back to the light at 27°C, the rate of glucose respiration shows a transient depression, accompanied by stimulated uptake of ammonium or nitrate ions (Müller-Stoll & Müntz 1971). The quantitative differences in the action of the uncoupler DNP (dinitrophenol) on the respiration of normal and cold-treated cells (Müntz 1965a) can be explained by lowered levels of ATP and other high-energy phosphates after storage in the cold (Müntz 1969).

The phosphorus requirement of *Scenedesmus* varies with temperature and consumption per unit weight of biomass is minimum at 25°C (Borchardt & Azad 1968, see also Pipes & Gotaas 1960). Such results obtained in cultures, should they be characteristic of algae in general, are significant not only in relation to the possible use of these organisms in removing phosphate, but also in relation to studies on the minimum phosphorus levels necessary for growth in natural environments (see Chapter 23, p. 636).

Nitrogen fixation and the excretion of nitrogenous products by blue-green algae are temperature dependent (Jones & Stewart 1969). Fogg and Than-Tun (1960) and Fogg and Stewart (1968) obtained $Q_{10}$ values for nitrogen fixation of about 6. Blue-green algae fix nitrogen in the Antarctic near 0°C (Fogg & Stewart 1968, Horne 1972) and thermal blue-green algae fix nitrogen at 46°C, a temperature which is below their absolute temperature maximum for growth (Stewart 1970). The fatty acids and other lipids of algae may vary with temperature. The total lipid content of *Anacystis* shows no significant changes between 26 and 41°C and while there are few qualitative changes between 26 and 35°C there is a drastic decrease in the ratio of unsaturated:saturated fatty acids (from $1 \cdot 01$ to $0 \cdot 68$) between 35 to 41°C. This is due mainly to an increased synthesis of palmitic acid and a suppression of hexadecenic acid (Holton *et al.* 1964). Similar

2A

effects have been reported for *Chlorella* (Harris & James 1969). Analytical
studies on a high-temperature strain of *Chlorella* have revealed a more complex
temperature effect (Patterson 1970). Here the degree of unsaturation was
maximal at 22°C and decreased at lower and at higher temperatures which also
stimulated the relative production of total fatty acids. There was a general
decrease in average chain length with increasing temperature. *Cyanidium* grown
at 55°C contains only half as much total lipid as cells adapted to 20°C (Kleinschmidt & McMahon 1970a). While the distribution of lipids into each of five
lipid classes was not influenced by temperature, the degree of unsaturation
decreased markedly with increasing temperature. Linolenic acid, comprising
30% of the total fatty acids at 20°C and preferentially bound to monogalactosyl-
diglyceride (Kleinschmidt & McMahon 1970b), was lacking in cells grown at
50°C (Kleinschmidt & McMahon 1970a).

## 4  HYDROGEN ION CONCENTRATION

### 4.1  *General*

The pH dependence of dissociation rates and the ionic state of polar inorganic
and organic compounds affects the availability of many algal nutrients such as
$CO_2$, iron (Stengel 1970) and organic acids (Cook 1965). pH also exerts an effect
on the electrical charge of the cell wall surface (Hegewald 1972), on ion transport
systems at the plasmalemma, and on the associated membrane potentials
(Bentrup 1971). Although changes in these properties alone may cause conspicuous changes in metabolic rates, some observations also point to a direct
influence of pH upon other metabolic activities. It should also be borne in mind
that most experiments are not carried out under static pH conditions and that
the hydrogen ion concentration of a synthetic nutrient solution is not necessarily
identical with the pH of the same medium gassed with, say, a $CO_2$-in-air mixture
(Soeder *et al.* 1964, 1966).

### 4.2  *pH requirements and pH tolerance*

The pH range at which algae occur is wide. Some thermophilic species can thrive
at extremely low pH (Brock 1969). The optimal pH range of *Zygogonium* sp.
extends down to pH 1·0 (Lynn & Brock 1969) and *Chlamydomonas acidophila*
has been isolated from peat bog water of pH 1·0 although they could not be
cultivated in the laboratory below pH 2·0 (Fott & McCarthy 1964). Certain algae
from alkaline waters have high pH optima. For example the optimum range for
*Spirulina platensis* from Lake Tchad extends from pH 8·5 to pH 11 (Clement &
van Landeghem 1971), and this alga tolerates pH 13·5 (Zarrouk 1966). Certain
algae also display a greater range of pH tolerance than required in the environments where they occur. For instance, *Staurastrum pingue* from a slightly acid

lake can be cultivated successfully between pH 4 and 11 (Schulle 1968). Kessler (1967a) has compared the acid tolerance of numerous *Chlorella* strains and has demonstrated species-specific differences which appear to be genetically determined. *Chlorella saccharophila*, with a lower pH limit of 2, shows the greatest acid tolerance and *C. minutissima*, with a lower pH limit of 5·5 shows the greatest acid sensitivity. Similar investigations have been carried out on *Ankistrodesmus*, *Scenedesmus* (Kessler 1967b) and *Golenkinia* (Ellis & Machlis 1968). Reynolds and Allen (1968) emphasized the role of pH as a factor determining the composition of freshwater phytoplankton communities and Merilaeinen (1967) and Patrick (1968) provide data on diatom communities.

### 4.3  *Effect of pH on various metabolic processes*

Pigment changes may occur as a result of variation in pH. Starosta (1970) studied *Ankistrodesmus* under carefully controlled pH conditions on a glucose medium in the dark and found that a substantial transformation of violaxanthin into zeaxanthin required a low external pH. An increased synthesis of secondary carotenoids in *Chlorella zofingiensis* exposed to low pH has been shown to be due to an acid inhibition of nitrate reductase and nitrite reductase so that nitrogen-deficiency resulted (Kessler & Czygan 1965). Carotenogenesis in streptomycin-bleached *Euglena* is favoured by an acid pH with an optimum at pH 6 for dark-grown cells and at pH 4 for light-grown cells (Dolphin 1970).

The rates at which polyphosphates and several organic phosphates are synthesized in *Ankistrodesmus* are also influenced markedly by pH (Ullrich 1972) and nitrogen fixation by blue-green algae is pH-dependent (see Fogg *et al.* 1973). pH may also affect the state and mobility of heavy metals. This is illustrated by the findings of Kanazawa and Kanazawa (1969) who observed an increase of $Cu^{2+}$ toxicity in *Chlorella* with decrease in pH. Similarly the action of organic inhibitors depends in many cases on pH. For example, maleic hydrazide is mutagenic in *Anacystis* at pH 5 but not at pH 8 (Gupta & Kumar 1970).

# 5  OSMOTIC EFFECTS

## 5.1  *General*

The clear-cut ecological differentiation of algae into halophilic or non-halophilic and stenohaline or euryhaline forms is reflected in their geographical distribution (Biebl 1962, den Hartog 1967) as well as in their physiology (Batterton & Van Baalen 1971, Gutknecht & Dainty 1968). Osmoregulation and osmoadaptation rely a great deal on the activity and the specificity of ionic pumps. Some organisms which tolerate extreme salinities do so by forming osmotically active organic substances in the cell and while the molecular bases of halophily and extreme osmotolerance are poorly understood, it seems that osmoregulation is more important than is the osmotic stability of the cytoplasmic enzymes. For example several enzymes extracted from *Dunaliella viridis* were found to be by no means as halophilic as those of the bacterium *Halobacterium* (Larsen 1967).

## 5.2   *Adaptation to changes of osmotic pressure*

On transfer of *Ochromonas* (Kauss 1967) and red algae (Kauss 1968) to higher salinities there is an increase in isofloridoside, an osmotically active α-galacto-glyceride. *Chlorella* responds to osmotic stress by the accumulation of oligo-saccharides (Dedio 1966, 1968). In *Monochrysis lutheri* the addition of 0·5 M NaCl to seawater medium causes an 80 to 90% increase of intracellular cyclo-hexane-tetrol within four hours. Dilution of the medium reduces the cyclitol concentration to a new level within 10 minutes (Craigie 1969).

Salinity-induced changes in *Dunaliella* from green to brown or red, are accompanied by a decrease in RNA content per cell and by an accumulation of protein (Mushak 1968). Another aspect of adaptation to high-salinity stress is a rapid loss of TCA-soluble organic phosphates into the medium, as observed by Antonyan and Pinevich (1967) in *Chlorella*. If synchronous *Chlorella* cells are transferred to a medium of higher salt concentration the formation of daughter cells is more strongly inhibited than is the synthesis of biomass. The result is a transitory increase of dry matter per cell for one or two days (Soeder *et al.* 1967). In the same system an adaptation to decreased salt concentration requires more than three days. These adaptation phenomena are accompanied by phase shifts of the rhythm of cell development which reach a new constant state after one cell cycle (Soeder 1966).

## 5.3   *Effect of osmotic pressure and salinity on various metabolic processes*

Many algae show an inhibition of photosynthesis after a transfer to a medium of higher salinity or of elevated osmotic pressure (see, e.g. Mironyuk & Einor 1968 for *Dunaliella*, and Batterton & Van Baalen 1971 for freshwater blue-green algae). The effect of increased NaCl concentration on photosynthetic rates of blue-green algae was more acute than the effects of other osmotic stresses (Batterton & Van Baalen 1971). In *Chlorella* the sensitivity of photosynthesis to an increase of osmotic pressure of the medium varies greatly with the developmental stage of the alga, the cells being most sensitive to osmotic shock prior to autospore release (Schmidbauer & Ried 1967).

The effect of osmotic stresses upon respiration of *Chlorella* also depends on cell age. Greatest stimulation of respiration is obtained with autospores, whereas respiration of mature autospore mother cells is inhibited by transfer to a more concentrated medium (Ried *et al.* 1962, Schmidbauer & Ried 1967). These also show a depression of blue-light excited fluorescence (Döhler & Ried 1963). The short-term effect of increased salt concentration in these experiments was identical to that obtained using iso-osmotic solutions of polyethylene glycol or sorbitol. Prolonged exposure to 0·4 molar polyethylene glycol was inhibitory (see also Greenway *et al.* 1968). Both the stimulatory and the inhibitory effects of osmotic stress on respiration are thought to be due to two things: firstly an indirect stimulation of ATP turnover similar to that caused by salt respiration, and secondly, to a direct inhibition of respiratory enzymes, perhaps those involved in glycolysis (Schmidbauer & Ried 1967).

Other short-term responses of *Chlorella* to osmotic stress are the already mentioned release of TCA-soluble organic phosphates (Antonyan & Pinevich 1967) and a transitory excretion of nitrite in the presence of nitrate (Soeder *et al.* 1962).

The change in colour from green to red or brown which is observed in *Dunaliella* after a transfer to higher salinity is due to a stimulated synthesis of carotenoids, and a decomposition of chlorophyll (Mironyuk & Einor 1968, Mushak 1968). The observation that ichthyotoxin production by *Prymnesium parvum* increased with salinity may be of practical significance (Padilla 1970). Marine algae from the lower littoral are severely damaged by exposure to hypotonic seawater solutions (Biebl 1962) and a 20% decrease of salinity is sufficient to cause a long-term loss of semipermeability in *Codium bursa* (Jacob 1961). Gessner (1969) provides evidence that the permeability to water is also impaired by hypotonic stress. Whether these effects are directly or indirectly linked to the inhibition of photosynthesis in hypotonic media is unclear (Gessner 1971).

# 6 MECHANICAL FACTORS

## 6.1 *Turbulence*

The effect of wave action and littoral turbulence on the vertical zonation of marine benthic algae has recently been studied by Friedman (1969) and by Gessner and Hammer (1968). Doty (1971) developed a simple method for quantitation of turbulence and demonstrated a positive influence of wave action upon photosynthetic biomass production. The importance of current velocity on distribution and production of the algal flora of running waters has been emphasized by Backhaus (1967) and McIntire (1968). The effects of current velocity upon these algae includes a stimulation of phosphate uptake and of respiration rate (Schumacher & Whitford 1965). A general stimulation of metabolic activities by turbulence is observed in algal suspensions (Müntz 1965b). However, shaking *Anabaena cylindrica* at a rate of 140 oscillations per minute inhibits growth, although the alga grows well at 90 oscillations per minute (Fogg & Than-Tun 1960). Growth of *Anabaena spiroides* is severely inhibited by a more than slight turbulence (Volk & Phinney 1968).

## 6.2 *Hydrostatic pressure*

Vidaver (1969) has provided a comprehensive review of the effects of hydrostatic pressure on algae. Unless sub-cellular structures are damaged by extreme pressure, the effect of pressure is to cause a reversible decrease of all metabolic activities. In general, it appears that hydrostatic pressure is not an ecologically important factor for algae. An exception is those blue-green algae which contain gas vacuoles and are able to adjust their position in the water column as a result of pressure changes in the vacuoles (Fogg & Walsby 1971, see also Jost & Jones 1970). Photosynthesis by dinoflagellates is reversibly inhibited at a pressure of up to 400 atm. and is more pressure-sensitive than is bioluminescence (Marchand 1968). Vidaver (1969) studied the effects of pressure on photosynthesis and transient phenomena at pressures up to 20,000 p.s.i. in several benthic marine algae and in *Ankistrodesmus*. He found that the algae showed small changes in the absorption spectra of their pigments at extreme pressure with the isolated phycobilins showing spectral shifts in the order of 4 to 5nm in a long-wavelength direction.

# 7 REFERENCES

ALLEN M.B. (1969) Structure, physiology, and biochemistry of the *Chrysophyceae*. *A. Rev. Microbiol.* **23**, 29–46.

ANTONYAN A.A. & PINEVICH V.V. (1967) Some characteristics of phosphorus metabolism in *Chlorella* due to action of different severe conditions of the medium. *Biol. Zh. Arm.* **20**, 28–36.

ASATO Y. & FOLSOME C.E. (1969) Mutagenesis of *Anacystis nidulans* by N-methyl-N-nitro-N-nitrosoguanidine and UV irradiation. *Mut. Res.* **8**, 531–6.

BACKHAUS D. (1967) Ökologische Untersuchungen an den Aufwuchsalgen der obersten Donau und ihrer Quellflüsse. I. Voruntersuchungen. *Arch. Hydrobiol. Suppl.* **30**, 364–399.

BAKER A.L., BROCK A.J. & KLEMER A.R. (1969) Some photosynthetic characteristics of a naturally occurring population of *Oscillatoria agardhii* Gomont. *Limnol. Oceanogr.* **14**, 327–33.

BARBER J. (1968a) Measurement of the membrane potential and evidence for active transport of ions in *Chlorella pyrenoidosa*. *Biochim. biophys. Acta*, **150**, 618–25.

BARBER J. (1968b) The efflux of potassium from *Chlorella pyrenoidosa*. *Biochim. biophys. Acta*, **163**, 531–8.

BATTERTON J.C. JR. & VAN BAALEN C. (1971) Growth responses of blue-green algae to sodium chloride concentration. *Arch. Mikrobiol.* **76**, 151–65.

BEALE S.I. & APPLEMAN D. (1971) Chlorophyll synthesis in *Chlorella*. Regulation by degree of light limitation of growth. *Pl. Physiol., Lancaster*, **47**, 230–5.

BECKER J.-D., DÖHLER G. & EGLE K. (1968) Die Wirkung monochromatischen Lichts auf die extrazelluläre Glykolsäure-Ausscheidung bei der Photosynthese von *Chlorella*. *Z. PflPhysiol.* **58**, 212–21.

BECKER W. (1972) Physiologische Untersuchungen zur Photosynthese von Algen unter extremen Temperaturbedingungen. Dissertation, Tübingen 1972.

BELCHER J.H. (1967) Reproduction and growth of the mixotrophic flagellate *Furcilla stigmatophora* (Skuja) Korsh. (*Volvocales*). *Arch. Mikrobiol.* **55**, 327–41.

BELCHER J.H. (1970) The resistance to desiccation and heat of the asexual cysts of some freshwater Prasinophyceae. *Br. Phycol. J.* **5**, 173–7.

BELYANIN V.N. & KOVROV B.G. (1968) Mathematical model of biosynthesis in a light-limited microalgae culture. *Dokl. Akad. Nauk. S.S.S.R.* **179**, 1463–6.

BENTRUP F.W. (1971) Elektrophysiologie der Zelle. *Fortschr. Bot.* **33**, 51–61.

BERMAN T. & RODHE W. (1971) Distribution and migration of *Peridinium* in Lake Kinneret. *Mitt. Int. Ver. Limnol.* **19**, 266–76.

BIEBL R. (1962) Physiological aspects of ecology: seaweeds. In *Physiology and Biochemistry of Algae*, ed. Lewin R.A. pp. 799–815. Academic Press, New York & London.

BIEBL R. (1969a) Untersuchungen zur Temperaturresistenz arktischer Süßwasseralgen im Raum von Barrow, Alaska. *Mikroskopie*, **25**, 3–6.

BIEBL R. (1969b) Studien zur Hitzeresistenz der Gezeitenalge *Chaetomorpha cannabina*. *Protoplasma*, **67**, 451–72.

BIEBL R. (1970) Vergleichende Untersuchungen zur Temperaturresistenz von Meeresalgen entlang der pazifischen Küste Nordamerikas. *Protoplasma*, **69**, 61–83.

BORCHARDT J.A. & AZAD H.S. (1968) Biological extraction of nutrients. *J. Water Poll. Contr. Fed.* **40**, 1739–54.

BROCK T.D. (1969) Microbial growth under extreme conditions. *Symp. Soc. gen. Microbiol.* **19**, 15–41.

BROWN D.L. & TREGUNNA E.B. (1967) Inhibition of respiration during photosynthesis by some algae. *Can. J. Bot.* **45**, 1135–43.

BROWN T.E., RICHARDSON F.L. & VAUGHN M.L. (1967) Development of red pigmentation in *Chlorococcum wimmeri* (Chlorophyta : Chlorococcales). *Phycologia*, **6**, 167–184.

BUETOW D.E. (1962) Differential effects of temperature on the growth of *Euglena gracilis*. *Expl. Cell Res.* **27**, 137–42.

BUNT J. S. (1965) Measurements of photosynthesis and respiration in a marine diatom (*Fragilaria sublinearis*) with the mass spectrometer and with carbon-14. *Nature, Lond.* **207**, 1373–5.

BUNT J.S., OWENS O. VAN H. & HOCH G. (1966) Exploratory studies on the physiology and ecology of a psychrophilic diatom. *J. Phycol*, **2**, 96–100.

CLAUSS H. (1968) Der Einfluß von Lichtqualität auf die Morphogenese von *Acetabularia mediterranea. Ber. dt. bot. Ges.* **80**, 755.

CLAUSS H. (1970) Der Einfluß von Rot- und Blaulicht auf die Hillaktivität von *Acetabularia*-Chloroplasten. *Planta*, **91**, 32–7.

CLEMENT G. & VAN LANDEGHEM H. (1971) *Spirulina*: ein günstiges Objekt für die Massenkultur von Mikroalgen. *Ber. dt. bot. Ges.* **83**, 559–65.

CONSTANTOPOULOS G. & BLOCH K. (1967) Effect of light intensity on the lipid composition of *Euglena gracilis*. *J. biol. Chem.* **242**, 3538–42.

COOK J.R. (1965) Influence of light on acetate utilization in green *Euglena*. *Pl. Cell Physiol., Tokyo*, **6**, 301–7.

COOK J.R. (1968) Photo-inhibition of cell division and growth in euglenoid flagellates. *J. Cell Physiol.* **71**, 177–84.

CRAIGIE J.S. (1969) Some salinity-induced changes in growth, pigments and cyclohexanetetrol content of *Monochrysis lutheri*. *J. Fish. Res. Board Can.* **26**, 2959–67.

DAVIES D.R., HOLT P.D. & PAPWORTH D.G. (1969) The survival curves of haploid and diploid *Chlamydomonas reinhardi* exposed to radiations of different LET. *Int. J. Radiat. Biol.* **15**, 75–87.

DEDIO H. (1966) Untersuchungen über den Aminosäure- und Kohlenhydratstoffwechsel bei Synchronkulturen von *Chlorella pyrenoidosa*. Dissertation, Universität Frankfurt am Main.

DEDIO H. (1968) Entwicklungsabhängiger Anstau von Oligosacchariden bei *Chlorella fusca*. *Ber. dt. bot. Ges.* **81**, 359–63.

DOEMEL W.N. & BROCK T.D. (1970) The upper temperature limit of *Cyanidium caldarium*. *Arch. Mikrobiol.* **72**, 326–32.

DÖHLER G. & RIED A. (1963) Über den Einfluß abgestufter Konzentrationen verschiedener Plasmolytica auf die Chloroplastenfluorescenz synchron kultivierter *Chlorella pyrenoidosa*. *Arch. Mikrobiol.* **46**, 190–216.

DOLPHIN W.D. (1970) Photoinduced carotenogenesis in chlorotic *Euglena gracilis*. *Pl. Physiol., Lancaster*, **46**, 685–91.

DOTY M.S. (1971) Measurement of water movement in reference to benthic algal growth. *Botanica mar.* **14**, 32–5.

ELLIS R.J. & MACHLIS L. (1968) Nutrition of the green alga *Golenkinia*. *Am. J. Bot.* **55**, 590–9.

ELSTER H.J. & MOTSCH B. (1966) Untersuchungen über das Phytoplankton und die organische Urproduktion in einigen Seen des Hochschwarzwalds, im Schluchsee und im Bodensee. *Arch. Hydrobiol. Suppl.* **28**, 291–376.

EPEL B.L. & BUTLER W.L. (1969) Cytochrome a₃. Destruction by light. *Science, N.Y.* **166**, 621–2.

EPEL B.L. & BUTLER W.L. (1970) Inhibition of respiration in *Prototheca zopfii* by light. *Pl. Physiol., Lancaster*, **45**, 728–34.

FAY P. (1970) Photostimulation of nitrogen fixation in *Anabaena cylindrica*. *Biochim. biophys. Acta*, **216**, 353–6.

FINDENEGG I. (1966) Die Bedeutung kurzwelliger Strahlung für die planktische Primärproduktion. *Verh. Int. Verein. Limnol.* **16**, 314–20.

FOGG G.E. (1962) Extracellular products. In *Physiology and Biochemistry of Algae*, ed. Lewin R.A. pp. 475–89. Academic Press, New York & London.

FOGG G.E. & HORNE A.J. (1968) The physiology of antarctic freshwater algae. *Antarctic Res.* 632–8.

FOGG G.E., NALEWAJKO C. & WATT W.D. (1965) Extracellular products of phytoplankton photosynthesis. *Proc. R. Soc., Ser. B.* **162**, 517–34.

FOGG G.E. & STEWART W.D.P. (1968) *In situ* determinations of biological nitrogen fixation in Antarctica. *Brit. Antarctic Surv. Bull.* **15**, 39–46.

FOGG G.E., STEWART W.D.P., FAY P. & WALSBY A.E. (1973) *The Blue-green Algae*. Academic Press, London & New York.

FOGG G.E. & THAN-TUN (1960) Interrelations of photosynthesis and assimilation of elementary nitrogen in a blue-green alga. *Proc. R. Soc. B.* **153**, 117–27.

FOGG G.E. & WALSBY A.E. (1971) Buoyancy regulation and the growth of planktonic blue-green algae. *Mitt. Internat. Verein. Limnol.* **19**, 182–8.

FOTT B. & McCARTHY A.J. (1964) Three acidophilic flagellates in pure culture. *J. Protozool.* **11**, 116–20.

FRANZISKET L. (1968) Zur Ökologie der Fadenalgen im Skelett lebender Riffkorallen. *Zool. Jb. (Physiol.)* **74**, 246–53.

FRIEDMANN I. (1969) Geographic and environmental factors controlling life history and morphology in *Prasiola stipitata* Suhr. *Öst. bot. Z.* **116**, 203–25.

GERHARDT B. & WIESSNER W. (1967) On the light-dependent reactivation of photosynthetic activity by manganese. *Biochem. biophys. Res. Commun.* **28**, 958–64.

GESSNER F. (1969) The osmotic regulations in *Valonia ventricosa*. A.J. Agardh. *Int. Rev. ges. Hydrobiol.* **54**, 529–32.

GESSNER F. (1971) Wasserpermeabilität und Photosynthese bei marinen Algen. *Botanica mar.* **14**, 27–31.

GESSNER F. & HAMMER L. (1968) The littoral algal vegetation on the east coast of Venezuela. *Int. Rev. ges. Hydrobiol.* **52**, 657–92.

GILET R. & TERRIER J. (1969) The effect of the process of restoration makes it possible to distinguish acute and subacute irradiations in *Chlorella*. *C.r. hebd. Séances Acad. Sci. Ser. D. Sci. Natur., Paris*, **269**, 383–5.

GOLDMAN C.R., MASON D.T. & WOOD B.J.B. (1963) Light injury and light inhibition in an antarctic freshwater phytoplankton community. *Limnol. Oceanogr.* **8**, 313–21.

GOVINDJEE R., RABINOWITCH E. & GOVINDJEE (1968) Maximum quantum yield and action spectrum of photosynthesis and fluorescence in *Chlorella*. *Biochim. biophys. Acta*, **162**, 539–44.

GRADMANN D. (1970) Einfluß von Licht, Temperatur und Außenmedium auf das elektrische Verhalten von *Acetabularia crenulata*. *Planta*, **93**, 323–53.

GRANT B.R. (1967) The action of light on nitrate and nitrite assimilation by the marine chlorophyte, *Dunaliella tertiolecta* Butcher. *J. gen. Microbiol.* **48**, 379–89.

GREENWAY H., HILLER R.G. & FLOWERS T. (1968) Respiratory inhibition in *Chlorella* produced by 'purified' polyethylene glycol 1540. *Science, N.Y.* **159**, 984–5.

GROSS R.E. & DUGGER W.M. (1969) Photoinhibition of growth of a yellow and a colorless form of *Chlamydomonas reinhardtii*. *Photochem. Photobiol.* **10**, 243–50.

GUPTA R.S. & KUMAR H.D. (1970) The effect of maleic hydrazide on growth and mutation of a blue-green alga. *Arch. Mikrobiol.* **70**, 330–9.

GUTKNECHT J. & DAINTY J. (1968) Ionic relations of marine algae. *A. Rev. Oceanogr. Mar. Biol.* **6**, 163–200.

HALLDAL P. (1968) Photosynthetic capacities and photosynthetic action spectra of endozoic algae of the massive *Favia*. *Biol. Bull. mar. biol. Lab., Woods Hole*, **134**, 411–24.

HALLDAL P. (1970) *Photobiology of microorganisms*. John Wiley, New York & London.

HALLDAL P. & FRENCH C.S. (1958) Algal growth in crossed gradients of light intensity and temperature. *Pl. Physiol., Lancaster*, **33**, 249–52.

HARRIS P. & JAMES A.T. (1969) The effect of low temperature on fatty acid biosynthesis in plants. *Biochem. J.* **112**, 325–30.

DEN HARTOG C. (1967) Brackish water as an environment for algae. *Blumea, Leiden*, **15**, 31–43.

HAUSCHILD A.H.W., NELSON C.D. & KROTKOV G. (1962) The effect of light quality on the products of photosynthesis in *Chlorella vulgaris*. *Can. J. Bot.* **40**, 179–89.

HAUSCHILD A.H.W., NELSON C.D. & KROTKOV G. (1964) Concurrent changes in the products and the rate of photosynthesis in *Chlorella vulgaris* in the presence of blue light. *Naturwissenschaften*, **51**, 274.

HEBER U. (1969) Biochemische Untersuchungen zu Frostwirkung und Frostschutz. *Ber. dt. bot. Ges.* **82**, 81–2.

HEGEWALD E. (1972) Untersuchungen zum Zeta-Potential von Planktonalgen. *Arch. Hydrobiol./Suppl.* **42**, 14–90.

HINDAK F. & KOMAREK J. (1968) Cultivation of the cryosestonic alga *Koliella tatrae* (Kol) Hind. *Biol. Plant.* **10**, 95–7.

HOARE D.S., HOARE S.L. & MOORE R.B. (1967) The photoassimilation of organic compounds by autotrophic blue-green algae. *J. gen. Microbiol.* **49**, 351–70.

HOGG J., WILLIAMS E.J. & JOHNSTON R.J. (1969) The membrane electrical parameters of *Nitella translucens*. *J. theoret. Biol.* **24**, 317–34.

HOLM-HANSEN O. (1963) Viability of blue-green and green algae after freezing. *Physiologia Pl.* **16**, 530–40.

HOLM-HANSEN O. (1967) Factors affecting the viability of lyophilized algae. *Cryobiology*, **4**, 17–23.

HOLM-HANSEN O. (1968) Ecology, physiology and biochemistry of blue-green algae. *A. Rev. Microbiol.* **22**, 47–70.

HOLTON R.W., BLECKER H.H. & ONORE M. (1964) Effect of growth temperature on the fatty acid composition of a blue-green alga. *Phytochemistry*, **3**, 595–602.

HORNE A.J. (1972) The ecology of nitrogen fixation on Signy Island, South Orkney Islands. *Brit. Antarc. Surv. Bull.* **27**, 893–902.

HULBURT E.M. & GUILLARD R.R.L. (1968) The relationship of the distribution of the diatom *Skeletonema tropicum* to temperature. *Ecology*, **49**, 337–9.

ICHIMURA S., NAGASAWA S. & TANAKA T. (1968) On the oxygen and chlorophyll maxima found in the metalimnion of a mesotrophic lake. *Bot. Mag., Tokyo*, **81**, 1–18.

IGNATIADES L. & SMAYDA T.J. (1970) Autecological studies on the marine diatom *Rhizosolenia fragilissima* Bergon. I. The influence of light, temperature, and salinity. *J. Phycol.* **6**, 332–9.

JACOB F. (1961) Zur Biologie von *Codium bursa* (L.) Agardh und seiner endophytischen Cyanophyceen. *Arch. Protistenk.* **105**, 345–406.

JAMES T.W. (1962) Metabolic aspects of continuous and synchronized cultures of flagellates. *Proc. 13th Int. Congr. Physiol. Sci., Leiden*, 1962, pp. 788–91.

JAVORNICKY P. (1966) Light as the main factor limiting the development of diatoms in Slapy Reservoir. *Verh. Int. Verein. Limnol.* **16**, 701–12.

JAVORNICKY P. (1970) On the utilization of light by fresh-water phytoplankton. *Arch. Hydrobiol., Suppl.* **39** (Algological Studies 2/3), 68–85.

JØRGENSEN E.G. (1964) Adaptation to different light intensities in the diatom, *Cyclotella meneghiniana* Kütz. *Physiologia Pl.* **17**, 136–45.

JØRGENSEN E.G. (1968) The adaptation of plankton algae. II. Aspects of the temperature adaptation of *Skeletonema costatum. Physiologia Pl.* **21**, 423–7.

JØRGENSEN E.G. (1969) The adaptation of plankton algae. IV. Light adaptation in different algal species. *Physiologia Pl.* **22**, 1307–15.

JONES K. & STEWART W.D.P. (1969) Nitrogen turnover in marine and brackish habitats. III. The production of extracellular nitrogen by *Calothrix scopulorum. J. mar. biol. Ass. U.K.* **49**, 475–88.

JOST M. & JONES D.D. (1970) Morphological parameters and macromolecular organization of gas vacuole membranes of *Microcystis aeruginosa* Kütz. emend. Elenkin. *Can. J. Microbiol.* **16**, 159–64.

KAJIMURA M. (1968) On fruiting season of *Caulerpa scalpelliformis* (R.Br.) Ag. var. *denticulata* (Decsn.). Weber van Bosse in the Oki Islands Shimane Prefecture. *Bull. Jap. Soc. Phycol.* **16**, 38–43.

KANAZAWA T. & KANAZAWA K. (1969) Specific inhibitory effect of copper on cellular division in *Chlorella. Pl. Cell Physiol., Tokyo*, **10**, 495–502.

KARLANDER E.P. & KRAUSS R.W. (1966) Responses of heterotrophic cultures of *Chlorella vulgaris* Beijerinck to darkness and light. I. Pigment and pH changes. *Pl. Physiol., Lancaster*, **41**, 1–6.

KAUSS H. (1967) Isofloridosid und Osmoregulation bei *Ochromonas malhamensis. Z. PflPhysiol.* **56**, 453–565.

KAUSS H. (1968) α-Galaktosylglyceride und Osmoregulation in Rotalgen. *Z. PflPhysiol.* **58**, 428–33.

KESSLER E. (1967a) Physiologische und biochemische Beiträge zur Taxonomie der Gattung *Chlorella*. III. Merkmale von 8 autotrophen Arten. *Arch. Mikrobiol.* **55**, 346–57.

KESSLER E. (1967b) Physiologische und biochemische Beiträge zur Taxonomie der Gattungen *Ankistrodesmus* und *Scenedesmus*. II. Säureresistenz. *Arch. Mikrobiol.* **58**, 270–4.

KESSLER E. & CZYGAN F.-C. (1965) *Chlorella zofingiensis*: Isolierung neuer Stämme und ihre physiologisch-biochemischen Eigenschaften. *Ber. dt. bot. Ges.* **78**, 342–7.

KHOJA T. & WHITTON B.A. (1971) Heterotrophic growth of blue-green algae. *Arch. Mikrobiol.* **79**, 280–2.

KIRK J.T.O. & READE J.A. (1970) The action spectrum of photosynthesis in *Euglena gracilis* at different stages of chloroplast development. *Aust. J. biol. Sci.* **23**, 33–41.

KLEINSCHMIDT M.G. & McMAHON V.A. (1970a) Effect of growth temperature on the lipid composition of *Cyanidium caldarium*. I. Class separation of lipids. *Pl. Physiol., Lancaster*, **46**, 286–9.

KLEINSCHMIDT M.G. & McMAHON V.A. (1970b) Effect of growth temperature on the lipid composition of *Cyanidium caldarium*. II. Glycolipid and phospholipid components. *Pl. Physiol., Lancaster*, **46**, 290–3.

KOK B. (1965) Photosynthesis: the path of energy. In *Plant Biochemistry*, eds. Bonner J. & Varner J.E. pp. 904–60. Academic Press, New York & London.

KOL E. (1968) *Kryobiologie. Biologie des Schnees und des Eises. I. Kryovegetation.* E. Schweizerbart'sche Verlagsbuchhandlung, Stuttgart.

KOWALLIK W. (1962) Über die Wirkung des blauen und des roten Spektralbereichs auf die Zusammensetzung und Zellteilung synchronisierter *Chlorellen. Planta*, **50**, 337–65.

KOWALLIK W. (1963) Die Zellteilung von *Chlorella* im Laufe einer Farblichtkultur. *Planta*, **60**, 100–8.

KOWALLIK W. (1965) Die Proteinproduktion von *Chlorella* im Licht verschiedener Wellenlängen. *Planta*, **64**, 191–200.

KOWALLIK W. (1967) Action spectrum for an enhancement of endogenous respiration by light in *Chlorella. Pl. Physiol., Lancaster*, **42**, 672–6.

KOWALLIK W. & GAFFRON H. (1966) Respiration induced by blue light. *Planta*, **69**, 92–5.

KOWALLIK U. & KOWALLIK W. (1969) Eine wellenlängeabhängige Atmungssteigerung während der Photosynthese bei *Chlorella. Planta*, **84**, 141–57.

KUZ'MENKO M.I. (1968) Assimilation of 1-$C^{14}$-alanine and 5-$C^{14}$-glutamic acid by certain blue-green algae. *Gidrobiol. Zh.* **4**, 41–9.

KYLIN A. (1966) The effect of light, carbon dioxide, and nitrogen nutrition on the incorporation of S from external sulphate into different S-containing fractions in *Scenedesmus*, with special reference to lipid S. *Physiologia Pl.* **19**, 883–7.

LANGE O.L. & METZNER H. (1965) Lichtabhängiger Kohlenstoff-Einbau in Flechten bei tiefen Temperaturen. *Naturwissenschaften*, **52**, 191.

LANGE W. (1970) Cyanophyta-bacteria systems: effects of added carbon compounds or phosphate on algal growth at low nutrient concentrations. *J. Phycol., N.Y.* **6**, 230–4.

LARSEN H. (1967) Biochemical aspects of extreme halophilism. *Adv. Microbial Physiol.* **1**, 97–132.

LAUDENBACH B. & PIRSON A. (1969) Über den Kohlenhydratumsatz in *Chlorella* unter dem Einfluß von blauem und rotem Licht. *Arch. Mikrobiol.* **67**, 226–42.

LEWIN R.A., ed. (1962) *Physiology and biochemistry of algae.* Academic Press, New York & London.

LORD J.M., CODD G.A. & MERRETT M.J. (1970) The effect of light quality on glycolate formation and excretion in algae. *Pl. Physiol., Lancaster*, **46**, 855–6.

LORENZEN H. (1963) Temperatureinflüsse auf *Chlorella pyrenoidosa* unter besonderer Berücksichtigung der Zellentwicklung. *Flora, Jena*, **153**, 554–92.

LUKNITSKAYA A.F. (1963) Effect of the rearing temperature on the thermostability of some algae. *Citologiya*, **5**, 135–41.

LUKNITSKAYA A.F. (1967) Do *Chlamydomonas* have a heat hardening capacity? *Citologiya*, **9**, 800–3.

LUND J.W.G. (1965) The ecology of freshwater phytoplankton. *Biol. Rev.* **40**, 231–93.

LYMAN H., EPSTEIN H.T. & SCHIFF J. (1960) Ultraviolet inactivation and photoreactivation of chloroplast development in *Euglena* without cell death. *J. Protozool.* **6**, 264–5.

LYNE R.L. & STEWART W.D.P. (1973) Emerson enhancement of carbon fixation but not of acetylene reduction (nitrogenase activity) in *Anabaena cylindrica. Planta*, **109**, 27–38.

LYNN R. & BROCK T.D. (1969) Notes on the ecology of a species of *Zygogonium* (Kütz.) in Yellowstone National Park. *J. Phycol.* **5**, 181–5.

LYNTOVA M.I., FEL'MAN N.L. & DROBYSHEV V.P. (1968) Changes in cellular thermoresistance of marine algae due to environmental temperature. *Citologiya*, **10**, 1538–45.

LYSEK G. & SIMONIS W. (1968) Substrataufnahme und Phosphatstoffwechsel bei *Ankistrodesmus braunii*. I. Beteiligung der Polyphosphate bei der Aufnahme von Glucose und 2-Desoxyglucose im Dunkeln und im Licht. *Planta*, **79**, 133–45.

MADDUX W.S. & JONES R.F. (1964) Some interactions of temperature, light intensity, and nutrient concentration during the continuous culture of *Nitzschia closterium* and *Tetraselmis* sp. *Limnol. Oceanogr.* **9**, 79–86.

MARCHAND C. (1968) Comparative effects of increased hydrostatic pressures and bio-luminescence, photosynthesis and alcoholic fermentation. *C.r. hebd. Séanc. Acad. Sci. Ser. D. Sci. Natur., Paris*, **267**, 2376–8.

MARGULIS L. (1968) Visible light: Mutagen or killer? *Science, N.Y.* **160**, 1255–6.

MATSUDA Y., KIKUCHI T. & ISHIDA M.R. (1971) Studies on chloroplast development in *Chlamydomonas reinhardtii*. I. Effect of brief illumination on chlorophyll synthesis. *Pl. Cell Physiol., Tokyo*, **12**, 127–35.

MATTHERN R.O., KOSTICK J.A. & OKADA I. (1969) Effect of total illumination upon continuous *Chlorella* production in a high intensity light system. *Biotechnol. Bioeng.* **11**, 863–74.

MCCOMBIE A.M. (1960) Actions and interactions of temperature, light intensity and nutrient concentration of the green alga, *Chlamydomonas reinhardi*. *J. Fish. Res. Bd. Can.* **17**, 871–94.

MCINTIRE C.D. (1968) Structural characteristics of benthic algal communities in laboratory streams. *Ecology*, **49**, 520–37.

MEFFERT M.E. (1971) Cultivation and growth of two planktonic *Oscillatoria* species. *Mitt. Internat. Ver. Limnol.* **19**, 189–205.

MERILAEINEN J. (1967) The diatom flora and the hydrogen-ion concentration of the water. *A. Bot. Fenn.* **4**, 51–8.

METZNER H. & LORENZEN H. (1960) Untersuchungen über den Photosynthese-Gaswechsel an vollsynchronen *Chlorella*-Kulturen. *Ber. dt. bot. Ges.* **73**, 410–17.

MIRONYUK V.L. & EINOR L.O. (1968) Oxygen exchange and the content of pigments in various forms of *Dunaliella salina* Teod. under conditions of increased NaCl content. *Gidrobiol. Zh.* **4**, 23–9.

MIYOSHI Y. & TSUBO Y. (1969) Permanent bleaching of *Euglena* by chloramphenicol. *Pl. Cell Physiol., Tokyo*, **10**, 221–5.

MORRIS I. & AHMED J. (1968) The effect of light on nitrate and nitrite assimilation by *Chlorella* and *Ankistrodesmus*. *Physiologia Pl.* **22**, 1166–74.

MÜLLER H. (1970) Wachstum und Phosphatbedarf von *Nitzschia actinastroides* (Lemm.) v. Goor in statischer und homokontinuierlicher Kultur unter Phosphatlimitierung. *Arch. Hydrobiol. Suppl.* **38**, 399–484.

MÜLLER-STOLL W.R. & MÜNTZ K. (1971) Über die Folgewirkungen einer Kältebehandlung im Dunkeln auf Wachstum und verschiedene Stoffwechselvorgänge bei *Chlorella pyrenoidosa*. *Arch. Hydrobiol. Suppl.* **39**, 206–25.

MÜNTZ K. (1965a) Die Wirkung von 2,4-Dinitrophenol auf den Dunkelstoffwechsel kältebehandelter Zellen von *Chlorella pyrenoidosa*. *Naturwissenschaften*, **52**, 646–7.

MÜNTZ K. (1965b) Vergleichende Untersuchungen über die Stickstoffund Glucoseaufnahme aus Nährlösungen in Abhängigkeit von den Durchmischungsbedingungen. *Z. allg. Mikrobiol.* **5**, 362–77.

MÜNTZ K. (1969) On growth and metabolism of *Chlorella pyrenoidosa* during cold temperatures in the dark. *Flora, Jena*, **160**, 139–57.

MUSHAK P.O. (1968) Content and state of nucleic acids in the alga *Dunaliella salina* Teod. depending on NaCl concentrations. *Ukr. Bot. Zh.* **25**, 91–5.

ODOEVSKAYA N.S., GERASIMENKO L.M. & GORYUNOVA S.V. (1969) Long-term storage of collection cultures of some algae species. *Mikrobiologiya*, **38**, 544–7.

OEQUIST G. (1969) Adaptations in pigment composition and photosynthesis by far red radiation in *Chlorella pyrenoidosa*. *Physiologia Pl.* **22**, 516–28.

OKU T., OKAYAMA S., AIGA I. & SASA T. (1968) Appearance of plastoquinone during the greening of bleached cells of *Chlorella protothecoides*. *Pl. Cell Physiol., Tokyo*, **9**, 599–602.

PAASCHE E. (1968) The effect of temperature, light intensity, and photoperiod on coccolith formation. *Limnol. Oceanogr.* **13**, 178–81.

PAASCHE E. (1969) Light-dependent coccolith formation in two forms of *Coccolithus pelagi-*

*cus.* With remarks on the $^{14}$C zero-thickness counting efficiency of coccolithophorids. *Arch. Mikrobiol.* **67**, 199–208.

PADILLA G.M. (1970) Growth and toxigenesis of the chrysomonad, *Prymnesium parvum* as a function of salinity. *J. Protozool.* **17**, 456–62.

PAREKH J.M., BHALALA J., TALREJA S.T. & DOSHI Y.A. (1969) Uptake of I-131 by marine algae and the factors influencing iodine enrichment by seaweeds from seawater. *Curr. Sci.* **38**, 268–9.

PATRICK R. (1968) The structure of diatom communities in similar ecological conditions. *Am. Nat.* **102**, 173–83.

PATTERSON G.W. (1970) Effect of culture temperature on fatty acids composition of *Chlorella sorokiniana. Lipids,* **5**, 597–600.

PETERFI S., NAGY-TOTH F. & BRUGOVITZKY E. (1969) Variata glucidelor si protidelor la *Scenedesmus* in diferite conditii de cultivare intensiva. *Contributti botanice,* **21**, 321–9.

PICKETT J.M. (1971) Effects of flashes of red or blue light on the composition of starved *Chlorella pyrenoidosa. Pl. Physiol., Lancaster* **47**, 226–9.

PINTNER I.J. & PROVASOLI L. (1968) Heterotrophy in subdued light of 3 *Chrysochromulina* species. *Bull. Misaki Mar. Biol. Inst. Kyoto Univ.* **12**, 25–31.

PIPES W.O. & GOTAAS H.B. (1960) Utilization of organic matter by *Chlorella* grown in sewage. *Appl. Microbiol.* **8**, 163–9.

PIRSON A., LORENZEN H. & KOEPPER A. (1959) A sensitive stage in synchronized cultures of *Chlorella. Pl. Physiol., Lancaster,* **34**, 353–5.

PIRSON A. & RUPPEL G. (1962) Über die Induktion einer Teilungshemmung in synchronen Kulturen von *Chlorella. Arch. Mikrobiol.* **42**, 299–309.

RAVEN J.A. (1968a) The linkage of light-stimulated Cl influx to K and Na influxes in *Hydrodictyon africanum. J. exp. Bot.* **19**, 233–53.

RAVEN J.A. (1968b) The mechanism of photosynthetic use of bicarbonate by *Hydrodictyon africanum. J. exp. Bot.* **19**, 193–206.

RAVEN J.A. (1969a) Action spectra for photosynthesis and light-stimulated ion transport processes in *Hydrodictyon africanum. New Phytol.* **68**, 45–62.

RAVEN J.A. (1969b) Effects of inhibitors on photosynthesis and the active influxes of K and Cl in *Hydrodictyon africanum. New Phytol.* **68**, 1089–113.

RAVEN J.A., MACROBBIE E.A.C. & NEUMANN J. (1969) The effect of Dio-9 on photosynthesis and ion transport in *Nitella, Tolypella,* and *Hydrodictyon. J. exp. Bot.* **20**, 221–35.

REYNOLDS C.S. & ALLEN S.E. (1968) Changes in the phytoplankton of Oak Mere following the introduction of base-rich water. *Br. phycol. Bull.* **3**, 451–62.

RIED A. (1967) Zuordnung von Übergangseffekten im $O_2$-Austausch von *Chlorella* zu verschiedenen Lichtreaktionen. *Ber. dt. bot. Ges.* **79**, 112–15.

RIED A. (1969a) Über die Wirkung des blauen Lichts auf den photosyntetischen Gasaustausch von *Chlorella. Planta,* **87**, 333–46.

RIED A. (1969b) Physiologische Aspekte der Vertikalzonierung von Algen des marinen Litorals. *Ber. dt. bot. Ges.* **82**, 127–41.

RIED A. (1970) Energetic aspects of the interaction between photosynthesis and respiration. In *Prediction and Measurement of Photosynthetic Productivity,* Centre for Agricultural Publishing and Documentation, Wageningen, pp. 231–46.

RIED A., MÜLLER I. & SOEDER C.J. (1962) Wirkung der Stoffwechselinhibitoren auf den respiratorischen Gaswechsel von *Chlorella pyrenoidosa. Vortr. Gesamtgeb. Bot. N.F.* **1**, 187–94.

RIETH A. (1967) Zur Frostresistenz von *Porphyridium cruentum* (Ag.) Naeg. Der Einfluß der Periodenlänge bei Kultur unter einem rhythmischen Wechsel von (Wärme + Licht)- mit (Frost + Dunkel). *Zeiten. Monatsber. Deut. Akd. Wiss., Berlin,* **9**, 725–30.

RIGANO C. (1971) Studies on nitrate reductase from *Cyanidium caldarium. Arch. Mikrobiol.* **76**, 265–76.

ROBINSON J.B. (1969a) Sulphate influx in characean cells. I. General characteristics. *J. exp. Bot.* **20**, 201–11.

ROBINSON J.B. (1969b) Sulphate influx in characean cells. II. Links with light and metabolism in *Chara australis. J. exp. Bot.* **20**, 212–20.

RODHE W. (1965) Standard correlations between pelagic photosynthesis and light. *Mem. Ist. Ital. Idrobiol. Marco de Marchi*, **18**, *Suppl.*, 365–81.

RYTHER J.H. & GUILLARD R.R.L. (1962) Studies of marine planktonic diatoms. III. Some effects of temperature on respiration of five species. *Can. J. Microbiol.* **8**, 447–53.

SARGENT D.F. & TAYLOR C.P.S. (1971) Proof of two distinct enhancement effects of blue light on oxygen uptake in *Chlorella. Nature, Lond.* **232**, 649–50.

SCHMID G.H. & SCHWARZE P. (1969) Blue light enhanced respiration in a colorless *Chlorella* mutant. *Hoppe-Seyler's Z. Physiol. Chem.* **350**, 1513–20.

SCHMIDBAUER A. & RIED A. (1967) Einfluß hypertonischer Medien auf den Stoffwechsel synchron kultivierter *Chlorella. Arch. Mikrobiol.* **58**, 275–95.

SCHÖLM H.E. (1968) Untersuchungen zur Hitze- und Frostresistenz einheimischer Süßwasseralgen. *Protoplasma*, **65**, 97–118.

SCHULLE H. (1968) Ökologische und physiologische Untersuchungen an Planktonalgen des Titisees. Dissertation, Freiburg i. Br.

SCHUMACHER G.J. & WHITFORD L.A. (1965) Respiration and $P^{32}$ uptake in various species of freshwater algae as affected by current. *J. Phycol.* **1**, 78–80.

SEMENENKO V.E., VLADIMIROVA M.G. & ORLEANSKAYA O. (1967) Physiological characteristics of *Chlorella* sp. during extremely high temperatures. I. Disassociation action of extreme temperatures on the cellular function of *Chlorella. Fiziol. Rast.* **14**, 612–25.

SENGER H. (1962) Über die Wirkung von Glucose und Kohlensäure auf das Wachstum synchroner *Chlorella*-Kulturen. *Vortr. Ges. geb. Bot.*, N.F. **1**, 205–16.

SENGER H. (1969) Charakterisierung einer Synchronkultur von *Scenedesmus obliquus*, ihrer potentiellen Photosyntheseleistung und des Photosynthese-Quotienten während des Entwicklungscyclus. *Planta*, **90**, 243–66.

SENGER H. (1970) Quantenausbeuten und unterschiedliches Verhalten der beiden Photosysteme und des Photosyntheseapparates während des Entwicklungsablaufes von *Scenedesmus obliquus* in Synchronkulturen. *Planta*, **92**, 327–46.

SENGER H. & BISHOP N.I. (1966) The light-dependent formation of nucleic acids in cultures of synchronized *Chlorella. Pl. Cell Physiol., Tokyo*, **7**, 441–56.

SENGER H. & SCHOSER G. (1966) Die spektralabhängige Teilungsinduktion in mixotrophen Synchronkulturen von *Chlorella. Z. PflPhysiol.* **54**, 308–20.

SETLIK I. (1968) Gas exchange measurements: Photosynthesis. *A. Rep. Lab. Algol. Trebon* 1967, 128–40.

SETLIK I., BERKOVA E. & KUBIN ST. (1969) Irradiation and temperature dependence of photosynthesis in some chlorococcal strains. *A. Rep. Lab. Algol. Trebon* 1968, 134–8.

SHELEF G., OSWALD W. & GOLUEKE C.G. (1969) The continuous culture of algal biomass on wastes. In *Continuous Cultivation of Microorganisms*, ed. Malek I. *et al.* pp. 601–29. Academic Press, New York & London.

SHNEYOUR A., BEN'SHAUL Y. & AVRON M. (1969) Structural changes in *Euglena gracilis* grown autotrophically in the light with 3, (3,4-dichlorophenyl)-1,1-dimethyl urea (DCMU). *Exp. Cell Res.* **58**, 1–9.

SIMMER J. (1969) Production physiology of algae: Conditions for optimum use of radiant energy in a photosynthetic system. *A. Rep. Lab. Algol. Trebon* 1968, 127–30.

SLOBODSKOI L.I., SID'KO F.YA., BELYANIN V.I., ALYPOV V.F. & BERESNEV G.F. (1969) Analytical expression of the effect of temperature on microalgae productivity. *Biofizika*, **14**, 196–9.

SMAYDA T.J. (1969) Experimental observations on the influence of temperature, light, and salinity on cell division of the marine diatom, *Detonula confervacea* (Cleve). Gran. *J. Phycol.* **5**, 150–7.

Soeder C.J. (1966) Wirkungen von Außenfaktoren auf die Rhythmik der synchronen Zellentwicklung von *Chlorella fusca* Shihira et Krauss. *Ber. dt. bot. Ges.* **79**, 138–46.

Soeder C.J., Müller I. & Ried A. (1962) Über den Einfluß von Salzkonzentration und Nitratangebot auf den respiratorischen Quotienten und das N-reduzierende System von vollsynchroner *Chlorella pyrenoidosa*. *Vortr. Gesamtgeb. Bot.*, *N.F.* **1**, 195–200.

Soeder C.J., Ried A. & Strotmann H. (1964) Hemmwirkung von $CO_2$ auf späte Stadien der Zellentwicklung von *Chlorella*. *Beitr. Biol. Pfl.* **40**, 159–71.

Soeder C.J., Schulze G. & Thiele D. (1967) Einfluß verschiedener Kulturbedingungen auf das Wachstum in Synchronkulturen von *Chlorella fusca* Shihira et Krauss. *Arch. Hydrobiol. Suppl.* **23**, 127–71.

Soeder C.J., Strotmann H. & Galloway R.A. (1966) Carbon dioxide induced delay of cellular development in two *Chlorella* species. *J. Phycol.* **2**, 117–20.

Sokawa Y. (1967a) Effect of light on the chlorophyll formation in the 'glucose bleached' cells of *Chlorella protothecoides*. *Pl. Cell Physiol.*, *Tokyo*, **8**, 495–508.

Sokawa Y. (1967b) Effects of light on the deoxyribonucleic acid formation and cellular division in *Chlorella protothecoides*. *Pl. Cell Physiol.*, *Tokyo*, **8**, 509–22.

Sokawa Y. & Hase E. (1968) Suppressive effect of light on the formation of DNA and on the increase of deoxythymidine monophosphate kinase in *Chlorella protothecoides*. *Pl. Cell Physiol.*, *Tokyo*, **9**, 461–6.

Sorokin C. (1959) Tabular comparative data for the low- and high-temperature strains of *Chlorella*. *Nature, Lond.* **184**, 613–14.

Sorokin C. (1960) Injury and recovery of photosynthesis. The capacity of cells of different developmental stages to regenerate their photosynthetic activity. *Physiologia Pl.* **13**, 20–35.

Sorokin C. & Krauss R.W. (1962) Effects of temperature and illuminance on *Chlorella* growth uncoupled from cell division. *Pl. Physiol.*, *Lancaster*, **37**, 37–42.

Spektorova L.V.V.I. (1967) Analysis of the potential productivity of *Chlorella* by way of synchronized cultures. *Bot. Zh.* **52**, 73–81.

Starosta B. (1970) Die Regulation der durch Stickstoffmangel ausgelösten Sekundär-carotinbildung bei *Ankistrodesmus*. Dissertation, Universität München, 112 pp.

Steemann Nielsen E. (1964) On a complication in marine productivity work due to the influence of ultraviolet light. *J. Cons. Int. Explor. Mer.* **24**, 130–5.

Steemann Nielsen E., Hansen V.K. & Jørgensen E.G. (1962) The adaptation to different light intensities in *Chlorella vulgaris* and the time dependence on transfer to a new light intensity. *Physiologia Pl.* 505–17.

Steemann Nielsen E. & Jørgensen E.G. (1968) The adaptation of plankton algae. I. General part. *Physiologia Pl.* **21**, 401–13.

Stengel E. (1970) Zustandsänderungen verschiedener Eisenverbindungen in Nährlösungen für Algen. *Arch. Hydrobiol. Suppl.* **38**, 151–69.

Stepanskii V.I. & Yaglova L.G. (1968) Temperature dependence of the membrane potential of *Nitella mucronata* cells. *Biol. Nauk.* **11**, 79–83.

Stewart W.D.P. (1970) Nitrogen fixation by blue-green algae in Yellowstone thermal areas. *Phycologia*, **9**, 261–9.

Strotmann H. (1967a) Untersuchungen zur lichtabhängigen Nitritreduktion von *Chlorella*. *Planta*, **77**, 32–48.

Strotmann H. (1967b) Blaulichteffekt auf die Nitritreduktion von *Chlorella*. *Planta*, **73**, 376–80.

Strotmann H. & Ried A. (1969) Polarographische Messung der Nitritreduktion von *Chlorella* in monochromatischem Licht. *Planta*, **85**, 250–69.

Talling J.F. (1971) The underwater light climate as a controlling factor in the production ecology of freshwater phytoplankton. *Mitt. Internat. Verein. Limnol.* **19**, 214–43.

Taniguchi M. (1969) An experiment to get the young culture of conchocelis of *Porphyra* in any season. *Bull. Jap. Soc. Scient. Fish.* **35**, 333–5.

TANNER W. (1969) Light-driven active uptake of 3-O-methylglucose via an inducible hexose uptake system of *Chlorella*. *Biochem. biophys. Res. Commun.* **36**, 278–83.

TANNER W., LOOS E., KLOB W. & KANDLER O. (1968) The quantum requirement for light-dependent anaerobic glucose assimilation by *Chlorella vulgaris. Z. PflPhysiol.* **59**, 301–3.

TERRY O. & EDMUNDS L.N. JR. (1969) Semi-continuous culture and monitoring system for temperature-synchronized *Euglena. Biotechnol. Bioeng.* **11**, 745–56.

THROM G. (1971) Einfluß von Hemmstoffen und des Redoxpotentials auf die lichtabhängige Änderung des Membranpotentials bei *Griffithsia.* setacea. *Z. PflPhysiol.* **64**, 281–96.

TRUKHIN N.V. (1968) Effect of the carbon supply on the response of *Chlorella pyrenoidosa* to blue and red light. *Fiziol. Rast.* **15**, 652–7.

TRUKHIN N.V. (1969) The utilization of light energy by *Chlorella* under laboratory culture conditions. *Tr. Inst. Biol. Vnutr. Vod. Akad. Nauk, SSSR.* **19**, 32–8.

ULLRICH W.R. (1972) Einfluß von $CO_2$ und pH auf die$^{32}$P- Markierung von Polyphosphaten und organischen Phosphaten bei *Scenedesmus* im Licht. *Planta*, **102**, 37–54.

VALLEE M. & JEANJEAN R. (1968) Le systeme de transport de $SO_4{}^{2-}$ chez *Chlorella pyrenoidosa* et sa regulation. I. Etude cinetique de la permeation. *Biochim. biophys. Acta*, **150**, 599–606.

VENKATARAMAN G.S., AMELUNXEN F. & LORENZEN H. (1969) Note to the fine structure of *Anacystis nidulans* during its synchronous growth. *Arch. Mikrobiol.* **69**, 370–2.

VIDAVER W. (1969) Hydrostatic pressure effects on photosynthesis. *Int. Rev. ges. Hydrobiol.* **54**, 697–747.

VOLK S.L. & PHINNEY H.K. (1968) Mineral requirements for the growth of *Anabaena spiroides in vitro. Can. J. Bot.* **46**, 619–30.

VOLLENWEIDER R.A. (1965) Calculation models of photosynthesis-depth curves and some implications regarding day rate estimates in primary production measurements. *Mem. Ist. Ital. Idrobiol. Marco de Marchi*, **18**, *Suppl.*, 425–57.

VOLLENWEIDER R.A. (1969) *A manual on methods for measuring primary production in aquatic environments.* IBP Handbook No. 12, Blackwell Scientific Publications, Oxford.

WATT W.D. (1966) Release of dissolved organic material from the cells of phytoplankton populations. *Proc. R. Soc. B., London*, **164**, 521–551.

WATT W.D. (1969) Extracellular release of organic matter from two freshwater diatoms. *Ann. Bot.* **33**, 427–437.

WATT W.D. & FOGG G.E. (1966) The kinetics of extracellular glycollate production by *Chlorella pyrenoidosa. J. exp. Bot.* **17**, 117–34.

WHITTON B.A. (1968) Effect of light and toxicity of various substances to *Anacystis nidulans. Pl. Cell Physiol., Tokyo*, **9**, 23–6.

WIESSNER W. (1969) Effect of autotrophic or photoheterotrophic growth conditions on *in vivo* absorption of visible light by green algae. *Photosynthetica, Praha*, **3**, 225–32.

WILSON B.W. & JAMES T.W. (1966) Energetics and the synchronized cell cycle. In *Cell Synchrony*, eds. Cameron I.L. & Padilla G.M. pp. 236–55. Academic Press, New York & London.

ZARROUK C. (1966) Contribution a l'étude d'une Cyanophycée. Influence de divers facteurs chimiques et physiques sur la croissance et la photosynthèse de *Spirulina maxima. Paris: Thèse Sciences.*

ZIEGLER R. & SCHANDERL S.H. (1969) Chlorophyll degradation and the kinetics of dephytylated derivates in a mutant of *Chlorella. Photosynthetica, Praha*, **3**, 45–54.

ZIMMERMANN U. (1969) Ökologische und physiologische Untersuchungen an der planktischen Blaualge *Oscillatoria rubescens* D.C. unter besonderer Berücksichtigung von Licht und Temperatur. *Schweiz. Z. Hydrol.* **31**, 1–58.

# CHAPTER 27

# VITAMINS AND GROWTH REGULATORS

## L. PROVASOLI

Haskins Laboratories,
Biology Department, Yale University,
New Haven, Connecticut, U.S.A.

and

## A. F. CARLUCCI

Institute of Marine Resources,
University of California,
La Jolla, California, U.S.A.

1    **Introduction**   742

2    **Vitamin requirements**   742
2.1   Patterns of requirements in algal groups   744
2.2   Variability of requirements 746
2.3   Specificity of the vitamin requirements   748
     (a) Thiamine   748
     (b) Vitamin $B_{12}$   749
2.4   Vitamin $B_{12}$ uptake and the binding factor   751
2.5   Bioassays   753
     (a) Seawater   753
     (b) Fresh water   754
     (c) Particulate material 756

3    **Vitamin and algal interactions** 756
3.1   Vitamin content of algae   756
3.2   Production of vitamins by algae 757
3.3   Utilization of algal produced vitamins by algae   757

4    **Ecology**   758
4.1   Occurrence of vitamins in waters 758
     (a) Seawater   758
     (b) Fresh water   759

4.2   Toxicity to marine phytoplankton   759
4.3   Probable role of vitamins   759

5    **Plant hormones and algal regulators**   762
5.1   Auxins   763
     (a) Occurrence   763
     (b) Effects of exogenous IAA   764
5.2   Cytokinins   765
     (a) Occurrence   765
     (b) Effects of exogenous cytokinins   766
5.3   Gibberellins   767
     (a) Occurrence   767
     (b) Effects of exogenous gibberellins   767
5.4   Unknown morphological regulators   768

6    **Acknowledgements**   771

7    **References**   771

8    **Appendix**   778

9    **Appendix references**   783

# 1  INTRODUCTION

Many algae require vitamins (auxotrophy); most higher plants do not. Apart from this difference and their need to live in water or moist environments, the nutrition of higher plants and algae is very similar.

The delay in recognition of auxotrophy in algae was mostly due to the success in the 1890s of Beijerinck, Molisch, and Miguel in culturing a few freshwater green algae and diatoms in inorganic media. We know now that these two algal groups are outstanding in comprising a large, perhaps a preponderant, number of species which do not require vitamins. The belief that algae are autotrophic was furthered by using mineral media in isolating algae, thus selecting against vitamin-requiring species. A few hundred cultures of autotrophic algae accumulated in laboratory collections in the intervening 40 years, and *Chlorella* became a favourite of plant physiologists as a convenient tool and a 'typical' alga!

Many algae cannot be grown on mineral media. Pringsheim pioneered the field, introducing soil-water and soil-extract cultures for growing the more exacting free-living algae (see Pringsheim 1937). It is not surprising that this addition proved so successful: it contains nitrates, ammonia, vitamins, an array of bacterial metabolites, and trace metals chelated by humic acids. The search for the growth factors present in soil extract and peptone resulted in the discovery that thiamine was needed for the colourless flagellates *Polytomella*, *Polytoma*, and *Chilomonas* (Lwoff & Dusi 1937a,b,c). Hutner (1936) found that animal proteins had a factor needed by *Euglena gracilis* which was not present in plant proteins. Liver concentrates used against pernicious anemia, were very active for *Euglena* (Provasoli *et al.* 1948); vitamin $B_{12}$ was finally recognized as the factor needed by *Euglena* (Hutner *et al.* 1949). *Euglena gracilis*, because of its great sensitivity became, with *Ochromonas malhamensis*, a $B_{12}$ bioassay organism for analysis of blood, urine and other natural materials (Baker & Frank 1968). The finding that soil extract was replaceable by trace metals and vitamins for several freshwater, brackish and marine algae (Provasoli & Pintner 1953) permitted the cultivation in artificial media of many algae requiring soil extract. Through the efforts of Droop, J. Lewin, R. Lewin, Pringsheim, Fries, Guillard and many others, numerous algae were found to be auxotrophic.

# 2  VITAMIN REQUIREMENTS

The subject has been reviewed often as work progressed (Provasoli 1958a, 1963, Lewin 1961, Droop 1962, Thomas 1968). An updated list of the vitamin requirements of single species of algae is given in the Appendix. It comprises freshwater and marine forms and some colourless (apochlorotic) species whose affinity with, and probable derivation from, photosynthetic species seems undisputed.

The algae stand out among microorganisms for needing only three vitamins. Vitamin $B_{12}$ and thiamine are required alone or in combination by the majority of the auxotrophic algae, and $B_{12}$ seems required more often than thiamine. Biotin so far has been shown to be necessary for a few chrysomonads and dino-flagellates and one euglenoid. The presence of bacteria, yeasts, and fungi in the same environments colonized by algae indicates that these environments have the variety of B vitamins needed by these heterotrophs; the survival of algal mutants requiring other vitamins besides $B_{12}$, thiamine and biotin would seem likely. Hence, the narrow spectrum of vitamins needed does not seem to be fortuitous.

Auxotrophy, or lack of it, does not relate to sources of energy used by the algae; it appears to be an independent loss of function, *sensu* Lwoff (1943). Obligate photoautotrophic species may need vitamins as do some *Synura* spp. and many marine algae; and obligate heterotrophs, such as algae which have lost their photosynthetic apparatus, may not need vitamins, i.e. *Polytoma uvella* and *P. obtusum*.

Since vitamins are present in all environments, auxotrophy does not strictly correlate with the trophic state of the environment. Thus autotrophs are found in sewage lagoons and auxotrophs in oligotrophic waters. However, the order of incidence of auxotrophs and autotrophs may be expected to differ either in number of individuals or species. Obviously, in environments perennially rich in vitamins the auxotrophs may thereby have an advantage in avoiding the necessity of synthesizing these vitamins.

An evaluation of the incidence of auxotrophy in pennate diatoms of marine littoral rock pools was carried out by Lewin (1972); 58% of the 55 clones isolated were auxotrophic. Since these clones (i.e. single colonies on agar) originated from single cells, the data show that in a rich environment the auxotrophs, numerically, compete favourably with the autotrophs. Data of this type, extended to other environments and seasons, are far more relevant ecologically than the usual data on numbers of cells of each species because single species may yield clones widely differing in their requirements (see below 2.2).

Incidence of auxotrophy in algal groups reveals differences. As evident (Table 27.1), members of the Cyanophyceae, Chlorophyceae, Xanthophyceae and Phaeophyceae have the least number of species requiring vitamins. The predominance of autotrophs in these algal groups seems undisputed, even taking into consideration that the information available may not be completely reliable and may be biased. Bias depends on the type of medium used for isolation. Mineral media were used for years, especially in isolating green algae, thus selecting for autotrophic species. To avoid bias, isolations should be done with media containing vitamins, and the clones obtained in pure cultures tested for autotrophy and need for single vitamins (at least three serial transfers in the same variable should be done to exhaust carry-over). Reliability of data on auxotrophy depends upon the precautions used to preserve chemical purity in compounding the media—a consideration which became conspicuous after the discovery and availability of vitamins (i.e. after the 1940s).

These precautions, now routinely applied in working with vitamins, are too often ignored in nutritional work with algae, but are largely unnecessary in collections. Thus algae maintained in ostensibly mineral media may not be autotrophic. Some of the data on blue-green algae may be misleading because of the notorious difficulty of obtaining cultures devoid of contaminants, which may provide vitamins for the algae.

Members of the Euglenophyceae, Cryptophyceae, Dinophyceae and Chryso-phyceae show, on the contrary, a clear predominance of auxotrophic species. These results are reliable, but may be biased because the interest in vitamin requirements may result in autotrophy going unreported. The morphologically more complex groups also show clear differences: the Phaeophyceae do not require, but some may be stimulated by vitamins; most Rhodophyceae require vitamin $B_{12}$.

## 2.1  Patterns of requirements in algal groups

While these cautionary remarks indicate that it is unwise at this time to attach importance to ratios between auto- and auxotrophy, the data seem representative despite the relatively small sample of approximately 380 species.

Each algal group shows a definite pattern for incidence of autotrophy and in the incidence of the vitamin required (Table 27.1). Some algal groups do not require biotin; among these the Xanthophyceae, Phaeophyceae, Cyanophyceae and Chlorophyceae stand out for having a predominance of autotrophic species. Among the diatoms, the auxotrophic species prevail, and in the red seaweeds (but not in the unicellular red alga *Porphyridium*) all are so far auxotrophic. The blue-green algae and the red seaweeds require only vitamin $B_{12}$, while the diatoms and the green algae may also need thiamine.

In the algal groups where auxotrophy dominates, biotin appears as a new requirement of low incidence. The sample size is far too small in the crypto-monads and euglenoids to consider it significant that biotin is required by only one species. The dinoflagellates have a majority of species requiring vitamin $B_{12}$ and the chrysomonads show a slight preference for thiamine, which seems to increase with the number of species studied (Pintner & Provasoli 1968).

These trends indicate that incidence of auxotrophy and of the vitamins required might serve as elements in an emerging physiological taxonomy. Since algal groups have varied and defined morphological tendencies (Smith 1933, Fritsch 1945) and the algal flagellates, physio-morphological tendencies towards animality (i.e. phagotrophy and loss of photosynthesis, Lwoff 1943), an attempt was made to relate auxotrophy or lack of it to these tendencies (Provasoli & Pintner 1953, Provasoli 1956). The incidence of auxotrophy is low in the Chlorophyceae, which have a weak tendency towards loss of photosynthesis (i.e. few apochlorotic species) and no tendency towards phagotrophy. The blue-green algae were not considered in these publications; however, they show a tendency to loss of photosynthetic apparatus and acquisition of auxotrophy. At

**Table 27.1.** Summary of vitamin requirements.

| Algal groups | Vit. B₁₂ | Thiamine | Biotin | B₁₂ Thiamine Biotin | B₁₂ Thiamine | B₁₂ Biotin | Thiamine Biotin | No req. | Vit. req. | B₁₂/Thiamine | Predom-inance |
|---|---|---|---|---|---|---|---|---|---|---|---|
| **(1) Do not require biotin** | | | | | | | | | | | |
| Cyanophyceae | 8 | 0 | 0 | 0 | 0 | 0 | 0 | 35 | 8 | 8/0 | B₁₂ |
| Rhodophyceae | 11 | 0 | 0 | 0 | 0 | 0 | 0 |  | 11 | 11/0 | B₁₂ |
| Bacillariophyceae | 35 | 5 | 8 | 0 | 0 | 0 | 0 | 23 | 48 | 43/13 | B₁₂ |
| Xanthophyceae | 0 | 0 | 0 | 0 | 0 | 0 | 0 | 7 | 0 | 0 | No |
| Phaeophyceae | 1 | 0 | 0 | 0 | 0 | 0 | 0 | 9* | 1 | 0 | No |
| Chlorophyceae | 38 | 13 | 0 | 0 | 10 | 0 | 0 | 106 | 61 | 48/23 | B₁₂ |
| **(2) Require biotin** | | | | | | | | | | | |
| Chrysophyceae and Haptophyceae | 2 | 8 | 0 | 2 | 12 | 2 | 1 | 1 | 27 | 18/23 | Thiamine |
| Cryptophyceae | 2 | 2 | 0 | 0 | 1 | 1 | 1 | 0 | 7 | 5/5 | No |
| Dinophyceae | 17 | 1 | 0 | 1** | 5 | 0 | 1 |  | 25 | 24/6 | B₁₂ |
| Euglenophyceae | 3 | 0 | 0 | 0 | 10 | 1 | 1 |  | 15 | 14/12 | No |
| Totals of the requirements | 172 | 82 | 14 | | | | | 185 | 203 | | |

\* Four of the 9 species are stimulated by vitamins  \*\* Crypthecodinium cohnii is highly stimulated by thiamine that time only few diatoms were represented; they seem to have a moderate preference for auxotrophy, yet only a few colourless species are known. High incidence of auxotrophy is found in the Chrysophyceae and Dinophyceae which have many apochlorotic and phagotrophic species.

The above considerations fit the hypothesis that the narrow spectrum of vitamins needed by and typifying the algae among microorganisms is not fortuitous but is in some way dependent on the morpho-physiological evolution of the algae.

This evolution proceeds even at present; euglenoids have numerous apochlorotic natural counterparts (*Astasia, Distigma*, etc.), and loss of photosynthesis in *Euglena gracilis* can be achieved artificially by applying heat (Pringsheim & Pringsheim 1952), antibiotics or other chemicals. Some Chrysophyceae form pseudopodia, are phagotrophic, and have many apochlorotic counterparts, some of which might have originated by an unequal division of species bearing a single chloroplast.

## 2.2   *Variability of requirements*

Variability in vitamin requirements by isolates (clones) of the same species was noted first by Lewin and Lewin (1960). They found that isolates of *Amphora coffaeiformis* could be either autotrophic, needed either thiamine or $B_{12}$, or both. When, besides vitamins, the physiology of each clone was characterized for ability to grow in darkness and utilization of carbon sources, the ten clones of *A. coffaeiformis* showed nine different physiological patterns. Similarly, six clones of *Nitzschia closterium* (*Phaeodactylum tricornutum*) yielded five different patterns. These findings of Lewin and Lewin cannot be attributed to faulty speciation; neither are they isolated cases. Of three clones of *Coccolithus huxleyi*, two need thiamine, and one $B_{12}$ (Guillard 1968). Unpublished data of Hargraves and Guillard show that each of five clones of *Bellerochea* sp. (*polymorpha*?) have differences in vitamin requirements, qualitative or quantitative, or related to $B_{12}$ specificity, and differences in tolerance of low salinity and high temperature. Clones of *Bellerochea* sp. (*spinifera*?) may need $B_{12}$ only, or $B_{12}$ and thiamine; *Fragilaria pinnata* has an autotrophic clone and one requiring $B_{12}$.

The clones of *F. pinnata* and *B.* sp. (*polymorpha*?), for example, differ also in their half-saturation constants (*K*) for nitrate uptake. These differences seem adaptations to environmental availability of nitrate: clones isolated from water poor in nitrate have a more efficient uptake system (Carpenter & Guillard 1971). Whether the variations in vitamin requirements are also related to environmental availability is unknown.

Variations in requirements within a clone have been observed when cultures have been maintained for years in culture collections. Media, frequency of transfer, temperatures, and light intensity are apt to vary and the cultures are not always kept in an optimal state, especially in private collections where research is the primary aim and maintenance of cultures secondary. Either adaptation or selection of mutants can then occur. A strain of *Polytoma obtusum* which utilized acetate and butyrate without a lag phase, after 12 years in culture in peptone media without fatty acids, could utilize butyric acid only after a long lag phase, and the readapted clone grew in successive transfers without lag (Provasoli 1938). This indicated that assimilation depended on production of adaptive enzymes; a subclone could be trained by repeated passages to utilize caproic acid. This induction of adaptive enzymes was confirmed by Cirillo (1957).

Permanence in culture of autotrophic clones may favour auxotrophic mutants as suggested by unpublished data kindly supplied by Guillard and Haines. Strains of *Melosira nummuloides*, isolated in 1954, 1956, and 1959 proved to be autotrophic. After 10 years' maintenance in a medium containing $B_{12}$, thiamine and biotin, the three clones grew without vitamins through nine months of transfers, but very slowly, and yields were moderate. Addition of thiamine eliminated the lag phase and yielded dense cultures. Fresh isolation of *M. nummuloides* from the same ponds gave autotrophic strains. The old cultures apparently are still synthesizing thiamine, but slowly.

Slow vitamin synthesis may also be found in freshly isolated clones. *Cryptheco-dinium* (*Gyrodinium*) *cohnii* cannot be grown without biotin; however, in the absence of thiamine, growth is slow and is about a 100-fold lower (Provasoli & Gold 1962). Subtler differentials in growth have been observed in many other species of algae and the effect of the added vitamin or other nutrients is termed 'stimulatory', indicating that different species or clones may have different rates of synthetic power—a low rate of synthesis may finally lead to permanent loss of synthetic power, i.e. appearance of an absolute need.

Laboratory maintenance does not, however, invariably entail loss of synthetic powers. Depending on the cultural conditions, it may favour less exacting mutants. Tassigny (1971) found that axenic cultures of *Closterium strigosum* had an absolute requirement for vitamin $B_{12}$ but that after being maintained in standard mineral media with soil extract (relying on the soil extract as a source of $B_{12}$) the clone became autotrophic. Bacterized cultures of *Closterium braunii* kept for 30 years require only $B_{12}$, while a clone freshly isolated from the same locality had a more complex requirement. *Raphidonema nivale*, when first tested in 1959, showed a definite requirement for thiamine. When the experiments were repeated twice in 1970 the clone no longer required thiamine; no information was given on the maintenance medium (Hoham 1971).

Undoubtedly, other unreported variations have happened in cultivated clones, but no physiological variations were reported in many strains which have been used for many years by researchers (i.e. *Chlorella* spp., *Euglena gracilis*, *Ochromonas malhamensis*, etc.). A suspected degeneration may afflict cultures of *O. danica* and *O. malhamensis* which were kept for 16 years in rich media (Klein 1972). As these algae have long been in routine use for vitamin assays, one may expect a thorough examination of this situation.

Such physiological variations among clones either inherent or acquired in culture may underlie many discrepancies in the literature which are not due to faulty identification. Since the physiological variations are apparently connected with localities and environments, extrapolation of data pertaining to a clone of any species to other clones of the same species but collected in other localities is risky.

## 2.3  *Specificity of the vitamin requirements*

The need for an exogenously indispensable molecule obviously points to an inability of the organism to synthesize the substance. This physiological impairment is seldom total; often the biosynthesis is only partly blocked at one step on the pathway, as in the familiar phenomenon of leaky mutants studied in microbial genetics. In such cases, the intermediate after the block, if taken up, replaces the vitamin.

### (a) *Thiamine*

The thiamine molecule is composed of a thiazole and a pyrimidine moiety. These moieties can often replace intact thiamine and indeed were the first growth factors identified for algal flagellates in 1937–38 by Lwoff and Dusi. The early literature was discussed by Lwoff (1943, 1947); these and later results were tabulated by Lewin (1961). The thiamine requirement is satisfied by the thiazole moiety alone in six species, by the pyrimidine moiety in four species; four species require both moieties, and only *Ochromonas malhamensis* (Hutner & Provasoli 1955), and a mutant of *Chlamydomonas reinhardtii* (Eversole 1956) require the intact thiamine molecule. For some species, homologues of the thiazole and pyrimidine moieties were found to be as active as the natural ones (see Lwoff 1947), indicating which structural modifications of the molecule are compatible with activity. In the last 10 years no further data have been added on moiety utilization and specificity—data that could be useful in differentiating the nutritional requirements of an algal species for ecological purposes. Renewed work seems necessary.

The hypothesis of Vishniac and Riley (1961) that scarcity and variability in concentration of thiamine in Long Island Sound may be due to non-biological destruction of thiamine has been confirmed by Gold *et al.* (1966). Disappearance of chemically analysable thiamine in seawater increases sharply between 10° and 30°C, and there is total destruction at 37°C in 24 hours at the alkaline pH of seawater. The same phenomenon but limited to the range 20 to 30°C was recorded using *Crypthecodinium cohnii* as bioassay organism. The discrepancy would be explained if *C. cohnii* utilizes some products of thiamine degradation. However, the degradation products of thiamine in seawater remain unidentified, and information is lacking on the ability of *C. cohnii* to utilize the moieties of thiamine.

## (b) Vitamin $B_{12}$

Vitamin $B_{12}$ has attracted more interest than thiamine. Vitamin $B_{12}$ ($\alpha$-5,6-dimethylbenzimidazolyl cyanocobalamin) is one of a family of cobalamins. The molecule is composed of two major portions, a large corrin nearly planar group, and a nucleotide side chain ($\alpha$-5,6-dimethylbenzimidazole-D-ribofuranose phosphate). The analogues can be subdivided into three groups: a nucleotide-like benzimidazole side chain (includes $B_{12}$), adenine or guanine-like side chains, and lack of a nucleotide side chain (factor B). Organisms needing cobalamins fall into three patterns of increasingly narrow specificity: one, the *Escherichia coli* type utilizes all analogues including factor B lacking the nucleotide chain; two, the *Lactobacillus leichmannii*-type utilizes all analogues except factor B (includes *Euglena gracilis*); three, the *mammalian* type utilizes only analogues with dimethylbenzimidazole nucleotide (includes *Ochromonas malhamensis*).

Algae, like other microorganisms, comprise species having these three specificity patterns. Interest in the $B_{12}$ specificity of the algae was stimulated by the finding that many marine bacteria produce and release cobalamins in the waters, but only a minority of them (15 to 30%) produce true vitamin $B_{12}$.

Thus algae having a narrow specificity (*mammalian* type) would be at a disadvantage. The early data of Droop *et al.* (1959) showed that the only two diatoms tested had the *E. coli* type of wide specificity. Hence, diatoms would be favoured, even if needing $B_{12}$, and their versatility of utilization would contribute to the conditions favouring the diatom spring blooms. Guillard (1968) surveyed 24 clones of centric diatoms and found that the three specificity patterns were represented in this group with the *E. coli* and *mammalian* specificities being favoured. The results of Guillard are particularly valuable ecologically because the analogues have been assayed at 4 ng $l^{-1}$, a level close to natural concentrations of $B_{12}$ in seawater. Droop *et al.* used a range of concentrations, but the lowest was 10 ng $l^{-1}$. Analogues were considered active by Guillard if 4 ng $l^{-1}$ gave at least 10% of the growth elicited by an equal amount of vitamin $B_{12}$. In a summary of the 74 species analysed (Table 27.2), 36 fall in the *mammalian* pattern, 15 in the *Lactobacillus* and 23 in the *E. coli* pattern. Utilization of *E. coli*-pattern is more diffuse in diatoms than other algal groups.

A survey on the specificity of nine species of Rhodophyceae shows that the red seaweeds do not seem to fall completely into classical patterns. This is partly due to the inclusion in the analogues tested of the cobalamins extracted by Neujahr (1956) from sewage sludge (factors $X_3$, $Z_1$, $Z_2$, $Z_3$) which, like factor B, do not have a nucleotide group. If the results on the red algae are analysed using only the usual analogues, *Antithamnion* sp., *A. glanduliferum*, *A. sarniense* (Tatewaki & Provasoli 1964), *Asterocytis ramosa* (Fries & Petterson 1968) and *Polysiphonia urceolata* (Fries 1964) would have a *mammalian* specificity; the conchocelis phase of *Porphyra tenera* (Iwasaki 1965) would have the *E. coli* specificity. *Goniotrichum alsidii* (*elegans*) (Fries 1960) is a special case; it utilizes factor A, but not pseudo $B_{12}$, so it does not fit completely into the *Lactobacillus*

**Table 27.2.** Algal Specificity toward vitamin $B_{12}$ analogues

| Algae | Specificity type* | No. |
|---|---|---|
| CYANOPHYCEAE | | |
| *Agmenellum quadruplicatum* (PR-6)[1], *Coccochloris elabens* (17a)[1], *Phormidium persicinum.* | C | 3 |
| CHLOROPHYCEAE AND PRASINOPHYCEAE | | |
| *Balticola droebakensis, B. buetschlii, Brachiomonas submarina, Chlamydomonas pulsatilla, Platymonas tetrathele, Pyramimonas inconstans, Stephanoptera pluvialis, Volvox globator, V. tertius.* | M | 9 |
| BACILLARIOPHYCEAE[2] | | |
| *Amphora perpusilla* (=*coffaeiformis*), *A. coffaeiformis, Bellerochea* sp. (F.8) (*spinifera?*), *Biddulphia* (?) sp., *Cyclotella caspia, Cylindrotheca closterium, C. pseudonana* (=*Cyclotella nana*), *Ditylum brightwellii, Nitzschia* sp., *N. frustulum, N. punctata, Skeletonema* sp. *S. costatum, Stephanopyxis turris, Thalassiosira pluvialis.* | C | 15 |
| *Chaetoceros pelagicus, C. simplex, Coscinodiscus asteromphalus, Fragilaria* sp. (F-3), *Nitzschia ovalis.* | L | 5 |
| *Achnanthes brevipes, Chaetoceros ceratosporum, C. lorenzianum, C. pseudocrinitus, Cylindrotheca pseudonana* (=*Cyclotella nana*), *C. closterium, Fragilaria* sp. (13-3) (*pinnata?*), *Rhizosolenia setigera, Skeletonema costatum.* | M | 9 |
| CHRYSOPHYCEAE | | |
| *Microglena arenicola, Ochromonas malhamensis, Synura caroliniana, S.petersenii.* | M | 4 |
| *Monochrysis lutheri.* | L | 1 |
| HAPTOPHYCEAE | | |
| *Cricosphaera* (*Syracosphaera*) *elongata, Isochrysis galbana, Prymnesium parvum.* | M | 3 |
| CRYPTOPHYCEAE | | |
| *Chroomonas marina, Cryptomonas ovata, Cyanophora paradoxa*[6], *Hemiselmis virescens.* | M | 4 |
| *Rhodomonas* sp. (several marine clones) | C | 1+ |
| EUGLENOPHYCEAE | | |
| *Euglena gracilis, E. viridis, Trachelomonas abrupta, T. pertii.* | L | 4 |
| *Phacus pyrum.* | M | 1 |
| *Eutreptiella* sp.[3] | C | 1 |
| DINOPHYCEAE | | |
| *Entomosigma* sp. (*Heterosigma akashivo*)[3], *Gyrodinium californicum, G. resplendens, G. uncatenum, Heterosigma inlandica*[3], *Woloszynskia limnetica.* | M | 6 |
| *Amphidinium carteri* (*Klebsii*), *A. rhynchocephalum*(?), *Exuviaella* sp.[3], *E. cassubica*[4], *Peridinium balticum*[4]. | L[5] | 5 |
| *Gymnodinium nelsoni*[3], *Peridinium hangoei*[3], *Polykrikos schwartzii.* | C | 3 |

* C, *E. coli*-type of specificity; L, *Lactobacillus*-type; M, *mammalian*-type. Most data were taken from Droop *et al.* (1959). (1) Results of van Baalen (1962). (2) The Bacillariophyceae data are from Guillard (1968), except for *Amphora perpusilla, Skeletonema costatum* (Droop *et al.* 1959), and *Stephanopyxis turris* (Pintner & Provasoli unpubl.). (3) Results of Iwasaki (1969, 1971a,b), Iwasaki

specificity pattern; several dinoflagellates have the same peculiarity (Table 27.2, note 4). However, *all* of these red algae utilized in different degrees (from 25 to 100% of the growth obtained with $B_{12}$) factors $Z_1$, $Z_2$, $Z_3$ and Ib (= factor β-ribosephosphate), which lack the nucleotide side chain; but do not utilize (except for *Porphyra tenera*) factor B which also lacks the nucleotide group. Utilization of the molecule without nucleotide typified, as noted, the *E. coli* specificity. Why the Neujahr compounds and not factor B are active for the red seaweeds is not known. Since factor Ib is utilized also by the red algae, the activity of the Neujahr factors could depend on the presence of other similar side groups of non-nucleotide type. *Asterocytis* differed from all the other red algae because of its narrow specificity; it utilized only three cobalamins: vitamin $B_{12}$, factor III and factor $Z_1$.

## 2.4    *Vitamin $B_{12}$ uptake and the binding factor*

In studies on the quantitative assay of vitamin $B_{12}$ with *Euglena gracilis* Kristensen (1955) found that variations in growth curves depended on the age of the *Euglena* culture used as inoculum. The effect was traced to the presence of a thermolabile inhibitor in the supernatant liquid of the cultures. The inhibitor combined with vitamin $B_{12}$, forming a non-dialysable complex which was not absorbed by wild-type *E. coli*. The titre of inhibitor increased with the age of the *Euglena* culture: 1 ml supernatant of a five-day old culture binds 2·2 ng $B_{12}$, and at 43 days it binds 17·3 ng $B_{12}$. Kristensen (1956) concluded that the $B_{12}$-binding factor of *Euglena* resembled in its properties the intrinsic factor, a thermolabile $B_{12}$-binding glycoprotein found in the gastric juice of mammals which is indispensable for the uptake of $B_{12}$ by the intestinal mucosa. The $B_{12}$ bound to the intrinsic factor and to the binder of *Euglena* is not available to the microorganisms used for vitamin $B_{12}$ assay. A similar undialysable $B_{12}$-binding factor was found by Ford *et al.* (1955) in extracts of *Ochromonas malhamensis* cells and in the supernatant of old cultures.

Droop (1966), using a chemostat to measure the parameters of the vitamin $B_{12}$ requirement in *Monochrysis lutheri*, found internal inconsistencies with the chemostat's operation that indicated the presence of factors unaccounted for by theory. The interfering factor was a $B_{12}$ binder produced by *Monochrysis*. Using supernatants of *Monochrysis*, of *Isochrysis galbana* (requiring $B_{12}$), and of *Phaeodactylum tricornutum* and testing them on the cells of the other species, he found that all three species produced the inhibitor, even *P. tricornutum*, which does not need vitamins (Droop 1968). Every $B_{12}$-requirer was inhibited by its own and by the heterologous filtrates. The inhibition is competitive, since addition of $B_{12}$ above the quantity necessary to satisfy the binding power restored growth. The inhibitory action was destroyed by autoclaving. Hence, excretion of, and

---

*et al.* (1968), Iwasaki & Sasada (1969)    (4) Results of Provasoli & McLaughlin (1955).    (5) The two *Amphidinium*, two *Exuviaella* and *P. balticum* utilize factor A, but not pseudo $B_{12}$.    (6) See note 21, p. 916.

sensitivity to the thermolabile inhibitor were not species specific. Also, it was not exclusively excreted under conditions of $B_{12}$ limitation as it appeared to be in the experiments of Kristensen (1955), since it was produced by *Monochrysis* under conditions of thiamine depletion and by a non-vitamin requirer, *P. tricornutum*. Using labelled vitamin $B_{12}$ and the Sephadex-gel filtration technique of Daisley (1961) for separating the bound from the unbound $B_{12}$, Droop confirmed that the inhibitor indeed bound vitamin $B_{12}$.

The mode of action of the inhibitor was clarified by using radioactive vitamin $B_{12}$ with *Phaeodactylum tricornutum* (Droop 1968). Even though this organism does not need $B_{12}$, it took up $B_{12}$ from fresh media at about the same rate as the $B_{12}$-requirers. In supernatants of old cultures the uptake of $B_{12}$ was blocked by the inhibitor, but growth was hardly repressed since *P. tricornutum* does not require $B_{12}$.

The inhibitor seemed to be released at a constant relative rate by the cells; excretion of the inhibitor parallels the apparently constant rate of excretion of polysaccharides by *Isochrysis* studied by Marker (1965). Ford (1958) also had found that the degree of binding in *Ochromonas* supernatants increased with absorbance of the culture (i.e. number of cells) and was not particularly associated with old cultures. One ml of a four-day-old dense culture bound 2 to 4 ng $B_{12}$, a value comparable to the one found by Kristensen. In the same paper Ford (1958) reported that pseudovitamin $B_{12}$ and factor A, which cannot substitute for $B_{12}$ in growth, are none the less taken up by *O. malhamensis* to about the same degree as vitamin $B_{12}$ and 'inhibited competitively the growth response to vitamin $B_{12}$ apparently by blocking a cell mechanism for binding the vitamin'. Apparently, the inability to utilize analogues is not due to a selective binding capacity but to the inherent inactivity of the molecule for the organism. Droop (1968) from his chemostat experiments with *Monochrysis* concludes that adsorption at the cell surface is initially very rapid as compared with the steady-state rate of uptake, but the capacity is limited since saturation of the cell surface follows a Langmuir isotherm. Similarly, Bradbeer (1971) proposed that the initial rapid phase of uptake of $B_{12}$ by *Ochromonas* represents the loading of a $B_{12}$ carrier at the cell surface and that the slower secondary phase consists of an energy-dependent release of $B_{12}$ from the carrier into the interior of the cell. However, Daisley (1970) found that antisera up to 50% final concentration did not agglutinate *Euglena* cells, and suggested that the binder may not be a component of the cell-wall.

Daisley (1970) confirmed previous work on *Ochromonas* and *Euglena*. The binder of *Euglena* is present in cell extracts and in supernatants; the $B_{12}$-binder complex is not dialysable, and is destroyed by autoclaving for 10 minutes at 10 lb pressure. Trichloracetic and perchloric acids precipitated the protein and liberated in dialysable form 80% of the vitamin $B_{12}$. Purification was done by gel filtration on Sephadex G-200 and the void volume containing the protein binder was fractionated in a DEAE-cellulose column. The cellular and extracellular binders are very similar but seem to be heterogeneous. The binding capacity of the purified material from the supernatant (i.e. extracellular) was higher than the cellular, but of much lower capacity than the commercial sample of porcine intrinsic factor. It is a glycoprotein [68 to 79% protein, 8 to 17% reducing sugars (xylose and fructose), 0·25 to 1·6% hexosamines] and is not absorbed firmly on zirconyl-phosphate gel as is the intrinsic factor.

It seems highly probable that the $B_{12}$-binding protein is involved in active uptake of vitamin $B_{12}$ by the cells and that when excess is produced it is excreted into the medium. Excretion of other algal metabolites is common (see Fogg 1966, Hellebust Chapter 30, p. 838); *O. danica* excretes macromolecules, even membranes, in the culture fluids (Aaronson 1971). The ecological implications will be discussed later.

## 2.5  *Bioassays*

A knowledge of dissolved vitamin $B_{12}$, thiamine, and biotin concentrations in marine and fresh waters is necessary for quantitative ecology. The extremely low concentrations of these vitamins in many samples require methods whose sensitivities can only be approached by bioassays.

### (a) *Seawater*

The earlier assays employing both bacteria and algae as test organisms have been adequately summarized (Belser 1963, Lewin 1961, Provasoli 1963). Incubations were usually from one to 21 days and vitamin concentrations were determined from terminal responses to vitamin levels. Measurements were made of cell numbers or of absorbance. Ryther and Guillard (1962) reported on a method for the direct assay of seawater samples for vitamin $B_{12}$. The response to vitamin by the test alga, *Cyclotella nana*, was measured optically. Before then seawater assays usually employed freshwater test organisms and the samples had to be pretreated, e.g. dialysed against freshwater, diluted to low salinities, or extracted with solvents such as phenol. Gold (1964) described a technique for $B_{12}$ bioassay, also using *C. nana*, in which [14]C incorporated during photosynthesis was used to measure the algal response to vitamin concentrations over a 24-hour period. Carlucci and Silbernagel (1966a) modified Gold's technique and described the most sensitive method for vitamin $B_{12}$ assay yet available. This latter procedure called for 48-hour incubation of the test alga with the sample, a longer period for allowing the cells to adapt to the variety of seawaters. Generally, different samples of seawater show varying degrees of inhibition of *C. nana* as compared with a charcoal-treated seawater standard. The equations for calculating the dissolved $B_{12}$ concentration in a sample also take into account the inhibition by the sample. It is therefore possible to estimate seawater toxicity to the alga. If facilities for radiocarbon work are not available, incubations are continued for about three more days after which time cell numbers are proportional to vitamin concentrations in the samples. The vitamin concentrations are calculated using the same equation as for the $^{14}C$ uptake assay. The method, in a step-by-step procedure, has recently been summarized (Strickland & Parsons 1968).

Thiamine is required by considerably fewer phytoplankton species than $B_{12}$ and has received less attention. However, the excellent review by Lewin (1961) has summarized the earlier methodology. Provasoli and Gold (1962) suggested that *Gyrodinium cohnii* could be used to bioassay seawater for thiamine. Vishniac (1961) described a thiamine assay which employed the phycomycete, 'Isolate S-3', as a test organism. Response to the vitamin was read turbidimetrically after a seven-day incubation.

Carlucci and Silbernagel (1966b) developed a procedure which measured thiamine concentrations in seawater samples by determining $^{14}C$ incorporation by cells of *Monochrysis lutheri*, a member of the Chrysophyceae. This assay is extremely sensitive and, as with the $B_{12}$ assay, gives an estimate of the sample inhibition to the test alga. Natarajan and Dugdale (1966) described a thiamine assay for seawater using the marine yeast *Cryptococcus albidus*.

Biotin is required for growth by only a few phytoplankton species. Lewin (1961) has summarized the earlier methods for the assays used to determine the vitamin in marine and freshwaters. Belser (1959) and Carlucci and Belser (1963) reported a biotin assay which employed biotin-requiring mutants of the marine bacterium, *Serratia marinorubra*, to assay biotin in seawater. The assays were not sensitive, however. Litchfield and Hood (1965) also used the biotin-requiring mutant of *S. marinorubra* to assay seawater for biotin. The lowest level of biotin which could be detected was 5 ng $l^{-1}$, also not a sensitive assay. Antia (1963) reported a biotin assay in which an unidentified marine bacterium was the test organism; method sensitivity was down to 3 ng $l^{-1}$. Provasoli and Gold (1962) described a biotin assay with a colourless heterotrophic dinoflagellate which could measure 1 to 10 ng $l^{-1}$ of $B_{12}$; precision was not very good. Carlucci and Silbernagel (1967) gave a technique in which $^{14}C$ incorporation by *Amphidinium carterae* was used to measure biotin concentrations in seawater. More recently, Ohwada (1972) reported a 48-hour assay using a marine bacterium, *Achromobacter* sp. which was sensitive to 0·1 ng $l^{-1}$ of biotin.

In summary, therefore, it appears that dissolved vitamin $B_{12}$, thiamine, and biotin concentrations in seawater samples can be determined with sensitive bioassays. The most sensitive methods are summarized in Table 27.3.

Many of the bioassay test organisms respond to vitamin analogues in addition to the complete vitamin. The analogue response should be known before an interpretation of sample vitamin concentrations is made. Sometimes it is advisable to assay the same sample with two or more test organisms in order to account for differential vitamin response noted for example, by Daisley and Fisher (1958).

(b) *Freshwater*

Lewin (1961) has summarized the methods available for dissolved vitamin $B_{12}$, thiamine, and biotin concentrations of freshwaters. Carlucci and Bowes (1972b)

Table 27.3. Summary of sensitive seawater vitamin assay methods[1]

| Vitamin | Test Organism | Measurement of Response | Amount of Vitamin Added to Internal Standard (ng l$^{-1}$) | Range (ng l$^{-1}$) | Reference |
|---|---|---|---|---|---|
| B$_{12}$ | Cyclotella nana[2] | $^{14}$C | 1 | 0·05 to 3 | Carlucci & Silbernagel (1966a) |
| Thiamine | Monochrysis lutheri[2] | $^{14}$C | 10 | 2 to 35 | Carlucci & Silbernagel (1966b) |
| Biotin | Achromobacter sp. | Cell density | 0·5 | 0·5 to 8 | Ohwada (1972) |

[1] All assays require 48 hours of incubation.
[2] Cell counts after five to seven days of incubation are proportional to vitamin concentrations and may be used to calculate vitamin concentrations.

adapted their $B_{12}$, thiamine, and biotin seawater assays for freshwater analyses by diluting samples $1:3$ (v./v) with charcoal-treated seawater. Response to vitamin concentrations was reflected in $^{14}CO_2$ uptake or in cell numbers as with the marine assays. The sensitivities of these assays were about $25\%$ of those of the marine assays, but these were still in ecologically important ranges. Ohwada *et al.* (1972) also adapted microbiological bioassays used for seawater to determine thiamine and biotin in freshwater.

### (c) *Particulate material*

The determinations of the vitamin $B_{12}$, thiamine, and biotin contents of particulate material such as phytoplankton and zooplankton are rather difficult. The major reason is the lack of adequate methodology (for earlier references see Provasoli 1963). Strohecker and Henning (1966) published a manual which contained a number of methods for vitamin analysis of particulate materials. Procedures for $B_{12}$, thiamine, and biotin analyses were included. The methods were generally insensitive and required large amounts of material for the analysis, e.g. gram to kilogram quantities. Ohwada and Taga (1972) and Ohwada *et al.* (1972) modified standard techniques to extract vitamins from freshwater and seawater particulate materials. Carlucci and Bowes (1972a) developed assays for the determination of particulate vitamin $B_{12}$, thiamine, and biotin. One extraction procedure could be used for all three vitamins. The extracts could then be assayed for the specific vitamin using standard vitamin bioassays.

## 3  VITAMIN AND ALGAL INTERACTIONS

### 3.1  *Vitamin content of algae*

The vitamin contents of many algae have been summarized (Kanazawa 1963, Provasoli 1963). Most of the data have been obtained for freshwater species. Table 27.4 gives the ranges which are found in several groups of algae. The

**Table 27.4.** Vitamin contents of algae (Kanazawa 1963)

| Algae | $B_{12}$ (ng g$^{-1}$) | Thiamine ($\mu$g g$^{-1}$) | Biotin (ng g$^{-1}$) |
|---|---|---|---|
| Chlorophyceae | 12–150 | 0·9–23 | 115–230 |
| Phaeophyceae | 3–76 | 0·27–1·10 | 126–282 |
| Rhodophyceae | 15–291 | 0·53–4·60 | 18–294 |

ranges represent values of at least five different species of algae. Within any one group there was considerable variation. More recently, Carlucci and Bowes (1972a) showed that the contents of required vitamin found in pure cultures of algae varied and were dependent upon the vitamin level in the external medium. Starved cultures contained less vitamin than those grown in high concentrations of vitamin. With non-essential vitamins, e.g. biotin in *Skeletonema costatum*, there was no difference between starved and non-starved cultures. It appears that phytoplankton organisms can synthesize non-essential vitamins at rates in step with cell division.

## 3.2 *Production of vitamins by algae*

The dissolved vitamins in the sea and in freshwaters have been thought to originate mainly from the activities of bacteria, both in the water column and in the sediments (Burkholder & Burkholder 1956, 1958, Burkholder 1963, Provasoli 1963). However, several algae had been reported to release vitamins in the culture medium: *Ochromonas malhamensis* (Ford & Goulden 1959), unidentified marine algae (Burkholder 1963), *Chlamydomonas* sp. (Nakamura & Gowans 1964), *Coccomyxa* sp. and *Chlorella pyrenoidosa* (Bednar & Holm-Hansen 1964) and *Ochromonas danica* (Aaronson *et al.* 1971). The observations were generally qualitative. Carlucci and Bowes (1970a), using sensitive bioassays, found that ecologically important phytoplankton species released vitamins into the culture medium during some portion of their growth. The results are summarized qualitatively in Table 27.5. *Skeletonema costatum* and *Stephanopyxis turris*

**Table 27.5.** Vitamin production by algae

| Algae | 2 ng $B_{12}$ l$^{-1}$ | | | 12 ng $B_{12}$ l$^{-1}$ | | |
|---|---|---|---|---|---|---|
| | $B_{12}$ | Thiamine | Biotin | $B_{12}$ | Thiamine | Biotin |
| *Skeletonema costatum* | U | + | + | U | + | + |
| *Stephanopyxis turris* | U | + | + | U | + | + |
| *Gonyaulax polyedra*[1] | | | | U | + | + |
| | 10 ng thiamine l$^{-1}$ | | | 120 ng thiamine l$^{-1}$ | | |
| *Coccolithus huxleyi* | — | U | + | + | U | + |

U = required vitamin taken up.    (1) Not tested with 2 ng l$^{-1}$.

(vitamin $B_{12}$-requirers) produced thiamine and biotin when growing with either 12 or 2 ng l$^{-1}$ of external $B_{12}$, and *Coccolithus huxleyi* (thiamine-requirer) produced vitamin $B_{12}$ and biotin with 120 ng l$^{-1}$ of thiamine, but only biotin with 10 ng l$^{-1}$ of thiamine. The amount of vitamin produced by an alga and the rate at which it was produced varied with the phytoplankton species, the concentration of the required vitamin, and incubation time. It was concluded that vitamins produced during early and exponential growth were due to cell excretions, and those produced during stationary growth resulted from both excretion and cell lysis.

## 3.3 *Utilization of algal produced vitamins by algae*

Carlucci and Bowes (1970b) have reported that vitamin-requiring algae can utilize the excreted vitamin when they are grown in the same culture vessel as

2B

the producer. The excretion of inhibitory materials was also observed. Vitamin utilization was most easily observed in cultures where two phytoplankton species were present. *Dunaliella tertiolecta* and *S. costatum* produced utilizable thiamine for *C. huxleyi*. *C. huxleyi* released utilizable vitamin $B_{12}$ for *C. nana*. *D. tertiolecta*, *Phaeodactylum tricornutum*, and *S. costatum* produced utilizable biotin for *A. carteri*. The amount of utilizable vitamin and rate at which it was released depended on the species present and conditions of incubation. In systems with more than two species, beneficial effects to utilizers were often observed for short durations during the growth period only. Droop (1968) has also reported syntrophic growth: '*Monochrysis* can be grown in a vitamin $B_{12}$-free medium in the presence of *Nannochloris oculata*, showing that the latter, which has no requirements for the vitamin, must excrete it'.

## 4  ECOLOGY

### 4.1  *Occurrence of vitamins in waters*

(a) *Seawater*

Reports have appeared on the distribution of vitamins in seawaters. Earlier work has been summarized (Lewin 1961, Provasoli 1963). More recently several reports have appeared: vitamin $B_{12}$ (Carlucci 1967, Natarajan 1967b, Carlucci & Silbernagel 1966c, 1967; Ohwada & Taga 1972); thiamine (Carlucci 1967, Natarajan 1967a,b; Ohwada & Taga 1972), and biotin (Litchfield & Hood 1965, Carlucci 1967, Natarajan 1967a, Ohwada 1972; Ohwada & Taga 1972). Table

**Table 27.6.** Average vitamin concentrations in Pacific Ocean waters, (ng $1^{-1}$) (Carlucci 1970)

| Water | Vitamin $B_{12}$ | Thiamine | Biotin |
|---|---|---|---|
| Scripps Institution Pier | 2·9 | 15 | 3·8 |
| Coastal | 1·6 | 9 | 2·6 |
| Central Pacific | 0·1 | 8 | 1·3 |

27.6 gives average vitamin concentrations for seawaters from the Pacific Ocean. The values of vitamin concentrations reported by the above workers for waters from other locations in the Pacific and Atlantic Oceans agree with those reported in Table 27.6. Unfortunately, few data are available for any one location and these are representative of only one sampling. In general, Arctic waters show the highest concentrations of vitamins (Carlucci unpubl.) and open-ocean waters the least. It appears that waters of intermediate depths contain higher levels than those below or above (Daisley & Fisher 1958); the upper waters, however, are most involved in biological activities and there levels probably fluctuate greatly during the year. The seasonal distributions of vitamins indicate that their concentrations vary in the same way as nitrogen and phosphorus (Menzel & Spaeth 1962, Carlucci unpubl.). Spring and late autumn maxima were observed. Ohwada and Taga (1972) found that particulate thiamine and biotin in surface water of the North Pacific were about 1% of their dissolved concentrations. In coastal waters particulate thiamine was 155% of the dissolved thiamine level, and particulate biotin was 54% of the dissolved biotin level.

### (b) Fresh water

Early studies on the distributions of vitamins were summarized by Lewin (1961). Subsequent work has been done by only a few investigators and mostly with $B_{12}$. Kashiwada et al. (1960) found up to 62·5 ng $l^{-1}$ of $B_{12}$ in Lake Ikeda in Japan with maximum concentrations generally being detected at 20 m. Tal (1962) observed $B_{12}$ concentrations up to 12 ng $l^{-1}$ in fish ponds in Israel. Daisley (1969) surveyed the waters of nine English lakes for vitamin $B_{12}$ concentrations over a 15-month period. Low vitamin $B_{12}$ concentrations (0·1 ng $l^{-1}$) were found in low productivity lakes and high concentrations (15 ng $l^{-1}$) were found in nutrient-rich lakes. Carlucci and Bowes (1972b) observed that the oligotrophic Lake Tahoe in California contained vitamin $B_{12}$, thiamine and biotin only in a few samples from the euphotic zone. Ohwada et al. (1972) found that seasonal cycles of vitamin $B_{12}$ varied from undetectable to about 6 ng $l^{-1}$; dissolved thiamine varied between 10 and about 350 ng $l^{-1}$; and biotin varied between 8 and about 80 ng $l^{-1}$; Particulate thiamine and biotin levels were less than the dissolved amounts. Particulate vitamin $B_{12}$ levels were higher than dissolved $B_{12}$ levels in spring when diatoms predominated.

### 4.2 Toxicity to marine phytoplankton

There are many examples from laboratory studies of seawater inhibition of algae (Johnston 1963a,b, Provasoli 1963, Droop 1968, Carlucci 1970). In addition to determining the dissolved vitamin concentrations in seawater the methods of Carlucci and Silbernagel (1966a,b, 1967) also measure seawater toxicity (as compared with a standard charcoal-treated seawater) to the test algae. In the 1967 study of Carlucci (1970) 30% of the samples were toxic. Water from May and June was especially inhibitory to C. nana, the $B_{12}$-assay organism, and was sporadically toxic during the rest of the period. Few samples were toxic to M. lutheri and A. carterae, the thiamine- and biotin-assay algae, respectively, during the first half of the study. From late June onwards, most samples were inhibitory.

### 4.3 Probable role of vitamins

Evaluating the role of vitamins in marine ecology is difficult. Only a few studies have been comprehensive enough to estimate the importance of vitamins in primary productivity and species succession. Menzel and Spaeth (1962) reported that a moderate diatom bloom occurred in the Sargasso Sea when $B_{12}$ concentrations were highest. When the level of $B_{12}$ dropped to barely detectable levels the diatoms also decreased. Most of the Sargasso Sea diatoms require $B_{12}$. *Coccolithus huxleyi*, a member of the Haptophyceae requiring thiamine, predominated during times when diatom numbers were low. Menzel and Spaeth (1962) did not believe that $B_{12}$ limited primary productivity but that it may have influenced the composition of the phytoplankton. Vishniac and Riley (1961) also showed a correlation between high concentrations of $B_{12}$ and diatoms. At the termination of the bloom vitamin $B_{12}$ had fallen to 4 ng $l^{-1}$, but this concentration was considered by Droop (1957) to be more than sufficient to produce very high populations of *Skeletonema costatum*, one of the two major components of the bloom. Consequently, Vishniac and Riley (1961) concluded that termination of the bloom was due to limiting nitrogen, not $B_{12}$. However, Wood (1962) found that levels of $B_{12}$ lower than 8 ng $l^{-1}$ reduced the growth rate of *Skeletonema* by 50% or more in chemostat cultures. Carlucci (1970) in an extensive study of the plankton off La Jolla, California tried to correlate vitamin $B_{12}$, thiamine, and biotin concentrations over a six-month period with phytoplankton composition and succession. He looked for a negative correlation (low vitamin concentrations, high phytoplankton standing stock) or a positive correlation (high vitamin concentrations, high phytoplankton standing stock). There was no statistical evidence that phytoplankton production as a whole was limited by the concentrations of vitamins, although, of course, individual species of phytoplankton may have been affected. In one instance a sudden growth of the red tide organism, *Gonyaulax polyedra*, could be correlated with the utilization of vitamin $B_{12}$. The many positive correlations between high vitamins in the waters and high standing stock of phytoplankton led Carlucci (1970) to the hypothesis, proven true, that phytoplankton may secrete vitamins.

However, none of these studies, though comprehensive, has been sufficiently detailed in the light of recent findings on the complexities of the cycle of vitamins in water. Any analysis attempting to unravel the role of vitamins in blooms and algal succession should consider the variability, sensitivity and specificity of the vitamin requirements of the most important algae. The vitamin producers and consumers in the system should be identified, and the occurrence of competitive and non-competitive inhibitors should also be considered. Awareness that some algal species may vary widely in their vitamin requirements and their physiological reactions to other ecological parameters no longer permits direct extrapolation of literature data on a species to the population found in the same or any other locality many years later.

Laboratory sensitivity data have little ecological value when used as a means of judging the limiting concentrations of a vitamin in natural situations. The art of bioassay methods, whose aim is to achieve the highest possible sensitivity of the organism, depends on improvements in media and on methods of evaluating precisely the growth response of the organism to minute doses of the variable. With such improvements the sensitivity of *Monochrysis lutheri* was increased by Droop (1957) from 5 ng $l^{-1}$ to 0·1 ng $l^{-1}$. However, in nature the organism is seldom operating under conditions of optimal and constant temperature, insolation or salinity and nurtured in optimal media (i.e. devoid of inhibitors and stresses).

Whether limiting concentrations affect growth rates or not is ecologically important, as suggested by Daisley (1957). Droop (1961) has elegantly shown that the growth rates of *Monochrysis* remain unaltered for a large range of $B_{12}$ concentrations (0·1 to 100 ng $l^{-1}$) but *Monochrysis* may be atypical. Division rate of *Ochromonas malhamensis* doubles in the sensitive range from 13 ng $l^{-1}$ (the limiting concentration) to 1,000 ng $l^{-1}$ (Ford 1958). Differences in division rates in other algae were also found to vary within the sensitive range of concentration of the vitamin: in *Skeletonema costatum* (Wood 1962) and in *Cylindrotheca nana* (= *Cyclotella nana*) with vitamin $B_{12}$, in *Monochrysis lutheri* with thiamine, and *Amphidinium carteri* with biotin (Carlucci & Silbernagel 1969). The comparison between *Ochromonas* and *Monochrysis* shows that algae differ, like other microorganisms and for other parameters, in their ability to capture needed molecules from dilute concentrations: *Monochrysis* from a 0·1 ng $l^{-1}$ solution, *Ochromonas* from 13 ng $l^{-1}$, a difference of two orders of magnitude. Several phytoplankton species differ widely in their sensitivity to $B_{12}$ (Table 2 in Provasoli 1958a). Even though the laboratory data cannot be extrapolated to natural situations they give a fair judgment on the *relative* degree of sensitivity of different organisms toward a vitamin, circumstantial evidence which may help in ecological evaluations.

The specificity of the requirements shows that clones within a species, may differ substantially in their ability to use either precursors or analogues of thiamine and $B_{12}$. These differences can also be ecologically relevant since we know that many marine bacteria produce analogues of $B_{12}$ and that degradation of vitamins may occur in seawater under the influence of temperature and solar radiation (Gold 1968, Carlucci *et al.* 1969). To take advantage of these physiological differences, natural waters should be analysed with algae having *mammalian* specificity to assess the quantity of $B_{12}$ and algae with *E. coli* specificity to measure total cobalamins. Algae able to utilize either the thiazole or the pyrimidine moieties of thiamine could be developed as bioassay organisms. For seawater, however, we lack an alga which can utilize only intact thiamine, to follow the degradation rate of thiamine.

The finding that algae may excrete substantial quantities of the vitamins which they are able to synthesize adds the algae to the roster of the vitamin producers in waters. We need to extend this finding over a larger number of ecologically important algae of fresh and saline waters because algal biomass is apparently far larger than bacterial biomass. As a consequence it is reasonable to postulate that the changes in the biological properties of waters during an algal bloom (removal of vitamins needed and release of other vitamins) may condition waters to favour algae requiring the vitamin released by the previous bloom. This selective preconditioning of the waters may be one of the factors operating in the seasonal succession of algal species (Provasoli 1971).

Excretion of toxic substances and of the $B_{12}$-binding factor are other elements in selective conditioning of waters. Production by algae of unidentified anti-algal factors has been reviewed by Lefèvre (1964) and production of bacterial antibiotics by Sieburth (1968). The $B_{12}$-binding factor may be one of the unidentified anti-algal factors of Lefèvre, some of which are also heat-labile and non-dialysable. The very little we know about the binding factor leads to a paradox. It is produced and released in the medium by a total of 10 auxotrophic species having varied vitamin requirements and by two marine autotrophs (adding to the results of Kristensen, Ford, Droop and our unpublished work). Is it produced by all algae; does the quantity produced vary? It binds $B_{12}$, and bound $B_{12}$ is unavailable non-specifically to all the algae tested so far. Yet free $B_{12}$ is present in waters as attested to by bioassays and blooms of $B_{12}$ requirers. We know, however, that the uptake of $B_{12}$ by $B_{12}$ requirers is very rapid and Droop's (1968) computations indicate that 90% of the available $B_{12}$ is taken up by *Monochrysis* in the first third of the logarithmic phase. The apparently logarithmic production of the $B_{12}$ binder occurs in the remainder of the logarithmic growth phase. Because of the timing of the two processes, the cells at first show a rapid luxury consumption of $B_{12}$ and the production of the binder affects only the second phase of uptake which is slow and steadily declining. As a result the specific growth rate depends upon the 'cell vitamin quota' and the effect of the binder, rather than directly on the medium concentration (Droop 1971). The ecological effects of the binder might be to partially slow down growth rates and consumption and to preserve some $B_{12}$ in unavailable form. If none of the algae is really able to utilize the bound $B_{12}$, the future utilization of this store of $B_{12}$ would depend on the rate of degradation of the binding complex by proteolytic microorganisms.

It is also important to appreciate that bacteria and other microorganisms are the other producers and consumers of vitamins in the water system (Burkholder 1963), and that measurements of vitamins in the waters reflect only the balance, *not the rate* at which the two processes are operating. Radioactive vitamin $B_{12}$ and cobalt could be used to identify the producers and consumers and to follow the turnover rates under natural conditions. The techniques have been used by M. Parker in a survey of the ecology of $B_{12}$ in Lake Washington (pers. comm.).

## 5 PLANT HORMONES AND ALGAL REGULATORS

Many observations hint that plant hormones or other growth regulators should operate in algae, especially in the multicellular, more organized seaweeds. Like higher plants, seaweeds show apical dominance, photoperiodism, orderly differentiation of complex specialized structures, etc. (see Chapter 28, p. 788).

Two main lines of research have been used to detect growth substances in algae: extraction, characterization by chromatography and plant bioassay; and also the physiological response of algae to applications of growth substances.

Conrad and Saltman (1962) reviewed critically the earlier literature which dealt mainly with auxins. Since then kinins and gibberellins have received increasing attention. Each category of substances will be treated separately here.

## 5.1 *Auxins*

(a) *Occurrence*

Early work was based on extraction and used the standard *Avena* curvature test to determine auxin activity. Indoleacetic acid (IAA) equivalents of the observed coleoptile activity ranging from 0·05 to 0·5 μg kg$^{-1}$ of fresh weight (f.w.) were reported by van Overbeek (1940) for *Macrocystis pyrifera*, while the corresponding value for *Bryopsis* was 80 μg kg$^{-1}$. Williams (1949) found auxin activity in extracts of *Laminaria agardhii*. Mowat (1965), with differential extractions and chromotographic separation, detected a variety of *Avena*-active substances of the auxin type in *Chlorella pyrenoidosa, Ochromonas malhamensis, Oscillatoria* sp., mixed marine phytoplankton (mostly diatoms), *Laminaria cloustonii* and *L. digitata* (approximate equivalents 1 to 20 μg (kg f.w.)$^{-1}$. Several indole compounds, not all of them biologically active, were located in the chromatograms by chemical test, but chemical identification was not carried out. Augier (1965) detected activity in extracts of the red seaweed *Botryocladia botryoides* and identified with six colour tests the presence in the extracts of IAA and tryptophan. Schiewer (1967a,b) found auxin activity in *Cladophora sericea, Pylaiella (Pilayella) littoralis, Nemalion multifidum, Furcellaria fastigiata, Ceramium rubrum, Enteromorpha prolifera* and *E. compressa*. Chromatography in different solvents and colour tests indicate that IAA was responsible for most activity. The low level of IAA activity in *Ceramium* and *Enteromorpha* was due apparently to inhibitors. Alkaline hydrolysis of the algae liberated large amounts of auxins.

Doubts on the above results are cast by the following reports: Booth (1958) has shown that sugars may be present in plant ether extracts. These sugars in chromatograms lay near to the position of IAA, gave a colour reaction with Ehrlich reagent similar to IAA and stimulated coleoptile growth. A control series with sugar added, and a careful pH control seem necessary precautions when using the coleoptile test. It is doubtful that these precautions were always used; the authors generally refer to a standard coleoptile test without giving details and Buggeln and Craigie (1971) were unable to detect the presence of endogenous, bound IAA in 10 marine seaweeds including species of *Fucus, Laminaria, Chondrus,* and *Acetabularia*. Using two-dimensional thin-layer chromatography, they found that the spot of exogenously added IAA was completely separated from other Ehrlich-positive spots, some of which were biologically active, and questioned the validity of previous findings based on colour reactions and biological activity. Dawes (1971) using other advanced techniques also failed to find IAA in *Caulerpa prolifera*, thus confirming the results of Buggeln and Craigie. In the absence of chemical evidence for the presence of

IAA in marine algae, the chromogenic reactions used, because of their incomplete specificity, are insufficient proof.

## (b) *Effects of exogenous IAA*

Work on the effects of exogenous IAA on algae was often undertaken as another way to detect whether IAA or other auxin-like substances were involved in regulating growth or morphogenesis of algae. Many of the early controversial reports on the effect of IAA and other growth-promoting substances on unicellular algae can be ignored because the growth substances had been added as alcoholic solutions. Bach and Fellig (1958) and Street *et al.* (1958) demonstrated that the ethanol solvent, not the growth substances, elicited the increased growth observed. Other reports are questionable because the growth substances were used at supra-hormonal levels (30 to 50 mg l$^{-1}$) and the increase of growth could be simply a nutrient effect (Leonian & Lilly 1937, Pratt 1937). Conrad and Saltman (1962) critically discuss this literature. IAA can indeed stimulate growth in several unicellular algae, and a great variability of response, or lack of it, was observed between different species and strains.

Bacteria-free cultures of two *Chlorella* spp., *Scenedesmus obliquus*, *Chlorococcum* sp. *Protococcus viridis*, and *Euglena gracilis* were not stimulated by IAA ranging from 1 μg to 10 mg l$^{-1}$ (Fernandez *et al.* 1968). Synchronous cultures of *Chlorella fusca* were similarly not stimulated by IAA up to 10 mg l$^{-1}$; pH-dependent inhibition intervened at higher concentrations (Lien *et al.* 1971).

Growth of some blue-green algae is apparently stimulated by IAA. Bunt (1961) observed that an unidentified *Nostoc* grew well in the presence of a *Caulobacter* and poorly without it; 1 or 2 drops of a 1 mg l$^{-1}$ solution of IAA replaced the bacterial effect. Ahmad and Winter (1968a) obtained a modest growth stimulation (50 to 70%) in the range of 0·17 μg to 1·7 mg l$^{-1}$ for several blue-green algae (including species of *Chlorogloea*, *Nostoc*, *Anabaena* and *Tolypothrix*) but the cultures were not axenic. A slight increase in growth (~20%) of *Nostoc punctiforme* was reported at 100 μg l$^{-1}$ IAA by Fernandez *et al.* (1968). Stimulation obtained by adding indole to cultures of *Chlorogloea*, *Nostoc*, and *Tolypothrix*, suggests that indole might be a precursor of IAA for these species (Ahmad & Winter 1968b). Further studies (Ahmad & Winter 1970) under aseptic conditions with *Chlorogloea fritschii* indicated that radioactive indole was indeed a precursor of IAA and that tryptamine was an intermediate in the synthesis. Tryptamine had been found to be an intermediate also in the conversion of tryptophan to IAA in *Chlorogloea* (Ahmad & Winter 1969), but was not detected by Schiewer (1967b) in the conversion of tryptophan by *Furcellaria fastigiata*.

Growth of the dinoflagellates *Gymnodinium breve* and *G. splendens* was not affected by IAA (W. Wilson personal communication), but a four-fold stimulation of growth was obtained by adding 50 μg l$^{-1}$ IAA to axenic cultures of another dinoflagellate, *Exuviaella* sp. (Iwasaki 1971b). This does not seem to be a specific effect of IAA since similar stimulation of growth was obtained with 2 mg l⁻¹ of kinetin, 0·4 mg l⁻¹ of gibberellic acid, and several purines and pyrimidines.

Modification of growth patterns seems to occur in the more organized seaweeds. Applications of a gradient of IAA to plants of *Bryopsis plumosa* grown in an inverted position (upside down) caused differentiation of rhizoids at the proximal end of most of the lateral branches. Control plants did not form basal rhizoids (Jacobs 1951).

Moss (1965) found that apices of *Fucus vesiculosus* from which the apical cell was removed developed three or four times the number of buds and branches produced in the controls when the anti-auxins 1-naphthoxyacetic acid and 2,3,5-triiodobenzoic acid were added at 10 mg l⁻¹. On the contrary, the auxins 3-indolylacetic, 2,4-dichlorophenoxyacetic and 1-naphthaleneacetic acids did not stimulate branch production. Chromatography of extracts of apical segments of *Fucus* showed that a mixture of stimulatory and inhibitory growth substances was present. These data support the conclusion of Moss that growth regulators are involved in apical dominance in *Fucus* (see also Chapter 28, p. 788).

Long-term thallus growth of the green algal coenocyte *Caulerpa sertularioides* was obtained either in the presence of a sand substrate rich in microorganisms, or when sap of *C. sertularioides* or *C. racemosa* was added to the seawater (2 to 10%, v/v). The sap could be replaced by 85 mg l⁻¹ of IAA (Mishra & Kefford 1969). The effect of these treatments was to promote the formation of upright shoots, resulting in extensive proliferation of blades, and in halving of rhizome elongation. Similar experiments have been carried out by Dawes (1971) with *Caulerpa prolifera* using artificial seawater. Blade proliferation was obtained with 0·17 mg l⁻¹ of IAA (5 to 10 new blades as compared with 1 or 2 blades in controls). At this concentration rhizome elongation was also obtained; inhibition of rhizome elongation, even death, resulted when IAA was higher than 1·7 mg l⁻¹. The retarded rhizome elongation found by Mishra and Kefford (1969) for *C. sertularioides* may be a response to the much higher concentrations of IAA used.

It is interesting that IAA stimulates growth modification in *Caulerpa* when no IAA could be detected by Dawes (1971) in this plant. A better yield of the *Conchocelis* phase of axenic *Porphyra tenera* was obtained with 20 μg l⁻¹ of IAA (Iwasaki 1965) and even larger yields (five times greater) with 0·4 mg l⁻¹ of gibberellic acid. Kinetin induced only longer filaments and became inhibitory at 0·5 mg l⁻¹.

## 5.2 *Cytokinins*

### (a) *Occurrence*

Reports on the occurrence of kinetin-like substances are scarce and based on extractions and bioassay. The chemical composition is unknown; we know only that the extracts mimic the action of kinetin in the bioassay. Bentley-Mowat and Reid (1968) calibrated and modified the radish leaf bioassay, confirmed that the response is not affected by IAA, but found that in 50% of the experiments $10^{-5}$ g ml$^{-1}$ gibberellic acid (GA$_3$) gives an activity approximately equivalent to $10^{-8}$ g ml$^{-1}$ of kinetin. When testing algal extracts for kinetin activity, GA$_3$ was also tested, and if found active, the experiment was discarded. Using this assay they found that extracts of axenic cultures of *Gymnodinium splendens* and *Phaeodactylum tricornutum* had kinetin activity; the approximate kinetin equivalents varying between 0·1 to 1·0 mg kg$^{-1}$ on an algal fresh weight basis, with a single report of 10 mg. Extracts of marine phytoplankton were also active. The activity of the algal extracts declined in a few weeks when stored at $-20°C$, while pure kinetin remained fully active for two years under the same conditions.

The brown alga *Laminaria digitata* contains cytokinin-like substances in the stipe and holdfast region; the two extracts differ in the $R_f$ region of highest kinetin activity and in the slope of the absorption peak at the region 266–275nm (Hussain & Boney 1969).

The content of cytokinin- and gibberellin-like substances in the motile and non-motile phases of *Cricosphaera elongata* and *C. carterae* differed (Hussain & Boney 1971). Cytokinin-like substances were more abundant in the motile phase cells and gibberellin-like substances similar to GA$_3$ and GA$_7$ predominated in the non-motile phase. Addition of kinetin to motile cells of *C. elongata* retarded settling; gibberellic acid favoured it.

Seawater taken from the *Fucus-Ascophyllum* zone supports much better growth of several seaweeds in culture than other types of seawater. A liquid-liquid extraction for 20 h at 130°C in ethyl acetate yielded an extract of good cytokinin-like activity in assays with tobacco callus cultures. Active extracts were obtained only from water collected in October, the time at which the seaweed community is at its peak (Pedersén & Fridborg 1972). The cytokinin was identified as 6-(3-methyl-2-butenylamino) purine (Pedersén 1973).

## (b) *Effects of exogenous cytokinins*

The rate of cell division of the colourless unicellular alga *Polytoma uvella* is increased by addition of 60 μg l$^{-1}$ of kinetin but the total yield is unaffected (Moewus 1959). *Phaeodactylum tricornutum* and *Gymnodinium splendens* respond to 0·1 to 10 mg l$^{-1}$ of kinetin with a slight increase in yield, while growth of *Nannochloris oculata* is not appreciably affected (Bentley-Mowat & Reid 1969).

Better results were obtained with seaweeds. *Ectocarpus fasciculatus* (with one infectant) and *Pylaiella* (*Pilayella*) *littoralis* (contaminated by diatoms and bacteria) need kinetin (2·4 to 4·8 mg l$^{-1}$) to preserve good growth rates and a normal morphology (Pedersén 1968). Kinetin does not increase growth of the axenic *Conchocelis* phase of *Porphyra tenera* (Iwasaki 1965), but increases leaf elongation and area sixfold at 1 mg l$^{-1}$ while inhibiting stem elongation of impure *Porphyra* fronds (Shimo & Nakatani 1964). Cap formation is accelerated in *Acetabularia mediterranea* by 21 μg l$^{-1}$ of kinetin and a larger percentage of caps are produced (Spencer 1968).

## 5.3 Gibberellins

### (a) Occurrence

Reports on the occurrence of gibberellin-like activity are scanty and are based, as usual, on bioassays of chromatographed extracts of algae. Gibberellin-like activity was found first in *Fucus vesiculosus* ($\sim$10 μg (kg f.w.)$^{-1}$; $R_f$ 0·3–0·4) and the activity was due to either gibberellic acid or gibberellin $A_1$ or $A_6$ (Radley 1961). The low activity may denote the presence of an inhibitor. Mowat (1965) found gibberellin-like activity in *Fucus spiralis* (probably due to $GA_1$ or $GA_3$) with a $GA_3$-equivalent of 2·0 to 20 μg (kg f.w.)$^{-1}$, and in the green unicellular *Tetraselmis* sp. (probably $GA_4$ or $GA_7$; $GA_3$ equivalents 10 to 60 μg kg$^{-1}$). Kato *et al.* (1962) were unable to demonstrate gibberellin-like activity in extracts of *Macrocystis pyrifera*, *Gonyaulax polyedra* and *Chlamydomonas reinhardtii*, while Jennings and McComb (1967) found activity in the red seaweed *Hypnea musciformis*. Extracts of *Enteromorpha prolifera* and *Ecklonia* gametophytes possess considerable activity (probably $GA_3$) (Jennings 1968).

### (b) Effects of exogenous gibberellins

Gibberellic acid elicited tubular elongations in axenic *Ulva lactuca* (Provasoli 1958b) at a concentration of 0·1 mg l$^{-1}$. Saono (1964) reported a slight increase in growth (approximately 20%) in *Chlorella vulgaris*, *C. pyrenoidosa*, *Scenedesmus obliquus* and *S. quadricauda*. The active concentrations were 1 to 10 mg l$^{-1}$ and higher concentrations were inhibitory. However, no growth stimulation was found by Fernandez *et al.* (1968) with *C. ellipsoidea*.

Concentrations of 34 μg l$^{-1}$ of gibberellic acids considerably shortened the lag phase (from 6 to 1 day) and increased total yields (40%) of the poisonous dinoflagellate *Gymnodinium breve* (Paster & Abbott 1970). However, in repeated experiments using gibberellic acid, $GA_4$ and $GA_7$, W. B. Wilson (pers. comm.) failed to obtain similar stimulations of axenic cultures of *G. breve* and *G. splendens*. Occasional stimulation was obtained in bacterized cultures. A 3-fold increase in growth of *Exuviella* sp. was obtained with 0·4 mg l$^{-1}$ (Iwasaki 1971b).

Growth rate and elongation of *Enteromorpha prolifera* were increased considerably (approximately 50%) by addition of 0·1 mg l$^{-1}$ of gibberellic acid (Jennings 1968) and *Enteromorpha* extracts showed a similar effect. Growth and elongation of gametophytes of *Ecklonia* were similarly stimulated, 10 μg l$^{-1}$ but higher concentrations were inhibitory.

A non-specific five-fold increase in growth of the *Conchocelis* phase of *Porphyra tenera* was obtained with 0·4 mg l$^{-1}$ of gibberellic acid (Iwasaki 1965) and increased growth (30%) of *Porphyra* fronds was obtained with 1 μg l$^{-1}$ by Kinoshita and Teramoto (1958).

*Desmarestia viridis* and *D. ligulata* were not affected by concentration ranges of 1 to 1,000 μg l$^{-1}$ of gibberellic acid (M. Tatewaki pers. comm.).

### 5.4 *Unknown morphological regulators*

The hypothesis that regulatory mechanisms should govern the orderly development of the complex structures and life-cycles of many multicellular algae seems logical and biologically sound.

The above reported research fails in our opinion to identify unequivocally the known land plant hormones as the active growth substances in algae. As noted in the auxin section there are doubts that the bioassays of the algal extracts always differentiated between true auxin effects and effects caused by other substances (Booth 1958). The identification of the active substances based on colour tests of spots on chromatograms may be misleading since the chromatographic tests were not specific for auxins (Buggeln & Craigie 1971). In other cases the $R_f$ of active spots do not correspond to any of the known auxins.

Undeniably, in many cases, applications of exogenous land plant hormones have affected the development or growth of the algae. But from this evidence we cannot be certain that algae normally employ such compounds to regulate their processes. Conversely, algae produce substances which have hormonal activity on land plants. Again these effects do not identify the substance. Land plants respond to a variety of quite different molecules with an auxin response; natural kinins differ chemically from kinetin, and many gibberellins have been described. Evidently we need chemical identification of the active purified fractions of algal origin. This might be quite rewarding since the paucity of genuine responses of algae to land plant hormones indicates that algae may use for similar purposes quite different chemicals, which are none the less active on land plants.

Work with axenic cultures of algae brings some support to this hypothesis. While several seaweeds conserve their morphological characters in culture with artificial media, a few others lose their normal morphology. Morphology of *Ulva lactuca*, when the accompanying natural microflora was removed, started to show signs of degeneration, thin finger-like short germlings were produced. These never grew more than 2 to 4 mm long, neither did they flatten, a stage preceding the formation of the typical expanded blade of the sea lettuce. Colonies of uniseriate branching filaments were also formed by the development of colourless rhizoids originating from single cells of the finger-like germling. These filaments became green and branched, forming small fungus-like colonies. Addition to an enriched aged seawater base (ASW III) of water solutions of indolacetic acid, kinetin and gibberellin were tried alone and in combination. IAA (50 μg l$^{-1}$) and kinetin (100 μg l$^{-1}$) induced more and longer finger-like germlings. The length of these filaments was increased dramatically by adding gibberellins to the IAA-kinetin base (optimum 0·1 mg l$^{-1}$); a flat blade (3 cm) was obtained by adding to the basal media 30 mg l$^{-1}$ of adenine and 0·2 mg l$^{-1}$ of kinetin. Adenine alone had no effect (Provasoli 1958b). These experiments could not be repeated using many other samples of seawater or mineral media, indicating that other factors, besides the addition of plant hormones were needed to restore the normal morphology of *Ulva*.

A few years later *Monostroma oxyspermum* was axenized by M. Tatewaki and lost its normal morphology (a leaf very similar externally to *Ulva*). The change in morphology was even more drastic; two types of loose cells were formed, rhizoidal cells with an extremely long rhizoidal part and groups of 1 to 3 roundish cells often contained in an extra common cell wall. As in the case of *Ulva*, additions of amino acids, nucleic acids, B vitamins, plant hormones, yeast and liver extracts, peptones, etc., did not restore normal morphology; neither did variations in the mineral nutrients and trace metals. Normal morphology of *M. oxyspermum* could be easily restored however by the filter-sterilized supernatants of cultures of two marine bacteria (among 20 isolates from seaweeds and seawater), of several axenic red algae and by the brown substances released into the medium by axenic *Sphacelaria* sp. or by freshly collected *Fucus* (Provasoli & Pintner 1964). None of these filtrates restores normal morphology in *Ulva*. Since the brown substances produced by *Fucus* and *Sphacelaria* proved similar and were characterized tentatively as condensed tannins (Craigie & McLachlan 1964), land plant tannins were tried but found inactive for *Monostroma*.

Little is known about algal tannins except that they are phenolics and many phenols have a hormonal effect on land plants presumably by either depressing or stimulating IAA peroxidase. Almost all of the available compounds known to be active were tried in the range ($10^1$ to $10^4$ μg $l^{-1}$) on the axenic *Monostroma* and *Ulva*. None restored normal morphology in *Monostroma* but ferulic acid, phenylalanine and other hydroxy and methoxy derivatives of cinnamic and benzoic acids induced in *Ulva* sporadically (20 to 40%) tubular filaments of solid appearance up to 2 cm long (Provasoli & Pintner 1966, Provasoli 1969). The sporadic nature of the results may be due to an uneven inoculum since the fungus-like colonies undergo various stages of growth and even produce swimming zoospores. Subsequently, two red algae *Polysiphonia urceolata* (?) and *Dasya pedicellata* (?) were brought into axenic culture and lost their normal morphology; again the usual array of nutrients and media modifications did not restore normal morphology.

Since the elimination of the microflora is responsible for morphological abnormalities, over 300 clones of marine bacteria were isolated from natural seaweeds and from infected cultures of algae showing normal morphology. Fortunately, the large number of bacterial isolates to be tested precluded the use of filtrates and the axenic cultures of the abnormal seaweeds were infected directly with single bacterial clones. Six additional clones were active on *M. oxyspermum*; more than 40 clones induced in *Ulva* a vigorous growth but of hollow tubular filaments indistinguishable to the unaided eye from *Enteromorpha*, and another twenty bacterial clones restored the erect and normal growth of *P. urceolata*, but none, even mixtures of ten clones at a time, restored the normal appearance of *Dasya pedicillata*.

The most active bacterial clones were grown in batches, and their filter-sterilized supernatants were tested for activity on the algae. While the supernatants of the *Monostroma* active clones were active for *Monostroma*, no supernatant was active on *Ulva* and *Polysiphonia*; syntrophic algal-bacterial growth is needed for the restoration of normal morphology. Specific bacteria are needed by each alga. Whenever a polyvalent activity was found, the bacterial isolate was not clonal and additional restreaks resulted in species specific clones.

Loss of morphology in *Grinnellia americana* occurred when most of the bacteria were eliminated. High concentrations of vitamin $B_{12}$ sometimes induce normal morphology and one bacterial strain isolated from *Grinnellia* cultures restored normal morphology five times out of seven (Tsukidate 1970).

A gradual decline of growth rate and appearance of abnormal pigmentation or morphology occurred in the cultures of *Nemalion helminthoides* and *Goniotrichum alsidii* (*elegans*) of Fries after several years of axenic cultivation. Normality in *Nemalion* and *Goniotrichum* could be restored by the addition to artficial media of almost equal quantities of seawater taken from the *Fucus-Ascophyllum* zone but not from seawater collected near shore. Addition of 40 mg $1^{-1}$ of casamino acids had similar beneficial effects. After fractionation, the activity of the casamino acids was localized in a peptide fraction. Chromatographic migration was partial but revealed traces of phenolic groups, and stimulation of *Goniotrichum* was also obtained from ultra-violet absorbing spots. A low molecular weight substance may be active alone or in combination with a peptide. Since the substances are insoluble in ether and chloroform, the presence of free auxin and kinins seems to be excluded (Fries 1970). The *Monostroma* factor is apparently also a small molecule, soluble in alcohols but insoluble in ether, chloroform and acetone.

Interesting progress was made by Fries with *Goniotrichum alsidii* (*elegans*) (see Fries 1972). *G. alsidii* isolated by Fries in 1960 has been maintained since then in modified $ASP_6$ + three vitamins; its pigmentation has become dark violet and instead of forming threads, it grew more like a callus. Addition to this medium of coumarin ($4 \cdot 10^{-6}$ to $4 \cdot 10^{-7}$ M), caffeic acid ($4 \cdot 10^{-5}$ to $4 \cdot 10^{-6}$ M) and ferulic acid ($10^{-7}$ M) stimulated growth and formation of long branched threads. Erratic results in early experiments were traced to variations of the previous history of the inoculum. Uniform results were obtained if *Goniotrichum* was grown for 10 days at under 9 hours light per day and then transferred to the medium with phenolics and grown under 18 hours light per day or in continuous light. The effects of the above phenolics do not seem to be connected with auxin activity since addition of various levels of IAA alone or combined with the phenolics did not modify the growth of *Goniotrichum*.

Ferulic acid also stimulated *Ulva* irregularly, but we have not experimented with light regimes. Coumarin, depending on the concentration used and the period of contact with the alga, either inhibits or stimulates growth of sporelings of the red algae, *Plumaria elegans*, *Antithamnion plumula* and *Polysiphonia brodiaei* (Boney 1967).

We have already reported the work of Pedersén (1968) on the restoration of morphology in impure *Ectocarpus fasciculatus* and *Pyliaella* (*Pilayella*) *littoralis* by kinetin. Despite the presence of microorganisms these results acquire ecological importance since the cytokinin present in seawater of the *Fucus–Ascophyllum* zone has been identified as 6-(-3 methyl-2-butenylamino) purine and the extracts are active on several seaweeds (Pedersén 1973). Perhaps cytokinins alone are responsible for these effects, but the results of Fries with *Goniotrichum* and ours with *Ulva* indicate that other substances may be necessary either alone or in combination with plant hormones for other seaweeds. The matter may be further complicated by the presence of inhibitors of auxins (Moss 1965) and of gibberellins (Jennings 1969, Hussain & Boney 1971).

# 6 ACKNOWLEDGEMENTS

We gratefully acknowledge our friends who generously gave information: Drs. M. R. Droop, L. Fries, R. R. L. Guillard, H. Iwasaki, J. C. Lewin, R. A. Lewin, R. E. Norris, R. C. Starr, M. Tatewaki, and W. H. Thomas. This study was supported in part by research grant GA–33480X of the National Science Foundation; in part by the California Cooperative Oceanic Fisheries Investigation sponsored by the State of California, and in part by the U.S. Atomic Energy Commission, Contract AT(11–1) GEN 10, P.A. 20.

# 7 REFERENCES

AARONSON S. (1971) The synthesis of extracellular macromolecules and membranes by a population of the phytoflagellate *Ochromonas danica*. *Limnol. Oceanog.* **16**, 1–9.

AARONSON S., DEANGELIS B., FRANK O. & BAKER H. (1971) Secretion of vitamins and amino acids into the environment by *Ochromonas danica*. *J. Phycol.* **7**, 215–18.

AHMAD M.R. & WINTER A. (1968a) Studies on the hormonal relationships of algae in pure culture. I. The effect of indole-3-acetic acid on the growth of blue-green and green algae. *Planta*, **78**, 277–86.

AHMAD M.R. & WINTER A. (1968b) Studies on the hormonal relationships of algae in pure culture. II. The effect of potential precursors on the growth of several freshwater blue-green algae. *Planta*, **81**, 16–27.

AHMAD M.R. & WINTER A. (1969) Studies on the hormonal relationships of algae in pure culture. III. Tryptamine is an intermediate in the conversion of tryptophan to indole-3-acetic acid by the blue-green alga *Chlorogloea fritschii*. *Planta*, **88**, 61–6.

AHMAD M.R. & WINTER A. (1970) Studies on the hormonal relationships in algae in pure culture. IV. The metabolism of indole-2-C$^{14}$ by the blue-green alga *Chlorogloea fritschii*. *Z. PflPhysiol.* **62**, 393–7.

ANTIA N.J. (1963) A microbiological assay for biotin in seawater. *Can. J. Microbiol.* **9**, 403–9.

AUGIER H. (1965) Les substances de croissance chez la Rhodophycée *Botryocladia botryoides* (Wulf.). Feldm. *C. r. hebd. Séanc. Acad. Sci., Paris*, **260**, 2304–6.

BACH M.K. & FELLIG J. (1958) Effect of ethanol and auxins on the growth of unicellular algae. *Nature, Lond.* **182**, 1359–60.

BAKER H. & FRANK H. (1968) *Clinical vitaminology: methods and interpretation.* Wiley-Interscience, New York.

BEDNAR T.W. & HOLM-HANSEN O. (1964) Biotin liberation by the lichen alga *Coccomyxa* sp. and by *Chlorella pyrenoidosa. Pl. Cell Physiol., Tokyo,* **5,** 297–303.

BELSER W.L. (1959) Bioassay of organic micronutrients in the sea. *Proc. natn. Acad. Sci., U.S.A.* **45,** 1535–42.

BELSER W.L. (1963) Bioassay of trace substances. In *The Sea,* Vol. 2 ed. Hill M.W. pp. 220–31. Interscience, New York.

BENTLEY-MOWAT J.A. & REID S.M. (1968) Investigations of the radish leaf bioassay for kinetins, and demonstration of kinetin-like substances in algae. *Ann. Bot.* **32,** 23–32.

BENTLEY-MOWAT J.A. & REID S.M. (1969) Effect of gibberellins, kinetin and other factors on the growth of unicellular algae in culture. *Botanica mar.* **12,** 185–93.

BONEY A.D. (1967) The effects of coumarin on the growth and viability of sporelings of red algae. *Planta,* **76,** 114–23.

BOOTH A. (1958) Non-hormone growth promotion shown by aqueous extracts. *J. exp. Bot.* **9,** 306–10.

BRADBEER C. (1971) Transport of vitamin $B_{12}$ in *Ochromonas malhamensis. Archs. Biochem. Biophys.* **144,** 184–92.

BUGGELN R.G. & CRAIGIE J.S. (1971) Evaluation of evidence for the presence of indole-3-acetic acid in marine algae. *Planta,* **97,** 173–8.

BUNT J.S. (1961) Blue-green algae. *Nature, Lond.* **192,** 1274–5.

BURKHOLDER P.R. (1963) Some nutritional relationships among microbes of sea sediments and water. In *Symposium on Marine Microbiology,* ed. Oppenheimer C.H. pp. 133–50. C. C. Thomas, Springfield, Ill.

BURKHOLDER P.R. & BURKHOLDER L.M. (1956) Vitamin $B_{12}$ in suspended solids and marsh muds collected along the coast of Georgia. *Limnol. Oceanog.* **1,** 202–8.

BURKHOLDER P.R. & BURKHOLDER L.M. (1958) Studies on B vitamins in relation to productivity of Bahia Forforescente, Puerto Rico. *Bull. Mar. Sci., Gulf Carib.* **8,** 201–23.

CARLUCCI A.F. (1967) Determination of vitamins in seawater. In *Chemical Environment in the Aquatic Habitat,* eds. Golterman H.L. & Clymo R.S. pp. 239–44. N. V. Noord-Hollandsche Uitgevers Maatschappij-Amsterdam.

CARLUCCI A.F. (1970) The ecology of the plankton off La Jolla, California in the period April through September, 1967. II. Vitamin $B_{12}$, thiamine and biotin. *Bull. Scripps Inst. Oceanog.* **17,** 23–30.

CARLUCCI A.F. & BELSER W.L. (1963) A method for the bioassay of metabolites in seawater with *Serratia marinorubra. Bacteriol. Proc.* **3.**

CARLUCCI A.F. & BOWES P.M. (1970a) Production of vitamin $B_{12}$, thiamine, and biotin by phytoplankton. *J. Phycol.* **6,** 351–7.

CARLUCCI A.F. & BOWES P.M. (1970b) Vitamin production and utilization by phytoplankton in mixed culture. *J. Phycol.* **6,** 393–400.

CARLUCCI A.F. & BOWES P.M. (1972a) Vitamin $B_{12}$, thiamine, and biotin contents of marine phytoplankton. *J. Phycol.* **8,** 133–7.

CARLUCCI A.F. & BOWES P.M. (1972b) Determination of vitamin $B_{12}$, thiamine and biotin in Lake Tahoe waters using modified marine bioassay techniques. *Limnol. Oceanog.* **17,** 774–7.

CARLUCCI A.F. & SILBERNAGEL S.B. (1966a) Bioassay of seawater. I. A $^{14}C$ uptake method for the determination of concentrations of vitamin $B_{12}$ in seawater. *Can. J. Microbiol.* **12,** 175–83.

CARLUCCI A.F. & SILBERNAGEL S.B. (1966b) Bioassay of seawater. II. Methods for the determination of concentrations of dissolved vitamin $B_{12}$ in seawater. *Can. J. Microbiol.* **12,** 1079–89.

CARLUCCI A.F. & SILBERNAGEL S.B. (1966c) Bioassay of seawater. III. Distribution of vitamin $B_{12}$ in the northeast Pacific Ocean. *Limnol. Oceanogr.* **11,** 642–6.

CARLUCCI A.F. & SILBERNAGEL S.B. (1967) Bioassay of seawater. IV. The determination of dissolved biotin in sea water using $^{14}$C uptake by cells of *Amphidinium carteri*. *Can. J. Microbiol.* **13**, 979–86.

CARLUCCI A.F. & SILBERNAGEL S.B. (1969) Effect of vitamin concentration on growth and development of vitamin-requiring algae. *J. Phycol.* **5**, 64–7.

CARLUCCI A.F., SILBERNAGEL S.B. & MCNALLY P.M. (1969) Influence of temperature and solar radiation on persistence of vitamin $B_{12}$, thiamine, and biotin in seawater. *J. Phycol.* **5**, 302–5.

CARPENTER E.J. & GUILLARD R.R.L. (1971) Intraspecific differences in nitrate half-saturation constants for three species of marine phytoplankton. *Ecology*, **52**, 183–5.

CIRILLO V.P. (1957) Long-term adaptation to fatty acids by the phytoflagellate, *Polytoma uvella*. *J. Protozool.* **4**, 60–2.

CONRAD H.M. & SALTMAN P. (1962) Growth substances. In *Physiology and biochemistry of algae*, ed. Lewin R.A. pp. 663–71. Academic Press, New York & London.

CRAIGIE J.S. & MCLACHLAN J. (1964) Excretion of colored ultraviolet-absorbing substances by marine algae. *Can. J. Bot.* **42**, 23–33.

DAISLEY K.W. (1957) Vitamin $B_{12}$ in marine ecology. *Nature, Lond.* **180**, 1042–3.

DAISLEY K.W. (1961) Gel filtration of seawater: separation of free and bound forms of vitamin $B_{12}$. *Nature, Lond.* **191**, 868–9.

DAISLEY K.W. (1969) Monthly survey of vitamin $B_{12}$ concentrations in some waters of the English Lake District. *Limnol. Oceanog.* **14**, 224–8.

DAISLEY K.W. (1970) The occurrence and nature of *Euglena gracilis* proteins that bind vitamin $B_{12}$. *Int. J. Biochem.* **1**, 561–74.

DAISLEY K.W. & FISHER L.R. (1958) Vertical distribution of vitamin $B_{12}$ in the sea. *J. mar. biol. Ass., U.K.* **37**, 683–6.

DAWES C.J. (1971) Indole-3-acetic acid in the green algal coenocyte *Caulerpa prolifera* (Chlorophyceae, Siphonales). *Phycologia*, **10**, 375–9.

DROOP M.R. (1957) Vitamin $B_{12}$ in marine ecology. *Nature, Lond.* **180**, 1041–2.

DROOP M.R. (1961) Vitamin $B_{12}$ and marine ecology: the response of *Monochrysis lutheri*. *J. mar. biol. Ass., U.K.* **41**, 69–76.

DROOP M.R. (1962) Organic micronutrients. In *Physiology and biochemistry of algae*, ed. Lewin R.A. pp. 141–54. Academic Press, New York & London.

DROOP M.R. (1966) Vitamin $B_{12}$ and marine ecology. III. An experiment with a chemostat. *J. mar. biol. Ass., U.K.* **46**, 659–71.

DROOP M.R. (1968) Vitamin $B_{12}$ and marine ecology. IV. The kinetics of uptake, growth and inhibition in *Monochrysis lutheri*. *J. mar. biol. Ass., U.K.* **48**, 689–733.

DROOP M.R. (1970) Vitamin $B_{12}$ and marine ecology. V. Continuous culture as an approach to nutritional kinetics. *Helgolander wiss. Meeresunters*, **20**, 629–36.

DROOP M.R., MCLAUGHLIN J.J.A., PINTNER I.J. & PROVASOLI L. (1959) Specificity of some protophytes toward vitamin $B_{12}$-like compounds. In *Int. Oceanog. Congress—Preprints*, *A.A.A.S.* 916–18.

EVERSOLE R.A. (1956) Biochemical mutants of *Chlamydomonas reinhardi*. *Am. J. Bot.* **43**, 404–7.

FERNANDEZ A., BALLONI W. & MATERASSI R. (1968) Sul comportamento di alcuni ceppi di microalghe nei confronti delle sostanze di crescita. *Agricoltura Italiana*, 281–6.

FOGG G.E. (1966) The extracellular products of algae. *Oceanogr. Mar. Biol. Ann. Rev.* **4**, 195–212.

FORD J.E. (1958) $B_{12}$-vitamins and the growth of *Ochromonas malhamensis*. *J. gen. Microbiol.* **19**, 161–72.

FORD J.E. & GOULDEN J.D.S. (1959) The influence of vitamin $B_{12}$ on the growth rate and cell composition of the flagellate *Ochromonas malhamensis*. *J. gen. Microbiol.* **20**, 267–76.

FORD J.E., GREGORY M.E. & HOLDSWORTH E.S. (1955) Uptake of $B_{12}$-vitamins in *Ochromonas malhamensis*. *Biochem. J.* **61**, xxiii.

FRIES L. (1960) The influence of different $B_{12}$ analogues on the growth of *Goniotrichum elegans*. *Physiologia Pl.* **13**, 264–75.

FRIES L. (1964) *Polysiphonia urceolata* in axenic culture. *Nature, Lond.* **202**, 110.

FRIES L. (1970) The influence of microamounts of organic substances other than vitamins on the growth of some red algae in axenic culture. *Br. Phycol. J.* **5**, 39–46.

FRIES L. (1972) The influence of phenolic compounds on the growth of *Goniotrichum elegans* (Chauv.) In: *Proc. VII Int. Seaweed Symp.* pp. 575–9. Tokyo University Press.

FRIES L. & PETTERSON H. (1968) On the physiology of the red alga *Asterocytis ramosa* in axenic culture. *Br. Phycol. Bull.* **3**, 417–22.

FRITSCH F.E. (1945) *The structure and reproduction of the algae*. Vols. I & II. Cambridge Univ. Press.

GOLD K. (1964) A microbiological assay for vitamin $B_{12}$ in seawater using radiocarbon. *Limnol. Oceanog.* **9**, 343–7.

GOLD K. (1968) Some factors affecting the stability of thiamine. *Limnol. Oceanog.* **13**, 185–8.

GOLD K., ROELS O.A. & BANK H. (1966) Temperature dependent destruction of thiamine in seawater. *Limnol. Oceanog.* **11**, 410–13.

GUILLARD R.R.L. (1968) $B_{12}$ specificity of marine centric diatoms. *J. Phycol.* **4**, 59–64.

HOHAM R. (1971) Laboratory and field studies on snow algae of the Pacific Northwest. Ph.D. Thesis, Univ. of Washington, Seattle.

HUSSAIN A. & BONEY A.D. (1969) Isolation of kinetin-like substances from *Laminaria digitata*. *Nature, Lond.* **223**, 504–5.

HUSSAIN A. & BONEY A.D. (1971) Plant growth substances associated with motile and nonmotile phases of two *Cricosphaera* species (Order Prymnesiales, Class Haptophyceae). *Botanica mar.* **14**, 17–21.

HUTNER S.H. (1936) The nutritional requirement of two species of *Euglena*. *Arch. Protist.* **88**, 93–106.

HUTNER S.H. & PROVASOLI L. (1955) Comparative biochemistry of flagellates. In *Biochemistry and Physiology of Protozoa*, eds. Hutner S.H. & Lwoff A., vol. **2**, 18–43. Academic Press, New York & London.

HUTNER S.L., PROVASOLI L., STOKSTAD E.L.R., HOFFMAN C.E., BELT M., FRANKLIN A.L. & JUKES T.H. (1949) Assay of antipernicious anemia factor with *Euglena*. *Soc. Exp. Biol. Med.* **70**, 118–20.

IWASAKI H. (1965) Nutritional studies of the edible seaweed *Porphyra tenera*. I. The influence of different $B_{12}$ analogues, plant hormones, purine and pyrimidines on the growth of *Conchocelis*. *Pl. Cell Physiol., Tokyo*, **6**, 325–36.

IWASAKI H. (1969) Studies on the red tide dinoflagellates. III. On *Peridinium hangoei* Schiller appeared on Gokasho Bay, Shima Peninsula. *Bull. Plankton Soc. Japan*, **16**, 132–9.

IWASAKI H. (1971a) Studies on the red tide dinoflagellates. V. On *Polykrikos schwartzi*, Butschli. *Bull. Jap. Sci. Fish.* **37**, 606–9.

IWASAKI H. (1971b) Studies on the red tide dinoflagellates. VI. *Eutreptiella* sp. and *Exuviaella* sp. *J. Ocean. Soc., Japan*, **27**, 152–7.

IWASAKI H. & SASADA K. (1969) Studies on the red tide dinoflagellates. II. On *Heterosigma inlandica* appeared in Gokasha Bay, Shima Peninsula. *Bull. Jap. Soc. Sci. Fish*, **35**, 943–7.

IWASAKI H., FUJIYAMA T. & YAMASHITA E. (1968) Studies on the red tide dinoflagellates. I. On *Entomosigma* sp. appeared in coastal area of Fukuyama. *J. Fish. & Anim. Husb., Hiroshima Univ.* **7**, 259–67.

JACOBS W.P. (1951) Studies on cell-differentiation: the role of auxin in algae, with particular reference to rhizoid-formation in *Bryopsis*. *Biol. Bull. mar. biol. Lab., Woods Hole*, **101**, 300–6.

JENNINGS R.C. (1968) Gibberellins as endogenous growth regulators in green and brown algae. *Planta*, **80**, 34–42.

JENNINGS R.C. (1969) Gibberellin antagonism by material from a brown alga. *New Phytol.* **68**, 683–8.

JENNINGS R.C. & McCOMB A.J. (1967) Gibberellins in the red alga *Hypnea musciformis* (Wulf.) Lamour. *Nature, Lond.* **215**, 872–3.

JOHNSTON R. (1963a) Antimetabolites as an aid to the study of phytoplankton nutrition. *J. mar. biol. Ass., U.K.* **43**, 409–25.

JOHNSTON R. (1963b) Seawater, the natural medium of phytoplankton. I. General features. *J. mar. biol. Ass., U.K.* **43**, 427–56.

KANAZAWA A. (1963) Vitamins in algae. *Bull. Jap. Soc. Sci. Fish.* **29**, 713–31.

KASHIWADA K., KAKIMOTO D. & KANAZAWA A. (1960) Studies of vitamin $B_{12}$ in natural water. *Rec. Oceanog. Work Jap.* **5**, 71–6.

KATO J., PURVES W.K. & PHINNEY B.O. (1962) Gibberellin-like substances in plants. *Nature, Lond.* **196**, 687–8.

KINOSHITA S. & TERAMOTO T. (1958) On the efficiency of gibberellin on the growth of *Porphyra*-frond. *Bull. Jap. Soc. Phyc.* **6**, 85–8.

KLEIN S. (1972) Conservation of glycerolated cultures of *Ochromonas danica* and *O. malhamensis* at – 10 C. *J. Protozool.* **19**, 140–3.

KRISTENSEN H.P.O. (1955) Investigations into the *Euglena gracilis* method for quantitative assay of vitamin $B_{12}$. *Acta physiol. scand.* **33**, 232–7.

KRISTENSEN H.P.O. (1956) A vitamin $B_{12}$-binding factor formed in cultures of *Euglena gracilis* var. *bacillaris*. *Acta physiol. scand.* **37**, 8–13.

LEFÈVRE M. (1964) Extracellular products of algae. In *Algae and Man*, ed. Jackson D.F. pp. 337–67. Plenum Press, N.Y.

LEONIAN L.H. & LILLY V.G. (1937) Is heterauxin a growth promoting substance? *Am. J. Bot.* **24**, 135–9.

LEWIN J.C. & LEWIN R.A. (1960) Auxotrophy and heterotrophy in marine littoral diatoms. *Can. J. Microbiol.* **6**, 127–34.

LEWIN R.A. (1961) Phytoflagellates and algae. In *Encyc. Plant Physiol.*, ed. Ruhland W., **14**, 401–17. Springer-Verlag, Berlin, Heidelberg & New York.

LEWIN R.A. (1972) Auxotrophy in marine littoral diatoms. In *VII Int. Seaweed Symposium*, pp. 316–18. Tokyo Univ. Press.

LIEN T., PETTERSEN R. & KNUTSEN G. (1971) Effects of indole-3-acetic acid and gibberellin on synchronous cultures of *Chlorella fusca*. *Physiologia Pl.* **24**, 185–90.

LITCHFIELD C.D. & HOOD D.W. (1965) Microbiological assay for organic compounds in sea-water. I. Quantitative assay procedure and biotin distribution. *Appl. Microbiol.* **13**, 886–94.

LWOFF A. (1943) *L'évolution physiologique. Étude des pertes de fonction chez les micro-organisms*. Hermann Co., Paris.

LWOFF A. (1947) Some aspects of the problem of growth factors for protozoa. *A. Rev. Microbiol.* **1**, 101–14.

LWOFF A. & DUSI H. (1937a) La pyrimidine et le thiazol, facteurs de croissance pour le flagellé *Polytomella coeca*. *C. r. hebd. Séanc. Acad. Sci., Paris*. **205**, 630.

LWOFF A. & DUSI H. (1937b) Le thiazol, facteur de croissance pour les flagelles *Polytoma caudatum* et *Chilomonas paramaecium*. *C. r. hebd. Séanc. Acad. Sci., Paris*, **205**, 756.

LWOFF A. & DUSI H. (1937c) Le thiazol, facteur de croissance pour le flagellé *Polytoma ocellatum*. *C. r. hebd. Séanc. Acad. Sci., Paris*, **205**, 882.

MARKER A.F. (1965) Extracellular carbohydrate liberation in the flagellates *Isochrysis galbana* and *Prymnesium parvum*. *J. mar. biol. Ass., U.K.* **45**, 755–72.

MENZEL D.W. & SPAETH J.P. (1962) Occurrence of vitamin $B_{12}$ in the Sargasso Sea. *Limnol. Oceanog.* **7**, 151–4.

MISHRA A.K. & KEFFORD N.P. (1969) Developmental studies on the coenocytic alga, *Caulerpa sertularioides*. *J. Phycol.* **5**, 103–9.

MOEWUS F. (1959) Stimulation of mitotic activity by benzidine and kinetin in *Polytoma uvella*. *Trans. Am. Microscop. Soc.* **78**, 295–304.

MOSS B. (1965) Apical dominance in *Fucus vesiculosus*. *New Phytol.* **64**, 387–92.

MOWAT J.A. (1965) A survey of results on the occurrence of auxins and gibberellins in algae. *Botanica mar.* **8**, 149–55.

NAKAMURA K. & GOWANS C.S. (1964) Nicotinic acid excreting mutants in *Chlamydomonas*. *Nature, Lond.* **202**, 826–7.

NATARAJAN K.V. (1967a) Distribution of thiamine, biotin and niacin in the sea. *Appl. Microbiol.* **16**, 366–9.

NATARAJAN K.V. (1967b) Distribution and significance of vitamin $B_{12}$ and thiamine in the Subarctic Pacific Ocean. *Limnol. Oceanog.* **15**, 655–7.

NATARAJAN K.V. & DUGDALE R.C. (1966) Bioassay and distribution of thiamine in the sea. *Limnol. Oceanog.* **11**, 621–9.

NEUJAHR H.Y. (1956) On vitamins in sewage sludge. IV. Isolation of new vitamin $B_{12}$-like factors. *Acta chem. scand.* **10**, 917–27.

OHWADA K. (1972) Bioassay of biotin and its distribution in the sea. *Marine Biology*, **14**, 10–17.

OHWADA K. & TAGA N. (1972) Distribution and seasonal variation of vitamin $B_{12}$, thiamine, and biotin in the sea. *Mar. Chem.* **1**, 61–73.

OHWADA K., OTSUHATA M. & TAGA N. (1972) Seasonal cycles of vitamin $B_{12}$, thiamine, and biotin in the surface water of Lake Tsukui. *Bull. Jap. Soc. Sci. Fish.* **38**, 817–23.

PASTER Z. & ABBOTT B.C. (1970) Gibberellic acid: a growth factor in the unicellular alga *Gymnodinium breve*. *Science, N.Y.* **169**, 600–1.

PEDERSÉN M. (1968) *Ectocarpus fasciculatus*: marine brown alga requiring kinetin. *Nature, Lond.* **218**, 776.

PEDERSÉN M. (1973) Identification of a cytokinin 6-(3-methyl-2-butenylamino) purine in seawater and the effects of cytokinins on brown algae. *Physiologia Pl.* **28**, 101–5.

PEDERSÉN M. & FRIDBORG G. (1972) Cytokinin-like activity in seawater from the *Fucus-Ascophyllum* zone. *Experientia*, **28**, 111–12.

PINTNER I.J. & PROVASOLI L. (1968) Heterotrophy in subdued light of 3 *Chrysochromulina* species. *Bull. Misaki Mar. Biol. Inst. Kyoto*, **2**, 25–31.

PRATT R. (1937) Influence of auxins on growth of *Chlorella vulgaris*. *Am. J. Bot.* **25**, 498–501.

PRINGSHEIM E.G. (1937) Assimilation of different organic substances by saprophytic flagellates. *Nature, Lond.* **139**, 196.

PRINGSHEIM E.G. & PRINGSHEIM O. (1952) Experimental elimination of chromatophores and eye-spot in *Euglena gracilis*. *New Phytol.* **51**, 65–76.

PROVASOLI L. (1938) Studi sulla nutrizione dei protozoi. *Boll. Zool. Agraria e Bachicolt*, **9**, 1–124.

PROVASOLI L. (1956) Alcune considerazioni sui caratteri morfologici e fisiologici della alge. (Reflections on the morphology and physiology of Algae.) *Boll. Zool. Agraria e Bachicolt.* **22**, 143–88.

PROVASOLI L. (1958a) Growth factors in unicellular marine algae. In *Perspectives in marine biology*, ed. Buzzati-Traverso, A.A. pp. 385–403. Univ. Calif. Press.

PROVASOLI L. (1958b) Effect of plant hormones on *Ulva*. *Biol. Bull. mar. biol. Lab., Woods Hole*, **114**, 375–84.

PROVASOLI L. (1963) Organic regulation of phytoplankton fertility. In *The Sea*, ed. Hill M.N. **2**, 165–219. Interscience, N.Y.

PROVASOLI L. (1969) Algal nutrition and eutrophication. In *Eutrophication*, Publ. No. 1700. Nat. Acad. Sci. Washington, pp. 589–90.

PROVASOLI L. (1971) Nutritional relationships of marine organisms. In *Fertility of the sea*, ed. Costlow J.D., pp. 369–82. Gordon & Breach, N.Y.

PROVASOLI L. & GOLD K. (1962) Nutrition of the American strain of *Gyrodinium cohnii*. *Arch. Microbiol.* **42**, 196–203.

PROVASOLI L. & MCLAUGHLIN J.J.A. (1955) Auxotrophy in some marine and brackish dinoflagellates. *J. Protozool.* **2** (*Suppl.*), 10.

PROVASOLI L. & PINTNER I.J. (1953) Ecological implications of *in vitro* nutritional requirements of algal flagellates. *Ann. N.Y. Acad. Sci.* **56**, 839–51.

PROVASOLI L. & PINTNER I.J. (1964) Symbiotic relationships between microorganisms and seaweeds. (*Abst.*) *Am. J. Bot.* **51**, 681.

PROVASOLI L. & PINTNER I.J. (1966) The effect of phenolic compounds on the morphology of *Ulva*. Proc. Abst. Papers. 11th Pacific Science Cong. In *Fisheries*, **7**, 23.

PROVASOLI L., HUTNER S.H. & SCHATZ A. (1948) Streptomycin-induced chlorophyll-less races of *Euglena. Proc. Soc. expt. Biol. Med.* **69**, 279–82.

RADLEY M. (1961) Gibberellin-like substances in plants. *Nature, Lond.* **191**, 684–5.

RYTHER J.G. & GUILLARD R.R.L. (1962) Studies of marine planktonic diatoms. II. Use of *Cyclotella nana* Hustedt for assays of vitamin $B_{12}$ in seawater. *Can. J. Microbiol.* **8**, 437–45.

SAONO S. (1964) Effect of gibberellic acid on the growth and multiplication of some soil microorganisms and unicellular green algae. *Nature, Lond.* **204**, 1328–9.

SCHIEWER V. (1967a) Auxinvorkommen und Auxinstoffwechsel bei mehrzelligen Ostseealgen. I. Zum Vorkommen von Indol-3-essigsaure. *Planta*, **74**, 313–23.

SCHIEWER V. (1967b) Auxinvorkommen und Auxinstoffwechsel bei mehrzelligen Ostseealgen. II. Zur Entstehung von Indol-3-essigsaure aus Tryptophan, unter Berucksichtigung des Einflusses der marinen Bakterienflora. *Planta*, **75**, 152–60.

SHIMO S. & NAKATANI S. (1964) The effect of kinetin on the growth of *Porphyra. J. Agricult. Lab. No. 5. Central Res. Lab. of Electricity*, 55–60.

SIEBURTH J.MCN. (1968) The influence of algal antibiosis on the ecology of marine microorganisms. In *Adv. Microbiol. Sea*, eds. Droop M.R. & Ferguson-Wood E.J. pp. 63–94. Academic Press, New York & London.

SMITH G.M. (1933) *The fresh-water algae of the United States.* McGraw Hill (1950, 2nd edit.).

SPENCER T. (1968) Effect of kinetin on the phosphatase enzymes of *Acetabularia. Nature, Lond.* **217**, 62–4.

STREET H.E., GRIFFITH D.J., THRESHER C.L. & OWENS M. (1958) Ethanol as a carbon source for the growth of *Chlorella vulgaris. Nature, Lond.* **182**, 1360–1.

STRICKLAND J.D.H. & PARSONS T.R. (1968) A practical handbook of seawater analysis. *Bull. Fish. Res. Bd. Can.* **167**, 311 pp.

STROHECKER R. & HENNING H.M. (1966) *Vitamin Assay.* Chemical Rubber Co., Cleveland, 360 pp.

TAL E. (1962) Preliminary study of fluctuations of vitamins $B_{12}$ content in fish ponds in the Berth-Shean Valley. *Bamidgeh*, **14**, 19–26.

TASSIGNY M. (1971) Observations sur les bésoins en vitamines des desmidiées (Chlorophycées-Zygnematales). *J. Phycol.* **7**, 213–15.

TATEWAKI M. & PROVASOLI L. (1964) Vitamin requirements of three species of *Antithamnion. Botanica mar.* **6**, 193–203.

THOMAS W.H. (1968) Nutrient requirements and utilization: Algae, Table 27, pp. 210–28. In *Metabolism*, eds. Altman P.L. & Ditmer D.S. *Fed. Am. Soc. Exp. Biol.*

TSUKIDATE J. (1970) Some notes on *Grinnellia americana* in culture. *Bull. Jap. Soc. Sci. Fish.* **36**, 1109–14.

VAN BAALEN C. (1962) Studies on marine blue-green algae. *Botanica mar.* **4**, 129–39.

VAN OVERBEEK J. (1940) Auxin in marine algae. *Pl. Physiol.*, Lancaster, **15**, 291–9.

VISHNIAC H.S. (1961) A biological assay for thiamine in seawater. *Limnol. Oceanog.* **6**, 31–5.

VISHNIAC H.S. & RILEY G.A. (1961) Cobalamin and thiamine in Long Island Sound: patterns of distribution and ecological significance. *Limnol. Oceanog.* **6**, 36–41.

WILLIAMS L.G. (1949) Growth regulating substances in *Laminaria agardhii. Science, N.Y.* **110**, 169–70.

WOOD E.A. (1962) An evaluation of the role of vitamin $B_{12}$ in the marine environment. Ph.D. Thesis, Yale Univ., New Haven, Conn.

# 8 APPENDIX TO CHAPTER 27

## VITAMIN REQUIREMENTS OF SOME ALGAL SPECIES

| Species | B₁₂ | Thiamine | Biotin* |
|---|---|---|---|
| **CYANOPHYCEAE** | | | |
| *Agmenellum quadruplicatum* (115), *Anabaena cylindrica* (3), *A. gelatinosa* (15), *A. naviculoides* (15), *A. variabilis* (59), *Anacystis marina* (115), *A. nidulans* (59), *Aphanizomenon flos-aquae* (32), *Calothrix brevissima* (116), *C. parietina* (3), *C. scopulorum* (105), *Chroococcus turgidus* (3), *Coccochloris peniocystis* (32), *Gloeocapsa dimidiata* (32), *G. membranina* (119), *Lyngbya aestuarii* (3, 115), *Microcoleus chthonoplastes* (115), *M. tenerrimus* (115), *M. vaginatus fuscus* (6), *Microcystis* (*Diplocystis*) *aeruginosa* (32), *Nostoc entophytum* (105), *N. muscorum* (3, 59), *Oscillatoria brevis* (105), *O. lutea* (6), *O. rubescens* (121), *Phormidium foveolarum* (3), *P. luridum* (3), *P. tenue* (32), *Plectonema nostocorum* (32), *P. notatum* (3), *P. terebrans* (115), *Schizothrix calcicola* (6), *S. calcicola glomerata* (6), *S. calcicola mucosa* (6), *S. calcicola spiralis* (6), *S. calcicola vermiformis* (6), *Synechococcus cedrorum* (3). | 0 | 0 | 0 |
| *Agmenellum quadruplicatum* (115), *Coccochloris elabens* (115), *Lyngbya lagerheimii* (115), *Oscillatoria amphibia* (115), *O. lutea auxotrophica* (6), *O. lutea contorta* (6), *O. subtilissima* (115), *Phormidium persicinum* (82). | R | 0 | 0 |
| **RHODOPHYCEAE** | | | |
| *Porphyridium purpureum* (= *cruentum*) (18) | 0 | 0 | 0 |
| *Antithamnion* sp. (111), *A. glanduliferum* (111), *A. spirographidis* (*sarniense*) (111), *Asterocytis ramosa* (31), *Bangia fuscopurpurea* (92), *Erythrotrichia carnea* (30), *Goniotrichum alsidii* (*elegans*) (28), *Nemalion helminthoides* (*multifidum*) (29), *Polysiphonia urceolata* (30), *Porphyra tenera* (Conchocelis phase) (50), *Rhodella maculata* (114) | R | 0 | 0 |
| **BACILLARIOPHYCEAE** | | | |
| *Amphora coffaeiformis* (64), *Asterionella formosa* (91), *Detonula confervacea* (36), *Fragilaria capucina* (91), *F.* sp. (*pinnata?*) (40), *Melosira* sp. (*nummuloides*) (38, 36), *Navicula corymbosa* (63, 64), *N. incerta* (64), *N. menisculus* (64), *N. pelliculosa* (60), *Nitzschia angularis affinis* (64), *N. curvilineata* (64), *N. filiformis* (64), *N. frustulum* (64), *N. hybridaeformis* (64), *N. laevis* (64), *N. lanceolata* (63, 64), *N. marginata* (64), *N. obtusa scalpelliformis* (64), *Phaeodactylum tricornutum* (46), *Stauroneis amphoroides* (64), *Tabellaria flocculosa* (91), *Thalassiosira nordenskioldii* (36). | 0 | 0 | 0 |

| | | |
|---|---|---|
| R | 0 | 0 |
| 0 | R | 0 |
| R | R | 0 |
| R | 0 | 0 |
| 0 | R | 0 |
| R | R | 0 |
| R | R | R |
| 0 | 0 | 0 |
| R | R | 0 |
| 0 | 0 | R |
| R | 0 | R |
| R | R | R |

*Achnanthes brevipes* (64), *Amphora coffaeiformis* (64) = *A. perpusilla* (48), *A. lineolata* (64), *Bellerochea* sp. (*polymorpha?*) (40), *B.* sp. (*spinifera?*) (40), *Chaetoceros ceratosporum* (37), *C. gracilis* (13, 113), *C. lorenzianus* (36), *C. pelagicus* (36), *C. pseudocrinitus* (37), *Coscinodiscus asteromphalus* (36), *Cyclotella* sp. (*cryptica*) (64), *C.* sp. (64) *Cylindrotheca closterium* (= *Nitzschia closterium*) (37), *C. pseudonana* (= *Cyclotella nana*) (39), *Ditylum brightwellii* (37), *Fragilaria* sp. (36), *F.* sp. (*pinnata?*) (40), *F.* sp. (*rotundissima?*) (40), *Licmophora hyalina* (68), *Melosira* sp. (36), *M. juergensii* (108), *Nitzschia frustulum* (64), *N. ovalis* (64), *N. punctata* (64), *N. seriata* (36), *Opephora* sp. (64), *Rhizosolenia setigera* (36), *Skeletonema costatum* (17), *Stephanopyxis turris* (91), *Synedra affinis* (64) (= *Fragilaria brevistriata*) (68), *Synedra* sp. (*rotundissima?*) (40), *Thalassiosira fluviatilis* (36), *T. nordenskioldii* (36)

*Amphora coffaeiformis* (64), *A. paludosa duplex* (64), *Bellerochea* sp. (*polymorpha?*) (40), *Cylindrotheca fusiformis* (= *Nitzschia closterium*) (62, 64), *Navicula pelliculosa* (40).

*Amphipleura rutilans* (64, 68), *Amphora coffaeiformis* (64), *Bellerochea* sp. (*polymorpha?*) (40), *B.* sp. (*spinifera?*) (40), *Cylindrotheca fusiformis* (= *Nitzschia closterium*) (64, 37), *Cyclotella caspia* (36, 37), *Nitzschia alba* (65), *N. leucosigma* (65), *N. putrida* (65).

CRYPTOPHYCEAE

*Cryptomonas ovata* (96), *Cyanophora paradoxa*[a] (96).
*Chilomonas paramecium* (70), *Rhodomonas lens* (91).
*Chroomonas salina* (5), *Hemiselmis virescens* (19),
*Rhodomonas ovalis* (55)

CHRYSOPHYCEAE

*Stichochrysis immobilis* (91)
*Microglena arenicola* (19), *Monochrysis lutheri* (19),
*Ochromonas minuta* (85), *Pavlova gyrans* (84),
*Syncrypta* (*Synochromonas*) *korschikoffii* (87)
*Synura sphagnicola* (85)
*Ochromonas danica* (43, 86)
*Synura caroliniana* (85), *S. petersenii* (85)
*Ochromonas malhamensis* (49, 86), *Poteriochromonas stipitata* (85)

Vitamin requirements—continued

| Species | $B_{12}$ | Thiamin | Biotin |
|---|---|---|---|
| **HAPTOPHYCEAE** | | | |
| *Coccolithus huxleyi* (36), *Hymenomonas carterae* (96) | R | 0 | 0 |
| *Apistonema aestuari* (=*Pontosphaera roscoffensis*) (85), *Coccolithus* sp. (85), *C. huxleyi* (36, 84), *Dicrateria inornata* (85), *Hymenomonas* sp. No. 156 (84), *Ochrosphaera neapolitana* (84), *Pleurochrysis scherffelii* (91), *Syracosphaera* sp. No. 181 (84) | 0 | R | 0 |
| *Chrysochromulina kappa* (85), *C. brevefilum* (85), *C. strobilus* (85), *Cricosphaera* (*Syracosphaera*) *elongata* (17), *Isochrysis galbana* (91) *Prymnesium parvum* (19) | R | R | 0 |
| **DINOPHYCEAE** | | | |
| *Peridinium cinctum ovoplanum* (12), *Symbiodinium microadriaticum*[b] (72, 73). | 0 | 0 | 0 |
| *Entomosigma* sp. (*Heterosigma akashivo*) (54), *Exuviaella* sp. (53), *E. cassubica* (94), *Glenodinium foliaceum* (19), *G. halli* (34), *G. monotis* (81), *Gonyaulax polyedra* (42, 112), *Gymnodinium splendens* (107), *Gyrodinium californicum* (95), *G. resplendens* (94, 95), *G. uncatenum* (94, 95), *Heterosigma inlandica* (56), *Peridinium balticum* (94), *P. chattoni* (94), *P. hangoei* (51), *P. trochoideum* (19), *Woloszynskia limnetica* (=*Peridinium* sp.) (91). | R | 0 | 0 |
| *Crypthecodinium* (*Gyrodinium*) *cohnii* (93). | 0 | 0 | R |
| *Cachonina niei* (69). | R | R | 0 |
| *Prorocentrum micans* (18, 57). | R | 0 | R |
| *Amphidinium carteri* (*Klebsi*) (71), *A. rhynchocephalum*? (71), *Gymnodinium breve* (2, 120), *Oxyrrhis marina* (19, 20), *Polykrikos schwartzii* (52). | R | R | R |
| **PHAEOPHYCEAE** | | | |
| *Desmarestia ligulata* (76), *D. viridis* (76), *Petalonia zoostericola* (110), *Scytosiphon lomentaria* (110), *Waerniella lucifuga* (18, 22). | 0 | 0 | 0 |
| *Ectocarpus confervoides* (9), *E. fasciculatus* (80), *E. parasitica* (18), *Litosiphon pusillus* (80). | 0 to stimulatory | 0 | 0 |
| *Pylaiella* (*Pilayella*) *littoralis* (80). | R | 0 | 0 |
| **XANTHOPHYCEAE AND EUSTIGMATOPHYCEAE** | | | |
| *Botrydiopsis intercedens* (8), *Bumilleriopsis brevis* (8), *Chlorellidium tetrabotrys* (8), *Monodus subterraneus* (74, 75), *Polyedriella helvetica* (8), *Tribonema aequale* (7), *T. minus* (8). | 0 | 0 | 0 |

EUGLENOPHYCEAE

Euglena anabaena minor (1).
Trachelomonas abrupta (97), T. pertii (91), Euglena spirogyra (122)
Euglena pisciformis (24).
Astasia klebsii (= A. longa, = A. chattoni) (1, 91), Euglena gracilis bacillaris (1), E. gracilis typica (1, 91), E. gracilis urophora (1, 91), E. klebsii (1), E. mutabilis (68), E. stellata (91), E. viridis (91), Peranema trichophorum (106),
Phacus pyrum (91).
Eutreptiella sp. (53).

CHLOROPHYCEAE, CHAROPHYCEAE AND PRASINOPHYCEAE

Astrephomene gubernaculifera (104), Bracteacoccus cinnabarinus (79), B. engadiensis (79), B. minor (79), B. terrestris (79), Chara aspera (27), C. globularis (27), C. zeilanica (27), Chlamydomonas agloeformis (1, 91). C. eugametos (117, 118), C. moewusii (1, 47, 91), C. pichinchae (44), C. reinhardtii (68, 91), Chlorella autotrophica (102), C. candida (102), C. ellipsoidea (102), C. emersonii (102), C. emersonii globosa (102), C. fusca (102), C. fusca vacuolata (102), C. infusionum (102), C. infusionum auxenophila (102), C. miniata (102), C. mutabilis (102), C. nocturna (102), C. photophila (102), C. pringsheimii (102), C. pyrenoidosa (Emerson) (100), C. regularis (102), C. regularis aprica (102), C. regularis imbricata (102), C. saccharophila (102), C. simplex (102), C. sorokiniana (102), C. vannielii (102), C. variabilis (102), C. vulgaris (91, 102), C. vulgaris luteoviridis (102), Chlorococcum aplanosporum (79), C. diplobionticum (79), C. echinozygotum (79), C. ellipsoideum (79), C. hypnosporum (79), C. macrostigmatum (79), C. minutum (79), C. multinucleatum (79), C. oleofaciens (79), C. perforatum (79), C. pinguideum (79), C. punctatum (79), C. scabellum (79), C. tetrasporum (79), C. vacuolatum (79), C. wimmeri (79), Chlorosarcinopsis auxotrophica (35), C. eremi (35), Cosmarium botrytis (109), C. impressulum (109), C. laeve (109), C. lundelli (109), C. meneghini (109), Cyanidium caldarium[c] (3, 4), Cylindrocystis brebissonii (44), Dictyochloris fragrans (79), Dunaliella primolecta (91), D. salina (33, 91), D. viridis (33), Eudorina elegans (25), Haematococcus pluvialis (91, 16), Lobomonas pyriformis (77), Micrasterias cruxmelitensis (109), Nannochloris atomus (91), N. oculata (19), Nautococcus pyriformis (79), Neochloris alveolaris (79), N. aquatica (79), N. gelatinosa (79), N. minuta (79), N. pseudoalveolaris (79), Pandorina morum (78), P. unicocca (98), Polytoma obtusum (1, 91), P. uvella (1, 91), Radiosphaera dissecta (79), Scenedesmus obliquus (91), Selenastrum minutum (66), Spongiochloris excentrica (79), S. lamellata (79), S. spongiosis (79), Spongiococcum alabamense (79), S. excentricum (79), S. multinucleatum (79), S. tetrasporum (79), Staurastrum gladiosum (109), S. paradoxum (109), S. sebaldii ornatum (109), Staurodesmus pachyrhynchus (109), Stephanoptera gracilis (33), Stichococcus cylindricus (91), Stigeoclonium aestivale (14), S. helveticum (14), S. subsecundum (14), S. tenue (14), S. pascheri (14), S. variabile (14), S. farctum (14).

Vitamin requirements—*continued*

| Species | $B_{12}$ | Thiamin | Biotin |
|---|---|---|---|
| *Balticola buetschlii* (21), *B. droebakensis* (21), *Chlamydomonas chlamydogama* (47), *C. mundana* (26), *C. mundana astigmata* (26), *C. pallens* (88), *C. pulsatilla* (21), *Chlorosarcinopsis sempervirens* (35), *Chlorosphaera consociata* (68), *Closterium braunii* Z 951/2 a (109), *C. macilentum* (109), *C. peracerosum* (109), *C. pusillum* (109), *C. siliqua* (109), *C. strigosum* (109), *C. turgidum* (109), *Cylindrocystis brebissonii* (109), *Cosmarium turpini* (58), *Desmidium swartzii* (109), *Docidium manubrium* (109), *Gonium pectorale* (104), *Haematococcus droebakensis* (16), *H. buetschlii* (16), *H. pluvialis* (89), *Hyalotheca dissiliens* (109), *Lobomonas rostrata* (66), *L. sphaerica* (123), *Nannochloris* sp. (13, 113), *Platydorina caudata* (41), *Platymonas tetrathele* (21), *Pleurotaenium trabecula* (109), *Stichococcus* sp. (*cylindricus?*) (67), *Volvox aureus* (90), *V. globator* (83), *V. tertius* (83, 90), *Volvulina steinii* (11), *V. pringsheimii* (11), *Xanthidium cristatum* (109). | R | 0 | 0 |
| *Chlorella protothecoides* (102), *C. protothecoides communis* (102), *C. protothecoides galactophila* (102), *C. protothecoides mannophila* (102), *Coelastrum morus* (68, 66), *Gloeocystis gigas* (68), *Gonium multicoccum* (99), *Polytoma caudatum* (70), *P. ocellatum* (70), *Polytomella caeca* (70), *Prototheca zopfii* (1, 91), *Raphidonema nivale* (44), *Selenastrum minutum* (68). | 0 | R | 0 |
| *Acetabularia mediterranea* (101), *Astrephomene gubernaculifera* (10), *Brachiomonas submarina* (21), *Eremosphaera viridis* (103), *Haematococcus capensis* typ. (89), *H. capensis borealis* (89), *H. zimbabwiensis* (89), *Pyramimonas inconstans* (91), *Stephanosphaera pluvialis* (21), *Volvox globator* (90). | R | R | 0 |

\* R = required, 0 = not required; *a* see p. 916, note 21; *b* see p. 916, note 24; *c*, but see p. 916, note 20.

# 9 APPENDIX REFERENCES

1 ALBRITTON E.C. (1954) *Standard values in nutrition and metabolism.* W.B. Saunders, Philadelphia.

2 ALDRICH D.V. (1962) Photoautotrophy in *Gymnodinium breve* Davis. *Science, N.Y.* **137**, 988–90.

3 ALLEN M.B. (1952) The cultivation of *Myxophyceae. Arch. Mikrobiol.* **17**, 34–53.

4 ALLEN M.B. (1959) Studies with *Cyanidium caldarium*, an anomalously pigmented chlorophyte. *Arch. Mikrobiol.* **32**, 270–7.

5 ANTIA N.J., CHENG J.Y. & TAYLOR F.J.R. (1969) The heterotrophic growth of a marine photosynthetic cryptomonad (*Chroomonas salina*). *Proc. VI Int. Seaweed Symp.* **6**, 17–29. Direccion General Pesca Maritima, Madrid.

6 BAKER A.F. & BOLD H.C. (1970) Phycological studies. X. Taxonomic studies on the *Oscillatoriaceae.* Univ. Texas Publ. No. 7004., 105 pp.

7 BELCHER J.H. & FOGG G.E. (1958) Studies on the growth of Xanthophyceae in pure culture. *Tribonema aequale. Arch. Mikrobiol.* **30**, 17–22.

8 BELCHER J.H. & MILLER J.D.A. (1960) Studies on the growth of Xanthophyceae in pure culture. IV. Nutritional types amongst the Xanthophyceae. *Arch. Mikrobiol.* **36**, 219–28.

9 BOALCH G.T. (1961) Studies on *Ectocarpus* in culture. II. Growth and nutrition of a bacteria-free culture. *J. mar. biol. Ass., U.K.* **41**, 287–304.

10 BROOKS A.E. (1965) Physiology and genetics of *Astrephomene gubernaculifera.* Ph.D. Thesis, Indiana University, Bloomington, Indiana.

11 CAREFOOT J.R. (1967) Nutrition of *Volvulina. J. Protozool.* **14**, 15–18.

12 CAREFOOT J.R. (1968) Culture and heterotrophy of the freshwater dinoflagellate *Peridinium cinctum* fa *ovoplanum. J. Phycol.* **4**, 129–31.

13 CARLUCCI A. unpubl.

14 COX E.R. & BOLD H.C. (1966) Phycological studies. VII. Taxonomic investigations of *Stigeoclonium.* Univ. Texas Publ. No. 6618, 167 pp.

15 DE P.K. (1939) The role of blue-green algae in nitrogen fixation in rice-fields. *Proc. R. Soc. Lond., B.* **127**, 121–39.

16 DROOP M.R. (1954) Studies on the phytoflagellate *Haematococcus pluvialis.* Thesis, Cambridge Univ., England.

17 DROOP M.R. (1955) A pelagic marine diatom requiring cobalamin. *J. mar. biol. Assoc., U.K.* **34**, 229–31.

18 DROOP M.R. (1957) Auxotrophy and organic compounds in the nutrition of marine phytoplankton. *J. gen. Microbiol.* **16**, 286–93.

19 DROOP M.R. (1958) Requirement for thiamine among some marine and supralittoral protista. *J. mar. biol. Ass., U.K.* **37**, 323–9.

20 DROOP M.R. (1959) Water-soluble factors in the nutrition of *Oxyrrhis marina. J. mar. biol. Ass., U.K.* **38**, 605–20.

21 DROOP M.R. (1961) *Haematococcus pluvialis* and its allies. III. Organic Nutrition. *Rev. Algol.* **5**, 247–59.

22 DROOP M.R. (1962) Organic nutrients. In *Physiology and Biochemistry of Algae*, ed. Lewin R.A. pp. 141–59. Academic Press, New York & London.

23 DROOP M.R., McLAUGHLIN J.J.A., PINTERN I.J. & PROVASOLI L. (1959) Specificity of some protophytes toward vitamin $B_{12}$-like compounds. *Preprints Intern. Oceanog. Congr.* 916–18. *A.A.A.S.*

24 DUSI H. (1939) La pyrimidine et le thiazol facteurs de croissance pour le flagellé à chlorophylle, *Euglena pisciformis. C. r. Séanc. Soc. Biol.* **130**, 419–21.

25  Dusi H. (1940) Culture bacteriologiquement pure et nutrition autotrophe d'*Eudorina elegans*. Rôle du fer pour la formation des colonies. *Ann. Inst. Pasteur*, **64**, 340–3.

26  Eppley R.W. & Maciasr F.M. (1962) Rapid growth of sewage lagoon *Chlamydomonas* with acetate. *Physiologia Pl.* **15**, 72–9.

27  Forsberg C. (1965) Nutritional studies of *Chara* in axenic culture. *Physiologia Pl.* **18**, 275–90.

28  Fries L. (1959) A marine red alga requiring vitamin $B_{12}$. *Nature, Lond.* **183**, 558–9.

29  Fries L. (1961) Vitamin requirements of *Nemalion multifidum*. *Experientia*, **17**, 75–6.

30  Fries L. (1964) *Polysiphonia urceolata* in axenic culture. *Nature, Lond.* **202**, 110.

31  Fries L. & Pettersson H. (1968) On the physiology of the red alga *Asterocytis ramosa* in axenic culture. *Br. Phycol. Bull.* **3**, 417–22.

32  Gerloff G.C., Fitzgerald G.P. & Skoog F. (1950) The isolation, purification, and culture of blue-green algae. *Am. J. Bot.* **37**, 216–18.

33  Gibor A. (1956) The culture of brine algae. *Biol. Bull. mar. biol. Lab., Woods Hole*, **111**, 223–9.

34  Gold K. (1964) Aspects of marine dinoflagellate nutrition measured by $C^{14}$ assimilation. *J. Protozool.* **11**, 85–9.

35  Groover R.D. & Bold H.C. (1969) Phycological studies. VIII. The taxonomy and comparative physiology of the Chlorosarcinales and certain other edaphic algae. *Univ. Texas Publ.* No. 6907, 165 pp.

36  Guillard R.R.L. (1963) Organic sources of nitrogen for marine centric diatoms. *Symp. Mar. Microbiol.*, ed. Oppenheimer C.H. pp. 93–104. Thomas Co., Springfield, Ill.

37  Guillard R.R.L. (1968) $B_{12}$ specificity of marine centric diatoms. *J. Phycol.* **4**, 59–64.

38  Guillard R.R.L. & Cassie V. (1963) Minimum cyanocobalamin requirements of some marine diatoms. *Limnol. Oceanog.* **8**, 161–5.

39  Guillard R.R.L. & Ryther J.H. (1962) Studies of marine planktonic diatoms. I. *Cyclotella nana* Hustedt, and *Detonula confervacea* (Cleve). *Can. J. Microbiol.* **8**, 229–39.

40  Hargraves P.E. & Guillard R.R.L. Personal communication.

41  Harris D.O. (1969) Nutrition of *Platydorina caudata*. *J. Phycol.* **5**, 205–10.

42  Haxo F.T. & Sweeney B.M. (1955) Bioluminescence in *Gonyaulax polyedra*. In *The Luminescence of Biological Systems*, ed. Johnson F.H. pp. 415–20. A.A.A.S.

43  Heinrich H.C. (1955) Der B–Vitamin-Bedarf der Chrysophyceen, *Ochromonas danica* nom. provis. Pringsheim und *Ochromonas malhamensis* Pringsheim. *Naturwissenschaften*, **14**, 418.

44  Hoham R. (1971) Laboratory and field studies on snow algae of the Pacific Northwest. Ph.D. Thesis, Univ. of Washington, Seattle.

45  Holz G.G. (1954) The oxidative metabolism of a cryptomonad flagellate, *Chilomonas paramecium*. *J. Protozool.* **1**, 114–20.

46  Hutner S.H. (1948) Essentiality of constituents of sea water for growth of a marine diatom. *Trans. N.Y. Acad. Sci.* **10**, 136–41.

47  Hutner S.H. & Provasoli L. (1951) The phytoflagellates. In *Biochemistry and Physiology of Protozoa*, ed. Lwoff A. **1**, 27–128. Academic Press, New York & London.

48  Hutner S.H. & Provasoli L. (1953) A pigmented marine diatom requiring vitamin $B_{12}$ and uracil. *News Bull. Phycol. Soc. Amer.* **6**, 7–8.

49  Hutner S.H., Provasoli L. & Filfus J. (1953) Nutrition of some phagotrophic fresh-water chrysomonads. *Ann. N.Y. Acad. Sci.* **56**, 852–62.

50  Iwasaki H. (1965) Nutritional studies of the edible seaweed *Porphyra tenera* I. The influence of different $B_{12}$ analogues, plant hormones, purines and pyrimidines on the growth of *Conchocelis*. *Pl. Cell Physiol., Tokyo*, **6**, 325–36.

51  Iwasaki H. (1969) Studies on the red tide dinoflagellates. III. On *Peridinium hangoei* Schiller appeared in Gokasho Bay, Shima Peninsula. *Bull. Plankton Soc., Japan*, **16**, 132–9.

52 IWASAKI H. (1971) Studies on the red tide dinoflagellates. V. On *Polykrikos schwartzi* Butschli. *Bull. Jap. Soc. Sci. Fish.* **37**, 606–9.

53 IWASAKI H. (1971) Studies on the red tide dinoflagellates. VI. *Eutreptiella* sp. and *Exuviaella* sp. *J. Ocean. Soc., Japan*, **27**, 152–7.

54 IWASAKI H., FUJIYAMA T. & YAMASHITA E. (1968) Studies on the red tide dinoflagellates. I. On *Entomosigma* sp. appeared in coastal area of Fukuyama. *J. Fac. Fish. & Anim. Husb. Hiroshima Univ.* **7**, 259–67.

55 IWASAKI H., OKAKA Y. & TANAKE S. (1969) Studies on the red tide dinoflagellates. IV. On *Rhodomonas ovalis* Nygaard appeared in coastal area of Fukuyama. *Bull. Plankton Soc., Japan*, **16**, 140–4.

56 IWASAKI H. & SASADA K. (1969) Studies on the red tide dinoflagellates. II. On *Heterosigma inlandica* appeared in Gokasho Bay, Shima Peninsula. *Bull. Jap. Soc. Sci. Fish.* **35**, 943–7.

57 KAIN J.M. & FOGG G.E. (1960) Studies on the growth of marine phytoplankton. III. *Prorocentrum micans* Ehrenberg. *J. mar. biol. Ass., U.K.* **39**, 33–50.

58 KORN R.W. (1969) Nutrition of *Cosmarium turpini*. *Physiologia Pl.* **22**, 1158–1165.

59 KRATZ W.A. & MYERS J. (1955) Nutrition and growth of several blue-green algae. *Am. J. Bot.* **42**, 282–7.

60 LEWIN J.C. (1953) Heterotrophy in diatoms. *J. gen. Microbiol.* **9**, 305–13.

61 LEWIN J.C. (1963) Heterotrophy in marine diatoms. In *Symp. Marine Microbiol.*, ed. Oppenheimer C.H. pp. 229–35. Thomas Co., Springfield, Ill.

62 LEWIN J.C. (1965) The thiamine requirement of a marine diatom. *Phycologia*, **4**, 142–4.

63 LEWIN J.C. & LEWIN R.A. (1959) Auxotrophy and heterotrophy in marine littoral diatoms. In *Preprints Int. Oceanog. Congress*, A.A.A.S. pp. 928–9.

64 LEWIN J.C. & LEWIN R.A. (1960) Auxotrophy and heterotrophy in marine littoral diatoms. *Can. J. Microbiol.* **6**, 127–34.

65 LEWIN J.C. & LEWIN R.A. (1967) Culture and nutrition of some apochlorotic diatoms of the genus *Nitzschia*. *J. gen. Microbiol.* **46**, 361–7.

66 LEWIN R.A. (1952) Vitamin requirements in the Chlorococcales. *Phycol. News Bull.* **5**, 21–2.

67 LEWIN R.A. (1954) A marine *Stichococcus* sp. which requires vitamin $B_{12}$. *J. gen. Microbiol.* **10**, 93–6.

68 LEWIN R.A. (1958) Vitamin-bezonoj de algoj, In *Sciencaj Studoj*, ed. Neergaard, P., pp. 187–92. Modersmaalet, Haderslev, Copenhagen.

69 LOEBLICH A.R. III (1969) Nutrition of the marine dinoflagellate *Cachonina niei*. *J. Protozool.* **16** (*Suppl.*), p. 20.

70 LWOFF A. & DUSI H. (1938) Culture de divers flagelles leucophytes en milieu synthetique. *C. r. Seanc. Soc. Biol.* **127**, 53–5.

71 MCLAUGHLIN J.J.A. & PROVASOLI L. (1957) Nutritional requirements and toxicity of two marine amphidinium. *J. Protozool.* **4** (*Suppl.*), p. 7.

72 MCLAUGHLIN J.J.A. & ZAHL P.A. (1959) Vitamin requirements in symbiotic algae. In *Preprints Int. Oceanog. Congress*, A.A.A.S. pp. 930–1.

73 MCLAUGHLIN J.J.A. & ZAHL P.A. (1959) Axenic zooxanthellae from various invertebrate hosts. *Ann. N.Y. Acad. Sci.* **77**, 55–72.

74 MILLER J.D.A. & FOGG G.E. (1957) Studies on the growth of Xanthophyceae in pure culture. I. The mineral nutrition of *Monodus subterraneus* Petersen. *Arch. Mikrobiol.* **28**, 1–17.

75 MILLER J.D.A. & FOGG G.E. (1958) Studies on the growth of Xanthophyceae in pure culture. II. The relations of *Monodus subterraneus* to organic substances. *Arch, Mikrobiol.* **30**, 1–16.

76 NAKAHARA H. (1972) Alternation of generations of some brown algae in unialgal and axenic cultures. Ph.D. Thesis, Hokkaido University, Sapporo, Japan.

77   OSTERUD K.L. (1946) Trophic potentialities of the green flagellate *Lobomonas piriformis* Pringsheim. *Physiol. Zool.* **19**, 19–34.

78   PALMER E.G. & STARR R.C. (1971) Nutrition of *Pandorina morum. J. Phycol.* **7**, 85–9.

79   PARKER B.C., BOLD H.C. & DEASON T.R. (1961) Facultative heterotrophy in some chlorococcacean algae. *Science, N.Y.* **133**, 761–3.

80   PEDERSÉN M. (1969) Marine brown algae requiring vitamin $B_{12}$. *Physiologia Pl.* **22**, 977–83.

81   PINCEMIN J.M. (1971) Action de facteurs physiques, chimiques et biotiques sur quelques dinoflagellés et diatomeés en culture. Ph.D. Thesis, Université D'Aix-Marseille.

82   PINTNER I.J. & PROVASOLI L. (1958) Artificial cultivation of a red-pigmented marine blue-green alga *Phormidium persicinum. J. gen. Microbiol.* **18**, 190–7.

83   PINTNER I.J. & PROVASOLI L. (1959) The nutrition of *Volvox globator* and *V. tertius. Proc. IX Int. Bot. Congr. Montreal, Abstr.* pp. 300–1.

84   PINTNER I.J. & PROVASOLI L. (1963) Nutritional characteristics of some chrysomonads. In *Symp. Marine Microbiol.*, ed. Oppenheimer C.H. pp. 114–21. Thomas Co., Springfield, Ill.

85   PINTNER I.J. & PROVASOLI L. (1968) Heterotrophy in subdued light of 3 *Chrysochromulina* species. *Bull. Misaki Mar. Biol. Inst. Kyoto Univ.* **12**, 25–31.

86   PRINGSHEIM E.G. (1955) Kleine Mitteilungen ueber Flagellaten und Algen. III. Uber *Ochromonas danica* n. sp. und andere Arten der Gattung. *Arch. Microbiol.* **23**, 181–92.

87   PRINGSHEIM E.G. (1958) Uber Mixotrophie bei Flagellaten. *Planta*, **52**, 405–30.

88   PRINGSHEIM E.G. (1962) *Chlamydomonas pallens*, a new organism proposed for assays of vitamin $B_{12}$. *Nature, Lond.* **195**, 604.

89   PRINGSHEIM E.G. (1966) Nutritional requirements of *Haematococcus pluvialis* and related species. *J. Phycol.* **2**, 1–7.

90   PRINGSHEIM E.G. (1970) Identification and cultivation of European *Volvox* sp. *Antonie van Leeuwenhoek*, **36**, 33–43.

91   PROVASOLI L. (1958) Nutrition and ecology of protozoa and algae. *A. Rev. Microbiol.* **12**, 279–308.

92   PROVASOLI L. (1964) Growing marine seaweeds. In *Proc. IV Int. Seaweed Symp.* ed. De Virville A.D. & Feldman J. pp. 9–17. Pergamon Press, Oxford, London, New York & Paris.

93   PROVASOLI L. & GOLD K. (1962) Nutrition of the American strain of *Gyrodinium cohni. Arch. Mikrobiol.* **42**, 196–203.

94   PROVASOLI L. & MCLAUGHLIN J.J.A. (1955) Auxotrophy in some marine and brackish dinoflagellates. *J. Protozool.* **2** (*Suppl.*), 10.

95   PROVASOLI L. & MCLAUGHLIN J.J.A. (1963) Limited heterotrophy of some photosynthetic dinoflagellates. In *Symp. Marine Microbiol.* ed. Oppenheimer C.H. pp. 105–13. Thomas Co. Springfield, Ill.

96   PROVASOLI L. & PINTNER I.J. (1953) Ecological implications of *in vitro* nutritional requirements of algal flagellates. *Ann. N.Y. Acad. Sci.* **56**, 839–51.

97   PROVASOLI L. & PINTNER I.J. (1955) Culture of *Trachelomonas* in chemically defined media. *Phycol. News Bull.* **8**, 7.

98   RAYBURN W.R. (1971) Morphology, nutrition and physiology of sexual reproduction of *Pandorina unicocca* sp. nov. Ph.D. Thesis, Indiana Univ., Bloomington, Ind.

99   SAITO S. (1972) Growth of *Gonium multicoccum* in synthetic media. *J. Phycol.* **8**, 169–75.

100  SAMEJIMA H. & MYERS J. (1958) On the heterotrophic growth of *Chlorella pyrenoidosa. J. gen. Microbiol.* **18**, 107–15.

101  SHEPARD D.C. (1969) Axenic culture of *Acetabularia* in synthetic medium. In *Methods in Cell Physiology IV*, pp. 46–69. Academic Press, New York & London.

102 SHIHIRA I. & KRAUSS R.W. (1965) *Chlorella*: physiology and taxonomy of 41 isolates. 97 pp. Port City Press, Baltimore, Md.

103 SMITH R.L. & BOLD H.C. (1966) Phycological studies VI. Investigations on the algal genera *Eremosphaera* and *Oocystis*. 121 pp. Univ. Texas Publ. no. 6612.

104 STEIN J.R. (1966) Growth and mating of *Gonium pectorale* (Volvocales) in defined media. *J. Phycol.* **2**, 27–8.

105 STEWART W.D.P. (1962) Fixation of elemental nitrogen by marine blue-green algae. *Ann. Botany, London, N.S.* **26**, 439–45.

106 STORM J. & HUTNER S.H. (1953) Nutrition of *Peranema*. *Ann. N.Y. Acad. Sci.* **56** 901–9.

107 SWEENEY B.M. (1954) *Gymnodinium splendens*, a marine dinoflagellate requiring vitamin $B_{12}$. *Am. J. Bot.* **41**, 821–4.

108 SWIFT M.J. & MCLAUGHLIN J.J.A. (1970) Some nutritional, physiological, and morphological studies on *Melosira juergensii* and *Heterocapsa kollmeriana*. *Ann. N.Y. Acad. Sci.* **175**, 577–600.

109 TASSIGNY M. (1971) Observations sur les bésoins en vitamines des Desmidiées (Chlorophycées-Zygnematales). *J. Phycol.* **7**, 213–15.

110 TATEWAKI M. Personal communication.

111 TATEWAKI M. & PROVASOLI L. (1964) Vitamin requirements of three species of *Antithamnion*. *Botanica mar.* **6**, 193–203.

112 THOMAS W.H. (1955) Heterotrophic nutrition and respiration of *Gonyaulax polyedra*. *J. Protozool.* **2** (*Suppl.*), 2–3.

113 THOMAS W.H. (1966) Effects of temperature and illuminance on cell division rates of three species of tropical oceanic phytoplankton. *J. Phycol.* **2**, 17–22, and personal communication.

114 TURNER M.F. (1970) A note on the nutrition of *Rhodella*. *Br. Phycol. J.* **5**, 15–18.

115 VAN BAALEN C. (1962) Studies on marine blue-green algae. *Botanica mar.* **4**, 129–39.

116 WATANABE A. (1951) Production in cultural solution of some amino acids by the atmospheric nitrogen-fixing blue-green algae. *Archs. Biochem. Biophys.* **34**, 50–5.

117 WETHERELL D.F. (1958) Obligate phototrophy in *Chlamydomonas eugametos*. *Physiologia Pl.* **11**, 260–74.

118 WETHERELL D.F. & KRAUSS R.W. (1957) X-Ray induced mutations in *Chlamydomonas eugametos*. *Am. J. Bot.* **44**, 609–19.

119 WILLIAMS A.E. & BURRIS R.H. (1952) Nitrogen fixation by blue-green algae and their nitrogenous composition. *Am. J. Bot.* **39**, 340–2.

120 WILSON W.B. (1966) The suitability of seawater for the survival and growth of *Gymnodinium breve* and some effects of phosphorus and nitrogen on its growth. Florida Board Conservation, Professional Papers, Series **7**, 42 pp.

121 STAUB R. (1961) Ernahrungsphysiologisch-autokologische untersuckungen an der planktischen Blaualge *Oscillatoria rubescens* DC. *Schweizerische Zeitschrift fur Hydrologie*, **23**, 82–198a.

122 LEEDALE G.F., MEEUSE B.J.D. & PRINGSHEIM E.G. (1965) Structure and physiology of *Euglena spirogyra*. III–VI. *Arch. Microbiol.* **50**, 133–55.

123 PRINGSHEIM E.G. (1964) Chlorophyllarme Algen II. *Lobomonas sphaerica* nov. spec. *Arch. Mikrobiol.* **46**, 227–37.

# CHAPTER 28

# MORPHOGENESIS

## BETTY MOSS

Department of Botany,
University of Newcastle upon Tyne,
Newcastle upon Tyne, U.K.

| 1 | Introduction 788 | 7 | Laminariales 794 |
|---|---|---|---|
| 2 | Cyanophyceae 789 | 8 | Fucales 799 |
| 3 | Flagellated unicells and colonies 790 | 9 | Other brown algae 801 |
| | | 10 | Rhodophyceae 803 |
| 4 | Non-flagellated unicells and colonies 791 | 11 | References 807 |
| 5 | Green siphonaceous algae 792 | | |
| 6 | Filamentous and parenchymatous Chlorophyceae 793 | | |

## 1 INTRODUCTION

Between the prokaryotic cells of blue-green algae and the eukaryotic giant plants of *Macrocystis* is a diversity of morphological structure and development unsurpassed in any other plant group. Early studies were concerned with descriptive morphology but the development of culture techniques brought with it the possibility of experimental work, first on unicellular forms, and latterly on more complex forms. As a result, an extensive literature on the effects of nutrients and physiological factors upon algal growth and morphogenesis has developed and this is covered in Chapters 22, p. 610; 27, p. 741; and 29, p. 814. The development of biochemical techniques, frequently associated with electron microscopy, has provided information on subjects such as silicification and calcification (Chapter 24, p. 655), walls and mucilages (Chapter 2, p. 40), chloroplasts (Chapter 4, p. 124) and cytoplasmic organelles (Chapter 3, p. 86). This chapter attempts a more general review of recent studies on growth and morphogenesis in algae. As many of the studies reported in the other chapters deal particularly with unicellular, simple filamentous, and colonial algae, these groups will be touched on only briefly here and particular attention will be given to the larger Phaeophyceae and Rhodophyceae.

## 2 CYANOPHYCEAE

The Cyanophyceae are unique amongst the algae in having a prokaryotic cellular organization. Recent studies on the cell walls and the external sheath, the plasmalemma, the photosynthetic thylakoids, the nucleoplasm and cellular inclusions are reviewed by Lang (1968) and Fogg et al. (1973). Unusual cellular inclusions are considered by Jensen and Bowen (1970). Detailed changes in ultrastructure associated with heterocyst development are discussed by Fay and Lang (1971) and Lang and Fay (1971).

Morphological plasticity induced by environmental factors has long been recognized among the blue-green algae. Drouet (1962), studying natural populations of Microcoleus vaginatus, found that cell size and shape as well as characters of the external sheath varied enormously. This one organism grew into forms which had been allocated to four other genera including nine different species. What were formerly thought of as distinct species Drouet now considers as ecophenes or ecological variants. Other studies such as Drouet and Daily (1956) for the unicellular Cyanophyceae, Fan (1956) for the genus Calothrix, Drouet (1968) for the Oscillatoriaceae and Forest and Kahn (1970) for various algae confirm the plasticity of morphological form in the blue-green algae, although one wonders whether Drouet and Daily (1956) and Drouet (1968) have been too sweeping in their reduction of numbers of species (see also Chapter 1, p. 30). More recently Stanier et al. (1971) have used the plane or planes of successive cell divisions and various biochemical characteristics such as GC ratios to separate the unicellular blue-green algae into eight typological groups. There is also taxonomic controversy due to the fact that unicells may produce filamentous mutants (Padan & Shilo 1969, Ingram & Van Baalen 1970, Kunisawa & Cohen-Bazire 1970).

In some instances, laboratory cultures have been combined with field studies to demonstrate the effect of known environmental factors upon morphology. Zehnder and Gorham (1960) using Microcystis and McLachan et al. (1963) using Aphanizomenon flos-aquae have shown that colony shape and size appeared to be due partly to heredity and partly to nutrition. When the latter species was kept in culture for prolonged periods with soil extract medium, it became partially or completely non-colonial. Species of Calothrix and Lyngbya exhibited great variability when cultured in different media (Pearson & Kingsbury 1966), especially in relation to twisting of the filaments, the presence or absence of constrictions at the cross walls, the abundance of heterocysts, pseudovacuoles

2C

and fragmentation of filaments. Rope-like twisting of the filaments was seen in liquid culture, but not when they were grown on agar. Changes in colour are often used as a taxonomic criterion in the blue-green algae and yet it is well known, for example, that under nitrogen-rich conditions the cells of blue-green algae are usually deep green in colour, whereas they are often orange under nitrogen-starved conditions. Fay *et al.* (1964) also noted changes in colour during the life cycle of *Chlorogloea fritschii* kept in quasi-synchronous culture. The small pale blue-green cells of the young filament developing from an endospore became more deeply blue-green and then yellowish-green as they matured before endospore formation was reached again. Heterocysts, the position and abundance of which are used sometimes as taxonomic characters are now known to vary in abundance depending on the level of ammonium-nitrogen in the environment and both Stein (1963) using *Tolypothrix tenuis* and Pearson and Kingsbury (1966) conclude that the heterocyst alone is not a good taxonomic criterion on which to separate species.

A photo-controlled developmental life cycle has been reported for *Nostoc muscorum* (Lazaroff & Vishniac 1961, 1964) and for *Nostoc commune* (Robinson & Miller 1970). In both species red light was found to promote the formation of motile trichomes while green light reversed the photoactivation. Robinson and Miller (1970) suggest the possibility of a phytochrome type of control associated with morphogenesis. The effects of IAA on growth and development of *Nostoc muscorum* have been determined by Ahmad and Winter (1970) and Ahmad (1971). Concentrations of IAA of $10^{-9}$ to $15^{-5}$M increased the growth rate e.g. after five days in $10^{-8}$M IAA the growth rate was increased by 77% (on a dry weight basis). Even after 24 days in all concentrations of IAA the increased growth rate continued.

## 3 FLAGELLATED UNICELLS AND COLONIES

Some groups of flagellated algae have received more attention than others. *Euglena* species, because of the relative ease with which they can be grown and handled in the laboratory, coupled with their 'plant-like' and 'animal-like' characteristics, are favourites for biological study. Details of the growth and morphology of *Euglena* have been brought together admirably by Leedale (1967) and Buetow (1968). At the other extreme, members of the Cryptophyceae have received little attention (Lucas 1970, Hibberd *et al.* 1971). In the Dinophyceae, ultrastructural details of the theca have advanced to the stage where Dodge and Crawford (1970) suggest that they could form the basis for some revision of the taxonomy of the class.

The Haptophyceae are distinguished from other flagellates by their ability to form surface scales or coccoliths. Details of their fine structure are considered by Black (1961) and the general biology of the group is reviewed by Paasche (1968). The motile stage of *Coccolithus pelagicus* shows extracellular mineralization of the organic matrix scale which is formed and subsequently mineralized in the Golgi cisternae before the coccoliths are released to the exterior of the shell (Pienaar 1969, Manton & Leedale 1969, Manton & Peterfi 1969). Since von Stosch (1955, 1958, 1967) first recorded a coccolithophorid with a benthic phase, the development of culture techniques has extended our knowledge of the group and cleared up many taxonomic problems (Parke 1961). Leadbeater (1970), for example, has demonstrated the connexion between the *Apistonema* thallus in the life cycle of von Stosch's original isolate of *Syracosphaera carterae*, and the coccolith bearing stages which are identical with *Hymenomenas carterae sensu* Manton & Peterfi (1969).

Comparatively few studies of ultrastructure have been made on the Volvocales (Pitelka 1963) although cell morphology has been described for *Chlamydomonas* (Sagar & Palade 1954, 1957), and Hobbs (1971) and Lang (1963) have found that the individual cells in a colony of *Eudorina* and *Volvox* are very similar. A biochemical approach to growth and gametogenesis has been made, e.g. by Kates and Jones (1964) who studied the control of gamete development and enzyme levels in synchronous culture of *Chlamydomonas*; by Jones *et al.* (1968) who investigated protein turnover during growth and differentiation, and by Jones (1971) who investigated periodic enzyme activity and synthesis during gamete differentiation. There is marked differentiation between somatic and reproductive cells in *Volvox*. Darden (1966) and Starr (1968, 1970) have studied differentiation in axenic culture and have shown that growth and reproduction of many species can be chemically controlled. The induction of sexual reproductive cells through the use of filtrate from sexual male cultures has also been demonstrated. These aspects are considered in more detail in Chapter 29, p. 814.

# 4 NON-FLAGELLATE UNICELLS AND COLONIES

Diatoms possess the unique capacity of forming morphologically complex shells composed of hydrated amorphous silica. A vast literature exists on purely descriptive morphology but we have little information on the mechanism whereby these intricately patterned shells are built. Lewin (1955, 1965) and Lewin and Chew (1968) have investigated silicon metabolism in culture experiments and ultrastructural studies have shown that a three-layered membrane encloses all siliceous material (Lewin & Guillard 1963, Busby & Lewin 1967). This membrane, they suggest, controls the transport of silicic acid into the membrane vesicle, followed by its subsequent polymerization as silica. These aspects are considered in detail in Chapter 24, p. 655.

The taxonomy of diatoms is based largely on the morphology of their shells. Careful culturing of clones, however, has shown that polymorphism can occur. For example, Holmes and Reimann (1966) showed that three morphological phases could be recognized during the life cycle of *Coscinodiscus concinnus*.

Shells of the post auxospore stage were typical of *C. concinnus*, shells of the following stage were typical of *C. granii*, while shells of the pre-auxospore stage were quite distinct from both of these. In a wild population of *Mastogloia*, Stoermer (1967) found that shells of two strikingly different morphological forms occurred. Total conversion of one form to the other took place at a single division, so that no intergradations between the two forms were found.

Polymorphism has long been noted among the desmids (Trainor *et al.* 1971). In field collections, many species vary both in shape and in ornamentation of their cell walls. These variations can take the form of cells intermediate in morphology between different varieties or even species. Experimental work on a clone of *Staurastrum polymorphum* Bréb. raised in culture (Brandham & Godward 1965) showed that temperature could affect the form of the cell. Low temperatures favoured the growth of the quadriflagellate form, whereas high temperatures favoured the tri-radiate form. The change from one form to the other was in this instance abrupt, occurring at a single cell division.

Colonial forms, such as *Scenedesmus*, have been the subject of several investigations into polymorphism (Trainor *et al.* 1971). In this genus it has been shown that nutrients in the medium may considerably affect morphogenetic development of the colony (Kylin & Das 1967, Trainor 1966, 1969, Trainor & Rowland 1968, Trainor & Roskosky 1967). With the addition of 1·5% yeast extract to a basal medium, cells of *S. dimorphus* were joined end-to-end in *Dactylococcus*-like chains. When cultures were transferred daily to different media containing ammonium ions buffered at 8·5 or above, unicellular populations were obtained. Both Swale (1967) and Fott (1968) have reported the presence of *Chodatella* stages in young cultures of *Scenedesmus*.

## 5  GREEN SIPHONACEOUS ALGAE

Since the classic experiments of Hämmerling (1932, 1934) on *Acetabularia* this alga has become a favourite organism for morphogenetic studies. Details of its growth, morphogenesis, and ultrastructure have been summarized recently by Puiseux-Dao (1970) and recent advances into various aspects of its biology are included in the proceedings of a symposium edited by Brachet and Bonotto (1969). Developing from Hämmerling's original hypothesis that morphogenetic substances were produced by the nucleus and stored in the cytoplasm, there are now many laboratories where investigations into control mechanisms, particularly those directing the synthesis of proteins in the cytoplasm of *Acetabularia*, are in progress (Schweiger 1970, Dillard 1970, Zetsche *et al.* 1970).

The ultrastructure of the growing tips of branches of *Bryopsis* (Burr & West 1970) and of several species of *Caulerpa* (Sabnis 1969, Dawes & Barilott 1969, Mishra 1969, Mishra & Kefford 1969) have indicated the presence of what they term a 'meristemplasm', devoid of chloroplasts but rich in other organelles such as mitochondria. Using time lapse photography, Dawes and Barilott (1969) showed that when *Caulerpa* sp. was kept under a 12 hours light:12 hours dark cycle, growth of the blade and movement of the chloroplasts at the blade tip were rhythmic, with the peak occurring 2 to 4 hours after initiation of the dark phase.

*Bryopsis* cultured in seawater does not produce rhizoids unless 100 mg l$^{-1}$ of IAA are added (Jacobs 1951) and Williams (1952) noted that 5 mg l$^{-1}$ IAA increased the length of apical segments of *Codium decorticatum*. Thimann and Beth (1959) reported a two-fold stimulation in the growth of rhizoids of *Acetabularia mediterranea* after incubation for 25 days with IAA at 10$^{-5}$M or 10$^{-6}$M. Chen and Jacobs (1966, 1968) found that rhizoids of *Caulerpa prolifera* were initiated in culture some 1 to 3 hours after the beginning of a light period, and that none appeared to be initiated in the dark. The pattern and rate of growth of *C. sertularioides* was studied in culture by Mishra and Kefford (1969). They found that at 26°C and a 12:12 hour light:dark cycle the thalli survived in untreated seawater up to two months. They recorded rates of rhizome, rhizoid and upright shoot elongation of 0·4, 0·81 and 0·54 cm day$^{-1}$ respectively. Although *C. sertularioides* is a coenocytic structure, there is no uniformity in the growth rate of its different parts. For continued growth beyond two months, certain additives were required in the medium. A substratum of sand rich in microorganisms, or the addition of 5 × 10$^{-4}$M of IAA or of expressed sap from thalli of *Caulerpa* sp., were equally successful in promoting continued growth and differentiation.

*Codium fragile* has been cultured from fertile gametes by Borden and Stein (1969). In half-strength Provasoli enrichment medium, the cultures were kept at 8 ± 2°C under a 16 hours light:8 hours dark cycle with the illumination at about 1000–1500 lx. Zygote development was variable and, under such culture conditions, they gave rise to branching filaments although utricles, necessary for the development of the adult thallus, were not formed. Similar developmental stages to those in culture were collected from the shore, and these authors suggest, like Délépine (1959), that there may always be a juvenile filamentous stage in the life cycle of this species.

Many of the Caulerpales secrete calcium carbonate but this is dealt with in Chapter 24, p. 662. The study of crystal growth of calcium carbonate in *Halimeda* has been described by Wilbur *et al.* (1969).

## 6  FILAMENTOUS AND PARENCHYMATOUS CHLOROPHYCEAE

These larger green algae are more difficult to culture in the laboratory than are the unicellular forms and this is a reason why they have not been the subject of so many experimental investigations. Studies on cell division in *Spirogyra* by Fowke and Pickett-Heaps (1969) show that the early formation of the cross wall is achieved by the annual ingrowth of a septum and that later the cross wall is completed by a phragmaplast containing Golgi vesicles, longitudinally-aligned micro-tubules and associated electron dense material. They suggest that *Spirogyra* may represent an intermediate stage in the evolution of the phragmaplast seen in higher plants. Such a phragmaplast has been seen in *Chara* (Pickett-Heaps 1967) and *Fritschiella* (McBride 1967). The ultrastructure of the unique method of cell division in *Oedogonium* has been described by Hill and Machlis (1968) and Pickett-Heaps and Fowke (1969).

In *Chara delicatula* light may control oospore germination for they germinate well in the light but not in the dark. Red light also induced germination, but blue light and white light had little effect (Takatori & Imahori 1971). The authors suggest the possibility that a phytochrome system may participate in the germination of these oospores.

Provasoli (1958) studied the effect of certain growth substances on the development of *Ulva*. Certain concentrations of either IAA or kinetin were found to stimulate both filament initiation and rhizoid growth. The optimum concentration for filament initiation was 0·05 mg l$^{-1}$ of IAA whereas 0·3 mg l$^{-1}$ was inhibitory. 0·01 mg l$^{-1}$ of kinetin encouraged rhizoid initiation without affecting the rate of elongation of the rhizoids. Eaton *et al.* (1966) examined both polarity and regeneration of *Enteromorpha* in different media and under different light intensities. From tubular segments cut from the plant, rhizoids were always produced from the basal end. When segments of thallus were immersed for eight hours in solutions of 0·005% and 0·001% of IAA, NAA, kinetin and gibberellic acid respectively, a marked increase in the length of the regenerated rhizoids was noted, with the maximum effect being caused by gibberellin. Details of other effects of exogenous organic substances and of unknown growth factors on *Enteromorpha*, *Monostroma* and *Ulva* are considered in Chapter 27, p. 767.

## 7 LAMINARIALES

The heteromorphic life cycle of the Laminariales provides two distinct phases of growth and morphological development. The fertilized egg gives rise to the new sporophyte generation, generally while it is still attached to the oogonium wall. A short celled filament is produced from which the basal cell grows out into an attaching rhizoid, while the remaining cells divide transversely and longitudinally. Kain (1969) has estimated that in four species which she studied, the time for a complete cell division in these young sporophytes is about two days at a temperature of 10°C and a saturating irradiance of approximately 1,000 lx. Thus a 100-celled sporophyte could result within 9 to 10 days of the first division of the zygote. The growth rate at 15 to 20°C was approximately the same as at 10°C. During further development of *Laminaria hyperborea* there is a straight line relationship between the logarithm of sporophyte length and the logarithm of the number of cells per sporophyte up to the 1,000-celled stage (Kain 1965). The growth rate in the sea of the larger, partly polystromatic, sporophytes was estimated as a 10% increase in length per day.

Laboratory culture experiments show that both light and temperature may affect the growth rate of young sporophytes. Under low light energies (13 μg cals cm$^{-2}$ sec$^{-1}$) sporophytes of *L. hyperborea*, *L. digitata*, *L. saccharina* and *Saccorhiza polyschides* did not develop above 19°C, while at 5°C development proceeded slowly (Fig. 28.1, Kain 1969). *Saccorhiza* grew fastest at 17°C while *L. saccharina* rarely formed sporophytes at this temperature. Sundene (1962b, 1963) gave an upper temperature limit of 10 to 12°C for sporophyte development

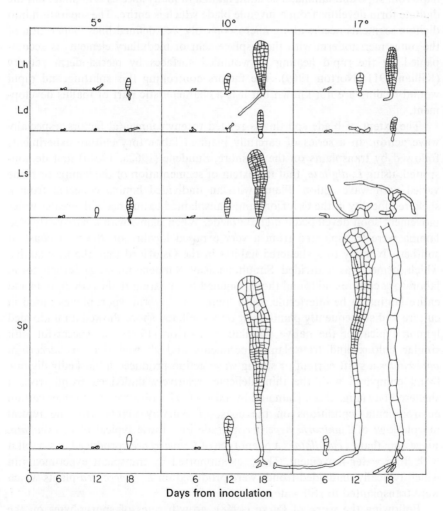

**Fig. 28.1.** Gametophytes or sporophytes after 6, 12 and 18 days in culture at 5°C, 10°C and 17°C. Lh, *L. hyperborea*; Ld, *L. digitata*; Ls, *L. saccharina*; Sp, *Saccorhiza*. Redrawn by S. Mercer after Kain (1969).

of *Chorda tomentosa. Alaria esculenta* showed good development up to 14°C with none above 20°C. The minimum irradiance necessary for growth of the young sporophytes, as measured by cell counts, is very low. Kain (1969) gives an approximate figure of 1 μg cal cm$^{-2}$ sec.$^{-1}$ and a saturation level of about 50 μg cal cm$^{-2}$ sec.$^{-1}$ at 10°C. At the same temperature she estimated that the rate of cell division at 20 μg cal cm$^{-2}$ sec.$^{-1}$ was half that in saturating irradiance. Information on the light requirements of the juvenile stages of *Macrocystis* is given by Anderson and North (1968).

In most species, as the sporophyte enlarges it differentiates into basal hapteron, stipe and lamina. The adult lamina in many species is digitate, but the digitate form develops from a juvenile blade which is entire. The separation into digitate segments occurs along microscopically well defined lines. Separation of the outer meristoderm with the displacement of medullary elements is accompanied by the rapid healing of wounded surfaces by meristoderm activity (Killian 1911, Norton 1969). The factors controlling this splitting and rapid wound healing are not known but it is a characteristic part of thallus development.

The extent of blade splitting is related to environmental factors, especially wave action. In a series of carefully planned laboratory culture experiments followed by transplants on the seashore, Sundene (1962a, 1964a) first demonstrated, using *L. digitata*, that the extent of segmentation of the lamina or blade varied with wave action. Plants with an undivided lamina collected from a sheltered locality in the Oslofjord and transplanted to an area with greater water movement soon developed a digitate blade. When plants with a narrow divided lamina were transplanted from a very exposed locality on the west coast of southern Norway to a sheltered habitat in the Oslofjord then the new lamina which formed was undivided. Simultaneously, Sundene raised gametophytes in laboratory culture and found those obtained from parent thalli with digitate and entire lamina to be interfertile. Both 'pure' and 'hybrid' sporophytes raised in culture and subsequently planted out on a sheltered shore showed an undivided lamina typical of the native population. Norton (1969) was successful with similar culture and transplant experiments and showed that in *Saccorhiza polyschides* a swift current or strong wave action produced a markedly digitate frond compared with the thin, delicate, relatively undivided frond from a sheltered bay (Fig. 28.2). Kain and Svendson (1971) observed that under certain environmental conditions on the coast of Norway plants with the typical morphology of *Laminaria hyperborea* grade into those typical of *L. cucullata*, suggesting that *L. cucullata* is a phenotype of *L. hyperborea* produced in a habitat with little water movement. This was supported by transplant experiments in which typically undivided fronds were produced on *L. hyperborea* plants which were transplanted to still water.

Following the work of Parke (1948), growth rates of sporophytes on the shore have been obtained for several species. On British shores the most rapid rate of growth of *Laminaria saccharina* occurs between March and June when an increase of 2·3 cm per day in the sub-littoral zone in Argyllshire has been recorded. Nicholson (1970) found an increase of as much as 2 cm per day in *Nereocystis* and Norton (1969) has recorded a maximum of 4·7 cm per week for *Saccorhiza*. Neushul and Haxo (1963) gave the maximum growth rates for *Macrocystis* in the laboratory and in parallel field studies, as a doubling in area every 16 to 20 days, and in length every 20 to 30 days.

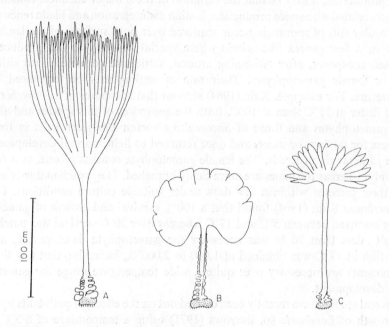

**Fig. 28.2.** Three forms of *Saccorhiza polyschides* from different habitats. *A*, with a continuous strong current, but no wave action; *B*, with little current or wave action, and *C*, with continuous turbulence. Redrawn by S. Mercer after Norton (1969).

All the species, so far studied extensively on the shore, show a marked seasonal periodicity in vegetative growth and reproduction. Growth is generally entirely vegetative during the first year, with the fronds reaching maturity after 12 to 16 months. The time at which spore production occurs is by no means synchronous for all Laminariales growing in one locality. For example, at Port Erin, Isle of Man, *L. hyperborea* and *Saccorhiza* sporulate in winter, while *L. digitata* spores mainly in the summer (Kain 1969). The old frond disintegrates soon after sporing. Simultaneously a new one is developed from the meristem at the base of the old frond so that at certain times of the year the fronds present a form with both the old disintegrating blade and the new blade growing rapidly above the stipe. A few genera, e.g. *Saccorhiza* and *Chorda*, appear to be monocarpic, and no renewal of the vegetative lamina occurs after sporing. Black

(1948, 1950) found marked seasonal variations in the major chemical constituents of the fronds of *L. digitata* and *L. saccharina*, while the composition of the stipes remained relatively constant. Mannitol was at its minimum value in the fronds in spring when laminaran was absent, but crude proteins, ash and alginic acid were then at their maximum. The percentage dry matter increased to a maximum in autumn. As it is the frond which shows seasonal periodicity in vegetative growth and reproduction, it may be that the variation of these major chemical constituents is correlated with spore production, lamina disintegration and blade renewal.

Irregular sori of sporangia occur scattered over both surfaces of the lamina except in a few genera like *Alaria* where special sporophylls are produced. Released zoospores, after swimming around, settle and germinate into either male or female gametophytes. Their rate of settling may be influenced by temperature. For example, Kain (1964) showed that zoospores of *L. hyperborea* settled faster at 17°C than at 10°C. Both the zoospores of this species and also their gametophytes and those of *Saccorhiza* (Norton 1969) can exist in total darkness for 60 days or more, and once returned to light, normal development of the gametangia proceeds. The female gametophyte consists of one, or a few cells only, whereas the males are more richly branched. They reach maturity and shed their gametes within a few days under suitable culture conditions. For *L. hyperborea* Kain (1964) found that a 100% survival and growth of gametophytes occurred between 5°C and 17°C whereas over 20°C survival was patchy. At 10°C, less than 26 lx was necessary for gametophyte development, and saturation at 17°C was obtained at 1,000 to 2,000 lx, indicating that low light requirements are necessary over quite a wide temperature range for gametophyte development.

Several studies have recently been carried out on the effects of pollutants upon the growth of *Laminaria* sp. Burrows (1971) using a temperature of 8·5°C, a daylength of 18 hours and a light intensity of 3,000 lx, measured the increase in frond area of young plants of *L. saccharina* in culture medium containing various pollutants. The growth rate was depressed in seawater from certain industrial areas compared with water from the open sea. The detergent 'Bluesyl' depressed the rate of growth, although even at high concentrations (10 p.p.m.) the plants continued to grow slowly. Kain and Hopkins (1971) tested the effects of a number of toxins upon the gametophytes and early stages in sporophyte development of *L. hyperborea*. They found that mercury, copper, zinc and the herbicide, 'Atrazine', were most toxic, whereas the herbicides, MCPA and 2,4D and detergents were less toxic. Sporophytes were produced after 8 to 10 days as usual in zinc concentrations of 0·25 and 0·5 p.p.m., but they grew very slowly, suggesting that the gametophytes were better able to withstand the toxin than were the sporophytes. The effects of similar substances upon the respiratory rate of fronds were also determined. It was generally found that concentrations ranging from one hundred to several thousand times greater were required to inhibit respiration of the mature lamina than were required to inhibit growth of the gametophytes in culture experiments.

MANTON I. & PETERFI L.S. (1969) Observations on the fine structure of coccoliths, scales and the protoplast of a freshwater coccolithophorid, *Hymenomonas roseola* Stein, with supplementary observations on the protoplast of *Cricophaera carterae*. *Proc. R. Soc. B.* **172**, 1–15.

McBRIDE G.E. (1967) Cytokinesis in the green alga *Fritschiella*. *Nature, Lond.* **216**, 131.

McLACHAN J., HAMMER U. & GORHAM P.R. (1963) Observations on the growth and colony habits of 10 strains of *Aphanizomenon flos-aquae*. *Phycologia*, **2**, 157–68.

MISHRA A. (1969) Fine structure of the growing point of the coenocytic alga, *Caulerpa sertularioides*. *Can. J. Bot.* **47**, 1599–603.

MISHRA A. & KEFFORD N. (1969) Developmental studies on the coenocytic alga *Caulerpa sertularoides*. *J. Phycol.* **5**, 103–9.

MOSS B. (1950) The anatomical structure and chemical composition of receptacles of *Fucus vesiculosus* from 3 contrasting habitats. *Ann. Bot.* **14**, 396–410.

MOSS B. (1952) Variations in chemical composition during the development of *Himanthalia elongata* (L.) S.F. Gray. *J. mar. biol. Ass., U.K.* **31**, 29–34.

MOSS B. (1963) Wound healing and regeneration in *Fucus vesiculosus* L. *Proc. IV Inter. Seaweed Symp.*, eds. Devirville A.D. & Feldman J. pp. 117–22. Pergamon Press, Oxford, London, New York & Paris.

MOSS B. (1965) Apical dominance in *Fucus vesiculosus*. *New Phytol.* **64**, 387–96.

MOSS B. (1967) The apical meristem of *Fucus*. *New Phytol.* **66**, 67–74.

MOSS B. (1968) The transition from vegetative to fertile tissue in *Fucus vesiculosus*. *Br. phycol. Bull.* **3**, 567–73.

MOSS B. (1969) Apical meristems and growth control in *Himanthalia elongata*. *New Phytol.* **68**, 387–97.

MOSS B. (1970) Meristems and growth control in *Ascophyllum nodosum* (L.) Le Jol. *New Phytol.* **69**, 253–60.

MOSS B. (1971) Meristems and morphogenesis in *Ascophyllum nodosum* ecad Mackaii (Cotton). *Br. phycol. J.* **6**, 187–93.

MOSS B. & LACEY A. (1963) The development of *Halidrys siliquosa* (L.) Lyng. *New Phytol.* **62**, 67–74.

MÜLLER D.G. (1961) Über jahres- und lunarperiodische Erscheingen bei einigen Braunalgen. *Botanica mar.* **4**, 140–55.

NAKAMURA Y. (1965) Development of zoospores in *Ralfsia*-like thallus with special reference to the life cycle of the Scytosiphonales. *Bot. Mag., Tokyo*, **78**, 109–10.

NAKAZAWA S. (1969) Notes on the Fucales. A review on the egg polarity. *Bull. Jap. Soc. Phytol.* **17**, 42–7.

NAYLOR M. (1949) Observations on the anatomy of *Durvillea antarctica* (Chamisso) Hariot. *New Phytol.* **13**, 285–308.

NEUSHUL M. & HAXO F.T. (1963) Studies on the giant kelp *Macrocystis*. I. Growth of young plants. *Am. J. Bot.* **50**, 349–53.

NICHOLSON N.L. (1970) Field studies on the giant kelp *Nereocystis*. *J. Phycol.* **6**, 177–82.

NIENBURG W. (1929) Zur Entwicklungsgeschichte der *Fucus*-Keimlinge. *Ber. dt. bot. Ges.* **47**, 527–9.

NIZAMUDDIN M. (1962) Studies on *Cystophyllum muricatum* (Turner). *New Phytol.* **61**, 233–43.

NIZAMUDDIN M. (1964) The anatomy and life history of *Cystophora, Acrocarpia* and *Gaulocystis* (Fucales). *Botanica mar.* **7**, 42–63.

NORRIS R.E. (1971) Development of the foliose thallus of *Weeksia fryeana* (Rhodophyceae). *Phycologia*, **10**, 205–13.

NORTON T.A. (1969) Growth form and environment in *Saccorhiza polyschides*. *J. mar. biol. Ass., U.K.* **49**, 1025–45.

OLTMANNS F. (1889) Beitrage zur Kenntnis der Fucaceen. *Bild. Bot.* **14**, 1889.

OLTMANNS F. (1904) *Morphologie und Biologie der Algen.* 1st edit. Jena.

PAASCHE E. (1968) Biology and physiology of coccolithophorids. *A. Rev. Microbiol.* **22**, 71–86.

PADAN E. & SHILO M. (1969) Short-trichome mutant of *Plectonema boryanum. J. Bact.* **97,** 975–6.

PAPENFUSS G. (1967) The history, morphology and taxonomy of *Hormophysa* (Fucales, Cystoseiraceae). *Phytomorph.* **17,** 42–7.

PARKE M. (1948) Studies on British Laminiaraceae. I. Growth in *Laminaria saccharina* (L.) Lamour. *J. mar. biol. Ass., U.K.* **27,** 651–709.

PARKE M. (1961) Some remarks concerning the class Chrysophyceae. *Br. phycol. Bull.* **2,** 47–55.

PEARSON J. & KINGSBURY J. (1966) Culturally induced variation in four morphologically diverse blue-green algae. *Am. J. Bot.* **53,** 192–200.

PEDERSON M. (1968) *Ectocarpus fasciculatus.* Marine brown algae requiring kinetin. *Nature, Lond.* **218,** 716.

PETERSON D.M. & TORREY J.G. (1968) Amino-acid incorporation in developing Fucus embryos. *Pl. Physiol., Lancaster,* **43,** 941–7.

PICKETT-HEAPS J.D. (1967) Ultrastructure and differentiation in *Chara* sp. II. Mitosis. *Aust. J. biol. Sci.* **20,** 883–94.

PICKETT-HEAPS J.D. & FOWKE L.C. (1969) Cell division in *Oedogonium.* I. Mitosis, cytokinesis and cell elongation. *Aust. J. biol. Sci.* **22,** 857–94.

PIENAAR R. (1969) The fine structure of *Hymenomonas carterae.* II. Observations on scale and coccolith production. *J. Phycol.* **5,** 321–31.

PITELKA D.R. (1963) *Electron microscopic structure of protozoa.* Pergamon Press, Oxford.

PROVASOLI I. (1958) Effect of plant hormones upon *Ulva. Biol. Bull. mar. biol. Lab., Woods Hole,* **114,** 375–84.

PUISEUX-DAO S. (1970) Acetabularia *and Cell Biology.* Logos Press, London.

RICHARDSON N. & DIXON P. (1968) Life history of *Bangia fuscopurpurea* (Dillw.) Lyngb. in culture. *Nature, Lond.* **218,** 496–7.

ROBINSON B.L. & MILLER J. (1970) Photomorphogenesis in the blue-green alga *Nostoc commune. Physiologia Pl.* **23,** 461–72.

RUSSELL G. (1970) Rhizoid production in *Dictyota dichotoma. Br. phycol. J.* **5,** 243–5.

SABNIS D.D. (1969) Observations on the ultrastructure of the coenocytic marine alga *Caulerpa prolifera* with particular reference to some unusual cytoplasmic components. *Phycologia,* **7,** 24–32.

SAGER R. & PALADE G.E. (1954) Chloroplast structure in yellow strains of *Chlamydomonas. Expl. Cell Res.* **7,** 584–8.

SAGER R. & PALADE G.E. (1957) Structure and development of the chloroplast in *Chlamydomonas. J. biophys. biochem. Cytol.* **3,** 463–87.

SEARLE S.R.B. (1968) Morphological studies of red algae of the order Gigartinales. *Univ. Calif. Pub.* **43,** 1–86.

SCHWEIGER H.G. (1970) Regulatory problems in *Acetabularia.* In *Biology of* Acetabularia, eds. Brachet J. & Bonotto S. pp. 3–13. Academic Press, New York & London.

SPARLING S. (1961) A report on the culture of some species of *Halosaccion, Rhodomenia* and *Fauchea. Am. J. Bot.* **48,** 493–99.

STANIER R.Y., KUNISAWA R., MANDEL M. & COHEN-BAZIRE G. (1971) Purification and properties of unicellular blue-green algae (order Chroococcales). *Bact. Rev.* **35,** 171–205.

STARR R.C. (1968) Cellular differentiation in *Volvox. Proc. natn. Acad. Sci., U.S.A.* **59,** 1082–8.

STARR R.C. (1970) Control of differentiation in *Volvox. Dev. Biol. Supplement,* **4,** 59–100.

STEIN J.R. (1963) Morphological variation of a *Tolypothrix* in culture. *Br. phycol. Bull.* **2,** 206–9.

STOERMER E.F. (1967) Polymorphism in *Mastogloia. J. Phycol.* **3,** 73–7.

STOSCH H.A. VON (1955) Ein morphologischer Phasenwechsel bie einer Coccolithophoride. *Naturwissenschaften,* **42,** 423.

STOSCH H.A. VON (1958) Der Geisselapparat einer Coccolithophoride. *Naturwissenschaften*, **45**, 140–1.

STOSCH H.A. VON (1967) Haptophyceae. In *Encyclopedia of Plant Physiology*, Berlin, **18**, 646–56.

SUBRAHMANYAN R. (1957) Observations on the anatomy, cytology and development of *Pelvetia canaliculata* Dene et Thur III. The liberation of reproductive bodies, fertilization and embryology. *J. Indian bot. Soc.* **36**, 373–95.

SUNDENE O. (1962a) Growth in the sea of *Laminaria digitata* sporophytes from culture. *Nytt. Mag. Bot.* **9**, 5–22.

SUNDENE O. (1962b) The implications of transplant and culture experiments on the growth and distribution of *Alaria esculenta*. *Nytt. Mag. Bot.* **9**, 155–74.

SUNDENE O. (1962c) Form variation in *Antithamnion plumula*. Experiments on Plymouth and Oslofjord strains in culture. *Nytt. Mag. Bot.* **7**, 181–7.

SUNDENE O. (1963) Reproduction and ecology of *Chorda tomentosa*. *Nytt. Mag. Bot.* **10**, 159–67.

SUNDENE O. (1964a) The ecology of *Laminaria digitata* in Norway in view of transplant experiments. *Nytt. Mag. Bot.* **11**, 83–107.

SUNDENE O. (1964b) *Antithamnion tenuissimum* (Hauck). Schiffner in culture. *Nytt. Mag. Bot.* **11**, 5–10.

SWALE E.M. (1967) A clone of *Scenedesmus* with *Chodatella* stages. *Br. phycol. Bull.* **3**, 281–93.

TAKATORI S. & IMAHORI K. (1971) Light reactions in the control of oospore germination of *Chara delicatula*. *Phycologia*, **10**, 221–8.

TATEWAKI M. (1966) Formation of a crustaceous sporophyte with unilocular sporangia in *Scytosiphon lomentaria*. *Phycologia*, **6**, 62–6.

THIMANN K.W. & BETH K. (1959) Action of auxins on *Acetabularia* and the effect of enucleation. *Nature, Lond.* **183**, 946–8.

TRAINOR F.R. (1966) A study of wall ornamentation in cultures of *Scenedesmus*. *Am. J. Bot.* **53**, 995–1000.

TRAINOR F. (1969) *Scenedesmus* morphogenesis. Trace elements and spine formation. *J. Phycol.* **5**, 185–90.

TRAINOR F.R. & ROSKOSKY F.G. (1967) Control of unicell formation in a soil *Scenedesmus*. *Can. J. Bot.* **45**, 1657–64.

TRAINOR F.R. & ROWLAND H.L. (1968) Control of colony and unicell formation in a synchronised *Scenedesmus*. *J. Phycol.* **4**, 310–17.

TRAINOR F.R., ROWLAND H.L., LYLIS J.C., WINTER P.A. & BONANOMI P.L. (1971) Some examples of polymorphism in algae. *Phycologia*, **16**, 113–19.

VALLANCE K.B. & COULT D.A. (1951) The composition of the gas in the vesicles of *Fucus vesiculosus*. *Ann. Bot.* **15**, 409–16.

WEST J.A. (1968) Morphology and reproduction of the red alga *Acrochaetium pectinatum* in culture. *J. Phycol.* **4**, 89–99.

WEST J.A. (1969) The life histories of *Rhodochorton purpureum* and *R. tenue* in culture. *J. Phycol.* **5**, 12–21.

WHITAKER D.M. (1940) Physical factors of growth. *Growth (Suppl.)*, pp. 75–90.

WILBUR K., COLINVAUX L.H. & WATABE N. (1969) Electron microscope study of calcification in the alga *Halimeda* (order Siphonales). *Phycologia*, **8**, 27–36.

WILLIAMS L.G. (1952) Effects of IAA on growth in *Codium*. *Am. J. Bot.* **39**, 107–9.

WYNNE M.J. (1969) Life history and systematic studies of some Pacific North American Phaeophyceae (Brown algae). *Univ. Calif. Publs. Bot.* **50**, 1–88.

ZEHNDER A. & GORHAM P.R. (1960) Factors influencing the growth of *Microcystis aeruginosa* Kutz. emend Elenkin. *Can. J. Microbiol.* **6**, 645–60.

ZETSCHE K., GREININGER G.E. & ANDERS J. (1970) Regulation of enzyme activity during morphogenesis of nucleate and anucleate cells of *Acetabularia*. In, *Biology of* Acetabularia, eds. Brachet J. & Bonotto S. pp. 87–111. Academic Press, New York & London.

# CHAPTER 29

# REPRODUCTION

## M. J. DRING

Department of Botany,
The Queen's University of Belfast,
Northern Ireland, U.K.

1 Introduction 814
2 Environmental factors affecting spore formation and liberation 815
2.1 Light 815
2.2 Temperature 818
2.3 Inorganic chemicals 821
2.4 Organic chemicals 823
2.5 Exposure 825
3 Endogenous rhythms of reproductive activity 826
3.1 Circadian rhythms 826
3.2 Lunar rhythms 826
3.3 Other rhythms 827

4 Physiology of sexuality 827
4.1 Hormonal control of gamete development 827
4.2 Chemotactic attraction of gametes 829
4.3 Pairing and fusion of gametes 830

5 Biochemistry of reproduction 830

6 References 831

## 1 INTRODUCTION

Reproduction is defined in this review as any process leading to an increase in the numbers of individuals of a species. This definition includes cell division in unicellular organisms, and this process will be referred to where useful parallels can be drawn with reproduction in multicellular organisms. The term *spore* will be used for any structure which effects reproduction, although an attempt will be made to indicate the role of each spore type discussed in the life-cycle of the species. Since such indications must necessarily be brief, the reader is referred to Ettl *et al.* (1967) for details. Previous reviews of reproductive physiology in algae have usually been restricted to asexual reproduction (e.g. Erben 1962) or sexual reproduction (e.g. Coleman 1962, Wiese 1969), although Lang (1965) has presented a synopsis of the whole field of the physiology of growth and development in algae.

Most studies of algal reproduction under different environmental conditions have been conducted by workers interested primarily in obtaining the complete life-cycle of selected species in culture, rather than in a physiological study of the reproductive process. The results obtained in such studies, therefore, appear incomplete to the physiologist, but they may contain indications of physiologically-interesting responses, and the collation of the results from a large number of such studies may reveal general patterns which would repay detailed physiological investigation. For these reasons, many studies of this type are discussed in this review, although the more limited number of investigations motivated by physiological interests inevitably figure more prominently.

## 2  ENVIRONMENTAL FACTORS AFFECTING SPORE FORMATION AND LIBERATION

### 2.1  *Light*

(a) *Light intensity*

There have been many studies in recent years of the rate of development of reproductive structures or of the intensity of reproductive development at different light intensities. Most of these show that the rate or intensity of reproduction increases with increasing light intensity, provided that other conditions (such as temperature and photoperiod) are suitable (e.g. *Hydrodictyon*, Neeb 1952; *Porphyra*, Kurogi *et al.* 1962). In addition, the results have often indicated that there is a threshold intensity below which reproduction will not occur, as for example in *Acetabularia* (Terborgh & Thimann 1965), the gametophytes of *Laminaria* (Kain 1964) and *Desmarestia* (Chapman & Burrows 1970), and both gametophytes and tetrasporophytes of *Polysiphonia* (Edwards 1970), or that reproduction is inhibited at high light intensities, as in tetrasporophytes of *Bonnemaisonia* (Chen *et al.* 1970a) and gametophytes of *Laminaria* (Hsiao & Druehl 1971). In some investigations, variations in light intensity have been combined with variations in daylength, and evidence has been obtained that the response was to the total incident energy of the radiation rather than to a specific intensity or daylength (e.g. *Acetabularia*, *Vaucheria*, *Ectocarpus*, various diatoms; see Dring 1970). Studies of several diatom species show that high light intensities maintained for long photoperiods have an inhibitory effect on reproduction (e.g. Bruckmayer-Berkenbusch 1955, Rozumek 1968).

It is generally assumed that the effects of light intensity on reproduction are correlated with changes in photosynthetic activity, but, except in two investigations (Neeb 1952, Kain 1964), the rate of photosynthesis has not been measured in such experiments, and a direct correlation has not been demonstrated. Indeed, some of the results obtained indicate that photosynthesis or growth was not saturated at energies which were saturating or inhibitory for reproduction (Kain 1964, Knaggs 1966a). Some other observations of the effects of light intensity on reproduction are best interpreted as a response to increased photosynthesis, in the absence of a full investigation. Examples include: *Volvox* (Starr 1970) and *Closterium* (Kies 1964), and also two reports of the differential effects of high and low light energies in *Ectocarpus*, where high energies favour unilocular, and low energies plurilocular, sporangia (Müller 1962) and *Derbesia* where sporangia are produced in strong light and vegetative propagules in weak light (Hustede 1964). The role of light in the control of gametogenesis in *Chlamydomonas* is less clear. This was the subject of intense investigation in the late 1950s, but there has been no recent work on this aspect of reproduction in *Chlamydomonas*, and so nothing can be added to the discussions of Coleman (1962) and Lang (1965). However, Ellis and Machlis (1968) have shown that sexual reproduction in *Golenkinia* is controlled by a similar system, in which light may have a specific qualitative role as well as a quantitative role mediated through the photosynthetic system. It seems possible that the control of egg liberation in *Fucus* is similar, since there appears to be a qualitative requirement for at least 8 hours of light but, above this threshold, the number of eggs released increases with the length of the irradiation period (Abe 1970).

Another type of response to light intensity which cannot be explained simply on the basis of enhanced photosynthesis is the response to a sudden increase in light intensity. This phenomenon has been most fully investigated in *Rhodochorton* (Knaggs 1966b), but has also been observed in some diatoms (von Stosch & Drebes 1964, Werner 1971c). In *Rhodochorton*, tetrasporangium formation was induced by transfer to a higher light intensity, after a period of adaptation to a lower intensity. Sporangium formation occurred after 30 days at intensities of 500, 4,000 or 8,000 erg cm$^{-2}$ sec$^{-1}$, provided that the alga had been growing for at least 30 days before the transfer at some lower intensity. The reproductive response could also be induced by 10 days at the higher light intensity, followed by a return to the original lower intensity. Knaggs' hypothesis that reproduction is induced by a change in light intensity does not, however, explain why plants brought in from the shore did not form spores in intensities higher than 500 erg cm$^{-2}$ sec$^{-1}$, in spite of good vegetative growth (Knaggs 1966a), and the system deserves more detailed, quantitative study.

### (b) *Light quality*

Although several morphogenetic effects of light have been reported for algal species (see Dring 1971), few specific effects of light on reproduction had been observed until recently. Cap formation in *Acetabularia* will not occur in the absence of blue light (Terborgh 1965, Richter & Kirschstein 1966), but this effect may be due to the stimulation of photosynthesis by blue light after its decline in red light to a low level (Terborgh 1966). The effects of light on the morphogenesis and reproduction of *Acetabularia* appear to involve a complex interaction between red and blue wavelengths, which has yet to be fully elucidated (see Clauss 1970, Puiseux-Dao 1970). The effect of red light on aplanospore formation in *Trebouxia* (the algal symbiont of *Cladonia cristatella*) is apparently more straightforward. Irradiation for one minute with a high intensity of red light was sufficient to induce a four-fold increase in spore production, and the effects of red light could be reversed by subsequent exposure to the same energy of far-red light (Giles 1970). This appears, therefore, to be another example of the activity of phytochrome in an alga, although a complete action spectrum has yet to be determined. The irradiation period in this investigation was considerably shorter than in other studies of algal phytochrome systems, but the total energy used ($6·7 \times 10^5$ erg cm$^{-2}$) is slightly higher than that required to produce a 50% response in other systems (Dring 1971).

Blue light has a specific effect on reproduction in a number of species. Female gametophytes of *Laminaria saccharina* remained vegetative indefinitely in red light, but 12 hours' irradiation with blue light was sufficient to induce egg production 10 days later in 50% of the gametophytes (Lüning & Dring 1972). There are also reports that blue light stimulates spore liberation in *Monostroma* (Shihira 1958) and *Nitophyllum* (Sagromsky 1961), and inhibits zoospore formation in *Protosiphon* (Durant *et al.* 1968). Cell division was induced by blue light in synchronous cultures of a *Chlorella* species (Senger & Schoser 1966), but was inhibited in *Prototheca* (Epel & Krauss 1966), *Chlamydomonas* and *Chlorella pyrenoidosa* (Carroll *et al.* 1970). These responses may be linked to the effects of blue light on respiration in some of these species (Kowallik 1970, Epel & Butler 1970). The formation or liberation of spores in several other algae has been shown to be stimulated by transfer from light to darkness (e.g. *Pelvetia*, Jaffe 1954) or from darkness to light (e.g. *Chlamydomonas*, Kates & Jones 1964a; *Enteromorpha*, Christie & Evans 1962; *Derbesia*, Ziegler & Kingsbury 1964; *Sargassum* and *Caulerpa*, Suto 1950; *Zonaria*, Liddle 1968), and, by analogy with the many similar responses which are specifically induced by blue light, it seems possible that some of these responses are also controlled by blue light. Recent reports that yellow light stimulates cell division or spore formation in species which are inhibited by blue light (Durant *et al.* 1968, Carroll *et al.* 1970), parallel the demonstration of a blue/yellow reversible effect of light on spore discharge in the fungus *Sphaerobolus* (Ingold 1969).

The transfer from darkness to light may also act as an entraining stimulus for endogenous rhythms, as has been shown for the circadian rhythm of sporulation in *Oedogonium* (Bühnemann 1955, Ruddat 1960). Interruption of a dark period by light with the intensity of moonlight was observed to synchronize the endogenous lunar rhythm of egg production in *Dictyota* (Bünning & Müller 1961), and, again, blue wavelengths appear to be the most effective in inducing this response (Vielhaben 1963, Kumke 1971).

## (c) *Light period*

The responses of algae to changes in daylength, including several reproductive responses, have recently been reviewed (Dring 1970), and a detailed account of the physiology of the photoperiodic control of reproduction in *Porphyra tenera* (Dring 1967a,b,c, Rentschler 1967) is included in this review. Richardson (1970) has described some of the details of a similar response in *Bangia*, but provides little direct evidence of the physiological similarity of this response to the response in *Porphyra*. There is evidence in two reports that the sexual reproduction of two brown algae may be under photoperiodic control. The gametophytes of *Desmarestia* were found to remain vegetative when grown in continuous light at 10°C, but one cycle of 10:$\overline{14}$ h (10 h light:14 h dark), followed by a return to continuous light, induced 13% of the branches to become fertile, and fertility was increased to 85% after 4 short-day cycles (Nakahara & Nakamura 1971). In *Laminaria longicruris*, sexual reproduction was more prolific in 5-hour days than in 12-hour days, but brief interruptions of long dark periods with red light reduced sexuality to the level recorded in long-day conditions (Byrne & Craigie 1971). Far-red light reversed the effects of red light, indicating that phytochrome is the photoreceptor pigment for this response, as in *Porphyra* and all flowering plants.

Some of the results of Rozumek (1968) suggest that a photoperiodic response may be involved in auxospore formation in the pennate diatom *Rhabdonema*, since the effects of a particular daylength depended on whether the light was given as a single photoperiod (e.g. 8:$\overline{16}$ h), or was divided into two shorter photoperiods (e.g. 4:$\overline{10}$:4:$\overline{6}$ h). Most of the other results, however, indicate that auxospore formation is controlled by the total light energy incident upon the cultures, and further work is necessary before the photoperiodic suspicions can be substantiated.

## 2.2 *Temperature*

### (a) *Constant temperature conditions*

Studies of the rate or intensity of reproductive development at different temperatures are easily and frequently conducted, but physiologically-complete results are obtained only if the temperature range investigated is wide enough to include both extremes, and if the differences between adjacent treatments are fairly small. Well-defined temperature optima for reproduction in two diatom species have been obtained by measurements at each of eight temperatures, ranging over 20°C, in steps of 2·5 to 3°C. Rozumek (1968) found that the optimal temperature for auxospore formation in *Rhabdonema* was 15 to 21°C, and Holmes (1966), working with *Coscinodiscus*, recorded a similar optimum for the production of male gametes, but a slightly higher optimum for auxospore production. The physiological significance of the effects of temperature on reproduction is difficult to assess, however, unless some other aspect of the alga's metabolism is measured simultaneously, and is correlated with reproductive activity. The results of Trainor and Burg (1965a) indicate that the optimal temperature for growth in *Scenedesmus* is higher than that for reproduction, and this observation suggests many physiological questions which could be pursued further. Another difficulty in the interpretation of temperature data results from the observations of Tseng and Chang (1956). They demonstrated that spore production in *Porphyra tenera* can be divided into at least three phases— sporangium formation, spore formation and spore release—and that each phase has a different temperature optimum. The results of Kurogi and Hirano (1956) with the same species, and, more recently, of Chen *et al.* (1970b) with *P. miniata* confirm these findings. Studies of 'spore production' at different temperatures are, therefore, liable to reveal only a compromise between the optima for several processes. Information on the quantitative variation of reproductive activity with temperature is also available for the following algae, but, since the observations on the effects of temperature were incidental to the main investigation in most of these studies, the results are rarely complete: *Cladophora* (Bellis 1968); *Derbesia* (Sears & Wilce 1970); *Enteromorpha* (Christie & Evans 1962); *Platydorina* (Harris & Starr 1969); *Protosiphon* (Durant *et al.* 1968); *Tetraspora* (Rhodes & Herndon 1967); *Petalonia* (Hsiao 1970).

Temperature may have qualitative effects on reproduction, in contrast to the purely quantitative effects discussed so far. In species in which different spore types are produced by the same thallus, the temperature conditions may influence the type of spore which is formed. For example, Müller (1962, 1963) has reported that *Ectocarpus* produced unilocular sporangia at low temperatures (below 16°C), and plurilocular sporangia at higher temperatures (above 13°C), whereas this situation appears to be reversed in *Elachista* (Kornmann 1962). Similar observations relate to the diatoms. In *Stephanopyxis*, resting spores were formed at low temperatures (3 to 15°), and gametes and auxospores at higher temperatures (21°C; von Stosch & Drebes 1964), and the results of Holmes (1966) indicate that the optimal temperature for the formation of male gametes in *Coscinodiscus* differs from that for the production of auxospores. The course of the developmental cycle may also be affected by temperature, as in *Urospora* (Kornmann 1961, 1970) and *Desmotrichum* (Rhodes 1970). A further effect of temperature has been reported by Stein (1966b), who observed that the ability of certain populations of *Gonium pectorale* to cross successfully was significantly affected by temperature.

## (b) *Sudden changes in temperature*

Klebs (1896) showed that sudden temperature changes, whether increases or decreases, could induce spore formation, and similar effects have recently been demonstrated in *Derbesia* (Sears & Wilce 1970), *Fucus* (Abe 1970) and *Coscinodiscus* (Werner 1971c). The only systematic studies of the effects of sudden temperature changes, however, appear to be those of Kurogi and co-workers on species of *Porphyra*, in which increases and decreases of temperature of different magnitudes were investigated over the whole range of culture temperatures. In *P. tenera*, temperature increases resulted in decreased spore liberation, whereas decreases resulted in increased liberation, the intensity of the effect being roughly proportional to the severity of the temperature change (Kurogi & Hirano 1956), but a similar investigation with *P. umbilicalis* (Kurogi *et al.* 1967) produced inconclusive results.

One well-established effect of a sudden change in temperature is to act as a 'Zeitgeber' (entraining stimulus) for an endogenous rhythm, and this effect has been demonstrated in *Oedogonium* (Ruddat 1960), *Vaucheria* (Rieth 1959) and *Derbesia* (Page & Sweeney 1968). Some of the other observed effects of sudden temperature changes may be attributable to their effects on an endogenous rhythm.

### (c) *Interactions between temperature and light*

There are few detailed studies of the interaction of temperature with light intensity. The two investigations of auxospore formation in diatoms, which have already been mentioned (Holmes 1966, Rozumek 1968) are of interest, since both studies indicate that at temperatures above or below the optimum, spore production is restricted to a narrower range of light energies. A curious inter-action between light energy and temperature has been reported for *Ectocarpus* (Müller 1962), since sporangial development was not influenced by light at 13° or 19°C (see p. 819), but, at 16°C unilocular sporangia were formed preferentially at high light energies, and plurilocular sporangia were formed at low light energies.

Interactions between photoperiod and temperature have also been investigated. The critical daylength of the short-day-plant *Porphyra tenera* was decreased by about 1 hour at 15°C, compared with 20°C and 25°C (Dring 1967c), but it was not clear whether this effect was due to the action of temperature on the photoperiodic process itself, or on some other phase of sporangium production. In the long-day species, *P. umbilicalis*, spore production was observed to occur in shorter daylengths at 1°C than at 5 to 15°C (Kurogi *et al.* 1967), and this effect can also be interpreted as a reduction in the critical daylength at low temperatures. In *Desmarestia* gametophytes, on the other hand, reproduction occurred in daylengths shorter than 10 hours at 18°C, 12 hours at 14°C and 14 hours at 10°C, and in all daylengths the fertility was higher at the lower temperatures (Nakahara & Nakamura 1971). In this alga, therefore, the critical daylength would seem to be increased at lower temperatures, but an alternative interpretation of these observations, which could also be applied to the effects of temperature in *P. umbilicalis*, is that photoperiodic control of reproduction is relaxed at low temperatures.

Recent observations on sexual reproduction in gametophytes of *Laminaria saccharina* (Lüning & Dring 1972) similarly indicate that light becomes less effective as a controlling factor at low temperatures. At 15°C, eggs are formed only in blue light, but at 10°C and 5°C, the gametophytes also become fertile in red light, although to a lesser extent than in blue light. Kain (1964) has also observed that gametophytes of *L. hyperborea* become fertile in lower light intensities at 10°C than at 17°C, and these observations confirm earlier reports that sexual reproduction can be induced by low temperature in several species of Laminariales (e.g. Schreiber 1930, Kemp & Cole 1961, Sundene 1963).

## 2.3 *Inorganic chemicals*

### (a) *Sudden changes of medium*

As reported by Klebs (1896), reproduction of some algal species can be induced by transfer to fresh medium, whereas other species respond to dilution of the medium, or transfer to distilled water. Evidence that gamete production is stimulated by transfer to distilled water or nitrogen-free medium is available for the following green algae: *Chlamydomonas* (except *C. moewusii*, Trainor 1959), *Cylindrocystis* and *Netrium* (Biebl, quoted by Harris & Starr 1969), *Golenkinia* (Ellis & Machlis 1968), *Oedogonium* (Hill & Machlis 1970), *Pandorina* (Coleman 1962), *Platydorina* (Harris & Starr 1969), *Scenedesmus* and *Coelastrum* (Trainor & Burg 1965a,b). In addition, asexual reproduction in *Spongiochloris* is inhibited by ammonium ions (McLean & Trainor 1965), and the results of Moner (1954) can be interpreted as evidence that colony formation in *Pediastrum* occurs more readily in depleted medium (see p. 825). The only indication that high nutrient concentrations may inhibit spore production in a non-green alga is provided by the work of Knaggs (1967) on *Rhodochorton*. Conversely, transfer to fresh medium appears to stimulate the spontaneous formation of male gametes in certain strains of *Volvox* (Kochert 1968), and release of gametes in *Chaetomorpha* (Köhler 1956). Sexual reproduction in several species of diatoms is enhanced by transfer to fresh medium (e.g. *Coscinodiscus asteromphalus*, Werner 1971c; *Stephanopyxis*, von Stosch & Drebes 1964) or inhibited by diluted medium (e.g. *Melosira*, Bruckmayer-Berkenbusch 1955; *Coscinodiscus concinnus*, Holmes 1966). Gametophytes of *Laminaria* have also been induced to form gametes by transfer to fresh medium (Kain 1964), and such treatment also enhances vegetative reproduction in the green algae *Protosiphon* (O'Kelley & Deason 1962), *Enteromorpha* (Lersten & Voth 1960) and *Tetraspora* (Rhodes & Herndon 1967). Some generalizations about the significance of a response either to an increase or to a decrease in the concentration of the medium have been suggested (e.g. Coleman 1962, Lang 1965), but the range of species which has been investigated seems too narrow for meaningful conclusions to be drawn at present.

As with sudden alterations of light intensity and of temperature, sudden changes of medium can also synchronize an endogenous rhythm of reproductive activity (*Derbesia*, Page & Kingsbury 1968).

### (b) *Specific inorganic chemicals*

*Nitrogen compounds.* Pringsheim (1918) observed that nitrogen-free culture medium was as effective as distilled water in inducing sexuality in *Cylindrocystis*,

2D

and Sager and Granick (1954) demonstrated that any nitrogen compound which can be assimilated and used for growth (nitrate or ammonium ions, urea, glutamine, etc.) will inhibit gamete production in *Chlamydomonas*, and, indeed, will cause the de-differentiation of gametes which have already been formed in nitrogen-free conditions. Nitrogen is now known to be the specific inhibitor of sexual reproduction in all species of the green algae listed above which produce gametes in distilled water or stale medium. There have, however, been few attempts to determine the precise role of nitrogen compounds in this inhibition. The effects of a range of nitrogen concentrations have been determined for *Chlamydomonas* (Tsubo 1956), *Golenkinia* (Ellis & Machlis 1968), *Platydorina* (Harris & Starr 1969) and *Oedogonium* (Hill & Machlis 1970). Some of these investigations indicate that a low concentration of nitrogen results in higher gamete production than does complete absence of the element, but these observations are probably due to the effects of nitrogen on growth of the alga before the onset of sexuality, rather than to its effects on reproduction. Using synchronous cultures of *Chlamydomonas*, Kates and Jones (1964a) showed that, when the cells were transferred to a nitrogen-free medium in the middle of a 12-hour light period, neither the synchronization or the timing of the cell division was affected. However, the cells divided to form four gametes instead of two vegetative cells. If a nitrogen source was added up to 9 hours after the transfer to nitrogen-free medium, the majority of cells underwent a vegetative division, but addition of nitrogen 12 hours after transfer permitted 30 to 40% of the cells to form gametes. The division of the cells occurred 15 hours after transfer, and so the addition of nitrogen at this stage had no inhibitory effect on gametogenesis. Similar results have been obtained with non-synchronous cultures of *Golenkinia* (Ellis & Machlis 1968). Kates and Jones (1964a) also followed the de-differentiation of gametes into vegetative cells in the presence of nitrogen, and found that loss of gametic activity was complete 10–14 hours after the addition of ammonium or nitrate ions. Some biochemical changes which accompany gametogenesis in *Chlamydomonas* have been investigated by Kates and Jones (1964b see p. 830), but there is still no clear picture of the mechanism by which nitrogen compounds inhibit sexuality in these algae.

The situation is further complicated by the demonstration that the inhibitory effect of nitrogen on vegetative reproduction in *Spongiochloris* is due to the ammonium ion alone (McLean & Trainor 1965). This suggests that a different type of nitrogen-inhibition must be operative in this alga. Since ammonium ions penetrate the cells, McLean and Trainor (1965) suggested that the primary effect was a change in the pH of the cell sap, leading to a shift in enzyme activity and thence to zoosporogenesis. Outside the Chlorophyceae, the only suggestion that high nitrogen concentrations may inhibit reproduction concerns tetraspore production in *Rhodochorton* (Knaggs 1967), but further experimentation is required to amplify the available evidence. In contrast to these many reports of the inhibitory effects of nitrogen on algal reproduction, O'Kelley and Deason (1962) have shown that nitrogen is essential for zoospore formation in *Proto-*

*siphon*, and that both ammonium and nitrate ions were more effective in fulfilling the requirement than was urea, although urea was a better source of nitrogen for the vegetative growth of the alga. This careful and thorough study of the reduction of zoospore production in stale cultures could serve as a model for investigations of the role of nutrients in reproduction.

*Other inorganic compounds.* Calcium plays an important role in algal reproduction. It appears to be essential for spore release in both *Protosiphon* (O'Kelley & Herndon 1961) and *Chlorococcum* (Gilbert & O'Kelley 1964), since the replacement of calcium by strontium inhibited the release, but not the formation of spores. The requirement for calcium in these algae was attributed to its function as a component of the pectic materials in the cell wall, and it was suggested that the normal mechanism of wall digestion could not operate in the absence of calcium. Zygospore formation in *Golenkinia* was also restricted to media with a high calcium content (Ellis & Machlis 1968), but the importance of calcium in this alga is probably related to its effects on the mating of gametes. These effects are discussed by Wiese (1969).

The role of carbon dioxide in the control of gamete formation and activity in *Chlamydomonas* is discussed by Coleman (1962), and other reports of the induction or enhancement of reproductive activity by $CO_2$ concern *Volvox* (Starr 1970), placoderm desmids (Starr 1955, Lippert 1969) and *Stigeoclonium* (Abbas & Godward 1963). As with increases in light intensity, these effects are generally assumed to be due to enhanced photosynthesis, but a direct correlation has not been demonstrated.

Sulphur, as sulphate, was essential for zoospore formation in *Protosiphon* (O'Kelley & Deason 1962), and gamete production in *Hydrodictyon* was inhibited by phosphorus deficiency (Neeb 1952), but Tsubo (1961) found that neither element was indispensible for sexual reproduction in *Chlamydomonas*. The production of male gametes and auxospores in *Cyclotella* appears to be induced by changes in salinity (Schultz & Trainor 1970), but experiments with *C. meneghiniana* indicate that the response can also be induced by artificial seawater lacking sodium chloride (Schultz & Trainor 1968). In another centric diatom, *Ditylum*, sexual reproduction is apparently enhanced by the absence of manganese (Steele 1965), and iodine has been implicated in zoospore production by the brown alga *Petalonia* (Hsiao 1969).

## 2.4   *Organic chemicals*

### (a) *Exogenous growth substances*

Although the usual range of growth substances has been tested on a wide variety of algae (Conrad & Saltman 1962), relatively few effects on reproduction have been reported. Sporangium production by the conchocelis-phase of *Porphyra* in short-day conditions was inhibited by $10^{-3}M$ indoleacetic acid (IAA), but, in long-day conditions, sporangium production was unaffected by IAA concentrations of between $10^{-7}$ and $10^{-3}$M (Dring 1967c). Gibberellic acid was found to have little effect on the response of the conchocelis in either long- or short-day conditions. Both of these observations agree with results obtained in investigations of photoperiodic responses in flowering plants. Aplanospore formation in *Trebouxia* was stimulated by $3 \times 10^{-4}$M IAA, as well as by red light (Giles 1970, see p. 817), but $6 \times 10^{-5}$M IAA was without effect. Since coumarate inhibits the response of this alga to red light, and coumarate activates IAA-oxidase in flowering plants, it was concluded that the primary effect of red light was to stimulate the synthesis of IAA, which in turn stimulates spore production.

Other recent reports of the effects of growth substances on algal reproduction include the stimulation of spore production in *Derbesia* by IAA (Hustede 1964), and the delay of gamete discharge in *Fucus* by naphthaleneacetic acid and kinetin (Moss 1967). In *Acetabularia*, cap formation is stimulated by $10^{-7}$M kinetin applied to small (7 mm) nucleate or enucleate cells, but kinetin has less effect on larger cells (Spencer 1968). In addition, $10^{-4}$M triiodobenzoic acid inhibited cap formation.

### (b) *Other exogenous organic compounds*

A wide variety of organic chemicals has been added to cultures of various algae, and the effects on reproduction have been duly reported. For example, asexual reproduction of *Porphyra* was promoted by purines (Teramoto & Kinoshita 1960); auxospore formation in *Melosira* was promoted by ascorbic acid, yeast extract and soil extract (Bruckmayer-Berkenbusch 1955), although unaffected by 2,4-dinitrophenol, iodoacetic acid or *o*-phenanthroline (Erben 1959); spermatogenesis was induced in *Coscinodiscus* by the addition of 40 mg $l^{-1}$ of methanol, combined with a suitable aeration treatment (Werner 1971c); and Stein (1966a) has reported on a variety of exogenous organic compounds which appear to be required for zygote formation by specific populations of *Gonium*. It is difficult to determine the specific activity of such compounds, however, and little physiologically meaningful information can be derived from these observations.

### (c) *Endogenous organic compounds*

The work of Rawitscher-Kunkel and Machlis (1962) on *Oedogonium*, and of Starr and co-workers on *Volvox* (e.g. Starr 1970), which is discussed in another section (see p. 827), has demonstrated that algae may produce organic substances which act as hormones and influence the sexual reproductive activity of other individuals of the same species. In certain strains of *Volvox aureus*, similar or identical hormones also affect asexual reproduction by inducing the production of parthenospores (Darden & Sayers 1969). Asexual reproduction of *V. carteri* is also affected by ultra-violet-irradiation of undivided gonidia. This treatment resulted in the formation of a high percentage of daughter colonies with one or more missing gonidia, although completely normal colonies were produced in the next generation (Kochert & Yates 1970). This effect was interpreted as evidence for the existence of an organic morphogenetic substance whose distribution in the gonidium determined the number and position of daughter colonies. Ultra-violet-irradiation of dividing cells may, however, produce a variety of non-heritable defects (e.g. in *Micrasterias*, Waris & Kallio 1964), and further work is required to substantiate this interpretation. Another, earlier report of the production of a 'swarming substance' in cultures of *Pediastrum* (Moner 1954) is also open to doubt. The transfer of young colonies to stale medium, in which older colonies had already produced swarmers, was shown to induce early swarmer production, and the effect of the old medium was shown not to be due to a change in pH. However, the possibility that swarmer production was inhibited in fresh medium (e.g. by high nitrogen concentrations; see pp. 821–2) was not convincingly eliminated, and the persistence of the inducing ability of stale medium after autoclaving also argues against the activity of an organic substance.

## 2.5  *Exposure*

There have been many ecological observations that algal species growing in the littoral zone release their spores shortly after re-immersion of the thalli by an incoming tide, and laboratory simulation of these conditions induces spore release, even in species which are seldom or never exposed to the air in nature (e.g. *Laminaria* spp., Schreiber 1930). There have, however, been few physiological investigations of the response to a period of exposure, although two possible mechanisms which have been suggested are that the sudden entry of water into partially-dehydrated cells results in the rupture of the sporangia, or that dehydration causes contraction of the whole thallus and the gradual extrusion of spores from conceptacles (e.g. *Pelvetia*, Subrahmanyan 1957) or sporangia. The recipe given by Pollock (1970) for obtaining *Fucus* eggs would seem to exclude both of these possibilities, however, since the re-immersion in seawater which triggered the release of gametes was preceded by 10 to 15 minutes in cold tap-water. Smith (1947) attempted to disentangle the effects of light and exposure on gamete release in *Ulva*, but his combined laboratory and field observations forced him to conclude that neither incipient drying nor illumination could account completely for the daily release of gametes in the field. A more systematic study of the influence of exposure on spore release in *Gloiopeltis* by Matsui (1969), indicates that, in one species (*G. tenax*), the sporangia mature some 12 hours before spore liberation occurs in submerged conditions, but a period of exposure causes an immediate release of spores from mature sporangia. In a second species (*G. furcata*), however, spore liberation appears to occur shortly after the maturation of the sporangia, and exposure delays maturation.

## 3  ENDOGENOUS RHYTHMS OF REPRODUCTIVE ACTIVITY

### 3.1  *Circadian rhythms*

Intensive studies of the diurnal rhythm of asexual reproduction in *Oedogonium* (Bühnemann 1955, Ruddat 1960) have contributed much to our understanding of endogenous circadian rhythms, and the significance of this work in relation to other work on endogenous rhythms in plants and animals has been discussed by Bünning (1967). However, few other diurnal rhythms of reproductive activity have been reported for algal species. Zoospore production in *Vaucheria* is also controlled by an endogenous circadian rhythm (Rieth 1959), which can be entrained by temperature changes or by light:dark cycles. Cell division in synchronous cultures of unicellular algae shows a regular diurnal periodicity (Lorenzen 1970), but in constant conditions of light and temperature such cultures soon become arhythmic and it is doubtful whether any endogenous oscillator is involved. In dinoflagellates, however, the rhythm of cell division does persist in continuous dim light (Sweeney & Hastings 1962). Diurnal rhythms of spore release have been reported, in the field and in the laboratory, for several species of red algae (see Boney 1965), but investigations of such rhythms in *Nitophyllum* (Sagromsky 1961) and *Gloiopeltis* (Matsui 1969) have shown that they are dependent upon a regular alternation of light and darkness, and do not persist in continuous conditions.

### 3.2  *Lunar rhythms*

There is one example of a lunar rhythm of reproductive activity in an alga—egg production in *Dictyota*—which has been fully investigated and documented (Bünning & Müller 1961, Müller 1962, Vielhaben 1963, Kumke 1971), and this, again, is fully discussed by Bünning (1967), but no other example has been satisfactorily investigated. Early observations of lunar rhythms of reproductive activity in littoral algae are listed by Sweeney and Hastings (1962), and other reports concern species of *Ulva* (Smith 1947, Chihara 1968, Subbaramaiah 1970), *Enteromorpha* (Christie & Evans 1962), *Fucus* (Kalähne *et al.* 1956), *Pelvetia* (Subrahmanyan 1957), *Hizikia* (Suto 1951), *Zonaria* (Liddle 1968) and *Porphyra* (see Boney 1965). Subbaramaiah (1970) observed that the appearance of swarmers in cultures of *Ulva fasciata*, maintained under constant laboratory conditions, coincided with the time of spring tides, and it proved impossible to induce swarmer formation in the cultures during neap tides. Kurogi *et al.* (1967) obtained evidence for a 28-day rhythm of spore production in laboratory cultures of *Porphyra umbilicalis* under long-day conditions. These cannot, however, be regarded as unequivocal demonstrations of the endogenous nature of these rhythms, since, in the first example, the data were apparently not quantitative and detailed results are not available, and, in the second, the observations were not continued for a sufficiently long period. A two-weekly rhythm of gamete release in *Pelvetia* was observed to coincide with the only tides which covered the plants on the shore (i.e. spring tides), and artificial exposure of plants for long periods in the laboratory, followed by re-immersion, was found to liberate gametes at other times (Subrahmanyan 1957). The periodicity of reproductive activity in *Pelvetia* was, therefore, attributed to the periodicity of the exposure pattern in nature, rather than to any endogenous rhythm, and it is possible that a similar interpretation applies to some of the other responses listed above.

### 3.3  *Other rhythms*

Page and Sweeney (1968) showed that the 4-5-day periodicity of gamete formation in *Derbesia* is controlled by an endogenous rhythm. The rhythm period was hardly affected by temperature or light intensity within the range permitting gamete formation, but increases of temperature or changes of medium could act as entraining stimuli. This appears to be the only example of a rhythm with such a period, although Beth (1962) has presented unconvincing field observations of a 4-5-day periodicity in gametangia production in another siphonaceous green alga, *Halimeda*.

## 4  PHYSIOLOGY OF SEXUALITY

Much of this topic is covered by a recent review of fertilization in algae (Wiese 1969), and the present review will, therefore, be confined to a discussion of more recent developments in investigations of the hormonal control of sexuality.

### 4.1  *Hormonal control of gamete development*

Rawitscher-Kunkel and Machlis (1962) have demonstrated that the complex sequence of events in the sexual reproduction of *Oedogonium* is controlled by four distinct hormones, produced by one or other of the sexes. Three of these are produced by the female, and act chemotactically, but one is produced by the dwarf male plants and stimulates the final division of the oogonial mother cell into an oogonium and a suffultory cell. The only other evidence for the production of hormones by one sex which induce the development of sexual cells in other individuals, either of the same or opposite sex, comes from work with *Volvox* (Starr 1970), although Rozumek (1968) observed that cultures of *Rhabdonema* only formed gametes when both sexes were cultured in the same vessel, but was unsuccessful in inducing gamete formation in one sex by filtrates of cultures from the other sex.

Induction systems have now been demonstrated in six species of *Volvox*, and four geographically-isolated strains of one of the species (*V. carteri*) have also been investigated (see Starr 1968, 1970, Darden 1970). This series of investigations of similar phenomena in several closely-related species and strains is valuable, since it has demonstrated that considerable physiological differences exist within the genus, and since it emphasizes the dangers of making wide generalizations based on physiological investigations of a single strain of a single species. Starr has so often stressed the differences between the systems, however, that an account of their similarities is overdue.

In all the strains investigated, the inducer is formed by colonies which have produced male gametes, and is apparently liberated at the same time as the gametes. Two parthenosporic strains of *V. aureus* and the female strain of *V. dissipatrix* have also been found to produce an inducer. The gonidia in asexual colonies which are exposed to the appropriate inducer will either differentiate so as to form sexual colonies, or will develop directly into reproductive cells (i.e. will be converted into eggs, sperm packets or parthenospores). Special conditions, such as high light intensities and high $CO_2$ levels may, however, be necessary for 100% induction of gonidia, or for the induction of particular types of reproductive cell (e.g. male colonies in the NB-3 strain of *V. carteri* var. *weismannia*). The inducers are species- and variety-specific (e.g. *V. carteri* var. *nagariensis* will not cross-induce with var. *weismannia*), but not entirely strain-specific (e.g. there is some cross-induction between Indian, Australian and Nebraskan strains of *V. carteri* var. *weismannia*). All of the inducers which have been studied in detail are active at very low concentrations (e.g. $3 \times 10^{-15}$M in *V. carteri* var. *nagariensis*), are heat labile and are inactivated by pronase. Estimates of the molecular weights of different inducers range from 10,000 to over 200,000, but these are all based on characterizations by different authors at different times and a comparative study of several inducers under identical conditions might well resolve some of these apparent differences. All of these results indicate that the inducers are proteins or polypeptides. More detailed studies of the male-inducing-substance from *V. aureus* (Darden & Sayers 1971) have provided little more specific information. Darden (1971) has demonstrated that sexual differentiation in *V. aureus* can also be induced by treatment with an extract of vegetative colonies combined with a histone preparation, but the activity of the extract again appears to depend on its protein content since the extract is totally inactivated by pronase.

The major differences between the inducers from different species are of two types. The different inducers can be divided into two distinct groups on the basis of their effect on the gonidia. The inducers of one group apparently influence the *development* of gonidia, converting them directly into reproductive cells (e.g. *V. aureus*, *V. gigas*, *V. rousseletii*), whereas the other inducers influence the *differentiation* of gonidia, so that they develop into sexual spheroids, containing eggs or sperm packets, instead of developing into asexual spheroids, containing only asexual gonidia (e.g. all strains of *V. carteri*, *V. obversus*). Although two distinct patterns of gonidial development have also been observed (see Starr 1970), the two patterns of inducer activity do not appear to correlate with these developmental patterns. Neither is it possible to correlate the pattern of inducer activity with the section of the genus to which particular species belong. Thus, *V. carteri* and *V. gigas* are both placed in the section Merillosphaera (Smith 1944), and both show the same pattern of gonidial development—the gonidia enlarging before the onset of cleavage, rather than between successive cleavages (Starr 1970)—but the inducers produced by *V. carteri* influence differentiation, whereas those produced by *V. gigas* affect development. It will probably be necessary to investigate even more strains in order to elucidate the significance of this difference between the two groups of inducers.

The other major difference between inducers that has been reported, concerns the period of exposure to the inducer which is necessary to effect induction. In *V. carteri* var. *weismannia* (Kochert 1968) and *V. rousseletii* (McCracken & Starr 1970), the minimum exposure times were found to be 12 and 15 hours, respectively, and continual exposure to the inducer was necessary in *V. gigas* (Vande Berg & Starr 1971). Parthenospore induction in the DS-strain of *V. aureus* also required long exposure to the inducer (more than 24 hours; Darden & Sayers 1969), but in the M5 strain of the same species, Darden (1966) reported 36% induction after exposure to the inducer for only 5 minutes. It seems unlikely that large protein molecules (a molecular weight of over 200,000 was reported for this strain) will penetrate the cells in so short a time, and the minimum exposure times reported for different inducers may be a function of the extent to which the inducer molecules are adsorbed onto the colonial matrix. Starr (1972) has recently proposed a working model for the action of the inducers, based on the operon theory of Jacob and Monod, and this could provide a basis for further experimentation on the biochemical and genetical aspects of the induction system in *Volvox*.

## 4.2 *Chemotactic attraction of gametes*

Evidence for the production of chemicals by the female (or '—') gametes or gametangia which attract the male (or '+') gametes is available for many species, principally from the Chlorophyceae and Phaeophyceae (see Machlis & Rawitscher-Kunkel 1967, Wiese 1969), but detailed physiological work, in which cell-free filtrates from female cultures have been shown to attract male gametes, has been mainly restricted to four genera: *Chlamydomonas* (Tsubo 1961), *Oedogonium* (Rawitscher-Kunkel & Machlis 1962), *Fucus* (Cook & Elvidge 1951) and *Ectocarpus* (Müller 1968). *Oedogonium* appears to be the only one of these genera in which the chemotactic responses have a high degree of specificity. Thus, filtrates from the oogonia of *O. cardiacum* did not attract the male gametes of *O. geniculatum* (Machlis & Rawitscher-Kunkel 1967), whereas filtrates from eggs of *Fucus serratus* and *F. vesiculosus* attracted the sperm of both species as well as that of *F. spiralis* (Cook & Elvidge 1951). This lack of specificity of the chemotactic response in *Fucus* is reflected in the wide range of simple organic compounds which exhibit a similar chemotactic activity to the natural substance. For example, n-hexane will attract the male gametes of *Fucus*, *Ascophyllum* (Cook & Elvidge 1951) and *Ectocarpus* (Müller 1968), and also the '+' gametes of *Chlamydomonas* (Tsubo 1961). Attempts to identify the chemical constituents of natural chemotactic substances have so far been unsuccessful, but Müller (1968) has used gas-liquid chromatography in his study of *Ectocarpus* and this technique is certainly more promising than the more primitive methods available to earlier workers (see also Müller & Jaenicke 1973).

### 4.3   *Pairing and fusion of gametes*

In several isogamous genera of the Chlorophyceae, gamete pairing occurs without chemotactic attraction, and random meeting of gametes of opposite mating type is followed by the adhesion of the flagellar tips. This adhesion is accomplished by an efficient agglutination reaction, which has been intensively studied in a number of *Chlamydomonas* strains by Wiese (1969). Unlike most of the chemotactic systems investigated, the agglutination reactions have a high degree of specificity for the sex, species and phase of development of the gametes, and for the region of the flagellae which will agglutinate. The agglutination substances can be isolated and characterized. They appear to be high-molecular weight glycoproteins, but the basis of their species- or sex-specificity has yet to be elucidated (see Wiese 1969). However, the recent demonstration that the phytohaemoglutinin, concanavalin A, has a differential effect on the agglutination of '+' and '—' gametes (Wiese & Shoemaker 1970) may mark a significant advance.

In most other algal species in which the fusion of gametes is not preceded either by a highly-specific chemotactic attraction, or by flagellar agglutination, it must be assumed that some species- and sex-specific mechanism initiates the adhesion and fusion of sexually-different gametes of the same species. The physical or chemical basis of this specificity could well be investigated in systems which have received exhaustive morphological study, such as the sexual reproduction of oogamous diatoms or of the trichogynous red algae.

## 5   BIOCHEMISTRY OF REPRODUCTION

Investigations of the biochemical changes which accompany reproduction in algae have so far been restricted to two systems. Kates and Jones, with various co-workers, have studied gametogenesis in synchronous cultures of *Chlamydomonas* (Kates & Jones 1964a), and Werner (1971a,b,c) has followed some of the biochemical changes which occur during the developmental cycle of *Coscinodiscus*.

In *Chlamydomonas*, the specific activities of certain enzymes in cells undergoing synchronous vegetative division were compared with those of cells which had been induced to form gametes by transfer to nitrogen-free medium (see

p. 822). The activity of alanine dehydrogenase was reduced in the cells forming gametes (Kates & Jones 1964b), but the implications of this observation do not appear to have been followed up. In a further study of gamete differentiation in an arginine-requiring mutant of *C. reinhardtii*, Jones *et al.* (1968) demonstrated that, although net protein synthesis stopped during gametogenesis, the turnover of proteins occurred at the same rate as in vegetatively dividing cells, and that an exogenous supply of arginine, which was essential for vegetative division in this mutant, was not required for gametogenesis. DNA and RNA were synthesized at the same rates during gametogenesis as in vegetative cell division, but analysis of the RNA fractions indicated that during gametogenesis, transfer-RNA is synthesized preferentially to ribosomal-RNA. The information obtained in these studies is somewhat disjointed as yet, but this is an extremely promising system for the study of the biochemistry of gametogenesis.

The developmental cycle of the centric diatom *Coscinodiscus* consists of a long series of vegetative cell divisions, during which the diameter of the valves decreases from about 200 μm to 80–90 μm. During this phase, the cells cannot be induced to form gametes, but, once the diameter has decreased to a critical size of about 90 μm, gametogenesis can be induced by a variety of environmental manipulations (Werner 1971a). The relative amounts of DNA, protein and potassium per unit surface area of the cells was found to change from 1:1:1 in cells of 170 μm diameter to about 5:2:0·5 in cells of 80 μm diameter. It was suggested, therefore, that sexual inducibility was a function of the relative sizes of the nucleus, cytoplasm and vacuole, or, more simply, that gametogenesis could only be induced when the surface/volume ratio was above a critical level. The extent to which different metabolic inhibitors suppressed cell division changes during development (Werner 1971b), and the rate of uptake of adenosine, AMP and ATP per unit surface area was about 50% lower in the smaller, inducible cells (Werner 1971c). A difficulty in the interpretation of these results is that only 5% of the population was induced to form gametes by the treatment used. This may mean either that a more effective environmental manipulation could be found to induce gametogenesis in this species, or that only a small proportion of the cells is really inducible even when the population consists wholly of small cells. If the second interpretation is correct, then measurement of the biochemical attributes of the whole population will tell us little about the specific changes accompanying the development of the capacity to undergo gametogenesis.

## 6 REFERENCES

ABBAS A. & GODWARD M.B.E. (1963) Effects of experimental culture in *Stigeoclonium*. *Br. phycol. Bull.* **2**, 281–2.

ABE M. (1970) A method of inducing egg liberation in *Fucus evanescens Bot. Mag., Tokyo*, **83**, 254–5.

BELLIS V.J. (1968) Unialgal cultures of *Cladophora glomerata* (L.) Kütz. I. Response to temperature. *J. Phycol.* **4**, 19–23.

BETH K. (1962) Reproductive phases in populations of *Halimeda tuna* in the Bay of Naples. *Pubbl. Staz. zool. Napoli*, **32** (*Suppl.*), 515–34.

BONEY A.D. (1965) Aspects of the biology of the seaweeds of economic importance. *Adv. mar. Biol.* **3**, 105–235.

BRUCKMAYER-BERKENBUSCH H. (1955) Die Beeinflussung der Auxosporenbildung von *Melosira nummuloides* durch Außenfaktoren. *Arch. Protistenk.* **100**, 183–211.

BÜHNEMANN F. (1955) Das endodiurnale System der Oedogoniumzelle. III. Über den Temperatureinfluß. *Z. Naturf.* **10b**, 305–10.

BÜNNING E. (1967) *The Physiological Clock*. Revised 2nd ed. Longmans/Springer-Verlag, New York.

BÜNNING E. & MÜLLER D. (1961) Wie messen Organismen lunare Zyklen? *Z. Naturf.* **16b**, 391–5.

BYRNE J.E. & CRAIGIE J.S. (1971) Evidence for a phytochrome mediated response of gametogenesis in the brown alga *Laminaria longicruris*. *Canad. Soc. Pl. Physiol.*, *Regional Meeting Abstracts*, Jan. 1971, p. 3.

CARROLL J.W., THOMAS J., DUNAWAY C. & O'KELLEY J.C. (1970) Light-induced synchronisation of algal species that divide preferentially in darkness. *Photochem. Photobiol.* **12**, 91–8.

CHAPMAN A.R.O. & BURROWS E.M. (1970) Experimental investigations into the controlling effects of light conditions on the development and growth of *Desmarestia aculeata* (L.) Lamour. *Phycologia*, **9**, 103–8.

CHEN L.C-M., EDELSTEIN T. & McLACHLAN J. (1970a) Vegetative development of the gametophyte of *Bonnemaisonia hamifera* from a filamentous state. *Can. J. Bot.* **48**, 523–5.

CHEN L.C-M., EDELSTEIN T., OGATA E. & McLACHLAN J. (1970b) The life-history of *Porphyra miniata*. *Can. J. Bot.* **48**, 385–9.

CHIHARA M. (1968) Field, culturing and taxonomic studies of *Ulva fenestrata* P. & R. and *Ulva scagelii* sp. nov. (Chlorophyceae) in British Columbia and Northern Washington. *Syesis*, **1**, 87–102.

CHRISTIE A.O. & EVANS L.V. (1962) Periodicity in the liberation of gametes and zoospores of *Enteromorpha intestinalis* Link. *Nature, Lond.* **193**, 193–4.

CLAUSS H. (1970) Effect of red and blue light on morphogenesis and metabolism of *Acetabularia mediterranea*. In *Biology of* Acetabularia, eds. Brachet J. & Bonotto S. pp. 177–91. Academic Press, New York & London.

COLEMAN A.W. (1962) Sexuality. In *Physiology and Biochemistry of Algae*, ed. Lewin R.A. pp. 711–29. Academic Press, New York & London.

CONRAD H.M. & SALTMAN P. (1962) Growth substances. In *Physiology and Biochemistry of Algae*, ed. Lewin R.A. pp. 663–71. Academic Press, New York & London.

COOK A.H. & ELVIDGE J.A. (1951) Fertilization in the Fucaceae: investigations on the nature of the chemotactic substance produced by eggs of *Fucus serratus* and *F. vesiculosus*. *Proc. R. Soc. B.* **138**, 97–114.

DARDEN W.H. (1966) Sexual differentiation in *Volvox aureus*. *J. Protozool.* **13**, 239–55.

DARDEN W.H. (1970) Hormonal control of sexuality in the genus *Volvox*. *Ann. N.Y. Acad. Sci.* **175**, 757–63.

DARDEN W.H. (1971) A new system of male induction in *Volvox aureus* M5. *Biochem. biophys. Res. Comm.* **45**, 1205–11.

DARDEN W.H. & SAYERS E.R. (1969) Parthenospore induction in *Volvox aureus* DS. *Microbios.* **1**, 171–6.

DARDEN W.H. & SAYERS E.R. (1971) The effect of selected chemical and physical agents on the male-inducing substance from *Volvox aureus* M5. *Microbios.* **3**, 209–14.

DRING M.J. (1967a) Effects of daylength on growth and reproduction of the *Conchocelis*-phase of *Porphyra tenera*. *J. mar. biol. Ass.*, *U.K.* **47**, 501–10.

DRING M.J. (1967b) Phytochrome in red alga, *Porphyra tenera*. *Nature, Lond.* **215**, 1411–12.

DRING M.J. (1967c) Photoperiodic studies on algae. Ph.D. Thesis, University of London.

DRING M.J. (1970) Photoperiodic effects in microorganisms. In *Photobiology of Micro-organisms*, ed. Halldal P. pp. 345–68. Wiley-Interscience, London, New York, Sydney & Toronto.

DRING M.J. (1971) Light quality and the photomorphogenesis of algae in marine environments. In *Fourth European Marine Biology Symposium*, ed. Crisp D.J. pp. 375–92. Cambridge University Press.

DURANT J.P., SPRATLING L. & O'KELLEY J.C. (1968) A study of light intensity, periodicity and wavelength on zoospore production by *Protosiphon botryoides* Klebs. *J. Phycol.* **4**, 356–62.

EDWARDS P. (1970) Field and cultural observations on the growth and reproduction of *Polysiphonia denudata* from Texas. *Br. phycol. J.* **5**, 145–53.

ELLIS R.J. & MACHLIS L. (1968) Control of sexuality in *Golenkinia*. *Am. J. Bot.* **55**, 600–10.

EPEL B.L. & BUTLER W.L. (1970) Inhibition of respiration in *Prototheca zopfii* by light. *Pl. Physiol., Lancaster*, **45**, 728–34.

EPEL B. & KRAUSS R.W. (1966) The inhibitory effect of light on growth of *Prototheca zopfii* Kruger. *Biochim. biophys. Acta*, **120**, 73–83.

ERBEN K. (1959) Untersuchungen über Auxosporen-*entwicklung* und Meioseauslösung an *Melosira nummuloides* (Dillw.) C.A. Agardh. *Arch. Protistenk.* **104**, 165–210.

ERBEN K. (1962) Sporulation. In *Physiology and Biochemistry of Algae*, ed. Lewin R.A. pp. 701–10. Academic Press, New York & London.

ETTL H., MÜLLER D.G., NEUMANN K., STOSCH H.A. VON & WEBER W. (1967) Vegetative Fortpflanzung, Parthogenese und Apogamie bei Algen. *Handb. PflPhysiol.* **18**, 597–776.

GILBERT W.A. & O'KELLEY J.C. (1964) The effects of replacement of calcium by strontium on the reproduction of *Chlorococcum echinozygotum*. *Am. J. Bot.* **51**, 866–9.

GILES K.L. (1970) The phytochrome system, phenolic compounds, and aplanospore formation in a lichenized strain of *Trebouxia*. *Can. J. Bot.* **48**, 1343–6.

HARRIS D.O. & STARR R.C. (1969) Life history and physiology of reproduction of *Platydorina caudata* Kofoid. *Arch. Protistenk.* **111**, 138–55.

HILL G.J.C. & MACHLIS L. (1970) Defined media for growth and gamete production by the green alga, *Oedogonium cardiacum*. *Pl. Physiol., Lancaster*, **46**, 224–6.

HOLMES R.W. (1966) Short-term temperature and light conditions associated with auxospore formation in the marine centric diatom, *Coscinodiscus concinnus* W. Smith. *Nature, Lond.* **209**, 217–18.

HSIAO S.I.C. (1969) Life-history and iodine nutrition of the marine brown alga, *Petalonia fascia* (O.F. Müll.) Kuntze. *Can. J. Bot.* **47**, 1611–16.

HSIAO S.I.C. (1970) Light and temperature effects on the growth, morphology and reproduction of *Petalonia fascia*. *Can. J. Bot.* **48**, 1359–61.

HSIAO S.I.C. & DRUEHL L.D. (1971) Environmental control of gametogenesis in *Laminaria saccharina*. I. The effects of light and culture media. *Can. J. Bot.* **49**, 1503–8.

HUSTEDE H. (1964) Entwicklungsphysiologische Untersuchungen über den Generationswechsel zwischen *Derbesia neglecta* Berth. und *Bryopsis halymeniae* Berth. *Botanica mar.* **6**, 134–42.

INGOLD C.T. (1969) Effect of blue and yellow light during the later developmental stages of *Sphaerobolus*. *Am. J. Bot.* **56**, 759–66.

JAFFE L. (1954) Stimulation of the discharge of gametangia from a brown alga by a change from light to darkness. *Nature, Lond.* **174**, 743.

JONES R.F., KATES J.R. & KELLER S.J. (1968) Protein turnover and macromolecular synthesis during growth and gametic differentiation in *Chlamydomonas reinhardtii*. *Biochim. biophys. Acta*, **157**, 589–98.

KAIN J.M. (1964) Aspects of the biology of *Laminaria hyperborea*. III. Survival and growth of gametophytes. *J. mar. biol. Ass., U.K.* **44**, 415–33.

KALÄHNE M., WARTENBERG A. & BAUCH R. (1956) Biologisch-ökologische Studien an der Gattung *Fucus*. III. Die rhythmische Bildung der Geschlechtszellen beim Blasentang. *Wiss. Z. Ernst. Moritz Arndt-Univ.*, *Greifswald*, **5/6**, 333–42.

KATES J.R. & JONES R.F. (1964a) The control of gametic differentiation in liquid cultures of *Chlamydomonas*. *J. cell. comp. Physiol.* **63**, 157–64.

KATES J.R. & JONES R.F. (1964b) Variation in alanine dehydrogenase and glutamate dehydrogenase during the synchronous development of *Chlamydomonas*. *Biochim. biophys. Acta*, **86**, 438–47.

KEMP L. & COLE K. (1961) Chromosomal alternation of generations in *Nereocystis luetkeana* (Mertens) Postels & Ruprecht. *Can. J. Bot.* **39**, 1711–24.

KIES L. (1964) Über die experimentelle Auslösung von Fortpflanzungsvorgängen und die Zygotenkeimung bei *Closterium acerosum* (Schrank) Ehrenbg. *Arch. Protistenk.* **107**, 331–50.

KLEBS G. (1896) *Die Bedingungen der Fortpflanzung bei einigen Algen und Pilzen*. Fischer, Jena.

KNAGGS F.W. (1966a) Role of light energy in the initiation and development of tetrasporangia on cultured specimens of a red algal species. *Nature, Lond.* **212**, 431.

KNAGGS F.W. (1966b) *Rhodochorton purpureum* (Lightf.) Rosenvinge. Observations on the relationship between reproduction and environment. I. The relationship between the energy of incident light and tetrasporangium production. *Nova Hedwigia*, **11**, 405–11.

KNAGGS F.W. (1967) *Rhodochorton floridulum* (Dillwn.) Näg. Observations on the relationship between reproduction and the environment. *Nova Hedwigia*, **14**, 31–8.

KOCHERT G. (1968) Differentiation of reproductive cells in *Volvox carteri*. *J. Protozool.* **15**, 438–52.

KOCHERT G. & YATES I. (1970) A UV-labile morphogenetic substance in *Volvox carteri*. *Devl. Biol.* **23**, 128–35.

KÖHLER K. (1956) Entwicklungsgeschichte, Geschlechtsbestimmung und Befruchtung bei *Chaetomorpha*. *Arch. Protistenk.* **101**, 223–68.

KORNMANN P. (1961) Über *Codiolum* und *Urospora*. *Helgoländer wiss. Meeresunters.* **8**, 42–57.

KORNMANN P. (1962) Plurilokuläre Sporangien bei *Elachista fucicola*. *Helgoländer wiss. Meeresunters.* **8**, 293–7.

KORNMANN P. (1970) Advances in marine phycology on the basis of cultivation. *Helgoländer wiss. Meeresunters.* **20**, 39–61.

KOWALLIK W. (1970) Light effects on carbohydrate and protein metabolism in algae. In *Photobiology of Microorganisms*, ed. Halldal P. pp. 165–85. Wiley-Interscience, London, New York, Sydney & Toronto.

KUMKE J. (1971) Spektrale Empfindlichkeitsverteilung der lichtinduzierten Oogon-Entleerung bei der Braunalge *Dictyota*. *Naturwissenschaften*, **58**, 571.

KUROGI M. & HIRANO K. (1956) Influences of water temperature on the growth, formation of monosporangia and monospore-liberation in the *Conchocelis*-phase of *Porphyra tenera* Kjellm. *Bull. Tohoku Reg. Fish. Res. Lab.* **8**, 45–61.

KUROGI M., AKIYAMA K. & SATO S. (1962) Influences of light on the growth and maturation of *Conchocelis*-thallus of *Porphyra*. I. Effect of photoperiod on the formation of monosporangia and liberation of monospores. *Bull. Tohoku Reg. Fish. Res. Lab.* **20**, 121–126.

KUROGI M., SATO S. & YOSHIDA T. (1967) Effect of water temperature on the liberation of monospores from the *Conchocelis* of *Porphyra umbilicalis* (L.) Kütz. *Bull. Tohoku Reg. Fish. Res. Lab.* **27**, 131–9.

LANG A. (1965) Physiology of growth and development in algae. A synopsis. *Handb. PflPhysiol.* **15/1**, 680–715.

LERSTEN N.R. & VOTH P.D. (1960) Experimental control of zoid discharge and rhizoid formation in the green alga *Enteromorpha*. *Bot. Gaz.* **122**, 33–45.

LIDDLE L.B. (1968) Reproduction in *Zonaria farlowii*. I. Gametogenesis, sporogenesis and embryology. *J. Phycol.* **4**, 298–305.

LIPPERT B.E. (1969) The effect of carbon dioxide on conjugation in *Closterium*. *XI Int. Bot. Congr. Abs.* p. 129.

LORENZEN H. (1970) Synchronous cultures. In *Photobiology of Microorganisms*, ed. Halldal P. pp. 187–212. Wiley-Interscience, London, New York, Sydney & Toronto.

LÜNING K. & DRING M.J. (1972) Reproduction induced by blue light in female gametophytes of *Laminaria saccharina*. *Planta*, **104**, 252–6.

MACHLIS L. & RAWITSCHER-KUNKEL E. (1967) Mechanisms of gametic approach in plants. In *Fertilization: Comparative Morphology, Biochemistry and Immunology*, eds. Metz C.B. & Monroy A. pp. 117–61. vol. 1. Academic Press, New York & London.

MATSUI T. (1969) Studies on the liberation and germination of spores in *Gloiopeltis tenax* (Turn.) J. Ag. and *G. furcata* Post. et Rupr. *J. Shimonoseki Univ. Fish.* **17**, 189–231.

MONER J.G. (1954) Evidence for a swarming substance which stimulates colony formation in the development of *Pediastrum duplex* Meyen. *Biol. Bull. mar. biol. Lab.*, *Woods Hole*, **107**, 236–46.

MOSS B. (1967) The culture of fertile tissue of *Fucus vesiculosus*. *Br. phycol. Bull.* **3**, 209–12.

MÜLLER D. (1962) Über jahres- und lunarperiodische Erscheinungen bei einigen Braunalgen. *Botanica mar.* **4**, 140–55.

MÜLLER D.G. (1963) Die Temperaturabhängigkeit der Sporangienbildung bei *Ectocarpus siliculosus* von verschiedenen Standorten. *Pubbl. Staz. zool. Napoli*, **33**, 310–14.

MÜLLER D.G. (1968) Versuche zur Charakterisierung eines Sexual-Lockstoffes bei der Braunalge *Ectocarpus siliculosus*. I. Methoden, Isolierung und gaschromatographischer Nachweis. *Planta*, **81**, 160–8.

MÜLLER D.G. & JAENICKE L. (1973) Fucoserraten, the female sex attractant of *Fucus serratus* L. (Phaeophyta). *F.E.B.S. Letters*, **30**, 137–9.

MCCRACKEN M.D. & STARR R.C. (1970) Induction and development of reproductive cells in the K-32 strains of *Volvox rousseletii*. *Arch. Protistenk.* **112**, 262–82.

MCLEAN R.J. & TRAINOR F.R. (1965) Zoospore inhibition in *Spongiochloris typica*. *J. Phycol.* **1**, 58–60.

NAKAHARA H. & NAKAMURA Y. (1971) The life-history of *Desmarestia tabacoides* Okamura. *Bot. Mag.*, *Tokyo*, **84**, 69–75.

NEEB O. (1952) *Hydrodictyon* als Objekt einer vergleichenden Untersuchung physiologische Größen. *Flora*, *Jena*, **139**, 39–95.

O'KELLEY J.C. & DEASON T.R. (1962) Effect of nitrogen, sulphur and other factors on zoospore production by *Protosiphon botryoides*. *Am. J. Bot.* **49**, 771–7.

O'KELLEY J.C. & HERNDON W.R. (1961) Alkaline earth elements and zoospore release and development in *Protosiphon botryoides*. *Am. J. Bot.* **48**, 796–802.

PAGE J.Z. & KINGSBURY J.M. (1968) Culture studies on the marine green alga *Halicystis parvula-Derbesia tenuissima*. II. Synchrony and periodicity in gamete formation and release. *Am. J. Bot.* **55**, 1–11.

PAGE J.Z. & SWEENEY B.M. (1968) Culture studies on the marine green alga *Halicystis parvula-Derbesia tenuissima*. III. Control of gamete formation by an endogenous rhythm. *J. Phycol.* **4**, 253–60.

POLLOCK E.G. (1970) Fertilisation in *Fucus*. *Planta*, **92**, 85–99.

PRINGSHEIM E.G. (1918) Die Kultur der Desmidiaceen. *Ber. dt. bot. Ges.* **36**, 482–5.

PUISEUX-DAO S. (1970) *Acetabularia and Cell Biology*. Logos Press, London.

RAWITSCHER-KUNKEL E. & MACHLIS L. (1962) The hormonal integration of sexual reproduction in *Oedogonium*. *Am. J. Bot.* **49**, 177–83.

RENTSCHLER H.-G.(1967) Photoperiodische Induktion der Monosporenbildung bei *Porphyra tenera* Kjellm. (Rhodophyta-Bangiophyceae). *Planta*, **76**, 65–74.

RHODES R.G. (1970) Relation of temperature to development of the macrothallus of *Desmotrichum undulatum*. *J. Phycol.* **6**, 312–14.

RHODES R.G. & HERNDON W.R. (1967) Relationship of temperature to zoospore production in *Tetraspora gelatinosa*. *J. Phycol*. **3**, 1–3.

RICHARDSON N. (1970) Studies on the photobiology of *Bangia fuscopurpurea*. *J. Phycol*. **6**, 215–19.

RICHTER G. & KIRSCHSTEIN M.J. (1966) Regeneration und Photosynthese-Leistung kern-haltinger Zell-Teilstücke von *Acetabularia* in blauer roter Strahlung. *Z. PflPhysiol*. **54**, 106–17.

RIETH A. (1959) Periodizität beim Ausschlüpfen der Schwärmsporen von *Vaucheria sessilis* de Candolle. *Flora, Jena*, **147**, 35–42.

ROZUMEK K.-E. (1968) Der Einfluß der Umweltfaktoren Licht und Temperatur auf die Ausbildung der Sexualstadien bei der pennaten Diatomee *Rhabdonema adriaticum* Kütz. *Beitr. Biol. Pfl*. **44**, 365–88.

RUDDAT M. (1960) Versuche zur Beeinflussung und Auslösung der endogenen Tages-rhythmik bei *Oedogonium cardiacum* Wittr. *Z. Bot*. **49**, 23–46.

SAGER R. & GRANICK S. (1954) Nutritional control of sexuality in *Chlamydomonas rein-hardi*. *J. gen. Physiol*. **37**, 729–42.

SAGROMSKY H. (1961) Durch Licht-Dunkel-Wechsel induzierter Rhythmus der Entleerung der Tetrasporangien von *Nitophyllum punctatum*. *Pubbl. Staz. zool. Napoli*, **32**, 29–40.

SCHREIBER E. (1930) Untersuchungen über Parthenogenesis, Geschlechtsbestimmung und Bastardierungsvermögen bei Laminarien. *Planta*, **12**, 331–53.

SCHULTZ M.E. & TRAINOR F.R. (1968) Production of male gametes and auxospores in the centric diatoms *Cyclotella meneghiniana* and *C. cryptica*. *J. Phycol*. **4**, 85–8.

SCHULTZ M.E. & TRAINOR F.R. (1970) Production of male gametes and auxospores in a polymorphic clone of the centric diatom *Cyclotella*. *Can. J. Bot*. **48**, 947–51.

SEARS J.R. & WILCE R.T. (1970) Reproduction and systematics of the marine alga *Derbesia* (Chlorophyceae) in New England. *J. Phycol*. **6**, 381–92.

SENGER H. & SCHOSER G. (1966) Die spektralabhängige Teilungsinduktion in mixotrophen Synchronkulturen von *Chlorella*. *Z. PflPhysiol*. **54**, 308–20.

SHIHIRA I. (1958) The effect of light on gamete liberation in *Monostroma*. *Bot. Mag., Tokyo*, **71**, 378–85.

SMITH G.M. (1944) A comparative study of the species of *Volvox*. *Trans. Amer. Microsc. Soc*. **63**, 265–310.

SMITH G.M. (1947) On the reproduction of some Pacific coast species of *Ulva*. *Am. J. Bot*. **34**, 80–7.

SPENCER T. (1968) Effect of kinetin on the phosphatase enzymes of *Acetabularia*. *Nature, Lond*. **217**, 62–4.

STARR R.C. (1955) Isolation of sexual strains of placoderm desmids. *Bull. Torrey bot. Club*. **82**, 261–5.

STARR R.C. (1968) Cellular differentiation in *Volvox*. *Proc. natn. Acad. Sci., U.S.A*. **59**, 1082–8.

STARR R.C. (1970) Control of differentiation in *Volvox*. *Developmental Biology Supplement*, **4**, 59–100.

STARR R.C. (1972) A working model for the control of differentiation during development of the embryo of *Volvox carteri* f. *nagariensis*. *Soc. bot. Fr., Mémoires*, 1972, pp. 175–82.

STEELE R.L. (1965) Induction of sexuality in two centric diatoms. *Bioscience*, **15**, 298.

STEIN J.R. (1966a) Growth and mating of *Gonium pectorale* (Volvocales) in defined media. *J. Phycol*. **2**, 23–8.

STEIN J.R. (1966b) Effect of temperature on sexual populations of *Gonium pectorale* (Vol-vocales). *Am. J. Bot*. **53**, 941–4.

STOSCH H.A.v. & DREBES G. (1964) Entwicklungsgeschichtliche Untersuchungen an zentrischen Diatomeen. IV. Die Planktondiatomee *Stephanopyxis turris*—ihre Behand-lung und Entwicklungsgeschichte. *Helgoländer wiss. Meeresunters*. **11**, 209–57.

SUBBARAMAIAH K. (1970) Growth and reproduction of *Ulva fasciata* Delele in nature and in culture. *Botanica mar*. **8**, 25–7.

SUBRAHMANYAN R. (1957) Observations on the anatomy, cytology, development of the reproductive structures, fertilization and embryology of *Pelvetia canaliculata* Dene. et Thur. III. The liberation of reproductive bodies, fertilization and embryology. *J. Indian bot. Soc.* **36**, 373–95.

SUNDENE O. (1963) Reproduction and ecology of *Chorda tomentosa*. *Nytt. Mag. Bot.* **10**, 159–67.

SUTO S. (1950) Studies on shedding, swimming and fixing of the spores of seaweeds. *Bull. Jap. Soc. scient. Fish.* **16**, 1–9.

SUTO S. (1951) On shedding of eggs, liberation of embryos, and their later fixing in *Hizikia fusiforme*. *Bull. Jap. Soc. scient. Fish.* **17**, 9–12.

SWEENEY B.M. & HASTINGS J.W. (1962) Rhythms. In *Physiology and Biochemistry of Algae*, ed. Lewin R.A. pp. 687–700. Academic Press, New York & London.

TERAMOTO K. & KINOSHITA S. (1960) On the effects of amino-acids and purines on the growth of *Porphyra*. *Bull. Jap. Soc. Phycol.* **8**, 90–5.

TERBORGH J. (1965) Effects of red and blue light on the growth and morphogenesis of *Acetabularia crenulata*. *Nature, Lond.* **207**, 1360–3.

TERBORGH J. (1966) Potentiation of photosynthetic oxygen evolution in red light by small quantities of monochromatic blue light. *Pl. Physiol., Lancaster*, **41**, 1401–10.

TERBORGH J.W. & THIMANN K.V. (1965) The control of development in *Acetabularia crenulata* by light. *Planta*, **64**, 241–53.

TRAINOR F.R. (1959) A comparative study of sexual reproduction in four species of *Chlamydomonas*. *Am. J. Bot.* **46**, 65–70.

TRAINOR F.R. & BURG C.A. (1965a) Motility in *Scenedesmus dimorphus*, *Scenedesmus obliquus* and *Coelastrum microporum*. *J. Phycol.* **1**, 15–18.

TRAINOR F.R. & BURG C.A. (1965b) *Scenedesmus obliquus* sexuality. *Science*, **148**, 1094–5.

TSENG C.K. & CHANG T.J. (1956) Conditions of *Porphyra* conchospores formation and discharge and discharge rhythm. *Acta bot. sinica*, **5**, 33–49.

TSUBO Y. (1956) Observations on sexual reproduction in a *Chlamydomonas*. *Bot. Mag., Tokyo*, **69**, 1–6.

TSUBO Y. (1961) Sexual reproduction of *Chlamydomonas* as affected by ionic balance in the medium. *Bot. Mag., Tokyo*, **74**, 442–8.

VANDE BERG W.J. & STARR R.C. (1971) Structure, reproduction and differentiation in *Volvox gigas* and *Volvox powersii*. *Arch. Protistenk.* **113**, 195–219.

VIELHABEN V. (1963) Zur Deutung des semilunaren Fortpflanzungszyklus von *Dictyota dichotoma*. *Z. Bot.* **51**, 156–73.

WARIS H. & KALLIO P. (1964) Morphogenesis in *Micrasterias*. *Adv. Morphogenesis*, **4**, 45–80.

WERNER D. (1971a) Der Entwicklungscyclus mit Sexualphase bei der marinen Diatomee *Coscinodiscus asteromphalus*. I. Kultur und Synchronisation von Entwicklungsstadien. *Arch. Mikrobiol.* **80**, 43–9.

WERNER D. (1971b) Der Entwicklungscyclus mit Sexualphase bei der marinen Diatomee *Coscinodiscus asteromphalus*. II. Oberflächenabhängige Differenzierung während der vegetativen Zellverkleinerung. *Arch. Mikrobiol.* **80**, 115–33.

WERNER D. (1971c) Der Entwicklungscyclus mit Sexualphase bei der marinen Diatomee *Coscinodiscus asteromphalus*. III. Differenzierung und Spermatogenese. *Arch. Mikrobiol.* **80**, 134–46.

WIESE L. (1969) Algae. In *Fertilization. Comparative Morphology, Biochemistry and Immunology*, vol. 2, eds. Metz C.B. & Monroy A. pp. 135–88. Academic Press, New York & London.

WIESE L. & SHOEMAKER D.W. (1970) On sexual agglutination and mating-type substances (gamones) in isogamous heterothallic Chlamydomonads. II. The effect of Concanavalin A upon the mating type reaction. *Biol. Bull. mar. biol. Lab., Woods Hole*, **138**, 88–95.

ZIEGLER J.R. & KINGSBURY J.M. (1964) Cultural studies on the marine green alga *Halicystis parvula-Derbesia tenuissima*. I. Normal and abnormal sexual and asexual reproduction. *Phycologia*, **4**, 105–16.

# CHAPTER 30

# EXTRACELLULAR PRODUCTS

## J. A. HELLEBUST

Department of Botany,
University of Toronto,
Toronto, Canada.

1   Introduction  838

2   Nature of extracellular products  839
2.1  Carbohydrates  839
2.2  Nitrogenous substances  841
2.3  Organic acids  842
2.4  Lipids  844
2.5  Phenolic substances  844
2.6  Organic phosphates  844
2.7  Volatile substances  844
2.8  Enzymes  845
2.9  Vitamins  845
2.10 Sex factors  846
2.11 Growth inhibitors and stimulators  846
2.12 Toxins  848

3   Characteristics of the release processes  848
3.1  General  848
3.2  Growth stage  849
3.3  Environmental and physiological factors  850

4   Ecological aspects  852
4.1  Primary productivity  852
4.2  Symbiosis  853
4.3  Succession  854

5   References  854

## 1  INTRODUCTION

The production of a great variety of extracellular substances by algae is now well established. It is also clear that such substances often play important roles in algal growth and physiology, as well as in aquatic food chains and ecosystems in general. Reviews on this subject have been published by Fogg (1962, 1966, 1971), and extensive discussions of relevant work in the field may also be found in some recent original publications (Chapman & Rae 1969, Gocke 1970, Weinmann 1970, Aaronson 1971).

The production of extracellular substances will be referred to as liberation, release, or excretion without attributing different meanings to these terms.

Possible mechanisms for the release of substances by algae will be discussed following a consideration of the nature of such products. Only a brief account will be given of the important ecological aspects of extracellular products.

Because of the increasingly voluminous literature on extracellular products some of the relevant work will be listed in tables, and only a portion of the available information will be discussed in any detail.

# 2 NATURE OF EXTRACELLULAR PRODUCTS

## 2.1 *Carbohydrates*

Simple and complex polysaccharides are liberated by a large number of taxonomically diverse algae (Table 30.1). The amounts released may represent a considerable fraction of the photoassimilated carbon of some algae during active growth; e.g. mucilaginous *Chlamydomonas* spp., 15–57% (Lewin 1956), *Porphyridium cruentum*, about 15% (Jones 1962), *Katodinium dorsalisulcum*, up to 90% of total polysaccharides produced (Prager *et al.* 1959), *Anabaena flos-aquae*, 28% (Moore & Tischer 1965), *Phaeocystis pouchetii*, 16 to 64% (Guillard & Hellebust 1971). Other algae produce extracellular polysaccharides mainly when they enter stationary growth phase; e.g. *Isochrysis galbana*, *Prymnesium parvum*, *Dunaliella tertiolecta*, *Rhodomonas* sp., *Pyramimonas* sp. (Guillard & Wangersky 1958, Marker 1965).

Common constituents of the extracellular polysaccharides, and heteropolysaccharides are: glucose, galactose, mannose, rhamnose, fucose, arabinose, xylose and uronic acids (Bishop *et al.* 1954, Lewin 1956, Jones 1962, Moore & Tischer 1964, Marker 1965, Guillard & Hellebust 1971).

Simple sugars and sugar alcohols are usually found in very small amounts in extracellular products of algae, with the following notable exceptions: glucose by blue-green lichen symbionts, ribitol, erythritol and sorbitol by green lichen symbionts, glucose and maltose by zoochlorellae, glycerol by zoozanthellae and *Dunaliella tertiolecta*, mannitol by *Olisthodiscus* sp., and *Pyramimonas* sp., and cyclohexanetetrol ($U_1$ in Hellebust 1965, co-chromatographed and found to be identical with cyclohexanetetrol contained in *Monochrysis lutheri* (Ramanathan *et al.* 1966, Craigie pers. comm.)) by *Monochrysis lutheri* and *Isochrysis galbana* (Table 30.1).

A variety of complex polysaccharides produced by green, brown and red seaweeds are released through the cell membranes to form the cell wall matrix (Percival & McDowell 1967). These substances may be considered extracellular products, and may account for part of the large quantities of polysaccharide materials released by seaweeds into the aquatic environment under various conditions (Sieburth 1969).

Other cases of release of polysaccharide material that at least initially remains part of algal cell walls or colony structures are numerous, and a few may be mentioned, for example, the mucilage tubes of *Amphipleura rutilans* (Lewin 1958), the exocellular sheath of *Phaeodactylum tricornutum* (Lewin *et al.* 1958), the mucilaginous stalks of *Gomphonema olivaceum* (Huntsman & Sloneker 1971), and chitan fibres of *Thalassiosira fluviatilis* and *Cyclotella cryptica* (McLachlan *et al.* 1965, Blackwell *et al.* 1967).

**Table 30.1.** Extracellular carbohydrates

| Family or group | Genera | Carbohydrate | References |
|---|---|---|---|
| Cyanophyceae | *Anabaena, Nostoc* | polysaccharides | Hough *et al.* (1952), Moore & Tischer (1965), Bishop *et al.* (1954) |
| | *Anabaena, Nostoc, Calothrix, Scytonema* (lichen symbionts) | simple sugars | Fogg (1952), Drew & Smith (1967), Richardson *et al.* (1968), Smith *et al.* (1969) |
| Chlorophyceae* | *Chlamydomonas, Scenedesmus, Platymonas, Palmella, Oocystis, Dunaliella, Pyramimonas* | polysaccharides | Lewin (1956), Allen (1956), Guillard & Wangersky (1958), Maksimova & Pimenova (1969), Weinmann (1970) |
| | *Chlorella, Scenedesmus,* | simple sugars | Tolbert & Zill (1967), |
| | Zoochlorellae | simple sugars | Muscatine (1965), Muscatine *et al.* (1967), Weinmann (1970) |
| | *Chlorella, Chlorococcum, Pyramimonas, Dunaliella. Trebouxia* and other lichen symbionts | sugar alcohols | Hellebust (1965), Richardson *et al.* (1968) |
| Bacillariophyceae | *Nitzschia, Synedra, Stephanodiscus, Skeletonema, Thalassiosira* | polysaccharides | Lewin (1955), Hellebust (1965), Tokuda (1969), Watt (1969) |
| Dinophyceae | *Katodinium, Gymnodinium* | polysaccharides | McLaughlin *et al.* (1960) Khailov *et al.* (1967) |
| Zooxanthellae | | glycerol | Muscatine (1967), Smith *et al.* (1969) |
| Xanthophyceae | *Olisthodiscus* | mannitol | Hellebust (1965) |
| Phaeophyceae | *Fucus, Ascophyllum, Laminaria* | polysaccharides heteropolysaccharides | Sieburth (1969) |
| Rhodophyceae | *Porphyridium, Chondrus* | polysaccharides heteropolysaccharides | Jones (1962), Sieburth (1969) |

* Including Prasinophyceae

## 2.2 *Nitrogenous substances*

Amino acids and peptides are very common in algal filtrates (Table 30.2), but in most cases represent only a small fraction of the total extracellular material

**Table 30.2.** Extracellular nitrogenous substances

| Family or group | Genera | Nitrogenous substances | References |
|---|---|---|---|
| Cyanophyceae | *Anabaena, Calothrix, Chlorogloea, Coccochloris, Lyngbya, Microcystis, Nostoc, Oscillatoria, Westiellopsis* | amino acids, peptides, proteins | Watanabe (1951), Fogg (1952), Stewart (1963, 1964), Fay *et al.* (1964), Hellebust (1965), Whitton (1965), Pattnaik (1966), Pearce & Carr (1967, 1969), Rzhanova (1967), Fogg & Pattnaik (1968), Jones & Stewart (1969) |
| Chlorophyceae* | *Scenedesmus* | glucosamine | Weinmann (1970) |
| | *Chlamydomonas* | nicotinic acid | Nakamura & Gowans (1964) |
| | *Chlamydomonas, Chlorella, Scenedesmus, Dunaliella, Pyramimonas* | amino acids, peptides | Allen (1956), Hellebust (1965), Sen (1966), Gocke (1970), Weinmann (1970) |
| | *Chlamydomonas, Volvox* | proteins | Förster *et al.* (1956), Darden (1966) |
| Euglenophyceae | *Euglena* | amino acids glycoproteins protoporphyrin | McCalla (1963) Daisley (1970) Dubash & Rege (1967) |
| Chrysophyceae | *Ochromonas* | amino acids, nucleic acids, proteins | Aaronson (1971), Aaronson *et al.* (1971) |
| Dinophyceae | *Exuviaella, Gymnodinium, Peridinium* | amino acids, peptides, protein | Hellebust (1965) Khailov *et al.* (1967) |
| Zooxanthellae | | nucleoside-polyphosphates | von Holt (1968) |
| Bacillariophyceae | *Chaetoceros, Cyclotella, Skeletonema, Stephanodiscus, Synedra, Rhizosolenia, Thalassiosira* | amino acids, peptides | Hellebust (1965), Watt (1969) |
| Rhodophyceae | *Porphyridium, Chondrus, Polysiphonia* | complex nitrogenous (protein) materials | Jones (1962), Sieburth (1969) |
| Phaeophyceae | *Ectocarpus Ascophyllum, Fucus, Laminaria* | complex nitrogenous (protein) materials | Fogg & Boalch (1958), Sieburth (1969) |

* Including Prasinophyceae

(Hellebust 1965, Watt 1966, 1969, Gocke 1970). Blue-green algae, on the other hand, liberate very large portions of their assimilated nitrogenous substances into the medium (Fogg 1952, 1966, Jones & Stewart 1969). *Anabaena cylindrica* releases about 30% of its organic nitrogen in rapidly growing cultures (Fogg 1952), and similar relative amounts of extracellular nitrogenous material have been reported for other nitrogen fixing blue-greens (14–42%, Watanabe 1951). *Calothrix scopulorum* releases 20–60% of its assimilated nitrogen, and, immediately after transfer to less favourable growth conditions the proportion released may be even higher (Jones & Stewart 1969). Data from experiments on the rate of $^{15}$N-labelling of intra- and extracellular nitrogenous compounds show that the release of organic nitrogen is not specifically associated with the $N_2$-fixing process (Stewart 1964, Jones & Stewart 1969).

A large proportion of the extracellular nitrogenous material released by blue-green algae is in the form of polypeptides, and only small amounts of free amino acids are usually found (Watanabe 1951, Fogg 1952, Stewart 1963, Whitton 1965, Jones & Stewart 1969). The extracellular peptides consist of about a dozen amino acids with glycine, glutamic acid, aspartic acid, alanine and serine predominating, and no, or trace amounts of basic amino acids (Whitton 1965, Jones & Stewart 1969).

Several algae (species of *Euglena*, *Monochrysis*, *Ochromonas*, *Skeletonema*) release high-molecular weight vitamin-$B_{12}$ binding substances (Kristensen 1956, Ford 1958, Droop 1968, 1970, Daisley 1970). Daisley demonstrated that the binding factor released by *Euglena gracilis* is a glycoprotein. The factor appears in the media of healthy, growing cells. Other extracellular proteins, or protein-containing substances with unique properties, such as toxins, growth inhibitors or promoters, reproductive factors and enzymes will be discussed elsewhere in this chapter.

### 2.3   *Organic acids*

Glycolic acid is the organic acid most commonly liberated by algae. Only some of the publications dealing with glycolic acid release by *Chlorella* are listed in Table 30.3 (see Fogg 1966, and Chapter 17, p. 474). Hellebust (1965) found glycolate in the filtrates of all but one of 23 marine planktonic algal species studied. However, only for four of these algae did glycolate constitute more than 10% of the total extracellular products. Watt (1969) found that glycolate was the major extracellular product of the freshwater diatom *Stephanodiscus hantzschii* but no glycolate was detected in the filtrate of another freshwater diatom, *Synedra acus*, grown under similar conditions. Glycolate has also been detected in experiments with natural freshwater (Fogg & Nalewajko 1964, Fogg & Watt 1965, Watt 1966) and marine (Hellebust 1965) phytoplankton. The release of glycolate by algae is favoured by conditions where $CO_2$ limits photosynthesis (Pritchard *et al.* 1962, Watt & Fogg 1966). The studies to date indicate that the amount of glycolate released under favourable conditions varies considerably for different species.

Only a few reports deal with the release of other organic acids (Table 30.3, see also Fogg 1962). Algae grown anaerobically liberate large quantities of fermentation products including lactate which is most common, acetate and formate.

**Table 30.3.** Extracellular organic acids

| Family or group | Genera | Organic acids | References |
|---|---|---|---|
| Cyanophyceae | *Coccochloris, Anacystis, Oscillatoria* | glycolate | Hellebust (1965), Döhler & Braun (1971), Cheng (1971) |
| Chlorophyceae* | *Chlorella, Chlorococcum, Dunaliella, Pyramimonas, Hormotila, Ankistrodesmus, Chlamydomonas, Scenedesmus,* | glycolate | Allen (1956), Tolbert & Zill (1957), Nalewajko et al. (1963), Whittingham & Pritchard (1963), Hellebust (1965), Watt & Fogg (1966), Muscatine (1967), Monahan & Monahan (1971), Weinmann (1970) |
| | Zoochlorellae | glycolate | |
| | *Chlorella, Scenedesmus* | formate, acetate lactate | Syrett & Wong (1963), Begum & Syrett (1970), Maksimova & Pimenova (1969), Hirt et al. (1971) |
| | *Chlamydomonas, Chlorella* | keto acids | Maksimova & Pimenova (1969), Collins & Kalnins (1967) |
| | *Ankistrodesmus* | mesotartrate isocitrate | Chang (1967) |
| | *Gloeomonas* | acid semialdehyde | Badour & Waygood (1971) |
| Xanthophyceae | *Olisothodiscus* | glycolate | Hellebust (1965) |
| Chrysophyceae | *Monochrysis* | glycolate | Hellebust (1965) |
| Haptophyceae | *Coccolithus, Cricosphaera, Isochrysis* | glycolate | Hellebust (1965) |
| Dinophyceae | *Exuviaella, Amphidinium, Peridinium, Gymnodinium* | glycolate | Hellebust (1965) |
| Bacillariophyceae | *Chaetoceros, Skeletonema, Cyclotella, Thalassiosira, Rhizosolenia, Stephanodiscus* | glycolate | Hellebust (1965), Watt (1969) |

* Including Prasinophyceae

## 2.4 *Lipids*

Little is known about the possible release of lipids by healthy algal cells. Chloroform extracts of media of growing marine phytoplankton cultures contained from 2·8 to 10·3% of the total extracellular materials in the chloroform-soluble fraction, indicating the liberation of small amounts of lipid compounds (Hellebust 1965). Growth-inhibiting substances released by *Chlamydomonas* and *Chlorella* cells, particularly in old cultures, appear to be unsaturated fatty acids (Spoehr *et al.* 1949, Proctor 1957, Scutt 1964), or peroxides of such acids. The marine diatom, *Asterionella japonica*, releases a fatty acid which inhibits bacterial growth (Aubert *et al.* 1970). *Ochromonas danica*, under good growth conditions, releases into the medium a variety of membranous structures of high lipid content (Aaronson 1971).

## 2.5 *Phenolic substances*

Large quantities of phenolic materials are frequently released by brown seaweeds such as *Fucus vesiculosus*, *Ascophyllum nodosum* and *Laminaria* spp. (Khailov 1963, Craigie & McLachlan 1964, Conover & Sieburth 1966, Sieburth & Jensen 1969, Sieburth 1969). Craigie and McLachlan (1964) characterized the material released by *Fucus vesiculosus* and other brown algae as polyphenols such as flavanols or catechin-type tannins. Sieburth and Jensen (1968, 1969) found the complex polyphenolic substances released by brown seaweeds indistinguishable from the common yellow-coloured humic substances (Gelbstoff) found in high concentrations in inshore seawater (Kalle 1966). Other reports of the release of yellow-coloured ultra-violet-absorbing substances by brown algae may also involve phenolic material (Fogg & Boalch 1958, Armstrong & Boalch 1960, Yentsch & Reichert 1962).

## 2.6 *Organic phosphates*

Only a few reports concerning the release of organic phosphates by healthy algae have been published (Kuenzler & Ketchum 1962, Antia *et al.* 1963, Johannes 1964, Strickland & Solórzano 1966). However, according to a recent study by Kuenzler (1970) on several species of marine planktonic algae, this may be of common occurrence, particularly when inorganic phosphorus is abundant. Unhealthy or dying algae, e.g. at the end of a phytoplankton bloom, obviously release considerable amounts of organic phosphates (Ansell *et al.* 1963, Watt & Hayes 1963, Holmes *et al.* 1967). A special case of production of extracellular organic phosphates is the release of nucleoside-polyphosphates by zooxanthellae (von Holt 1968).

## 2.7 *Volatile substances*

Special odours are often associated with algal blooms indicating that many algae release volatile compounds. Armstrong and Boalch (1960) detected ultraviolet-absorbing volatile compounds released by several marine planktonic and macro-algae during healthy growth. Volatile organic substances, such as dimethyl sulphide, are frequently liberated by unicellular marine algae (Ishida 1968, Kadota & Ishida 1968).

The volatile compounds produced by some common odour-causing algae in freshwater reservoirs have been studied by Collins and co-workers. *Chlamydomonas globosa* releases formaldehyde, acetaldehyde and methyl ethyl ketone (Collins & Bean 1963). *Synura petersenii* releases acetaldehyde, furfuraldehyde, *n*-haptanal, acetone, valeraldehyde, ethyl acetate, ethyl alcohol and acetic acid (Collins & Kalnins 1965). Similar volatiles are also released by *Cryptomonas ovata* (Collins & Kalnins 1966). Volatile sex factors are produced by some algae (see below, p. 846).

### 2.8 *Enzymes*

*Ochromonas variabilis*, a member of the Chrysophyceae liberates enzymes which hydrolyse sucrose, starch, and proteins (Pringsheim 1952). Another species, *O. danica*, releases both acid and alkaline phosphatases even in the presence of inorganic phosphate (Aaronson 1971). Proteolytic enzymes are liberated by the colourless diatom *Nitzschia putrida* (Pringsheim 1951) and several species of green soil algae produce extracellular starch-degrading enzymes (Bischoff & Bold 1963). Another soil alga, *Monodus subterraneus*, a member of the Xanthophyceae, is able to hydrolyse glutamine extracellularly by releasing a glutaminase (Belmont & Miller 1965). The pennate diatoms *Nitzschia filiformis* and *N. frustulum* produce extracellular enzymes which depolymerize agar (Lewin & Lewin 1960).

Other extracellular enzymes secreted through the cell membrane, but which are chiefly associated with the cell wall, are acid and alkaline phosphatases produced by various algae in the absence of inorganic phosphates (Kuenzler & Perras 1965), and cellulases which digest the cell walls of *Scenedesmus obliquus*, and probably other green algae when autospores are released (Burczyk *et al.* 1970). Evidence for the release of cell wall degrading enzymes into the medium by *Chlamydomonas reinhardtii* has also been obtained (Schlösser 1966, Claes 1971).

### 2.9 *Vitamins*

Several species of marine planktonic algae produce extracellular vitamins. *Dunaliella tertiolecta* and *Skeletonema costatum* release thiamine and *Cyclotella nana*, *Phaeodactylum tricornutum* and *S. costatum* release biotin (Carlucci & Bowes 1970a,b). Extracellular production of vitamins occurs during exponential growth and during stationary phase conditions (Carlucci & Bowes 1970a). Burkholder (1963) obtained evidence for extracellular production of thiamine by several seaweeds, as well as by the dinoflagellate *Gonyaulax tamerensis*.

High concentrations of dissolved vitamin $B_{12}$, thiamine, and biotin have been associated with high phytoplankton biomass indicating that in nature phytoplankton may be responsible for the production of these vitamins (Carlucci 1970). Experiments on the production and utilization of vitamins by algae in mixed cultures frequently indicate that certain species release toxic substances. Carlucci and Bowes (1970b) suggest that these substances may in some cases be similar to the vitamin-binding factors described by Droop (1968, 1970) and Daisley (1970).

The following vitamins have been detected in media of autotrophically growing axenic cultures of the freshwater alga *Ochromonas danica*: ascorbate, vitamin $B_6$, $N^5$-methyltetrahydrofolate, tetrahydrofolate polyglutamates, nicotinic acid, pantothenate, riboflavin and vitamin E (Aaronson *et al.* 1971). *Chlorella pyrenoidosa* and the green, lichen alga, *Coccomyxa* sp. liberate biotin (Bednar & Holm-Hansen 1964). A special case of rapid excretion is that of nicotinic acid by a *Chlamydomonas eugametos* mutant (Nakamura & Gowans 1964) (see also Chapter 27, p. 741).

### 2.10  *Sex factors*

Sperm-attracting substances are produced extracellularly by several genera of green algae (Pascher 1931a,b, Hoffman 1960). However, Ellis and Machlis (1968) could find no evidence for the release of substances which chemotactically attracted the sperm of a *Golenkinia* sp. *Fucus* eggs liberate a volatile substance which attracts sperms (Cook & Elvidge 1951), and similarly, female gametes of another brown alga, *Ectocarpus siliculosus*, release a highly volatile sexual attractant for male gametes (Müller 1968).

A female sex hormone (gamone) liberated by *Chlamydomonas eugametos* gametes causes agglutination of gametes of opposite mating types (Förster *et al.* 1956). The substance is a glycoprotein of very high molecular weight. Cells of both mating types of *C. moewusii* liberate a substance into the medium in the light which increases mating activities of cells in the dark (Tsubo 1961). Gametes of *Chlamydomonas reinhardtii* produce an extracellular lytic factor which serves to lyse cell walls of the gametes prior to protoplasmic fusion (Claes 1971). The factor is heat-labile and is probably an enzyme. The colonial green alga, *Volvox aureus*, releases a large, relatively heat-stable substance which induces differentiation of male colonies (Darden 1966). Proteolytic enzymes destroy the activity of the substance, indicating that it consists at least partly of protein (see also Chapter 27, p. 741; and Chapter 29, p. 814).

### 2.11  *Growth inhibitors and stimulators*

Many algae release substances which inhibit either their own growth, that of other species, or both. The extensive early literature concerning such compounds has been reviewed by Fogg (1962) and Lefèvre (1964), and will not be discussed here. General aspects of the influence of algal inhibitors in the ecology of marine ecosystems have been discussed by Sieburth (1968).

Reports on the formation of growth-inhibiting peroxides of unsaturated fatty acids (chlorellin) by *Chlorella* (Spoehr *et al.* 1949) have been confirmed although they are not found in filtrates of growing cultures (Scutt 1964). Peroxides of polyunsaturated fatty acids also accumulate in stationary phase cells of *Ochromonas danica* but there is no evidence that they accumulate in the medium (Aaronson & Bensky 1967), although this is implied in a study of a *Nitzschia* sp. by Badour and Gergis (1965).

Yellow polyphenolic substances liberated by *Fucus vesiculosus* are inhibitory to some unicellular algae, including *Skeletonema costatum* (Craigie & McLachlan 1964). However, some phenolic substances liberated by red and brown seaweeds and some unicellular algae appear to be necessary for normal development and completion of life cycles of *Ulva* and *Monostroma* (Provasoli 1965). Several seaweeds produce extracellular antibiotic substances (Conover & Sieburth 1964, Sieburth 1968).

Unicellular algae which are reported to produce extracellular antibacterial substances include *Isochrysis galbana* (Bruce & Duff 1967), *Asterionella japonica* (Aubert *et al.* 1970), and *Pandorina morum* (Harris 1971a). The production of extracellular antibacterial substances (fatty acid and nucleosides) by *A. japonica* appears to be inhibited by the release of a protein by *Prorocentrum micans* in mixed culture (Aubert 1971b). Small amounts of acrylic acid, an antibacterial substance present in *Phaeocystis pouchetii* (Sieburth 1959), are released into the medium by growing cultures of this alga (Guillard & Hellebust 1971).

A comprehensive study of the production of extracellular autoinhibitors and mutually inhibiting substances by members of the Volvocaceae indicates a complex system of growth inhibitors (Harris 1971b). For example, the autoinhibitory substance produced by stationary phase cultures of *Platydorina caudata* appears to be a protein which is specific for this species (Harris 1970) and yet culture filtrates from *Volvulina pringsheimii* and *Eudorina california* are inhibitory to most members of this family (Harris 1971a).

*Cricosphaera elongata*, during its motile phase, produces an extracellular substance, inhibitory to wheat coleoptiles, and with properties similar to abscisic acid (Hussain & Boney unpubl., see Boney 1970). *Peridinium polonicum*, a freshwater dinoflagellate, liberates a toxin which inhibits the growth of the green algae *Scenedesmus obliquus* and *Dunaliella tertiolecta* (Nozawa 1970). A recent example of chemical warfare by two unrelated marine planktonic algae is that of the inhibition of *Skeletonema costatum* by *Olisthodiscus luteus* (Pratt 1966).

Nalewajko *et al.* (1963) reported that glycolic acid, a common algal extracellular product, enhances growth of a planktonic strain of *Chlorella pyrenoidosa* by eliminating the lag phase. A recent study by Droop (1966) demonstrates that salts of several weak organic acids, including glycolate, eliminate the lag phase of algal cells in unbuffered media by increasing the capacity of the media for $CO_2$. The autostimulatory effects of *Hormotila blennista* filtrates are probably due, in part at least, to the liberation of weak acids by this alga (Monahan & Trainor 1970). *Enteromorpha linza* liberates organic compounds which stimulate the growth of this and another *Enteromorpha* species (Berglund 1969).

## 2.12   *Toxins*

Substances toxic to animals are produced by several algae, and in some cases at least are released into the medium. *Prymnesium parvum* when grown photo-autotrophically produces proteinaceous extracellular toxins (Shilo *et al*. 1953) which are released only during the late exponential growth phase (Shilo 1967). Toxins produced by blue-green algae, e.g. *Microcystis aeruginosa* and *Apha-nizomenon flos-aquae*, are usually endotoxins, being released only after cell lysis (see Shilo 1967, Gentile & Maloney 1969). However, the 'very fast-death' factor produced by *Anabaena flos-aquae* is often released into the surrounding water during growth (Gorham *et al*. 1964).

McLaughlin and Provasoli (1957) reported that the dinoflagellates *Amphi-dinium klebsi* and *A. rhynchocephalum* release substances toxic to fish. Similarly, the toxin produced by *Gymnodinium veneficum* has been described as an exotoxin (Abbot & Ballentine 1957). The freshwater dinoflagellate, *Peridinium polonicum*, produces an extracellular toxin which kills fish (Nozawa 1970). Other reports on toxin-producing dinoflagellates, e.g. *Gymnodinium breve* (Starr 1958) and *Gonyaulax catenella* (Burke *et al*. 1960) indicate that only healthy cultures pro-duce endotoxins.

# 3   CHARACTERISTICS OF THE RELEASE PROCESSES

## 3.1   *General*

The release of simple substances, such as sugars, amino acids and organic acids by healthy algal cells probably occurs chiefly by diffusion through the cell plasmalemma. The rate of such release will, therefore, depend on the concentra-tion gradient of the substance across the membrane, and the permeability constant of the membrane for the substance (Stadelmann 1962, 1969). Data on relative amounts of intra- and extracellular simple molecules support such an assumption. The relative amounts of extracellular amino acids produced by *Scenedesmus quadricauda* reflect those of the intracellular amino acids, with aspartic and glutamic acids predominating (Gocke 1970). Similarly, the most abundant extracellular products of *Olisthodiscus* sp. and *Pyramimonas* sp. (mannitol), *Dunaliella tertiolecta* (glycerol), and a marine *Chlorella* sp. (proline) are also the most abundant intracellular metabolites in these algae (Hellebust 1965). However, there appears to be a considerable amount of selectivity in the release processes of metabolites by symbiotic algae *in situ* (see p. 853), but such selectivity is probably based on differential membrane permeability for cell metabolites. There appears to be a good correlation between the proportion of cell material in the ethanol-soluble fraction (simple molecules mainly) and the amount of organic carbon released by plankton algae (Wallen & Geen 1971). Active excretion of small substances is also possible, but no convincing evidence for such processes in algae is available.

Large molecules, such as polysaccharides, proteins and polyphenolic substances, are probably excreted by more complex processes such as the fusion of intracellular vesicles containing the macromolecules with the plasmalemma resulting in extracellular release (McCully 1968, Whaley *et al.* 1972). The production of slime-tracks by motile pennate diatoms (Hopkins 1967), and extracellular mucilage during movement of blue-green algae (Walsby 1968) represent special cases of release of macromolecules.

Both simple and complex substances may, in addition, be liberated by healthy cells during vegetative reproductive processes such as autospore release (Schlösser 1966), or sexual reproductive processes such as gamete fusion (Claes 1971). The loss of cell contents in general may also occur through autolysis, or any kind of cell lysis, particularly during stationary phase growth in old cultures. Some of the extracellular products discussed here undoubtedly are due to such cell breakdown (e.g. Marker 1965). We are, however, mainly concerned with the release of substances by healthy cells, although it is often difficult to be certain that dying cells are absent even in rapidly growing cultures (Guillard & Wangersky 1958, Aaronson 1971).

The rate of production of extracellular products by the above processes depends on physiological and environmental factors affecting membrane permeability, on intracellular concentrations of simple metabolites, and on the ability of complex substances intended for excretion to form cell wall substances, gelatinous capsules, etc., which may subsequently be liberated in part into the medium (Watt 1969, Guillard & Hellebust 1971).

## 3.2 *Growth stage*

Cells in lag and stationary growth phases generally release more organic carbon than do exponentially growing cells (Guillard & Wangersky 1958, Nalewajko *et al.* 1963, Marker 1965). Autolysis may account for much of the extracellular products derived from lag and stationary phase cells. However, normal release by living cells of high concentrations of intracellular metabolites and products, not removed in normal biosynthetic processes due to poor growth conditions, is also possible (Hellebust 1967, Jones & Stewart 1969). Large quantities of polysaccharides may be liberated by cells which are actively metabolizing, but which are unable to divide (McLaughlin *et al.* 1960).

Exponentially growing cells release into the medium from less than 1% to over 50% of their photoassimilated carbon, depending on the species and growth conditions (Hellebust 1965, Nalewajko 1966, Gocke 1970, Guillard & Hellebust 1971). The rate of excretion of the simple metabolite glycolic acid (see Chapter 17, p. 474) is much higher in rapidly growing cultures than in older cultures (Nelson & Tolbert 1969). However, exponentially growing *Thalassiosira fluviatilis* cells release only about 5% of the photoassimilated carbon, while the proportion released increases to over 20% when the cells enter stationary phase (Hellebust 1967). A few studies have been made of excretion by cells at different stages of their life cycle. *Ankistrodesmus braunii* releases less glycolate during cell division than at other stages of its life cycles in synchronous culture (Gimmler *et al.* 1969). However, *Scenedesmus obliquus*, releases glycolate only at the stage of cell division (Hess *et al.* 1967). *Phaeocystis pouchetii* liberates 15 to 64% of its photoassimilated carbon when growing as gelatinous colonies, while the proportion released during its motile, single-cell stage, is only about 3% (Guillard & Hellebust 1971).

### 3.3  *Environmental and physiological factors*

In general, any environmental condition which inhibits cell multiplication, but still permits photoassimilation to continue, results in the release of high proportions of the photoassimilate. Wallen and Geen (1971) found good correlations between the proportion of photoassimilated carbon in the ethanol-soluble fraction of the cells and the proportion released by the cells. Blue-green light, which predominates near the bottom of the euphotic zone, favours low proportions of alcohol-soluble photosynthate and low excretion rates.

High light intensities often result in release of large proportions of photosynthate by algae, possibly due to membrane damage (Fogg *et al.* 1965, Hellebust 1965, Nalewajko 1966, Bellin & Ronayne 1968). Nalewajko (1966) found good correlation between relative inhibition of photosynthesis and percentage release of photoassimilate at high light intensities. The proportion of the photoassimilated carbon released is relatively unaffected by light intensities over an intermediate range where little or no inhibition of photosynthesis takes place (Hellebust 1965, Nalewajko 1966).

Relatively small proportions of cellular carbon are lost by algal cells during dark periods following periods of photoassimilation (Hellebust 1965). This may be due to lower intracellular concentrations of simple metabolites in cells kept in the dark. However, cells fixing $^{14}CO_2$ in the dark, presumably into organic acids, release high proportions of such $^{14}C$-labelled compounds (Fogg *et al.* 1965, Watt 1969).

Under conditions where $CO_2$ limits the rate of photosynthesis, e.g. high light intensities, low $CO_2$ concentrations and high pH, glycolic acid is often a major extracellular product (Tolbert & Zill 1957, Whittingham & Pritchard 1963, Watt & Fogg 1966). However, *Chlamydomonas reinhardtii*, previously grown with air (0·03% $CO_2$) does not excrete significant amounts of glycolate, while cells previously grown with 1% $CO_2$-enriched air do (Nelson & Tolbert 1969) (see Chapter 17, p. 474).

Low pH favours high rates of excretion by isolated zoochlorellae (Muscatine 1967) and *Haematococcus pluvialis* (McLachlan & Craigie 1965). The high excretion rate of *H. pluvialis* may be due to the high proportion of the photosynthate found in the water-soluble fraction at this pH, rather than on a pH effect on membrane permeability.

Little is known about the effects of temperature on the production of extracellular compounds by algae. Rapid changes of environmental conditions, including temperature, often result in high rates of extracellular release (Jones & Stewart 1969, Sieburth 1969). Low temperatures, which permit growth of *Katodinium dorsalisulcum*, decrease the relative amount of extracellular carbohydrate produced by this alga (McLaughlin *et al.* 1960).

Under anaerobic conditions, in the presence of glucose, organic acids formed by fermentation processes are released into the medium by several algae (e.g. Syrett & Wong 1963, Kowallik 1969). Light reduces such extracellular organic acid production, apparently by inhibiting glycolysis (Hirt *et al.* 1971).

Craigie (1969) demonstrated that the concentration of cyclohexanetetrol increases in *Monochrysis lutheri* cells with increasing salinities. When cells grown at high salinities are transferred to low salinity media, cyclohexanetetrol is rapidly released. High salinities favour the production of proline in a marine *Chlorella* sp. and, probably as a consequence of this, the release of this amino acid (W. Bell & J. A. Hellebust unpubl.).

Under conditions of mineral deficiency, which inhibit cell division but permit continued photosynthesis, large amounts of capsular polysaccharides are liberated by *Navicula pelliculosa* (Lewin 1955).

Some special cases of release of metabolites which accumulate in cells due to mutations, presumably involving loss of control of specific metabolic pathways, may be mentioned. A temperature shock mutant of *Cyclotella cryptica* releases large amounts of gelatinous products resulting in slime colonies (Werner 1970). Mutants of *Chlamydomonas eugametos* lacking end-product control of the pathway leading to nicotinic acid release this vitamin into the medium (Nakamura & Gowans 1964).

Hood *et al.* (1969) suggest that the rapid production of extracellular peptides and amino acids by blue-green algae may be a consequence of failure to control amino acid biosynthesis through end-product regulation.

The synthesis of chlorophyll in *Euglena* is blocked in the dark at an intermediate stage beyond which further reactions can proceed only in the presence of light. Dubash and Rege (1967) demonstrated that excretion of the chlorophyll precursor, protoporphyrin IX, is about 13-fold higher in dark-grown cells of *E. gracilis* than in light-grown cells. However, prolonged dark-adaptation limits this porphyrin excretion to about two times that in light.

Excretion of metabolites may result from the addition of substances which interfere with metabolic pathways. An example of this is the release of δ-aminolevulinic acid when levulinic acid, which inhibits chlorophyll synthesis, is added to the medium (Beale 1970).

Another special case of extracellular production is the release of a portion of the carbon skeletons of $^{14}$C-amino acids taken up by marine planktonic algae (Stephens & North 1971). This is probably due to the ability of such cells to concentrate extracellular amino acids over a hundred-fold by special uptake processes (Hellebust 1970), and the deamination of the accumulated amino acids resulting in high intracellular concentrations of the corresponding organic acids, some of which are then released.

## 4  ECOLOGICAL ASPECTS

### 4.1  *Primary productivity*

There is good evidence that healthy, actively growing phytoplankton species release a considerable proportion of their photoassimilated carbon into the aquatic environment (Fogg 1958, Fogg *et al*. 1965, Hellebust 1965, Horne *et al*. 1969, Anderson & Zeutschel 1970, Samuel *et al*. 1971, Thomas 1971, Choi 1972, Saunders 1972). Accurate estimates of the relative amounts of photoassimilates lost by planktonic algae under various environmental conditions are therefore very important in determining the primary productivity of aquatic ecosystems. Fogg and co-workers (Fogg 1958, 1966, 1971, Fogg *et al*. 1965) have discussed this problem in some detail, and only a brief consideration of recent work will be presented here.

In general, higher excretion rates have been observed with samples of natural phytoplankton than with laboratory cultures of such algae (Fogg *et al*. 1965, Hellebust 1965, Watt 1966, Nalewajko & Marin 1969, Takahashi *et al*. 1971). One cannot always be certain, however, that this reflects with any accuracy what the algae do *in situ* because the process of incubating plankton samples in bottles for various lengths of time may itself be detrimental to natural populations and result in high excretion rates.

There are also methodological problems. Arthur and Rigler (1967) found that when increasing volumes of lake water from $^{14}$C-productivity samples were filtered according to standard productivity procedures, the amount of radioactivity retained on the membrane filter per unit volume decreased. They attributed this to increasing damage to algal cells during vacuum filtration. Other workers have compared filtration and centrifugation techniques for determining excretion rates, as well as the nature of extracellular and intracellular products, and concluded that little cell damage occurs during careful filtration procedures (Fogg *et al*. 1965, Hellebust 1965, Marker 1965, Guillard & Hellebust 1971). Nalewajko and Lean (1972) have shown that the observations by Arthur and Rigler (1967) can be explained by a saturable process of adsorption of $^{14}$C-bicarbonate and $^{14}$C-organic compounds in the algal medium on the membrane filter. These effects are apparently only important in samples from low productivity areas.

Fogg *et al*. (1965) studied the excretion rates for *in situ* lake- or seawater samples and found that 7 to 50% of the photoassimilate was released. Over a wide range of light intensities the liberation of extracellular products was proportional to the amount of photoassimilated carbon in the cells. However, at intensities sufficiently high to inhibit photosynthesis unusually high release rates were found. Excretion rates are also very high towards the end of a phytoplankton bloom (Hellebust 1965). Comparative studies of extracellular production by plankton samples indicate that such release is proportionally higher in oligo-trophic than eutrophic areas (Anderson & Zeutschel 1970, Thomas 1971, Saunders 1972).

Recent studies of extracellular release by seaweeds show that a very large proportion of the cellular organic carbon may be released under normal condi-tions (Kroes 1970). Khailov and Burkalova (1969) estimated that 23 to 39% of the photoassimilated carbon is released by various seaweeds. Sieburth (1969) estimated that about 40% of the net carbon fixed by *Fucus vesiculosus* is released as various extracellular products. However, Majak *et al.* (1966) and Craigie *et al.* (1966) found that only 0 to 4·4% of the photoassimilated carbon is released by some green and red seaweeds.

It should be noted that all these excretion studies with plankton algae or seaweeds have been carried out with non-axenic samples, and at least some of the variations in results obtained may be due to different levels of bacterial activities resulting in consumption of extracellular products (Khailov 1968, Sieburth 1969). Sieburth (1969) tried to avoid this problem by using rapid-flow systems to remove extracellular products quickly for analysis, or, by the addition of a bactericidal substance. Anderson and Zeutschel (1970) added the antibiotics penicillin and streptomycin to their phytoplankton samples in an attempt to prevent bacterial activity but such a technique is not very successful in short-term experiments (Hellebust unpubl.).

## 4.2 *Symbiosis*

Autoradiographic studies of transfer of $^{14}$C-photosynthate from algal symbionts to their fungal or invertebrate partners demonstrate that a large portion of photoassimilated carbon is released by the algal cells and assimilated by the heterotrophic symbionts (see Smith *et al.* 1969). Furthermore, recently isolated algal symbionts have very high excretion rates (Smith & Drew 1965, Muscatine *et al.* 1967). The very high excretion rates probably represent a special adaptation of the algae to their symbiotic mode of life. Algal symbionts often release less of their photosynthate after prolonged culture than immediately following isolation from their symbiotic environment (Follman 1960, Richardson *et al.* 1968). Symbiotic algae usually excrete highly specific compounds such as glycerol, erythritol, ribitol, mannitol, sorbitol, glucose and maltose (see Table 30.1).

The heterotrophic symbiont may regulate both the metabolism and the membrane permeability of the algal symbiont. Symbiotic zooxanthellae release mainly glycerol, while isolated cultured zooxanthellae release mainly insoluble mucoid materials (Smith *et al.* 1969). Also, while free glucose is the main photo-synthetic product of *Nostoc* in the lichen *Peltigera polydactyla*, cultured *Nostoc* cells do not contain detectable amounts of free glucose (Drew & Smith 1967). The addition of host homogenate to zooxanthellae isolated from the mollusc *Tridacna crocea* increased about sixteen-fold the excretion of glycerol by the algal

2E

cells (Muscatine 1967). Similarly, lichen substances, such as usnic and lecanoric acids, may increase the permeability of algal cells and, thus, regulate the rate of transfer of photosynthate from algal to fungal symbiont (Follmann 1960, Follmann & Villagran 1965, Kinraide & Ahmadjian 1970).

The release of carbohydrates by algal symbionts appears to be selective. The major intracellular soluble metabolite in zooxanthellae from the coelenterate *Pocillopora damicornis* and the mollusc *Tridacna crocea* is glucose, and only small amounts of glycerol are found, while these algae release chiefly glycerol when living in the host tissue (L. Muscatine & E. Chernichiari unpubl., see Smith *et al.* 1969). Algal symbionts in lichens release very specific sugars or sugar alcohols from a variety of soluble photosynthetic products contained in the cells (Smith *et al.* 1969).

### 4.3  *Succession*

The production of extracellular growth inhibiting and promoting substances by a large number of algae in culture (see p. 846) indicates that such substances may play important roles in the succession of species commonly observed in aquatic ecosystems. The considerable differences in abilities of samples taken from different water bodies to support growth of selected algal assay organisms agree with such an assumption (Johnston 1963, 1964, Smayda 1970).

The interactions of *Chlamydomonas globosa* and *Chlorococcum ellipsoideum* was studied by Kroes (1971) in a filter culture system which only allowed exchange of media. Under these conditions, *C. ellipsoideum* inhibits *C. globosa* but not vice versa. Pratt (1966) compared the interactions of *Skeletonema costatum* and *Olisthodiscus luteus* in mixed cultures with the succession of these species in Narragansett Bay, Rhode Island, and concluded that the mechanisms regulating growth rates and cell densities in cultures probably also operate in nature. However, the large difference in cell densities between culture experiments and those found in nature, even under bloom conditions, makes such extrapolations rather uncertain.

The release of vitamins (Carlucci & Bowes 1970a,b) and vitamin binding factors (Droop 1968), which affect the growth of many planktonic species at extremely low concentrations may also be important in determining species succession. Aubert (1971a) believes that a variety of substances released by marine organisms, including algae, play important roles in determining species composition and succession (cf. Lucas 1947, 1958).

## 5  REFERENCES

AARONSON S. (1971) The synthesis of extracellular macromolecules and membranes by a population of the phytoflagellate *Ochromonas danica*. *Limnol. Oceanogr.* **16**, 1–9.
AARONSON S. & BENSKY B. (1967) Effect of aging of a cell population on lipids and drug resistance in *Ochromonas danica*. *J. Protozool.* **14**, 76–8.

AARONSON S., DEANGELIS B., FRANK O. & BAKER H. (1971) Secretion of vitamins and amino acids into the environment by *Ochromonas danica*. *J. Phycol.* **7**, 215–18.

ABBOT B.C. & BALLANTINE D. (1957) The toxin from *Gymnodinium veneficum* Ballantine. *J. mar. biol. Ass., U.K.* **36**, 169–89.

ALLEN M.B. (1956) Excretion of organic compounds by *Chlamydomonas*. *Arch. Mikrobiol.* **24**, 163–8.

ANDERSON G.C. & ZEUTSCHEL R.P. (1970) Release of dissolved organic matter by marine phytoplankton in coastal and off-shore areas of the Northeast Pacific Ocean. *Limnol. Oceanogr.* **15**, 402–7.

ANSELL A.D., RAYMONT J.E.G., LANDER K.F., CROWLEY E. & SHACKLEY P. (1963) Studies on the mass culture of *Phaeodactylum*. II. The growth of *Phaeodactylum* and other species in outdoor tanks. *Limnol. Oceanogr.* **8**, 184–206.

ANTIA N.J., MCALLISTER C.D., PARSONS T.R., STEPHENS K. & STRICKLAND J.D.H. (1963) Further measurements of primary productivity using a large-volume plastic sphere. *Limnol. Oceanogr.* **8**, 166–83.

ARMSTRONG F.A. & BOALCH G.T. (1960) Volatile organic matter in algal culture media and sea water. *Nature, Lond.* **185**, 761–2.

ARTHUR C.R. & RIGLER F.H. (1967) A possible source of error in the $^{14}$C method of measuring primary productivity. *Limnol. Oceanogr.* **12**, 121–4.

AUBERT M. (1971a) Télémediateurs chemiques et équilibre biologique océanique. Première Partie. Théorie générale. *Rev. Int. Oceanogr. Méd.* **21**, 5–16.

AUBERT M. (1971b) Télémediateurs chemiques et équilibre biologique océanique. Deuxieme Partie. Nature chemique de l'inhibiteur de la synthèse d'un antibiotique produit par un diatomée. *Rev. Int. Oceanogr. Méd.* **21**, 17–22.

AUBERT M., PESANDO D. & GAUTHIER M. (1970) Phénomènes d'antibiose d'origine phytoplanktonique en milieu marine. Substances antibactériènne produites par un diatomée *Asterionella japonica* (Cleve). *Rev. Int. Oceanogr. Méd.* **18–19**, 69–76.

BADOUR S.S. & GERGIS M.S. (1965) Cell division and fat accumulation in *Nitzschia* sp. grown in continuously illuminated mass culture. *Arch. Mikrobiol.* **51**, 94–102.

BADOUR S.S. & WAYGOOD E.R. (1971) Excretion of an acid semialdehyde by *Gloeomonas*. *Phytochemistry*, **10**, 967–76.

BEALE S.L. (1970) The biosynthesis of δ-aminolevulinic acid in *Chlorella*. *Pl. Physiol., Lancaster*, **45**, 504–6.

BEDNAR T.W. & HOLM-HANSEN O. (1964) Biotin liberation by the lichen alga *Coccomyxa* sp. and by *Chlorella pyrenoidosa*. *Pl. Cell Physiol., Tokyo*, **5**, 297–303.

BEGUM F. & SYRETT P.J. (1970) Fermentation of glucose by *Chlorella*. *Arch. Mikrobiol.* **72**, 344–52.

BELLIN J.S. & RONAYNE M.E. (1968) Effects of photodynamic action on the cell membrane of *Euglena*. *Physiologia Pl.* **21**, 1060–6.

BELMONT L. & MILLER J.D.A. (1965) The utilization of glutamine by algae. *J. exp. Bot.* **16**, 318–24.

BERGLUND H. (1969) Stimulation of growth of two marine algae by organic substances excreted by *Enteromorpha linza* in unialgal and axenic cultures. *Physiologia Pl.* **22**, 1069–73.

BISCHOFF H. & BOLD H. (1963) Some algae from Enchanted Rock and related algal species. *Phycol. Stud.* **4**, Univ. Texas Publ. 6318.

BISHOP, C.T., ADAMS G.A. & HUGHES E.O. (1954) A polysaccharide from the blue-green alga *Anabaena cylindrica*. *Can. J. Chem.* **32**, 999–1004.

BLACKWELL J., PARKER K.D. & RUDALL K.M. (1967) Chitin fibres of the diatoms *Thalassiosira fluviatilis* and *Cyclotella cryptica*. *J. molec. Biol.* **28**, 383–5.

BONEY A.D. (1970) Scale-bearing phytoflagellates. An interim review. *Oceanogr. Mar. Biol. Ann. Rev.* **8**, 251–305.

BRUCE D.L. & DUFF D.C.B. (1967) The identification of two antibacterial products of the marine planktonic alga *Isochrysis galbana*. *J. gen. Microbiol.* **48**, 293–8.

BURCZYK J., GRZYBECK H., BANAS J. & BANAS E. (1970) Presence of cellulase in the algae *Scenedesmus* and *Chlorella*. *Expl. Cell. Res.* **63**, 451–3.

BURKE J.M., MARCHISOTTO J., McLAUGHLIN J.J.A. & PROVASOLI L. (1960) Analysis of the toxin produced by *Gonyaulax catenella* in axenic culture. *Ann. N.Y. Acad. Sci.* **90**, 837–42.

BURKHOLDER P.R. (1963) Some nutritional relationships among microbes of sea sediments and waters. In *Symposium on Marine Microbiology*, ed. Oppenheimer C.H. pp. 133–50. C.C. Thomas, Springfield.

CARLUCCI A.F. (1970) The ecology of the plankton off La Jolla, California in the period April through September, 1967. II. Vitamin $B_{12}$, thiamine and biotin. *Bull. Scripps Inst. Oceanogr.* **17**, 23–31.

CARLUCCI A.F. & BOWES P.M. (1970a) Production of vitamin $B_{12}$, thiamine, and biotin by phytoplankton. *J. Phycol.* **6**, 351–7.

CARLUCCI A.F. & BOWES P.M. (1970b) Vitamin production and utilization by phytoplankton in mixed culture. *J. Phycol.* **6**, 393–400.

CHANG W. (1967) *Excretion of organic acids during photosynthesis by synchronized algae.* Ph.D. Thesis, Michigan State University, East Lansing, Michigan.

CHAPMAN G. & RAE A.C. (1969) Excretion of photosynthate by a benthic diatom. *Mar. Biol.* **3**, 341–51.

CHENG KAR-HING (1971) *The excretion, uptake and oxidation of glycolic acid by* Oscillatoria. M.Sc. Thesis, York University, Toronto, Ont.

CHOI I.C. (1972) Phytoplankton production and extracellular release of dissolved organic carbon in the Western North Atlantic. *Deep-Sea Research,* **19**, 731–5.

CLAES H. (1971) Autolyse der Zellwand bei den Gameten von *Chlamydomonas reinhardii*. *Arch. Mikrobiol.* **78**, 180–8.

COLLINS R.P. & BEAN G.H. (1963) Volatile constituents obtained from *Chlamydomonas globosa*. The carbonyl fraction. *Phycologia,* **3**, 55–9.

COLLINS R.P. & KALNINS K. (1965) Volatile constituents of *Synura petersenii*. I. The carbonyl fraction. *Lloydia,* **28**, 48–52.

COLLINS R.P. & KALNINS K. (1966) Carbonyl compounds produced by *Cryptomonas ovata* var. *pulustris*. *J. Protozool.* **13**, 435–7.

COLLINS R.P. & KALNINS K. (1967) Keto acids produced by *Chlamydomonas reinhardti*. *Can. J. Microbiol.* **13**, 995–9.

CONOVER J.T. & SIEBURTH J. McN. (1964) Effect of *Sargassum* distribution on its epibiota and antibacterial activity. *Botanica mar.* **6**, 147–57.

CONOVER J.J. & SIEBURTH J.McN. (1966) Effects of tannins excreted from Phaeophyta on planktonic animal survival in tide pools. In *Proc. V intern. Seaweed Symp.*, Halifax, 1965, eds. Young E.G. & McLachlan J.L. pp. 99–100. Pergamon Press, Oxford, London, New York & Paris.

COOK A.H. & ELVIDGE J.A. (1951) Fertilization in the Fucaceae. Investigations on the nature of the chemotactic substance produced by eggs of *Fucus serratus* and *F. vesiculosus*. *Proc. R. Soc. B.* **138**, 97–114.

CRAIGIE J.S. (1969) Some salinity-induced changes in growth, pigments and cyclohexane-tetrol content of *Monochrysis lutheri*. *J. Fish. Res. Bd. Can.* **26**, 2959–67.

CRAIGIE J.S. & McLACHLAN J. (1964) Excretion of coloured ultraviolet absorbing substances by marine algae. *Can. J. Bot.* **42**, 23–33.

CRAIGIE J.S., McLACHLAN J., MAJAK W., ACKMAN R.C. & TOCHER C.S. (1966) Photosynthesis in algae. II. Green algae with special reference to *Dunaliella* spp. and *Tetraselmis* spp. *Can. J. Bot.* **44**, 1247–54.

DAISLEY K.W. (1970) The occurrence and nature of *Euglena gracilis* proteins that bind vitamin $B_{12}$. *Int. J. Biochem.* **1**, 561–74.

DARDEN JR. W.H. (1966) Sexual differentiation in *Volvox aureus*. *J. Protozool.* **13**, 239–255.

DÖHLER G. & BRAUN F. (1971) Untersuchungen der Beziehung zwischen extracelluläre Glykolsäure-Ausscheidung und der photosynthetischen $CO_2$-Aufnahme bei der Blaualge *Anacystis nidulans. Planta*, **98**, 357–61.

DREW E.A. & SMITH D.C. (1967) Studies in the physiology of lichens. VIII. The physiology of the *Nostoc* symbiont of *Peltigera polydactyla* compared with cultured and free-living forms. *New Phytol.* **66**, 379–88.

DROOP M.R. (1966) Organic acids and bases and the lag phase in *Nannochloris oculata. J. mar. biol. Assoc.*, *U.K.* **46**, 673–8.

DROOP M.R. (1968) Vitamin $B_{12}$ and marine ecology. IV. The kinetics of uptake, growth and inhibition in *Monochrysis lutheri. J. mar. biol. Ass.*, *U.K.* **48**, 689–733.

DROOP M.R. (1970) Vitamin $B_{12}$ and marine ecology. V. Continuous culture as an approach to nutritional kinetics. *Helgoländer Wiss. Meeresunters.* **20**, 629–36.

DUBASH P.J. & REGE D.V. (1967) Excretion of protoporphyrin IX by *Euglena. Biochem. biophys. Acta*, **141**, 209–11.

ELLIS R.J. & MACHLIS L. (1968) Control of sexuality in *Golenkinia. Am. J. Bot.* **55**, 600–10.

FAY P., KUMAR H.D. & FOGG G.E. (1964) Cellular factors affecting nitrogen fixation in the blue-green alga *Chlorogloea fritschii. J. gen. Microbiol.* **35**, 351–60.

FOGG G.E. (1952) The production of extracellular nitrogenous substances by a blue-green alga. *Proc. R. Soc. B.* **139**, 372–97.

FOGG G.E. (1958) Extracellular products of phytoplankton and the estimation of primary production. *Rapp. Cons. Explor. Mer.* **144**, 56–60.

FOGG G.E. (1962) Extracellular products. In *Physiology and Biochemistry of Algae*, ed. Lewin R.A. pp. 475–89. Academic Press, New York & London.

FOGG G.E. (1966) The extracellular products of algae. In *Oceanogr. Mar. Biol. Ann. Rev.* **4**, 195–212.

FOGG G.E. (1971) Extracellular products of algae in fresh water. *Arch. Hydrobiol.* **5**, 1–25.

FOGG G.E. & BOALCH G.T. (1958) Extracellular products in pure cultures of a brown alga. *Nature, Lond.* **181**, 789–90.

FOGG G.E. & NALEWAJKO C. (1964) Glycollic acid as an extracellular product of phytoplankton *Verh. Int. Verein. Theor. Angew. Limnol.* **15**, 806–10.

FOGG G.E., NALEWAJKO C. & WATT W.D. (1965) Extracellular products of phytoplankton photosynthesis. *Proc. R. Soc. B.* **162**, 517–34.

FOGG G.E. & PATTNAIK H. (1968) The release of extracellular nitrogenous products by *Westiellopsis prolifica* Janet. *Phykos*, **5**, 58–67,

FOGG G.E. & WATT W.D. (1965) The kinetics of release of extracellular products of photosynthesis by phytoplankton. *Mem. 1st Ital. Idrobiol.* **18** (*Suppl.*), 165–74.

FOLLMAN G. (1960) Die Durchlässigkeitseigenschaften der Protoplasten von Phycobionten aus *Cladonia fuscata* (Huds.) Schrad. *Naturwissenschaften*, **17**, 405–6.

FOLLMAN G. & VILLAGRAN V. (1965) Flechtenstoffe and Zellpermiabilität. *Z. Naturf. B.* **20**, 723.

FORD J.E. (1958) $B_{12}$-vitamins and growth of the flagellate *Ochromonas malhamensis. J. gen. Microbiol.* **19**, 161–72.

FÖRSTER H., WIESE L. & BRAUNITZER G. (1956) Über das agglutinierend wirkende Gynogamone von *Chlamydomonas eugametos. Z. Naturf.* **11b**, 315–17.

GENTILE J.H. & MALONEY T.E. (1969) Toxicity and environmental requirements of a strain of *Aphanizomenon flos-aquae* (L.) Ralfs. *Can. J. Microbiol.* **15**, 165–73.

GIMMLER H., ULLRICH W., DOMANSKY-KADEN J. & URBACH W. (1969) Excretion of glycolate during synchronous culture of *Ankistrodesmus* in the presence of disalicylidene-propanediamine or hydroxypyridinemethanesulfonate. *Pl. Cell Physiol., Tokyo*, **10**, 103–12.

GOCKE K. (1970) Untersuchungen über Abgabe und Aufnahme von Aminosäuren und Polypeptiden durch Planktonorganismen. *Arch. Hydrobiol.* **67**, 285–367.

GORHAM P.R., MCLACHLAN J., HAMMER U.T. & KIM W.K. (1964) Isolation and culture of toxic strains of *Anabaena flos-aquae* (Lyngb.) de Bréb. *Verh. Int. Ver. Limnol.* **15**, 796–804.

GUILLARD R.R.L. & HELLEBUST J.A. (1971) Growth and the production of extracellular substances by two strains of *Phaeocystis poucheti. J. Phycol.* **7**, 330–8.

GUILLARD R.R.L. & WANGERSKY P.J. (1958) The production of extracellular carbohydrates by some marine flagellates. *Limnol. Oceanogr.* **3**, 449–54.

HARRIS D.O. (1970) An autoinhibitory substance produced by *Platydorina caudata* Kofoid. *Pl. Physiol., Lancaster.* **45**, 210–14.

HARRIS D.O. (1971a) Growth inhibitors produced by the green algae Volvocaceae. *Arch. Mikrobiol.* **76**, 47–50.

HARRIS D.O. (1971b) A model system for the study of algal growth inhibitors. *Arch. Protistenk.* **113**, 230–4.

HELLEBUST J.A. (1965) Excretion of some organic compounds by marine phytoplankton. *Limnol. Oceanogr.* **10**, 192–206.

HELLEBUST J.A. (1967) Excretion of organic compounds by cultured and natural populations of marine phytoplankton. In *Estuaries*, ed. Lauff G.H. pp. 361–6. *Am. Assoc. Adv. Sci. Publ.* No. 83, Washington, D.C.

HELLEBUST J.A. (1970) Uptake and assimilation of organic substances by marine phytoplankters. In *Organic matter in natural waters*, ed. Hood D.W. Sympos. University of Alaska, Fairbanks, Alaska. 1968. *Inst. Mar. Sci. Publ.* **1**, 225–56.

HESS J.L., TOLBERT N.E. & PIKE L.M. (1967) Glyollate biosynthesis by *Scenedesmus* and *Chlorella* in the presence or absence of $NaHCO_3$. *Planta,* **74**, 278–85.

HIRT G., TANNER W. & KANDLER O. (1971) Effect of light on the rate of glycolysis in *Scenedesmus obliquus. Pl. Physiol., Lancaster,* **47**, 841–3.

HOFFMAN L.R. (1960) Chemotaxis of *Oedogonium* species. *S. West. Nat.* **5**, 111–16.

HOLMES R.W., WILLIAMS P.M. & EPPLEY R.W. (1967) Red water in La Jolla Bay, 1964–1966. *Limnol. Oceanogr.* **12**, 503–12.

HOOD W., LEAVER A.G. & CARR N.G. (1969) Extracellular nitrogen and the control of arginine biosynthesis in *Anabaena variabilis. Biochem. J.* **114**, 12P.

HOPKINS J.T. (1967) The diatom trail. *J. Quekett Microscop. Club,* **30**, 209–17.

HORNE A.J., FOGG G.E. & EAGLE D.J. (1969) Studies *in situ* of the primary products of an area of inshore Antarctic Sea. *J. mar. biol. Ass., U.K.* **49**, 393–405.

HOUGH L., JONES J.K.M. & WADMAN W.H. (1952) An investigation of the polysaccharide components of certain fresh-water algae. *J. Chem. Soc.,* 3393–9.

HUNTSMAN S.A. & SLONEKER J.H. (1971) An exocellular polysaccharide from the diatom *Gomphonema olivaceum. J. Phycol.* **7**, 261–4.

ISHIDA Y. (1968) Physiological studies on evolution of dimethylsulfide from unicellular marine algae. *Mem. Coll. Agr. Kyoto Univ.* **94**, 47–82.

JOHNSTON R. (1963) Sea water, the natural medium for phytoplankton. Part 1. General aspects. *J. mar. biol. Ass., U.K.* **43**, 427–56.

JOHNSTON R. (1964) Sea water, the natural medium for phytoplankton. Part II. Trace metals and chelation. *J. mar. biol. Ass., U.K.* **44**, 87–110.

JOHANNES R.E. (1964) Uptake and release of dissolved organic phosphorous by representatives of a coastal marine ecosystem. *Limnol. Oceanogr.* **9**, 224–34.

JONES K. & STEWART W.D.P. (1969) Nitrogen turnover in marine and brackish habitats. III. The production of extracellular nitrogen by *Calothrix scopulorum. J. mar. biol. Ass., U.K.* **49**, 475–88.

JONES R.F. (1962) Extracellular mucilage of the red alga *Porphyridium cruentum. J. Cell. Comp. Physiol.* **60**, 61–4.

KADOTA H. & ISHIDA Y. (1968) Evolution of volatile sulfur compounds from unicellular marine algae. *Bull. Misaki Mar. Biol. Inst., Kyoto Univ.* **12**, 35–48.

KALLE K. (1966) The problem of the Gelbstoff in the sea. *Oceanogr. Mar. Biol. Ann. Rev.* **4**, 91–104.

KHAILOV K.M. (1963) Some unknown organic substances in sea water. *Dokl. Akad. Nauk. S.S.S.R.* **147**, 1355–7.

KHAILOV K.M. (1968) Extracellular microbial hydrolysis of polysaccharides dissolved in sea water. *Microbiology*, **37**, 424–7.

KHAILOV K.M. & BURKALOVA Z.P. (1969) Release of dissolved organic matter by marine seaweeds and distribution of their total organic products to inshore communities. *Limnol. Oceanogr.* **14**, 521–7.

KHAILOV K.M. & BURLAKOVA Z.P. & LANSKAYA L.A. (1967) Exogenous metabolites in cultures of *Gymnodinium Kovalevskii*. *Microbiology*, **36**, 498–502.

KINRAIDE W.T.B. & AHMADJIAN V. (1970) The effects of usnic acid on the physiology of two cultured species of the lichen alga *Trebouxia* Puym. *Lichenologist*, **4**, 234–47.

KOWALLIK W. (1969) Eine förderne Wirkung von Blaulicht auf die Säureproduktion anaerob gehalterner *Chlorellen*. *Planta*, **87**, 372–84.

KRISTENSEN H.P.O. (1956) A vitamin $B_{12}$-binding factor formed in culture of *Euglena gracilis* var. *bacillaris*. *Acta physiol. scand.* **37**, 8–13.

KROES H.W. (1970) Excretion of mucilage and yellow-brown substances by some brown algae from the intertidal zone. *Botanica mar.* **13**, 107–10.

KROES H.W. (1971) Growth interactions between *Chlamydomonas globosa* Snow and *Chlorococcum ellipsoideum* Deason & Bold under different experimental conditions, with special attention to the role of pH. *Limnol. Oceanogr.* **16**, 869–79.

KUENZLER E.J. (1970) Dissolved organic phosphorus excretion by marine phytoplankton. *J. Phycol.* **6**, 7–13.

KUENZLER E.J. & KETCHUM B.H. (1962) Rate of phosphorus uptake by *Phaeodactylum tricornutum*. *Biol. Bull. mar. biol. Lab.*, *Woods Hole*, **123**, 134–45.

KUENZLER E.J. & PERRAS J.P. (1965) Phosphatases of marine algae. *Biol. Bull. mar. biol. Lab.*, *Woods Hole*, **128**, 271–84.

LEFÈVRE M. (1964) Extracellular products of algae. In *Algae and Man*, ed. Jackson D.F. pp. 337–67. Plenum Press, New York.

LEWIN J.C. (1955) The capsule of the diatom *Navicula pelliculosa*. *J. gen. Microbiol.* **13**, 162–9.

LEWIN J.C. & LEWIN R.A. (1960) Auxotrophy and heterotrophy in marine littoral diatoms. *Can. J. Microbiol.* **6**, 127–34.

LEWIN J.C., LEWIN R.A. & PHILPOTT D.E. (1958) Observations on *Phaeodactylum tricornutum*. *J. gen. Microbiol.* **18**, 418–26.

LEWIN R.A. (1956) Extracellular polysaccharides of green algae. *Can. J. Microbiol.* **2**, 665–72.

LEWIN R.A. (1958) The mucilage tubes of *Amphipleura rutilans*. *Limnol. Oceanogr.* **3**, 111–13.

LUCAS C.E. (1947) The ecological effects of external metabolites. *Biol. Rev.* **22**, 270–95.

LUCAS C.E. (1958) External metabolites and productivity. *Rapp. P. -v. Réun. Int. Explor. Mer.* **144**, 155–8.

MAJAK W., CRAIGIE J.S. & McLACHLAN J. (1966) Photosynthesis in algae. I. Accumulation products in the Rhodophyceae. *Can. J. Bot.* **44**, 541–9.

MAKSIMOVA I.V. & PIMENOVA M.N. (1969) Liberation of organic acids by green unicellular algae. *Microbiology*, **38**, 64–70.

MARKER A.F.H. (1965) Extracellular carbohydrate liberation in the flagellates *Isochrysis galbana* and *Prymnesium parvum*. *J. mar. biol. Ass.*, *U.K.* **45**, 755–72.

McCALLA D.R. (1963) Accumulation of extracellular amino acids by *Euglena gracilis*. *J. Protozool.* **10**, 491–5.

McCULLY M.E. (1968) Histological studies of the genus *Fucus*. III. Fine structure and possible functions of the epidermal cells of the vegetative thallus. *J. Cell Sci.* **3**, 1–16.

McLACHLAN J. & CRAIGIE J.S. (1965) Effects of carboxylic acids on growth and photosynthesis of *Haematococcus pluvialis*. *Can. J. Bot.* **43**, 1449–56.

McLACHLAN J., McINNES A.G. & FALK M. (1965) Studies on the chitan (chitin: poly-N-acetylglucosamine) fibres of the diatom *Thallasiosira fluviatilis* Hustedt. *Can. J. Bot.* **43**, 707–13.

McLAUGHLIN J.J.A. & PROVASOLI L. (1957) Nutritional requirements and toxicity of two marine *Amphidinium. J. Protozool.* **4** (*Suppl.*), 7.

McLAUGHLIN J.J.A., ZAHL P.A., NOVAK A., MARCHISOTTO J. & PRAGER J. (1960) Mass cultivation of some phytoplanktons. *Ann. N.Y. Acad. Sci.* **90**, 856–65.

MONAHAN T.J. & MONAHAN F.R. (1971) Stimulatory properties of filtrate from the green alga *Hormotila blennista*. II. Fractionation of filtrate. *J. Phycol.* **7**, 170–6.

MONAHAN T.J. & TRAINOR F.R. (1970) Stimulatory properties of filtrate from the green alga *Hormotila blennista*. I. Description. *J. Phycol.* **6**, 263–9.

MOORE B.G. & TISCHER R.G. (1964) Extracellular polysaccharides of algae: effects on life-support systems. *Science, N.Y.* **145**, 586–7.

MOORE B.G. & TISCHER R.G. (1965) Biosynthesis of extracellular polysaccharides by the blue-green alga *Anabaena flos-aquae. Can. J. Microbiol.* **11**, 877–85.

MÜLLER D.G. (1968) Ein leicht flüchtiges Gyno-Gamon der Braunalge *Ectocarpus siliculosus. Naturwissenschaften*, **54**, 496–7.

MUSCATINE L. (1965) Symbiosis of hydra and algae. III. Extracellular products of algae. *Comp. Biochem. Physiol.* **16**, 77–92.

MUSCATINE L. (1967) Glycerol excretion by symbiotic algae from corals and *Tridacna* and its control by the host. *Science, N.Y.* **156**, 516–19.

MUSCATINE L., KARAKASHIAN S.J. & KARAKASHIAN M.W. (1967) Soluble extracellular products of algae symbiotic with a ciliate, a sponge and a mutant hydra. *Comp. Biochem. Physiol.* **20**, 1–12.

NAKAMURA K. & GOWANS C.S. (1964) Nicotinic acid-excreting mutants in *Chlamydomonas. Nature, Lond.* **202**, 826–7.

NALEWAJKO C. (1966) Photosynthesis and excretion in various planktonic algae. *Limnol. Oceanogr.* **11**, 1–10.

NALEWAJKO C., CHOWDHURI N. & FOGG G.E. (1963) Excretion of glycollic acid and the growth of a planktonic *Chlorella*. In *Studies on Microalgae and Photosynthetic Bacteria. Jap. Soc. Plant Physiol.*, pp. 171–83, Univ. of Tokyo Press, Tokyo.

NALEWAJKO C. & LEAN D. (1972) Retention of dissolved compounds by membrane filters as an error in the $^{14}$C method of primary production measurements. *J. Phycol.* **8**, 37–43.

NALEWAJKO C. & MARIN L. (1969) Extracellular production in relation to growth of four planktonic algae and of phytoplankton populations from Lake Ontario. *Can. J. Bot.* **47**, 405–13.

NELSON E.B. & TOLBERT N.E. (1969) The regulation of glycolate metabolism in *Chlamydomonas reinhardtii. Biochim. biophys. Acta*, **184**, 263–70.

NOZAWA K. (1970) The effect of *Peridinium* toxin on other algae. *Bull. Misaki Mar. Biol. Inst.*, No. 12, pp. 21–4.

PASCHER A. (1931a) Über Gruppenbildung und Geschlechtswechsel bei den Gameten einer Chlamydomonadine (*Chlamydomonas paupera*). *Jb. wiss. Bot.* **75**, 551–80.

PASCHER A. (1931b) Über einen neuen einzelligen Organismus mit Eibefruchtung. *Beih. bot. Zbl.* **48**, 446–80.

PATTNAIK H. (1966) Studies on nitrogen fixation by *Westiellopsis prolifica* Janet. *Ann. Bot.* **30**, 231–8.

PEARCE J. & CARR N.G. (1967) The metabolism of acetate by the blue-green algae *Anabaena variabilis* and *Anacystis nidulans. J. gen. Micribiol.* **49**, 301–13.

PEARCE J. & CARR N.G. (1969) The incorporation and metabolism of glucose by *Anabaena variabilis. J. gen. Microbiol.* **54**, 451–62.

PERCIVAL E. & McDOWELL R.H. (1967) *Chemistry and enzymology of marine algal polysaccharides*. Academic Press, London & New York.

PRAGER J.C., BURKE J.M., MARCHISOTTO, J. MCLAUGHLIN J.J.A. (1959) Mass culture of a tropical dinoflagellate and the chromatographic analysis of extracellular polysaccharides *J. Protozool.* **6** (*Suppl.*), 19–20.

PRATT D.M. (1966) Competition between *Skeletonema costatum* and *Olisthodiscus luteus* in Naragansett Bay and in culture. *Limnol. Oceanogr.* **11**, 447–55.

PRINGSHEIM E.G. (1951) Über farblose diatomeen. *Arch. Mikrobiol.* **16**, 18–27.

PRINGSHEIM E.G. (1952) On the nutrition of *Ochromonas .Quart. J. Microsc. Sci.* **93**, 71–96.

PRITCHARD G.G., GRIFFIN W.J. & WHITTINGHAM C.P. (1962) The effect of carbon dioxide concentration, light intensity and iso-nicotinyl hydrazide on the photosynthetic production of glycolic acid by *Chlorella. J. exp. Bot.* **13**, 176–84.

PROCTOR V.W. (1957) Some controlling factors in the distribution of *Haematococcus pluvialis. Ecology,* **38**, 457–62.

PROVASOLI L. (1965) Nutritional aspects of algal growth. *Proc. Can. Soc. Pl. Physiol.* **6**, 126–7.

RAMANATHAN J.D., CRAIGIE J.S., MCLACHLAN J., SMITH D.G. & MCINNES A.G. (1966) The occurrence of D-(+)-14/25-cyclohexanetetrol in *Monochrysis lutheri* Droop. *Tetrahedron Letters,* **14**, 1527–31.

RICHARDSON D.H.S., HILL D.J. & SMITH D.C. (1968) Lichen physiology. XI. The role of the alga in determining the pattern of carbohydrate movement between lichen symbionts. *New Phytol.* **67**, 469–86.

RZHANOVA G.N. (1967) Extracellular nitrogen-containing compounds of two nitrogen-fixing species of blue-green algae. *Microbiology,* **36**, 536–40.

SAMUEL S., SHAH N.H. & FOGG G.E. (1971) Liberation of extracellular products of photosynthesis by tropical phytoplankton. *J. mar. biol. Ass., U.K.* **51**, 793–8.

SAUNDERS G.W. (1972) The kinetics of extracellular release of soluble organic matter by plankton. *Verh. Ver. Int. Limnol.* **18**, 140–6.

SCHLÖSSER U. (1966) Enzymatisch gesteurte Freisetzung von Zoosporen bei *Chlamydomonas reinhardtii* Dangeard in Synchronkultur. *Arch. Mikrobiol.* **54**, 129–59.

SCUTT J.E. (1964) Autoinhibitor production by *Chlorella vulgaris. Am. J. Bot.* **51**, 581–4.

SEN N. (1966) Growth of a planktonic *Chlorella* and formation of glycine. *J. Ind. Bot. Soc.* **45**, 175–87.

SHILO M. (1967) Formation and mode of action of algal toxins. *Bact. Rev.* **31**, 180–93.

SHILO M., ASCHNER M. & SHILO M. (1953) The general properties of the exotoxin of the phytoflagellate *Prymnesium parvum. Bull. Res. Council, Israel,* **2**, 446.

SIEBURTH J.McN. (1959) Acrylic acid, an 'antibiotic' principle in *Phaeocystis* blooms in Antarctic waters. *Science, N.Y.* **132**, 676–7.

SIEBURTH J.McN. (1968) The influence of algal antibiosis on the ecology of marine microorganisms. *Adv. Microbiol. Sea,* **1**, 63–94.

SIEBURTH J.McN. (1969) Studies on algal substances in the sea. III. The production of extracellular organic matter by littoral marine algae. *J. Exp. Mar. Biol. Ecol.* **3**. 290–309.

SIEBURTH J.McN. & JENSEN A. (1968) Studies on algal substances in the sea. I. Gelbstoff (humic material) in terrestrial and marine waters. *J. Exp. Mar. Biol. Ecol.* **2**, 174–89.

SIEBURTH J.McN. & JENSEN A. (1969) Studies on algal substances in the sea. II. The formation of Gelbstoff (humic material) by phaeophyte exudates. *J. Exp. Mar. Biol. Ecol.* **3**, 275–89.

SMAYDA T.J. (1970) Growth potential bioassay of water masses using diatom cultures: Phosphorescent Bay (Puerto Rico) and Caribbean waters. *Helgoländer Wiss. Meeresunters.* **20**, 172–94.

SMITH D.C. & DREW E.A. (1965) Studies in the physiology of lichens. V. Translocation from the algal layer to the medulla in *Peltigera polydactyla. New Phytol.* **64**, 109–200.

SMITH D.C., MUSCATINE L. & LEWIS D.H. (1969) Carbohydrate movement from autotrophs to heterotrophs in parasitic and mutualistic symbiosis. *Biol. Rev.* **44**, 17–90.

SPOEHR H.A., SMITH J.H.C., STRAIN H.H., MILNER H.W. & HARDIN G.J. (1949) Fatty acid antibacterials from plants. *Carnegie Inst. Wash. Publ. No.* **586**, 1–67.

STADELMANN E.J. (1962) Permeability. In *Physiology and Biochemistry of Algae*, ed. Lewin R.A. pp. 493–528. Academic Press, New York & London.

STADELMANN E.J. (1969) Permeability of the plant cell. *A. Rev. Pl. Physiol.* **20**, 585–606.

STARR T.J. (1958) Notes on a toxin from *Gymnodinium breve*. *Texas Rep. Biol. Med.* **16**, 500–7.

STEPHENS G.C. & NORTH B.B. (1971) Extrusion of carbon accompanying uptake of amino acids by marine phytoplankters. *Limnol. Oceanogr.* **16**, 752–7.

STEWART W.D.P. (1963) Liberation of extracellular nitrogen by two nitrogen-fixing blue-green algae. *Nature, Lond.* **200**, 1020–1.

STEWART W.D.P. (1964) Nitrogen fixation by Myxophyceae from marine environments. *J. gen. Microbiol.* **36**, 415–22.

STRICKLAND J.D.H. & SOLÓRZANO L. (1966) Determination of monoesterase hydrolyzable phosphate and phosphomonoesterase activity in sea water. In *Some Contemporary Studies in Marine Science*, ed. Barnes H. pp. 665–74, Allen & Unwin, London.

SYRETT P.J. & WONG H.A. (1963) The fermentation of glucose by *Chlorella vulgaris*. *Biochem. J.* **89**, 308–15.

TAKAHASHI M., SHIMURA S., YAMAGUCHI Y. & FUJITA Y. (1971) Photoinhibition of phytoplankton photosynthesis as a function of exposure time. *J. Oceanogr. Soc. Japan*, **27**, 43–50.

THOMAS J.P. (1971) Release of dissolved organic matter from natural populations of marine phytoplankton. *Mar. Biol.* **11**, 311–23.

TOKUDA H. (1969) Excretion of carbohydrate by a marine pennate diatom, *Nitzschia closterium*. *Rec. Oceanogr. Wks. Japan*, **10**, 109–22.

TOLBERT N.E. & ZILL L.P. (1957) Excretion of glycolic acid by *Chlorella* during photosynthesis. In *Research in Photosynthesis*, ed. Gaffron H. *et al.* pp. 228–31. Interscience, New York.

TSUBO Y. (1961) Effect of the supernatant of illuminated cultures on dark mating in *Chlamydomonas moewusii* var. *rotunda*. *Bot. Mag.*, *Tokyo*, **74**, 519–23.

VON HOLT C. (1968) Uptake of glycine and release of nucleoside polyphosphates by zooxanthellae. *Comp. Biochem. Physiol.* **26**, 1071–9.

WALLEN D.G. & GEEN G.H. (1971) The value of the photosynthate in natural phytoplankton populations in relation to light quality. *Mar. Biol.* **10**, 157–68.

WALSBY A.E. (1968) Mucilage secretion and the movement of blue-green algae. *Protoplasma*, **65**, 223–38.

WATANABE A. (1951) Production in cultural solution of some amino acids by the atmospheric nitrogen fixing blue-green algae. *Archs. Biochem. Biophys.* **34**, 50–5.

WATT W.D. (1966) Release of dissolved organic material from the cells of phytoplankton populations. *Proc. R. Soc. B.* **164**, 521–51.

WATT W.D. (1969) Extracellular release of organic matter from two freshwater diatoms. *Ann. Bot.* **33**, 427–37.

WATT W.D. & FOGG G.E. (1966) The kinetics of extracellular glycollate production by *Chlorella pyrenoidosa*. *J. exp. Bot.* **17**, 117–34.

WATT W.D. & HAYES F.R. (1963) Tracer study of the phosphorus cycle in sea water. *Limnol. Oceanogr.* **8**, 276–85.

WEINMANN G. (1970) Gelöste Kohlenhydrate und andere organische Stoffe in natürlichen Gewässern und in Kulturen von *Scenedesmus quadricauda*. *Arch. Hydrobiol.* (*Suppl.*) **37**, 164–242.

WERNER D. (1970) Isolierung einer Gallertkolonien bildenden Mutante von *Cyclotella cryptica* (Diatomea) nach Temperaturschockbehandlung. *Int. Rev. Ges. Hydrobiol.* **55**, 403–7.

WHALEY W.G., DAUWALDER M. & KEPHART J.E. (1972) Golgi apparatus: Influence on cell surfaces. *Science, N.Y.* **175**, 596–9.

WHITTINGHAM C.P. & PRITCHARD G.G. (1963) The production of glycollate during photosynthesis in *Chlorella*. *Proc. R. Soc., B.* **157**, 366–80.

WHITTON B.A. (1965) Extracellular products of blue-green algae. *J. gen. Microbiol.* **40**, 1–11.

YENTSCH C.S. & REICHERT C.A. (1962) The interrelationship between water-soluble yellow substances and chloroplastic pigments in marine algae. *Botanica mar.* **3**, 65–74.

# CHAPTER 31

# MOVEMENTS

## W. NULTSCH

Department of Botany,
University of Marburg,
Marburg, Germany.

1    Introduction   864

2    Mechanisms of movement   865
2.1  Flagellar movement   865
2.2  Amoeboid movement   868
2.3  Metabolic movement   869
2.4  Gliding movement   869
     (a) Cyanophyceae   870
     (b) Diatoms   871
     (c) Euglenophyceae   872
     (d) Desmids   872
     (e) Rhodophyceae   872

3    Influence of external factors   872
3.1  Light
     (a) Photokinesis   873
     (b) Photo-phobotaxis   875
     (c) Photo-topotaxis   878

3.2  Chemical agents   880
3.3  Temperature   882
3.4  Other factors   882
     (a) Galvanotaxis   882
     (b) Geotaxis   883
     (c) Thigmotaxis   883

4    Intracellular movements   883
4.1  Cytoplasmic streaming   883
4.2  Movement of nuclei   884
4.3  Movement of chloroplasts   884

5    References   887

## 1  INTRODUCTION

Movement is widespread among the algae. Motile vegetative cells occur in all groups with the exception of brown algae, in which motility is restricted to gametes and zoospores. Movement can be observed in single cells as well as in colonies and filaments. Moreover, in many algae intracellular movements occur. Active or passive movements occur but this chapter deals only with active movements. Several different mechanisms are developed, and in some organisms visible organelles exist which are responsible for the movement, while in other forms no corresponding organelles or structures have been found. The speed of movement varies widely dependent on the different mechanisms and some examples are given in Table 31.1. The speed of movement also depends, of course, on other internal and external factors. External factors can also influence the direction of movement and induce particular reactions in the movement behaviour, such as the so-called 'stop-responses' (see p. 875). Recent reviews which include studies on movement in algae are Bendix (1960), Clayton (1964), Haupt (1966), Hand and Davenport (1970), Haupt and Schönbohm (1970) and Nultsch (1970).

**Table 31.1.** Speed of movement in several algae

| | Speed ($\mu$m sec$^{-1}$) | | |
| Family/organism | Average rate | Maximum rate | Reference |
| --- | --- | --- | --- |
| Cyanophyceae | | | |
| *Phormidium autumnale* | 2.2 | 2.8 | Nultsch (1961) |
| *Anabaena variabilis* | 0.3 | 0.5 | Drews (1959), Nultsch & Hellmann (1972) |
| | | | |
| Euglenophyceae | | | |
| *Euglena gracilis* | 100 | 180 | Wolken & Shin (1958) |
| | | | |
| Bacillariophyceae | | | |
| *Bacillaria paradoxa* | 7–8 | — | Jarosch (1958) |
| *Nitzschia palea* | 8–10 | 14 | Nultsch (1956) |
| | | | |
| Chlorophyceae | | | |
| *Chlamydomonas reinhardtii* | 110 | 230 | Nultsch (unpubl.) |
| *Micrasterias denticulata* | 0.5 | 1 | Neuscheler (1967a,b) |
| | | | |
| Rhodophyceae | | | |
| *Porphyridium cruentum* | 1 | 2 | Geitler (1944), Vischer (1955) |

# 2 MECHANISMS OF MOVEMENT

According to Doflein and Reichenow (1949) the following types of locomotion can be distinguished in unicellular organisms: 1. movements caused by swinging organelles, such as flagella and cilia; 2. amoeboid movements; 3. movements due to active contractions of contractile elements inside the cells; 4. gliding movements. These will now be considered briefly.

## 2.1 *Flagellar movement*

Flagellated species, or developmental stages such as gametes or zoospores, occur in all groups of algae with the exception of the blue-green algae and the red algae. Flagellar motion may result from the action of a single flagellum or pairs or bundles of flagella.

The ultrastructure of flagella, and cilia is uniform in eukaryotic organisms

(Brokaw 1962). As shown in Fig. 31.1, they consist of nine peripheral double fibrils, surrounding two central fibrils which appear as single tubes, and a common wall is absent. Each peripheral fibril possesses two 'arms', which arise from one side and point, if the observer is looking from the flagellar base towards the tip, in a clockwise direction (Afzelius 1959). The diameters of the double fibrils are 20 to 25 µm, to 35 to 37 µm and the arms are about 15 µm long and 5 µm thick. However, arms are not easily seen on the peripheral fibrils (Ringo 1967a), and sometimes may be absent.

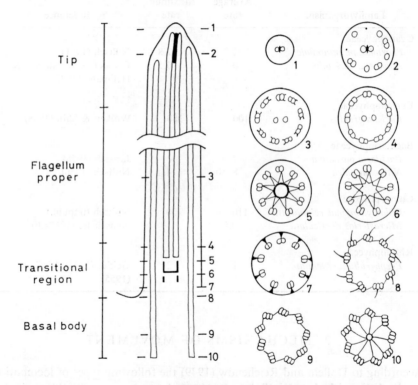

**Fig. 31.1.** Schematic drawing of an idealized longitudinal section through the flagellum and the basal body of *Chlamydomonas reinhardtii*. The four regions of the flagellum are designated, and 10 typical cross sections are shown for the numbered points marked along the flagellum. Cross section 3 shows the typical 9 + 2 arrangement (modified from Ringo 1967a).

The peripheral subfibrils appear to be tubular (Fig. 31.2) and are composed of almost spherical subunits, about 4·5 µm in diameter and consisting of a protein with a molecular weight of 40,000. The subunits are arranged in longitudinal arrays so that one tube shows 13 subunits when viewed in cross section and shares three or four subunits with the adjacent tube in the common wall (Ringo 1967b). The outer boundary of the flagellum is a semipermeable unit membrane, which is continuous with the cell membrane. The fibrils, however, run into the cytoplasm, where they originate from a complicated structure, called the basal body, blepharoplast or kinetosome. This basal body consists of nine triplet fibrils, which are connected with the adjacent fibrils (Fig. 31.1). At its proximal end, thin filaments run from the triplet to the centre, forming a cart-wheel-like structure. The distal end of the basal body is the doublet-to-triplet transition region, in which stellate patterns occur. The basal body, however, seems to be concerned with the synthesis of the flagellum rather than with the control of flagellar activity. Sometimes flagella show hair-like protrusions from the surface and additional structures of minor interest have been reported (DuPraw 1969). (For further details see Chapter 3, p. 87.)

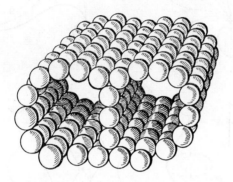

**Fig. 31.2.** Schematic drawing of a peripheral doublet fibril consisting of spherical subunits, without the 'arms'.

In spite of the uniform structure of the flagella, there is a great diversity in the ways the flagella beat (Pohl 1962). Some examples are given in Fig. 31.3. The *pull* or, in biflagellated forms, *breast stroke* type is shown in Fig. 31.3A; the propeller type in Fig. 31.3B; the undulatory type in which sinusoidal waves pass along the flagellum from the base to the tip in Fig. 31.3C. The pattern of flagellar movement may change in the same individual, depending on internal and external factors.

The molecular mechanisms of flagellar movement are not yet known. With the aid of the 'glycerol model' technique Hoffmann-Berling (1955, 1962) showed that glycerol-extracted flagella of sperms and trypanosomes display rhythmic move-ments on exposure to ATP. Tibbs (1958) has demonstrated the existence of ATPase activity in isolated flagella of *Polytoma uvella*. In the same organism, Brokaw (1961) increased the frequency of beating in isolated and reactivated flagella by adding exogenous ATP. Gibbons (1963, 1967) found later in *Tetra-hymena* that the ATPase activity is associated with a protein, the so-called dynein, which constitutes the arms of the peripheral double fibrils. In general therefore we may assume that during flagellar movement the contractile fibrils undergo an alternation of contraction and relaxation, resulting in unilateral bending, rotation or undulation of the flagellum, and that the energy is supplied as ATP.

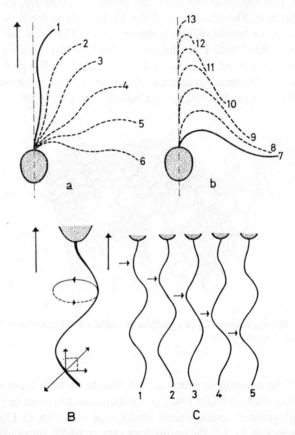

**Fig. 31.3.** Patterns of flagellar movement in algae. A. 'pull' type; a. power stroke, b. return stroke: B. 'propeller' type; C. 'undulatory' type, with a wave (arrow) running over the flagellum. Successive positions are numbered. (After Krijgsman & Metzner, from Pohl 1962.)

## 2.2 Amoeboid movement

Amoeboid movements are frequent in flagellates, e.g. in the Rhizochloridaceae, while in most other groups of algae amoeboid stages scarcely occur. Only in zoospores and gametes of a few species, such as *Aphanochaete pascheri*, *Bumilleria stigeoclonium* and *Vaucheria geminata*, has amoeboid movement been observed (Drews & Nultsch 1962).

Amoeboid movement is characterized by the formation of pseudopodia.

These are transient protuberances of cytoplasm, which vary in size and shape and which originate spontaneously at any site of the cell surface. They may be broad and lobed (*lobopodia*), thin and filamentous (*filopodia*) or connected to neighbouring threads, forming anastomoses (*rhizopodia*). The mechanism of amoeboid movement has not been investigated in algae but it is probably not very different from that of animal amoebae (Drews & Nultsch 1962, Abe 1964, Goldacre 1964, Kamiya 1964, Allen 1968, 1970). A theoretical explanation of amoeboid movement mechanism is given by Jarosch (1971).

### 2.3   *Metabolic movement*

Metabolic movements are defined as transient changes in the shape of elastic cells. They are mostly undulatory, and do not result in an overall change in cell shape. In the latter respect they are quite distinct from amoeboid movements (Drews & Nultsch 1962). They are common in flagellates such as *Euglena*. One or more waves of contraction (Fig. 31.4) can run over the cell surface within 10

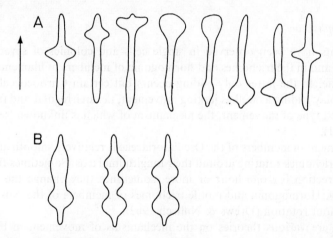

**Fig. 31.4.**   Metabolic movements in *Euglena deses*. A. Single wave, running over the cell in direction of the arrow. B. Two or three waves, running simultaneously. (After Kamiya, after Drews & Nultsch 1962.)

to 20 seconds. They are not strictly locomotions but can result in locomotion under certain circumstances. The mechanism of these movements is not yet clear but according to Diskus (1956) they are caused by a contractile layer of the periplast.

### 2.4   *Gliding movement*

Gliding movement is characterized by 'negative' criteria: the absence of definitive organelles of movement, and the lack of visible changes in cell shape. It occurs in several groups of algae, such as some blue-green algae, all diatoms which possess true raphes, several desmids and in the monospores and spermatia of red algae. Gliding movement can sometimes be observed in members of the Euglenophyceae (Jarosch 1962) and in other flagellated forms (Brokaw 1962).

The speed of gliding is generally very low compared with the flagellar movement and there are also considerable differences between the groups which display gliding movement. For example, the average speed of diatoms is about 10 times as fast as that of the other groups (Table 31.1). The different speeds of gliding are probably due to different movement mechanisms (see below). In many cases correlations between gliding movement and secretion of mucilage exist. On the one hand movement can be caused by a unidirectional secretion of mucilage and on the other hand the mucilage may only smooth the surface of the substrate and serve as lubricant. The chemical composition of the mucilage depends on the species and, probably, on the physiological conditions. Generally, the mucilages are polysaccharides consisting of various sugars, especially uronic acids (see O'Colla 1962).

## (a) *Cyanophyceae*

Movements have been observed in single cells and colonies of several Chroococcales and in the trichomes and hormogonia of members of filamentous blue-green algae. In the latter only gliding occurs but certain Chroococcales sometimes display gliding as well as jerking movement, i.e. an irregular and frequently interrupted type of movement, the mechanism of which is unknown (see Stanier *et al.* 1971).

Movement in members of the Oscillatoriaceae is relatively smooth and steady with the trichomes rotating around their longitudinal axis. Sometimes they move in one direction for one hour or more, sometimes they change the direction frequently. Hormogonia and motile trichomes of members of the Nostocaceae, glide without rotation (Drews & Nultsch 1962).

There are various theories on the mechanisms of movement in blue-green algae: a unilateral secretion of mucilage (Fechner 1915); propulsion of the secreted mucilage by protoplasmic fibrils fixed at the cell surface (Jarosch 1958, 1962); changes in surface tension (Burkholder 1934); contraction waves running over the surface of the trichomes (Schmid 1923); osmotic (Hansgirg 1883) and electro-osmotic processes (Toman 1955). Since all blue-green algae investigated secrete mucilage during movement and glide only when in contact with a solid substrate, the secretion of mucilage seems to be a prerequisite of movement, although the mechanism remains unclear.

Recently Halfen and Castenholz (1970) studying *Oscillatoria princeps* observed fibrils 5 to 8 μm wide on the surface of the trichomes. These were arranged in a parallel array and were obliquely aligned creating a helix with a pitch of about 60°. These workers suggest that gliding is the result of unidirectional waves travelling on the cell surface which act against a solid substrate and are produced by a lateral deformation of these fibrils. Movement in blue-green algae is discussed in detail by Fogg *et al.* (1973).

## (b) *Diatoms*

Diatoms display an alternating 'forward' and 'backward' movement, rather like a shunting engine. If no external factors act unilaterally, the direction of movement is random in cells which lack morphological polarity. However, in diatoms, such as *Gomphonema*, that are asymmetrical about their transverse axis, movement in the direction of the narrowest end lasts longer than the backward movement (Drum & Hopkins 1966). Species of some genera, such as *Navicula*, *Pinnularia* and *Pleurosigma*, move in more or less straight paths, while in *Amphora* the paths may be straight or curved. In *Nitzschia* movement is curved or circular, and the radii of the curves or the curvatures of circles may vary (Nultsch 1956).

Only diatoms with raphes are motile and the raphe appears to function by transmitting the motive force from the protoplast, rather than acting as an 'organelle' which causes motility. For example, contact between the raphe and a solid substrate is a prerequisite for movement. Thus, during locomotion the cells normally adhere to the surface of the substrate at their posterior ends (Drum & Hopkins 1966). In plasmolysed cells movement continues until the protoplast withdraws from the raphe system and especially from the terminal nodule (Höfler 1940, Nultsch 1962d, Drum & Hopkins 1966).

Further evidence for the involvement of the raphe comes from studies on particle transport along them. This can be observed when diatoms are immersed in suspensions of carmine or indigo or when detritus particles are adjacent to the raphes. This particle transport was first observed by Ehrenberg (1838) and is an important base of the protoplasm streaming theory of Müller (1889) which suggests that locomotion of diatoms is the result of the motive forces which the cytoplasm, protruding out of the raphe, generates on the cell surface, and its direction is the resultant of these forces. Martens (1940) has reported that velocity and direction of the particle transport are not related to velocity and direction of cell locomotion. This finding, however, is not unequivocal, because Nultsch (1962d) observed that a backward particle transport in forward moving cells is much more frequent than is transport in the forward direction, a finding confirmed by Drum and Hopkins (1966). The force responsible for particle transport is certainly sufficient to drive the cells, because the mass of the transported particles can be greater than that of the diatoms moving them. However, there are arguments against the theory of protoplasmic streaming and these are discussed in detail by Nultsch (1962d) and Drews and Nultsch (1962).

Jarosch (1958, 1960, 1962) proposed an alternative hypothesis which suggests that movement is the result of submicroscopic waves in extramembranous protein fibrils, which cause the shifting of secreted mucilage or a gliding of the cell over the substratum. More recently, Drum and Hopkins (1966) suggested that diatom movement is caused by any stimulus, 'which initiates contraction of locomotor fibrils and concurrent release of crystalloid body material at either the central or terminal pore; the secreted material is forced out of the cell and into the raphe systems where it streams until striking an object', which is large enough to function as a substrate. Both fibrils and crystalloid bodies have been found in electron micrographs and seem to be restricted to motile pennate diatoms. However the chemical nature of the material is unknown; further investigations in this area are required.

### (c) *Euglenophyceae*

Under certain conditions, *Euglena* displays gliding movements (Drews & Nultsch 1962) which, unlike metabolic movement, is not accompanied by visible deformations of the cells. The speed of gliding in *Euglena* is 2 to 3 $\mu$ sec$^{-1}$ (Günther 1928, Jarosch 1958), which is about a hundredth of the speed of flagellar movement of the same organism. The mechanism of the movement is not known but Mainx (1928) suggested that it was due to a unidirectional secretion of mucilage and Günther (1928) postulated that it is caused by invisible contractions of the periplast.

### (d) *Desmids*

Movement of desmids, first described by Ehrenberg (1838), is due to an unidirectional secretion of mucilage through terminal cell pores (Klebs 1885, 1886) and the propulsive force may result from an increase in the volume of the mucilage due to swelling. Mucilage can be demonstrated by staining or by negative contrast with Indian ink. In some desmids pores occur over the whole cell wall (Drawert & Mix 1961), and mucilage is probably secreted over the entire cell surface, functioning perhaps as a lubricant there.

### (e) *Rhodophyceae*

Gliding movements occur in some red algae, such as *Porphyridium cruentum*, and in spermatia, monospores and tetraspores of several species (Drews & Nultsch 1962). Secretion of mucilage, demonstrated by Geitler (1944, 1953) and Pringsheim (1968b) is obviously not sufficient to explain the mechanism of movement. Changes in the shape of the cells during movement have been observed by Geitler (1944) and Vischer (1955).

## 3  INFLUENCE OF EXTERNAL FACTORS

Irrespective of its mechanism, movement of algae can be affected by several external factors, such as light, temperature, gravity, chemicals and electric and mechanical stimuli. Furthermore, movement can be influenced by the same stimulus in different ways, resulting in quite distinct types of responses. Two main types can be distinguished: *kinesis*, i.e. an effect on the speed of movement, and *taxis*, i.e. a reaction that leads to a distinct pattern in spatial arrangement or distribution of the organisms. Taxes can be subdivided into two types, which are called *topotaxis* and *phobotaxis*. Topotactic responses result in a movement towards the source of the stimulus (*positive*) or away from it (*negative*), while phobotactic responses are caused by a temporal decrease (positive) or increase (negative) in the intensity of the stimulus, $\pm dI/dt$.

The mechanisms of the reactions are different in the various classes of algae and depend on the mechanism of movement, the speed of movement and other factors. Furthermore, a topotactic response can be the result of either a steering act or of trial and error motion. On the other hand, a stimulus is received by an individual when it passes a steep spatial gradient of the stimulus so that a phobotactic response can be caused this way too. Thus, the behaviour of the organisms during the reaction can vary considerably from family to family or even from genus to genus.

## 3.1   *Light*

Light is one of the most important ecological factors affecting movement of almost all motile algal species investigated so far (Haupt 1959c), and all the reaction types mentioned above have been observed. Responses may be obtained not only to visible light but also to ultra-violet and infra-red radiation.

### (a) *Photokinesis*

Negative photokinesis in algae was first observed by Strasburger (1878) in motile stages of *Botrydium granulatum* and other species. In his experiments, the swarmers became immotile in light, while they continued to move in the dark. More extensive investigations were carried out by Bolte (1920), who observed both positive and negative photokinesis. Most green flagellates showed positive reactions. They became immotile in the dark, but resumed their movement after a short irradiation. Negative photokinesis occurred only in a few species of *Phacus*, *Lepocinclis* and *Chlamydomonas*. Later, Luntz (1931a,b, 1932) found in some Volvocales, e.g. in *Eudorina*, that at lower intensities the photokinetic effect is a function of the light quantity, while under continuous illumination it depends only on light intensity. Some further incidental observations of several authors, reviewed by Nultsch (1970), revealed contradictory results.

In the last decade photokinesis has been investigated by Nultsch and co-workers in blue-green algae (Nultsch 1962a, Nultsch & Hellmann 1972), in diatoms (Nultsch 1971a) and in the red alga, *Porphyridium cruentum* (Nultsch unpubl.). In all these forms, which are motile also in the dark, light has a positive photokinetic effect. As shown in Fig. 31.5, movement is accelerated by

light, until an optimum is reached. At higher light intensities the speed of movement decreases. On this basis Nultsch (1970) redefined photokinesis as 'an
acceleration of movement by light or, in organisms which are immotile in the
dark, even an excitation of movement by light'. Consequently, negative photokinesis is a light-induced decrease of the speed, compared with the speed of the
dark movement, eventually resulting in immotility.

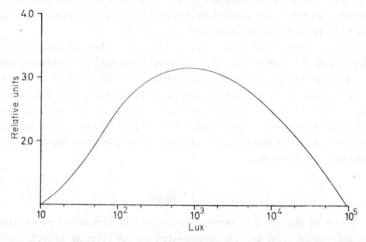

**Fig. 31.5.** Photokinesis, in relative units, of *Anabaena variabilis* at different intensities of white light. (After Nultsch & Hellmann 1972.)

Action spectra studies on blue-green algae reveal that the photokinetically
active light is absorbed by photosynthetic pigments. However, there are remarkable differences in the spectral sensitivity of members of the Oscillatoriaceae on
the one hand and members of the Nostocaceae on the other (Nultsch 1962c,
Nultsch & Hellmann 1972). As shown in Fig. 31.6, the action spectrum of
*Phormidium ambiguum*, and of other *Phormidium* species, resembles the absorption spectrum of chlorophyll *a*, while radiation absorbed by carotenoids and
biliproteins is scarcely effective, if at all. On the contrary, in *Anabaena variabilis*
the maximum of photokinetic activity coincides with the absorption maximum
of C-phycocyanin. While blue light is ineffective, red light absorbed by chlorophyll *a* is very active, although no distinct peak can be observed. From these
results Nultsch concluded that, in species of the genus *Phormidium*, the photokinetic effect is the result of an additional ATP supply from cyclic photophosphorylation with photosystem I, while in *Anabaena variabilis* the photo-induced
acceleration of movement is due to an energy supply from a phosphorylation
process of the pseudocyclic type which is mediated both by photosystem I
and photosystem II. A response at wavelengths up to 800 nm in *Phormidium* may
also be due to photosystem I activity. These suggestions have been confirmed by
experiments using photosynthetic inhibitors (Nultsch 1965, Nultsch & Jeeji-Bai

1966), uncouplers (Nultsch 1966, 1967, 1968b) and redox-systems (Nultsch 1968a). Similar results demonstrating correlations between photokinesis and photophosphorylation processes have been obtained with the diatom *Nitzschia communis* (Nultsch 1971a) and with the red alga *Porphyridium cruentum* (Nultsch & Dillenburger unpubl.). The photokinetic action spectrum of *Euglena* measured by Wolken and Shin (1958) can be interpreted in the same way. Thus it seems

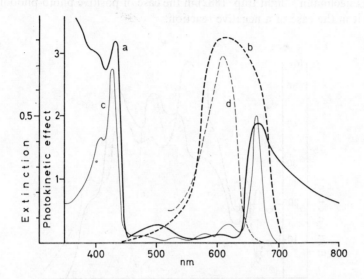

**Fig. 31.6.** Photokinetic action spectra, in relative units, of *Phormidium ambiguum* (a) and *Anabaena variabilis* (b). Absorption spectrum of chlorophyll *a* (c) and C-phycocyanin (d). (After Nultsch & Hellmann 1972.)

that photokinesis in algae is the result of an additional energy supply from photophosphorylation processes, while movement in the dark is maintained by ATP supplied from oxidative phosphorylation (Nultsch 1970).

### (b) *Photo-phobotaxis*

Photo-phobotactic reactions, variously called *phobic responses*, *shock reactions*, *stop responses* or *motor responses*, are caused by sudden changes in light intensity. Photo-phobotaxis is widespread in algae, but the mode of the phobic responses, i.e. the behaviour of the individuals after stimulation, is different in the various groups and depends on the morphology of the organism and the mechanism of movement. In slowly moving organisms, such as diatoms and blue-green algae, it is simply a stop, sometimes preceded by a slow-down, and normally followed by a return. In *Euglena* the phobic response results in a sharp turn of the cell, due to an abrupt sideward beating of the flagellum to the ventral side (Haupt 1959c). Similar responses have been observed in dinoflagellates (Metzner 1929).

In *Chlamydomonas* the backward swimming is caused by a sudden forward beating of the flagella (Ringo 1967a). At low light intensities, the phobic responses are mostly positive, i.e. they are caused by a decrease in light intensity, provided that it exceeds a distinct threshold value, the so-called 'discrimination threshold'. Conversely, negative responses usually occur at high light intensities. If a light field is projected by a slit onto a preparation of algae, they gather in the light field (Engelmann's 'light trap' 1882) in the case of positive photo-phobotaxis, or leave it in the case of a negative reaction.

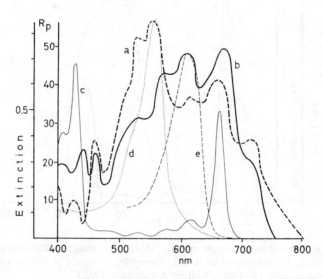

**Fig. 31.7.** Photo-phobotactic action spectra (Rp) in relative units of *Phormidium uncinatum* (a) and *Phormidium ambiguum* (b). Absorption spectrum of chlorophyll *a* (c), C-phycoerythrin (d) and C-phycocyanin (e). (After Nultsch 1962c).

Photo-phobotactic action spectra of blue-green algae have been measured by Nultsch (1962b,c) in some species of *Phormidium*. As shown in Fig. 31.7, species which contain C-phycoerythrin, such as *Phormidium uncinatum*, display maximum phobotactic activity at about 560nm, while in forms which only contain C-phycocyanin, such as *Phormidium ambiguum*, this maximum shifts to 615nm. Moreover, in both species a second maximum exists at about 680nm, which coincides with the red absorption maximum of chlorophyll *a in vivo*, while the effect of blue-light is greater than its absorption by chlorophyll *a*. Because of the striking similarity between these photo-phobotactic and some photosynthetic action spectra in several blue-green algae (Duysens 1952, Haxo & Norris 1953, Nultsch & Richter 1963), it has been suggested that photo-phobotactic responses are caused by sudden changes in the rate of photosynthesis. This has been confirmed by experiments using photosynthetic inhibitors (Nultsch 1965,

Nultsch & Jeeji-Bai 1966), uncouplers (Nultsch 1966, 1967, 1968b) and redox-systems (Nultsch 1968a). Based on the results of these investigations Nultsch (1970) has concluded that in blue-green algae photo-phobotaxis is coupled with the photosynthetic electron transport chain. He suggested that sudden changes in the steady state of the electron transport, provided that they exceed a distinct threshold value, are transformed into bioelectric potential changes, e.g. action potentials, which are quickly conducted by the cytoplasmic membrane to the locomotor apparatus, causing a phobic response. However, in other algae such as flagellates (Bünning & Schneiderhöhn 1956, Diehn 1969) and diatoms (Nultsch 1971a) only radiation of shorter wavelengths is photo-phobotactically active, in particular the near ultra-violet, and visible light up to 500 or 550 nm, as shown in Fig. 31.8. Consequently, in these forms correlations between photo-phobotaxis and photosynthesis can be excluded. This striking difference is

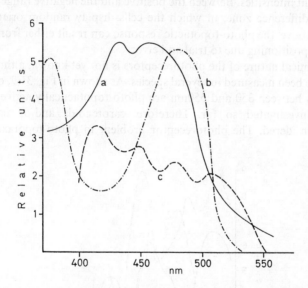

**Fig. 31.8.** Action spectra, in relative units, of positive photo-phobotaxis of *Nitzschia communis* (a) and *Euglena gracilis* (b) and of negative photo-phobotaxis of *Euglena gracilis* (c). (After Nultsch 1971a, Diehn 1969, Bünning & Schneider-höhn 1956.)

probably due to the existence of a chloroplast envelope in the eukaryotic algae, which separates the thylakoids from the surrounding cytoplasm and in this way prevents the immediate transmission of the potential changes to the locomotor apparatus.

As the action spectra of photo-phobotaxis and photosynthesis are essentially identical in purple bacteria as well (Manten 1948, Thomas & Nijenhuis 1950, Duysens 1952, Clayton 1953a,b), we may conclude that two different photo-phobotactic reaction mechanisms exist: firstly, the prokaryotic or photosynthetic type, which occurs in blue-green algae and photosynthetic bacteria. Here, photophobotactic responses are due to sudden changes in the steady state of photosynthetic electron transport caused by sudden changes in light intensity. Secondly, there is the eukaryotic type, where only radiation of wavelengths shorter than 550nm is active, so that only yellow pigments act as photoreceptors.

### (c) *Photo-topotaxis*

Photo-topotactic reactions, observed first by Treviranus (1817), are very common in algae. In photo-topotaxis the response is positive, when the organisms move towards the light source, and negative when they move away from it. Normally, positive reactions are observed at lower light intensities and negative reactions at higher light intensities. Between the positive and the negative range there is the so-called indifference zone, in which the cells display random orientation. As mentioned above, the photo-topotactic response can result either from a steering act or from positioning due to trial and error.

The chemical nature of the photoreceptors is not yet known, although action spectra have been measured in several species. As shown in Fig. 31.9, only shorter wavelengths between 350 and 550nm are photo-topotactically active in most of the algae investigated so far. Therefore, carotenoids and/or flavoproteins must be considered. The photoreceptor problem in photo-topotaxis has been

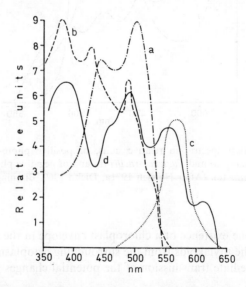

**Fig. 31.9.** Action spectra of positive photo-topotaxis, in relative units, of *Chlamydomonas reinhardtii* (a), *Nitzschia communis* (b), *Prorocentrum micans* (c) and *Phormidium autumnale* (d). (After Nultsch *et al.* 1971, Nultsch 1971a, Halldal 1958, Nultsch 1961.)

compared very often with that in phototropism. However, the maximum activity at 503nm in *Chlamydomonas* (Nultsch *et al.* 1971) and the extension of the action spectrum to longer wavelengths in diatoms (Nultsch 1971a) and in *Prorocentrum* (Halldal 1958) favours the possible photoreceptor role of carotenoids, because the range about 500nm is poorly absorbed by most of the flavoproteins. Finally, because of the strong effectiveness of light absorbed by C-phycoerythrin and C-phycocyanin in phototaxis of blue-green algae (Nultsch 1961, 1962c) it must be concluded that even biliproteins can participate in the perception of the phototactic stimulus.

In *Euglena*, the photo-topotactic orientation is thought to be the result of numerous successive steering acts, each of them caused by a slight sideward beating of the flagellum due to a short shading of the photoreceptor by the stigma. The photoreceptor is considered to be a flavoprotein located in the paraflagellar body (Vavra 1956), while the stigma contains large amounts of carotenoids. Periodical shading occurs during rotation of the cells around their long axis until the flagellated end points to the light source. However, Halldal (1958) has emphasized that the type of phototactic orientation exhibited by *Euglena* is unique and is not typical of all flagellates. There are many which exhibit a perfect phototactic orientation although they lack a stigma, and there are others, such as *Chlamydomonas* (Ringo 1967a) which do not rotate as they move. In both the latter cases periodical shading of the photoreceptor by the stigma is impossible. According to Halldal (1963), the direction of movement in flagellates under unilateral illumination is regulated by asymmetrical beating of the flagella, whenever the photoreceptor is shaded by any part of the cell. This continues until the flagellated end of the cells points to the light source, a position where, for morphological reasons, the paraflagellar body is not shaded. In *Volvox aureus* photo-topotactic orientation results from unequal flagellar activity at either side of the colony, due to flagellar activity ceasing in those cells nearest the light source (Huth 1970a,b, Hand & Haupt 1971).

In the Cyanophyceae several different types of response are observed. The trichomes of the Nostocaceae, which do not rotate during movement, bend their tips to the light source when they are exposed to unilateral irradiation (Drews 1959) and this is probably a true steering act. Blue-green algae of the family Oscillatoriaceae, and diatoms as well, display an alternating 'back and forth' movement. The orientation of their long axis is random and under unilateral irradiation, no active adjustment of the direction of movement occurs. However, filaments with their long axis at an angle to the light beam, and exhibiting a positive response, show prolonged movements towards the light and shorter movements away from it. In this way, the organisms approach the light source by degrees. This 'trial-and-error' type of movement is not a 'steering mechanism' in the strict sense of the term. A peculiar type of topotactic movement occurs in certain members of the Chroococcales (Pringsheim 1968a, Stanier *et al.* 1971), in which whole colonies glide towards the light source or away from it. A similar behaviour has been observed in *Porphyridium cruentum* where single cells display

photo-topotactic orientation (Pringsheim 1968b). The photo-topotactic movement of desmids has been investigated by Neuscheler (1967a,b). *Micrasterias*
cells orientate their long axes parallel to the light beam in the case of a positive
reaction and perpendicular to it in the case of phototactic indifference.

### 3.2  *Chemical agents*

*Chemotaxis* is the effect of chemicals on the movement of microorganisms.
Chemotactic activity is usually measured by a capillary method (Halldal 1962,
Ziegler 1962), in which a glass capillary filled with a solution of the chemical is
introduced into a suspension of the test organism and attraction or repulsion of
the organism is then observed. We really have to distinguish between topotactic
and phobotactic responses in chemotaxis, too, although both the reactions result
in a movement towards the source of the stimulus in the case of a positive
reaction and *vice versa*. However, in the case of topotaxis movement is a result
of a steering act, while it results from a series of successive responses in the case
of phobotaxis (Fig. 31.10). Positive and negative chemotaxis have been observed
in numerous algae (Table 31.2), and there is no real difference in the chemotactic
behaviour of coloured and colourless forms (Ziegler 1962). In blue-green algae
and diatoms only phobic responses take place. In flagellates both reaction types
have been reported to occur. In the early investigations of Pfeffer (1888) all

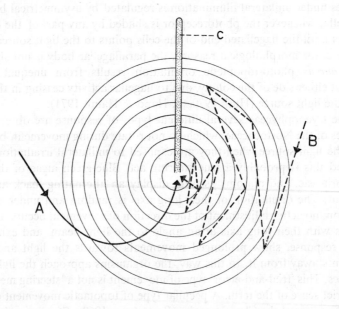

**Fig. 31.10.** Schematic diagram of topic (A) and phobic (B) chemotactic movement. C is a capillary containing the chemotactic agent. The concentration gradient
is shown by the concentric rings. (After Nultsch 1971b.)

species displayed topotactic responses when the long axis of the cell was orientated parallel to the gradient. Links (1955) stated that phobochemotaxis is the most common if not the only type of chemotaxis.

**Table 31.2.** Occurrence of chemotaxis in algae

| Family/organism | Reference |
|---|---|
| Cyanophyceae | |
| *Oscillatoria, Phormidium* | Fechner 1915, Schmid 1923. |
| Euglenophyceae | |
| *Euglena, Astasia, Trachelomonas,* | Engelmann 1882, Pfeffer 1888, Aderhold |
| *Phacus* | 1888, Frank 1904, Mainx 1928. |
| Cryptophyceae | |
| *Chilomonas, Cryptomonas* | Pfeffer 1888, Garrey 1900, Galiano 1921. |
| Dinophyceae | |
| *Glenodinium* | Pfeffer 1888 |
| Bacillariophyceae | |
| *Navicula, Pinnularia, Nitzschia* | Engelmann 1882, Aderhold 1888, Nultsch 1956. |
| Chlorophyceae | |
| *Chlamydomonas, Polytoma,* | Pfeffer 1888, Frank 1904, Pringsheim |
| *Pandorina* | 1921, Pringsheim & Mainx 1926. |
| Some colourless genera | |
| *Bodo, Monas, Tetramitus,* | Pfeffer 1888. |
| *Trepomonas, Hexamitus* | |

Various substances have been examined, e.g. ions, organic acids, alcohols, esters, aldehydes, sugars, and amino acids, and many of them were reported to be chemotactically active. However, the results were often contradictory, and no unequivocal correlations between chemical structure and chemotactic activity could be detected. For example, Pringsheim and Mainx (1926) found that within homologous series compounds consisting of 4-carbon atoms showed the strongest chemotactic activity in *Polytoma uvella*. On the other hand, Links (1955), investigating the chemotactic action of fatty acids ($C_4$ to $C_{16}$), did not find any correlation between effectiveness and chain length. To determine how many chemoreceptors exist for a particular species, the effect of an attractant is investigated in the presence of another chemotactic agent and if they compete with each other, it is generally agreed that the same chemoreceptor is involved. In this way, at least eight different chemoreceptors have been detected in *Polytoma*, which are sensitive to oxygen (*aerotaxis*), acids, alkalies, aliphatic compounds, aromatic compounds, ammonium salts, calcium salts and thiosulphate.

The chemical nature of algal chemoreceptors is not yet known. Adler (1969), who investigated chemotaxis in bacteria, found that they are not enzymes that catalyse the metabolism of the attractants or parts of permeases and related transport systems. Further attempts to identify them are necessary.

### 3.3 *Temperature*

Orientation of motile microorganisms in a given temperature gradient is called *thermotaxis*. As in other taxes, topic and phobic responses are possible, but the mode of the response has not been investigated in most organisms (Haupt 1962a). As shown in Table 31.3, thermotaxis has been observed in several algae.

**Table 31.3.** Occurrence of thermotaxis in algae

| Family/organism | Reference |
| --- | --- |
| Cyanophyceae | |
| *Oscillatoria* | Mendelssohn 1902b,c. |
| Euglenophyceae | |
| *Euglena, Phacus, Carteria* | Franzé 1893, Günther 1928, Reimers 1928. |
| Bacillariophyceae | |
| *Pinnularia, Nitzschia, Navicula* | Mendelssohn 1902a, Reimers 1928. |
| Chlorophyceae | |
| *Haematococcus, Volvox* | Reimers 1928. |

Some display positive reactions, e.g. *Navicula radiosa*, which moves from colder areas to 28 to 30°C (Reimers 1928), while others, e.g. *Haematococcus*, react negatively and move from warmer areas to 5 to 10°C.

### 3.4 *Other factors*

In addition to the taxes mentioned above, tactic reactions can also be induced by other stimuli, e.g. *galvanotaxis* by electric stimuli, *geotaxis* by gravity and *thigmotaxis* by contact with solid bodies.

### (a) *Galvanotaxis*

Galvanotaxis is an active orientation in an electric current, and should not be confused with a passive electrophoretic transport of particles. In the case of a cathodic ( −ve) reaction the individuals move towards the cathode, while in the case of anodic (+ve) reaction they move to the anode. Both reactions occur in algae (Umrath 1959) and both can occur in the same species, depending on the physiological state of the organism and on the environmental factors. For example in *Polytoma uvella* and other species Gebauer (1930) found anodic responses at high partial pressures of $O_2$ and cathodic responses at low ones and, at an intermediate $O_2$ pressure, anodic reactions at high pH and *vice versa*. The mode of galvanotaxis has been reported to depend on light conditions such as pre-illumination and phototactic disposition during the experiment (Terry 1906, Bancroft 1907, Mast 1927). The mechanism of galvanotactic orientation has scarcely been investigated. According to Bancroft (1915) *Euglena* cells adjust their position by sideward beating of the flagellum until the flagellated end points to the cathode.

### (b) *Geotaxis*

Geotaxis means an active orientation of microorganisms in the field of gravity. Our knowledge of geotaxis is mainly based on occasional observations, reviewed by Haupt (1962b). Since geotaxis can easily be simulated by other gradients in the medium, e.g. in $O_2$ and $CO_2$ concentration, many reports of geotaxis are equivocal. Indeed Brinkmann (1968) has questioned the existence of geotaxis in *Euglena* and in other unicellular algae.

### (c) *Thigmotaxis*

Negative thigmotaxis has been reported by Wagner (1934) for *Nitzschia putrida*. However, many algae return or change their direction of movement if they are mechanically prevented from moving forward. It is questionable, therefore, whether a particular term is necessary to describe this type of behaviour.

## 4 INTRACELLULAR MOVEMENTS

Intracellular movements can be autonomous or caused by external factors. They may concern the whole cytoplasm, with the exception of the cortical regions, or be restricted to single organelles, such as nuclei and chloroplasts. Movement of these organelles can be passive or active. Earlier reviews include those of Haupt (1962c) and Haupt and Schönbohm (1970).

### 4.1 *Cytoplasmic streaming*

Cytoplasmic streaming has been observed in numerous algae, e.g. in *Chara*, *Nitella*, *Spirogyra*, *Acetabularia*, *Bryopsis*, *Caulerpa* and *Codium* (Haupt 1962c, Kamiya 1959, 1962, Küster 1956). In *Spirogyra* and some desmids 'agitation' and 'circulation' types occur, in which the direction and the speed of streaming in the cytoplasmic layer lining the cell wall as well as in the strands crossing the vacuole is random and unsteady. In the internodal cells of *Chara* and *Nitella* we find the 'rotation' type. Here the cytoplasm streams in a constant direction on one side of the cell and in the opposite direction on the other. On the side walls the moving cytoplasm is separated by a quiescent zone. Rotation takes place only in the endoplasm, while the ectoplasm and the cytoplasmic membranes do not move. Unlike some other algae and higher plants, such as *Vallisneria*, the chloroplasts are not displaced by the streaming endoplasm, because they stick to the ectoplasm. *Acetabularia*, *Bryopsis* and *Codium* show 'multistriate streaming' (Kamiya 1962). In this type, streaming is restricted to definite cytoplasmic tracks, the direction of which either coincides with the long axis of the cell, or deviates from it as it follows a helical course. Streaming in neighbouring strands, which are separated from each other by quiescent zones, is very often, but not always, in opposite directions. Chloroplasts in the strands may be displaced passively.

The motive force for cytoplasmic streaming is an active, energy consuming shearing force generated at the boundary between the ectoplasm and the endoplasm (Hayashi 1964). However, we do not know how the active shearing force is produced at this boundary.

### 4.2  *Movement of nuclei*

Migration of nuclei in algae is mostly associated with the fertilization process (Haupt 1962c), e.g. in *Spirogyra* where the nuclei pass intercellular bridges, or in red algae where the male nucleus passes down the trichogyne. In *Acetabularia* thousands of nuclei migrate from the rhizoid through the stalk into the cap at the beginning of cyst formation. It is uncertain whether these movements are active or passive.

### 4.3  *Movement of chloroplasts*

In addition to the passive displacements of chloroplasts by cytoplasmic streaming mentioned above, active movements of chloroplasts caused by light or other factors occur in numerous algae (Table 31.4). One can distinguish two extreme positions in relation to light: a 'high intensity arrangement' and a 'low intensity arrangement', because they are induced by irradiation of higher or lower intensities and the movements leading to these positions are called high and low intensity movements respectively (Haupt 1959a, Haupt & Schönbohm 1970).

The arrangements of the chloroplasts vary depending on the incident light, and as a general rule we can say that they move to the least irradiated parts of the cells in strong light and *vice versa*. The final arrangement, of course, depends on the morphology of the cell and the chloroplasts. Several distinct types, described in detail by Senn (1908), exist and some of these are shown in Fig. 31.11. In several algae, such as *Dictyota* (Nultsch & Pfau unpubl.), any position between these extreme arrangements can be taken by the chloroplasts at medium intensities, but in other genera, such as *Mougeotia*, there is a distinct light intensity which separates the ranges of 'high' and 'low' intensities. Its magnitude depends on the species in question as well as on internal and external factors.

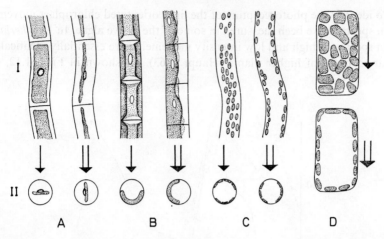

**Fig. 31.11.** Chloroplast arrangements in some algae, showing surface views (*I*) and cross sections (*II*) (with the exception of D). Single arrow: low light intensity; double arrow: high light intensity. A. *Mougeotia*; B. *Hormidium*; C. *Vaucheria*; D. *Dictyota*. (Modified from Haupt 1963, Senn 1908, Nultsch & Pfau unpubl.)

In addition to the above arrangements, there exists in several algae a particular 'dark arrangement'. For example, in *Vaucheria* the distribution of chloroplasts in the cytoplasm is random in the dark and in some centric diatoms (Table 31.4) chloroplast displacement, although caused by light, is not related to the light direction (Senn 1919).

**Table 31.4.** Occurrence of chloroplast displacement in algae

| Family/organism | Reference |
|---|---|
| Chrysophyceae | Molisch 1901. |
| *Chromulina* | |
| Bacillariophyceae | Lüders 1854, Karsten 1905. |
| *Achnanthes, Biddulphia, Coscino-* | |
| *discus* and other centric diatoms | |
| Chlorophyceae | |
| *Eremosphaera, Ulothrix, Hormidium,* | Moore 1901, Senn 1908, |
| *Ulva, Bryopsis, Acetabularia,* | Winkler 1900, De Bary & Strasburger 1877. |
| *Mougeotia, Mesotaenium* | |
| Xanthophyceae | Stahl 1880. |
| *Vaucheria* | |

2F

To identify the photoreceptors of the light orientated chloroplast movement, action spectra have been measured in some of the above algae. In *Vaucheria*, the action spectra of high and low intensity movements are essentially identical and similar to those of higher plants (Haupt 1963). As shown in Fig. 31.12, only

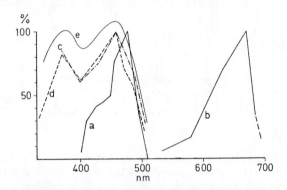

**Fig. 31.12.** Action spectra, in terms of relative quantum efficiency, of chloroplast orientation movements. *Mougeotia* high (a) and low (b) intensity movements, *Vaucheria* high (c) and low (d) intensity movements. For comparison the absorption spectrum of riboflavin is given (e). (Modified from Haupt 1959b, 1963, Schön-bohm 1963.)

ultra-violet, violet and blue light is active. The action spectra resemble the absorption spectrum of riboflavin and it has been suggested that flavins are the photoreceptors of both reactions in these organisms (Haupt & Schönbohm 1970). On the other hand, in *Mougeotia* and in *Mesotaenium* the action spectra of high and low intensity movements are quite different from each other. While the high intensity movement is also caused by light of shorter wavelengths (Schönbohm 1963), the photoreceptor of the low intensity movement, which is sensitive to red light (Fig. 31.12), is phytochrome (Haupt 1959b, Haupt & Schönbohm 1970).

It has been shown by Fischer-Arnold (1963) in *Vaucheria* and by Bock and Haupt (1961) in *Mougeotia*, that the photoreceptor molecules are not localized in the chloroplasts themselves, but in the cytoplasm. Moreover, in polarized light the response depends on the electric vector in relation to the long axis of the cell: if it is orientated parallel to the long axis, the high intensity chloroplast arrangement occurs, while perpendicular orientation results in the low intensity arrangement (see Fig. 31.13). From these and other results it has been concluded by Haupt and coworkers that the photoreceptor molecules are 'localized in the cortical cytoplasm, perhaps associated with the plasmalemma', and that their orientation is dichroitic and parallel to the cell surface as demonstrated in Fig. 31.13. This is true for both blue and red light responses. The mechanism of chloroplast movement is an open question (Zurzycki 1962). It has been shown by Schönbohm (1969, 1972b) that the chloroplast movement in *Mougeotia* is

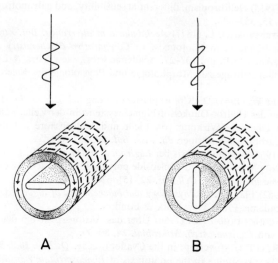

**A**             **B**

**Fig. 31.13.** Absorption of polarized light by oriented photoreceptors in *Mougeotia*. Heavy dashes represent photoreceptors which can absorb the polarized light due to their orientation, while thin ones represent such receptors which cannot. The resulting absorption is indicated by dots near the periphery in the cytoplasmic layer; the absorption gradient is shown by arrows. The electrical vector of the light is indicated by waves. (After Haupt & Schönbohm 1970.)

inhibited by SH-group blocking agents such as PCMB (*p*-chloromercuribenzoate), and the inhibitory effect is completely reversed by the addition of cysteine. This favours the concept that chloroplast displacement is brought about by the contractile protein fibrils detected by Schönbohm (1972a).

# 5 REFERENCES

ABÉ T.H. (1964) Mechanisms of ameboid movement based on dynamic organization: Morphophysiological study of ameboid movement. IV. In *Primitive Motile Systems in Cell Biology*, eds. Allen R.D. & Kamiya N. pp. 221–36. Academic Press, New York & London.

ADERHOLD R. (1888) Beitrag zur Kenntnis richtender Kräfte bei der Bewegung niederer Organismen. *Jena. Z. Naturw.* **22** (N.F. 15), 310–42.

ADLER J. (1969) Chemoreceptors in bacteria. *Science, N.Y.* **166**, 1588–97.

AFZELIUS B. (1959) Electron microscopy of the sperm tail. *J. biophys. biochem. Cytol.* **5**, 269–78.

ALLEN R.D. (1968) Differences of a fundamental nature among several types of amoeboid movement. In *Aspects of cell motility*, XXIII Symp. S.E.B. pp. 151–68. Cambridge University Press.

ALLEN R.D. (1970) Comparative aspects of amoeboid movement. *Acta Protozoologica*, **7**, 291–9.

BANCROFT F.W. (1907) The mechanism of the galvanotropic orientation in *Volvox. J. exp. Zool.* **4**, 157–63.

BANCROFT F.W. (1915) Heliotropism, differential sensibility, and galvanotropism in *Euglena*. *J. exp. Zool.* **15**, 383–428.

DE BARY A. & STRASBURGER E. (1877) *Acetabularia mediterranea. Bot. Ztg.* **35**, 713–58.

BENDIX S. (1960) Pigments in Phototaxis. In *Comparative Biochemistry of Photoreactive Systems*, ed. Allen M.B. pp. 107–27. Academic Press, New York & London.

BERTHOLD G. (1882) Beiträge zur Morphologie und Physiologie der Meeresalgen. *Jb. wiss. Bot.* **13**, 559–717.

BOCK G. & HAUPT W. (1961) Die Chloroplastendrehung bei *Mougeotia*. III. Die Frage der Lokalisierung des Hellrot-Dunkelrot-Pigmentsystems in der Zelle. *Planta*, **57**, 518–30.

BOLTE E. (1920) Über die Wirkung von Licht und Kohlensäure auf die Beweglichkeit grüner und farbloser Schwärmer. *Jb. wiss. Bot.* **59**, 287–324.

BRINKMANN K. (1968) Keine Geotaxis bei *Euglena. Z. PflPhysiol.* **59**, 12–16.

BROKAW C.J. (1961) Movement and nucleoside polyphosphatase activity of isolated flagella from *Polytoma uvella. Expl. Cell Res.* **22**, 151–62.

BROKAW C.J. (1962) Flagella. In *Physiology and biochemistry of algae*, ed. Lewin R.A. pp. 595–601. Academic Press, New York & London.

BÜNNING E. & SCHNEIDERHÖHN G. (1956) Über das Aktionsspektrum der phototaktischen Reaktionen von *Euglena. Arch. Mikrobiol.* **24**, 80–90.

BURKHOLDER P.R. (1934) Movement in the Cyanophyceae. *Q. Rev. Biol.* **9**, 438–59.

CLAYTON R.K. (1953a) Studies in the phototaxis of *Rhodospirillum rubrum*. I. Action spectrum, growth in green light and Weber Law adherence. *Arch. Mikrobiol.* **19**, 107–24.

CLAYTON R.K. (1953b) Studies in the phototaxis of *Rhodospirillum rubrum*. II. The relation between phototaxis and photosynthesis. *Arch Mikrobiol.* **19**, 125–40.

CLAYTON R.K. (1964) Phototaxis in microorganisms. In *Photophysiology*, vol. II, ed. Giese A.C. pp. 51–77. Academic Press, New York & London.

DIEHN B. (1969) Action spectra of the phototactic response in *Euglena. Biochim. biophys. Acta*, **177**, 136–43.

DISKUS A. (1956) Färbestudien an den Schleimkörperchen und Schleimausscheidungen einiger Euglenen. *Protoplasma*, **45**, 460–77.

DOFLEIN G. & REICHENOW E. (1949) Lehrbuch der Protozoenkunde, Teil I: Allgemeine Naturgeschichte der Protozoen. Aufl., Jena, 6.

DRAWERT H. & MIX M. (1961) Licht- und elektronenmikroskopische Untersuchungen an Desmidiaceen. II. Mitt. Hüllgallerte und Schleimbildung bei *Micrasterias, Pleurotaenium* und *Hyalotheca. Planta*, **56**, 237–61.

DREWS G. (1959) Beiträge zur Kenntnis der phototaktischen Reaktionen der Cyanophyceen. *Arch. Protistenk.* **104**, 389–430.

DREWS G. & NULTSCH W. (1962) Spezielle Bewegungsmechanismen von Einzellern (Bakterien, Algen). In *Handbuch der Pflanzenphysiologie*, vol. **17**/2, ed. Ruhland W. pp. 876–919. Springer Verlag, Berlin, Göttingen, Heidelberg.

DRUM R.W. & HOPKINS J.T. (1966) Diatom locomotion: An explanation. *Protoplasma*, **62**, 1–33.

DUPRAW E.J. (1969) *Cell and Molecular Biology*. Academic Press, New York & London.

DUYSENS L.N.M. (1952) Transfer of exitation energy in photosynthesis. Thesis, Utrecht.

EHRENBERG C.G. (1838). *Die Infusionsthierchen als vollkommene Organismen*. Leopold Voss, Leipzig.

ENGELMANN T.W. (1882) Über Licht- und Farbenperception niederster Organismen. *Pflügers Arch. ges. Physiol.* **29**, 387–400.

FECHNER R. (1915) Die Chemotaxis der Oscillarien und ihre Bewegungserscheinungen überhaupt. *Z. Bot.* **7**, 289–364.

FISCHER-ARNOLD G. (1963) Untersuchungen über die Chloroplastenbewegung bei *Vaucheria sessilis. Protoplasma*, **56**, 495–520.

FOGG G.E., STEWART W.D.P., FAY P. & WALSBY A.E. (1973) *The Blue-green Algae*. Academic Press, New York & London.

FRANK TH. (1904) Cultur und chemische Reizerscheinungen der *Chlamydomonas tingens*. *Bot. Ztg.* **62**, 153–87.

FRANZÉ R. (1893) Zur Morphologie und Physiologie der Stigmata der Mastigophoren. *Z. wiss. Zool.* **56**, 138–64.

GALIANO F.E. (1921) Sur les réactions chemiotactiques du flagellé *Chilomonas*. *Z. C. r. hebd. Séanc. Akad. Sci., Paris*, **172**, 776–9.

GARREY W.E. (1900) The effect of ions upon the aggregation of flagellated infusoria. *Am. J. Physiol.* **3**, 291–315.

GEBAUER H. (1930) Zur Kenntnis der Galvanotaxis von *Polytoma uvella* und einigen anderen Volvocineen. *Beitr. Biol. Pfl.* **18**, 463–98.

GEITLER L. (1944) Furchungsteilung, Locomotion, Plasmolyse und Ökologie der Bangiacee *Porphyridium cruentum. Flora, Jena*, N.F. **37**, 300–33.

GEITLER L. (1953) Das Bewegungsverhalten von *Porphyridium cruentum* am natürlichen Standort. *Öst. bot. Z.* **100**, 672–6.

GIBBONS I.R. (1963) Studies on the protein components of cilia from *Tetrahymena pyriformis. Proc. natn. Acad. Sci., U.S.A.* **50**, 1002–10.

GIBBONS I.R. (1967) The organization of cilia and Flagella. In *Molecular Organisation and Biological Function*, ed. Allen J.M. pp. 211–37. Harper & Row, New York.

GOLDACRE R.J. (1964) On the mechanism and control of amoeboid movement. In *Primitive Motile Systems in Cell Biology*, eds. Allen R.D. & Kamiya N. pp. 237–55. Academic Press, New York & London.

GÜNTHER F. (1928) Über den Bau und die Lebensweise der Euglenen, besonders der Arten *E. terricola, geniculata, proxima, sanguinea* und *luccus* nov. spec. *Arch. Protistenk*, **60**, 511–90.

HALFEN L.N. & CASTENHOLZ R.W. (1970) Gliding in a blue-green alga: a possible mechanism. *Nature, Lond.* **225**, 1163–5.

HALLDAL P. (1958) Action spectra of phototaxis and related problems in Volvocales, *Ulva*-gametes and Dinophyceae. *Physiologia Pl.* **11**, 118–53.

HALLDAL P. (1962) Taxes. In *Physiology and Biochemistry of Algae*, ed. Lewin R.A. pp. 583–93. Academic Press, New York & London.

HALLDAL P. (1963) Zur Frage des Photoreceptors bei der Topophototaxis der Flagellaten. *Ber. dt. bot. Ges.* **76**, 323–7.

HAND W.G. & DAVENPORT D. (1970) The experimental analysis of phototaxis and photokinesis in flagellates. In *Photobiology of Microorganisms*, ed. Halldal P. pp. 253–82. Wiley-Interscience, London, New York, Sydney & Toronto.

HAND W.G. & HAUPT W. (1971) Flagellar activity of the colony members of *Volvox aureus* Ehrbg. during light stimulation. *J. Protozool.* **18**, 361–4.

HANSGIRG A. (1883) Bemerkungen über die Bewegungen der Oscillatoriaceen. *Bot. Ztg.* **41**, 831–43.

HAUPT W. (1959a) Chloroplastenbewegung. In *Handbuch der Pflanzenphysiologie*, vol. 17/1, ed. Ruhland W. pp. 278–317. Springer Verlag, Berlin, Göttingen, Heidelberg.

HAUPT W. (1959b) Die Chloroplastendrehung bei *Mougeotia*. I. Über den quantitativen und qualitativen Lichtbedarf der Schwachlichtbewegung. *Planta*, **53**, 484–501.

HAUPT W. (1959c) Die Phototaxis der Algen. In *Handbuch der Pflanzenphysiologie*, vol. 17/1, ed. Ruhland W. pp. 318–70. Springer Verlag, Berlin, Göttingen, Heidelberg.

HAUPT W. (1962a) Thermotaxis. In *Handbuch der Pflanzenphysiologie*, vol. 17/2, ed. Ruhland W. pp. 29–33. Springer Verlag, Berlin, Göttingen, Heidelberg.

HAUPT W. (1962b) Geotaxis. In *Handbuch der Pflanzenphysiologie*, vol. 17/2, ed. Ruhland W. pp. 390–5. Springer Verlag, Berlin, Göttingen, Heidelberg.

HAUPT W. (1962c) Intracellular Movements. In *Physiology and Biochemistry of Algae*, ed. Lewin R.A. pp. 567–72. Academic Press, New York & London.

HAUPT W. (1963) Photoreceptorprobleme der Chloroplastenbewegung. *Ber. dt. bot. Ges.* **76**, 313–22.

HAUPT W. (1966) Phototaxis in plants. *Int. Rev. Cytol.* **19**, 267–99.

HAUPT W. & SCHÖNBOHM W. (1970) Light-oriented chloroplasts movements. In *Photobiology of Microorganisms*, ed. Halldal P. pp. 283–307. Wiley-Interscience, London, New York, Sydney, Toronto.

HAXO F.T. & NORRIS P.S. (1953) Photosynthetic activity of phycobilins in some red and blue-green algae. *Biol. Bull. mar. biol. Lab., Woods Hole*, **105**, 374–81.

HAYASHI T. (1964) Role of the cortical gel layer in cytoplasmic streaming. In *Primitive Motile Systems in Cell Biology*, eds. Allen R.D. & Kamiya N. pp. 19–29. Academic Press, New York & London.

HÖFLER K. (1940) Aus der Protoplasmatic der Diatomeen. *Ber. dt. bot. Ges.* **58**, 97–120.

HOFFMANN-BERLING H. (1955) Geißelmodelle und ATP. *Biochim. biophys. Acta*, **16**, 146–154.

HOFFMANN-BERLING H. (1962) Die innere Mechanik der Geißelbewegung. In *Handbuch der Pflanzenphysiologie*, vol. 17/2, ed. Ruhland W. pp. 836–42. Springer Verlag, Berlin, Göttingen, Heidelberg.

HUTH K. (1970a) Bewegung und Orientierung bei *Volvox aureus* Ehrenbg. I. Mechanismus der phototaktischen Reaktion. *Z. PflPhysiol.* **62**, 436–50.

HUTH K. (1970b) Bewegung und Orientierung bei *Volvox aureus* Ehrbg. II. Richtungsabweichung bei taktischen Reaktionen. *Z. PflPhysiol.* **63**, 344–51.

JAROSCH R. (1958) Zur Gleitbewegung der niederen Organismen. *Protoplasma*, **50**, 277–89.

JAROSCH R. (1960) Die Dynamik im Characeen-Protoplasma. *Phyton*, **15**, 43–66.

JAROSCH R. (1962) Gliding. In *Physiology and Biochemistry of Algae*, ed. Lewin R.A. pp. 573–81, Academic Press, New York & London.

JAROSCH R. (1971) Vergleichende Studien zur amöboiden Beweglichkeit. *Protoplasma*, **72**, 79–100.

KAMIYA N. (1959) Protoplasmic streaming. In *Protoplasmatologia—Handbuch der Protoplasmaforschung*, vol. VIII, 3a, eds. Heilbrunn L.V. & Weber F. Springer, Wien.

KAMIYA N. (1962) Protoplasmic streaming. In *Handbuch der Pflanzenphysiologie*, vol. 17/2, ed. Ruhland W. pp. 979–1035. Springer Verlag, Berlin, Göttingen, Heidelberg.

KAMIYA N. (1964) The motive force of endoplasmatic streaming in the ameba. In *Primitive Motile Systems in Cell Biology*, eds. Allen R.D. & Kamiya N. pp. 257–77. Academic Press, New York & London.

KARSTEN G. (1905) Das Phytoplankton des antartkischen Meeres. In *Wiss. Ergebnisse der deutschen Tiefsee-Expedition*, Band II, 16, ed. Chun C.

KLEBS G. (1885) Über Bewegung und Schleimbildung der Desmidiaceen. *Biol. Zbl.* **5**, 353–67.

KLEBS G. (1886) Über die Organisation der Gallerte bei einigen Algen und Flagellaten. *Unters. Bot. Inst. Tübingen*, **2**, 333–417.

KÜSTER E. (1905) Über den Einfluß von Lösungen verschiedener Konzentrationen auf die Orientierungsbewegungen der Chromatophoren. *Ber. dt. bot. Ges.* **23**, 254–6.

KÜSTER E. (1956) *Die Pflanzenzelle. 3rd ed.* Fischer, Jena.

LINKS J. (1955) I. A hypothesis for the mechanism of (phobo-) chemotaxis. II. The carotenoids, steroids and fatty acids of *Polytoma uvella*. Thesis, Leiden.

LÜDERS J.E. (1854) Beobachtungen über die Organisation, Teilung und Kopulation der Diatomeen. *Bot. Ztg.* **20**, 41–69.

LUNTZ A. (1931a) Untersuchungen über die Phototaxis. I. Die absoluten Schwellenwerte und die relative Wirksamkeit von Spektralfarben bei grünen und farblosen Einzelligen. *Z. vergl. Physiol.* **14**, 68–92.

LUNTZ A. (1931b) II. Lichtintensität und Schwimmgeschwindigkeit bei *Eudorina elegans*. *Z. vergl. Physiol.* **15**, 652–78.

LUNTZ A. (1932) III. Die Umkehr der Reaktionsrichtung bei starken Lichtintensitäten und ihre Bedeutung für eine allgemeine Theorie der photischen Reizwirkung. *Z. vergl. Physiol.* **16**, 204–17.

MANTEN A. (1948) Phototaxis in the purple bacterium, and the relation between phototaxis and photosynthesis. *Antonie van Leeuwenhoeck*, **14**, 65–86.

MAINX F. (1928) Beiträge zur Morphologie und Physiologie der Eugleninen. I. Morphologische Beobachtungen, Methoden und Erfolge der Reinkultur. II. Untersuchungen über die Ernährungs- und Reizphysiologie. *Arch. Protistenk.* **60**, 305–414.

MARTENS P. (1940) La locomotion des diatomées. *Cellule*, **48**, 279–305.

MAST S.O. (1927) Response to electricity in *Volvox* and the nature of galvanic stimulation. *Z. vergl. Physiol.* **5**, 739–61.

MENDELSSOHN M. (1902a) Recherches sur la thermotaxie des organismes unicellulaires. *J. Physiol. Path. gén.* **4**, 393–409.

MENDELSSOHN M. (1902b) Recherches sur l'interference de la thermotaxie avec d'autres tactismes et sur le mécanisme du mouvement thermotactique. *J. Physiol. Path. gén.* **4**, 475–88.

MENDELSSOHN M. (1902c) Le nature et le rôle biologique de la thermotaxie. *J. Physiol. Path. gén.* **4**, 489–96.

METZNER P. (1929) Bewegungsstudien an Peridineen. *Z. Bot.* **22**, 225–65.

MOLISCH H. (1901) Über den Goldglanz von *Chromophyton Rosanoffii* Woronin. *Sitzungsber. d. kais. Akad. d. Wissenschft.* in Wien, *mathem-naturw. Klasse*, **110**, Abt. **1**, 354–78.

MOORE G.T. (1901) New or little known unicellular Algae. II. *Eremosphaera* and *Excentrosphaera*. *Bot. Gaz.* **32**, 309–24.

MÜLLER O. (1889) Durchbrechungen der Zellwand in ihrer Beziehung zur Ortsbewegung der Bacillariaceen. *Ber. dt. bot. Ges.* **7**, 169–80.

NEUSCHELER W. (1967a) Bewegung und Orientierung bei *Micrasterias denticulata* Bréb. im Licht. I. Zur Bewegungs- und Orientierungsweise. *Z. PflPhysiol.* **57**, 46–59.

NEUSCHELER W. (1967b) Bewegung und Orientierung bei *Micrasterias denticulata* Bréb. im Licht. II. Photokinesis und Phototaxis. *Z. PflPhysiol.* **57**, 151–71.

NULTSCH W. (1956) Studien über die Phototaxis der Diatomeen. *Arch. Protistenk.* **101**, 1–68.

NULTSCH W. (1961) Der Einfluß des Lichtes auf die Bewegung der Cyanophyceen.I.Phototaxis von *Phormidium autumnale*. *Planta*, **56**, 632–47.

NULTSCH W. (1962a) Der Einfluss des Lichtes auf die Bewegung der Cyanophyceen. II. Photokinesis bei *Phormidium autumnale*. *Planta*, **57**, 613–23.

NULTSCH W. (1962b) Der Einfluss des Lichtes auf die Bewegung der Cyanophyceen. III. Photo-phobotaxis von *Phormidium uncinatum*. *Planta*, **58**, 647–63.

NULTSCH W. (1962c) Phototaktische Aktionsspektren von Cyanophyceen. *Ber. dt. bot. Ges.* **75**, 443–53.

NULTSCH W. (1962d) Über das Bewegungsverhalten der Diatomeen. *Planta*, **58**, 22–33.

NULTSCH W. (1965) Light reaction systems in Cyanophyceae. *Photochem. Photobiol.* **4**, 613–19.

NULTSCH W. (1966) Über den Antagonismus von Atebrin und Flavinnucleotiden im Bewegungs- und Lichtreaktionsverhalten von *Phormidium uncinatum*. *Arch. Mikrobiol.* **55**, 187–9.

NULTSCH W. (1967) Untersuchungen über den Einfluss von Entkopplern auf die Bewegungsaktivität und das phototaktische Reaktionsverhalten blaugrüner Algen. *Z. PflPhysiol.* **56**, 1–11.

NULTSCH W. (1968a) Einfluss von Redox-Systemen auf die Bewegungsaktivität und das phototaktische Reaktionsverhalten von *Phormidium uncinatum*. *Arch. Mikrobiol.* **63**, 295–320.

NULTSCH W. (1968b) Effect of desaspidin and DCMU on photokinesis of blue-green algae. *Photochem. Photobiol.* **10**, 119–23.

NULTSCH W. (1970) Photomotion of microorganisms and its interactions with photosynthesis. In *Photobiology of Microorganisms*, ed. Halldal P. pp. 213–49. Wiley Interscience, London, New York, Sydney & Toronto.

NULTSCH W. (1971a) Phototactic and photokinetic action spectra of the diatom *Nitzschia communis*. *Photochem. Photobiol.* **14**, 705–12.

NULTSCH W. (1971b) *General Botany*. Academic Press, New York & London (translation of the 3rd German edition). Georg Thieme Verlag, Stuttgart 1968.

NULTSCH W. & RICHTER G. (1963) Aktionsspektren des photosynthetischen $^{14}CO_2$-Einbaus von *Phormidium uncinatum*. *Arch. Mikrobiol.* **47**, 207–13.

NULTSCH W. & JEEJI-BAI (1966) Untersuchungen über den Einfluss von Photosynthese-Hemmstoffen auf das phototaktische und photokinetische Reaktionsverhalten blaugrüner Algen. *Z. PflPhysiol.* **54**, 84–98.

NULTSCH W., THROM G. & RIMSCHA I. (1971) Phototaktische Untersuchungen an *Chlamydomonas reinhardii* Dangeard in homokontinuierlicher Kultur. *Arch. Mikrobiol.* **80**, 351–69.

NULTSCH W. & HELLMANN W. (1972) Untersuchungen zur Photokinesis von *Anabaena variabilis* Kützing. *Arch. Mikrobiol.* **82**, 76–90.

O'COLLA P.S. (1962) Mucilages. In *Physiology and Biochemistry of Algae*, ed. Lewin R.A. pp. 337–56. Academic Press, New York & London.

PFEFFER W. (1888) Über chemotaktische Bewegungen von Bakterien, Flagellaten und Volvocineen. *Unters. Bot. Inst. Tübingen*, **2**, 582–661.

POHL R. (1962) Die äußere Mechanik der Geißelbewegung. In *Handbuch der Pflanzenphysiologie*, vol. 17/2, ed. Ruhland W. pp. 843–75. Springer Verlag, Berlin, Göttingen, Heidelberg.

PRINGSHEIM E.G. (1921) Zur Physiologie saprophytischer Flagellaten. *Beitr. allgem. Bot.* **2**, 88–137.

PRINGSHEIM E.G. (1968a) Cyanophyceen-Studien. *Arch. Mikrobiol.* **63**, 331–55.

PRINGSHEIM E.G. (1968b) Kleine Mitteilungen über Flagellaten und Algen. XV. Zur Kenntnis der Gattung *Porphyridium* (Rhodophyceae). *Arch. Mikrobiol.* **61**, 169–80.

PRINGSHEIM E.G. & MAINX F. (1926) Untersuchungen an *Polytoma uvella* Ehrbg., insbesondere über Beziehungen zwischen chemotaktischer Reizwirkung und chemischer Konstitution. *Planta*, **1**, 583–623.

REIMERS H. (1928) Über die Thermotaxis niederer Organismen. *Jb. wiss. Bot.* **67**, 242–90.

RINGO D.L. (1967a) Flagellar motion and fine structure of the flagellar apparatus in *Chlamydomonas*. *J. Cell Biol.* **33**, 543–71.

RINGO D.L. (1967b) The arrangement of subunits in flagellar fibers. *J. Ultrastruct. Res.* **17**, 266–77.

SCHMID G. (1923) Das Reizverhalten künstlicher Teilstücke, die Kontraktilität und das osmotische Verhalten von *Oscillatoria jenensis*. *Jb. wiss. Bot.* **62**, 328–419.

SCHIMPER A.F.W. (1885) Untersuchungen über die Chlorophyllkörner und die ihnen homologen Gebilde. *Jb. wiss. Bot.* **16**, 1–247.

SCHÖNBOHM E. (1963) Untersuchungen über die Starklichtbewegung des *Mougeotia*-Chloroplasten. *Z. Bot.* **51**, 233–76.

SCHÖNBOHM E. (1969) Die Hemmung der lichtinduzierten Bewegung des *Mougeotia*-Chloroplasten durch p-Chlormercuribenzoat. (Versuche zur Mechanik der Chloroplasten bewegung.) *Z. PflPhysiol.* **61**, 250–60.

SCHÖNBOHM E. (1972a) Experiments on the mechanism of chloroplasts movement in light-oriented chloroplast arrangement. *Acta Protozool.* **11**, 211–24.

SCHÖNBOHM E. (1972b) Die Wirkung von SH-Blockern sowie von Licht und Dunkel auf die Verankerung der *Mougeotia*-Chloroplasten im cytoplasmatischen Wandbelag. II. Zur Mechanik der Chloroplastenbewegung. *Z. PflPhysiol.* **66**, 113–32.

SENN G. (1908) *Die Gestalts- und Lageveränderung der Pflanzen-Chromatophoren*. Wilhelm Engelmann, Leipzig.

SENN G. (1919) Weitere Untersuchungen über Gestalts- und Lageveränderungen der Chromatophoren. *Z. Bot.* **11**, 81–141.

STAHL E. (1880) Über den Einfluß von Richtung und Stärke der Beleuchtung auf einige Bewegungserscheinungen im Pflanzenreiche. *Bot. Ztg.* **38**, 297–413.

STANIER R.Y., KUNISAWA R., MANDEL M. & COHEN-BAZIRE G. (1971) Purification and properties of unicellular blue-green algae (order Chroococcales). *Bact. Rev.* **35**, 171–205.

STRASBURGER E. (1878) *Wirkung des Lichtes und der Wärme auf Schwärmsporen.* Fischer Verlag, Jena.

TERRY O.P. (1906) Galvanotropism of *Volvox. Am. J. Physiol.* **15**, 235–43.

THOMAS J.B. & NIJENHUIS L.E. (1950) On the relation between phototaxis and photosynthesis in *Rhodospirillum rubrum. Biochim. biophys. Acta,* **6**, 317–24.

TIBBS J. (1958) The properties of algal and sperm flagella, obtained by sedimentation. *Biochim. biophys. Acta,* **28**, 636–7.

TOMAN M. (1955) Theorie der Oscillatorienbewegung. *Arb. d. II. Sektion d. Slowakischen Wiss. Acad. Biol. Serie,* **1**, 5–23.

TREVIRANUS L.G. (1817) Vermischte Schriften anatomischen und physiologischen Inhalts. **2**, 71–92.

UMRATH K. (1959) Galvanotaxis. In *Handbuch der Pflanzenphysiologie,* vol. 17/1, ed. Ruhland W. pp. 164–7. Springer Verlag, Berlin, Göttingen, Heidelberg.

VÁVRA J. (1956) Ist der Photoreceptor eine unabhängige Organelle der Eugleniden? *Arch. Mikrobiol.* **25**, 223–5.

VISCHER W. (1955) *Porphyridium cruentum* (Ag) Naeg. und die Bewegung seiner Monosporen. Ber. Schweiz. Bot. Ges. **65**, 459–74.

WAGNER J. (1934) Beiträge zur Kenntnis der *Nitzschia putrida* Benecke, insbesondere ihrer Bewegung. *Arch. Protistenk.* **82**, 86–113.

WINKLER H. (1900) Über Polarität, Regeneration und Heteromorphose bei *Bryopsis. Jb. wiss. Bot.* **35**, 449–69.

WOLKEN J.J. & SHIN E. (1958) Photomotion in *Euglena gracilis.* I. Photokinesis. II. Phototaxis. *J. Protozool.* **5**, 39–46.

ZIEGLER H. (1962) Chemotaxis. In *Handbuch der Pflanzenphysiologie,* vol. 17/2, ed. Ruhland W. pp. 484–532. Springer Verlag, Berlin, Göttingen, Heidelberg.

ZURZYCKI J. (1962) The mechanism of the movements of plastids. In *Handbuch der Pflanzenphysiologie,* vol. 17/2, ed. Ruhland W. pp. 940–78. Springer Verlag, Berlin, Göttingen, Heidelberg.

# CHAPTER 32

# SYNCHRONOUS CULTURES

## H. LORENZEN and M. HESSE

Pflanzenphysiologisches Institut der Universität,
34 Göttingen, Untere Karspüle 2, Germany.

1 Introduction 894

2 Methods used to synchronize cultures of unicellular algae 895

3 Requirements of a permanent synchronous culture 897

4 Some developmental and physiological aspects studied using synchronous cultures 899

5 Influence of circadian rhythms 901

6 Synchronous cultures and high productivity 904

7 References 905

## 1  INTRODUCTION

Synchronized cultures have been used in cell research for over 20 years and a variety of more or less successful methods are now available. A great deal of attention has been paid to synchronous cultures of bacteria and of the ciliate, *Tetrahymena*, and to the synchronous growth of photosynthetic unicellular organisms. In this chapter we shall outline briefly the principles of synchronous cultures, as they apply to algae and attempt to provide some general views on how to synchronize unicellular algae, how to handle synchronous cultures, and how to obtain maximal rates of growth. We shall consider in detail only the more recent papers of general interest. The older literature is reviewed by Kuhl and Lorenzen (1964); Lorenzen (1970); Pirson and Lorenzen (1966) and Tamiya (1966).

## 2 METHODS USED TO SYNCHRONIZE CULTURES OF UNICELLULAR ALGAE

A knowledge of the physiological state of the organism used is an important preliminary for successful synchrony. For example, the growth conditions should guarantee rapid cell divisions and the shortest possible generation time, which is genetically fixed, must be ascertained. The range of size, shape, or of physiological activities, at these optimum conditions only manifests what we may call type variability (see Czurda 1935). The total or maximum variability is significantly greater than type variability; at extreme conditions the cells are still growing but are not enabled for further divisions (irreversible variability; giant cells).

Starting from such a knowledge there are various ways of obtaining synchronous cultures. Few workers deal in detail, however, with the methods of synchronous culture which they use. A list of the more important types of culture, the organisms used, and some of the authors who have used them is given in Table 32.1. It depends on the specific research topic under investigation which method of synchronization is best, but it is always necessary to check that the treatment is producing standardized cells which are not subject to an inhibition of growth or to environmental stress.

During log phase growth some unicellular algae are always in a state of cell division. That is, the cell number is steadily increasing at a high rate. In synchronous culture, on the other hand, cell numbers are fairly constant for a long time phase and then there is a short release when the cell numbers actually increase (Fig. 32.1A). If log phase cultures are placed after dilution in fresh medium with half or considerably more of the volume renewed there is no lag phase and the culture continues to grow at an optimum rate.

One may differentiate between cultures synchronized completely, partially, or in groups (Lorenzen 1957). The most important are completely synchronized cultures, in which all the cells (or more than 98%) take part in the division burst at the same time. This burst in a good synchronous culture should be complete in less than 10% of the time of the cell cycle. Most algae, however, require a rather longer time (30 to 50% of the cycle) to double in cell number, e.g. *Euglena* (Cook 1971) although the colourless *Polytomella agilis*, growing in a complex organic medium, has a doubling time of less than 2 hours at 25°C after growth (substance production) for 22 hours at 9°C (Cantor & Koltz 1971). Meeker (1964) has obtained similar results in studies on *Astasia*. Complete synchrony seldom occurs under natural conditions and there, induced synchrony, as a result of favourable growth conditions, is more usual. Actually, if the cells are enabled to grow permanently in their shortest possible generation time, they are placed in the best of all possible worlds and neither pampered nor mishandled.

The demand for large quantities of synchronized cells cannot be met using conventional laboratory type cultures which usually provide not more than

**Table 32.1.** A summary of some methods used to synchronize unicellular algae

| | Treatment | Alga | Reference |
|---|---|---|---|
| 1. | Starvation until cell division ceases, e.g. by lack of vitamin, micro-nutrient or macro-nutrient; LD or LL; inorganic medium | *Euglena* *Navicula* | Shlyk & Sukhover (1968) Darley & Volcani (1971) |
| 2. | Temperature changes (i.e. alteration of a sub-optimum temperature which prevents cell division with an optimum one); LD or LL, 2 cycles only; inorganic medium | *Chlamydomonas* *Dunaliella* | Rooney et al. (1971) Wegmann & Metzner (1971) |
| 3. | Temperature changes which enable the cell cycle to proceed; LL with serial dilution, many cycles; inorganic medium | *Anacystis* *Euglena* | Lorenzen & Venkataraman (1969) Terry & Edmunds (1970) |
| 4. | Temperature changes which enable the cell cycle to proceed, LL with serial dilution, many cycles; organic medium | *Euglena* | Padilla & Bragg (1968) |
| 5. | LD alterations, few cycles, inorganic medium | *Chlorella ellipsoidea* *Scenedesmus* *Chlamydomonas* | Tamiya et al. (1953) Komarek et al. (1968) Bernstein (1960) Kates & Jones (1967) |
| 6. | LD alterations with serial dilution; many cycles; inorganic medium | *Chlorella* *Chlamydomonas* | Lorenzen (1957) Tamiya et al. (1961) Schlösser (1966) |
| 7. | LD alterations with serial dilution; many cycles; organic medium. | *Chlorella* | Senger (1965) |
| 8. | LD alterations with continuous dilution; inorganic medium | *Chlorella* *Scenedesmus* | Pfau et al. (1971) Bongers (1958) |
| 9. | LD presynchronization followed by a LL with 2–3 serial dilutions; inorganic medium | *Chlorella* 7-11-05 | Baker & Schmidt (1964) |
| 10. | LL alternating fluorescent white light (in air) with yellow/red light (in air + added $CO_2$); (3 cycles); inorganic medium | *Chlamydomonas* *Chlorella* *Protosiphon* | Carroll et al. (1970) Carroll et al. (1970) Carroll et al. (1970) |

5 1 day$^{-1}$ (about 2 to 3 gm dry weight of cells). To overcome this problem Dalmon and Gilet (1969) used a volume of 15 1, but only 90% of the cells took part in every cell cycle burst. The greatest trouble using such large volumes is the unavoidable reduction in light per cell which results from mutual shading in large containers.

In addition to the methods listed in Table 32.1 different types of pretreatments, in particular centrifugation or filtration, have been used to isolate populations of distinct size from cultures (selective harvesting). Such procedures, however, affect the physiological state of the cells and, of course, cannot be used in studies with permanently synchronized cultures. An exception to this generalization is the case of diatoms. Here it is possible to select cells of certain diameter (large, medium, small) after a few weeks of growth in sufficient quantity for analytical and physiological work (Werner 1971) and one can study physiological differences associated with the characteristic decrease in cell size during many mitotic cycles.

It must be noted, furthermore, that starting experiments with synchronous cells and then changing to growth conditions which do not maintain further synchronous cycling, gives more specific information than that gained just using batch cultures (see Guerin-Dumartrait *et al.* 1970). In such *Chlamydomonas* cells gametogenesis is inducible and during its course (about 18 hours) about 90% of the cytoplasmic, and later of the plastidic ribosomes degrade as does the bulk of the RNA. Most of the intermediates are then used for DNA-synthesis connected with sexual differentiation (Siersma & Chiang 1971).

## 3  REQUIREMENTS OF A PERMANENT SYNCHRONOUS CULTURE

A good synchronous culture should provide cell material which, from generation to generation, is homogenous, not only with respect to cell number, but also with respect to other features such as dry weight, cell constituents etc. If this is not the case the cells differ in successive cycles (see Bishop & Senger 1971, using *Scenedesmus*, but check Senger 1970). Examples of good synchronous cultures are those obtained with our method by Galling (1970) where the changes of free and membrane-bound polysomes were repetitive in *Chlorella*, and those of *Chlamydomonas* obtained by Siersma and Chiang (1971) where even the amount of 70 S-chloroplast RNA and of 79 S-cytoplasmic RNA was the same from one cycle to the other. That is, we must ensure that the starting cells of each succeeding cycle are the same, both morphologically and physiologically. In this way we can measure the behaviour of millions of cells and receive information on how the standardized (normal) cell growing in its shortest possible generation time is reacting.

Figure 32.1 illustrates the method and gives some results obtained with such permanent synchronous cultures. Figure 32.1A shows that *Chlorella* cells

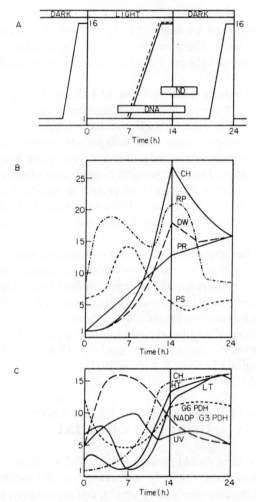

**Fig. 32.1.** Growth and metabolic interactions in a permanent, semi-continuous (dilution after every cycle), synchronous culture of *Chlorella*, 211-8b. LD, 14:10; temperature 32°C; light intensity, 15,000lx; gas phase, 1·5% $CO_2$ in air; inorganic medium; pH, 6·1–6·5.

A.   Time course of variation in cell number during the LD cycle. DNA represents the time of synthesis of the bulk of DNA; ND represents the time of 4 nuclear divisions. The broken line during the light phase shows the later cell number produced, if samples are darkened each hour beginning after 7 hours of light (induction of cell division).

B.   Time course of relative variations in some major cell components and metabolic activities. CH, carbohydrates; PR, protein; DW, dry weight; RP, respiration; PS, photosynthesis ($O_2$ evolution).

C.   Time courses of relative variations in total chlorophyll (CH); specific activities of glucose-6-phosphate dehydrogenase (G6PDH) and NADP-glucose-3-phosphate dehydrogenase (NADP G-3-PDH), and resistance against high temperature shock (20 min at 45°C) (HT); low temperature shock (2 hours at 4·5°C) (LT), and ultraviolet light (UV). The temperature shock data are taken from Lorenzen (1963), the UV data from Taube (1970).

acquire the ability for division without further increase in biomass before DNA-synthesis is complete (see also Spektorov & Nikolskaya 1971). If the cells are darkened after different lengths of light it is found that the induction of cell division (broken line during light phase in Fig. 32.1A) begins after 7 hours of light. After 12 hours of light there is no further increase in the later number of autospores produced. It should be noted that the nuclear divisions in the 14:10 hours light:dark period (LD) commences after 12 hours of illumination.

The importance of the shortest possible generation time for synchronous cultures has been mentioned above. If development is not restricted to the shortest possible generation time (about 20 hours in the case of *Chlorella* and cannot be shortened by changes of light intensity or duration, temperature, adding of organic nutrients and so on) there is evidence that the events during the cell cycle are influenced at different rates (Lorenzen & Schleif 1966). Thus, it is not recommendable to give the length of generation times in arbitrary units (e.g. 0·0, 0·1 . . . 0·9, 1·0). When data are presented in this way we really can deduce nothing about whether the cells were growing under optimum conditions or not. We also have to recognize that all our culture conditions, even the optimum ones, are geared to the overall cell cycle and that we really do not know the specific optimum requirements of the different developmental stages. It may be possible, nevertheless, to shorten the cycle by applying short-time changes of external conditions during cell development, e.g. higher amount of nutrients, differences in light intensity or higher temperature etc. at different stages. There is really no great advantage in this, however, and processes of adaptation certainly influence metabolic activities to different and unknown degrees.

A word must also be said about the term 'unbalanced growth' which leads to cells of abnormal composition and which is regarded by some authors as a characteristic of all synchronous cultures. This is simply not the case. Semicontinuous synchronous cultures growing in a genetically fixed shortest generation time provide ideal conditions of balanced growth, and cultures which can run optimally for months. It is also found generally that synchronous growth allows better preservation of cell organelles, although Galling (1970) reported a lower number of polysomes in synchronous cultures compared with cells from batch cultures.

## 4 SOME DEVELOPMENTAL AND PHYSIOLOGICAL ASPECTS STUDIED USING SYNCHRONOUS CULTURES

The development of a variety of organelles has been studied in synchronous culture. Atkinson *et al.* (1971) studying *Chlorella*, detected centrioles (diameter 200 nm) in all developmental stages and these duplicated only during the course of autospore formation. Microtubules on the other hand were only transitory, appearing before the first mitosis (after 10 hours of light) and were not detected following cytokinesis. In *Chlamydomonas*, microtubules take part in chloroplast division and cytokinesis. The cytoplasmic organelles duplicate within 4 hours at the middle of the 12:12 LD (Gould & Jones 1968). Cyclic changes of thylakoid membranes have also been observed in *Chlamydomonas* by Schor *et al.* (1970). Chloroplast and nuclear division commence before cytokinesis but the cycle finishes with the separation and release of autospores.

Studies on polypeptide synthesis by Erben (1971) using a cell-free system from synchronous *Chlorella* cells show a splitting of the cycle into three different parts (0 to 7 hours, 8 to 16 hours, and the rest of time). Transcription and translation of chloroplast components in synchronous *Chlamydomonas* are discussed by Armstrong *et al.* (1971).

The problems of DNA synthesis during the life cycle are of interest. In Fig. 32.1A the time course of DNA synthesis in synchronous *Chlorella* is shown. This, of course, represents only the bulk synthesis and there is some evidence that the plastid-DNA is synthesized earlier. Neomycin (0·5 mg ml$^{-1}$) reduces the DNA content in *Chlorella* by about 25% and, according to Galling and Wolff (1972), the RNA-content is lowered about 15%. No single fraction of the nucleic acids was specially affected by this treatment. Wanka and his group obtained evidence that a simultaneous synthesis of specific proteins required for DNA replication occurs and these can be separated by use of specific inhibitors (Wanka *et al.* 1972). Variations in DNA in synchronous cultures of *Chlorella* and *Porphyridium* have been considered by Gense-Vimon (1971). Evidence was obtained for at least three types of DNA. The satellite DNA increases slowly throughout the cycle at a low rate (Wanka *et al.* 1970), the major component only between 10 to 18 hours (Fig. 32.1) and, according to Schönherr and Wanka (1971), the DNA polymerase increases almost continuously. The deoxyribonuclease is highly active only during the phase of maximum DNA-synthesis in *Chlorella* (Schönherr *et al.* 1970) and in *Euglena* (Walther & Edmunds 1970).

Nitrate reduction variations have been studied in synchronous culture by Kanazawa *et al.* (1970). It was found that the autospores released in the dark utilize small amounts of nitrite and nitrate. Glycolate metabolism has been investigated in synchronous cultures of *Euglena* (Codd & Merrett 1971), *Ankistrodesmus* (Chang & Tolbert 1970), and *Gloeomonas* (Badour & Waygood 1971).

In Fig. 32.1C the course of two enzymes with either a maximum or a minimum of specific activity in the middle of the light phase is given. Schmidt's group checked many enzymes and two recent papers deal with the continuous inducibility of isocitrate lyase (Baechtel *et al.* 1970) and the regulation of ribulose-1,5-diphosphate carboxylase (Molloy & Schmidt 1970). Isocitrate lyase activity increases parallel to DNA synthesis. The same enzyme, which is present throughout the life cycle and which is involved in chlorophyll synthesis, has been partially purified from photoautotrophically and photoheterotrophically grown *Gloeomonas* synchronized with a LD of 12:12 hours by Foo *et al.* (1971). There is no evidence so far for isoenzymes of isocitrate lyase in *Gloeomonas*. McCullough and John (1972) obtained variable results with *Chlorella* in so far as they found a temporary control of the *de novo* synthesis of this enzyme, cells 12 hours old being unable to synthesize the enzyme before a lag of 8 hours was finished.

Senger and Bishop (1969) observed, with *Scenedesmus obliquus*, highest $O_2$ production after 8 hours in the light phase and this was connected with the lowest quantum requirement and the highest enhancement effect. They obtained a similar time course to Lorenzen (1959) with *Chlorella* where the absolute difference of photosynthetic capacity during one cycle with a 16-fold increase in cell number varied only by a factor of 2 (see also Fig. 32.1B). In *Scenedesmus* there is no increase in pigment content during the dark phase although in *Chlorella* up to 15% of the pigment is synthesized in the dark.

Using continuous illumination during 2 cycles after LD-synchronization with the high temperature strain of *Chlorella* 7–11–05, Reitz *et al.* (1967) were able to show that the same fatty acids (mostly saturated ones of C16 and C18) were present when the alga was grown with or without added glucose. However, with glucose the overall level was lower. The synthesis of single fatty acids occurred at different times and did not parallel photosynthetic activity. Nelson *et al.* (1969) found a minimum of glycolate synthesis during the release of daughter cells by synchronous *Scenedesmus* cells.

Cells of *Chlorella* are most sensitive to radiation before DNA synthesis starts (Mampl *et al.* 1971), but the most sensitive stages against heat or cold shocks are a few hours later (see Fig. 32.1C). DNA synthesis and changes in radiosensitivity in *Chlamydomonas reinhardtii* were reported by Cechacek and Hillova (1970). As found by Arnaud (1970) the synthesis of m-RNA takes place at the beginning of the *Chlorella* cycle in the light, but that of ribosomal RNA occurs with a maximum at the beginning of the dark time. Cattolico and Jones (1972) isolated, using a sonication procedure, a stable high molecular ribosomal RNA with high levels of 23-S rRNA.

The synthesis of protein is seen in Fig. 32.1B. Here the first hours of the light phase are very important, but protein is synthesized throughout the cycle, even in the dark. The levels of free amino acids and protein amino acids in the course of the *Chlorella* cycle show spectacular differences. There is no periodism by the protein amino acids as a percentage of total cellular nitrogen but there are tremendous changes in the free amino acids. Most of the peptide amino acids are incorporated in low molecular weight peptides (Hare & Schmidt 1970).

Despite claims of effects caused by phytohormones, there is no evidence that these have any effect on *Chlorella*, a finding which has recently been confirmed for IES and gibberellic acid by Lien *et al.* (1971).

## 5 INFLUENCE OF CIRCADIAN RHYTHMS

To achieve good synchrony and high productivity there must be a favourable interaction between cell cycling and circadian rhythm. The importance of the biological clock in this connexion has been shown particularly in *Euglena*

(Edmunds & Funch 1969, Terry & Edmunds 1970) and the general problems have been discussed by Ehret and Wille (1970). There is no evidence of circadian rhythms in the prokaryotic blue-green algae.

A generation time of about 24 hours is not in itself proof that a circadian oscillator occurs in a particular organism (Mitchell 1971) and some unicellular algae do not show a circadian oscillation in continuous light. The easiest way to test whether there is an endogenous circadian rhythm or not is to check for internal oscillations in metabolic activities under constant conditions in non-dividing cells (Sweeney 1969).

A generation time of about 24 hours in unicellular algae may result from the division cycle itself being the endogenous clock, or else the cell cycle is controlled by the same clock as the circadian rhythm. There is little evidence either way and we really do not know what is responsible for the induction of synchrony. Certainly, the pace-maker is influenced by some parameter other than energy supply. Mihara and Hase (1971) discussed in this context the early start of protein synthesis in the chloroplasts of Chlamydomonas.

The liberation of daughter cells in Chlorella is also affected by an endogenous rhythm which coincides with the beginning of the light time. If biomass production reaches a certain level and if the conditions allow a previous mitosis, the cells hatch 20 hours after receiving light. If synchronized cells growing under unfavourable conditions are used, the rest of the population has to wait for the next 'gate', exactly 24 hours later, before hatching occurs. In the latter case the number of daughter cells produced may be higher and the offspring may be of a different size to those produced from cells grown under optimum conditions. Thus, there is some evidence that the mechanism of synchronization can be explained on the basis that the alga is unable to measure time when it is placed in darkness for some hours and that the cells are phased again by the next light time.

Autospores of Chlorella can be maintained after a complete LD cycle of 14:10 hours at a constant temperature of 32°C and with continuous bubbling with $1 \cdot 5\%$ $CO_2$ in air in total darkness. Hesse (1971) sampled every 3 hours and exposed these cells to another LD-cycle. Biomass in these synchronous LD-cycles was then measured as cell number, chlorophyll content, crude protein, carbohydrates and dry weight. Each factor was then plotted against the duration of the additional dark phase (the waiting time) after the last complete LD-cycle. The results (Fig. 32.2A) show that the algae exhibit a rhythmical pattern of productivity in relation to the length of the waiting time. This circadian rhythm is truly endogenous due to the constant conditions of the waiting time (in the dark) and because extreme values for biomass production always correspond to the duration of these constant conditions at the start of the LD-cycles.

Further evidence of the endogenous character of such a rhythm comes from the finding that the circadian rhythm is the same after synchronizing in a light-dark (LD)-cycle of 14:10 or of 16:16 hours as in non-synchronous cultures kept in the dark after a light-light (LL) treatment. The last example also shows that the endogenous rhythm is independent of the cell age.

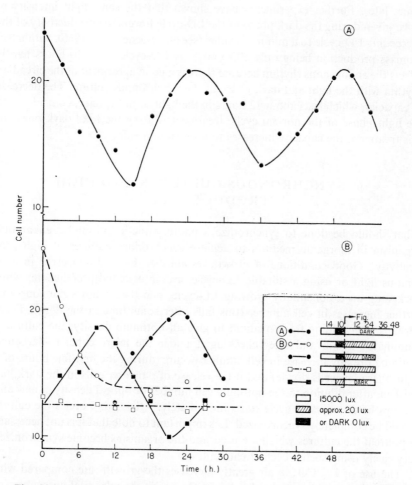

**Fig. 32.2.** Variations in cell number in a permanent synchronous culture of *Chlorella*. LD of 14:10 (15,000–0 lx) and temperature, 32°C.

A. After one cycle the culture was diluted with fresh medium to a standard cell number (1 = $1 \cdot 56 \times 10^6$ cells ml$^{-1}$) and kept in the dark (waiting time). Every 3 hours a sample was exposed to a complete LD. The resulting number of autospores produced after this LD are plotted.

B. Details as in A, but the light regime was varied as shown in inset.

The beginning of the dark phase acts in all these experiments as an inducer of the rhythm (*Zeitgeber*). If the dark phase during the waiting time is reset by a small quantity of light (e.g. less than 20 lx) the rhythm fades out within 12 hours. Transferring the algae from this low light to total darkness again induces the endogenous circadian rhythm (Fig. 32.2B). The phase of maximum productivity occurs after 9 to 12 hours in darkness, that of minimum productivity 12 hours later. Further experiments have shown that the same light intensity of 20 lx, given during the dark phase of the LD-cycle lowers the productivity of the succeeding LD-cycle to a minimum value (see also Soeder *et al.* 1966), with mean biomass production being only 50 to 60%. A LD-cycle of 14:10 hours hardly affects the endogenous rhythm because there is a good agreement of the circadian rhythm with the light and dark phase of the synchronous culture. The decrease of capacity, which lasts about 12 hours in the light or in the dark, coincides with the light phase of the normal cyclic treatment; during the total dark phase of this treatment the capacity increases to a maximum value.

## 6 SYNCHRONOUS CULTURES AND HIGH PRODUCTIVITY

What should be done to synchronize a photosynthetic unicellular eukaryotic organism in inorganic medium to achieve synchronous cultures of high productivity? Good conditions of growth, established by trial and error in continuous light or using a suitable LD-cycle, are an essential prerequisite. After darkening an exponentially growing LL-cycle, one then follows the course of further increases in cell number (this falls after some time in the dark). Then, perhaps after adding fresh medium to give an optimum density, the culture is illuminated again, and after checking it under the microscope, re-darken it again before any increase in cell number occurs (during the first few hours of a light phase few cells divide) and if the release of daughter cells in the dark had finished, after dilution to the optimum cell number or optical density, a new and perhaps better adjusted cycle starts immediately. After 3 to 5 cycles the culture should be completely synchronized. It is important to note that it is not necessary to pretreat the cultures which are used and the organisms become synchronized very easily using such a cyclic treatment.

The use of 1% $CO_2$ in air greatly increases the growth rate compared with air grown cultures, but higher levels of $CO_2$ inhibit the release of autospores or zoospores in unicellular algae. Also in *Chlorella*, the start of the light phase is particularly sensitive to $CO_2$, high $CO_2$ levels resulting in a lower number of daughter cells (Grünberg & Galloway 1970).

During the dark phase the cells become activated in some way to produce more $O_2$ per unit biomass during the next light phase compared with LL grown cultures. This increase in capacity does not occur if low light levels are provided during the dark phase. Thus, by ensuring complete darkness, one can obtain high productivity. On the other hand, it is also possible to achieve low productivity, if this is desirable, e.g. if one wishes cells with a high degree of resistance to some factor. The effects of adding small quantities of non-photosynthetic light during the dark phase can also cause a fading of the endogenous circadian rhythm (Hesse 1971). This explains why there is a lower productivity per hour of light with increasing light time, or with an unfavourable ratio of light to dark phases. *Vice versa* darkening for at least 10 hours results in cells showing high productivity during the next light phase.

# 7 REFERENCES

ARMSTRONG J.J., SURZYCKI S.J., MOLL B. & LEVINE R.P. (1971) Genetic transcription and translation specifying chloroplast components in *Chlamydomonas reinhardi*. *Biochemistry*, **10**, 692–701.

ARNAUD M. (1970) Ribonucleic acid metabolism during the cell cycle of *Chlorella*. *Bull. Inform. Sci. Tech.* **145**, 57–9.

ATKINSON A.W., GUNNING B.E.S., JOHN P.C.L. & McCULLOUGH W. (1971) Centrioles and microtubules in *Chlorella*. *Nature, New Biology*, **234**, 24–5.

BADOUR S.S. & WAYGOOD E.R. (1971) Glyoxylate carboxy-lyase activity in the unicellular green alga *Gloeomonas* sp. *Biochim. biophys. Acta*, **242**, 493–9.

BAECHTEL F.S., HOPKINS H.A. & SCHMIDT R.R. (1970) Continuous inducibility of isocitrate lyase during the cell cycle of the eucaryote *Chlorella*. *Biochim. biophys. Acta*, **217**, 216–19.

BAKER A.L. & SCHMIDT R.R. (1964) Polyphosphate metabolism during nuclear division in synchronously growing *Chlorella*. *Biochim. biophys. Acta*, **82**, 624–6.

BERNSTEIN E. (1960) Synchronous division in *Chlamydomonas moewusii*. *Science N.Y.* **131**, 1528–9.

BISHOP N.I. & SENGER H. (1971) Preparation and photosynthetic properties of synchronous cultures of *Scenedesmus*. In: *Methods in Enzymology XXIII*A, ed. San Pietro A. pp. 53–66. Academic Press, New York & London.

BONGERS L.H.J. (1958) Changes in photosynthetic activity during algal growth and multiplication. *Medel. Landbouwhogesch. Wageningen*, **58**, 1–10.

CANTOR M.H. & KOLTZ J. (1971) Synchronous growth of *Polytomella agilis*. *Experientia*, **27**, 801–3.

CARROLL J.W., THOMAS J., DUNAWAY C. & O'KELLEY J.C. (1970) Light-induced synchronization of algal species that divide preferentially in darkness. *Photochem. Photobiol.* **12**, 91–8.

CATTOLICO R.A. & JONES R.F. (1972) Isolation of stable ribosomal RNA from whole cells of *Chlamydomonas reinhardtii*. *Biochim. biophys. Acta*, **269**, 259–64.

CECHACEK Z. & HILLOVA J. (1970) DNA synthesis and changes in radiosensitivity during the cell cycle of vegetative culture of *Chlamydomonas reinhardtii*. *Int. J. Radiat. Biol.* **17**, 127–34.

CHANG W.H. & TOLBERT N.E. (1970) Excretion of glycolate, mesotartrate and isocitrate lactone by synchronized cultures of *Ankistrodesmus braunii*. *Pl. Physiol., Lancaster*, **46**, 377–85.

CODD G.A. & MERRETT M.J. (1971) The regulation of glycolate metabolism in division synchronized cultures of *Euglena*. *Pl. Physiol., Lancaster*, **47**, 640–3.

COOK J.R. (1971) Synchronous Cultures: *Euglena*. In: *Methods in Enzymology, XXIII*A, ed. San Pietro A. pp. 74–8. Academic Press, New York & London.

CZURDA V. (1935) Über die Begriffe 'Typus' und 'Variabilität' in der Systematik der Algen. *Beih. Bot. Cbl.* **53**, *Abt. A*. 158–71.

DALMON J. & GILET R. (1969) Realization of synchronous axenic and dense *Chlorella* cultures. *Folio Microbiologica*, **14**, 97–103.

DARLEY W.M. & VOLCANI R.E. (1971) Synchronized Cultures: Diatoms. In: *Methods in Enzymology, XXIII*A, ed. San Pietro A. pp. 85–96. Academic Press, New York & London.

EDMUNDS L.N. & FUNCH R.R. (1969) Circadian rhythm of cell division in *Euglena*: effects of a random illumination regime. *Science N.Y.* **165**, 500–2.

EHRET C.F. & WILLE J.J. (1970) The photobiology of circadian rhythms in protozoa and other eukaryotic microorganisms. In: *Photobiology of Microorganisms*, ed. Halldal P. pp. 369–416. Wiley Interscience, London, New York, Sydney & Toronto.

ERBEN K. (1971) Ein Leistungsvergleich zellfreier Polypeptidsynthese aus unterschiedlichen Entwicklungsstadien von *Chlorella*. *Z. Pflanzenphysiol.* **64**, 446–61.

FOO S.K., BADOUR S.S. & WAYGOOD E.R. (1971) Regulation of isocitrate lyase from autotrophic and photoheterotrophic cultures of *Gloeomonas* sp. *Canad. J. Bot.* **49**, 1647–1653.

GALLING G. (1970) Ribosomes in unicellular green algae. II. Dynamic alterations in the polysome patterns of *Chlorella*. *Cytobiologie*, **2**, 359–75.

GALLING G. & WOLFF E. (1972) Die Wirkung von Neomycin auf Zellteilung, Wachstum und Nucleinsäuren synchronisierter Chlorellen. *Arch. Mikrobiol.* **81**, 295–304.

GENSE-VIMON M-T. (1971) Occurrence of the G + C rich satellite DNA in unicellular algae. *Biochem. biophys. Res. Commun.* **42**, 347–52.

GOULD R.B. & JONES R.F. (1968) Electron-microscopic observations of microtubules during the synchronized vegetative cell cycle of *Chlamydomonas reinhardtii*. *Anat. Rec.* **160**, 355.

GRÜNBERG H. & GALLOWAY R. (1970) Some effects of high carbon dioxide content on autospore release and number in *Chlorella*. *J. Phycol.*, **6**, 222–4.

GUERIN-DUMARTRAIT E., MIHARA S. & MOYSE A. (1970) Composition de *Chlorella pyrenoidosa*, structure des cellules et de laurs lamelles chloroplastiques, en fonction de la carence en azote et de la lévee de carence. *Canad. J. Botany*, **48**, 1147–54.

HARE T.A. & SCHMIDT R.R. (1970) Nitrogen metabolism during synchronous growth of *Chlorella*. II. Free-, peptide-, and protein-amino acid distribution. *J. Cell Physiol.* **75**, 73–82.

HESSE M. (1971) Endogene Schwankungen der Produktionsfähigkeit bei *Chlorella* und ihre Beeinflussung durch Licht sehr niedriger Intensität und durch tiefe Temperatur. Dissertation, Göttingen, 1–84.

KANAZAWA T., KANAZAWA K., KIRK M.R. & BASSHAM J.A. (1970) Difference in nitrate reduction in 'light' and 'dark' stages of synchronously grown *Chlorella pyrenoidosa* and resultant metabolic changes. *Pl. Cell Physiol., Tokyo*, **11**, 445–52.

KATES J.R. & JONES R.F. (1967) Periodic increases in enzyme activity in synchronized cultures of *Chlamydomonas reinhardtii*. *Biochim. biophys. Acta*, **145**, 153–8.

KOMAREK J., RUZICKA J. & SIMMER J. (1968) Synchronization of the cultures of *Scenedesmus quadricauda* by optimalizing the length of the light period. *Biol. Plant., Praha*, **10**, 177–89.

KUHL A. & LORENZEN H. (1964) Handling and culturing of *Chlorella*. In: *Methods in Cell Physiology*, ed. Prescott D.M., Bd. 1, 159–87. Academic Press, New York & London.

LIEN T., PETTERSEN R. & KNUTSEN G. (1971) Effects of indole-3-acetic acid and gibberellin on synchronous cultures of *Chlorella fusca*. *Physiologia Pl.* **24**, 185–90.

LORENZEN H. (1957) Synchrone Zellteilungen von *Chlorella* bei verschiedenen Licht-Dunkel-Wechseln. *Flora, Jena*, **144**, 473–96.

LORENZEN H. (1959) Die photosynthetische Sauerstoffproduktion von *Chlorella* bei langfristig intermittierender Belichtung. *Flora, Jena*, **147**, 382–404.

LORENZEN H. (1963) Temperatureinflüsse auf *Chlorella pyrenoidosa* unter besonderer Berücksichtigung der Zellentwicklung. *Flora, Jena*, **153**, 554–92.

LORENZEN H. (1970) Synchronous cultures. In: *Photobiology of Microorganisms*, ed. Halldal P. pp. 187–212. Wiley Interscience, London, New York, Sydney & Toronto.

LORENZEN H. & SCHLEIF J. (1966) Zur Bedeutung der kürzest möglichen Generationsdauer in Synchronkulturen von *Chlorella*. *Flora, Abt. A.* **156**, 673–83.

LORENZEN H. & VENKATARAMAN G.S. (1969) Synchronous cell divisions in *Anacystis nidulans* Richter. *Arch. Mikrobiol.* **67**, 251–5.

McCULLOUGH W. & JOHN P.C.L. (1972) A temporal control of the *de novo* synthesis of isocitrate lyase during the cell cycle of the eucaryote *Chlorella pyrenoidosa. Biochim. biophys. Acta*, **296**, 287–96.

MAMPL Q., ALTMANN H. & BIEBL R. (1971) Unterschiede und Beeinflussung der Erholungskapazität von *Chlorella* zellen in verschiedenen Stadien des Zellzyklus. *Radiat. Bot.* **11**, 201–7.

MEEKER G.L. (1964) The effect of β-mercaptoethanol on synchronously dividing populations of *Astasia longa. Expl. Cell Res.* **35**, 1–8.

MIHARA S. & HASE E. (1971) Studies on the vegetative life-cycle of *Chlamydomonas reinhardi* Dangeard in synchronous culture. II. Effects of chloramphenicol and cycloheximide on the length of cell cycle. *Pl. Cell Physiol., Tokyo*, **12**, 237–41.

MITCHELL J.L.A. (1971) Photoinduced division synchrony in permanently bleached *Euglena gracilis. Planta*, **100**, 244–57.

MOLLOY G.R. & SCHMIDT R.R. (1970) Studies on the regulation of ribulose-1, 5-diphosphate carboxylase synthesis during the cell cycle of the eucaryote *Chlorella. Biochem. biophys. Res. Commun.* **40**, 1125–33.

NELSON E.B., TOLBERT N.E. & HESS J.L. (1969) Glycolate stimulation of oxygen evolution during photosynthesis. *Pl. Physiol., Lancaster,* **44**, 55–9.

PADILLA G.M. & BRAGG R.J. (1968) Responses of the flagellates *Prymnesium parvum* and *Euglena gracilis* to light, dark, and synchrony-inducing temperature cycles. *J. Cell. Biol.* **39**, 101a.

PFAU J., WERTHMÜLLER K. & SENGER H. (1971) Permanent automatic synchronization of micro-algae achieved by photoelectrically controlled dilution. *Arch. Mikrobiol.* **75**, 338–45.

PIRSON A. & LORENZEN H. (1966) Synchronized dividing algae. *A. Rev. Pl. Physiol.* **17**, 439–58.

REITZ R.C., HAMILTON J.G. & COLE F.E. (1967) Fatty acid concentrations in synchronous cultures of *Chlorella pyrenoidosa*, grown in the presence and absence of glucose. *Lipids*, **2**, 381–9.

ROONEY D.W., YEN B.C. & MIKITA D.J. (1971) Synchronization of *Chlamydomonas* division with intermittent hypothermia. *Expl. Cell Res.* **65**, 94–8.

SCHLÖSSER U. (1966) Enzymatisch gesteuerte Freisetzung von Zoosporen bei *Chlamydomonas reinhardii* Dangeard in Synchronkultur. *Arch. Mikrobiol.* **54**, 129–59.

SCHÖNHERR O.T. & WANKA F. (1971) An investigation of DNA polymerase in synchronously growing *Chlorella* cells. *Biochim. biophys. Acta*, **232**, 83–93.

SCHÖNHERR O.T., WANKA F. & KUYPER C.M.A. (1970) Periodic change of deoxyribonuclease activity in synchronous cultures of *Chlorella. Biochim. biophys. Acta*, **224**, 74–9.

SCHOR S.L., SIEKEWITZ P. & PALADE G.E. (1970) Cyclic changes in thylakoid membranes of synchronized *Chlamydomonas reinhardi. Proc. natn. Acad. Sci. U.S.A.* **66**, 174–80.

SENGER H. (1965) Die teilungsinduzierende Wirkung des Lichtes in mixotrophen Synchronkulturen von *Chlorella. Arch. Mikrobiol.* **51**, 307–22.

SENGER H. (1970) Charakterisierung einer Synchronkultur von *Scenedesmus obliquus*, ihrer potentiellen Photosyntheseleistung und des Photosynthese-Quotienten während des Entwicklungs-Cyclus. *Planta*, **90**, 243–66.

SENGER H. & Bishop N.R. (1969) Changes in the photosynthetic apparatus during the synchronous life-cycle of *Scenedesmus obliquus*. In: *Progress in Photosynthesis Research*, **1**, ed. Metzner, H. pp. 425–34. H. Laupp, Jr., Tübingen.

SHLYK A.A. & SUKHOVER L.K. (1968) Fractionation of metabolically heterogeneous pigments of *Euglena. Dokl. Akad. Nauk. S.S.S.R.* **181**, 1274–7 (Russ.).

SIERSMA P.W. & CHIANG K.S. (1971) Conservation and degradation of cytoplasmic and chloroplast ribosomes in *Chlamydomonas reinhardii. J. molec. Biol.* **58**, 167–85.

SOEDER C.J., SCHULZE G. & THIELE D. (1966) A new type of 'light inhibition', occurring in synchronous cultures of *Chlorella*. *Verh. int. Ver. Limnol.* **16**, 1595–1601.

SPEKTOROV C.S. & NIKOLSKAYA T.V. (1971) On the correlation between DNA synthesis and autoformation in *Chlorella pyrenoidosa* Pringsh 82 T. *Fiziol. Rast.* **18**, 754–9.

SWEENEY B.M. (1969) The cell division cycle. *Rhythmic phenomena in plants*, 104–16. Academic Press, New York & London.

TAMIYA H. (1966) Synchronous cultures of algae. *A. Rev. Pl. Physiol.* **17**, 1–26.

TAMIYA H. IWAMURA T., SHIBATA K., HASE E. & NIHEI T. (1953) Correlation between photosynthesis and light-independent metabolism in the growth of *Chlorella*. *Biochim. biophys. Acta*, **12**, 23–40.

TAMIYA H., MORIMURA Y., YOKOTA M. & KUNIEDA R. (1961) Mode of nuclear division in synchronous cultures of *Chlorella*: comparison of various methods of synchronization. *Pl. Cell Physiol., Tokyo*, **2**, 383–403.

TAUBE Ö. (1970) Ultraviolet inactivation and photoreactivation in synchronized *Chlorella*. *Physiologia Pl.* **23**, 755–61.

TERRY O.W. & EDMUNDS L.N. (1970) Rhythmic settling induced by temperature cycles in continuously-stirred autotrophic cultures of *Euglena gracilis* (Z strain). *Planta*, **93**, 128–42.

WALTHER W.G. & EDMUNDS L.N. (1970) Periodic increase in deoxyribonuclease activity during the cell cycle in synchronized *Euglena*. *J. Cell Biol.* **46**, 613–17.

WANKA P., JOOSTEN H.F. & DE GRIP W.J. (1970) Composition and synthesis of DNA in synchronously growing cells of *Chlorella pyrenoidosa*. *Arch. Mikrobiol.* **75**, 25–36.

WANKA F., MOORS J. & KRIJZER F.N.C.M. (1972) Dissociation of nuclear DNA replication from concomitant protein synthesis in synchronous cultures of *Chlorella*. *Biochim. biophys. Acta*, **269**, 153–161.

WEGMANN K. & METZNER H. (1971) Synchronization of *Dunaliella* cultures. *Arch. Mikrobiol.* **78**, 360–7.

WERNER D. (1971) Der Entwicklungscyclus mit Sexualphase bei der marinen Diatomee *Coscinodiscus asteromphalus*. I. Kultur und Synchronization von Entwicklungsstadien. *Arch. Mikrobiol.* **80**, 43–9.

# A NOTE ON TAXONOMY

## W. D. P. STEWART

Department of Biological Sciences,
University of Dundee
Scotland, U.K.

One of the problems in compiling a book of this type is that the contributors are not algal taxonomists. Nevertheless, it is very important that the systematic position of those algae mentioned in it is standardized and I have attempted to do this to the generic level. Each major group has been given the rank of a class thus:

1 Cyanophyceae
2 Rhodophyceae
3 Cryptophyceae
4 Dinophyceae
5 Haptophyceae
6 Chrysophyceae
7 Xanthophyceae
8 Eustigmatophyceae
9 Phaeophyceae
10 Prasinophyceae
11 Bacillariophyceae
12 Chlorophyceae
13 Euglenophyceae
14 Charophyceae

I have then included each algal genus (with a few exceptions) in an order within one of the above classes. Interested readers can obtain additional information by consulting the taxonomic works which have been followed thus: Parke and Dixon (1968) have been followed for the Cyanophyceae, Rhodophyceae, Cryptophyceae, Dinophyceae, Haptophyceae, Chrysophyceae, Phaeophyceae and Prasinophyceae. I have included the class Eustigmatophyceae and separated those members of it from the Xanthophyceae as recommended by Hibberd and Leedale (1970). Apart from that the Xanthophyceae are treated as in Parke and Dixon (1968). The classification of the Bacillariophyceae follows Hendey (1974), the Chlorophyceae, Round (1963) and the Euglenophyceae are classified according to Leedale (1967). The Charophyceae contains one order, the Charales, as recommended by Wood and Imahori (1965). I have not included the Rhaphido-phyceae (Chloromonadophyceae) because of its uncertain taxonomic position (see Chapter 5, and Guillard and Lorenzen, 1972).

In general the generic and specific names used by the contributors in the text have not been changed. Changes would lead to confusion among physiologists who know and recognize their 'physiological entities' by these names, even although they may not be strictly correct from a taxonomic point of view. There has been one exception. That is *Chlamydomonas reinhardtii* (*C. reinhardi*, *C. reinhartii*, etc.) which in the original manuscripts appeared in five different spellings. According to Dr. P.C. Silva (*pers. comm.* to Dr. D. Hibberd) there are two taxonomically acceptable ways of spelling this name: *Chlamydomonas reinhardtii* and *C. reinhardii*, and I have standardized on the former, except in Chapter 14 where the author preferred to use the spelling to which he was accustomed.

The detailed taxonomic arrangement used is as follows:

## Class CYANOPHYCEAE[1]*

Order Chroococcales[2]: *Agmenellum, Anacystis,*[3] *Aphanocapsa, Chroococcus, Coccochloris, Gloeocapsa, Merismopedia, Microcystis, Synechococcus, Synechocystis*

Order Nostocales[4]: *Anabaena, Anabaenopsis, Aphanizomenon, Arthrospira, Calothrix, Chlorogloea,*[5] *Cylindrospermum, Fremyella, Gloeotrichia, Hormothamnion, Hydrocoleum, Lyngbya, Microchaete, Microcoleus, Nostoc* (= *Amorphonostoc*), *Oscillatoria, Phormidium, Plectonema, Rivularia, Schizothrix, Scytonema, Spirulina, Tolypothrix, Trichodesmium*

Order Stigonematales: *Chlorogloeopsis,*[6] *Hapalosiphon, Mastigocladus, Westiellopsis*

## Class RHODOPHYCEAE[6a]

Order Nemaliales: *Acanthopeltis, Acrochaetium, Batrachospermum, Bonnemaisonia, Chaetangium, Chantransia, Delisia, Falkenbergia,*[7] *Galaxaura, Gelidium, Lemanea, Nemalion, Pseudogloeophloea, Pterocladia, Rhodochorton, Sacheria, Sirodotia, Suhria, Thorea, Trailliella*[8]

Order Gigartinales: *Agardhiella, Ahnfeltia, Anatheca, Chondrus, Cystoclonium, Eucheuma, Furcellaria, Gigartina, Gracilaria, Gymnogongrus, Hypnea, Iridaea, Iridophycus, Meristotheca, Nizymenia, Phyllophora, Plocamium, Rhodoglossum, Stictosporum*

* Notes to which superior figures refer appear on pp. 915–16.

Order Cryptonemiales: *Aeodes, Bossiella, Constantinea, Corallina, Cyrtymenia* (= *Phyllymenia*), *Dilsea, Dumontia, Endocladia, Gloiopeltis, Grateloupia, Holmsella, Joculator, Lithothamnion, Pachymenia, Phymatolithon, Polyides, Serraticardia, Tichocarpus, Weeksia*

Order Rhodymeniales: *Botryocladia, Chylocladia, Coeloseira, Halosaccion, Lomentaria, Rhodymenia*

Order Ceramiales: *Acanthophora, Antithamnion, Bornetia, Bostrychia, Brongniartella, Callithamnion, Campylaephora, Ceramium, Chondria, Dasya, Delesseria, Erythrocystis, Gonimophyllum, Griffithsia, Grinnellia, Janczewskia, Laingia, Laurencia, Lenormandia, Membranoptera, Nitophyllum, Phycodrys, Plumaria, Polycoryne, Polyneura, Polysiphonia, Ptilota, Rhodomela, Rytiphlaea*

Order Porphyridiales: *Porphyridium, Rhodella, Rhodosorus*

Order Goniotrichales: *Asterocytis, Goniotrichum*

Order Bangiales: *Bangia, Compsopogon, Erythrotrichia, Porphyra, Smithora*

## Class CRYPTOPHYCEAE

Order Cryptomonadales: *Chilomonas, Chroomonas, Cryptomonas, Hemiselmis, Rhodomonas*

## Class DINOPHYCEAE

Order Prorocentrales: *Exuviaella, Prorocentrum*

Order Peridiniales: *Amphidinium, Aureodinium, Cachonina, Entomosigma, Glenodinium, Gonyaulax, Gymnodinium, Gyrodinium, Heterosigma, Katodinium, Oxyrrhis, Peridinium, Polykrikos, Woloszynskia*

Order Phytodiniales: *Crypthecodinium*

## Class HAPTOPHYCEAE

Order Isochrysidales: *Dicrateria, Isochrysis*

Order Prymnesiales: *Apistonema, Chrysochromulina, Coccolithus, Cricosphaera, Crystallolithus, Hymenomonas, Ochrosphaera, Phaeocystis, Pleurochrysis, Prymnesium, Syracosphaera*

## Class CHRYSOPHYCEAE

Order Ochromonadales: *Mallomonas, Microglena, Monas, Ochromonas, Paraphysomonas, Pavlova, Poteriochromonas, Syncrypta, Synura*

Order Phaeothamniales: *Phaeosaccion*

Order Chromulinales: *Chromulina, Gloeochrysis,*[9] *Hydrurus, Monochrysis, Pascherella, Pseudopedinella, Sphaleromantis*

Order Thallochrysidales: *Thallochrysis*

Order Phaeoplacales: *Stichochrysis*[10]

## Class XANTHOPHYCEAE

Order Heterochloridales: *Olisthodiscus*

Order Mischococcales: *Botrydiopsis, Chlorellidium, Monodus, Ophiocytium*

Order Tribonematales: *Bumilleria, Bumilleriopsis, Heterococcus, Tribonema*

Order Vaucheriales: *Botrydium, Vaucheria*

## Class EUSTIGMATOPHYCEAE[11]

*Pleurochloris,*[12] *Polyedriella,*[13] *Vischeria*[14]

## Class PHAEOPHYCEAE

Order Ectocarpales: *Chordaria, Compsonema, Ectocarpus, Elachista, Giffordia, Levringiella, Mesogloia, Myrionema, Pylaiella* (= *Pilayella*), *Ralfsia, Spermatochnus, Stilophora, Streblonema, Waerniella*

Order Dictyosiphonales: *Desmotrichum, Dictyosiphon, Ishige, Litosiphon*

Order Scytosiphonales: *Petalonia, Scytosiphon*

Order Cutleriales: *Cutleria* (*Aglaozonia*[15])

Order Desmarestiales: *Desmarestia*

Order Laminariales: *Alaria, Arthrothamnus, Chorda, Cymathaera, Ecklonia, Egregia, Eisenia, Laminaria, Macrocystis, Nereocystis, Saccorhiza*

Order Sphacelariales: *Sphacelaria*

Order Dictyotales: *Dictyopteris, Dictyota, Padina, Zonaria*

Order Fucales: *Ascophyllum, Bifurcaria, Cystophora, Cystophyllum, Cystoseira,*[16] *Durvillea, Fucus, Halidrys, Himanthalia, Hormophysa, Hormosira, Pelvetia, Sargassum* (incl. *Hizikia), Turbinaria*

## Class PRASINOPHYCEAE

Order Pyramimonadales: *Asteromonas, Heteromastix, Micromonas, Platymonas, Prasinocladus, Pyramimonas, Stephanoptera, Tetraselmis*

Order Halosphaerales: *Pterosperma*

Order Pedinomonadales: *Mesostigma*[10]

## Class BACILLARIOPHYCEAE

Order Bacillariales: *Achnanthes, Amphipleura, Amphiprora, Amphora, Asterionella, Bacillaria, Bellerochea, Biddulphia, Chaetoceros, Coscinodiscus, Cyclotella, Cylindrotheca, Detonula, Ditylum, Fragilaria, Gomphonema, Hantzschia, Lauderia, Licmophora, Lithodesmium, Mastogloia, Melosira, Navicula, Nitzschia, Opephora, Phaeodactylum, Pinnularia, Pleurosigma, Rhabdonema, Rhizosolenia, Skeletonema, Stauroneis, Stephanodiscus, Stephanopyxis, Synedra, Tabellaria, Thalassiosira*

## Class CHLOROPHYCEAE[17]

Order Volvocales: *Astrephomene, Balticola,*[18] *Brachiomonas, Carteria,*[19] *Chlamydobotrys, Chlamydomonas, Chlorogonium, Diplostauron, Dunaliella, Eudorina, Furcilla, Gloeomonas, Gonium, Haematococcus, Hyalogonium, Lobomonas, Pandorina, Platydorina, Pleodorina, Polytoma, Polytomella, Pyrobotrys, Stephanosphaera, Volvox, Volvulina*

Order Tetrasporales: *Gloeocystis, Hormotila, Nautococcus, Palmella, Tetraspora*

Order Chlorococcales: *Ankistrodesmus, Botryococcus, Bracteacoccus, Chlorella,*

*Chlorococcum, Chodatella, Codiolum, Coelastrum, Crucigenia, Dictyochloris, Eremosphaera, Franceia, Golenkinia, Hydrodictyon, Keratococcus (= Ourococcus), Kirchneriella, Myrmecia, Neochloris, Oocystis, Ostreobium, Pediastrum, Protococcus, Protosiphon, Prototheca, Pseudochlorococcum, Radiosphaera, Rhaphidium, Scenedesmus, Selanastrum, Spongiochloris, Spongiococcum, Tetracystis, Trebouxia*

Order Chlorosphaerales: *Chlorosarcinopsis, Chlorosphaera, Coccomyxa, Hyalococcus, Nannochloris*

Order Ulotrichales: *Hormidium, Klebsormidium, Koliella, Raphidonema, Stichococcus, Ulothrix*

Order Ulvales: *Enteromorpha, Monostroma, Ulva*

Order Prasiolales: *Prasiola*

Order Microsporales: *Microspora*

Order Chaetophorales: *Aphanochaete, Draparnaldia, Fritschiella, Stigeoclonium, Trentepohlia*

Order Cladophorales: *Chaetomorpha, Cladophora, Pithopora, Rhizoclonium*

Order Acrosiphoniales: *Acrosiphonia, Spongomorpha, Urospora*

Order Sphaeropleales: *Sphaeroplea*

Order Oedogoniales: *Bulbochaete, Oedogonium*

Order Siphonocladales: *Anadyomene, Apjohnia, Blastophysa, Boodlea, Cladophoropsis, Dictyosphaeria, Microdictyon, Siphonocladus, Struvea, Valonia, Valoniopsis*

Order Dasycladales: *Acetabularia* (inc. *Acicularia*), *Batophora, Bornetella, Cymopolia, Dasycladus, Halicoryne, Neomeris*

Order Derbesiales: *Derbesia* (inc. *Halicystis*)

Order Codiales: *Avrainvillea, Bryopsis, Codium, Tydemania*

Order Caulerpales: *Caulerpa, Chlorodesmis, Halimeda, Penicillus, Udotea*

Order Dichotomosiphonales: *Dichotomosiphon, Pseudodichotomosiphon*

Order Mesotaeniales: *Cylindrocystis, Mesotaenium, Netrium*

Order Desmidiales: *Closterium, Cosmarium, Desmidium, Docidium, Hyalotheca, Micrasterias, Pleurotaenium, Staurastrum, Staurodesmus, Xanthidium*

Order Zygnematales: *Mougeotia, Spirogyra, Zygogonium*

## Class EUGLENOPHYCEAE

Order Euglenales: *Astasia, Euglena, Lepocinclis, Phacus, Trachelomonas*

Order Eutreptiales: *Distigma, Eutreptiella*

Order Heteronematales: *Peranema*

## Class CHAROPHYCEAE

Order Charales: *Chara, Lamprothamnion, Nitella, Nitellopsis, Tolypella*

## GENERA OF UNCERTAIN TAXONOMIC POSITION

*Cyanidium,*[20] *Cyanophora,*[21] *Glaucocystis,*[22] *Gonyostomum,*[23] *Symbiodinium,*[24] *Vacuolaria,*[23] *Zoochlorella*[24]

## NOTES

1  The filamentous *Myxosarcina, Beggiatoa, Leucothrix* and *Vitreoscilla* are considered by some to be colourless members of the Cyanophyceae but their taxonomic position is uncertain.

2  Stanier *et al.* (1971) have classified unicellular blue-green algae into various typological groups based on physiological, biochemical and morphological factors.

3  Stanier *et al.* (1971) consider that the alga known as *Anacystis nidulans* should be classified as a strain of *Synechococcus elongatus.*

4  Kenyon *et al.* (1972) have classified various Nostocales into typological groups based on physiological, biochemical and morphological factors.

5  Placed in the Chroococcales by Parke and Dixon (1968) but some species at least are now known to be members of the Nostocales.

6  See Kenyon *et al.* (1972).

6a  Also see Dixon (1973).

7   *Falkenbergia rufolanosa* is the tetrasporic stage of *Asparagopsis armata* according to Parke and Dixon (1968).

8   *Trailliella intricata* is the tetrasporic stage of *Bonnemaisonia hamifera* according to Parke and Dixon (1968).

9   *Gleochrysis litoralis* and *G. maritima* = *Ruttnera litoralis* and *R. maritima* (Ochromonadales) respectively according to Parke and Dixon (1968).

10   According to M. Parke (pers. comm.).

11   This class was separated from the Xanthophyceae by Hibberd and Leedale (1970). As yet there has been no separation of the Eustigmatophyceae into orders.

12   As yet only two species of *Pleurochloris* (*P. commutata* and *P. magna*) can be placed with certainty in this class (Hibberd & Leedale, 1970).

13   As yet only one species of *Polyedriella* (*P. helvetica*) can be placed with certainty in this class (Hibberd & Leedale, 1970).

14   As yet only two species of *Vischeria* (*V. punctata* and *V. stellata*) can be placed with certainty in this class. (Hibberd & Leedale, 1970).

15   *Aglaozonia parvula* is the diploid stage of *Cutleria multifida* (Parke & Dixon, 1968).

16   For a discussion of the taxonomy and synonomy of *Cystoseira*, see Roberts (1970) and references cited therein.

17   The arrangement of the orders follows Round (1963) who places them in the sub-phylum Chlorophytina of the phylum Chlorophyta, except for the order Acrosiphoniales which I have separated from the Cladophorales (see Round, 1971). Round (1971) has also suggested some other modifications to his 1963 scheme and his interesting papers should be consulted.

18   *Balticola* is included in the genus *Haematococcus* by Silva (1962).

19   *Carteria ovata* is the diploid motile stage of *Chlamydomonas variabilis* according to Silva (1962).

20   The taxonomic position of *Cyanidium caldarium*, a much used experimental tool, is uncertain. The consensus of opinion places it in the Rhodophyceae (see Schnepf & Brown, 1971).

21   *Cyanophora paradoxa* is a symbiotic association of a cryptomonad and an organism similar to a blue-green alga (see Hall & Claus, 1963 and Robinson & Preston, 1971).

22   *Glaucocystis nostochinearum* is a symbiotic association of an apochlorotic unicellular green alga and an organism similar to a blue-green alga (see Hall & Claus, 1967 and Robinson & Preston, 1971).

23   *Gonyostomum* and *Vacuolaria* are placed by some in the Rhaphidophyceae (see Guillard & Lorenzen, 1972 and p. 163).

24   The taxonomic position of these zoochlorellae is uncertain.

## REFERENCES

DIXON P.S. (1973) *Biology of the Rhodophyta.* Oliver & Boyd, Edinburgh.
GUILLARD R.R.L. & LORENZEN C.J. (1972) Yellow-green algae with chlorophyllide *c*. *J. Phycol.*, **8**, 10–11.

HALL W.T. & CLAUS G. (1963) Ultrastructural studies on the blue-green algal symbiont in *Cyanophora paradoxa* Korschikoff. *J. Cell Biol.*, **19**, 551–63.

HALL W.T. & CLAUS G. (1967) Ultrastructural studies on the cyanelles of *Glaucocystis nostochinearum* Itzigsohn. *J. Phycol.*, **3**, 37–51.

HENDEY N.I. (1974) A revised check list of British marine diatoms. *J. Mar. biol. Ass. U.K.* **54**, 277–300.

HIBBERD D.J. & LEEDALE G.F. (1970) Eustigmatophyceae—a new algal class with unique organization of the motile cell. *Nature, Lond.*, **224**, 758–60.

KENYON C.N., RIPPKA R. & STANIER R.Y. (1972) Fatty acid composition and physiological properties of some filamentous blue-green algae. *Arch. Mikrobiol.*, **83**, 216–36.

LEEDALE G.F. (1967) *Euglenoid flagellates*. Prentice-Hall, Englewood Cliffs, N.J.

PARKE M. & DIXON P.S. (1968) Check-list of British marine algae—second revision. *J. mar. biol. Ass. U.K.*, **48**, 783–832.

ROBERTS M. (1970) Studies on marine algae of the British Isles, 8. *Cystoseira tamariscifolia* (Hudson) Papenfuss. *Br. phycol. J.* **5**, 201–10.

ROBINSON D.G. & PRESTON R.D. (1971) Studies on the fine structure of *Glaucocystis nostochinearum* Itzigs. II. Membrane morphology and taxonomy. *Br. phycol. J.*, **6**, 113–28.

ROUND F.E. (1963) The taxonomy of the Chlorophyta. *Br. phycol. Bull.* **2**, 224–35.

ROUND F.E. (1971) The taxonomy of the Chlorophyta II. *Br. phycol. J.* **6**, 235–64.

SCHNEPF E. & BROWN JR. R.M. (1971) On relationships between endosymbiosis and the origin of plastids and mitochondria. In *Origin and continuity of cell organelles.* **2**, ed. Reinert, J. & Ursprung, H. pp. 299–322, Springer-Verlag, Berlin.

SILVA P.C. (1962) Classification of algae. In *Physiology and biochemistry of algae*, ed. Lewin, R.A., pp. 827–37, Academic Press, London and New York.

STANIER R.Y., KUNISAWA R., MANDEL M. & COHEN-BAZIRE G. (1971) Purification and properties of unicellular blue-green algae (Order Chroococcales). *Bact. Rev.* **35**, 171–205.

WOOD R.D. & IMAHORI K. (1965) *A revision of the Characeae, First Part. Monograph of the Characeae.* Weinheim.

2G

# SPECIES INDEX

Numbers in italics refer to the taxonomic appendix.

*Acanthopeltis*  63, *910*
 *japonica*
  agaroses of  12
  sterols of  272
*Acanthophora*  *911*
 *spicifera*  183
*Acetabularia*  *914*
  absence of IAA from  763
  cell wall of  43, 53
  effect of light quality on  720
  ion uptake by  699, 701–3, 719
  mannan of  13
  metaphosphates of  645
  morphogenesis of  792
  movement in  883–4
  nucleic acids of  297–301
  protein synthesis of  317
  reproduction of  815, 816, 824
  storage products of  216
  ultrastructure of  97, 101, 142
 *clavata*  180
 *crenulata*
  carotenoids of  180
  cell walls of  54, 74
  ion fluxes in  701–3
  storage products of  216, 217, 224
 *mediterranea*
  carotenoids of  180
  DNA of  299
  effect of growth substances on  766,
   793
  glycolate biosynthesis in  485
  ion fluxes in  698, 701–3
  storage products in  210, 217
  vitamin requirements of  782
 *mobii*  180
 *wettsteinii*  180
*Achnanthes*  885, *913*
 *brevipes*
  heterotrophic growth of  537
  utilization of vitamin B$_{12}$ analogues by
   750
  vitamin requirements of  779
*Achromobacter*  754, 755
*Acicularia*  *914*
 *schenkii*  180
*Acrochaetium*  23, *910*
 *pectinatum*  805

*Acrosiphonia*  *914*
 *arcta*  180
 *centralis* ($=$ *Spongomorpha arcta*)  74
 *sonderi*  180
*Acrosiphoniales*  *914*
  carotenoids of  180
*Aeodes*  63, *910*
 *orbitosa*  69
 *ulvoidea*  69
*Agardhiella*  63, *910*
 *tenera*  68
*Aglaozonia*  803, *912, 916*
*Agmenellum*  *910*
 *quadruplicatum*
  heterotrophic growth of  537
  utilization of vitamin B$_{12}$ analogues by
   750
  vitamin requirements of  778
*Ahnfeltia*  63, *910*
 *plicata*  218, 222
 *stellata*  272
*Alaria*  *913*
  laminaran in  213
  sporangia in  798
 *esculenta*
  alginates of  11
  effect of light and temperature on
   development of  796
*Amorphonostoc*  75, *910*
*Amphidinium*  *911*
  carotenoids of  187
  extracellular products of  843
  thylakoid arrangement in  141
 *carteri*
  fatty acids of  248
  vitamin metabolism in  750, 754, 758,
   761, 780
 *klebsi*  848
 *rhynchocephalum*  750, 780, 848
*Amphipleura*  142, *913*
 *pellucida*  657
 *rutilans*
  heterotrophic growth of  537
  mucilage of  840
  polysaccharides of  75
  vitamin requirements of  779
*Amphiprora*  *913*
 *paludosa*  537

919

*Amphora*  871, *913*
  *coffaeiformis*
    heterotrophic growth of  534–7
    vitamin metabolism of  746, 750, 778,
      779
  *lineolata*
    heterotrophic growth of  537
    vitamin requirements of  779
  *paludosa* var. *duplex*  779
  *perpusilla* (= *coffaeiformis*)  750, 779
*Anabaena*  910
  anaerobic growth of  467
  carotenoids of  190
  DNA hybridization in  285
  effect of IAA on  764
  electron transport in  400
  enzymes of  466, 562–3, 586–8, 642
  extracellular products of  840, 841
  fatty acids of  241, 244
  heterocyst production in  564, 571
  heterotrophic growth of  535, 537, 548
  inorganic nutrition of  612, 621
  nitrogen metabolism in  562–4, 571,
    586–8, 621
  oxygen uptake in  491
  phosphorus metabolism in  639, 642,
    647
  photophosphorylation in  405, 407
  photosystems of  368-9
  polysaccharides of  75
  respiration in  521
  steroids of  18
  *aerulosa* var. *oscillatorioides*  190
  *cylindrica*
    carotenoids of  190–2
    effect of turbulence on growth of  729
    extracellular nitrogen from  842
    fatty acid metabolism of 241, 242, 257
    glutamine synthetase in  564, 594
    heterocysts of  564–7
    heterotrophic growth of  537
    hydrogenase in  457, 562
    inorganic nutrition of  617, 624
    nitrogen fixation by  561–3, 569
    nitrogen metabolism of  462, 586, 587
      589, 591, 592, 594, 617
    phycocyanins of  197
    photorespiration by  476, 570
    ribosomes of  284
    vitamin requirements of  778
  *flos-aquae*
    carotenoids of  190
    effect of high $O_2$ tension on  509
    extracellular products of  839
    fatty acids of  241, 242
    heterotrophic growth of  537, 548
    inorganic nutrition of  624
    nitrogenase activity and heterocyst
      numbers in  567

    nitrogen fixation by  568, 572
    production of toxins by  848
    storage products of  218
  *gelatinosa*  778
  *naviculoides*  778
  *oscillarioides*  621
  *spiroides*  729
  *variabilis*
    carotenoids of  190
    DNA hybridization in  285
    growth of on combined nitrogen  571
    heterotrophic growth of  537, 543, 548
    inorganic nutrition in  615, 619, 624
    movement in  865, 874, 875
    oxidative phosphorylation by  522
    photosystems of  368, 369, 401
    phycocyanins of  197
    pyruvate metabolism in  562
    respiration in  508, 521
    ribosomes of  284
    vitamin requirements of  778
*Anabaenopsis*  *910*
  *circularis*
    bacterial decomposition of  573
    nitrogen fixation in dark by  569
*Anacystis*  *910*
  absorption spectra of  358, 363, 364
  acetyl CoA carboxylase in  448
  ATP synthesis in  410
  carotenoids of  192
  DNA of  284
  effect of DCMU on  549
  effect of light on growth of  718, 720
  effect of temperature on growth of
    723, 725
  extracellular products of  843
  glycolate production in  495
  inorganic nutrition of  612, 617, 618
  light absorption by  351, 369, 371, 373
  oxygen uptake in  490
  photosynthesis in  368, 397, 400, 403,
    438
  phytoflavin from  563
  protein synthesis in  602
  steroids of  18
  synchronous cultures of  896
  *marina*
    fatty acids of  242
    heterotrophic growth of  537
    taxonomy of  29
    vitamin requirements of  778
  *montana*
    ribosomes of  601
    RNA polymerase of  284
  *nidulans* (see also *Synechococcus elon-*
    *gatus*)  *915*
    absorption spectra of  360, 364  370
    aldolase in  20
    antibiotic sensitivity of  285

carotenoids of 188, 189, 191
chlorophyll of 4, 169–71
chromatic adaptation of 377–9
fatty acid metabolism of 242, 244, 257, 259
heterotrophic growth of 537, 543, 547, 548
hydrogenase, absence in 457
inorganic nutrition of 616, 618, 624
metaphosphates in 645
nucleic acid metabolism in 282–5
photorespiration in 476
photosystems of 368
phycocyanins of 197, 198
quinones of 17
respiration of 521
taxonomy of 30
uric acid utilization by 575
vitamin requirements of 778
*Anadyomene* 13, *914*
*stellata* 180
*Anatheca* 63, *910*
*dentata* 66
*Ankistrodesmus* *913*
ATP synthesis in 410
effect of hydrostatic pressure on 730
effect of manganese deficiency on 618
effect of pH on 727
effect of phosphorus deficiency on 640
extracellular products of 843
glycolate excretion by 487, 494, 496
glycolate metabolism in 900
hydrogenase in 457, 466
photophosphorylation in 412, 440, 644
polyphosphate formation in 648, 649
taxonomy of 27, 29
utilization of amino acids as nitrogen source by 595
*braunii*
carotenoids of 183
control of nitrate assimilation in 589, 591, 592
glycolate metabolism in 480, 488, 850
inorganic nutrition in 613, 616
nitrate reductase activity in 585
nitrite reductase activity in 587, 588
photoassimilation of organic compounds in 548
photorespiration in 476
protein synthesis in 600
*falcatus* 535
*Antithamnion* *911*
genophore segregation in 150
vitamin metabolism in 622, 759, 778
*cruciatum* 805
*glanduliferum* 749, 778
*plumula*
effect of coumarin on 770, 806

effect of light and temperature on 805
*sarniense* 749
*spirographidis* (= *sarniense*) 778
*subulatum* 149–50
*Aphanizomenon* *910*
*flos-aquae*
carotenoids of 187, 189, 190
morphological plasticity in 789
nitrogen fixation by 569
release of toxins by 848
vitamin requirements of 778
*Aphanocapsa* *910*
fatty acids of 243
phycocyanins of 197, 198
*Aphanochaete* *914*
*pascheri* 868
*Apistonema* *911*
life cycle of 663, 791
structure of cell wall of 664
*aestuari* (= *Pontosphaera roscoffensis*) 780
*carterae* 274
*Apjohnia* 13, *914*
*laetevirens* 75
*Arthrospira* 187, 189, 190, *910*
*maxima* 198
*Arthrothamnus* 213, *913*
*Ascophyllum* *913*
alginic acids of 58
chemotaxis in 830
extracellular products of 840, 841
flagella of 88
growth and development in 799–801
growth factors from 769, 770
*nodosum*
alginates of 11
carotenoids of 185
cellulose of 13
extracellular products of 844
fucose in 72, 73
hydrogenase in 457
mannitol in 220, 225
osmoregulation in 225
polyols in 221
*Astasia* *915*
chemotaxis in 881
cytochromes in 616
effect of temperature on growth of 723
evolution of 746
respiration of 506
synchronous culture of 895
*klebsii* (= *A. longa*, = *A. chattoni*) 781
*longa*
DNA of 286
fatty acid metabolism of 251, 253
heterotrophic growth of 534, 536, 541, 543, 551
oxidative phosphorylation of 518–19

*Astasia*
  *longa—cont.*
    respiration in   509, 512
    vitamin requirements of   781
  *ocellata*   192
*Asterionella*   913
  *formosa*
    phosphorus storage in   639
    respiration of   509
    vitamin requirements of   778
  *japonica*
    extracellular products of   844, 847
    fatty acids of   249
*Asterocytis*   622, 751, *911*
  *ramosa*   749, 778
*Asteromonas*   913
  *gracilis*   96
  *propulsa*   181
*Astrephomene*   913
  *gubernaculifera*
    heterotrophic growth in   534
    vitamin requirements of   781, 782
Athiorhodaceae   256
*Aureodinium*   141, *911*
*Avena*   763
*Avrainvillea*   13, *914*
  *nigricans*   180
  *rawsoni*   180
*Azolla*   575
*Azotobacter*   259


*Bacillaria*   913
  *paradoxa*   865
Bacillariales   913
Bacillariophyceae (see also diatoms)   *909, 913*
  alkaline phosphatase in   642
  ATP levels in   638
  carbohydrates of   75, 214, 218
  carotenoids of   4–6, 152, 185, 361
  chlorophylls of   161–4, 358
  chloroplasts of   133, 137, 140, 148–51
  cyclitols in   224
  DNA of   302
  extracellular products of   840, 841,
    843
  fatty acids of   249–50
  hydrogenase, absence of in   458
  lipids of   140
  movement in, 865, 881, 882, 885
  phylogeny of   261
  pigments of   153
  soluble metabolites in   9
  sterols of   276
  surface structures of   90
  thylakoid arrangement of   142
  urease in   597
  vitamin requirements of   745 750,
    778–9

*Bacillus*
  *subtilis*
    ammonia release by   574
    gene arrangement in   336
*Balticola*   913, *916*
  *buetschlii*
    heterotrophic growth of   537
    vitamin metabolism in   750, 782
  *droebakensis*
    apochlorosis and   551
    heterotrophic growth of   537
    vitamin requirements of   750, 782
*Bangia*   911
  chloroplast membranes of   130, 133
  effect of daylength on reproduction of
    818
  phycoerythrins of   8
  requirement for vitamin $B_{12}$ in   622
  thylakoid arrangement in   141
  *fuscopurpurea*
    absence of starch from   207
    cell wall polysaccharides   53
    chloroplast of   132
    effect of daylength on   806
    mannans of   12
    vitamin requirements of   778
Bangiaceae   63
Bangiales   *911*
  biliproteins of   356
  enzymes of   20
  intercellular plugs in   14
  mannans of   12
  phycoerythrins of   8
  pigment adaptation in   377
  sterols of   272
  taxonomy of   804
  xylans of   55
Bangiophycidae   164, 356, 377, 804
*Batophora*   914
  cell wall development in   46
  cell wall polysaccharides of   53
  mannan in   13
  *oerstedi*
    carotenoids of   180
    inulin-like storage products of   216
    polysaccharides of   10
*Batrachospermum*   910
  chloroplast development in   148
  chloroplast movement in   885
  phycoerythrins of   8
  polysaccharides of   14
  respiration of   509
  storage products of   217, 218, 222
  thylakoids of   141
  *virgatum*   196
*Beggiatoa*   915
  classification of   30
  fatty acid metabolism in   251
  heterotrophic growth of   534

*Bellerochea* *913*
  (*polymorpha*?)   746, 779
  (*spinifera*?)   746, 750, 779
*Biddulphia* *913*
  chloroplast movement in   885
  flagella of   87, 88
  storage products in   214
  vitamin metabolism in   750
  *sinensis*   249
*Bifurcaria* *913*
  *bifurcata*
    cellulose in   14
    polysaccharides of   73
*Blasia*   575
*Blastophysa*   13, *914*
  *rhizopus*   180
*Bodo*   881
*Boehemeria*
  *nivea*   50
*Bonnemaisonia*   910, *916*
  effect of light on reproduction of   815
  iodine in   23
  phycoerythrins of   8
  *hamifera*   8
*Boodlea*   *914*
  *coacta*   180
  *kaeneana*   180
*Bornetella*   *914*
  *sphaerica*   180
*Bornetia*   *911*
  *secundiflora*   218
*Bossiella*   *910*
  *orbigniana*   669
*Bostrychia*   *911*
  *scorpioides*   218, 222
*Botrydiopsis*   *912*
  *alpina*   185
  *intercedens*
    heterotrophic growth of   536
    vitamin requirements of   780
*Botrydium*   *912*
  carotenoids of   4
  chlorophyll of   163
  thylakoid arrangement in   142
  *granulatum*
    dictyosome doubling in   104
    photokinesis in   873
    sterols of   274
*Botryocladia*   *911*
  *botryoides*   763
*Botryococcus*   *913*
  *braunii*   17
*Brachiomonas*   *913*
  *submarina*
    heterotrophic growth of   534, 539, 543, 547, 548, 550, 551
    vitamin metabolism in   750, 782
  *submarina* var. *pulsifera*   543

*Bracteacoccus*   457, *913*
  *cinnabarinus*
    heterotrophic growth of   536
    vitamin requirements of   781
  *engadiensis*
    heterotrophic growth of   535
    vitamin requirements of   781
  *minor*
    heterotrophic growth of   535
    vitamin requirements of   781
  *terrestris*
    heterotrophic growth of   536
    vitamin requirements of   781
*Brongniartella*   *911*
  *byssoides*   805
*Bryopsis*   *914*
  carotenoids of   6, 180
  chloroplast movement in   885
  cytoplasmic streaming in   883, 884
  IAA in   763, 793
  nuclear pores in   96
  proteins of   20
  thylakoid arrangement in   142
  ultrastructure of   792
  xylan of   13
  *corticulans*   180
  *hypnoides*   180
  *muscosa*   180
  *plumosa*   765
*Bulbochaete*   95, *914*
  *hiloensis*   110
*Bumilleria*   *912*
  flagella of   88
  nuclear envelope of   96
  *sicula*   142
  *stigeoclonium*   868
*Bumilleriopsis*   *912*
  *brevis*
    heterotrophic growth of   536, 542
    vitamin requirements of   780

*Cachonina*   *911*
  *niei*   780
*Callithamnion*   *911*
  *tetricum*   218
*Callitriche*
  *aquatica*   625
*Calothrix*   *910*
  extracellular products of   840, 841
  fatty acids of   244
  heterotrophic growth of   535, 537, 548
  morphological placticity in   789
  nitrogen fixation by   572, 573
  polysaccharides of   75
  storage products of   218
  *brevissima*
    bacterial decomposition of   573
    vitamin requirements of   778

*Calothrix—cont.*
  *parietina*
    carotenoids of　191
    heterotrophic growth of　537
    vitamin requirements of　778
  *scopulorum*
    carotenoids of　191
    extracellular nitrogen from　573, 842
    nitrogen fixation by　572
    phycocyanins of　197
    vitamin requirements of　778
*Campylaephora　911*
  *hypnaeoides*　12
*Carteria　913, 916*
  polyols in　221
  thermotaxis in　882
  thylakoid arrangement in　142
  *mediterranea*　537
*Caulerpa　914*
  caulerpin and caulerpacin from　23
  cytoplasmic streaming in　883
  effect of light on　817
  hydrogenase and　457
  pigments of　154
  polysaccharides of　73, 74
  reproduction in　817
  temperature requirements of　721
  ultrastructure of　792
  xylan of　13
  *cupressoides*　181
  *distichophylla*　181
  *filiformis*
    amylopectin in　210
    carotenoids of　181
    xylans of　55
  *lentillifera*　181
  *prolifera*
    carotenoids of　178, 181
    effect of light on　793
    IAA and　763, 765
  *racemosa*
    carotenoids of　181
    effect of IAA on　765
    xylan of　57
  *sedoides*　128
  *serrulata*　181
  *sertularioides*
    carotenoids of　181
    effect of IAA on　765
    growth and development of　793
*Caulerpales*　181, *914*
  carotenoids of　181
  secretion of calcium carbonate by　793
*Caulobacter*　764
*Cavicularia*　575
*Ceramiaceae*　63
*Ceramiales　911*
  halogens in　23
  phycoerythrins of　8

    sterols of　273
    storage products of　219, 220
*Ceramium　911*
  galactans of　63
  iodine in　625
  mitochondria of　101
  production of adhesive in　110
  *boydenii*
    agaroses of　12
    porphyran of　66
  *rubrum*
    absorption spectra of　195, 355
    auxin activity in　763
    fatty acids of　245
    sterols of　273
    storage products of　218, 222
*Chaetangium　910*
  *fastigiatum*　57
*Chaetoceros*　841, 843, *913*
  *ceratosporum*
    heterotrophic growth of　537
    vitamin metabolism in　750, 779
  *gracilis*　779
  *lorenzianum*　750, 779
  *pelagicus*
    heterotrophic growth of　537
    vitamin metabolism in　750, 779
  *pseudocrinitus*
    heterotrophic growth of　537
    vitamin metabolism in　750, 779
  *septentrionale*　249
  *simplex*　750
*Chaetomorpha　914*
  carotenoids of　180
  cellulose in　13, 48, 51, 52
  cell walls of　44–6
  effect of fresh medium on reproduction
    in　821
  fructosides in　217
  ion fluxes in　697, 703, 704
  polysaccharides of　73
  proteins of　19
  *area*　180
  *antennina*　180
  *crassa*　270
  *darwinii*　698, 703, 704
  *linum*　180
  *melagonium*
    carotenoids of　180
    cellulose of　49, 50
    cell wall of　44
  *princeps*　44
  *rupestris*　210
*Chaetophorales　914*
*Chantransia*　885, *910*
*Chara　915*
  cytoplasmic streaming in　883
  dictyosomes of　103, 104
  effect of calcium levels on　614

effect of temperature on membrane potential of 724
ion uptake in 688, 695, 701
microtubules in 97
nuclear division in 100
optimum phosphorus levels for 639
photophosphorylation in 440
phragmaplast in 794
plasmodesmata in 95
thylakoid arrangement in 142
*aspera* 781
*australis*
  chloride fluxes in 703
  starches of 210
*braunii* 693
*ceratophylla* 181
*corallina*
  ion uptake in 683–9, 691, 694
  photophosphorylation in 405, 412
*delicatula* 794
*fibrosa* 146
*foetida* 688–9
*globularis*
  flux measurements in 683
  iodine in 625
  vitamin requirements of 781
*vulgaris* 278
*zeilanica* 781
Charales 909, 915
Charophyceae 909, 915
  carotenoids of 5
  chlorophylls of 161–4
  deposition of calcite in 662
  polysaccharides of 210
  sterols of 278
  thylakoids of 142–3
  vitamin requirements of 781–2
*Chilomonas* 911
  carbon metabolism in 506
  chemotaxis in 881
  combustion ratios of 546
  respiration in 506
  thiamine requirement of 742
*paramecium*
  fatty acids of 247
  heterotrophic growth of 533, 534, 543
  trichocysts of 89
  vitamin requirements of 779
*Chlamydobotrys* 913
  acetate utilization in 547, 549
  photophosphorylation in 412
*stellata*
  heterotrophic growth of 537, 547, 548
  production of chlorophyll *b* by 3
*Chlamydomonas* 913
  anaerobic growth of 466, 467
  carbon metabolism in 506
  cellulose in 90

chloroplasts of 125, 142, 143, 147, 148, 366
contractile vacuoles in 91
effect of ultraviolet light on 718
extracellular products of 839–41, 843, 844
fatty acids of 252
flagella of 87
glycolate metabolism in 477, 479, 480, 486, 488, 494, 496
hydrogen evolution by 20
hydroxypyruvate reductase in 483
inorganic nutrition of 612, 614, 618
isoagglutinins of 24
morphogenesis of 791
movement of 829–30, 873, 876, 879, 881
nuclear and cytoplasmic inheritance in 316–38
nucleic acid metabolism in 287, 290–302
oxygen uptake in 490–1
photoassimilation in 547
photophosphorylation in 403, 407, 408
photoreduction in 400
photorespiration in 492
pigments of 167, 373, 551, 720
polyols in 221
protein synthesis in 317, 598, 902
reproduction in 816, 817, 821–3, 829–830
respiration in 506, 516
surface antigens of 25
synchronous growth of 896–900
taxonomy of 29
temperature adaptation in 724
vitamin production by 757
*acidophila* 726
*agloeformis*
  carotenoids of 178
  vitamin requirements of 781
*chlamydogama* 782
*debaryana* 457
*dorsocentralis* 534, 536, 548
*dysosmos*
  heterotrophic growth in 534, 548, 550
  hydrogenase in 457
  isocitrate lyase activity in 508
*eugametos*
  chlorophyll synthesis in 169, 170
  extracellular products of 846, 851
  genetic analysis in 319
  heterotrophic growth of 537
  hydrogenase in 457
  mutants of 323
  vitamin requirements of 781
*globosa*
  extracellular products of 845

*Chlamydomonas*
  *globasa—cont.*
    heterotrophic growth of  548
    inhibition of, by algal products  854
  *gloeopara*  584
  *humicola*  537, 548
  *intermedia*  457
  *microsphaerella*  584
  *moewusii*
    chemotaxis of mating types  846
    heterotrophic growth of  537, 549
    hydrogenase in  457
    reproduction of  820
    vitamin requirements of  781
  *monoica*  534, 536
  *mundana*
    heterotrophic growth of  537, 543, 547, 548
    vitamin requirements of  782
  *mundana* var. *astigmata*
    heterotrophic growth of  543
    vitamin requirements of  782
  *pallens*  622, 782
  *peterfi*  584
  *pichinchae*  781
  *pseudococcum*  536, 548
  *pseudogloe*
    heterotrophic growth of  536, 548
    hydrogenase, absence in  458
  *pulchra*  534, 536, 548
  *pulsatilla*
    apochlorosis and  551
    heterotrophic growth of  537, 543
    photoassimilation of  547, 548, 550
    vitamin metabolism in  750, 782
  *reinhardtii* (*reinhardi*)  910
    arginine metabolism in  597, 598
    chlorophyll synthesis in  167–9
    extracellular products of  845, 846, 850
    fatty acid metabolism in  251, 252
    flagella of  87, 96
    gamete differentiation in  831
    gibberellins in  767
    glycolate metabolism in  481, 850
    heterotrophic growth of  534, 542, 548
    hydrogenase in  457
    inorganic nutrition of  613
    life cycle of  318
    mitochondria of  101
    movement of  865, 878
    mutant strains of  424–31, 439, 515
    nitrogen metabolism of  587, 589, 590, 592, 594
    nuclear division in  99
    nuclear genetics of  320, 322
    nucleic acid metabolism of  291–5, 900
    protein synthesis in  600, 601
    pyrenoid division in  146
    respiration in  476, 511, 512, 514

    starch in  212
    thylakoids of  143, 258
    vitamin requirements of  748, 781
  *stellata*  169
  *subglobosa*  534
  *spreta*  534
*Chlorella*  913
  anaerobic growth of  406, 467, 716
  ATP measurements in  638
  carbon dioxide fixation in  437, 438, 446, 447
  catalase in  482
  chloroplasts of  125, 142, 366
  effect of IAA on  764
  effects of light on  379, 717, 718, 720, 721, 817
  effects of temperature on  722, 723, 725, 726
  electron transport in  397, 400, 404, 406
  extracellular products of  716, 840–4, 847, 848, 851
  fatty acid metabolism in  251, 254, 259, 545, 900
  glycolate metabolism in  477, 478, 480, 483–9, 495, 496, 716
  glyoxylate cycle in  545
  hydrogen metabolism in  20, 456, 466
  inorganic nutrition in  612–24
  ion uptake by  540, 707, 719, 725
  light absorption by  351, 358, 361, 362, 366, 371, 373, 379
  lysine in  20
  microbodies in  91, 484
  nitrogen metabolism in  448, 585–90, 594–8
  nucleic acid metabolism in  284, 294–297, 899–900
  osmoregulation in  225, 728, 729
  oxygen uptake in  476, 490
  pH range of  727
  phosphorus metabolism in  639–40, 643, 645, 647
  photosynthesis in  476
  photorespiration in  492
  phototrophy in  531, 533, 537, 539, 546–9
  pigments of  167, 168, 170, 358, 361, 362, 551
  protein synthesis in  600, 899, 900
  reproduction in  817
  respiration in  492, 506, 512, 515
  sporopollenin in  90
  sterols of  269
  storage products of  211, 224
  symbiotic growth of  25
  synchronous culture of  896–900
  taxonomy of  26, 27, 28, 269
  urea assimilation in  448

vitamin requirements of 747
*autotrophica*
  heterotrophic growth of 537
  vitamin requirements of 781
*candida*
  heterotrophic growth of 535
  sterols of 269
  vitamin requirements of 781
*ellipsoidea*
  characteristics of 28
  chlorophyll synthesis in 171
  DNA of 295, 296
  effect of gibberellin on 767
  heterotrophic growth of 535, 536, 546
  inorganic nutrition of 613, 615, 619
  pentose phosphate cycle in 507
  phototrophy in 548
  pyrenoid division in 146
  sterols of 269
  storage products of 224
  synchronous culture of 896
  taxonomy of 27, 28
  urea metabolism in 596
  vitamin requirements of 781
*emersonii*
  heterotrophic growth of 535
  sterols of 269
  vitamin requirements of 781
*emersonii* var. *globosa*
  heterotrophic growth of 535
  vitamin requirements of 781
*fusca*
  characteristics of 28
  effect of IAA on 764
  effect of temperature change on 723, 724
  fatty acid metabolism in 255
  heterotrophic growth of 535, 716
  hydrogenase in 457, 465
  inorganic nutrition of 620, 621
  nitrate metabolism in 586, 588–90
  phosphate metabolism in 643
  sterols of 269
  storage products in 223, 224
  vitamin requirements of 781
*fusca* var. *rubescens* 28
*fusca* var. *vacuolata*
  heterotrophy in 535
  nitrate metabolism in 585, 589, 591, 592
  urea metabolism in 596
  vitamin requirements of 781
*glucotropha* 269
*homosphaera*
  characteristics of 28
  hydrogenase in 457
*infusionum*
  heterotrophic growth of 537
  vitamin requirements of 781

*infusionum* var. *auxenophila*
  heterotrophic growth of 537
  vitamin requirements of 781
*kessleri*
  characteristics of 28
  hydrogenase in 457
*luteoviridis*
  characteristics of 28
  chlorophyll synthesis in 170
  heterotrophic growth of 537
  hydrogenase in 458
*miniata*
  heterotrophic growth of 535
  sterols of 269
  vitamin requirements of 781
*minutissima*
  characteristics of 28
  hydrogenase in 458
  pH range of 727
*mutabilis*
  heterotrophic growth of 536
  vitamin requirements of 781
*nocturna*
  heterotrophic growth of 536
  sterols of 269
  vitamin requirements of 781
*photophila*
  heterotrophic growth of 537
  vitamin requirements of 781
*pringsheimii*
  heterotrophic growth of 535
  vitamin requirements of 781
*prototbecoides*
  characteristics of 28
  chlorophyll synthesis in 167–70
  DNA of 295, 296, 718
  effects of light on 718, 720
  glucose metabolism in 509, 536, 552
  hydrogenase, absence in 458
  vitamin requirements of 782
*prototbecoides* var. *communis*
  sterols of 269
  vitamin requirements of 782
*prototbecoides* var. *galactophila* 782
*prototbecoides* var. *mannophila*
  sterols of 269
  vitamin requirements of 782
*pyrenoidosa*
  absorption spectra of 362, 369–71
  carotenoids of 178
  cell wall of 90
  chlorophyll of 169, 356, 359
  effect of growth substances on 763, 767
  effects of light on 371, 512, 592, 817
  extracellular products of 757, 846, 847
  fatty acids of 252, 256

*Chlorella*
  *pyrenoidosa—cont.*
    glycolate metabolism in   481, 488
    heterotrophy in   536, 537, 539, 542,
      546–9
    inorganic nutrition of   612, 613, 616,
      618–20, 623
    ion uptake in   706
    nitrogen metabolism in   585, 586, 590,
      592–4, 596, 601, 602
    nuclear division in   99
    nucleic acid metabolism in   294–7
    phosphorus metabolism in   640
    photochemical systems in   365, 370
    photorespiration in   476
    protein of   20, 602
    respiration of   507, 508, 512
    storage products of   210, 211, 224
    urea metabolism in   596
    vitamin metabolism in   757, 781, 846
  *regularis*
    heterotrophic growth of   536
    vitamin requirements of   781
  *regularis* var. *aprica*   781
  *regularis* var. *imbricata*   781
  *rubescens*   535
  *saccharophila*
    characteristics of   28
    heterotrophic growth of   535, 537
    hydrogenase, absence in   458
    inorganic nutrition of   617
    pH range of   727
    sterols of   269
    vitamin requirements of   781
  *simplex*
    heterotrophic growth of   535
    sterols of   269
    vitamin requirements of   781
  *sorokiniana*
    heterotrophic growth of   535
    sterols of   269
    vitamin requirements of   781
  *vannielii*
    heterotrophic growth of   535
    sterols of   269
    vitamin requirements of   781
  *variabilis*
    heterotrophic growth of   537
    vitamin requirements of   781
  *variegata*   169
  *vulgaris*
    carotenoids of   178
    characteristics of   28
    chlorophylls of   170, 359
    effect of gibberellic acid on   767
    effects of light on   171, 512, 717
    fatty acid metabolism in   252, 254,
      257, 260
    heterotrophy in   536, 537, 542, 548

    hydrogenase in   457, 458
    inorganic nutrition of   612, 613, 616,
      617, 619, 623
    light adaptation in   717
    nitrogen metabolism in   585, 594, 595
    polyphosphate synthesis by   647
    respiration in   476, 488, 508, 512
    sterols of   269
    storage products of   211, 224
    taxonomy of   27, 28
    vitamin requirements of   781
  *vulgaris* var. *luteoviridis*   781
  *vulgaris* var. *tertia*   28
  *zofingiensis*
    characteristics of   28
    effect of low pH on   727
    heterotrophic growth of   537
    hydrogenase, absence in   458
*Chlorellidium*   *912*
  *tetrabotrys*
    heterotrophic growth of   534, 536
    vitamin requirements of   780
Chlorococcaceae   269–70
Chlorococcales   *913*
  active glucose transport in   539
  carbon source in   532
  chemotrophy in   531
  sterols of,   269–70
  taxonomy of   26–9
*Chlorococcum*   *914*
  effect of IAA on   764
  extracellular products of   840, 843
  hydrogenase, absence in   458, 466
  phototrophy in   531, 537
  proteins of   20
  reproduction of   823
  *aplanosporum*   781
  *diplobionticum*   781
  *echinozygotum*
    inorganic nutrition of   615
    vitamin requirements of   781
  *ellipsoideum*
    production of inhibitor by   854
    vitamin requirements of   781
  *hypnosporum*   781
  *mactostigmatum*   781
  *minutum*   781
  *multinucleatum*   781
  *oleofaciens*   781
  *perforatum*   781
  *pinguideum*   781
  *punctatum*   781
  *scabellum*   781
  *tetrasporum*   781
  *vacuolatum*
    hydrogenase in   457
    vitamin requirements of   781
  *wimmeri*
    effect of light on pigments of   720

vitamin requirements of 781
*Chlorodesmis* 13, *914*
  *comosa* 180
*Chlorogloea* *910*
  effect of IAA on 764
  extracellular products of 841
  fatty acids of 243
  *fritschii*
    effect of IAA on 764
    fatty acids of 241, 242
    heterotrophic growth of 535, 536, 542, 543
    life cycle of 790
    nitrogen fixation by 569
    PHB in 23
    plastoglobuli in 147
*Chlorogloeopsis* *910*
  fatty acids of 244
  heterotrophic growth of 535, 548
*Chlorogonium* *913*
  glyoxysomes in 485
  microbodies in 484
  *elongatum* 508, 534
  *euchlorum* 534, 535, 548
Chlorophyceae *909, 913*
  anaerobic growth in 467
  auxotrophy in 743–5
  carbohydrates of 9, 13–15, 55, 73, 145, 210, 217, 219
  carotenoids of 4–6, 152, 154, 178–83, 192, 361
  chemotaxis in 829, 881
  chemotrophy in 531
  chlorophyll of 3, 161–5, 358
  chloroplasts of 126, 131, 133, 145, 154, 885
  chromatic adaptation of 379
  cyclitols in 223, 224
  deposition of aragonite in 662, 668
  DNA of 22, 298, 301
  effect of potassium deficiency on 613
  enzymes of 457–8, 533
  extracellular products of 840, 841, 843
  eyespots in 147
  fatty acids of 251–3
  flagella of 87
  gametogenesis in 822, 830
  hydrocarbons of 17
  iron in 617
  lipids of 16–19
  manganese in 618
  mannitol in 220
  measurement of ATP in 638
  morphogenesis in 790–4
  movement in 865, 881, 882, 885
  phosphorus levels for growth of 639
  phylogeny of 261
  pigments of 154, 365
  polyols of 221

proteins of 20
pyrenoid of 144
rRNA of 304
silicification in 656
sterols of 269–70
sulphur metabolism in 612
thermotaxis in 882
thylakoids of 142, 154
urea assimilation by 595, 597
vitamin metabolism in 750, 756, 781–782
*Chlorosarcinopsis* *914*
  *auxotrophica* 781
  *eremi* 781
  *sempervirens* 782
*Chlorosphaera* *914*
  *consociata* 782
Chlorosphaerales *914*
*Chodatella* *914*
  polymorphism in 792
  taxonomy of 27
*Chondria* 23, *911*
  *dasyphylla* 273
*Chondrus* *910*
  absence of IAA from 763
  carrageenan of 14
  extracellular products of 840, 841
  galactans of 63
  *crispus*
    carrageenan from 67
    sterols of 270, 272, 273
    storage products in 218
  *giganteus* 272
  *ocellatus* 272
*Chorda* *913*
  lamina renewal in 797
  thylakoid arrangement in 142
  *filum* 11
  *tomentosa* 796
*Chordaria* 11, *912*
  *flagelliformis* 11
Chordariales 11, 213
*Chromatium* 256
*Chromulina* *912*
  effect of light on heterotrophic growth of 716
  chloroplast movement in 885
  *pusilla* 101
Chromulinales *912*
Chroococcaceae 191
Chroococcales *910*
  carotenoids of 191–2
  fatty acids of 19, 260
  gliding movements in 870, 879
*Chroococcus* *910*
  carotenoids of 191
  inorganic nutrition of 624
  lipids of 19
  *minutus* 8

*Chrooococcus—cont.*
 *turgidus* 537, 778
*Chroomonas* 141, *911*
 *marina* 750
 *salina*
  chemotrophy in 531
  storage products in 221
  vitamin requirements of 779
*Chrysochromulina* *911*
 haptonemata of 89
 scale production in 105
 surface scales of 90, 664
 thylakoids of 142
 *brevifilum*
  mitochondrial multiplication in 101
  vitamin requirements of 780
 *chiton*
  haptonemata of 89
  scales of 90, 105–7, 665
 *kappa* 780
 *strobilus* 780
Chrysophyceae *909, 912*
 alkaline phosphatase in 642
 ATP measurement in 638
 auxotrophy in 744–6, 779
 carbohydrates of 9, 145, 218
 carotenoids of 4–6, 152, 185
 chemotaxis in 881
 chemotrophy in 531
 chlorophyll of 161–4, 358
 chloroplasts of 131, 133, 137, 140,
  150–1, 885
 cyclitols in 224
 DNA of 299
 extracellular products of 841, 843, 845
 eyespot of 147
 flagella of 87, 89
 glycerol production in 222
 hydrocarbons of 17
 lipids and fatty acids of 16, 140, 245–
  248
 paramylon in 215
 phagotrophy in 538
 phylogeny of 261
 polyols in 221
 silicon in 623, 656
 sterols of 274
 storage products of 145, 214
 thermotaxis in 882
 thylakoids of 142
 trichocysts of 89
 vitamin metabolism in 750, 779
*Chylocladia* 885, *911*
*Cladonia*
 *cristatella* 817
 *impexa* 269
*Cladophora* *914*
 carotenoids of 178, 180
 cellulose in 13, 48, 51

cell wall of 43–5
 effect of temperature on 819
 polysaccharides of 73
 reproduction of 819
 respiration of 509
 storage products of 217, 221
 *crispata* 180
 *fascicularis* 180
 *graminea* 180
 *laetevirens* 217
 *lehmanniana* 180
 *membranacea* 180
 *prolifera*
  carotenoids of 180
  cell wall of 44
 *rupestris*
  carotenoids of 180
  cell wall of 44
  storage products of 210, 211, 217
 *sericea* 763
 *trichotoma* 180
Cladophorales *914*
 carbohydrates of 9, 216, 217
 carotenoids of 180
 cell wall of 44
 sterols of 270
*Cladophoropsis* 13, *914*
 *herpestica* 180
 *zollingeri* 180
*Closterium* *915*
 calcium sulphate crystals in 662
 effect of light on reproduction of 816
 thylakoid arrangement in 142
 *braunii* 747, 782
 *macilentum* 782
 *peracerosum* 782
 *pusillum* 782
 *siliqua* 782
 *stringosum* 747, 782
 *turgidum* 782
*Clostridium* 562
 *ethylicum* 562
 *kluyveri* 562
 *pasteurianum* 562
*Coccochloris* 841, 843, *910*
 *elabens*
  carotenoids of 191
  heterotrophic growth of 537
  vitamin metabolism of 750, 778
 *peniocystis*
  carotenoids of 192
  respiration in 521
  vitamin requirements of 778
*Coccolithus* *911*
 coccoliths of 90, 105, 720, 725
 extracellular products of 843
 scale production in 105
 thylakoid arrangement in 142
 vitamin requirements of 780

*huxleyi*
  calcification of 663–7
  fatty acids of 246
  mannitol in 220
  nitrate reductase of 587
  vitamin metabolism of 746, 757–80
*pelagicus* 663–6, 790
*Coccomyxa* *914*
  extracellular products of 846
  hydrogenase, absence in 458
  inorganic nutrition of 614
  storage products in 221
  vitamin metabolism in 757, 846
Codiales 180, *914*
*Codiolum* 15, *914*
*Codium 914*
  carbon dioxide fixation in 439
  carotenoids in 6
  cell wall of 53, 54, 73
  cytoplasmic streaming in 883, 884
  mannan of 13
  pigments of 154
  proteins of 19
*bursa*
  carotenoids of 180
  effect of low salinity on 729
*coronatum* 180
*decorticatum* 793
*dimorphum* 180
*duthieae* 180
*elongatum* 180
*fragile*
  carotenoids of 178, 180
  fatty acids of 253
  growth and development of 793
  polysaccharides of 54, 74, 210
*lucasii* 180
*muelleri* 180
*spongiosum* 180
*tomentosum* 180
*Coelastrum* *914*
  effect of fresh medium on reproduction
    of 821
  hydrogenase in 457
*morus* 782
*proboscideum* 600
*Coelosira* *911*
*pacifica* 272
Collema 575
  *tuniforme* 575
*Compsonema* *912*
  *sporangiiferum* 803
*Compsopogon* *911*
  *oishii* 8
*Constantinea* *910*
  *subulifera* 208
*Convoluta*
  *convoluta* 660
  *roscoffensis* 25

*Corallina* *910*
  *officinalis*
    galactan of 66
    sterols of 272
    storage products of 207, 218, 222
*Coscinodiscus* *913*
  cell wall of 657
  chloroplast movement in 885
  reproduction in 818, 819, 824, 830–1
  storage products in 214
  *asteromphalus*
    effect of fresh medium on reproduction
      of 821
    heterotrophic growth of 537
    vitamin metabolism in 750, 779
  *concinnus*
    effect of depleted medium on repro-
      duction of 821
    polymorphism in 791
  *granii* 792
*Cosmarium* *915*
  *botrytis*
    genetic control in 319
    vitamin requirements of 781
  *impressulum* 781
  *laeve* 781
  *lundelli* 781
  *meneghini* 781
  *turpini* 782
*Cricosphaera* *911*
  coccoliths of 90, 105
  extracellular products of 843
  *carterae*
    calcification in 662
    fatty acids of 246
    growth substances in 766
  *elongata*
    fatty acids of 246
    growth substances in 766
    heterotrophic growth of 537
    production of inhibitor by 847
    vitamin metabolism in 750, 780
*Crucigenia* *914*
  *apiculata* 457
*Crypthecodinium* *911*
  *cohnii* 745, 747, 748, 780
*Cryptococcus*
  *albidus* 754
Cryptomonadales *911*
*Cryptomonas* *911*
  chemotaxis in 881
  chloroplast envelope in 133
  ecological distribution of 717
  fatty acids of 247
  flagella of 88
  thylakoids of 141
  trichocysts of 89
  xanthophylls of 187
  *appendiculata* 247

*Cryptomonas—cont.*
  *maculata*  247
  *ovata*
    carotenoids of  187
    extracellular products of  845
    fatty acids of  247
    vitamin metabolism in  750, 779
  *stigmatica*  135
*Cryptonemiales 910*
    deposition of calcite in  662
    phycoerythrins of  8
    sterols of  272
*Cryptophyceae 909, 911*
    auxotrophy in  744, 745, 779
    biliproteins of  6–9, 194, 196
    carotenoids of  5, 152, 154, 187, 361
    chemotaxis in  881
    chlorophyll of  4, 161–4, 358
    chloroplasts of  133, 137, 141, 154
    eyespot in  147
    glycerol production in  222
    growth studies of  790
    heterotrophic growth of  533
    hydrocarbons of  17
    lipids and fatty acids of  16, 245
    phylogeny of  261
    starch storage in  145
    sugars in  220
    thylakoids of  130, 154
    trichocysts of  89, 91, 104
    vitamin metabolism in  750, 779
*Crystallolithus 911*
    coccoliths of  90
    dictyosomes of  103
  *hyalinus*  664–5
*Cutleria 803, 912*
*Cutleriales 912*
*Cyanidium 915, 916*
    absorption spectra of  364
    chlorophyll of  164
    lipids of  245, 726
    proteins of  19
    taxonomy of  26, 152
    temperature requirement of  722, 724, 726
  *caldarium*
    anaerobic growth in  467
    chloroplasts of  144
    fatty acids of  244, 245
    heterotrophic growth of  536
    pigments of  153, 164, 167, 183, 197, 200, 365
    sterols of  273
    storage products of  209
    vitamin requirements of  781
*Cyanophora 915, 916*
  *paradoxa*
    carotenoids of  192
    vitamin metabolism in  750, 779

Cyanophyceae (see also blue-green algae)
  *909, 910*
    ATP measurements in  638
    auxotrophy in  743–5, 750, 778
    biliproteins of  6–9, 194, 196, 356, 364–5
    carotenoids of  5, 152, 153, 187–92
    chemotaxis in  881
    chemotrophy in  531
    chlorophyll of  153, 162–4, 170
    citrulline production in  447
    cytochromes in  617
    deposition of calcite in  662
    DNA hybridization in  21
    extracellular products of  840, 841, 843
    glycolate metabolism in  477, 481, 488
    hydrocarbons of  17
    hydrogenase in  457
    in relation to red algae  9
    lipids and fatty acids of  16–19, 241–4, 256–60
    morphogenesis of  789–90
    movement in  865, 870–1, 879, 881, 882
    nucleic acid metabolism in  282–5
    organic nitrogen utilization by  595
    phosphorylases of  209
    photosynthetic apparatus of  126–8, 153
    phototaxis in  879
    pigment adaptation in  377–9
    polyols of  221
    polysaccharides of  75, 208–10, 218
    proteins of  20
    ribosomes of  22
    RNA of  22–3
    taxonomy of  29–32
    thermotaxis in  882
    ultrastructure of  86
    vitamin requirements of  778
*Cyclotella 913*
    effect of salinity on reproduction of  823
    extracellular products of  841, 843
    heterotrophic growth of  535, 542
    light adaptation in  717
    vitamin requirements of  779
  *caspia*
    heterotrophic growth of  537
    vitamin metabolism in  750, 779
  *cryptica*
    cell wall of  657
    chitin in  75, 90
    extracellular products of  840, 851
    fatty acids of  249
    inorganic nutrition of  623, 657–62
    vitamin requirements of  779
  *meneghiniana*  823
  *nana*
    extracellular vitamins from  845
    heterotrophic growth of  537

nitrate reductase of 587
sterols of 276
storage products of 218
vitamin metabolism of 750, 753, 755, 758, 761, 779
*Cylindrocystis* 821, *915*
  *brebissonnii* 781, 782
*Cylindrospermum* *910*
  carotenoids of 190
  hydrogenase, absence in 457
  nitrogenase in 573
*Cylindrotheca* 623, *913*
  *closterium* 750, 779
  *fusiformis*
    DNA synthesis in 302
    heterotrophic growth of 534, 542, 544
    inorganic nutrition of 623, 657, 659, 661
    vitamin requirements of 779
  *nana* (= *Cyclotella nana*) 761
  *pseudonana* (= *Cyclotella nana*) 750, 779
  *tricorfusiformis* 249
*Cymathaera* 213, *913*
*Cymopolia* 53, *914*
  *barbata* 180
*Cyrtymenia* *910*
  *sparsa* 272
*Cystoclonium* *910*
  *purpureum*
    storage products of 218, 222
    translocation in 94
*Cystophora* 799, *913*
*Cystophyllum* 799, *913*
*Cystoseira* *913*, *916*
  branching in 800
  flagella of 87
  storage products in 213
  *abrotanifolia* 11
  *barbata* 11
  *stricta* 213

*Dasya* *911*
  *pedicellata*
    in axenic conditions 769
    sterols of 273
*Dasycladales* *914*
  carotenoids of 180
  fructosans in 216
*Dasycladus* *914*
  cell wall of 46, 53
  polysaccharides of 13, 53
  *clavaeformis* 180
*Delesseria* *911*
  chloroplasts of 134
  translocation in 94
  *sanguinea*
    storage products of 218
    translocation in 94

*Delisia* *910*
  *pulchra* 8
*Derbesia* *914*
  carotenoids of 180
  mannan of 13, 15
  pigments of 154
  proteins of 19
  reproduction of 816, 817, 819–21, 824, 827
  *lamourouxii* 180
  *tenuissima* 180
  *vaucheriaeformis* 180
Derbesiales 180, *914*
*Desmarestia* *912*
  inorganic nutrition of 612
  reproduction in 815, 818, 820
  *aculeata*
    alginates of 11
    growth and development of 803
    storage products in 224
  *ligulata*
    effect of gibberellic acid on 767
    vitamin requirements of 780
  *viridis*
    effect of gibberellic acid on 767
    vitamin requirements of 780
Desmarestiales 11, *912*
Desmidiales *915*
*Desmidium* *915*
  *swartzii* 782
*Desmotrichum* 819, *912*
*Detonula* *913*
  *confervacea* 778
*Dichotomosiphon* *914*
  carotenoids of 181
  xylan of 13
  *tuberosus* 179, 181
Dichotomosiphonales 181, *914*
*Dicrateria* *911*
  *inornata*
    fatty acids of 246
    vitamin requirements of 780
*Dictyochloris* 535, *914*
  *fragrans* 781
*Dictyopteris* 213, *913*
  *polypodioides* 11
*Dictyosiphon* *912*
  *foeniculaceus* 11
Dictyosiphonales 11, *912*
*Dictyosphaeria* *914*
  cellulose in 13
  hydrogenase, absence in 458
  *cavernosa* 180
  *favulosa* 180
  *versluysii* 180
*Dictyota* *913*
  apical dominance in 802
  cell wall of 92–4
  chloroplasts of 137, 884, 885

*Dictyota—cont.*
   egg production in   817, 826
   flagella of   88
   inorganic nutrition of   621, 623
   polysaccharide production in   107
  *dichotoma*
   alginates of   11
   effect of molybdenum on   621
*Dictyotales*   *913*
   alginates of   11
   nuclear division in   100
*Dictyuchus*   88
*Dilsea*   *910*
  *carnosa*   272
  *edulis*
   carrageenan of   70
   starch in   207
*Dinophyceae*   *909, 911*
   auxototrophy in   744, 745, 746, 750
   carotenoids of   5, 152, 154, 186–7, 361
   chemotaxis in   881
   chemotrophy in   531
   chlorophylls of   161–4, 358
   extracellular products of   840, 841, 843
   fatty acids of   248, 250
   measurement of ATP levels in   638
   phagotrophy in   538
   phylogeny of   261
   pigments of   154
   pyrenoid of   145
   starch reserves in   212
   thylakoid arrangement in   141, 154
   ultrastructure studies of   790
   vitamin requirements of   780
   wall of   90
*Diplostauron*   *913*
  *elegans*   534, 548
*Distigma*   746, *915*
*Ditylum*   823, *913*
  *brightwellii*
   fatty acids of   249
   inorganic nutrition of   618, 659
   nitrate reductase activity in   586
   vitamin metabolism in   750, 779
*Docidium*   *915*
  *manubrium*   782
*Draparnaldia*   88, *914*
*Dumontia*   *910*
  *incrassata*
   carrageenan of   70
   storage products of   218
*Dunaliella*   *913*
   apochlorosis and   551
   effect of light on   407
   extracellular products of   840, 841, 843
   osmoregulation in   225, 728, 729
   polyols in   221
   production of glycerol in   9
   synchronous cultures of   896

  *primolecta*
   fatty acids of   252
   heterotrophic growth of   537
   vitamin requirements of   781
  *salina*
   carotenoids of   181
   heterotrophic growth of   537
   vitamin requirements of   781
  *tertiolecta*
   extracellular products of   758, 839,
    845, 848
   fatty acids of   252
   inhibition of, by algal products   847
   inorganic nutrition of   584, 624
   nitrogen metabolism in   584, 587
   vitamin production by   758
  *viridis*
   heterotrophic growth of   537
   osmoregulation in   728
   vitamin requirements of   781
*Durvillea*   801, *913*

*Ecklonia*   *913*
   gibberellins from   767
   laminaran in   213
  *radiata*   213
*Ectocarpales*   *912*
   alginates of   11
   laminaran in   212
   morphogenesis in   801–2
*Ectocarpus*   *912*
   alginates of   11
   chemotaxis in   829, 830
   chloroplasts of   137
   extracellular products of   110, 814
   reproduction of   815, 816, 819, 820
   vitamin $B_{12}$ requirement of   622
  *confervoides*
   alginates of   11
   morphogenesis of   801–2
   vitamin requirements of   780
  *fasciculatus*
   effect of iodine deficiency on   625
   effect of low temperatures on   802
   effect of kinetin on   766, 770, 802
   vitamin requirements of   780
  *parasitica*   780
  *siliculosus*
   chemotaxis of gametes of   846
   morphogenesis of   802
   pheromones of   24
*Egregia*   *913*
   chloroplast of   137
   intercellular connections in   94
   mitochondrial DNA in   103
  *menziesii*   101
*Eisenia*   438, *913*
  *bicyclis*   9, 213, 220, 221

*Elachista*  819, *912*
*Endocladia*  63, *910*
Endocladiaceae  63
*Enteromorpha*  *914*
    copper toxicity of  619
    effect of growth substances on  767, 794
    fatty acids of  253
    protein formation in  110, 111
    reproduction of  817, 819, 821, 826
    sulphated polysaccharides of  14
    stimulation of growth by algal products
        847
    *compressa*
        auxin activity in  763
        fatty acids of  253
        polysaccharides of  74, 210
    *intestinalis*  269, 270
    *linza*
        production of autostimulants by  847
        sterols of  270
    *prolifera*  763, 767
*Entomosigma*  750, 780, *911*
*Eremosphaera*  885, *914*
    *viridis*  782
*Erythrocystis*  25, *911*
*Erythrotrichia*  *911*
    *carnea*  778
*Escherischia*
    *coli*
        effect of streptomycin on  328
        gene arrangement in  336
        nucleic acid metabolism in  284–6
        -type specificity for vitamins  749, 750
*Eucheuma*  63, *910*
    *spinosum*  68
*Eudorina*  *913*
    cell morphology of  791
    DNA repair in  303
    nuclear and cytoplasmic inheritance in
        316, 319
    photokinesis in  873
    *california*  847
    *elegans*
        heterotrophic growth of  537
        vitamin requirements of  781
*Euglena*  *915*
    absorption spectra of  356, 358
    chloroplasts of  125, 137, 142, 148,
        356, 358
    effects of physico-chemical factors on
        718, 720, 723, 725, 727, 875–83
    enzymes of  20, 91, 438, 457, 508, 511,
        642
    extracellular products of  841, 842
    fatty acid metabolism in  255, 448, 545
    glycolate metabolism in  478, 480, 481,
        488, 494, 899
    growth and morphology of  90, 718,
        720, 723, 790

    heterotrophic growth of  540, 716
    inorganic nutrition of  258, 612, 613,
        615, 617, 620, 621
    movement in  869, 872, 875, 879, 881–
        883
    mutagenesis in  325
    nucleic acid metabolism in  21, 285–
        290, 303
    pigments of  167–9, 187, 551, 727
    phosphorus metabolism in  91, 639,
        640, 642
    photosynthesis in  407, 438
    protein synthesis in  317, 600
    respiration in  476–94, 506–20, 725
    sterols of  276–8
    storage products of  215–17, 223
    synchronous cultures of  895, 896, 899–
        901
    vitamin metabolism of  742, 752
    *anabaenae*  537
    *anabaenae* var. *minor*  781
    *deses*
        heterotrophic growth of  537
        metabolic movements in  869
    *gracilis*
        acid phosphatase in  641
        active transport in  540
        chlorophyll synthesis in  167, 169, 170,
            171
        effect of IAA on  764
        evolution of  746
        extracellular products of  842, 851
        fatty acid metabolism in  251, 253–5,
            258
        heterotrophic growth of  531–4, 540,
            542, 547, 551
        hydrogenase in  457
        inorganic nutrition in  615–22
        lysine in  20
        movement in  865, 877
        nitrogen sources for  595
        nucleic acid metabolism in  98, 103,
            287
        photorespiration in  484
        photosystems in  366
        protein synthesis in  600
        ribosomes of  289
        sterols in  276, 277
        storage products in  219
        vitamin metabolism in  622, 742, 747
            749–51, 842
    *gracilis* var. *bacillaris*
        enzymes of  215, 507, 523
        cytochromes in  616
        heterotrophic growth of  533, 536,
            540, 542, 548
        respiration in  507, 509–12, 517–18, 523
        vitamin requirements of  622, 781
    *gracilis* var. *saccharophila*  533, 536

*Euglena—cont.*
  *gracilis* var. *typica* 781
  *gracilis* var. *urophora*
    heterotrophic growth of 533, 534
    vitamin requirements of 781
  *granulata* 91
  *heliorubescens* 187
  *klebsii*
    heterotrophic growth of 537
    vitamin requirements of 781
  *mutabilis* 781
  *pisciformis*
    heterotrophic growth of 537
    vitamin requirements of 781
  *spirogyra* 781
  *stellata*
    heterotrophic growth of 537
    vitamin requirements of 781
  *viridis*
    heterotrophic growth of 537
    vitamin requirements of 750, 781
Euglenales *915*
Euglenophyceae *909, 915*
  auxotrophy in 744, 745
  carotenoids of 4–6, 152, 154, 187
  chemotaxis in 882
  chemotrophy in 531
  chlorophyll of 4, 154, 161–4, 358
  chloroplasts of 126, 129, 131, 133
  extracellular products of 841
  flagella of 87
  hydrogenase in 457
  lipids and fatty acids of 16, 251–3
  measurement of ATP in 638
  movement in 865, 870, 872, 881, 882
  organic nitrogen sources for 595
  paramylon formation in 215
  pellicle of 90
  phagotrophy in 538
  phylogeny of 261
  polysaccharide storage in 145, 146
  pyrenoid of 145
  sterols of 276
  thermotaxis in 882
  thylakoids of 142, 154
  vitamin metabolism in 750, 781
Eustigmatophyceae *909, 912, 916*
  carotenoids of 152, 154, 183–5
  chlorophyll of 154, 161–4
  chloroplasts of 131
  thylakoid arrangement in 142, 154
  vitamin requirements of 780
Eutreptiales *915*
*Eutreptiella* 750, 781, *915*
*Exuviaella* *911*
  effect of growth substances on 764, 767
  extracellular products of 841, 843
  vitamin metabolism in 750, 780

  *cassubica* 750, 780
*Falkenbergia* 23, *910, 916*
*Favia* 181
  *pallida* 379
*Flexibacter* 192
Florideophycidae
  biliproteins of 356
  enzymes of 20
  intercellullar plugs in 14
  thallus construction in 804
*Fragilaria* 750, 779, *913*
  *capucina*
    temperature requirements of 721
    vitamin requirements of 778
  *crotonensis*
    respiration of 509
    temperature requirements of 721
  *pinnata* 749
  (*pinnata?*) 750, 778, 779
  (*rotundissima?*) 779
  *sublinearis* 722
*Franceia* 27, *914*
*Fremyella* 18, *910*
  *diplosiphon*
    absorption spectra of 195, 199, 200
    pigments of 196–8
    sterols of 278
*Fritschiella* 794, *914*
Fucaceae 73
Fucales *913*
  alginates of 11
  laminaran in 212
  morphogenesis of 799–801
*Fucus* *913*
  carbon dioxide compensation point of 492
  carotenoids of 185
  chemotaxis in 829, 830, 846
  chloroplasts of 125, 142
  control of egg liberation in 816, 819, 824, 825
  extracellular products of 840, 841
  growth factors of 763, 769, 770
  growth and development of 799–800
  ion uptake in 707
  lipids of 257
  lunar rhythms in 826
  organelles of 88, 96
  osmoregulation in 225
  polysaccharide production in 58, 107
  storage products in 213, 217, 220, 221, 225
  *ceranoides* 221
  *evanescens*
    intercellular connections in 94
    sterols of 274
  *platycarpus* 250
  *serratus*
    alginates of 11

chemotaxis in 829
fatty acids of 250
storage products of 213, 221
*spiralis*
  chemotaxis in 829
  cyclitols in 224
  gibberellins in 767
*vesiculosus*
  alginates of 11
  cellulose in 13
  chemotaxis of 829
  extracellular products of 844, 847
  fatty acids of 250
  growth and development of 799–800
  growth substances and 765, 767
  nuclear division in 100
  storage products of 72, 73, 214, 221
*virsoides* 219–21
*Furcellaria* 63, *910*
  *fastigiata*
    auxin activity in 763, 764
Furcellariaceae 63
  carrageenans of 69
  growth rates of 804
  sterols of 272
  storage products of 218, 224
*Furcilla* 722, *913*
  *stigmatophora* 534, 548


*Galaxaura* *910*
  *fastigiata* 8
Gelidiaceae 8, 63
Gelidiales 8
*Gelidium* *910*
  as source of agar 64
  galactans of 63
  *amansii*
    agars of 12, 64, 66
    cellulose in 14
    sterols of 272
  *cartilagineum* 224
  *japonicum*
    agaroses of 12
    sterols of 272
  *subcostatum*
    agaroses of 12
    sterols of 272
*Giffordia* 142, *912*
*Gigartina* 63, *910*
  *stellata*
    carrageenans of 71
    sterols of 272
    storage products of 218, 222
Gigartinales 8, 272, *910*
Gigartinaceae 63
*Glaucocystis* 915, *916*
  cell wall of 44, 45
  contractile vacuoles in 92

taxonomy of 26
*nostochinearum* 192
*Glenodinium* 881, *911*
  *foliaceum*
    carotenoids of 187
    nucleus of 98
    pigments of 154
    vitamin requirements of 780
  *halli* 780
  *monotis* 780
*Gloeocapsa* *910*
  fatty acids and lipids of 243, 258
  hydrogenase, absence in 457
  nitrogen fixation by 567
  *alpicola* 521
  *dimidiata* 778
  *membranina* 778
  *polydermatica* 537
*Gloeochrysis* 912, *916*
  *maritima* 274
*Gloeocystis* *913*
  *gigas*
    inorganic N metabolism in 584
    vitamin requirements of 782
  *maxima* 584
*Gloeomonas* *913*
  enzymes in 485
  extracellular products of 843
  glycolate metabolism in 900
  glyoxysomes in 485
*Gloeotrichia* 564, *910*
*Gloiopeltis* *910*
  effect of exposure on spore release in
    825
  endogenous rhythms in 826
  galactans of 63
  *furcata*
    funoran of 65
    spore release in 825
    sterols of 272
  *tenax* 825
*Golenkinia* *914*
  chemotaxis of sperm of 846
  control of reproduction in 816, 821–
    823
  effect of acetate on chlorophyll
    synthesis in 169
  flagella of 87
  pH range of 727
*Gomphonema* *913*
  movement in 871
  thylakoid arrangement in 142
  *olivaceum* 14, 75, 840
  *parvulum* 657
*Gonimophyllum* *911*
  *skottsbergii* 101
Goniotrichales *911*
*Goniotrichum* *911*
  effect of growth factors on 770, 771

*Goniotrichum—cont.*
    vitamin requirements of  622
  *alsidii (elegans)*
    effect of axenic conditions on  770
    vitamin metabolism in  749, 778
*Gonium  913*
    glyoxysomes in  485
    reproduction in  824
  *multicoccum*
    heterotrophic growth of  537
    vitamin requirements of  782
  *octonarium*  534
  *pectorale*
    heterotrophic growth of  537
    inorganic nutrition of  614, 616
    reproduction in  819
    sexual compatability in  24
    vitamin requirements of  782
  *quadratum*
    isocitrate lyase activity in  508
    heterotrophic growth of  534
  *sacculiferum*  537, 548
  *sociale*  537
*Gonyaulax*  141, *911*
  *catenella*  848
  *polyedra*
    bioluminescent organelles in  523
    fatty acids of  248
    gibberellins and  767
    heterotrophic growth of  537
    nitrate reductase in  587
    vitamin metabolism in  757, 760, 780
  *tamerensis*  845
*Gonyostomum*  *915, 916*
  *semen*  163
*Gossypium*  53
*Gracilaria*  *910*
    as source of agar  64
    galactans of  63
    translocation in  94
  *edulis*  183
  *lichenoides*  183
  *sjoestedtii*  183
  *verrucosa*
    agaroses of  12
    sterols of  272
*Gracilariaceae*  63
*Grateloupia*  63, *910*
  *elliptica*
    porphyrans of  66
    sterols of  272
*Grateloupiaceae*  63, 69
*Griffithsia*  *911*
    ion uptake by  697–701, 703, 719
    nucleic acids of  22, 301, 303
  *flabelliformis*  697–701
  *monile*  697–701
  *pacifica*  14, 94
  *pulvinata*  697–701

  *setacea*  700
  *teges*  697–701
*Grinnellia*  *911*
  *americana*
    effect of axenic conditions on  770
    sterols of  273
*Gymnodinium*  *911*
    carotenoids of  187
    extracellular products of  840, 841, 843
    thylakoid arrangement in  141
  *breve*
    effect of growth substances on  764, 767
    heterotrophic growth of  531, 537
    toxin production by  848
    vitamin requirements of  780
  *foliacium*  537
  *splendens*
    effect on growth substances on  764, 766, 767
    vitamin requirements of  780
  *veneficum*
    pigments of  154, 187
    toxin production by  848
*Gymnogongrus*  63, *910*
*Gyrodinium*  551, *911*
  *californicum*  750, 780
  *cohnii*
    fatty acids of  248
    heterotrophic growth of  536, 543
    in assay for thiamine  754
    nuclear division in  97
  *resplendens*  750, 780
  *uncatenum*  750, 780

*Haematococcus*  *913*
    carotenoids of  183
    inorganic N metabolism in  584
    thermotaxis in  882
  *buetschii*  782
  *capensis*  782
  *capensis* var. *borealis*  782
  *droebakensis*  782
  *pluvialis*
    apochlorosis in  551
    carotenoids of  181
    effect of pH on excretion by  850
    heterotrophic growth of  534, 536, 537, 539, 542, 544, 548, 550
    vitamin requirements of  781, 782
  *zimbabwiensis*  782
*Halicoryne*  13, *914*
*Halicystis*  *914*
    cell wall of  15
    ion uptake in  705
    inorganic nutrition of  612
  *osterhoutii*  705

*ovalis*
    carotenoids of 180
    ion uptake in 698, 705
*Halidrys* *913*
    flagella of 87
    growth and development of 799–800
    *siliquosa* 11
*Halimeda* *914*
    calcification in 668, 669, 793
    carotenoids of 6
    endogenous rhythms in 827
    xylan of 13
    *discoidea* 181
    *opuntia* 181
    *tuna* 181
*Halobacterium* 728
*Halosaccion* *911*
    *glandiforma* 183
    *ramentaceum* 272
Halosphaerales *913*
*Hantzschia* 657, *913*
*Hapalosiphon* *910*
    *fontinalis* var. *globosus* 621
    *laminosus*
        fatty acids of 242
        taxonomy of 29
Haptophyceae *909*, *911*
    ATP measurements in 638
    carotenoids of 4–6, 152, 153, 185
    chemotrophy in 531
    chlorophyll of 153, 161–4, 358
    chloroplasts of 131, 133 137
    extracellular products of 843
    fatty acids of 245–8
    haptonemata of 89
    phylogeny of 261
    pyrenoid of 145, 146
    scales on 90
    sterols of 274
    thylakoids of 142
    ultrastructure of 790
    urease in 597
    vitamin metabolism in 745, 750, 780
*Hartmannella*
    *rhysodes* 251
*Hemiselmis* *911*
    thylakoids of 141
    trichocysts of 89
    *brunnescens* 247
    *rufescens* 247
    *virescens*
        fatty acids of 247
        heterotrophic growth of 537
        vitamin metabolism in 750, 779
    *viridis* 187
Heterochloridales *912*
*Heterococcus* 88, *912*
*Heteromastix* *913*
    cartoenoids of 181

surface structures of 89, 90
    *rotunda* 251, 253
Heteronematales *915*
*Heterosigma* *911*
    *akashivo* 750, 780
    *inlandica* 750, 780
*Hexamitus* 881
*Himanthalia* *913*
    flagellar spines in 88
    growth and development of 779–801
    pores in walls of 94
    *elongata* 11
    *lorea* 14, 73
*Hizikia* 826, *913*
*Holmsella* 94, *910*
*Hormidium* 885, *914*
    *flaccidum* 458
*Hormophysa* 801, *913*
*Hormosira* 800, *913*
*Hormothamnion* *910*
    *enteromorphoides* 190
*Hormotila* 843, *913*
    *blennista* 847
*Hyalococcus* 221, *914*
*Hyalogonium* *913*
    *klebsii* 535
*Hyalotheca* *915*
    *dissiliens* 782
*Hydra* 25
*Hydrocoleum* 190, *910*
*Hydrodictyon* *914*
    effects of light on 692, 815
    effect of phosphorus deficiency on 823
    ion uptake in 687–92
    nuclear division in 99
    phosphorylase activity in 210, 212
    *africanum* 683–7, 719
    *reticulatum* 270
*Hydrurus* 509, *912*
    *foetidus* 214
*Hymenomonas* *911*
    acid phosphatase in 91
    calcification in 105, 663, 665–7
    photosynthesis in 439
    vitamin requirements of 780
    *carterae*
        calcification of 662–7
        life cycle of 791
        vitamin requirements of 780
    *huxleyii* 185
    *roseola* 662, 663
*Hypnea* *910*
    galactans of 63
    hyrdogenase, absence in 458
    *japonica* 272, 273
    *musciformis* 767
    *specifera* 69
Hypneaceae 63

*Iridaea*   63, *910*
  *laminaroides*
    cellulose in   14
    sterols of   272
*Iridophycus*   *910*
  *cornucopiae*   272
  *flaccidum*
    osmoregulation in   225
    storage products of   218
*Ishige*   *912*
  *okamurai*   9, 213
Isochrysidales   *911*
*Isochrysis*   *911*
    carotenoids of   185
    extracellular products of   843
    vitamin B$_{12}$ binding factor from   752
  *galbana*
    carotenoids of   185
    extracellular products of   839, 847
    fatty acids of   246
    heterotrophic growth of   537
    storage products of   214
    vitamin metabolism in   750, 751, 780


*Janczewskia*   *911*
  *gardneri*   100
*Joculator*   *910*
  *maximus*   207


*Katodinium*   840, *911*
  *dorsalisulcum*   839, 851
*Keratococcus*   466, *914*
  *bicaudatus*   458
*Kirchneriella*   *914*
  *lunaris*   99
*Klebsiella*
  *pneumoniae*   563
*Klebsormidium*   *914*
    catalase in   91
    nuclear division in   100
*Koliella*   *914*
  *tatrae*   722


*Lactobacillus*   749, 750
  *leichmannii*   749
*Laingia*   *911*
  *pacifica*   66
*Laminaria*   *913*
    absence of IAA from   763
    alginic acids of   58
    effect of pollutants on   798
    extracellular products of   840, 841,
      844
    polysaccharides of   51, 107-8, 213
    pyrenoids of   145
    reproduction in   815, 821

    thylakoids of   142
  *aghardhii*   763
  *cloustonii*
    auxin activity in   763
    cyclitols in   224
  *cucullata*   796
  *digitata*
    alginates of   11
    auxin activity in   763
    cellulose in   13
    cytokinins in   766
    growth and development of   795-8
    laminaran in   213
    sterols of   275
    translocation in   95
    zinc uptake in   620
  *faeroensis*   275
  *hyperborea*
    alginates of   11
    cellulose in   13
    growth and development in   794-8
    laminaran in   212
    reproduction in   820
    secretory cell of   108
    translocation in   95
  *longicruris*   818
  *saccharina*
    alginates of   11
    growth and development in   795-8
    laminaran in   213
    reproduction in   817, 820
    secretory cells in   108, 109
    translocation in   95
Laminariales   *913*
    alginates of   11
    effect of temperature on reproduction
      of   821
    laminaran in   212
    morphogenesis in   794-8
    sieve tubes of   94
    trumpet hyphae in   95
*Lamprothamnium*   *915*
  *succinctum*   684, 685
*Lauderia*   *913*
  *borealis*   249
*Laurencia*   *911*
    biochemical recognition in   25
    galactans of   63
    halogenated compounds in   23
    intercellular connections in   92
    thylakoid arrangement in   141
  *pinnatifida*
    porphyrans of   65
    sterols of   273
    xylan of   57
  *spectabilis*   208
*Lemanea*   *910*
  *fluviatilis*   218, 222
  *nodosa*   218

*Lenormandia* 911
  *prolifera* 183
*Lepocinclis* 873, *915*
*Leptogium*
  *lichenoides* 575
*Leucothrix* 915
  *mucor* 521
*Levringiella* 912
  *gardneri* 100
*Lichina* 575
*Licmophora* 660, 669, *913*
  *hyalina* 779
*Lithodesmium* 913
  flagella of 87
  thylakoid arrangement in 142
  *undulatum* 98
*Lithothamnion* 218, 222, *911*
*Litosiphon* 622, *912*
  *pusillus*
    effect of kinetin on 802
    vitamin requirements of 780
*Lobomonas* 913
  *pyriformis* 781
  *rostrata* 782
  *sphaerica* 782
*Lomentaria* 911
  chloroplast development in 148
  intercellular connections in 92
  thylakoid arrangement in 141
*Lyngbya* 910
  acetylene reduction by 568
  DNA renaturation in 284
  extracellular products of 841
  fatty acids of 243
  heterotrophic growth of 535, 537, 548
  morphology of 789
  taxonomy of 29, 30
  *aestuarii*
    heterotrophic growth of 537
    vitamin requirements of 778
  *confervoides* 8
  *lagerheimii*
    fatty acids of 242
    heterotrophic growth of 537
    vitamin requirements of 778

*Macrocystis* 913
  growth of 796, 797
  inositols of 23
  mannitol in 220
  sieve tubes in 94, 95
  *pyrifera*
    growth substances and 763, 767
    storage products in 221, 224
*Mallomonas* 912
  *epithalattia* 537
Mastigocladaceae 191

*Mastigocladus* 910
  carotenoids of 191
  nitrogen fixation by 516
  *laminosus*
    fatty acids of 242
    nitrogen fixation by 561, 573
    phycocyanins of 197
    taxonomy of 29
*Mastogloia* 792, *913*
*Melosira* 913
  reproduction in 821, 824
  vitamin requirements of 779
  *juergensii* 779
  *lineata* 537
  *nummuloides*
    heterotrophic growth of 535, 537
    vitamin requirements of 747, 778
*Membranoptera* 96, *911*
  *platyphylla* 100
Merillosphaera 829
*Merismopedia* 910
  *punctata* 191
*Meristotheca* 910
  *papulosa* 273
*Mesogloia* 11, *912*
  *vermiculata* 11
*Mesostigma* 89, 90, 105, *913*
Mesotaeniales *915*
*Mesotaenium* 885, 886, *915*
*Micrasterias* 915
  catalase in 91
  effect of ultraviolet irradiation on 825
  Golgi bodies of 103, 104, 112, 113
  microbodies in 484
  movement of 880
  nuclear pores in 96
  *cruxmelitensis* 781
  *denticulata* 865
*Microchaete* 910
  fatty acids of 244
  heterotrophic growth of 537, 548
*Micrococcus*
  *radiodurans* 284
*Microcoleus* 910
  *chthonoplastes*
    heterotrophic growth of 537
    vitamin requirements of 778
  *paludosus* 190
  *tenerrimus* 778
  *vaginatus*
    carotenoids of 190
    morphology of 789
  *vaginatus* var. *fuscus* 778
*Microcystis* 910
  extracellular products of 841
  fatty acids of 243
  inorganic nutrition of 616, 618
  morphology of 789

*Microcystis—cont.*
  *aeruginosa*
    carotenoids of  191
    effect of boron on  623
    toxins from  848
    vitamin requirements of  778
*Microdictyon*  *914*
  *setchellianum*  180
*Microglena*  *912*
  *arenicola*
    heterotrophic growth of  537
    vitamin metabolism in  750, 779
*Micromonas*  *913*
  mitochondria of  101
  polyols in  221
  scales of  90
*Microspora*  *914*
  lack of plasmodesmata in  95
  mitosis in  100
Microsporales  *914*
Mischococcales  *912*
*Monas*  881, *912*
*Monochrysis*  *912*
  extracellular products of  20, 843
  vitamin metabolism of  752, 762, 842
  *lutheri*
    extracellular products of  839, 851
    fatty acids of  246
    heterotrophic growth of  537
    osmoregulation in  225, 728
    storage products in  220, 221, 223–5
    vitamin metabolism in  750, 751, 754, 755, 760, 761, 779
*Monodus*  549, *912*
  *subterraneus*
    cell wall polysaccharides of  75
    extracellular enzymes from  845
    fatty acid metabolism in  250, 254
    heterotrophic growth of  537
    storage products in  213, 219, 222
    vitamin requirements of  780
*Monostroma*  *914*
  effect of axenic conditions on  769
  effect of growth substances on  794, 847
  effect of light quality on reproduction of  817
  *nitidum*  270
  *oxyspermum*  769
*Mougeotia*  625, 884–7, *915*
*Myrionema*  802, *912*
*Myrmecia*  221, *914*
*Myxosarcina*  *915*
  *chroococcoides*  242, 259


*Nannochloris*  *914*
  phototrophy in  531
  storage products of  220, 221

  vitamin requirements of  782
  *atomus*  781
  *oculata*
    effect of kinetin on  766
    heterotrophic growth of  537, 539
    vitamin metabolism in  758, 781
*Nautococcus*  *913*
  *pyriformis*  781
*Navicula*  *913*
  movement in  871, 881, 882
  synchronous culture of  896
  *corymbosa*  778
  *incerta*
    heterotrophic growth of  535
    vitamin requirements of  778
  *menisculus*
    heterotrophic growth of  537
    vitamin requirements of  778
  *pelliculosa*
    carotenoids of  185
    fatty acids of  249
    heterotrophic growth of  535, 544, 548
    polysaccharides of  75, 851
    respiration of  513
    silicification of  513, 623, 657–61
    sterols of  276
    vitamin requirements of  778, 779
  *radiosa*  882
Nemaliales  8, 272, 662, *910*
*Nemalion*  622, *910*
  *helminthoides*  770, 778
  *multifidum*
    auxin activity in  763
    carotenoids of  183
    vitamin requirements of  778
*Neochloris*  535, *914*
  *alveolaris*
    heterotrophic growth of  536
    vitamin requirements of  781
  *aquatica*  781
  *gelatinosa*  781
  *minuta*  781
  *pseudoalveolaris*  781
*Neomeris*  53, *914*
  *annulata*  180
*Nereocystis*  *913*
  growth of  797
  sieve tubes in  94, 95
*Netrium*  821, *915*
*Neurospora*  320, 335
*Nitella*  *915*
  carbon dioxide compensation point in  492
  cell wall of  43, 46
  chloroplast division in  148
  cytoplasmic streaming in  883
  DNA of  301
  inorganic nutrition of  614
  ion transport in  687–8, 691–4, 697

membrane potential of 692, 693, 724
microtubules in 97
plasmodesmata in 95
proteins of 19
*clavata* 683, 685, 689–691
*flexilis*
ion uptake in 684, 685
sterols of 278
*opaca* 278
*syncarpa* 181
*translucens*
ion transport in 683–91, 694
phosphorus uptake in 640, 642
photoassimilation in 548
polysaccharides of 75, 210
*Nitellopsis* 915
*obtusa* 683, 685
*Nitophyllum* 911
effect of light quality on reproduction of 817
endogenous rhythms of 826
*Nitzschia* 913
apochlorosis and 551
extracellular products of 840, 847
hydrogenase, absence in 458
movements in 871, 881, 882
thylakoid arrangement in 142
vitamin metabolism in 750
*actinastroides* 722
*alba*
calcification and silicification in 657–660
heterotrophic growth of 534, 536
taxonomy of 25
vitamin requirements of 779
*angularis*
fatty acids of 249
heterotrophic growth of 535
*angularis* var. *affinis* 778
*closterium*
carotenoids of 185
heterotrophic growth of 536, 537, 542
phosphorus metabolism in 639
sterols of 276
vitamin requirements of 746, 779
*closterium* forma *minutissima* 531
*communis* 875, 877, 878
*curvilineata*
heterotrophic growth of 534, 535
vitamin requirements of 778
*filiformis*
extracellular enzymes from 845
heterotrophic growth of 535
vitamin requirements of 778
*frustulum*
extracellular enzymes from 845
heterotrophic growth of 535–7
vitamin metabolism of 750, 778, 780

*hybridaeformis*
heterotrophic growth of 537
vitamin requirements of 778
*laevis*
heterotrophic growth of 535, 536
vitamin requirements of 778
*lanceolata* 778
*leucosigma*
heterotrophic growth of 534, 536
vitamin requirements in 779
*marginata*
heterotrophic growth of 535, 542
vitamin requirements of 778
*obtusa* 537
*obtusa* var. *scalpelliformis* 778
*ovalis*
heterotrophic growth of 537
*palea*
effect of copper on 619
movement of 865
*punctata*
heterotrophic growth of 535
vitamin metabolism in 750, 779
*putrida*
extracellular enzymes from 845
heterotrophic growth of 534, 536
thigmotaxis in 883
vitamin requirements of 779
*seriata* 779
*tennuissima* 536
*thermalis*
fatty acids of 249
heterotrophic growth of 535
*Nizymenia* 804, *910*
*Nostoc* *910*
anaerobic growth of 467
carotenoids of 190
citrulline production by 564
DNA hybridization in 285
effect of IAA on 764
extracellular products of 840, 841, 853
ferredoxin in 617
polysaccharides of 75
storage products of 218
symbioses with 575
temperature requirements of 722
*commune*
carotenoids of 190, 191
life cycle of 790
nitrogen fixation by 573
storage products in 218, 221
*entophytum*
nitrogen fixation by 572
vitamin requirements of 778
*muscorum*
absorption spectra of 355
carbamoyl phosphate synthetase in 594
DNA hybridization in 285

*Nostoc*
  *muscorum—cont.*
    fatty acids of 242
    α-granules of 209
    growth and development of 790
    heterotrophic growth of 535, 537, 543, 548
    hydrogenase in 457
    inorganic nutrition of 623, 624
    lyophilization of 722
    nitrogen fixation in 573
    pigments of 190, 192, 197
    vitamin requirements of 778
  *punctiforme*
    effect of IAA on 764
    heterotrophic growth of 535
    nitrogen fixation in 573
    phycocyanin in 197
  *sphaericum* 521
Nostocaceae
  carotenoids of 190, 870, 879
Nostocales *910, 915*
  fatty acids of 19
  taxonomy of 30


Ochromonadales *912*
*Ochromonas* *912*
  acid phosphatase in 91
  chloroplast of 133, 142
  contractile vacuoles in 91
  extracellular products of 841, 842
  flagella of 88, 96, 104
  mitochondrial ribosomes in 103
  mitosis in 99
  osmoadaptation in 728
  phagotrophy in 538
  pigments of 153, 163–4, 167, 358
  respiration in 506
  storage products of 214
  sulpholipids in 240
  vitamin $B_{12}$ assay with 622, 742
  *danica*
    carotenoids of 185
    chloroplast of 133, 304
    extracellular products of 753, 844–7
    fatty acid metabolism in 246, 248, 254, 257
    heterotrophic growth of 535, 538, 548
    methionine metabolism in 599
    sterols of 274
    sulpholipids in 612
    vitamin metabolism in 747, 757, 779
  *malhamensis*
    anaerobic growth of 467
    auxin activity in 763
    enzyme distribution in 515
    fatty acid metabolism in 246, 248, 508

    heterotrophic growth of 531, 535, 538, 548, 551
    osmoregulation in 225
    propionate metabolism in 448
    sterols of 274
    storage products of 214, 215, 218, 220, 221
    sulpholipids in 612
    urea production by 597
    vitamin metabolism in 622, 747–52, 757, 761, 779
  *minuta* 779
  *sociabilis* 274
  *variabilis* 845
*Ochrosphaera* *911*
  *neapolitana* 780
Oedogoniales *914*
*Oedogonium* *914*
  cell division in 794
  Golgi bodies of 113
  mitochondria of 101
  nuclear division in 99, 100
  organic substances from 824
  plasmodesmata of 95
  pyrenoid multiplication in 146
  reproduction in 817, 819, 821, 822, 826, 827, 829
  rhythms in 819, 826
  sterols of 270
  wall rupture in 104
  *cardiacum*
    chemotaxis of 829
    chloroplast of 146
  *geniculatum* 829
*Olisthodiscus* 839, 840, 843, 848, *912*
  *luteus*
    flagella of 88
    production of inhibitor by 847, 854
*Oocystis* *914*
  cell wall of 44, 45, 48
  extracellular products of 840
  inorganic nutrition of 621
  *marssonii* 458
  *naegelii* 535
  *polymorpha* 269
*Opephora* 779, *913*
*Ophiocytium* *912*
  chlorophyll of 163
  thylakoid arrangement in 142
*Oscillatoria* *910*
  auxin activity in 763
  carotenoids of 191
  ecological distribution of 717
  extracellular products of 841, 843
  fatty acids of 243
  heterotrophy of 535, 537
  hydrogenase, absence in 457
  inorganic nutrition of 624

lipids of 19
movement in 881, 882
nitrogen fixation in 568
storage products of 209
taxonomy of 30
*agardhii* 190, 191, 197, 198
*amoena*
carotenoids of 190
storage products of 208
*amphibia*
heterotrophic growth of 537
storage products of 218, 221
vitamin requirements of 778
*brevis* 778
*chalybia* 208
*irrigua* 8
*limosa*
inorganic nutrition of 614
pigments of 7, 190
*lutea* 778
*lutea* var. *auxotrophica* 778
*lutea* var. *contorta* 778
*princeps*
heterotrophic growth of 537
movement in 870
phycoerythrins of 8
polysaccharides of 209
*rubescens*
carotenoids of 184, 190, 192
light adaptation in 717
vitamin requirements of 778
*subtilissima*
heterotrophic growth of 537
vitamin requirements of 778
*tenuis* 190
*terebriformis* 7
*williamsii* 242
Oscillatoriaceae
carotenoids of 190
morphological plasticity in 789
movement in 870, 879
*Ostreobium* 914
chromatic adaptation in 379
pigments of 6, 181
*Oxyrrhis* 538, 911
*marina*
heterotrophic growth of 533, 534,
538, 543, 544
vitamin requirement of 544, 780

*Pachymenia* 63, 911
*carnosa* 69, 70
*Padina* 885, 913
*pavonia* 14, 73
*Palmella* 840, 913
*Pandorina* 913
chemotaxis in 881
reproduction in 821

*morum*
antibacterial substances from 847
heterotrophic growth of 537
inorganic nutrition of 614
syngen pairs in 24
vitamin requirements of 781
*unicocca* 781
*Paraphysomonas* 103, 912
*Pascherella* 912
*tetras* 537
*Pavlova* 912
*gyrans* 779
*mesolychnon* 215
*Pediastrum* 914
reproduction in 821, 825
silicon in 656
Pedinomonadales 913
*Peltigera* 575
*aphthosa* 575
*canina* 575
*polydactyla* 575, 853
*rufescens* 575
*Pelvetia* 913
growth and development of 799, 800
polysaccharide synthesis in 107, 108
reproduction in 817, 825–7
*canaliculata*
alginates of 11
cellulose of 13
mannitol in 220
sterols of 275
*wrightii* 72, 73
*Penicillus* 13, 43, 914
*capitatus* 180
*dumetosus* 668
*Peranema* 538, 915
*trichophorum* 781
Peridiniales 662, 911
*Peridinium* 911
extracellular products of 841, 843
thylakoid arrangement in 141
*balticum* 780
*chattoni* 780
*cinctum* var. *ovoplanum* 780
*hangoei* 750, 780
*polonicum* 847, 848
*triquetrum* 248
*trochoideum*
calcite in 662
fatty acids of 248
vitamin requirements of 780
*westii* 146
*Petalonia* 819, 823, 912
*fascia* 802
*zoostericola* 780
*Phacus* 915
chloroplasts of 138, 142
movement in 873, 881, 882
*pyrum* 750, 781

*Phaeocystis*   23, *911*
  *pouchetii*   839, 847, 850
*Phaeodactylum*   *913*
  cell wall of   669
  chlorophyll of   368
  fractionation of   366
  *nutum*   249
  *tricornutum*
    cell wall of   657
    chlorophylls of   366, 368
    extracellular products of   840, 845
    growth factors and   766
    heterotrophic growth of   531, 537
    phosphorus metabolism in   639
    polysaccharides in   75
    storage products of   214, 218, 224
    vitamin metabolism in   751, 752, 758,
      778
Phaeophyceae   *909, 912*
  alginic acid in   58
  aragonite in   662
  auxotrophy in   743–5
  carotenoids of   4–6, 152–3, 185,
    361
  cellulose in   13–14
  chemotaxis in   829
  chemotrophy and   531
  chlorophyll of   4, 153, 161–4, 358
  chloroplasts of   131, 133, 137, 148,
    150–1, 885
  chromatic adaptation in   379
  cyclitols in   223, 224
  DNA of   298
  extracellular products of   840, 841
  eyespot of   147
  fatty acids of   250–1
  flagella of   87–8
  fucoidins in   10
  hydrocarbons of   17
  hydrogenase in   457, 458
  inositols of   23
  laminaran of   213
  lipids of   16–19
  mannitol in   220–2
  nuclear envelope of   96
  phylogeny of   261
  proteins of   20
  pyrenoid of   145
  silicification in   656
  soluble metabolites of   9
  sterols of   274
  sugars of   15
  sulphated polysaccharides of   14–15,
    72
  sulpholipids in   612
  thylakoids of   142
  vitamin metabolism in   622, 756,
    780
Phaeoplacales   *912*

*Phaeosaccion*   *912*
  *collinsii*   214
Phaeothamniales   *912*
*Phormidium*   *910*
  absorption spectra of   363
  carotenoids of   191
  nitrogen fixation in   568
  photokinesis in   874
  steroids of   18
  taxonomy of   29–30
  *ambiguum*   874–6, 881
  *autumnale*
    carotenoids of   190
    movements of   865, 878
    phycobilins of   7
  *ectocarpi*
    carotenoids of   188, 190
    phycoerythrins of   8
  *foveolarum*
    carotenoids of   188, 190
    heterotrophic growth of   535
    vitamin requirements of   778
  *luridum*
    heterotrophic growth of   537
    photosystems of   369
    pigments of   190, 197
    respiration in   521
    ribosomes of   284, 285
    vitamin requirements of   778
  *luridum* var. *olivaceae*   278
  *persicinum*
    absorption spectra of   355
    pigments of   188, 190, 191, 196
    vitamin metabolism in   750, 778
  *tenue*   778
  *uncinatum*   876
*Phycodrys*   *911*
  *rubens*   218, 222
  *sinuosa*   183
*Phyllophora*   *910*
  *membranifolia*   218, 272
Phyllophoraceae   63
*Phyllymenia*   63, *910*
  *cornea*   69
*Phymatolithon*   *911*
  *compactum*   218, 222
Phytodiniales   *911*
*Pinnularia*   *913*
  dictyosomes of   103
  movements of   871, 881, 882
*Pinus*   51
*Pithopora*   509, *914*
*Platydorina*   819, 821, 822, *913*
  *caudata*
    autoinhibitor from   847
    vitamin requirements of   782
*Platymonas*   *913*
  enzymes of   20
  extracellular products of   840

theca of 90
thylakoids of 142
*convolutae* 25
*subcordiformis* 90
*tetrathele*
  theca of 90, 105
  vitamin metabolism in 750, 782
*Plectonema 910*
  anaerobic growth of 467
  fatty acids of 244
  heterotrophic growth of 537, 548
  nitrogenase activity in 563, 568
  taxonomy of 29–30
*boryanum*
  chlorophyll synthesis in 170
  DNA of 21, 283
  nitrogen fixation by 561, 568
  phycocyanins of 197, 198
  polyphosphate bodies in 645
*nostocorum* 778
*notatum*
  heterotrophic growth of 535
  vitamin requirements of 778
*terebrans*
  fatty acids of 242
  vitamin requirements of 778
*Pleodorina 913*
  *californica* 537
  *illinoiensis* 537
*Pleurochloris 912, 916*
  *commutata* 185
*Pleurochrysis,* 664, 665, *911,*
  *scherffelii*
    cellulose synthesis in 48
    Golgi in 113
    microtubules in 97
    scales of 90, 105, 663, 664
    vitamin requirements of 780
*Pleurosigma* 871, *913*
*Pleurotaenium 915*
  *trabecula* 782
*Plocamium 910*
  *coccineum* 245
  *pacificum* 208
  *vulgare* 272
*Plumaria 911*
  *elegans*
    effect of coumarin on 770
    growth of 805–7
*Pocillopora*
  *damicornis* 854
*Polycoryne 911*
  *gardneri* 100
*Polyedriella* 142, *912, 916*
  *helvetica*
    heterotrophic growth of 537
    vitamin requirements of 780
*Polyides 911*
  *caprinus* 272

*rotundus*
  sterols of 272
  storage products of 219, 222
*Polykrikos 911*
  *schwartzii* 750, 780
*Polyneura* 492, *911*
  *hilliae* 805
*Polysiphonia 911*
  extracellular products of 841
  halogens in 23, 624
  mitochondrial DNA in 103
  reproduction in 815
  vitamin requirements of 622
*boldii* 805, 806
*brodiaei* 770
*fastigiata*
  porphyran of 66
  storage products in 222, 224
*lanosa*
  fishy odour of 23
  sterols of 273
*nigrescens* 222, 273
*subtillissima* 273
*urceolata*
  requirement for iodine in 625
  vitamin metabolism in 749, 769, 778
*Polytoma 913*
  apochlorosis and 551
  chemotaxis in 881
  heterotrophy in 532
  respiration in 506, 519–21
  starch in 211
  vitamin requirements of 742
*caudatum* 782
*obtusum*
  alterations in, with prolonged culture
    747
  heterotrophic growth of 534, 541
  storage products in 508
  vitamin requirements of 743, 781
*ocellatum*
  heterotrophic growth of 534
  vitamin requirements of 782
*uvella*
  carotenoids of 192
  effect of kinetin on 766
  fatty acid metabolism in 254
  heterotrophic growth of 532–4
  movement in 867, 881, 883
  respiration in 509, 519
  starch in 210, 211
  vitamin requirements of 743, 781
*Polytomella 913*
  apochlorosis and 551
  glyoxysomes in 485
  respiration in 506, 519–21, 617
  vitamin requirements of 742
*agilis*
  respiration in 507, 508, 512, 514, 519

*Polytomella*
  *agilis—cont.*
    starch production in   211
    synchronous culture of   895
  *caeca*
    glyoxylate cycle in   508, 545
    heterotrophic growth of   533, 534,
      539–41, 545
    microbodies in   484
    respiration in   508, 510, 511, 515, 519,
      520, 522
    vitamin requirements of   782
*Porphyra   911*
  galactans of   63
  morphology of   43, 803
  reproduction of   815, 819, 823, 824
  temperature tolerance of   722
  *laciniata*   197, 198
  *leucosticta*
    growth of   805, 806
    storage products of   219
  *linearis*   219
  *miniata*
    reproduction of   819
    storage products of   219
  *naiadum*   207
  *perforata*
    osmoregulation in   225
    storage products of   207, 219, 224
  *purpurea*   272
  *suborbiculata*   8
  *tenera*
    chloroplast DNA in   304
    effect of growth substances on   765–7
    inorganic nutrition of   612, 619, 622,
      624, 625
    morphology of   803, 806
    reproduction in   818–20
    vitamin metabolism in   622, 749, 778
  *umbilicalis*
    hydrogenase in   457
    lunar rhythms in   826
    growth of   805
    polysaccharides of   12, 53, 57, 65, 71,
      207
    reproduction of   820
    storage products of   207, 219, 224
  *yezoensis*
    photosystems of   363, 368
    sulphite metabolism in   612
*Porphyridiales*   272, *911*
*Porphyridium   911*
  anaerobic growth of   467
  extracellular products of   840, 841
  light absorption by   351, 364, 371,
    373
  mucilage of   89
  non-motility of   87
  nucleic acids of   22, 303, 899

photosynthetic electron flow in   397,
  403, 408
  storage products of   219, 223, 224
  temperature tolerance of   722
  thylakoids of   141
  *aerugineum*   183, 196
  *aeruginosa*   245
  *cruentum*
    extracellular products of   839
    fatty acid metabolism in   245, 251, 254
    heterotrophic growth of   537
    hydrogenase in   457
    movement of   865, 872, 873, 875, 879
    nitrogen sources for   595
    pigments of   8, 183, 366, 369, 371, 377
    sterols of   270, 272, 273
    storage products in   224
    thylakoids of   128
    vitamin requirements of   778
  *purpureum*   778
*Poteriochromonas   125, 912*
  *stipitata*
    fatty acids of   246
    vitamin requirements of   779
*Prasinocladus   221, 913*
  *marinus*   96
*Prasinophyceae   909, 913*
  association of enzymes and plastids in
    533
  carotenoids of   4–6, 152, 154, 178, 181
  chlorophylls of   3, 154, 161–4, 358
  extracellular products of   840, 841, 842
  eyespots in   147
  fatty acids of   251–3
  flagella of   89
  mannitol in   220–2
  nuclear membrane of   96
  phylogeny of   261
  polyols in   221
  pyrenoid of   144
  soluble metabolites in   9
  starch storage in   145, 212
  surface scales of   90
  temperature tolerance of   722
  thylakoid arrangement in   142, 154
  trichocysts of   89
  urease in   597
  vitamin metabolism in   750, 781–2
*Prasiolales   914*
*Prasiola   914*
  *crispa*   722
  *japonica*   12
*Prorocentrales   911*
*Prorocentrum   911*
  phototopotaxis in   879
  thylakoids of   141, 148
  *micans*
    carotenoids of   186
    extracellular protein from   847

fatty acids of 248
heterotrophic growth of 537
nuclear pores of 96
phototopotaxis in 878
vitamin requirements of 780
*Protococcus 914*
*viridis* 764
*Protosiphon 914*
inorganic nutrition of 614, 615
reproduction in 817, 819, 821–3
synchronous culture of 896
*botryoides*
carotenoids of 181, 183
hydrogenase, absence in 458
inorganic nutrition of 613
*Prototheca 914*
apochlorosis and 551
cytochromes in 617
effects of light on 718, 817
taxonomy of 25
*chlorelloides* 25, 26
*ciferrii* 532, 533
*zopfii*
glycogen in 210
heterotrophic growth of 535, 536
nitrogen sources for 595
respiration in 507–17, 544, 617
taxonomy of 25–6
vitamin metabolism in 544, 782
Prymnesiales *911*
*Prymnesium 911*
heterotrophic growth of 544
mitosis in 99
scales of 90, 664
storage products in 214, 217
thylakoids of 142
*parvum*
carotenoids of 185
division of 98
extracellular products of 729, 839, 848
fatty acids of 246
haptonemata of 89
heterotrophy in 531, 537, 540
scale production in 105
storage products in 218, 221, 222, 224
vitamin metabolism in 622, 750, 780
*Pseudochlorococcum* 468, *914*
*typicum* 29
*Pseudodichotomosiphon* 13, *914*
*Pseudogloeophloea 910*
intercellular connections in 92
thylakoids of 141
*Pseudopedinella* 246, *912*
*Pterocladia 910*
effect of cobalt on 621
galactans of 63
*capillacea* 621
*tenuis*
agaroses of 12

sterols of 272
*Pterosperma* 181, *913*
*Ptilota 911*
*serrata* 273
*Pylaiella* (= *Pilayella*) *912*
*littoralis*
growth factors and 763, 766, 770
vitamin requirements of 622, 780
Pyramimonadales *913*
*Pyramimonas 913*
extracellular products of 839, 840, 841, 843, 848
scales of 89, 90
storage products in 221
thylakoids of 142
*grossii* 89
*inconstans* 750, 782
*Pyrobotrys* 412, 547, *913*
*Pythium* 101

*Radiosphaera 914*
*dissecta*
heterotrophic growth of 537
vitamin requirements of 781
*Ralfsia* 802, *912*
*Raphidonema 914*
*nivale* 747, 782
*Rhabdonema* 818, 827, *913*
Rhaphidophyceae 161–4, 358, *910*
*Rhaphidium* 457, 466, *914*
Rhizochloridaceae 868
*Rhizoclonium 914*
cellulose in 13
storage products in 217
*implexum* 180
*tortuosum* 180
*Rhizosolenia* 841, 843, *913*
*fragilissima* 722
*setigera*
respiration of 725
vitamin metabolism in 750, 779
*Rhodella 911*
mucilage of 89
non-motility of 87
thylakoid arrangement in 141
ultrastructure of 102, 103
*maculata*
heterotrophic growth of 537
nuclear envelope of 96
vitamin requirements of 778
*Rhodochorton* 816, 821, 822, *910*
*floridulum*
absorption spectrum of 195
xylan of 57
*purpureum* 806
*rothii* 196
*Rhodoglossum* 63, *910*
*pulcherum* 272

2H

*Rhodomela* 23, *911*
  *confervoides* 270, 273
  *larix*
    mitotic index of 805
    porphyran of 66
    sterols of 273
    storage products of 219, 222
  *subfusca* 245
Rhodomelaceae 219
*Rhodomonas* *911*
  carotenoids of 187
  extracellular products of 839
  thylakoid arrangement in 141
  vitamin metabolism in 750
  *lens*
    fatty acids of 247
    vitamin requirements of 779
  *ovalis* 779
Rhodophyceae *909, 910*
  argaroses in 10, 12
  aragonite in 662, 668
  ATP measurements in 638
  biliproteins in 6–9, 153, 194–6, 356, 364
  calcite in 662
  carbohydrates in 12–15, 55, 63–72, 146, 208, 217, 218
  carotenoids of 5, 6, 152, 153, 183, 361
  chemotrophy and 531
  cellulose in 14, 51
  chlorophyll of 4, 161–4, 358
  chloroplasts of 126, 129, 133, 141, 885
  compared with blue-green algae 9
  cyclitols in 223, 224
  DNA of 302
  extracellular products of 840, 841
  fatty acids and lipids of 16–19, 244–245
  hydrocarbons of 17
  hydrogenase in 457–8
  inositols of 23
  ion transport in 697
  morphogenesis in 803–7
  movement of 865, 872, 885
  nitrogen sources for 595
  phylogeny of 261
  pigment adaptation in 377–9
  polyols in 220, 222
  proteins of 20
  RNA of 22, 303
  sterols of 270–4
  sulphite reduction in 612
  taurine in 612
  thylakoids in 141, 143
  vitamin metabolism in 621, 744–5, 749, 756, 778
*Rhodopseudomonas*
  *spheroides* 482
*Rhodosorus* *911*

  *marinus*
    effect of blue light on 443
    polysaccharide storage in 146
*Rhodospirillum*
  *rubrum* 486
*Rhodymenia* *911*
  cellulose of 51
  intercellular connections in 92, 93
  xylan of 57
  *palmata*
    sterols of 270, 272, 273
    storage products of 207, 219
    xylan of 56, 57
  *palmata* var. *mollis* 804
  *pertusa*
    cellulose in 14
    storage products in 207, 208
Rhodymeniales 8, 272, *911*
*Rivularia* 75, 219, *910*
  *bullata* 218, 224
Rivulariaceae 191
*Rytiphlaea* *911*
  *tinctoria* 273


*Saccorhiza* *913*
  laminaran in 213
  growth of 797–8
  *polyschides* 795–7
*Sacheria* *910*
  *fluviatilis* 219
*Salmonella* 30
*Saprochaete* 26
*Saprolegnia* 88
*Saprospira*
  *grandis*
    carotenoids of 192
    cytochromes from 521
*Sargassum* *913*
  branching patterns in 800
  carbon dioxide compensation point of 492
  hydrogenase, absence in 458
  laminaran in 213
  reproduction of 817
  *linifolium* 11
  *natans* 221
  *pallidum* 72
  *ringgoldianum* 274
  *wightii* 221
*Scenedesmus* *914*
  anaerobic growth of 467
  chlorophyll of 164, 167, 356, 359
  effects of light on 373, 720
  extracellular products of 840, 841, 843
  heterotrophic growth of 535
  hydrogen metabolism in 20, 457, 466
  inorganic nutrition of 613, 617, 618, 621, 641

nitrogen metabolism in   585, 587, 594
oxyhydrogen reaction in   438
pH range of   727
phosphorus metabolism in   412, 639,
    640, 648
photosystems of   367, 373
polymorphism in   792
pyrenoid of   146
reproduction in   818, 821
respiration in   476–81, 483, 488, 490,
    492, 496, 546
synchronous culture of   896
taxonomy of   27, 29
temperature requirements of   722–5,
    818
thylakoids of   142, 143
*acuminatus*   537, 548
*acutiformis*   537, 548
*brasiliensis*   181
*costulatus*   535, 542, 551
*dimorphus*
    heterotrophic growth of   536
    polymorphism in   792
*obliquus*
    absorption spectra of   366
    carotenoids of   178
    chlorophyll synthesis in   167, 171
    effect of growth factors on   764, 767
    electron transport in   424–31
    extracellular products of   845, 850
    fatty acid metabolism in   252, 256
    heterotrophic growth of   536, 548
    inhibition of   847
    inorganic nutrition of   611, 619, 624
    photosynthesis in   901
    sterols of   269
    vitamin requirements of   781
*obtusiusculus*   535
*quadricauda*
    effect of gibberellic acid on   767
    extracellular products of   848
    fatty acids of   252
    heterotrophic growth of   536, 539, 548
    phosphorus sources for growth of   643
    protein synthesis in   600
*Schizothrix*   192, *910*
    *calcicola*   778
*Scytonema*   910
    extracellular products of   840
    storage products in   218
    *hoffmanii*   192
Scytonemataceae   191
*Scytosiphon*   912
    flagella of   87
    growth and development of   802–3
    thylakoid arrangement of   142
    *lomentaria*
        alginates of   11
        vitamin requirements of   780

Scytosiphonales   802–3, *912*
*Selenastrum*   914
    *gracile*   457
    *minutum*   781, 782
*Serratia*
    *marinorubra*   754
*Serraticardia*   911
    *maxima*   208, 219, 222, 224
Siphonocladales   178, 180, *914*
*Siphonocladus*   13, *914*
*Sirodotia*   910
    *suecica*   8
*Skeletonema*   913
    extracellular products of   840–3
    vitamin metabolism in   750
    *costatum*
        effect of temperature on growth of
            723, 724
        fatty acids of   249
        heterotrophic growth of   537
        inhibition of   847, 854
        storage products of   214, 218
        vitamin metabolism of   621, 750, 757–
            758, 760, 761, 779, 845
    *tropicum*   721
*Smithora*   141, *911*
    *naiadum*   8, 110
Solieraceae   63
*Spermatochnus*   11, *912*
    *paradoxus*   11
*Sphacelaria*   913
    chloroplasts of   130, 137, 139, 142,
        144, 149–51
    genophore of   303
    growth factors from   769
    *bipinnata*   11
Sphacelariales   11, *913*
*Sphaerobolus*   817
*Sphaeroplea*   914
    *annulina*   180
    *cambrica*   180
Sphaeropleales   180, *914*
*Sphaleromantis*   912
    carotenoids of   185
    flagellar scales of   89
*Spirogyra*   915
    cell division in   793
    cytoplasmic streaming in   883
    fatty acids of   253
    nuclear division in   100
    nuclear movement in   884
    *setiformis*   209–11
*Spirulina*   910
    fatty acids of   19, 241, 244, 260
    heterotrophic growth of   537
    *geitleri*
        carotenoids of   5
        fatty acids of   260
    *maxima*   278

*Spirulina—cont.*
  *platensis*
    carotenoids of  6
    fatty acids of  242, 260
    pH range of  726
*Spongiochloris*  *914*
  heterotrophic growth of  535
  reproduction of  821, 822
  *excentrica*  781
  *lamellata*  781
  *spongiosis*  781
*Spongiococcum*  *914*
  cellulose in  13
  heterotrophic growth of  535
  *alabamense*  781
  *excentricum*  781
  *multinucleatum*  781
  *tetrasporum*  781
*Spongomorpha*  13, 15, *914*
  *coalita*  180
*Staurastrum*  *915*
  *gladiosum*  781
  *paradoxum*  781
  *pingue*  726-7
  *polymorphium*  792
  *sebaldii* var. *ornatum*  781
*Staurodesmus*  *915*
  *pachyrhynchus*  781
*Stauroneis*  *913*
  *amphoroides*
    heterotrophic growth of  537
    vitamin requirements of  778
*Stephanodiscus*  *913*
  cell wall of  658
  extracellular products of  840, 841, 843
  *hantzschii*  842
*Stephanoptera*  *913*
  *gracilis*
    heterotrophic growth of  537
    vitamin requirements of  781
  *pluvialis*  750
*Stephanopyxis*  819, 821, *913*
  *turris*  750, 757, 779
*Stephanosphaera*  *913*
  *pluvialis*
    heterotrophic growth of  534
    vitamin requirements of  782
*Stereocaulon*  575
*Stichochrysis*  *912*
  *immobilis*  779
*Stichococcus*  *914*
  chlorophylls of  3, 359
  storage products of  9, 220, 221
  *bacillaris*
    heterotrophic growth of  535
    hydrogenase, absence in  458
    inorganic nutrition of  620
  *cylindricus*  781, 782
  *diplosphaera*  535

*Stictosporum*  804, *910*
*Stigeoclonium*  *914*
  contractile vacuoles in  91
  flagella of  88
  nuclear division in  **100**
  plasmodesmata in  95
  reproduction in  823
  *aestivale*  781
  *farctum*  781
  *helveticum*  781
  *subsecundum*  781
  *pascheri*  781
  *tenue*  781
  *variabile*  781
Stigonematales  *910*
*Stilophora*  *912*
  *rhizodes*  221
*Streblonema*  *912*
  *anomalum*  802
*Struvea*  *914*
  carotenoids of  180
  cellulose in  13
*Suhria*  63, *910*
*Symbiodinium*  *915*
  *microadriaticum*  780
*Syncrypta*  *912*
  (= *Synochromonas*) *korschikoffii*  779
*Synechococcus*  *910*
  absorption spectra of  363
  fatty acids of  243
  photosystems of  368
  taxonomy of  30, 31
  *cedrorum*
    heterotrophic growth of  537
    nitrogen sources for  595
    taxonomy of  29–30
    vitamin requirements of  778
  *elongatus* (see also *Anacystis nidulans*)
    *915*
    carotenoids of  191
    hydrogenase in  457
  *lividus*  197
  *salina*  625
*Synechocystis*  457, *910*
*Synedra*  840, 841, *913*
  *acus*  842
  *affinis* (= *Fragilaria brevistriata*)  779
  *rotundissima*  779
*Synura*  *912*
  cell wall in  658
  chlorophyll of  163
  flagella of  88
  thylakoid arrangement in  142
  vitamin requirements of  743
  *caroliniana*  750, 779
  *petersenii*
    extracellular products of  845
    morphogenesis of  656
    sterols of  274

vitamin requirements of 779
*sphagnicola* 779
*Syracosphaera* 780, *911*
  *carterae*
    calcification in 662
    life cycle of 791

*Tabellaria* *913*
  *flocculosa* 778
*Tetracystis* *914*
  proteins of 20
  pyrenoids of 146
*Tetrahymena*
  ATPase in 867
  mitochondria of 101
  synchronous culture of 894
*Tetramitus* 881
*Tetraselmis* *913*
  gibberellins in 767
  heterotrophic growth of 535
  storage products of 221
  *carteriiformis* 537
  *chui* 535
  *subcordiformis* 535
  *tetrathele* 535, 537
*Tetraspora* 819, 821, *913*
Tetrasporales *913*
*Thalassiosira* 840, 841, 843 *913*
  *fluviatilis*
    chitin in 75, 90
    extracellular products of 840, 849
    heterotrophic growth of 537
    respiration in 725
    storage products in 220
    vitamin requirements of 779
  *nordenskioldii* 778, 779
  *pluvialis* 750
Thallochrysidales *912*
*Thallochrysis* *912*
  *litoralis* 274
*Thorea* *910*
  *okadai* 8
*Tichocarpus* *911*
  *crinitus*
    carrageenan of 70
    sterols of 272
*Tolypella* 694, *915*
  *intricata* 683–5, 688
*Tolypothrix* *910*
  apochlorosis and 552
  effect of IAA on 764
  photosynthetic enzymes in 438
  *tenuis*
    bacterial decomposition of 573
    heterocysts in 790
    heterotrophic growth of 535, 547, 548
    nitrogen fixation by 574
    pigments of 7, 191, 196, 200, 377

storage products in 218
*Trachelomonas* 881, *915*
  *abrupta* 750, 781
  *pertii* 750, 781
*Traillella* *910, 916*
  *intricata* 219, 222
*Trebouxia* *914*
  aplanospore development in 817, 824
  extracellular products of 840
  polyols in 221
  sterols of 269
  thylakoid arrangement in 142
  *erici* 645
*Trentepohlia* 221, *914*
  *aurea* 181
*Trepomonas* 881
*Tribonema* *912*
  carotenoids of 4
  chlorophyll of 163
  flagella of 88
  plasmodesmata, absence in 95
  thylakoids of 142
  *aequale*
    heterotrophic growth of 536
    vitamin requirements of 780
  *bombycinum* 164, 358
  *minus*
    chloroplast of 136
    heterotrophic growth of 533, 535
    vitamin requirements of 780
Tribonematales *912*
*Trichodesmium* 568, 569, *910*
  *erythraeum* 19, 242
*Tridacna*
  *crocea* 853–4
*Turbinaria* *913*
  branching pattern in 800
  hydrogenase, absence in 458
  *conoides* 221
*Tydemania* 662, *914*

*Udotea* 13, *914*
  *flabellum* 181
  *petiolata* 181
*Ulmus* 51
*Ulothrix* 95, 100, 885, *914*
Ulotrichales 270, *914*
*Ulva* *914*
  acid phosphatase in 104
  chloroplast movement in 885
  effect of axenic conditions on 769, 771
  effect of growth substances on 794, 847
  inorganic nutrition of 621, 623, 624
  ion transport in 104
  pigments of 181, 373
  polysaccharides of 14, 51
  reproduction in 825, 826

*Ulva—cont.*
  *fasciata*
    hydrogenase in 457
    lunar rhythms in 826
  *lactuca*
    amino acid synthesis in 593
    carotenoids of 181
    effect of growth substances on 767, 768
    hydrogenase in 457
    inorganic nutrition of 621
    maltose in 219
    polysaccharides of 74
    sterols of 269, 270
  *mutabilis* 99
  *pertusa*
    sterols of 270
    storage products in 210, 211
  *rigida* 178
Ulvales 10, *914*
*Urospora* *914*
    cellulose in 13, 15
    reproduction in 819
    storage products in 216, 217


*Vacuolaria* 88, 91, 92, 104, *915*, *916*
  *virescens* 163
*Vallisneria* 884
*Valonia* *914*
    cellulose of 13, 49, 51, 53
    cell walls of 44–6
    ion uptake in 704–5
    sulphate and 612
  *fastigiata* 180
  *macrophysa*
    carotenoids of 180
    salt influx into 705
  *utricularis* 180
  *ventricosa*
    cell wall of 45
    ion uptake in 698, 704–5
*Valoniopsis* *914*
    cellulose of 13
    ion uptake in 698, 706
  *pachynema* 180
*Vaucheria* *912*
    cellulose in 14
    chloroplast movement in 885, 886
    nuclear division in 100
    pigments of 3–4, 163
    reproduction in 815
    rhythms in 819, 826
    thylakoids of 142
  *geminata* 868
  *hamata*
    chlorophylls of 164
    sterols of 274
  *sessilis* 250

*Vaucheriales* *912*
*Vischeria* 142, *912*, *916*
*Vitreoscilla* 521, *915*
Volvocales *913*
    chemotrophy in 531
    cryptic species in 24–5
    glycerol in 222
    photoheterotrophy in 716
    photokinesis in 873
    production of growth inhibitors by 847
    ultrastructure and cell morphology of 791
*Volvox* *913*
    chloroplasts of 142, 146
    extracellular products of 841
    genetics of 319
    isoagglutinins in 24
    morphology of 791
    plasmodesmata in 95
    reproduction in 816, 821, 823, 824, 827
    thermotaxis in 882
  *aureus*
    parthenospores of 824
    phototopotaxis in 879
    sex factors from 828, 829, 846
    vitamin requirements of 782
  *carteri* 824, 827–9
  *carteri* var. *nagariensis* 828
  *carteri* var. *weismannia* 828, 829
  *dissipatrix* 828
  *gigas* 828, 829
  *globator* 750, 782
  *obversus* 828
  *rousseletii* 828, 829
  *tertius* 750, 782
*Volvulina* *913*
  *pringsheimii*
    inhibitors from 847
    vitamin requirements of 782
  *steinii*
    heterotrophic growth of 534
    vitamin requirements of 782


*Waerniella* *912*
  *lucifuga*
    heterotrophic growth of 437
    vitamin requirements of 780
*Weeksia* *911*
  *fryeana* 804
*Westiellopsis* 841, *910*
  *prolifica* 563
*Woloszynskia* *911*
    nuclear division in 97
    pigments of 154
    thylakoids of 141
  *limnetica* 750, 780

*Xanthidium* 915
  *cristatum* 782
Xanthophyceae 909, 912
  auxotrophy in 743–5
  carotenoids of 4–6, 152, 153, 183–5
  cellulose in 14
  chlorophylls of 153, 161–4, 358
  chloroplasts of 131, 133, 137, 140, 148, 150, 151
  compared with other algae 3–4
  extracellular products of 840, 843, 845
  flagella of 87
  eyespot of 147
  heterotrophy in 535
  hydrocarbons of 17
  lipids and fatty acids of 16, 40, 250–1

  nucleus of 96
  phylogeny of 261
  polyols of 222
  silicification in 656
  sterols of 274
  storage products in 213
  thylakoids of 142
  urease in 597
  vitamin requirements of 780


*Zonaria* 817, 826, 913
*Zoochlorella* 219, 915
*Zostera*
  *marina* 14
Zygnematales 915
*Zygogonium* 717, 726, 915

# SUBJECT INDEX

absorption of light   346–79
absorption spectra
   effect of light intensity and quality on
     373–9
   of biliproteins   194–5, 198, 355, 363–
     365
   of blue-green algae   360, 363, 369, 378
   of carotenoids   355, 357, 361, 362
   of chlorophyll   162–6, 352, 353, 356–63,
     366, 547
   of chloroplasts   356, 357
   of crude extracts   199
   of photosystems I and II   367, 369, 371
   of polysaccharides   209, 210
   uses of   347
acetaldehyde   845
acetate
   assimilation of   532–52
   as source of energy   509–10, 513
   effect of, on chlorophyll synthesis   169
   extracellular production of   843
   in glyoxylate cycle   485, 486, 507, 508,
     545
   in lipid metabolism   255
   reduction of   461
   requiring mutants   320, 322, 323, 326–8
     335–6
   toxicity of   539
acetate flagellates
   acid tolerance in   538–40
   and the 'glucose block'   541
   effect of pH on   511
   inability of, to use glucose   507, 532
   requirement for thiamine of   544
   use of simple acids by   532
acetate kinase   540
acetic acid
   as carbon source   533
   extracellular production of   845
   penetration of cell membranes by   539
acetone
   effect of, on cytochrome   426
   extracellular production of   845
acetyl CoA
   in arginine metabolism   597
   in fatty acid synthesis   448
   in glyoxysomes   485
   in metabolite synthesis   442, 445
acetyl CoA carboxylase   448

acetylene reduction
   as measure of nitrogenase activity   561–
     571, 720
   by heterocysts   565
   inhibition of   570
N-acetylglutamate   597
acetylglutamate kinase (ATP:N-acetyl
   glutamate-5-phosphotransferase)   597
N-acetylglutamate-5-phosphotransferase
   598
acetylglutamate synthetase (acetyl CoA:L-
   glutamate N-acetyl transferase)   597
N-acetyl glutamic semialdehyde   597
N-acetylglutamyl phosphate   597
acetylglutamyl phosphate reductase (N-
   acetyl L-glutamate semialdehyde:
   NADP oxidoreductase)   597
N-acetyl ornithine   597
acetylornithine-glutamate transacetylase
   (N-acetyl-ornithine L-glutamate N-
   acetyltransferase)   597
acetylornithine-transaminase (N-acetyl
   L-ornithine:2 oxoglutarate aminotransferase)   597
acid phosphatase
   effect of phosphorus on activity of   641
   extracellular production of   845
   in Golgi apparatus   104
   in unicellular algae   91
acrylic acid
   occurrence of   23
   production of, by algae   847
actidione (see cycloheximide)
actinomycetes   601
actinomycin D
   as inhibitor of hydrogenase   464
   as inhibitor of nitrate assimilation   590
   in nucleic acid metabolism   294
action spectra
   in relation to movement   874–9
   of chloroplast orientation   886
   of coccolith formation   666
   of ion uptake   687, 688, 719
   of nitrogen fixation   563
   of oxygen evolution   371
   of photoheterotrophy   716
   of photoinhibition   718
   of photoreduction   460
   of photorespiration   495

of photosystems 398, 409, 719
of protein synthesis 721
of redox reactions 397, 703
of regreening 718, 720
of respiration 412, 495
active transport 677–707
ATP for 550
effect of light on 719
effect of temperature on 724–5
of carbohydrates 539
of iodine 625
occurrence of 697–707
vacuolar fluxes and 696
adaptation
of photosynthetic pigments 377–9
to temperature 724
to variations in light quality 716–17
adenine
as nitrogen source 595
requiring mutants 323
adenosine
uptake in cell division 831
adenosine diphosphate (ADP)
in arginine metabolism 597
in ATP synthesis 348, 405, 428, 644
in photophosphorylation 392, 393, 405
in photosynthesis 374
in respiration 510
in urea metabolism 596
adenosine diphosphoglucose
in starch storage 208
utilization of 210, 211
adenosine diphosphoglucose : α-1, 4-glucan
transferase 211
adenosine diphosphoglucose : α-1, 4-glucan
α-4-glucosyltransferase 208
adenosine diphosphoglucose pyrophosphorylase 211
adenosine diphosphoglucose transferase
211
adenosine diphosphoglucosyl transferase
209
adenosine monophosphate (AMP)
as phosphorus source 643
in arginine metabolism 597
in cell cycle 831
in enzyme activity 209
in polysaccharide synthesis 215
in respiration 511
adenosine-3′-phosphate-5′-phosphosulphate
612
adenosine triphosphatase 428, 429, 615,
659, 670, 687, 831, 867
adenosine triphosphate (ATP)
effect of osmotic stress on 729
effect of phosphorus on level of 641
effect of temperature on level of 725
in active transport 539, 687–9, 700 705–
706

in amino acid synthesis 593, 597
in $O_2$ reduction 436, 439–40
in glucose metabolism 549–50
in glycolate metabolism 483
inhibition of 522
in movement 867, 868, 874
in nitrate metabolism 592–3
in nitrogen fixation 561, 562, 565
in photophosphorylation 392–412
in photosynthesis 348, 374, 402, 449
in polyphosphate synthesis 648
in sulphate metabolism 612
in urea metabolism 596
measurement of, in vivo 638
synthesis of 394, 405, 409, 410, 428, 522,
621, 644
adenosine triphosphate : urea amido-lyase
596, 597
S-adenosyl methionine 612
S-adenosyl methionine-magnesium-
protoporphyrin methyltransferase 615
adenylates 409
adhesive 110
adonirubin ester 154
adonixanthin ester 154
ADP (see adenosine diphosphate)
aeodan 69
aerotaxis 881
agar 10, 12, 63–5
agarobiose (4-O-β-D-galactopyranosyl-
3, 6-anhydro-L-galactose)
in porphyran 66
structure of 64
agaropectin 64
agarose 10, 12, 63–5, 72
agglutination 830
ALA (see δ-aminolevulinic acid)
alanine
as carbon source 536
extracellular production of 842
synthesis of 442, 593
L-α-alanine 594
alanine dehydrogenase
effect of nitrogen source on 594
function of 593
in reproducing cells 831
alcohols
as carbon source 533
tolerance of 538–40
aldolase 20, 220, 547
alginates 58–62
alginic acid
composition of 10, 11
in cell walls 58–62, 72
occurrence of 10, 797
alkaline phosphatase
effect of phosphorus on activity of 641–
642
extracellular production of 845

allantoic acid 597
allantoin 595, 597
allophanate 596
allophanate amidohydrolase 596
allophycocyanin
    absorption spectrum of 195, 355, 364,
        365
    characteristics of 198, 354, 356
    occurrence of 7, 26, 194, 354, 364
C-allophycocyanin 153
alloxanthin 5, 152, 154, 186, 187
allulose 217
altrose 218–19
amides
    as nitrogen source 595
    formation of 593–4
amination 593
amines
    as nitrogen source 595
    as taxonomic feature 23–4
    in osmoregulation 225
amino acids
    as carbon sources 533
    as nitrogen sources 595, 597–9
    extracellular production of 841–2, 848,
        851
    in cell walls 658
    in osmoregulation 225
    in synchronous cultures 901
    in taxonomy 21
    metabolism of 436, 442, 593, 599, 615,
        620
    requiring mutants 323
    translocation of 94, 95
amino acid decarboxylase 20
amino acyl synthetase 317
α-amino adipic acid 20, 599
p-amino-benzoate requiring mutants 320,
    322, 323
δ-aminolevulinic acid (ALA)
    effect of glucose on 552
    occurrence of 594
    origin of 167–70
    release of 851
δ-aminolevulinic acid synthetase 168
2-amino-purine 319
amino-transferase 496
amino triazine 459
3 amino-1, 2, 4-triazole 806
amitrole 209, 211
ammonia
    and metabolite synthesis 442, 447
    and nitrogen fixation 563, 564, 571–2
    and urea metabolism 596
    release of, from blue-green algae 574
    requiring mutants 320, 322, 323
ammonium molybdate 621
ammonium-N
    as nitrogen source for growth 584–93

assimilation of 593–9
    effect of, on arginine synthesis 599
    effect of, on reproduction 821–3
    effect of, on sucrose synthesis 217
    effect of temperature on uptake of 725
    in control of nitrate reduction 588–93
    in urea metabolism 596
ammonium sulphate 574
amoeboid movement 868–9
AMP (see adenosine monophosphate)
amylase 208, 211, 538
amylopectin 207, 208, 210, 211
α-amylase 211
amylose 208, 210–12
α-amylolysis 209, 210
β-amylolysis 207, 210
α-amyrin 273
β-amyrin 273, 275, 276
amytal 516–17, 521–2
anaerobic growth
    effect of, on extracellular products 851
    features of 467–8
anaphase 99, 100
anastomoses 869
3,6-anhydro-α-D-galactopyranose 41
1,4-anhydro-α-L-galactopyranose 64
3,6-anhydro-α-L-galactopyranose 41, 64
3,6-anhydro-α-D-galactose 67
3,6-anhydro-D-galactose 2-sulphate 71
3,6-anhydro-L-galactose 10, 12, 64–70
3,6-anhydro-4-O-β-galactopyranosyl-D-
    galactitol 69
3,6-anhydro-2-O-methyl-L-galactose 65,
    66
Antarctic
    algae of 722
    nitrogen fixation in 725
antheraxanthin 5, 152–4, 183, 185,
    187
antibiotics
    effect of light on toxicity of 720
    effect of, on growth 511
    effect of, on ribosomes 147, 285, 296
    production of, by algae 24, 847
    resistant mutants 323–4, 328
antibodies 400
antimetabolite resistant mutants 323
antimycin A 403, 511, 512, 516–22, 687
aphanicin 187
aphanin 187
aphanizophyll 153–4, 187–91
apical dominance
    in algae 762
    growth regulators in 765
apical meristem 799, 800, 804
aplanospore
    effect of IAA on 824
    effect of light quality on development of
        816, 824

apochlorosis 73
1,4-L-arabinobiose 551–2
1,5-L-arabinobiose 73
arabinogalactan 75
α-L-arabinopyranose 41
3-O-α-L-arabinopyranosyl-L-arabinose 74
arabinose 54, 75, 105, 223, 839
D-arabinose 74
L-arabinose 10, 15, 73, 74
L-arabinose 3-sulphate 73
arachidonic acid 18, 238, 241, 248, 254, 258
aragonite 662, 668
Arctic
    algae of 722
    vitamins in waters of 758
arginase (L-arginine ureohydrolase) 597
arginine
    in reproduction 831
    requiring mutants 320, 322, 323
    synthesis of 442, 446, 449, 594,
        597–9
    utilization of 597–9
arginine deiminase (L-arginine iminohydro-
        lase) 598, 599
arginine desmidase (arginine deiminase)
        598
argininosuccinate 597
argininosuccinate lyase (L-argininosuccinate
        arginine lyase) 598
argininosuccinate synthetase (L-citrulline:
        L-aspartate ligase) 598
arginyl-tRNA 293, 598
arginyl-tRNA synthetase 293
arsenate
    effect of, on active ion uptake 688
    inhibition of nitrite reductase by 587,
        592, 593
    inhibition of phosphorus uptake by 640
ascophyllan 72
ascorbate
    extracellular production of 846
    metabolism of 563
ascorbate-NNN′N′-tetramethyl-p-phenylene
        diamine (ascorbate-TMPD) 522
ascorbic acid 824
asexual reproduction 814, 824, 826
asparagine 536, 595
aspartate
    as carbon source 536
    in arginine metabolism 597
    production of 442, 444, 492, 593
aspartic acid
    effect of, on silicic acid uptake 658
    extracellular production of 842, 848
D-aspartic acid 595
assimilation (see also specific compounds)
    oxidative 544–6
    phototrophic 546–9
astacin 152, 187, 188

astaxanthin 152, 154, 181, 182, 187, 641
atebrin 522
ATP (see adenosine triphosphate)
'Atrazine' 798
attachment point of chloroplast DNA
        334–6, 338
autoradiography
    of chloroplast DNA 299, 304
    of photosynthate dispersal 569–70
    of polysaccharide production 107–10
    of protein production 110
    of theca formation 105
autospore 613, 900, 904
autotrophic growth 435–43, 448, 449, 507
        616
auxins 763–5
auxospores
    effect of daylength on formation of 818,
        820
    effect of organic compounds on forma-
        tion of 824
    effect of salinity on formation of 823
    effect of temperature on formation of
        819, 820
auxotrophy 320, 323, 538, 742–62
azide 516, 517, 521, 522


bacteria
    chemotaxis in 882
    decomposition of algae by 573
    growth factors from 768–71
    growth saturation in 544
    hydrogenase in 464–6
    nucleic acid hybridization in 285
    photosynthetic 126–8, 244, 256
    phototaxis in 878
    ribosomes of 601
bacteriochlorophyll 30, 128
bacteriopheophytin 360
basal bodies 96, 866, 867
base ratios 281, 284, 286, 294–6, 299–301,
        303–4
benzimidazole 622
benzoquinone 396, 397
p-benzoquinone 428
benzyl viologen 461, 586, 587
bicarbonate
    active uptake of 689
    role of, in coccolith formation 666
biliproteins 6–9, 194–200, 356, 879
binding factor of vitamin $B_{12}$ 751–3, 761–
        762, 842, 854
bioassay
    for phosphorus 641
    for vitamins 753–6, 760
biotin
    algal production of 757–8, 845, 846
    bioassay for 753–6

biotin—*cont.*
  content of algae  756-7
  ecology of  758-62
  requirement by algae  435, 438, 743-5,
    747, 778-82
biotin-dependent carboxylase  435
bipyridilum  398
birefringence of cell walls  53, 55
blepharoplast  87, 867
blue drop phenomenon  347
blue-green algae (see also Cyanophyceae)
  arginine biosynthesis in  598
  auxotrophy in  744
  effect of hydrostatic pressure on  730
  effect of IAA on  764
  effect of lysozyme on  515
  effect of osmotic pressure on  728
  effect of oxygen level on  509
  effect of pH on  726-7
  effect of temperature on  722, 725
  enzymes of  209, 446
  extracellular products of  839-42, 848-
    851
  ferredoxin in  617
  growth of, in the dark  550
  inorganic metabolism in  611-14, 618,
    620-5
  in symbiosis  575
  movement of,
    gliding  870
    in response to chemicals  880
    in response to light  873-9
  nitrogen fixation by  560-75, 620
  oxidative phosphorylation in  520-
    522
  pentose phosphate cycle in  507
  photoheterotrophy in  716
  photosynthesis in  126-8, 399,
    512
  respiration in  512, 550
  ribosomes of  601
  storage products of  217, 222
  toxicity of  848
bonding
  between thylakoids  144
  in cell walls  43, 49, 54, 55, 57, 60,
    70
boron  611, 622-3
borosilicates  660
brassicasterol  270-1, 273, 274, 276
bromide  695
bromine  621
buoyant density of DNA  286, 287, 290-1,
    294-6, 299
butenyl cycloheptadiene  24
butyrate  507, 509, 542-3
*n*-butyrate  543
butyric acid  533
butyryl-CoA  508

C-550
  absorption changes and  376
  in photosynthetic electron transport  395,
    407
  occurrence of  373, 374, 402, 426, 429
calcification  92, 662-70
calcite  662-7
calcium
  as inorganic nutrient  611, 614-15, 669
  effect of deficiency of  659, 660
  effect of phosphate on uptake of  641
  role of, in reproduction  823
calcium carbonate  662-70, 690
callose  48, 94
calorhodin-α  192
calorhodin-β  192
caloxanthin  189-91
Calvin-Benson cycle
  inactivation of  507
  metabolism of  435-43
  occurrence of  349
campesterol  267, 270-4
canavanine-resistant mutants  320, 322, 323
canthaxanthin  152-4, 181-2, 187, 190-2
carbamycin-resistant mutants  326, 327, 336
carbamoyl phosphate
  in metabolite synthesis  442
  synthesis of  446-7, 449, 593, 594, 597
carbamoyl phosphate synthetase (ATP:
    carbamate phosphotransferase)  442,
    447, 593, 594
carbamycin  328-9
carbohydrate
  active transport of  539
  assimilation of  550
  as source of carbon  533
  as taxonomic feature  9-15
  effect of magnesium on metabolism of
    615
  effect of phosphorus on levels of  640
  extracellular production of  839-40
  in cell walls  658
  in synchronous cultures  898
  production of  436
  storage of  206-20
  transfer of, in symbioses  575
carbon
  as inorganic nutrient  611
  extracellular release of  716, 849, 850-3
  heterotrophic use of  530-52
  in photosynthesis  402, 441
  in respiration  506-10, 513-14
  sources of  435
carbon dioxide
  catalytic roles of  448, 449
  chemosynthetic reduction of  462
  compensation point of  492
  effect of, on fatty acid metabolism  255,
    448

effect of, on glycolate production 477, 487
effect of, on ion uptake 719
effect of, on nitrate assimilation 591–592
effect of, on phosphorus metabolism 641, 648
effect of, on photorespiration 491–3
effect of, on pigmentation 379
effect of, on reproduction 823, 828
fixation of 125, 348, 392, 393, 434–49, 460
growth rates on 542–3
in amino acid synthesis 593
release of 476
carbonic acid anhydrase
effect of $CO_2$ on 493
inhibition of 666
occurrence of 435, 437, 448
carbon monoxide 459, 465, 514–16, 518, 521
carbon reduction cycle
course of 435–43
in glycolate pathway 475
NADP as reductant for 400
carbonyl cyanide m-chlorophenyl hydrazone (CCCP) 463, 593, 640, 688, 689, 700–702, 706–7
carboxydismutase 448
carboxylase (see specific enzymes)
β-carboxylation 444–6, 449
cardiolipin 239, 256
carnitine 255
carnitine palmityl transferase 255
α-carotene 5, 6, 153–4, 177–9, 183, 184, 187, 192, 354, 361
β-carotene 4, 5, 6, 26, 152-4, 177–9, 181, 183, 185, 187, 190–2, 346, 354–5, 361–2, 368, 373
γ-carotene 154, 177–8, 181, 189, 192
δ-carotene 178
ε-carotene 153, 154, 178, 179, 185, 187, 361
carotenoids
absorption spectra of 353–5, 363
as taxonomic feature 4–6, 152–5
chromatic adaptation of 379
distribution of 28, 90, 147, 152–5, 176–194
effect of light on synthesis of 720
effect of pH on 727
effect of salinity on 729
in light absorption 346–7, 351, 361–3, 368
in movement 878, 879
in photosystems I and II 370, 373
nature and nomenclature of 176–8
carrabiose (4-O-β-D-galactopyranosyl-D-galactose) 68

carrabiose dimethyl acetal 68
carrageenan 10, 12, 63, 66–72
λ-carrageenan 68
ι-carrageenan 68, 70, 71
ϰ-carrageenan 67–71
μ-carrageenan 68–9
cartwheel pattern
of centriole 100
of flagellar tubules 87
catalase
in microbodies 91, 484, 485, 522
properties of 481–2
cation exchange pump 686, 687
caulerpacin 23
caulerpin 23
CCCP (see carbonyl cyanide m-chlorophenyl hydrazone)
$C_4$-dicarboxylic acid cycle 492
cell division
biochemistry of 612, 831
effect of light quality on 817
effect of selenomethionine on 613
effect of temperature on 723–4
endogenous rhythms of 826
extracellular production at 850
in synchronous cultures 895, 897
in various algae 793–6
cellobiose 48, 49
cell surfaces 24–5
cellulase 845
cellulose (β-1,4-glucan)
as taxonomic feature 4, 12–14
in cell wall 15, 42, 43, 44, 46, 47, 90
in scales 90, 105, 107, 665
structure of 48–53
cell wall
disruption of 515
effect of strontium on 615
mutations of 323
organization of 42–8
polysaccharides of 9–15, 40–76, 90, 92
proteins of 20–1
types of 42–8, 657–8, 660–1, 789
centriole
development of 899
structure and function of 96–100
centromere
in mapping genes 321–2
in nuclear division 99, 100
centrosome 98
cephalin 239
cephalodia 575
cerebroside 18
cesium 641
chain form
of alginates 60
of carrageenan 70–2
of cellulose molecule 49, 50, 53
of mannan 53, 54

chain form—*cont.*
  of xylan  55
chair form
  in carrageenan  71
  in mannan  54
  in polymannuronic acid  60, 62
  in polyguluronic acid  60, 62
chalinasterol  270, 271, 277
chemotrophy  530–2, 541–4
chemoreceptors  881–2
chemotaxis
  features of  880–2
  in sexual reproduction  827, 829–30, 846
D-*chiro*-inositol  223–4
chitan  840
chitin (1,4-2-acetamido-2-deoxy-β-D-
    glucopyranose)  26, 75, 90
chloramphenicol
  effect of, on DNA synthesis  287
  effect of, on growth and respiration  511
  effect of, on protein synthesis  600
  effect of, on ribosomes  147, 168, 285
  reduction by H₂ in dark  461
chlorellin  847
chloretone  459, 465
chloride
  effect of deficiency of  459, 460
  effect of light on uptake of  719
  effect of phosphate on uptake of  641
  fluxes  683–707
  requirement for growth  611, 621
*m*-chlorobenzhydrazamic acid  523
chlorobium vesicles  128
*p*-chloromercuribenzoate (PCMB)  211
    887
4-chloro-2-methyl phenoxyacetic acid
    (MCPA)  798
3′(*p*-chlorophenyl)1′, 1′ dimethylurea
    (CMU)  446, 447, 563
chlorophyllase  718
chlorophyll (see also specific types below)
    161–71
  analysis of  164–6
  as taxonomic feature  3–4, 153–4, 185
  characteristics of  354
  content of  169–71, 258, 615–17, 898
  distribution of  161–4, 372, 399, 532, 551
  effect of germanium on  662
  effect of iron deficiency on  169, 616, 617
  effect of light on  168, 170–1, 258, 718,
    720
  effect of magnesium deficiency on  615
  effect of manganese deficiency on  258,
    617
  effect of rifampicin on  168, 302
  effect of salinity on  729
  effect of silicon deficiency on  623
  in light absorption  346, 356–60
  in plastoglobuli  147

site of  125–8
structure of  161–4
synthesis of  167–9, 900
chlorophyll *a*
  adaptation of  377–9, 717
  analysis of  164–6
  as taxonomic feature  3–4, 29, 31, 153–4
  characteristics of  354
  distribution of  161–4, 358, 364
  effect of iron deficiency on  616
  fluorescence of  349–51, 353, 357, 693
  forms of  349, 356–60
  in heterocysts  566–7
  in light absorption  346–79
  in movement  874–6
  in photosystems I and II  365–79, 396,
    399
  proportions of  359
  spectral bands of  165–6, 352, 365, 874–
    875
  structure of  161–4
  synthesis of  167–70
chlorophyll *a₄₃₆*  370
chlorophyll *a₆₆₀*  349, 358, 359, 370
chlorophyll *a₆₇₀*  349, 356, 358, 359, 366,
    368–70
chlorophyll *a₆₈₀*  349, 356, 358, 359, 366,
    369, 370
chlorophyll *a₆₈₅*  358, 359, 369, 370
chlorophyll *a₆₉₀*  349, 358, 359, 369, 370
chlorophyll *a₆₉₅*  349, 358, 366, 369, 370
chlorophyll *a₇₀₅*  349, 356, 358, 359, 369,
    370
chlorophyll *a₇₂₀*  717
chlorophyll *b*
  as taxonomic feature  3–4, 153–4
  characteristics of  354
  estimation of  165–6
  in light absorption  346, 348, 351
  in photosystems I and II  365–73
  occurrence of  358, 161–4
  spectral bands of  165–6, 352, 353, 357
    362
  structure of  162
  synthesis of  167–9
chlorophyll *b₆₄₀*  359
chlorophyll *b₆₅₀*  359, 370
chlorophyll *c*
  adaptation of  379
  characteristics of  354
  estimation of  165–6
  occurrence of  4, 153–4, 161–4, 354, 358,
    368, 370
  spectral bands of  165–6, 352, 353
  structure of  162
chlorophyll *d*
  characteristics of  354
  estimation of  165–6
  occurrence of  4, 153, 161–4, 354, 358

spectral bands of 165–6, 352, 353
structure of 162
chlorophyll *e* 164, 358
chlorophyllide 358
chloroplast
  absorption spectra of 356, 357, 366
  active transport in 683, 693, 703
  autonomy of 303–5
  cytochrome of 426
  development of 168–9, 551
  division of 148–51, 900
  DNA of 21–2, 125, 134, 148–51, 287–304, 325, 334–40, 718
  effect of light on 720
  envelope 125, 128–40
  evolution of 9, 19–23, 32, 129, 155, 259–261, 282, 303–4
  genes in 315
  glycolate metabolism in 475, 478, 482–486
  lipids and fatty acids of 19, 216, 254–8
  membranes of 258, 401, 407, 426–30, 683, 693, 703
  mitochondria and 511
  movement of 884–7
  mutations of 323, 328
  nitrogen metabolism in 591
  photosynthesis in 399, 439
  pigments of 152–5, 399
  protein synthesis in 600–2, 902
  ribosomes of 22–3, 125, 289–303, 317, 328, 429
  role of, in theca production 105
  RNA of 285, 288–97, 317
  starch in 208
  structure of 28, 92, 93, 102, 103, 125–51, 427
chlorpromazine 562
cholesta-5,7-dienol 277
cholestane 267, 268
cholest-7-enol 277
cholest-4-en-3-one 271, 273
cholesterol 17, 18, 266–8, 270, 272–3, 277–8
choline 23, 239, 257, 622
chondrillasta-5,7-dienol 277, 278
chondrillast-7-enol 277
chondrillasterol 27, 268, 269, 276–8
chromate 461
chromatic adaptation 377–9
chromatophore 532, 551
chromoprotein 197–8
chromosome
  duplication and segregation at meiosis 361
  in nuclear division 97–101
chrysolaminaran 214–15, 532
chrysomonads (see also Chrysophyceae)
  chlorophylls of 4

vitamin requirements of 744
cilia 865, 866
circadian rhythms
  in relation to photosynthesis 440–1
  of sporulation 817, 826
  synchronous cultures and 901–4
cisternae
  enzymes of 112
  of chloroplasts 140
  of Golgi complex 103, 104, 112, 113
  scale production in 105, 107, 664
citrate
  in metabolite synthesis 442
  mitochondrial oxidation of 517
  toxicity with manganese 618
citrulline
  as nitrogen source 595
  formation of 437, 442, 447, 598
  in amino acid synthesis 594, 597
  in nitrogen fixation 564
  metabolism of 597–9
classification (see taxonomy)
cleosine 326, 327, 329, 336
clionasterol 269–71, 274, 277, 278
CMU (see 3′(*p*-chlorophenyl)1′,1′dimethy-lurea)
CoA (see coenzyme A)
cobalamin 749–51, 761
cobalt 611, 621–2
coccoliths
  effect of light on formation of 720
  effect of temperature on formation of, 725
  production of 105, 790
  structure of 90, 662–7
coccolithosome 664
coccolithophorids
  calcification in 662–7
  life cycle and taxonomy of 691
coenzyme A (CoA)
  acetylization of 544
  association of, with iron containing acids 617
  in arginine metabolism 597
  in lipid metabolism 255
coenzyme $Q_7$ 516
coenzyme $Q_9$ 509
colchicine
  effect of, on microtubules 97
  inhibition of diatom mitosis by 659, 660
collagen 669
complementation studies 316
concanavalin-A
  in gamete fusion 830
  reaction of, with starch 207
contractile vacuoles
  in osmoregulation 91–2, 104, 112
  mutations of 323
copper
  as inorganic nutrient 611, 619

copper—*cont.*
  effect of deficiency of 459
  effect of, on $O_2$ uptake 515
  toxicity of 619, 720, 727, 798
coproporphyrinogen 616
coumarin 770, 806
crassulacian acid metabolism 491
cristae 101–3
crocoxanthin 154, 186, 187
crossing over of nuclear genes 320–3
crustaxanthin ester 154, 181, 183
cryophilic algae 721, 722
cryptomonads (see also Cryptophyceae)
  biliproteins of 356
  chemotrophy of 531
  lipids of 16
  starches of 212
  taxonomy of 29
cryptoxanthin 190–1
α-cryptoxanthin 183
β-cryptoxanthin 6, 153–4, 183
β-cryptoxanthin-5′,6′-monoepoxide 153–4
β-cryptoxanthin-5′,6′,5′,6′-diepoxide 153–154
crystalline inclusions 125, 146
CTP (see cytidine triphosphate)
cuticle 43, 53
cyanate 590
cyanelles 192
cyanide
  inhibitory effects of 459, 465, 481, 514, 516, 518, 520–2, 562, 587
  reduction of, by nitrogenase 563
  stimulatory effects of, on respiration 515, 516
cyanocobalamin (see also Vitamin $B_{12}$) 622
cycads 575
cyclitols
  as storage products 206, 223–5
  in osmoadaptation 728
  structure of 223
cycloartenol 266–70, 276
cyclohexane tetrol
  extracellular production of 839, 851
  in osmoadaptation 728
cycloheximide 147, 168, 285, 287, 296, 297, 320, 322, 323, 430, 590, 600
1,2,3,4/5-cyclohexanepentol 223–4
1,4/2,5-cyclohexanetetrol 223, 225
D(+)-1,4/2,5-cyclohexanetetrol 223–4
cysteine
  as nitrogen source 595
  as sulphur source 612
  synthesis of 599
cytidine triphosphate (CTP) 602
cytochrome (see also specific types below)
  effect of blue light on 718
  in mutants 426, 518
  role of, in photosynthesis 348

cytochrome *a* (see also specific types below) 516–20, 616–17
cytochrome $a_3$ 512, 516, 517, 519–21
cytochrome $a_{593}$ 518
cytochrome $a_{605}$ 518
cytochrome $a_{607}$ 518
cytochrome $a_{609}$ 517
cytochrome *b* (see also specific types below) 511, 516–23, 616, 617
cytochrome $b_{555}$ 516
cytochrome $b_{558}$ 518
cytochrome $b_{559}$ ($b_3$) 373, 374, 395, 402, 425, 426, 429, 512, 516
cytochrome $b_{561}$ 517, 518
cytochrome $b_{563}$–$_{564}$($b_6$) 373, 374, 402, 403, 408, 511–12, 516–18
cytochrome $b_{568}$ 518
cytochrome *c* (see also specific types below 509, 516–22, 616, 617
cytochrome $c_1$ 519
cytochrome $c_{549}$ 516
cytochrome $c_{551}$ 512, 516, 518
cytochrome $c_{552}$ 518
cytochrome $c_{553}$ 429
cytochrome $c_{555}$ 518
cytochrome $c_{556}$ 517, 518
cytochrome *c* oxidase 508, 509, 522
cytochrome *c* reductase 522
cytochrome *f* 373–4, 395–401, 407, 512
cytochrome $f_{558}$ 373–4
cytochrome *o* 521
cytochrome oxidase 511, 517, 519, 523
cytohets 332, 335, 339
cytokinesis
  cell wall formation and 659
  effect of silicon starvation on 661
  microtubules in 900
  mutations of 323
cytokinins 765–6, 771
cytoplasm
  division of 99, 613
  protein synthesis in 602–3
  streaming of 883–4
cytoplasmic
  genes 315–17, 326–7, 338–40
  inheritance 325–8
cytosine 22
cytosol 408


2,4 D (see 2,4 dinitrophenol)
DAPA (see α-ε-diaminopimelic acid)
daylength, effect of, on reproduction 817–818, 820–1, 823
DCCD (see dicyclohexyl carbodiimide)
DCMU (see 3′(3,4-dichlorophenyl) 1′,1′dimethyl urea)
DCPIP (see 2,6-dichloro-phenolindophenol)
n-decanoic acid 19

decarboxylation 435
degree of polymerization (DP)
    of cellulose in walls 53
    of mannan in walls 54
    of xylan 56
22-dehydrocholesterol 271, 273
dehydrogenase 401
7-dehydroporiferasterol 274, 275
7-dehydrositosterol 274
deoxycholate 366, 368
2-deoxy-*myo*-inositol 223–4
deoxyri bonuclease (DNase) 101, 102, 103, 287, 296, 302, 900
deoxyribonucleic acid (DNA)
    acid insolubility of 637
    as taxonomic feature 21–2, 282
    base ratios of 31, 282
    effect of light on 718, 721
    effect of nutrient deficiency on 620–3, 661
    effect of temperature on 723
    in centrioles 97
    in dinoflagellates 98
    in protein synthesis 602
    in pyrenoids 144
    in silicification 660, 669
    metabolism of 286–303, 831, 897–900
    of blue-green algae 282–4
    of chloroplasts 125, 134, 148–51, 286–303, 317, 334–40, 718
    of mitochondria 101–3, 286–8, 290, 294, 299
    replication of 287, 291
    satellite 287, 290–2, 294–7
    segregation of 333–4
deoxyribonucleic acid polymerase 283, 293, 296, 301, 661, 662, 899
deoxyribotides 622
desiccation
    effect of, on nitrogen fixation 573, 575
    effect of, on spore release 825
desaspidin 547, 688
desmids
    calcium sulphate in 662
    cytoplasmic streaming in 883
    movement of 870, 872, 880
desmosterol 270–5
detergent
    as phosphorus source 643
    toxicity of 798
deuterium oxide 256
diadinoxanthin 5, 152–4, 183–7
α-ε-diaminopimelic acid (DAPA)
    as nitrogen source 595
    in cell wall of blue-green algae 20
    in lysine synthesis 599
diaphorase 518, 586
diastase 208

diatoms (see also Bacillariophyceae)
    acetate assimilation by 533
    auxin activity in 763
    auxotrophy in 744
    blooms of 759–62
    carotenes of 361
    cell selection in 897
    chemotrophy in 531
    chlorophylls of 4, 163
    chloroplast movement in 885
    chromatic adaptation in 379
    effects of chemicals on movement of 880
    effects of light on 379, 815, 816, 820, 875, 877, 879
    effects of pH on 727
    effects of temperature on 724, 819, 820
    extracellular products of 842, 844, 845, 849
    hydrocarbons of 17
    inorganic nutrition of 611, 616, 618, 623–4, 791
    lipids of 16
    movement of 870–2, 875, 877, 879, 880, 885
    phosphorus levels for 639
    polysaccharides of 14, 145, 214
    pyrenoid of 144
    surface structures of 90, 656–61
    taxonomy of 791
    vitamin metabolism in 621, 749, 759–60
diatoxanthin 5, 152–4, 183–7
dibromothymoquinone 460
2,6-dichloro-phenolindophenol (DCPIP) 397–8, 408, 461, 462, 475, 481, 522, 563
2,4-dichlorophenoxyacetic acid 573
3′(3,4-dichlorophenyl) 1′,1′ dimethyl urea (DCMU) 374–6, 398, 409–10, 443, 463, 547, 549, 570, 591, 640, 688, 700
dichroism 55, 59
dictyosomes
    acid phosphatase in 91
    in scale production 107
dicyclohexyl carbodiimide (DCCD) 688, 707
di-enoic acid 251, 259
diepoxyneoxanthin 187, 188
digalactosyl diglyceride 18, 19, 240, 254, 256, 257, 259
digitonin 368
22-dihydrobrassicasterol 271, 273, 274, 277
22-dihydrochondrillasterol 268, 269
5,6-dihydroergosterol 273
24,25-dihydrolanosterol 270, 271
dihydroxyacetone-phosphate 222, 436
2,7-dihydroxynaphthalene 486
diimide 561
di-iodotyrosine 23, 625

α-5,6-dimethylbenzimidazoyl cyanocoba-
　　lamin (see Vitamin B$_{12}$)
α-5,6-dimethylbenzimidazole-D-
　　ribofuranose phosphate　749
4α,24-dimethyl-5α-cholesta-8(9)-en-3β-ol
　　276
N,N-dimethyl formamide　164–5
dimethyl propiothetin　23
4,4-dimethyl sterols　269, 270, 276
2,4-dinitrophenol (DNP)　459–64, 517,
　　521, 550, 587, 592, 593, 688, 705,
　　725, 798
dinoflagellates
　calcareous cysts in　662
　carbon assimilation in　533
　carotenoids of　4–6
　chlorophylls of　4
　chloroplast of　129, 131, 137, 152
　chromatic adaptation in　379
　effect of hydrostatic pressure on　730
　effect of IAA on　764
　endogenous rhythms in　826
　flagella of　87
　lipids of　16
　movement of, in response to light　875
　nuclear division in　97
　nuclear membrane of　96
　osmoregulation in　92
　pyrenoid of　144, 145, 146
　toxin production by　848
　trichocysts of　89
　vitamin requirements of　744
dinoxanthin　154, 185, 186
Dio-9　688
γ,δ-dioxovaleric acid　594
diphenylamine　401, 519, 521
diphosphatidyl glycerol　239
diploid nuclei　324
α,α′-dipyridyl　523
diquat　586, 587
disalicylidine propanediamine (DSPD)
　496
dissociation constants　539
dithionite　461, 519, 561–2, 565, 586
dithiothreitol
　effect of, on cell wall　46
　requirement of nitrate reductase　587
DNA (see deoxyribonucleic acid)
DNP (see 2,4-dinitrophenol)
1,14,5-docosanediol 1,14 disulphate　257
docosohexaenoic acid　16
DP (see degree of polymerization)
drug resistance
　in mutant isolation　325
　markers　292
　mutants showing　320
DSPD (see disalicylidine propanediamine)
dulcitol (galactitol)　221–2
dynein　867

echinenone　5, 152–4, 181, 182, 190–2, 368
ecology
　of extracellular production　852–4
　of nitrogen fixation　570–5
　of phosphates　643, 649
　of vitamins　753, 758–62
ecophenes　789
EDTA (see ethylenediamine tetraacetic acid)
egg
　chemotaxes of, in Phaeophyceae　829
　control of release of, in Phaeophyceae
　　816, 817, 820, 825, 826
　development of, in Phaeophyceae　799
electrical properties　692–3, 697–707
electric stimuli, movement in response to
　882–3
electrogenic pump　680–2, 690, 693, 699,
　702, 705
electron acceptor
　in mutants　424, 425
　in photosynthesis　391–412
　nitrate as　410
　sulphate as　410
electron donor
　artificial　426, 586
　cytochrome as　516
　for nitrate reductase　585–7
　for nitrite reductase　587–8
　for nitrogen fixation　561, 562
　water as　126, 405, 424, 425
electron spin resonance (ESR) studies　616,
　618
electron transport
　cyclic　374, 392–412
　effect of manganese deficiency on photo-
　　synthetic　618
　in mitochondria　515–22
　in mutants　424–31
　in photosynthesis　348, 370, 391–412
　microsomal　522
　non-cyclic　374, 392–412
　rhodoquinone in　509
electron paramagnetic resonance (EPR)
　spectra　519, 523, 618
elementary fibrils　51
Emerson enhancement effect　347, 348,
　370, 393–8, 404, 563, 716, 900
emisson spectra　350, 666
endogenous rhythms　817, 820, 821, 826–7
endoplasmic reticulum
　and Golgi complex　103, 106, 107, 110,
　　111
　and nucleus　96
　around pit-connection　92, 93
　in flagellar development　88
　in nuclear division　99, 100
　in plasmodesmata　94, 95
　in polysaccharide secretion　108, 109
　in sieve elements　94–5

occurrence of  87
of chloroplasts  131, 133–40, 141
endotoxin  848
energetics
    of nitrate assimilation  592–3
    of photosynthetic carbon reduction
        439–40
energy coupling  687–8
energy transfer
    between chlorophylls  360, 370
    from biliproteins  365
    from carotenes to chlorophyll *a*  362–
        363
enzymes (see also specific enzymes)
    as taxonomic feature  20–1
    branching  209
    effect of light intensity on  717
    extracellular production of  845
    heat stability of  722
    in fatty acid metabolism  255
    in gametogenesis  791
    in Golgi apparatus  104, 113
    in mannitol synthesis  220–2
    in nitrate reduction  584–8, 590–1
    in paramylon metabolism  215
    in polysaccharide metabolism  210, 211,
        533, 540, 541
    in pyrenoid  145
    iron-containing  617
    osmotic stability of  728, 729
    photorespiratory  484–6
    photosynthetic  435, 436, 438–9, 717
    potassium as activator of  613
    Q enzymes  208, 209, 509, 516
    respiratory  507–22, 545, 729
episterol  277
EPR (see electron paramagnetic resonance)
ergosta-5,22-dienol  277
5α-erogsta-8,14-dien-3β-ol  269, 271
ergost-5-enol  268, 269, 272, 273
ergost-7-enol  268, 269, 277
ergostenone  27
ergosterol  17, 267–9, 272–6, 509, 538
erythritol  221–2, 839, 853
erythromycin
    effect of, on ribosomes  147, 285
    resistant mutants  320, 323, 326, 327
        335, 336
erythrose 4-phosphate  436
ESR (see electron spin resonance)
ester sulphate  72, 73, 75
ethanol
    as carbon source  538
    assimilation of  533
    effect of, on chlorophyll  169
    effect of, on glyoxylate cycle  511
    extracellular production of  845
    in electron transport  520
    mitochondrial reduction by  519

specific growth rates on  543
ethanolamine  239
ethionine  688
ethyl acetate  845
24-ethylcholest-7-enol  277, 278
ethylenediamine tetraacetic acid(EDTA)
    619
ethylene glycol  212
ethylene-methane sulphonate  319
24-ethylidene lophenol  268, 269
euglenarhodone  187
eukaryotic algae (see also specific algae)
    chloroplasts of  393
    hydrogenase in  466
    lipids of  16–19
    photophobotaxis in  877–8
    photosynthetic apparatus of  125
    ribosomes of  22, 601
    taxonomy of  25–9
euryhaline algae  624, 727
evolution
    of chloroplasts  9, 19, 21–3, 32, 129, 155,
        259–61, 282, 303–4
    of ribosomes  601
exchange diffusion  677–82, 691, 703
exocytosis  107, 110
exotoxin  848
exposure, effect of on spore formation  825
extracellular products
    ecological aspects of  852–4
    nature of  839–48
    release processes of  848–52
extraction-replacement studies  393, 398,
        401, 403
eyespot
    carotenoids in  187
    description of  125
    mutations of  323
    structure and function of  147


factor A  622, 752
factor B  622, 749, 751
factor G  622
factor H  622
factor I  622
factor 1  751
factor III  751
factor X₃  749
factor Z  622, 749, 751
factor Z₂  749, 751
factor Z₃  622, 749, 751
FAD (see flavin adenine dinucleotide)
fats
    accumulation of, under phosphorus
        deficiency  640
    photoassimilation and  549
fatty acids (see also lipids)
    as carbon source  533

fatty acids—*cont.*
　assimilation of　544–6, 550
　as taxonomic feature　16, 19
　chemistry of　237–8, 251–6
　composition of　240–51
　distribution of　31, 241–53, 259
　effect of light on　720
　effect of temperature on　725, 726
　oxidation of　510
　phylogeny of　259–61
　respiration of　549, 550
　synthesis of　448, 900
　tolerance of algae to　538–40
fatty acid synthetase　255
FCCP (see *p*-trifluoromethoxy carbonyl
　　cyanide phenyl hydrazone)
feedback inhibition
　of arginine synthesis　598
　of nitrate assimilation　588–593
fenestration　104, 112, 113
fermentation　463, 851
ferredoxin
　antibodies to　400
　characteristics of　395
　functions of　373–4, 400–5, 425, 440, 617
　　461, 462, 466, 561–3, 586–8, 612,
ferredoxin *b*　400
ferredoxin-NADP oxidoreductase
　　(ferredoxin-NADH reductase)　373,
　　374, 400, 403, 562
ferredoxin-nitrite reductase　588
ferredoxin reducing substance (FRS)　373,
　　374
ferricyanide　397, 405, 461–2
ferrocytochrome *c*　522
ferrous sulphate　616
fertilization　318, 327
ferulic acid　769, 770
fibres
　in choloroplasts　144
　in flagella　87–9
　in haptonemata　89
fibrils　866–8
filopodia　869
flagella
　agglutination of　614
　in gamete fusion　830
　microtubules in　96–7
　movement by　865–8, 879
　types of　3, 87–90
　ultrastructure of　866–7
flavacin　189, 191
flavanol　844
flavins　464, 886
flavin adenine dinucleotide (FAD)
　in nitrogen metabolism　585–8
　reduction of　461
flavin mononucleotide (FMN)　405, 461,
　　585–8

flavin mononucleotide-nitrate reductase
　　590
flavodoxin　400
flavoprotein　400, 408, 519, 523, 617, 878,
　　879
flavoprotein dehydrogenase　511
flexibacteria　30, 192
flexixanthin　192, 193
Flimmergeissel (hairy flagella)　88, 104
floridoside　218–20, 223
fluoride
　effect of, on $O_2$ consumption　515
　inhibition of phosphatase by　642
　toxicity of　619
fluorescence
　of algae　466
　of biliproteins　356, 365
　of chlorophylls　349–53, 360, 362
　of flavoproteins　519
　of photosystems I and II　369, 370, 376,
　　460
*p*-fluorophenyl alanine (PFA)　642
FMN (see flavin mononucleotide)
formaldehyde　845
formate
　extracellular production of　843
　reduction　461
formazan　565
N-formyl-methionyl tRNA　317
four carbon acid cycle　436
freshwater algae (see also phytoplankton)
　extracellular products of　839–48
　ion uptake by　683–97
　nitrogen fixation by　571–4
　vitamins and　742, 754–6, 759
FRS (see ferredoxin reducing substance)
fructosans　216–17
fructose　216, 218–19, 535
D-fructose　15
fructose-1,6-diphosphate　436
fructose-6-phosphate　222, 436
fructose diphosphate aldolase　436
fructose-1,6-diphosphate-1-phosphatase
　　436, 438, 442
frustules　90
fucan sulphate　73
fucoidin　10, 72, 73, 107
α-L-fucopyranose　41
4-O-α-L-fucopyranosyl-D-xylose　73
fucose　839
L-fucose　10, 15, 72, 73, 75
fucosterol　17, 267, 270–1, 273–5, 278
fucoxanthin　4, 5, 152–4, 185–7, 347, 354,
　　361, 362
fumarase　447
fumarate　461, 511, 549, 597
funoran　63, 65
furcelleran　63, 69
furfuraldehyde　845

galactan 63–72
galactitol 221–2
galactoaraban 19
galactoarabinoxylan 73–4
β-1,3-galactobiose 73
β-1,6-galactobiose 73
galactofucan 73
D-galactofuranose 74
galactolipids 19, 258
α-L-galactopyranose 41
β-D-galactopyranose 41, 64
3-O-α-D-galactopyranosyl-D-galactose 68, 69
3-O-β-D-galactopyranosyl-D-galactose 74
3-O-α-L-galactopyranosyl-D-galactose 66
4-O-β-D-galactopyranosyl-D-galactose 66, 68–70
4-O-β-D-galactopyranosyl-L-galactose 66
4-O-β-D-galactopyranosyl-L-galactose 6-sulphate 65
O-β-D-galactopyranosyl-(1→4)-O-α-L-galactopyranosyl-(1→3)-D-galactose 66
O-β-D-galactopyranosyl-(1→4)-O-α-D-galactopyranosyl-(1→3)-D-galactose 68
O-α-D-galactopyranosyl-(1→3)-O-β-D-galactopyranosyl-(1→4)-D-galactose 68, 70
O-α-D-galactopyranosyl-(1→3)-O-β-D-galactopyranosyl-(1→4)-O-β-D-galactopyranosyl-(1→4)-D-galactose 68
O-β-D-galactopyranosyl-(1→4)-O-α-L-galactopyranosyl-(1→3)-O-D-galactopyranosyl-(1→4)-L-galactose 66
4-O-β-D-galactopyranosyl-2-O-methyl-D-galactose 70
3-O-(α-L-galactopyranosyl 6-sulphate)-D-galactose 65
3-O-(α-L-galactopyranosyl 6-sulphate)-6-O-methyl-D-galactose 65
2-O-β-D-galactopyranosyl-D-threitol 70
galactose (see also specific compounds below)
    as carbon source 535
    extracellular production of 839
    in cell walls 54, 72, 75
    in lipids 258
    in scales 90, 105, 665
    storage of 218–19
D-galactose 10, 12, 15, 63–6, 69, 72–5
1,3-D-galactose 75
1,3-β-D-galactose 65, 66, 68
1,6-D-galactose 75
L-galactose 10, 12, 15, 63, 65, 66
1,4-α-L-galactose 66
1,4-α-D-galactose 2,6-disulphate 68,71

D-galactose 2-sulphate 69
galactose 4-sulphate 66, 69
1,3-β-D-galactose 4-sulphate 67, 75
D-galactose 4-sulphate 69, 74
galactose 6-sulphate 66
1,4-galactose 6-sulphate 70
D-galactose 6-sulphate 69, 73, 74
1,3-D-galactose 6-sulphate 75
1,4-α-D-galactose 6-sulphate 67
L-galactose 6-sulphate 10, 65, 71
α-galactoside 220
galactosyldiglyceride 618
galactosylglyceride 617
D-galacturonic acid 15, 66, 105
galvanotaxis 882
gametes
    amoeboid movement in 868
    chemotaxis in 829–30
    control of 791
    effect of temperature on formation of 819
    flagella of 865
    inhibition of, by combined nitrogen 822–3
    lunar rhythms of release of 827
    pairing and fusion of 830
gametogenesis
    biochemistry of 830–1
    control of 319, 791, 816, 821–3, 897
gametophyte
    control of 821
    effect of light on 815, 817, 818, 820
    effect of toxins on 798
    morphology of, under various conditions 796, 798
gamone 846
gas bladders 800, 801
gas vacuoles 730
GDP-glucose (see guanosine diphospho-glucose)
gelatin 27, 29
genes
    cytoplasmic 315–17, 326–7
    linkage and mapping of 320–3
    mutations of 319–20, 326–7, 426, 429
    nuclear 315–25
genetic lesion 426, 428 616
genome 169, 285
genophore
    chloroplast 303
    definition of 125
    distribution of 141, 142
    structure of 129, 134, 136, 138, 140
    transmission of, in chloroplasts 148–51
gentamycin 464
gentio-oligosaccharides 213
geotaxis 883
germanic acid 659, 661
germanium
    as inhibitor 661–2
    utilization of 623, 624

giant cells 683–706
gibberellic acid
    effect of, on growth 764, 765, 767,
        794
    effect of, on reproduction 824
    occurrence of 766–7
gibberellins 767, 768, 794
gliding 869–72
glucan 58, 208, 209
α-(1,4)-glucan 207–12
β-(1,3)-glucan 9, 206, 212–16
β-(1,3)-glucanase 215, 216
β-(1,3)-glucan glucohydrolase 215
α-glucan phosphorylase 210
β-(1,3)-glucan phosphorylase 215, 216
β-(1,3)-glucan synthetase 215
glucofructans 217
L-gluco-heptulose 223
glucokinase 541
gluconeogenesis 438, 447, 485, 489
β-D-glucopyranose 41
O-D-glucopyranosyluronic acid-(1→3)-O-
    D-mannopyranosyl-(1→2)-O-D-
    mannose 75
3-O-β-D-glucopyranosyluronic acid-L-
    fucose 73
O-D-glucopyranosyluronic acid-(1→4)-L-
    rhamnopyranosyl-(1→3)-D-
    glucopyranosyluronic acid-(1→3)-D-
    xylose 74
4-O-β-D-glucopyranosyluronic acid-L-
    rhamnose 74
O-D-glucopyranosyluronic acid-(1→4)-L-
    rhamnose 2-sulphate 74
3-O-D-glucopyranosyluronic acid-D-xylose
    74
4-O-D-glucopyranosyluronic acid-D-xylose
    74
glucosamine 841
glucose
    active transport of 539
    assimilation of 209, 223, 483, 506, 509,
        532–8, 540–52
    block 540–1
    effect of, on chlorophyll synthesis 169–
        170
    effect of, on fermentation 463
    effect of, on phosphate metabolism 647
    extracellular production of 839, 853,
        854
    respiration of 640, 716, 725
D-glucose 15, 51, 55, 75, 223
glucose-1-phosphate 210, 215, 541
glucose-6-phosphate 562, 643
glucose-6-phosphate dehydrogenase 507,
    562, 898
glucose phosphate isomerase 220
β-(1→3)-glucosyl hydrolase 215
6-O-D-glucosylmaltotriose 217

β-(1→3)-glucosyl phosphorylase 215
glucosyl transferase 208, 215
β-(1→3)-glucosyl transferase 215
D-glucuronic acid 10, 15, 70, 72, 74, 75
glucuronoxylofucan 73
D-glucuronoxylorhamnans 74–5
glutamate
    as carbon source 517
    effect of vitamin $B_{12}$ on 622
    formation of 593
    in arginine metabolism 597
    in metabolite synthesis 442
    in nitrogen fixation 563
L-glutamate 517
glutamate-glyoxylate amino-transferase
    475, 483, 484
glutamate-hydroxypyruvate amino-
    transferase 475
glutamate-oxaloacetate transaminase 724
glutamate synthetase 564
glutamic acid
    as nitrogen source 595
    effect of, on uptake of silicic acid 658
    extracellular production of 842, 848
    in nitrogen fixation 563
D-glutamic acid 595
L-glutamic acid 594
glutamic dehydrogenase (L-glutamate:
    NAD(P) oxidoreductase) 593, 594
glutaminase 845
glutamine
    as carbon source 536
    as source of ammonia 447
    effect of, on uptake of silicic acid 658
    inhibition of gametes by 822
    in nitrogen fixation 563
    synthesis 594
D-glutamine 595
glutamine synthetase 564, 594
γ-glutamyl hydroxamic acid 585
glutamyl-tRNA 289
glutathione 587
glyceraldehyde-3-phosphate 436, 700
glycerate 476
glycerate dehydrogenase 475
glycerate-1,3-diphosphate 436
glycerate kinase 475
glycerol
    algal production of 9, 839, 840, 848,
        853, 854
    growth on 531, 535, 540, 667
    in osmoregulation 225
    storage of 221–2
    structure of 239
D-glycerol(1→1)-O-α-D-galactopyranose
    220
L-glycerol-(1→1)-O-α-D-galactopyranose
    220
α-glycerophosphate 220, 222, 643

glycine
  metabolism of 442, 461, 476–80, 535,
    536, 618
  production of 483, 488, 489, 842
glycine oxidase 475
glycogen 26, 208, 210, 533
glycolate: DCPIP oxidoreductase 475,
  481
glycolate
  assimilation 488
  effect of light on production of 495–6,
    716, 721
  effect of manganese on 618
  enzymes of 484–6
  excretion of 480, 486–96, 716, 721, 842–
    843, 850
  in metabolite synthesis 442
  pathway 474–83, 488, 489, 491, 900
  synthesis of 478–9, 489, 493, 901
glycolate dehydrogenase 481, 482, 484,
  485, 491, 493, 496
glycolate-glyoxylate shuttle 482–3, 489
glycolate oxidase 91, 475, 476, 480–4, 492,
  495
glycolate reductase 475
glycolic acid
  assimilation of 488
  production of 549, 842, 843, 847, 849,
    850
glycolysis 506–8, 851
glycoprotein 90, 104, 107, 110, 753
glycosaminoglycan 71, 72
glycosidic linkages 42
O-glycosyl-1′-hydroxyl-1′,2′-dihydro-3′,4′-
  didehydro-α-carotene 192, 193
glycyl-glycine 595
glycyl-serine 595
glyoxylate
  in photoassimilation 549
  production of 479–81
  reduction of 482–3
glyoxylate cycle 445, 485–6, 506–8, 511,
  545
glyoxylate reductase 482
glyoxysomes 91, 484-6
Golgi bodies
  acid phosphatase in 91, 104
  coccolith development from 664, 669
  flagellar development from 88, 104
  in sieve elements 95
  occurrence of 86, 92, 93, 103–13, 794
  polysaccharide synthesis in 48, 105,
    107–13
  protein secretion by 110
  scale production by 105–7
grana 126, 130, 131, 136, 141, 142, 368,
  371
α-granules 208, 209
gravity 883

growth
  effect of hydrostatic pressure on 730
  effect of inorganic compounds on 612,
    616, 619, 620
  effect of light on 715–21
  effect of pH on 726–7
  effect of temperature on 721–6
  effect of turbulence on 729
  effect of vitamin $B_{12}$ on 622
  heterotrophic 530–52
  inhibition of 510–11, 619
  kinetics of 541–4
  morphogenesis and 708–807
  of cell walls 42–8
  osmotic effects of 727–9
  rates of 542–3
  regulation of 762–71
  synchronous 894–904
growth factors
  discovery of 742
  effect of, on reproduction 823–4
  effect of, on sporelings 806
  extracellular release of 846–7
  in algal control 762–71
GTP (see guanosine triphosphate)
guanine 22
guanosine diphosphoglucose (GDP-glucose)
  48, 208
guanosine triphosphate (GTP) 602
α-L-gulopyranuronic acid 41
gulose 218–19
L-guluronic acid 15, 62, 72
α-L-guluronic acid 60–1
1,4-L-guluronic acid 10, 11, 58, 59


haemoprotein 519
hairs 3, 88
haliclonasterol 270, 271
halogenated compounds 23
halophilic algae 727
haplotrophs 532–8
n-haptanal 845
hapteron 796
haptonemata 87, 89
Hatch and Slack pathway 3, 437
heavy metals (see also specific metals) 540
helical structures 42–6, 55–7, 60, 70–2, 146
helium 463
hemicellulose 57, 90
n-heneicosahexaene 16–17
1-heptacosene 27
n-heptadecane 17
4(n-heptyl)hydroxyquinoline N-oxide
  (HOQNO) 521
herbicides 209, 459, 573
heterocysts
  effect of nitrogen source on 789, 790
  in taxonomy 790

heterocysts—*cont.*
　lipids of　256, 258–9
　nitrogen fixation in　561, 564–7, 571, 575
　thylakoids of　126
heterokont flagella　87, 88
heterotrophy　444–9, 507, 530–52, 615, 667,
　　716
heteroxanthin　153–4, 183, 186
4,7,10,13 hexadecatetraenoic acid　258
hexadecenic acid　725
*n*-hexane　830
hexokinase　541
hexose (see also specific compounds)
　production of　442, 443, 447
　utilization of　532–3, 535
hexosedophosphatase　220
hexosekinase　541
hexosephosphate　541, 544, 545
higher plants
　boron in　623
　electron transport in　399–403
　lipids of　18
　ribosomes of　22
Hill and Bendall hypothesis　374, 396–7
Hill reaction　259, 260, 371, 463, 567, 615,
　　618, 619, 621
histidine　595
homocysteine　612
HOQNO (see 4(*n*-heptyl)hydroxyquinoline
　　N-oxide)
hormogonia　870
hormones
　and algal regulation　762–71, 824–5
　and sexuality　827, 846
hyaluronate　72
hybridization　21–2, 284–8
hydrazine　561
hydrocarbons　16–18
hydrogen
　active transport of　689, 696, 702
　bonding　43, 49, 54, 55, 57, 60, 70, 144
　metabolism of　396, 456–66, 540, 611, 617,
　　619
hydrogenase　3, 27–9, 440, 456–7, 561–2,
　　617
hydrogen peroxide　479, 485, 496
hydrogen sulphide　459, 461, 465
hydrolase　91, 215, 515
hydrostatic pressure　730
β-hydroxybutyrate　517
1′-hydroxy-1′,2′-dehydro-γ-carotene　192,
　　193
3-hydroxy-echinenone　154, 187, 188
hydroxylamine　459–61, 463, 465, 585
hydroxylamine reductase　587
hydroxylysine　54
4-hydroxymyxoxanthophyll　189
hydroxyproline
　as nitrogen source　595

in cell walls　20, 54, 90, 658, 665, 669
β-hydroxypropionate　510, 520
α-hydroxypyridine methane sulphonate
　　496
hydroxypyruvate　476
hydroxypyruvate reductase　483
8-hydroxyquinoline　508, 523
hydroxysulphonate　483
hyponitrite reductase　587
hypoxanthine　615


IAA (see indoleacetic acid)
ichthyotoxin　729
idose　218–19
indole　764
indoleacetic acid (IAA)　763–5, 768, 790,
　　793–4, 823–4
indoleacetic acid peroxidase　769
3-indolylacetic-2,4-dichlorophenoxy-acetic
　　acid　765
induction
　of alkaline phosphatase　642
　of gamete production　827–9, 831
　of nitrate assimilating enzymes　589, 590
inheritance
　cytoplasmic　315–17, 325–38
　nuclear　315–25
inhibitor/inhibition (see also specific com-
　　pounds)
　effect of pH on　727
　extracellular production of　846
　feedback　589–90, 598
　of cell division　831
　of $CO_2$ reduction　459
　of electron flow　403, 515–23
　of growth　717–18, 800
　of growth factors　771
　of hydrogenase　464, 465
　of oxygen evolution　459
　of photophosphorylation　547
　of photorespiration　496–7
　of photosynthesis　666
　of protein synthesis　600
　of respiration　510–11
　of TCA cycle　507
　of vitamin uptake　751–3
inorganic chemicals (see specific compounds)
inorganic nutrients　610–25
inositol　23, 239, 257
intercellular
　connections　92–5
　plug　14, 92–4
inulin　10, 216
iodine
　as inorganic nutrient　611, 625
　effect of, on reproduction　823
　effect of temperature on uptake of　725
　occurrence of　23

iodoacetamide 459
ionic pumps 727
ion uptake (see also specific ions) 641,
 676–707
ion interactions 59, 62
ion transport
 effect of light on 718–20
 effect of pH on 726
 effect of temperature on 724–5
 in Golgi apparatus 104
iron
 affinity of vanadium for 625
 as nutrient 611, 616–17
 effect of, on copper toxicity 619
 non-haem 519, 523
 requirement for 169, 459, 465, 466, 561,
  588
isoagglutinins 24–5
isocitrate
 metabolism of 486, 517, 522, 545
 production of 843
isocitrate lyase
 activity and purification of 900
 inducibility of 900
 in glyoxylate cycle 485, 486, 508
 photoassimilation and 547
 reported absence in Euglena 545
 suppression of 549
isocryptoxanthin 189, 191
isoenzymes 208, 209, 466
isofloridoside 218–20, 225, 728
isofloridoside phosphate 220
28-isofucosterol 267, 270, 271, 273,
 278
isogamy 318
isokont flagella 87
isoleucyl-tRNA 289
isomerism
 of fatty acid molecules 238
 of sterols 267
isonicotinyl hydrazide 217, 496
isoprenoid units 176, 177
isopropanol 461, 467
isozeaxanthin 190–1

kanamycin 320, 323, 326, 327
α-ketoglutarate (2-oxoglutarate) 442, 507,
 516–20, 563, 593, 597
4-keto-3′-hydroxy-β-carotene 187, 188,
 190, 191
4-ketomyxol-2′-methyl pentoside 190–1
ketose 127
kinase (see specific enzymes)
kinesis 873
kinetin 764–6, 768, 771, 794, 824
kinetosome 867
Krebs' cycle (see tricarboxylic acid
 cycle)

lactate
 extracellular production of 843
 growth on 542–3
 metabolism of 461, 463, 517, 518
L-lactate 518
D-lactate 518
D-lactate dehydrogenase 481
lactic acid 532
lactic dehydrogenase 620
4-O-lactyl-β-D-fructofuranosyl-α-D-
 glucopyranoside 217
lamina 796–7
laminaran
 chemistry of 9, 48, 213, 220
 distribution of 9, 16, 145, 231
 storage of 212–14
laminaribiose 214
laminaribiose phosphorylase 215
laminarinase 214
1-O-β-laminaritetraosyl mannitol 212
laminitol 23, 223–4
lanosterol 266–8, 270
lauric acid 178, 533
lecanoric acid 854
lecithin 18, 239, 538
L-leucanthemitol 223–4
leucine
 incorporation of 799
 translocation of 94
L-leucine 595
leucophytes 532
leucoplasts 15, 551
leucosin 145
lichen
 activity of, at low temperatures 722
 extracellular products of 839, 840, 853–
  854
 nitrogen fixation by 575
life cycles 316, 440–3, 643, 849–50, 899–901
light
 absorption 346–90
 and culture conditions 895–8
 effects (see next main entry below)
 emission 346–90
 flashing 393
 photosynthesis and 346–79, 394–5, 404,
  429, 441, 443
 ultraviolet (see ultraviolet irradiation)
light effects on
 ATP 550, 688
 biliproteins 199
 calcification 665–8
 chlorophyll 168, 170–1, 258, 616, 718, 720
 chloroplasts 884–7
 circadian rhythms 902–4
 copper toxicity 619
 cytochromes 397, 616
 DNA synthesis 291, 294, 296
 extracellular products 850–2

light effects on—*cont.*
  glycerol synthesis  222
  glycolate synthesis  477
  growth  542, 543, 715–21, 795–8, 802, 805
  hydrogenase  465
  inulin accumulation  217
  ion uptake  686, 687, 689, 691, 700, 701,
    702, 706
  lipid composition  258
  manganese uptake  618
  movements  873–80, 884–7
  nitrate assimilation  591–2
  nitrogen fixation  569–71
  oxygen uptake  490–1
  paramylon utilization  216
  pH  427–8
  photoreduction  458, 460, 463
  photorespiration  495–6
  phosphorus metabolism  639–41, 647–8
  reproduction  790, 794, 815–18, 820–1,
    828
  respiration  511–12
lincomycin  147, 286
linkage
  of chloroplast genes  325, 329, 336, 338–
    340
  of cytoplasmic genes  329–40
  of nuclear genes  320–3
linoleic acid  251, 254, 258
linolenate  241, 245, 251, 255
α-linolenate  241, 251, 258
linolenic acid  726
α-linolenic acid  238, 248, 250, 251, 254,
    259–61
γ-linolenic acid  238, 248, 250, 251, 254,
    258, 260, 261
lipases (see also specific enzymes)  426, 538
lipids (see also fatty acids)  236–65
  as taxonomic feature  15–19
  effect of light on  720
  effect of nutrient deficiencies on  552,
    618, 623
  effect of temperature on  725–6
  extracellular production of  844
  formation of  436, 442, 448, 449, 552
  in cell walls  658
  in chloroplasts  136, 140, 147
  in polysaccharide synthesis  48
  storage of  216
lipoprotein  90, 104
liverworts  575
lobopodia  869
loroxanthin  154, 178, 179
luciferin-luciferase assay  638
lunar rhythm  817, 826–7
lutein  4, 5, 6, 153–4, 178, 179, 181, 183,
    185, 187, 192, 354, 363, 368
lutein-5,6-epoxide  154, 179, 182
lycopene  154, 177, 181, 189

lyophilization  722
lysine
  as nitrogen source  595
  in diatom cell walls  658
  occurrence of  20
  synthesis  599
lysosomes  90–1
lysophosphatidyl ethanolamine  256
lysozyme  515, 565
lyxose  218–19

macronutrients  611–15
magnesium
  as nutrient  611, 614, 615, 660
  requirement for chlorophyll  169
  requirement for nitrogenase  562
magnesium chloride  612
magnesium sulphate  587
malate
  as carbon source  517, 545
  in β-carboxylation  444
  in electron transport  520
  in glyoxylate cycle  485, 492, 545
  in metabolite synthesis  442
  oxidation of  516, 517
  production of  437, 511
  reduction of  461
L-malate  517, 518
malate synthetase  485, 508, 511, 545
maleic hydrazide  727
malonic semialdehyde  508, 510, 520
malonyl CoA  448
maltase  211
maltose
  extracellular production of  839, 853
  storage of  217–19
maltotetraose  217
maltotriose  217
*mammalian* type—vitamin specificity  749,
    750, 761
manganese
  as inorganic nutrient  611, 616–18
  effects of deficiency of  258, 459, 466, 496
  effect of, on nitrogen fixation  621
  effect of, on reproduction  823
  in photosynthesis  402, 459, 466, 618
mannan  12–13, 53–4
  β-1,3-D-mannan  75
  β-1,4-mannan  12, 13
mannan triacetate  60
mannitol
  as carbon source  535, 544
  extracellular production of  839, 840,
    848, 853
  in osmoregulation  225
  in polysaccharides  9, 15, 212, 213, 219–
    222, 798
  translocation of  94, 95

1-mannitol acetate  226
1,6-mannitol di-β-glucoside  220
1-mannitol-β-glucoside  220
mannitol-1-phosphatase  220
β-D-mannopyranose  41
β-D-mannopyranuronic acid  41
mannose  212, 839
D-mannose  15, 51, 53, 54, 72, 74, 75
β-1,4-D-mannose  12, 13, 42, 54
4-O-β-D-mannosyl-L-gulose  58
D-mannuronic acid  15, 58, 62, 72
β-D-mannuronic acid  60–1
β-1,4-D-mannuronic acid  10, 11, 58
mapping
    of cytoplasmic genes  329–38
    of nuclear genes  320–3
marine algae (see also individual genera)
    cobalamin production by  749
    extracellular products of  839–48
    ion uptake by  697–706
    nitrogen fixation by  571, 572
    vitamin requirements of  743
MCPA (see 4-chloro-2-methyl phenoxy-
        acetic acid)
medulla  94, 95
meiosis  290–2, 315–24, 330–8
Mendel's laws  315
membrane
    active transport across (see active trans-
        port)
    cell (see plasmalemma)
    of chloroplast (see chloroplast)
    of flagella  87, 88, 867, 868
    of Golgi body  104, 105, 112, 113
    of haptonemata  89
    of intercellular plug  92–4
    of lysosomes  91
    of microbodies  91, 92
    of mitochondria  101–2, 514, 520
    of nuclei  95–6
    of photosynthetic apparatus  125–51
    of pyrenoid  144
    permeability of, to ATP  550
membrane potential
    and absorption changes  375, 376
    effect of light on  719
    effect of pH on  726
    effect of temperature on  724–5
menadione  459, 522
menaquinones  17
mercury  720, 798, 806
meristemplasm  792
meristoderm  796
mesokaryotic nucleus  98
mesotartrate  843
metaphosphates  654, 646
methanol  824
methionine
    as nitrogen source  595

as sulphur source  612
    effect of vitamin B₁₂ on  622
D-methionine  599, 612
L-methionine  599, 612
methionine sulphoximine  320, 322, 323,
        332
6-O-methyl agarobiose  66
β-methyl aspartate  622
6-(3-methyl-2-butenylamine) purine  766,
        771
2-methyl-4-chlorophenoxyacetic acid  573
5-methyl cytosine  286, 622
24-methyl agnosterol  275, 276
methylene blue  461, 462
24-methylene cholesterol  270, 271, 273, 275
24-methylene cycloartenol  268, 269, 276
24-methylene lanosterol  275, 276
24-methylene lophenol  268, 269, 276
methyl ethyl ketone  845
2-O-methyl-4-O-(β-D-galactopyranosyl)-D-
        galactose  70
3-O-(2-O-methyl-α-D-galactopyranosyl)-D-
        galactose  70
4-O-(6-O-methyl-β-D-galactopyranosyl)-D-
        galactose  70
4-O-(6-O-methyl-β-D-galactopyranosyl)-L-
        galactose 6-sulphate  65
2-O-methyl-D-galactose  69, 70
3-O-methyl galactose  66
4-O-methyl-D-galactose  66
6-O-methyl-D-galactose  12, 65, 69, 70
2-O-methyl-L-galactose  66
4-O-methyl-L-galactose  66, 69
methylmalonyl CoA  448, 508
2-O-methyl-4-O-(6-O-methyl-β-D-
        galactopyranosyl)-D-galactose  70
6-O-methyl-(2-O-methyl-D-
        galactopyranosyl)-D-galactose  70
N'methyl-N-nitro-N-nitroso-guanidine
        (NTG)  319, 326–8
4-methyl sterols  269, 270, 276
N⁵-methyltetrahydrofolate  846
methyl viologen  398, 461, 466, 586, 587,
        612
2-O-methyl-D-xylose  74
2-O-methyl-L-xylose  73
4-α-methyl zymosterol  275, 276
metaphase  98–100
MFA (see monofluoroacetate)
microaerophilic nitrogen fixation  564–75
microbodies
    enzymes of  484–6
    in unicellular algae  91
microfibrils  42–6, 51–6, 90, 107
micronone  181
micronutrients  538, 616–21
microsome  522
microtubules
    development of  899

microtubules—*cont.*
  in chloroplasts 146
  in nuclear division 96–100
  of flagella 87, 88, 96
  structure and function of 96–7
mineralization
  calcification 662–70
  silicification 655–62
mitochondria
  active transport in 688, 700
  DNA of 101–3, 286–8, 290, 294–300
  effect of starvation on 513–14
  genes of 315, 329
  enzymes of 484–6, 509–10
  in meristemplasm 792
  in sieve elements 94
  isolation of 514–22
  membranes of 514, 520
  morphological changes in 510–12
  occurrence of 86
  polysomes of 288
  ribosomes of 290, 317
  RNA of 317
  serine production in 483
  structure and function of 101–3, 108,
    109
mitosis 97–8, 338, 623, 657–61
molybdenum
  and nitrogenase 561, 567
  and nitrogen fixation 572, 574, 620, 621
  and nitrate reductase activity 586
  as inorganic nutrient 611, 620–1
monodoxanthin 154, 187, 188
Monod model 541–4
monoenic acid 251
monofluoroacetate (MFA) 507
monogalactosyl diglyceride 16, 18, 19, 240,
  256–9
mono-iodotyrosine 23, 625
monospores 872
morphogenesis 707, 789–807
Mössbauer spectra 617
movement
  amoeboid 868–9
  effect of chemicals on 880–2
  effect of electric stimuli on 882–3
  effect of gravity on 883
  effect of light on 873–80
  effect of temperature on 882
  flagellar 87–9, 865–8
  gliding 69–70
  intracellular 883–7
  metabolic 869
  speed of 865
mucilage 89, 94, 840, 849, 870–2
mucopolysaccharide 24, 104, 112
mutagenesis
  of cytoplasmic genes 325–8
  of nuclear genes 319–20

mutants
  acetate 320, 322, 323, 326–8, 335
  adenine 323
  amino acid 323
  *p*-amino-benzoate 320–3
  ammonia 320, 323
  antibiotic and drug resistant 320–8,
    335–7
  arginine 320–3
  auxotrophic 320–3
  chloroplast 323, 328, 430
  cytoplasmic 325–9
  genes and 314–40, 426, 429
  nuclear 319–24
  photosynthetic 393, 394, 424–31, 518
  ribosomal 328, 337
  structural 323
  temperature sensitive 320–8, 336
  vitamin requiring 320–3
mutations (see mutants)
mutatochrome 153–4, 189–91
*myo*-inositol 223–4, 239
myristic acid 258, 533
mytilitol 223–4
myxol-2′-glucoside 188
myxol-2′-O-methyl-methylpentoside 190–
  191
myxol-2′-methylpentoside 188, 189
myxorhodin-α 191
myxorhodin-β 191
myxoxanthin 187, 361
myxoxanthophyll 5, 6, 153, 187–92, 361,
  368, 567

NAA (see naphthalene acetic acid)
NAD and related compounds (see
    nicotinamide adenine dinucleotide)
NADH and related compounds (see
    nicotinamide adenine dinucleotide
    (reduced form))
NADP and related compounds (see
    nicotinamide adenine dinucleotide
    phosphate)
NADPH and related compounds (see
    nicotinamide adenine dinucleotide
    phosphate (reduced form))
naphthaleneacetic acid (NAA) 765, 794,
  800, 824
1-naphthoxyacetic acid 765, 800
neamine 327, 328, 335, 336
neoagarobiose (3-O-(3,6-anhydro-α-L-
    galactopyranosyl)-D-galactose) 64
neoagarohexose 65
neoagarotetrose 65
neocarrabiose sulphate (3-O-(3,6-anhydro-
    α-D-galactopyranosyl)-D-galactose 4-
    sulphate) 68
neomycin 900

neoxanthin 5, 152–4, 178, 179, 183–4, 187
Neujahr factors 751
neurotoxic compounds 24
nicotinamide 320, 322, 323
nicotinamide adenine dinucleotide (NAD)
    225, 475
nicotinamide adenine dinucleotide-
    glutamate dehydrogenase 594
nicotinamide adenine dinucleotide-
    hydroxypyruvate reductase 475, 483
nicotinamide adenine dinucleotide (reduced
    form) (NADH) 462, 475, 511, 516–22,
    547, 585–7
nicotinamide adenine dinucleotide (reduced
    form) dehydrogenase 520
nicotinamide adenine dinucleotide (reduced
    form) diaphorase 590
nicotinamide adenine dinucleotide (reduced
    form) hydroxypyruvate reductase 482
nicotinamide adenine dinucleotide (reduced
    form)-nitrate reductase 588, 621
nicotinamide adenine dinucleotide (reduced
    form) oxidase 511, 516–22
nicotinamide adenine dinucleotide
    phosphate (NADP) 348, 373–5, 392–
    412, 424–7, 461, 462, 475, 562, 597, 612,
    617, 618
nicotinamide adenine dinucleotide
    phosphate-glucose-3-phosphate
    dehydrogenase 898
nicotinamide adenine dinucleotide
    phosphate-glutamate dehydrogenase
    594
nicotinamide adenine dinucleotide
    phosphate reductase 588
nicotinamide adenine dinucleotide
    phosphate (reduced form) (NADPH)
    225, 348–9, 392–412, 436, 439, 449, 475,
    483, 507, 522, 585–8, 597
nicotinamide adenine dinucleotide
    phosphate (reduced form)-glyoxylate
    reductase 475, 479, 482
nicotinic acid 841, 846, 851
nitrate
    as nitrogen source 584–93
    effect of deficiency of 613
    effect of light on uptake of 719
    effect of, on gamete production 822
    effect of, on nitrogen fixation 571–2
    effect of, on polyphosphate synthesis 674
    effect of temperature on uptake of 725
    reduction of 27, 28, 461, 550–2, 584–8,
        621, 900
nitrate reductase (NAD(P)H : nitrate
    oxidoreductase)
    characteristics and function of 585–7
    control of 589–93
    effect of pH on 727
    effect of sodium deficiency on 624

heat stability of 722
nitrite
    as nitrogen source 584–93
    effect of light on uptake of 719
    effect of sodium deficiency on 624
    in synchronous cultures 899
    reduction of 461, 462, 617
    release under osmotic stress 729
nitrite reductase (NAD(P)H : nitrite
    oxidoreductase)
    characteristics and function of 585, 587–
        588
    effect of pH on 727
    in nitrite reduction 462, 464
    regulation of 589, 590
nitrogen
    assimilation of 583–99
    effect of mineral deficiency on metabolism
        of 615, 621
    effect of, on carboxylation 445, 446
    effect of, on mating 319
    effect of, on photorespiration 493
    effect of, on photosynthesis 443
    effect of, on pigmentation 169, 509, 552,
        790
    effect of, on RNA synthesis 293
    extracellular production of 841–2
    fixation (see nitrogen fixation)
    hydrogen production under 463
    metabolism of inorganic 583–93, 611
    metabolism of organic 593–9
    transfer of fixed 573–4
nitrogenase
    and nitrogen fixation 560–76
    effect of phosphorus levels on activity of
        639, 641
    occurrence in algae 466
nitrogen fixation 560–76
    biochemistry of 561–4
    effect of inorganic elements on 621, 624
    effect of light on 720
    effect of pH on 727
    effect of temperature on 725–6
    lipids and 258
    photosynthesis and 569–70
    physiological ecology of 570–4
    site of 128, 564–7
nitrophenol 461
non-reciprocal exchange 338
nostoxanthin 190–1, 193
NTG (see N′methyl-N′-nitro-N-nitroso-
    guanidine)
nucleic acids (see also DNA and RNA)
    comparative aspects of 303–4
    effect of ultraviolet light on 718
    extracellular production of 841
    metabolism of 281–304
    monomers for 446
    taxonomic use of 21–3

nucleolus  98, 100, 102, 103
nucleoplasm  789
nucleoside diphosphate sugars  48, 215
nucleoside phosphatase  113
nucleoside polyphosphates  841, 844
nucleoside triphosphates  659
nucleotides  613
nucleus
    and chloroplasts  133
    and Golgi bodies  108–10
    division of the (see also meiosis and
        mitosis)  97–103, 613
    DNA of the  286–303
    envelope of the  95–103
    genes of the  315–25
    morphogenic substances in the  792
    movement of the  884
    occurrence of  86, 92–4
    of *Acetabularia*  297–303
    structure of  95–6
nutrients
    effect of, on chlorophyll  169–70
    effect of, on algal development  789, 792
    effect of, on photosynthetic rate  441,
        443
    inorganic  610–25
    in synchronous cultures  896
    mineral, and nitrogen fixation  572


obtusifoliol  275, 276
oleandomycin  326, 327, 336
olefins  17
oleic acid  241, 244, 245, 251, 254, 259
oligosaccharides  216, 218–19, 728
olive oil  538
oogonia  827, 829
operon theory  829
organelles  86–123
organic acids  842–3, 848, 851
organic compounds
    effect of, on algal reproduction  823–5
    effect of, on algal growth  530–52
ornithine  442, 595, 597
ornithine transcarbamoylase (carbamoyl-
        phosphate: L-ornithine carbamoyl
        transferase)  598
*O*-ornithyl phosphatidyl glycerol  256
orotic acid  288, 615
orthophosphate  637, 643–5, 647
orthophosphate  643–8
oscillaxanthin  5, 6, 153–4, 187–91
oscillol-2,2′-di(O-methyl) methyl pentoside
        190–1
osmoadaptation  727, 728
osmoregulation  91, 92, 104, 225, 727
osmotic pressure
    effect of, on metabolism  728–9
    effect of, on photosynthetic products  443

ouabaine  625, 686, 687
oxalate  702
oxaloacetate
    as substrate  545
    in metabolite synthesis  442, 444
    production of  485
oxidative phosphorylation (see phosphory-
        lation)
oximes  585
24-oxocholesterol  275
2-oxoglutarate (see α-ketoglutarate)
oxygen
    effect of, on chemotaxis  881
    effect of, on glycolate production  477,
        479, 487, 490
    effect of, on nitrogenase  562, 568, 569
    effect of, on photorespiration  492–3, 496
    effect of, on photosynthesis  441, 443,
        476
    effect of, on phosphorus uptake  640
    effect of, on polyphosphate formation
        648
    effect of, on respiration  508–9
    evolution  348, 373, 392–412
    in primeval atmosphere  467
    reduction  461
    uptake  490–1, 510–14, 725
oxyhydrogen reaction  437, 438, 440
oxytrophs  532–7

P416  522
P430
    characteristics of  395
    distribution of  373–4, 399
    function of  376–7, 407
P450  522
P680–690
    characteristics of  375–7, 395
    distribution of  399
    function of  350, 370, 373–9, 407
P700
    characteristics of  395
    distribution of  356, 368, 369, 398, 567
    function of  350, 370–5, 399, 407, 408
P750  358, 360
paddy fields  574
palmitic acid
    metabolism of  254, 255
    occurrence of  250
    production of  725
    structure of  238
palmitoleic acid  238, 250
panose  217
pantothenate  846
paraflagellar body  879
paramylon  145, 215–16, 443
parthenospore  824, 828, 829
passive diffusion  677–82, 691–3, 706

PChl. (see protochlorophyllide)
PCP (see pentachlorophenol)
pectin 14, 105, 107, 614
pedigree analysis 331
Peitschengeissel (whiplash flagella) 88
pellicle 90
pentachlorophenol (PCP) 593
1-pentacosene 27
n-pentadecane 17
D-2,3,5/4,6-pentahydroxycyclo-hexanone 223–4
pentose phosphate pathway 447, 506–8, 550
peptides 841–2, 900
perichloroplastic matrix 137
perichloroplastic space 137
peridinin 4, 152, 154, 186, 187, 361
periplast 90, 869, 872
permeability 538–40, 848
permease 540
peroxisomes
    enzymes of 480–6, 489, 511, 522
    fractionation of 515
    glycolate pathway in 476
    occurrence of 91, 522
    phylogeny of 497
PFA (see p-fluorophenyl alanine)
pH
    effect of, on coccolith formation 667
    effect of, on copper toxicity 619
    effect of, on extracellular product release 850
    effect of, on glucose metabolism 540
    effect of, on glycolate metabolism 477, 480–1, 487, 494–5
    effect of, on ion uptake 689, 690
    effect of, on membrane potential 692, 700
    effect of, on nitrogen fixation 572
    effect of, on photosynthetic rate 440–1
    effect of, on polyphosphate formation 648
    effect of, on respiration 511
    in taxonomy 27
    regulation of intracellular 446, 449
    requirements and tolerance 28, 538–40, 726–7
phagotrophic algae 533, 538, 744, 746
PHB (see poly-β-hydroxybutyrate)
O-phenanthroline 459, 460, 463, 465
phenazine methosulphate (PMS) 402, 405 408, 428, 461, 509
phenolic substances 844, 847
phenosafranine 461
phenotype 323–9
phenotypic effects 317, 339
phenylalanine
    as nitrogen source 595
    effect of, on morphology 769

L-phenylalanine 594
phenylalanyl-tRNA 289
phenylurethane 459, 463
pheophytin 165, 166
pheromones 24
phlorizin 522, 688
phobotaxis 873
phoenicopterone 154, 181, 183, 192
phosphatase
    effect of phosphorus on activity of 641–642
    extracellular production of 845
    in phosphorus metabolism 643
phosphate
    active uptake of 688–9
    effect of current velocity on uptake of 729
    effect of magnesium deficiency on 615
    extracellular production of organic 844
    loss in high salinity 728, 729
    metabolism of 637–49
    requirement for photosynthesis 459
    requirement of nitrate reductase for 587
    turnover during photosynthesis 439–40
phosphatidic acid 239
phosphatidyl choline 18, 239, 254, 256, 257
phosphatidyl ethanolamine 18, 256
phosphatidyl glycerol 18, 254 256, 257, 260
phosphatidyl inositol 18, 239
phosphoenolpyruvate 442–5, 485
phosphoenolpyruvate carboxykinase 445
phosphoenolpyruvate carboxylase 435, 442–6, 449
phosphoglucomutase 541
6-phosphogluconate dehydrogenase 507
3-phosphoglyceraldehyde dehydrogenase 436
3-phosphoglycerate
    in $CO_2$ fixation cycle 436, 442
    in glycolate pathway 476, 478
3-phosphoglycerate kinase 436
phosphoglycerate phosphatase 475, 480, 483
phosphoglycolate 478–9, 488
phosphoglycolate phosphatase
    function and properties of 480
    in chloroplasts 479
    in glycolate excretion 487, 494
    in photorespiration 475, 483
phospholipids 239, 256, 260
phosphopentose epimerase 426
phosphoribose isomerase 436
phosphoribulokinase 436
phosphorus
    as inorganic nutrient 611
    effect of temperature on requirement for 725
    in relation to nitrogen fixation 572

phosphorus—*cont.*
  metabolism 636–49
  requirement for gamete production 823
phosphorylase 208, 209
phosphorylation
  and movement 874, 875
  cyclic 460, 463
  non-cyclic 700
  oxidative 407–10, 440, 509–11, 515–22, 541, 563, 641, 644, 689
  photophosphorylation (see separate entry below)
  requirement of, for chlorine 621
  substrate 644
phosphotriose isomerase 436
photoautotrophy 530
photoheterotrophy 716
photoinhibition (see also light effects on) 717–18
photokinesis 873–5
photo-organotrophy 530–2, 541–4, 549–50
photoperiodism 762
photophobotaxis 875–8
photophosphorylation 391–412
  active transport and 687
  as source of ATP 563
  $CO_2$ as catalyst for 448
  cyclic 392–412, 440
  effect of manganese deficiency on 618
  effect of phosphorus level on 641
  in heterocysts 567
  in mutants 427–9
  mechanisms of 644
  movement and 874–5
  photoassimilation and 547–9
  polyphosphate synthesis and 647, 648
  respiration and 550
photoreceptors
  for chloroplast orientation 886
  eyespots as 147
  nature of 873–80
photoreduction
  and manganese deficiency 618
  copper in 619
  inhibition of 459
  occurrence and mechanism of 458–61, 563
  of benzoquinone 396
photorespiration 474–97, 511–12, 570
photosynthesis
  as source of ATP 550
  bacterial 547
  $CO_2$ fixation in 435–43
  effect of inorganic ions on 614, 616, 618, 619, 623, 624
  effect of hydrostatic pressure on 730
  effect of light intensity on 715, 717, 718
  effect of osmotic pressure on 728
  effect of phosphorus levels on 640

  effect of temperatures on 722–4
  electron flow in 391–412, 424–31
  evolution of 126
  fatty acids and 260
  glycolate excretion during 486
  in coccolith formation 666
  inhibition of 476
  in synchronous cultures 898
  light absorption and 346–79
  mutations of 323, 424–31
  nitrogen metabolism and 569–70, 590
  phylogeny and 259
  phosphorylation and 391–412, 427–9
  polyphosphate synthesis and 647–8
  products of 208, 222, 223
  reproductive rate and 815–16
  respiration and 511
  site of 125–54
  sulpholipids and 257
photosynthetic carbon reduction cycle (see Calvin-Benson cycle)
photosynthetic lamellae (see thylakoids)
photosynthetic pyridine nucleotide reductase (PPNR) 617
photosystem I
  characteristics of 394
  chemical components of 398–403
  chlorophyll of 396
  coupling sites 407–8
  effect of inorganic ions on 618, 619
  effect of, on $O_2$ uptake 512
  energy transfer in 363
  in heterocysts 567
  ion uptake and 687, 719
  light absorption and 347–79
  movement and 874
  mutants and 426, 428
  nitrogen fixation and 720
  oxyhydrogen reaction and 440
  photoheterotrophy and 716
  photoinhibition and 718
  photophosphorylation and 393–412
  photoreduction and 460
  regulation of 411–12
  separation of 365–9
  site of 143, 368
photosystem II
  absence of, from heterocysts 566, 567
  acetate assimilation and 169, 547
  characteristics of 394
  chemical components of 398–403
  chlorophyll of 163, 396
  $CO_2$ as catalyst for 448
  coupling sites 407, 408
  effect of manganese deficiency on 618
  energy transfer in 363
  fluorescence yield of 404
  ion uptake and 691, 719
  light absorption and 347–79

movement and 874
mutations affecting 425–7, 429
oxyhydrogen reaction and 440
photophosphorylation and 393–412
photoreduction and 460, 464
photorespiration and 489
separation of 366–9
site of 143, 368
phototaxis 625
phototopotaxis 878–80
phototropisin 878
phragmaplast 794
phycobilin
　absorption spectra of 353–5
　as taxonomic feature 7–9, 153–4
　effect of hydrostatic pressure on 730
　effect of light on synthesis of 199
　in light absorption 346, 351, 363–5
　in photosystems 370
　occurrence of 4
　site of 128
phycobiliprotein granules (phycobilisomes)
　129, 134, 140, 141, 196, 368
phycobionts 222
phycocyanin (see also specific types below)
　absorption spectra of 355, 363, 364
　adaptation of 377–9
　as index of nitrogen content 564
　characteristics of 354
　fluorescence 365
　in chloroplasts 141
　in heterocysts 567
　in photosystems I and II 369, 370
　occurrence of 7, 31, 154, 196, 354, 363
　structures of 197–9
　types of 197–9
C-phycocyanin
　absorption spectra of 195, 355
　in relation to movement 874–6, 879
　occurrence of 7–9, 26, 153, 194, 347
　physical properties of 356
　structure and function of 197–8
R-phycocyanin
　absorption spectra of 198
　occurrence of 7–9, 153, 194, 199
　physical properties of 356
　structure and function of 199
phycocyanobilin 7, 193, 194, 197, 199
phycoerythrin (see also specific types below)
　absorption spectra of 355, 364–5
　adaptation of 377
　as taxonomic feature 8
　characteristics of 354
　in Synechococcus 31
　occurrence of 7, 141, 154, 354, 363
　types of 194–7
B-phycoerythrin 194, 195, 197, 356
C-phycoerythrin
　absorption spectra of 195, 355

in relation to movement 876, 879
occurrence of 7–9, 153, 194
physical properties of 356
structure and functions of 196
R-phycoerythrin
　absorption spectra of 195, 197, 355
　occurrence of 7–9, 153, 194, 347
　physical properties of 356
　structure and function of 197
phycoerythrobilin 7, 193, 194, 196, 197, 199
phycourobilin 194, 197
phylloquinone (see also vitamin $K_1$) 17
phyllymenan 69
physico-chemical factors affecting growth 714–30
phytane 18
phytochrome
　functions of 168, 790, 794, 817, 818, 886
　occurrence and structure of, 7, 193, 194
phytoene 177
phytoflavin (flavodoxin) 400, 563
phytoplankton
　effect of light on vertical distribution of 717
　effect of pH on 727
　extracellular products of 842–8, 851–853
　nitrate reductase in 587
　nitrogen fixation by 570–4
　optimum phosphorus levels for 639
　vitamin metabolism of 754, 757, 760, 761
Piericidin A 519, 520
pigments (see also specific pigments)
　adaptation of 377–9
　as taxonomic feature 3–9
　effect of light on 720–1
　effect of pH on 727
　in solution 353–6
　of chloroplasts 152–5
　photosynthetic 346–79, 399
　respiratory 516
　species lacking photosynthetic 530, 551
plasmalemma
　in cellulose production 107
　in ion transport 683, 686, 690, 693, 696–706
　in extracellular release 848
　in wall formation 657, 658
　occurrence and structure of 89, 94, 95, 103, 107, 126–8, 789
plasmodesmata 94, 95
plastids 124–55
　association with enzymes 533
　location of cytochromes in 616
　occurrence of 86
　origin of 9, 19, 21–3, 32, 120, 153, 259–261, 282, 284, 302–4

plastocyanin
  characteristics of 395
  copper requirement for 619
  in mutants 425
  in photosynthesis 373, 374
  photo-oxidation of 400–1
  reduction of 407
plastoglobuli 125, 147
plastoquinones (PQ)
  as taxonomic feature 17
  effect of light intensity on 373, 720
  in photosynthesis 374, 407, 408
  photo-oxidation of 400, 401
  requirement for 403
  site of 147
PMS (see phenazine methosulphate)
polar granules 31
polar lipids 238, 255
polarity 799
pollutants 798
polyenoic acid 251, 259
polyglucans 208, 209, 214
polyglucan phosphorylase 211
polyguluronic acid 58, 60–2
poly-β-hydroxybutyrate (PHB) 23, 259
polymannuronic acid 58, 60, 62
polymorphism 657, 791, 792
polynucleotide phosphorylase 285
polyols 220–2, 225
polypeptides
  of chloroplast membranes 426, 431
  synthesis studies 900
polyphosphate
  as storage product 642–3
  effect of acid on 637
  effect of pH on 727
  metabolism 645–9
  structure of 645, 646
  synthesis and function of 647–9
polyphosphate-glucose phosphotransferase
    649
polysaccharides (see also specific com-
    pounds)
  as taxonomic feature 9–15
  enzymes of 533
  extracellular production of 839–40, 849,
    851
  in cell walls 40–76, 92
  in Golgi apparatus 104, 105, 112
  in mucilage 89
  storage of 144–6, 207–25, 533
  sulphated 63–75, 92, 107–8, 612
polysomes 288–90, 897, 900
polytomaxanthin 192
polyuridine 75
polyuronide alginic acid 107
pores 92, 93, 96
poriferasterol 267–70, 274, 277, 278
porphyran 10, 12, 63, 65–6

porphyrins 169, 436, 442, 614
potassium
  as nutrient 611, 613–14
  effect of light on uptake of 719
  effect of temperature on uptake of 724–
    725
  effect of, on copper toxicity 619
  transport of 683–7, 689, 691–2, 696–
    707
potassium cyanide 508, 516, 519, 521
PPNR (see photosynthetic pyridine
    nucleotide reductase)
PQ (see plastoquinones)
primary production
  effect of light on 715
  role of extracellular products in 852–3
  role of vitamins in 759
pristane 17, 18
prokaryotic organisms (general)
  ATP in 393
  nitrogenase in 466
  photophobotaxis in 876–8
  photosynthetic apparatus in 125–8
  ribosomes of 601
proline 442, 848, 851
pronase 46
prophase 97–100
propionate 448, 507–10, 549, 622
propionyl CoA 448
proplastids 133, 148–50, 285–6, 551
prostaglandins 241
protein
  as taxonomic feature 19–21
  effect of light quality on 720
  effect of nutrient deficiencies on 615,
    622, 623
  effect of salinity on 728
  extracellular production of 841, 842,
    849
  histone 21, 98
  in cell walls 43, 46, 54, 90
  in flagella 87, 88
  in Golgi apparatus 110
  in microtubules 97
  in pyrenoid 146
  in scales 665
  in synchronous cultures 898, 901
  sulphur in 612
  synthesis 599–602
    cytoplasmic 792
    DNA and 296
    inhibitors of 464, 511
    in plastids 147, 317, 902
    magnesium for 615
    of monomers for 446
  turnover during growth 791, 798,
    799
protochlorophyll 3, 6, 168
protochlorophyllide (PChl.) 167, 168

proton gradient   522
proton movement   427, 428
protoporphyrin
    extracellular production of   841, 851
    iron requirement for   616
protozoa
    algal carotenoids in   192
    evolution of   532
    mitochondrial preparations of   516
    ribosomes of   601
pseudopodia   746, 868–9
pseudo-vitamin $B_{12}$   622, 752
pthiocol   459, 465
puromycin   464, 590
putrescene   595
pyrenoid(s)
    cap   145, 146
    early work on   125
    in theca production   105
    multiplication of   146
    occurrence of   28, 87, 96, 103, 125, 429
    structure and function of   144–6
pyrenoxanthin   154, 178, 179
pyridine-adenine dinucleotides (see also
        NAD(H) and NADP(H))   401
pyridine haemochromogen   516
pyridine nucleotide (see also NAD(H)
        and NADP(H))   522
pyridine nucleotide oxidases   522
pyrimidines   436, 442, 446, 449
pyrosulphite   461
pyruvate
    and ammonium assimilation   593
    and dark respiration   516, 517, 549
    and fermentation   463
    and nitrogen fixation   562
    and oxidative assimilation   544, 545
    growth rates on   543
    reduction of by $H_2$   461
pyruvate dehydrogenase   562
pyruvate-ferredoxin oxido-reductase   562
pyruvate kinase   507
pyruvic acid
    as carbon source   532
    in β-carboxylation   444
    in metabolite synthesis   442
pyruvic acid carboxylase   544
$Q_{10}$
    of ion uptake   723, 725
    of nitrogen fixation   573, 725
Q enzyme   208–11
quantosomes   430–1
quantum efficiency
    of acetate photoassimilation   547
    of cation transport system   687
    of fluorescence   353
    of photosystem II   351
    red drop in   397
    red rise in   398

quantum requirement
    of photophosphorylation dependent
        processes   406
    of photosynthesis and photoreduction
        460
    of photosystems I and II   398
    of synchronous cultures   900
quantum yield
    of cyclic photophosphorylation   411
    of oxygen evolution   358
    of photo-oxidation of cytochrome f   401
    of photosynthesis   347, 393, 460
quinacrine   688
quinone   371, 461, 462

radiation
    effects of, on gene recombination   324–5
    sensitivity of Chlorella to   901
raphe   871
reciprocal exchange   338
recombination   324, 330, 338
red drop phenomenon   347, 397 687
red rise phenomenon   398
redox cofactor   402
redox potential
    effect of light on   404
    in relation to iron uptake   616
    of cytochromes   401, 402, 426, 511
    of photosynthetic processes   374, 397
    of plastocyanin   401
regulation
    of alkaline phosphatase   642
    of algae, by plant hormones   762–71
    of amino acid synthesis   594, 598
    of chlorophyll synthesis   168–9
    of nitrate assimilation   588–93
    of products of photosynthetic carbon
        reduction cycle   441–3
    of rate of autotrophic $CO_2$ fixation   440–
        441
repression
    of arginine synthesising enzymes   598
    of nitrate reducing enzymes   588–93
reproduction   292–0, 814–31
respiration
    cytochromes in   616
    dark   505–23
    effect of iron and manganese on   616
    effect of light on   511–12, 718, 719
    effect of osmotic stress on   728–9
    effect of pH on   511
    effect of phosphorus levels on   640, 641
    effect of temperature on   511, 725
    effect of thiamine on   544
    effect of turbulence on 729
    endogenous   510, 512–16, 521
    inhibition of   507, 508, 515–23, 619, 718,
        798

respiration—*con*
  in synchronous cultures  898
  nitrogen fixation and  569–70
  photorespiration (see photorespiration)
  photosynthesis and  511
  reproduction and  817
  site of  128
respiratory quotient  507, 512, 546
α-L-rhamnopyranose  41
O-L-rhamnopyranosyl-(1→4)-O-D-
    xylopyranosyl-(1→3)-D-glucose  74
rhamnose
  extracellular production of  839
  storage of  218, 219
L-rhamnose  10, 15, 74, 75
1,2-L-rhamnose  75
L-rhamnose 2-sulphate  74
rhizoids  43, 44, 793, 794, 799
rhizopodia  869
rhodoquinone  509
rhythms  817, 826–7, 901–4
ribitol  221–2, 839, 853
riboflavin  846, 886
*ribo*-hexulose  217
ribonuclease (RNase)  284, 288, 301, 303
ribonucleic acid (RNA) (see also specific
    types below)
  acid resistance of  637
  as taxonomic feature  22–3
  effect of light quality on  721
  effect of nutrient deficiencies on  615,
    616, 620, 622, 623, 661
  effect of salinity on  728
  in chloroplasts  147, 285, 317
  in gametogenesis  831, 897
  in nucleus  95, 96
  in prokaryotic and eukaryotic organisms
    601
  in protein synthesis  601–2
  in pyrenoids  145
  polyphosphate and  647
  synthesis of  284, 282–7, 301–3, 897–900
ribonucleic acid-messenger (mRNA)  289,
    297
ribonucleic acid-ribosomal (rRNA)  284-5,
    288–90, 293, 297, 301, 303
ribonucleic acid-transfer (tRNA)  293, 297
ribonucleic acid polymerase  284, 293, 302
ribonucleic acid synthetase  289
ribose
  in scales  90, 665
  storage of  218–19
ribose-5-phosphate  436
ribosomes
  chloroplast  125, 130–3, 135, 137, 138,
    147–8, 303, 600
  degradation of  897
  effect of zinc deficiency on  620

  in protein synthesis  600–2
  mitochondrial  103
  mutations of  328, 337
  types and properties of  22–3, 87, 284–5,
    288–90, 292–4, 296–7, 301–2, 601
ribotides  622
ribulose diphosphate carboxylase  435,
    436, 438, 449, 475
ribulose diphosphate oxygenase  475
ribulose-1,5-diphosphate  436
ribulose-1,5-diphosphate carboxylase  429,
    615
ribulose-5,5-diphosphate carboxylase  900
ribulose diphosphoric acid (RuDP)  475,
    477, 478, 488, 496
ribulose diphosphoric acid carboxylase
    478, 479, 489, 494
ribulose diphosphoric acid oxygenase  478,
    479, 494
ribulose-5-phosphate (Ru-5-P)  436
ribulose-5-phosphate kinase  479
rifampicin  168, 293, 294, 301, 303, 317
rifamycin  284, 290
RNA and related compounds (see ribo-
    nucleic acid)
mRNA (see ribonucleic acid-messenger)
rRNA (see ribonucleic acid-ribosomal)
tRNA (see ribonucleic acid-transfer)
rotenone  516, 517, 519–22
rubimedin  402
rubidium
  effect of phosphate on uptake of  641
  replacement of potassium by  613
RuDP and related compounds (see ribulose
    diphosphoric acid)
Ru-5-P and related compounds (see
    ribulose-5-phosphate)


salicylaldoxime  463
salinity
  effect of, on extracellular products  851
  effect of, on nitrogen fixation  572
  effect of, on pigmentation  379
  effect of, on reproduction  823
  tolerance of  727–9
saproxanthin  192, 193
sargasterol  274, 275
saringosterol  275
satellite DNA (see deoxyribonucleic acid)
scales
  composition of  107
  development of  105–7
  flagellar  88, 90
  surface  90
scintillons  523
*scyllo*-inositol  223–4
seawater
  growth factors in  766, 770, 771

humic substances in   844
  toxicity of   759, 761
  vitamins in   753–4, 757–9
sedimentation coefficient   601
sedoheptulose-1,7-diphosphate   436
sedoheptulose-1,7-diphosphate-1-
    phosphatase   436, 438
sedoheptulose-7-phosphate   436
segregation   315–16, 330–8
selenium   612
selenomethionine   613
serine
  effect of manganese deficiency on   618
  effect of vitamin $B_{12}$ on   622
  in glycolate pathway   476–80
  in metabolite synthesis   442
  occurrence and structure of   239
  production   483, 488, 489, 842
D-serine   595
serine-glyoxylate amino transferase   475,
    483
serine hydroxymethyl transferase   475, 476,
    483, 491
sex factors   846
sexual reproduction   814, 816, 818, 820–4
sexuality
  in algae   319
  physiology of   827–30
sheath   840
sieve plates   94, 95
sieve tubes   94, 95
silica   90, 655–62
silicalemma   658
silicic acid
  effect of germanium on metabolism of
    661–2
  requirement in diatoms   623
  uptake of   658–61
silicification   655–62, 669
silicoflagellates   656
silicon
  as inorganic nutrient   611, 623–4, 660–1
  effect of deficiency of   514, 660
  effect of germanium on uptake of   661–2
  requirement in DNA synthesis   303
siphonein   6, 154, 178–81
siphonaxanthin   6, 154, 178–82
sitosterol   17, 18, 267, 270, 271, 273, 274,
    278
soda cellulose   60
sodium
  as inorganic nutrient   611, 613, 624
  effect of light on uptake of   719
  transport of   683–7, 689, 691–2, 696–707
sodium chloride (see also salinity)   621
sodium dimethyl dithiocarbamate   619
sodium mannoglycerate   218–19
sorbitol
  extracellular production of   829, 853

storage of   221–2
sorbose   218–19
Soret band   164, 358, 375
soxhlet extraction   260
specific extinction coefficients   352
spectinomycin
  inhibitory effects of   168, 293, 294, 430
  resistant mutants   320, 323, 326, 327, 336
sperms   829, 846, 867
spermatia   872
spinasterol   270, 271
spindle in nuclear division   97–101
spines   88
spiral growth   46
spiromycin
  effect of, on ribosomes   147, 285
  resistant mutants of Chlamydomonas
    326, 327, 336
sporangia
  effect of growth factors on production of
    823
  effect of light on development of   816, 820
  effect of temperature on development of
    819, 820
  of Laminariales   798
spore(s)
  as reproductive structures   814–31
  factors affecting formation and liberation
    of   815–25
  rhythms of release of   826–7
sporophylls   213, 798
sporophyte   794–8
sporopollenin   90
squalene   18, 266, 268
squalene-1,2-oxide   266, 268
starch
  chlorophycean   210–12
  cyanophycean   145
  floridean   145, 207–8
  granules   92, 93, 102, 103
  in chloroplasts   135, 137, 144, 145
  in theca formation   105
  ingestion of, by phagotrophs   538
  myxophycean   208–10
  storage forms of   206–12, 532
  role of   9–10, 212
starch phosphorylase   541
stearic acid   238, 254
stellate pattern   87
stenohaline algae   727
stereo-specificity   254
steroids
  as taxonomic feature   17–18
  chemistry of   237
sterols   266–80
sterol glycoside   18
stigma   879
5,24,28-stigmasta-diene-3β,7α-diol   275
5α-stigmasta-8,14-dien-3β-ol   269, 271

3,5,24,28-stigmastatrien-7-one   275
stigmasterol   270, 271, 274
stipe   796
storage products   206–25, 238
streptomycin
 as mutagen   321, 325–8
 bleaching by   286, 551, 612
 dependence on   335, 336
 resistance, cytoplasmic inheritance of
  329, 336, 337
 resistant mutants   320, 323
stroma
 and coupling sites of phosphorylation
  407
 connected to nucleoplasm   140
 definition of   125
 generation of NADPH in   401
 microtubules in   146
 photosystems sited in   368, 371
 release of ATP into   408
strontium
 effect of phosphate on uptake of   641
 substitution of, for calcium   614–15
succinate
 in glyoxylate cycle   486, 545
 in respiration   516–22, 540, 545, 549
succinate dehydrogenase   509, 518, 522
succinic acid   540, 549
succinoxidase   510–11, 516–20
succinyl CoA   442
sucrose
 as carbon source   535
 production in green algae   9–10
 specific growth rates on   542–3
 storage of   217–19, 225
sugar(s) (see also specific compounds)
 active transport of   539
 as carbon source   532–8
 as storage products   217–20
 extracellular production of   839–40,
  848
 in diatom cell walls   658
 in osmoregulation   225
 residues in plant polysaccharides   15
sulphate
 active uptake of   689
 as source of sulphur   611, 612
 effect of deficiency of   613
 effect of light on uptake of   719
 effect of phosphorus on uptake of   641
 effect of temperature on uptake of   725
 reduction by $H_2$   461
 requirement in reproduction   823
sulphate ester   15
sulphated polysaccharides   63–75, 107, 108
sulphite
 reducing system   612
 reduction of, by $H_2$   461, 462
sulphite reductase   612

sulpholipid
 as source of sulphur   612
 metabolism of   254, 259
 plant   240, 256, 257
sulphonate   481, 496
sulphonium compounds   612
sulphonoglycolipid   257
sulphoquinovosyl diglyceride   18, 260
sulphur
 as inorganic nutrient   611–13
 effect of deficiency of   613
 requirement for silicic acid uptake   658
 requirement for silicon utilization   624
sulphuric acid   64, 612
symbionts
 biochemical recognition of   25
 transfer between   853–4
symbiosis
 extracellular products in   853–4
 nitrogen fixation in   575
symbiotic hypothesis of plastid evolution
  9, 19, 21–3, 32, 129, 155, 259–61, 282,
  284, 302–4
synaptonemal complexes   100
synchronous cultures   894–904
 chlorophyll synthesis in   171
 circadian rhythms and   901–3
 development and physiology of   899–901
 photorespiration in   494
 requirements of   897–8
syngen pairs   24–5


tannins   769, 844
tartrate   642
taurine   612
taxonomy
 biliproteins in   7–9, 194–5
 biochemical   1–32
 fatty acids in   16, 259–61
 hydrogenase in   466
 morphological features and   789–90
 sterols in   18, 269
TCA (see tricarboxylic acid)
tellurite   521
telophase   99
temperature effects on
 chlorophyll synthesis   170–1, 551
 coccolith formation   667
 excretion   850, 851
 growth and development   721–6, 792,
  795–6, 798, 802, 805, 818–21
 ion uptake   686, 701, 702, 706
 movement   882
 nitrogen fixation   573
 photochemical reactions   398, 463
 pigmentation   379
 respiration   511
 synchronous cultures   896, 898

temperature sensitive mutants 320, 323, 326–8, 336
terpenoid 442
tetracycline 147
tetrad analysis 320–3, 329–38
tetrahydrofolate polyglutamate 846
tetrametaphosphate 646
tetramethylthiuram disulphide 619
tetra-nitro blue tetrazolium 521
1,3,4,6-tetra-O-methylfructose 217
2,3,4,6-tetra-O-methylglucose 217
tetraspores 872
tetrasporophytes 815
tetraterpene 176
tetrathionate 461
tetrol 225
theca 90, 105
thermophilic algae 170–1, 721, 722
thermotaxis 882
thiamine
 algal production of 757–8, 845
 bioassay for 753–6
 content of algae 756–7
 ecology of 758–62
 effect of, on respiration 544
 requirement for 28, 320, 322, 323, 544, 742–8, 778–82
thiamine pyrophosphate 544
thiamine pyrophosphate-glycoaldehyde 479
thigmotaxis 883
thiocyanate 509, 523
thiosulphate 461, 612
D-threonine 595
thylakoids
 cyclic changes of 900
 freeze etching of 366
 hydrogen pump of 406–7
 photophosphorylation and 393
 photosystems on 366–8, 408, 411
 structure and functions of 125–44, 153, 154, 789
thymine 622
tocopherol 17
α-tocopherolquinone 17
toluidine blue 461
tonoplast 698–706
topotaxis 873
toxins (see also specific compounds)
 effects of, on growth and development 798, 806
 release of, by algae 846, 848
transaldolase 507
transaminase 594
transamination 593, 594
trans-3-hexadecenoic acid 18, 238, 254 257–60
transformation 284
transketolase 436, 507

translocation
 in chloroplasts 145
 products of 220
 route in red algae 94–5
transphosphorylation 428
trehalose 218–19
trehalose phosphorylase 219
tricarboxylic acid (TCA) 539
tricarboxylic acid cycle 506–8, 520
 β-carboxylation and the 445, 449
 effect of blue light on the, 443, 445, 446
 oxidation of organic acids and the 544, 550
 reductive 436, 438
trichocyst 87, 89, 91, 104
p-trifluoromethoxy carbonyl cyanide phenyl hydrazone (FCCP) 521–2
triglycerides 238–40
trihydroxy-β-carotene 190–1
2,3,5-tri-iodobenzoic acid 765, 800, 824
trimetaphosphate 646
3,4,6-tri-O-methylfructose 217
triose phosphate isomerase 480
triphenyl tetrazolium chloride (TTC) 461, 565
triterpene 266, 276
trumpet hyphae 95
trypanosome 867
trypsin 515
tryptamine 764
tryptophan 595, 763
TTC (see triphenyl tetrazolium chloride)
turbulence, effects of 729

ubiquinones 17, 538
UDP and related compounds (see uridine diphosphate)
ultraviolet irradiation
 as mutagen 319, 324, 327, 330
 bleaching of Euglena by 286
 effect of, on growth 717–18
 effect of, on reproduction 824–5
 resistance to 898
unicellular forms 87–92, 97
uracil
 effect of magnesium deficiency on 615
 incorporation into DNA 294
urea
 assimilation of 448, 595
 inhibition of gametes by 822–3
 metabolism of 596–7
urea : $CO_2$ ligase 596
urease 596–7
uric acid 595, 597
uridine diphosphogalactose (UDP galactose) 220
uridine diphosphoglucose (UDP glucose)
 in polysaccharide synthesis 48

uridine diphosphoglucose—*cont.*
　in starch storage　208
　utilization of　211
uridine diphosphoglucose
　　pyrophosphorylase　215
uridine diphosphoglucosyl transferase　209
uridine triphosphate (UTP)　602
uronic acids
　extracellular production of　839
　in agar　64
　in alginic acid　58
　in cell walls　47, 658
　in fucoidan　72, 73
　in mucilages　870
　in porphyran　66
　in thecae　90
usnic acid　854
UTP (see uridine triphosphate)

vaccenic acid　244, 259
vacuolar sap　612
vacuoles
　in ion transport　683, 693–7
　occurrence of　87, 92, 93
valeraldehyde　845
vanadium　611, 624–5
vaucheriaxanthin　4–6, 153–4, 183–6
vesicles　94
violaxanthin　5, 152–4, 178–9, 183, 185,
　　727
viologen dyes　405
vitamins (see also specific types)
　bioassays for　753–6
　content of algae　756–7
　ecology of　758–62
　production by algae　757–8, 845–6, 854
　requiring mutants　323
　requirements of algae　742–56, 778–82
vitamin B$_6$
　dependence of phosphorylase formation
　　on　209, 211
　extracellular production of　846
vitamin B$_{12}$ (α-5,6-dimethyl benzimidazolyl
　　cyanocobalamin)
　algal production of　757–8, 845
　analogues of　749–51, 754
　bioassay for　753–6
　content of algae　756–7
　ecology of　758–62
　effect of, on morphology　770
　requirement for　31, 508, 611, 621, 622,
　　742–51, 778–82
　uptake of　751–3
vitamin E　846
vitamin K$_1$ (see also phylloquinone)
　as photosynthetic inhibitor　459
　requirement for　405
volatile substances　844–5

water
　in photosynthesis　348, 392, 396, 398, 407
　production of　511
　transport and ion fluxes　705
wavelength
　and photosynthetic pigments　396
　effect of, on products of photosynthesis
　　443
　of light absorbed by algae　346–79

xanthine　595
xanthophylls
　as taxonomic feature　4–6,
　characteristics of　354
　effect of silicon deficiency on　623
　in photosystem II　370, 373
　occurrence of　176–94, 361, 368
X-ray
　mutants of *Scenedesmus*　459
　studies of cell walls　43–4, 48–55, 58–62,
　　70
　studies on fructosans　216
　studies on starch grains　210
xylan
　as taxonomic feature　12–13, 15
　in cell walls　53, 55–8, 75
β-1,3-xylan　12, 43
β-1,4-D-xylan　60
xylogalactan　14–15
xylomannan　12, 53
β-D-xylopyranose　41
3-O-β-D-xylopyranosyl-L-fucose　72, 73
xylose
　extracellular production of　839
　storage of　218–19
D-xylose　10, 12, 15, 51, 55, 66, 69–75
1,3-D-xylose　42, 55, 75
1,4-D-xylose　57, 75
D-xylose 2-sulphate　74
xylose-5-phosphate　436

zeaxanthin　5, 26, 152–4, 178, 183–7, 190–
　　192, 368, 727
zinc
　as inorganic nutrient　611, 620
　role of　615
　toxicity of　798, 807
zoochlorellae
　extracellular products of　839, 840, 843
　　850
　in *Hydra*　25
zoospore
　amoeboid movement in　868
　effect of endogenous rhythms on
　　production of 826
　effect of nutrient deficiency on　613, 822–
　　823

effect of light quality on production of 817
flagella of   865
inhibition of release of   904
role in life cycle   798
zooxanthellae
  DNA of   22
  extracellular products of   840, 841, 844, 853, 854
  glycerol in   222, 839
  role of, in calcification   669

zygospore
  change in respiration on germination of 514
  effect of calcium on production of   823
  in *Chlamydomonas* life cycle   291
zygote
  effect of organic compounds on formation of   824
  in *Chlamydomonas* life cycle   316, 318
  irradiation of   330
zymogen   110